Algebra

Beginning and Intermediate

Richard N. Aufmann
Palomar College, California

Joanne S. Lockwood
Plymouth State University, New Hampshire

Houghton Mifflin Company Boston New York

Publisher: Jack Shira
Senior Sponsoring Editor: Lynn Cox
Senior Development Editor: Dawn Nuttall
Editorial Associate: Melissa Parkin
Senior Project Editor: Audrey Bryant
Editorial Assistant: Sarah Cleary
Senior Production/Design Coordinator: Carol Merrigan
Manufacturing Manager: Florence Cadran
Senior Marketing Manager: Ben Rivera
Marketing Associate: Lisa Lawler

Cover Photographer: Cover image © Anthony Johnson/Getty Images

PHOTOGRAPH CREDITS: p. xxxiv David Young-Wolff/PhotoEdit Inc.; p. 60 Lester V. Bergman/CORBIS; p. 61 CORBIS; p. 71 Frozen Images/The Image Works, Inc.; p. 73 Mary Steinbacher/PhotoEdit, Inc.; p. 74 Journal Courier-Steve Warnowski/The Image Works; p. 117 AP/Wide World Photos; p. 123 AP/Wide World Photos; P. 194 Barbara Stitzer/PhotoEdit, Inc.; p. 243 Eric Fowke/PhotoEdit, Inc.; p. 251 Bettmann/CORBIS; p. 276 Jonathan Nourak/PhotoEdit, Inc.; p. 312 CORBIS; p. 319 Michael Newman/PhotoEdit, Inc.; p. 321 Liaison Photo/Getty Images; P. 322 Michael Newman/PhotoEdit Inc.; p. 340 Bettman/CORBIS; p. 374 Reuters NewMedia, Inc./CORBIS; p. 375 top Jeff Greenberg/PhotoEdit, Inc.; p. 375 bottom Liaison Photo/Getty Images; p. 376 Bettman/CORBIS;p. 388 AFP/CORBIS; p. 396 Spencer Grant/PhotoEdit, Inc.; p. 446 David Barber/PhotoEdit, Inc.; p. 465 AP/Wide World Photos; p. 466 AP/Wide World Photos; p. 478 Bill Aron/PhotoEdit, Inc.; P. 479 top Gunter Marx Photography/CORBIS; p. 479 bottom Vicki Silbert/PhotoEdit, Inc.; P. 480 Lee Snider/CORBIS; p. 482 Michael S. Yamashita/CORBIS; p. 489 AP/Wide World Photos; p. 490 Copyright Cameraphoto Arte, Venice/Art Resource, NY; p. 498 Jonathan Nourak/PhotoEdit, Inc.; p. 504 Bonnie Kamin/PhotoEdit, Inc.; p. 508 Gianni Dagli Orti/CORBIS; p. 538 Karl Weatherly/CORBIS; p. 553 Kevin Schafer/CORBIS; 562 Mark Richards/PhotoEdit, Inc.; p. 583 Kevin Schafer/CORBIS; p. 593 Jonathan Nourak/PhotoEdit, Inc.; p. 634 Jeff Greenberg/PhotoEdit, Inc.; p. 684 Michael Newman/PhotoEdit, Inc.; p. 698 Michael Newman/PhotoEdit, Inc.; p. 743 Getty Images; p. 745 NOVASTOCK/PhotoEdit, Inc.; p. 746 Topham/The Image Works.

Text Credits: p. 31 "How did you find your most recent job?" graphic copyright © Knight Ridder/Tribune Media Services. Reprinted with permission; p. 583 chart reprinted by permission of the author, Dr. Barry Noon; p. 747 G. Polya, *How to Solve It; A New Aspect of Mathematical Method.* Copyright © 1945, renewed 1973 by Princeton University Press. Reprinted by permission of Princeton University Press; p. 747 reprinted with permission from *Student Math Notes.* Copyright © 1991 by the National Council of Teachers of Mathematics. All rights reserved; pp. 748-750 Copyright © 2003 Fair Isaac Corporation. Reprinted with permission.

Custom Publishing Editor: Luis Rosas
Custom Publishing Production Manager: Tina Kozik
Project Coordinator: Andrea Wagner

Printed in the United States of America.

ISBN-13: 978-0-618-68259-1
ISBN-10: 0-618-68259-7
N-05475

3 4 5 6 7 8 9 – CM – 07 06

Houghton Mifflin
Custom Publishing

222 Berkeley Street • Boston, MA 02116

Address all correspondence and order information to the above address.

2 Solving Equations and Inequalities 75

Contents

Preface

Algebra: Beginning and Intermediate examines the fundamental ideas of algebra. The text has been designed not only to meet the needs of the traditional college student, but also to serve the needs of returning students whose mathematical proficiency may have declined during years away from formal education.

Careful attention has been given to implementing the standards suggested by AMATYC and NCTM. Each chapter opens by illustrating and referencing a mathematical application within the chapter. At the end of each section there are "Applying the Concepts" exercises that include writing, synthesis, critical thinking, and challenge problems. At the end of each chapter there is a "Focus on Problem Solving" that introduces students to various problem-solving strategies. This is followed by "Projects and Group Activities," which can be used for cooperative learning activities.

Instructional Features

Interactive Approach

Algebra: Beginning and Intermediate uses an interactive style that provides students with the opportunity to try a skill as it is presented. Each section is divided into objectives, and every objective contains one or more sets of matched-pair examples. The example in each pair is worked out; the problem is for the student to work. By solving this problem, the student practices concepts as they are presented in the text. There are *complete* worked-out solutions to these problems in an appendix at the back of the book. By comparing their solutions to model solutions, students are able to obtain immediate feedback on and reinforcement of the concepts.

Emphasis on Problem-Solving Strategies

Algebra: Beginning and Intermediate features a carefully sequenced approach to application problems that emphasizes proven strategies to solve problems. Students are encouraged to develop their own strategies, draw diagrams, and to write their strategies as part of the solution to each application problem. Model strategies are presented as guides for students to follow as they attempt the Problem corresponding to each numbered Example in the text. Having students provide strategies is a natural way to incorporate writing into the math curriculum.

Emphasis on Applications

The traditional approach to teaching algebra covers only the straightforward manipulation of numbers and variables and thereby fails to teach students the practical value of algebra. By contrast, *Algebra: Beginning and Intermediate* contains an extensive collection of contemporary application problems. Wherever appropriate, the last objective of a section presents applications that require the student to use the skills covered in that section to solve practical problems. This carefully integrated applied approach generates awareness of the value of algebra as a real-life tool.

Completely Integrated Learning System Organized by Objective

Each chapter begins with a list of the learning objectives included within that chapter. Each objective is then restated in the chapter to remind the student of the current topic of discussion. The same objectives that organize the text are used as the structure for exercises and the testing programs.

Acknowledgments

The authors would like to thank the following people who have reviewed this manuscript and provided many valuable suggestions.

Edith Hays, *Texas Woman's University, TX*
Feliks Kostanyan, *DeVry University, NY*
Christine Mirbaha, *The Community College of Baltimore County-Dundalk Campus, MD*
Earl T. Odom, Jr., *Bishop State Community College, AL*
Angelia Reynolds, *Gulf Coast Community College, FL*
Lauri Semarne
Barry S. Spieler, *Birmingham-Southern College, AL*
Elise Price, *Tarrant County College-Southeast Campus, TX*
Danny Whited, *Virginia Intermont College, VA*
David White, *The Victoria College, TX*

Chapter Opening Features

NEW! Chapter Opener

New, motivating chapter opener photos and captions have been added, illustrating and referencing a specific application from the chapter. A list of the applications in the chapter is also provided.

The web icon at the bottom of the second page lets students know of additional online resources at **math.college.hmco.com/students.**

Objective-Specific Approach

Each chapter begins with a list of learning objectives which form the framework for a complete learning system. The objectives are woven throughout the text (i.e. Exercises, Prep Tests, Chapter Review Exercises, Chapter Tests, Cumulative Review Exercises) as well as through the print and multimedia ancillaries. This results in a seamless learning system delivered in one consistent voice.

Leonardo DaVinci designed the first parachute in 1485. People have been jumping out of planes ever since. The highest parachute jump ever made was one by U.S. Air Force Captain Joseph W. Kittinger, Jr. in 1960. Kittinger, who was performing high-altitude escape experiments for the Air Force, jumped from 102,800 feet above sea level—over 19 miles up! His free fall lasted more then 4.5 minutes and he reached a velocity of 614 mph. Factors that affect velocity include the initial altitude of the jump, the size of the parachute, and the weight and body position of the diver. The velocity of a falling object can be modeled by a quadratic equation, as shown in the **Project on pages 436–438.**

Applications

Architecture, *p. 430*
Carpentry, *p. 444*
Consecutive integer problems, *pp. 442, 444*
Construction, *pp. 432, 433*
Consumerism, *p. 444*
Discount, *p. 444*
Food science, *p. 427*
Geometry, *pp. 403, 418, 419, 430, 432, 433, 435, 442, 444*
Investments, *p. 444*
Motion pictures, *p. 442*
Number problems, *pp. 426, 427, 432*
Physics, *pp. 433, 437, 438*
Prime numbers, *p. 436*
Space science, *p. 432*
Sports, *pp. 438, 441*
Symmetry, *p. 435*
Temperature, *p. 444*

Objectives

Section 6.1
1 Factor a monomial from a polynomial
2 Factor by grouping

Section 6.2
1 Factor trinomials of the form $x^2 + bx + c$
2 Factor completely

Section 6.3
1 Factor trinomials of the form $ax^2 + bx + c$ using trial factors
2 Factor trinomials of the form $ax^2 + bx + c$ by grouping

Section 6.4
1 Factor the difference of two perfect squares and factor perfect-square trinomials
2 Factor the sum or the difference of two cubes
3 Factor trinomials that are quadratic in form
4 Factor completely

Section 6.5
1 Solve equations by factoring
2 Application problems

6
Factoring

Need help? For online student resources, such as section quizzes, visit this web site: math.college.hmco.com

397

pages 396–397

76 CHAPTER 2 Solving Equations and Inequalities

Prep Test

1. Subtract: $8 - 12$
2. Multiply: $-\frac{3}{4}\left(-\frac{4}{3}\right)$
3. Multiply: $-\frac{5}{8}(16)$
4. Write 90% as a decimal.
5. Write 0.75 as a percent.
6. Evaluate $3x^2 - 4x - 1$ when $x = -4$.
7. Simplify: $3x - 5 + 7x$
8. Simplify: $8x - 9 - 8x$
9. Simplify: $6x - 3(6 - x)$

Go Figure

How can you cut a donut into 8 equal pieces with three cuts of the knife?

page 76

NEW! Prep Test and Go Figure

Prep Tests occur at the beginning of each chapter and test students on *previously covered* concepts that are required in the coming chapter. Answers are provided in the Answer Appendix. Objective references are also provided if a student needs to review specific concepts.

The **Go Figure** problem that follows the *Prep Test* is a playful puzzle problem designed to engage students in problem solving.

Aufmann Interactive Method (AIM)

An Interactive Approach

Algebra: Beginning and Intermediate uses an interactive style that provides a student with an opportunity to try a skill as it is presented. Each section is divided into objectives, and every objective contains one or more sets of matched-pair examples. The first example in each set is worked out; the second example, called "Problem," is for the student to work. By solving this problem, the student actively practices concepts as they are presented in the text.

There are *complete*, *worked-out* solutions to these problems in an appendix. By comparing their solution to the solution in the appendix, students obtain immediate feedback on, and reinforcement of, the concept.

Example 2 Multiply: $(2b^3 - b + 1)(2b + 3)$

Solution

$$\begin{array}{r} 2b^3 - b + 1 \\ 2b + 3 \\ \hline 6b^3 \qquad - 3b + 3 \\ 4b^4 \qquad - 2b^2 + 2b \qquad \\ \hline 4b^4 + 6b^3 - 2b^2 - b + 3 \end{array}$$

- Multiply $2b^3 - b + 1$ by 3.
- Multiply $2b^3 - b + 1$ by $2b$. Arrange the terms in descending order.
- Add the terms in each column.

Problem 2 Multiply: $(2y^3 + 2y^2 - 3)(3y - 1)$

Solution See page S18.

Example 3 Multiply: $(4a^3 - 5a - 2)(3a - 2)$

Solution

$$\begin{array}{r} 4a^3 - 5a - 2 \\ 3a - 2 \\ \hline -8a^3 \qquad + 10a + 4 \\ 12a^4 \qquad - 15a^2 - 6a \qquad \\ \hline 12a^4 - 8a^3 - 15a^2 + 4a + 4 \end{array}$$

- Multiply $4a^3 - 5a - 2$ by -2.
- Multiply $4a^3 - 5a - 2$ by $3a$.
- Add the terms in each column.

Problem 3 Multiply: $(3x^3 - 2x^2 + x - 3)(2x + 5)$

Solution See page S18.

page 358

SECTION 5.3

Problem 1

A. $(-2y + 3)(-4y) = 8y^2 - 12y$

B. $-a^2(3a^2 + 2a - 7) = -3a^4 - 2a^3 + 7a^2$

Problem 2

$$\begin{array}{r} 2y^3 + 2y^2 - 3 \\ 3y - 1 \\ \hline -2y^3 - 2y^2 \qquad + 3 \\ 6y^4 + 6y^3 \qquad - 9y \qquad \\ \hline 6y^4 + 4y^3 - 2y^2 - 9y + 3 \end{array}$$

Problem 3

$$\begin{array}{r} 3x^3 - 2x^2 + x - 3 \\ 2x + 5 \\ \hline 15x^3 - 10x^2 + 5x - 15 \\ 6x^4 - 4x^3 + 2x^2 - 6x \qquad \\ \hline 6x^4 + 11x^3 - 8x^2 - x - 15 \end{array}$$

Problem 4

$(4y - 5)(3y - 3)$
$= 12y^2 - 12y - 15y + 15$
$= 12y^2 - 27y + 15$

Problem 5

$(3a + 2b)(3a - 5b)$
$= 9a^2 - 15ab + 6ab - 10b^2$
$= 9a^2 - 9ab - 10b^2$

Problem 6

$(2a + 5c)(2a - 5c) = (2a)^2 - (5c)^2$
$= 4a^2 - 25c^2$

Problem 7

$(3x + 2y)^2 = (3x)^2 + 2(3x)(2y) + (2y)^2$
$= 9x^2 + 12xy + 4y^2$

Problem 8

$(6x - y)^2 = (6x)^2 + 2(6x)(-y) + (-y)^2$
$= 36x^2 - 12xy + y^2$

Problem 9

Strategy To find the area in terms of x, replace the variables L and W in the equation $A = LW$ with the given values, and solve for A.

page S18

AIM for Success

Welcome to *Algebra: Beginning and Intermediate*. As you begin this course, we know two important facts: (1) We want you to succeed. (2) You want to succeed. To do that requires an effort from each of us. For the next few pages, we are going to show you what is required of you to achieve that success and how you can use the features of this text to be successful.

Motivation

One of the most important keys to success is motivation. We can try to motivate you by offering interesting or important ways mathematics can benefit you. But, in the end, the motivation must come from you. On the first day of class, it is easy to be motivated. Eight weeks into the term, it is harder to keep that motivation.

To stay motivated, there must be outcomes from this course that are worth your time, money, and energy. List some reasons you are taking this course. Do not make a mental list—actually write them out.

> **Take Note**
> Motivation alone will not lead to success. For instance, suppose a person who cannot swim is placed in a boat, taken out to the middle of a lake, and then thrown overboard. That person has a lot of motivation but there is a high likelihood the person will drown without some help. Motivation gives us the desire to learn but is not the same as learning.

Although we hope that one of the reasons you listed was an interest in mathematics, we know that many of you are taking this course because it is required to graduate, it is a prerequisite for a course you must take, or because it is required for your major. Although you may not agree that this course is necessary, it is! If you are motivated to graduate or complete the requirements for your major, then use that motivation to succeed in this course. Do not become distracted from your goal to complete your education!

Commitment

To be successful, you must make a committment to succeed. This means devoting time to math so that you achieve a better understanding of the subject.

List some activities (sports, hobbies, talents such as dance, art, or music) that you enjoy and at which you would like to become better.

ACTIVITY	*TIME SPENT*	*TIME WISHED SPENT*

Thinking about these activities, put the number of hours that you spend each week practicing these activities next to the activity. Next to that number, indicate the number of hours per week you would like to spend on these activities.

Whether you listed surfing or sailing, aerobics or restoring cars, or any other activity you enjoy, note how many hours a week you spend doing it. To succeed in math, you must be willing to commit the same amount of time. Success requires some sacrifice.

The "I Can't Do Math" Syndrome

There may be things you cannot do, such as lift a two-ton boulder. You can, however, do math. It is much easier than lifting the two-ton boulder. When

xxv

page xxv

NEW! AIM for Success Student Preface

This new student 'how to use this book' preface explains what is required of a student to be successful and how this text has been designed to foster student success, including the Aufmann Interactive Method (AIM). *AIM for Success* can be used as a lesson on the first day of class or as a project for students to complete to strengthen their study skills. There are suggestions for teaching this lesson in the *Instructor's Resource Manual.*

Problem Solving

Focus on Problem Solving

At the end of each chapter is a Focus on Problem Solving feature which introduces the student to various successful problem-solving strategies. Strategies such as drawing a diagram, applying solutions to other problems, working backwards, inductive reasoning, and trial and error are some of the techniques that are demonstrated.

Focus on Problem Solving 263

Focus on Problem Solving

Find a Pattern Polya's four recommended problem-solving steps are stated below.

1. Understand the problem.
2. Devise a plan.
3. Carry out the plan.
4. Review your solution

Point of Interest

Carl Friedrich Gauss (1777–1855) has been called the "Prince of Mathematicians" by some historians. He applied his genius to many areas of mathematics and science. A unit of magnetism, the gauss, is named in his honor. Some electronic equipment (televisions, for instance) contain a degausser that controls magnetic fields.

One of the several ways of devising a plan is first to try to find a pattern. Karl Friedrich Gauss supposedly used this method to solve a problem that was given to his math class when he was in elementary school. As the story goes, his teacher wanted to grade some papers while the class worked on a math problem. The problem given to the class was to find the sum

$$1 + 2 + 3 + 4 + \cdots + 100$$

Gauss quickly solved the problem by seeing a pattern. Here is what he saw.

$$1 + 2 + 3 + 4 + \cdots + 97 + 98 + 99 + 100$$

(101, 101, 101, 101)

Note that

$$1 + 100 = 101$$
$$2 + 99 = 101$$
$$3 + 98 = 101$$
$$4 + 97 = 101$$

Gauss noted that there were 50 sums of 101. Therefore, the sum of the first 100 natural numbers is

$$1 + 2 + 3 + 4 + \cdots + 97 + 98 + 99 + 100 = 50(101) = 5050$$

Try to solve the following problems by finding a pattern.

1. Find the sum $2 + 4 + 6 + \cdots + 96 + 98 + 100$.
2. Find the sum $1 + 3 + 5 + \cdots + 97 + 99 + 101$.
3. Find another method of finding the sum $1 + 3 + 5 + \cdots + 97 + 99 + 101$ given in the previous exercise.
4. Find the sum $\frac{1}{1\cdot 2} + \frac{1}{2\cdot 3} + \frac{1}{3\cdot 4} + \cdots + \frac{1}{49\cdot 50}$.
   Hint: $\frac{1}{1\cdot 2} = \frac{1}{2}$, $\frac{1}{1\cdot 2} + \frac{1}{2\cdot 3} = \frac{2}{3}$, $\frac{1}{1\cdot 2} + \frac{1}{2\cdot 3} + \frac{1}{3\cdot 4} = \frac{3}{4}$
5. The following problem shows that checking a few cases does not always result in a conjecture that is true for *all* cases. Select any two points on a circle and draw a *chord*, the line connecting the points (see the drawing in the left margin). The chord divides the circle into two regions. Now select three different points and draw chords connecting each of the three points with every other point. The chords divide the circle into four regions. Now select four points and connect each of the points with every other point. Make a conjecture about the relationship between the number of regions and the number of points on the circle. Does your conjecture work for five points? six points?

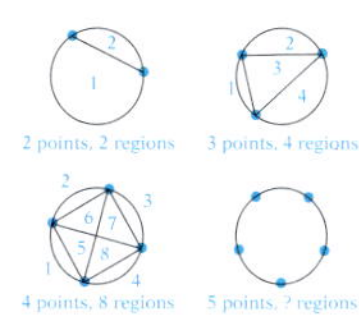

page 263

536 CHAPTER 8 Rational Exponents and Radicals

2 Application problems

A right triangle contains one 90° angle. The side opposite the 90° angle is called the **hypotenuse.** The other two sides are called **legs.**

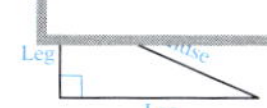

Pythagoras, a Greek mathematician, is credited with the discovery that the square of the hypotenuse of a right triangle is equal to the sum of the squares of the two legs. This is called the Pythagorean Theorem.

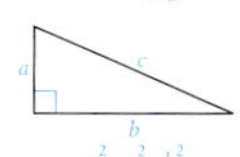

$c^2 = a^2 + b^2$

Pythagorean Theorem

The square of the hypotenuse, c, of a right triangle is equal to the sum of the squares of the two legs, a and b.

$$c^2 = a^2 + b^2$$

Example 3 A ladder 20 ft long is leaning against a building. How high on the building will the ladder reach when the bottom of the ladder is 8 ft from the building? Round to the nearest tenth.

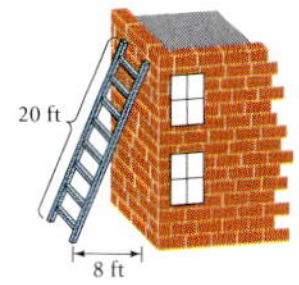

Strategy To find the distance, use the Pythagorean Theorem. The hypotenuse is the length of the ladder. One leg is the distance from the bottom of the ladder to the base of the building. The distance along the building from the ground to the top of the ladder is the unknown leg.

Solution

$$c^2 = a^2 + b^2$$
$$20^2 = 8^2 + b^2$$
$$400 = 64 + b^2$$
$$336 = b^2$$
$$(336)^{\frac{1}{2}} = (b^2)^{\frac{1}{2}}$$
$$\sqrt{336} = b$$
$$18.3 \approx b$$

The distance is 18.3 ft.

Problem 3 Find the diagonal of a rectangle that is 6 cm in length and 3 cm in width. Round to the nearest tenth.

Solution See page S28.

page 536

Problem-Solving Strategies

The text features a carefully developed approach to problem solving that emphasizes the importance of *strategy* when solving problems. Students are encouraged to develop their own strategies—to draw diagrams, to write out the solution steps in words—as part of their solution to a problem. In each case, model strategies are presented as guides for students to follow as they attempt the parallel Problem. Having students provide strategies is a natural way to incorporate writing into the math curriculum.

Real Data and Applications

Applications

One way to motivate an interest in mathematics is through applications. Wherever appropriate, the last objective of a section presents applications that require the student to use problem-solving strategies, along with the skills covered in that section, to solve practical problems. This carefully integrated applied approach generates student awareness of the value of algebra as a real-life tool.

Applications are taken from many disciplines including agriculture, business, carpentry, chemistry, construction, Earth science, education, manufacturing, nutrition, real estate, and sports.

SECTION 2.1 Introduction to Equations **83**

Point of Interest

As noted at the right, 172,841 of the dogs or litters registered were Labrador retrievers. Listed below are the next most popular breeds and their registrations.
Golden retrievers: 66,300
German shepherds: 57,660
Dachshunds: 54,773
Beagles: 52,026

Example 9 During a recent year, nearly 1.2 million dogs or litters were registered with the American Kennel Club. The most popular breed was the Labrador retriever, with 172,841 registered. What percent of the registrations were Labrador retrievers? Round to the nearest tenth of a percent. (*Source:* American Kennel Club)

Strategy To find the percent, solve the basic percent equation using B = 1.2 million = 1,200,000 and A = 172,841. The percent is unknown.

Solution

$$PB = A$$
$$P(1{,}200{,}000) = 172{,}841$$
$$\frac{P(1{,}200{,}000)}{1{,}200{,}000} = \frac{172{,}841}{1{,}200{,}000}$$
$$P \approx 0.144$$
$$P \approx 14.4\%$$

Approximately 14.4% of the registrations were Labrador retrievers.

Problem 9 A student correctly answered 72 of the 80 questions on an exam. What percent of the questions were answered correctly?

Solution See page S5.

Example 10 According to the Centers for Disease Control and Prevention, 30.8% of the adult population of Kentucky smokes. How many adults in Kentucky smoke? Use a figure of 3,000,000 for the number of adults in Kentucky.

Strategy To find the number of adults who smoke, solve the basic percent equation using B = 3,000,000 and P = 30.8% = 0.308. The amount is unknown.

Solution

$$PB = A$$
$$0.308(3{,}000{,}000) = A$$
$$924{,}000 = A$$

Approximately 924,000 adults in Kentucky smoke.

Problem 10 The price of a digital camcorder is $895. A 6% sales tax is added to the price. How much is the sales tax?

Solution See page S5.

The simple interest that an investment earns is given by the equation $I = Prt$, where I is the simple interest, P is the principal, or amount invested, r is the simple interest rate, and t is the time.

Take Note

A $1500 investment has an annual simple interest rate of 7%. Find the simple interest earned on the investment after 18 months.

page 83

SECTION 3.1 The Rectangular Coordinate System **209**

34. **Wealth** The table shows the number of U.S. households with a net worth of $1 million or more every three years from 1989 to 2001. Also shown is the millionaire households as a percent of all households in the United States. (*Source:* Edward Wolff, professor of economics, New York University; Federal Reserve) Graph the scatter diagram for these data.

Millionaire households (in millions), x	3.0	3.1	3.4	4.1	5.0
Percent of all households, y	3.2	3.2	3.4	4.0	4.7

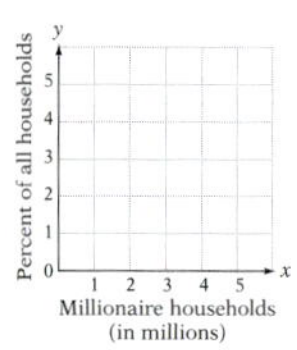

35. **Telecommunications** As shown in the table below, the percent of households with telephones has increased over the years. Draw a scatter diagram for these data. (*Source:* U.S. Census Bureau/Federal Communications Commission)

Year, x	'20	'30	'40	'50	'60	'70	'80	'90	'00
Percent, y	35	41	37	62	78	90	93	95	97

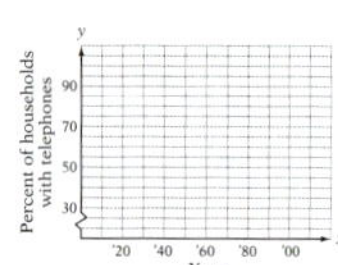

4 **36.** **Temperature** On November 5 in midstate New Hampshire, the temperature at 6 A.M. was 34°F. At 2 P.M. on the same day, the temperature was 58°F. Find the average rate of change in temperature per hour.

37. **Demography** The table at the right shows the population of Texas each decade from 1850 to 2000. (*Source:* U.S. Bureau of the Census)

a. Find the average annual rate of change in the population of Texas from 1900 to 2000.

b. Was the average annual rate of change in the population from 1900 to 1950 greater than or less than the average annual rate of change in the population from 1950 to 2000?

c. Was the average annual rate of change in the population of Texas from 1980 to 1990 greater than or less than the average annual rate of change in the population of California from 1980 to 1990? (See page 132 for the average annual rate of change in the population of California from 1980 to 1990.)

d. During which decade was the average annual rate of change the least? What was the average annual rate of change during the decade?

Population of Texas

Year	Population (in millions)
1850	0.2
1860	0.6
1870	0.8
1880	1.6
1890	2.2
1900	3.0
1910	3.9
1920	4.7
1930	5.8
1940	6.4
1950	7.7
1960	9.6
1970	11.2
1980	14.2
1990	17.0
2000	20.9

38. **Cremation** In the United States in 1990, there were 2.15 million deaths and 367,975 cremations. In 1999, there were 2.35 million deaths and 595,617 cremations. (*Source:* Cremation Association of North America)

a. In the 1990s, what was the average annual rate of change in the number of deaths in the United States? Round to the nearest thousand.

b. In the 1990s, what was the average annual rate of change in the number of cremations in the United States? Round to the nearest thousand.

c. In the 1990s, which increased at a greater rate, the number of deaths or the number of cremations?

page 209

Real Data

Real data examples and exercises, identified by an icon, ask students to analyze and solve problems taken from actual situations. Students are often required to work with tables, graphs, and charts drawn from a variety of disciplines.

Student Pedagogy

Icons

The icons at each objective head remind students of the many and varied additional resources available for each objective.

Graphing Calculator Boxes

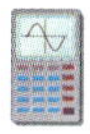

Graphing calculators are incorporated as an optional feature at appropriate places throughout the text. This material is designated by the graphing calculator icon.

Key terms, in bold, emphasize important terms. The key terms are also provided in a **Glossary** at the back of the text.

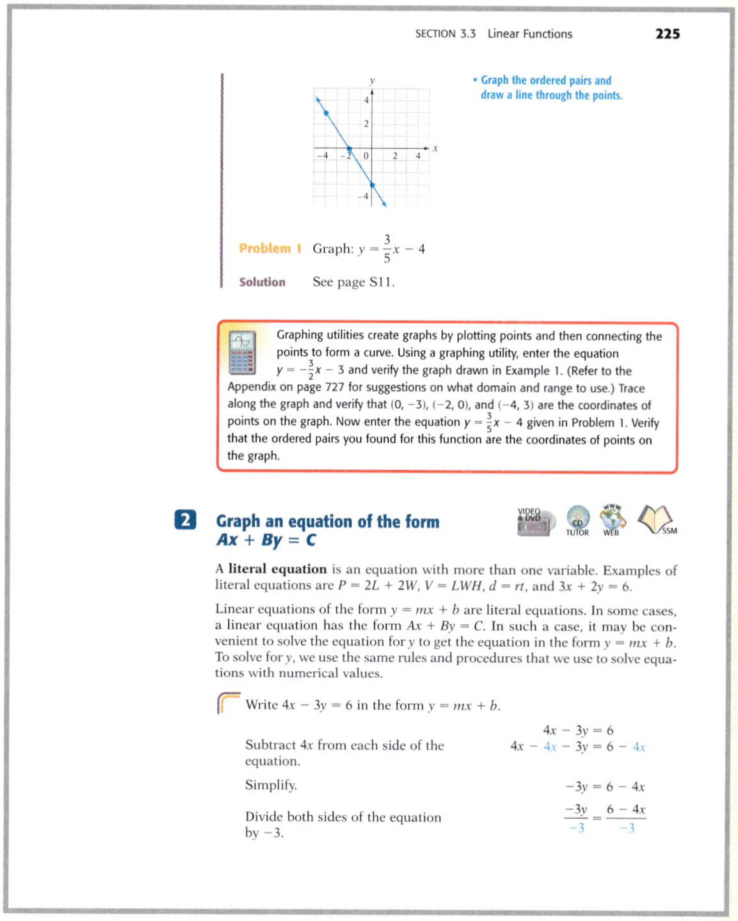

SECTION 3.3 Linear Functions 225

• Graph the ordered pairs and draw a line through the points.

Problem 1 Graph: $y = \frac{3}{5}x - 4$

Solution See page S11.

Graphing utilities create graphs by plotting points and then connecting the points to form a curve. Using a graphing utility, enter the equation $y = -\frac{3}{2}x - 3$ and verify the graph drawn in Example 1. (Refer to the Appendix on page 727 for suggestions on what domain and range to use.) Trace along the graph and verify that (0, −3), (−2, 0), and (−4, 3) are the coordinates of points on the graph. Now enter the equation $y = \frac{3}{5}x - 4$ given in Problem 1. Verify that the ordered pairs you found for this function are the coordinates of points on the graph.

2 Graph an equation of the form $Ax + By = C$

A **literal equation** is an equation with more than one variable. Examples of literal equations are $P = 2L + 2W$, $V = LWH$, $d = rt$, and $3x + 2y = 6$.

Linear equations of the form $y = mx + b$ are literal equations. In some cases, a linear equation has the form $Ax + By = C$. In such a case, it may be convenient to solve the equation for y to get the equation in the form $y = mx + b$. To solve for y, we use the same rules and procedures that we use to solve equations with numerical values.

Write $4x - 3y = 6$ in the form $y = mx + b$.

$4x - 3y = 6$

Subtract $4x$ from each side of the equation. $4x - 4x - 3y = 6 - 4x$

Simplify. $-3y = 6 - 4x$

Divide both sides of the equation by −3. $\frac{-3y}{-3} = \frac{6 - 4x}{-3}$

page 225

Take Note

These margin notes alert students to a point requiring special attention or are used to amplify the concept under discussion.

Point of Interest

These margin notes contain interesting sidelights about mathematics, its history, or its application.

Instructional Examples

Colored brackets indicate an instructional example that usually has explanatory remarks to the left of the solution. These examples are used to provide additional examples of important concepts as well as examine important techniques or principles that are often used in the solution of other types of problems.

Key Concepts are presented in green boxes in order to highlight these important concepts and to provide for easy reference.

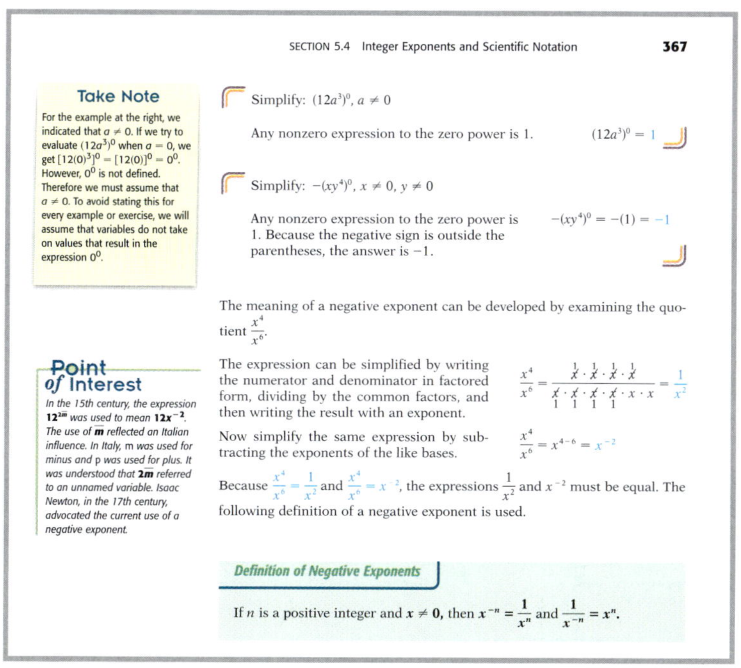

SECTION 5.4 Integer Exponents and Scientific Notation 367

Take Note

For the example at the right, we indicated that $a \neq 0$. If we try to evaluate $(12a^3)^0$ when $a = 0$, we get $[12(0)^3]^0 = [12(0)]^0 = 0^0$. However, 0^0 is not defined. Therefore we must assume that $a \neq 0$. To avoid stating this for every example or exercise, we will assume that variables do not take on values that result in the expression 0^0.

Simplify: $(12a^3)^0$, $a \neq 0$

Any nonzero expression to the zero power is 1. $(12a^3)^0 = 1$

Simplify: $-(xy^4)^0$, $x \neq 0$, $y \neq 0$

Any nonzero expression to the zero power is 1. Because the negative sign is outside the parentheses, the answer is −1. $-(xy^4)^0 = -(1) = -1$

The meaning of a negative exponent can be developed by examining the quotient $\frac{x^4}{x^6}$.

Point of Interest

In the 15th century, the expression $12^{2\bar{m}}$ was used to mean $12x^{-2}$. The use of $\bar{m}$ reflected an Italian influence. In Italy, m was used for minus and p was used for plus. It was understood that $2\bar{m}$ referred to an unnamed variable. Isaac Newton, in the 17th century, advocated the current use of a negative exponent.

The expression can be simplified by writing the numerator and denominator in factored form, dividing by the common factors, and then writing the result with an exponent. $\frac{x^4}{x^6} = \frac{x \cdot x \cdot x \cdot x}{x \cdot x \cdot x \cdot x \cdot x \cdot x} = \frac{1}{x^2}$

Now simplify the same expression by subtracting the exponents of the like bases. $\frac{x^4}{x^6} = x^{4-6} = x^{-2}$

Because $\frac{x^4}{x^6} = \frac{1}{x^2}$ and $\frac{x^4}{x^6} = x^{-2}$, the expressions $\frac{1}{x^2}$ and x^{-2} must be equal. The following definition of a negative exponent is used.

Definition of Negative Exponents

If n is a positive integer and $x \neq 0$, then $x^{-n} = \frac{1}{x^n}$ and $\frac{1}{x^{-n}} = x^n$.

page 367

Exercises and Projects

Exercises

The exercise sets of *Algebra: Beginning and Intermediate* emphasize skill building, skill maintenance, and applications. Icons identify appropriate writing, data analysis, and calculator exercises.

Before each exercise set are **Concept Review** exercises designed to test a student's understanding of the section material. Used as oral exercises, these can lead to interesting class discussions.

Included in each exercise set are **Applying Concepts** that present extensions of topics, require analysis, or offer challenge problems. The writing exercises ask students to explain answers, write about a topic in the section, or research and report on a related topic.

32. Number Problem Find two numbers whose difference is 14 and whose product is a minimum.

33. Ranching A rancher has 200 ft of fencing to build a rectangular corral alongside an existing fence. Determine the dimensions of the corral that will maximize the enclosed area.

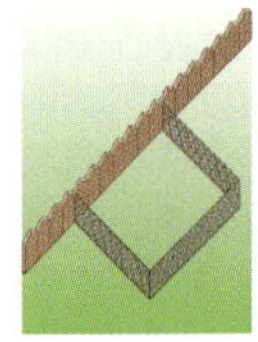

9.7 Applying Concepts

34. Show that the minimum value of

$$S(x) = (2 - x)^2 + (5 - x)^2 + (4 - x)^2 + (7 - x)^2$$

occurs when x is the average of the numbers 2, 5, 4, and 7.

Use a graphing utility to find the minimum or maximum value of the function. Round to the nearest tenth. See the Graphing Calculator Appendix for assistance.

35. $f(x) = x^4 - 2x^2 + 4$

36. $f(x) = x^4 + 2x^3 + 1$

37. $f(x) = -x^6 + x^4 - x^3 + x$

38. $f(x) = -x^8 + x^6 - x^4 + 5x^2 + 7$

39. On the basis of Exercises 35–38, make a conjecture about the relationship between the sign of the leading coefficient of a polynomial function of even degree and whether that function will have a minimum or a maximum value.

Solve.

40. Traffic Control Traffic engineers try to determine the effect a traffic light has at an intersection. By gathering data about the intersection, engineers can determine the approximate number of cars that enter the intersection in the horizontal direction and those that enter in the vertical direction. The engineers would also collect information on the time it takes a stopped car to regain the normal posted speed limit. One model of this situation is $T = \left(\frac{H + V}{2}\right)R^2 + (0.08H - 1.08V)R + 0.58V$, where H is the number of cars arriving at the intersection from the horizontal direction, V is the number of cars arriving at the intersection from the vertical direction, and R is the percent of time the light is red in the horizontal direction. T is the total delay time for all cars and is measured as the number of times the traffic light changes from red to green and back to red.

a. Graph this equation for $H = 100$, $V = 150$, and $0 \le R \le 1$.

b. Write a sentence that explains why the graph is drawn only for $0 \le R \le 1$.

c. What percent of the time should the traffic light remain red in the horizontal direction to minimize T? Round to the nearest whole percent.

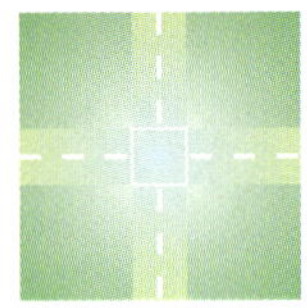

page 620

Projects & Group Activities

Population and Land Allocation*

Rhode Island

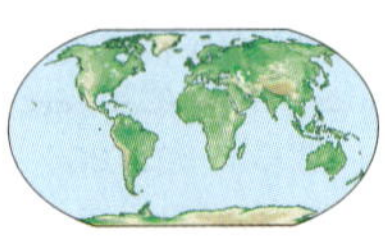

In this project, you are asked to determine hypothetical land allocation for the world's population today. Use the figure 6×10^9 for the current world population and the figure 2.9×10^8 for the current U.S. population. One square mile is approximately 2.8×10^7 ft^2.

1. If every person in the world moved to Texas and each person were given an equal amount of land, how many square feet of land would each person have? The area of Texas is 2.619×10^5 mi^2.
2. If every person in the United States moved to Rhode Island and each person were given an equal amount of land, how many square feet of land would each person have? The area of Rhode Island is 1.0×10^3 mi^2. Round to the nearest whole number.
3. Suppose every person in the world were given a plot of land the size of a two-car garage (22 ft × 22 ft).
   a. How many people would fit in a square mile? Round to the nearest hundred.
   b. How many square miles would be required to accommodate the entire world population? Round to the nearest hundred.
4. If the total land area of Earth were divided equally, how many acres of land would each person be allocated? Use a figure of 5.7×10^7 mi^2 for the land area of Earth. One acre is 43,560 ft^2. Round to the nearest tenth.
5. If every person on Earth were given a plot of land the size of a two-car garage, what would be the carrying capacity of Earth? Round to the nearest hundred billion.

*Source: **www.infoplease.com**

page 388

Projects and Group Activities

The Projects and Group Activities feature at the end of each chapter can be used as extra credit or for cooperative learning activities. The projects cover various aspects of mathematics, including the use of calculators, collecting data from the internet, data analysis, and extended applications.

End of Chapter

Chapter Summary

At the end of each chapter there is a Chapter Summary that includes *Key Words* and *Essential Rules and Procedures*, with accompanying examples, that were covered in the chapter. These chapter summaries provide a single point of reference as the student prepares for a test. Each concept references the objective and the page number from the lesson where the concept is introduced.

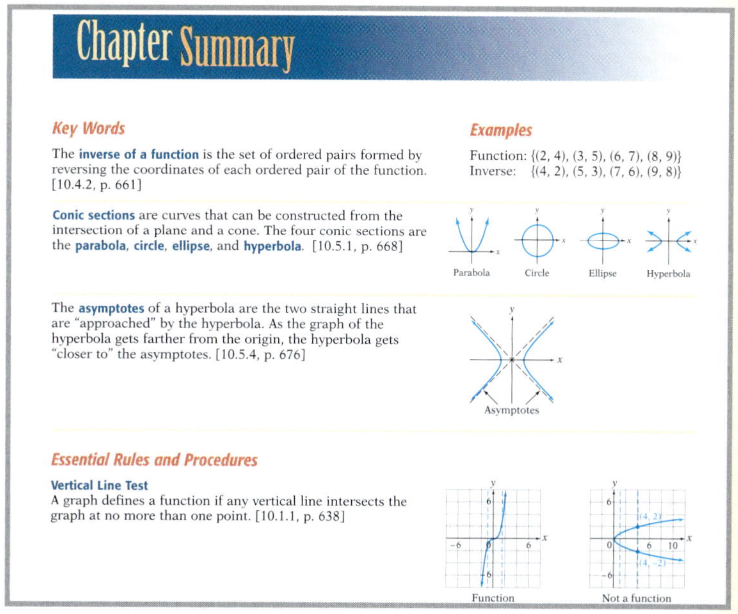

Chapter Summary

Key Words

The **inverse of a function** is the set of ordered pairs formed by reversing the coordinates of each ordered pair of the function. [10.4.2, p. 661]

Conic sections are curves that can be constructed from the intersection of a plane and a cone. The four conic sections are the **parabola**, **circle**, **ellipse**, and **hyperbola**. [10.5.1, p. 668]

The **asymptotes** of a hyperbola are the two straight lines that are "approached" by the hyperbola. As the graph of the hyperbola gets farther from the origin, the hyperbola gets "closer to" the asymptotes. [10.5.4, p. 676]

Examples

Function: {(2, 4), (3, 5), (6, 7), (8, 9)}
Inverse: {(4, 2), (5, 3), (7, 6), (9, 8)}

Parabola Circle Ellipse Hyperbola

Asymptotes

Essential Rules and Procedures

Vertical Line Test
A graph defines a function if any vertical line intersects the graph at no more than one point. [10.1.1, p. 638]

Function Not a function

page 687

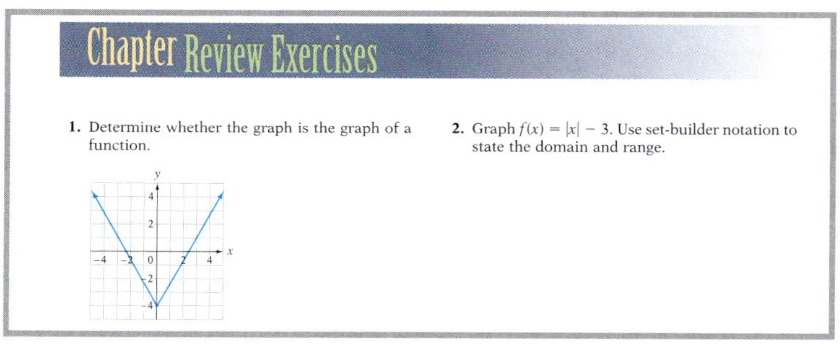

Chapter Review Exercises

1. Determine whether the graph is the graph of a function.

2. Graph $f(x) = |x| - 3$. Use set-builder notation to state the domain and range.

page 690

Chapter Review Exercises

Review exercises are found at the end of each chapter. These exercises are selected to help the student integrate all of the topics presented in the chapter.

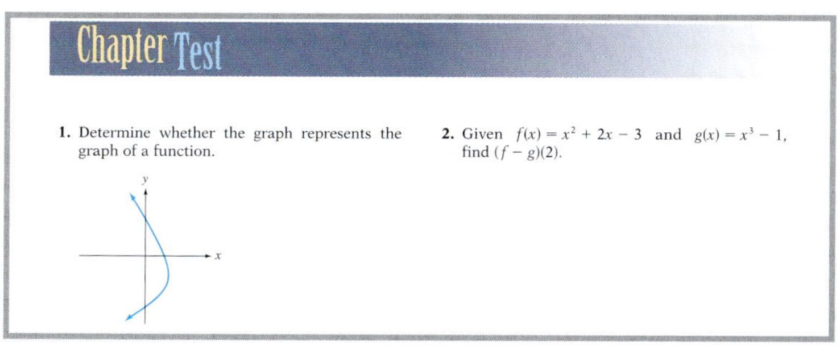

Chapter Test

1. Determine whether the graph represents the graph of a function.

2. Given $f(x) = x^2 + 2x - 3$ and $g(x) = x^3 - 1$, find $(f - g)(2)$.

page 693

Chapter Test

The Chapter Test exercises are designed to simulate a possible test of the material in the chapter.

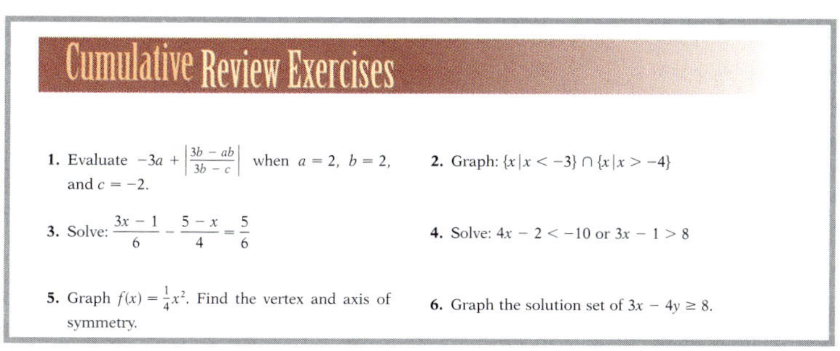

Cumulative Review Exercises

1. Evaluate $-3a + \left|\frac{3b - ab}{3b - c}\right|$ when $a = 2$, $b = 2$, and $c = -2$.

2. Graph: $\{x \mid x < -3\} \cap \{x \mid x > -4\}$

3. Solve: $\frac{3x - 1}{6} - \frac{5 - x}{4} = \frac{5}{6}$

4. Solve: $4x - 2 < -10$ or $3x - 1 > 8$

5. Graph $f(x) = \frac{1}{4}x^2$. Find the vertex and axis of symmetry.

6. Graph the solution set of $3x - 4y \geq 8$.

page 695

Cumulative Review Exercises

Cumulative Review Exercises, which appear at the end of each chapter (beginning with Chapter 2), help students maintain skills learned in previous chapters.

The answers to all Chapter Review Exercises, all Chapter Test exercises, and all Cumulative Review Exercises are given in the Answer Section. Along with the answer, there is a reference to the objective that pertains to each exercise.

Instructor Resources

Algebra: Beginning and Intermediate has a complete set of teaching aids for the instructor.

Instructor's Annotated Edition This edition contains a replica of the student text and additional items just for the instructor. Answers to all exercises are provided.

Instructor's Solutions Manual The *Instructor's Solutions Manual* contains worked-out solutions for all exercises in the text.

Instructor's Resource Manual with Testing This resource includes ready-to-use printed Chapter Tests, cumulative tests, and final exams, as well as a printout of one example of each of the algorithmic items in *HM Testing*. A lesson plan for the *AIM for Success* and suggested course sequences are also provided.

HM ClassPrep with HM Testing CD-ROM *HM ClassPrep* contains a multitude of text-specific reources for instructors to use to enhance the classroom experience. These resources can be easily accessed by chapter or resource type and can also link you to the text's website. *HM Testing* is our computerized test generator and contains a database of algorithmic test items as well as providing **online testing** and **gradebook** functions.

Instructor Text-specific Website The resources available on the *Class Prep CD* are also available on the instructor website at math.college.hmco.com/instructors. Appropriate items are password protected. Instructors also have access to the student part of the text's web site.

Student Resources

Student Solutions Manual The *Student Solutions Manual* contains complete solutions to all odd-numbered exercises in the text.

Math Study Skills Workbook *by Paul D. Nolting* This workbook is designed to reinforce skills and minimize frustration for students in any math class, lab, or study skills course. It offers a wealth of study tips and sound advice on note taking, time mangement, and reducing math anxiety. In addition, numerous opportunities for self-assessment enable students to track their own progress.

HM Eduspace® online learning environment *Eduspace* is a text-specific online learning environment that combines an algorithmic tutorial program with homework capabilities. Specific content is available 24 hours a day to help students further understand the textbook.

SMARTHINKING™ live, online tutoring Houghton Mifflin has partnered with SMARTHINKING to provide an easy-to-use and effective online tutorial service. **Whiteboard Simulations** and **Practice Area** promote real-time visual interaction.

Three levels of service are offered:

- **Text-specific Tutoring** provides real-time, one-on-one instruction with a specially qualified 'e-structor.'

- **Questions Any Time** allows students to submit questions to the tutor outside the scheduled hours and receive a reply within 24 hours.
- **Independent Study Resources** connect students with around-the-clock access to additional educational services, including interactive websites, diagnostic tests, and Frequently Asked Questions posed to SMARTHINKING e-structors.

HM mathSpace® Student Tutorial CD-ROM This tutorial algorithmically generates excercises and provides step-by-step solutions.

Houghton Mifflin Instructional Videos and DVDs This first edition offers text-specific videos and DVDs, hosted by Dana Mosely, covering all sections of the text and providing a valuable resource for further instruction and review. Next to every objective head, the 

serves as a reminder that the objective is covered in a video/DVD lesson.

Student Text-specific website Online student resources can be found at this text's website at math.college.hmco.com/students.

AIM for Success

Welcome to *Algebra: Beginning and Intermediate*. As you begin this course, we know two important facts: (1) We want you to succeed. (2) You want to succeed. To do that requires an effort from each of us. For the next few pages, we are going to show you what is required of you to achieve that success and how you can use the features of this text to be successful.

Motivation

One of the most important keys to success is motivation. We can try to motivate you by offering interesting or important ways mathematics can benefit you. But, in the end, the motivation must come from you. On the first day of class, it is easy to be motivated. Eight weeks into the term, it is harder to keep that motivation.

To stay motivated, there must be outcomes from this course that are worth your time, money, and energy. List some reasons you are taking this course. Do not make a mental list—actually write them out.

> **Take Note**
>
> Motivation alone will not lead to success. For instance, suppose a person who cannot swim is placed in a boat, taken out to the middle of a lake, and then thrown overboard. That person has a lot of motivation but there is a high likelihood the person will drown without some help. Motivation gives us the desire to learn but is not the same as learning.

____

____

____

Although we hope that one of the reasons you listed was an interest in mathematics, we know that many of you are taking this course because it is required to graduate, it is a prerequisite for a course you must take, or because it is required for your major. Although you may not agree that this course is necessary, it is! If you are motivated to graduate or complete the requirements for your major, then use that motivation to succeed in this course. Do not become distracted from your goal to complete your education!

Commitment

To be successful, you must make a committment to succeed. This means devoting time to math so that you achieve a better understanding of the subject.

List some activities (sports, hobbies, talents such as dance, art, or music) that you enjoy and at which you would like to become better.

ACTIVITY	*TIME SPENT*	*TIME WISHED SPENT*

Thinking about these activities, put the number of hours that you spend each week practicing these activities next to the activity. Next to that number, indicate the number of hours per week you would like to spend on these activities.

Whether you listed surfing or sailing, aerobics or restoring cars, or any other activity you enjoy, note how many hours a week you spend doing it. To succeed in math, you must be willing to commit the same amount of time. Success requires some sacrifice.

The "I Can't Do Math" Syndrome

There may be things you cannot do, such as lift a two-ton boulder. You can, however, do math. It is much easier than lifting the two-ton boulder. When you first learned the activities you listed above, you probably could not do them well. With practice, you got better. With practice, you will be better at math. Stay focused, motivated, and committed to success.

It is difficult for us to emphasize how important it is to overcome the "I Can't Do Math" Syndrome. If you listen to interviews of very successful athletes after a particularly bad performance, you will note that they focus on the positive aspect of what they did, not the negative. Sports psychologists encourage athletes to always be positive—to have a "Can Do" attitude. Develop this attitude toward math.

Strategies for Success

Textbook Reconnaissance Right now, do a 15-minute "textbook reconnaissance" of this book. Here's how:

First, read the table of contents. Do it in three minutes or less. Next, look through the entire book, page by page. Move quickly. Scan titles, look at pictures, notice diagrams.

A textbook reconnaissance shows you where a course is going. It gives you the big picture. That's useful because brains work best when going from the general to the specific. Getting the big picture before you start makes details easier to recall and understand later on.

Your textbook reconnaissance will work even better if, as you scan, you look for ideas or topics that are interesting to you. List three facts, topics, or problems that you found interesting during your textbook reconnaissance.

____

____

____

The idea behind this technique is simple: It's easier to work at learning material if you know it's going to be useful to you.

Not all the topics in this book will be "interesting" to you. But that is true of any subject. Surfers find that on some days the waves are better than others, musicians find some music more appealing than other music, computer gamers find some computer games more interesting than others, car enthusiasts find some cars more exciting than others. Some car enthusiasts would rather have a completely restored 1957 Chevrolet than a new Ferrari.

Know the Course Requirements To do your best in this course, you must know exactly what your instructor requires. Course requirements may be stated in a *syllabus*, which is a printed outline of the main topics of the course, or they may be presented orally. When they are listed in a syllabus or on other printed pages, keep them in a safe place. When they are presented orally, make sure to take complete notes. In either case, it is important that you understand them completely and follow them exactly. Be sure you know the answer to the following questions.

1. What is your instructor's name?
2. Where is your instructor's office?
3. At what times does your instructor hold office hours?
4. Besides the textbook, what other materials does your instructor require?
5. What is your instructor's attendance policy?
6. If you must be absent from a class meeting, what should you do before returning to class? What should you do when you return to class?
7. What is the instructor's policy regarding collection or grading of homework assignments?

8. What options are available if you are having difficulty with an assignment? Is there a math tutoring center?
9. Is there a math lab at your school? Where is it located? What hours is it open?
10. What is the instructor's policy if you miss a quiz?
11. What is the instructor's policy if you miss an exam?
12. Where can you get help when studying for an exam?

Remember: Your instructor wants to see you succeed. If you need help, ask! Do not fall behind. If you are running a race and fall behind by 100 yards, you may be able to catch up but it will require more effort than had you not fallen behind.

Take Note

Besides time management, there must be realistic ideas of how much time is available. There are very few people who can *successfully* work full-time and go to school full-time. If you work 40 hours a week, take 15 units, spend the recommended study time given at the right, and sleep 8 hours a day, you will use over 80% of the available hours in a week. That leaves less than 20% of the hours in a week for family, friends, eating, recreation, and other activities.

Time Management We know that there are demands on your time. Family, work, friends, and entertainment all compete for your time. We do not want to see you receive poor job evaluations because you are studying math. However, it is also true that we do not want to see you receive poor math test scores because you devoted too much time to work. When several competing and important tasks require your time and energy, the only way to manage the stress of being successful at both is to manage your time efficiently.

Instructors often advise students to spend twice the amount of time outside of class studying as they spend in the classroom. Time management is important if you are to accomplish this goal and succeed in school. The following activity is intended to help you structure your time more efficiently.

List the name of each course you are taking this term, the number of class hours each course meets, and the number of hours you should spend studying each subject outside of class. Then fill in a weekly schedule like the one below. Begin by writing in the hours spent in your classes, the hours spent at work (if you have a job), and any other commitments that are not flexible with respect to the time that you do them. Then begin to write down commitments that are more flexible, including hours spent studying. Remember to reserve time for activities such as meals and exercise. You should also schedule free time.

	Monday	Tuesday	Wednesday	Thursday	Friday	Saturday	Sunday
7–8 a.m.							
8–9 a.m.							
9–10 a.m.							
10–11 a.m.							
11–12 p.m.							
12–1 p.m.							
1–2 p.m.							
2–3 p.m.							
3–4 p.m.							
4–5 p.m.							
5–6 p.m.							
6–7 p.m.							
7–8 p.m.							
8–9 p.m.							
9–10 p.m.							
10–11 p.m.							
11–12 a.m.							

We know that many of you must work. If that is the case, realize that working 10 hours a week at a part-time job is equivalent to taking a three-unit class. If you must work, consider letting your education progress at a slower rate to allow you to be successful at both work and school. There is no rule that says you must finish school in a certain time frame.

Schedule Study Time As we encouraged you to do by filling out the time management form above, schedule a certain time to study. You should think of this time the way you would the time for work or class—that is, reasons for missing study time should be as compelling as reasons for missing work or class. "I just didn't feel like it" is not a good reason to miss your scheduled study time.

Although this may seem like an obvious exercise, list a few reasons you might want to study.

____

____

____

Of course we have no way of knowing the reasons you listed, but from our experience one reason given quite frequently is "To pass the course." There is nothing wrong with that reason. If that is the most important reason for you to study, then use it to stay focused.

One method of keeping to a study schedule is to form a ***study group***. Look for people who are committed to learning, who pay attention in class, and who are punctual. Ask them to join your group. Choose people with similar educational goals but different methods of learning. You can gain insight from seeing the material from a new perspective. Limit groups to four or five people; larger groups are unwieldy.

There are many ways to conduct a study group. Begin with the following suggestions and see what works best for your group.

1. Test each other by asking questions. Each group member might bring two or three sample test questions to each meeting.
2. Practice teaching each other. Many of us who are teachers learned a lot about our subject when we had to explain it to someone else.
3. Compare class notes. You might ask other students about material in your notes that is difficult for you to understand.
4. Brainstorm test questions.
5. Set an agenda for each meeting. Set approximate time limits for each agenda item and determine a quitting time.

And finally, probably the most important aspect of studying is that it should be done in relatively small chunks. If you can only study three hours a week for this course (probably not enough for most people), do it in blocks of one hour on three separate days, preferably after class. Three hours of studying on a Sunday is not as productive as three hours of paced study.

Text Features That Promote Success

There are 11 chapters in this text. Each chapter is divided into sections, and each section is subdivided into learning objectives. Each learning objective is labeled with a number from 1 to 5.

Preparing for a Chapter Before you begin a new chapter, you should take some time to review previously learned skills. There are two ways to do this. The first is to complete the ***Cumulative Review***, which occurs after every chapter (except Chapter 1). For instance, turn to page 394. The questions in this review are taken from the previous chapters. The answers for all these exercises can be found on page A24. Turn to that page now and locate the answers for the Chapter 5 Cumulative Review. After the answer to the first exercise, which is 4, you will see the objective reference [1.4.2]. This means that this question was taken from Chapter 1, Section 4, Objective 2. If you missed this question, you should return to that objective and restudy the material.

A second way of preparing for a new chapter is to complete the ***Prep Test***. This test focuses on the particular skills that will be required for the new chapter. Turn to page 342 to see a Prep Test. The answers for the Prep Test are the first set of answers in the answer section for a chapter. Turn to page A21 to see the answers for the Chapter 5 Prep Test. Note that an objective reference is given for each question. If you answer a question incorrectly, restudy the objective from which the question was taken.

Before the class meeting in which your professor begins a new section, you should read each objective statement for that section. Next, browse through the objective material, being sure to note each word in bold type. These words indicate important concepts that you must know in order to learn the material. Do not worry about trying to understand all the material. Your professor is there to assist you with that endeavor. The purpose of browsing through the material is so that your brain will be prepared to accept and organize the new information when it is presented to you.

Turn to page 2. Write down the title of the first objective in Section 1.1. Under the title of the objective, write down the words in the objective that are in bold print. It is not necessary for you to understand the meaning of these words. You are in this class to learn their meaning.

_______________ _______________ _______________ _______________

_______________ _______________ _______________ _______________

_______________ _______________ _______________ _______________

_______________ _______________ _______________ _______________

_______________ _______________ _______________ _______________

Math is Not a Spectator Sport To learn mathematics you must be an active participant. Listening and watching your professor do mathematics is not enough. Mathematics requires that you interact with the lesson you are studying. If you filled in the blanks above, you were being interactive. There are other ways this textbook has been designed to help you be an active learner.

Instructional Examples Colored brackets indicate an instructional example that usually has explanatory remarks to the left of the solution. These examples are used for two purposes. The first is to provide additional examples of important concepts. Secondly, these examples examine important techniques or principles that are often used in the solution of other types of problems.

$$\frac{2}{5}x - 3 = -7$$
$$\frac{2}{5}x - 3 + 3 = -7 + 3$$
$$\frac{2}{5}x = -4$$
$$\frac{5}{2} \cdot \frac{2}{5}x = \frac{5}{2}(-4)$$
$$x = -10$$

The solution is −10.

Solve: $\frac{2}{5}x - 3 = -7$

	$\frac{2}{5}x - 3 = -7$
Add 3 to each side of the equation.	$\frac{2}{5}x - 3 + 3 = -7 + 3$
Simplify.	$\frac{2}{5}x = -4$
Multiply each side of the equation by the reciprocal of the coefficient $\frac{2}{5}$.	$\frac{5}{2} \cdot \frac{2}{5}x = \frac{5}{2}(-4)$
Simplify. Now the equation is in the form *variable = constant*.	$x = -10$

Check $\frac{2}{5}x - 3 = -7$

$$\frac{2}{5}(-10) - 3 \;\Big|\; -7$$
$$-4 - 3 \;\Big|\; -7$$
$$-7 = -7$$

Write the solution. The solution is −10.

page 93

When you complete the example, get a clean sheet of paper. Write down the problem and then try to complete the solution without referring to your notes or the book. When you can do that, move on to the next part of the objective.

Leaf through the book now and write down the page numbers of two other occurrences of an instructional example.

Example/Problem Pairs One of the key instructional features of this text is Example/Problem pairs. Note that each Example is completely worked out and the "Problem" following the example is not. Study the worked-out example carefully by working through each step. Then work the Problem. If you get stuck, refer to the page number following the Problem which directs you to the page on which the Problem is solved—a complete worked-out solution is provided. Try to use the given solution to get a hint for the step you are stuck on. Then try to complete your solution.

Example 2 Solve: $5 = 9 - 2x$

Solution

$5 = 9 - 2x$	
$5 - 9 = 9 - 9 - 2x$	• Subtract 9 from each side of the equation.
$-4 = -2x$	• Simplify.
$\frac{-4}{-2} = \frac{-2x}{-2}$	• Divide each side of the equation by −2.
$2 = x$	• Simplify.
The solution is 2.	• Write the solution.

Problem 2 Solve: $11 = 11 + 3x$

Solution See page S5.

Problem 2:

$$11 = 11 + 3x$$
$$11 - 11 = 11 - 11 + 3x$$
$$0 = 3x$$
$$\frac{0}{3} = \frac{3x}{3}$$
$$0 = x$$

The solution is 0.

page 94

When you have completed your solution, check your work against the solution we provided. (Turn to page S5 to see the solution of Problem 2.) Be aware that frequently there is more than one way to solve a problem. Your answer, however, should be the same as the given answer. If you have any question as to whether your method will "always work," check with your instructor or with someone in the math center.

Browse through the textbook and write down the page numbers where two other Example/Problem pairs occur.

Remember: Be an active participant in your learning process. When you are sitting in class watching and listening to an explanation, you may think that you understand. However, until you actually try to do it, you will have no confirmation of the new knowledge or skill. Most of us have had the experience of sitting in class thinking we knew how to do something only to get home and realize that we didn't.

Take Note

There is a strong connection between reading and being a successful student in math or in any other subject. If you have difficulty reading, consider taking a reading course. Reading is much like other skills. There are certain things you can learn that will make you a better reader.

Word Problems Word problems are difficult because we must read the problem, determine the quantity we must find, think of a method to do that, and then actually solve the problem. In short, we must formulate a *strategy* to solve the problem and then devise a *solution*.

Note in the Example/Problem pair below that part of every word problem is a strategy and part is a solution. The strategy is a written description of how we will solve the problem. In the corresponding Problem, you are asked to formulate a strategy. Do not skip this step, and be sure to write it out.

Example 9 During a recent year, nearly 1.2 million dogs or litters were registered with the American Kennel Club. The most popular breed was the Labrador retriever, with 172,841 registered. What percent of the registrations were Labrador retrievers? Round to the nearest tenth of a percent. (*Source:* American Kennel Club)

Strategy To find the percent, solve the basic percent equation using $B = 1.2$ million $= 1{,}200{,}000$ and $A = 172{,}841$. The percent is unknown.

Solution

$$PB = A$$
$$P(1{,}200{,}000) = 172{,}841$$
$$\frac{P(1{,}200{,}000)}{1{,}200{,}000} = \frac{172{,}841}{1{,}200{,}000}$$
$$P \approx 0.144$$
$$P \approx 14.4\%$$

Approximately 14.4% of the registrations were Labrador retrievers.

Problem 9 A student correctly answered 72 of the 80 questions on an exam. What percent of the questions were answered correctly?

Solution See page S5.

page 83

Problem 9:

Strategy:

To find the percent of questions answers correctly, solve the basic percent equation using $B = 80$ and $A = 72$.

Solution:

$$PB = A$$
$$P(80) = 72$$
$$80P = 72$$
$$\frac{80P}{80} = \frac{72}{80}$$
$$P = 0.9$$

90% of the questions were answered correctly.

Take Note

If a rule has more than one part, be sure to make a notation to that effect.

Rule Boxes Pay special attention to rules placed in boxes. These rules give you the reasons certain types of problems are solved the way they are. When you see a rule, try to rewrite the rule in your own words.

Multiplying each side of an equation by the same number (NOT 0) will not change the solution.

Multiplication Property of Equations

Each side of an equation can be multiplied by the same nonzero number without changing the solution of the equation.

page 79

Find and write down two page numbers on which there are examples of rule boxes.

Chapter Exercises When you have completed studying an objective, do the exercises in the exercise set that correspond with that objective. The exercises are labeled with the same number as the objective. Math is a subject that needs to be learned in small sections and practiced continually in order to be mastered. Doing all of the exercises in each exercise set will help you master the problem-solving techniques necessary for success. As you work through the exercises for an objective, check your answers to the odd-numbered exercises with those in the back of the book.

Preparing for a Test There are important features of this text that can be used to prepare for a test.

- Chapter Summary
- Chapter Review Exercises
- Chapter Test

After completing a chapter, read the Chapter Summary. (See page 390 for the Chapter 5 Summary.) This summary highlights the important topics covered in the chapter. The page number following each topic refers you to the page in the text on which you can find more information about the concept.

Following the Chapter Summary are Chapter Review Exercises (see page 391) and a Chapter Test (see page 393). Doing the review exercises is an important way of testing your understanding of the chapter. The answer to each review exercise is given at the back of the book, along with its objective reference. After checking your answers, restudy any objective from which a question you missed was taken. It may be helpful to retry some of the exercises for that objective to reinforce your problem-solving techniques.

The Chapter Test should be used to prepare for an exam. We suggest that you try the Chapter Test a few days before your actual exam. Take the test in a quiet place and try to complete the test in the same amount of time you will be allowed for your exam. When taking the Chapter Test, practice the strategies of successful test takers: (1) scan the entire test to get a feel for the questions; (2) read the directions carefully; (3) work the problems that are easiest for you first; and perhaps most importantly, (4) try to stay calm.

When you have completed the Chapter Test, check your answers. If you missed a question, review the material in that objective and rework some of the exercises from that objective. This will strengthen your ability to perform the skills in that objective.

Your career goal here. →

Is it difficult to be successful? YES! Successful music groups, artists, professional athletes, chefs, and ______________ have to work very hard to achieve their goals. They focus on their goals and ignore distractions. The things we ask you to do to achieve success take time and commitment. We are confident that if you follow our suggestions, you will succeed.

Algebra: Beginning and Intermediate

At the beginning of a Little League season, a schedule for all the teams in the league is created. Every team must play every other team at least once, and every team should play the same number of games. A pattern can be used to set up a schedule that meets both of these requirements. In the **Project on page 65,** *a pattern is used to determine the number of games to be scheduled in a softball league.*

Applications

1 Real Numbers and Variable Expressions

Objectives

Section 1.1

1. Order relations
2. Opposites and absolute value

Section 1.2

1. Add integers
2. Subtract integers
3. Multiply integers
4. Divide integers
5. Application problems

Section 1.3

1. Write rational numbers as decimals
2. Add and subtract rational numbers
3. Multiply and divide rational numbers
4. Convert among percents, fractions, and decimals

Section 1.4

1. Exponential expressions
2. The Order of Operations Agreement

Section 1.5

1. Evaluate variable expressions
2. The Properties of Real Numbers
3. Simplify variable expressions using the Properties of Addition
4. Simplify variable expressions using the Properties of Multiplication
5. Simplify variable expressions using the Distributive Property
6. Simplify general variable expressions

Section 1.6

1. Translate a verbal expression into a variable expression
2. Translate a verbal expression into a variable expression and then simplify the resulting expression
3. Translate application problems

WWW WEB **Need help? For online student resources, such as section quizzes, visit this web site: math.college.hmco.com**

PrepTest

1. What is 127.1649 rounded to the nearest hundredth?

2. Add: 49,147 + 596

3. Subtract: 5004 − 487

4. Multiply: 407 × 28

5. Divide: 456 ÷ 19

6. What is the smallest number into which both 8 and 12 divide evenly?

7. What is the greatest number that divides evenly into both 16 and 20?

8. Without using 1, write 21 as a product of two whole numbers.

9. Represent the shaded portion of the figure as a fraction.

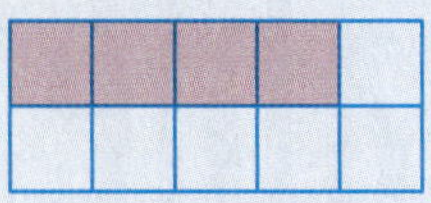

Go Figure

If you multiply the first 20 natural numbers ($1 \cdot 2 \cdot 3 \cdot 4 \cdot 5 \cdot \cdots \cdot 17 \cdot 18 \cdot 19 \cdot 20$), how many zeros will be at the end of the number?

1.1 Introduction to Integers

1 Order relations

VIDEO & DVD CD TUTOR WEB SSM

It seems to be a human characteristic to group similar items. For instance, a botanist places plants with similar characteristics in groups called species. Nutritionists classify foods according to food groups; for example, pasta, crackers, and rice are among the foods in the bread group.

Mathematicians place objects with similar properties in groups called sets. A **set** is a collection of objects. The objects in a set are called **elements** of the set.

The **roster method** of writing sets encloses a list of the elements in braces. The set of sections within an orchestra is written {brass, percussion, strings, woodwinds}. When the elements of a set are listed, each element is listed only once. For instance, if the list of numbers 1, 2, 3, 2, 3 were placed in a set, the set would be {1, 2, 3}.

The numbers that we use to count objects, such as the number of students in a classroom or the number of people living in an apartment house, are the natural numbers.

Natural numbers = {1, 2, 3, 4, 5, 6, 7, 8, 9, 10, . . .}

The three dots mean that the list of natural numbers continues on and on and that there is no largest natural number.

Point of Interest

The Alexandrian astronomer Ptolemy began using omicron, o, the first letter of the Greek word that means "nothing," as the symbol for zero in A.D. 150. It was not until the 13th century, however, that Fibonacci introduced 0 to the Western world as a placeholder so that we could distinguish, for example, 45 from 405.

The natural numbers do not include zero. The number zero is used for describing data such as the number of people who have run a two-minute mile and the number of college students at Providence College who are under the age of 10. The set of whole numbers includes the natural numbers and zero.

$$\textbf{Whole numbers} = \{0, 1, 2, 3, 4, 5, 6, 7, \ldots\}$$

The whole numbers do not provide all the numbers that are useful in applications. For instance, a meteorologist also needs numbers below zero.

$$\textbf{Integers} = \{\ldots, -5, -4, -3, -2, -1, 0, 1, 2, 3, 4, 5, \ldots\}$$

Each integer can be shown on a number line. The integers to the left of zero on the number line are called **negative integers.** The integers to the right of zero are called **positive integers,** or natural numbers. Zero is neither a positive nor a negative integer.

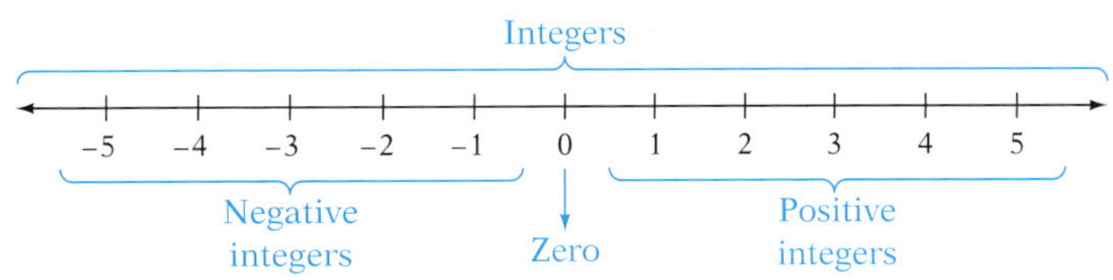

The **graph of an integer** is shown by placing a heavy dot on the number line directly above the number. The graphs of −3 and 4 are shown on the number line below.

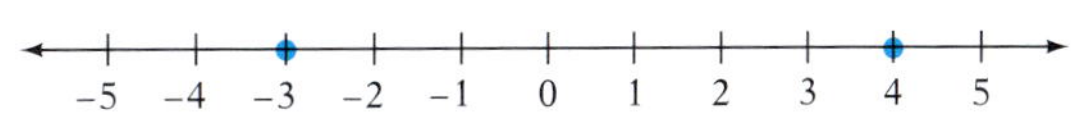

Consider the following sentences.

The quarterback threw the football, and the receiver caught *it*.

A student purchased a computer and used *it* to write English and history papers.

In the first sentence, *it* is used to mean the football; in the second sentence, *it* means the computer. In language, the word *it* can stand for many different objects. Similarly, in mathematics, a letter of the alphabet can be used to stand for a number. Such a letter is called a **variable.** Variables are used in the following definition of inequality symbols.

Definition of Inequality Symbols

If a and b are two numbers and a is to the left of b on the number line, then a **is less than** b. This is written $\boldsymbol{a < b}$.

If a and b are two numbers and a is to the right of b on the number line, then a **is greater than** b. This is written $\boldsymbol{a > b}$.

Point of Interest

The symbols for "is less than" and "is greater than" were introduced by Thomas Harriot around 1630. Before that, ⊏ and ⊐ were used for $<$ and $>$, respectively.

Negative 4 is less than negative 1.

$-4 < -1$

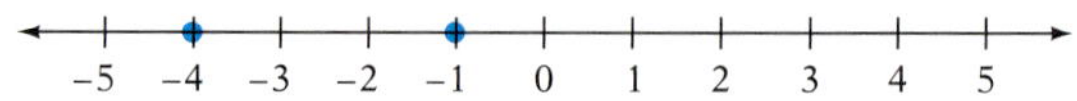

5 is greater than 0.

$5 > 0$

There are also inequality symbols for **is less than or equal to** ($\leq$) and **is greater than or equal to** ($\geq$).

$7 \leq 15$ 7 is less than or equal to 15. This is true because $7 < 15$.

$6 \leq 6$ 6 is less than or equal to 6. This is true because $6 = 6$.

Example 1 Use the roster method to write the set of negative integers greater than or equal to -6.

Solution $A = \{-6, -5, -4, -3, -2, -1\}$ • **A set is designated by a capital letter.**

Problem 1 Use the roster method to write the set of positive integers less than 5.

Solution See page S1.

Example 2 Given $A = \{-6, -2, 0\}$, which elements of set A are less than or equal to -2?

Solution $-6 < -2$ • **Find the order relation between each element of set *A* and -2.**

$-2 = -2$

$0 > -2$

The elements -6 and -2 are less than or equal to -2.

Problem 2 Given $B = \{-5, -1, 5\}$, which elements of set B are greater than -1?

Solution See page S1.

2 Opposites and absolute value

Two numbers that are the same distance from zero on the number line but are on opposite sides of zero are **opposite numbers,** or **opposites.** The opposite of a number is also called its **additive inverse.**

The opposite of 5 is -5.

The opposite of -5 is 5.

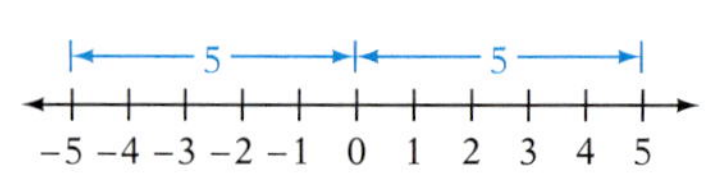

The negative sign can be read "the opposite of."

$-(2) = -2$ The opposite of 2 is -2.

$-(-2) = 2$ The opposite of -2 is 2.

Example 3 Find the opposite number. **A.** 6 **B.** -51

Solution **A.** -6 **B.** 51

Problem 3 Find the opposite number. **A.** -9 **B.** 62

Solution See page S1.

The **absolute value** of a number is its distance from zero on the number line. Therefore, the absolute value of a number is a positive number or zero. The symbol for absolute value is two vertical bars, $|\ |$.

The distance from 0 to 3 is 3. Therefore, the absolute value of 3 is 3.

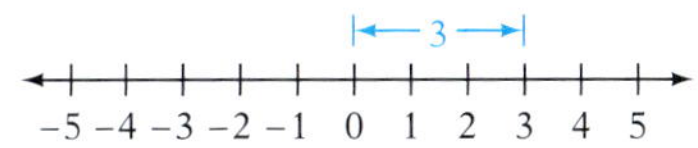

$$|3| = 3$$

The distance from 0 to -3 is 3. Therefore, the absolute value of -3 is 3.

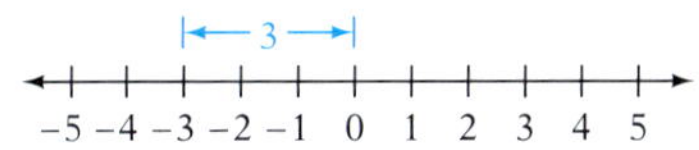

$$|-3| = 3$$

Point of Interest

The definition of absolute value given in the box is written in what is called rhetorical style. That is, it is written without the use of variables. This is how all mathematics was written prior to the Renaissance. During that period from the 14th to the 16th century, the idea of expressing a variable symbolically was developed. In terms of that symbolism, the definition of absolute value is

$$|x| = \begin{cases} x, & x > 0 \\ 0, & x = 0 \\ -x, & x < 0 \end{cases}$$

Absolute Value

The absolute value of a positive number is the number itself. The absolute value of zero is zero. The absolute value of a negative number is the opposite of the negative number.

Example 4 Evaluate. **A.** $|-4|$ **B.** $-|-10|$

Solution **A.** $|-4| = 4$

B. $-|-10| = -10$

- The absolute value sign does not affect the negative sign in front of the absolute value sign. You can read $-|-10|$ as "the opposite of the absolute value of negative 10."

Problem 4 Evaluate. **A.** $|-5|$ **B.** $-|-9|$

Solution See page S1.

1.1 Concept Review

Determine whether the statement is always true, sometimes true, or never true.

1. The absolute value of a number is positive.
2. The absolute value of a number is negative.
3. If x is an integer, then $|x| < -3$.
4. If x is an integer, then $|x| > -2$.
5. The opposite of a number is a positive number.
6. The set of positive integers is $\{0, 1, 2, 3, 4, \ldots\}$.

7. If a is an integer, then $a \leq a$.

8. If a and b are integers and $a > b$, then $a \geq b$.

9. If x is a negative integer, then $|x| = -x$.

10. Suppose a and b are integers and $a < b$. Then $|a| < |b|$.

1.1 Exercises

1 **1.** How do the whole numbers differ from the natural numbers?

2. Explain the difference between the symbols $<$ and $\leq$.

Place the correct symbol, $<$ or $>$, between the two numbers.

3. $-2 \quad -5$ **4.** $-6 \quad -1$ **5.** $-16 \quad 1$ **6.** $-2 \quad 13$

7. $3 \quad -7$ **8.** $5 \quad -6$ **9.** $0 \quad -3$ **10.** $8 \quad 0$

11. $-42 \quad 27$ **12.** $-36 \quad 49$ **13.** $21 \quad -34$ **14.** $53 \quad -46$

15. $-27 \quad -39$ **16.** $-51 \quad -20$ **17.** $-131 \quad 101$ **18.** $127 \quad -150$

Use the roster method to write the set.

19. the natural numbers less than 9

20. the natural numbers less than or equal to 6

21. the positive integers less than or equal to 8

22. the positive integers less than 4

23. the negative integers greater than -7

24. the negative integers greater than or equal to -5

Solve.

25. Given $A = \{-7, 0, 2, 5\}$, which elements of set A are greater than 2?

26. Given $B = \{-8, 0, 7, 15\}$, which elements of set B are greater than 7?

27. Given $D = \{-23, -18, -8, 0\}$, which elements of set D are less than -8?

28. Given $C = \{-33, -24, -10, 0\}$, which elements of set C are less than -10?

29. Given $E = \{-35, -13, 21, 37\}$, which elements of set E are greater than -10?

30. Given $F = \{-27, -14, 14, 27\}$, which elements of set F are greater than -15?

31. Given $B = \{-52, -46, 0, 39, 58\}$, which elements of set B are less than or equal to 0?

32. Given $A = \{-12, -9, 0, 12, 34\}$, which elements of set A are greater than or equal to 0?

33. Given $C = \{-23, -17, 0, 4, 29\}$, which elements of set C are greater than or equal to -17?

34. Given $D = \{-31, -12, 0, 11, 45\}$, which elements of set D are less than or equal to -12?

35. Given that set A is the positive integers less than 10, which elements of set A are greater than or equal to 5?

36. Given that set B is the positive integers less than or equal to 12, which elements of set B are greater than 6?

37. Given that set D is the negative integers greater than or equal to -10, which elements of set D are less than -4?

38. Given that set C is the negative integers greater than -8, which elements of set C are less than or equal to -3?

2 **39.** What is the additive inverse of a number?

40. What is the absolute value of a number?

Find the opposite number.

41. 22 **42.** 45 **43.** -31 **44.** -88

45. -168 **46.** -97 **47.** 630 **48.** 450

Evaluate.

49. $-(-18)$ **50.** $-(-30)$ **51.** $-(49)$ **52.** $-(67)$

53. $|16|$ **54.** $|19|$ **55.** $|-12|$ **56.** $|-22|$

57. $-|29|$ **58.** $-|20|$ **59.** $-|-14|$ **60.** $-|-18|$

61. $-|0|$ **62.** $|-30|$ **63.** $-|34|$ **64.** $-|-45|$

Solve.

65. Given $A = \{-8, -5, -2, 1, 3\}$, find
a. the opposite of each element of set A.
b. the absolute value of each element of set A.

66. Given $B = \{-11, -7, -3, 1, 5\}$, find
a. the opposite of each element of set B.
b. the absolute value of each element of set B.

Place the correct symbol, $<$ or $>$, between the two numbers.

67. $|-83| \quad |58|$ **68.** $|22| \quad |-19|$ **69.** $|43| \quad |-52|$ **70.** $|-71| \quad |-92|$

71. $|-68| \quad |-42|$ **72.** $|12| \quad |-31|$ **73.** $|-45| \quad |-61|$ **74.** $|-28| \quad |43|$

1.1 Applying Concepts

Write the given numbers in order from least to greatest.

75. $|-5|, 6, -|-8|, -19$

76. $-4, |-15|, -|-7|, 0$

77. $-(-3), -22, |-25|, |-14|$

78. $-|-26|, -(-8), |-17|, -(5)$

A meteorologist may report a wind-chill temperature. This is the equivalent temperature, including the effects of wind and temperature, that a person would feel in calm air conditions. The table that follows gives the wind-chill temperature for various wind speeds and temperatures. For instance, when the temperature is 5°F and the wind is blowing at 15 mph, the wind-chill temperature is −25°F. Use this table for Exercises 79 and 80. (*Source: Information Please Almanac*)

Wind-Chill Factors

Wind Speed (in mph)	*Thermometer Reading (in degrees Fahrenheit)*																
	35	**30**	**25**	**20**	**15**	**10**	**5**	**0**	**−5**	**−10**	**−15**	**−20**	**−25**	**−30**	**−35**	**−40**	**−45**
5	33	27	21	19	12	7	0	−5	−10	−15	−21	−26	−31	−36	−42	−47	−52
10	22	16	10	3	−3	−9	−15	−22	−27	−34	−40	−46	−52	−58	−64	−71	−77
15	16	9	2	−5	−11	−18	−25	−31	−38	−45	−51	−58	−65	−72	−78	−85	−92
20	12	4	−3	−10	−17	−24	−31	−39	−46	−53	−60	−67	−74	−81	−88	−95	−103
25	8	1	−7	−15	−22	−29	−36	−44	−51	−59	−66	−74	−81	−88	−96	−103	−110
30	6	−2	−10	−18	−25	−33	−41	−49	−56	−64	−71	−79	−86	−93	−101	−109	−116
35	4	−4	−12	−20	−27	−35	−43	−52	−58	−67	−74	−82	−89	−97	−105	−113	−120
40	3	−5	−13	−21	−29	−37	−45	−53	−60	−69	−76	−84	−92	−100	−107	−115	−123
45	2	−6	−14	−22	−30	−38	−46	−54	−62	−70	−78	−85	−93	−102	−109	−117	−125

79. **Meteorology** Which of the following weather conditions feels colder?

a. a temperature of 5°F with a 20 mph wind or a temperature of 10°F with a 15 mph wind

b. a temperature of −25°F with a 10 mph wind or a temperature of −15°F with a 20 mph wind

80. **Meteorology** Which of the following weather conditions feels warmer?

a. a temperature of 25°F with a 25 mph wind or a temperature of 10°F with a 10 mph wind

b. a temperature of −5°F with a 10 mph wind or a temperature of −15°F with a 5 mph wind

Complete.

81. On the number line, the two points that are four units from 0 are ______ and ______.

82. On the number line, the two points that are six units from 0 are ______ and ______.

83. On the number line, the two points that are seven units from 4 are ______ and ______.

84. On the number line, the two points that are five units from −3 are ______ and ______.

85. If a is a positive number, then $-a$ is a ______ number.

86. If a is a negative number, then $-a$ is a ______ number.

87. An integer that is its own opposite is ______.

88. True or False. If $a \geq 0$, then $|a| = a$.

89. True or False. If $a \leq 0$, then $|a| = -a$.

90. Give examples of some games that use negative integers in the scoring.

91. Give some examples of English words that are used as variables.

92. Write an essay describing your interpretation of the data presented in the table at the right.

The Twenty-First Century: An Easier Ladder to Climb
Number of 30- to 34-year-olds competing for jobs

	1992	*1999*	*2005*
Men	10.3 million	9.2 million	8.5 million
Women	8.3 million	7.9 million	7.8 million
Total	18.6 million	17.1 million	16.3 million

1.2 Operations with Integers

1 Add integers

VIDEO & DVD CD TUTOR WEB SSM

A number can be represented anywhere along the number line by an arrow. A positive number is represented by an arrow pointing to the right, and a negative number is represented by an arrow pointing to the left. The size of the number is represented by the length of the arrow.

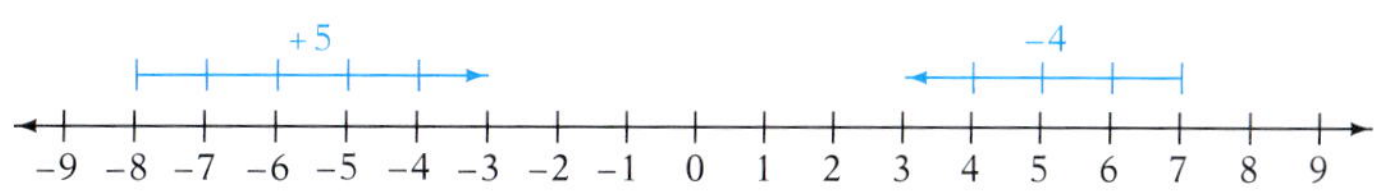

Addition is the process of finding the total of two numbers. The numbers being added are called **addends.** The total is called the **sum.** Addition of integers can be shown on the number line. To add integers, find the point on the number line corresponding to the first addend. At that point, draw an arrow representing the second addend. The sum is the number directly below the tip of the arrow.

$4 + 2 = 6$

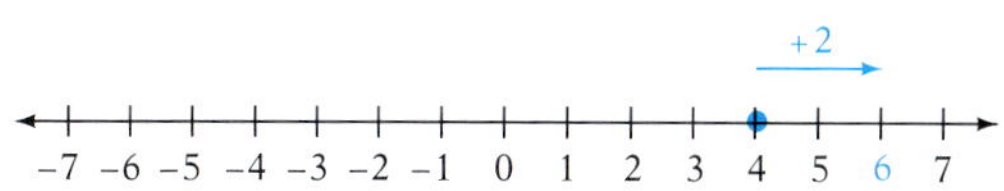

$-4 + (-2) = -6$

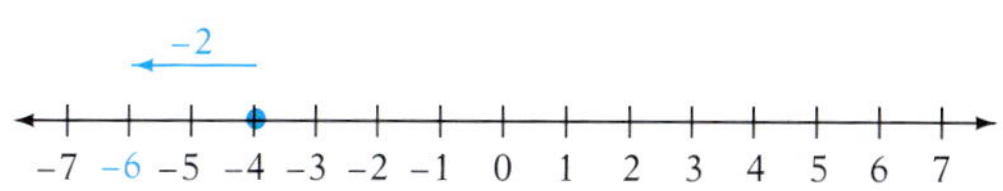

$-4 + 2 = -2$

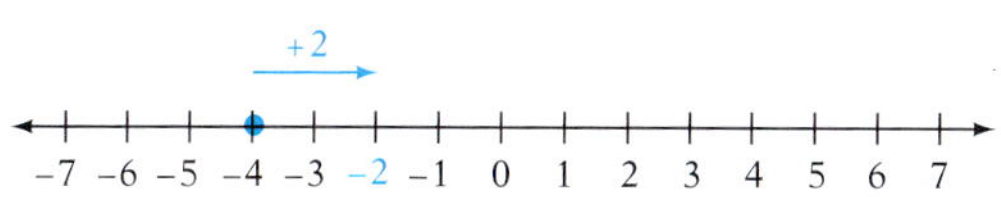

$4 + (-2) = 2$

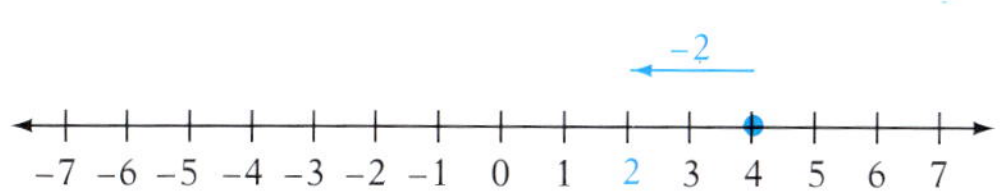

The pattern for addition shown on the preceding number lines is summarized in the following rules for adding integers.

Addition of Integers

Integers with the same sign

To add two numbers with the same sign, add the absolute values of the numbers. Then attach the sign of the addends.

$2 + 8 = 10$

$-2 + (-8) = -10$

Integers with different signs

To add two numbers with different signs, find the absolute value of each number. Then subtract the smaller of these absolute values from the larger one. Attach the sign of the number with the larger absolute value.

$-2 + 8 = 6$

$2 + (-8) = -6$

Example 1 Add. **A.** $162 + (-247)$ **B.** $-14 + (-47)$ **C.** $-4 + (-6) + (-8) + 9$

Solution

A. $162 + (-247) = -85$

- The signs are different. Subtract the absolute values of the numbers $(247 - 162)$. Attach the sign of the number with the larger absolute value.

B. $-14 + (-47) = -61$

- The signs are the same. Add the absolute values of the numbers $(14 + 47)$. Attach the sign of the addends.

C. $-4 + (-6) + (-8) + 9$
$= -10 + (-8) + 9$
$= -18 + 9$
$= -9$

- To add more than two numbers, add the first two numbers. Then add the sum to the third number. Continue until all the numbers have been added.

Problem 1 Add.

A. $-162 + 98$ **B.** $-154 + (-37)$ **C.** $-36 + 17 + (-21)$

Solution See page S1.

2 Subtract integers

VIDEO & DVD CD TUTOR WEB SSM

Subtraction is the process of finding the difference between two numbers. Subtraction of an integer is defined as addition of the opposite integer.

Subtract 8 − 3 by using addition of the opposite.

Subtraction → Addition of the Opposite

$$8 - (+3) = 8 + (-3) = 5$$

(+3) and (−3): Opposites

To subtract one integer from another, add the opposite of the second integer to the first integer.

first number	−	second number	=	first number	+	the opposite of the second number		
40	−	60	=	40	+	(−60)	=	−20
−40	−	60	=	−40	+	(−60)	=	−100
−40	−	(−60)	=	−40	+	60	=	20
40	−	(−60)	=	40	+	60	=	100

Example 2 Subtract: $-12 - 8$

Solution $-12 - 8 = -12 + (-8)$ • Rewrite subtraction as addition of the opposite.

$= -20$ • Add.

Problem 2 Subtract: $-8 - 14$

Solution See page S1.

When subtraction occurs several times in a problem, rewrite each subtraction as addition of the opposite. Then add.

Example 3 Subtract: $-8 - 30 - (-12) - 7 - (-14)$

Solution $-8 - 30 - (-12) - 7 - (-14)$

$= -8 + (-30) + 12 + (-7) + 14$ • Rewrite each subtraction as addition of the opposite.

$= -38 + 12 + (-7) + 14$ • Add the first two numbers. Then add the sum to the third number. Continue until all the numbers have been added.

$= -26 + (-7) + 14$

$= -33 + 14$

$= -19$

Problem 3 Subtract: $4 - (-3) - 12 - (-7) - 20$

Solution See page S1.

3 Multiply integers

VIDEO & DVD CD TUTOR WEB SSM

Point of Interest

The cross × was first used as a symbol for multiplication in 1631 in a book titled The Key to Mathematics. *In that same year, another book,* Practice of the Analytical Art, *advocated the use of a dot to indicate multiplication.*

Multiplication is the process of finding the product of two numbers.

Several different symbols are used to indicate multiplication. The numbers being multiplied are called **factors;** for instance, 3 and 2 are factors in each of the examples at the right. The result is called the **product.** Note that when parentheses are used and there is no arithmetic operation symbol, the operation is multiplication.

$3 \times 2 = 6$
$3 \cdot 2 = 6$
$(3)(2) = 6$
$3(2) = 6$
$(3)2 = 6$

When 5 is multiplied by a sequence of decreasing integers, each product decreases by 5.

$(5)(3) = 15$
$(5)(2) = 10$
$(5)(1) = 5$
$(5)(0) = 0$

The pattern developed can be continued so that 5 is multiplied by a sequence of negative numbers. The resulting products must be negative in order to maintain the pattern of decreasing by 5.

$(5)(-1) = -5$
$(5)(-2) = -10$
$(5)(-3) = -15$
$(5)(-4) = -20$

This illustrates that the product of a positive number and a negative number is negative.

When -5 is multiplied by a sequence of decreasing integers, each product increases by 5.

$(-5)(3) = -15$
$(-5)(2) = -10$
$(-5)(1) = -5$
$(-5)(0) = 0$

The pattern developed can be continued so that -5 is multiplied by a sequence of negative numbers. The resulting products must be positive in order to maintain the pattern of increasing by 5.

$(-5)(-1) = 5$
$(-5)(-2) = 10$
$(-5)(-3) = 15$
$(-5)(-4) = 20$

This illustrates that the product of two negative numbers is positive.

This pattern for multiplication is summarized in the following rules for multiplying integers.

Multiplication of Integers

Integers with the same sign
To multiply two numbers with the same sign, multiply the absolute values of the numbers. The product is positive.

$4 \cdot 8 = 32$
$(-4)(-8) = 32$

Integers with different signs
To multiply two numbers with different signs, multiply the absolute values of the numbers. The product is negative.

$-4 \cdot 8 = -32$
$(4)(-8) = -32$

Example 4 Multiply. **A.** $-42 \cdot 62$ **B.** $2(-3)(-5)(-7)$

Solution **A.** $-42 \cdot 62$
$= -2604$

• The signs are different. The product is negative.

B. $2(-3)(-5)(-7)$
$= -6(-5)(-7)$
$= 30(-7)$
$= -210$

• To multiply more than two numbers, multiply the first two numbers. Then multiply the product by the third number. Continue until all the numbers have been multiplied.

Problem 4 Multiply. **A.** $-38 \cdot 51$ **B.** $-7(-8)(9)(-2)$

Solution See page S1.

4 Divide integers

VIDEO & DVD CD TUTOR WEB SSM

For every division problem there is a related multiplication problem.

Division: $\frac{8}{2} = 4$ Related multiplication: $4 \cdot 2 = 8$

This fact can be used to illustrate the rules for dividing signed numbers.

The quotient of two numbers with the same sign is positive.

$\frac{12}{3} = 4$ because $4 \cdot 3 = 12$.

$\frac{-12}{-3} = 4$ because $4(-3) = -12$.

The quotient of two numbers with different signs is negative.

$\frac{12}{-3} = -4$ because $-4(-3) = 12$.

$\frac{-12}{3} = -4$ because $-4 \cdot 3 = -12$.

Division of Integers

Integers with the same sign

To divide two numbers with the same sign, divide the absolute values of the numbers. The quotient is positive.

Integers with different signs

To divide two numbers with different signs, divide the absolute values of the numbers. The quotient is negative.

Note that $\frac{-12}{3} = -4$, $\frac{12}{-3} = -4$, and $-\frac{12}{3} = -4$. This suggests the following rule.

If a and b are two integers, and $b \neq 0$, then $\frac{a}{-b} = \frac{-a}{b} = -\frac{a}{b}$.

Read $b \neq 0$ as "b is not equal to 0." The reason why the denominator must not be equal to 0 is explained in the following discussion of zero and one in division.

Zero and One in Division

Zero divided by any number other than zero is zero.	$\frac{0}{a} = 0, a \neq 0$	because $0 \cdot a = 0$.
Division by zero is not defined.	$\frac{4}{0} = ?$	$? \times 0 = 4$ There is no number whose product with zero is 4.
Any number other than zero divided by itself is 1.	$\frac{a}{a} = 1, a \neq 0$	because $1 \cdot a = a$.
Any number divided by 1 is the number.	$\frac{a}{1} = a$	because $a \cdot 1 = a$.

Example 5 Divide. **A.** $(-120) \div (-8)$ **B.** $\frac{95}{-5}$ **C.** $-\frac{-81}{3}$

Solution **A.** $(-120) \div (-8) = 15$

B. $\frac{95}{-5} = -19$

C. $-\frac{-81}{3} = -(-27) = 27$

Problem 5 Divide. **A.** $(-135) \div (-9)$ **B.** $\frac{84}{-6}$ **C.** $-\frac{36}{-12}$

Solution See page S1.

5 Application problems

VIDEO & DVD CD TUTOR WEB SSM

In many courses, your course grade depends on the *average* of all your test scores. You compute the average by calculating the sum of all your test scores and then dividing that result by the number of tests. Statisticians call this average an **arithmetic mean.** Besides its application to finding the average of your test scores, the arithmetic mean is used in many other situations.

Stock market analysts calculate the **moving average** of a stock. This is the arithmetic mean of the changes in the value of a stock for a given number of days. To illustrate the procedure, we will calculate the 5-day moving average of a stock. In actual practice, a stock market analyst may use 15 days, 30 days, or some other number.

The table below shows the amount of increase or decrease, in cents, in the closing price of a stock for a 10-day period.

Day 1	*Day 2*	*Day 3*	*Day 4*	*Day 5*	*Day 6*	*Day 7*	*Day 8*	*Day 9*	*Day 10*
+50	−175	+225	0	−275	−75	−50	+50	−475	−50

To calculate the 5-day moving average of this stock, determine the average of the stock for days 1 through 5, days 2 through 6, days 3 through 7, and so on.

Days 1–5	*Days 2–6*	*Days 3–7*	*Days 4–8*	*Days 5–9*	*Days 6–10*
+50	−175	+225	0	−275	−75
−175	+225	0	−275	−75	−50
+225	0	−275	−75	−50	+50
0	−275	−75	−50	+50	−475
−275	−75	−50	+50	−475	−50
Sum = −175	Sum = −300	Sum = −175	Sum = −350	Sum = −825	Sum = −600
$\text{Av} = \frac{-175}{5} = -35$	$\text{Av} = \frac{-300}{5} = -60$	$\text{Av} = \frac{-175}{5} = -35$	$\text{Av} = \frac{-350}{5} = -70$	$\text{Av} = \frac{-825}{5} = -165$	$\text{Av} = \frac{-600}{5} = -120$

The 5-day moving average is the list of means: −35, −60, −35, −70, −165, and −120. If the numbers in the list tend to increase, the price of the stock is showing an upward trend; if they decrease, the price of the stock is showing a downward trend. For the stock shown here, the price of the stock is showing a downward trend. These trends help analysts determine whether to recommend a stock.

Example 6 The daily low temperatures, in degrees Celsius, during one week were recorded as follows: −8°, 2°, 0°, −7°, 1°, 6°, −1°. Find the average daily low temperature for the week.

Strategy To find the average daily low temperature:

- ▶ Add the seven temperature readings.
- ▶ Divide the sum by 7.

Solution

$$
\begin{aligned}
&-8 + 2 + 0 + (-7) + 1 + 6 + (-1)\\
&\quad = -6 + 0 + (-7) + 1 + 6 + (-1)\\
&\quad = -6 + (-7) + 1 + 6 + (-1)\\
&\quad = -13 + 1 + 6 + (-1)\\
&\quad = -12 + 6 + (-1)\\
&\quad = -6 + (-1)\\
&\quad = -7
\end{aligned}
$$

$$-7 \div 7 = -1$$

The average daily low temperature was −1°C.

Problem 6 The daily high temperatures, in degrees Celsius, during one week were recorded as follows: −5°, −6°, 3°, 0°, −4°, −7°, −2°. Find the average daily high temperature for the week.

Solution See page S1.

1.2 Concept Review

Determine whether the statement is always true, sometimes true, or never true.

1. The sum of two integers is larger than either of the integers being added.
2. The sum of two nonzero integers with the same sign is positive.
3. The quotient of two integers with different signs is negative.
4. To find the opposite of a number, multiply the number by −1.
5. To find the opposite of an integer, divide the integer by −1.
6. To perform the subtraction $a - b$, add the opposite of a to b.
7. $4(-8) = -4$
8. If x and y are two integers and $y = 0$, then $\frac{x}{y} = 0$.

9. If x and y are two different integers, then $x - y = y - x$.

10. If x is an integer and $4x = 0$, then $x = 0$.

1.2 Exercises

1 **1.** Explain how to add two integers with the same sign.

2. Explain how to add two integers with different signs.

Add.

3. $-3 + (-8)$ **4.** $-12 + (-1)$ **5.** $-4 + (-5)$ **6.** $-12 + (-12)$

7. $6 + (-9)$ **8.** $4 + (-9)$ **9.** $-6 + 7$ **10.** $-12 + 6$

11. $2 + (-3) + (-4)$ **12.** $7 + (-2) + (-8)$ **13.** $-3 + (-12) + (-15)$

14. $9 + (-6) + (-16)$ **15.** $-17 + (-3) + 29$ **16.** $13 + 62 + (-38)$

17. $-3 + (-8) + 12$ **18.** $-27 + (-42) + (-18)$

19. $13 + (-22) + 4 + (-5)$ **20.** $-14 + (-3) + 7 + (-6)$

21. $-22 + 10 + 2 + (-18)$ **22.** $-6 + (-8) + 13 + (-4)$

23. $-126 + (-247) + (-358) + 339$ **24.** $-651 + (-239) + 524 + 487$

2 **25.** Explain the meaning of the words *minus* and *negative*.

26. Explain how to rewrite $6 - (-9)$ as addition of the opposite.

Subtract.

27. $16 - 8$ **28.** $12 - 3$ **29.** $7 - 14$

30. $7 - (-2)$ **31.** $3 - (-4)$ **32.** $-6 - (-3)$

33. $-4 - (-2)$ **34.** $6 - (-12)$ **35.** $-12 - 16$

36. $-4 - 3 - 2$ **37.** $4 - 5 - 12$ **38.** $12 - (-7) - 8$

39. $-12 - (-3) - (-15)$ **40.** $4 - 12 - (-8)$ **41.** $13 - 7 - 15$

42. $-6 + 19 - (-31)$ **43.** $-30 - (-65) - 29 - 4$ **44.** $42 - (-82) - 65 - 7$

45. $-16 - 47 - 63 - 12$

46. $42 - (-30) - 65 - (-11)$

47. $-47 - (-67) - 13 - 15$

48. $-18 - 49 - (-84) - 27$

3 **49.** Explain how to multiply two integers with the same sign.

50. Explain how to multiply two integers with different signs.

51. Name the operation in each expression. Justify your answer.
a. $8(-7)$ **b.** $8 - 7$ **c.** $8 - (-7)$ **d.** $-xy$ **e.** $x(-y)$ **f.** $-x - y$

52. Name the operation in each expression. Justify your answer.
a. $(4)(-6)$ **b.** $4 - (6)$ **c.** $4 - (-6)$ **d.** $-ab$ **e.** $a(-b)$ **f.** $-a - b$

Multiply.

53. $14 \cdot 3$

54. $62 \cdot 9$

55. $5(-4)$

56. $4(-7)$

57. $-8(2)$

58. $-9(3)$

59. $(-5)(-5)$

60. $(-3)(-6)$

61. $(-7)(0)$

62. $-32 \cdot 4$

63. $-24 \cdot 3$

64. $19(-7)$

65. $6(-17)$

66. $-8(-26)$

67. $-4(-35)$

68. $-5(23)$

69. $5 \cdot 7(-2)$

70. $8(-6)(-1)$

71. $(-9)(-9)(2)$

72. $-8(-7)(-4)$

73. $-5(8)(-3)$

74. $(-6)(5)(7)$

75. $-1(4)(-9)$

76. $6(-3)(-2)$

77. $4(-4) \cdot 6(-2)$

78. $-5 \cdot 9(-7) \cdot 3$

79. $-9(4) \cdot 3(1)$

80. $8(8)(-5)(-4)$

81. $-6(-5)(12)(0)$

82. $7(9) \cdot 10(-1)$

4 Write the related multiplication problem.

83. $\dfrac{-36}{-12} = 3$

84. $\dfrac{28}{-7} = -4$

85. $\dfrac{-55}{11} = -5$

86. $\dfrac{-20}{-10} = 2$

Divide.

87. $12 \div (-6)$

88. $18 \div (-3)$

89. $(-72) \div (-9)$

90. $(-64) \div (-8)$

91. $0 \div (-6)$

92. $-49 \div 0$

93. $45 \div (-5)$

94. $-24 \div 4$

95. $-36 \div 4$

96. $-56 \div 7$

97. $-81 \div (-9)$

98. $-40 \div (-5)$

99. $72 \div (-3)$

100. $44 \div (-4)$

101. $-60 \div 5$

102. $144 \div 9$

103. $78 \div (-6)$

104. $84 \div (-7)$

105. $-72 \div 4$

106. $-80 \div 5$

107. $-114 \div (-6)$

108. $-128 \div 4$

109. $-130 \div (-5)$

110. $(-280) \div 8$

111. $-132 \div (-12)$

112. $-156 \div (-13)$

113. $-182 \div 14$

114. $-144 \div 12$

115. $143 \div 11$

116. $168 \div 14$

5 **117.** **Temperature** Find the temperature after a rise of 9°C from −6°C.

118. **Temperature** Find the temperature after a rise of 7°C from −18°C.

119. **Temperature** The high temperature for the day was 10°C. The low temperature was −4°C. Find the difference between the high and low temperatures for the day.

120. **Temperature** The low temperature for the day was −2°C. The high temperature was 11°C. Find the difference between the high and low temperatures for the day.

121. **Chemistry** The temperature at which mercury boils is 360°C. Mercury freezes at −39°C. Find the difference between the temperature at which mercury boils and the temperature at which it freezes.

122. **Chemistry** The temperature at which radon boils is −62°C. Radon freezes at −71°C. Find the difference between the temperature at which radon boils and the temperature at which it freezes.

The elevation, or height, of places on Earth is measured in relation to sea level, or the average level of the ocean's surface. The following table shows height above sea level as a positive number and depth below sea level as a negative number. (*Source: Information Please Almanac*)

Continent	*Highest Elevation (in meters)*		*Lowest Elevation (in meters)*	
Africa	Mt. Kilimanjaro	5895	Qattara Depression	−133
Asia	Mt. Everest	8848	Dead Sea	−400
Europe	Mt. Elbrus	5634	Caspian Sea	−28
North America	Mt. McKinley	6194	Death Valley	−86
South America	Mt. Aconcagua	6960	Salinas Grandes	−40

123. **Geography** Use the table to find the difference in elevation between Mt. Elbrus and the Caspian Sea.

124. **Geography** Use the table to find the difference in elevation between Mt. Aconcagua and Salinas Grandes.

125. **Geography** Use the table to find the difference in elevation between Mt. Kilimanjaro and the Qattara Depression.

126. **Geography** Use the table to find the difference in elevation between Mt. McKinley and Death Valley.

127. **Geography** Use the table to find the difference in elevation between Mt. Everest and the Dead Sea.

128. **Temperature** The daily low temperatures, in degrees Celsius, during one week were recorded as follows: 4°, −5°, 8°, 0°, −9°, −11°, −8°. Find the average daily low temperature for the week.

129. **Temperature** The daily high temperatures, in degrees Celsius, during one week were recorded as follows: −8°, −9°, 6°, 7°, −2°, −14°, −1°. Find the average daily high temperature for the week.

130. **Temperature** On January 22, 1943, the temperature at Spearfish, South Dakota, rose from −4°F to 45°F in two minutes. How many degrees did the temperature rise during those two minutes?

131. **Temperature** In a 24-hour period in January of 1916, the temperature in Browning, Montana, dropped from 44°F to −56°F. How many degrees did the temperature drop during that time?

The table at the right shows the average temperatures at different cruising altitudes for airplanes. Use the table for Exercises 132 and 133.

Cruising Altitude	*Average Temperature*
12,000 ft	16°F
20,000 ft	−12°F
30,000 ft	−48°F
40,000 ft	−70°F
50,000 ft	−70°F

132. **Aviation** What is the difference between the average temperatures at 12,000 ft and at 40,000 ft?

133. **Aviation** How much colder is the average temperature at 30,000 ft than at 20,000 ft?

134. **Education** To discourage random guessing on a multiple-choice exam, a professor assigns 5 points for a correct answer, −2 points for an incorrect answer, and 0 points for leaving the question blank. What is the score for a student who had 20 correct answers, had 13 incorrect answers, and left 7 questions blank?

135. **Education** To discourage random guessing on a multiple-choice exam, a professor assigns 7 points for a correct answer, −3 points for an incorrect answer, and −1 point for leaving the question blank. What is the score for a student who had 17 correct answers, had 8 incorrect answers, and left 2 questions blank?

136. **The Stock Market** The value of a share of Wal-Mart stock on August 10, 2001, was $53.60. The table below shows the amount of increase or decrease, to the nearest 10 cents, from the August 10 closing price of the stock for a 10-day period. Calculate the 5-day moving average for this stock.

Day 1	*Day 2*	*Day 3*	*Day 4*	*Day 5*	*Day 6*	*Day 7*	*Day 8*	*Day 9*	*Day 10*
−140	+20	−40	−30	−90	+80	−170	+30	−50	−120

137. **The Stock Market** The value of a share of Disney stock on August 10, 2001, was \$27.40. The table below shows the amount of increase or decrease, to the nearest 10 cents, from the August 10 closing price of the stock for a 10-day period. Calculate the 5-day moving average for this stock.

Day 1	*Day 2*	*Day 3*	*Day 4*	*Day 5*	*Day 6*	*Day 7*	*Day 8*	*Day 9*	*Day 10*
−20	−20	−50	−20	−10	+30	−10	+30	0	+50

1.2 Applying Concepts

Simplify.

138. $|-7 + 12|$

139. $|13 - (-4)|$

140. $|-13 - (-2)|$

141. $|18 - 21|$

Assuming the pattern is continued, find the next three numbers in the pattern.

142. $-7, -11, -15, -19, \ldots$

143. $16, 11, 6, 1, \ldots$

144. $7, -14, 28, -56, \ldots$

145. $1024, -256, 64, \ldots$

Solve.

146. 32,844 is divisible by 3. By rearranging the digits, find the largest possible number that is still divisible by 3.

147. 4563 is not divisible by 4. By rearranging the digits, find the largest possible number that is divisible by 4.

148. How many three-digit numbers of the form 8__4 are divisible by 3?

In each exercise, determine which statement is false.

149. **a.** $|3 + 4| = |3| + |4|$ **b.** $|3 - 4| = |3| - |4|$ **c.** $|4 + 3| = |4| + |3|$ **d.** $|4 - 3| = |4| - |3|$

150. **a.** $|5 + 2| = |5| + |2|$ **b.** $|5 - 2| = |5| - |2|$ **c.** $|2 + 5| = |2| + |5|$ **d.** $|2 - 5| = |2| - |5|$

Determine which statement is true for all real numbers.

151. **a.** $|x + y| \le |x| + |y|$ **b.** $|x + y| = |x| + |y|$ **c.** $|x + y| \ge |x| + |y|$

152. **a.** $||x| - |y|| \le |x| - |y|$ **b.** $||x| - |y|| = |x| - |y|$ **c.** $||x| - |y|| \ge |x| - |y|$

153. Is the difference between two integers always smaller than either one of the numbers in the difference? If not, give an example for which the difference between two integers is greater than either integer.

154. If $-4x$ equals a positive integer, is x a positive or a negative integer? Explain your answer.

155. Explain why division by zero is not allowed.

1.3 Rational Numbers

1 Write rational numbers as decimals

VIDEO & DVD CD TUTOR WEB SSM

Point of Interest

As early as A.D. 630 the Hindu mathematician Brahmagupta wrote a fraction as one number over another separated by a space. The Arab mathematician al Hassar (around A.D. 1050) was the first to show a fraction with the horizontal bar separating the numerator and denominator.

A **rational number** is the quotient of two integers. Therefore, a rational number is a number that can be written in the form $\frac{a}{b}$, where a and b are integers, and b is not zero. A rational number written in this way is commonly called a **fraction.**

$\frac{a}{b}$ $a \leftarrow$ an integer, $b \leftarrow$ a nonzero integer

$\left.\frac{2}{3}, \frac{-4}{9}, \frac{18}{-5}, \frac{4}{1}\right\}$ Rational numbers

Because an integer can be written as the quotient of the integer and 1, every integer is a rational number.

$5 = \frac{5}{1}$ $\quad -3 = \frac{-3}{1}$

A number written in **decimal notation** is also a rational number.

three-tenths $\quad 0.3 = \frac{3}{10}$

thirty-five hundredths $\quad 0.35 = \frac{35}{100}$

negative four-tenths $\quad -0.4 = \frac{-4}{10}$

A rational number written as a fraction can be written in decimal notation.

Take Note

The fraction bar can be read "divided by."

$\frac{5}{8} = 5 \div 8$

Example 1 Write $\frac{5}{8}$ as a decimal.

Solution

```
    0.625   ← This is called a terminating decimal.
 8)5.000
  -4 8
     20
    -16
      40
     -40
       0   ← The remainder is zero.
```

$\frac{5}{8} = 0.625$

Take Note

Dividing the numerator by the denominator results in a remainder of 0. The decimal 0.625 is a terminating decimal.

Problem 1 Write $\frac{4}{25}$ as a decimal.

Solution See page S1.

Example 2 Write $\frac{4}{11}$ as a decimal.

Solution

$$\begin{array}{r} 0.3636\ldots \\ 11\overline{)4.0000} \\ \underline{-3\,3} \\ 70 \\ \underline{-66} \\ 40 \\ \underline{-33} \\ 70 \\ \underline{-66} \\ 4 \end{array}$$

← This is called a **repeating decimal.**

← The remainder is never zero.

$\frac{4}{11} = 0.\overline{36}$ ← The bar over the digits 3 and 6 is used to show that these digits repeat.

Take Note

No matter how far we carry out the division, the remainder is never zero. The decimal $0.\overline{36}$ is a repeating decimal.

Problem 2 Write $\frac{4}{9}$ as a decimal. Place a bar over the repeating digits of the decimal.

Solution See page S1.

Rational numbers can be written as fractions, such as $-\frac{6}{7}$ or $\frac{8}{3}$, where the numerator and denominator are integers. But every rational number also can be written as a repeating decimal (such as 0.25767676...) or a terminating decimal (such as 1.73). This was illustrated in Examples 1 and 2.

Numbers that cannot be written either as a repeating decimal or as a terminating decimal are called **irrational numbers.** For example, 2.45445444544445... is an irrational number. Two other examples are $\sqrt{2}$ and π.

$$\sqrt{2} = 1.414213562\ldots \qquad \pi = 3.141592654\ldots$$

The three dots mean that the digits continue on and on without ever repeating or terminating. Although we cannot write a decimal that is exactly equal to $\sqrt{2}$ or to π, we can give approximations of these numbers. The symbol $\approx$ is read "is approximately equal to." Shown below are $\sqrt{2}$ rounded to the nearest thousandth and π rounded to the nearest hundredth.

$$\sqrt{2} \approx 1.414 \qquad \pi \approx 3.14$$

The rational numbers and the irrational numbers taken together are called the **real numbers.**

2 Add and subtract rational numbers

VIDEO & DVD CD TUTOR WEB SSM

To add or subtract fractions, first rewrite each fraction as an equivalent fraction with a common denominator, using the least common multiple (LCM) of the denominators as the common denominator. Then add the numerators, and place the sum over the common denominator. Write the answer in simplest form.

> **Take Note**
>
> You can find the LCM by multiplying the denominators and then dividing by the greatest *common factor* of the two denominators. In the case of 6 and 10, $6 \cdot 10 = 60$. Now divide by 2, the greatest common factor of 6 and 10.
>
> $$60 \div 2 = 30$$
>
> Alternatively, you can use as the common denominator the product of the denominators, which in this case is 60. Write each fraction with a denominator of 60. Add the fractions. Then simplify the sum.
>
> $$-\frac{5}{6} + \frac{3}{10} = -\frac{50}{60} + \frac{18}{60} = \frac{-50 + 18}{60} = \frac{-32}{60} = -\frac{8}{15}$$

The sign rules for adding integers apply to addition of rational numbers.

Example 3 Add: $-\frac{5}{6} + \frac{3}{10}$

Solution Prime factorization of 6 and 10:

$6 = 2 \cdot 3 \qquad 10 = 2 \cdot 5$

$\text{LCM} = 2 \cdot 3 \cdot 5 = 30$

• Find the LCM of the denominators 6 and 10. The LCM of denominators is sometimes called the **least common denominator** (LCD).

$$-\frac{5}{6} + \frac{3}{10} = -\frac{25}{30} + \frac{9}{30}$$

• Rewrite the fractions as equivalent fractions, using the LCM of the denominators as the common denominator.

$$= \frac{-25 + 9}{30} = \frac{-16}{30}$$

• Add the numerators, and place the sum over the common denominator.

$$= -\frac{8}{15}$$

• Write the answer in simplest form.

Problem 3 Subtract: $\frac{5}{9} - \frac{11}{12}$

Solution See page S1.

Example 4 Simplify: $-\frac{3}{4} + \frac{1}{6} - \frac{5}{8}$

Solution

$$-\frac{3}{4} + \frac{1}{6} - \frac{5}{8} = -\frac{18}{24} + \frac{4}{24} - \frac{15}{24}$$

• The LCM of 4, 6, and 8 is 24.

$$= \frac{-18}{24} + \frac{4}{24} + \frac{-15}{24} = \frac{-18 + 4 + (-15)}{24} = \frac{-29}{24} = -\frac{29}{24}$$

Problem 4 Simplify: $-\frac{7}{8} - \frac{5}{6} + \frac{1}{2}$

Solution See page S2.

To add or subtract decimals, write the numbers so that the decimal points are in a vertical line. Then proceed as in the addition or subtraction of integers. Write the decimal point in the answer directly below the decimal points in the problem.

Example 5 Add: $14.02 + 137.6 + 9.852$

Solution

$$\begin{array}{r} 14.02 \\ 137.6 \\ +\ 9.852 \\ \hline 161.472 \end{array}$$

- Write the decimals so that the decimal points are in a vertical line.
- Write the decimal point in the sum directly below the decimal points in the problem.

Problem 5 Add: $3.097 + 4.9 + 3.09$

Solution See page S2.

Example 6 Add: $-114.039 + 84.76$

Solution

$$\begin{array}{r} 114.039 \\ -\ 84.76 \\ \hline 29.279 \end{array}$$

- The signs are different. Subtract the absolute value of the number with the smaller absolute value from the absolute value of the number with the larger absolute value.

$$-114.039 + 84.76 = -29.279$$

- Attach the sign of the number with the larger absolute value.

Problem 6 Subtract: $16.127 - 67.91$

Solution See page S2.

3 Multiply and divide rational numbers

VIDEO & DVD CD TUTOR WEB SSM

The sign rules for multiplying and dividing integers apply to multiplication and division of rational numbers. The product of two fractions is the product of the numerators divided by the product of the denominators.

Example 7 Multiply: $\frac{3}{8} \cdot \frac{12}{17}$

Solution

$$\frac{3}{8} \cdot \frac{12}{17} = \frac{3 \cdot 12}{8 \cdot 17}$$

- Multiply the numerators. Multiply the denominators.

$$= \frac{3 \cdot \overset{1}{\cancel{2}} \cdot \overset{1}{\cancel{2}} \cdot 3}{2 \cdot \underset{1}{\cancel{2}} \cdot \underset{1}{\cancel{2}} \cdot 17}$$

- Write the prime factorization of each factor. Divide by the common factors.

$$= \frac{9}{34}$$

- Multiply the numbers remaining in the numerator. Multiply the numbers remaining in the denominator.

Problem 7 Multiply: $-\frac{7}{12} \cdot \frac{9}{14}$

Solution See page S2.

The **reciprocal** of a fraction is the fraction with the numerator and denominator interchanged. For instance, the reciprocal of $\frac{2}{3}$ is $\frac{3}{2}$, and the reciprocal of $-\frac{5}{4}$ is $-\frac{4}{5}$. To divide fractions, multiply the dividend by the reciprocal of the divisor.

Take Note

The method for dividing fractions is sometimes stated "To divide fractions, invert the divisor and multiply." Inverting the divisor means writing its reciprocal.

Example 8 Divide: $\frac{3}{10} \div \left(-\frac{18}{25}\right)$

Solution

$$\frac{3}{10} \div \left(-\frac{18}{25}\right) = -\left(\frac{3}{10} \div \frac{18}{25}\right)$$

• The signs are different. The quotient is negative.

$$= -\left(\frac{3}{10} \cdot \frac{25}{18}\right)$$

• Change division to multiplication. Invert the divisor.

$$= -\left(\frac{3 \cdot 25}{10 \cdot 18}\right)$$

• Multiply the numerators. Multiply the denominators.

$$= -\left(\frac{\overset{1}{\cancel{3}} \cdot \overset{1}{\cancel{5}} \cdot 5}{2 \cdot \underset{1}{\cancel{5}} \cdot 2 \cdot \underset{1}{\cancel{3}} \cdot 3}\right)$$

$$= -\frac{5}{12}$$

Problem 8 Divide: $-\frac{3}{8} \div \left(-\frac{5}{12}\right)$

Solution See page S2.

To multiply decimals, multiply as in the multiplication of whole numbers. Write the decimal point in the product so that the number of decimal places in the product equals the sum of the decimal places in the factors.

Example 9 Multiply: $(-6.89)(0.00035)$

Solution

```
     6.89    2 decimal places
× 0.00035    5 decimal places
     3445
    2067
0.0024115    7 decimal places
```

• Multiply the absolute values.

$(-6.89)(0.00035) = -0.0024115$

• The signs are different. The product is negative.

Problem 9 Multiply: $(-5.44)(3.8)$

Solution See page S2.

To divide decimals, move the decimal point in the divisor to make it a whole number. Move the decimal point in the dividend the same number of places to the right. Place the decimal point in the quotient directly over the decimal point in the dividend. Then divide as in the division of whole numbers.

Example 10 Divide: $1.32 \div 0.27$. Round to the nearest tenth.

Solution $0.27.\overline{)1.32.}$

• Move the decimal point two places to the right in the divisor and in the dividend. Place the decimal point in the quotient.

Take Note

Moving the decimal point in the numerator and denominator is the same as multiplying the numerator and the denominator by the same number. For the problem at the right, we have

$$1.32 \div 0.27 = \frac{1.32}{0.27} = \frac{1.32}{0.27} \cdot \frac{100}{100} = \frac{132}{27}$$

$$\begin{array}{r} 4.88 \approx 4.9 \\ 27.\overline{)132.00} \\ \underline{-108} \\ 24\,0 \\ \underline{-21\,6} \\ 2\,40 \\ \underline{-2\,16} \\ 24 \end{array}$$

- The symbol ≈ is used to indicate that the quotient is an approximate value that has been rounded off.

Problem 10 Divide $-0.394 \div 1.7$. Round to the nearest hundredth.

Solution See page S2.

4 Convert among percents, fractions, and decimals

VIDEO & DVD CD TUTOR WEB SSM

"A population growth rate of 3%," "a manufacturer's discount of 25%," and "an 8% increase in pay" are typical examples of the many ways in which percent is used in applied problems. **Percent** means "parts of 100." Thus 27% means 27 parts of 100.

In applied problems involving a percent, it is usually necessary either to rewrite the percent as a fraction or a decimal, or to rewrite a fraction or a decimal as a percent.

To write 27% as a fraction, remove the percent sign and multiply by $\frac{1}{100}$.

$$27\% = 27\left(\frac{1}{100}\right) = \frac{27}{100}$$

To write a percent as a decimal, remove the percent sign and multiply by 0.01.

To write 33% as a decimal, remove the percent sign and multiply by 0.01.

$$33\% = 33(0.01) = 0.33$$

Move the decimal point two places to the left. Then remove the percent sign.

Write 100% as a decimal.

$$100\% = 100(0.01) = 1$$

Example 11 Write 130% as a fraction and as a decimal.

Solution

$$130\% = 130\left(\frac{1}{100}\right) = \frac{130}{100} = 1\frac{3}{10}$$

- To write a percent as a fraction, remove the percent sign and multiply by $\frac{1}{100}$.

$$130\% = 130(0.01) = 1.30$$

- To write a percent as a decimal, remove the percent sign and multiply by 0.01.

Problem 11 Write 125% as a fraction and as a decimal.

Solution See page S2.

Example 12 Write $33\frac{1}{3}\%$ as a fraction.

Solution $33\frac{1}{3}\% = 33\frac{1}{3}\left(\frac{1}{100}\right) = \frac{100}{3}\left(\frac{1}{100}\right)$

$= \frac{1}{3}$

• Write the mixed number $33\frac{1}{3}$ as the improper fraction $\frac{100}{3}$.

Problem 12 Write $16\frac{2}{3}\%$ as a fraction.

Solution See page S2.

Example 13 Write 0.25% as a decimal.

Solution $0.25\% = 0.25(0.01) = 0.0025$

• Remove the percent sign and multiply by 0.01.

Problem 13 Write 6.08% as a decimal.

Solution See page S2.

A fraction or decimal can be written as a percent by multiplying by 100%. Recall that $100\% = 1$, and multiplying a number by 1 does not change the number.

To write $\frac{5}{8}$ as a percent, multiply by 100%.

$$\frac{5}{8} = \frac{5}{8}(100\%) = \frac{500}{8}\% = 62.5\% \text{ or } 62\frac{1}{2}\%$$

To write 0.82 as a percent, multiply by 100%.

$$0.82 = 0.82(100\%) = 82\%$$

Move the decimal point two places to the right. Then write the percent sign.

Example 14 Write as a percent. **A.** 0.027 **B.** 1.34

Solution **A.** $0.027 = 0.027(100\%) = 2.7\%$

• To write a decimal as a percent, multiply by 100%.

B. $1.34 = 1.34(100\%) = 134\%$

Problem 14 Write as a percent. **A.** 0.043 **B.** 2.57

Solution See page S2.

Example 15 Write $\frac{5}{6}$ as a percent. Round to the nearest tenth of a percent.

Solution $\frac{5}{6} = \frac{5}{6}(100\%) = \frac{500}{6}\% \approx 83.3\%$

• To write a fraction as a percent, multiply by 100%.

Problem 15 Write $\frac{5}{9}$ as a percent. Round to the nearest tenth of a percent.

Solution See page S2.

Example 16 Write $\frac{7}{16}$ as a percent. Write the remainder in fractional form.

Solution $\frac{7}{16} = \frac{7}{16}(100\%) = \frac{700}{16}\% = 43\frac{3}{4}\%$ • Multiply the fraction by 100%.

Problem 16 Write $\frac{9}{16}$ as a percent. Write the remainder in fractional form.

Solution See page S2.

1.3 Concept Review

Determine whether the statement is always true, sometimes true, or never true.

1. To multiply two fractions, you must first rewrite the fractions as equivalent fractions with a common denominator.
2. The number $\frac{\pi}{3}$ is an example of a rational number.
3. A rational number can be written as a terminating decimal.
4. The number 2.585585558... is an example of a repeating decimal.
5. An irrational number is a real number.
6. 37%, 0.37, and $\frac{37}{100}$ are three numbers that have the same value.
7. To write a decimal as a percent, multiply the decimal by $\frac{1}{100}$.
8. If a, b, c, and d are natural numbers, then $\frac{a}{b} + \frac{c}{d} = \frac{a+c}{b+d}$.
9. -12 is an example of a number that is both an integer and a rational number.
10. $\frac{7}{9}$ is an example of a number that is both a rational number and an irrational number.

1.3 Exercises

1 Write as a decimal. Place a bar over the repeating digits of a repeating decimal.

1. $\frac{1}{3}$
2. $\frac{2}{3}$
3. $\frac{1}{4}$
4. $\frac{3}{4}$
5. $\frac{2}{5}$
6. $\frac{4}{5}$
7. $\frac{1}{6}$
8. $\frac{5}{6}$
9. $\frac{1}{8}$
10. $\frac{7}{8}$

11. $\frac{2}{9}$ **12.** $\frac{8}{9}$ **13.** $\frac{5}{11}$ **14.** $\frac{10}{11}$ **15.** $\frac{7}{12}$

16. $\frac{11}{12}$ **17.** $\frac{4}{15}$ **18.** $\frac{8}{15}$ **19.** $\frac{7}{16}$ **20.** $\frac{15}{16}$

21. $\frac{6}{25}$ **22.** $\frac{14}{25}$ **23.** $\frac{9}{40}$ **24.** $\frac{21}{40}$ **25.** $\frac{15}{22}$

26. $\frac{19}{22}$ **27.** $\frac{11}{24}$ **28.** $\frac{19}{24}$ **29.** $\frac{5}{33}$ **30.** $\frac{25}{33}$

31. $\frac{3}{37}$ **32.** $\frac{14}{37}$

2 Simplify.

33. $\frac{2}{3} + \frac{5}{12}$ **34.** $\frac{1}{2} + \frac{3}{8}$ **35.** $\frac{5}{8} - \frac{5}{6}$ **36.** $\frac{1}{9} - \frac{5}{27}$

37. $-\frac{5}{12} - \frac{3}{8}$ **38.** $-\frac{5}{6} - \frac{5}{9}$ **39.** $-\frac{6}{13} + \frac{17}{26}$ **40.** $-\frac{7}{12} + \frac{5}{8}$

41. $-\frac{5}{8} - \left(-\frac{11}{12}\right)$ **42.** $\frac{1}{3} + \frac{5}{6} - \frac{2}{9}$ **43.** $\frac{1}{2} - \frac{2}{3} + \frac{1}{6}$ **44.** $-\frac{3}{8} - \frac{5}{12} - \frac{3}{16}$

45. $-\frac{5}{16} + \frac{3}{4} - \frac{7}{8}$ **46.** $\frac{1}{2} - \frac{3}{8} - \left(-\frac{1}{4}\right)$ **47.** $\frac{3}{4} - \left(-\frac{7}{12}\right) - \frac{7}{8}$ **48.** $\frac{1}{3} - \frac{1}{4} - \frac{1}{5}$

49. $\frac{2}{3} - \frac{1}{2} + \frac{5}{6}$ **50.** $\frac{5}{16} + \frac{1}{8} - \frac{1}{2}$ **51.** $\frac{5}{8} - \left(-\frac{5}{12}\right) + \frac{1}{3}$ **52.** $\frac{1}{8} - \frac{11}{12} + \frac{1}{2}$

53. $-\frac{7}{9} + \frac{14}{15} + \frac{8}{21}$ **54.** $1.09 + 6.2$ **55.** $-32.1 - 6.7$ **56.** $5.13 - 8.179$

57. $-13.092 + 6.9$ **58.** $2.54 - 3.6$ **59.** $5.43 + 7.925$ **60.** $-16.92 - 6.925$

61. $-3.87 + 8.546$ **62.** $6.9027 - 17.692$

63. $2.09 - 6.72 - 5.4$ **64.** $16.4 + 3.09 - 7.93$

65. $-18.39 + 4.9 - 23.7$ **66.** $19 - (-3.72) - 82.75$

67. $-3.07 - (-2.97) - 17.4$ **68.** $-3.09 - 4.6 - 27.3$

69. $317.09 - 46.902 + 583.0714$ **70.** $71.0235 - 86.0974 + 254.309$

3 Simplify.

71. $\frac{1}{2}\left(-\frac{3}{4}\right)$ **72.** $-\frac{2}{9}\left(-\frac{3}{14}\right)$ **73.** $\left(-\frac{3}{8}\right)\left(-\frac{4}{15}\right)$

74. $\frac{5}{8}\left(-\frac{7}{12}\right)\frac{16}{25}$

75. $\left(\frac{1}{2}\right)\left(-\frac{3}{4}\right)\left(-\frac{5}{8}\right)$

76. $\left(\frac{5}{12}\right)\left(-\frac{8}{15}\right)\left(-\frac{1}{3}\right)$

77. $\frac{3}{8} \div \frac{1}{4}$

78. $\frac{5}{6} \div \left(-\frac{3}{4}\right)$

79. $-\frac{5}{12} \div \frac{15}{32}$

80. $\frac{1}{8} \div \left(-\frac{5}{12}\right)$

81. $-\frac{4}{9} \div \left(-\frac{2}{3}\right)$

82. $-\frac{6}{11} \div \frac{4}{9}$

83. $(1.2)(3.47)$

84. $(-0.8)(6.2)$

85. $(-1.89)(-2.3)$

86. $(6.9)(-4.2)$

87. $(1.06)(-3.8)$

88. $(-2.7)(-3.5)$

89. $(1.2)(-0.5)(3.7)$

90. $(-2.4)(6.1)(0.9)$

91. $(-0.8)(3.006)(-5.1)$

Simplify. Round to the nearest hundredth.

92. $-24.7 \div 0.09$

93. $-1.27 \div (-1.7)$

94. $9.07 \div (-3.5)$

95. $-354.2086 \div 0.1719$

96. $-2658.3109 \div (-0.0473)$

97. $(-3.92) \div (-45.008)$

4

98. **a.** Explain how to convert a fraction to a percent.
b. Explain how to convert a percent to a fraction.

99. **a.** Explain how to convert a decimal to a percent.
b. Explain how to convert a percent to a decimal.

100. Explain why multiplying a number by 100% does not change the value of the number.

Write as a fraction and as a decimal.

101. 75%

102. 40%

103. 50%

104. 10%

105. 64%

106. 88%

107. 175%

108. 160%

109. 19%

110. 87%

111. 5%

112. 2%

113. 450%

114. 380%

115. 8%

116. 4%

Write as a fraction.

117. $11\frac{1}{9}\%$

118. $37\frac{1}{2}\%$

119. $31\frac{1}{4}\%$

120. $66\frac{2}{3}\%$

121. $\frac{1}{2}\%$

122. $5\frac{3}{4}\%$

123. $6\frac{1}{4}\%$

124. $83\frac{1}{3}\%$

Write as a decimal.

125. 7.3%

126. 9.1%

127. 15.8%

128. 0.3%

129. 9.15%

130. 121.2%

131. 18.23%

132. 0.15%

Write as a percent.

133. 0.15 **134.** 0.37 **135.** 0.05 **136.** 0.02

137. 0.175 **138.** 0.125 **139.** 1.15 **140.** 2.142

141. 0.008 **142.** 0.004 **143.** 0.065 **144.** 0.083

Write as a percent. Round to the nearest tenth of a percent.

145. $\frac{27}{50}$ **146.** $\frac{83}{100}$ **147.** $\frac{1}{3}$ **148.** $\frac{3}{8}$

149. $\frac{4}{9}$ **150.** $\frac{9}{20}$ **151.** $2\frac{1}{2}$ **152.** $1\frac{2}{7}$

Write as a percent. Write the remainder in fractional form.

153. $\frac{3}{8}$ **154.** $\frac{7}{16}$ **155.** $\frac{5}{14}$ **156.** $\frac{4}{7}$

157. $1\frac{1}{4}$ **158.** $2\frac{5}{8}$ **159.** $1\frac{5}{9}$ **160.** $1\frac{13}{16}$

Employment The graph at the right shows the responses to a survey that asked respondents, "How did you find your most recent job?" Use the graph for Exercises 161 to 163.

161. What fraction of the respondents found their most recent jobs on the Internet?

162. What fraction of the respondents found their most recent jobs through a referral?

163. Did more or fewer than one-quarter of the respondents find their most recent jobs through a newspaper ad?

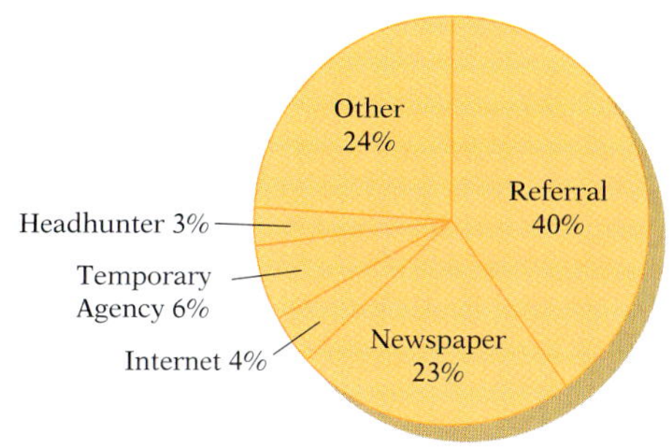

How Did You Find Your Most Recent Job?

1.3 Applying Concepts

Classify each of the following numbers as a natural number, an integer, a positive integer, a negative integer, a rational number, an irrational number, or a real number. List all that apply.

164. -1 **165.** 28

166. $-\frac{9}{34}$ **167.** -7.707

168. $5.2\overline{6}$ **169.** 0.171771777...

Solve.

170. Find the average of $\frac{5}{8}$ and $\frac{3}{4}$.

171. **The Postal Service** Postage for first-class mail is \$.37 for the first ounce or fraction of an ounce and \$.23 for each additional ounce or fraction of an ounce. Find the cost of mailing, by first-class mail, a letter that weighs $4\frac{1}{2}$ oz.

Government The table at the right shows the surplus or deficit, in billions of dollars, for the federal budget for selected years from 1945 through 2000. A negative sign (−) indicates a deficit. Use this table for Exercises 172 to 177. (*Source:* U.S. Office of Management and Budget)

Year	*Federal Budget Surplus or Deficit (in billions of dollars)*
1945	−47.553
1950	−3.119
1955	−2.993
1960	0.301
1965	−1.411
1970	−2.842
1975	−53.242
1980	−73.835
1985	−212.334
1990	−221.194
1995	−163.899
1996	−107.450
1997	−21.940
1998	69.246
1999	79.263
2000	117.305

172. In which year listed was the deficit lowest?

173. Find the difference between the deficits in the years 1980 and 1985.

174. Calculate the difference between the surplus in 1960 and the deficit in 1955.

175. How many times greater was the deficit in 1985 than in 1975? Round to the nearest whole number.

176. What was the average deficit per quarter, in millions of dollars, for the year 1970?

177. Find the average surplus or deficit for the years 1995 through 2000. Round to the nearest million.

178. Use a calculator to determine the decimal representations of $\frac{17}{99}$, $\frac{45}{99}$, and $\frac{73}{99}$. Make a conjecture as to the decimal representation of $\frac{83}{99}$. Does your conjecture work for $\frac{33}{99}$? What about $\frac{1}{99}$?

179. If the same positive integer is added to both the numerator and the denominator of $\frac{2}{5}$, is the new fraction less than, equal to, or greater than $\frac{2}{5}$?

180. A magic square is one in which the numbers in every row, column, and diagonal sum to the same number. Complete the magic square at the right.

$\frac{2}{3}$		
	$\frac{1}{6}$	$\frac{5}{6}$
		$-\frac{1}{3}$

181. Let x represent the price of a car. If the sales tax is 6% of the price, express the total of the price of the car and the sales tax in terms of x.

182. Let x represent the price of a suit. If the suit is on sale at a discount rate of 30%, express the price of the suit after the discount in terms of x.

183. The price of a pen was 60¢. During a storewide sale, the price was reduced to a different whole number of cents, and the entire stock was sold for \$54.59. Find the price of the pen during the sale.

184. Find three natural numbers a, b, and c such that $\frac{1}{a} + \frac{1}{b} + \frac{1}{c}$ is a natural number.

185. If a and b are rational numbers and $a < b$, is it always possible to find a rational number c such that $a < c < b$? If not, explain why. If so, show how to find one.

186. In your own words, define **a.** a rational number, **b.** an irrational number, and **c.** a real number.

187. Explain why you "invert and multiply" when dividing a fraction by a fraction.

188. Explain why you need a common denominator when adding two fractions and why you don't need a common denominator when multiplying two fractions.

1.4 Exponents and the Order of Operations Agreement

1 Exponential expressions

VIDEO & DVD CD TUTOR WEB SSM

Point of Interest

René Descartes (1596–1650) was the first mathematician to use exponential notation extensively as it is used today. However, for some unknown reason, he always used xx for x^2.

Repeated multiplication of the same factor can be written using an exponent.

$$2 \cdot 2 \cdot 2 \cdot 2 \cdot 2 = 2^5 \leftarrow \text{exponent (2 is the base)} \qquad a \cdot a \cdot a \cdot a = a^4 \leftarrow \text{exponent (}a\text{ is the base)}$$

The **exponent** indicates how many times the factor, called the **base,** occurs in the multiplication. The multiplication $2 \cdot 2 \cdot 2 \cdot 2 \cdot 2$ is in **factored form.** The exponential expression 2^5 is in **exponential form.**

2^1 is read "the first power of two" or just "two." → Usually the exponent 1 is not written.

2^2 is read "the second power of two" or "two squared."

2^3 is read "the third power of two" or "two cubed."

2^4 is read "the fourth power of two."

2^5 is read "the fifth power of two."

a^5 is read "the fifth power of a."

There is a geometric interpretation of the first three natural-number powers.

To evaluate an exponential expression, write each factor as many times as indicated by the exponent. Then multiply.

$$3^5 = 3 \cdot 3 \cdot 3 \cdot 3 \cdot 3 = 243$$

$$2^3 \cdot 3^2 = (2 \cdot 2 \cdot 2) \cdot (3 \cdot 3) = 8 \cdot 9 = 72$$

Take Note

The -4 is squared only when the negative sign is *inside* the parentheses. In $(-4)^2$, we are squaring -4; in -4^2, we are finding the opposite of 4^2.

Example 1 Evaluate $(-4)^2$ and -4^2.

Solution $(-4)^2 = (-4)(-4) = 16$

$-4^2 = -(4 \cdot 4) = -16$

Problem 1 Evaluate $(-5)^3$ and -5^3.

Solution See page S2.

Take Note

The product of an even number of negative factors is positive. The product of an odd number of negative factors is negative.

Example 2 Evaluate $(-2)^4$ and $(-2)^5$.

Solution

$$\begin{aligned}(-2)^4 &= (-2)(-2)(-2)(-2)\\ &= 4(-2)(-2)\\ &= -8(-2)\\ &= 16\end{aligned}$$

$$\begin{aligned}(-2)^5 &= (-2)(-2)(-2)(-2)(-2)\\ &= 4(-2)(-2)(-2)\\ &= -8(-2)(-2)\\ &= 16(-2)\\ &= -32\end{aligned}$$

Problem 2 Evaluate $(-3)^3$ and $(-3)^4$.

Solution See page S2.

Example 3 Evaluate $(-3)^2 \cdot 2^3$ and $\left(-\frac{2}{3}\right)^3$.

Solution $(-3)^2 \cdot 2^3 = (-3)(-3) \cdot (2)(2)(2) = 9 \cdot 8 = 72$

$$\left(-\frac{2}{3}\right)^3 = \left(-\frac{2}{3}\right)\left(-\frac{2}{3}\right)\left(-\frac{2}{3}\right) = -\frac{2 \cdot 2 \cdot 2}{3 \cdot 3 \cdot 3} = -\frac{8}{27}$$

Problem 3 Evaluate $(3^3)(-2)^3$ and $\left(-\frac{2}{5}\right)^2$.

Solution See page S2.

2 The Order of Operations Agreement

VIDEO & DVD | CD TUTOR | WEB | SSM

Evaluate $2 + 3 \cdot 5$.

There are two arithmetic operations, addition and multiplication, in this problem. The operations could be performed in different orders.

Add first.	$\underbrace{2 + 3} \cdot 5$	Multiply first.	$2 + \underbrace{3 \cdot 5}$
Then multiply.	$\underbrace{5 \cdot 5}_{25}$	Then add.	$\underbrace{2 + 15}_{17}$

In order to prevent there being more than one answer to the same problem, an Order of Operations Agreement has been established.

The Order of Operations Agreement

Step 1 Perform operations inside grouping symbols. Grouping symbols include parentheses ( ), brackets [ ], absolute value symbols | |, and the fraction bar.

Step 2 Simplify exponential expressions.

Step 3 Do multiplication and division as they occur from left to right.

Step 4 Do addition and subtraction as they occur from left to right.

Example 4 Simplify: $12 - 24(8 - 5) \div 2^2$

Solution

$12 - 24(8 - 5) \div 2^2$

$= 12 - 24(3) \div 2^2$ • Perform operations inside grouping symbols.

$= 12 - 24(3) \div 4$ • Simplify exponential expressions.

$= 12 - 72 \div 4$ • Do multiplication and division as they occur from left to right.

$= 12 - 18$ • Do addition and subtraction as they occur from left to right.

$= -6$

Problem 4 Simplify: $36 \div (8 - 5)^2 - (-3)^2 \cdot 2$

Solution See page S2.

One or more of the steps shown in Example 4 may not be needed to simplify an expression. In that case, proceed to the next step in the Order of Operations Agreement.

Example 5 Simplify: $\frac{4+8}{2+1} - |3-1| + 2$

Solution

$$\frac{4+8}{2+1} - |3-1| + 2$$

$= \frac{12}{3} - |2| + 2$ • Perform operations inside grouping symbols (above and below the fraction bar and inside the absolute value symbols).

$= \frac{12}{3} - 2 + 2$ • Find the absolute value of 2.

$= 4 - 2 + 2$ • Do multiplication and division as they occur from left to right.

$= 2 + 2$ • Do addition and subtraction as they occur from left to right.

$= 4$

Problem 5 Simplify: $27 \div 3^2 + (-3)^2 \cdot 4$

Solution See page S2.

When an expression has grouping symbols inside grouping symbols, first perform the operations inside the *inner* grouping symbols by following Steps 2, 3, and 4 of the Order of Operations Agreement. Then perform the operations inside the *outer* grouping symbols by following Steps 2, 3, and 4 in sequence.

Example 6 Simplify: $6 \div [4 - (6 - 8)] + 2^2$

Solution

$6 \div [4 - (6 - 8)] + 2^2$

$= 6 \div [4 - (-2)] + 2^2$ • Perform operations inside inner grouping symbols.

$= 6 \div 6 + 2^2$ • Perform operations inside outer grouping symbols.

$= 6 \div 6 + 4$ • Simplify exponential expressions.

$= 1 + 4$ • Do multiplication and division.

$= 5$ • Do addition and subtraction.

Problem 6 Simplify: $4 - 3[4 - 2(6 - 3)] \div 2$

Solution See page S2.

1.4 Concept Review

Determine whether the statement is always true, sometimes true, or never true.

1. $(-5)^2$, -5^2, and $-(5)^2$ all represent the same number.

2. By the Order of Operations Agreement, addition is performed before division.

3. In the expression 3^8, 8 is the base.

4. The expression 9^4 is in exponential form.

5. To evaluate the expression $6 + 7 \cdot 10$ means to determine what one number it equals.

6. When using the Order of Operations Agreement, we consider the absolute value symbol a grouping symbol.

7. Using the Order of Operations Agreement, it is possible to get more than one correct answer to a problem.

8. The Order of Operations Agreement is used for natural numbers, integers, rational numbers, and real numbers.

9. The expression $(2^3)^2$ equals $2^{(3^2)}$.

10. Suppose a and b are positive integers. Then $a^b = b^a$.

1.4 Exercises

1 Evaluate.

1. 6^2
2. 7^4
3. -7^2
4. -4^3
5. $(-3)^2$
6. $(-2)^3$
7. $(-3)^4$
8. $(-5)^3$
9. $\left(\frac{1}{2}\right)^2$
10. $\left(-\frac{3}{4}\right)^3$
11. $(0.3)^2$
12. $(1.5)^3$
13. $\left(\frac{2}{3}\right)^2 \cdot 3^3$
14. $\left(-\frac{1}{2}\right)^3 \cdot 8$
15. $(0.3)^3 \cdot 2^3$
16. $(0.5)^2 \cdot 3^3$
17. $(-3) \cdot 2^2$
18. $(-5) \cdot 3^4$
19. $(-2) \cdot (-2)^3$
20. $(-2) \cdot (-2)^2$
21. $2^3 \cdot 3^3 \cdot (-4)$
22. $(-3)^3 \cdot 5^2 \cdot 10$
23. $(-7) \cdot 4^2 \cdot 3^2$
24. $(-2) \cdot 2^3 \cdot (-3)^2$
25. $\left(\frac{2}{3}\right)^2 \cdot \frac{1}{4} \cdot 3^3$
26. $\left(\frac{3}{4}\right)^2 \cdot (-4) \cdot 2^3$
27. $8^2 \cdot (-3)^5 \cdot 5$

2 28. Why do we need an Order of Operations Agreement?

29. Describe each step in the Order of Operations Agreement.

Simplify by using the Order of Operations Agreement.

30. $4 - 8 \div 2$
31. $2^2 \cdot 3 - 3$
32. $2(3 - 4) - (-3)^2$
33. $16 - 32 \div 2^3$
34. $24 - 18 \div 3 + 2$
35. $8 - (-3)^2 - (-2)$

36. $16 + 15 \div (-5) - 2$

37. $14 - 2^2 - |4 - 7|$

38. $3 - 2[8 - (3 - 2)]$

39. $-2^2 + 4[16 \div (3 - 5)]$

40. $6 + \dfrac{16 - 4}{2^2 + 2} - 2$

41. $24 \div \dfrac{3^2}{8 - 5} - (-5)$

42. $96 \div 2[12 + (6 - 2)] - 3^3$

43. $4 \cdot [16 - (7 - 1)] \div 10$

44. $16 \div 2 - 4^2 - (-3)^2$

45. $18 \div |9 - 2^3| + (-3)$

46. $16 - 3(8 - 3)^2 \div 5$

47. $4(-8) \div [2(7 - 3)^2]$

48. $\dfrac{(-10) + (-2)}{6^2 - 30} \div |2 - 4|$

49. $16 - 4 \cdot \dfrac{3^3 - 7}{2^3 + 2} - (-2)^2$

50. $(0.2)^2 \cdot (-0.5) + 1.72$

51. $0.3(1.7 - 4.8) + (1.2)^2$

52. $(1.8)^2 - 2.52 \div 1.8$

53. $(1.65 - 1.05)^2 \div 0.4 + 0.8$

54. $\dfrac{3}{8} \div \left|\dfrac{5}{6} + \dfrac{2}{3}\right|$

55. $\left(\dfrac{3}{4}\right)^2 - \left(\dfrac{1}{2}\right)^3 \div \dfrac{3}{5}$

1.4 Applying Concepts

56. Using the Order of Operations Agreement, describe how to simplify Exercise 46.

Place the correct symbol, < or >, between the two numbers.

57. $(0.9)^3 \quad 1^5$

58. $(-3)^3 \quad (-2)^5$

59. $(-1.1)^2 \quad (0.9)^2$

Simplify.

60. $1^2 + 2^2 + 3^2 + 4^2$

61. $1^3 + 2^3 + 3^3 + 4^3$

62. $(-1)^3 + (-2)^3 + (-3)^3 + (-4)^3$

63. $(-2)^2 + (-4)^2 + (-6)^2 + (-8)^2$

Solve.

64. Find a rational number, r, that satisfies the condition.
a. $r^2 < r$ **b.** $r^2 = r$ **c.** $r^2 > r$

65. **Computers** A computer can do 6,000,000 additions in 1 s. To the nearest second, how many seconds will it take the computer to do 10^8 additions?

66. The sum of two natural numbers is 41. Each of the two numbers is the square of a natural number. Find the two numbers.

Determine the ones digit when the expression is evaluated.

67. 34^{202}

68. 23^{502}

69. 27^{622}

70. Prepare a report on the Kelvin scale. Include in your report an explanation of how to convert between the Kelvin scale and the Celsius scale.

1.5 Variable Expressions

1 Evaluate variable expressions

VIDEO & DVD · CD TUTOR · WEB · SSM

> **Point of Interest**
>
> *Historical manuscripts indicate that mathematics is at least 4000 years old. Yet it was only 400 years ago that mathematicians started using variables to stand for numbers. The idea that a letter can stand for some number was a critical turning point in mathematics.*

Often we discuss a quantity without knowing its exact value, such as the price of gold next month, the cost of a new automobile next year, or the tuition for next semester. In algebra, a letter of the alphabet is used to stand for a quantity that is unknown or one that can change, or *vary*. The letter is called a **variable.** An expression that contains one or more variables is called a **variable expression.**

A variable expression is shown at the right. The expression can be rewritten by writing subtraction as the addition of the opposite.

$$3x^2 - 5y + 2xy - x - 7$$

$$3x^2 + (-5y) + 2xy + (-x) + (-7)$$

Note that the expression has five addends. The **terms** of a variable expression are the addends of the expression. The expression has five terms.

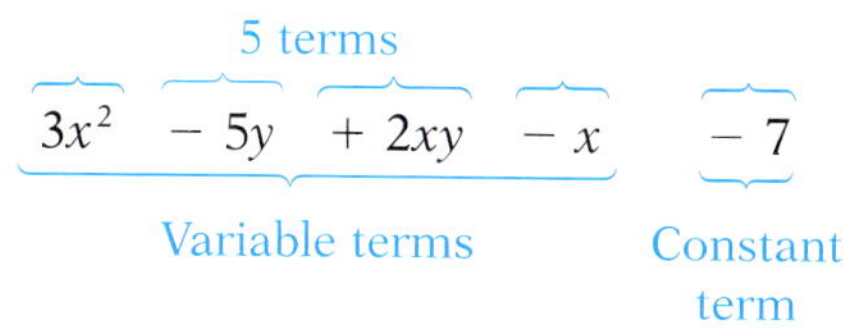

The terms $3x^2$, $-5y$, $2xy$, and $-x$ are **variable terms.**

The term -7 is a **constant term,** or simply a **constant.**

Each variable term is composed of a **numerical coefficient** and a **variable part** (the variable or variables and their exponents).

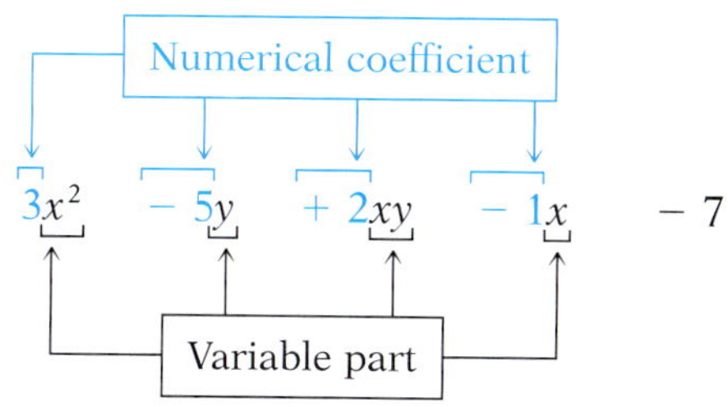

When the numerical coefficient is 1 or −1, the 1 is usually not written ($x = 1x$ and $-x = -1x$).

Example 1 Name the variable terms of the expression $2a^2 - 5a + 7$.

Solution $2a^2, -5a$

Problem 1 Name the constant term of the expression $6n^2 + 3n - 4$.

Solution See page S3.

Variable expressions occur naturally in science. In a physics lab, a student may discover that a weight of 1 pound will stretch a spring $\frac{1}{2}$ inch. A weight of 2 pounds will stretch the spring 1 inch. By experimenting, the student can discover that the distance the spring will stretch is found by multiplying the weight, in pounds, by $\frac{1}{2}$. By letting W represent the number of pounds attached to the spring, the distance, in inches, the spring stretches can be represented by the variable expression $\frac{1}{2}W$.

With a weight of W pounds, the spring will stretch $\frac{1}{2} \cdot W = \frac{1}{2}W$ inches.

With a weight of 10 pounds, the spring will stretch $\frac{1}{2} \cdot 10 = 5$ inches.

With a weight of 3 pounds, the spring will stretch $\frac{1}{2} \cdot 3 = 1\frac{1}{2}$ inches.

Replacing the variable or variables in a variable expression with numbers and then simplifying the resulting numerical expression is called **evaluating the variable expression.**

Example 2 Evaluate $ab - b^2$ when $a = 2$ and $b = -3$.

Solution

$ab - b^2$

$2(-3) - (-3)^2$ • Replace each variable in the expression with the number it represents.

$= 2(-3) - 9$ • Use the Order of Operations Agreement to simplify the resulting numerical expression.

$= -6 - 9$

$= -15$

Problem 2 Evaluate $2xy + y^2$ when $x = -4$ and $y = 2$.

Solution See page S3.

Example 3 Evaluate $\frac{a^2 - b^2}{a - b}$ when $a = 3$ and $b = -4$.

Solution

$\frac{a^2 - b^2}{a - b}$

$\frac{(3)^2 - (-4)^2}{3 - (-4)}$ • Replace each variable in the expression with the number it represents.

$= \frac{9 - 16}{3 - (-4)}$ • Use the Order of Operations Agreement to simplify the resulting numerical expression.

$= \frac{-7}{7}$

$= -1$

Problem 3 Evaluate $\frac{a^2 + b^2}{a + b}$ when $a = 5$ and $b = -3$.

Solution See page S3.

Example 4 Evaluate $x^2 - 3(x - y) - z^2$ when $x = 2$, $y = -1$, and $z = 3$.

Solution

$x^2 - 3(x - y) - z^2$

$(2)^2 - 3[2 - (-1)] - (3)^2$ • Replace each variable in the expression with the number it represents.

$= (2)^2 - 3(3) - (3)^2$ • Use the Order of Operations Agreement to simplify the resulting numerical expression.

$= 4 - 3(3) - 9$

$= 4 - 9 - 9$

$= -5 - 9$

$= -14$

Problem 4 Evaluate $x^3 - 2(x + y) + z^2$ when $x = 2$, $y = -4$, and $z = -3$.

Solution See page S3.

Example 5 The diameter of the base of a right circular cylinder is 5 cm. The height of the cylinder is 8.5 cm. Find the volume of the cylinder. Round to the nearest tenth.

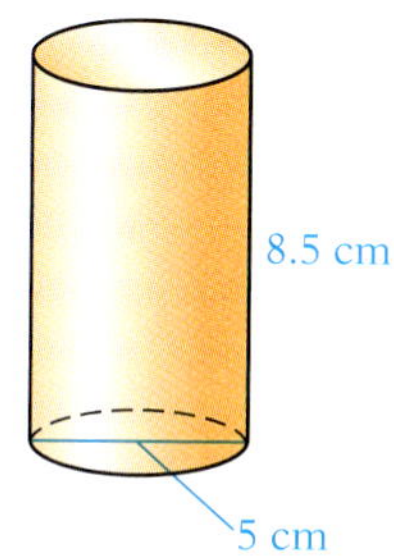

Take Note

For your reference, geometric formulas are printed inside the front cover of this textbook.

Solution

$V = \pi r^2 h$ • Use the formula for the volume of a right circular cylinder.

$V = \pi(2.5)^2(8.5)$ • $r = \frac{1}{2}d = \frac{1}{2}(5) = 2.5$

$V = \pi(6.25)(8.5)$ • Use the π key on your calculator to enter the value of π.

$V \approx 166.9$

The volume is approximately 166.9 cm^3.

Problem 5 The diameter of the base of a right circular cone is 9 cm. The height of the cone is 9.5 cm. Find the volume of the cone. Round to the nearest tenth.

Solution See page S3.

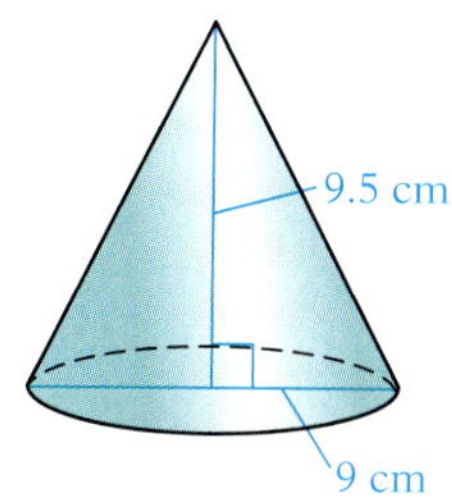

A graphing calculator can be used to evaluate variable expressions. When the value of each variable is stored in the calculator's memory and a variable expression is then entered on the calculator, the calculator evaluates that variable expression for the values of the variables stored in its memory. See the Appendix for a description of keystroking procedures.

2 The Properties of Real Numbers

VIDEO & DVD · CD TUTOR · WEB · SSM

The Properties of the Real Numbers describe the way operations on numbers can be performed. Here are some of the Properties of the Real Numbers and an example of each.

The Commutative Property of Addition

If a and b are real numbers, then $a + b = b + a$.

Two terms can be added in either order; the sum is the same.

$$4 + 3 = 7 \quad \text{and} \quad 3 + 4 = 7$$

The Commutative Property of Multiplication

If a and b are real numbers, then $a \cdot b = b \cdot a$.

Two factors can be multiplied in either order; the product is the same.

$$(5)(-2) = -10 \quad \text{and} \quad (-2)(5) = -10$$

The Associative Property of Addition

If a, b, and c are real numbers, then $(a + b) + c = a + (b + c)$.

When three or more terms are added, the terms can be grouped (with parentheses, for example) in any order; the sum is the same.

$$2 + (3 + 4) = 2 + 7 = 9 \quad \text{and} \quad (2 + 3) + 4 = 5 + 4 = 9$$

The Associative Property of Multiplication

If a, b, and c are real numbers, then $(a \cdot b) \cdot c = a \cdot (b \cdot c)$.

When three or more factors are multiplied, the factors can be grouped in any order; the product is the same.

$$(2 \cdot 3) \cdot 4 = 6 \cdot 4 = 24 \quad \text{and} \quad 2 \cdot (3 \cdot 4) = 2 \cdot 12 = 24$$

The Addition Property of Zero

If a is a real number, then $a + 0 = 0 + a = a$.

The sum of a term and zero is the term.

$$4 + 0 = 4 \qquad 0 + 4 = 4$$

The Multiplication Property of Zero

If a is a real number, then $a \cdot 0 = 0 \cdot a = 0$.

The product of a term and zero is zero.

$$(5)(0) = 0 \qquad (0)(5) = 0$$

The Multiplication Property of One

If a is a real number, then $a \cdot 1 = 1 \cdot a = a$.

The product of a term and 1 is the term.

$$6 \cdot 1 = 6 \qquad 1 \cdot 6 = 6$$

The Inverse Property of Addition

If a is a real number, then $a + (-a) = (-a) + a = 0$.

The sum of a number and its opposite is zero.
The opposite of a number is called its **additive inverse.**

$$8 + (-8) = 0 \qquad (-8) + 8 = 0$$

The Inverse Property of Multiplication

If a is a real number, and $a \neq 0$, then $a \cdot \frac{1}{a} = \frac{1}{a} \cdot a = 1$.

The product of a number and its reciprocal is 1.

$\frac{1}{a}$ is the **reciprocal** of a. $\frac{1}{a}$ is also called the **multiplicative inverse** of a.

$$7 \cdot \frac{1}{7} = 1 \qquad \frac{1}{7} \cdot 7 = 1$$

The Distributive Property

If a, b, and c are real numbers, then $a(b + c) = ab + ac$ or $(b + c)a = ba + ca$.

By the Distributive Property, the term outside the parentheses is multiplied by each term inside the parentheses.

$$\begin{aligned} 2(3 + 4) &= 2 \cdot 3 + 2 \cdot 4 \\ 2 \cdot 7 &= 6 + 8 \\ 14 &= 14 \end{aligned} \qquad \begin{aligned} (4 + 5)2 &= 4 \cdot 2 + 5 \cdot 2 \\ 9 \cdot 2 &= 8 + 10 \\ 18 &= 18 \end{aligned}$$

Example 6 Complete the statement by using the Commutative Property of Multiplication.

$(6)(5) = (?)(6)$

Solution $(6)(5) = (5)(6)$ • The Commutative Property of Multiplication states that $a \cdot b = b \cdot a$.

Problem 6 Complete the statement by using the Inverse Property of Addition.

$7 + ? = 0$

Solution See page S3.

Example 7 Identify the property that justifies the statement.

$2(8 + 5) = 16 + 10$

Solution The Distributive Property

• **The Distributive Property states that $a(b + c) = ab + ac$.**

Problem 7 Identify the property that justifies the statement.

$5 + (13 + 7) = (5 + 13) + 7$

Solution See page S3.

3 Simplify variable expressions using the Properties of Addition

VIDEO & DVD CD TUTOR WEB SSM

Like terms of a variable expression are terms with the same variable part. (Because $x^2 = x \cdot x$, x^2 and x are not like terms.)

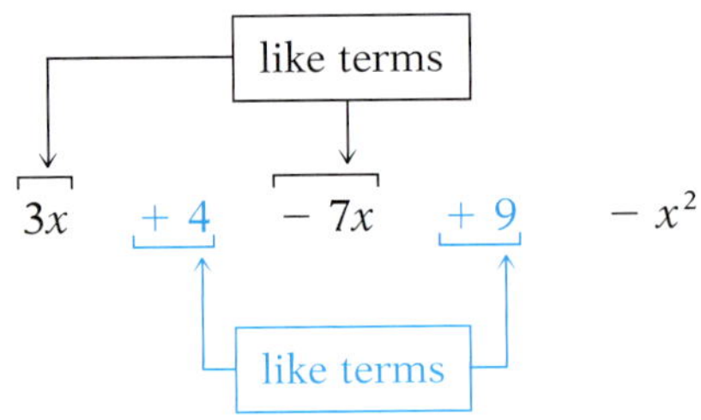

Constant terms are like terms. 4 and 9 are like terms.

To **combine** like terms, use the Distributive Property $ba + ca = (b + c)a$ to add the coefficients.

$$\begin{aligned} 2x + 3x &= (2 + 3)x \\ &= 5x \end{aligned}$$

Example 8 Simplify. **A.** $-2y + 3y$ **B.** $5x - 11x$

Solution

A. $-2y + 3y$

$= (-2 + 3)y$ • **Use the Distributive Property $ba + ca = (b + c)a$.**

$= 1y$ • **Add the coefficients.**

$= y$ • **Use the Multiplication Property of One.**

B. $5x - 11x$

$= [5 + (-11)]x$ • **Use the Distributive Property $ba + ca = (b + c)a$.**

$= -6x$ • **Add the coefficients.**

Problem 8 Simplify. **A.** $9x + 6x$ **B.** $-4y - 7y$

Solution See page S3.

In simplifying more complicated expressions, use the Properties of Addition.

The Commutative Property of Addition can be used when adding two like terms. The terms can be added in either order. The sum is the same.

$$2x + (-4x) = -4x + 2x$$
$$[2 + (-4)]x = (-4 + 2)x$$
$$-2x = -2x$$

The Associative Property of Addition is used when adding three or more terms. The terms can be grouped in any order. The sum is the same.

$$3x + 5x + 9x = (3x + 5x) + 9x = 3x + (5x + 9x)$$
$$8x + 9x = 3x + 14x$$
$$17x = 17x$$

By the Addition Property of Zero, the sum of a term and zero is the term.

$$5x + 0 = 0 + 5x = 5x$$

By the Inverse Property of Addition, the sum of a term and its additive inverse is zero.

$$7x + (-7x) = -7x + 7x = 0$$

Example 9 Simplify. **A.** $8x + 3y - 8x$ **B.** $4x^2 + 5x - 6x^2 - 2x$

Solution **A.** $8x + 3y - 8x$

$= 3y + 8x - 8x$ • Use the Commutative Property of Addition to rearrange the terms.

$= 3y + (8x - 8x)$ • Use the Associative Property of Addition to group like terms.

$= 3y + 0$ • Use the Inverse Property of Addition.

$= 3y$ • Use the Addition Property of Zero.

B. $4x^2 + 5x - 6x^2 - 2x$

$= 4x^2 - 6x^2 + 5x - 2x$ • Use the Commutative Property of Addition to rearrange the terms.

$= (4x^2 - 6x^2) + (5x - 2x)$ • Use the Associative Property of Addition to group like terms.

$= -2x^2 + 3x$ • Combine like terms.

Problem 9 Simplify. **A.** $3a - 2b + 5a$ **B.** $x^2 - 7 + 9x^2 - 14$

Solution See page S3.

4 Simplify variable expressions using the Properties of Multiplication

VIDEO & DVD CD TUTOR WEB SSM

The Properties of Multiplication are used in simplifying variable expressions.

The Associative Property is used when multiplying three or more factors. $2(3x) = (2 \cdot 3)x = 6x$

The Commutative Property can be used to change the order in which factors are multiplied. $(3x) \cdot 2 = 2 \cdot (3x) = 6x$

By the Multiplication Property of One, the product of a term and 1 is the term. $(8x)(1) = (1)(8x) = 8x$

By the Inverse Property of Multiplication, the product of a term and its reciprocal is 1. $5x \cdot \frac{1}{5x} = \frac{1}{5x} \cdot 5x = 1, x \neq 0$

Example 10 Simplify. **A.** $2(-x)$ **B.** $\frac{3}{2}\left(\frac{2x}{3}\right)$ **C.** $(16x)2$

Solution

A. $2(-x) = 2(-1 \cdot x)$ • $-x = -1x = -1 \cdot x$

$= [2 \cdot (-1)]x$ • Use the Associative Property of Multiplication to group factors.

$= -2x$ • Multiply.

B. $\frac{3}{2}\left(\frac{2x}{3}\right) = \frac{3}{2}\left(\frac{2}{3}x\right)$ • Note that $\frac{2x}{3} = \frac{2}{3} \cdot \frac{x}{1} = \frac{2}{3}x$.

$= \left(\frac{3}{2} \cdot \frac{2}{3}\right)x$ • Use the Associative Property of Multiplication to group factors.

$= 1x$ • Use the Inverse Property of Multiplication.

$= x$ • Use the Multiplication Property of One.

C. $(16x)2 = 2(16x)$ • Use the Commutative Property of Multiplication to rearrange factors.

$= (2 \cdot 16)x$ • Use the Associative Property of Multiplication to group factors.

$= 32x$ • Multiply.

Problem 10 Simplify. **A.** $-7(-2a)$ **B.** $-\frac{5}{6}(-30y^2)$ **C.** $(-5x)(-2)$

Solution See page S3.

5 Simplify variable expressions using the Distributive Property

VIDEO & DVD CD TUTOR WEB SSM

The Distributive Property is used to remove parentheses from a variable expression.

$3(2x - 5)$
$= 3(2x) - 3(5)$
$= 6x - 15$

An extension of the Distributive Property is used when an expression contains more than two terms.

$4(x^2 + 6x - 1)$
$= 4(x^2) + 4(6x) - 4(1)$
$= 4x^2 + 24x - 4$

Example 11 Simplify. **A.** $-3(5 + x)$ **B.** $-(2x - 4)$ **C.** $(2y - 6)2$ **D.** $5(3x + 7y - z)$

Solution

A. $-3(5 + x)$

$= -3(5) + (-3)x$ • Use the Distributive Property.

$= -15 - 3x$ • Multiply.

B. $-(2x - 4)$
$= -1(2x - 4)$ • Just as $-x = -1x$, $-(2x - 4) = -1(2x - 4)$.
$= -1(2x) - (-1)(4)$ • Use the Distributive Property.
$= -2x + 4$ • Note: When a negative sign immediately precedes the parentheses, remove the parentheses and change the sign of *each* term inside the parentheses.

C. $(2y - 6)2$
$= (2y)(2) - (6)(2)$ • Use the Distributive Property $(b + c)a = ba + ca$.
$= 4y - 12$

D. $5(3x + 7y - z)$
$= 5(3x) + 5(7y) - 5(z)$ • Use the Distributive Property.
$= 15x + 35y - 5z$

Problem 11 Simplify. **A.** $7(4 + 2y)$ **B.** $-(5x - 12)$ **C.** $(3a - 1)5$ **D.** $-3(6a^2 - 8a + 9)$

Solution See page S3.

6 Simplify general variable expressions

When simplifying variable expressions, use the Distributive Property to remove parentheses and brackets used as grouping symbols.

Example 12 Simplify: $4(x - y) - 2(-3x + 6y)$

Solution $4(x - y) - 2(-3x + 6y)$
$= 4x - 4y + 6x - 12y$ • Use the Distributive Property to remove parentheses.
$= 10x - 16y$ • Combine like terms.

Problem 12 Simplify: $7(x - 2y) - 3(-x - 2y)$

Solution See page S3.

Example 13 Simplify: $2x - 3[2x - 3(x + 7)]$

Solution $2x - 3[2x - 3(x + 7)]$
$= 2x - 3[2x - 3x - 21]$ • Use the Distributive Property to remove the inner grouping symbols.
$= 2x - 3[-x - 21]$ • Combine like terms inside the grouping symbols.
$= 2x + 3x + 63$ • Use the Distributive Property to remove the brackets.
$= 5x + 63$ • Combine like terms.

Problem 13 Simplify: $3y - 2[x - 4(2 - 3y)]$

Solution See page S3.

1.5 Concept Review

Determine whether the statement is always true, sometimes true, or never true.

1. The Associative Property of Addition states that two terms can be added in either order and the sum will be the same.
2. When three numbers are multiplied, the numbers can be grouped in any order and the product will be the same.
3. The Multiplication Property of One states that multiplying a number by one does not change the number.
4. The sum of a number and its additive inverse is zero.
5. The product of a number and its multiplicative inverse is one.
6. The terms x and x^2 are like terms because both have a coefficient of 1.
7. $3(x + 4) = 3x + 4$ is an example of the correct application of the Distributive Property.
8. To add like terms, add the coefficients; the variable part remains unchanged.

1.5 Exercises

1 1. What is the meaning of the phrase "evaluate a variable expression"?

Evaluate the variable expression when $a = 2$, $b = 3$, and $c = -4$.

2. $a^2 - b^2$
3. $b^2 - c^2$
4. $(a + b)^2$
5. $a^2 + b^2$
6. $2a - (c + a)^2$
7. $(b - a)^2 + 4c$
8. $b^2 - \frac{ac}{8}$
9. $\frac{5ab}{6} - 3cb$
10. $(b - 2a)^2 + bc$

Evaluate the variable expression when $a = -2$, $b = 4$, $c = -1$, and $d = 3$.

11. $(b + d)^2 - 4a$
12. $(d - a)^2 - 3c$
13. $(d - a)^2 \div 5$
14. $(b - c)^2 \div 5$
15. $b^2 - 2b + 4$
16. $a^2 - 5a - 6$

17. $\frac{bd}{a} \div c$

18. $\frac{2ac}{b} \div (-c)$

19. $2(b + c) - 2a$

20. $(b + c)^2 + (a + d)^2$

21. $4ac + (2a)^2$

22. $3dc - (4c)^2$

Evaluate the variable expression when $a = 2.7$, $b = -1.6$, and $c = -0.8$.

23. $c^2 - ab$

24. $(a + b)^2 - c$

25. $\frac{b^3}{c} - 4a$

Solve. Round to the nearest tenth.

26. **Geometry** Find the volume of a sphere that has a radius of 8.5 cm.

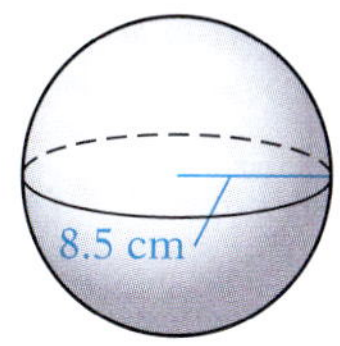

27. **Geometry** Find the volume of a right circular cylinder that has a radius of 1.25 in. and a height of 5.25 in.

28. **Geometry** The radius of the base of a right circular cylinder is 3.75 ft. The height of the cylinder is 9.5 ft. Find the surface area of the cylinder.

29. **Geometry** The length of one base of a trapezoid is 17.5 cm, and the length of the other base is 10.25 cm. The height is 6.75 cm. What is the area of the trapezoid?

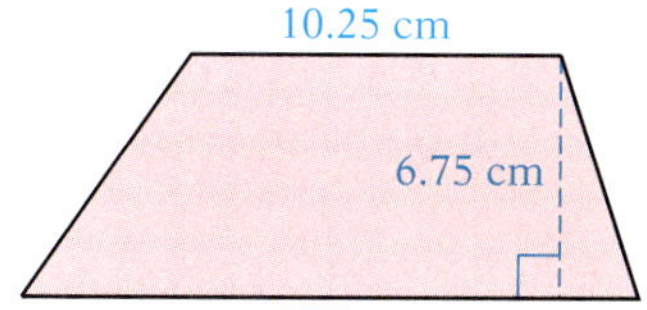

30. **Geometry** A right circular cone has a height of 2.75 in. The diameter of the base is 1 in. Find the volume of the cone.

31. **Geometry** A right circular cylinder has a height of 12.6 m. The diameter of the base is 7 m. Find the volume of the cylinder.

32. The value of z is the value of $a^2 - 2a$ when $a = -3$. Find the value of z^2.

33. The value of a is the value of $3x^2 - 4x - 5$ when $x = -2$. Find the value of $3a - 4$.

34. The value of c is the value of $a^2 + b^2$ when $a = 2$ and $b = -2$. Find the value of $c^2 - 4$.

2 Use the given property to complete the statement.

35. The Commutative Property of Multiplication
$2 \cdot 5 = 5 \cdot ?$

36. The Commutative Property of Addition
$9 + 17 = ? + 9$

37. The Associative Property of Multiplication
$(4 \cdot 5) \cdot 6 = 4 \cdot (? \cdot 6)$

38. The Associative Property of Addition
$(4 + 5) + 6 = ? + (5 + 6)$

39. The Distributive Property
$2(4 + 3) = 8 + ?$

40. The Addition Property of Zero
$? + 0 = -7$

41. The Inverse Property of Addition
$8 + ? = 0$

42. The Inverse Property of Multiplication
$\frac{1}{-5}(-5) = ?$

Identify the property that justifies the statement.

43. $-7 + 7 = 0$

44. $(-8)\left(-\frac{1}{8}\right) = 1$

45. $23 + 19 = 19 + 23$

46. $-21 + 0 = -21$

47. $3 \cdot 5 = 5 \cdot 3$

48. $-32(0) = 0$

49. $(4 \cdot 3) \cdot 5 = 4 \cdot (3 \cdot 5)$

50. $\frac{1}{4}(1) = \frac{1}{4}$

51. What are *like terms*? Give an example of two like terms. Give an example of two terms that are not like terms.

52. Explain the meaning of the phrase "simplify a variable expression."

53. Which of the following are like terms?
$3x, 3x^2, 5x, 5xy$

54. Which of the following are like terms?
$-7a, -7b, -4a, -7a^2$

3 Simplify.

55. $-12a + 17a$

56. $-3a + 12a$

57. $5ab - 7ab$

58. $9ab - 3ab$

59. $-12xy + 17xy$

60. $-15xy + 3xy$

61. $-3ab + 3ab$

62. $-7ab + 7ab$

63. $-\frac{1}{2}x - \frac{1}{3}x$

64. $-\frac{2}{5}y + \frac{3}{10}y$

65. $\frac{3}{8}x^2 - \frac{5}{12}x^2$

66. $\frac{2}{3}y^2 - \frac{4}{9}y^2$

67. $3x + 5x + 3x$

68. $8x + 5x + 7x$

69. $5a - 3a + 5a$

70. $10a - 17a + 3a$

71. $-5x^2 - 12x^2 + 3x^2$

72. $-y^2 - 8y^2 + 7y^2$

73. $7x - 8x + 3y$

74. $8y - 10x + 8x$

75. $7x - 3y + 10x$

76. $8y + 8x - 8y$

77. $3a - 7b - 5a + b$

78. $-5b + 7a - 7b + 12a$

79. $3x - 8y - 10x + 4x$

80. $3y - 12x - 7y + 2y$

81. $x^2 - 7x - 5x^2 + 5x$

82. $3x^2 + 5x - 10x^2 - 10x$

4 Simplify.

83. $4(3x)$

84. $12(5x)$

85. $-3(7a)$

86. $-2(5a)$

87. $-2(-3y)$

88. $-5(-6y)$

89. $(4x)2$

90. $(6x)12$

91. $(3a)(-2)$

92. $(7a)(-4)$

93. $(-3b)(-4)$

94. $(-12b)(-9)$

95. $-5(3x^2)$ **96.** $-8(7x^2)$ **97.** $\frac{1}{3}(3x^2)$ **98.** $\frac{1}{5}(5a)$

99. $\frac{1}{8}(8x)$ **100.** $-\frac{1}{4}(-4a)$ **101.** $-\frac{1}{7}(-7n)$ **102.** $\left(\frac{3x}{4}\right)\left(\frac{4}{3}\right)$

103. $\frac{12x}{5}\left(\frac{5}{12}\right)$ **104.** $(-6y)\left(-\frac{1}{6}\right)$ **105.** $(-10n)\left(-\frac{1}{10}\right)$ **106.** $\frac{1}{3}(9x)$

107. $\frac{1}{7}(14x)$ **108.** $-\frac{1}{5}(10x)$ **109.** $-\frac{1}{8}(16x)$ **110.** $-\frac{2}{3}(12a^2)$

111. $-\frac{5}{8}(24a^2)$ **112.** $-\frac{1}{2}(-16y)$ **113.** $-\frac{3}{4}(-8y)$ **114.** $(16y)\left(\frac{1}{4}\right)$

5 Simplify.

115. $-(x + 2)$ **116.** $-(x + 7)$ **117.** $2(4x - 3)$ **118.** $5(2x - 7)$

119. $-2(a + 7)$ **120.** $-5(a + 16)$ **121.** $-3(2y - 8)$ **122.** $-5(3y - 7)$

123. $(5 - 3b)7$ **124.** $(10 - 7b)2$ **125.** $-3(3 - 5x)$ **126.** $-5(7 - 10x)$

127. $3(5x^2 + 2x)$ **128.** $6(3x^2 + 2x)$ **129.** $-2(-y + 9)$ **130.** $-5(-2x + 7)$

131. $(-3x - 6)5$ **132.** $(-2x + 7)7$ **133.** $2(-3x^2 - 14)$ **134.** $5(-6x^2 - 3)$

135. $-3(2y^2 - 7)$ **136.** $-8(3y^2 - 12)$ **137.** $-(6a^2 - 7b^2)$

138. $3(x^2 + 2x - 6)$ **139.** $4(x^2 - 3x + 5)$ **140.** $-2(y^2 - 2y + 4)$

141. $-3(y^2 - 3y - 7)$ **142.** $2(-a^2 - 2a + 3)$ **143.** $4(-3a^2 - 5a + 7)$

6 Simplify.

144. $4x - 2(3x + 8)$ **145.** $6a - (5a + 7)$ **146.** $9 - 3(4y + 6)$

147. $10 - (11x - 3)$ **148.** $5n - (7 - 2n)$ **149.** $8 - (12 + 4y)$

150. $3(x + 2) - 5(x - 7)$ **151.** $2(x - 4) - 4(x + 2)$ **152.** $12(y - 2) + 3(7 - 3y)$

153. $6(2y - 7) - 3(3 - 2y)$ **154.** $3(a - b) - 4(a + b)$ **155.** $2(a + 2b) - (a - 3b)$

156. $4[x - 2(x - 3)]$ **157.** $2[x + 2(x + 7)]$ **158.** $-2[3x + 2(4 - x)]$

159. $-5[2x + 3(5 - x)]$ **160.** $-3[2x - (x + 7)]$ **161.** $-2[3x - (5x - 2)]$

162. $2x - 3[x - 2(4 - x)]$ **163.** $-7x + 3[x - 7(3 - 2x)]$ **164.** $-5x - 2[2x - 4(x + 7)] - 6$

165. $4a - 2[2b - (b - 2a)] + 3b$ **166.** $2x + 3(x - 2y) + 5(3x - 7y)$ **167.** $5y - 2(y - 3x) + 2(7x - y)$

1.5 Applying Concepts

Simplify.

168. $C - 0.7C$

169. $\frac{1}{3}(3x + y) - \frac{2}{3}(6x - y)$

170. $-\frac{1}{4}[2x + 2(y - 6y)]$

Complete.

171. **a.** A number that has no reciprocal is ______ .

b. A number that is its own reciprocal is ______ .

172. Determine whether the statement is true or false. If the statement is false, give an example that illustrates that it is false.

a. Division is a commutative operation.
b. Division is an associative operation.
c. Subtraction is an associative operation.
d. Subtraction is a commutative operation.
e. Addition is a commutative operation.

173. Give examples of two operations that occur in everyday experience that are not commutative (for example, putting on socks and then shoes).

1.6 Translating Verbal Expressions into Variable Expressions

1 Translate a verbal expression into a variable expression

VIDEO & DVD CD TUTOR WEB SSM

One of the major skills required in applied mathematics is the ability to translate a verbal expression into a variable expression. This requires recognizing the verbal phrases that translate into mathematical operations. Here is a partial list of the phrases used to indicate the different mathematical operations.

Addition	added to	6 added to y	$y + 6$
	more than	8 more than x	$x + 8$
	the sum of	the sum of x and z	$x + z$
	increased by	t increased by 9	$t + 9$
	the total of	the total of 5 and y	$5 + y$
	plus	b plus 17	$b + 17$
Subtraction	minus	x minus 2	$x - 2$
	less than	7 less than t	$t - 7$
	less	7 less t	$7 - t$
	subtracted from	5 subtracted from d	$d - 5$
	decreased by	m decreased by 3	$m - 3$
	the difference between	the difference between y and 4	$y - 4$

Point of Interest

The way in which expressions are symbolized has changed over time. Here are some expressions as they may have appeared in the early 16th century.

R p. 9 *for x + 9. The symbol* **R** *was used for a variable to the first power. The symbol* **p.** *was used for plus.*

R m. 3 *for x − 3. Again, the symbol* **R** *was used for the variable. The symbol* **m.** *was used for minus.*

The square of a variable was designated by **Q** *and the cube was designated by* **C.** *The expression* $x^3 + x^2$ *was written* **C p. Q.**

Multiplication	times	10 times t	$10t$
	of	one half of x	$\frac{1}{2}x$
	the product of	the product of y and z	yz
	multiplied by	y multiplied by 11	$11y$
	twice	twice n	$2n$
Division	divided by	x divided by 12	$\frac{x}{12}$
	the quotient of	the quotient of y and z	$\frac{y}{z}$
	the ratio of	the ratio of t to 9	$\frac{t}{9}$
Power	the square of	the square of x	x^2
	the cube of	the cube of a	a^3

Translating a phrase that contains the word *sum, difference, product,* or *quotient* can sometimes cause a problem. In the examples at the right, note where the operation symbol is placed.

the *sum* of x *and* y (+)	$x + y$
the *difference* between x *and* y (−)	$x - y$
the *product* of x *and* y (·)	$x \cdot y$
the *quotient* of x *and* y (÷)	$\frac{x}{y}$

Example 1 Translate into a variable expression.

A. the total of five times b and c

B. the quotient of eight less than n and fourteen

C. thirteen more than the sum of seven and the square of x

Solution

A. the total of five times b and c

$5b + c$

- Identify words that indicate mathematical operations.
- Use the operations to write the variable expression.

B. the quotient of eight less than n and fourteen

$\frac{n - 8}{14}$

- Identify words that indicate mathematical operations.
- Use the operations to write the variable expression.

C. thirteen more than the sum of seven and the square of x

$(7 + x^2) + 13$

Problem 1 Translate into a variable expression.

A. eighteen less than the cube of x

B. y decreased by the sum of z and nine

C. the difference between the square of q and the sum of r and t

Solution See page S3.

In most applications that involve translating phrases into variable expressions, the variable to be used is not given. To translate these phrases, a variable must be assigned to an unknown quantity before the variable expression can be written.

> **Take Note**
>
> The expression $n(6 + n^3)$ must have parentheses. If we write $n \cdot 6 + n^3$, then by the Order of Operations Agreement, only the 6 is multiplied by n, but we want n to be multiplied by the *total* of 6 and n^3.

Example 2 Translate "a number multiplied by the total of six and the cube of the number" into a variable expression.

Solution the unknown number: n

- **Assign a variable to one of the unknown quantities.**

the cube of the number: n^3
the total of six and the cube of the number: $6 + n^3$

- **Use the assigned variable to write an expression for any other unknown quantity.**

$n(6 + n^3)$

- **Use the assigned variable to write the variable expression.**

Problem 2 Translate "a number added to the product of five and the square of the number" into a variable expression.

Solution See page S3.

Example 3 Translate "the quotient of twice a number and the difference between the number and twenty" into a variable expression.

Solution the unknown number: n
twice the number: $2n$
the difference between the number and twenty: $n - 20$

$$\frac{2n}{n - 20}$$

Problem 3 Translate "the product of three and the sum of seven and twice a number" into a variable expression.

Solution See page S3.

2 Translate a verbal expression into a variable expression and then simplify the resulting expression

After translating a verbal expression into a variable expression, simplify the variable expression by using the Properties of the Real Numbers.

Example 4 Translate and simplify "the total of four times an unknown number and twice the difference between the number and eight."

Solution the unknown number: n

- **Assign a variable to one of the unknown quantities.**

four times the unknown number: $4n$
twice the difference between the number and eight: $2(n - 8)$

- **Use the assigned variable to write an expression for any other unknown quantity.**

$4n + 2(n - 8)$ • Use the assigned variable to write the variable expression.

$= 4n + 2n - 16$ • Simplify the variable expression.

$= 6n - 16$

Problem 4 Translate and simplify "a number minus the difference between twice the number and seventeen."

Solution See page S4.

Example 5 Translate and simplify "the difference between five-eighths of a number and two-thirds of the same number."

Solution the unknown number: n

five-eighths of the number: $\frac{5}{8}n$

two-thirds of the number: $\frac{2}{3}n$

• Assign a variable to one of the unknown quantities.
• Use the assigned variable to write an expression for any other unknown quantity.

$$\frac{5}{8}n - \frac{2}{3}n$$
$$= \frac{15}{24}n - \frac{16}{24}n$$
$$= -\frac{1}{24}n$$

• Use the assigned variable to write the variable expression.
• Simplify the variable expression.

Problem 5 Translate and simplify "the sum of three-fourths of a number and one-fifth of the same number."

Solution See page S4.

3 Translate application problems

VIDEO & DVD CD TUTOR WEB SSM

Many of the applications of mathematics require that you identify an unknown quantity, assign a variable to that quantity, and then attempt to express another unknown quantity in terms of that variable.

Suppose we know that the sum of two numbers is 10 and that one of the two numbers is 4. We can find the other number by subtracting 4 from 10.

one number: 4
other number: $10 - 4 = 6$
The two numbers are 4 and 6.

Now suppose we know that the sum of two numbers is 10, we don't know either number, and we want to express *both* numbers in terms of the *same* variable.

Let one number be x. Again, we can find the other number by subtracting x from 10.

one number: x
other number: $10 - x$
The two numbers are x and $10 - x$.

Note that the sum of x and $10 - x$ is 10.

$$x + (10 - x) = x + 10 - x = 10$$

Example 6 The length of a swimming pool is 20 ft longer than the width. Express the length of the pool in terms of the width.

Solution the width of the pool: W • Assign a variable to the width of the pool.

the length is 20 more than the width: $W + 20$ • Express the length of the pool in terms of W.

> **Take Note**
> Any variable can be used. For example, if the width is y, then the length is $y + 20$.

Problem 6 An older computer takes twice as long to process a set of data as does a newer model. Express the amount of time it takes the older computer to process the data in terms of the amount of time it takes the newer model.

Solution See page S4.

Example 7 An investor divided \$5000 between two accounts, one a mutual fund and the other a money market fund. Use one variable to express the amounts invested in each account.

Solution the amount invested in the mutual fund: x • Assign a variable to the amount invested in one account.

the amount invested in the money market fund: $5000 - x$ • Express the amount invested in the other account in terms of x.

> **Take Note**
> It is also correct to assign the variable to the amount in the money market fund. Then the amount in the mutual fund is $5000 - x$.

Problem 7 A guitar string 6 ft long was cut into two pieces. Use one variable to express the lengths of the two pieces.

Solution See page S4.

1.6 Concept Review

Determine whether the statement is always true, sometimes true, or never true.

1. "Five less than n" can be translated as "$5 - n$."
2. A variable expression contains an equals sign.
3. If the sum of two numbers is 12 and one of the two numbers is x, then the other number can be expressed as $x - 12$.
4. The words *total* and *times* both indicate multiplication.
5. The words *quotient* and *ratio* both indicate division.
6. The expressions $7y - 8$ and $(7y) - 8$ are equivalent.
7. The phrase "five times the sum of x and y" and the phrase "the sum of five times x and y" yield the same variable expression.

1.6 Exercises

1 Translate into a variable expression.

1. d less than nineteen

2. the sum of six and c

3. r decreased by twelve

4. w increased by fifty-five

5. a multiplied by twenty-eight

6. y added to sixteen

7. five times the difference between n and seven

8. thirty less than the square of b

9. y less the product of three and y

10. the sum of four-fifths of m and eighteen

11. the product of negative six and b

12. nine increased by the quotient of t and five

13. four divided by the difference between p and six

14. the product of seven and the total of r and eight

15. the quotient of nine less than x and twice x

16. the product of a and the sum of a and thirteen

17. twenty-one less than the product of s and negative four

18. fourteen more than one-half of the square of z

19. the ratio of eight more than d to d

20. the total of nine times the cube of m and the square of m

21. three-eighths of the sum of t and fifteen

22. s decreased by the quotient of s and two

23. w increased by the quotient of seven and w

24. the difference between the square of c and the total of c and fourteen

25. d increased by the difference between sixteen times d and three

26. the product of eight and the total of b and five

27. a number divided by nineteen

28. thirteen less a number

29. forty more than a number

30. three-sevenths of a number

31. the square of the difference between a number and ninety

32. the quotient of twice a number and five

33. the sum of four-ninths of a number and twenty

34. eight subtracted from the product of fifteen and a number

35. the product of a number and ten more than the number

36. six less than the total of a number and the cube of the number

37. fourteen added to the product of seven and a number

38. the quotient of three and the total of four and a number

39. the quotient of twelve and the sum of a number and two

40. eleven plus one-half of a number

41. the ratio of two and the sum of a number and one

42. a number multiplied by the difference between twice the number and nine

43. the difference between sixty and the quotient of a number and fifty

44. the product of nine less than a number and the number

45. the sum of the square of a number and three times the number

46. the quotient of seven more than twice a number and the number

47. the sum of three more than a number and the cube of the number

48. a number decreased by the difference between the cube of the number and ten

49. the square of a number decreased by one-fourth of the number

50. four less than seven times the square of a number

51. twice a number decreased by the quotient of seven and the number

52. eighty decreased by the product of thirteen and a number

53. the cube of a number decreased by the product of twelve and the number

54. the quotient of five and the sum of a number and nineteen

2 Translate into a variable expression. Then simplify the expression.

55. a number increased by the total of the number and ten

56. a number added to the product of five and the number

57. a number decreased by the difference between nine and the number

58. eight more than the sum of a number and eleven

59. the difference between one-fifth of a number and three-eighths of the number

60. a number minus the sum of the number and fourteen

61. four more than the total of a number and nine

62. the sum of one-eighth of a number and one-twelfth of the number

63. twice the sum of three times a number and forty

64. the sum of a number divided by two and the number

65. seven times the product of five and a number

66. sixteen multiplied by one-fourth of a number

67. the total of seventeen times a number and twice the number

68. the difference between nine times a number and twice the number

69. a number plus the product of the number and twelve

70. nineteen more than the difference between a number and five

71. three times the sum of the square of a number and four

72. a number subtracted from the product of the number and seven

73. three-fourths of the sum of sixteen times a number and four

74. the difference between fourteen times a number and the product of the number and seven

75. sixteen decreased by the sum of a number and nine

76. eleven subtracted from the difference between eight and a number

77. five more than the quotient of four times a number and two

78. twenty minus the sum of four-ninths of a number and three

79. six times the total of a number and eight

80. four times the sum of a number and twenty

81. seven minus the sum of a number and two

82. three less than the sum of a number and ten

83. one-third of the sum of a number and six times the number

84. twice the quotient of four times a number and eight

85. the total of eight increased by the cube of a number and twice the cube of the number

86. the sum of five more than the square of a number and twice the square of the number

87. twelve more than a number added to the difference between the number and six

88. a number plus four added to the difference between three and twice the number

89. the sum of a number and nine added to the difference between the number and twenty

90. seven increased by a number added to twice the difference between the number and two

91. fourteen plus the product of three less than a number and ten

92. a number plus the product of the number minus five and seven

3 Write a variable expression.

93. The sum of two numbers is 18. Express the two numbers in terms of the same variable.

94. The sum of two numbers is 20. Express the two numbers in terms of the same variable.

95. **Genetics** The human genome contains 11,000 more genes than the roundworm genome. Express the number of genes in the human genome in terms of the number of genes in the roundworm genome. (*Source:* Celera, *USA Today* research)

96. WEB **Telecommunications** In 1947, phone companies began using area codes. According to information found at **555-1212.com**, there were 151 more area codes at the beginning of 2001 than there were in 1947. Express the number of area codes in 2001 in terms of the number of area codes in 1947.

97. **Space Exploration** A survey in *USA Today* reported that almost three-fourths of Americans think that money should be spent on exploration of Mars. Express the number of Americans who think that money should be spent on exploration of Mars in terms of the total number of Americans.

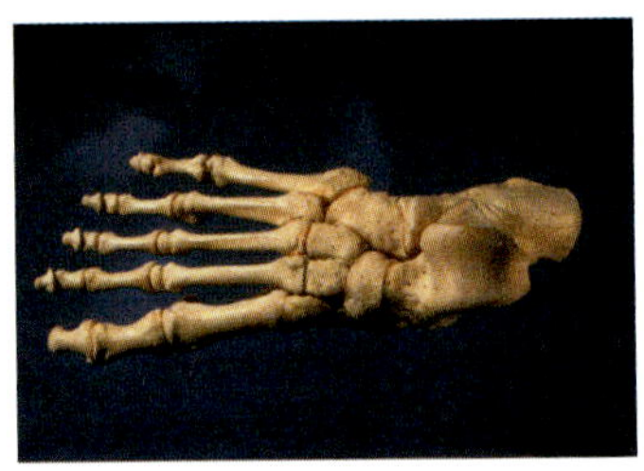

98. **Biology** According to the American Podiatric Medical Association, the bones in your foot account for one-fourth of all bones in your body. Express the number of bones in your foot in terms of the number of bones in your body.

99. **Sports** In football, the number of points awarded for a touchdown is three times the number of points awarded for a safety. Express the number of points awarded for a touchdown in terms of the number of points awarded for a safety.

100. **Cost of Living** A cost-of-living calculator at **HOMECARE.com** shows that a person living in Manhattan would need twice the salary of a person living in San Diego. Express the salary needed to live in Manhattan in terms of the salary needed to live in San Diego.

101. **Astronomy** The planet Saturn has 9 more moons than Jupiter. Express the number of moons of Saturn in terms of the number of moons of Jupiter. (*Source:* NASA)

102. **Earth Science** According to the National Oceanic and Atmospheric Administration, the number of tornadoes in 1990 was 235 more than the number of tornadoes in 2000. Express the number of tornadoes in 1990 in terms of the number of tornadoes in 2000.

103. **Agriculture** In a recent year, Alabama produced one-half the number of pounds of pecans that Texas produced in the same year. Express the amount of pecans produced in Alabama in terms of the amount produced in Texas. (*Source:* National Agricultural Statistics Service)

104. **Taxes** According to the Internal Revenue Service, it takes about one-fifth the time to fill out Schedule B (interest and dividends) than it does to complete Schedule A (itemized deductions). Express the amount of time it takes to fill out Schedule B in terms of the time it takes to fill out Schedule A.

105. **Demography** According to the U.S. Bureau of the Census, the world population in the year 2050 is expected to be twice the world population in 1980. Express the expected world population in 2050 in terms of the world population in 1980.

106. **Geometry** The length of a rectangle is 5 m more than twice the width. Express the length of the rectangle in terms of the width.

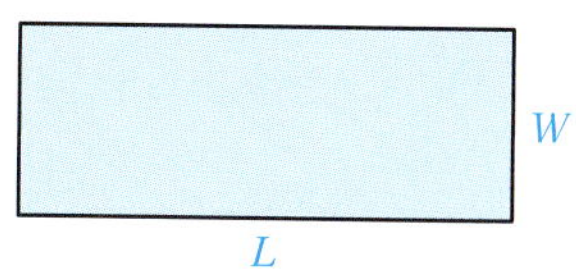

107. **Geometry** In a triangle, the measure of the smallest angle is 10 degrees less than one-half the measure of the largest angle. Express the measure of the smallest angle in terms of the measure of the largest angle.

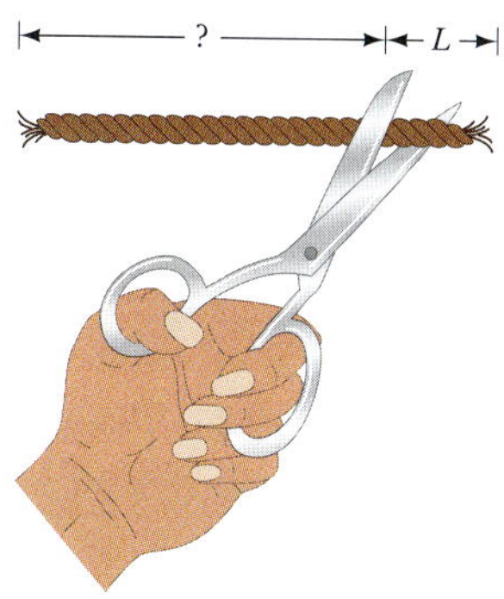

108. **Sports** A halyard 12 ft long is cut into two pieces. Use the same variable to express the lengths of the two pieces.

109. **Coins** A coin bank contains thirty-five coins in nickels and dimes. Use the same variable to express the number of nickels and the number of dimes in the coin bank.

110. **Travel** Two cars are traveling in opposite directions and at different rates. Two hours later the cars are 200 mi apart. Express the distance traveled by the faster car in terms of the distance traveled by the slower car.

111. **Wages** An employee is paid \$720 per week plus \$27 for each hour of overtime worked. Express the employee's weekly pay in terms of the number of hours of overtime worked.

112. **Billing** An auto repair bill is \$138 for parts and \$45 for each hour of labor. Express the amount of the repair bill in terms of the number of hours of labor.

113. **Consumerism** The cost of a cellular phone service is \$19.95 per month plus \$.08 per minute for phone calls. Express the monthly cost of the phone service in terms of the number of minutes of phone calls.

1.6 Applying Concepts

114. **Chemistry** The chemical formula for glucose (sugar) is $C_6H_{12}O_6$. This formula means that there are twelve hydrogen atoms, six carbon atoms, and six oxygen atoms in each molecule of glucose. If x represents the number of atoms of oxygen in a pound of sugar, express the number of hydrogen atoms in the pound of sugar in terms of x.

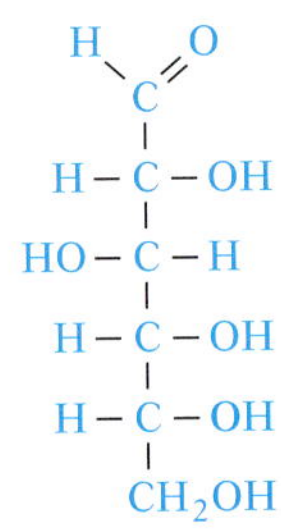

115. **Metalwork** A wire whose length is given as x inches is bent into a square. Express the length of a side of the square in terms of x.

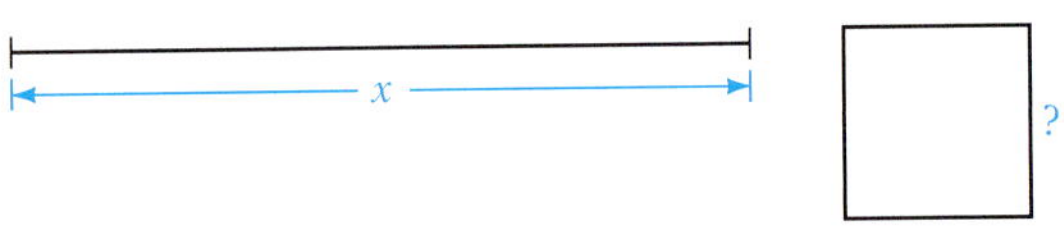

116. **Pulleys** A block-and-tackle system is designed so that pulling one end of a rope five feet will move a weight on the other end a distance of three feet. If x represents the distance the rope is pulled, express the distance the weight moves in terms of x.

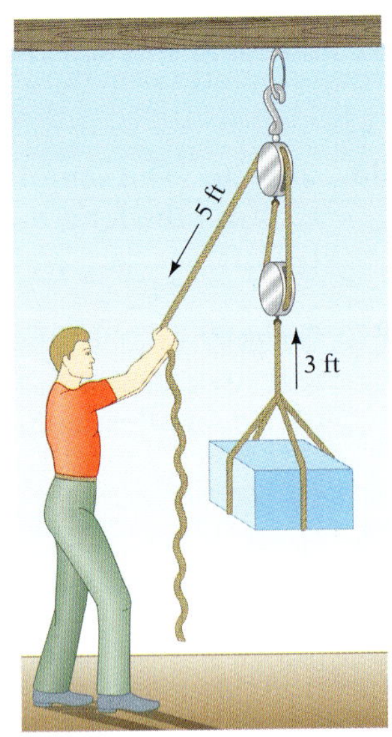

117. Translate the expressions $5x + 8$ and $5(x + 8)$ into phrases.

118. In your own words, explain how variables are used.

119. Explain the similarities and differences between the expressions "the difference between x and 5" and "5 less than x."

Focus on Problem Solving

From Concrete to Abstract

In your study of algebra, you will find that the problems are less concrete than those you studied in arithmetic. Problems that are concrete provide information pertaining to a specific instance. Algebra is more abstract. Abstract problems are theoretical; they are stated without reference to a specific instance. Let's look at an example of an abstract problem.

How many minutes are there in h hours?

A strategy that can be used to solve this problem is to begin by substituting a number for the variable.

How many minutes are there in 5 hours?

You know that there are 60 minutes in 1 hour. To find the number of minutes in 5 hours, multiply 5 by 60.

$60 \cdot 5 = 300$ There are 300 minutes in 5 hours.

Use the same procedure to find the number of minutes in h hours: multiply h by 60.

$60 \cdot h = 60h$ There are $60h$ minutes in h hours.

This problem might be taken a step further:

If you walk one mile in x minutes, how far can you walk in h hours?

Consider the same problem using numbers in place of the variables.

If you walk one mile in 20 minutes, how far can you walk in 3 hours?

To solve this problem, you need to calculate the number of minutes in 3 hours (multiply 3 by 60), and divide the result by the number of minutes it takes to walk one mile (20 minutes).

$\frac{60 \cdot 3}{20} = \frac{180}{20} = 9$ If you walk one mile in 20 minutes, you can walk 9 miles in 3 hours.

Use the same procedure to solve the related abstract problem. Calculate the number of minutes in h hours (multiply h by 60), and divide the result by the number of minutes it takes to walk one mile (x minutes).

$\frac{60 \cdot h}{x} = \frac{60h}{x}$ If you walk one mile in x minutes, you can walk $\frac{60h}{x}$ miles in h hours.

At the heart of the study of algebra is the use of variables. It is the variables in the problems above that make them abstract. But it is variables that allow us to generalize situations and state rules about mathematics.

Try each of the following problems.

1. How many hours are there in d days?
2. You earn d dollars an hour. What are your wages for working h hours?

3. If p is the price of one share of stock, how many shares can you purchase with d dollars?

4. A company pays a television station d dollars to air a commercial lasting s seconds. What is the cost per second?

5. After every v videotape rentals, you are entitled to one free rental. You have rented t tapes, where $t < v$. How many more do you need to rent before you are entitled to a free rental?

6. Your car gets g miles per gallon. How many gallons of gasoline does your car consume traveling t miles?

7. If you drink j ounces of juice each day, how many days will q quarts of the juice last?

8. A TV station has m minutes of commercials each hour. How many ads lasting s seconds each can be sold for each hour of programming?

9. A factory worker can assemble p products in m minutes. How many products can the factory worker assemble in h hours?

10. If one candy bar costs n nickels, how many candy bars can be purchased with q quarters?

Projects & Group Activities

Investigation into Operations with Even and Odd Integers

An **even integer** is an integer that is divisible by 2. For example, 6, −12, and 480 are even integers. An **odd integer** is an integer that is not evenly divisible by 2. For example, 3, −25, and 971 are odd integers.

Complete each statement with the word *even* or *odd*.

1. If k is an odd integer, then $k + 1$ is an ______ integer.
2. If k is an odd integer, then $k - 2$ is an ______ integer.
3. If n is an integer, then $2n$ is an ______ integer.
4. If m and n are even integers, then $m - n$ is an ______ integer.
5. If m and n are even integers, then mn is an ______ integer.
6. If m and n are odd integers, then $m + n$ is an ______ integer.
7. If m and n are odd integers, then $m - n$ is an ______ integer.
8. If m and n are odd integers, then mn is an ______ integer.
9. If m is an even integer and n is an odd integer, then $m - n$ is an ______ integer.
10. If m is an even integer and n is an odd integer, then $m + n$ is an ______ integer.

Applications of Patterns in Mathematics

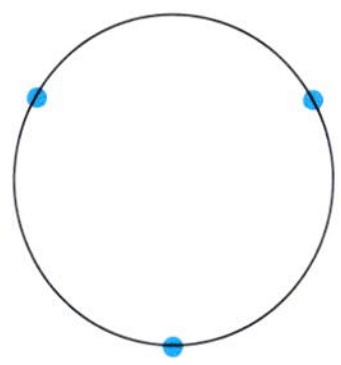

For the circle shown at the left, use a straight line to connect each dot on the circle with every other dot on the circle. How many different straight lines are there?

Follow the same procedure for each of the circles shown below. How many different straight lines are there for each?

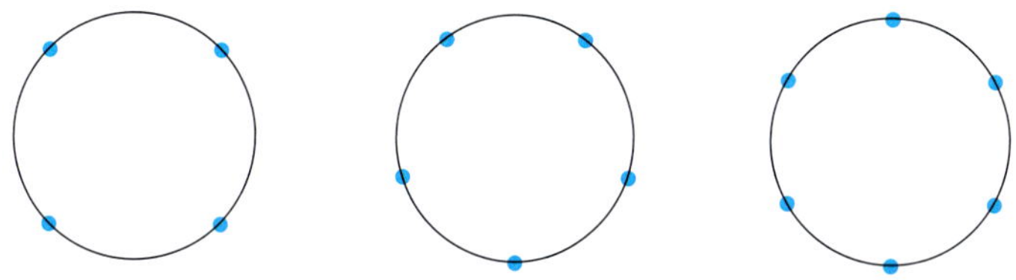

1. Find a pattern to describe the number of dots on a circle and the corresponding number of different lines drawn.

2. Use the pattern from Exercise 1 to determine the number of different lines that would be drawn in a circle with 7 dots and in a circle with 8 dots.

3. You are arranging a tennis tournament with nine players. How many singles matches will be played among the nine players if each player plays each of the other players once?

Now consider **triangular numbers,** which were studied by ancient Greek mathematicians. The numbers 1, 3, 6, 10, 15, 21 are the first six triangular numbers. Note in the following diagram that a triangle can be formed using the number of dots that correspond to a triangular number.

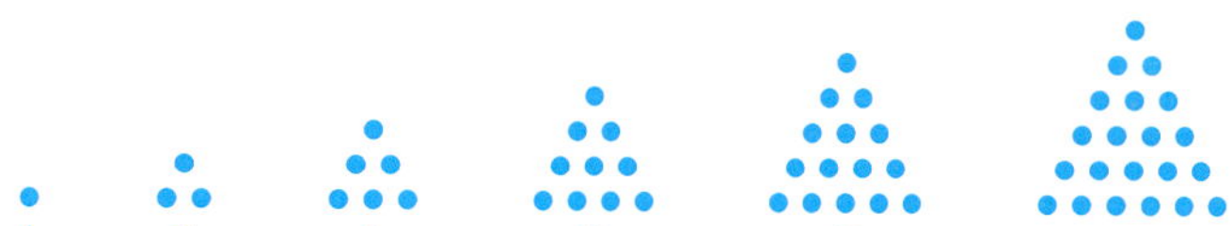

Observe that the number of dots in a row is one more than the number in the row above. The total number of dots can be found by addition.

$$1 = 1 \qquad 1 + 2 = 3$$
$$1 + 2 + 3 = 6 \qquad 1 + 2 + 3 + 4 = 10$$
$$1 + 2 + 3 + 4 + 5 = 15 \qquad 1 + 2 + 3 + 4 + 5 + 6 = 21$$

4. Use this pattern to find the seventh triangular number and the eighth triangular number. Check your answers by drawing the corresponding triangles of dots.

5. Discuss the relationship between triangular numbers and the pattern describing the correspondence between the number of dots on a circle and the number of different lines drawn (see Exercise 1).

6. Suppose you are in charge of scheduling softball games for a league. There are 10 teams in the league. Use the pattern of triangular numbers to determine the number of games that must be scheduled.

Chapter Summary

Key Words	Examples
A **set** is a collection of objects. The objects in the set are called the **elements** of the set. The **roster method** of writing sets encloses a list of the elements in braces. [1.1.1, p. 2]*	The set of even natural numbers less than 10 is written $A = \{2, 4, 6, 8\}$. The elements in this set are the numbers 2, 4, 6, and 8.
The set of **natural numbers** is $\{1, 2, 3, 4, 5, 6, 7, \ldots\}$. The set of **integers** is $\{\ldots, -4, -3, -2, -1, 0, 1, 2, 3, 4, \ldots\}$. [1.1.1, pp. 2–3]	
A number a **is less than** another number b, written $a < b$, if a is to the left of b on the number line. A number a **is greater than** another number b, written $a > b$, if a is to the right of b on the number line. The symbol $\le$ means **is less than or equal to**. The symbol $\ge$ means **is greater than or equal to**. [1.1.1, pp. 3–4]	$-9 < 7$ $4 > -5$
Two numbers that are the same distance from zero on the number line but on opposite sides of zero are **opposite numbers**, or **opposites**. The opposite of a number is also called its **additive inverse**. [1.1.2, p. 4]	5 and -5 are opposites.
The **absolute value** of a number is its distance from zero on the number line. The absolute value of a number is positive or zero. [1.1.2, p. 5]	$\lvert -12 \rvert = 12$ $-\lvert 7 \rvert = -7$
A **rational number** is a number that can be written in the form $\frac{a}{b}$, where a and b are integers and $b \neq 0$. A rational number written in this form is commonly called a **fraction**. [1.3.1, p. 21]	$\frac{3}{8}, \frac{-9}{11}$, and $\frac{4}{1}$ are rational numbers.
An **irrational number** is a number that has a decimal representation that never terminates or repeats. [1.3.1, p. 22]	π, $\sqrt{2}$, and 2.171171117111117... are irrational numbers.
The rational numbers and the irrational numbers taken together are called the **real numbers**. [1.3.1, p. 22]	$\frac{3}{8}, \frac{-9}{11}, \frac{4}{1}, \pi, \sqrt{2}$, and 2.171171117111117... are real numbers.
Percent means "parts of 100." [1.3.4, p. 26]	63% means 63 of 100 equal parts.
An expression of the form a^n is in **exponential form**, where a is the **base** and n is the **exponent**. [1.4.1, p. 33]	6^3 is an exponential expression in which 6 is the base and 3 is the exponent.

*The numbers in brackets refer to the objective where the Key Word or Essential Rule or Procedure is first introduced. For example, the reference [1.1.1] stands for Chapter 1, Section 1, Objective 1. This notation will be used in all Chapter Summaries throughout the text.

A **variable** is a letter that is used to stand for a quantity that is unknown or that can change. A **variable expression** is an expression that contains one or more variables. [1.5.1, p. 39]	$5x - 4y + 7z$ is a variable expression. It contains the variables x, y, and z.
The **terms of a variable expression** are the addends of the expression. Each term is a **variable term** or a **constant term**. [1.5.1, p. 39]	The expression $4a^2 - 6b^3 + 7$ has three terms: $4a^2$, $-6b^3$, and 7. $4a^2$ and $-6b^3$ are variable terms. 7 is a constant term.
A variable term is composed of a **numerical coefficient** and a **variable part**. [1.5.1, p. 39]	For the expression $8p^4r$, 8 is the coefficient and p^4r is the variable part.
Replacing the variables in a variable expression with numbers and then simplifying the numerical expression is called **evaluating the variable expression**. [1.5.1, p. 40]	To evaluate $a^2 - 2b$ when $a = -3$ and $b = -5$, simplify the expression $(-3)^2 - 2(-5)$.
The **additive inverse** of a number is the opposite of the number [1.5.2, p. 43]	The additive inverse of 8 is −8. The additive inverse of −15 is 15.
The **multiplicative inverse** of a number is the reciprocal of the number. [1.5.2, p. 43]	The multiplicative inverse of $\frac{3}{8}$ is $\frac{8}{3}$.
Like terms of a variable expression are the terms with the same variable part. [1.5.3, p. 44]	For the expression $2st - 3t + 9s - 11st$, the terms $2st$ and $-11st$ are like terms.

Essential Rules and Procedures

To add two integers with the same sign, add the absolute values of the numbers. Then attach the sign of the addends. [1.2.1, p. 10]	$9 + 3 = 12$ $-9 + (-3) = -12$
To add two integers with different signs, find the absolute value of each number. Then subtract the smaller of these absolute values from the larger one. Attach the sign of the number with the greater absolute value. [1.2.1, p. 10]	$9 + (-3) = 6$ $-9 + 3 = -6$
To subtract one integer from another, add the opposite of the second integer to the first integer. [1.2.2, p. 11]	$6 - 11 = 6 + (-11) = -5$ $-2 - (-8) = -2 + 8 = 6$
To multiply two integers with the same sign, multiply the absolute values of the numbers. The product is positive. [1.2.3, p. 12]	$3(5) = 15$ $(-3)(-5) = 15$
To multiply two integers with different signs, multiply the absolute values of the numbers. The product is negative. [1.2.3, p. 12]	$-3(5) = -15$ $3(-5) = -15$
To divide two integers with the same sign, divide the absolute values of the numbers. The quotient is positive. [1.2.4, p. 13]	$14 \div 2 = 7$ $-14 \div (-2) = 7$
To divide two integers with different signs, divide the absolute values of the numbers. The quotient is negative. [1.2.4, p. 13]	$14 \div (-2) = -7$ $-14 \div 2 = -7$

Zero and One in Division [1.2.4, p. 13]

$\frac{0}{a} = 0, a \neq 0$

Division by zero is undefined.

$\frac{a}{a} = 1, a \neq 0$

$\frac{a}{1} = a$

$\frac{0}{3} = 0$

$\frac{9}{0}$ is undefined.

$\frac{-2}{-2} = 1$

$\frac{-6}{1} = -6$

To convert a percent to a decimal, remove the percent sign and multiply by 0.01. [1.3.4, p. 26]

$85\% = 85(0.01) = 0.85$

To convert a percent to a fraction, remove the percent sign and multiply by $\frac{1}{100}$. [1.3.4, p. 26]

$25\% = 25 \cdot \frac{1}{100} = \frac{25}{100} = \frac{1}{4}$

To convert a decimal or a fraction to a percent, multiply by 100%. [1.3.4, p. 27]

$0.5 = 0.5(100\%) = 50\%$

$\frac{3}{8} = \frac{3}{8} \cdot 100\% = \frac{300}{8}\% = 37.5\%$

Order of Operations Agreement [1.4.2, p. 35]

Step 1 Perform operations inside grouping symbols.
Step 2 Simplify exponential expressions.
Step 3 Do multiplication and division as they occur from left to right.
Step 4 Do addition and subtraction as they occur from left to right.

$$\begin{aligned} &16 \div (-2)^3 + (7 - 12) \\ &= 16 \div (-2)^3 + (-5) \\ &= 16 \div (-8) + (-5) \\ &= -2 + (-5) \\ &= -7 \end{aligned}$$

The Commutative Property of Addition
If a and b are real numbers, then $a + b = b + a$. [1.5.2, p. 41]

$5 + 2 = 7$ and $2 + 5 = 7$

The Commutative Property of Multiplication
If a and b are real numbers, then $ab = ba$. [1.5.2, p. 42]

$6(-3) = -18$ and $-3(6) = -18$

The Associative Property of Addition
If a, b, and c are real numbers, then $(a + b) + c = a + (b + c)$. [1.5.2, p. 42]

$-1 + (4 + 7) = -1 + 11 = 10$
$(-1 + 4) + 7 = 3 + 7 = 10$

The Associative Property of Multiplication
If a, b, and c are real numbers, then $(ab)c = a(bc)$. [1.5.2, p. 42]

$(-2 \cdot 5) \cdot 3 = -10 \cdot 3 = -30$
$-2 \cdot (5 \cdot 3) = -2 \cdot 15 = -30$

The Addition Property of Zero
If a is a real number, then $a + 0 = 0 + a = a$. [1.5.2, p. 42]

$9 + 0 = 9 \quad 0 + 9 = 9$

The Multiplication Property of Zero
If a is a real number, then $a \cdot 0 = 0 \cdot a = 0$. [1.5.2, p. 42]

$-8(0) = 0 \quad 0(-8) = 0$

The Multiplication Property of One
If a is a real number, then $1 \cdot a = a \cdot 1 = a$. [1.5.2, p. 43]

$7 \cdot 1 = 7 \qquad 1 \cdot 7 = 7$

The Inverse Property of Addition
If a is a real number, then $a + (-a) = (-a) + a = 0$. [1.5.2, p. 43]

$4 + (-4) = 0 \qquad -4 + 4 = 0$

The Inverse Property of Multiplication
If a is a real number and $a \neq 0$, then $a \cdot \frac{1}{a} = \frac{1}{a} \cdot a = 1$. [1.5.2, p. 43]

$6 \cdot \frac{1}{6} = 1 \qquad \frac{1}{6} \cdot 6 = 1$

The Distributive Property
If a, b, and c are real numbers, then $a(b + c) = ab + ac$. [1.5.2, p. 43]

$$\begin{aligned} 5(2 + 3) &= 5 \cdot 2 + 5 \cdot 3 \\ &= 10 + 15 \\ &= 25 \end{aligned}$$

Chapter Review Exercises

1. Use the roster method to write the set of natural numbers less than 7.

2. Write $\frac{5}{8}$ as a percent.

3. Evaluate: $-|-4|$

4. Subtract: $16 - (-30) - 42$

5. Divide: $-561 \div (-33)$

6. Write $\frac{7}{9}$ as a decimal. Place a bar over the repeating digits of the decimal.

7. Simplify: $(6.02)(-0.89)$

8. Simplify: $\frac{-10 + 2}{2 + (-4)} \div 2 + 6$

9. Write 0.672 as a percent.

10. Write $79\frac{1}{2}\%$ as a fraction.

11. Multiply: $(-5)(-6)(3)$

12. Subtract: $6.039 - 12.92$

13. Given $A = \{-5, -3, 0\}$, which elements of set A are less than or equal to -3?

14. Write 7% as a decimal.

15. Evaluate: $\frac{3}{4} \cdot (4)^2$

16. Place the correct symbol, < or >, between the two numbers.

    $-2 \quad -40$

17. Add: $-\frac{2}{5} + \frac{7}{15}$

18. Evaluate: $(-3^3) \cdot 2^2$

19. Divide: $-\frac{7}{16} \div \frac{3}{8}$

20. Simplify: $2^3 \div 4 - 2(2 - 7)$

21. Write $2\frac{8}{9}$ as a percent. Write the remainder in fractional form.

22. Given $C = \{-12, -8, -1, 7\}$, find

a. the opposite of each element of set C.
b. the absolute value of each element of set C.

23. Divide: $0.2654 \div (-0.023)$
Round to the nearest tenth.

24. Simplify: $(7 - 2)^2 - 5 - 3 \cdot 4$

25. Add: $-12 + 8 + (-4)$

26. Add: $-\frac{5}{8} + \frac{1}{6}$

27. Add: $-1.329 + 4.89$

28. Subtract: $-5 - 22 - (-13) - 19 - (-6)$

29. Multiply: $\left(\frac{1}{3}\right)\left(-\frac{4}{5}\right)\left(\frac{3}{8}\right)$

30. Place the correct symbol, $<$ or $>$, between the two numbers.

$-43 \quad -34$

31. Simplify: $-7y^2 + 6y^2 - (-2y^2)$

32. Simplify: $(12x)\left(\frac{1}{4}\right)$

33. Simplify: $5(2x + 4) - 3(x - 6)$

34. Evaluate $a^2 - 3b$ when $a = 2$ and $b = -4$.

35. Complete the statement by using the Inverse Property of Addition.
$-9 + ? = 0$

36. Simplify: $3[2x - 3(x - 2y)] + 3y$

37. Simplify: $5(3 - 7b)$

38. Simplify: $2x + 3[4 - (3x - 7)]$

39. Identify the property that justifies the statement.
$-4(3) = 3(-4)$

40. Simplify: $3(8 - 2x)$

41. Evaluate $\frac{-2ab}{2b - a}$ when $a = -4$ and $b = 6$.

42. Simplify: $(-3)(-12y)$

43. Simplify: $4(3x - 2) - 7(x + 5)$

44. Simplify: $(16x)\left(\frac{1}{8}\right)$

45. Simplify: $-3(2x^2 - 7y^2)$

46. Evaluate $3(a - c) - 2ab$ when $a = 2$, $b = 3$, and $c = -4$.

47. Simplify: $2x - 3(x - 2)$

48. Simplify: $2a - (-3b) - 7a - 5b$

49. Simplify: $-5(2x^2 - 3x + 6)$

50. Simplify: $3x - 7y - 12x$

51. Simplify: $3x + (-12y) - 5x - (-7y)$

52. Simplify: $\left(-\frac{5}{6}\right)(-36b)$

53. Complete the statement by using the Distributive Property.
$(6 + 3)7 = 42 + ?$

54. Simplify: $4x^2 + 9x - 6x^2 - 5x$

55. Translate "b decreased by the product of seven and b" into a variable expression.

56. Translate "ten divided by the difference between y and two" into a variable expression.

57. Find the temperature after a rise of 14°C from −6°C.

58. The daily low temperatures, in degrees Celsius, for a three-day period were recorded as follows: −8°, 7°, −5°. Find the average low temperature for the three-day period.

59. Use the table to find the difference between the record high temperature and the record low temperature for January in Bismarck, North Dakota.

Bismarck, North Dakota

	Jan	*Feb*	*Mar*	*Apr*	*May*	*Jun*	*Jul*	*Aug*	*Sep*	*Oct*	*Nov*	*Dec*
Record High	62°F (1981)	69°F (1992)	80°F (1967)	93°F (1992)	96°F (1980)	107°F (1988)	109°F (1973)	107°F (1973)	105°F (1959)	95°F (1963)	75°F (1978)	65°F (1979)
Record Low	−44°F (1950)	−43°F (1994)	−28°F (1995)	−12°F (1975)	15°F (1967)	30°F (1969)	35°F (1971)	33°F (1988)	11°F (1974)	−10°F (1991)	−30°F (1985)	−40°F (1983)

Source: **www.weather.com**

60. Find the temperature after a rise of 7°C from −13°C.

61. The temperature on the surface of the planet Venus is 480°C. The temperature on the surface of the planet Pluto is −234°C. Find the difference between the surface temperatures on Venus and Pluto.

62. Translate and simplify "eight times the quotient of twice a number and sixteen."

63. Translate and simplify "the product of four and the sum of two and five times a number."

64. The length of the base of a triangle is 15 in. more than the height of the triangle. Express the length of the base of the triangle in terms of the height of the triangle.

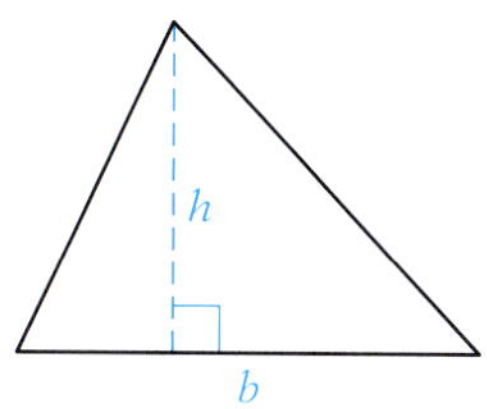

65. A coffee merchant made 20 lb of a blend of coffee using only mocha java beans and espresso beans. Use the same variable to express the amounts of mocha java beans and espresso beans in the coffee blend.

Chapter Test

1. Write 55% as a fraction.

2. Given $B = \{-8, -6, -4, -2\}$, which elements of set B are less than -5?

3. Subtract: $-9 - (-6)$

4. Write $\frac{3}{20}$ as a decimal.

5. Multiply: $\frac{3}{4}\left(-\frac{2}{21}\right)$

6. Divide: $-75 \div 5$

7. Evaluate: $\left(-\frac{2}{3}\right)^3 \cdot 3^2$

8. Add: $-7 + (-3) + 12$

9. Use the roster method to write the set of positive integers less than or equal to 6.

10. Write 1.59 as a percent.

11. Evaluate: $|-29|$

12. Place the correct symbol, < or >, between the two numbers.

$$-47 \quad -68$$

13. Subtract: $-\frac{4}{9} - \frac{5}{6}$

14. Divide: $-\frac{5}{8} \div \left(-\frac{3}{4}\right)$

15. Write $\frac{3}{13}$ as a percent. Round to the nearest tenth of a percent.

16. Write 6.2% as a decimal.

17. Subtract: $13 - (-5) - 4$

18. Write $\frac{13}{30}$ as a decimal. Place a bar over the repeating digits of the decimal.

19. Multiply: $(-0.9)(2.7)$

20. Add: $15 + (-8) + (-19)$

21. Given $A = \{-17, -6, 5, 9\}$, find

a. the opposite of each element of set A.
b. the absolute value of each element of set A.

22. Add: $-18.354 + 6.97$

23. Multiply: $-4(8)(-5)$

24. Simplify: $(9y)4$

25. Simplify: $7x + 5y - 3x - 8y$

26. Simplify: $8n - (6 - 2n)$

27. Identify the property that justifies the statement.
$\frac{3}{8}(1) = \frac{3}{8}$

28. Simplify: $-4(-x + 10)$

29. Simplify: $\frac{2}{3}x^2 - \frac{7}{12}x^2$

30. Simplify: $(-10x)\left(-\frac{2}{5}\right)$

31. Simplify: $(-4y^2 + 8)6$

32. Complete the statement by using the Inverse Property of Addition.
$-19 + ? = 0$

33. Evaluate $\frac{-3ab}{2a + b}$ when $a = -1$ and $b = 4$.

34. Simplify: $5(x + y) - 8(x - y)$

35. Simplify: $3(x^2 - 5x + 4)$

36. Evaluate $4(b - a) + bc$ when $a = 2$, $b = -3$, and $c = 4$.

37. Simplify: $6x - 3(y - 7x) + 2(5x - y)$

38. Simplify: $9(-4) \div [2(8 - 5)^2]$

39. Find the temperature after a rise of 12°C from −8°C.

40. Use the table to find the average of the record low temperatures in Fairbanks, Alaska, for the first four months of the year.

Fairbanks, Alaska

	Jan	Feb	Mar	Apr	May	Jun	Jul	Aug	Sep	Oct	Nov	Dec
Record High	50°F (1981)	47°F (1987)	56°F (1994)	74°F (1960)	89°F (1960)	96°F (1969)	94°F (1975)	93°F (1994)	84°F (1957)	65°F (1969)	49°F (1997)	44°F (1985)
Record Low	−61°F (1969)	−58°F (1993)	−49°F (1956)	−24°F (1986)	−1°F (1964)	30°F (1950)	35°F (1959)	27°F (1987)	3°F (1992)	−27°F (1975)	−46°F (1990)	−62°F (1961)

Source: **www.weather.com**

41. Translate "the difference between the sum of a and b and the square of b" into a variable expression.

42. Translate and simplify "twenty times the sum of a number and nine."

43. Translate and simplify "two more than a number added to the difference between the number and three."

44. The distance from Neptune to the sun is thirty times the distance from Earth to the sun. Express the distance from Neptune to the sun in terms of the distance from Earth to the sun.

45. A nine-foot board is cut into two pieces of different lengths. Use the same variable to express the lengths of the two pieces.

Without electricians, like the one shown here, we'd find it difficult to get through our day-to-day lives. Their knowledge of the relationship between voltage, current, and resistance makes them experts on ensuring that homes and buildings have electricity. We rely on electricity for power, which provides energy to turn on a television or a light, as well as a computer. The **Project on pages 179–181** *demonstrates how key elements of electricity are measured.*

Applications

2 Solving Equations and Inequalities

Objectives

Section 2.1

1. Determine whether a given number is a solution of an equation
2. Solve equations of the form $x + a = b$
3. Solve equations of the form $ax = b$
4. Applications of percent

Section 2.2

1. Solve equations of the form $ax + b = c$
2. Solve equations of the form $ax + b = cx + d$
3. Solve equations containing parentheses
4. Application problems

Section 2.3

1. Translate a sentence into an equation and solve
2. Consecutive integer problems
3. Value mixture problems
4. Uniform motion problems

Section 2.4

1. Perimeter problems
2. Problems involving angles formed by intersecting lines
3. Problems involving the angles of a triangle

Section 2.5

1. Markup problems
2. Discount problems

Section 2.6

1. Investment problems
2. Percent mixture problems

Section 2.7

1. Set operations and interval notation
2. Solve inequalities in one variable
3. Solve compound inequalities
4. Application problems

Section 2.8

1. Absolute value equations
2. Absolute value inequalities
3. Application problems

WEB

Need help? For online student resources, such as section quizzes, visit this web site: math.college.hmco.com

Prep Test

1. Subtract: $8 - 12$
2. Multiply: $-\frac{3}{4}\left(-\frac{4}{3}\right)$
3. Multiply: $-\frac{5}{8}(16)$
4. Write 90% as a decimal.
5. Write 0.75 as a percent.
6. Evaluate $3x^2 - 4x - 1$ when $x = -4$.
7. Simplify: $3x - 5 + 7x$
8. Simplify: $8x - 9 - 8x$
9. Simplify: $6x - 3(6 - x)$

Go Figure

How can you cut a donut into 8 equal pieces with three cuts of the knife?

2.1 Introduction to Equations

1 Determine whether a given number is a solution of an equation

VIDEO & DVD CD TUTOR WEB SSM

Point of Interest

One of the most famous equations ever stated is $E = mc^2$. *This equation, stated by Albert Einstein, shows that there is a relationship between mass* m *and energy* E.

An **equation** expresses the equality of two mathematical expressions. The expressions can be either numerical or variable expressions.

$$\left.\begin{aligned} 9 + 3 &= 12 \\ 3x - 2 &= 10 \\ y^2 + 4 &= 2y - 1 \\ z &= 2 \end{aligned}\right\} \text{Equations}$$

The equation at the right is true if the variable is replaced by 5.

$x + 8 = 13$

$5 + 8 = 13$ A true equation

The equation $x + 8 = 13$ is false if the variable is replaced by 7.

$7 + 8 = 13$ A false equation

A **solution** of an equation is a number that, when substituted for the variable, results in a true equation. 5 is a solution of the equation $x + 8 = 13$. 7 is not a solution of the equation $x + 8 = 13$.

Example 1 Is -3 a solution of the equation $4x + 16 = x^2 - 5$?

Solution

$4x + 16$	$= x^2 - 5$
$4(-3) + 16$	$(-3)^2 - 5$
$-12 + 16$	$9 - 5$

$4 = 4$

- Replace the variable by the given number, -3.
- Evaluate the numerical expressions using the Order of Operations Agreement.
- Compare the results. If the results are equal, the given number is a solution. If the results are not equal, the given number is not a solution.

Yes, -3 is a solution of the equation $4x + 16 = x^2 - 5$.

Problem 1 Is $\frac{1}{4}$ a solution of $5 - 4x = 8x + 2$?

Solution See page S4.

Example 2 Is -4 a solution of $4 + 5x = x^2 - 2x$?

Solution

$4 + 5x$	$= x^2 - 2x$
$4 + 5(-4)$	$(-4)^2 - 2(-4)$
$4 + (-20)$	$16 - (-8)$

$-16 \neq 24$

No, -4 is not a solution of the equation $4 + 5x = x^2 - 2x$.

- Replace the variable by the given number, -4.
- Evaluate the numerical expressions using the Order of Operations Agreement.
- Compare the results. If the results are equal, the given number is a solution. If the results are not equal, the given number is not a solution.

Problem 2 Is 5 a solution of $10x - x^2 = 3x - 10$?

Solution See page S4.

2 Solve equations of the form $x + a = b$

VIDEO & DVD CD TUTOR WEB SSM

To **solve an equation** means to find a solution of the equation. The simplest equation to solve is an equation of the form **variable = constant**, because the constant is the solution.

If $x = 5$, then 5 is the solution of the equation because $5 = 5$ is a true equation.

> **Take Note**
>
> Think of an equation as a balance scale. If the weights added to each side of the equation are not the same, the pans no longer balance.

The solution of the equation shown at the right is 7.

$$x + 2 = 9 \qquad 7 + 2 = 9$$

Note that if 4 is added to each side of the equation, the solution is still 7.

$$x + 2 + 4 = 9 + 4$$
$$x + 6 = 13 \qquad 7 + 6 = 13$$

If -5 is added to each side of the equation, the solution is still 7.

$$x + 2 + (-5) = 9 + (-5)$$
$$x - 3 = 4 \qquad 7 - 3 = 4$$

This illustrates the Addition Property of Equations.

Addition Property of Equations

The same number or variable term can be added to each side of an equation without changing the solution of the equation.

This property is used in solving equations. Note the effect of adding, to each side of the equation $x + 2 = 9$, the opposite of the constant term 2. After each side of the equation is simplified, the equation is in the form variable = constant. The solution is the constant.

$$x + 2 = 9$$
$$x + 2 + (-2) = 9 + (-2)$$
$$x + 0 = 7$$
$$x = 7$$

variable = constant

The solution is 7.

In solving an equation, the goal is to rewrite the given equation in the form variable = constant. The Addition Property of Equations can be used to rewrite an equation in this form. The Addition Property of Equations is used to **remove a term** from one side of an equation **by adding the opposite of that term** to each side of the equation.

Example 3 Solve and check: $y - 6 = 9$

Solution

$$y - 6 = 9$$
• The goal is to rewrite the equation in the form *variable = constant.*

$$y - 6 + 6 = 9 + 6$$
• Add the opposite of the constant term −6 to each side of the equation (the Addition Property of Equations).

$$y + 0 = 15$$
• Simplify using the Inverse Property of Addition.

$$y = 15$$
• Simplify using the Addition Property of Zero. Now the equation is in the form *variable = constant.*

Check

$$\begin{array}{c|c} y - 6 & 9 \\ \hline 15 - 6 & 9 \\ 9 & 9 \end{array}$$

$9 = 9$ • This is a true equation. The solution checks.

The solution is 15. • Write the solution.

Problem 3 Solve and check: $x - \frac{1}{3} = -\frac{3}{4}$

Solution See page S4.

Because subtraction is defined in terms of addition, the Addition Property of Equations makes it possible to subtract the same number from each side of an equation without changing the solution of the equation.

Solve: $y + \frac{1}{2} = \frac{5}{4}$

The goal is to rewrite the equation in the form *variable = constant*.

$$y + \frac{1}{2} = \frac{5}{4}$$

Add the opposite of the constant term $\frac{1}{2}$ to each side of the equation. This is equivalent to subtracting $\frac{1}{2}$ from each side of the equation.

$$y + \frac{1}{2} - \frac{1}{2} = \frac{5}{4} - \frac{1}{2}$$

$$y + 0 = \frac{5}{4} - \frac{2}{4}$$

$$y = \frac{3}{4}$$

The solution $\frac{3}{4}$ checks. The solution is $\frac{3}{4}$.

Example 4 Solve: $\frac{1}{2} = x + \frac{2}{3}$

Solution

$$\frac{1}{2} = x + \frac{2}{3}$$

$$\frac{1}{2} - \frac{2}{3} = x + \frac{2}{3} - \frac{2}{3}$$ • Subtract $\frac{2}{3}$ from each side of the equation.

$$-\frac{1}{6} = x$$ • Simplify each side of the equation.

The solution is $-\frac{1}{6}$.

Problem 4 Solve: $-8 = 5 + x$

Solution See page S4.

Note from the solution to Example 4 that an equation can be rewritten in the form *constant* = *variable*. Whether the equation is written in the form *variable* = *constant* or in the form *constant* = *variable*, the solution is the constant.

3 Solve equations of the form $ax = b$

The solution of the equation shown at the right is 3.	$2x = 6$	$2 \cdot 3 = 6$
Note that if each side of the equation is multiplied by 5, the solution is still 3.	$5 \cdot 2x = 5 \cdot 6$ $10x = 30$	$10 \cdot 3 = 30$
If each side is multiplied by −4, the solution is still 3.	$(-4) \cdot 2x = (-4) \cdot 6$ $-8x = -24$	$-8 \cdot 3 = -24$

This illustrates the Multiplication Property of Equations.

Multiplication Property of Equations

Each side of an equation can be multiplied by the same nonzero number without changing the solution of the equation.

This property is used in solving equations. Note the effect of multiplying each side of the equation $2x = 6$ by the reciprocal of the coefficient 2. After each side of the equation is simplified, the equation is in the form variable = constant. The solution is the constant.

$$2x = 6$$

$$\frac{1}{2} \cdot 2x = \frac{1}{2} \cdot 6$$

$$1x = 3$$

$$x = 3$$

variable = constant

The solution is 3.

In solving an equation, the goal is to rewrite the given equation in the form variable = constant. The Multiplication Property of Equations can be used to rewrite an equation in this form. The Multiplication Property of Equations is used to **remove a coefficient** from a variable term in an equation **by multiplying each side of the equation by the reciprocal of the coefficient.**

Example 5 Solve: $\frac{3x}{4} = -9$

Solution

$\frac{3x}{4} = -9$ • $\frac{3x}{4} = \frac{3}{4}x$

$\frac{4}{3} \cdot \frac{3}{4}x = \frac{4}{3}(-9)$ • Multiply each side of the equation by the reciprocal of the coefficient $\frac{3}{4}$ (the Multiplication Property of Equations).

$1x = -12$ • Simplify using the Inverse Property of Multiplication.

$x = -12$ • Simplify using the Multiplication Property of One. Now the equation is in the form *variable = constant.*

The solution is −12. • Write the solution.

Problem 5 Solve: $-\frac{2x}{5} = 6$

Solution See page S4.

Because division is defined in terms of multiplication, the Multiplication Property of Equations makes it possible to divide each side of an equation by the same number without changing the solution of the equation.

Solve: $8x = 16$

The goal is to rewrite the equation in the form *variable = constant*. $8x = 16$

Multiply each side of the equation by the reciprocal of 8. This is equivalent to dividing each side by 8.

$$\frac{8x}{8} = \frac{16}{8}$$
$$x = 2$$

The solution 2 checks. The solution is 2.

When using the Multiplication Property of Equations to solve an equation, multiply each side of the equation by the reciprocal of the coefficient when the coefficient is a fraction. Divide each side of the equation by the coefficient when the coefficient is an integer or a decimal.

Example 6 Solve and check: $4x = 6$

Solution

$4x = 6$

$\frac{4x}{4} = \frac{6}{4}$ • Divide each side of the equation by 4, the coefficient of x.

$x = \frac{3}{2}$ • Simplify each side of the equation.

Check

$$\begin{array}{c|c} \multicolumn{2}{c}{4x = 6} \\ \hline 4\left(\frac{3}{2}\right) & 6 \\ \multicolumn{2}{c}{6 = 6} \end{array}$$

• This is a true equation. The solution checks.

The solution is $\frac{3}{2}$.

Problem 6 Solve and check: $6x = 10$

Solution See page S4.

Before using one of the Properties of Equations, check to see whether one or both sides of the equation can be simplified. In Example 7, like terms appear on the left side of the equation. The first step in solving this equation is to combine the like terms so that there is only one variable term on the left side of the equation.

Example 7 Solve: $5x - 9x = 12$

Solution

$$5x - 9x = 12$$

$$-4x = 12 \quad \text{• Combine like terms.}$$

$$\frac{-4x}{-4} = \frac{12}{-4} \quad \text{• Divide each side of the equation by } -4.$$

$$x = -3$$

The solution is -3.

Problem 7 Solve: $4x - 8x = 16$

Solution See page S4.

4 Applications of percent

VIDEO & DVD | CD TUTOR | WEB | SSM

The solution of a problem that involves a percent requires solving the basic percent equation.

The Basic Percent Equation

$$\text{Percent} \cdot \text{Base} = \text{Amount}$$
$$P \cdot B = A$$

To translate a problem involving a percent into an equation, remember that the word *of* translates into "multiply" and the word *is* translates into "=". The base usually follows the word *of*.

20% of what number is 30?

Given: $P = 20\% = 0.20$
$A = 30$
Unknown: Base

$$PB = A$$
$$(0.20)B = 30$$
$$\frac{0.20B}{0.20} = \frac{30}{0.20}$$
$$B = 150$$

The number is 150.

What percent of 40 is 30?

Given: $B = 40$
$A = 30$
Unknown: Percent

$$PB = A$$
$$P(40) = 30$$
$$40P = 30$$
$$\frac{40P}{40} = \frac{30}{40}$$
$$P = \frac{3}{4}$$

We must write the fraction as a percent in order to answer the question.

$$P = 75\%$$

30 is 75% of 40.

Find 25% of 200.

Given: $P = 25\% = 0.25$
$B = 200$
Unknown: Amount

$$PB = A$$
$$0.25(200) = A$$
$$50 = A$$

25% of 200 is 50.

In most cases, we write a percent as a decimal before solving the basic percent equation. However, some percents are more easily written as a fraction. For example,

$$33\frac{1}{3}\% = \frac{1}{3} \qquad 66\frac{2}{3}\% = \frac{2}{3} \qquad 16\frac{2}{3}\% = \frac{1}{6} \qquad 83\frac{1}{3}\% = \frac{5}{6}$$

Example 8 12 is $33\frac{1}{3}\%$ of what number?

Solution

$$PB = A \qquad \bullet\ 33\frac{1}{3}\% = \frac{1}{3}$$
$$\frac{1}{3}B = 12$$
$$3 \cdot \frac{1}{3}B = 3 \cdot 12$$
$$B = 36$$

The number is 36.

Problem 8 27 is what percent of 60?

Solution See page S4.

Example 9 During a recent year, nearly 1.2 million dogs or litters were registered with the American Kennel Club. The most popular breed was the Labrador retriever, with 172,841 registered. What percent of the registrations were Labrador retrievers? Round to the nearest tenth of a percent. (*Source:* American Kennel Club)

Strategy To find the percent, solve the basic percent equation using $B = 1.2$ million $= 1{,}200{,}000$ and $A = 172{,}841$. The percent is unknown.

Solution

$$\begin{aligned} PB &= A \\ P(1{,}200{,}000) &= 172{,}841 \\ \frac{P(1{,}200{,}000)}{1{,}200{,}000} &= \frac{172{,}841}{1{,}200{,}000} \\ P &\approx 0.144 \\ P &\approx 14.4\% \end{aligned}$$

Approximately 14.4% of the registrations were Labrador retrievers.

Problem 9 A student correctly answered 72 of the 80 questions on an exam. What percent of the questions were answered correctly?

Solution See page S5.

Point of Interest

As noted at the right, 172,841 of the dogs or litters registered were Labrador retrievers. Listed below are the next most popular breeds and their registrations.

Golden retrievers: 66,300
German shepherds: 57,660
Dachshunds: 54,773
Beagles: 52,026

Example 10 According to the Centers for Disease Control and Prevention, 30.8% of the adult population of Kentucky smokes. How many adults in Kentucky smoke? Use a figure of 3,000,000 for the number of adults in Kentucky.

Strategy To find the number of adults who smoke, solve the basic percent equation using $B = 3{,}000{,}000$ and $P = 30.8\% = 0.308$. The amount is unknown.

Solution

$$\begin{aligned} PB &= A \\ 0.308(3{,}000{,}000) &= A \\ 924{,}000 &= A \end{aligned}$$

Approximately 924,000 adults in Kentucky smoke.

Problem 10 The price of a digital camcorder is \$895. A 6% sales tax is added to the price. How much is the sales tax?

Solution See page S5.

The simple interest that an investment earns is given by the equation $I = Prt$, where I is the simple interest, P is the principal, or amount invested, r is the simple interest rate, and t is the time.

A \$1500 investment has an annual simple interest rate of 7%. Find the simple interest earned on the investment after 18 months.

The time is given in months, but the interest rate is an annual rate. Therefore, we must convert 18 months to years.

18 months = 1.5 years

Take Note

$$\begin{aligned} 18 \text{ months} &= \frac{18}{12} \text{ years} \\ &= 1.5 \text{ years} \end{aligned}$$

To find the interest earned, solve the equation $I = Prt$ for I using $P = 1500$, $r = 0.07$, and $t = 1.5$.

$$I = Prt$$
$$I = 1500(0.07)(1.5)$$
$$I = 157.5$$

The interest earned is \$157.50.

Example 11 In April, Marshall Wardell was charged \$8.72 in interest on an unpaid credit card balance of \$545. Find the annual interest rate for this credit card.

Strategy The interest is \$8.72. Therefore, $I = 8.72$. The unpaid balance is \$545. This is the principal on which interest is calculated. Therefore, $P = 545$. The time is 1 month. Because the *annual* interest rate must be found and the time is given as 1 month, write 1 month as $\frac{1}{12}$ year. Therefore, $t = \frac{1}{12}$. To find the interest rate, solve the equation $I = Prt$ for r.

Solution

$$I = Prt$$
$$8.72 = 545r\left(\frac{1}{12}\right) \qquad \bullet\ I = 8.72,\ P = 545,\ t = \frac{1}{12}$$
$$8.72 = \frac{545}{12}r$$
$$\frac{12}{545}(8.72) = \frac{12}{545}\left(\frac{545}{12}r\right)$$
$$0.192 = r$$

The annual interest rate is 19.2%.

Problem 11 Clarissa Adams purchased a \$1000 municipal bond that earns an annual simple interest rate of 6.4%. How much must she deposit into a bank account that earns 8% annual simple interest so that the interest earned from each account after one year is the same?

Solution See page S5.

Take Note

Municipal bonds are frequently sold by government agencies to pay for roads, aqueducts, and other essential public needs. An investor who purchases one of these bonds is paid interest on the cost of the bond for a certain number of years. When that time period expires, the cost of the bond is returned to the investor.

The amount of a substance in a solution can be given as a percent of the total solution. For instance, if a certain fruit juice drink is advertised as containing 27% cranberry juice, then 27% of the contents of the bottle must be cranberry juice.

Solving problems involving mixtures is based on the percent mixture equation $Q = Ar$, where Q is the quantity of a substance in the solution, A is the amount of the solution, and r is the percent concentration of the substance.

The formula for a perfume requires that the concentration of jasmine be 1.2% of the total amount of perfume. How many ounces of jasmine are in a 2-ounce bottle of this perfume?

Point of Interest

In the jewelry industry, the amount of gold in a piece of jewelry is measured in karats. *Pure gold is 24 karats. A necklace that is 18 karats is* $\frac{18}{24} = 0.75 = 75\%$ gold.

The amount of perfume is 2 oz. Therefore, $A = 2$. The percent concentration is 1.2%, so $r = 0.012$. To find the number of ounces of jasmine, solve the equation $Q = Ar$ for Q.

$$Q = Ar$$
$$Q = 2(0.012)$$
$$Q = 0.024$$

The bottle contains 0.024 oz of jasmine.

Example 12 To make a certain color of blue, 4 oz of cyan must be contained in 1 gal of paint. What is the percent concentration of cyan in the paint?

Strategy The amount of cyan is given in ounces and the amount of paint is given in gallons; we must convert ounces to gallons or gallons to ounces. For this problem, we will convert gallons to ounces: 1 gal = 128 oz.

Solution To find the percent concentration of cyan, solve the equation $Q = Ar$ for r using $Q = 4$ and $A = 128$.

$$Q = Ar$$
$$4 = 128r$$
$$\frac{4}{128} = \frac{128r}{128}$$
$$0.03125 = r$$

The percent concentration of cyan is 3.125%.

Problem 12 The concentration of sugar in Choco-Pops cereal is 25%. If a bowl of this cereal contains 2 oz of sugar, how many ounces of cereal are in the bowl?

Solution See page S5.

2.1 Concept Review

Determine whether the statement is always true, sometimes true, or never true.

1. Both sides of an equation can be multiplied by the same number without changing the solution of the equation.

2. Both sides of an equation can be divided by the same number without changing the solution of the equation.

3. For an equation of the form $ax = b$, $a \neq 0$, multiplying both sides of the equation by the reciprocal of a will result in an equation of the form $x = constant$.

4. Use the Multiplication Property of Equations to remove a term from one side of an equation.

5. Adding a negative 3 to each side of an equation yields the same result as subtracting 3 from each side of the equation.

6. In the basic percent equation, the amount follows the word *of*.

7. An equation contains an equals sign.

2.1 Exercises

1 **1.** What is the difference between an equation and an expression?

2. How can you determine whether a number is a solution of an equation?

3. Is 4 a solution of $2x = 8$?

4. Is 3 a solution of $y + 4 = 7$?

5. Is -1 a solution of $2b - 1 = 3$?

6. Is -2 a solution of $3a - 4 = 10$?

7. Is 1 a solution of $4 - 2m = 3$?

8. Is 2 a solution $7 - 3n = 2$?

9. Is 5 a solution of $2x + 5 = 3x$?

10. Is 4 a solution of $3y - 4 = 2y$?

11. Is 0 a solution of $4a + 5 = 3a + 5$?

12. Is 0 a solution of $4 - 3b = 4 - 5b$?

13. Is -2 a solution of $4 - 2n = n + 10$?

14. Is -3 a solution of $5 - m = 2 - 2m$?

15. Is 3 a solution of $z^2 + 1 = 4 + 3z$?

16. Is 2 a solution of $2x^2 - 1 = 4x - 1$?

17. Is -1 a solution of $y^2 - 1 = 4y + 3$?

18. Is -2 a solution of $m^2 - 4 = m + 3$?

19. Is 4 a solution of $x(x + 1) = x^2 + 5$?

20. Is 3 a solution of $2a(a - 1) = 3a + 3$?

21. Is $-\frac{1}{4}$ a solution of $8t + 1 = -1$?

22. Is $\frac{1}{2}$ a solution of $4y + 1 = 3$?

23. Is $\frac{2}{5}$ a solution of $5m + 1 = 10m - 3$?

24. Is $\frac{3}{4}$ a solution of $8x - 1 = 12x + 3$?

25. Is 2.1 a solution of $x^2 - 4x = x + 1.89$?

26. Is 1.5 a solution of $c^2 - 3c = 4c - 8.25$?

2 **27.** How is the Addition Property of Equations used to solve an equation?

28. Explain why the goal in solving an equation is to write the equation in the form *variable* = *constant*.

29. Without solving the equation $x + \frac{13}{15} = -\frac{21}{43}$, determine whether x is less than or greater than $-\frac{21}{43}$. Explain your answer.

30. Without solving the equation $x - \frac{11}{16} = \frac{19}{24}$, determine whether x is less than or greater than $\frac{19}{24}$. Explain your answer.

Solve and check.

31. $x + 5 = 7$
32. $y + 3 = 9$
33. $b - 4 = 11$
34. $z - 6 = 10$
35. $2 + a = 8$
36. $5 + x = 12$
37. $m + 9 = 3$
38. $t + 12 = 10$
39. $n - 5 = -2$
40. $x - 6 = -5$
41. $b + 7 = 7$
42. $y - 5 = -5$
43. $a - 3 = -5$
44. $x - 6 = -3$
45. $z + 9 = 2$
46. $n + 11 = 1$
47. $10 + m = 3$
48. $8 + x = 5$
49. $9 + x = -3$
50. $10 + y = -4$
51. $b - 5 = -3$
52. $t - 6 = -4$
53. $2 = x + 7$
54. $-8 = n + 1$
55. $4 = m - 11$
56. $-6 = y - 5$
57. $12 = 3 + w$
58. $-9 = 5 + x$
59. $4 = -10 + b$
60. $-7 = -2 + x$
61. $m + \frac{2}{3} = -\frac{1}{3}$
62. $c + \frac{3}{4} = -\frac{1}{4}$
63. $x - \frac{1}{2} = \frac{1}{2}$
64. $x - \frac{2}{5} = \frac{3}{5}$
65. $\frac{5}{8} + y = \frac{1}{8}$
66. $\frac{4}{9} + a = -\frac{2}{9}$
67. $m + \frac{1}{2} = -\frac{1}{4}$
68. $b + \frac{1}{6} = -\frac{1}{3}$
69. $x + \frac{2}{3} = \frac{3}{4}$
70. $n + \frac{2}{5} = \frac{2}{3}$
71. $-\frac{5}{6} = x - \frac{1}{4}$
72. $-\frac{1}{4} = c - \frac{2}{3}$
73. $-\frac{1}{21} = m + \frac{2}{3}$
74. $\frac{5}{9} = b - \frac{1}{3}$
75. $\frac{5}{12} = n + \frac{3}{4}$
76. $d + 1.3619 = 2.0148$
77. $w + 2.932 = 4.801$
78. $-0.813 + x = -1.096$
79. $-1.926 + t = -1.042$

3 80. How is the Multiplication Property of Equations used to solve an equation?

81. Why, when the Multiplication Property of Equations is used, must the number that multiplies each side of the equation not be zero?

82. Without solving the equation $-\frac{15}{41}x = -\frac{23}{25}$, determine whether x is less than or greater than zero. Explain your answer.

83. Without solving the equation $\frac{5}{28}x = -\frac{3}{44}$, determine whether x is less than or greater than zero. Explain your answer.

Solve and check.

84. $5x = 15$
85. $4y = 28$
86. $3b = -12$
87. $2a = -14$

88. $-3x = 6$
89. $-5m = 20$
90. $-3x = -27$
91. $-6n = -30$

92. $20 = 4c$
93. $18 = 2t$
94. $-32 = 8w$
95. $-56 = 7x$

96. $8d = 0$
97. $-5x = 0$
98. $36 = 9z$
99. $35 = -5x$

100. $-64 = 8a$
101. $-32 = -4y$
102. $-42 = 6t$
103. $-12m = -144$

104. $\frac{x}{3} = 2$
105. $\frac{x}{4} = 3$
106. $-\frac{y}{2} = 5$
107. $-\frac{b}{3} = 6$

108. $\frac{n}{7} = -4$
109. $\frac{t}{6} = -3$
110. $\frac{2}{5}x = 12$
111. $-\frac{4}{3}c = -8$

112. $\frac{5}{6}y = -20$
113. $-\frac{2}{3}d = 8$
114. $-\frac{3}{5}m = 12$
115. $\frac{2n}{3} = 2$

116. $\frac{5x}{6} = -10$
117. $\frac{-3z}{8} = 9$
118. $\frac{-4x}{5} = -12$
119. $-6 = -\frac{2}{3}y$

120. $-15 = -\frac{3}{5}x$
121. $\frac{2}{9} = \frac{2}{3}y$
122. $-\frac{6}{7} = -\frac{3}{4}b$
123. $-\frac{2}{5}m = -\frac{6}{7}$

124. $5x + 2x = 14$
125. $3n + 2n = 20$
126. $7d - 4d = 9$
127. $10y - 3y = 21$

128. $2x - 5x = 9$
129. $\frac{x}{1.4} = 3.2$
130. $\frac{z}{2.9} = -7.8$
131. $3.4a = 7.004$

132. $2.3m = 2.415$
133. $-3.7x = 7.77$
134. $-1.6m = 5.44$

4 **135.** Employee A had an annual salary of \$22,000, Employee B had an annual salary of \$28,000, and Employee C had an annual salary of \$26,000 before each employee was given a 5% raise. Which of the three employees now has the highest annual salary? Explain how you arrived at your answer.

136. Each of three employees earned an annual salary of \$25,000 before Employee A was given a 3% raise, Employee B was given a 6% raise, and Employee C was given a 4.5% raise. Which of the three employees now has the highest annual salary? Explain how you arrived at your answer.

Solve.

137. 12 is what percent of 50?
138. What percent of 125 is 50?
139. Find 18% of 40.

140. What is 25% of 60?

141. 12% of what is 48?

142. 45% of what is 9?

143. What is $33\frac{1}{3}\%$ of 27?

144. Find $16\frac{2}{3}\%$ of 30.

145. What percent of 12 is 3?

146. 10 is what percent of 15?

147. 60% of what is 3?

148. 75% of what is 6?

149. 12 is what percent of 6?

150. 20 is what percent of 16?

151. $5\frac{1}{4}\%$ of what is 21?

152. $37\frac{1}{2}\%$ of what is 15?

153. Find 15.4% of 50.

154. What is 18.5% of 46?

155. Without solving an equation, determine whether 40% of 80 is less than, equal to, or greater than 80% of 40.

156. Without solving an equation, determine whether $\frac{1}{4}\%$ of 80 is less than, equal to, or greater than 25% of 80.

157. **Construction** The Arthur Ashe Tennis Stadium in New York has a seating capacity of 22,500 people. Of these seats, 1.11% are wheelchair-accessible seats. How many seats in the stadium are reserved for wheelchair accessibility? Round to the nearest whole number.

158. **e-Commerce** The graph at the right, based on data from Forrester Research, shows the percent share of the global e-commerce market for various regions of the world. The total projected revenue from e-commerce business in 2004 is $7500 billion. How many billion dollars more is Western Europe's projected share than Latin America's projected share?

Projected Regional Shares of Global E-Commerce, 2004

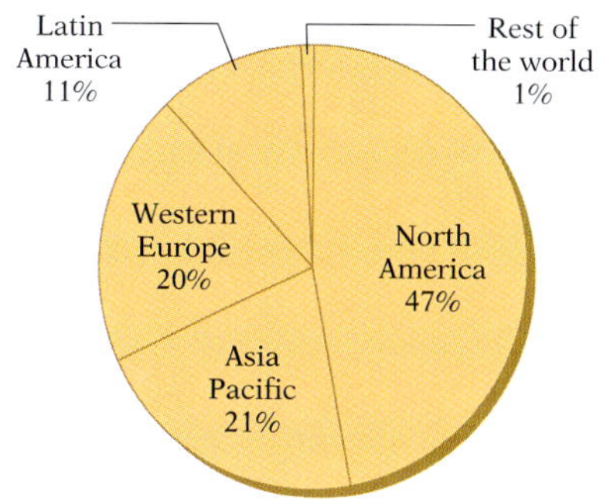

159. **Natural Resources** On average, a person uses 13.2 gal of water per day for showering. This is 17.8% of the total amount of water used per person per day in the average single-family home. Find the total amount of water used per person per day in the average single-family home. Round to the nearest whole number. (*Source:* American Water Works Association)

160. **Business** During a recent year, 276 billion product coupons were issued by manufacturers. Shoppers redeemed 4.8 billion of these coupons. What percent of the coupons issued were redeemed by customers? Round to the nearest tenth of a percent. (*Source:* NCH NuWorld Consumer Behavior Study, America Coupon Council)

161. **Travel** According to the Travel Industry Association of America, a typical traveler's total vacation bill is $1442. Of that amount, $735 is paid in cash. To the nearest percent, what percent of a typical traveler's vacation bill is paid in cash?

162. **Taxes** In a recent year, 238 U.S. airports collected $1.1 billion in passenger taxes. Of this amount, $88 million was spent on noise reduction. What percent of the passenger taxes collected was spent on noise reduction?

163. **The Internet** The circle graph at the right shows the responses to a question about downloading music from Napster. How many people were surveyed? (*Source:* Taylor Nelson Sofres Intersearch)

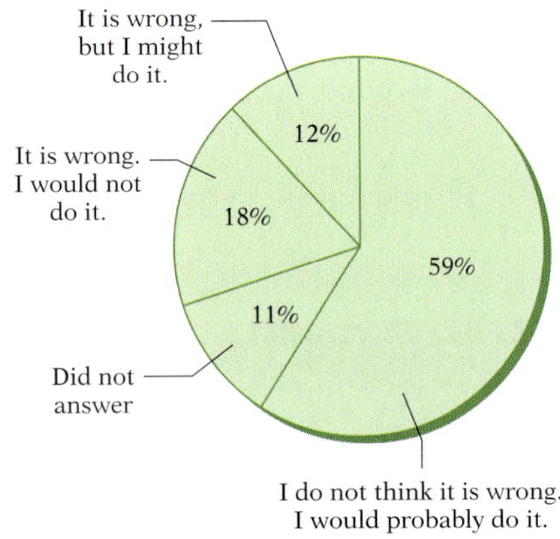

164. **EPA** In January 2001, the Environmental Protection Agency (EPA) required all 54,000 community water systems in the United States to reduce the amount of arsenic in drinking water to 40 parts per billion from 50 parts per billion. What percent decrease does this represent?

165. **Safety** Recently the National Safety Council collected data on the leading causes of accidental death. The findings revealed that for people age 20, 30 died from a fall, 47 from fire, 200 from drowning, and 1950 from motor vehicle accidents. What percent of the accidental deaths were not attributed to motor vehicle accidents? Round to the nearest percent.

166. **Health** According to *Health* magazine, the average American has increased his or her daily consumption of calories from 18 years ago by 11.6%. If the average daily consumption was 1970 calories 18 years ago, what is the average daily consumption today? Round to the nearest whole number.

167. **Energy** The Energy Information Administration reports that if every U.S. household switched 4 h of lighting per day from incandescent bulbs to compact fluorescent bulbs, we would save 31.7 billion kilowatt-hours of electricity a year, or 33% of the total electricity used for home lighting. What is the total electricity used for home lighting in this country? Round to the nearest tenth of a billion.

168. **Investment** If Miranda Perry invests $2500 in an account that earns an 8% annual simple interest rate, how much interest will Miranda have earned after 9 months?

169. **Investment** Kachina Caron invests $1200 in a simple interest account and earns $72 in 8 months. What is the annual simple interest rate?

170. **Investment** How much money must Andrea invest for two years in an account that earns an annual simple interest rate of 8% if she wants to earn $300 from the investment?

171. **Investment** Sal Boxer divided a gift of $3000 into two different accounts. He placed $1000 in one account that earned an annual simple interest rate of 7.5%. The remaining money was placed in an account that earned an annual simple interest rate of 8.25%. How much interest did Sal earn from the two accounts after one year?

172. **Investment** If Americo invests \$2500 at an 8% annual simple interest rate and Octavia invests \$3000 at a 7% annual simple interest rate, which of the two will earn the greater amount of interest in one year?

173. **Investment** An investor places \$1000 in an account that earns 9% annual simple interest and \$1000 in an account that earns 6% annual simple interest. If each investment is left in the account for the same period of time, is the interest rate on the combined investment less than 6%, between 6% and 9%, or greater than 9%?

174. **Mixture Problem** The concentration of platinum in a necklace is 15%. The necklace weighs 12 g. Find the amount of platinum in the necklace.

175. **Mixture Problem** A 250-milliliter solution of a fabric dye contains 5 ml of hydrogen peroxide. What is the percent concentration of the hydrogen peroxide?

176. **Mixture Problem** A carpet is made of a blend of wool and other fibers. If the concentration of wool in the carpet is 75% and the carpet weighs 175 lb, how much wool is in the carpet?

177. **Mixture Problem** Apple Dan's 32-ounce apple-flavored fruit drink contains 8 oz of apple juice. Forty ounces of a generic brand of apple-flavored fruit drink contains 9 oz of apple juice. Which of the two brands has the greater concentration of apple juice?

178. **Mixture Problem** Bakers use simple syrup in many of their recipes. Simple syrup is made by combining 500 g of sugar with 500 g of water and mixing it well until the sugar dissolves. What is the percent concentration of sugar in simple syrup?

179. **Mixture Problem** A pharmacist has 50 g of a topical cream that contains 75% glycerin. How many grams of the cream are not glycerin?

180. **Mixture Problem** A chemist has 100 ml of a solution that is 9% acetic acid. If the chemist adds 50 ml of pure water to this solution, what is the percent concentration of the resulting mixture?

181. **Mixture Problem** A 500-gram salt and water solution contains 50 g of salt. This mixture is left in the open air and 100 g of water evaporates from the solution. What is the percent concentration of salt in the remaining solution?

2.1 Applying Concepts

Solve.

182. $\frac{2m + m}{5} = -9$

183. $\frac{3y - 8y}{7} = 15$

184. $\frac{1}{\frac{1}{x}} = 5$

185. $\dfrac{1}{\frac{1}{x}} + 8 = -19$

186. $\dfrac{4}{\frac{3}{b}} = 8$

187. $\dfrac{5}{\frac{7}{a}} - \dfrac{3}{\frac{7}{a}} = 6$

188. **Geometry** Find the value of x in the diagram shown at the right.

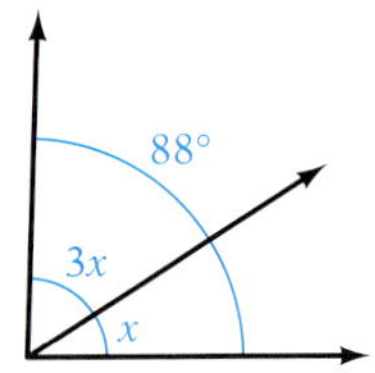

189. **Consumerism** The total cost for a dinner was \$97.52. This included a 15% tip calculated on the cost of the dinner after a 6% sales tax. Find the cost of dinner before the tip and tax.

190. **Business** A retailer decides to increase the original price of each item in the store by 10%. After the price increase, the retailer notices a significant drop in sales and so decides to reduce the current price of each item in the store by 10%. Are the prices back to the original prices? If not, are the prices lower or higher than the original prices?

191. If a quantity increases by 100%, how many times its original value is the new value?

192. Make up an equation of the form $x + a = b$ that has 2 as a solution.

193. Make up an equation of the form $ax = b$ that has -2 as a solution.

194. **Monthly Mortgage Payments** Suppose you have a \$100,000 30-year mortgage. What is the difference between the monthly payment if the interest rate on your loan is 7.75% and the monthly payment if the interest rate is 7.25%? (*Hint:* You will need to discuss the question with an agent at a bank that provides mortgage services, discuss the question with a real estate agent, or find and use the formula for determining the monthly payment on a mortgage.)

195. Write an explanation of how to use the tax rate schedule shown below. Calculate the income tax on a single taxpayer's taxable income of \$75,000.

2001 Tax Rate Schedule

If taxable income is more than	*but not more than*	*Tax is*	*of the amount over*
For single taxpayers			
0	\$27,050	15%	0
\$27,050	\$65,550	\$4,057.50 plus 27.5%	\$27,050
\$65,550	\$136,750	\$14,645.00 plus 30.5%	\$65,550
\$136,750	\$297,350	\$36,361.00 plus 35.5%	\$136,750
\$297,350	—	\$93,374.00 plus 39.1%	\$297,350
For married taxpayers, filing jointly			
0	\$45,200	15%	0
\$45,200	\$109,250	\$6,780.00 plus 27.5%	\$45,200
\$109,250	\$166,500	\$24,393.75 plus 30.5%	\$109,250
\$166,500	\$297,350	\$41,855.00 plus 35.5%	\$166,500
\$297,350	—	\$88,306.75 plus 39.1%	\$297,350

2.2 General Equations

1 Solve equations of the form $ax + b = c$

VIDEO & DVD CD TUTOR WEB SSM

In solving an equation of the form $ax + b = c$, the goal is to rewrite the equation in the form *variable* = *constant*. This requires applying both the Addition and the Multiplication Properties of Equations.

Solve: $\frac{2}{5}x - 3 = -7$

	$\frac{2}{5}x - 3 = -7$
Add 3 to each side of the equation.	$\frac{2}{5}x - 3 + 3 = -7 + 3$
Simplify.	$\frac{2}{5}x = -4$
Multiply each side of the equation by the reciprocal of the coefficient $\frac{2}{5}$.	$\frac{5}{2} \cdot \frac{2}{5}x = \frac{5}{2}(-4)$
Simplify. Now the equation is in the form *variable* = *constant*.	$x = -10$

Check $\frac{2}{5}x - 3 = -7$

$\frac{2}{5}(-10) - 3$	-7
$-4 - 3$	-7

$-7 = -7$

Write the solution. The solution is -10.

Example 1 Solve: $3x - 7 = -5$

Solution

$$3x - 7 = -5$$

$$3x - 7 + 7 = -5 + 7$$ • Add 7 to each side of the equation.

$$3x = 2$$ • Simplify.

$$\frac{3x}{3} = \frac{2}{3}$$ • Divide each side of the equation by 3.

$$x = \frac{2}{3}$$ • Simplify. Now the equation is in the form *variable* = *constant*.

The solution is $\frac{2}{3}$. • Write the solution.

Problem 1 Solve: $5x + 7 = 10$

Solution See page S5.

Example 2 Solve: $5 = 9 - 2x$

Solution

$$5 = 9 - 2x$$
$$5 - 9 = 9 - 9 - 2x$$ • Subtract 9 from each side of the equation.
$$-4 = -2x$$ • Simplify.
$$\frac{-4}{-2} = \frac{-2x}{-2}$$ • Divide each side of the equation by -2.
$$2 = x$$ • Simplify.

The solution is 2. • Write the solution.

Problem 2 Solve: $11 = 11 + 3x$

Solution See page S5.

Note in Example 4 in the Introduction to Equations section that the original equation, $\frac{1}{2} = x + \frac{2}{3}$, contained fractions with denominators of 2 and 3. The least common multiple of 2 and 3 is 6. The least common multiple has the property that both 2 and 3 will divide evenly into it. Therefore, if both sides of the equation are multiplied by 6, both of the denominators will divide evenly into 6. The result is an equation that does not contain any fractions. Multiplying an equation that contains fractions by the LCM of the denominators is called **clearing denominators.** It is an alternative method of solving an equation that contains fractions.

Solve: $\frac{1}{2} = x + \frac{2}{3}$

$$\frac{1}{2} = x + \frac{2}{3}$$

Multiply both sides of the equation by 6, the LCM of the denominators.

$$6\left(\frac{1}{2}\right) = 6\left(x + \frac{2}{3}\right)$$

Simplify each side of the equation. Use the Distributive Property on the right side of the equation.

$$3 = 6(x) + 6\left(\frac{2}{3}\right)$$

Note that multiplying both sides of the equation by the LCM of the denominators eliminated the fractions.

$$3 = 6x + 4$$

Solve the resulting equation.

$$3 - 4 = 6x + 4 - 4$$
$$-1 = 6x$$
$$\frac{-1}{6} = \frac{6x}{6}$$
$$-\frac{1}{6} = x$$

The solution is $-\frac{1}{6}$.

Example 3 Solve: $\frac{2}{3} + \frac{1}{4}x = -\frac{1}{3}$

Solution

$$\frac{2}{3} + \frac{1}{4}x = -\frac{1}{3}$$

• The equation contains fractions. Find the LCM of the denominators.

$$12\left(\frac{2}{3} + \frac{1}{4}x\right) = 12\left(-\frac{1}{3}\right)$$

• The LCM of 3 and 4 is 12. Multiply each side of the equation by 12.

$$12\left(\frac{2}{3}\right) + 12\left(\frac{1}{4}x\right) = -4$$

• Use the Distributive Property to multiply the left side of the equation by 12.

$$8 + 3x = -4$$

• The equation now contains no fractions.

$$8 - 8 + 3x = -4 - 8$$
$$3x = -12$$
$$\frac{3x}{3} = \frac{-12}{3}$$
$$x = -4$$

The solution is −4.

Problem 3 Solve: $\frac{5}{2} - \frac{2}{3}x = \frac{1}{2}$

Solution See page S5.

2 Solve equations of the form $ax + b = cx + d$

VIDEO & DVD CD TUTOR WEB SSM

In solving an equation of the form $ax + b = cx + d$, the goal is to rewrite the equation in the form *variable* = *constant*. Begin by rewriting the equation so that there is only one variable term in the equation. Then rewrite the equation so that there is only one constant term.

Solve: $4x - 5 = 6x + 11$

	$4x - 5 = 6x + 11$
Subtract $6x$ from each side of the equation.	$4x - 6x - 5 = 6x - 6x + 11$
Simplify. Now there is only one variable term in the equation.	$-2x - 5 = 11$
Add 5 to each side of the equation.	$-2x - 5 + 5 = 11 + 5$
Simplify. Now there is only one constant term in the equation.	$-2x = 16$
Divide each side of the equation by −2.	$\frac{-2x}{-2} = \frac{16}{-2}$
Simplify. Now the equation is in the form *variable* = *constant*.	$x = -8$

Check

$$\begin{array}{r|l} 4x - 5 = & 6x + 11 \\ \hline 4(-8) - 5 & 6(-8) + 11 \\ -32 - 5 & -48 + 11 \\ -37 = & -37 \end{array}$$

Write the solution. The solution is -8.

Example 4 Solve: $4x - 3 = 8x - 7$

Solution

$$4x - 3 = 8x - 7$$

$$4x - 8x - 3 = 8x - 8x - 7$$

- Subtract $8x$ from each side of the equation.

$$-4x - 3 = -7$$

- Simplify. Now there is only one variable term in the equation.

$$-4x - 3 + 3 = -7 + 3$$

- Add 3 to each side of the equation.

$$-4x = -4$$

- Simplify. Now there is only one constant term in the equation.

$$\frac{-4x}{-4} = \frac{-4}{-4}$$

- Divide each side of the equation by -4.

$$x = 1$$

- Simplify. Now the equation is in the form *variable* = *constant*.

The solution is 1.

- Write the solution.

Problem 4 Solve: $5x + 4 = 6 + 10x$

Solution See page S5.

3 Solve equations containing parentheses

VIDEO & DVD CD TUTOR WEB SSM

When an equation contains parentheses, one of the steps in solving the equation requires the use of the Distributive Property. The Distributive Property is used to remove parentheses from a variable expression.

$$a(b + c) = ab + ac$$

Solve: $4 + 5(2x - 3) = 3(4x - 1)$

$$4 + 5(2x - 3) = 3(4x - 1)$$

Use the Distributive Property to remove parentheses.

$$4 + 10x - 15 = 12x - 3$$

Simplify.

$$10x - 11 = 12x - 3$$

Subtract $12x$ from each side of the equation.

$$10x - 12x - 11 = 12x - 12x - 3$$

Simplify. Now there is only one variable term in the equation.

$$-2x - 11 = -3$$

Add 11 to each side of the equation.

$$-2x - 11 + 11 = -3 + 11$$

Simplify. Now there is only one constant term in the equation. $-2x = 8$

Divide each side of the equation by -2. $\dfrac{-2x}{-2} = \dfrac{8}{-2}$

Simplify. Now the equation is in the form *variable* = *constant*. $x = -4$

Check $4 + 5(2x - 3) = 3(4x - 1)$

$4 + 5(2x - 3)$	$3(4x - 1)$
$4 + 5[2(-4) - 3]$	$3[4(-4) - 1]$
$4 + 5(-8 - 3)$	$3(-16 - 1)$
$4 + 5(-11)$	$3(-17)$
$4 - 55$	-51
$-51 =$	-51

Write the solution. The solution is -4.

In Chapter 2, we discussed the use of a graphing calculator to evaluate variable expressions. The same procedure can be used to check the solution of an equation. Consider the example above. After we divide both sides of the equation by -2, the solution appears to be -4. To check this solution, store the value of x, -4, in the calculator. Evaluate the expression on the left side of the original equation: $4 + 5(2x - 3)$. The result is -51. Now evaluate the expression on the right side of the original equation: $3(4x - 1)$. The result is -51. Because the results are equal, the solution -4 checks. See the Appendix for a description of keystroking procedures for different models of graphing calculators.

Example 5 Solve: $3x - 4(2 - x) = 3(x - 2) - 4$

Solution

$3x - 4(2 - x) = 3(x - 2) - 4$

$3x - 8 + 4x = 3x - 6 - 4$ • Use the Distributive Property to remove parentheses.

$7x - 8 = 3x - 10$ • Simplify.

$7x - 3x - 8 = 3x - 3x - 10$ • Subtract $3x$ from each side of the equation.

$4x - 8 = -10$

$4x - 8 + 8 = -10 + 8$ • Add 8 to each side of the equation.

$4x = -2$

$\dfrac{4x}{4} = \dfrac{-2}{4}$ • Divide each side of the equation by 4.

$x = -\dfrac{1}{2}$ • The equation is in the form *variable* = *constant*.

The solution is $-\dfrac{1}{2}$.

Problem 5 Solve: $5x - 4(3 - 2x) = 2(3x - 2) + 6$

Solution See page S6.

Example 6 Solve: $3[2 - 4(2x - 1)] = 4x - 10$

Solution

$$\begin{aligned}
3[2 - 4(2x - 1)] &= 4x - 10 \\
3[2 - 8x + 4] &= 4x - 10 && \text{• Use the Distributive Property to remove the parentheses.} \\
3[6 - 8x] &= 4x - 10 && \text{• Simplify inside the brackets.} \\
18 - 24x &= 4x - 10 && \text{• Use the Distributive Property to remove the brackets.} \\
18 - 24x - 4x &= 4x - 4x - 10 && \text{• Subtract } 4x \text{ from each side of the equation.} \\
18 - 28x &= -10 \\
18 - 18 - 28x &= -10 - 18 && \text{• Subtract 18 from each side of the equation.} \\
-28x &= -28 \\
\frac{-28x}{-28} &= \frac{-28}{-28} && \text{• Divide each side of the equation by } -28. \\
x &= 1
\end{aligned}$$

The solution is 1.

Problem 6 Solve: $-2[3x - 5(2x - 3)] = 3x - 8$

Solution See page S6.

4 Application problems

Example 7 A company uses the equation $V = C - 6000t$ to determine the depreciated value V, after t years, of a milling machine that originally cost C dollars. If a milling machine originally cost \$50,000, in how many years will the depreciated value of the machine be \$38,000?

Strategy To find the number of years, replace C with 50,000 and V with 38,000 in the given equation, and solve for t.

Solution

$$\begin{aligned}
V &= C - 6000t \\
38{,}000 &= 50{,}000 - 6000t \\
38{,}000 - 50{,}000 &= 50{,}000 - 50{,}000 - 6000t \\
-12{,}000 &= -6000t \\
\frac{-12{,}000}{-6000} &= \frac{-6000t}{-6000} \\
2 &= t
\end{aligned}$$

The depreciated value of the machine will be \$38,000 in 2 years.

Problem 7 The value V of an investment of \$7500 at an annual simple interest rate of 6% is given by the equation $V = 450t + 7500$, where t is the amount of time, in years, that the money is invested. In how many years will the value of a \$7500 investment be \$10,200?

Solution See page S6.

Take Note

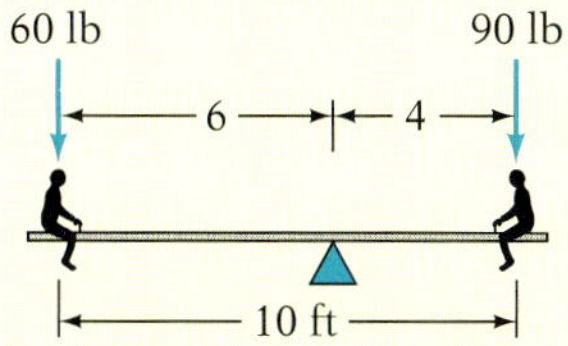

This system balances because

$$F_1x = F_2(d - x)$$
$$60(6) = 90(10 - 6)$$
$$60(6) = 90(4)$$
$$360 = 360$$

A lever system is shown below. It consists of a lever, or bar; a fulcrum; and two forces, F_1 and F_2. The distance d represents the length of the lever, x represents the distance from F_1 to the fulcrum, and $d - x$ represents the distance from F_2 to the fulcrum.

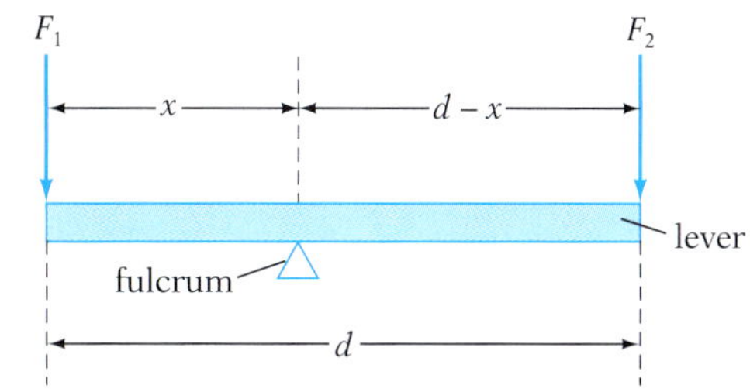

A principle of physics states that when the lever system balances,

$$F_1x = F_2(d - x)$$

Example 8 A lever is 10 ft long. A force of 100 lb is applied to one end of the lever, and a force of 400 lb is applied to the other end. When the system balances, how far is the fulcrum from the 100-pound force?

Strategy Draw a diagram of the situation.

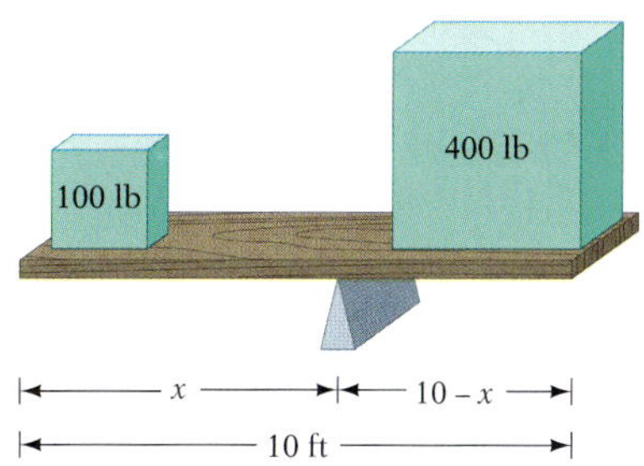

The lever is 10 ft long, so $d = 10$. One force is 100 lb, so $F_1 = 100$. The other force is 400 lb, so $F_2 = 400$. To find the distance of the fulcrum from the 100-pound force, replace the variables F_1, F_2, and d in the lever-system equation with the given values, and solve for x.

Solution

$$\begin{aligned}
F_1x &= F_2(d - x) \\
100x &= 400(10 - x) \\
100x &= 4000 - 400x \\
100x + 400x &= 4000 - 400x + 400x \\
500x &= 4000 \\
\frac{500x}{500} &= \frac{4000}{500} \\
x &= 8
\end{aligned}$$

The fulcrum is 8 ft from the 100-pound force.

Problem 8 A lever is 14 ft long. At a distance of 6 ft from the fulcrum, a force of 40 lb is applied. How large a force must be applied to the other end of the lever so that the system will balance?

Solution See page S6.

2.2 Concept Review

Determine whether the statement is always true, sometimes true, or never true.

1. The same variable term can be added to both sides of an equation without changing the solution of the equation.
2. The same variable term can be subtracted from both sides of an equation without changing the solution of the equation.
3. An equation of the form $ax + b = c$ cannot be solved if a is a negative number.
4. The solution of the equation $\frac{x}{3} = 0$ is 0.
5. The solution of the equation $\frac{x}{0} = 3$ is 0.
6. In solving an equation of the form $ax + b = cx + d$, the goal is to rewrite the equation in the form *variable* = *constant*.
7. In solving an equation of the form $ax + b = cx + d$, subtracting cx from each side of the equation results in an equation with only one variable term in it.

2.2 Exercises

1 Solve and check.

1. $3x + 1 = 10$ **2.** $4y + 3 = 11$ **3.** $2a - 5 = 7$ **4.** $5m - 6 = 9$

5. $5 = 4x + 9$ **6.** $2 = 5b + 12$ **7.** $13 = 9 + 4z$ **8.** $7 - c = 9$

9. $2 - x = 11$ **10.** $4 - 3w = -2$ **11.** $5 - 6x = -13$ **12.** $8 - 3t = 2$

13. $-5d + 3 = -12$ **14.** $-8x - 3 = -19$ **15.** $-7n - 4 = -25$ **16.** $-12x + 30 = -6$

17. $-13 = -11y + 9$ **18.** $2 = 7 - 5a$ **19.** $3 = 11 - 4n$ **20.** $-35 = -6b + 1$

21. $-8x + 3 = -29$ **22.** $-3m - 21 = 0$ **23.** $7x - 3 = 3$ **24.** $8y + 3 = 7$

25. $6a + 5 = 9$ **26.** $3m + 4 = 11$ **27.** $11 = 15 + 4n$ **28.** $4 = 2 - 3c$

29. $9 - 4x = 6$ **30.** $7 - 8z = 0$ **31.** $1 - 3x = 0$ **32.** $6y - 5 = -7$

33. $8b - 3 = -9$ **34.** $5 - 6m = 2$ **35.** $7 - 9a = 4$ **36.** $9 = -12c + 5$

37. $10 = -18x + 7$ **38.** $2y + \frac{1}{3} = \frac{7}{3}$ **39.** $3x - \frac{5}{6} = \frac{13}{6}$ **40.** $5y + \frac{3}{7} = \frac{3}{7}$

41. $9x + \frac{4}{5} = \frac{4}{5}$ **42.** $4 = 7 - 2w$ **43.** $7 = 9 - 5a$ **44.** $-6y + 5 = 13$

45. $-4x + 3 = 9$ **46.** $\frac{1}{2}a - 3 = 1$ **47.** $\frac{1}{3}m - 1 = 5$ **48.** $\frac{2}{5}y + 4 = 6$

49. $\frac{3}{4}n + 7 = 13$ **50.** $-\frac{2}{3}x + 1 = 7$ **51.** $-\frac{3}{8}b + 4 = 10$ **52.** $\frac{x}{4} - 6 = 1$

53. $\frac{y}{5} - 2 = 3$ **54.** $\frac{2x}{3} - 1 = 5$ **55.** $\frac{3c}{7} - 1 = 8$ **56.** $4 - \frac{3}{4}z = -2$

57. $3 - \frac{4}{5}w = -9$ **58.** $5 + \frac{2}{3}y = 3$ **59.** $17 + \frac{5}{8}x = 7$ **60.** $17 = 7 - \frac{5}{6}t$

61. $\frac{2}{3} = y - \frac{1}{2}$ **62.** $\frac{3}{4} = \frac{1}{12}x + 2$ **63.** $\frac{3}{8} = \frac{5}{12} - \frac{1}{3}b$ **64.** $\frac{2}{3} = \frac{3}{4} - \frac{1}{2}y$

65. $7 = \frac{2x}{5} + 4$ **66.** $5 - \frac{4c}{7} = 8$ **67.** $7 - \frac{5}{9}y = 9$

68. $6a + 3 + 2a = 11$ **69.** $5y + 9 + 2y = 23$ **70.** $7x - 4 - 2x = 6$

71. $11z - 3 - 7z = 9$ **72.** $2x - 6x + 1 = 9$ **73.** $b - 8b + 1 = -6$

74. $-1 = 5m + 7 - m$ **75.** $8 = 4n - 6 + 3n$ **76.** $0.15y + 0.025 = -0.074$

77. $1.2x - 3.44 = 1.3$ **78.** $3.5 = 3.5 + 0.076x$ **79.** $-6.5 = 4.3y - 3.06$

Solve.

80. If $2x - 3 = 7$, evaluate $3x + 4$. **81.** If $3x + 5 = -4$, evaluate $2x - 5$.

82. If $4 - 5x = -1$, evaluate $x^2 - 3x + 1$. **83.** If $2 - 3x = 11$, evaluate $x^2 + 2x - 3$.

2 Solve and check.

84. $8x + 5 = 4x + 13$ **85.** $6y + 2 = y + 17$ **86.** $7m + 4 = 6m + 7$

87. $11n + 3 = 10n + 11$ **88.** $5x - 4 = 2x + 5$ **89.** $9a - 10 = 3a + 2$

90. $12y - 4 = 9y - 7$

91. $13b - 1 = 4b - 19$

92. $15x - 2 = 4x - 13$

93. $7a - 5 = 2a - 20$

94. $3x + 1 = 11 - 2x$

95. $n - 2 = 6 - 3n$

96. $2x - 3 = -11 - 2x$

97. $4y - 2 = -16 - 3y$

98. $2b + 3 = 5b + 12$

99. $m + 4 = 3m + 8$

100. $4x - 7 = 5x + 1$

101. $6d - 2 = 7d + 5$

102. $4y - 8 = y - 8$

103. $5a + 7 = 2a + 7$

104. $6 - 5x = 8 - 3x$

105. $10 - 4n = 16 - n$

106. $2x - 4 = 6x$

107. $2b - 10 = 7b$

108. $8m = 3m + 20$

109. $9y = 5y + 16$

110. $-x - 4 = -3x - 16$

111. $8 - 4x = 18 - 5x$

112. $6 - 10a = 8 - 9a$

113. $5 - 7m = 2 - 6m$

114. $8b + 5 = 5b + 7$

115. $6y - 1 = 2y + 2$

116. $7x - 8 = x - 3$

117. $10x - 3 = 3x - 1$

118. $5n + 3 = 2n + 1$

119. $8a - 2 = 4a - 5$

120. $\frac{1}{5}d = \frac{1}{2}d + 3$

121. $\frac{3}{4}x = \frac{1}{12}x + 2$

122. $\frac{2}{3}a = \frac{1}{5}a + 7$

123. $\frac{4}{5}c - 7 = \frac{1}{10}c$

124. $\frac{1}{2}y = 10 - \frac{1}{3}y$

125. $\frac{2}{3}b = 2 - \frac{5}{6}b$

126. $8.7y = 3.9y + 9.6$

127. $4.5x - 5.4 = 2.7x$

128. $5.6x = 7.2x - 6.4$

Solve.

129. If $5x = 3x - 8$, evaluate $4x + 2$.

130. If $7x + 3 = 5x - 7$, evaluate $3x - 2$.

131. If $2 - 6a = 5 - 3a$, evaluate $4a^2 - 2a + 1$.

132. If $1 - 5c = 4 - 4c$, evaluate $3c^2 - 4c + 2$.

3 Solve and check.

133. $5x + 2(x + 1) = 23$

134. $6y + 2(2y + 3) = 16$

135. $9n - 3(2n - 1) = 15$

136. $12x - 2(4x - 6) = 28$

137. $7a - (3a - 4) = 12$

138. $9m - 4(2m - 3) = 11$

139. $5(3 - 2y) + 4y = 3$

140. $4(1 - 3x) + 7x = 9$

141. $10x + 1 = 2(3x + 5) - 1$

142. $5y - 3 = 7 + 4(y - 2)$

143. $4 - 3a = 7 - 2(2a + 5)$

144. $9 - 5x = 12 - (6x + 7)$

145. $3y - 7 = 5(2y - 3) + 4$

146. $2a - 5 = 4(3a + 1) - 2$

147. $5 - (9 - 6x) = 2x - 2$

148. $7 - (5 - 8x) = 4x + 3$

149. $3[2 - 4(y - 1)] = 3(2y + 8)$

150. $5[2 - (2x - 4)] = 2(5 - 3x)$

151. $3a + 2[2 + 3(a - 1)] = 2(3a + 4)$

152. $5 + 3[1 + 2(2x - 3)] = 6(x + 5)$

153. $-2[4 - (3b + 2)] = 5 - 2(3b + 6)$

154. $-4[x - 2(2x - 3)] + 1 = 2x - 3$

155. $0.3x - 2(1.6x) - 8 = 3(1.9x - 4.1)$

156. $0.56 - 0.4(2.1y + 3) = 0.2(2y + 6.1)$

Solve.

157. If $4 - 3a = 7 - 2(2a + 5)$, evaluate $a^2 + 7a$.

158. If $9 - 5x = 12 - (6x + 7)$, evaluate $x^2 - 3x - 2$.

159. If $2z - 5 = 3(4z + 5)$, evaluate $\frac{z^2}{z - 2}$.

160. If $3n - 7 = 5(2n + 7)$, evaluate $\frac{n^2}{2n - 6}$.

4 Black ice is an ice covering on roads that is especially difficult to see and therefore extremely dangerous for motorists. The distance that a car traveling 30 mph will slide after its brakes are applied is related to the outside temperature by the formula $C = \frac{1}{4}D - 45$, where C is the Celsius temperature and D is the distance in feet that the car will slide.

161. Physics Determine the distance a car will slide on black ice when the outside temperature is -3°C.

162. Physics Determine the distance a car will slide on black ice when the outside temperature is -11°C.

The pressure at a certain depth in the ocean can be approximated by the equation $P = \frac{1}{2}D + 15$, where P is the pressure in pounds per square inch, and D is the depth in feet.

163. Physics Find the depth of a diver when the pressure on the diver is 35 lb/in^2.

164. Physics Find the depth of a diver when the pressure on the diver is 45 lb/in^2.

The distance s, in feet, that an object will fall in t seconds is given by $s = 16t^2 + vt$, where v is the initial velocity of the object in feet per second.

165. Physics Find the initial velocity of an object that falls 80 ft in 2 s.

166. Physics Find the initial velocity of an object that falls 144 ft in 3 s.

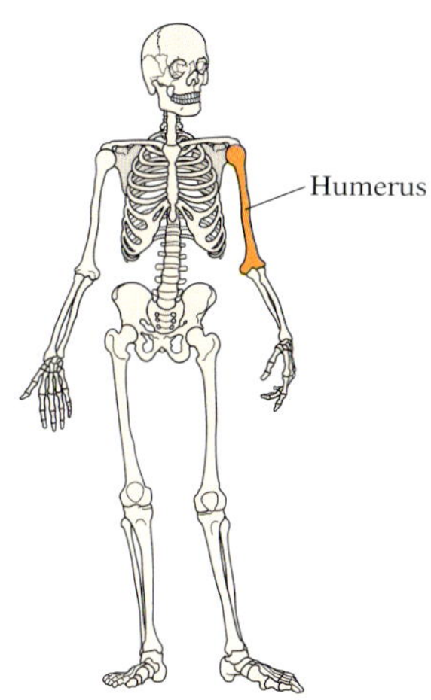

Anthropologists can approximate the height of a primate from the size of its humerus (the bone extending from the shoulder to the elbow) by using the equation $H = 1.2L + 27.8$, where L is the length of the humerus and H is the height of the primate, both in inches.

167. Anthropology An anthropologist estimates the height of a primate to be 66 in. What is the approximate length of the humerus of this primate? Round to the nearest tenth.

168. **Anthropology** The height of a primate is estimated to be 62 in. Find the approximate length of the humerus of this primate.

The fare, F, to be charged a customer by a taxi company is calculated using the formula $F = 1.50 + 0.95(m - 1)$, where m is the number of miles traveled.

169. **Business** A customer is charged \$6.25. How many miles was the passenger driven?

170. **Business** A passenger is charged \$9.10. Find the number of miles the customer was driven.

A person's accurate typing speed can be approximated by the equation $S = \frac{W - 5e}{10}$, where S is the accurate typing speed in words per minute, W is the number of words typed in ten minutes, and e is the number of errors made.

171. **Employment** After taking a 10-minute typing test, a job candidate was told that he had an accurate speed of 35 words per minute. He had typed a total of 390 words. How many errors did he make?

172. **Employment** A job applicant took a 10-minute typing test and was told that she had an accurate speed of 37 words per minute. If she had typed a total of 400 words, how many errors did she make?

A telephone company estimates that the number N of phone calls made per day between two cities of populations P_1 and P_2 that are d miles apart is given by the equation $N = \frac{2.51P_1P_2}{d^2}$.

173. **Business** Estimate the population P_2, given that P_1 is 48,000, the number of phone calls is 1,100,000, and the distance between the cities is 75 mi. Round to the nearest thousand.

174. **Business** Estimate the population P_1, given that P_2 is 125,000, the number of phone calls is 2,500,000, and the distance between the cities is 50 mi. Round to the nearest thousand.

To determine the break-even point, or the number of units that must be sold so that no profit or loss occurs, an economist uses the equation $Px = Cx + F$, where P is the selling price per unit, x is the number of units sold, C is the cost to make each unit, and F is the fixed cost.

175. **Business** An economist has determined that the selling price per unit for a television is \$250. The cost to make one television is \$165, and the fixed cost is \$29,750. Find the break-even point.

176. **Business** A manufacturing engineer determines that the cost per unit for a compact disc is \$3.35 and that the fixed cost is \$6180. The selling price for the compact disc is \$8.50. Find the break-even point.

Use the lever-system equation $F_1x = F_2(d - x)$.

177. In the lever-system equation $F_1x = F_2(d - x)$, what does x represent? What does d represent? What does $d - x$ represent?

178. Suppose an 80-pound child is sitting at one end of a 7-foot seesaw, and a 60-pound child is sitting at the other end. The 80-pound child is 3 feet from the fulcrum. Use the lever-system equation to explain why the seesaw balances.

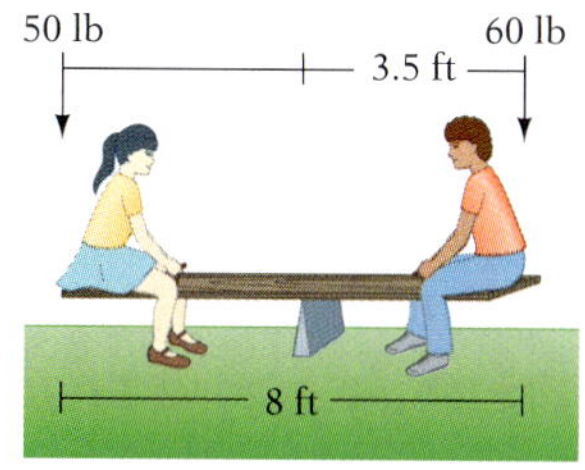

179. **Lever Systems** Two children are sitting 8 ft apart on a seesaw. One child weighs 60 lb and the second child weighs 50 lb. The fulcrum is 3.5 ft from the child weighing 60 lb. Is the seesaw balanced?

180. **Lever Systems** An adult and a child are on a seesaw that is 14 ft long. The adult weighs 175 lb and the child weighs 70 lb. How many feet from the child must the fulcrum be placed so that the seesaw balances?

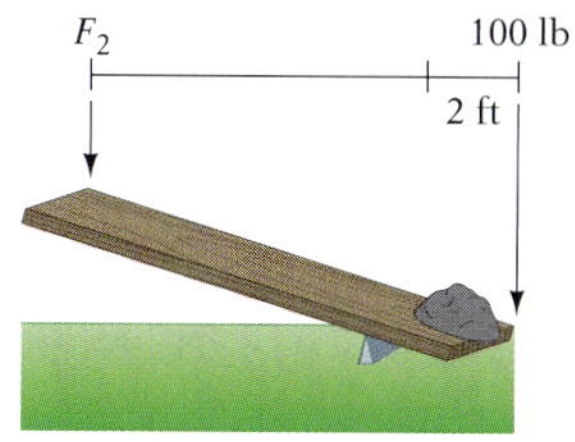

181. **Lever Systems** A lever 10 ft long is used to balance a 100-pound rock. The fulcrum is placed 2 ft from the rock. What force must be applied to the other end of the lever to balance the rock?

182. **Lever Systems** A 50-pound weight is applied to the left end of a seesaw that is 10 ft long. The fulcrum is 4 ft from the 50-pound weight. A weight of 30 lb is applied to the right end of the seesaw. Is the 30-pound weight adequate to balance the seesaw?

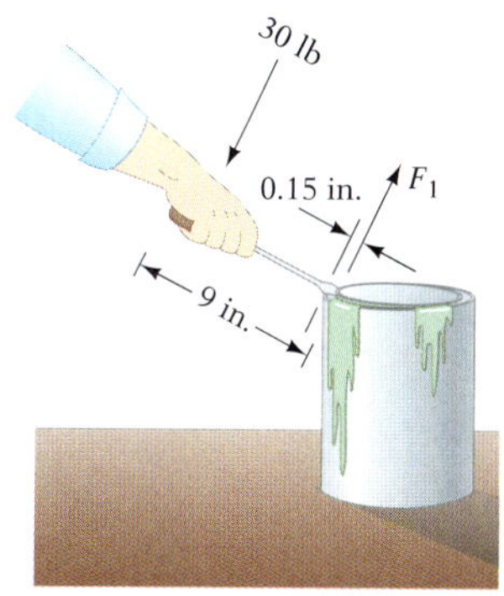

183. **Lever Systems** In preparation for a stunt, two acrobats are standing on a plank that is 18 ft long. One acrobat weighs 128 lb and the second acrobat weighs 160 lb. How far from the 128-pound acrobat must the fulcrum be placed so that the acrobats are balanced on the plank?

184. **Lever Systems** A screwdriver 9 in. long is used as a lever to open a can of paint. The tip of the screwdriver is placed under the lip of the lid with the fulcrum 0.15 in. from the lip. A force of 30 lb is applied to the other end of the screwdriver. Find the force on the lip of the lid.

2.2 Applying Concepts

Solve. If the equation has no solution, write "No solution."

185. $3(2x - 1) - (6x - 4) = -9$

186. $\frac{1}{5}(25 - 10b) + 4 = \frac{1}{3}(9b - 15) - 6$

187. $3[4(w + 2) - (w + 1)] = 5(2 + w)$

188. $\frac{2(5x - 6) - 3(x - 4)}{7} = x + 2$

189. Solve for x: $32{,}166 \div x = 518$ remainder 50

190. One-half of a certain number equals two-thirds of the same number. Find the number.

191. Write an equation of the form $ax + b = cx + d$ that has 4 as a solution.

192. If a, b, and c are real numbers, is it always possible to solve the equation $ax + b = c$? If not, what values of a, b, and/or c must be excluded?

193. Does the sentence "Solve $3x - 4(x - 1)$" make sense? Why or why not?

194. The equation $x = x + 1$ has no solution, whereas the solution of the equation $2x + 1 = 1$ is zero. Is there a difference between no solution and a solution of zero? Explain your answer.

195. Explain in your own words the steps you would take to solve the equation $\frac{3}{4}x - 7 = 2$. State the Property of Real Numbers or the Property of Equations that you used at each step.

196. Explain what is wrong with the following demonstration, which suggests that $2 = 3$.

$$2x + 5 = 3x + 5$$

$$2x + 5 - 5 = 3x + 5 - 5 \quad \text{• Subtract 5 from each side of the equation.}$$

$$2x = 3x$$

$$\frac{2x}{x} = \frac{3x}{x} \quad \text{• Divide each side of the equation by } x.$$

$$2 = 3$$

2.3 Application Problems

1 Translate a sentence into an equation and solve

VIDEO & DVD | CD TUTOR | WEB | SSM

An equation states that two mathematical expressions are equal. Therefore, to translate a sentence into an equation requires recognizing the words or phrases that mean "equals." Besides "equals," some of these phrases are "is," "is equal to," "amounts to," and "represents."

Once the sentence is translated into an equation, the equation can be solved by rewriting the equation in the form *variable* = *constant*.

Example 1 Translate "two times the sum of a number and eight equals the sum of four times the number and six" into an equation and solve.

Solution the unknown number: n

- **Assign a variable to the unknown quantity.**

two times the sum of a number and eight	equals	the sum of four times the number and six

- **Find two verbal expressions for the same value.**

$$2(n + 8) = 4n + 6$$

- **Write a mathematical expression for each verbal expression. Write the equals sign.**
- **Solve the equation.**

$$2n + 16 = 4n + 6$$
$$2n - 4n + 16 = 4n - 4n + 6$$
$$-2n + 16 = 6$$
$$-2n + 16 - 16 = 6 - 16$$
$$-2n = -10$$
$$\frac{-2n}{-2} = \frac{-10}{-2}$$
$$n = 5$$

The number is 5.

Problem 1 Translate "nine less than twice a number is five times the sum of the number and twelve" into an equation and solve.

Solution See page S6.

Example 2 The temperature of the sun on the Kelvin scale is 6500 K. This is 4740° less than the temperature on the Fahrenheit scale. Find the Fahrenheit temperature.

Strategy To find the Fahrenheit temperature, write and solve an equation using F to represent the Fahrenheit temperature.

Solution

6500	is	4740° less than the Fahrenheit temperature

$$6500 = F - 4740$$
$$6500 + 4740 = F - 4740 + 4740$$
$$11{,}240 = F$$

The temperature of the sun is 11,240°F.

Problem 2 A molecule of octane gas has eight carbon atoms. This represents twice the number of carbon atoms in a butane gas molecule. Find the number of carbon atoms in a butane molecule.

Solution See page S6.

> **Take Note**
>
> We could also let x represent the length of the longer piece and $10 - x$ represent the length of the shorter piece. Then our equation would be
>
> $$3(10 - x) = 2x$$
>
> Show that this equation results in the same solutions.

Example 3 A board 10 ft long is cut into two pieces. Three times the length of the shorter piece is twice the length of the longer piece. Find the length of each piece.

Strategy Draw a diagram. To find the length of each piece, write and solve an equation using x to represent the length of the shorter piece and $10 - x$ to represent the length of the longer piece.

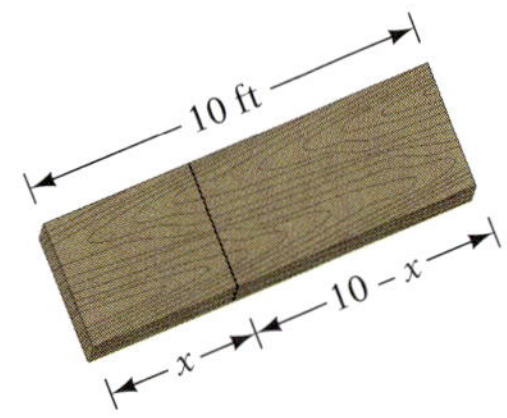

Solution

three times the length of the shorter piece	is	twice the length of the longer piece

$$3x = 2(10 - x)$$
$$3x = 20 - 2x$$
$$3x + 2x = 20 - 2x + 2x$$
$$5x = 20$$
$$\frac{5x}{5} = \frac{20}{5}$$
$$x = 4$$

• **The length of the shorter piece is 4 ft.**

$$10 - x = 10 - 4 = 6$$

• **Substitute the value of x into the variable expression for the longer piece and evaluate.**

The length of the shorter piece is 4 ft. The length of the longer piece is 6 ft.

Problem 3 A company manufactures 160 bicycles per day. Four times the number of 3-speed bicycles made each day equals 30 less than the number of 10-speed bicycles made each day. Find the number of 10-speed bicycles manufactured each day.

Solution See page S6.

2 Consecutive integer problems

VIDEO & DVD CD TUTOR WEB SSM

Recall that the integers are the numbers $\ldots, -4, -3, -2, -1, 0, 1, 2, 3, 4, \ldots$.

An **even integer** is an integer that is divisible by 2. Examples of even integers are -8, 0, and 22.

An **odd integer** is an integer that is not divisible by 2. Examples of odd integers are -17, 1, and 39.

Consecutive integers are integers that follow one another in order. Examples of consecutive integers are shown at the right. (Assume the variable n represents an integer.)

11, 12, 13
$-8, -7, -6$
$n, n + 1, n + 2$

Examples of **consecutive even integers** are shown at the right. (Assume the variable n represents an even integer.)

24, 26, 28
$-10, -8, -6$
$n, n + 2, n + 4$

Examples of **consecutive odd integers** are shown at the right. (Assume the variable n represents an odd integer.)

19, 21, 23,
$-1, 1, 3$
$n, n + 2, n + 4$

Solve: The sum of three consecutive odd integers is 51. Find the integers.

STRATEGY ***for solving a consecutive integer problem***

▶ Let a variable represent one of the integers. Express each of the other integers in terms of that variable. Remember that consecutive integers differ by 1. Consecutive even or consecutive odd integers differ by 2.

First odd integer: n
Second odd integer: $n + 2$
Third odd integer: $n + 4$

▶ Determine the relationship among the integers.

The sum of the three odd integers is 51.

$$n + (n + 2) + (n + 4) = 51$$
$$3n + 6 = 51$$
$$3n = 45$$
$$n = 15$$

$$n + 2 = 15 + 2 = 17$$
$$n + 4 = 15 + 4 = 19$$

• **Substitute the value of *n* into the variable expressions for the second and third integers.**

The three consecutive odd integers are 15, 17, and 19.

Example 4 Find three consecutive even integers such that three times the second integer is six more than the sum of the first and third integers.

Strategy

▶ First even integer: n
Second even integer: $n + 2$
Third even integer: $n + 4$

▶ Three times the second integer equals six more than the sum of the first and third integers.

Solution

$$3(n + 2) = n + (n + 4) + 6$$
$$3n + 6 = 2n + 10$$
$$n + 6 = 10$$
$$n = 4$$

• **The first even integer is 4.**

$$n + 2 = 4 + 2 = 6$$
$$n + 4 = 4 + 4 = 8$$

• **Substitute the value of *n* into the variable expressions for the second and third integers.**

The three consecutive even integers are 4, 6, and 8.

Problem 4 Find three consecutive integers whose sum is -12.

Solution See page S7.

3 Value mixture problems

VIDEO & DVD CD TUTOR WEB SSM

A **value mixture problem** involves combining two ingredients that have different prices into a single blend. For example, a coffee manufacturer may blend two types of coffee into a single blend.

A solution of a value mixture problem is based on the equation $AC = V$, where A is the amount of the ingredient, C is the cost per unit of the ingredient, and V is the value of the ingredient.

The value of 12 lb of coffee costing \$5.25 per pound is

$$V = AC$$
$$V = 12(\$5.25)$$
$$V = \$63$$

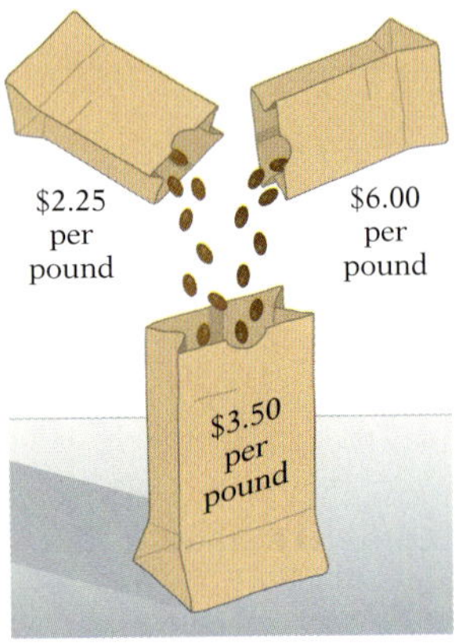

Solve: How many pounds of peanuts that cost \$2.25 per pound must be mixed with 40 lb of cashews that cost \$6.00 per pound to make a mixture that costs \$3.50 per pound?

STRATEGY for solving a value mixture problem

- ▶ For each ingredient in the mixture, write a numerical or variable expression for the amount of the ingredient used, the unit cost of the ingredient, and the value of the amount used. For the mixture, write a numerical or variable expression for the amount, the unit cost of the mixture, and the value of the amount. The results can be recorded in a table.

Amount of peanuts: x

	Amount, A	·	Unit cost, C	=	Value, V
Peanuts	x	·	2.25	=	$2.25x$
Cashews	40	·	6.00	=	$6.00(40)$
Mixture	$x + 40$	·	3.50	=	$3.50(x + 40)$

- ▶ Determine how the values of the individual ingredients are related. Use the fact that the sum of the values of these ingredients is equal to the value of the mixture.

The sum of the values of the peanuts and the cashews is equal to the value of the mixture.

$$2.25x + 6.00(40) = 3.50(x + 40)$$
$$2.25x + 240 = 3.50x + 140$$
$$-1.25x + 240 = 140$$
$$-1.25x = -100$$
$$x = 80$$

The mixture must contain 80 lb of peanuts.

Example 5 How many ounces of a gold alloy that costs \$320 an ounce must be mixed with 100 oz of an alloy that costs \$100 an ounce to make a mixture that costs \$160 an ounce?

Strategy ▶ Ounces of the \$320 gold alloy: x

	Amount	Cost	Value
\$320 alloy	x	320	$320x$
\$100 alloy	100	100	100(100)
Mixture	$x + 100$	160	$160(x + 100)$

▶ The sum of the values before mixing equals the value after mixing.

Solution

$$\begin{aligned} 320x + 100(100) &= 160(x + 100) \\ 320x + 10{,}000 &= 160x + 16{,}000 \\ 160x + 10{,}000 &= 16{,}000 \\ 160x &= 6000 \\ x &= 37.5 \end{aligned}$$

The mixture must contain 37.5 oz of the \$320 gold alloy.

Problem 5 A butcher combined hamburger that costs \$4.00 per pound with hamburger that costs \$2.80 per pound. How many pounds of each were used to make a 75-pound mixture that costs \$3.20 per pound?

Solution See page S7.

4 Uniform motion problems

VIDEO & DVD CD TUTOR WEB SSM

Any object that travels at a constant speed in a straight line is said to be in *uniform motion*. **Uniform motion** means that the speed and direction of an object do not change. For instance, a train traveling at a constant speed of 50 mph on a straight track is in uniform motion.

The solution of a uniform motion problem is based on the equation $d = rt$, where d is the distance traveled, r is the rate of travel, and t is the time spent traveling. For instance, suppose a train travels for 2 h at an average speed of 45 mph. Because the time (2 h) and the rate (45 mph) are known, we can find the distance traveled by solving the equation $d = rt$ for d.

$$\begin{aligned} d &= rt \\ d &= 45(2) \qquad \bullet\ r = 45, t = 2 \\ d &= 90 \end{aligned}$$

The train travels a distance of 90 mi.

A chef leaves a restaurant and drives to his home, which is 16 mi away. If it takes him 20 min to get home, what is the chef's average rate of speed?

Take Note

$20 \text{ min} = \frac{20}{60}\text{h}$

$= \frac{1}{3}\text{h}$

Because the answer must be in miles per *hour* and the given time is *minutes*, convert 20 min to $\frac{1}{3}$ h.

To find the rate of speed, solve the equation $d = rt$ for r using $d = 16$ and $t = \frac{1}{3}$.

$$\begin{aligned} d &= rt \\ 16 &= r\left(\frac{1}{3}\right) \\ 16 &= \frac{1}{3}r \\ (3)16 &= (3)\frac{1}{3}r \\ 48 &= r \end{aligned}$$

The average rate of speed is 48 mph.

Take Note

The abbreviation ft/s means feet per second.

If two objects are moving in opposite directions, then the rate at which the distance between them is increasing is the sum of the speeds of the two objects. For instance, in the diagram at the right, two joggers start from the same point and run in opposite directions. The distance between them is changing at a rate of 21 ft/s.

Similarly, if two objects are moving toward each other, the distance between them is decreasing at a rate that is equal to the sum of the speeds. The rate at which the two cyclists at the right are approaching one another is 35 mph.

Two cars start from the same point and move in opposite directions. The car moving west is traveling 50 mph, and the car moving east is traveling 65 mph. In how many hours will the cars be 230 mi apart?

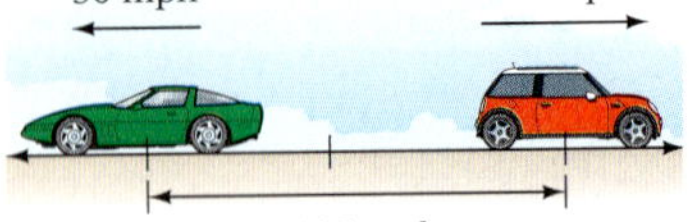

The cars are moving in opposite directions, so the rate at which the distance between them is changing is the sum of the rates of the cars. Thus $r = 115$.

$$50 \text{ mph} + 65 \text{ mph} = 115 \text{ mph}$$

The distance is 230 mi. To find the time, solve $d = rt$ for t.

$$\begin{aligned} d &= rt \\ 230 &= 115t \\ \frac{230}{115} &= \frac{115t}{115} \\ 2 &= t \end{aligned}$$

The time is 2 h.

Point of Interest

The jet stream that generally flows from west to east across the United States affects the time it takes a plane to fly from Los Angeles to New York. For instance, on a typical day, the flying time from New York City to Los Angeles is about 40 min longer than the trip from Los Angeles to New York City.

The typical cruising speed of a Boeing 777 jet is 525 mph. However, wind affects the speed of the plane. For example, when a plane is flying from California to New York, the wind is usually in the direction of the plane's flight, thereby increasing the effective speed of the plane. If the speed of the wind is 50 mph, then the effective speed of the plane is the sum of the plane's speed and the wind's speed: 525 mph + 50 mph = 575 mph.

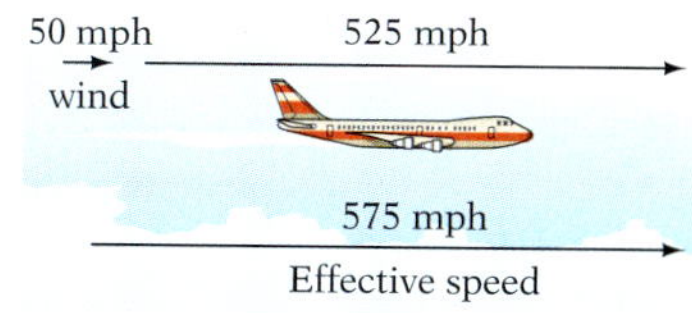

If the plane is traveling from New York to California, the wind is generally in the opposite direction of the plane's flight, so it decreases the speed of the plane. If the speed of the wind is 50 mph, then the effective speed of the plane is the difference between the plane's speed and the wind's speed: 525 mph − 50 mph = 475 mph.

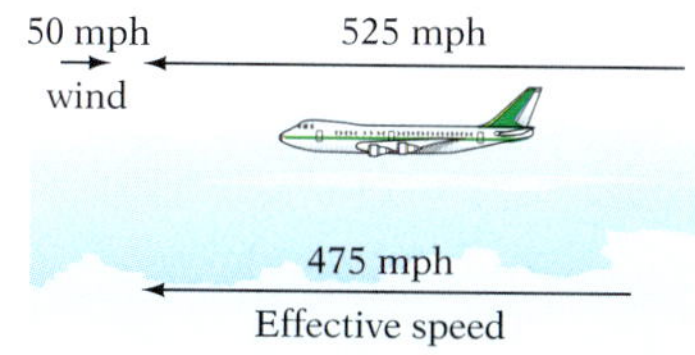

There are other situations in which these concepts may be applied.

Example 6 A Mississippi River sightseeing tour takes tourists from the tour dock on a 30-mile trip down the river to a Civil War historical site, where they spend 1 h walking around the grounds. The tour then returns to the tour dock. If the speed of the boat is set to travel at 16 mph in calm water and the rate of the current is 4 mph, find the total time of the trip.

Strategy To determine the total time:

- Find the time spent traveling down the river by solving $d = rt$ for t. The distance is 30 mi. Therefore, $d = 30$.

  Because the boat is traveling with the current, the rate of the boat is the sum of the boat's speed in calm water and the speed of the current: 16 mph + 4 mph = 20 mph. Thus $r = 20$ mph.

- Find the time spent returning to the tour station by solving $d = rt$ for t. The distance is 30 mi. Therefore, $d = 30$. Because the boat is traveling against the current, the rate of the boat is the difference between the boat's speed in calm water and the speed of the current: 16 mph − 4 mph = 12 mph. Thus $r = 12$ mph.

- Add the time spent traveling down the river, the time spent at the historical site (1 h), and the time returning.

Solution

$$d = rt$$
$$30 = 20t \quad \bullet\ d = 30,\ r = 16 + 4 = 20$$
$$\frac{30}{20} = \frac{20t}{20}$$
$$1.5 = t$$

The time spent traveling down the river is 1.5 h.

$$d = rt$$
$$30 = 12t \qquad \bullet\ d = 30, r = 16 - 4 = 12$$
$$\frac{30}{12} = \frac{12t}{12}$$
$$2.5 = t$$

The time spent returning to the tour station is 2.5 h.

Total time = 1.5 + 1 + 2.5 = 5

The total time of the trip is 5 h.

Problem 6 A plane that can normally travel at 175 mph in calm air is flying into a headwind of 35 mph. How long will it take the plane to fly 350 mi?

Solution See page S7.

For more difficult uniform motion problems, it may be helpful to record known information in a table. This is illustrated in the following example.

Solve: An executive has an appointment 785 mi from the office. The executive takes a helicopter from the office to the airport and a plane from the airport to the business appointment. The helicopter averages 70 mph, and the plane averages 500 mph. The total time spent traveling is 2 h. Find the distance from the executive's office to the airport.

STRATEGY ***for solving a uniform motion problem***

▶ For each object, write a numerical or variable expression for the distance, rate, and time. The results can be recorded in a table.

The total time of travel is 2 h.

Unknown time in the helicopter: t
Time in the plane: $2 - t$

	Rate, r	·	Time, t	=	Distance, d
Helicopter	70	·	t	=	$70t$
Plane	500	·	$2 - t$	=	$500(2 - t)$

▶ Determine how the distances traveled by the individual objects are related. For example, the total distance traveled by both objects may be known, or it may be known that the two objects traveled the same distance.

The total distance traveled is 785 mi.

$$70t + 500(2 - t) = 785$$
$$70t + 1000 - 500t = 785$$
$$-430t + 1000 = 785$$
$$-430t = -215$$
$$t = 0.5$$

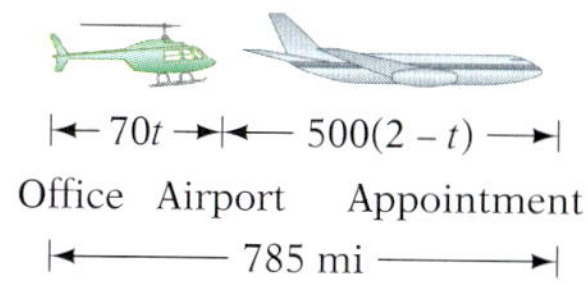

The time spent traveling from the office to the airport in the helicopter is 0.5 h. To find the distance between these two points, substitute 70 for r and 0.5 for t into the equation $rt = d$.

$$rt = d$$
$$70(0.5) = d$$
$$35 = d$$

The distance from the office to the airport is 35 mi.

Example 7 A long-distance runner started a course running at an average speed of 6 mph. One and one-half hours later, a cyclist traveled the same course at an average speed of 12 mph. How long after the runner started did the cyclist overtake the runner?

Strategy

▶ Unknown time for the cyclist: t
Time for the runner: $t + 1.5$

	Rate	Time	Distance
Runner	6	$t + 1.5$	$6(t + 1.5)$
Cyclist	12	t	$12t$

▶ The runner and the cyclist traveled the same distance.

Solution

$$6(t + 1.5) = 12t$$
$$6t + 9 = 12t$$
$$9 = 6t$$
$$\frac{3}{2} = t$$

• **The cyclist traveled for 1.5 h.**

$$t + 1.5 = 1.5 + 1.5$$
$$= 3$$

• **Substitute the value of *t* into the variable expression for the runner's time.**

The cyclist overtook the runner 3 h after the runner started.

Problem 7 Two small planes start from the same point and fly in opposite directions. The first plane is flying 30 mph faster than the second plane. In 4 h the planes are 1160 mi apart. Find the rate of each plane.

Solution See page S7.

2.3 Concept Review

Determine whether the statement is always true, sometimes true, or never true.

1. When translating a sentence into an equation, we can use any variable to represent an unknown number.

2. If a rope 19 in. long is cut into two pieces and one of the pieces has length x inches, then the length of the other piece can be expressed as $(x - 19)$ inches.

3. An even integer is a multiple of 2.

4. When $10d$ is used to represent the value of d dimes, 10 is written as the coefficient of d because it is the number of cents in one dime.

5. Given the consecutive odd integers -5 and -3, the next consecutive odd integer is -1.

6. If the first of three consecutive odd integers is n, then the second and third consecutive odd integers are represented as $n + 1$ and $n + 3$.

7. You have a total of 20 coins in nickels and dimes. If n represents the number of nickels you have, then $n - 20$ represents the number of dimes you have.

8. A gold alloy costing \$300 per ounce was mixed with a gold alloy costing \$200 per ounce. The cost of the mixture was \$250 per ounce.

9. Two planes are flying in opposite directions. One plane is flying at a rate of x mph, and the other plane is flying at a rate of y mph. The distance between the planes is increasing at a rate of $(x + y)$ mph.

2.3 Exercises

1 Translate into an equation and solve.

1. The difference between a number and fifteen is seven. Find the number.

2. The sum of five and a number is three. Find the number.

3. The product of seven and a number is negative twenty-one. Find the number.

4. The quotient of a number and four is two. Find the number.

5. Four less than three times a number is five. Find the number.

6. The difference between five and twice a number is one. Find the number.

7. Four times the sum of twice a number and three is twelve. Find the number.

8. Twenty-one is three times the difference between four times a number and five. Find the number.

9. Twelve is six times the difference between a number and three. Find the number.

10. The difference between six times a number and four times the number is negative fourteen. Find the number.

11. Twice the difference between a number and twenty-five is three times the number. Find the number.

12. Four times a number is three times the difference between thirty-five and the number. Find the number.

13. The sum of two numbers is twenty. Three times the smaller is equal to two times the larger. Find the two numbers.

14. The sum of two numbers is fifteen. One less than three times the smaller is equal to the larger. Find the two numbers.

15. The sum of two numbers is eighteen. The total of three times the smaller and twice the larger is forty-four. Find the two numbers.

16. The sum of two numbers is two. The difference between eight and twice the smaller number is two less than four times the larger. Find the two numbers.

Write an equation and solve.

17. **Depreciation** As a result of depreciation, the value of a car is now \$9600. This is three-fifths of its original value. Find the original value of the car.

18. **Structures** The length of the Royal Gorge Bridge in Colorado is 320 m. This is one-fourth the length of the Golden Gate Bridge. Find the length of the Golden Gate Bridge.

19. **Nutrition** One slice of cheese pizza contains 290 calories. A medium-size orange has one-fifth that number of calories. How many calories are in a medium-size orange?

20. **History** John D. Rockefeller died in 1937. At the time of his death, Rockefeller had accumulated a wealth of \$1,400 million, which was equal to one-sixty-fifth of the gross national product of the United States at that time. What was the U.S. gross national product in 1937? (*Source: The Wealthy 100: A Ranking of the Richest Americans, Past and Present*)

21. **Agriculture** A soil supplement that weighs 24 lb contains iron, potassium, and a mulch. There is five times as much mulch as iron and twice as much potassium as iron. Find the amount of mulch in the soil supplement.

22. **Commissions** A real estate agent sold two homes and received commissions totaling \$6000. The agent's commission on one home was one and one-half times the commission on the second home. Find the agent's commission on each home.

23. **Safety** Loudness, or the intensity of sound, is measured in decibels. The sound level of a television is about 70 decibels, which is considered a safe hearing level. A food blender runs at 20 decibels higher than a TV, and a jet engine's decibel reading is 40 less than twice that of a blender. At this level, exposure can cause hearing loss. Find the intensity of the sound of a jet engine.

Point of Interest

The low-frequency pulses or whistles made by blue whales have been measured at up to 188 decibels, making them the loudest sounds produced by a living organism.

24. **Education** A university employs a total of 600 teaching assistants and research assistants. There are three times as many teaching assistants as research assistants. Find the number of research assistants employed by the university.

25. **Geometry** Greek architects considered a rectangle whose length was approximately 1.6 times its width to be the most visually appealing. Find the length and width of a rectangle constructed in this manner if the sum of the length and width is 130 ft.

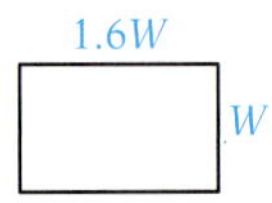

26. **Geography** Greenland, the largest island in the world, is 21 times larger than Iceland. The combined area of Greenland and Iceland is 880,000 mi^2. Find the area of Greenland.

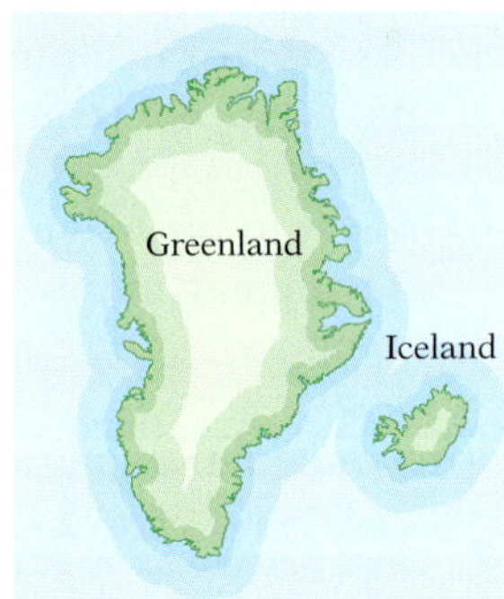

27. **Consumerism** The cost to replace a water pump in a sports car was \$600. This included \$375 for the water pump and \$45 per hour for labor. How many hours of labor were required to replace the water pump?

28. **Utilities** The cost of electricity in a certain city is \$.08 for each of the first 300 kWh (kilowatt-hours) and \$.13 for each kWh over 300 kWh. Find the number of kilowatt-hours used by a family that receives a \$51.95 electric bill.

29. **Business** A technical information hotline charges a customer \$9.00 plus \$.50 per minute to answer questions about software. How many minutes did a customer who received a bill for \$14.50 use this service?

30. **Metalwork** A wire 12 ft long is cut into two pieces. Each piece is bent into the shape of a square. The perimeter of the larger square is twice the perimeter of the smaller square. Find the perimeter of the larger square.

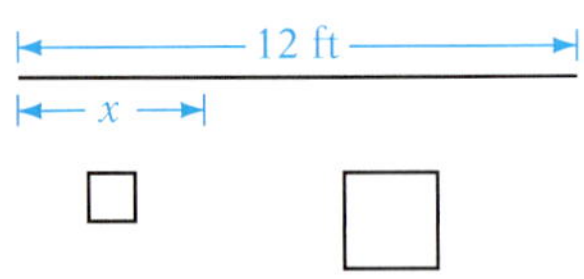

2 Consecutive Integer Problems

31. The sum of three consecutive integers is 54. Find the integers.

32. The sum of three consecutive integers is 75. Find the integers.

33. The sum of three consecutive even integers is 84. Find the integers.

34. The sum of three consecutive even integers is 48. Find the integers.

35. The sum of three consecutive odd integers is 57. Find the integers.

36. The sum of three consecutive odd integers is 81. Find the integers.

37. Find two consecutive even integers such that five times the first integer is equal to four times the second integer.

38. Find two consecutive even integers such that six times the first integer equals three times the second integer.

39. Nine times the first of two consecutive odd integers equals seven times the second. Find the integers.

40. Five times the first of two consecutive odd integers is three times the second. Find the integers.

41. Find three consecutive integers whose sum is negative twenty-four.

42. Find three consecutive even integers whose sum is negative twelve.

43. Three times the smallest of three consecutive even integers is two more than twice the largest. Find the integers.

44. Twice the smallest of three consecutive odd integers is five more than the largest. Find the integers.

45. Find three consecutive odd integers such that three times the middle integer is six more than the sum of the first and third integers.

46. Find three consecutive odd integers such that four times the middle integer is equal to two less than the sum of the first and third integers.

3 Value Mixture Problems

47. Use the equation Amount · Unit cost = Value to explain how to represent the value of x quarts of juice that cost \$1.40 per quart.

48. Suppose you are mixing peanuts that cost \$4 per pound with raisins that cost \$2 per pound. In the context of this situation, explain the meaning of the statement "The sum of the values of the individual ingredients is equal to the value of the mixture."

49. A restaurant chef mixes 20 lb of snow peas costing \$1.99 a pound with 14 lb of petite onions costing \$1.19 a pound to make a vegetable medley for the evening meal. Find the cost per pound of the mixture.

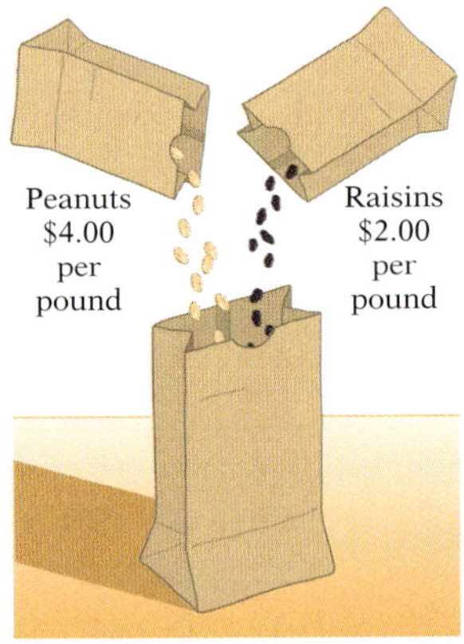

50. A coffee merchant combines coffee costing \$5.50 per pound with coffee costing \$3.00 per pound. How many pounds of each should be used to make 40 lb of a blend costing \$4.00 per pound?

51. Adult tickets for a play cost \$5.00 and children's tickets cost \$2.00. For one performance, 460 tickets were sold. Receipts for the performance were \$1880. Find the number of adult tickets sold.

52. The tickets for a local production of Gilbert and Sullivan's *H.M.S. Pinafore* cost \$6.00 for adults and \$3.00 for children and seniors. The total receipts for 505 tickets were \$1977. Find the number of adult tickets sold.

53. Fifty liters of pure maple syrup that costs \$9.50 per liter are mixed with imitation maple syrup that costs \$4.00 per liter. How much imitation maple syrup is needed to make a mixture that costs \$5.00 per liter?

54. To make a flour mix, a miller combined soybeans that cost \$8.50 per bushel with wheat that costs \$4.50 per bushel. How many bushels of each were used to make a mixture of 800 bushels that cost \$5.50 per bushel?

55. A hiking instructor is preparing a trail mix for a group of young hikers. She mixes nuts that cost \$3.99 a pound with pretzels that cost \$1.29 a pound to make a 20-pound mixture that costs \$2.37 a pound. Find the number of pounds of nuts used.

56. A silversmith combined pure silver that costs \$5.20 an ounce with 50 oz of a silver alloy that costs \$2.80 an ounce. How many ounces of the pure silver were used to make an alloy of silver that costs \$4.40 an ounce?

57. A tea mixture was made from 30 lb of tea that costs \$6.00 per pound and 70 lb of tea that costs \$3.20 per pound. Find the cost per pound of the tea mixture.

58. Find the cost per ounce of salad dressing mixture made from 64 oz of olive oil that costs \$8.29 and 20 oz of vinegar that costs \$1.99. Round to the nearest cent.

59. A fruitstand owner combined cranberry juice that costs \$4.20 per gallon with 50 gal of apple juice that costs \$2.10 per gallon. How much cranberry juice was used to make cranapple juice that costs \$3.00 per gallon?

60. Walnuts that cost \$4.05 per kilogram were mixed with cashews that cost \$7.25 per kilogram. How many kilograms of each were used to make a 50-kilogram mixture that costs \$6.25 per kilogram? Round to the nearest tenth.

4 Uniform Motion Problems

61. It takes a student 30 min to drive from school to work, a distance of 20 mi. At what average rate of speed does the student drive?

62. A cyclist rides at a rate of 16 mph for 45 min. How far does the cyclist travel in that time?

63. The typical cruising speed for a Boeing 747-400 airplane is 30 mph faster than that for a Boeing 737-800. If a Boeing 747-400 can make a trip of 1680 mi in 3 h, how long would it take a Boeing 737-800 to make the same trip? Round to the nearest tenth.

64. Some hotels in Las Vegas use moving sidewalks to transport guests between various places. Suppose a hotel guest decides to get on a moving sidewalk that is 195 ft long. If the guest walks at a rate of 5 ft/s and the moving sidewalk moves at a rate of 7 ft/s, how many seconds will it take the guest to walk from one end of the moving sidewalk to the other end?

65. Two cyclists start at 1:00 P.M. from opposite ends of a 54-mile race course. The average rate of speed of the first cyclist is 17 mph, and the average rate of speed of the second cyclist is 19 mph. At what time will the two cyclists meet?

66. At 9:00 A.M., Katrina starts down an exercise path and walks at a rate of 4 mph. At 9:30 A.M., Carla leaves from the same place and begins chasing Katrina, catching her at 10:00 A.M. Find Carla's jogging rate.

67. A bicyclist traveling at 18 mph overtakes an in-line skater who is traveling at 10 mph and had a 0.5-hour head start. How far from the starting point does the bicyclist overtake the in-line skater?

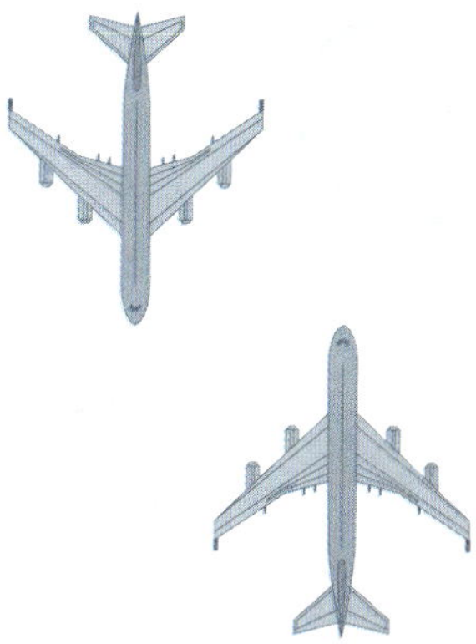

68. A helicopter traveling 120 mph overtakes a speeding car traveling at 90 mph. The car had a 0.5-hour head start. How far from the starting point did the helicopter overtake the car?

69. Two planes are 1380 mi apart and traveling toward each other. One plane is traveling 80 mph faster than the other plane. The planes meet in 1.5 h. Find the speed of each plane.

70. Two jet skiers leave the same dock at the same time and travel in opposite directions. One skier is traveling 14 mph slower than the other skier. In half an hour the skiers are 48 mi apart. Find the rate of the slower skier.

71. A ferry leaves a harbor and travels to a resort island at an average speed of 18 mph. On the return trip, the ferry travels at an average speed of 12 mph because of fog. The total time for the trip is 6 h. How far is the island from the harbor?

72. A commuter plane provides transportation from an international airport to surrounding cities. One commuter plane averaged 210 mph flying to a city and 140 mph returning to the international airport. The total flying time was 4 h. Find the distance between the two airports.

73. Two planes start from the same point and fly in opposite directions. The first plane is flying 50 mph slower than the second plane. In 2.5 h, the planes are 1400 mi apart. Find the rate of each plane.

74. Two hikers start from the same point and hike in opposite directions around a lake whose shoreline is 13 mi long. One hiker walks 0.5 mph faster than the other hiker. How fast did each hiker walk if they meet in 2 h?

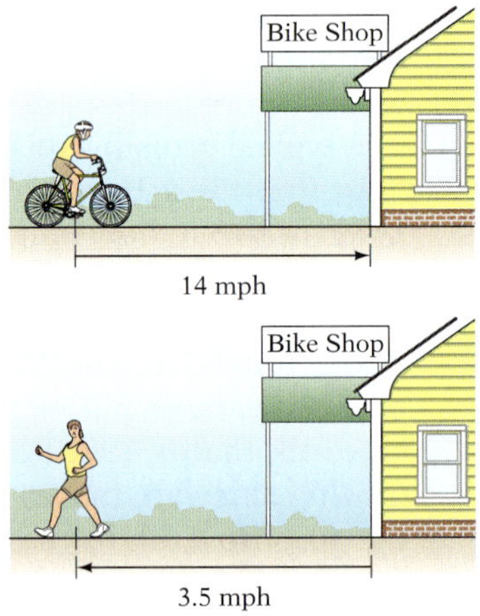

75. A student rode a bicycle to the repair shop and then walked home. The student averaged 14 mph riding to the shop and 3.5 mph walking home. The round trip took one hour. How far is it between the student's home and the bicycle shop?

76. An express train leaves Grand Central Station 1 h after a freight train leaves the same station. The express train is traveling 15 mph faster than the freight train. Find the rate at which each train is traveling if the express train overtakes the freight train in 3 h.

77. At noon a train leaves Washington, D.C., headed for Pittsburgh, Pennsylvania, a distance of 260 mi. The train travels at a speed of 60 mph. At 1 P.M. a second train leaves Pittsburgh headed for Washington, D.C., traveling at 40 mph. How long after the train leaves Pittsburgh will the two trains pass each other?

78. A plane leaves an airport at 3 P.M. At 4 P.M. another plane leaves the same airport traveling in the same direction at a speed 150 mph faster than that of the first plane. Four hours after the first plane takes off, the second plane is 250 mi ahead of the first plane. How far did the second plane travel?

2.3 Applying Concepts

79. The amount of liquid in a container triples every minute. The container becomes completely filled at 3:40 P.M. What fractional part of the container is filled at 3:39 P.M.?

80. **Wages** Four employees are paid at four consecutive levels on a wage scale. The difference between any two consecutive levels is \$320 per month. The average of the four employees' monthly wages is \$3880. What is the monthly wage of the highest-paid employee?

81. Find four consecutive odd integers whose sum is -48.

82. Find three consecutive odd integers such that the sum of the first and third integers is twice the second integer.

83. Find four consecutive integers such that the sum of the first and fourth integers equals the sum of the second and third integers.

84. In 1999, astronomers confirmed the existence of three planets orbiting a star other than the sun. The planets are approximately the size of Jupiter and orbit the star Upsilon Andromeda, which is approximately 260 trillion miles from Earth. How many years after leaving Earth would it take a spacecraft traveling 18 million miles per hour to reach this star? Round to the nearest hundred. (18 million miles per hour is about 1000 times faster than current spacecraft can travel.)

85. If a parade 2 mi long is proceeding at 3 mph, how long will it take a runner, jogging at 6 mph, to travel from the end of the parade to the start of the parade?

86. Two cars are headed directly toward each other at rates of 40 mph and 60 mph. How many miles apart are they 2 min before impact?

87. The following problem appears in a math text written around A.D. 1200. Two birds start flying from the tops of two towers 50 ft apart at the same time and at the same rate. One tower is 30 ft high, and the other tower is 40 ft high. The birds reach a grass seed on the ground at exactly the same time. How far is the grass seed from the 40-foot tower?

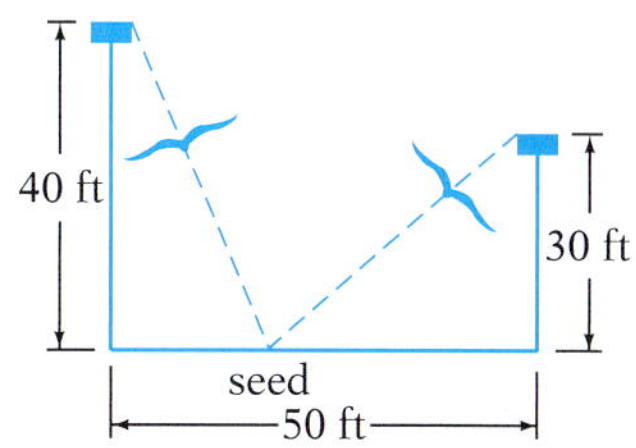

88. A car travels at an average speed of 30 mph for 1 mi. Is it possible for it to increase its speed during the next mile so that its average speed for the 2 mi is 60 mph?

89. In December of 1986, pilots Dick Rutan and Jeana Yeager flew the *Voyager* in the first nonstop, nonrefueled flight around the world. They flew east from Edwards Air Force Base in California on December 14, traveled 24,986.727 mi around the world, and returned to Edwards 216 h 3 min 44 s after their departure.

a. On what date did Rutan and Yeager land at Edwards Air Force Base after their flight?

b. What was their average speed in miles per hour? Round to the nearest whole number.

c. Find the circumference of Earth in a reference almanac, and then calculate the approximate distance above Earth that the flight was flown. Round to the nearest whole number.

90. A student jogs 1 mi at a rate of 8 mph and jogs back at a rate of 6 mph. Does it seem reasonable that the average rate is 7 mph? Why or why not? Support your answer.

2.4 Geometry Problems

1 Perimeter problems

VIDEO & DVD CD TUTOR WEB SSM

The **perimeter** of a plane geometric figure is a measure of the distance around the figure. Perimeter is used, for example, in buying fencing for a lawn and in determining how much baseboard is needed for a room.

The perimeter of a triangle is the sum of the lengths of the three sides. Therefore, if a, b, and c represent the lengths of the sides of a triangle, the perimeter, P, of the triangle is given by $P = a + b + c$.

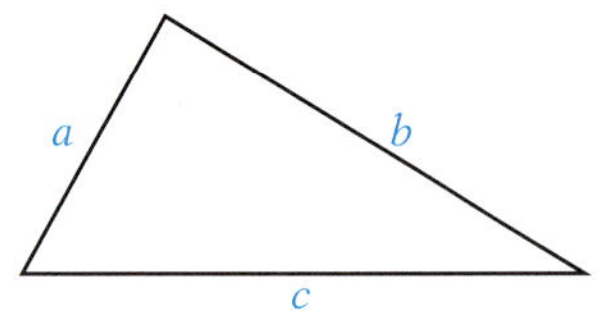

$P = a + b + c$

Two special types of triangles are shown below. An **isosceles triangle** has two sides of equal length. The three sides of an **equilateral triangle** are of equal length.

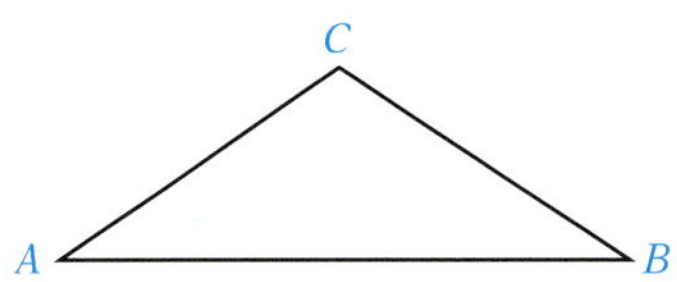

Isosceles triangle

$AC = BC$

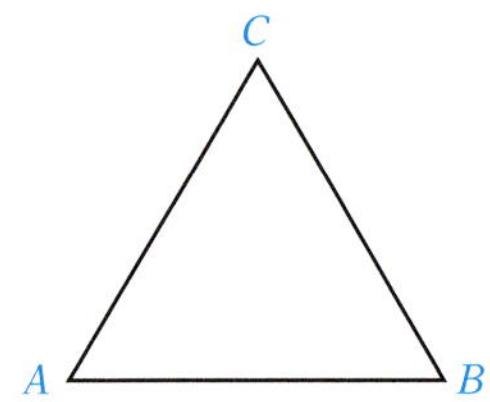

Equilateral triangle

$AB = AC = BC$

Point of Interest

Leonardo da Vinci painted the Mona Lisa on a rectangular canvas whose height was approximately 1.6 times its width. Rectangles with these proportions, called **golden rectangles,** *were used extensively in Renaissance art.*

The perimeter of a rectangle is the sum of the lengths of the four sides. Let L represent the length and W represent the width of a rectangle. Then the perimeter, P, of the rectangle is given by $P = L + W + L + W$. Combine like terms and the formula is $P = 2L + 2W$.

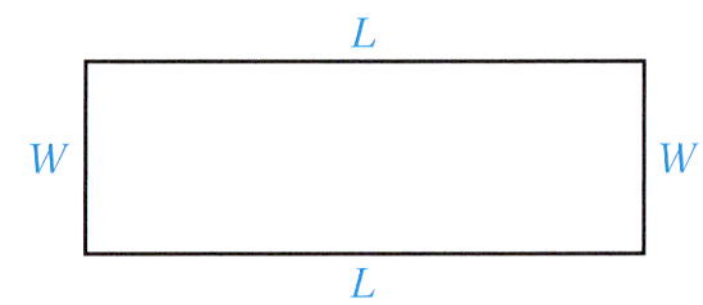

$P = 2L + 2W$

A square is a rectangle in which each side has the same length. Let s represent the length of each side of a square. Then the perimeter, P, of the square is given by $P = s + s + s + s$. Combine like terms and the formula is $P = 4s$.

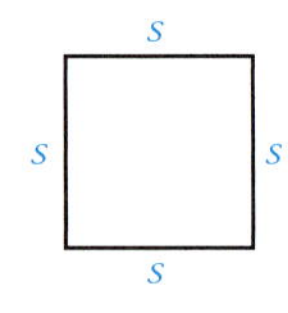

$P = 4s$

Formulas for Perimeters of Geometric Figures

Perimeter of a triangle	$P = a + b + c$
Perimeter of a rectangle	$P = 2L + 2W$
Perimeter of a square	$P = 4s$

The perimeter of a rectangle is 26 ft. The length of the rectangle is 1 ft more than twice the width. Find the width and length of the rectangle.

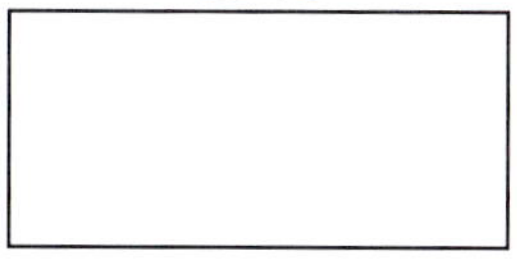

Strategy

Let a variable represent the width. Width: W

Represent the length in terms of that variable. Length: $2W + 1$

Use the formula for the perimeter of a rectangle.

Solution

	$P = 2L + 2W$
$P = 26$. Substitute $2W + 1$ for L.	$26 = 2(2W + 1) + 2W$
Use the Distributive Property.	$26 = 4W + 2 + 2W$
Combine like terms.	$26 = 6W + 2$
Subtract 2 from each side of the equation.	$24 = 6W$
Divide each side of the equation by 6.	$4 = W$
Find the length of the rectangle by substituting 4 for W in $2W + 1$.	$L = 2W + 1$ $= 2(4) + 1 = 8 + 1 = 9$

The width is 4 ft. The length is 9 ft.

Example 1 The perimeter of an isosceles triangle is 25 ft. The length of the third side is 2 ft less than the length of one of the equal sides. Find the measures of the three sides of the triangle.

Strategy

- Each equal side: x
  The third side: $x - 2$
- Use the equation for the perimeter of a triangle.

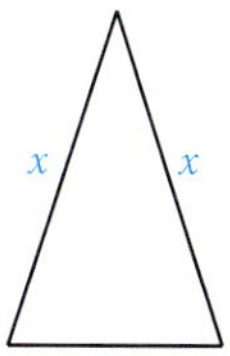

Solution

$$P = a + b + c$$
$$25 = x + x + (x - 2)$$
$$25 = 3x - 2$$
$$27 = 3x$$
$$9 = x$$

$x - 2 = 9 - 2 = 7$ • **Substitute the value of x into the variable expression for the length of the third side.**

Each of the equal sides measures 9 ft.
The third side measures 7 ft.

Problem 1 A carpenter is designing a square patio with a perimeter of 52 ft. What is the length of each side?

Solution See page S7.

2 Problems involving angles formed by intersecting lines

VIDEO & DVD CD TUTOR WEB SSM

A unit used to measure angles is the **degree**. The symbol for degree is °. ∠ is the symbol for angle.

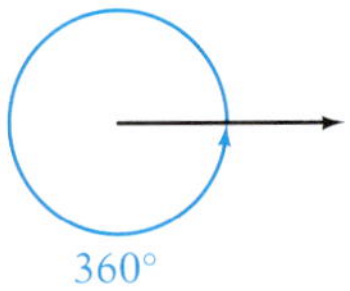

One complete revolution is 360°. It is probable that the Babylonians chose 360° for a circle because they knew that there are 365 days in one year, and 360 is the closest number to 365 that is divisible by many numbers.

A 90° angle is called a **right angle.** The symbol ∟ represents a right angle. Angle C ($\angle C$) is a right angle.

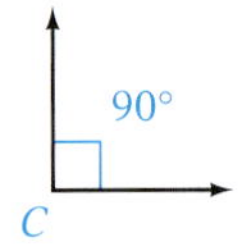

A 180° angle is called a **straight angle.** The angle at the right is a straight angle.

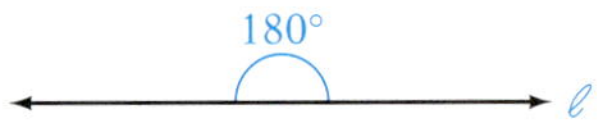

An **acute angle** is an angle whose measure is between 0° and 90°. $\angle A$ at the right is an acute angle.

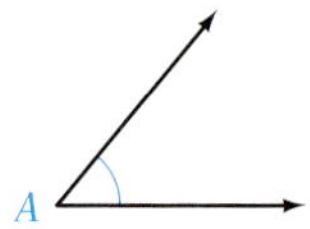

An **obtuse angle** is an angle whose measure is between 90° and 180°. $\angle B$ at the right is an obtuse angle.

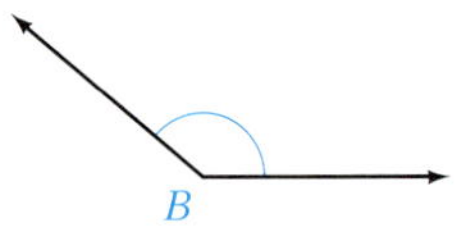

Given the diagram at the left, find x.

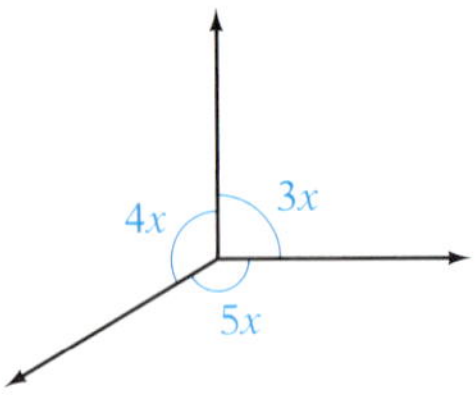

Strategy The sum of the measures of the three angles is 360°. To find x, write an equation and solve for x.

Solution

$$\begin{aligned} 3x + 4x + 5x &= 360^\circ \\ 12x &= 360^\circ \\ x &= 30^\circ \end{aligned}$$

The measure of x is 30°.

Complementary angles are two angles whose measures have the sum 90°.

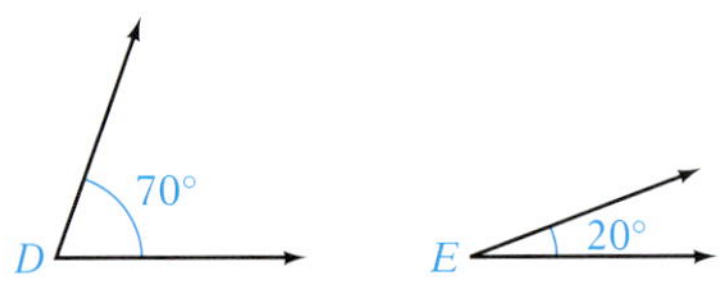

$\angle D + \angle E = 70^\circ + 20^\circ = 90^\circ$

$\angle D$ and $\angle E$ are complementary angles.

Supplementary angles are two angles whose measures have the sum 180°.

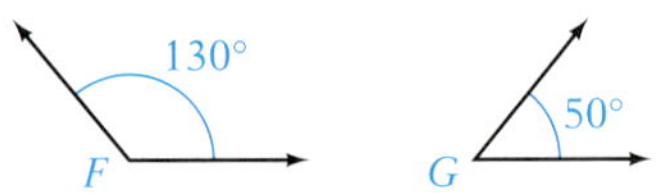

$\angle F + \angle G = 130^\circ + 50^\circ = 180^\circ$

$\angle F$ and $\angle G$ are supplementary angles.

Example 2 Find the complement of a 39° angle.

Strategy To find the complement, let x represent the complement of a 39° angle. Use the fact that complementary angles are two angles whose sum is 90° to write an equation. Solve for x.

Solution $39° + x = 90°$
$x = 51°$ • Subtract 39 from each side of the equation.

The complement of a 39° angle is a 51° angle.

Problem 2 Find the supplement of a 107° angle.

Solution See page S7.

Parallel lines never meet. The distance between them is always the same. The symbol $\|$ means "is parallel to." In the figure at the right, $\ell_1 \| \ell_2$.

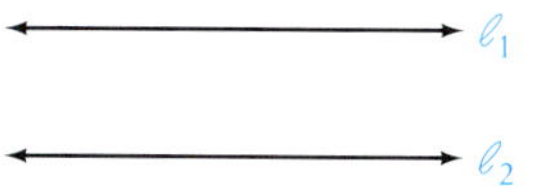

Perpendicular lines are intersecting lines that form right angles. The symbol $\perp$ means "is perpendicular to." In the figure at the right, $p \perp q$.

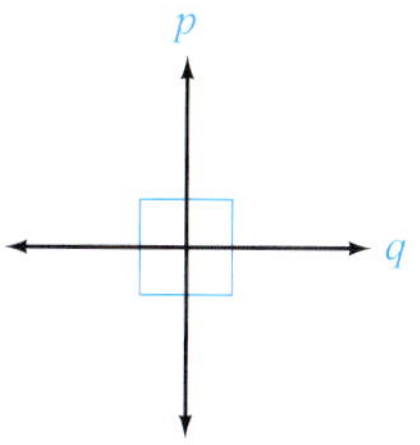

Four angles are formed by the intersection of two lines. If the two lines are perpendicular, then each of the four angles is a right angle. If the two lines are not perpendicular, then two of the angles formed are acute angles and two of the angles formed are obtuse angles. The two acute angles are always opposite each other, and the two obtuse angles are always opposite each other.

In the figure at the right, $\angle w$ and $\angle y$ are acute angles, and $\angle x$ and $\angle z$ are obtuse angles.

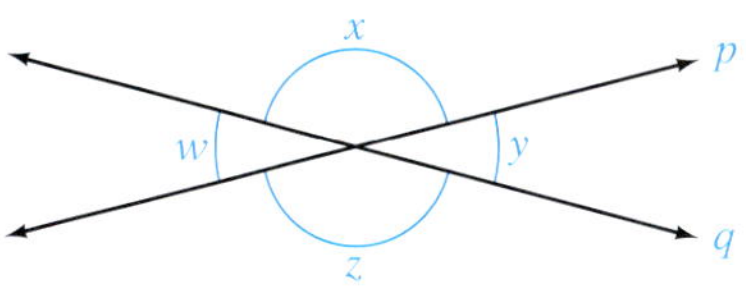

Two angles that are on opposite sides of the intersection of two lines are called **vertical angles.** Vertical angles have the same measure. $\angle w$ and $\angle y$ are vertical angles. $\angle x$ and $\angle z$ are vertical angles.

Vertical angles have the same measure.

$$\angle w = \angle y$$
$$\angle x = \angle z$$

Two angles that share a common side are called **adjacent angles.** For the figure shown above, $\angle x$ and $\angle y$ are adjacent angles, as are $\angle y$ and $\angle z$, $\angle z$ and $\angle w$, and $\angle w$ and $\angle x$. Adjacent angles of intersecting lines are supplementary angles.

Adjacent angles of intersecting lines are supplementary angles.

$$\angle x + \angle y = 180°$$
$$\angle y + \angle z = 180°$$
$$\angle z + \angle w = 180°$$
$$\angle w + \angle x = 180°$$

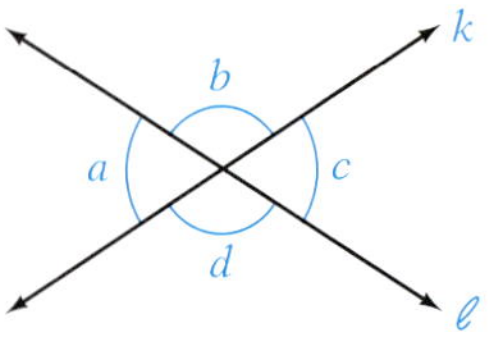

In the diagram at the left, $\angle b = 115°$. Find the measures of angles a, c, and d.

$\angle a$ is supplementary to $\angle b$ because $\angle a$ and $\angle b$ are adjacent angles of intersecting lines.

$$\angle a + \angle b = 180°$$
$$\angle a + 115° = 180°$$
$$\angle a = 65°$$

$\angle c = \angle a$ because $\angle c$ and $\angle a$ are vertical angles. $\quad \angle c = 65°$

$\angle d = \angle b$ because $\angle d$ and $\angle b$ are vertical angles. $\quad \angle d = 115°$

Example 3 Find x.

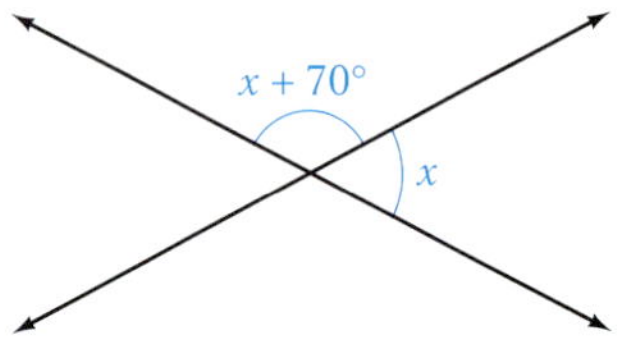

Strategy The angles labeled are adjacent angles of intersecting lines and are therefore supplementary angles. To find x, write an equation and solve for x.

Solution

$$x + (x + 70°) = 180°$$
$$2x + 70° = 180°$$
$$2x = 110°$$
$$x = 55°$$

Problem 3 Find x.

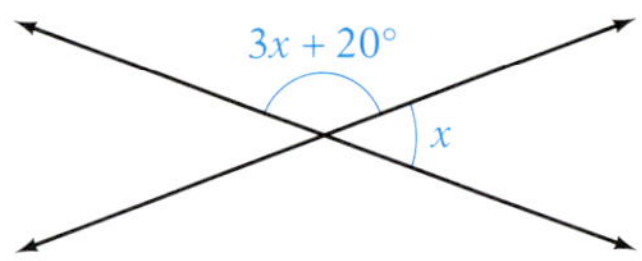

Solution See page S7.

A line that intersects two other lines at different points is called a **transversal**. If the lines cut by a transversal t are parallel lines and the transversal is not perpendicular to the parallel lines, all four acute angles have the same measure and all four obtuse angles have the same measure. In the figure at the right,

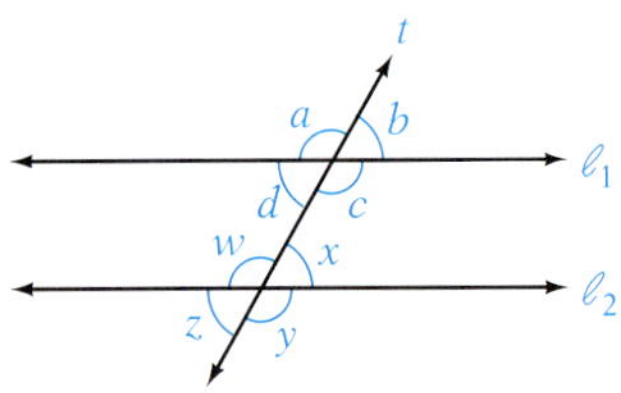

$$\angle b = \angle d = \angle x = \angle z$$
$$\angle a = \angle c = \angle w = \angle y$$

Alternate interior angles are two nonadjacent angles that are on opposite sides of the transversal and between the lines. In the figure above, $\angle c$ and $\angle w$ are alternate interior angles, and $\angle d$ and $\angle x$ are alternate interior angles. Alternate interior angles have the same measure.

Alternate interior angles have the same measure.

$$\angle c = \angle w$$
$$\angle d = \angle x$$

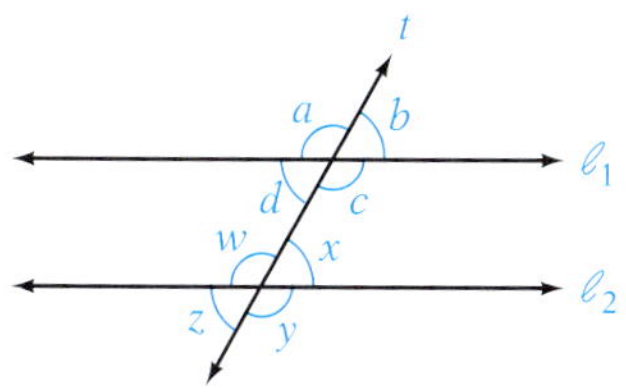

Alternate exterior angles are two nonadjacent angles that are on opposite sides of the transversal and outside the parallel lines. In the figure at the left, $\angle a$ and $\angle y$ are alternate exterior angles, and $\angle b$ and $\angle z$ are alternate exterior angles. Alternate exterior angles have the same measure.

Alternate exterior angles have the same measure.

$$\angle a = \angle y$$
$$\angle b = \angle z$$

Corresponding angles are two angles that are on the same side of the transversal and are both acute angles or are both obtuse angles. In the figure above, the following pairs of angles are corresponding angles: $\angle a$ and $\angle w$, $\angle d$ and $\angle z$, $\angle b$ and $\angle x$, and $\angle c$ and $\angle y$. Corresponding angles have the same measure.

Corresponding angles have the same measure.

$$\angle a = \angle w$$
$$\angle d = \angle z$$
$$\angle b = \angle x$$
$$\angle c = \angle y$$

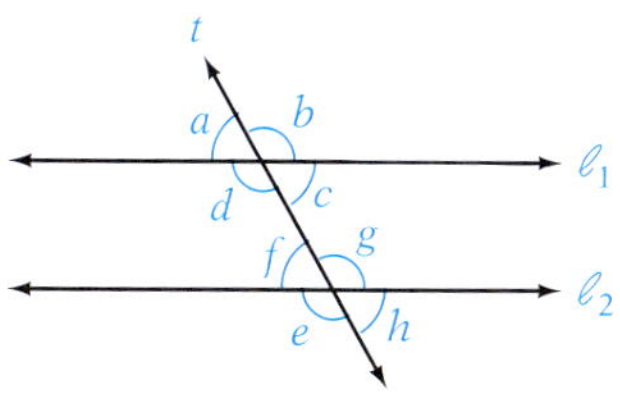

In the diagram at the left, $\ell_1 \parallel \ell_2$ and $\angle f = 58°$. Find the measures of $\angle a$, $\angle c$, and $\angle d$.

$\angle a$ and $\angle f$ are corresponding angles. $\qquad \angle a = \angle f = 58°$

$\angle c$ and $\angle f$ are alternate interior angles. $\qquad \angle c = \angle f = 58°$

$\angle d$ is supplementary to $\angle a$.

$$\angle d + \angle a = 180°$$
$$\angle d + 58° = 180°$$
$$\angle d = 122°$$

Example 4 Given $\ell_1 \parallel \ell_2$, find x.

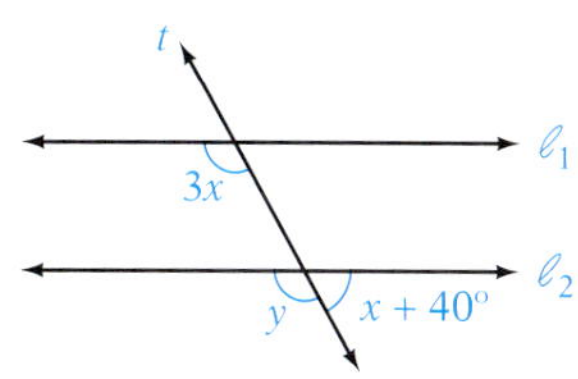

Strategy $3x = y$ because corresponding angles have the same measure. $y + (x + 40°) = 180°$ because adjacent angles of intersecting lines are supplementary angles.

Substitute $3x$ for y and solve for x.

Solution

$$y + (x + 40°) = 180°$$
$$3x + (x + 40°) = 180°$$
$$4x + 40° = 180°$$
$$4x = 140°$$
$$x = 35°$$

Problem 4 Given $\ell_1 \parallel \ell_2$, find x.

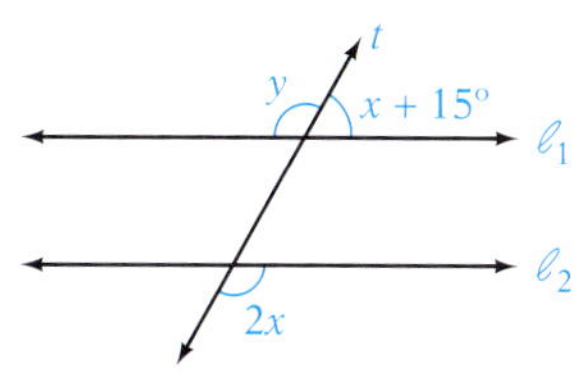

Solution See pages S7–S8.

3 Problems involving the angles of a triangle

If the lines cut by a transversal are not parallel lines, the three lines will intersect at three points. In the figure at the right, the transversal t intersects lines p and q. The three lines intersect at points A, B, and C. The geometric figure formed by line segments AB, BC, and AC is a **triangle.**

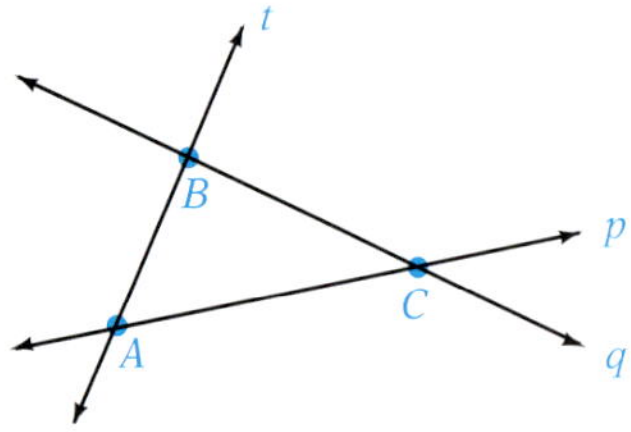

The angles within the region enclosed by the triangle are called **interior angles.** In the figure at the right, angles a, b, and c are interior angles. The sum of the measures of the interior angles is 180°.

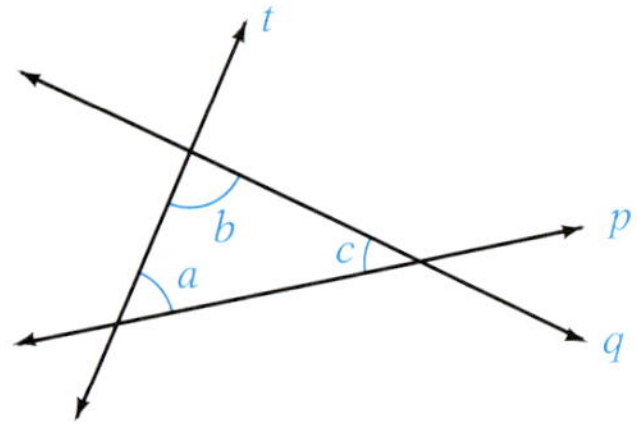

$$\angle a + \angle b + \angle c = 180°$$

The Sum of the Measures of the Interior Angles of a Triangle

The sum of the measures of the interior angles of a triangle is 180°.

An angle adjacent to an interior angle is an **exterior angle.** In the figure at the right, angles m and n are exterior angles for angle a. The sum of the measures of an interior and an exterior angle is 180°.

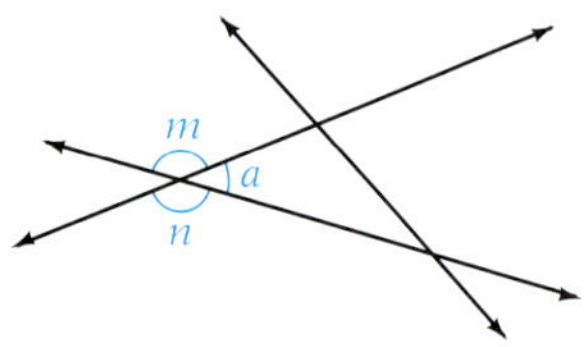

$$\angle a + \angle m = 180°$$
$$\angle a + \angle n = 180°$$

Given that $\angle c = 40°$ and $\angle e = 60°$, find the measure of $\angle d$.

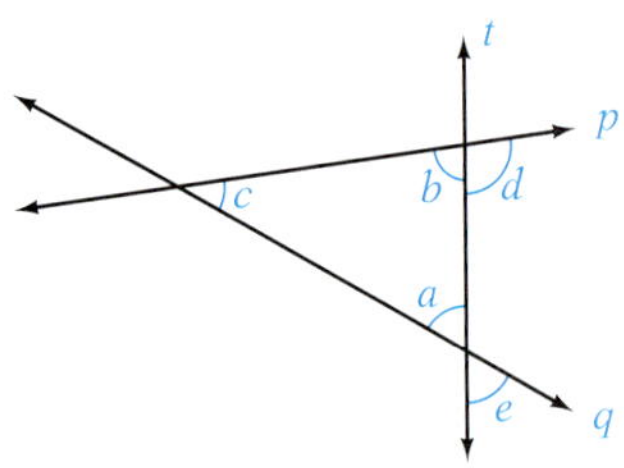

$\angle a$ and $\angle e$ are vertical angles. $\quad \angle a = \angle e = 60°$

The sum of the interior angles is 180°.

$$\begin{aligned} \angle c + \angle a + \angle b &= 180° \\ 40° + 60° + \angle b &= 180° \\ 100° + \angle b &= 180° \\ \angle b &= 80° \end{aligned}$$

$\angle b$ and $\angle d$ are supplementary angles.

$$\angle b + \angle d = 180°$$
$$80° + \angle d = 180°$$
$$\angle d = 100°$$

Example 5 Given that $\angle a = 45°$ and $\angle x = 100°$, find the measures of angles b, c, and y.

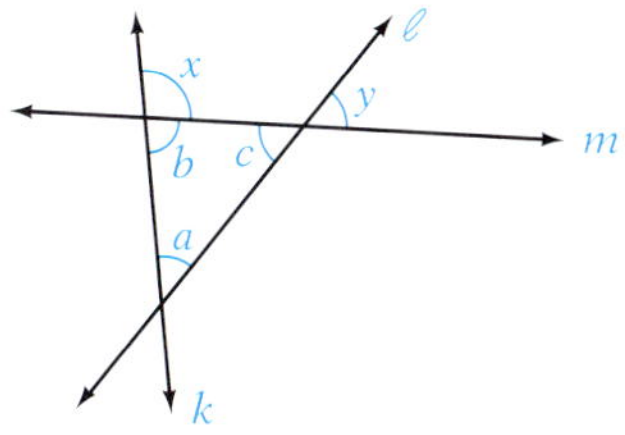

Strategy

- ▶ To find the measure of $\angle b$, use the fact that $\angle b$ and $\angle x$ are supplementary angles.
- ▶ To find the measure of $\angle c$, use the fact that the sum of the measures of the interior angles of a triangle is 180°.
- ▶ To find the measure of $\angle y$, use the fact that $\angle c$ and $\angle y$ are vertical angles.

Solution

$$\angle b + \angle x = 180°$$
$$\angle b + 100° = 180°$$
$$\angle b = 80°$$

$$\angle a + \angle b + \angle c = 180°$$
$$45° + 80° + \angle c = 180°$$
$$125° + \angle c = 180°$$
$$\angle c = 55°$$

$$\angle y = \angle c = 55°$$

Problem 5 Given that $\angle y = 55°$, find the measures of angles a, b, and d.

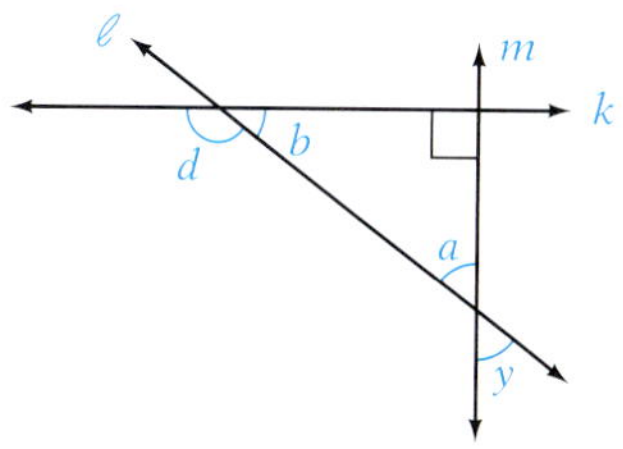

Solution See page S8.

Example 6 Two angles of a triangle measure 43° and 86°. Find the measure of the third angle.

Strategy To find the measure of the third angle, use the fact that the sum of the measures of the interior angles of a triangle is 180°. Write an equation using x to represent the measure of the third angle. Solve the equation for x.

Solution

$$x + 43° + 86° = 180°$$
$$x + 129° = 180°$$
$$x = 51°$$

The measure of the third angle is 51°.

Problem 6 One angle of a triangle is a right angle, and one angle measures 27°. Find the measure of the third angle.

Solution See page S8.

2.4 Concept Review

Determine whether the statement is always true, sometimes true, or never true.

1. The formula for the perimeter of a rectangle is $P = 2L + 2W$, where L represents the length and W represents the width of a rectangle.

2. In the formula for the perimeter of a triangle, $P = a + b + c$, the variables a, b, and c represent the measures of the three angles of a triangle.

3. If two angles of a triangle measure 52° and 73°, then the third angle of the triangle measures 55°.

4. An isosceles triangle has two sides of equal measure and two angles of equal measure.

5. The perimeter of a geometric figure is a measure of the area of the figure.

6. In a right triangle, the sum of the measures of the two acute angles is 90°.

2.4 Exercises

1 Geometry Problems

1. What is the difference between an isosceles triangle and an equilateral triangle?

2. Which of the following units can be used to measure perimeter?
   a. feet
   b. square miles
   c. meters
   d. ounces
   e. cubic inches
   f. centimeters

3. In an isosceles triangle, the third side is 50% of the length of one of the equal sides. Find the length of each side when the perimeter is 125 ft.

4. In an isosceles triangle, the length of one of the equal sides is three times the length of the third side. The perimeter is 21 m. Find the length of each side.

5. The perimeter of a rectangle is 42 m. The length of the rectangle is 3 m less than twice the width. Find the length and width of the rectangle.

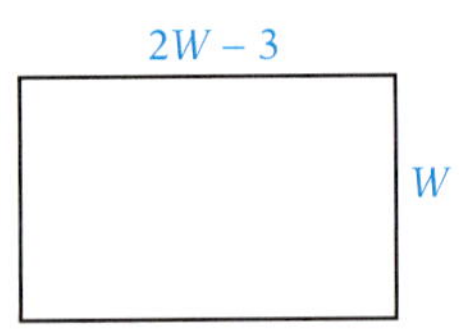

6. The width of a rectangle is 25% of the length. The perimeter is 250 cm. Find the length and width of the rectangle.

7. The perimeter of a rectangle is 120 ft. The length of the rectangle is twice the width. Find the length and width of the rectangle.

8. The perimeter of a rectangle is 50 m. The width of the rectangle is 5 m less than the length. Find the length and width of the rectangle.

9. The perimeter of a triangle is 110 cm. One side is twice the second side. The third side is 30 cm more than the second side. Find the length of each side.

10. The perimeter of a triangle is 33 ft. One side of the triangle is 1 ft longer than the second side. The third side is 2 ft longer than the second side. Find the length of each side.

11. The width of the rectangular foundation of a building is 30% of the length. The perimeter of the foundation is 338 ft. Find the length and width of the foundation.

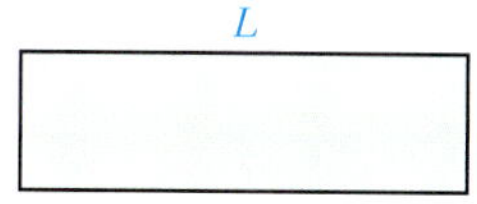

12. The perimeter of a rectangular playground is 440 ft. If the width is 100 ft, what is the length of the playground?

13. A rectangular vegetable garden has a perimeter of 64 ft. The length of the garden is 20 ft. What is the width of the garden?

14. Each of two sides of a triangular banner measures 18 in. If the perimeter of the banner is 46 in., what is the length of the third side of the banner?

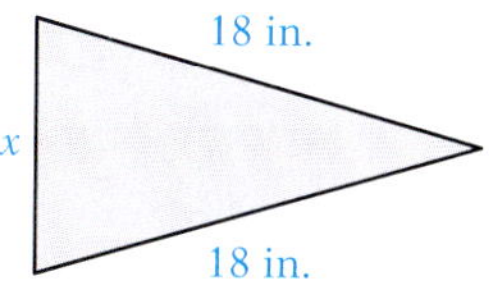

15. The perimeter of a square picture frame is 48 in. Find the length of each side of the frame.

16. A square rug has a perimeter of 32 ft. Find the length of each side of the rug.

2 Geometry Problems

17. Name the number of degrees in each of the following.
 a. a right angle
 b. a straight angle
 c. one complete revolution
 d. an acute angle
 e. an obtuse angle
 f. complementary angles
 g. supplementary angles

18. Determine whether enough information is given to solve the problem.
 a. Find the sum of the measures of the two acute angles in a right triangle.
 b. Find the measures of the two acute angles in a right triangle.

19. Find the complement of a 28° angle.

20. Find the complement of a 46° angle.

21. Find the supplement of a 73° angle.

22. Find the supplement of a 119° angle.

Find the measure of $\angle x$.

23.

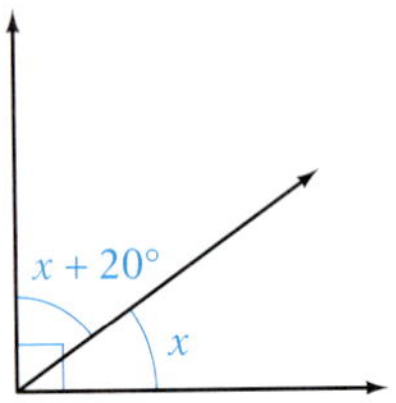

24.

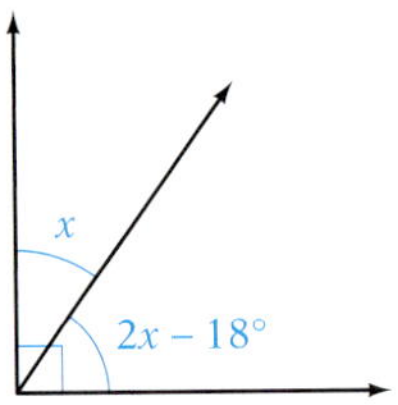

25.

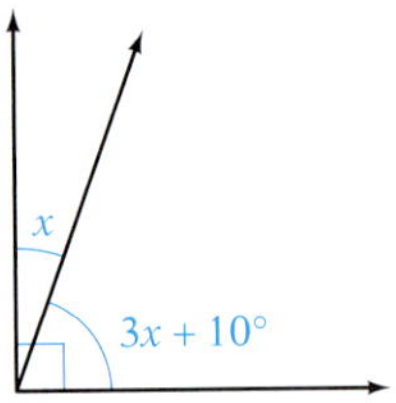

Find the measure of $\angle a$.

26.

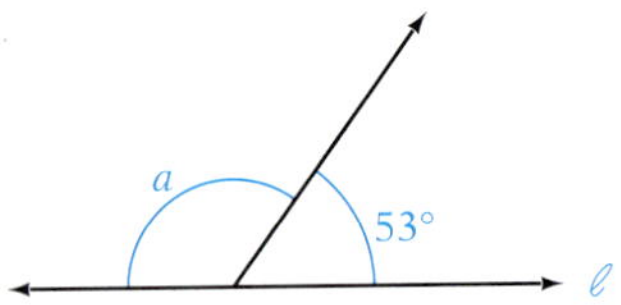

27.

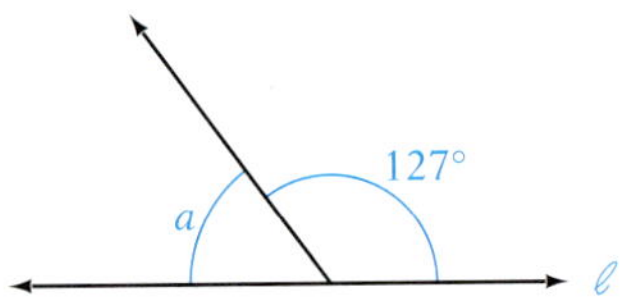

28.

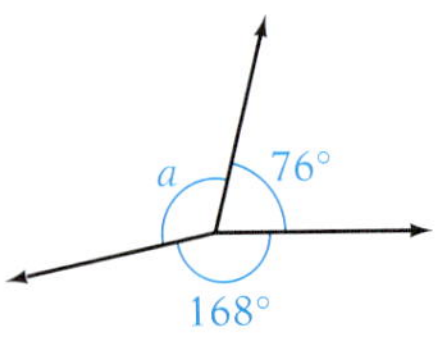

29.

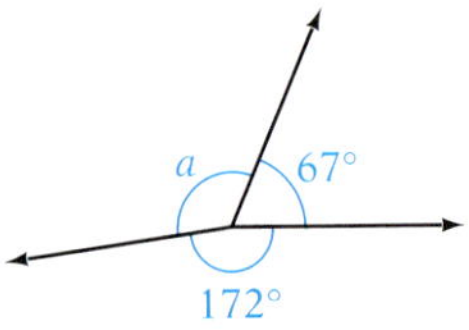

Find x.

30.

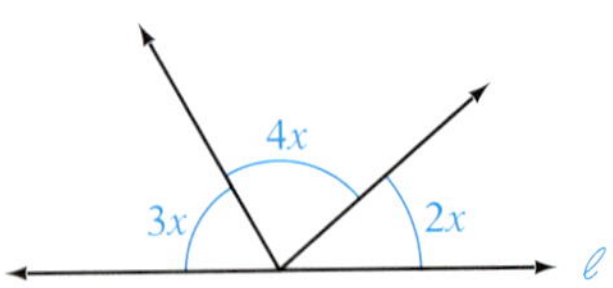

31.

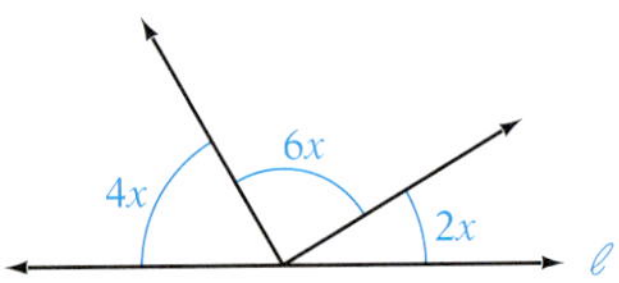

32.

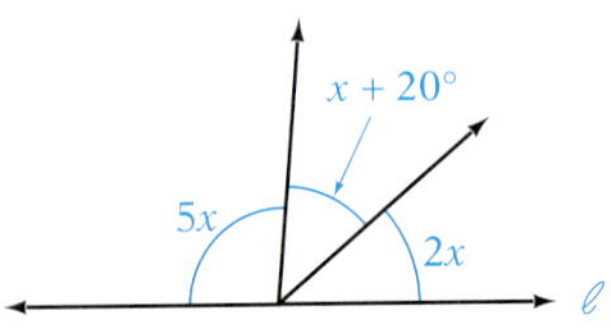

33.

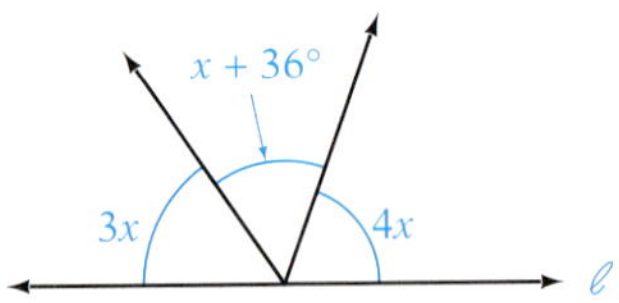

34.

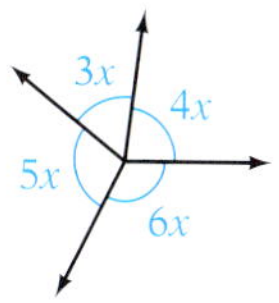

35.

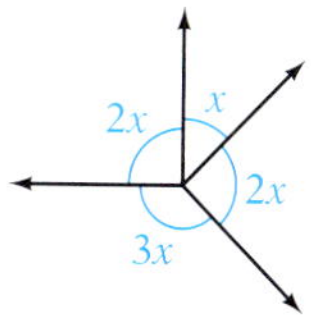

Find the measure of $\angle x$.

36.

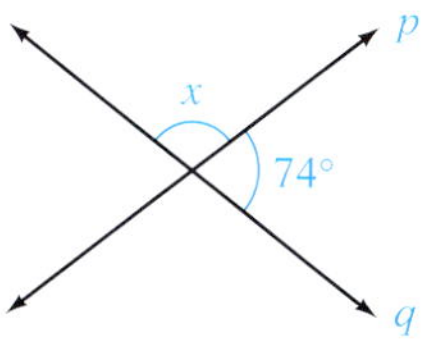

37.

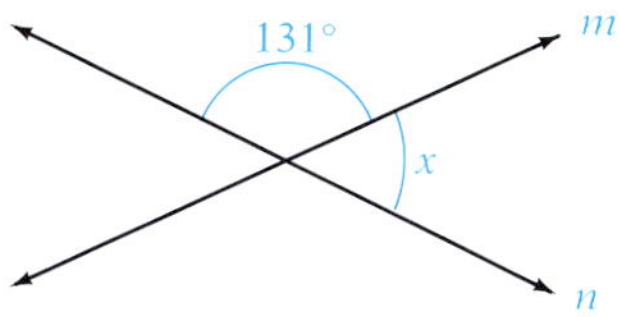

Find x.

38.

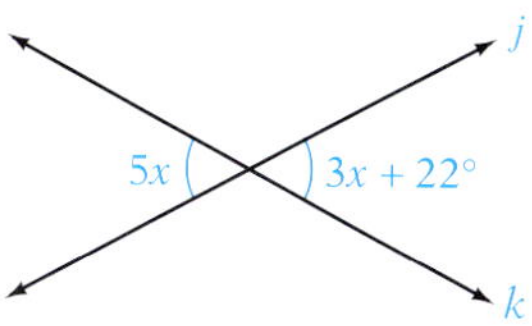

39.

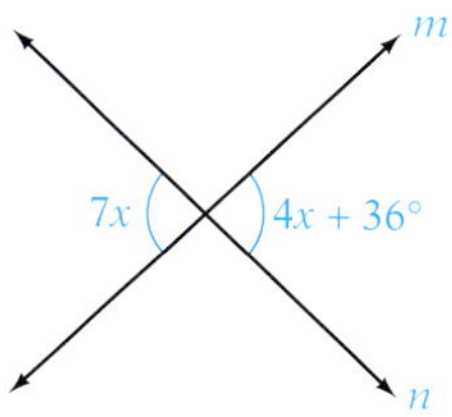

Given that $\ell_1 \| \ell_2$, find the measures of angles a and b.

40.

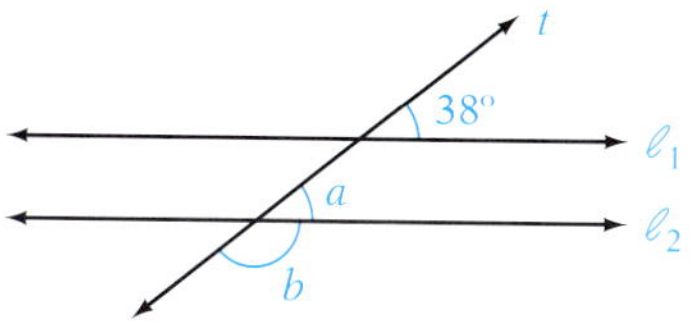

41.

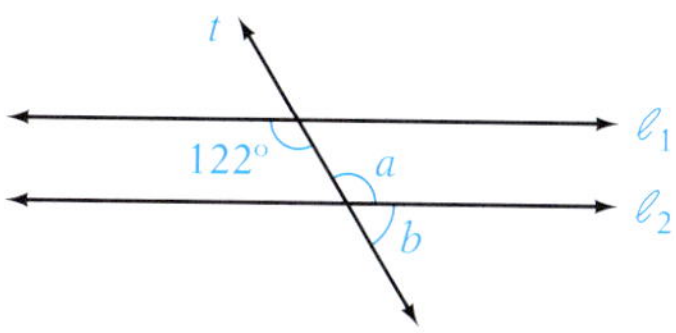

42.

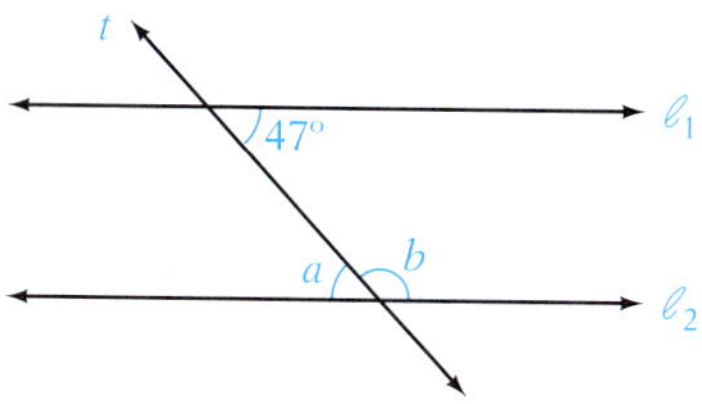

43.

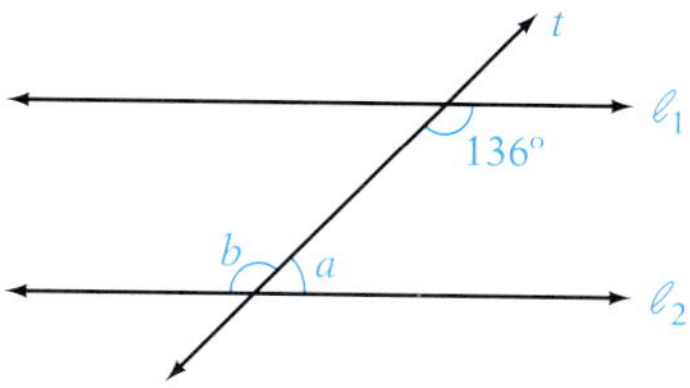

Given that $\ell_1 \| \ell_2$, find x.

44.

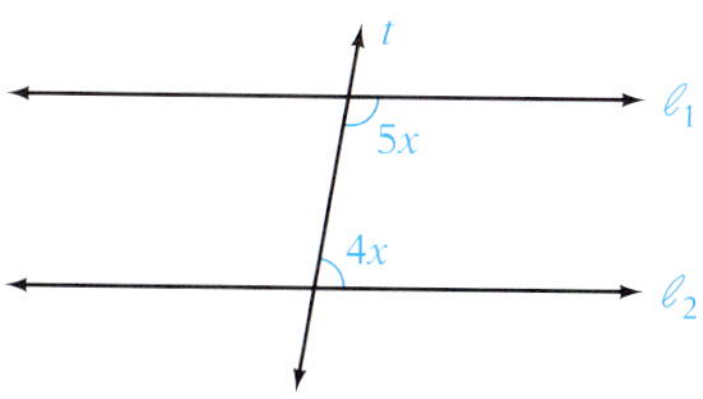

45.

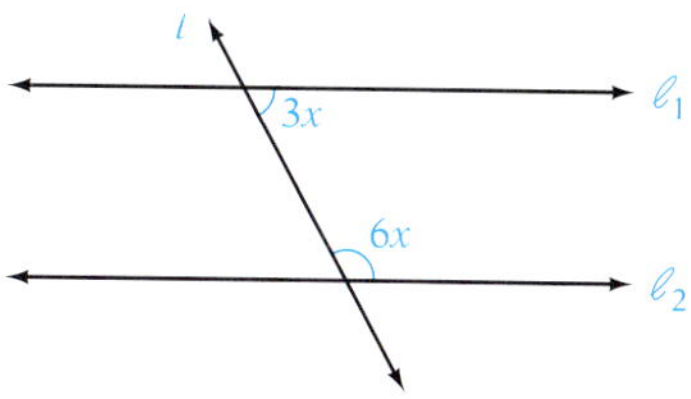

46.

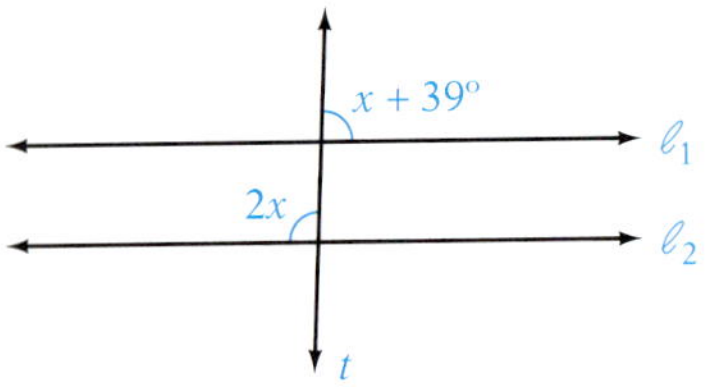

47.

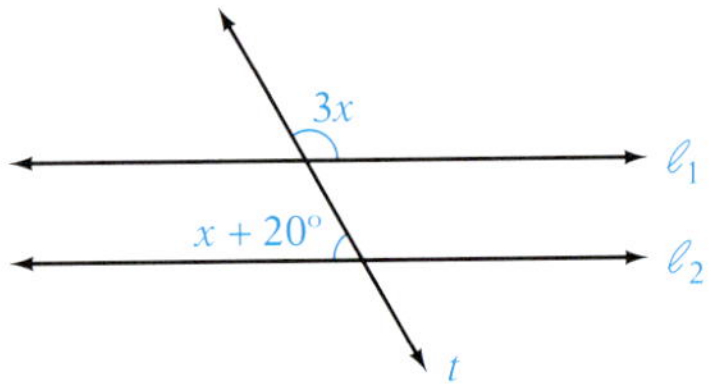

48. Given that $\angle a = 51°$, find the measure of $\angle b$.

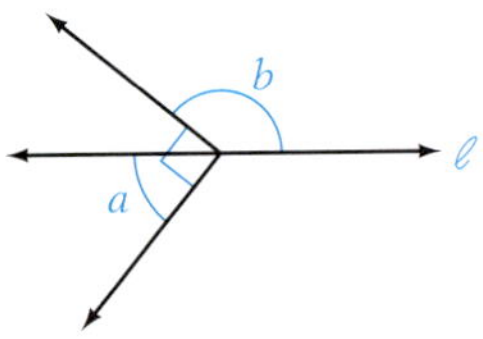

49. Given that $\angle a = 38°$, find the measure of $\angle b$.

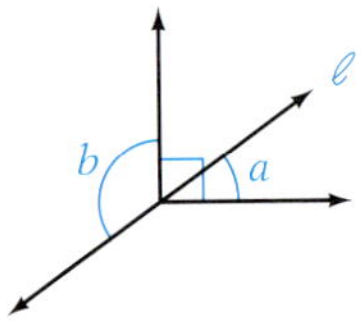

3 Geometry Problems

50. Given that $\angle a = 95°$ and $\angle b = 70°$, find the measures of angles x and y.

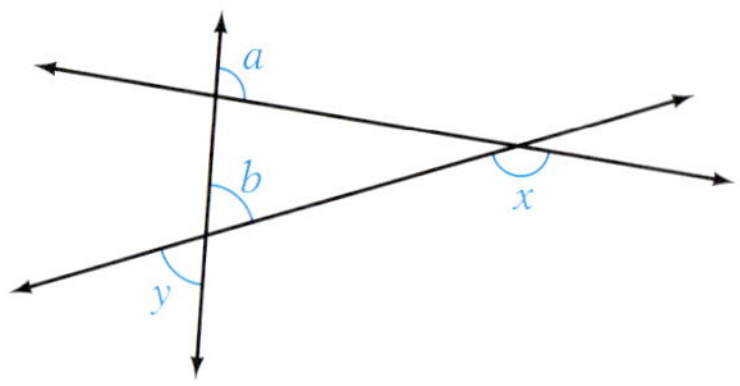

51. Given that $\angle a = 35°$ and $\angle b = 55°$, find the measures of angles x and y.

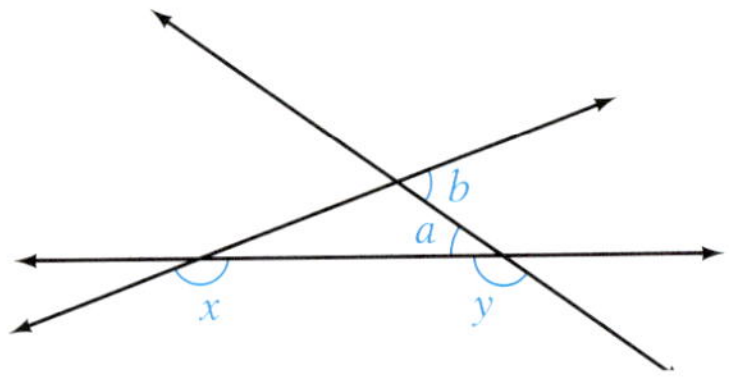

52. Given that $\angle y = 45°$, find the measures of angles a and b.

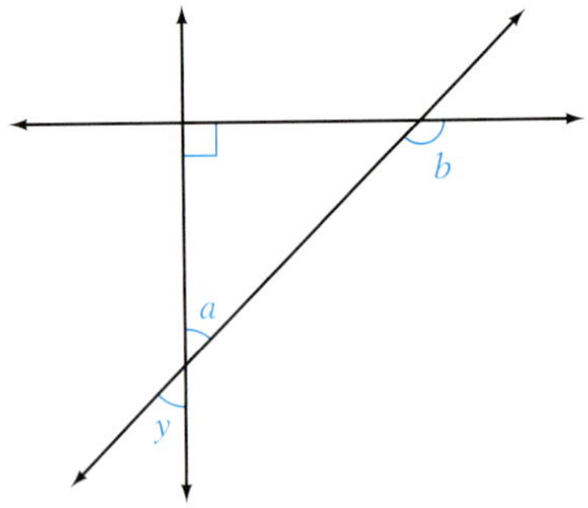

53. Given that $\angle y = 130°$, find the measures of angles a and b.

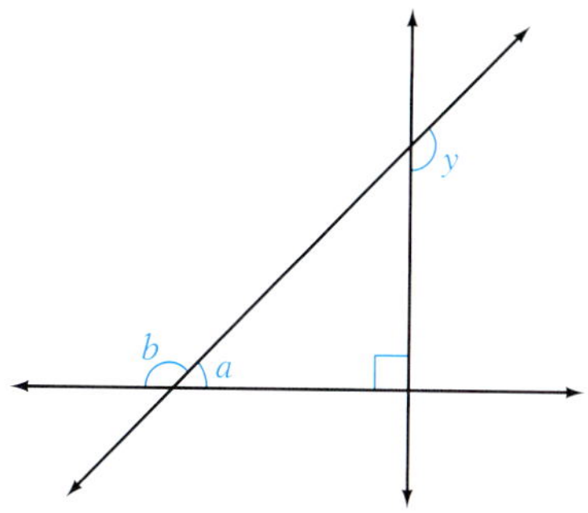

54. Given that $AO \perp OB$, express in terms of x the number of degrees in $\angle BOC$.

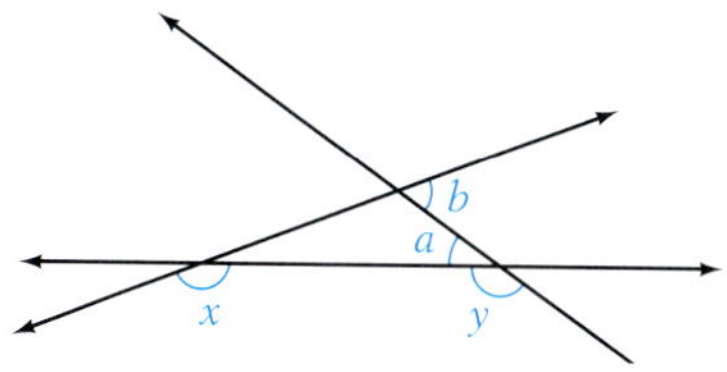

55. Given that $AO \perp OB$, express in terms of x the number of degrees in $\angle AOC$.

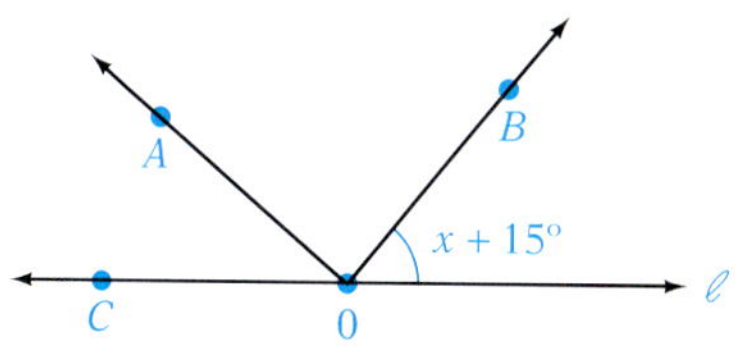

56. One angle in a triangle is a right angle, and one angle measures 30°. What is the measure of the third angle?

57. A triangle has a 45° angle and a right angle. Find the measure of the third angle.

58. Two angles of a triangle measure 42° and 103°. Find the measure of the third angle.

59. Two angles of a triangle measure 62° and 45°. Find the measure of the third angle.

60. A triangle has a 13° angle and a 65° angle. What is the measure of the third angle?

61. A triangle has a 105° angle and a 32° angle. What is the measure of the third angle?

62. In an isosceles triangle, one angle is three times the measure of one of the equal angles. Find the measure of each angle.

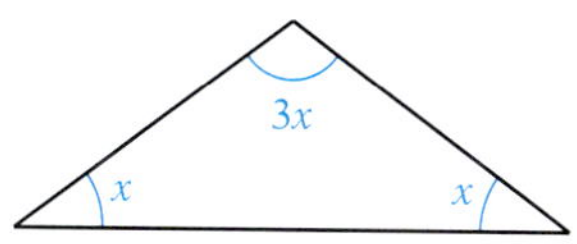

63. In an isosceles triangle, one angle is 10° less than three times the measure of one of the equal angles. Find the measure of each angle.

64. In an isosceles triangle, one angle is 16° more than twice the measure of one of the equal angles. Find the measure of each angle.

65. In a triangle, one angle is twice the measure of the second angle. The third angle is three times the measure of the second angle. Find the measure of each angle.

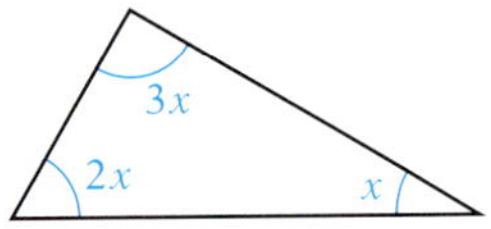

2.4 Applying Concepts

66. A rectangle and an equilateral triangle have the same perimeter. The length of the rectangle is three times the width. Each side of the triangle is 8 cm. Find the length and width of the rectangle.

67. The length of a rectangle is 1 cm more than twice the width. If the length of the rectangle is decreased by 2 cm and the width is decreased by 1 cm, the perimeter is 20 cm. Find the length and width of the original rectangle.

68. For the figure at the right, find the sum of the measures of angles x, y, and z.

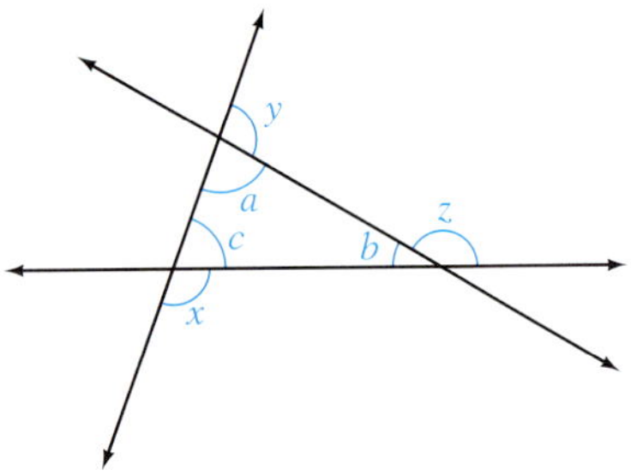

69. For the figure at the right, explain why $\angle a + \angle b = \angle x$. Write a rule that describes the relationship between an exterior angle of a triangle and the opposite interior angles. Use the rule to write an equation involving angles a, c, and z.

70. The width of a rectangle is $8x$. The perimeter is $48x$. Find the length of the rectangle in terms of the variable x.

71. Write a report stating the arguments of the proponents of changing to the metric system in the United States or a report stating the arguments of those opposed to switching from the U.S. Customary System of measurement to the metric system.

72. Prepare a report on the use of geometric form in architecture.

2.5 Markup and Discount Problems

1 Markup problems

VIDEO & DVD CD TUTOR WEB SSM

Cost is the price that a business pays for a product. **Selling price** is the price for which a business sells a product to a customer. The difference between selling price and cost is called **markup.** Markup is added to a retailer's cost to cover the expenses of operating a business. Markup is usually expressed as a percent of the retailer's cost. This percent is called the **markup rate.**

The basic markup equations used by a business are

$$\begin{aligned} \text{Selling price} &= \text{Cost} + \text{Markup} \\ S &= C + M \end{aligned}$$

$$\begin{aligned} \text{Markup} &= \text{Markup rate} \cdot \text{Cost} \\ M &= r \cdot C \end{aligned}$$

By substituting $r \cdot C$ for M in the first equation, we can also write selling price as

$$\begin{aligned} S &= C + M \\ S &= C + (r \cdot C) \\ S &= C + rC \end{aligned}$$

The equation $S = C + rC$ is the equation used to solve the markup problems in this section.

Example 1 The manager of a clothing store buys a suit for \$180 and sells the suit for \$252. Find the markup rate.

Strategy Given: $C = \$180$
$S = \$252$
Unknown markup rate: r
Use the equation $S = C + rC$.

Solution

$$S = C + rC$$

$$252 = 180 + 180r$$ • Substitute the values of C and S into the equation.

$$252 - 180 = 180 - 180 + 180r$$ • Subtract 180 from each side of the equation.

$$72 = 180r$$

$$\frac{72}{180} = \frac{180r}{180}$$ • Divide each side of the equation by 180.

$$0.4 = r$$ • The decimal must be changed to a percent.

The markup rate is 40%.

Problem 1 The cost to the manager of a sporting goods store for a tennis racket is \$120. The selling price of the racket is \$180. Find the markup rate.

Solution See page S8.

Example 2 The manager of a furniture store uses a markup rate of 45% on all items. The selling price of a chair is \$232. Find the cost of the chair.

Strategy Given: $r = 45\% = 0.45$
$S = \$232$
Unknown cost: C
Use the equation $S = C + rC$.

Solution

$$S = C + rC$$

$$232 = C + 0.45C$$

$$232 = 1.45C$$ • $C + 0.45C = 1C + 0.45C = (1 + 0.45)C$

$$160 = C$$

The cost of the chair is \$160.

Problem 2 A hardware store employee uses a markup rate of 40% on all items. The selling price of a lawnmower is \$266. Find the cost.

Solution See page S8.

2 Discount problems

VIDEO & DVD CD TUTOR WEB SSM

Discount is the amount by which a retailer reduces the regular price of a product for a promotional sale. Discount is usually expressed as a percent of the regular price. This percent is called the **discount rate.**

The basic discount equations used by a business are

$$\text{Sale price} = \text{Regular price} - \text{Discount}$$
$$S = R - D$$

$$\text{Discount} = \text{Discount rate} \cdot \text{Regular price}$$
$$D = r \cdot R$$

By substituting $r \cdot R$ for D in the first equation, we can also write sale price as

$$S = R - D$$
$$S = R - (r \cdot R)$$
$$S = R - rR$$

The equation $S = R - rR$ is the equation used to solve the discount problems in this section.

Example 3 In a garden supply store, the regular price of a 100-foot garden hose is \$48. During an "after-summer sale," the hose is being sold for \$36. Find the discount rate.

Strategy Given: $R = \$48$
$S = \$36$
Unknown discount rate: r
Use the equation $S = R - rR$.

Solution

$$S = R - rR$$
$$36 = 48 - 48r$$
- Substitute the values of R and S into the equation.

$$36 - 48 = 48 - 48 - 48r$$
- Subtract 48 from each side of the equation.

$$-12 = -48r$$
$$\frac{-12}{-48} = \frac{-48r}{-48}$$
- Divide each side of the equation by −48.

$$0.25 = r$$
- The decimal must be changed to a percent.

The discount rate is 25%.

Problem 3 A case of motor oil that regularly sells for \$39.80 is on sale for \$29.85. What is the discount rate?

Solution See page S8.

Example 4 The sale price for a chemical sprayer is \$27.30. This price is 35% off the regular price. Find the regular price.

Strategy Given: $S = \$27.30$

$r = 35\% = 0.35$

Unknown regular price: R

Use the equation $S = R - rR$.

Solution

$$S = R - rR$$
$$27.30 = R - 0.35R$$
$$27.30 = 0.65R \qquad \bullet\ R - 0.35R = 1R - 0.35R = (1 - 0.35)R$$
$$42 = R$$

The regular price is \$42.00.

Problem 4 The sale price for a telephone is \$43.50. This price is 25% off the regular price. Find the regular price.

Solution See page S8.

2.5 Concept Review

Determine whether the statement is always true, sometimes true, or never true.

1. In the markup equation $S = C + rC$, the variable r represents the markup rate.

2. Markup is an amount of money, whereas markup rate is a percent.

3. In the discount equation $S = R - rR$, the variable R represents the discount rate.

4. In the markup equation, S represents sale price, and in the discount equation, S represents selling price.

5. $R - 0.45R = 1R - 0.45R = (1 - 0.45)R = 0.55R$ is an example of the application of the Distributive Property.

2.5 Exercises

1 Markup Problems

1. Explain the difference between the cost and the selling price of a product.

2. Explain the difference between markup and markup rate.

3. A computer software retailer uses a markup rate of 40%. Find the selling price of a computer game that costs the retailer \$40.

4. A car dealer advertises a 5% markup over cost. Find the selling price of a car that costs the dealer $26,000.

5. An electronics store uses a markup rate of 58%. Find the selling price of a camcorder that costs the store owner $358.

6. The cost to a landscape architect for a 25-gallon tree is $65. Find the selling price of the tree if the markup rate used by the architect is 30%.

7. A set of golf clubs costing $360 is sold for $630. Find the markup rate on the set of golf clubs.

8. A jeweler purchases a diamond ring for $2399.50. The selling price is $4799. Find the markup rate.

9. A freezer costing $360 is sold for $520. Find the markup rate. Round to the nearest tenth of a percent.

10. A grocer purchases a can of fruit juice for $1.96. The selling price of the fruit juice is $2.45. Find the markup rate.

11. A watch costing $98 is sold for $156.80. Find the markup rate on the watch.

12. A sofa costing $320 is sold for $479. Find the markup rate. Round to the nearest tenth of a percent.

13. The selling price of a compact disc player is $168. The markup rate used by the seller is 40%. Find the cost of the compact disc player.

14. A manufacturer of exercise equipment uses a markup rate of 45%. One of the manufacturer's treadmills has a selling price of $580. Find the cost of the treadmill.

15. A markup rate of 40% was used on a basketball with a selling price of $82.60. Find the cost of the basketball.

16. A digitally recorded compact disc has a selling price of $18.90. The markup rate is 40%. Find the cost of the CD.

17. A markup rate of 25% is used on a computer that has a selling price of $2187.50. Find the cost of the computer.

18. The selling price of a portable CD player is $132. The markup rate used by the retailer is 60%. Find the cost to the retailer.

2 Discount Problems

19. Explain the meaning of sale price and discount.

20. Explain the difference between discount and discount rate.

21. A tennis racket that regularly sells for \$135 is on sale for 25% off the regular price. Find the sale price.

22. A fax machine that regularly sells for \$219.99 is on sale for $33\frac{1}{3}\%$ off the regular price. Find the sale price.

23. A supplier of electrical equipment offers a 5% discount for a purchase that is paid for within 30 days. A transformer regularly sells for \$230. Find the discount price of a transformer that is paid for 10 days after the purchase.

24. A clothing wholesaler offers a discount of 10% per shirt when 10 to 20 shirts are purchased and a discount of 15% per shirt when 21 to 50 shirts are purchased. A shirt regularly sells for \$27. Find the sale price per shirt when 35 shirts are purchased.

25. A car stereo system that regularly sells for \$425 is on sale for \$318.75. Find the discount rate.

26. A pair of roller blades that regularly sells for \$99.99 is on sale for \$79.99. Find the discount rate. Round to the nearest percent.

27. An oak bedroom set with a regular price of \$1295 is on sale for \$995. Find the discount rate. Round to the nearest tenth of a percent.

28. A stereo set with a regular price of \$495 is on sale for \$395. Find the markdown rate. Round to the nearest tenth of a percent.

29. A digital camera with a regular price of \$325 is on sale for \$201.50. Find the markdown rate.

30. A luggage set with a regular price of \$178 is on sale for \$103.24. Find the discount rate.

31. The sale price of a free-weight home gym is \$408, which is 20% off the regular price. Find the regular price.

32. The sale price of a toboggan is \$77, which is 30% off the regular price. Find the regular price.

33. A mechanic's tool set is on sale for \$180 after a markdown of 40% off the regular price. Find the regular price.

34. The sale price of a modem is \$126, which is 30% off the regular price. Find the regular price.

35. A telescope is on sale for \$165 after a markdown of 40% off the regular price. Find the regular price.

36. An exercise bike is on sale for \$390, having been marked down 25% off the regular price. Find the regular price.

2.5 Applying Concepts

37. A pair of shoes that now sells for \$63 has been marked up 40%. Find the markup on the pair of shoes.

38. The sale price of a motorcycle helmet is 25% off the regular price. The discount is \$70. Find the sale price.

39. A refrigerator selling for \$770 has a markup of \$220. Find the markup rate.

40. The sale price of a copy machine is \$765 after a discount of \$135. Find the discount rate.

41. The manager of a camera store uses a markup rate of 30%. Find the cost of a camera selling for \$299.

42. The sale price of a television is \$180. Find the regular price if the sale price was computed by taking $\frac{1}{3}$ off the regular price followed by an additional 25% discount on the reduced price.

43. A customer buys four tires, three at the regular price and one for 20% off the regular price. The four tires cost \$304. What was the regular price of a tire?

44. A lamp, originally priced at under \$100, was on sale for 25% off the regular price. When the regular price, a whole number of dollars, was discounted, the discounted price was also a whole number of dollars. Find the largest possible number of dollars in the regular price of the lamp.

45. A used car is on sale for a discount of 20% off the regular price of \$8500. An additional 10% discount on the sale price was offered. Is the result a 30% discount? What is the single discount that would give the same sale price?

46. Write a report on series trade discounts. Explain how to convert a series discount to a single-discount equivalent.

2.6 Applications: Problems Involving Percent

1 Investment problems

VIDEO & DVD CD TUTOR WEB SSM

The annual simple interest that an investment earns is given by the equation $Pr = I$, where P is the principal, or the amount invested, r is the simple interest rate, and I is the simple interest. The solution of an investment problem is based on this equation.

The annual interest rate on a \$3000 investment is 9%. The annual simple interest earned on the investment is

$$I = Pr$$
$$I = \$3000(0.09)$$
$$I = \$270$$

Solve: You have a total of \$8000 invested in two simple interest accounts. On one account, a money market fund, the annual simple interest rate is 11.5%. On the second account, a bond fund, the annual simple interest rate is 9.75%. The total annual interest earned by the two accounts is \$823.75. How much do you have invested in each account?

STRATEGY *for solving a problem involving money deposited in two simple interest accounts*

▶ For each amount invested, use the equation $Pr = I$. Write a numerical or variable expression for the principal, the interest rate, and the interest earned. The results can be recorded in a table.

The total amount invested is \$8000.

Amount invested at 11.5%: x
Amount invested at 9.75%: $8000 - x$

	Principal, P	$\cdot$	Interest rate, r	$=$	Interest earned, I
Amount at 11.5%	x	$\cdot$	0.115	$=$	$0.115x$
Amount at 9.75%	$8000 - x$	$\cdot$	0.0975	$=$	$0.0975(8000 - x)$

▶ Determine how the amounts of interest earned on the individual investments are related. For example, the total interest earned by both accounts may be known, or it may be known that the interest earned on one account is equal to the interest earned on the other account.

The total annual interest earned is \$823.75.

$$
\begin{aligned}
0.115x + 0.0975(8000 - x) &= 823.75 \\
0.115x + 780 - 0.0975x &= 823.75 \\
0.0175x + 780 &= 823.75 \\
0.0175x &= 43.75 \\
x &= 2500
\end{aligned}
$$

The amount invested at 9.75% is $8000 - x$.
Replace x with 2500 and evaluate.

$$8000 - x = 8000 - 2500 = 5500$$

The amount invested at 11.5% is \$2500.
The amount invested at 9.75% is \$5500.

Example 1 An investment of \$4000 is made at an annual simple interest rate of 4.9%. How much additional money must be invested at an annual simple interest rate of 7.4% so that the total interest earned is 6.4% of the total investment?

Strategy

▶ Additional amount to be invested at 7.4%: x

	Principal	Rate	Interest
Amount at 4.9%	4000	0.049	$0.049(4000)$
Amount at 7.4%	x	0.074	$0.074x$
Amount at 6.4%	$4000 + x$	0.064	$0.064(4000 + x)$

▶ The sum of the interest earned by the two investments equals the interest earned by the total investment.

Solution

$$
\begin{aligned}
0.049(4000) + 0.074x &= 0.064(4000 + x) \\
196 + 0.074x &= 256 + 0.064x \\
196 + 0.01x &= 256 \\
0.01x &= 60 \\
x &= 6000
\end{aligned}
$$

\$6000 must be invested at an annual simple interest rate of 7.4%.

Problem 1 An investment of \$3500 is made at an annual simple interest rate of 5.2%. How much additional money must be invested at an annual simple interest rate of 7.5% so that the total interest earned is \$575?

Solution See page S8.

2 Percent mixture problems

VIDEO & DVD CD TUTOR WEB SSM

The amount of a substance in a solution or alloy can be given as a percent of the total solution or alloy. For example, in a 10% hydrogen peroxide solution, 10% of the total solution is hydrogen peroxide. The remaining 90% is water.

> **Take Note**
>
> As stated at the right, in a 10% hydrogen peroxide solution, 10% of the total solution is hydrogen peroxide. Here we are referring to weight: The solution is 10% hydrogen peroxide by weight. For example, if the solution weighs 100 g, then the hydrogen peroxide in the solution weighs 10 g and the water weighs 90% of 100 g, or 90 g.

The solution of a percent mixture problem is based on the equation $Ar = Q$, where A is the amount of solution or alloy, r is the percent of concentration, and Q is the quantity of a substance in the solution or alloy.

The number of grams of silver in 50 g of a 40% silver alloy is

$$Q = Ar$$
$$Q = 50(0.40)$$
$$Q = 20$$

Solve: A chemist mixes an 11% acid solution with a 4% acid solution. How many milliliters of each solution should the chemist use to make a 700-milliliter solution that is 6% acid?

STRATEGY ***for solving a percent mixture problem***

▶ For each solution, use the equation $Ar = Q$. Write a numerical or variable expression for the amount of solution, the percent of concentration, and the quantity of the substance in the solution. The results can be recorded in a table.

The total amount of solution is 700 ml.

Amount of 11% solution: x
Amount of 4% solution: $700 - x$

	Amount of solution, A	$\cdot$	Percent of concentration, r	$=$	Quantity of substance, Q
11% solution	x	$\cdot$	0.11	$=$	$0.11x$
4% solution	$700 - x$	$\cdot$	0.04	$=$	$0.04(700 - x)$
6% solution	700	$\cdot$	0.06	$=$	$0.06(700)$

▶ Determine how the quantities of the substance in the individual solutions are related. Use the fact that the sum of the quantities of the substances being mixed is equal to the quantity of the substance after mixing.

The sum of the amounts of acid in the 11% solution and the 4% solution is equal to the amount of acid in the 6% solution.

$$0.11x + 0.04(700 - x) = 0.06(700)$$
$$0.11x + 28 - 0.04x = 42$$
$$0.07x + 28 = 42$$
$$0.07x = 14$$
$$x = 200$$

The amount of 4% solution is $700 - x$. Replace x by 200 and evaluate.

$$700 - x = 700 - 200 = 500$$

The chemist should use 200 ml of the 11% solution and 500 ml of the 4% solution.

Example 2 How many grams of pure acid must be added to 60 g of an 8% acid solution to make a 20% acid solution?

Strategy ▶ Grams of pure acid: x

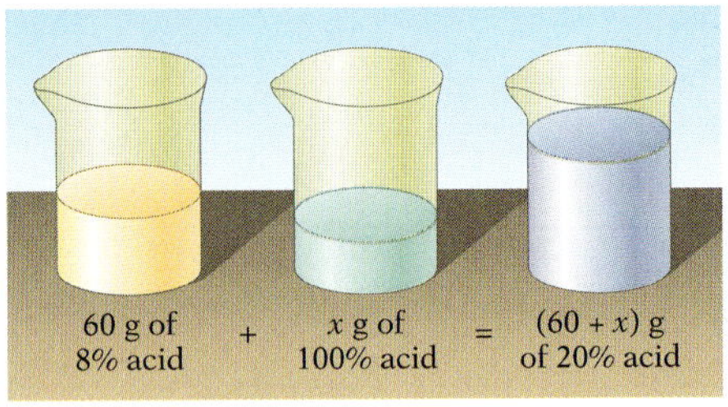

	Amount	Percent	Quantity
Pure acid (100%)	x	1.00	x
8%	60	0.08	0.08(60)
20%	$x + 60$	0.20	$0.20(x + 60)$

▶ The sum of the quantities before mixing equals the quantity after mixing.

Solution

$$\begin{aligned} x + 0.08(60) &= 0.20(x + 60) \\ x + 4.8 &= 0.20x + 12 \\ 0.8x + 4.8 &= 12 \\ 0.8x &= 7.2 \\ x &= 9 \end{aligned}$$

To make the 20% acid solution, 9 g of pure acid must be used.

Problem 2 A butcher has some hamburger that is 22% fat and some that is 12% fat. How many pounds of each should be mixed to make 80 lb of hamburger that is 18% fat?

Solution See pages S8–S9.

2.6 Concept Review

Determine whether the following statements are always true, sometimes true, or never true.

1. \$10,000 is invested in two accounts. If x is invested in one account, then $x - 10{,}000$ is invested in the other account.

2. The equation $Q = Ar$, where A is the amount of solution or alloy, r is the percent concentration, and Q is the quantity of a substance in the solution or alloy, is used in solving percent mixture problems.

3. If one solution has a 5% concentration and another solution has a 20% concentration, the mixture will have a 25% concentration.

4. An investment of $8000 is divided equally between two accounts with one account paying 6% and the other account paying 8%. The interest received is equivalent to 7% of the total investment.

2.6 Exercises

1 Investment Problems

1. Explain the meaning of each variable in the equation $I = Pr$. Give an example of how this equation is used.

2. Suppose $5000 is invested in two simple interest accounts. On one account the annual simple interest rate is 6%, and on the other the annual simple interest rate is 7%. The total annual interest earned on the two accounts is $330. In the context of this situation, explain each term of the equation

$$0.06x + 0.07(5000 - x) = 330$$

3. Joseph Abruzzio decides to divide a gift of $5000 into two different accounts. He deposits $2000 in an IRA account that earns an annual simple interest rate of 5.5%. The remaining money is placed in an account that earns an annual simple interest rate of 7.25%. How much interest will Joseph earn in one year from the two accounts?

4. Kristi invests $1500 at a 7.25% annual simple interest rate, and her sister Kari invests $2000 at a 6.75% annual simple interest rate. Which of the two will earn the greater amount of interest after one year? How much greater?

5. Deon Brown purchases a municipal bond for $2000. The bond earns an annual simple interest rate of 6.4%. How much must Deon invest in an account that earns 8% annual simple interest so that both investments will earn the same amount of interest in one year?

6. A $5000 investment at an annual simple interest rate of 5.2% earned as much interest after one year as another investment in an account that earned 6.5% annual simple interest. How much was invested at 6.5%?

7. Two investments earn a total annual income of $2825. One investment is in a 6.75% annual simple interest certificate of deposit. The other is in a 7.25% tax-free annual simple interest account. The total amount invested is $40,000. Find the amount invested in the certificate of deposit.

8. Two investments earn an annual income of $765. One investment earns an annual simple interest rate of 8.5%, and the other investment earns an annual simple interest rate of 10.2%. The total amount invested is $8000. How much is invested in each account?

9. An investment club invested \$5000 at an annual simple interest rate of 8.4%. How much additional money must be invested at an annual simple interest rate of 10.5% so that the total interest earned will be 9% of the total investment?

10. Two investments earn an annual income of \$465. One investment is a 5.5% tax-free annual simple interest account, and the other is a 4.5% annual simple interest certificate of deposit. The total amount invested is \$9600. How much is invested in each account?

11. Two investments earn an annual income of \$575. One investment earns an annual simple interest rate of 8.5%, and the other investment earns an annual simple interest rate of 6.4%. The total amount invested is \$8000. How much is invested in each account?

12. An investment club invested \$6000 at an annual simple interest rate of 4.0%. How much additional money must be invested at an annual simple interest rate of 6.5% so that the total annual interest earned will be 5% of the total investment?

13. Dee Pinckney made an investment of \$6000 at an annual simple interest rate of 5.5%. How much additional money must she invest at an annual simple interest rate of 10% in order for the total annual interest earned to be 7% of the investment?

14. An account executive deposited \$42,000 into two simple interest accounts. On the tax-free account the annual simple interest rate is 3.5%, and on the money market fund the annual simple interest rate is 4.5%. How much should be invested in each account so that both accounts earn the same annual interest?

15. An investment club invested \$13,600 into two simple interest accounts. On one account, the annual simple interest rate is 4.2%. On the other, the annual simple interest rate is 6%. How much should be invested in each account so that both accounts earn the same annual interest?

16. Orlando Salavarrio, a financial planner, recommended that 25% of a client's investment be placed in a 4% annual simple interest tax-free account, that 40% be placed in 6% certificates of deposit, and that the remainder be placed in a 9% high-risk investment. The total interest earned from the investments would be \$6550. Find the total amount to be invested.

2 Percent Mixture Problems

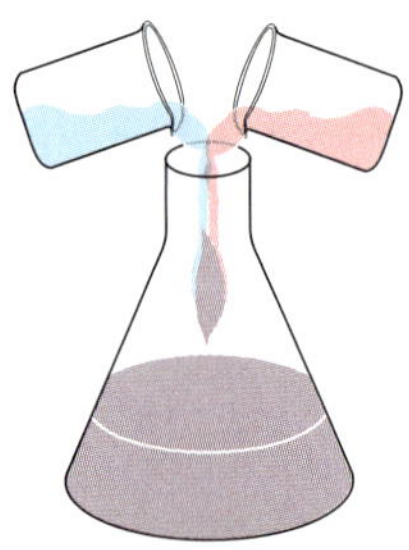

17. Suppose you are mixing a 6% acid solution with a 12% acid solution. In the context of this situation, explain the meaning of the statement "The sum of the quantities of the substances being mixed is equal to the quantity of the substance after mixing."

18. Suppose a goldsmith mixes 25 g of a 75% gold alloy with 40 g of a 50% gold alloy. In the context of this situation, explain each term of the equation

$$25(0.75) + 40(0.50) = 65x$$

19. A 32-ounce bottle of Orange Ade contains 8 oz of orange juice. At the same percent concentration, how much orange juice is contained in a 40-ounce bottle of Orange Ade?

20. A chemist has 4 L of a solution that contains 0.36 L of acetic acid. The chemist needs to prepare 6 L of a solution with the same percent concentration of acetic acid. How much acidic acid must the 6 L of the solution contain?

21. A 750-milliliter solution is 4% hydrogen peroxide. How much more hydrogen peroxide is in a 850-milliliter solution that is 5% hydrogen peroxide?

22. One 8-ounce bar of a silver alloy is 30% silver. The percent concentration of silver in a second bar, which weighs 12 oz, is 35%. Find the total amount of silver in the two bars.

23. A jeweler mixed 15 g of a 60% silver alloy with 45 g of a 20% silver alloy. What is the percent concentration of silver in the resulting alloy?

24. A goldsmith mixed 10 g of a 50% gold alloy with 40 g of a 15% gold alloy. What is the percent concentration of the resulting alloy?

25. A silversmith mixed 25 g of a 70% silver alloy with 50 g of a 15% silver alloy. What is the percent concentration of the resulting alloy?

26. A chemist mixed 100 ml of an 8% saline solution with 60 ml of a 5% saline solution. Find the percent concentration of the resulting mixture.

27. How many pounds of a 12% aluminum alloy must be mixed with 400 lb of a 30% aluminum alloy to make a 20% aluminum alloy?

28. How many pounds of a 20% copper alloy must be mixed with 600 lb of a 30% copper alloy to make a 27.5% copper alloy?

29. A hospital staff member mixed a 65% disinfectant solution with a 15% disinfectant solution. How many liters of each were used to make 50 L of a 40% disinfectant solution?

30. A butcher has some hamburger that is 20% fat and some hamburger that is 12% fat. How many pounds of each should be mixed to make 80 lb of hamburger that is 17% fat?

31. How many quarts of water must be added to 5 qt of an 80% antifreeze solution to make a 50% antifreeze solution?

32. Rubbing alcohol is typically diluted with water to 70% strength. If you need 3.5 oz of 45% rubbing alcohol, how many ounces of 70% rubbing alcohol and how much water should you combine?

33. How many ounces of pure water must be added to 60 oz of a 7.5% salt solution to make a 5% salt solution?

34. How much water must be evaporated from 10 gal of a 12% sugar solution in order to obtain a 15% sugar solution?

35. Many fruit drinks are actually only 5% real fruit juice. If you let 2 oz of water evaporate from 12 oz of a drink that is 5% fruit juice, what is the percent concentration of the result?

36. A student mixed 50 ml of a 3% hydrogen peroxide solution with 30 ml of a 12% hydrogen peroxide solution. Find the percent concentration of the resulting mixture. Round to the nearest tenth of a percent.

37. Eighty pounds of a 54% copper alloy is mixed with 200 lb of a 22% copper alloy. Find the percent concentration of the resulting mixture. Round to the nearest tenth of a percent.

38. A druggist mixed 100 cc (cubic centimeters) of a 15% alcohol solution with 50 cc of pure alcohol. Find the percent concentration of the resulting mixture. Round to the nearest tenth of a percent.

2.6 Applying Concepts

39. **Investment Problem** A financial manager invested 25% of a client's money in bonds paying 9% annual simple interest, 30% in an 8% annual simple interest account, and the remainder in 9.5% corporate bonds. Find the amount invested in each if the total annual interest earned is $1785.

40. **Percent Mixture Problem** A silversmith mixed 90 g of a 40% silver alloy with 120 g of a 60% silver alloy. Find the percent concentration of the resulting alloy. Round to the nearest tenth of a percent.

41. **Value Mixture Problem** Find the cost per pound of a tea mixture made from 50 lb of tea that costs $5.50 per pound and 75 lb of tea that costs $4.40 per pound.

42. **Percent Mixture Problem** How many kilograms of water must be evaporated from 75 kg of a 15% salt solution to produce a 20% salt solution?

43. **Percent Mixture Problem** A chemist added 20 g of pure acid to a beaker containing an unknown number of grams of pure water. The resulting solution was 25% acid. How many grams of water were in the beaker before the acid was added?

44. **Percent Mixture Problem** A radiator contains 6 L of a 25% antifreeze solution. How much should be drained and replaced with pure antifreeze to produce a 50% antifreeze solution?

45. **Consumer Price Index** The Consumer Price Index (CPI), or the "cost-of-living index," measures the change in the price of consumer goods and services. The graph at the right shows the percent increase in the CPI for the years 1990 through 2000.

a. During which year shown was the percent increase in the CPI the greatest?

b. During which year was the cost of consumer goods and services highest?

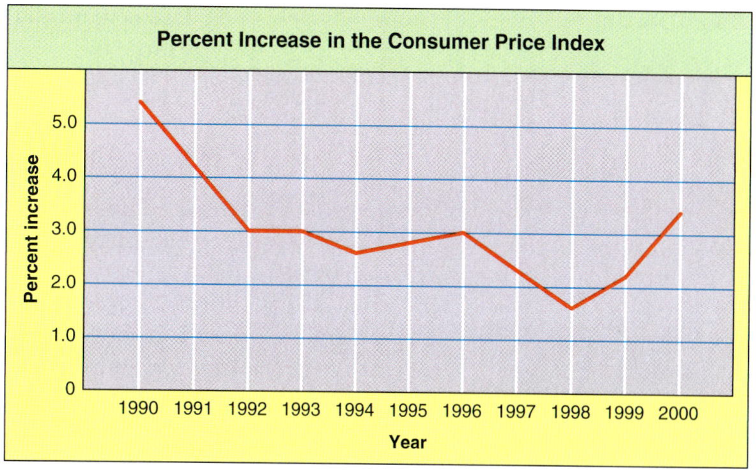

Source: Bureau of Labor Statistics

46. Write a report on series trade discounts. Explain how to convert a series discount to a single-discount equivalent.

2.7 Inequalities in One Variable

1 Set operations and interval notation

VIDEO & DVD CD TUTOR WEB SSM

The *roster method* of writing a set encloses the list of the elements of the set in braces. The set of whole numbers, written as $\{0, 1, 2, 3, 4, \ldots\}$, and the set of natural numbers, written as $\{1, 2, 3, 4, \ldots\}$, are **infinite sets.** The pattern of numbers continues without end. It is impossible to list all the elements of an infinite set.

The set of even natural numbers less than 10 is written $\{2, 4, 6, 8\}$. This is an example of a **finite set;** all the elements of the set can be listed.

The **empty set,** or **null set,** is the set that contains no elements. The symbol $\varnothing$ or $\{\,\}$ is used to represent the empty set.

A second method of representing a set is **set-builder notation.** Set-builder notation can be used to describe almost any set, but it is especially useful when writing infinite sets. In set-builder notation, the set of integers greater than -3 is written

$$\{x \mid x > -3, x \in \text{integers}\}$$

and is read "the set of all x such that x is greater than -3 and x is an element of the integers." This is an infinite set. It is impossible to list all the elements of the set, but the set can be described by using set-builder notation.

The set of real numbers less than 5 is written

$$\{x \mid x < 5, x \in \text{real numbers}\}$$

and is read "the set of all x such that x is less than 5 and x is an element of the real numbers."

Example 1 Use set-builder notation to write the set of integers greater than -8.

Solution $\{x \mid x > -8, x \in \text{integers}\}$

Problem 1 Use set-builder notation to write the set of real numbers less than 7.

Solution See page S9.

Just as operations such as addition and multiplication are performed on real numbers, operations are performed on sets. Two operations performed on sets are *union* and *intersection*.

Point of Interest

The symbols $\in$, $\cup$, *and* $\cap$ *were first used by Giuseppe Peano, in* Arithmetices Principia, Nova Exposita (The Principle of Mathematics, a New Method of Exposition), *published in 1889. The purpose of this book was to deduce the principles of mathematics from pure logic.*

The **union of two sets,** written $A \cup B$, is the set of all elements that belong to either A **or** B. In set-builder notation, this is written

$$A \cup B = \{x \mid x \in A \quad \text{or} \quad x \in B\}$$

Given $A = \{2, 3, 4\}$ and $B = \{0, 1, 2, 3\}$, the union of A and B contains all the elements that belong to either A or B. The elements that belong to both sets are listed only once.

$$A \cup B = \{0, 1, 2, 3, 4\}$$

The **intersection of two sets,** written $A \cap B$, is the set of all elements that are common to both A **and** B. In set-builder notation, this is written

$$A \cap B = \{x \mid x \in A \quad \text{and} \quad x \in B\}$$

Given $A = \{2, 3, 4\}$ and $B = \{0, 1, 2, 3\}$, the intersection of A and B contains all the elements that are common to both A and B.

$$A \cap B = \{2, 3\}$$

Example 2 Find $C \cup D$ given $C = \{1, 5, 9, 13, 17\}$ and $D = \{3, 5, 7, 9, 11\}$.

Solution $C \cup D = \{1, 3, 5, 7, 9, 11, 13, 17\}$

Problem 2 Find $A \cup C$ given $A = \{-2, -1, 0, 1, 2\}$ and $C = \{-5, -1, 0, 1, 5\}$.

Solution See page S9.

Example 3 Find $A \cap B$ given $A = \{x \mid x \in \text{natural numbers}\}$ and $B = \{x \mid x \in \text{negative integers}\}$.

Solution There are no natural numbers that are also negative integers. $A \cap B = \varnothing$

Problem 3 Find $E \cap F$ given $E = \{x \mid x \in \text{odd integers}\}$ and $F = \{x \mid x \in \text{even integers}\}$.

Solution See page S9.

Set-builder notation and the inequality symbols $>$, $<$, $\geq$, and $\leq$ are used to describe infinite sets of real numbers. These sets can be graphed on the real number line.

The graph of $\{x \mid x > -2, x \in \text{real numbers}\}$ is shown below. The set is the real numbers greater than -2. The parenthesis on the graph indicates that -2 is not included in the set.

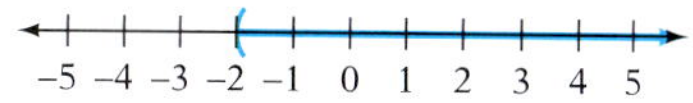

The graph of $\{x \mid x \geq -2, x \in \text{real numbers}\}$ is shown below. The set is the real numbers greater than or equal to -2. The bracket at -2 indicates that -2 is included in the set.

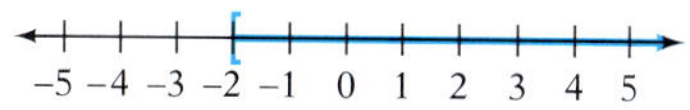

For the remainder of this section, all variables will represent real numbers. Using this convention, the above set is written $\{x \mid x \geq -2\}$.

Example 4 Graph: $\{x \mid x \leq 3\}$

Solution

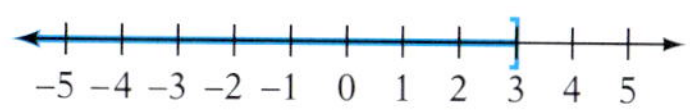

• **The set is the real numbers less than or equal to 3. Draw a right bracket at 3, and darken the number line to the left of 3.**

Problem 4 Graph: $\{x \mid x > -3\}$

Solution See page S9.

The union of two sets is the set of all elements belonging to either one or the other of the two sets. The set $\{x \mid x \leq -1\} \cup \{x \mid x > 3\}$ is the set of real numbers that are either less than or equal to -1 or greater than 3.

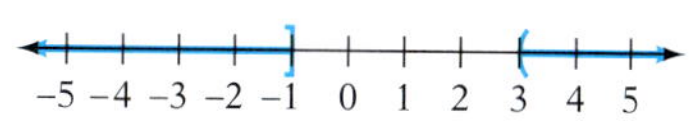

The set is written $\{x \mid x \leq -1 \text{ or } x > 3\}$.

The set $\{x \mid x > 2\} \cup \{x \mid x > 4\}$ is the set of real numbers that are either greater than 2 or greater than 4.

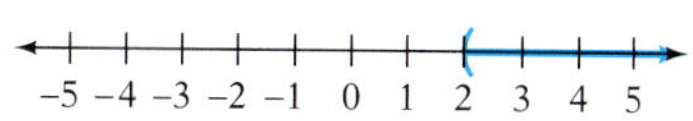

The set is written $\{x \mid x > 2\}$.

The intersection of two sets is the set that contains the elements common to both sets. The set $\{x \mid x > -2\} \cap \{x \mid x < 5\}$ is the set of real numbers that are greater than -2 and less than 5.

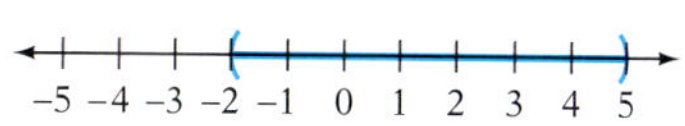

The set can be written $\{x \mid x > -2 \text{ and } x < 5\}$. However, it is more commonly written $\{x \mid -2 < x < 5\}$. This set is read "the set of all x such that x is greater than -2 and less than 5."

The set $\{x \mid x < 4\} \cap \{x \mid x < 5\}$ is the set of real numbers that are less than 4 and less than 5.

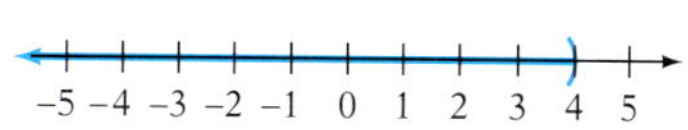

The set is written $\{x \mid x < 4\}$.

Example 5 Graph: $\{x \mid x < 0\} \cap \{x \mid x > -3\}$

Solution The set is $\{x \mid -3 < x < 0\}$.

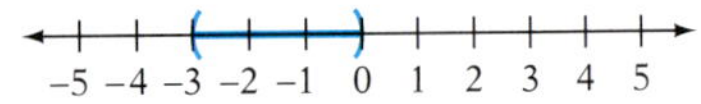

Problem 5 Graph: $\{x \mid x \geq 1\} \cup \{x \mid x \leq -3\}$

Solution See page S9.

Some sets can also be expressed using **interval notation.** For example, the interval notation $(-3, 2]$ indicates the interval of all real numbers greater than -3 and less than or equal to 2. As on the graph of a set, the left parenthesis indicates that -3 is not included in the set. The right bracket indicates that 2 is included in the set.

An interval is said to be **closed** if it includes both endpoints; it is **open** if it does not include either endpoint. An interval is **half-open** if one endpoint is included and the other is not. In each example given below, -3 and 2 are the **endpoints** of the interval. In each case, the set notation, the interval notation, and the graph of the set are shown.

$\{x \mid -3 < x < 2\}$ $(-3, 2)$ Open interval

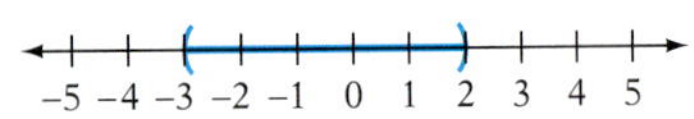

$\{x \mid -3 \leq x \leq 2\}$ $[-3, 2]$ Closed interval

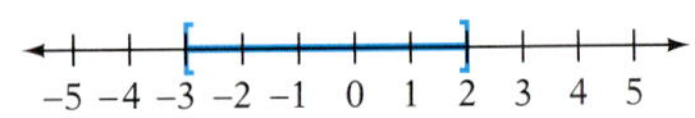

$\{x \mid -3 \leq x < 2\}$ $[-3, 2)$ Half-open interval

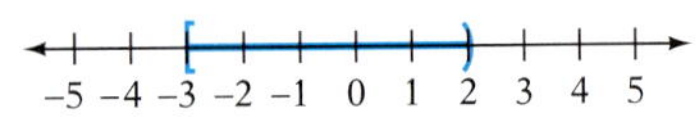

$\{x \mid -3 < x \leq 2\}$ $(-3, 2]$ Half-open interval

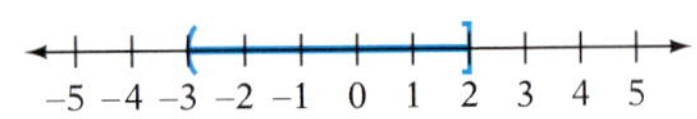

To indicate an interval that extends forever in one or both directions using interval notation, we use the **infinity symbol** ∞ or the **negative infinity symbol** $-\infty$. The infinity symbol is not a number; it is simply used as a notation to indicate that the interval is unlimited. In interval notation, a parenthesis is always used to the right of an infinity symbol or to the left of a negative infinity symbol, as shown in the following examples.

$\{x \mid x > 1\}$ $(1, \infty)$

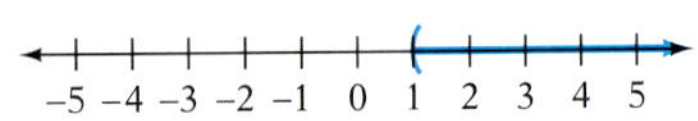

$\{x \mid x \geq 1\}$ $[1, \infty)$

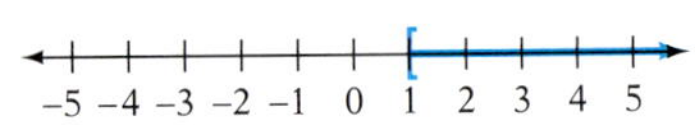

$\{x \mid x < 1\}$ $(-\infty, 1)$

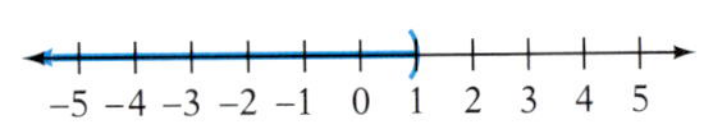

$\{x \mid x \leq 1\}$ $(-\infty, 1]$

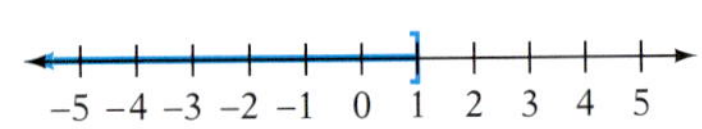

$\{x \mid -\infty < x < \infty\}$ $(-\infty, \infty)$

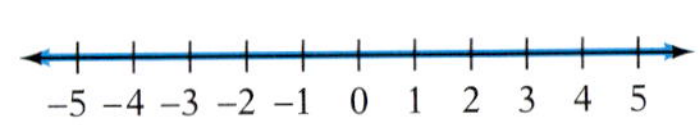

Example 6 Write $\{x \mid 0 < x \leq 5\}$ using interval notation.

Solution $(0, 5]$ • **The set is the real numbers greater than 0 and less than or equal to 5.**

Problem 6 Write $\{x \mid -8 \leq x < -1\}$ using interval notation.

Solution See page S9.

Example 7 Write $(-\infty, 9]$ using set-builder notation.

Solution $\{x \mid x \leq 9\}$ • The set is the real numbers less than or equal to 9.

Problem 7 Write $(-12, \infty)$ using set-builder notation.

Solution See page S9.

2 Solve inequalities in one variable

The **solution set of an inequality** is a set of numbers, each element of which, when substituted for the variable, results in a true inequality.

The inequality at the right is true if the variable is replaced by 3, −1.98, or $\frac{2}{3}$.

$$x - 1 < 4$$
$$3 - 1 < 4$$
$$-1.98 - 1 < 4$$
$$\frac{2}{3} - 1 < 4$$

There are many values of the variable x that will make the inequality $x - 1 < 4$ true. The solution set of the inequality is any number less than 5. The solution set can be written in set-builder notation as $\{x \mid x < 5\}$.

The graph of the solution set of $x - 1 < 4$ is shown at the right.

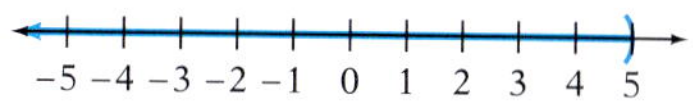

In solving an inequality, use the Addition and Multiplication Properties of Inequalities to rewrite the inequality in the form *variable* < *constant* or *variable* > *constant*.

The Addition Property of Inequalities

If $a > b$ and c is a real number, then the inequalities $a > b$ and $a + c > b + c$ have the same solution set.

If $a < b$ and c is a real number, then the inequalities $a < b$ and $a + c < b + c$ have the same solution set.

The Addition Property of Inequalities states that the same number can be added to each side of an inequality without changing the solution set of the inequality. This property is also true for an inequality that contains the symbol ≤ or ≥.

The Addition Property of Inequalities is used to remove a term from one side of an inequality by adding the additive inverse of that term to each side of the inequality. Because subtraction is defined in terms of addition, the same number can be subtracted from each side of an inequality without changing the solution set of the inequality.

Solve: $3x - 4 < 2x - 1$

$$3x - 4 < 2x - 1$$

Subtract $2x$ from each side of the inequality. $\quad x - 4 < -1$

Add 4 to each side of the inequality. $\quad x < 3$

Write the solution set. The solution set can be written in either set-builder notation or interval notation. $\{x \mid x < 3\}$

$(-\infty, 3)$

The Multiplication Property of Inequalities is used to remove a coefficient from one side of an inequality so that the inequality can be written in the form *variable* < *constant* or *variable* > *constant*.

The Multiplication Property of Inequalities

Rule 1

If $a > b$ and $c > 0$, then the inequalities $a > b$ and $ac > bc$ have the same solution set.

If $a < b$ and $c > 0$, then the inequalities $a < b$ and $ac < bc$ have the same solution set.

Rule 2

If $a > b$ and $c < 0$, then the inequalities $a > b$ and $ac < bc$ have the same solution set.

If $a < b$ and $c < 0$, then the inequalities $a < b$ and $ac > bc$ have the same solution set.

Take Note

$c > 0$ means c is a positive number. Note that the inequality symbols are not changed.

$c < 0$ means c is a negative number. Note that the inequality symbols are reversed.

Rule 1 states that when each side of an inequality is multiplied by a positive number, the inequality symbol remains the same. However, Rule 2 states that when each side of an inequality is multiplied by a negative number, the inequality symbol must be reversed.

Here are some examples of this property.

Rule 1		Rule 2	
$3 > 2$	$2 < 5$	$3 > 2$	$2 < 5$
$3(4) > 2(4)$	$2(4) < 5(4)$	$3(-4) < 2(-4)$	$2(-4) > 5(-4)$
$12 > 8$	$8 < 20$	$-12 < -8$	$-8 > -20$

Because division is defined in terms of multiplication, when each side of an inequality is divided by a positive number, the inequality symbol remains the same. When each side of an inequality is divided by a negative number, the inequality symbol must be reversed.

The Multiplication Property of Inequalities is also true for the symbol ≤ or ≥.

Solve: $-3x > 9$

Write the solution set in interval notation. $-3x > 9$

Divide each side of the inequality by the coefficient -3, and reverse the inequality symbol. Simplify.

$$\frac{-3x}{-3} < \frac{9}{-3}$$

$$x < -3$$

Write the solution set. $(-\infty, -3)$

Take Note

Each side of the inequality is divided *by* a negative number; the inequality symbol must be reversed.

Example 8 Solve: $x + 3 > 4x + 6$
Write the solution set in set-builder notation.

Solution

$x + 3 > 4x + 6$	
$-3x + 3 > 6$	• Subtract $4x$ from each side of the inequality.
$-3x > 3$	• Subtract 3 from each side of the inequality.
$x < -1$	• Divide each side of the inequality by -3, and reverse the inequality symbol.
$\{x \mid x < -1\}$	• Write the solution set.

Problem 8 Solve: $2x - 1 < 6x + 7$
Write the solution set in set-builder notation.

Solution See page S9.

When an inequality contains parentheses, the first step in solving the inequality is to use the Distributive Property to remove the parentheses.

Example 9 Solve: $5(x - 2) \geq 9x - 3(2x - 4)$
Write the solution set in set-builder notation.

Solution

$5(x - 2) \geq 9x - 3(2x - 4)$	
$5x - 10 \geq 9x - 6x + 12$	• Use the Distributive Property to remove parentheses.
$5x - 10 \geq 3x + 12$	• Simplify.
$2x - 10 \geq 12$	• Subtract $3x$ from each side of the inequality.
$2x \geq 22$	• Add 10 to each side of the inequality.
$x \geq 11$	• Divide each side of the inequality by 2.
$\{x \mid x \geq 11\}$	

Problem 9 Solve: $5x - 2 \leq 4 - 3(x - 2)$
Write the solution set in interval notation.

Solution See page S9.

3 Solve compound inequalities

A **compound inequality** is formed by joining two inequalities with a connective word such as "and" or "or." The inequalities shown below are compound inequalities.

$$2x < 4 \quad \text{and} \quad 3x - 2 > -8$$
$$2x + 3 > 5 \quad \text{or} \quad x + 2 < 5$$

The solution set of a compound inequality with the connective word *and* is the set of all elements common to the solution sets of both inequalities. Therefore, it is the *intersection* of the solution sets of the two inequalities.

Solve: $2x < 6$ and $3x + 2 > -4$
Write the solution set in interval notation.

Solve each inequality.

$$2x < 6 \quad \text{and} \quad 3x + 2 > -4$$
$$x < 3 \qquad\qquad 3x > -6$$
$$x > -2$$

$$\{x \mid x < 3\} \qquad \{x \mid x > -2\}$$

Find the intersection of the solution sets. $\{x \mid x < 3\} \cap \{x \mid x > -2\} = (-2, 3)$

Solve: $-3 < 2x + 1 < 5$
Write the solution set in set-builder notation.

This inequality is equivalent to the compound inequality $-3 < 2x + 1$ and $2x + 1 < 5$.

$$-3 < 2x + 1 < 5$$

$$\begin{array}{lcr} -3 < 2x + 1 & \text{and} & 2x + 1 < 5 \\ -4 < 2x & & 2x < 4 \\ -2 < x & & x < 2 \\ \{x \mid x > -2\} & & \{x \mid x < 2\} \end{array}$$

Find the intersection of the solution sets. $\{x \mid x > -2\} \cap \{x \mid x < 2\} = \{x \mid -2 < x < 2\}$

There is an alternative method for solving the inequality in the last example.

$$-3 < 2x + 1 < 5$$

Subtract 1 from each of the three parts of the inequality.

$$-3 - 1 < 2x + 1 - 1 < 5 - 1$$
$$-4 < 2x < 4$$

Divide each of the three parts of the inequality by the coefficient 2.

$$\frac{-4}{2} < \frac{2x}{2} < \frac{4}{2}$$
$$-2 < x < 2$$
$$\{x \mid -2 < x < 2\}$$

The solution set of a compound inequality with the connective word *or* is the *union* of the solution sets of the two inequalities.

Solve: $2x + 3 > 7$ or $4x - 1 < 3$
Write the solution set in set-builder notation.

Solve each inequality.

$$\begin{array}{rcr} 2x + 3 > 7 & \text{or} & 4x - 1 < 3 \\ 2x > 4 & & 4x < 4 \\ x > 2 & & x < 1 \\ \{x \mid x > 2\} & & \{x \mid x < 1\} \end{array}$$

Find the union of the solution sets. $\{x \mid x > 2\} \cup \{x \mid x < 1\} = \{x \mid x > 2 \text{ or } x < 1\}$

Example 10 Solve: $1 < 3x - 5 < 4$
Write the solution set in interval notation.

Solution

$$1 < 3x - 5 < 4$$

$$1 + 5 < 3x - 5 + 5 < 4 + 5$$ • Add 5 to each of the three parts of the inequality.

$$6 < 3x < 9$$ • Simplify.

$$\frac{6}{3} < \frac{3x}{3} < \frac{9}{3}$$ • Divide each of the three parts of the inequality by 3.

$$2 < x < 3$$

$(2, 3)$ • Write the solution set.

Problem 10 Solve: $-2 \le 5x + 3 \le 13$
Write the solution set in interval notation.

Solution See page S9.

Example 11 Solve: $11 - 2x > -3$ and $7 - 3x < 4$
Write the solution set in set-builder notation.

Solution

$$\begin{aligned} 11 - 2x &> -3 \\ -2x &> -14 \\ x &< 7 \end{aligned} \quad \text{and} \quad \begin{aligned} 7 - 3x &< 4 \\ -3x &< -3 \\ x &> 1 \end{aligned}$$

• Solve each inequality.

$\{x | x < 7\}$ $\{x | x > 1\}$

$\{x | x < 7\} \cap \{x | x > 1\} = \{x | 1 < x < 7\}$

• Find the intersection of the solution sets.

Problem 11 Solve: $5 - 4x > 1$ and $6 - 5x < 11$
Write the solution set in interval notation.

Solution See page S9.

Example 12 Solve: $3 - 4x > 7$ or $4x + 5 < 9$
Write the solution set in set-builder notation.

Solution

$$\begin{aligned} 3 - 4x &> 7 \\ -4x &> 4 \\ x &< -1 \end{aligned} \quad \text{or} \quad \begin{aligned} 4x + 5 &< 9 \\ 4x &< 4 \\ x &< 1 \end{aligned}$$

• Solve each inequality.

$\{x | x < -1\}$ $\{x | x < 1\}$

$\{x | x < -1\} \cup \{x | x < 1\} = \{x | x < 1\}$

• Find the union of the solution sets.

Problem 12 Solve: $2 - 3x > 11$ or $5 + 2x > 7$
Write the solution set in set-builder notation.

Solution See page S9.

4 Application problems

Example 13 Company A rents cars for \$6 a day and 14¢ for every mile driven. Company B rents cars for \$12 a day and 8¢ for every mile driven. You want to rent a car for 5 days. How many miles can you drive a Company A car during the 5 days if it is to cost less than a Company B car?

Strategy To find the number of miles, write and solve an inequality using N to represent the number of miles.

Solution Cost of Company A car < cost of Company B car

$$\begin{aligned} 6(5) + 0.14N &< 12(5) + 0.08N \\ 30 + 0.14N &< 60 + 0.08N \\ 30 + 0.06N &< 60 \\ 0.06N &< 30 \\ N &< 500 \end{aligned}$$

It is less expensive to rent from Company A if the car is driven less than 500 mi.

Problem 13 The base of a triangle is 12 in., and the height is $(x + 2)$ in. Express as an integer the maximum height of the triangle when the area is less than 50 in^2.

Solution See page S9.

Example 14 Find three consecutive odd integers whose sum is between 27 and 51.

Strategy To find the three consecutive odd integers, write and solve a compound inequality using x to represent the first odd integer.

Solution

$$\text{Lower limit of the sum} < \text{sum} < \text{upper limit of the sum}$$

$$27 < x + (x + 2) + (x + 4) < 51$$
$$27 < 3x + 6 < 51$$
$$27 - 6 < 3x + 6 - 6 < 51 - 6$$
$$21 < 3x < 45$$
$$\frac{21}{3} < \frac{3x}{3} < \frac{45}{3}$$
$$7 < x < 15$$

The three integers are 9, 11, and 13; or 11, 13, and 15; or 13, 15, and 17.

Problem 14 An average score of 80 to 89 in a history course receives a B grade. A student has grades of 72, 94, 83, and 70 on four exams. Find the range of scores on the fifth exam that will give the student a B for the course

Solution See page S9.

2.7 Concept Review

Determine whether the following statements are always true, sometimes true, or never true.

1. If $a > b$, then $-a < -b$.
2. If $a < b$ and $a \neq 0$, $b \neq 0$, then $\frac{1}{a} < \frac{1}{b}$.
3. When dividing both sides of an inequality by an integer, we must reverse the inequality symbol.
4. When subtracting the same real number from both sides of an inequality, we must reverse the inequality symbol.
5. Suppose $a < 1$. Then $a^2 < a$.
6. Suppose $a < b < 0$ and $c < d < 0$. Then $ac > bd$.

2.7 Exercises

1 **1.** Explain the difference between the union of two sets and the intersection of two sets.

2. Explain the difference between $\{x \mid x < 5\}$ and $\{x \mid x \leq 5\}$.

Use set-builder notation to write the set.

3. the integers greater than 4

4. the integers less than -2

5. the real numbers greater than or equal to -2

6. the real numbers less than or equal to 2

7. the real numbers between 0 and 1

8. the real numbers between -2 and 5

9. the real numbers between 1 and 4, inclusive

10. the real numbers between 0 and 2, inclusive

Find $A \cup B$.

11. $A = \{1, 4, 9\}, B = \{2, 4, 6\}$

12. $A = \{-1, 0, 1\}, B = \{0, 1, 2\}$

13. $A = \{2, 3, 5, 8\}, B = \{9, 10\}$

14. $A = \{1, 3, 5, 7\}, B = \{2, 4, 6, 8\}$

15. $A = \{-4, -2, 0, 2, 4\}, B = \{0, 4, 8\}$

16. $A = \{-3, -2, -1\}, B = \{-2, -1, 0, 1\}$

17. $A = \{1, 2, 3, 4, 5\}, B = \{3, 4, 5\}$

18. $A = \{2, 4\}, B = \{0, 1, 2, 3, 4, 5\}$

Find $A \cap B$.

19. $A = \{6, 12, 18\}, B = \{3, 6, 9\}$

20. $A = \{-4, 0, 4\}, B = \{-2, 0, 2\}$

21. $A = \{1, 5, 10, 20\}, B = \{5, 10, 15, 20\}$

22. $A = \{1, 3, 5, 7, 9\}, B = \{1, 9\}$

23. $A = \{1, 2, 4, 8\}, B = \{3, 5, 6, 7\}$

24. $A = \{-3, -2, -1, 0\}, B = \{1, 2, 3, 4\}$

25. $A = \{2, 4, 6, 8, 10\}, B = \{4, 6\}$

26. $A = \{-9, -5, 0, 7\}, B = \{-7, -5, 0, 5, 7\}$

Graph.

27. $\{x \mid -1 < x < 5\}$

28. $\{x \mid 1 < x < 3\}$

29. $\{x \mid 0 \leq x \leq 3\}$

30. $\{x \mid -1 \leq x \leq 1\}$

31. $\{x \mid x < 2\}$

32. $\{x \mid x < -1\}$

33. $\{x \mid x \geq 1\}$

34. $\{x \mid x \leq -2\}$

35. $\{x|x > 1\} \cup \{x|x < -1\}$

36. $\{x|x \le 2\} \cup \{x|x > 4\}$

37. $\{x|x \le 2\} \cap \{x|x \ge 0\}$

38. $\{x|x > -1\} \cap \{x|x \le 4\}$

39. $\{x|x > 1\} \cap \{x|x \ge -2\}$

40. $\{x|x < 4\} \cap \{x|x \le 0\}$

41. $\{x|x > 2\} \cup \{x|x > 1\}$

42. $\{x|x < -2\} \cup \{x|x < -4\}$

Write each interval in set-builder notation.

43. $(0, 8)$

44. $(-2, 4)$

45. $[-5, 7]$

46. $[3, 4]$

47. $[-3, 6)$

48. $(4, 5]$

49. $(-\infty, 4]$

50. $(-\infty, -2)$

51. $(5, \infty)$

52. $[-2, \infty)$

Write each set of real numbers in interval notation.

53. $\{x|-2 < x < 4\}$

54. $\{x|0 < x < 3\}$

55. $\{x|-1 \le x \le 5\}$

56. $\{x|0 \le x \le 3\}$

57. $\{x|x < 1\}$

58. $\{x|x \le 6\}$

59. $\{x|-2 \le x < 6\}$

60. $\{x|x \ge 3\}$

61. $\{x|x \in \text{real numbers}\}$

62. $\{x|x > -1\}$

2 63. State the Addition Property of Inequalities and give numerical examples of its use.

64. State the Multiplication Property of Inequalities and give numerical examples of its use.

65. Which numbers are solutions of the inequality $x + 7 \le -3$?
a. -17 **b.** 8 **c.** -10 **d.** 0

66. Which numbers are solutions of the inequality $2x - 1 > 5$?
a. 6 **b.** -4 **c.** 3 **d.** 5

Solve. Write the solution set in set-builder notation.

67. $x - 3 < 2$

68. $x + 4 \ge 2$

69. $4x \le 8$

70. $6x > 12$

71. $-2x > 8$

72. $-3x \le -9$

73. $3x - 1 > 2x + 2$

74. $5x + 2 \ge 4x - 1$

75. $2x - 1 > 7$

76. $4x + 3 \le -1$

77. $6x + 3 > 4x - 1$

78. $7x + 4 < 2x - 6$

79. $8x + 1 \ge 2x + 13$

80. $5x - 4 < 2x + 5$

81. $7 - 2x \ge 1$

82. $3 - 5x \le 18$

83. $4x - 2 < x - 11$

84. $6x + 5 \ge x - 10$

Solve. Write the solution set in interval notation.

85. $x + 7 \geq 4x - 8$

86. $3x + 1 \leq 7x - 15$

87. $6 - 2(x - 4) \leq 2x + 10$

88. $4(2x - 1) > 3x - 2(3x - 5)$

89. $2(1 - 3x) - 4 > 10 + 3(1 - x)$

90. $2 - 5(x + 1) \geq 3(x - 1) - 8$

91. $\frac{3}{5}x - 2 < \frac{3}{10} - x$

92. $\frac{5}{6}x - \frac{1}{6} \leq x - 4$

93. $\frac{1}{3}x - \frac{3}{2} \geq \frac{7}{6} - \frac{2}{3}x$

94. $\frac{7}{12}x - \frac{3}{2} < \frac{2}{3}x + \frac{5}{6}$

95. $\frac{1}{2}x - \frac{3}{4} > \frac{7}{4}x - 2$

96. $\frac{2 - x}{4} - \frac{3}{8} \geq \frac{2}{5}x$

97. $2 - 2(7 - 2x) < 3(3 - x)$

98. $3 + 2(x + 5) \geq x + 5(x + 1) + 1$

99. **a.** Which set operation is used when a compound inequality is combined with *or*?
b. Which set operation is used when a compound inequality is combined with *and*?

100. Explain why writing $-3 < x > 4$ does not make sense.

3 Solve. Write the solution set in interval notation.

101. $3x < 6$ and $x + 2 > 1$

102. $x - 3 \leq 1$ and $2x \geq -4$

103. $x + 2 \geq 5$ or $3x \leq 3$

104. $2x < 6$ or $x - 4 > 1$

105. $-2x > -8$ and $-3x < 6$

106. $\frac{1}{2}x > -2$ and $5x < 10$

107. $\frac{1}{3}x < -1$ or $2x > 0$

108. $\frac{2}{3}x > 4$ or $2x < -8$

109. $x + 4 \geq 5$ and $2x \geq 6$

110. $3x < -9$ and $x - 2 < 2$

111. $-5x > 10$ and $x + 1 > 6$

112. $7x < 14$ and $1 - x < 4$

113. $2x - 3 > 1$ and $3x - 1 < 2$

114. $4x + 1 < 5$ and $4x + 7 > -1$

115. $3x + 7 < 10$ or $2x - 1 > 5$

116. $6x - 2 < -14$ or $5x + 1 > 11$

Solve. Write the solution set in set-builder notation.

117. $-5 < 3x + 4 < 16$

118. $5 < 4x - 3 < 21$

119. $0 < 2x - 6 < 4$

120. $-2 < 3x + 7 < 1$

121. $4x - 1 > 11$ or $4x - 1 \leq -11$

122. $3x - 5 > 10$ or $3x - 5 < -10$

123. $2x + 3 \geq 5$ and $3x - 1 > 11$

124. $6x - 2 < 5$ or $7x - 5 < 16$

125. $9x - 2 < 7$ and $3x - 5 > 10$

126. $8x + 2 \leq -14$ and $4x - 2 > 10$

127. $3x - 11 < 4$ or $4x + 9 \geq 1$

128. $5x + 12 \geq 2$ or $7x - 1 \leq 13$

129. $3 - 2x > 7$ and $5x + 2 > -18$

130. $1 - 3x < 16$ and $1 - 3x > -16$

131. $5 - 4x > 21$ or $7x - 2 > 19$

132. $6x + 5 < -1$ or $1 - 2x < 7$

133. $3 - 7x \leq 31$ and $5 - 4x > 1$

134. $9 - x \geq 7$ and $9 - 2x < 3$

135. $\frac{2}{3}x - 4 > 5$ or $x + \frac{1}{2} < 3$

136. $\frac{5}{8}x + 2 < -3$ or $2 - \frac{3}{5}x < -7$

137. $-\frac{3}{8} \leq 1 - \frac{1}{4}x \leq \frac{7}{2}$

138. $-2 \leq \frac{2}{3}x - 1 \leq 3$

4 **139.** **Integer Problem** Five times the difference between a number and two is greater than the quotient of two times the number and three. Find the smallest integer that will satisfy the inequality.

140. **Integer Problem** Two times the difference between a number and eight is less than or equal to five times the sum of the number and four. Find the smallest number that will satisfy the inequality.

141. **Geometry** The length of a rectangle is 2 ft more than four times the width. Express as an integer the maximum width of the rectangle when the perimeter is less than 34 ft.

142. **Geometry** The length of a rectangle is 5 cm less than twice the width. Express as an integer the maximum width of the rectangle when the perimeter is less than 60 cm.

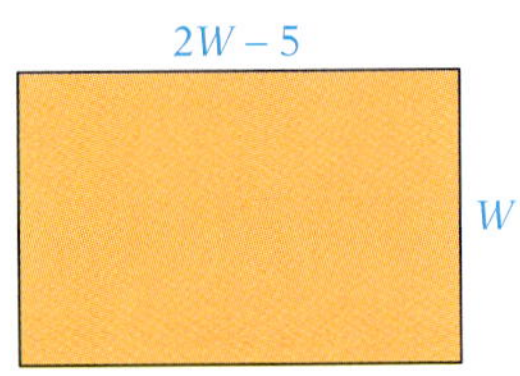

143. **Integer Problem** Find four consecutive integers whose sum is between 62 and 78.

144. **Integer Problem** Find three consecutive even integers whose sum is between 30 and 52.

145. **Geometry** One side of a triangle is 1 in. longer than the second side. The third side is 2 in. longer than the second side. Find, to the nearest whole number, the length of the second side of the triangle if the perimeter is more than 15 in. and less than 25 in.

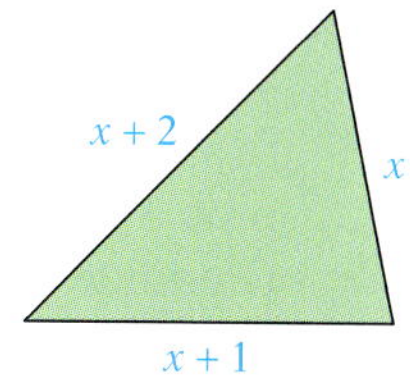

146. **Geometry** The length of a rectangle is 4 ft more than twice the width. Find, to the nearest whole number, the width of the rectangle if the perimeter is more than 28 ft and less than 40 ft.

147. **Consumerism** A cellular phone company offers its customers a rate of $99 for up to 200 min per month of cellular phone time, or a rate of $35 per month plus $.40 for each minute of cellular phone time. For how many minutes per month can a customer who chooses the second option use a cellular phone before the charges exceed those of the first option?

148. **Consumerism** In 2002, the computer service America Online offered its customers the option of paying $23.90 per month for unlimited use. Another option was a rate of $4.95 per month with 3 free hours plus $2.50 per hour thereafter. How many hours per month can you use this second option if it is to cost you less than the first option? Round to the nearest whole number.

149. **Consumerism** AirTouch advertises local paging service for $6.95 per month for up to 400 pages, and $.10 per page thereafter. A competitor advertises paging service for $3.95 per month for up to 400 pages and $.15 per page thereafter. For what number of pages per month is the AirTouch plan less expensive?

150. **Consumerism** Suppose PayRite Rental Cars rents compact cars for $32 per day with unlimited mileage, and Otto Rentals offers compact cars for $19.99 per day but charges $.19 for each mile beyond 100 mi driven per day. You want to rent a car for one week. How many miles can you drive during the week if Otto Rental is to be less expensive than PayRite?

151. **Consumerism** During a weekday, to call a city 40 mi away from a certain pay phone costs $.70 for the first 3 min and $.15 for each additional minute. If you use a calling card, there is a $.35 fee and then the rates are $.196 for the first minute and $.126 for each additional minute. How many minutes must a call last if it is to be cheaper to pay with coins rather than a calling card?

152. **Consumerism** Heritage National Bank offers two different checking accounts. The first charges $3 per month, and $.50 per check after the first 10 checks. The second account charges $8 per month with unlimited check writing. How many checks can be written per month if the first account is to be less expensive than the second account?

153. **Consumerism** Glendale Federal Bank offers a checking account to small businesses. The charge is \$8 per month plus \$.12 per check after the first 100 checks. A competitor is offering an account for \$5 per month plus \$.15 per check after the first 100 checks. If a business chooses the Glendale Federal Bank account, how many checks does the business write monthly if it is assumed that this account will cost less than the competitor's account?

154. **Commissions** George Stoia earns \$1000 per month plus 5% commission on the amount of sales. George's goal is to earn a minimum of \$3200 per month. What amount of sales will enable George to earn \$3200 or more per month?

155. **Automobiles** A new car will average at least 22 mpg for city driving and at most 27.5 mpg for highway driving. Find the range of miles that the car can travel on a full tank (19.5 gal) of gasoline.

156. **Education** An average score of 90 or above in an English course receives an A grade. A student has grades of 85, 88, 90, and 98 on four tests. Find the range of scores on the fifth test that will give the student an A grade.

2.7 Applying Concepts

Use the roster method to list the set of positive integers that are solutions of the inequality.

157. $8x - 7 < 2x + 9$

158. $2x + 9 \geq 5x - 4$

159. $5 + 3(2 + x) > 8 + 4(x - 1)$

160. $6 + 4(2 - x) > 7 + 3(x + 5)$

161. $-3x < 15$ and $x + 2 < 7$

162. $3x - 2 > 1$ and $2x - 3 < 5$

163. $-4 \leq 3x + 8 < 16$

164. $5 < 7x - 3 \leq 24$

165. **Temperature** The relationship between Celsius temperature and Fahrenheit temperature is given by the formula $F = \frac{9}{5}C + 32$. If the temperature is between 77°F and 86°F, what is the temperature range in degrees Celsius?

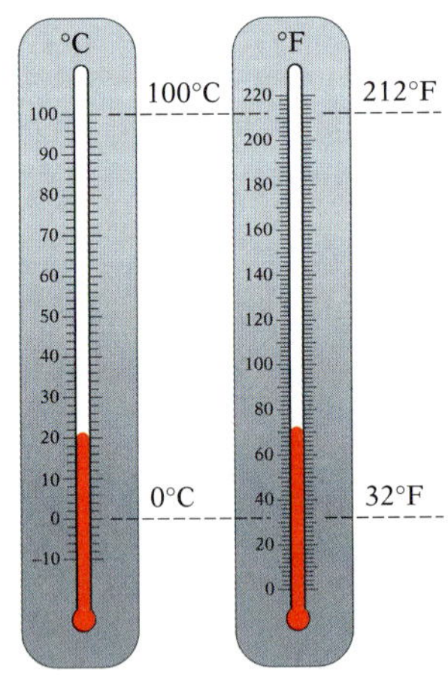

166. **Integer Problem** The average of two negative integers is less than or equal to −15. The smaller integer is 7 less than the larger integer. Find the greatest possible value for the smaller integer.

167. **Consumerism** The charges for a long-distance telephone call are \$1.56 for the first three minutes and \$.52 for each additional minute or fraction of a minute. What is the largest whole number of minutes a call can last if it is to cost you less than \$5.40?

2.8 Absolute Value Equations and Inequalities

1 Absolute value equations

VIDEO & DVD CD TUTOR WEB SSM

The **absolute value** of a number is its distance from zero on the number line. Distance is always a positive number or zero. Therefore, the absolute value of a number is always a positive number or zero.

The distance from 0 to 3 or from 0 to -3 is 3 units.

$|3| = 3 \qquad |-3| = 3$

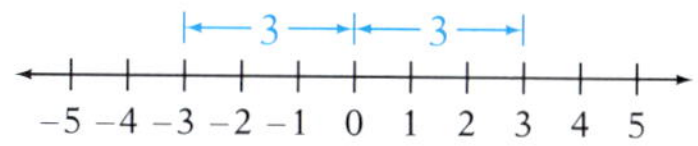

An equation containing an absolute value symbol is called an **absolute value equation.**

$|x| = 3$
$|x + 2| = 8$
$|3x - 4| = 5x - 9$

} Absolute value equations

Absolute Value Equations

If $a \geq 0$ and $|x| = a$, then $x = a$ or $x = -a$.

Given $|x| = 3$, then $x = 3$ or $x = -3$.

Solve: $|x + 2| = 8$

Remove the absolute value sign, and rewrite as two equations.

$|x + 2| = 8$

$x + 2 = 8 \qquad x + 2 = -8$

Solve each equation.

$x = 6 \qquad x = -10$

Check:

$$\begin{array}{c|c} |x + 2| & 8 \\ \hline |6 + 2| & 8 \\ |8| & \\ 8 = 8 & \end{array} \qquad \begin{array}{c|c} |x + 2| & 8 \\ \hline |-10 + 2| & 8 \\ |-8| & \\ 8 = 8 & \end{array}$$

Write the solution.

The solutions are 6 and -10.

Example 1 Solve.
A. $|x| = 15$ **B.** $|2 - x| = 12$
C. $|2x| = -4$ **D.** $3 - |2x - 4| = -5$

Solution **A.** $|x| = 15$

$x = 15 \qquad x = -15$

• **Remove the absolute value sign, and rewrite as two equations.**

The solutions are 15 and -15.

Take Note

Remember to check the solutions. For Part B:

$\lvert 2 - x\rvert = 12$	
$\lvert 2 - (-10)\rvert$	12
$\lvert 2 + 10\rvert$	12
$\lvert 12\rvert$	12
$12 = 12$	

$\lvert 2 - x\rvert = 12$	
$\lvert 2 - 14\rvert$	12
$\lvert -12\rvert$	12
$12 = 12$	

Take Note

$3 - \lvert 2x - 4\rvert = -5$	
$3 - \lvert 2(6) - 4\rvert$	-5
$3 - \lvert 12 - 4\rvert$	-5
$3 - \lvert 8\rvert$	-5
$3 - 8$	-5
$-5 = -5$	

$3 - \lvert 2x - 4\rvert = -5$	
$3 - \lvert 2(-2) - 4\rvert$	-5
$3 - \lvert -4 - 4\rvert$	-5
$3 - \lvert -8\rvert$	-5
$3 - 8$	-5
$-5 = -5$	

B. $|2 - x| = 12$

$2 - x = 12 \qquad 2 - x = -12$ • Remove the absolute value sign, and rewrite as two equations.

$-x = 10 \qquad -x = -14$ • Solve each equation.

$x = -10 \qquad x = 14$

The solutions are -10 and 14.

C. $|2x| = -4$ • The absolute value of a number must be nonnegative.

There is no solution to the equation.

D. $3 - |2x - 4| = -5$

$-|2x - 4| = -8$ • Solve for the absolute value.

$|2x - 4| = 8$ • Multiply each side of the equation by -1.

$2x - 4 = 8 \qquad 2x - 4 = -8$ • Remove the absolute value sign, and rewrite as two equations.

$2x = 12 \qquad 2x = -4$ • Solve each equation.

$x = 6 \qquad x = -2$

The solutions are 6 and -2.

Problem 1 Solve. **A.** $|x| = 25$ **B.** $|2x - 3| = 5$ **C.** $|x - 3| = -2$ **D.** $5 - |3x + 5| = 3$

Solution See page S10.

2 Absolute value inequalities

Recall that absolute value represents the distance between two points. For example, the solutions of the absolute value equation $|x - 1| = 3$ are the numbers whose distance from 1 is 3. Therefore, the solutions are -2 and 4.

The solutions of the **absolute value inequality** $|x - 1| < 3$ are the numbers whose distance from 1 is *less than* 3. Therefore, the solutions are the numbers greater than -2 and less than 4. The solution set is $\{x \mid -2 < x < 4\}$. *Note:* In this text, solutions to absolute value inequalities will always be written in set-builder notation.

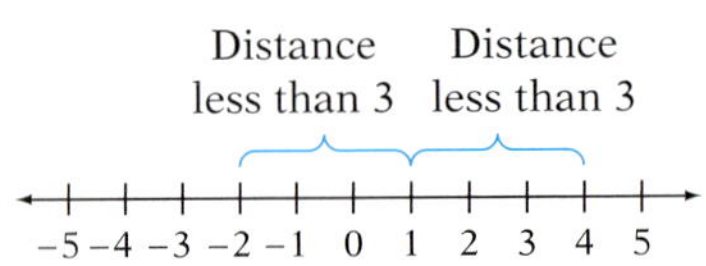

Absolute Value Inequalities of the Form $|ax + b| < c$

To solve an absolute value inequality of the form $|ax + b| < c$, solve the equivalent compound inequality $-c < ax + b < c$.

The compound inequality $-c < ax + b < c$ is equivalent to the intersection of the two inequalities $-c < ax + b$ and $ax + b < c$. The intersection of these two inequalities is shown on the following number line.

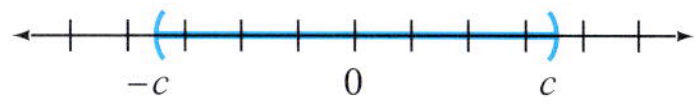

From this number line, we see that the solution of $|ax + b| < c$ is all the numbers for which $ax + b$ is greater than $-c$ and less than c.

Example 2 Solve: $|4x - 3| < 5$

Solution $|4x - 3| < 5$

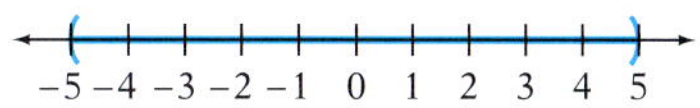

• The distance from zero is less than 5 units. $4x - 3$ is between -5 and 5.

$$-5 < 4x - 3 < 5$$

• Solve the equivalent compound inequality.

$$-5 + 3 < 4x - 3 + 3 < 5 + 3$$

• Add 3 to each of the three parts of the inequality.

$$-2 < 4x < 8$$

$$\frac{-2}{4} < \frac{4x}{4} < \frac{8}{4}$$

• Divide each of three parts of the inequality by 4.

$$-\frac{1}{2} < x < 2$$

$$\left\{x \,\middle|\, -\frac{1}{2} < x < 2\right\}$$

• Write the solution set.

Problem 2 Solve: $|3x + 2| < 8$

Solution See page S10.

Example 3 Solve: $|x - 3| < 0$

Solution $|x - 3| < 0$ • The absolute value of a number must be nonnegative.

$\varnothing$ • The solution set is the empty set.

Problem 3 Solve: $|3x - 7| < 0$

Solution See page S10.

The solutions of the absolute value inequality $|x + 1| > 2$ are the numbers whose distance from -1 is *greater than* 2. Therefore, the solutions are the numbers less than -3 or greater than 1. The solution set is $\{x \mid x < -3 \text{ or } x > 1\}$.

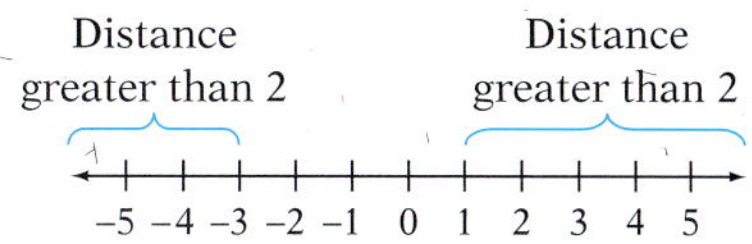

Absolute Value Inequalities of the Form $|ax + b| > c$

To solve an absolute value inequality of the form $|ax + b| > c$, solve the equivalent compound inequality $ax + b < -c$ or $ax + b > c$.

The solution of the compound inequality $ax + b < -c$ or $ax + b > c$ is shown on the following number line.

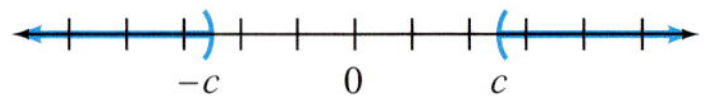

From this number line we see that the solution of $|ax + b| > c$ is the union of the numbers less than $-c$ and greater than c.

> **Take Note**
>
> Carefully observe the difference between the solution method for $|ax + b| > c$ shown here and that for $|ax + b| < c$ shown in Example 2.

Example 4 Solve: $|2x - 1| > 7$

Solution $|2x - 1| > 7$

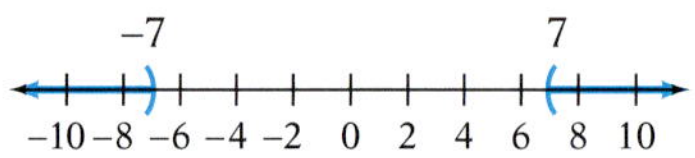

• **The distance from zero is greater than 7. $2x - 1$ is less than -7 or greater than 7.**

$$2x - 1 < -7 \quad \text{or} \quad 2x - 1 > 7$$
$$2x < -6 \qquad\qquad 2x > 8$$
$$x < -3 \qquad\qquad x > 4$$

• **Solve the equivalent compound inequality.**

$$\{x \mid x < -3\} \qquad \{x \mid x > 4\}$$

$$\{x \mid x < -3\} \cup \{x \mid x > 4\}$$
$$= \{x \mid x < -3 \text{ or } x > 4\}$$

• **Find the union of the solution sets.**

Problem 4 Solve: $|5x + 3| > 8$

Solution See page S10.

3 Application problems

VIDEO & DVD CD TUTOR WEB SSM

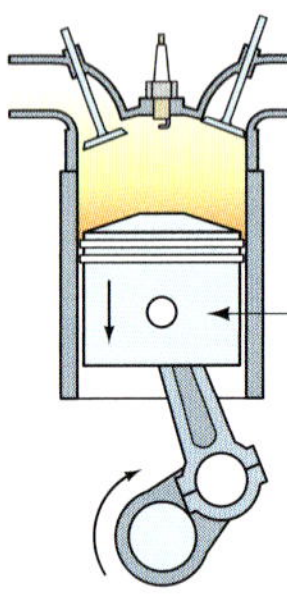

The **tolerance of a component,** or part, is the acceptable amount by which the component may vary from a given measurement. For example, the diameter of a piston may vary from the given measurement of 9 cm by 0.001 cm. This is written as 9 cm ± 0.001 cm, which is read "9 centimeters plus or minus 0.001 centimeter." The maximum diameter, or **upper limit,** of the piston is 9 cm + 0.001 cm = 9.001 cm. The minimum diameter, or **lower limit,** of the piston is 9 cm − 0.001 cm = 8.999 cm.

The lower and upper limits of the diameter could also be found by solving the absolute value inequality $|d - 9| \le 0.001$, where d is the diameter of the piston.

$$|d - 9| \le 0.001$$

$$-0.001 \le d - 9 \le 0.001$$
$$-0.001 + 9 \le d - 9 + 9 \le 0.001 + 9$$
$$8.999 \le d \le 9.001$$

The lower and upper limits of the diameter are 8.999 cm and 9.001 cm.

Example 5 A doctor has prescribed 2 cc of medication for a patient. The tolerance is 0.03 cc. Find the lower and upper limits of the amount of medication to be given.

Strategy Let p represent the prescribed amount of medication, T the tolerance, and m the given amount of medication. Solve the absolute value inequality $|m - p| \le T$ for m.

Solution

$|m - p| \le T$

$|m - 2| \le 0.03$ • Substitute the values of p and T into the inequality.

$-0.03 \le m - 2 \le 0.03$ • Solve the equivalent compound inequality.

$1.97 \le m \le 2.03$

The lower and upper limits of the amount of medication to be given to the patient are 1.97 cc and 2.03 cc.

Problem 5 A machinist must make a bushing that has a tolerance of 0.003 in. The diameter of the bushing is 2.55 in. Find the lower and upper limits of the diameter of the bushing.

Solution See page S10.

2.8 Concept Review

Determine whether the following statements are always true, sometimes true, or never true.

1. If $a > 0$ and $|x| = a$, then $x = a$ or $x = -a$.
2. The absolute value inequality $|x| < a$, $a > 0$, is equivalent to the compound inequality $x > a$ or $x < -a$.
3. If $|x + 3| = 12$, then $|x + 3| = 12$ or $|x + 3| = -12$.
4. An absolute value equation has two solutions.
5. If $|x + b| < c$, then $x + b < -c$ or $x + b > c$.

2.8 Exercises

1

1. Is 2 a solution of $|x - 8| = 6$?
2. Is -2 a solution of $|2x - 5| = 9$?
3. Is -1 a solution of $|3x - 4| = 7$?
4. Is 1 a solution of $|6x - 1| = -5$?
5. Explain why, if $a \ge 0$ and $|x| = a$, $x = a$ or $x = -a$.

6. Explain how to solve an absolute value equation.

Solve.

7. $|x| = 7$
8. $|a| = 2$
9. $|-t| = 3$
10. $|-a| = 7$
11. $|-t| = -3$
12. $|-y| = -2$
13. $|x + 2| = 3$
14. $|x + 5| = 2$
15. $|y - 5| = 3$
16. $|y - 8| = 4$
17. $|a - 2| = 0$
18. $|a + 7| = 0$
19. $|x - 2| = -4$
20. $|x + 8| = -2$
21. $|2x - 5| = 4$
22. $|4 - 3x| = 4$
23. $|2 - 5x| = 2$
24. $|2x - 3| = 0$
25. $|5x + 5| = 0$
26. $|3x - 2| = -4$
27. $|2x + 5| = -2$
28. $|x - 2| - 2 = 3$
29. $|x - 9| - 3 = 2$
30. $|3a + 2| - 4 = 4$
31. $|8 - y| - 3 = 1$
32. $|2x - 3| + 3 = 3$
33. $|4x - 7| - 5 = -5$
34. $|2x - 3| + 4 = -4$
35. $|3x - 2| + 1 = -1$
36. $|6x - 5| - 2 = 4$
37. $|4b + 3| - 2 = 7$
38. $|3t + 2| + 3 = 4$
39. $|5x - 2| + 5 = 7$
40. $3 - |x - 4| = 5$
41. $2 - |x - 5| = 4$
42. $|2x - 8| + 12 = 2$
43. $|3x - 4| + 8 = 3$
44. $2 + |3x - 4| = 5$
45. $5 + |2x + 1| = 8$
46. $5 - |2x + 1| = 5$
47. $3 - |5x + 3| = 3$
48. $8 - |1 - 3x| = -1$

2 49. Explain how to solve an absolute value inequality of the form $|ax + b| < c$.

50. Explain how to solve an absolute value inequality of the form $|ax + b| > c$.

Solve.

51. $|x| > 3$
52. $|x| < 5$
53. $|x + 1| > 2$
54. $|x - 2| > 1$
55. $|x - 5| \leq 1$
56. $|x - 4| \leq 3$
57. $|2 - x| \geq 3$
58. $|3 - x| \geq 2$
59. $|2x + 1| < 5$
60. $|3x - 2| < 4$
61. $|5x + 2| > 12$
62. $|7x - 1| > 13$
63. $|4x - 3| \leq -2$
64. $|5x + 1| \leq -4$
65. $|2x + 7| > -5$
66. $|3x - 1| > -4$

67. $|4 - 3x| \geq 5$

68. $|7 - 2x| > 9$

69. $|5 - 4x| \leq 13$

70. $|3 - 7x| < 17$

71. $|6 - 3x| \leq 0$

72. $|10 - 5x| \geq 0$

73. $|2 - 9x| > 20$

74. $|5x - 1| < 16$

3 75. **Mechanics** The diameter of a bushing is 1.75 in. The bushing has a tolerance of 0.008 in. Find the lower and upper limits of the diameter of the bushing.

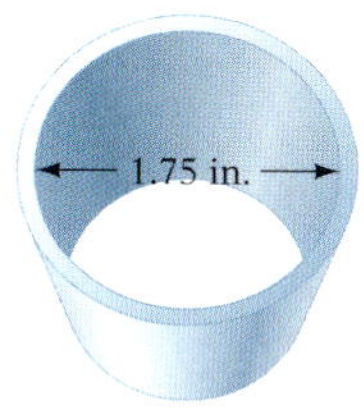

76. **Mechanics** A machinist must make a bushing that has a tolerance of 0.004 in. The diameter of the bushing is 3.48 in. Find the lower and upper limits of the diameter of the bushing.

77. **Medicine** A doctor has prescribed 2.5 cc of medication for a patient. The tolerance is 0.2 cc. Find the lower and upper limits of the amount of medication to be given.

78. **Automobiles** The diameter of a piston for an automobile is $3\frac{5}{16}$ in. with a tolerance of $\frac{1}{64}$ in. Find the lower and upper limits of the diameter of the piston.

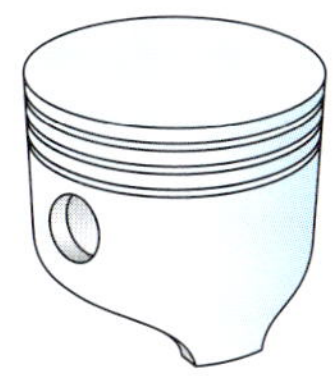

79. **Computers** A power strip is utilized on a computer to prevent the loss of programming by electrical surges. The power strip is designed to allow 110 volts plus or minus 16.5 volts. Find the lower and upper limits of voltage to the computer.

80. **Appliances** An electric motor is designed to run on 220 volts plus or minus 25 volts. Find the lower and upper limits of voltage on which the motor will run.

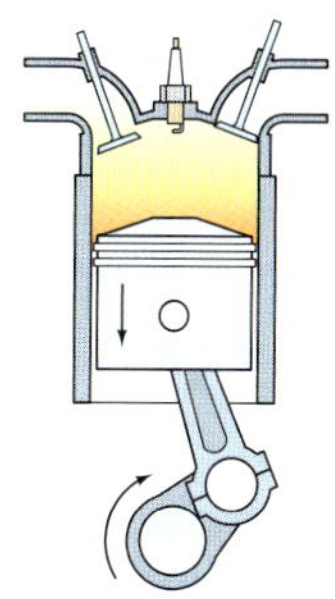

81. **Automobiles** A piston rod for an automobile is $10\frac{3}{8}$ in. with a tolerance of $\frac{1}{32}$ in. Find the lower and upper limits of the length of the piston rod.

The tolerance of the resistors used in electronics is given as a percent of the resistance, which is measured in ohms. For example, if a 20,000-ohm resistor has a tolerance of 4%, then the tolerance is $0.04(20{,}000) = 800$ ohms. The lower and upper limits of the resistor can be found by solving the absolute value inequality $|M - 20{,}000| \leq 800$, where M represents the amount of ohms.

82. **Electronics** Find the lower and upper limits of a 29,000-ohm resistor with a 2% tolerance.

83. **Electronics** Find the lower and upper limits of a 15,000-ohm resistor with a 10% tolerance.

84. **Electronics** Find the lower and upper limits of a 25,000-ohm resistor with a 5% tolerance.

85. **Electronics** Find the lower and upper limits of a 56-ohm resistor with a 5% tolerance.

2.8 Applying Concepts

Solve.

86. $\left|\frac{2x-5}{3}\right| = 7$

87. $\left|\frac{3x-2}{4}\right| + 5 = 6$

88. $\left|\frac{4x-2}{3}\right| > 6$

89. $\left|\frac{2x-1}{5}\right| \leq 3$

For what values of the variable is the equation true? Write the solution set in set-builder notation.

90. $|x + 3| = x + 3$

91. $|y + 6| = y + 6$

92. $|a - 4| = 4 - a$

93. $|b - 7| = 7 - b$

94. For real numbers x and y, which of the following is always true?
a. $|x + y| = |x| + |y|$ **b.** $|x + y| \leq |x| + |y|$ **c.** $|x + y| \geq |x| + |y|$

95. Write an absolute value inequality to represent all real numbers within 5 units of 2.

96. Write an absolute value inequality to represent all real numbers within k units of j. (Assume $k > 0$.)

97. Replace the question mark with $\leq$, $\geq$, or $=$.
a. $|x + y| \,?\, |x| + |y|$ **b.** $|x - y| \,?\, |x| - |y|$ **c.** $||x| - |y|| \,?\, |x| + |y|$
d. $\left|\frac{x}{y}\right| \,?\, \frac{|x|}{|y|}, y \neq 0$ **e.** $|xy| \,?\, |x||y|$

Focus on Problem Solving

Understand the Problem

The first of the four steps that Polya advocated to solve problems is to *understand the problem.* This aspect of problem solving is frequently not given enough attention. There are various exercises that you can try to achieve a good understanding of a problem. Some of these are listed here:

- Try to restate the problem in your own words.
- Determine what is known about this type of problem.
- Determine what information is given.
- Determine what information is unknown.
- Determine whether any of the information given is unnecessary.
- Determine the goal.

To illustrate this aspect of problem solving, consider the following famous ancient riddle.

As I was going to St. Ives,
I met a man with seven wives;
Each wife had seven sacks,
Each sack had seven cats,
Each cat had seven kits:
Kits, cats, sacks, and wives,
How many were going to St. Ives?

Point of Interest

Try this brain teaser: You have two U.S. coins that add up to \$.55. One is not a nickel. What are the two coins?

To answer the question in the poem, we will ask and answer some of the questions listed above.

1. What is the goal?

   The goal is to determine how many were going to St. Ives. (We know this from reading the last line of the poem.)

2. What information is necessary and what information is unnecessary?

   The first line indicates that the poet was going to St. Ives. The next five lines describe a man the poet *met* on the way. This information is irrelevant.

The answer to the question, then, is 1. Only the poet was going to St. Ives.

There are many other examples of the importance, in problem solving, of recognizing *irrelevant information*. One that frequently makes the college circuit is posed in the form of a test. The first line of a 100-question test states, "Read the entire test before you begin." The last line of the test reads, "Choose any *one* question to answer." Many people ignore the information given in the first line and just begin the test, only to find out much later that they did a lot more work than necessary.

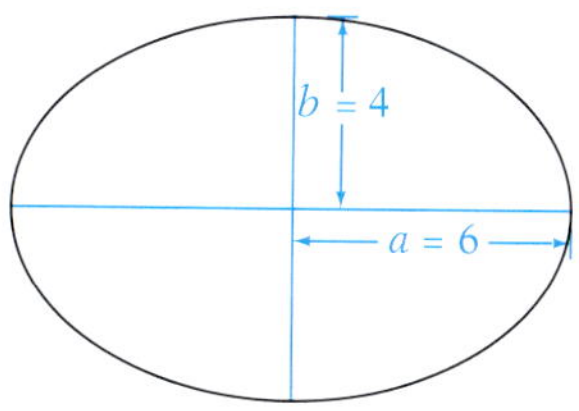

To illustrate another aspect of Polya's first step in the problem-solving process, consider the problem of finding the area of the oval-shaped region (called an *ellipse*) shown in the diagram at the left. This problem can be solved by doing some research to determine *what information is known about this type of problem*. Mathematicians have found a formula for the area of an ellipse. That formula is $A = \pi ab$, where a and b are as shown in the diagram. Therefore, $A = \pi(6)(4) = 24\pi \approx 75.40$ square units. Without the formula, this problem is difficult to solve. With the formula, it is fairly easy.

For each of the following problems, examine the problem in terms of the first step in Polya's problem-solving method. *Do not solve the problem.*

1. Johanna spent one-third of her allowance on a book. She then spent \$5.00 for a sandwich and iced tea. The cost of the iced tea was one-fifth the cost of the sandwich. Find the cost of the iced tea.

2. A flight from Los Angeles to Boston took 6 h. What was the average speed of the plane?

3. A major league baseball is approximately 5 in. in diameter and is covered with cowhide. Approximately how much cowhide is used to cover 10 baseballs?

4. How many donuts are in seven baker's dozen?

5. The smallest prime number is 2. Twice the difference between the eighth and the seventh prime numbers is two more than the smallest prime number. How large is the smallest prime number?

Projects & Group Activities

Venn Diagrams

Diagrams can be very useful when we are working with sets. The set diagrams shown below are called **Venn diagrams.**

In the Venn diagram at the right, the rectangle represents the set U, and the circles A and B represent subsets of set U. The common area of the two circles represents the intersection of A and B.

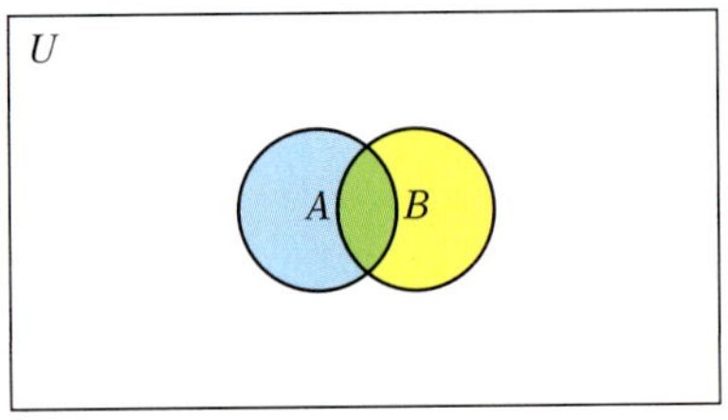

This Venn diagram represents the set $U = \{1, 2, 3, 4, 5, 6, 7, 8\}$. The sets $A = \{2, 3, 4, 5\}$ and $B = \{4, 5, 6, 7\}$ are subsets of U. Note that 1 and 8 are in set U but not in A or B. The numbers 4 and 5 are in both set A and set B.

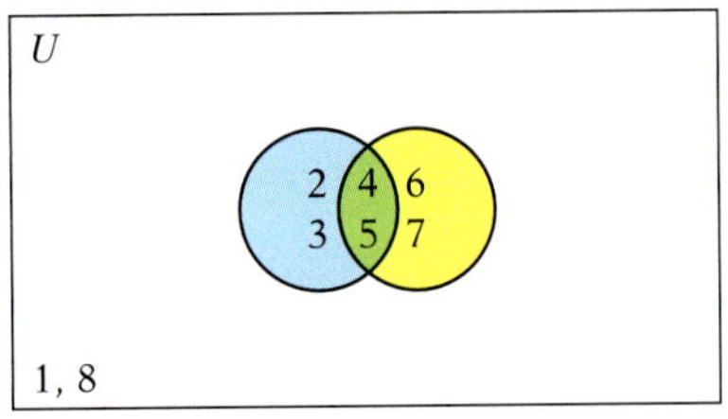

Sixty-five students at a small college enrolled in the following courses.

39 enrolled in English.
26 enrolled in mathematics.
35 enrolled in history.
19 enrolled in English and history.
11 enrolled in English and mathematics.
9 enrolled in mathematics and history.
2 enrolled in all three courses.

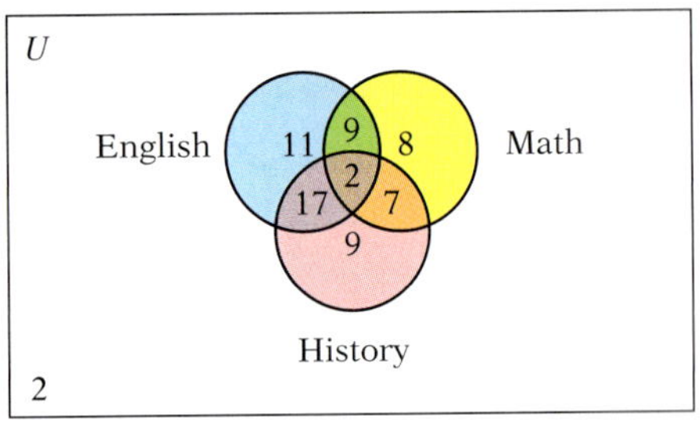

The Venn diagram is shown above. The diagram was drawn by placing the 2 students in the intersection of all three courses. Because 9 students took history and mathematics and 2 students are already in the intersection of history and mathematics, 7 more students are placed in the intersection of history and mathematics. Continue in this manner until all information is used.

Complete the following.

1. Use the Venn diagram above that shows course enrollments.
   a. How many students enrolled only in English and mathematics?
   b. How many students enrolled in English but did not enroll in mathematics or history?
   c. How many students did not enroll in any of the three courses?

2. Sketch the Venn diagram for $U = \{1, 2, 3, 4, 5, 6, 7, 8\}$, $A = \{2, 3, 4, 5, 6\}$, $B = \{4, 5, 6, 7, 8\}$, and $C = \{4, 5\}$.

3. A busload of 50 scouts stopped at a fast-food restaurant and ordered hamburgers. The scouts could have pickles, and/or tomatoes, and/or lettuce on their hamburgers.

31 ordered pickles.
36 ordered tomatoes.
31 ordered lettuce.
21 ordered pickles and tomatoes.
24 ordered tomatoes and lettuce.
22 ordered pickles and lettuce.
17 ordered all three.

a. How many scouts ordered a hamburger with pickles only?
b. How many scouts ordered a hamburger with lettuce and tomatoes without the pickles?
c. How many scouts ordered a hamburger without pickles, tomatoes, or lettuce?

Electricity

Since the Industrial Revolution at the turn of the century, technology has been a far-reaching and ever-changing phenomenon. Most of the technological advances that have been made, however, could not have been accomplished without electricity, and mathematics plays an integral part in the science of electricity.

Point of Interest

Ohm's law is named after Georg Simon Ohm (1789–1854), a German physicist whose work contributed to mathematics, acoustics, and the measurement of electrical resistance.

Central to the study of electricity is **Ohm's law,** one part of which states that $V = IR$, where V is voltage, I is current, and R is resistance.

The word *electricity* comes from the same root word as the word *electron*. Electrons are tiny particles in atoms. Each electron has an electric charge, and this is the fundamental cause of electricity. In order to move, electrons need a source of energy—for example, light, heat, pressure, or a chemical reaction. This is where voltage comes in. Basically, **voltage** is a measure of the amount of energy in a flow of electricity. Voltage is measured in units called **volts.**

Point of Interest

The ampere is named after André Marie Ampère (1775–1836), a French physicist and mathematician who formulated Ampère's law, a mathematical description of the magnetic field produced by a current-carrying conductor.

Current is a measure of how many electrons pass a given point in a fixed amount of time in a flow of electricity. This means that current is a measure of the strength of the flow of electrons through a wire—that is, their speed. Picture a faucet: The more you turn the handle, the more water you get. Similarly, the more current in a wire, the stronger the flow of electricity. Current is measured in **amperes,** often simply called **amps.** A current of 1 ampere would be sufficient to light the bulb in a flashlight.

Resistance is a measure of the amount of resistance, or opposition, to the flow of electricity. You might think of resistance as friction. The more friction, the slower the speed. Resistance is measured in **ohms.**

Watts measure the **power** in an electrical system, measuring both the strength and the speed of the flow of electrons. The power in an electrical system is equal to the voltage times the current, written $P = VI$, where P is the power measured in watts, V is voltage measured in volts, and I is current measured in amperes.

In an electrical circuit, a source of electricity, such as a battery or generator, drives electrons through a wire to the part of the machine that produces the output. Because we cannot see electrons, it may help to think of an electrical system as similar to water flowing through a pipe.

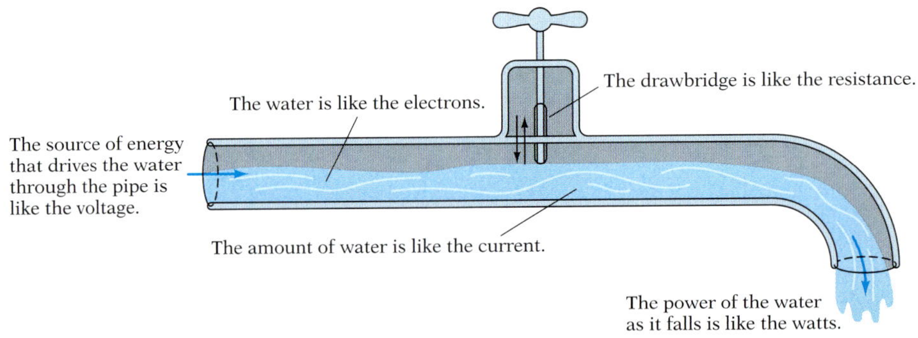

Here are the formulas introduced above, along with the meaning of each variable and the unit in which it is measured:

Formulas for Voltage and Power

$V = IR$

Voltage (volts) = Current (amperes) × Resistance (ohms)

$P = VI$

Power (watts) = Voltage (volts) × Current (amperes)

How many amperes of current pass through a wire in which the voltage is 90 volts and the resistance is 5 ohms?

$V = IR$

$90 = I \cdot 5$ • *V* is the voltage, measured in volts. *R* is the resistance, measured in ohms.

$\frac{90}{5} = \frac{I \cdot 5}{5}$ • Solve for *I*, the current, measured in amperes.

$18 = I$

18 amperes of current pass through the wire.

How much voltage does a 120-watt light bulb with a current of 20 amperes require?

$P = VI$

$120 = V \cdot 20$ • *P* is the power, measured in watts. *I* is the current, measured in amperes.

$\frac{120}{20} = \frac{V \cdot 20}{20}$ • Solve for *V*, the voltage, measured in volts.

$6 = V$

The light bulb requires 6 volts.

Find the power in an electrical system that has a voltage of 120 volts and resistance of 150 ohms.

First solve the formula $V = IR$ for I. This will give us the amperes. Then solve the formula $P = VI$ for P. This will give us the watts.

$$
\begin{aligned}
V &= IR \\
120 &= I \cdot 150 \\
\frac{120}{150} &= \frac{I \cdot 150}{150} \\
0.8 &= I
\end{aligned}
\qquad
\begin{aligned}
P &= VI \\
P &= 120(0.8) \\
P &= 96
\end{aligned}
$$

The power in the electrical system is 96 watts.

Solve.

1. How many volts pass through a wire in which the current is 20 amperes and the resistance is 100 ohms?

2. Find the voltage in a system in which the current is 4.5 amperes and the resistance is 150 ohms.

3. How many amperes of current pass through a wire in which the voltage is 100 volts and the resistance is 10 ohms?

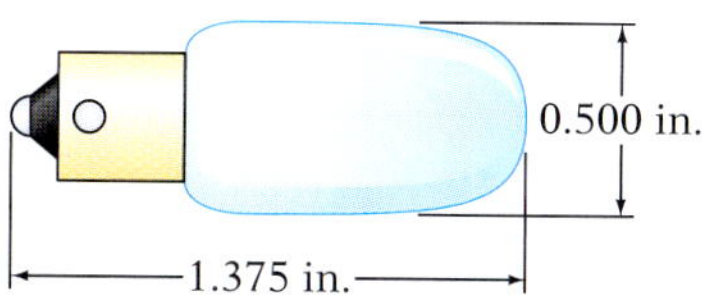

4. A lamp has a resistance of 70 ohms. How much current flows through the lamp when it is connected to a 115-volt circuit? Round to the nearest hundredth.

5. What is the resistance of a semiconductor that passes 0.12 ampere of current when 0.48 volt is applied to it?

6. Find the resistance when the current is 120 amperes and the voltage is 1.5 volts.

7. Determine the power in a light bulb when the current is 10 amperes and the voltage is 12 volts.

8. Find the power in a hand-held dryer that operates from a voltage of 115 volts and draws 2.175 amperes of current.

9. How much current flows through a 110-volt, 850-watt lamp? Round to the nearest hundredth.

10. A heating element in a clothes dryer is rated at 4500 watts. How much current is used by the dryer when the voltage is 240 volts?

11. A miniature lamp pulls 0.08 ampere of current while lighting a 0.5-watt bulb. What voltage battery is needed for the lamp?

12. The power of a car sound system that pulls 15 amperes is 180 watts. Find the voltage of the system.

13. Find the power of a lamp that has a voltage of 160 volts and resistance of 80 ohms.

14. Find the power in an electrical system that has a voltage of 105 volts and resistance of 70 ohms.

Chapter Summary

Key Words

An **equation** expresses the equality of two mathematical expressions. [2.1.1, p. 76]

Example: $5(3x - 2) = 4x + 7$ is an equation.

A **solution of an equation** is a number that, when substituted for the variable, results in a true equation. [2.1.1, p. 76]

Example: 1 is the solution of the equation $6x - 4 = 2$ because $6(1) - 4 = 2$ is a true equation.

To **solve** an equation means to find a solution of the equation. The goal is to rewrite the equation in the form **variable** = **constant**, because the constant is the solution. [2.1.2, p. 77]

Example: The equation $x = -9$ is in the form *variable* = *constant*. The constant, -9, is the solution of the equation.

Consecutive integers follow one another in order. [2.3.2, p. 108]

Example: 11, 12, 13 are consecutive integers. $-9, -8, -7$ are consecutive integers.

The **perimeter** of a geometric figure is a measure of the distance around the figure. [2.4.1, p. 124]

Example: The perimeter of a triangle is the sum of the lengths of the three sides. The perimeter of a rectangle is the sum of the lengths of the four sides.

An angle is measured in **degrees**. A 90° angle is a **right angle**. A 180° angle is a **straight angle**. One complete revolution is 360°. An **acute angle** is an angle whose measure is between 0° and 90°. An **obtuse angle** is an angle whose measure is between 90° and 180°. An **isosceles triangle** has two sides of equal measure. The three sides of an **equilateral triangle** are of equal measure. **Complementary angles** are two angles whose measures have the sum 90°. **Supplementary angles** are two angles whose measures have the sum 180°. [2.4.1, 2.4.2, pp. 124, 126]

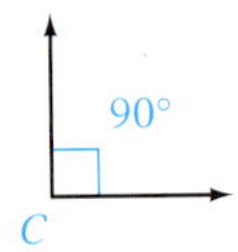

Parallel lines never meet; the distance between them is always the same. **Perpendicular lines** are intersecting lines that form right angles. Two angles that are on opposite sides of the intersection of two lines are **vertical angles;** vertical angles have the same measure. Two angles that share a common side are **adjacent angles;** adjacent angles of intersecting lines are supplementary angles. [2.4.2, p. 127]

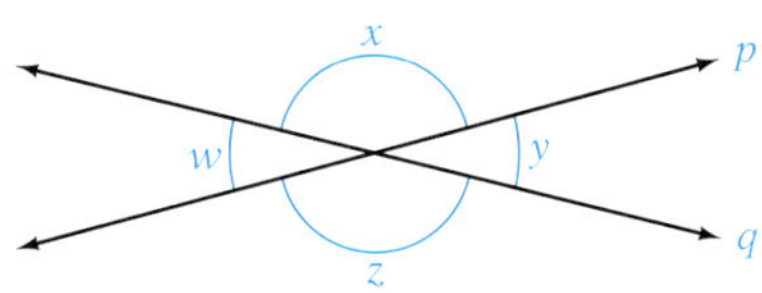

A line that intersects two other lines at two different points is a **transversal**. If the lines cut by a transversal are parallel lines, then pairs of equal angles are formed: **alternate interior angles**, **alternate exterior angles**, and **corresponding angles**. [2.4.2, pp. 128–129]	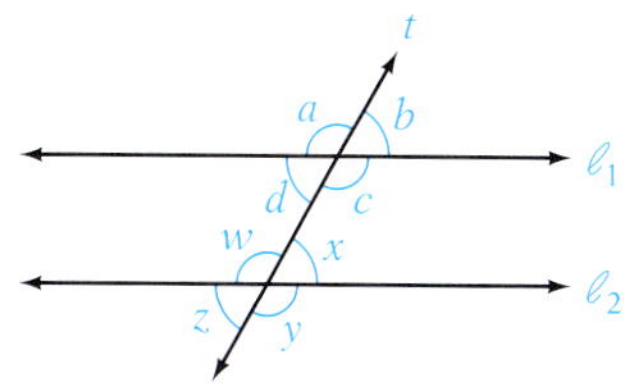 $\angle b = \angle d = \angle x = \angle z$ $\angle a = \angle c = \angle w = \angle y$
Cost is the price that a business pays for a product. **Selling price** is the price for which a business sells a product to a customer. **Markup** is the difference between selling price and cost. **Markup rate** is the markup expressed as a percent of the retailer's cost. [2.5.1, p. 138]	If a business pays \$100 for a product and sells that product for \$140, then the cost of the product is \$100, the selling price is \$140, and the markup is $\$140 - \$100 = \$40$.
Discount is the amount by which a retailer reduces the regular price of a product. **Discount rate** is the discount expressed as a percent of the regular price. [2.5.2, p. 140]	The regular price of a product is \$100. The product is now on sale for \$75. The discount on the product is $\$100 - \$75 = \$25$.
The **empty set**, or **null set**, is the set that contains no elements. [2.7.1, p. 153]	The empty set is written as $\varnothing$ or $\{\ \}$.
A **finite set** is a set in which the elements can be counted. [2.7.1, p. 153]	$A = \{1, 2, 3, 4\}$
An **infinite set** is a set in which it is impossible to list all the elements. [2.7.1, p. 153]	$A = \{1, 2, 3, 4, \ldots\}$
The **union** of two sets, written $A \cup B$, is the set that contains all the elements of A and all the elements of B. The elements that are in both set A and set B are listed only once. [2.7.1, p. 154]	If $A = \{2, 3, 4\}$ and $B = \{4, 5, 6\}$, then $A \cup B = \{2, 3, 4, 5, 6\}$.
The **intersection** of two sets, written $A \cap B$, is the set that contains the elements that are common to both A and B. [2.7.1, p. 154]	If $A = \{2, 3, 4\}$ and $B = \{4, 5, 6\}$, then $A \cap B = \{4\}$.
Interval notation is an alternative method of representing a set. An interval is **closed** if it includes both **endpoints**. An interval is **open** if it does not include either endpoint. An interval is **half-open** if one endpoint is included and the other endpoint is not included. A bracket is used to indicate a closed interval, and a parenthesis is used to indicate an open interval. [2.7.1, p. 156]	Closed interval Open interval Half-open interval 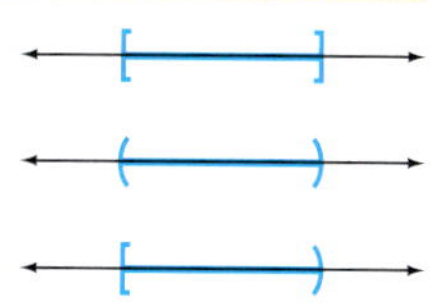
The **solution set of an inequality** is a set of numbers, each element of which, when substituted in the inequality, results in a true inequality. [2.7.2, p. 157]	Any number greater than 4 is a solution of the inequality $x > 4$.

A **compound inequality** is formed by joining two inequalities with a connective word such as "and" or "or." [2.7.3, p. 159]

$3x > 6$ and $2x + 5 < 7$

$2x + 2 < 3$ or $x + 2 > 4$

An **absolute value equation** is an equation that contains an absolute value symbol. [2.8.1, p. 169]

$|x - 2| = 3$

An **absolute value inequality** is an inequality that contains an absolute value symbol. [2.8.2, p. 170]

$|x - 4| < 5 \qquad |2x - 3| > 6$

Essential Rules and Procedures

The Addition Property of Equations
The same number or variable term can be added to each side of an equation without changing the solution of the equation. [2.1.2, p. 77]

$$\begin{aligned} x + 12 &= -19 \\ x + 12 - 12 &= -19 - 12 \\ x &= -31 \end{aligned}$$

The Multiplication Property of Equations
Each side of an equation can be multiplied by the same nonzero number without changing the solution of the equation. [2.1.3, p. 79]

$$\begin{aligned} -6x &= 24 \\ \frac{-6x}{-6} &= \frac{24}{-6} \\ x &= -4 \end{aligned}$$

To solve an equation of the form $ax + b = c$ or an equation of the form $ax + b = cx + d$, use both the Addition and the Multiplication Properties of Equations. [2.2.1 and 2.2.2, pp. 93, 95]

$$\begin{aligned} 7x - 4 &= 3x + 16 \\ 7x - 3x - 4 &= 3x - 3x + 16 \\ 4x - 4 &= 16 \\ 4x - 4 + 4 &= 16 + 4 \\ 4x &= 20 \\ \frac{4x}{4} &= \frac{20}{4} \\ x &= 5 \end{aligned}$$

To solve an equation containing parentheses, use the Distributive Property to remove parentheses. [2.2.3, p. 96]

$$\begin{aligned} 3(2x - 1) &= 15 \\ 6x - 3 &= 15 \\ 6x - 3 + 3 &= 15 + 3 \\ 6x &= 18 \\ \frac{6x}{6} &= \frac{18}{6} \\ x &= 3 \end{aligned}$$

The Basic Percent Equation
Percent · Base = Amount
$P \cdot B = A$
[2.1.4, p. 81]

40% of what number is 16?

$$\begin{aligned} PB &= A \\ 0.40B &= 16 \\ \frac{0.40B}{0.40} &= \frac{16}{0.40} \\ B &= 40 \end{aligned}$$

The number is 40.

Consecutive Integers
$n, n + 1, n + 2, \ldots$ [2.3.2, p. 108]

The sum of three consecutive integers is 135.

$$n + (n + 1) + (n + 2) = 135$$

Consecutive Even or Consecutive Odd Integers
$n, n + 2, n + 4, \ldots$ [2.3.2, p. 108]

The sum of three consecutive odd integers is 135.

$$n + (n + 2) + (n + 4) = 135$$

Value Mixture Equation
$V = AC$ [2.3.3, p. 110]

An herbalist has 30 oz of herbs costing \$2 per ounce. How many ounces of herbs costing \$1 per ounce should be mixed with the 30 oz to produce a mixture costing \$1.60 per ounce?

$$30(2) + 1x = 1.60(30 + x)$$

Uniform Motion Equation
$d = rt$ [2.3.4, p. 111]

A boat traveled from a harbor to an island at an average speed of 20 mph. The average speed on the return trip was 15 mph. The total trip took 3.5 h. How long did it take to travel to the island?

$$20t = 15(3.5 - t)$$

Formulas for Perimeter
Triangle: $P = a + b + c$
Rectangle: $P = 2L + 2W$
Square: $P = 4s$
[2.4.1, p. 124]

The perimeter of a rectangle is 108 m. The length is 6 m more than three times the width. Find the width of the rectangle.

$$108 = 2(3W + 6) + 2W$$

The perimeter of an isosceles triangle is 34 ft. The length of the third side is 5 ft less than the length of one of the equal sides. Find the length of one of the equal sides.

$$34 = x + x + (x - 5)$$

Sum of the Angles of a Triangle
The sum of the measures of the interior angles of a triangle is 180°.
$\angle A + \angle B + \angle C = 180°$
The sum of an interior and corresponding exterior angle is 180°.
[2.4.3, p. 130]

In a right triangle, the measure of one acute angle is 12° more than the measure of the smallest angle. Find the measure of the smallest angle.

$$x + (x + 12) + 90 = 180$$

Basic Markup Equation
$S = C + rC$ [2.5.1, p. 139]

The manager of a department buys a pasta maker machine for \$70 and sells the machine for \$98. Find the markup rate.

$$98 = 70 + 70r$$

Basic Discount Equation
$S = R - rR$ [2.5.2, p. 140]

The sale price for a leather bomber jacket is \$95. This price is 24% off the regular price. Find the regular price.

$$95 = R - 0.24R$$

Annual Simple Interest Equation
$I = Pr$ [2.1.4, 2.6.1, pp. 83, 145]

An investment of \$4000 is made at an annual simple interest rate of 5%. How much additional money must be invested at an annual simple interest rate of 6.5% so that the total interest earned is \$720?

$$0.05(4000) + 0.065x = 720$$

Percent Mixture Problems
$Q = Ar$ [2.1.4, 2.6.2, pp. 84, 147]

A silversmith mixed 120 oz of an 80% silver alloy with 240 oz of a 30% silver alloy. Find the percent concentration of the resulting silver alloy.

$$0.80(120) + 0.30(240) = x(360)$$

The Addition Property of Inequalities
If $a > b$, then $a + c > b + c$.
The property is also true for $a < b$. [2.7.2, p. 157]

$$x + 3 > -2$$
$$x + 3 - 3 > -2 - 3$$
$$x > -5$$

The Multiplication Property of Inequalities [2.7.2, p. 158]

Rule 1 If $a > b$ and $c > 0$, then $ac > bc$.
If $a < b$ and $c > 0$, then $ac < bc$.

$$3x > 2$$
$$\left(\frac{1}{3}\right)(3x) > \left(\frac{1}{3}\right)2$$
$$x > \frac{2}{3}$$

Rule 2 If $a > b$ and $c < 0$, then $ac < bc$.
If $a < b$ and $c < 0$, then $ac > bc$.

$$-2x < 5$$
$$\frac{-2x}{-2} > \frac{5}{-2}$$
$$x > -\frac{5}{2}$$

Solutions of Absolute Value Equations
If $a > 0$ and $|x| = a$, then $x = a$ or $x = -a$.
If $a = 0$ and $|x| = a$, then $x = 0$. [2.8.1, p. 169]

$$|x - 3| = 7$$
$$x - 3 = 7 \qquad x - 3 = -7$$
$$x = 10 \qquad x = -4$$

Absolute Value Inequalities of the Form $|ax + b| < c$
To solve an absolute value inequality of the form $|ax + b| < c$, $c > 0$, solve the equivalent compound inequality $-c < ax + b < c$. [2.8.2, p. 170]

$$|x - 5| < 9$$
$$-9 < x - 5 < 9$$
$$-9 + 5 < x - 5 + 5 < 9 + 5$$
$$-5 < x < 14$$

Absolute Value Inequalities of the Form $|ax + b| < c$
To solve an absolute value inequality of the form $|ax + b| > c$, solve the equivalent compound inequality $ax + b < -c$ or $ax + b > c$. [2.8.2, p. 171]

$$|x - 5| > 9$$
$$x - 5 < -9 \quad \text{or} \quad x - 5 > 9$$
$$x < -4 \quad \text{or} \quad x > 14$$

Chapter Review Exercises

1. Is 3 a solution of $5x - 2 = 4x + 5$?

2. Solve: $x - 4 = 16$

3. Solve: $8x = -56$

4. Solve: $5x - 6 = 29$

5. Solve: $5x + 3 = 10x - 17$

6. Solve: $3(5x + 2) + 2 = 10x + 5[x - (3x - 1)]$

7. What is 81% of 500?

8. 18 is 72% of what number?

9. 27 is what percent of 40?

10. Solve: $\dfrac{2x - 3}{3} + 2 = \dfrac{2 - 3x}{5}$

11. Find $A \cap B$ given $A = \{0, 1, 2, 3\}$ and $B = \{2, 3, 4, 5\}$.

12. Graph: $\{x \mid x < 1\}$

13. Solve: $3x - 7 > -2$
Write the solution set in interval notation.

14. Solve: $2x - 9 < 8x + 15$
Write the solution set in interval notation.

15. Solve: $-5 < 4x - 1 < 7$
Write the solution set in interval notation.

16. Solve: $5x - 2 > 8$ or $3x + 2 < -4$
Write the solution set in interval notation.

17. Solve: $3x < 4$ and $x + 2 > -1$
Write the solution set in set-builder notation.

18. Solve: $3x - 2 > -4$ or $7x - 5 < 3x + 3$
Write the solution set in set-builder notation.

19. Solve: $|2x - 3| = 8$

20. Solve: $|5x + 8| = 0$

21. Solve: $6 + |3x - 3| = 2$

22. Solve: $|2x - 5| \le 3$

23. Solve: $|4x - 5| \ge 3$

24. Solve: $|5x - 4| < -2$

25. A lever is 12 ft long. At a distance of 2 ft from the fulcrum, a force of 120 lb is applied. How large a force must be applied to the other end of the lever so that the system will balance? Use the lever-system equation $F_1x = F_2(d - x)$.

26. In Central America and Mexico, 1184 plants and animals are at risk of extinction. This represents approximately 10.7% of all the species at risk of extinction on Earth. Approximately how many plants and animals are at risk of extinction on Earth? (*Source:* World Conservation Union)

27. The pressure at a certain depth in the ocean can be approximated by the equation $P = 15 + \frac{1}{2}D$, where P is the pressure in pounds per square inch and D is the depth in feet. Use this equation to find the depth when the pressure is 55 lb/in^2.

28. Translate "seven less than twice a number is equal to the number" into an equation and solve.

29. A piano wire is 35 in. long. A note can be produced by dividing this wire into two parts so that three times the length of the shorter piece is twice the length of the longer piece. Find the length of the shorter piece.

30. The sum of two numbers is twenty-one. Three times the smaller number is two less than twice the larger number. Find the two numbers.

31. A ceiling fan that regularly sells for $90 is on sale for $60. Find the discount rate.

32. In an isosceles triangle, the measure of one of the two equal angles is 15° more than the measure of the third angle. Find the measure of each angle.

33. The manager of a sporting goods store buys packages of cleats for $5.25 each and sells them for $9.45 each. Find the markup rate.

34. A dairy owner mixes 5 gal of cream that is 30% butterfat with 8 gal of milk that is 4% butterfat. Find the percent concentration of butterfat in the resulting mixture.

35. The length of a rectangle is four times the width. The perimeter is 200 ft. Find the length and width of the rectangle.

36. The sum of three consecutive odd integers is −45. Find the integers.

37. Given that $\ell_1 \| \ell_2$, find the measures of angles a and b.

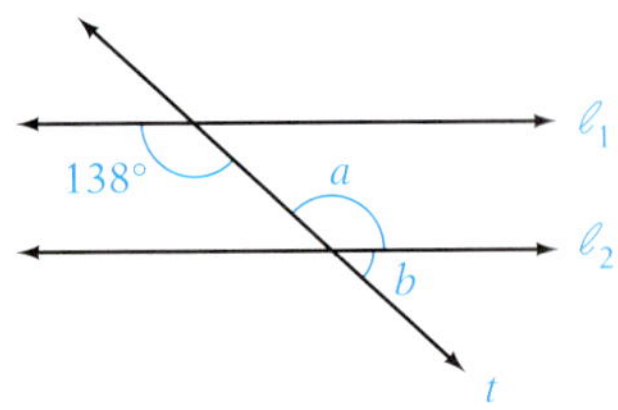

38. A doctor has prescribed 2 cc of medication for a patient. The tolerance is 0.25 cc. Find the lower and upper limits of the amount of medication to be given.

39. Find three consecutive integers such that five times the middle integer is twice the sum of the other two integers.

40. A silversmith combines 40 oz of pure silver that costs $8.00 per ounce with 200 oz of a silver alloy that costs $3.50 per ounce. Find the cost per ounce of the mixture.

41. A grocer mixed apple juice that costs $3.20 per gallon with 40 gal of cranberry juice that costs $5.50 per gallon. How much apple juice was used to make cranapple juice that costs $4.20 per gallon?

42. Two planes are 1680 mi apart and traveling toward each other. One plane is traveling 80 mph faster than the other plane. The planes meet in 1.75 h. Find the speed of each plane.

43. Two investments earn an annual income of \$635. One investment is earning 10.5% annual simple interest, and the other investment is earning 6.4% annual simple interest. The total investment is \$8000. Find the amount invested in each account.

44. An alloy containing 30% tin is mixed with an alloy containing 70% tin. How many pounds of each were used to make 500 lb of an alloy containing 40% tin?

45. An average score of 80 to 90 in a psychology class receives a B grade. A student has grades of 92, 66, 72, and 88 on four tests. Find the range of scores on the fifth test that will give the student a B for the course.

Chapter Test

1. Solve: $\frac{3}{4}x = -9$

2. Solve: $6 - 5x = 5x + 11$

3. Solve: $3x - 5 = -14$

4. Is -2 a solution of $x^2 - 3x = 2x - 6$?

5. Solve: $5x - 2(4x - 3) = 6x + 9$

6. Solve: $x - 3 = -8$

7. 30% of what is 12?

8. Solve: $4 - 3(x + 2) < 2(2x + 3) - 1$
Write the solution set in interval notation.

9. Solve: $3x - 2 > 4$ or $4 - 5x < 14$
Write the solution set in set-builder notation.

10. Solve: $|3 - 5x| = 12$

11. Solve: $|2x - 1| > 3$

12. Translate "the difference between three times a number and fifteen is twenty-seven" into an equation and solve.

13. The manager of a sports shop uses a markup rate of 50%. The selling price for a set of golf clubs is \$300. Find the cost of the golf clubs.

14. A financial manager has determined that the cost per unit for a calculator is \$15 and that the fixed costs per month are \$2000. Find the number of calculators produced during a month in which the total cost was \$5000. Use the equation $T = U \cdot N + F$, where T is the total cost, U is the cost per unit, N is the number of units produced, and F is the fixed cost.

15. Find three consecutive odd integers such that three times the first integer is one less than the sum of the second and third integers.

16. In a triangle, the first angle is 15° more than the second angle. The third angle is three times the second angle. Find the measure of each angle.

17. The width of a rectangle is 12 ft. The length is $(3x + 5)$ ft. Express as an integer the minimum length of the rectangle if the area is greater than 276 ft^2.

18. A file cabinet that normally sells for $99 is on sale for 20% off. Find the sale price.

19. Suppose an obtuse angle of two intersecting lines is 135°. What are the measures of the three other angles?

20. Agency A rents cars for $12 per day and 10¢ for every mile driven. Agency B rents cars for $24 per day with unlimited mileage. How many miles per day can you drive an Agency A car if it is to cost you less than an Agency B car?

21. A doctor prescribed 3 cc of medication for a patient. The tolerance is 0.1 cc. Find the lower and upper limits of the amount of medication to be given.

22. A butcher combines 100 lb of hamburger that costs $2.60 per pound with 60 lb of hamburger that costs $4.20 per pound. Find the cost of the hamburger mixture.

23. A jogger runs a distance at a speed of 8 mph and returns the same distance running at a speed of 6 mph. Find the total distance that the jogger runs if the total time running is 1 h and 45 min.

24. An investment of $12,000 is deposited into two simple interest accounts. On one account, the annual simple interest rate is 7.8%. On the other, the annual simple interest rate is 9%. The total interest earned for one year is $1020. How much was invested in each account?

25. How many ounces of pure water must be added to 60 oz of an 8% salt solution to make a 3% salt solution?

Cumulative Review Exercises

1. Subtract: $-6 - (-20) - 8$

2. Multiply: $(-2)(-6)(-4)$

3. Subtract: $-\frac{5}{6} - \left(-\frac{7}{16}\right)$

4. Divide: $-2\frac{1}{3} \div 1\frac{1}{6}$

5. Simplify: $-4^2 \cdot \left(-\frac{3}{2}\right)^3$

6. Simplify: $25 - 3 \cdot \frac{(5-2)^2}{2^3+1} - (-2)$

7. Evaluate $3(a - c) - 2ab$ when $a = 2$, $b = 3$, and $c = -4$.

8. Simplify: $3x - 8x + (-12x)$

9. Simplify: $2a - (-3b) - 7a - 5b$

10. Simplify: $(16x)\left(\frac{1}{8}\right)$

11. Simplify: $-4(-9y)$

12. Simplify: $-2(-x^2 - 3x + 2)$

13. Simplify: $-2(x - 3) + 2(4 - x)$

14. Simplify: $-3[2x - 4(x - 3)] + 2$

15. Use the roster method to write the set of negative integers greater than -8.

16. Write $\frac{7}{8}$ as a percent. Write the remainder in fractional form.

17. Write 342% as a decimal.

18. Write $62\frac{1}{2}\%$ as a fraction.

19. Is -3 a solution of $x^2 + 6x + 9 = x + 3$?

20. Solve: $x - 4 = -9$

21. Solve: $\frac{3}{5}x = -15$

22. Solve: $13 - 9x = -14$

23. Solve: $5x - 8 = 12x + 13$

24. Solve: $8x - 3(4x - 5) = -2x - 11$

25. Solve: $-\frac{3}{4}x > \frac{2}{3}$

26. Solve: $5x - 4 \geq 4x + 8$

27. Solve: $3x + 17 < 5x - 1$

28. Translate "the difference between eight and the quotient of a number and twelve" into a variable expression.

29. Translate and simplify "the sum of a number and two more than the number."

30. A lever is 25 ft long. At a distance of 12 ft from the fulcrum, a force of 26 lb is applied. How large a force must be applied to the other end of the lever so that the system will balance?

31. Three times the sum of the first and third of three consecutive odd integers is fifteen more than the second integer. Find the first integer.

32. Tickets for a school play sold for $2.25 for each adult and $.75 for each child. The total receipts for 75 tickets were $128.25. Find the number of adult tickets sold.

33. Two planes are 1400 mi apart and traveling toward each other. One plane is traveling 120 mph faster than the other plane. The planes meet in 2.5 h. Find the speed of the faster plane.

34. How many liters of a 12% acid solution must be mixed with 4 L of a 5% acid solution to make an 8% acid solution?

35. An investment advisor invested $10,000 in two accounts. One investment earned 9.8% annual simple interest, and the other investment earned 12.8% annual simple interest. The amount of interest earned in one year was $1085. How much was invested in the 9.8% account?

Thermometers, such as the one shown above, give us a reading of the air temperature. The wind-chill index combines the air temperature and the wind speed to indicate how cold the wind makes the air "feel." The greater the wind speed, the colder the air feels.

In the **Project on pages 265–266,** *you will be working with functions that determine the wind-chill index given the air temperature and the wind speed. These functions are linear functions, the topic of this chapter.*

Applications

3 Linear Functions and Inequalities in Two Variables

Objectives

Section 3.1

1. Points on a rectangular coordinate system
2. Find the length and midpoint of a line segment
3. Graph a scatter diagram
4. Average rate of change

Section 3.2

1. Evaluate a function

Section 3.3

1. Graph a linear function
2. Graph an equation of the form $Ax + By = C$
3. Application problems

Section 3.4

1. Find the slope of a line given two points
2. Graph a line given a point and the slope

Section 3.5

1. Find the equation of a line given a point and the slope
2. Find the equation of a line given two points
3. Application problems

Section 3.6

1. Find equations of parallel and perpendicular lines

Section 3.7

1. Graph the solution set of an inequality in two variables

WWW WEB **Need help? For online student resources, such as section quizzes, visit this web site: math.college.hmco.com**

Prep Test

1. Simplify: $-4(x - 3)$
2. Simplify: $\sqrt{(-6)^2 + (-8)^2}$
3. Simplify: $\dfrac{3 - (-5)}{2 - 6}$
4. Evaluate: $-2x + 5$ for $x = -3$.
5. Evaluate $\dfrac{2r}{r - 1}$ for $r = 5$.
6. Evaluate $2p^3 - 3p + 4$ for $p = -1$.
7. Evaluate $\dfrac{x_1 + x_2}{2}$ for $x_1 = 7$ and $x_2 = -5$.
8. Given $3x - 4y = 12$, find the value of x when $y = 0$.

Go Figure

If $\boxed{5} = 4$ and $\textcircled{5} = 6$ and $y = x - 1$, which of the following has the largest value?

$\boxed{x}$ $\quad$ $\textcircled{x}$ $\quad$ $\boxed{y}$ $\quad$ $\textcircled{y}$

3.1 The Rectangular Coordinate System

1 Points on a rectangular coordinate system

VIDEO & DVD CD TUTOR WEB SSM

Point of Interest

*A rectangular coordinate system is also called a **Cartesian** coordinate system in honor of Descartes.*

Before the 15th century, geometry and algebra were considered separate branches of mathematics. That changed when René Descartes, a French mathematician who lived from 1596 to 1650, founded **analytic geometry.** In this geometry, a *coordinate system* is used to study relationships between variables.

A **rectangular coordinate system** is formed by two number lines, one horizontal and one vertical, that intersect at the zero point of each line. The point of intersection is called the **origin.** The two lines are called **coordinate axes,** or simply **axes.**

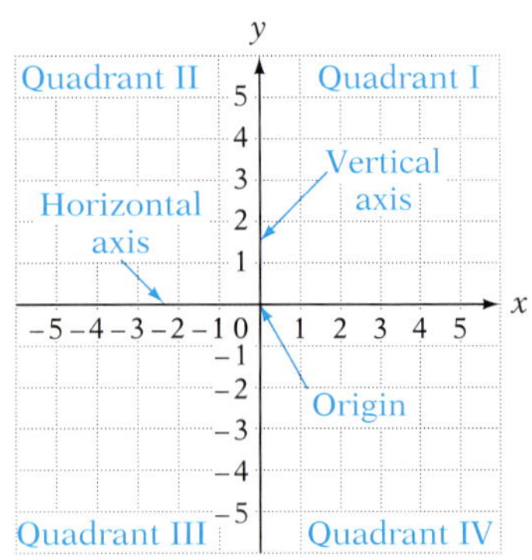

The axes determine a **plane,** which can be thought of as a large, flat sheet of paper. The two axes divide the plane into four regions called **quadrants.** The quadrants are numbered counterclockwise from I to IV.

Each point in the plane can be identified by a pair of numbers called an **ordered pair.** The first number of the pair measures a horizontal distance and is called the **abscissa.** The second number of the pair measures a vertical distance and is called the **ordinate.** The **coordinates** of the point are the numbers in the ordered pair associated with the point. The abscissa is also called the **first coordinate,** or ***x*-coordinate,** of the ordered pair, and the ordinate is also called the **second coordinate,** or ***y*-coordinate,** of the ordered pair.

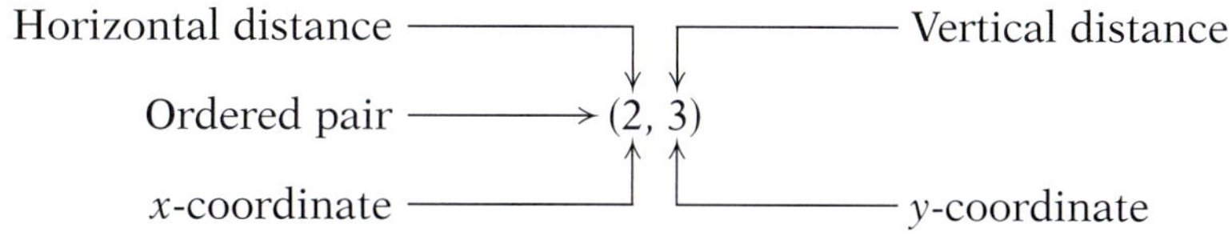

Point of Interest

Gottfried Leibniz introduced the words abscissa *and* ordinate. Abscissa *is from Latin meaning "to cut off." Originally, Leibniz used* abscissa linea, *"cut off a line." The root of* ordinate *is a Latin word used to suggest the sense of order.*

When drawing a rectangular coordinate system, we often label the horizontal axis x and the vertical axis y. In this case, the coordinate system is called an ***xy*-coordinate system.** To graph or plot a point in the xy-coordinate system, place a dot at the location given by the ordered pair. The **graph of an ordered pair** is the dot drawn at the coordinates of the point in the xy-coordinate system. The points whose coordinates are $(3, 4)$ and $(-2.5, -3)$ are graphed in the figure at the right.

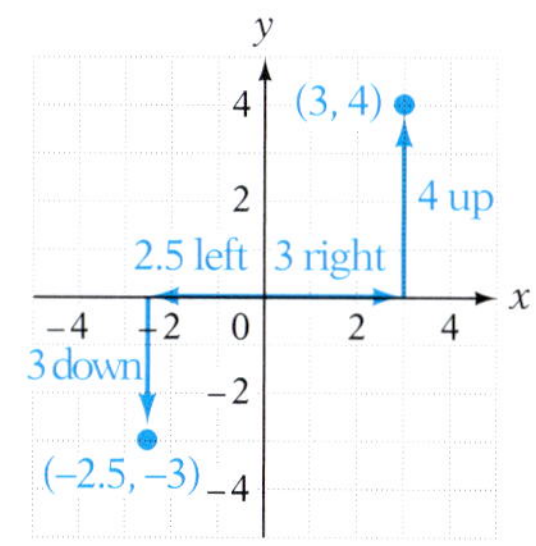

Take Note

The *ordered pair* is an important concept. Remember: There are two numbers (*pair*), and the *order* in which they are given is important.

The points whose coordinates are $(3, -1)$ and $(-1, 3)$ are graphed at the right. Note that the graphs are in different locations. The *order* of the coordinates of an ordered pair is important.

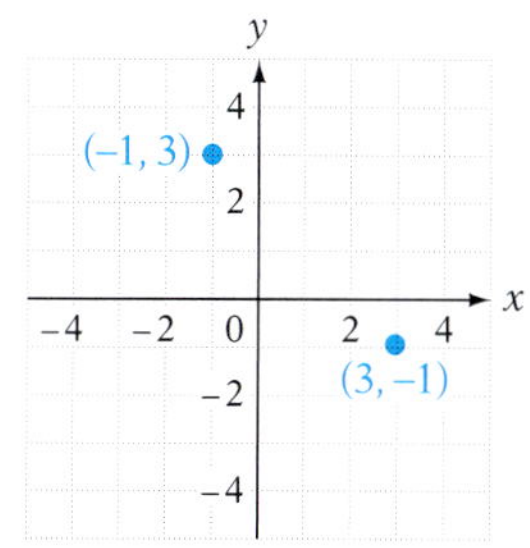

The xy-coordinate system is used to graph equations in *two variables*. Examples of equations in two variables are shown at the right.

$$y = 3x + 7$$
$$y = x^2 - 4x + 3$$
$$x^2 + y^2 = 25$$
$$y = \frac{y}{y^2 + 4}$$

A **solution of an equation in two variables** is an ordered pair (x, y) whose coordinates make the equation a true statement.

Is the ordered pair $(-3, 7)$ a solution of the equation $y = -2x + 1$?

Replace x by -3 and replace y by 7. Then simplify.

$$\begin{array}{c|l} y & = -2x + 1 \\ \hline 7 & -2(-3) + 1 \\ 7 & 6 + 1 \end{array}$$
$$7 = 7$$

Compare the results. If the resulting equation is a true statement, the ordered pair is a solution of the equation. If it is not a true statement, the ordered pair is not a solution of the equation.

Yes the ordered pair $(-3, 7)$ is a solution of the equation.

Besides the ordered pair $(-3, 7)$, there are many other ordered-pair solutions of the equation $y = -2x + 1$. For example, $(-5, 11)$, $(0, 1)$, $\left(-\frac{3}{2}, 4\right)$ and $(4, -7)$ are also solutions of the equation.

In general, an equation in two variables has an infinite number of solutions. By choosing any value of x and substituting that value into the equation, we can calculate a corresponding value of y. The resulting ordered-pair solution (x, y) of the equation can be graphed in a rectangular coordinate system.

Graph the solutions (x, y) of $y = x^2 - 1$ when x equals $-2, -1, 0, 1,$ and 2.

Substitute each value of x into the equation and solve for y. Then graph the resulting ordered pairs. It is convenient to record the ordered-pair solutions in a table similar to the one shown below.

x	$y = x^2 - 1$	y	(x, y)
-2	$y = (-2)^2 - 1$	3	$(-2, 3)$
-1	$y = (-1)^2 - 1$	0	$(-1, 0)$
0	$y = 0^2 - 1$	-1	$(0, -1)$
1	$y = 1^2 - 1$	0	$(1, 0)$
2	$y = 2^2 - 1$	3	$(2, 3)$

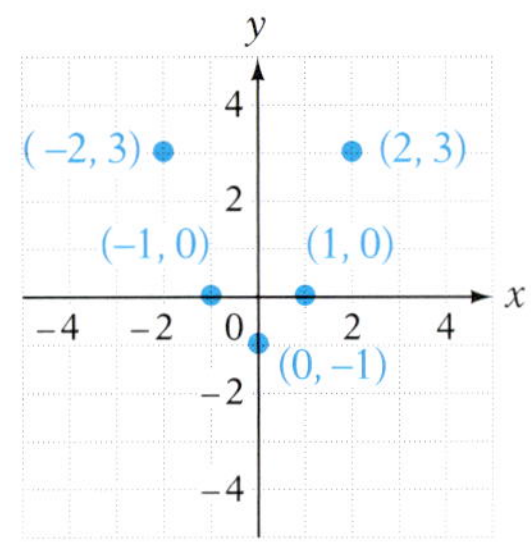

Example 1 Determine the ordered-pair solution of $y = \frac{x}{x - 2}$ that corresponds to $x = 4$.

Solution

$$y = \frac{x}{x-2}$$

$$y = \frac{4}{4-2} \quad \text{• Replace } x \text{ by 4 and solve for } y.$$

$$y = 2$$

The ordered-pair solution is $(4, 2)$.

Problem 1 Determine the ordered-pair solution of $y = \frac{3x}{x + 1}$ that corresponds to $x = -2$.

Solution See page S10.

Example 2 Graph the ordered-pair solutions of $y = x^2 - x$ when $x = -1$, 0, 1, and 2.

Solution

x	y
-1	2
0	0
1	0
2	2

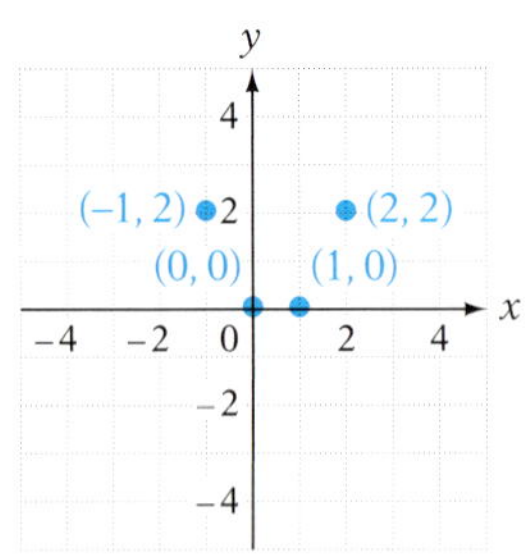

Problem 2 Graph the ordered-pair solutions of $y = |x + 1|$ when $x = -3, -2, -1, 0,$ and 1.

Solution See page S10.

2 Find the length and midpoint of a line segment

VIDEO & DVD CD TUTOR WEB SSM

Point of Interest

Although Pythagoras (c. 550 B.C.) is given credit for this theorem, a clay tablet called Plimpton 322 shows that the Babylonians knew the theorem 1000 years before Pythagoras!

The distance between two points in an xy-coordinate system can be calculated by using the Pythagorean Theorem.

Pythagorean Theorem

If a and b are the lengths of the legs of a right triangle and c is the length of the hypotenuse, then $a^2 + b^2 = c^2$.

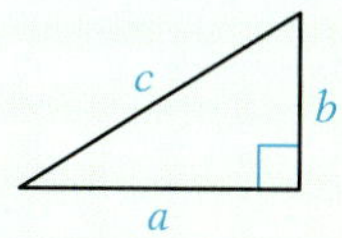

Consider the two points and the right triangle shown at the right. The vertical distance between $P_1(x_1, y_1)$ and $P_2(x_2, y_2)$ is $|y_2 - y_1|$. The horizontal distance between the points $P_1(x_1, y_1)$ and $P_2(x_2, y_2)$ is $|x_2 - x_1|$.

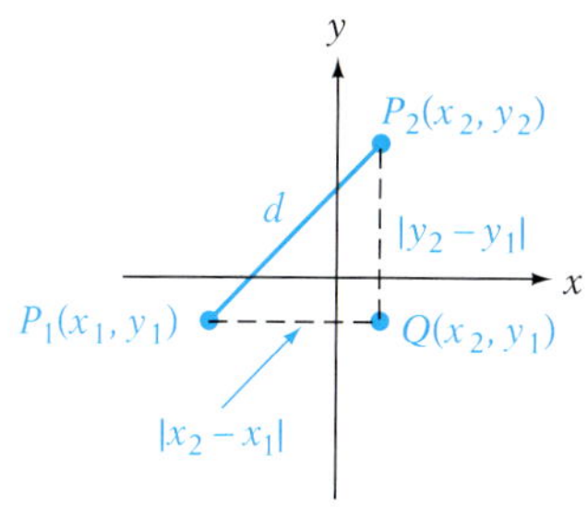

The quantity d^2 is calculated by applying the Pythagorean Theorem to the right triangle.

$$d^2 = |x_2 - x_1|^2 + |y_2 - y_1|^2$$

Because the square of a number is always nonnegative, the absolute value signs are not necessary.

$$d^2 = (x_2 - x_1)^2 + (y_2 - y_1)^2$$

The distance d is the square root of d^2.

$$d = \sqrt{(x_2 - x_1)^2 + (y_2 - y_1)^2}$$

The Distance Formula

If $P_1(x_1, y_1)$ and $P_2(x_2, y_2)$ are two points in the plane, then the distance d between the two points is given by

$$d = \sqrt{(x_2 - x_1)^2 + (y_2 - y_1)^2}$$

Example 3 Find the exact distance between the points whose coordinates are $(-3, 2)$ and $(4, -1)$.

Solution

$$d = \sqrt{(x_2 - x_1)^2 + (y_2 - y_1)^2}$$ • Use the distance formula.

$$= \sqrt{[4 - (-3)]^2 + (-1 - 2)^2}$$ • $(x_1, y_1) = (-3, 2)$ and $(x_2, y_2) = (4, -1)$.

$$= \sqrt{7^2 + (-3)^2} = \sqrt{49 + 9}$$
$$= \sqrt{58}$$

The distance between the points is $\sqrt{58}$.

Problem 3 Find the exact distance between the points whose coordinates are (5, −2) and (−4, 3).

Solution See page S10.

The **midpoint of a line segment** is equidistant from its endpoints. The coordinates of the midpoint of the line segment P_1P_2 are (x_m, y_m). The intersection of the horizontal line segment through P_1 and the vertical line segment through P_2 is Q, with coordinates (x_2, y_1).

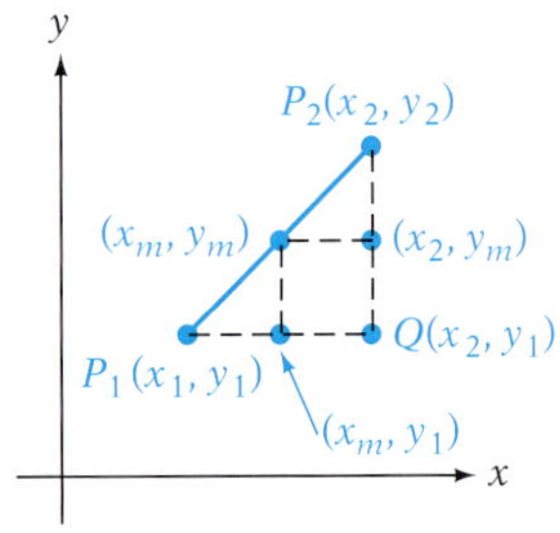

The x-coordinate x_m of the midpoint of the line segment P_1P_2 is the same as the x-coordinate of the midpoint of the line segment P_1Q. It is the average of the x-coordinates of the points P_1 and P_2.

$$x_m = \frac{x_1 + x_2}{2}$$

Similarly, the y-coordinate y_m of the midpoint of the line segment P_1P_2 is the same as the y-coordinate of the midpoint of the line segment P_2Q. It is the average of the y-coordinates of the points P_1 and P_2.

$$y_m = \frac{y_1 + y_2}{2}$$

The Midpoint Formula

If $P_1(x_1, y_1)$ and $P_2(x_2, y_2)$ are the endpoints of a line segment, then the coordinates of the midpoint (x_m, y_m) of the line segment are given by

$$x_m = \frac{x_1 + x_2}{2} \quad \text{and} \quad y_m = \frac{y_1 + y_2}{2}$$

Example 4 Find the coordinates of the midpoint of the line segment with endpoints (−5, 4) and (−3, 7).

Solution

$$x_m = \frac{x_1 + x_2}{2} \qquad y_m = \frac{y_1 + y_2}{2}$$

• **Use the midpoint formula.**

$$= \frac{-5 + (-3)}{2} \qquad = \frac{4 + 7}{2}$$

• **$(x_1, y_1) = (-5, 4)$ and $(x_2, y_2) = (-3, 7)$.**

$$= -4 \qquad = \frac{11}{2}$$

The midpoint is $\left(-4, \frac{11}{2}\right)$.

Problem 4 Find the coordinates of the midpoint of the line segment with endpoints $(-3, -5)$ and $(-2, 3)$.

Solution See page S10.

3 Graph a scatter diagram

VIDEO & DVD CD TUTOR WEB SSM

Discovering a relationship between two variables is an important task in the study of mathematics. These relationships occur in many forms and in a wide variety of applications. Here are some examples.

- A botanist wants to know the relationship between the number of bushels of wheat yielded per acre and the amount of watering per acre.
- An environmental scientist wants to know the relationship between the incidence of skin cancer and the amount of ozone in the atmosphere.
- A business analyst wants to know the relationship between the price of a product and the number of products that are sold at that price.

A researcher may investigate the relationship between two variables by means of *regression analysis,* which is a branch of statistics. The study of the relationship between two variables may begin with a **scatter diagram,** which is a graph of the ordered pairs of the known data.

The following table shows data collected by a university registrar comparing the grade-point averages (GPAs) of graduating high school seniors and their scores on a national test.

GPA, x	3.25	3.00	3.00	3.50	3.50	2.75	2.50	2.50	2.00	2.00	1.50
Score, y	1200	1200	1000	1500	1100	1000	1000	900	800	900	700

The scatter diagram for these data is shown at the right. Each ordered pair represents the GPA and the test score for a student. For example, the ordered pair (2.75, 1000) indicates that one student with a GPA of 2.75 had a test score of 1000.

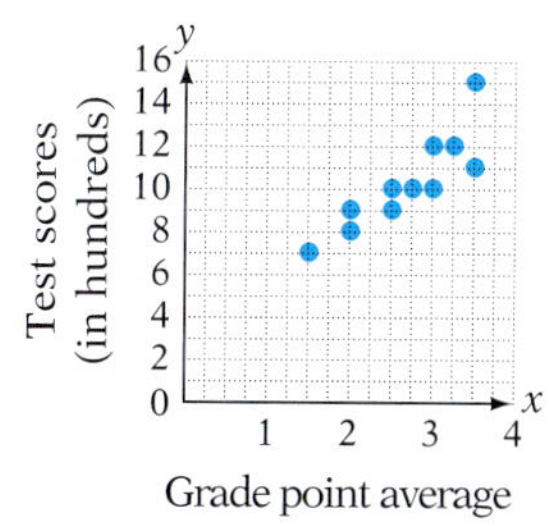

Example 5 The scatter diagram on the next page shows the amount of money, in billions of dollars, that venture capitalists invested in technology-based companies in the first quarter of each year from 1996 to 2000.

A. How much was invested in technology-based companies in the first quarter of 1998?

B. In the first quarter of which year was the amount invested greater than \$10 billion?

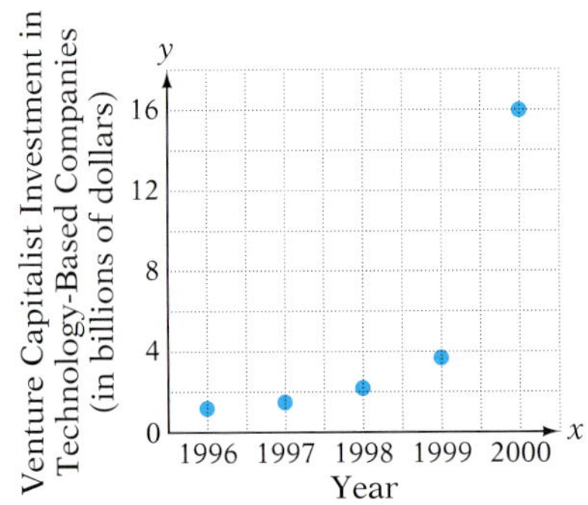

Source: PricewaterhouseCoopers

Strategy **A.** Read the graph. Find 1998 on the horizontal axis. Follow that mark up vertically to determine the amount invested in technology-based companies in the first quarter of that year.

B. Find the horizontal line that represents $10 billion. Determine the year for which the amount invested in the first quarter is above that line.

Solution **A.** In the first quarter of 1998, venture capitalists invested $2 billion in technology-based companies.

B. The amount invested in the first quarter was greater than $10 billion in 2000.

Problem 5 The table below shows the estimated number of adults, in millions, age 65 and older who will have some form of vision loss. Graph the scatter diagram for these data.

Year, x	2000	2010	2020	2030
Adults age 65 and older with vision loss (in millions), y	7.3	8.3	11.2	15.0

Source: Lighthouse National Survey on Vision Loss

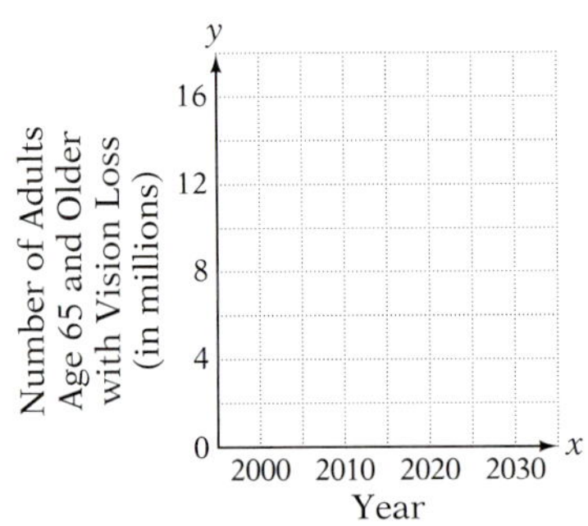

Solution See page S11.

4 Average rate of change

VIDEO & DVD CD TUTOR WEB SSM

Population of California	
Year	Population (in millions)
1850	0.1
1860	0.4
1870	0.6
1880	0.9
1890	1.2
1900	1.5
1910	2.4
1920	3.4
1930	5.7
1940	6.9
1950	10.6
1960	15.7
1970	20.0
1980	23.7
1990	29.8
2000	33.9

The table at the left shows the population of California each decade from 1850 to 2000. (*Source:* U.S. Bureau of the Census). These data are graphed in the scatter diagram below.

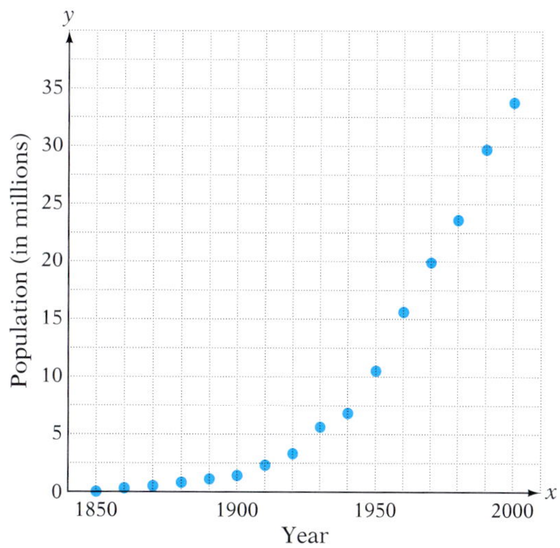

One way to describe how the population of California has changed over time is to calculate the change in population for a given time period. For example, we can find the change in population from 1990 to 2000.

$$\begin{aligned}\text{Population in 2000} - \text{population in 1990} &= 33.9 \text{ million} - 29.8 \text{ million}\\ &= 4.1 \text{ million}\end{aligned}$$

The amount of change in the population, 4.1 million, does not describe how rapid the growth in population was. To describe the rate of growth, or the *rate of change,* we must consider the number of years over which that growth took place. To determine how rapidly the population grew from 1990 to 2000, divide the change in population by the number of years during which that growth took place.

$$\begin{aligned}\frac{\text{Change in population}}{\text{Change in years}} &= \frac{\text{population in 2000} - \text{population in 1990}}{2000 - 1990}\\ &= \frac{33.9 - 29.8}{2000 - 1990}\\ &= \frac{4.1}{10} = 0.41\end{aligned}$$

The average rate of change was 0.41 million, or 410,000, people per year. This means that *on average,* from 1990 to 2000, the population of California increased 410,000 people per year.

Consider the two points on the graph that correspond to these years.

$$(1990, 29.8) \quad \text{and} \quad (2000, 33.9)$$

Note that **the average rate of change is calculated by dividing the difference between the y values of the two points by the difference between the x values of the two points.**

Average Rate of Change

The average rate of change of y with respect to x is $\frac{\text{change in } y}{\text{change in } x}$.

To calculate the average rate of change, think

$$\frac{y \text{ value of second point} - y \text{ value of first point}}{x \text{ value of second point} - x \text{ value of first point}}$$

Find the average annual rate of change in the population of California from 1980 to 1990.

In 1980, the population was 23.7 million: (1980, 23.7)
In 1990, the population was 29.8 million: (1990, 29.8)

$$\begin{aligned}\text{Average rate of change} &= \frac{\text{change in } y}{\text{change in } x}\\ &= \frac{\text{population in 1990} - \text{population in 1980}}{1990 - 1980}\\ &= \frac{29.8 - 23.7}{1990 - 1980}\\ &= \frac{6.1}{10} = 0.61\end{aligned}$$

The average rate of change in the population was 0.61 million, or 610,000, people per year.

Take Note

The rate of change in the population of California is not constant. It varies for different time periods.

From 1990 to 2000, the average rate of change was 410,000 people per year, whereas from 1980 to 1990, the average rate of change was 610,000 people per year. Therefore, the average rate of growth from 1980 to 1990 was greater than the average rate of growth from 1990 to 2000.

Example 6 Find the average annual rate of change in California's population from 1850 to 1950.

Solution In 1850, the population was 0.1 million: (1850, 0.1)
In 1950, the population was 10.6 million: (1950, 10.6)

$$\begin{aligned}\text{Average rate of change} &= \frac{\text{change in } y}{\text{change in } x}\\ &= \frac{\text{population in 1950} - \text{population in 1850}}{1950 - 1850}\\ &= \frac{10.6 - 0.1}{1950 - 1850}\\ &= \frac{10.5}{100} = 0.105\end{aligned}$$

The average rate of change in the population was 105,000 people per year.

Note that if the ordinates and abscissas are subtracted in the reverse order, the result is the same.

$$\text{Average rate of change} = \frac{\text{population in 1850} - \text{population in 1950}}{1850 - 1950}$$

$$= \frac{0.1 - 10.6}{1850 - 1950} = \frac{-10.5}{-100} = 0.105$$

Problem 6 Find the average annual rate of change in California's population from 1900 to 2000.

Solution See page S11.

The table below shows the number of women competing in triathlons in the United States in 1990, 1995, and 2000. (*Source:* USA Triathlon) Use these data for Example 7 and Problem 7.

Year	1990	1995	2000
Women competing in triathlons	26,000	39,100	72,000

Example 7 Find the average annual rate of change in the number of women competing in triathlons in the United States between 1990 and 2000.

Solution In 1990, the number of women competing was 26,000: (1990, 26,000)

In 2000, the number of women competing was 72,000: (2000, 72,000)

$$\text{Average rate of change} = \frac{\text{change in } y}{\text{change in } x}$$

$$= \frac{\text{women competing in 2000} - \text{women competing in 1990}}{2000 - 1990}$$

$$= \frac{72{,}000 - 26{,}000}{2000 - 1990}$$

$$= \frac{46{,}000}{10} = 4600$$

The average rate of change was 4600 women per year.

Problem 7 **A.** Find the average annual rate of change in the number of women competing in triathlons in the United States between 1990 and 1995.

B. Was this more than or less than the average annual rate of change in the number of women competing in triathlons in the United States between 1995 and 2000?

Solution See page S11.

3.1 Concept Review

Determine whether the following statements are always true, sometimes true, or never true.

1. An equation in two variables has an infinite number of solutions.
2. A rectangular coordinate system has two real number axes.
3. In the graph of an ordered pair, the first number is the y-coordinate and the second number is the x-coordinate.
4. The solution of an equation with two variables is an ordered pair.
5. The point $(-2, -4)$ is in the fourth quadrant.
6. In quadrant III, the y-coordinate of the point is negative.

3.1 Exercises

1 **1.** Describe a rectangular coordinate system.

2. Discuss the sign of a and the sign of b as they pertain to the quadrant in which the ordered pair (a, b) is graphed.

3. Graph the ordered pairs $(0, -1)$, $(2, 0)$, $(3, 2)$, and $(-1, 4)$.

4. Graph the ordered pairs $(-1, -3)$, $(0, -4)$, $(0, 4)$, and $(3, -2)$.

5. Find the coordinates of each point.

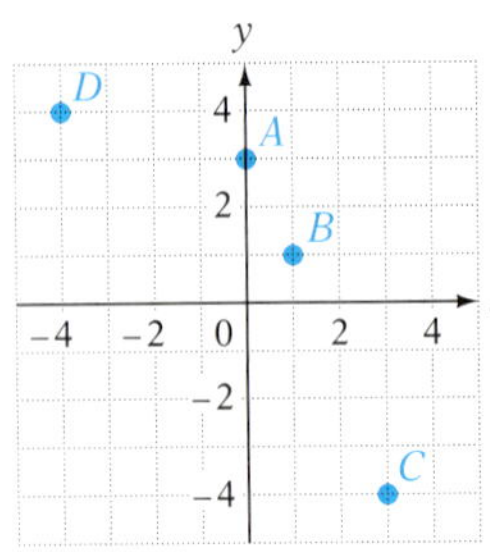

6. Find the coordinates of each point.

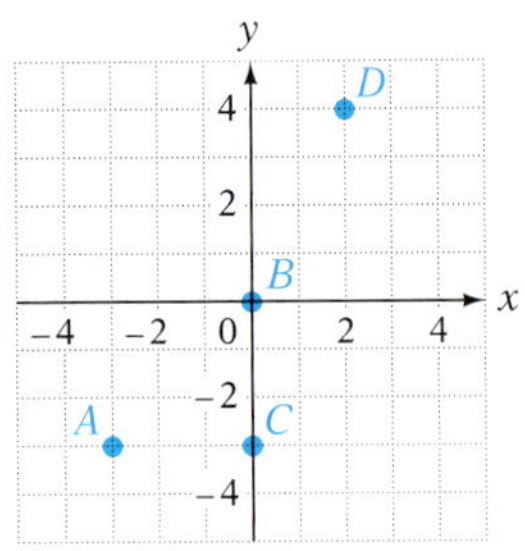

7. Draw a line through all points with an abscissa of 2.

8. Draw a line through all points with an abscissa of -3.

9. Draw a line through all points with an ordinate of -3.

10. Draw a line through all points with an ordinate of 4.

11. Graph the ordered-pair solutions of $y = x^2$ when $x = -2, -1, 0, 1$, and 2.

12. Graph the ordered-pair solutions of $y = -x^2 + 1$ when $x = -2, -1, 0, 1$, and 2.

13. Graph the ordered-pair solutions of $y = |x + 1|$ when $x = -5, -3, 0, 3$, and 5.

14. Graph the ordered-pair solutions of $y = -2|x|$ when $x = -3, -1, 0, 1$, and 3.

15. Graph the ordered-pair solutions of $y = -x^2 + 2$ when $x = -2, -1, 0, 1$, and 2.

16. Graph the ordered-pair solutions of $y = -x^2 + 4$ when $x = -3, -1, 0, 1$, and 3.

17. Graph the ordered-pair solutions of $y = x^3 - 2$ when $x = -1, 0, 1$, and 2.

18. Graph the ordered-pair solutions of $y = -x^3 + 1$ when $x = -1, 0, 1$, and $\frac{3}{2}$.

2 19. Explain how to find the distance between two points in the plane.

20. What is the midpoint of a line segment and how is it calculated?

Find the length and midpoint of the line segment between the given points.

21. $P_1(3, 5)$ and $P_2(4, 1)$

22. $P_1(-2, 3)$ and $P_2(5, -1)$

23. $P_1(0, 3)$ and $P_2(-2, 4)$

24. $P_1(6, -1)$ and $P_2(-3, -2)$

25. $P_1(-3, -5)$ and $P_2(2, -4)$

26. $P_1(-7, -5)$ and $P_2(-2, -1)$

27. $P_1(5, -2)$ and $P_2(-1, 5)$

28. $P_1(3, -5)$ and $P_2(6, 0)$

29. $P_1(5, -5)$ and $P_2(2, -5)$

3 Solve.

30. **Chemistry** The amount of a substance that can be dissolved in a fixed amount of water usually increases as the temperature of the water increases. Cerium selenate, however, does not behave in this manner. The graph at the right shows the number of grams of cerium selenate that will dissolve in 100 mg of water for various temperatures, in degrees Celsius.

a. Determine the temperature at which 25 g of cerium selenate will dissolve.

b. Determine the number of grams of cerium selenate that will dissolve when the temperature is 80°C.

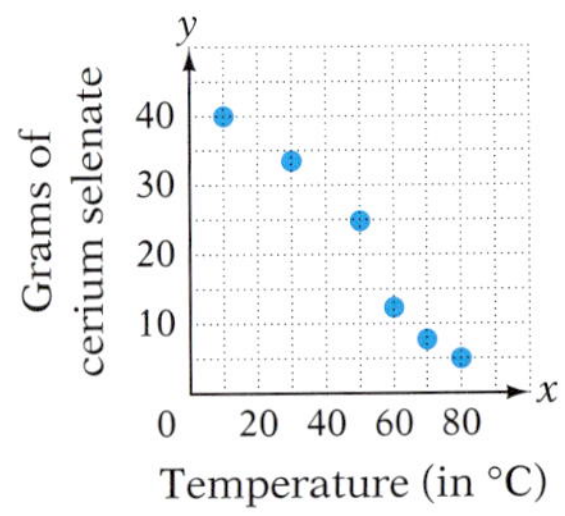

31. **Nutrition** Sodium content in foods is a measure of the salt content in the food. The scatter diagram at the right shows the approximate amount of sodium and the number of calories in various sandwiches offered at a McDonald's restaurant. (*Source:* Gourmet Connection website/Convence Labs)

a. How many calories are in a hamburger?

b. How many milligrams of sodium are in a Big Mac?

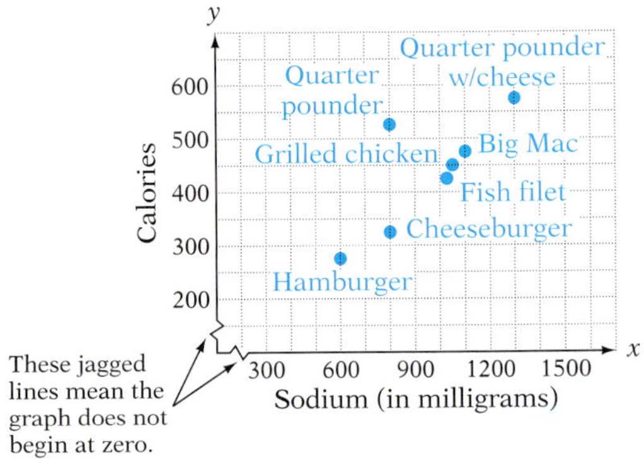

32. **Automotive Technology** When water freezes, it expands. This expansion is a potential problem for cars, because the freezing water can cause the engine block of a car to crack. Ethylene glycol is mixed with water to produce an antifreeze that helps prevent the water in the car from freezing. The scatter diagram at the right shows the changes in the freezing point of a solution of water and ethylene glycol that occur as the percent of ethylene glycol is increased.

a. At what temperature will a 45% mixture of ethylene glycol and water freeze?

b. What percent of ethylene glycol is required to protect the car to temperatures as low as −30°F?

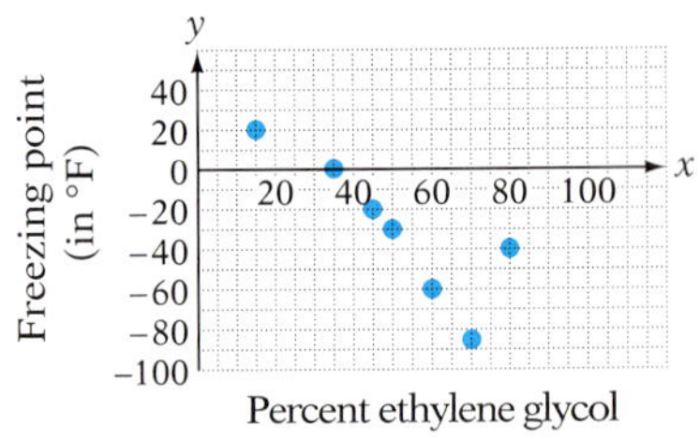

33. **Agriculture** The scatter diagram to the right shows the number of bushels of fresh apples that were in storage in the United States on May 1 of each year from 1997 to 2000. A bushel is about 42 pounds.

a. How many bushels of apples were in storage in 1997?

b. In which year was the number of bushels first greater than 12 million?

c. Looking at the trend of the data, would you expect the number of bushels of apples in storage in 2001 to be greater than or less than the number of bushels of apples in storage in 2000?

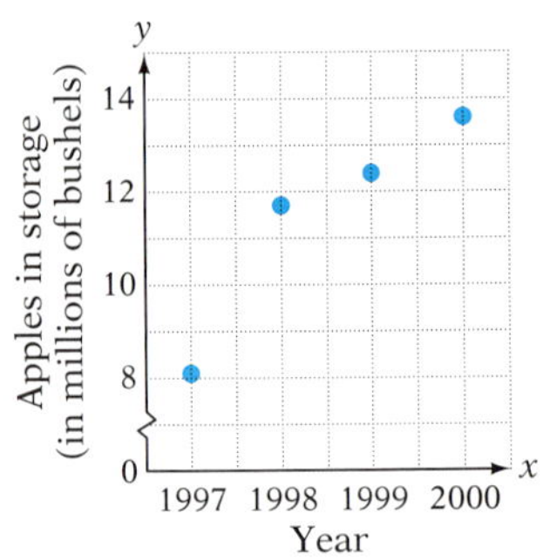

Source: U.S. Apple Association

34. **Wealth** The table shows the number of U.S. households with a net worth of \$1 million or more every three years from 1989 to 2001. Also shown is the millionaire households as a percent of all households in the United States. (*Source:* Edward Wolff, professor of economics, New York University; Federal Reserve) Graph the scatter diagram for these data.

Millionaire households (in millions), x	3.0	3.1	3.4	4.1	5.0
Percent of all households, y	3.2	3.2	3.4	4.0	4.7

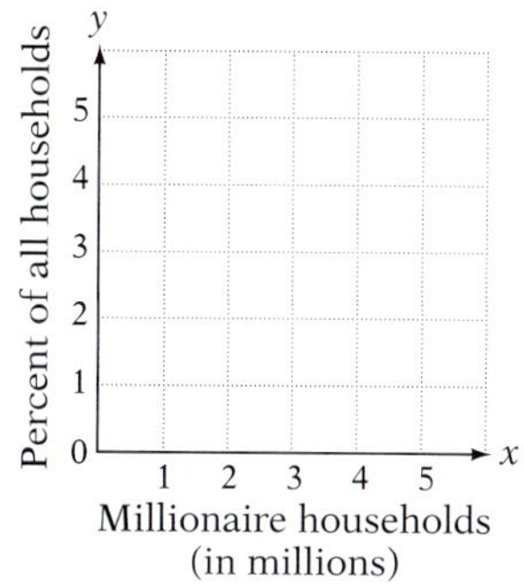

35. **Telecommunications** As shown in the table below, the percent of households with telephones has increased over the years. Draw a scatter diagram for these data. (*Source:* U.S. Census Bureau/Federal Communications Commission)

Year, x	'20	'30	'40	'50	'60	'70	'80	'90	'00
Percent, y	35	41	37	62	78	90	93	95	97

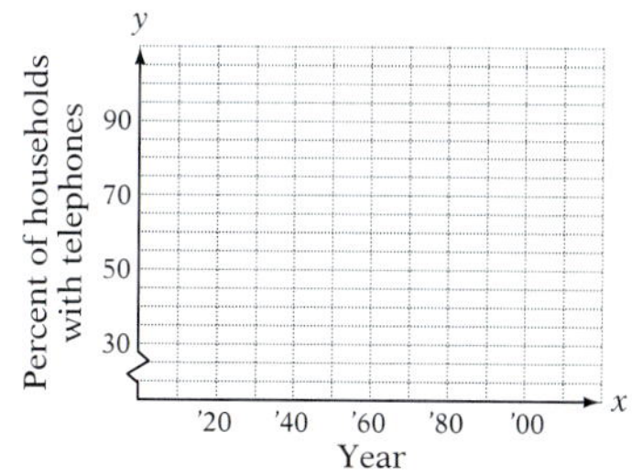

4 **36.** **Temperature** On November 5 in midstate New Hampshire, the temperature at 6 A.M. was 34°F. At 2 P.M. on the same day, the temperature was 58°F. Find the average rate of change in temperature per hour.

37. **Demography** The table at the right shows the population of Texas each decade from 1850 to 2000. (*Source:* U.S. Bureau of the Census)

a. Find the average annual rate of change in the population of Texas from 1900 to 2000.

b. Was the average annual rate of change in the population from 1900 to 1950 greater than or less than the average annual rate of change in the population from 1950 to 2000?

c. Was the average annual rate of change in the population of Texas from 1980 to 1990 greater than or less than the average annual rate of change in the population of California from 1980 to 1990? (See page 132 for the average annual rate of change in the population of California from 1980 to 1990.)

d. During which decade was the average annual rate of change the least? What was the average annual rate of change during the decade?

Population of Texas

Year	*Population (in millions)*
1850	0.2
1860	0.6
1870	0.8
1880	1.6
1890	2.2
1900	3.0
1910	3.9
1920	4.7
1930	5.8
1940	6.4
1950	7.7
1960	9.6
1970	11.2
1980	14.2
1990	17.0
2000	20.9

38. **Cremation** In the United States in 1990, there were 2.15 million deaths and 367,975 cremations. In 1999, there were 2.35 million deaths and 595,617 cremations. (*Source:* Cremation Association of North America)

a. In the 1990s, what was the average annual rate of change in the number of deaths in the United States? Round to the nearest thousand.

b. In the 1990s, what was the average annual rate of change in the number of cremations in the United States? Round to the nearest thousand.

c. In the 1990s, which increased at a greater rate, the number of deaths or the number of cremations?

39. **Traffic Fatalities** The number of pedestrians killed by motor vehicles in 1990 was 6482; in 2000 it was 4727. (*Source:* National Highway Transportation Safety Administration)

a. Find the average annual rate of change in the number of pedestrian fatalities from 1990 to 2000. (*Note:* A negative average rate of change denotes a decrease.)

b. Why can the answer to part a be considered encouraging?

40. **Energy Prices** The table at the right shows the average price per gallon of gasoline in the United States each month from January 2000 to April 2001. (*Source:* American Petroleum Institute)

a. Find the average rate of change per month in the price of gasoline from January 2000 to April 2001. Round to the nearest tenth of a cent.

b. Which was greater, the average rate of change per month from January to April 2000 or the average rate of change per month from January to April 2001? How much greater?

Month	*Price per Gallon*
January 2000	\$1.30
February	\$1.37
March	\$1.54
April	\$1.51
May	\$1.50
June	\$1.62
July	\$1.59
August	\$1.51
September	\$1.58
October	\$1.56
November	\$1.56
December	\$1.49
January 2001	\$1.47
February	\$1.48
March	\$1.45
April	\$1.53

41. **Trademarks** The table below shows the number of applications for trademark registration in the United States in 1990, 1995, and 2000. (*Source:* International Trademark Association) One reason for the large number of applications is the increase in popularity of the Internet and the desire to register domain names and web sites.

a. Find the average annual rate of change in the number of applications from 1990 to 2000.

b. How much greater was the average annual rate of change in the number of applications from 1995 to 2000 than from 1990 to 1995?

Year	1990	1995	2000
Applications (in thousands)	120	175	375

3.1 Applying Concepts

42. Graph the ordered pairs $\left(x, \frac{1}{x}\right)$, where $x \in \left\{-2, -1, -\frac{1}{2}, -\frac{1}{3}, \frac{1}{3}, \frac{1}{2}, 1, 2\right\}$.

43. Graph the ordered pairs (x, x^2), where $x \in \{-2, -1, 0, 1, 2\}$.

44. Draw a line passing through every point whose ordinate is the additive inverse of its abscissa.

45. Draw a line passing through every point whose abscissa equals its ordinate.

46. Describe the graph of all the ordered pairs (x, y) that are 5 units from the origin.

47. Consider two distinct fixed points in the plane. Describe the graph of all the points (x, y) that are equidistant from these fixed points.

3.2 Introduction to Functions

1 Evaluate a function

VIDEO & DVD CD TUTOR WEB SSM

In mathematics and its application, there are many times when it is necessary to investigate a relationship between two quantities. Here is a financial application: Consider a person who is planning to finance the purchase of a car. If the current interest rate for a five-year loan is 9%, the equation that describes the relationship between the amount that is borrowed B and the monthly payment P is $P = 0.020758B$.

For each amount B the purchaser may borrow, there is a certain monthly payment P. The relationship between the amount borrowed and the payment can be recorded as ordered pairs, where the first coordinate is the amount borrowed and the second coordinate is the monthly payment. Some of these ordered pairs are shown at the right.

$$0.020758B = P$$

(5000, 103.79)
(6000, 124.55)
(7000, 145.31)
(8000, 166.07)

A relationship between two quantities is not always given by an equation. The table at the right describes a grading scale that defines a relationship between a score on a test and a letter grade. For each score, the table assigns only one letter grade. The ordered pair (84, B) indicates that a score of 84 receives a letter grade of B.

Score	Grade
90–100	A
80–89	B
70–79	C
60–69	D
0–59	F

The graph at the right also shows a relationship between two quantities. It is a graph of the viscosity V of SAE 40 motor oil at various temperatures T. Ordered pairs can be approximated from the graph. The ordered pair (120, 250) indicates that the viscosity of the oil at 120°F is 250 units.

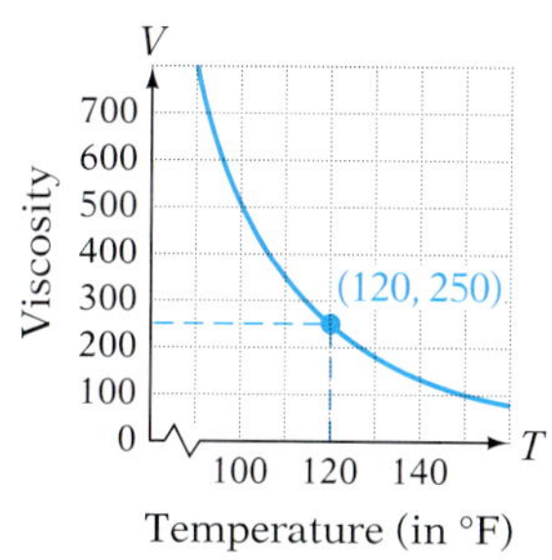

In each of these examples, there is a rule (an equation, a table, or a graph) that determines a certain set of ordered pairs.

Definition of Relation

A **relation** is a set of ordered pairs.

Here are some of the ordered pairs for the relations just described.

Relation	Some of the Ordered Pairs of the Relation
Car payment	(2500, 51.90), (3750, 77.85), (4396, 91.26)
Grading scale	(78, C), (98, A), (70, C), (81, B), (94, A)
Oil viscosity	(100, 500), (120, 250), (130, 175), (150,100)

These three relations actually represent a special type of relation called a function. Functions play an important role in mathematics and its application.

Point of Interest

Leonhard Euler (1707–1783) was the first to give a definition of function: "A function of a variable quantity is an analytic expression composed in any way whatsoever of the variable quantity and numbers or constant quantities."

Definition of Function

A **function** is a relation in which no two ordered pairs have the same first coordinate and different second coordinates.

The **domain** of a function is the set of the first coordinates of all the ordered pairs of the function. The **range** is the set of the second coordinates of all the ordered pairs of the function.

For the function defined by the ordered pairs

$$\{(2, 3), (4, 5), (6, 7), (8, 9)\}$$

the domain is $\{2, 4, 6, 8\}$ and the range is $\{3, 5, 7, 9\}$.

Example 1 Find the domain and range of the function $\{(2, 3), (4, 6), (6, 8), (10, 6)\}$.

Solution The domain is $\{2, 4, 6, 10\}$.
- The domain of the function is the set of the first components in the ordered pairs.

The range is $\{3, 6, 8\}$.
- The range of the function is the set of the second components in the ordered pairs.

Problem 1 Find the domain and range of the function $\{(-1, 5), (3, 5), (4, 5), (6, 5)\}$.

Solution See page S11.

For each element of the domain of a function there is a corresponding element in the range of the function. A possible diagram for the function of Example 1 is

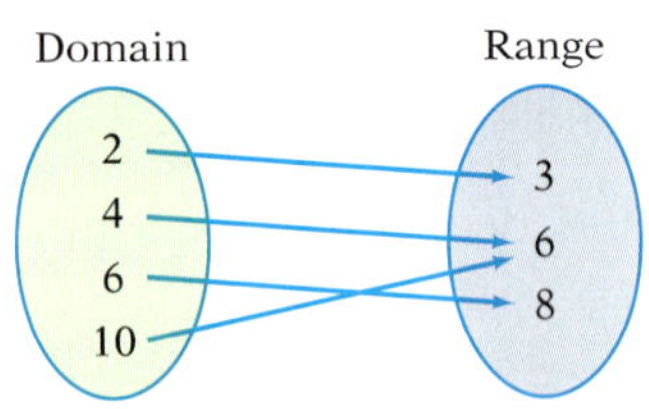

$\{(2, 3), (4, 6), (6, 8), (10, 6)\}$

Functions defined by tables or graphs, such as those described at the beginning of this section, have important applications. However, a major focus of this text is functions defined by equations in two variables.

The **square function,** which pairs each real number with its square, can be defined by the equation

$$y = x^2$$

This equation states that for a given value of x in the domain, the value of y in the range is the square of x. For instance, if $x = 6$, then $y = 36$ and if $x = -7$, then $y = 49$. Because the value of y *depends* on the value of x, y is called the **dependent variable** and x is called the **independent variable.**

> **Take Note**
> A pictorial representation of the square function is shown at the right. The function acts as a machine that changes a number from the domain into the square of that number.

A function can be thought of as a rule that pairs one number with another number. For instance, the square function pairs a number with its square. The ordered pairs for the values shown at the right are $(-5, 25)$, $\left(\frac{3}{5}, \frac{9}{25}\right)$, $(0, 0)$, and $(3, 9)$. For this function, the second coordinate is the square of the first coordinate. If we let x represent the first coordinate, then the second coordinate is x^2 and we have the ordered pair (x, x^2).

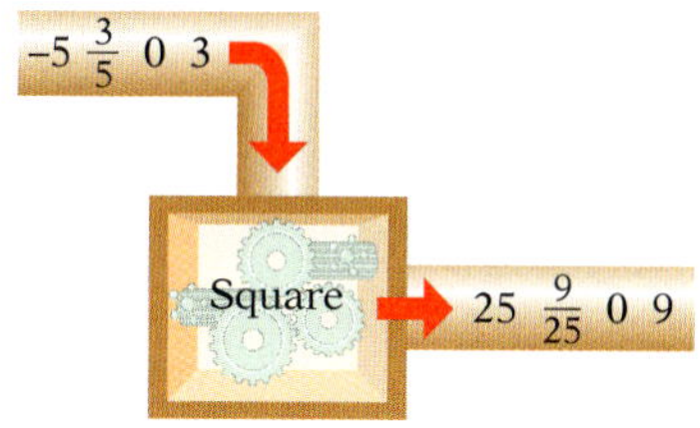

A function cannot have two ordered pairs with *different* second coordinates and the same first coordinate. However, a function can contain ordered pairs with the *same* second coordinate. For instance, the square function has the ordered pairs $(-3, 9)$ and $(3, 9)$; the second coordinates are the same but the first coordinates are different.

The double function pairs a number with twice that number. The ordered pairs for the values shown at the right are $(-5, -10)$, $\left(\frac{3}{5}, \frac{6}{5}\right)$, $(0, 0)$, and $(3, 6)$. For this function, the second coordinate is twice the first coordinate. If we let x represent the first coordinate, then the second coordinate is $2x$ and we have the ordered pair $(x, 2x)$.

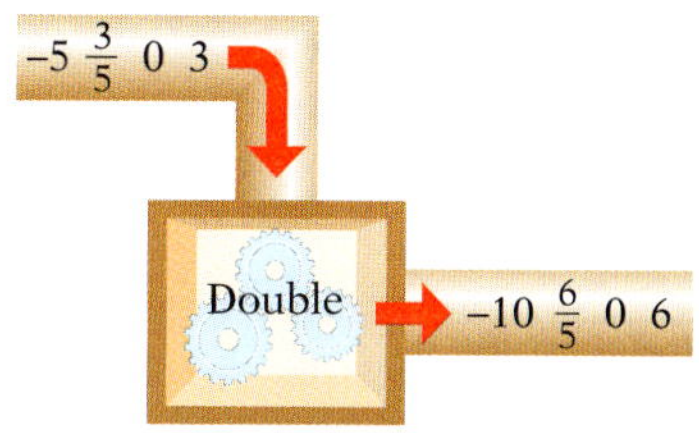

Not every equation in two variables defines a function. For instance, consider

$$y^2 = x^2 + 9$$

This equation does not define a function. Because

$$5^2 = 4^2 + 9 \qquad \text{and} \qquad (-5)^2 = 4^2 + 9$$

the ordered pairs $(4, 5)$ and $(4, -5)$ are both solutions of the equation. Consequently, there are two ordered pairs that have the same first coordinate, 4, but *different* second coordinates, 5 and -5. Therefore, the equation does not define a function. Other ordered pairs for this equation are $(0, -3)$, $(0, 3)$,

$(\sqrt{7}, -4)$, and $(\sqrt{7}, 4)$. A graphical representation of these ordered pairs is shown below.

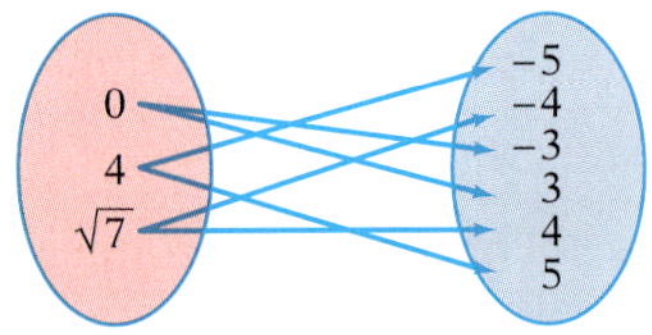

Note from this graphical representation that an element from the domain has two arrows to different elements in the range. Any time this occurs, the situation does not represent a function. However, this diagram does represent a relation. The relation for the values shown is $\{(0, -3), (0, 3), (4, -5), (4, 5), (\sqrt{7}, -4), (\sqrt{7}, 4)\}$.

The phrase "y is a function of x," or a similar phrase with different variables, is used to describe those equations in two variables that define functions. To emphasize that the equation represents a function, **functional notation** is used.

Just as x is commonly used as a variable to represent a number, the letter f is commonly used to name a function. The square function is written in functional notation as follows:

> **Take Note**
>
> The dependent variable y and $f(x)$ can be used interchangeably.

This is the value of the function. It is the number that is paired with x.

$$f(x) = x^2$$

The name of the function is f.

This is an algebraic expression that defines the relationship between the dependent and independent variables.

The symbol $f(x)$ is read "the *value* of f at x" or "f of x."

It is important to note that $f(x)$ does not mean f times x. The symbol $f(x)$ is the **value of the function** and represents the value of the dependent variable for a given value of the independent variable. We often write $y = f(x)$ to emphasize the relationship between the independent variable, x, and the dependent variable, y. Remember that y and $f(x)$ are different symbols for the same number.

The letters used to represent a function are somewhat arbitrary. All of the following equations represent the same function.

$$\left.\begin{aligned} f(x) &= x^2 \\ s(t) &= t^2 \\ P(v) &= v^2 \end{aligned}\right\} \text{These equations represent the square function.}$$

The process of determining $f(x)$ for a given value of x is called **evaluating the function.** For instance, to evaluate $f(x) = x^2$ when $x = 4$, replace x by 4 and simplify.

$$f(x) = x^2$$
$$f(4) = 4^2 = 16$$

The *value* of the function is 16 when $x = 4$. An ordered pair of the function is (4, 16).

Evaluate $g(t) = 3t^2 - 5t + 1$ when $t = -2$.

Replace t by -2 and then simplify.

$$\begin{aligned} g(t) &= 3t^2 - 5t + 1 \\ g(-2) &= 3(-2)^2 - 5(-2) + 1 \\ &= 3(4) - 5(-2) + 1 \\ &= 12 + 10 + 1 = 23 \end{aligned}$$

When t is -2, the value of the function is 23.
Therefore an ordered pair of the function is $(-2, 23)$.

It is possible to evaluate a function for a variable expression.

Evaluate $P(z) = 3z - 7$ when $z = 3 + h$.

Replace z by $3 + h$, and then simplify.

$$\begin{aligned} P(z) &= 3z - 7 \\ P(3 + h) &= 3(3 + h) - 7 \\ &= 9 + 3h - 7 \\ &= 3h + 2 \end{aligned}$$

When z is $3 + h$, the value of the function is $3h + 2$.
Therefore, an ordered pair of the function is $(3 + h, 3h + 2)$.

Example 2 Given $s(t) = 3t^2 - 4t + 5$, find $s(-4)$.

Solution

$$\begin{aligned} s(t) &= 3t^2 - 4t + 5 \\ s(-4) &= 3(-4)^2 - 4(-4) + 5 \\ &= 3(16) - 4(-4) + 5 \\ &= 48 + 16 + 5 \\ &= 69 \end{aligned}$$

- **Replace t by -4.**
- **Use the Order of Operations Agreement to simplify the numerical expression.**

Problem 2 Evaluate $G(x) = \dfrac{2x}{x + 4}$ when $x = -2$.

Solution See page S11.

There are several ways to use a graphing calculator to evaluate a function. The screens at the right show one way that will work on many calculators. See the Graphing Calculator Appendix for other methods of evaluating a function.

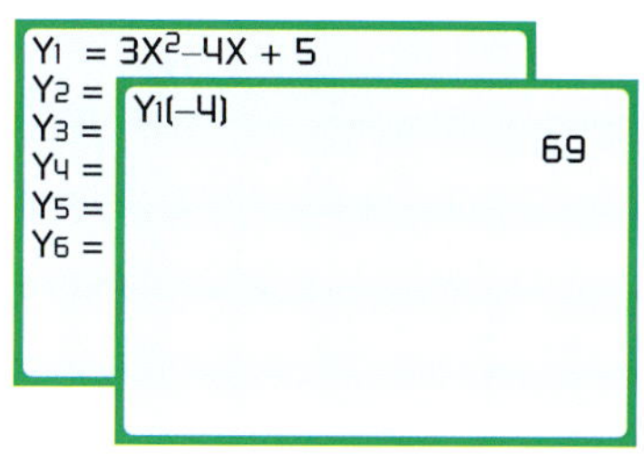

The range of a function contains all the elements that result from applying the function to each element of the domain. If the domain contains an infinite number of elements, then it may be difficult to find the range. However, if the domain has only a finite number of elements, then the range can be found by evaluating the function for each element in the domain.

Example 3 Find the range of $f(x) = x^3 + x$ if the domain is $\{-2, -1, 0, 1, 2\}$.

Solution

$$\begin{aligned} f(x) &= x^3 + x \\ f(-2) &= (-2)^3 + (-2) = -10 \\ f(-1) &= (-1)^3 + (-1) = -2 \\ f(0) &= 0^3 + 0 = 0 \\ f(1) &= 1^3 + 1 = 2 \\ f(2) &= 2^3 + 2 = 10 \end{aligned}$$

• **Replace x by each member of the domain. The range includes the values of $f(-2)$, $f(-1)$, $f(0)$, $f(1)$, and $f(2)$.**

The range is $\{-10, -2, 0, 2, 10\}$.

Problem 3 Find the range of $f(x) = x^2 - x - 2$ if the domain is $\{-3, -2, -1, 0, 1, 2\}$.

Solution See page S11.

The TABLE feature of a calculator can be used to determine range values of a function for selected values in the domain.

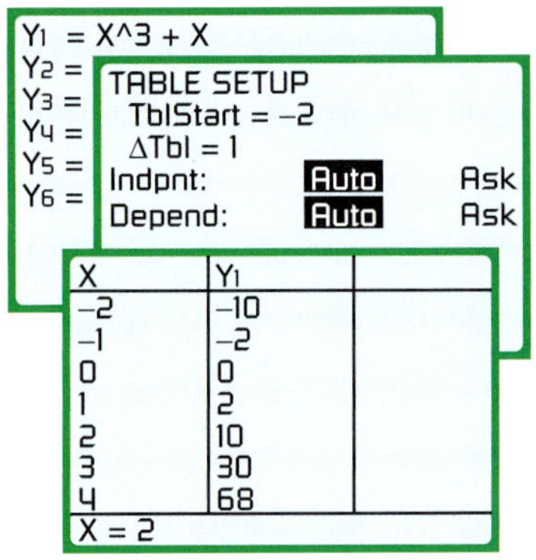

When a function is represented by an equation, the domain of the function is all real numbers for which the value of the function is a real number. For instance:

- The domain of $f(x) = x^2$ is all real numbers, because the square of every real number is a real number.
- The domain of $g(x) = \frac{1}{x-2}$ is all real numbers except 2, because when $x = 2$, $g(2) = \frac{1}{2-2} = \frac{1}{0}$, which is not a real number.

The domain of the grading-scale function is the set of whole numbers from 0 to 100. This is written $\{x \mid 0 \le x \le 100, x \in \text{whole numbers}\}$ in set-builder notation. The range is {A, B, C, D, F}.

Score	Grade
90–100	A
80–89	B
70–79	C
60–69	D
0–59	F

Example 4 What values, if any, are excluded from the domain of $f(x) = 2x^2 - 7x + 1$?

Solution Because the value of $2x^2 - 7x + 1$ is a real number for any value of x, the domain of the function is all real numbers.

No values are excluded from the domain of $f(x) = 2x^2 - 7x + 1$.

Problem 4 What values, if any, are excluded from the domain of $f(x) = \frac{2}{x - 5}$?

Solution See page S11.

3.2 Concept Review

Determine whether the following statements are always true, sometimes true, or never true.

1. A set of ordered pairs is called a function.
2. A function is a relation.
3. The domain of a function is the set of real numbers.
4. Zero is excluded from the domain of a function.
5. $f(a)$ represents a value in the range of a function when a is a value in the domain of the function.
6. Two ordered pairs of a function can have the same first coordinate.

3.2 Exercises

1

1. In your own words, explain what a function is.

2. What is the domain of a function? What is the range of a function?

3. Does the diagram below represent a function? Explain your answer.

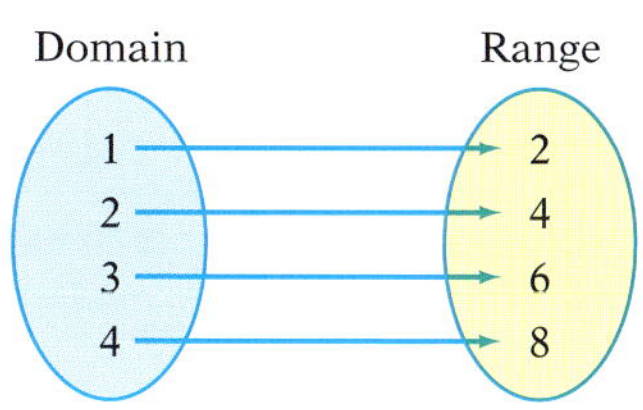

4. Does the diagram below represent a function? Explain your answer.

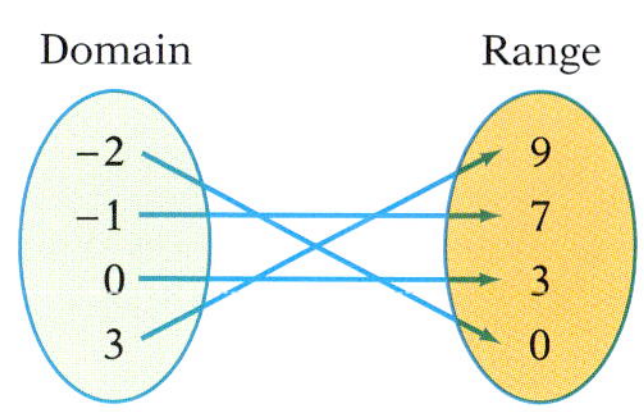

5. Does the diagram below represent a function? Explain your answer.

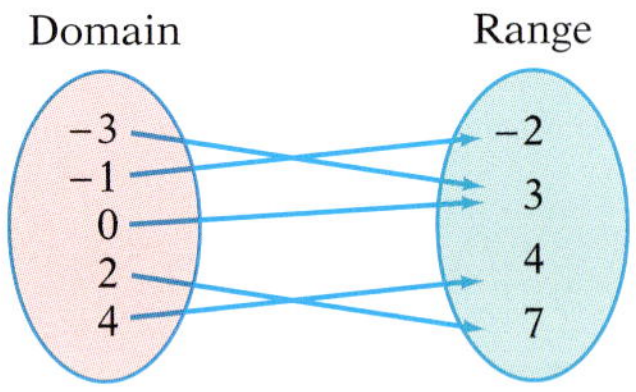

6. Does the diagram below represent a function? Explain your answer.

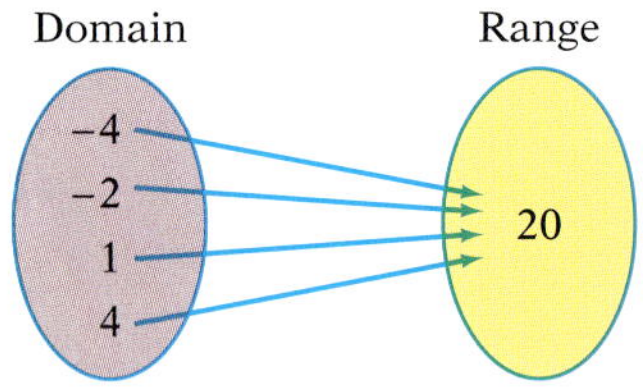

7. Does the diagram below represent a function? Explain your answer.

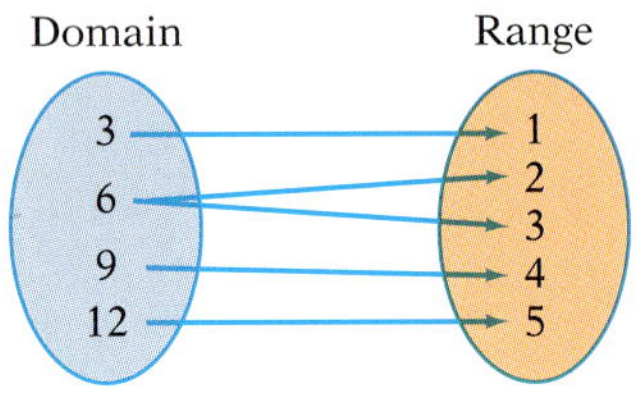

8. Does the diagram below represent a function? Explain your answer.

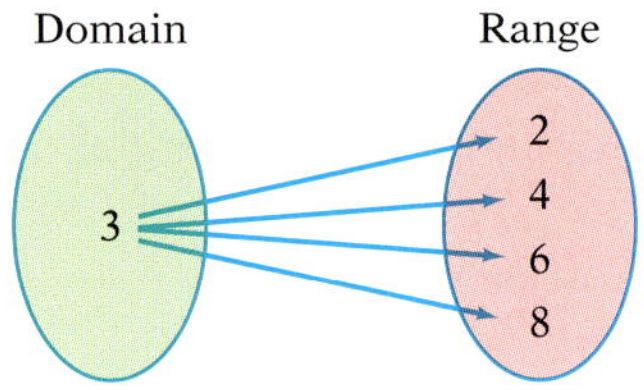

State whether each of the following relations is a function.

9. $\{(0, 0), (2, 4), (3, 6), (4, 8), (5, 10)\}$

10. $\{(1, 3), (3, 5), (5, 7), (7, 9)\}$

11. $\{(-2, -1), (-4, -5), (0, -1), (3, 5)\}$

12. $\{(-3, -1), (-1, -1), (0, 1), (2, 6)\}$

13. $\{(-2, 3), (-1, 3), (0, -3), (1, 3), (2, 3)\}$

14. $\{(0, 0), (1, 0), (2, 0), (3, 0), (4, 0)\}$

15. $\{(1, 1), (4, 2), (9, 3), (1, -1), (4, -2)\}$

16. $\{(3, 1), (3, 2), (3, 3), (3, 4)\}$

Solve.

17. **Shipping** The table at the right shows the cost to send an overnight package using United Parcel Service.
a. Does this table define a function?
b. Given $x = 2.75$ lb, find y.

Weight in pounds (x)	*Cost (y)*
$0 < x \le 1$	\$24.05
$1 < x \le 2$	\$26.83
$2 < x \le 3$	\$29.62
$3 < x \le 4$	\$32.40
$4 < x \le 5$	\$35.18

18. **Shipping** The table at the right shows the cost to send an "Express Mail" package using the U.S. Postal Service.
a. Does this table define a function?
b. Given $x = 0.5$ lb, find y.

Weight in pounds (x)	*Cost (y)*
$0 < x \le 0.5$	\$10.70
$0.5 < x \le 1$	\$14.90
$1 < x \le 2$	\$14.90
$2 < x \le 3$	\$18.10
$3 < x \le 4$	\$21.25

19. What does it mean to evaluate a function? Explain how to evaluate $f(x) = 3x$ when $x = 2$.

20. What is the value of a function?

Given $f(x) = 5x - 4$, evaluate:

21. $f(3)$ **22.** $f(-2)$ **23.** $f(0)$ **24.** $f(-1)$

Given $G(t) = 4 - 3t$, evaluate:

25. $G(0)$ **26.** $G(-3)$ **27.** $G(-2)$ **28.** $G(4)$

Given $q(r) = r^2 - 4$, evaluate:

29. $q(3)$ **30.** $q(4)$ **31.** $q(-2)$ **32.** $q(-5)$

Given $F(x) = x^2 + 3x - 4$, evaluate:

33. $F(4)$ **34.** $F(-4)$ **35.** $F(-3)$ **36.** $F(-6)$

Given $H(p) = \dfrac{3p}{p + 2}$, evaluate:

37. $H(1)$ **38.** $H(-3)$ **39.** $H(t)$ **40.** $H(v)$

Given $s(t) = t^3 - 3t + 4$, evaluate:

41. $s(-1)$ **42.** $s(2)$ **43.** $s(a)$ **44.** $s(w)$

45. **Business** Game Engineering has just completed the programming and testing for a new computer game. The cost to manufacture and package the game depends on the number of units Game Engineering plans to sell. The table at the right shows the cost per game for manufacturing various quantities. Evaluate this function when
a. $x = 7000$ **b.** $x = 20{,}000$

Number of Games Manufactured	*Cost to Manufacture One Game*
$0 < x \le 2500$	\$6.00
$2500 < x \le 5000$	\$5.50
$5000 < x \le 10{,}000$	\$4.75
$10{,}000 < x \le 20{,}000$	\$4.00
$20{,}000 < x \le 40{,}000$	\$3.00

46. **Pricing** Airport administrators have a tendency to price airport parking at a rate that discourages people from using the parking lot for long periods of time. The rate structure for an airport is given in the table at the right. Evaluate this function when
a. $t = 2.5$ h **b.** $t = 7$ h

Hours Parked	*Cost*
$0 < t \le 1$	\$1.00
$1 < t \le 2$	\$3.00
$2 < t \le 4$	\$6.50
$4 < t \le 7$	\$10.00
$7 < t \le 12$	\$14.00

47. **Real Estate** A real estate appraiser charges a fee that depends on the estimated value, V, of the property. A table giving the fees charged for various estimated values of the real estate appears at the right. Evaluate this function when
a. $V = \$5{,}000{,}000$ **b.** $V = \$767{,}000$

Value of Property	*Appraisal Fee*
$V < 100{,}000$	\$350
$100{,}000 \le V < 500{,}000$	\$525
$500{,}000 \le V < 1{,}000{,}000$	\$950
$1{,}000{,}000 \le V < 5{,}000{,}000$	\$2500
$5{,}000{,}000 \le V < 10{,}000{,}000$	\$3000

48. **Shipping** The cost to mail a priority overnight package by Federal Express depends on the weight, w, of the package. A table of the costs for selected weights is given at the right. Evaluate this function when
a. $w = 2$ lb 3 oz **b.** $w = 1.9$ lb

Weight (lb)	*Cost*
$0 < w < 1$	\$24.00
$1 \le w < 2$	\$26.75
$2 \le w < 3$	\$29.25
$3 \le w < 4$	\$32.00
$4 \le w < 5$	\$34.00

Find the domain and range of the function.

49. $\{(1, 1), (2, 4), (3, 7), (4, 10), (5, 13)\}$

50. $\{(2, 6), (4, 18), (6, 38), (8, 66), (10, 102)\}$

51. $\{(0, 1), (2, 2), (4, 3), (6, 4)\}$

52. $\{(0, 1), (1, 2), (4, 3), (9, 4)\}$

53. $\{(1, 0), (3, 0), (5, 0), (7, 0), (9, 0)\}$

54. $\{(-2, -4), (2, 4), (-1, 1), (1, 1), (-3, 9), (3, 9)\}$

55. $\{(0, 0), (1, 1), (-1, 1), (2, 2), (-2, 2)\}$

56. $\{(0, -5), (5, 0), (10, 5), (15, 10)\}$

What values, if any, are excluded from the domain of the function?

57. $f(x) = \dfrac{1}{x - 1}$

58. $g(x) = \dfrac{1}{x + 4}$

59. $h(x) = \dfrac{x + 3}{x + 8}$

60. $F(x) = \dfrac{2x - 5}{x - 4}$

61. $f(x) = 3x + 2$

62. $g(x) = 4 - 2x$

63. $G(x) = x^2 + 1$

64. $H(x) = \dfrac{1}{2}x^2$

65. $f(x) = \dfrac{x - 1}{x}$

66. $g(x) = \dfrac{2x + 5}{7}$

67. $H(x) = x^2 - x + 1$

68. $f(x) = 3x^2 + x + 4$

69. $f(x) = \dfrac{2x - 5}{3}$

70. $g(x) = \dfrac{3 - 5x}{5}$

71. $H(x) = \dfrac{x - 2}{x + 2}$

72. $h(x) = \dfrac{3 - x}{6 - x}$

73. $f(x) = \dfrac{x - 2}{2}$

74. $G(x) = \dfrac{2}{x - 2}$

Find the range of the function defined by each equation.

75. $f(x) = 4x - 3$; domain $= \{0, 1, 2, 3\}$

76. $G(x) = 3 - 5x$; domain $= \{-2, -1, 0, 1, 2\}$

77. $g(x) = 5x - 8$; domain $= \{-3, -1, 0, 1, 3\}$

78. $h(x) = 3x - 7$; domain $= \{-4, -2, 0, 2, 4\}$

79. $h(x) = x^2$; domain $= \{-2, -1, 0, 1, 2\}$

80. $H(x) = 1 - x^2$; domain $= \{-2, -1, 0, 1, 2\}$

81. $f(x) = 2x^2 - 2x + 2$; domain $= \{-4, -2, 0, 4\}$

82. $G(x) = -2x^2 + 5x - 2$; domain $= \{-3, -1, 0, 1, 3\}$

83. $H(x) = \dfrac{5}{1 - x}$; domain $= \{-2, 0, 2\}$

84. $g(x) = \dfrac{4}{4 - x}$; domain $= \{-5, 0, 3\}$

85. $f(x) = \dfrac{2}{x - 4}$; domain $= \{-2, 0, 2, 6\}$

86. $g(x) = \dfrac{x}{3 - x}$; domain $= \{-2, -1, 0, 1, 2\}$

87. $H(x) = 2 - 3x - x^2$; domain $= \{-5, 0, 5\}$

88. $G(x) = 4 - 3x - x^3$; domain $= \{-3, 0, 3\}$

3.2 Applying Concepts

89. Given $P(x) = 4x + 7$, write $P(-2 + h) - P(-2)$ in simplest form.

90. Given $G(t) = 9 - 2t$, write $G(-3 + h) - G(-3)$ in simplest form.

91. Given $f(z) = 8 - 3z$, write $f(-3 + h) - f(-3)$ in simplest form.

92. Given $M(n) = 5 - 2n$, write $M(-6 + h) - M(-6)$ in simplest form.

93. **Automotive Technology** The distance, s (in feet), that a car will skid on a certain road surface after the brakes are applied is a function of the car's velocity, v (in miles per hour). The function can be approximated by $s = f(v) = 0.017v^2$. How far will a car skid after its brakes are applied if it is traveling 60 mph?

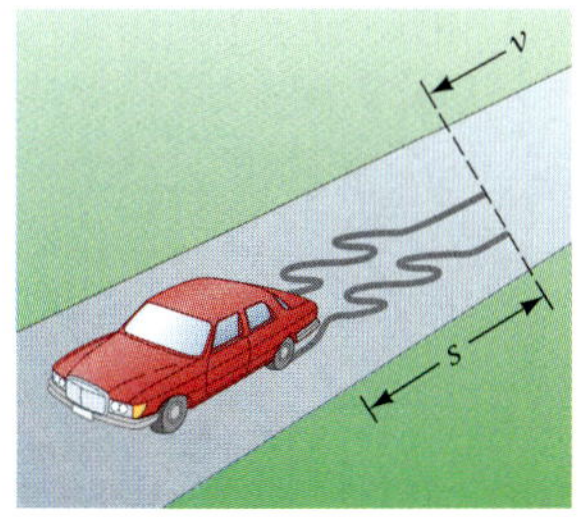

94. **Energy** The power a windmill can generate is a function of the velocity of the wind. The function can be approximated by $P = f(v) = 0.015v^3$, where P is the power in watts and v is the velocity of the wind in meters per second. How much power will be produced by a windmill when the velocity of the wind is 15 m/s?

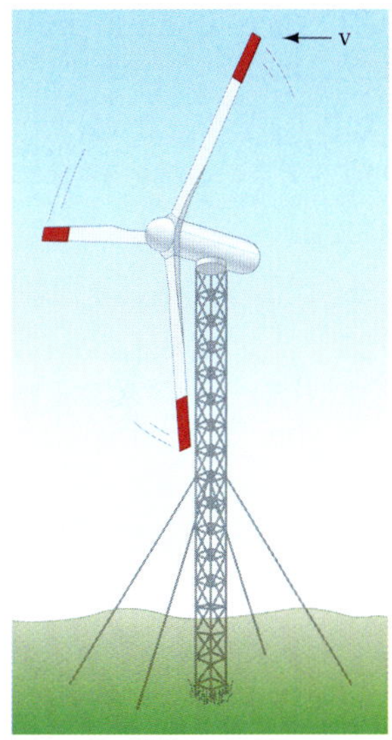

95. Find the set of ordered pairs (x, y) determined by the equation $y = x^3$, where $x \in \{-2, -1, 0, 1, 2\}$. Does the set of ordered pairs define a function? Why or why not?

96. Find the set of ordered pairs (x, y) determined by the equation $|y| = x$, where $x \in \{0, 1, 2, 3\}$. Does the set of ordered pairs define a function? Why or why not?

Each of the following graphs defines a function. Evaluate the function by estimating the ordinate (which is the value of the function) for the given value of t.

97. **Physics** The graph in Figure 1 shows the speed, v (in feet per second), at which a parachutist is falling during the first 20 s after jumping out of a plane. Estimate the speed at which the parachutist is falling when:

a. $t = 5$ s **b.** $t = 15$ s

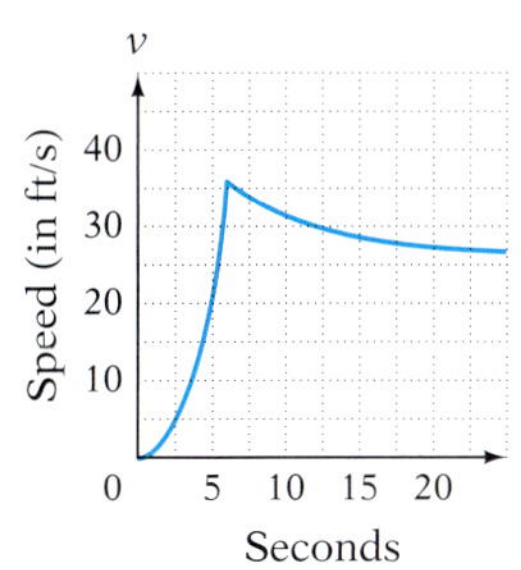

Figure 1

98. **Psychology** The graph in Figure 2 shows what an industrial psychologist has determined to be the average percent score, P, for an employee taking a performance test t weeks after training begins. Estimate the score an employee would receive on this test when:

a. $t = 4$ weeks **b.** $t = 10$ weeks

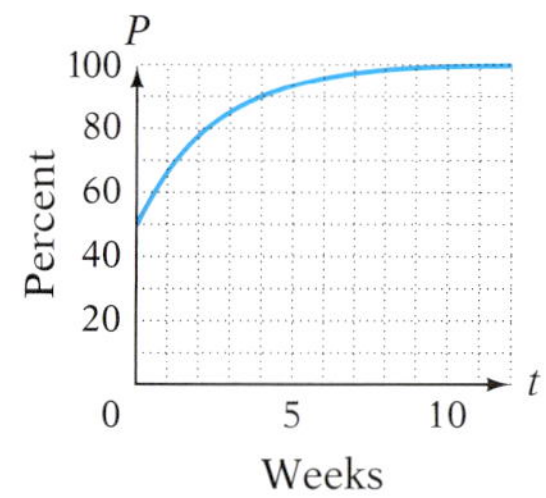

Figure 2

99. **Physics** The graph in Figure 3 shows the temperature, T (in degrees Fahrenheit), of a can of cola t hours after it is placed in a refrigerator. Use the graph to estimate the temperature of the cola when:

a. $t = 10$ h **b.** $t = 15$ h

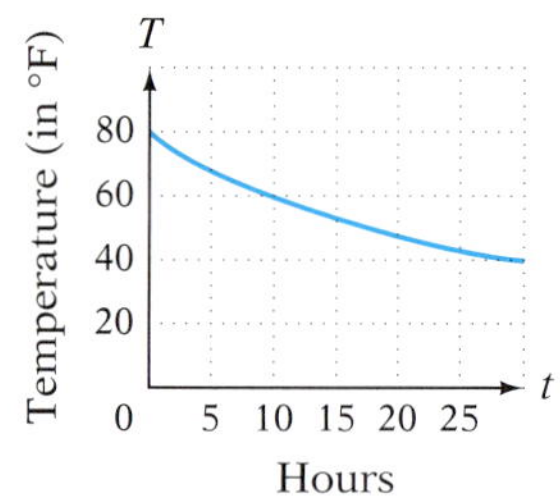

Figure 3

100. **Health Science** The graph in Figure 4 shows the decrease in the heart rate, r (in beats per minute), of a runner t minutes after the completion of a race. Use the graph to estimate the heart rate of a runner when:

a. $t = 5$ min **b.** $t = 20$ min

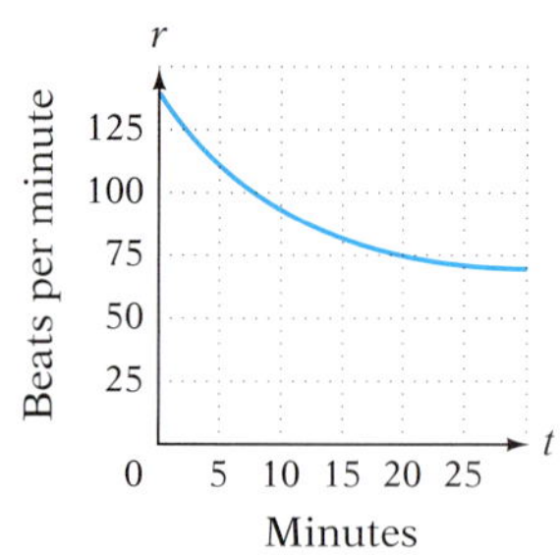

Figure 4

101. **Government** The table below gives the 50 states in the United States in terms of the Post Office's two-letter abbreviations. The first number following the name of each state is the figure provided by the U.S. Census Bureau for the population of the state in 2000; these figures are in millions and are rounded to the nearest hundred thousand. The second number following the name of each state is the number of representatives that state has in the House of Representatives.

a. Draw a scatter diagram of the data.

b. Explain the relationship between a state's population and the number of representatives that the state has in the House of Representatives.

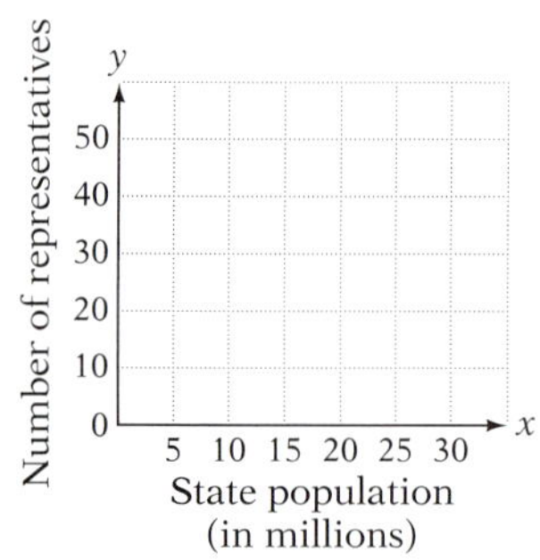

AL: 4.5, 7	HI: 1.2, 2	MA: 6.4, 10	NM: 1.8, 3	SD: 0.8, 1
AK: 0.6, 1	ID: 1.3, 2	MI: 10.0, 15	NY: 19.0, 29	TN: 5.7, 9
AZ: 5.1, 8	IL: 12.4, 19	MN: 4.9, 8	NC: 8.1, 13	TX: 20.9, 32
AR: 2.7, 4	IN: 6.1, 9	MS: 2.9, 4	ND: 0.6, 1	UT: 2.2, 3
CA: 33.9, 53	IA: 2.9, 5	MO: 5.6, 9	OH: 11.4, 18	VT: 0.6, 1
CO: 4.3, 7	KS: 2.7, 4	MT: 1.9, 1	OK: 3.5, 5	VA: 7.1, 11
CT: 3.4, 5	KY: 4.0, 6	NE: 1.7, 3	OR: 3.4, 5	WA: 5.9, 9
DE: 0.8, 1	LA: 4.5, 7	NV: 2.0, 3	PA: 12.3, 19	WV: 1.8, 3
FL: 16.0, 25	ME: 1.3, 2	NH: 1.2, 2	RI: 1.0, 2	WI: 5.4, 8
GA: 8.2, 13	MD: 5.3, 8	NJ: 8.4, 13	SC: 4.0, 6	WY: 0.5, 1

102. Write a report on the longitude-latitude coordinate system used to locate a position on Earth. Include in your report the definition of *nautical mile* and the coordinates of the city in which your school is located.

103. Investigate the relationship between two variables by collecting data, graphing the scatter diagram for these data, and discussing the results. For example, for each person in your classes who commutes to class by car, find out how many miles he or she commutes and the total number of miles on the odometer of the car he or she drives to class. Discuss whether you think there is a relationship between the two variables.

104. Give a real-world example of a relation that is not a function. Is it possible to give an example of a function that is not a relation? If so, give one. If not, explain why it is not possible.

3.3 Linear Functions

1 Graph a linear function

VIDEO & DVD CD TUTOR WEB SSM

Recall that the ordered pairs of a function can be written as $(x, f(x))$ or (x, y). The **graph of a function** is a graph of the ordered pairs (x, y) that belong to the function. Certain functions have characteristic graphs. A function that can be written in the form $f(x) = mx + b$ (or $y = mx + b$) is called a **linear function** because its graph is a straight line.

Examples of linear functions are shown at the right. Note that the exponent on each variable is 1.

$$f(x) = 2x + 5 \qquad (m = 2, b = 5)$$
$$P(t) = 3t - 2 \qquad (m = 3, b = -2)$$
$$y = -2x \qquad (m = -2, b = 0)$$
$$y = -\frac{2}{3}x + 1 \qquad \left(m = -\frac{2}{3}, b = 1\right)$$
$$g(z) = z - 2 \qquad (m = 1, b = -2)$$

The equation $y = x^2 + 4x + 3$ is not a linear function because it includes a term with a variable squared. The equation $f(x) = \frac{3}{x - 2}$ is not a linear function because a variable occurs in the denominator. Another example of an equation that is not a linear function is $y = \sqrt{x} + 4$; this equation contains a variable within a radical and so is not a linear function.

Consider $f(x) = 2x + 1$. Evaluating the linear function when $x = -3, -2, -1, 0, 1,$ and 2 produces some of the ordered pairs of the function. It is convenient to record the results in a table similar to the one at the right. The graph of the ordered pairs is shown in Figure 1 on the next page.

x	$f(x) = 2x + 1$	y	(x, y)
-3	$2(-3) + 1$	-5	$(-3, -5)$
-2	$2(-2) + 1$	-3	$(-2, -3)$
-1	$2(-1) + 1$	-1	$(-1, -1)$
0	$2(0) + 1$	1	$(0, 1)$
1	$2(1) + 1$	3	$(1, 3)$
2	$2(2) + 1$	5	$(2, 5)$

Evaluating the function when x is not an integer produces more ordered pairs to graph, such as $\left(-\frac{5}{2}, -4\right)$ and $\left(\frac{3}{2}, 4\right)$, as shown in Figure 2 below. Evaluating the function for still other values of x would result in more and more ordered pairs to graph. The result would be so many dots that the graph would look like the straight line shown in Figure 3, which is the graph of $f(x) = 2x + 1$.

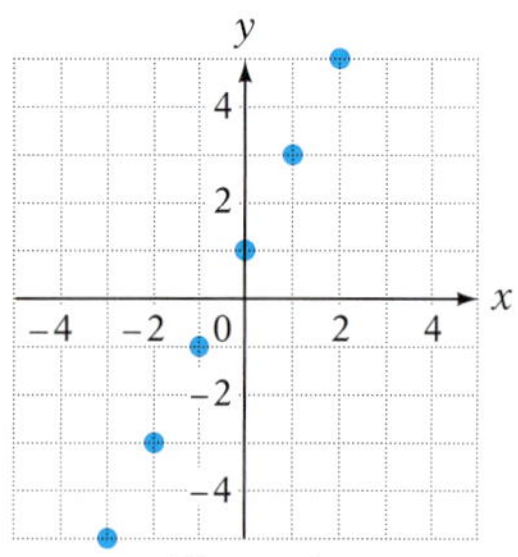

Figure 1

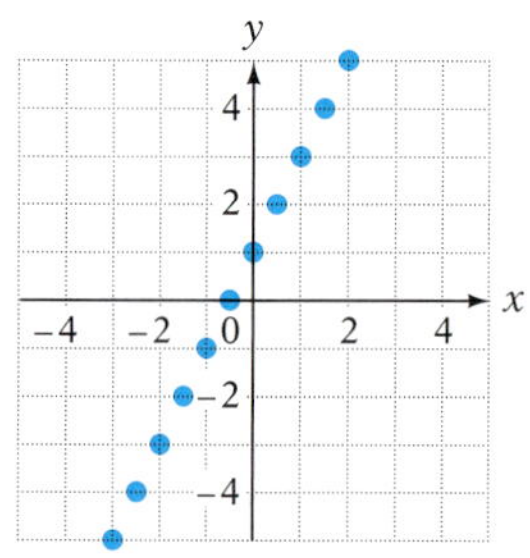

Figure 2

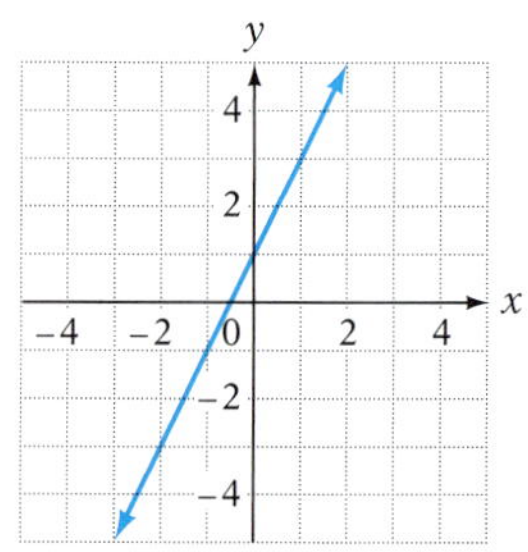

Figure 3

No matter what value of x is chosen, $2x + 1$ is a real number. This means the domain of $f(x) = 2x + 1$ is all real numbers. Therefore, we can use any real number when evaluating the function. Normally, however, values such as π or $\sqrt{5}$ are not used because it is difficult to graph the resulting ordered pairs.

Note from the graph of $f(x) = 2x + 1$ shown at the right that $(-1.5, -2)$ and $(3, 7)$ are the coordinates of points on the graph and that $f(-1.5) = -2$ and $f(3) = 7$. Note also that the point whose coordinates are $(2, 1)$ is not a point on the graph and that $f(2) \neq 1$. Every point on the graph is an ordered pair that belongs to the function, and every ordered pair that belongs to the function corresponds to a point on the graph.

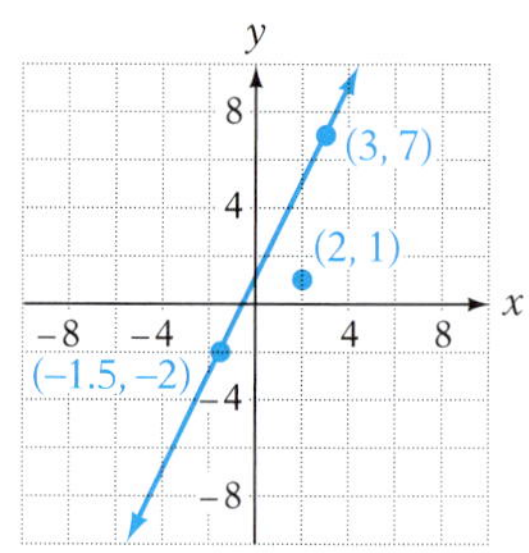

Whether an equation is written as $f(x) = mx + b$ or $y = mx + b$, the equation represents a linear function, and the graph of the equation is a straight line.

Because the graph of a linear function is a straight line, and a straight line is determined by two points, the graph of a linear function can be drawn by finding only two of the ordered pairs of the functions. However, it is recommended that you find at least *three* ordered pairs to ensure accuracy.

Take Note

When the coefficient of x is a fraction, choose values of x that are multiples of the denominator of the fraction. This will result in coordinates that are integers.

Example 1 Graph: $f(x) = -\frac{3}{2}x - 3$

Solution

x	$y = f(x)$
0	−3
−2	0
−4	3

• Find at least three ordered pairs. When the coefficient of x is a fraction, choose values of x that will simplify the evaluations. The ordered pairs can be displayed in a table.

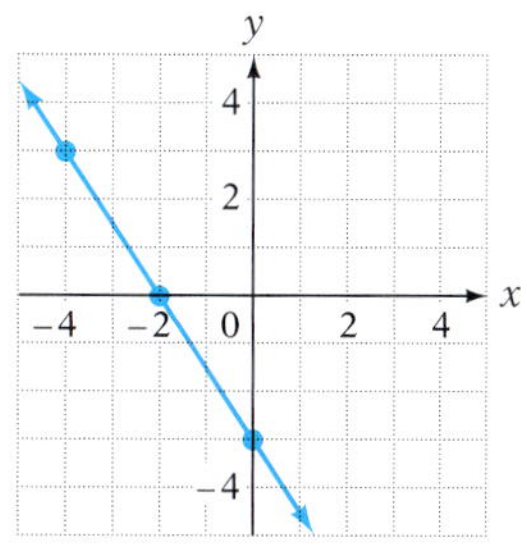

• Graph the ordered pairs and draw a line through the points.

Problem 1 Graph: $y = \frac{3}{5}x - 4$

Solution See page S11.

Graphing utilities create graphs by plotting points and then connecting the points to form a curve. Using a graphing utility, enter the equation $y = -\frac{3}{2}x - 3$ and verify the graph drawn in Example 1. (Refer to the Appendix on page 727 for suggestions on what domain and range to use.) Trace along the graph and verify that $(0, -3)$, $(-2, 0)$, and $(-4, 3)$ are the coordinates of points on the graph. Now enter the equation $y = \frac{3}{5}x - 4$ given in Problem 1. Verify that the ordered pairs you found for this function are the coordinates of points on the graph.

2 Graph an equation of the form $Ax + By = C$

VIDEO & DVD CD TUTOR WEB SSM

A **literal equation** is an equation with more than one variable. Examples of literal equations are $P = 2L + 2W$, $V = LWH$, $d = rt$, and $3x + 2y = 6$.

Linear equations of the form $y = mx + b$ are literal equations. In some cases, a linear equation has the form $Ax + By = C$. In such a case, it may be convenient to solve the equation for y to get the equation in the form $y = mx + b$. To solve for y, we use the same rules and procedures that we use to solve equations with numerical values.

Write $4x - 3y = 6$ in the form $y = mx + b$.

$$4x - 3y = 6$$

Subtract $4x$ from each side of the equation.

$$4x - 4x - 3y = 6 - 4x$$

Simplify.

$$-3y = 6 - 4x$$

Divide both sides of the equation by -3.

$$\frac{-3y}{-3} = \frac{6 - 4x}{-3}$$

Simplify. $y = \dfrac{6 - 4x}{-3}$

Divide each term in the numerator by the denominator. $y = \dfrac{6}{-3} - \dfrac{4x}{-3}$

Simplify. $y = -2 + \dfrac{4}{3}x$

Write the equation in the form $y = mx + b$. $y = \dfrac{4}{3}x - 2$

We will show two methods of graphing an equation of the form $Ax + By = C$. In the first method, we solve the equation for y and then follow the same procedure used for graphing an equation of the form $y = mx + b$.

Example 2 Graph: $3x + 2y = 6$

Solution

$$3x + 2y = 6$$
$$2y = -3x + 6$$
$$y = -\frac{3}{2}x + 3$$

• **Solve the equation for y.**

x	y
0	3
2	0
4	−3

• **Find at least three solutions.**

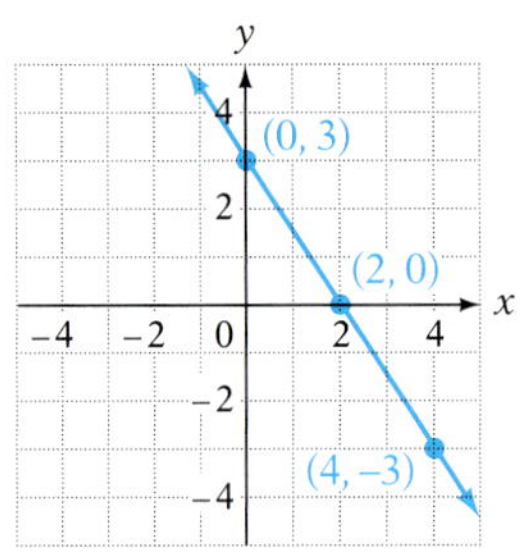

• **Graph the ordered pairs in a rectangular coordinate system. Draw a straight line through the points.**

Problem 2 Graph: $-3x + 2y = 4$

Solution See pages S11–S12.

Use a graphing utility to graph the equation $3x + 2y = 6$. First solve the equation for y and then enter the equation $y = -\frac{3}{2}x + 3$. Next trace along the graph and verify that (0, 3), (2, 0), and (4, −3) are the coordinates of points on the graph. Follow the same procedure for Problem 2.

An equation in which one of the variables is missing has a graph that is either a horizontal or a vertical line.

The equation $y = -2$ can be written

$$0 \cdot x + y = -2$$

Because $0 \cdot x = 0$ for any value of x, y is -2 for every value of x.

Some of the possible ordered-pair solutions of $y = -2$ are given in the following table. The graph is shown at the right.

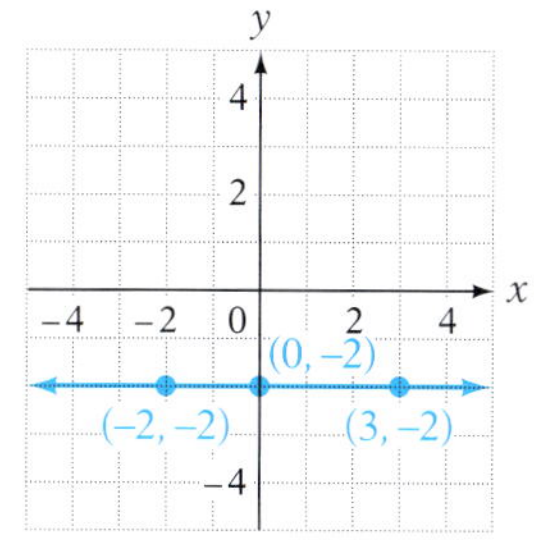

x	y
-2	-2
0	-2
3	-2

The equation $y = -2$ represents a function. Some of the ordered pairs of this function are $(-2, -2)$, $(0, -2)$, and $(3, -2)$. In functional notation we would write $f(x) = -2$. This function is an example of a constant function. No matter what value of x is selected, $f(x) = -2$.

Definition of the Constant Function

A function given by $f(x) = b$, where b is a constant, is a **constant function.** The graph of a constant function is a horizontal line passing through $(0, b)$.

Graph: $y + 4 = 0$

Solve for y.

$$y + 4 = 0$$
$$y = -4$$

The graph of $y = -4$ is a line passing through $(0, -4)$.

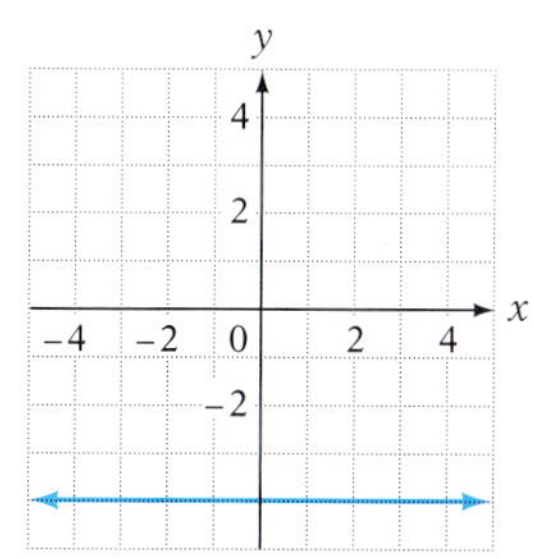

For each value in the domain of a constant function, the value of the function is the same (that is, it is constant). For instance, if $f(x) = 4$, then $f(2) = 4$, $f(3) = 4$, $f(\sqrt{3}) = 4$, $f(\pi) = 4$, and so on. The value of $f(x)$ is 4 for all values of x.

Evaluate $P(t) = -7$ when $t = 6$.

The value of the constant function is the same for all values of the variable.

$$P(t) = -7$$
$$P(6) = -7$$

For the equation $y = -2$, the coefficient of x is zero. For the equation $x = 2$, the coefficient of y is zero. For instance, the equation $x = 2$ can be written

$$x + 0 \cdot y = 2$$

No matter what value of y is chosen, $0 \cdot y = 0$ and therefore x is always 2.

Some of the possible ordered-pair solutions are given in the following table. The graph is shown at the right.

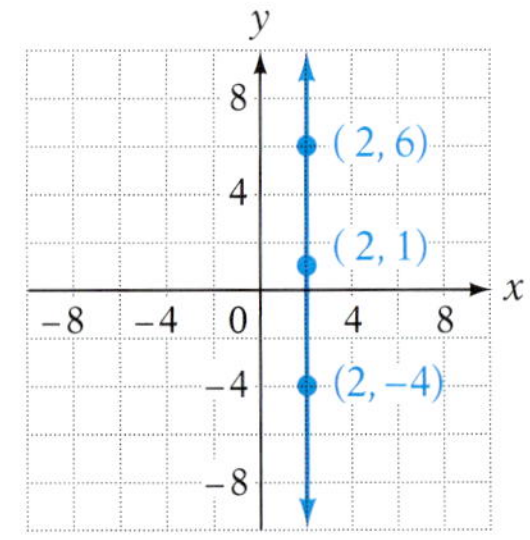

x	y
2	6
2	1
2	−4

Graph of $x = a$

The graph of $x = a$ is a vertical line passing through the point $(a, 0)$.

Take Note

The equation $y = b$ represents a function. The equation $x = a$ does *not* represent a function. Remember, not all equations represent functions.

Recall that a function is a set of ordered pairs in which no two ordered pairs have the same first coordinate and different second coordinates. Because $(2, 6)$, $(2, 1)$, and $(2, -4)$ are ordered pairs belonging to the equation $x = 2$, this equation does not represent a function, and the graph is not the graph of a function.

Example 3 Graph: $x = -4$

Solution

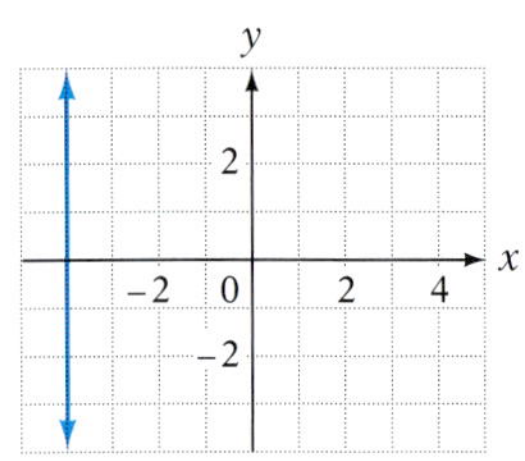

• **The graph of an equation of the form $x = a$ is a vertical line passing through the point whose coordinates are $(a, 0)$.**

Problem 3 Graph: $y - 3 = 0$

Solution See page S12.

A second method of graphing straight lines uses the *intercepts* of the graph.

The graph of the equation $x - 2y = 4$ is shown at the right. The graph crosses the x-axis at the point $(4, 0)$. This point is called the ***x*-intercept.** The graph also crosses the y-axis at the point $(0, -2)$. This point is called the ***y*-intercept.**

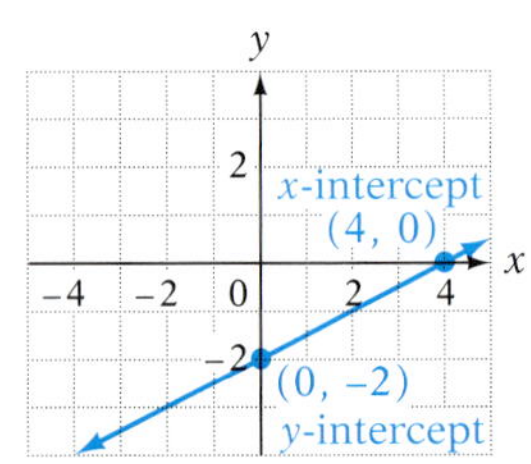

Take Note

The x-intercept occurs when $y = 0$. The y-intercept occurs when $x = 0$.

Find the x- and y-intercepts of the graph of the equation $3x + 4y = -12$.

To find the x-intercept, let $y = 0$.

$$3x + 4y = -12$$
$$3x + 4(0) = -12$$

(Any point on the x-axis has x-coordinate 0.)

$$3x = -12$$
$$x = -4$$

The x-intercept is $(-4, 0)$.

To find the y-intercept, let $x = 0$.
(Any point on the y-axis has x-coordinate 0.)

$$3x + 4y = -12$$
$$3(0) + 4y = -12$$
$$4y = -12$$
$$y = -3$$

The y-intercept is $(0, -3)$.

Find the y-intercept of the graph of the equation $y = \frac{2}{3}x + 7$.

$$y = \frac{2}{3}x + 7$$

To find the y-intercept, let $x = 0$.

$$y = \frac{2}{3}(0) + 7$$
$$y = 7$$

The y-intercept is $(0, 7)$.

Note that the y-coordinate of the y-intercept $(0, 7)$ has the same value as the constant term of $y = \frac{2}{3}x + 7$. As shown below, this is always true.

To find the y-intercept, let $x = 0$.

$$y = mx + b$$
$$y = m(0) + b$$
$$y = b$$

Thus the y-intercept is $(0, b)$.

y-Intercept of a Straight Line

For any equation of the form $y = mx + b$, the y-intercept is $(0, b)$.

A linear equation can be graphed by finding the x- and y-intercepts and then drawing a line through the two points.

Example 4 Graph: $4x - y = 4$ by using the x- and y-intercepts.

Solution x-intercept:

$$4x - y = 4$$
$$4x - 0 = 4 \quad \text{• To find the } x\text{-intercept, let } y = 0.$$
$$4x = 4$$
$$x = 1$$

The x-intercept is $(1, 0)$.

y-intercept:

$$4x - y = 4$$
$$4(0) - y = 4 \quad \text{• To find the } y\text{-intercept, let } x = 0.$$
$$-y = 4$$
$$y = -4$$

The y-intercept is $(0, -4)$.

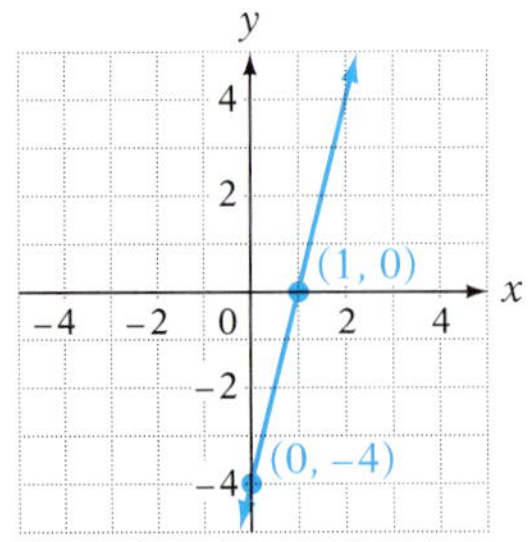

• **Graph the points (1, 0) and (0, −4). Draw a line through the two points.**

Problem 4 Graph: $3x - y = 2$ by using the x- and y-intercepts.

Solution See page S12.

3 Application problems

VIDEO & DVD CD TUTOR WEB SSM

There are a variety of applications of linear functions.

On the basis of data from *The Joy of Cooking,* the daily caloric allowance for a woman can be approximated by the equation $C = -7.5A + 2187.5$, where C is the caloric intake and A is the age of the woman.

A. Graph this equation for $25 \leq A \leq 75$.

B. The point whose coordinates are (45, 1850) is on the graph. Write a sentence that describes the meaning of this ordered pair.

A.

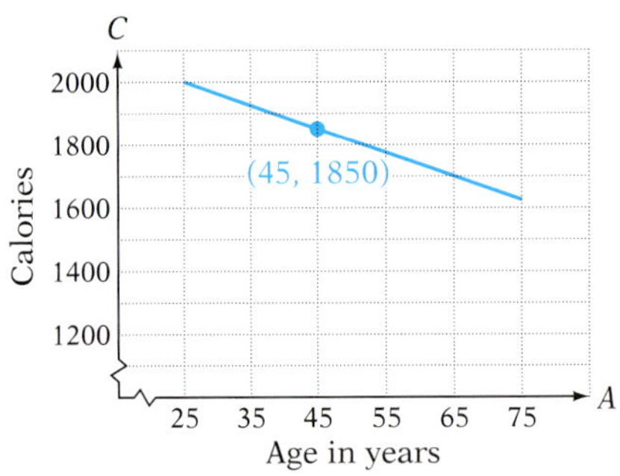

B. The ordered pair (45, 1850) means that the caloric allowance of a 45-year-old woman is 1850 calories per day.

Example 5 An electronics technician charges \$45 plus \$1 per minute to repair defective wiring in a home or apartment. The equation that describes the total cost, C (in dollars), to have defective wiring repaired is given by the equation $C = t + 45$, where t is the number of minutes the technician works. Graph this equation for $0 \leq t \leq 60$. The point whose coordinates are (15, 60) is on the graph. Write a sentence that describes the meaning of this ordered pair.

Solution

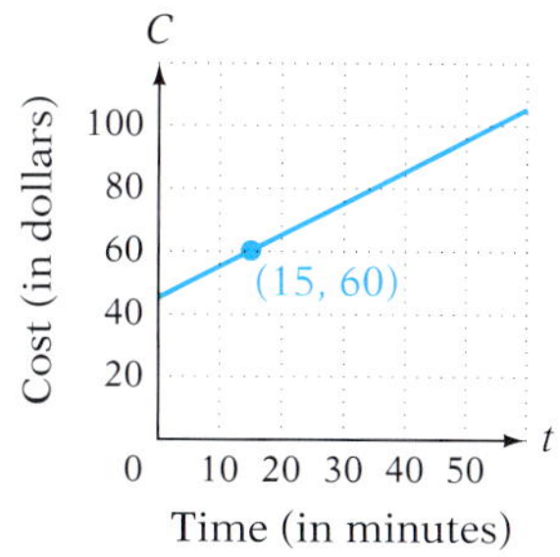

The ordered pair (15, 60) means that it costs \$60 for the technician to work 15 min.

Problem 5 The height, h (in inches), of a person and the length, L (in inches), of that person's stride while walking are related. The equation $h = \frac{3}{4}L + 50$ approximates the relationship. Graph this equation for $15 \le L \le 40$. The point whose coordinates are (32, 74) is on the graph. Write a sentence that describes the meaning of this ordered pair.

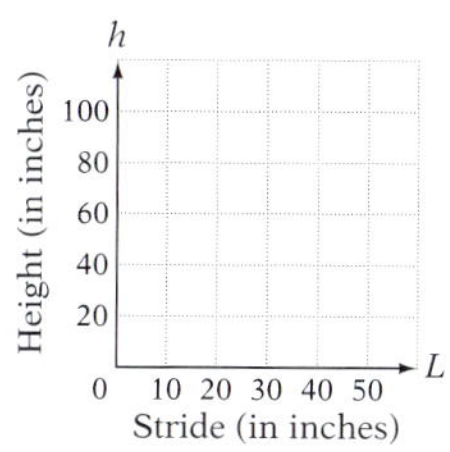

Solution See page S12.

3.3 Concept Review

Determine whether the following statements are always true, sometimes true, or never true.

1. The graph of $y = \frac{1}{x} + 2$ is the graph of a linear function.

2. The graph of $y - 3 = 0$ is a vertical line.

3. The graph of a straight line crosses the x-axis when $y = 0$.

4. The graph of a linear equation is a straight line.

5. The point at which a line crosses the y-axis is the x-intercept.

6. $xy + 2 = 0$ is an example of a linear equation.

3.3 Exercises

1 1. Explain how to graph a linear function by plotting points.

2. Give one example of a linear function and one example of a nonlinear function.

Graph.

3. $y = 3x - 4$

4. $y = -2x + 3$

5. $y = -\frac{2}{3}x$

6. $y = \frac{3}{2}x$

7. $y = \frac{2}{3}x - 4$

8. $y = \frac{3}{4}x + 2$

9. $y = -\frac{1}{3}x + 2$

10. $y = -\frac{3}{2}x - 3$

11. $y = \frac{3}{5}x - 1$

2 12. Explain how to find the x-intercept of the graph of an equation.

13. Explain how to find the y-intercept of the graph of an equation.

Graph

14. $2x - y = 3$

15. $2x + y = -3$

16. $2x + 5y = 10$

17. $x - 4y = 8$

18. $y = -2$

19. $x - 3y = 0$

20. $2x - 3y = 12$

21. $3x - y = -2$

22. $3x - 2y = 8$

Find the x- and y-intercepts and graph.

23. $x - 2y = -4$

24. $3x + y = 3$

25. $2x - 3y = 9$

26. $4x - 2y = 5$

27. $2x - y = 4$

28. $2x + y = 3$

29. $3x + 2y = 5$

30. $4x - 3y = 8$

31. $3x + 2y = 4$

32. $2x - 3y = 4$

33. $3x - 5y = 9$

34. $4x - 3y = 6$

3 Solve.

35. **Amusement Parks** The Rattler roller coaster at Fiesta in San Antonio, Texas, has a maximum speed of approximately 110 ft/s. The equation that describes the total number of feet traveled by the roller coaster in t seconds at this speed is given by $D = 110t$. Use the coordinate axes at the right to graph this equation for $0 \le t \le 10$. The point (5, 550) is on the graph. Write a sentence that describes the meaning of this ordered pair.

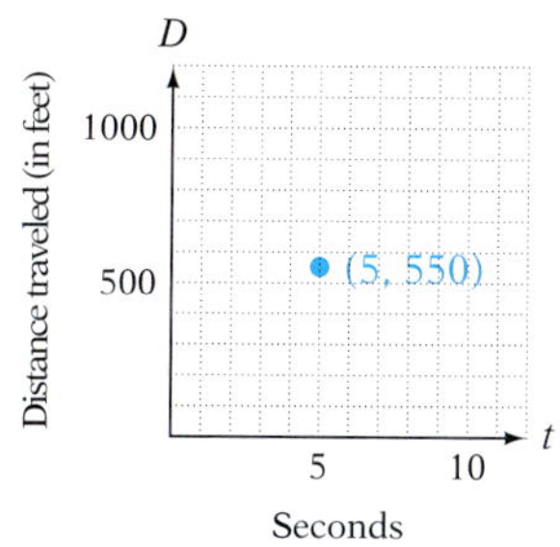

36. **Sports** A tennis pro is paid \$40 an hour for a private lesson by members of a condominium group. The equation that describes the total earnings, E (in dollars), received by the pro is $E = 40t$, where t is the total number of hours worked. Use the coordinate axes at the right to graph this equation for $0 \le t \le 30$. The point (25, 1000) is on the graph. Write a sentence that describes the meaning of this ordered pair.

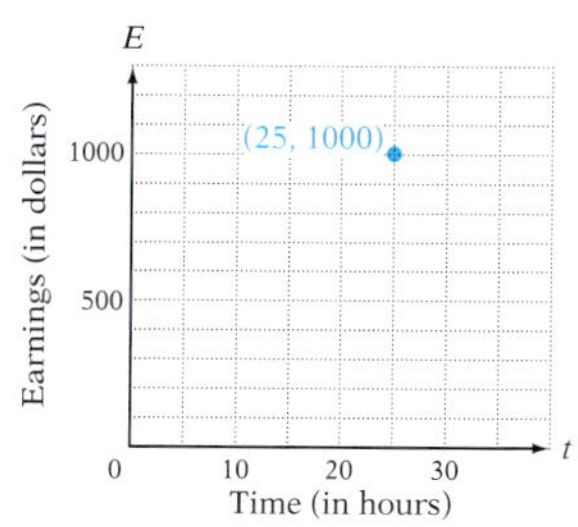

37. **Real Estate** A realtor receives \$400 per month plus 6% commission on sales. The equation that describes the total monthly income, I (in dollars), of the realtor is $I = 0.06s + 400$, where s is the amount of sales. Use the coordinate axes at the right to graph this equation for $0 \le s \le 150{,}000$. The point (60,000, 4000) is on the graph. Write a sentence that describes the meaning of this ordered pair.

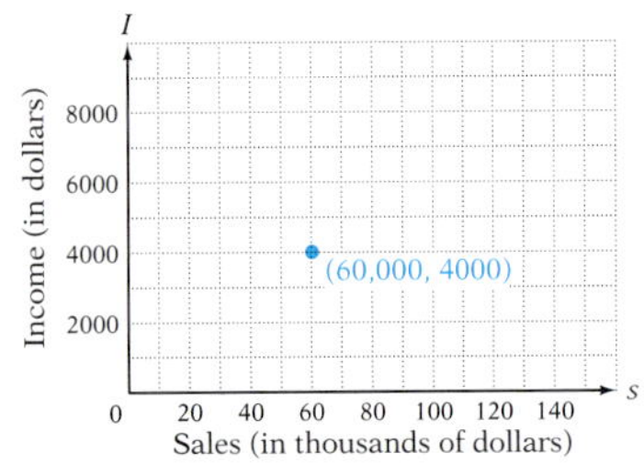

38. **Human Resources** An electronics company pays line workers \$9 per hour plus \$.05 for each transistor produced. The equation that describes the hourly wage, W (in dollars), is $W = 0.05n + 9$, where n is the number of transistors produced during any given hour. Use the coordinate axes at the right to graph this equation for $0 \le n \le 20$. The point (16, 9.80) is on the graph. Write a sentence that describes the meaning of this ordered pair.

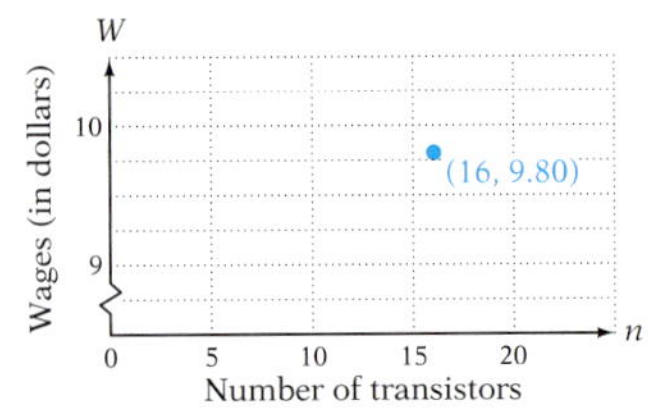

39. **Food Science** A caterer charges a flat rate of \$500 plus \$.95 for each hot appetizer. The equation that describes the cost, C (in dollars), of catering a dinner party is $C = 0.95n + 500$, where n is the number of hot appetizers. Use the coordinate axes at the right to graph this equation for $0 \le n \le 300$. The point (120, 614) is on the graph. Write a sentence that describes the meaning of this ordered pair.

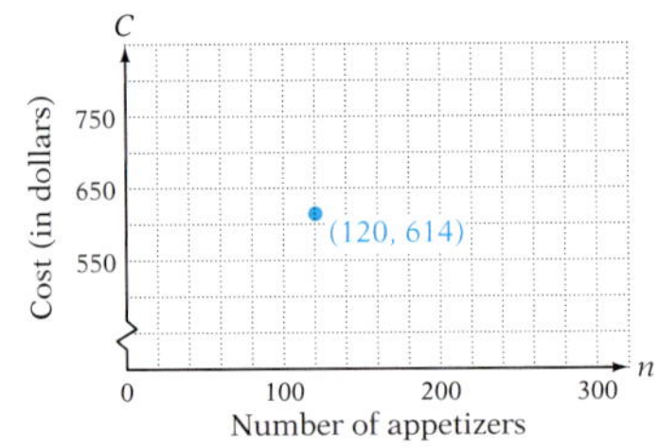

40. **Environmental Science** A tree service charges \$60 plus \$35 for each tree removed. The equation that describes the cost, C (in dollars), is $C = 35n + 60$, where n is the number of trees removed. Use the coordinate axes at the right to graph this equation for $0 \le n \le 100$. The point (50, 1810) is on the graph. Write a sentence that describes the meaning of this ordered pair.

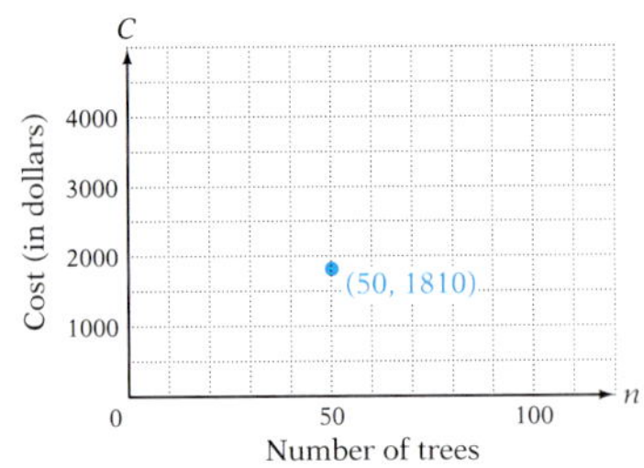

3.3 Applying Concepts

41. **Business** The sale price of an item is a function, s, of the original price, p, where $s(p) = 0.80p$. If an item's original price is \$200, what is the sale price of the item?

42. **Business** The markup of an item is a function, m, of its cost, c, where $m(c) = 0.25c$. If the cost of an item is \$150, what is the markup of the item?

Use a graphing utility to draw the graph of each of the following equations.

43. $f(x) = 1.2x + 2.3$

44. $f(x) = 2.4x + 0.5$

45. $f(x) = \dfrac{2x}{3} - \dfrac{5}{3}$

46. $y = -\frac{3x}{4} - \frac{5}{2}$

47. $3x - y = 4$

48. $2x + y = 3$

49. Explain the relationship between the zero of a linear function and the x-intercept of the graph of that function.

50. There is a relationship between the number of times a cricket chirps and the air temperature. The function $f(C) = 7C - 30$, where C is the temperature in degrees Celsius, can be used to approximate the number of times per minute that a cricket chirps. Discuss the domain and range of this function, and explain the significance of its x-intercept.

51. Explain what the graph of an equation represents.

52. Explain why you cannot graph the equation $4x + 3y = 0$ by using just its intercepts.

3.4 Slope of a Straight Line

1 Find the slope of a line given two points

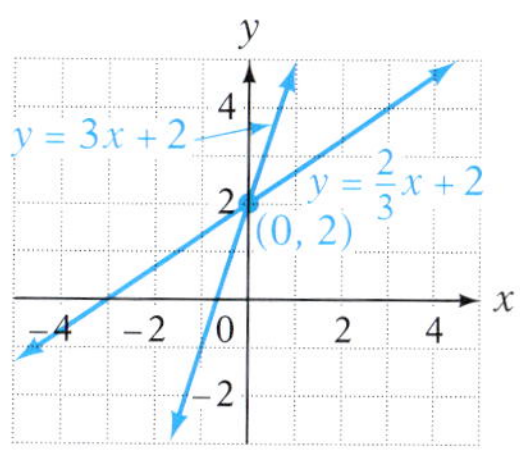

The graphs of $y = 3x + 2$ and $y = \frac{2}{3}x + 2$ are shown at the left. Each graph crosses the y-axis at the point $(0, 2)$, but the graphs have different slants. The **slope** of a line is a measure of the slant of the line. The symbol for slope is m.

The slope of a line containing two points is the ratio of the change in the y values between the two points to the change in the x values. The line containing the points whose coordinates are $(-1, -3)$ and $(5, 2)$ is shown below.

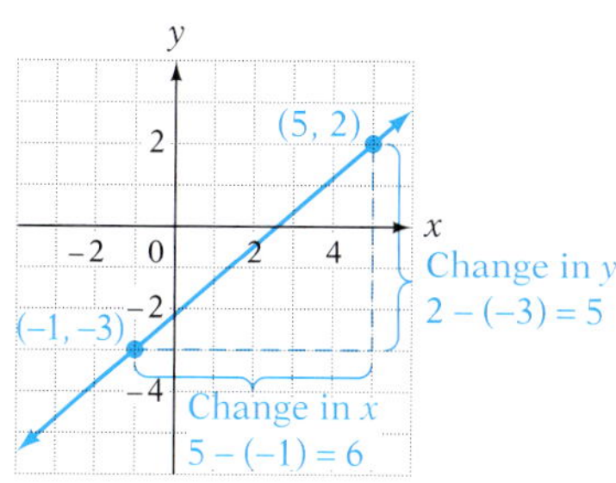

The change in the y values is the difference between the y-coordinates of the two points.

$$\text{Change in } y = 2 - (-3) = 5$$

The change in the x values is the difference between the x-coordinates of the two points.

$$\text{Change in } x = 5 - (-1) = 6$$

The slope of the line between the two points is the ratio of the change in y to the change in x.

$$\text{Slope} = m = \frac{\text{change in } y}{\text{change in } x} = \frac{5}{6} \qquad m = \frac{2 - (-3)}{5 - (-1)} = \frac{5}{6}$$

In general, if $P_1(x_1, y_1)$ and $P_2(x_2, y_2)$ are two points on a line, then

$$\text{Change in } y = y_2 - y_1 \qquad \text{Change in } x = x_2 - x_1$$

Using these ideas, we can state a formula for slope.

Slope Formula

The slope of the line containing the two points $P_1(x_1, y_1)$ and $P_2(x_2, y_2)$ is given by

$$m = \frac{y_2 - y_1}{x_2 - x_1}, \quad x_1 \neq x_2$$

Frequently, the Greek letter Δ is used to designate the change in a variable. Using this notation, we can write the equations for the change in y and the change in x as follows:

$$\text{Change in } y = \Delta y = y_2 - y_1 \qquad \text{Change in } x = \Delta x = x_2 - x_1$$

With this notation, the slope formula is written $m = \frac{\Delta y}{\Delta x}$.

In Section 1, Objective 4, of this chapter, we calculated the average rate of change in the population of California for different intervals of time. In each case, the average rate of change was different. Note that the graph of the population data did not lie on a straight line.

The graph of an equation of the form $y = mx + b$ is a straight line, and the average rate of change is constant. We can calculate the average rate of change between any two points on the line, and the rate of change will always be the same. For a linear equation, the rate of change is referred to as slope.

Find the slope of the line containing the points whose coordinates are $(-2, 0)$ and $(4, 5)$.

Let $P_1 = (-2, 0)$ and $P_2 = (4, 5)$. (It does not matter which point is named P_1 and which P_2; the slope will be the same.)

$$m = \frac{y_2 - y_1}{x_2 - x_1} = \frac{5 - 0}{4 - (-2)} = \frac{5}{6}$$

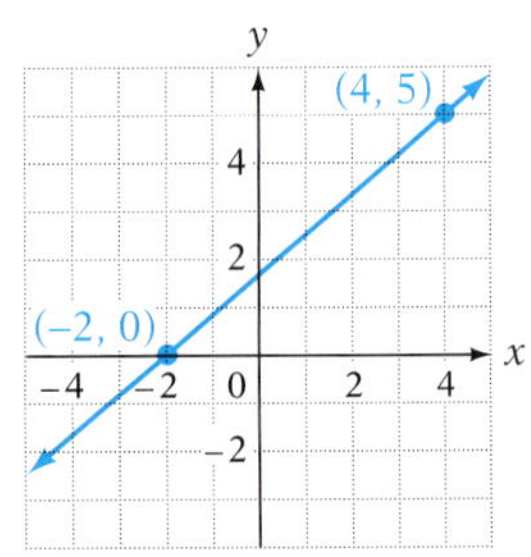

Positive slope

A line that slants upward to the right always has a **positive slope.**

Find the slope of the line containing the points whose coordinates are $(-3, 4)$ and $(4, 2)$.

Let $P_1 = (-3, 4)$ and $P_2 = (4, 2)$.

$$m = \frac{y_2 - y_1}{x_2 - x_1} = \frac{2 - 4}{4 - (-3)} = \frac{-2}{7} = -\frac{2}{7}$$

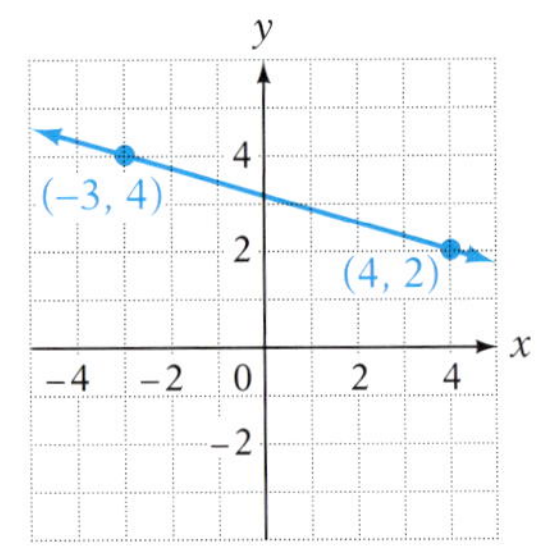

Negative slope

A line that slants downward to the right always has a **negative slope.**

Find the slope of the line containing the points whose coordinates are $(-2, 2)$ and $(4, 2)$.

Let $P_1 = (-2, 2)$ and $P_2 = (4, 2)$.

$$m = \frac{y_2 - y_1}{x_2 - x_1} = \frac{2 - 2}{4 - (-2)} = \frac{0}{6} = 0$$

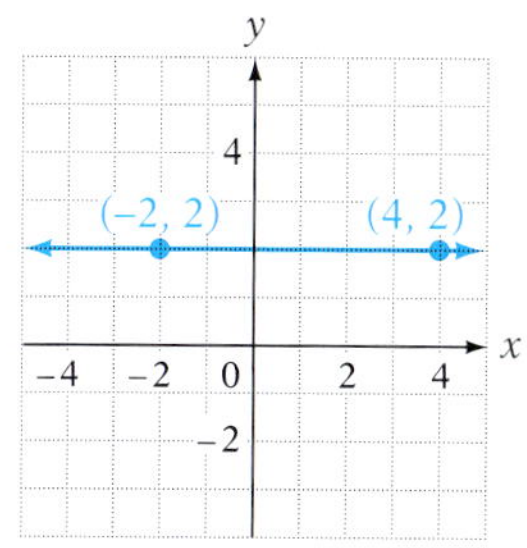

Zero slope

The horizontal line has **zero slope.**

Find the slope of the line containing the points whose coordinates are $(1, -2)$ and $(1, 3)$.

Let $P_1 = (1, -2)$ and $P_2 = (1, 3)$.

$$m = \frac{y_2 - y_1}{x_2 - x_1} = \frac{3 - (-2)}{1 - 1} = \frac{5}{0} \quad \text{Not a real number}$$

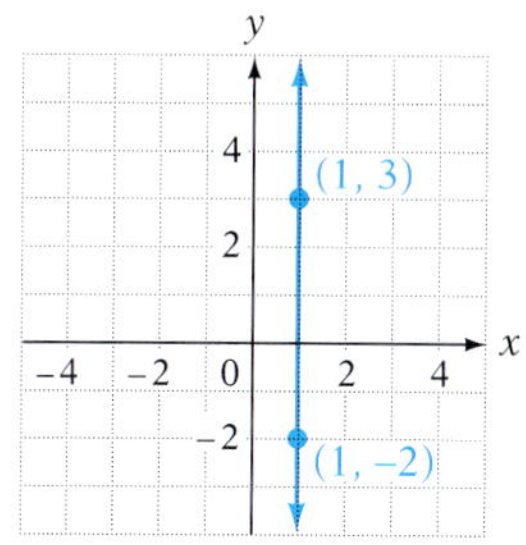

Undefined slope

The slope of a vertical line is **undefined.**

Example 1 Find the slope of the line containing the points whose coordinates are $(2, -5)$ and $(-4, 2)$.

Solution
$$m = \frac{y_2 - y_1}{x_2 - x_1} = \frac{2 - (-5)}{-4 - 2} = \frac{7}{-6} = -\frac{7}{6}$$

• Let $P_1 = (2, -5)$ and $P_2 = (-4, 2)$.

The slope is $-\frac{7}{6}$.

Problem 1 Find the slope of the line containing the points whose coordinates are $(4, -3)$ and $(2, 7)$.

Solution See page S12.

There are many applications of slope. Here are two examples.

The first record for the one-mile run was recorded in 1865 in England. Richard Webster ran the mile in 4 min 36.5 s. His average speed was approximately 19 feet per second.

The graph at the right shows the distance Webster ran during that run. From the graph, note that after 60 s (1 min) he had traveled 1140 ft and that after 180 s (3 min) he had traveled 3420 ft.

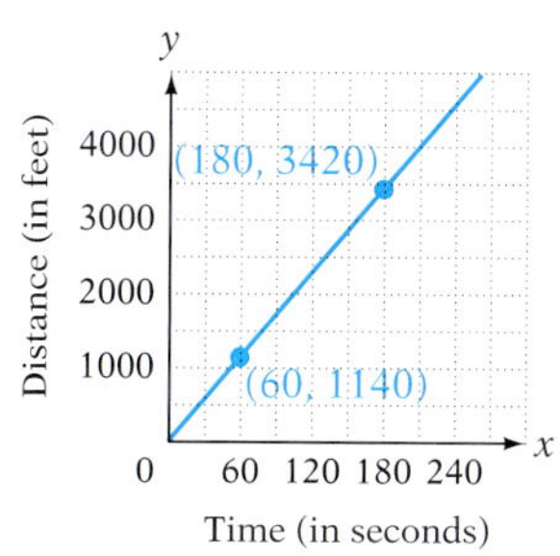

Let the point $(60, 1140)$ be (x_1, y_1) and the point $(180, 3420)$ be (x_2, y_2). The slope of the line between these two points is

$$m = \frac{y_2 - y_1}{x_2 - x_1} = \frac{3420 - 1140}{180 - 60} = \frac{2280}{120} = 19$$

Point of Interest

One of the motivations for the discovery of calculus was to solve a more complicated version of the distance-rate problem at the right.

You may be familiar with twirling a ball on the end of a string. If you release the string, the ball flies off in a path as shown below.

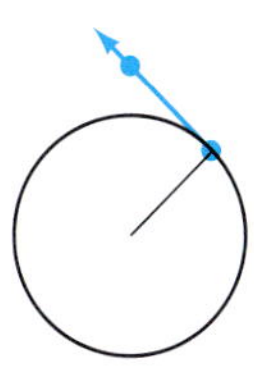

The question that mathematicians tried to answer was essentially "What is the slope (relative to the horizontal) of the line represented by the arrow?"

Answering questions similar to this led to the development of one aspect of calculus.

Note that the slope of the line is the same as Webster's average speed, 19 feet per second.

Average speed is related to slope.

The following example is drawn from economics.

As a result of cheaper digital networks and more competition, the cost of wireless phone use is decreasing. The average cost per minute in 1999 was 28¢. At the time of this writing, the average cost is projected to be 20¢ in 2003. (*Source:* The Strategic Group)

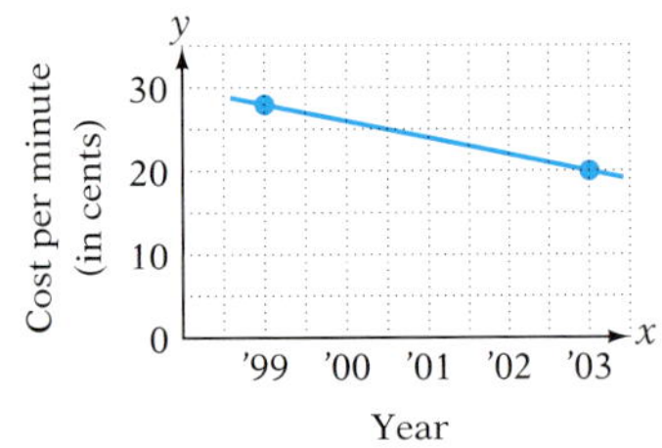

Let the point (1999, 28) be (x_1, y_1) and the point (2003, 20) be (x_2, y_2). The slope of the line between these two points is

$$m = \frac{y_2 - y_1}{x_2 - y_1} = \frac{20 - 28}{2003 - 1999} = \frac{-8}{4} = -2$$

Note that if we interpret negative slope as decreasing cost, then the slope of the line represents the rate at which the per-minute cost of wireless phone use is decreasing, 2¢ per year.

In general, any quantity that is expressed by using the word *per* is represented mathematically as slope. In the first example, the slope was 19 feet per second. In the second example, the slope was −2¢ per year.

Example 2 The graph shows the relationship between the cost of an item and the sales tax. Find the slope of the line between the two points shown on the graph. Write a sentence that states the meaning of the slope.

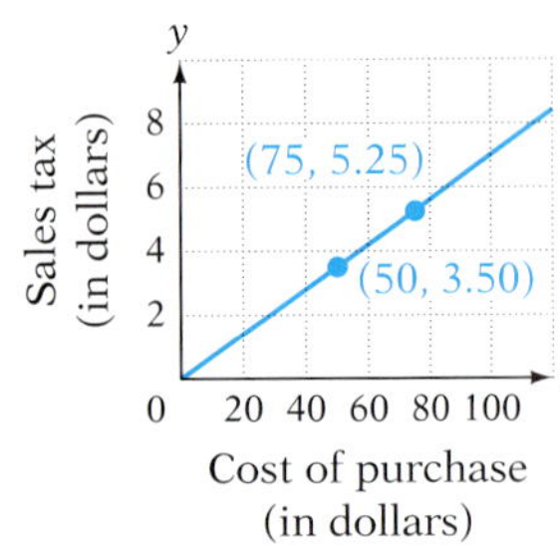

Solution $m = \frac{5.25 - 3.50}{75 - 50} = \frac{1.75}{25} = 0.07$

A slope of 0.07 means that the sales tax is \$.07 per dollar.

Problem 2 The graph shows the decrease in the value of a printing press for a period of six years. Find the slope of the line between the two points shown on the graph. Write a sentence that states the meaning of the slope.

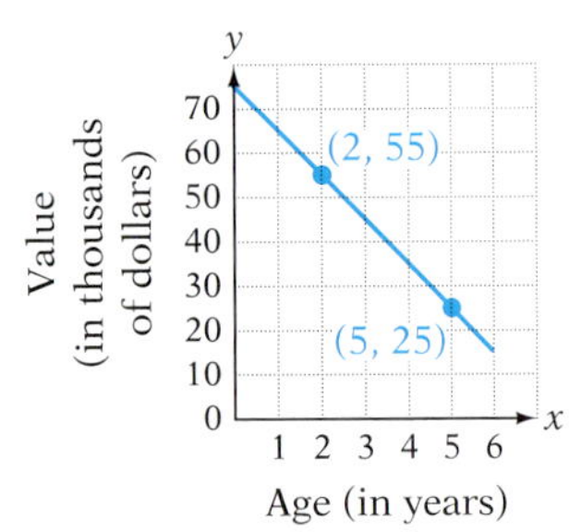

Solution See page S12.

2 Graph a line given a point and the slope

The graph of the equation $y = -\frac{3}{4}x + 4$ is shown at the right. The points (−4, 7) and (4, 1) are on the graph. The slope of the line is

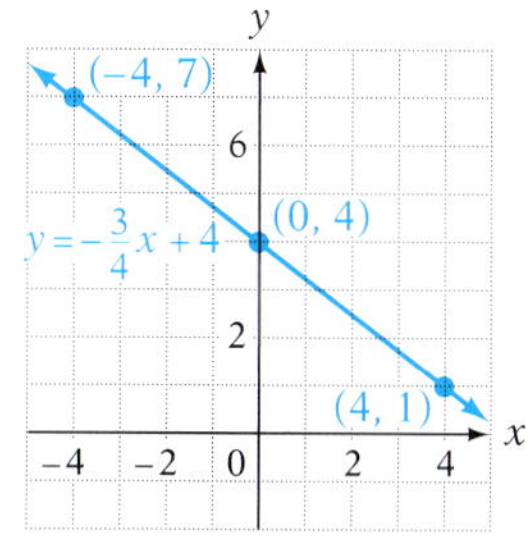

$$m = \frac{7 - 1}{-4 - 4} = \frac{6}{-8} = -\frac{3}{4}$$

Note that the slope of the line has the same value as the coefficient of x. The y-intercept is (0, 4).

Slope-Intercept Form of a Straight Line

The equation $y = mx + b$ is called the **slope-intercept form** of a straight line. The slope of the line is m, the coefficient is x. The y-intercept is $(0, b)$.

When the equation of a straight line is in the form $y = mx + b$, the graph can be drawn by using the slope and the y-intercept. First locate the y-intercept. Use the slope to find a second point on the line. Then draw a line through the two points.

Graph $y = \frac{5}{3}x - 4$ by using the slope and the y-intercept.

The slope is the coefficient of x.

$$m = \frac{5}{3} = \frac{\text{change in } y}{\text{change in } x}$$

The y-intercept is (0, −4).

Beginning at the y-intercept (0, −4), move right 3 units (change in x) and then up 5 units (change in y).

The point whose coordinates (3, 1) is a second point on the graph. Draw a line through the points (0, −4) and (3, 1).

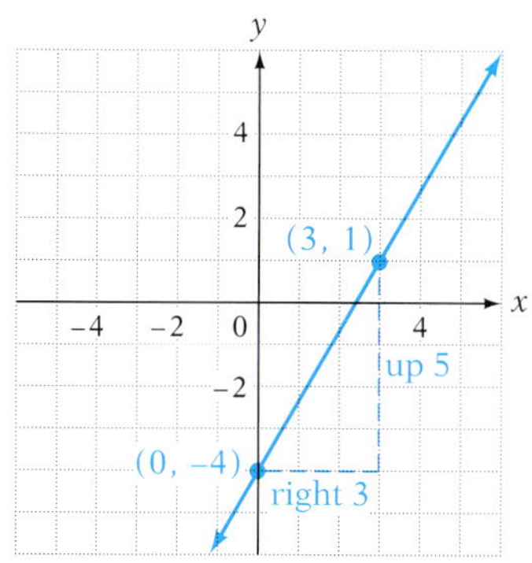

Take Note

When graphing a line by using its slope and y-intercept, *always* start at the y-intercept.

Graph $x + 2y = 4$ by using the slope and the y-intercept.

Solve the equation for y.

$$\begin{aligned} x + 2y &= 4 \\ 2y &= -x + 4 \\ y &= -\frac{1}{2}x + 2 \end{aligned}$$

The slope is $m = -\frac{1}{2} = \frac{-1}{2}$ and the y-intercept is (0, 2).

Beginning at the y-intercept (0, 2), move right 2 units (change in x) and then down 1 unit (change in y).

The point whose coordinates are (2, 1) is a second point on the graph. Draw a line through the points (0, 2) and (2, 1).

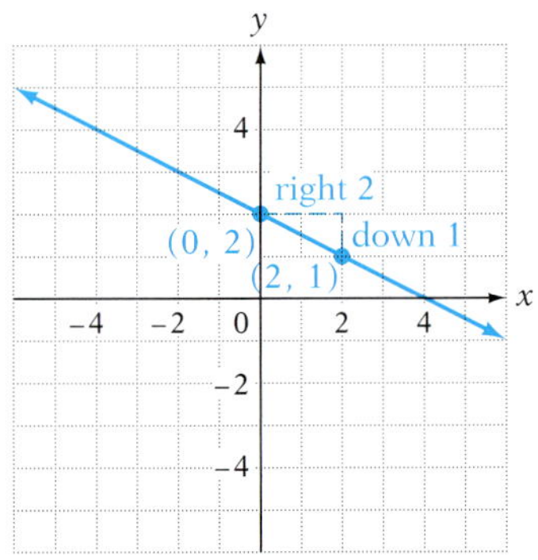

Example 3 Graph: $y = -\frac{3}{2}x + 4$ by using the slope and the y-intercept.

Solution y-intercept = (0, 4) • Locate the y-intercept.

$m = -\frac{3}{2} = \frac{-3}{2}$ • $m = \frac{\text{change in } y}{\text{change in } x}$

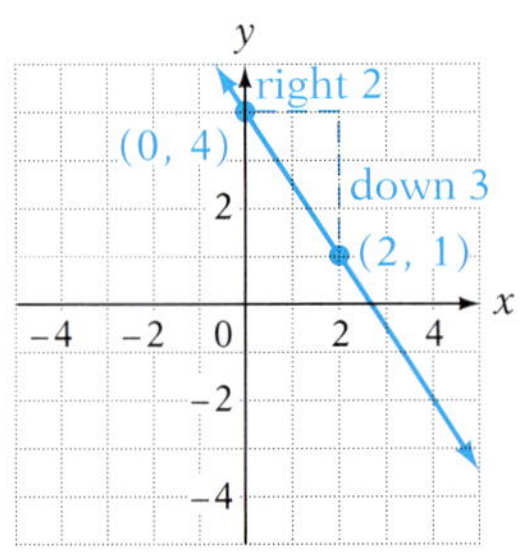

- Beginning at the y-intercept, (0, 4), move right 2 units and then down 3 units. (2, 1) is a second point on the graph.
- Draw a line through the points (0, 4) and (2, 1).

Problem 3 Graph: $2x + 3y = 6$ by using the slope and the y-intercept.

Solution See page S12.

The graph of a line can be drawn when any point on the line and the slope of the line are given.

Graph the line that passes through the point (−4, −4) and has slope 2.

When the slope is an integer, write it as a fraction with denominator 1.

$$m = 2 = \frac{2}{1} = \frac{\text{change in } y}{\text{change in } x}$$

Locate (−4, −4) in the coordinate plane. Beginning at that point, move right 1 unit (change in x) and then up 2 units (change in y).

The point whose coordinates are (−3, −2) is a second point on the graph. Draw a line through the points (−4, −4) and (−3, −2).

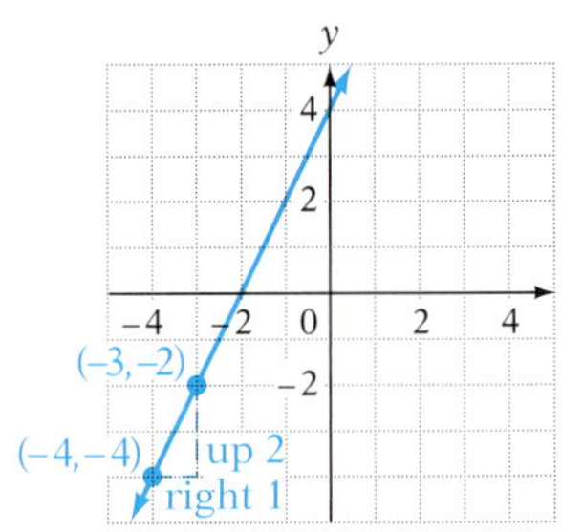

Take Note

This example differs from the others in that a point other than the y-intercept is used. In this case, start at the given point.

Example 4 Graph the line that passes through the point $(-2, 3)$ and has slope $= -\frac{4}{3}$.

Solution

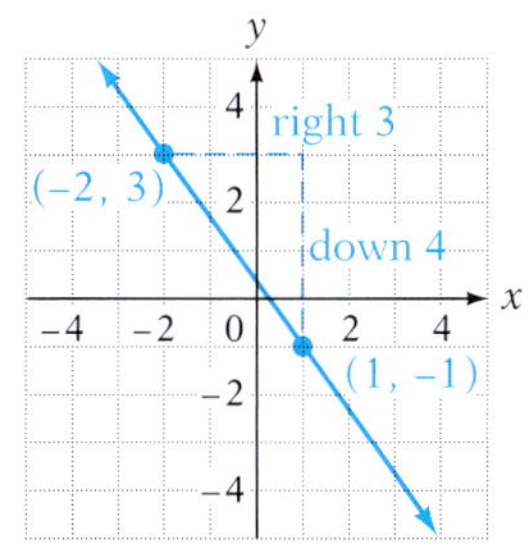

- $(x_1, y_1) = (-2, 3)$;

$$m = -\frac{4}{3} = \frac{-4}{3} = \frac{\text{change in } y}{\text{change in } x}$$

- Beginning at the point $(-2, 3)$, move right 3 units and then down 4 units. $(1, -1)$ is a second point on the line.
- Draw a line through the points $(-2, 3)$ and $(1, -1)$.

Problem 4 Graph the line that passes through the point $(-3, -2)$ and has slope 3.

Solution See page S12.

3.4 Concept Review

Determine whether the following statements are always true, sometimes true, or never true.

1. The slope of a line that slants downward to the left is positive.
2. The y-intercept of a line is the point at which the line crosses the y-axis.
3. The slope of a vertical line is zero.
4. A line whose slope is undefined is parallel to the y-axis.

3.4 Exercises

1 **1.** Explain in your own words how to find the slope of a line given two points on the line.

2. What is the difference between a line that has positive slope and one that has negative slope? Can a line have zero slope? If so, explain what its graph would look like.

Find the slope of the line containing the points P_1 and P_2.

3. $P_1(1, 3)$, $P_2(3, 1)$ **4.** $P_1(2, 3)$, $P_2(5, 1)$ **5.** $P_1(-1, 4)$, $P_2(2, 5)$

6. $P_1(3, -2)$, $P_2(1, 4)$ **7.** $P_1(-1, 3)$, $P_2(-4, 5)$ **8.** $P_1(-1, -2)$, $P_2(-3, 2)$

9. $P_1(0, 3)$, $P_2(4, 0)$

10. $P_1(-2, 0)$, $P_2(0, 3)$

11. $P_1(2, 4)$, $P_2(2, -2)$

12. $P_1(4, 1)$, $P_2(4, -3)$

13. $P_1(2, 5)$, $P_2(-3, -2)$

14. $P_1(4, 1)$, $P_2(-1, -2)$

15. $P_1(2, 3)$, $P_2(-1, 3)$

16. $P_1(3, 4)$, $P_2(0, 4)$

17. $P_1(0, 4)$, $P_2(-2, 5)$

18. $P_1(-2, 3)$, $P_2(-2, 5)$

19. $P_1(-3, -1)$, $P_2(-3, 4)$

20. $P_1(-2, -5)$, $P_2(-4, -1)$

21. **Uniform Motion** The graph below shows the relationship between the distance traveled by a motorist and the time of travel. Find the slope of the line between the two points shown on the graph. Write a sentence that states the meaning of the slope.

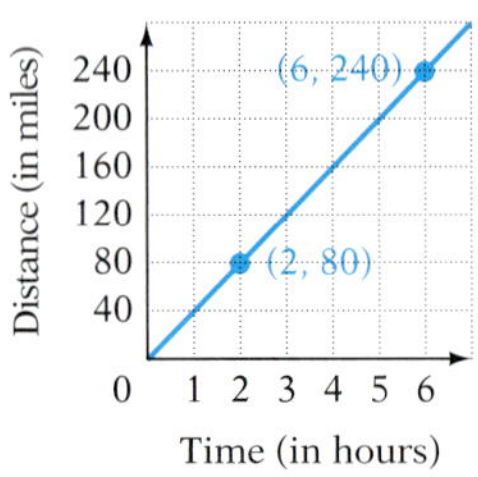

22. **Aeronautics** The graph below shows how the altitude of an airplane above the runway changes after take-off. Find the slope of the line. Write a sentence that states the meaning of the slope.

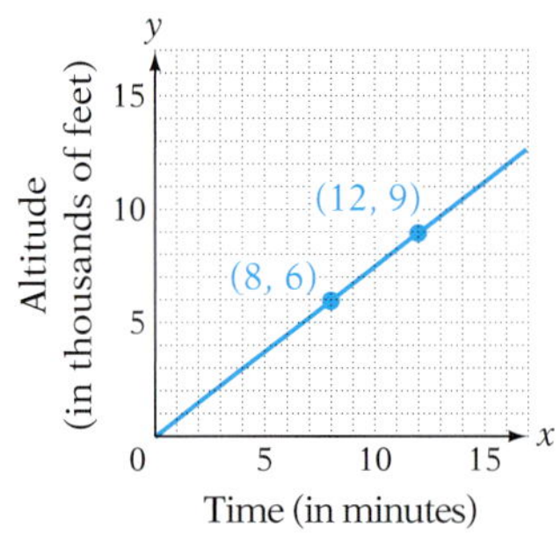

23. **Automotive Technology** The graph below shows how the amount of gas in the tank of a car decreases as the car is driven. Find the slope of the line. Write a sentence that states the meaning of the slope.

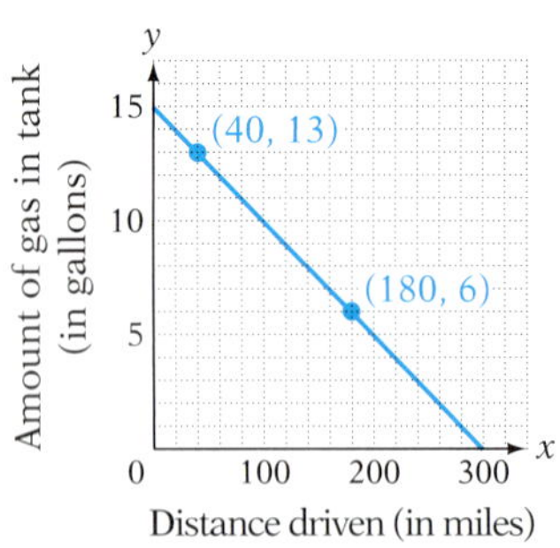

24. **Meteorology** The troposphere extends from the surface of Earth to an elevation of approximately 11 km. The graph below shows the decrease in temperature of the troposphere as altitude increases. Find the slope of the line. Write a sentence that states the meaning of the slope.

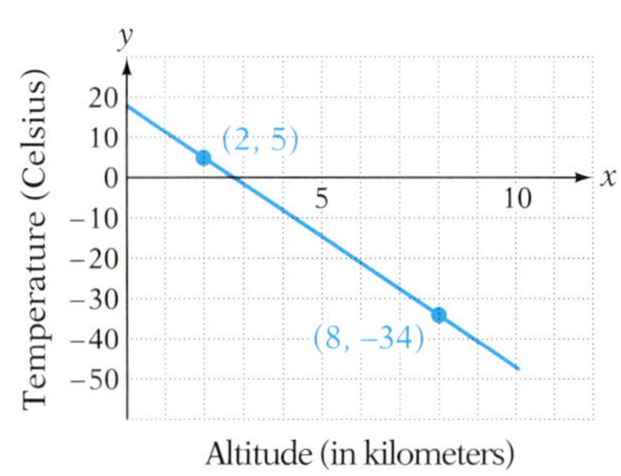

25. **Sports** Lois and Tanya start from the same place on a jogging course. Lois is jogging at 9 km/h, and Tanya is jogging at 6 km/h. The graphs below show the total distance traveled by each jogger and the total distance between Lois and Tanya. Which lines represent which distances?

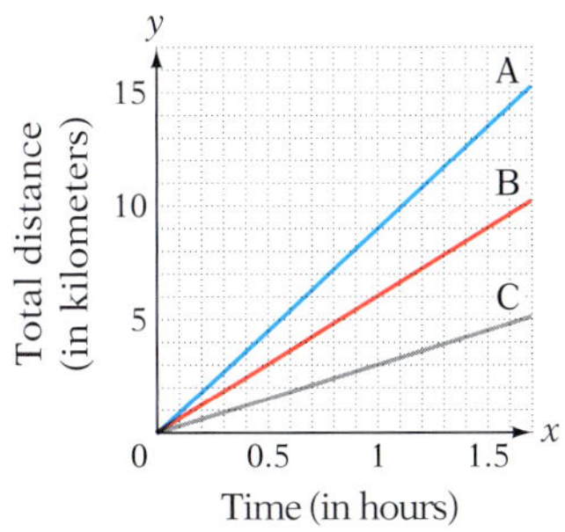

26. **Chemistry** A chemist is filling two cans from a faucet that releases water at a constant rate. Can 1 has a diameter of 20 mm, and can 2 has a diameter of 30 mm. The depth of the water in each can is measured at 5-second intervals. The graph of the results is shown below. On the graph, which line represents the depth of the water for which can?

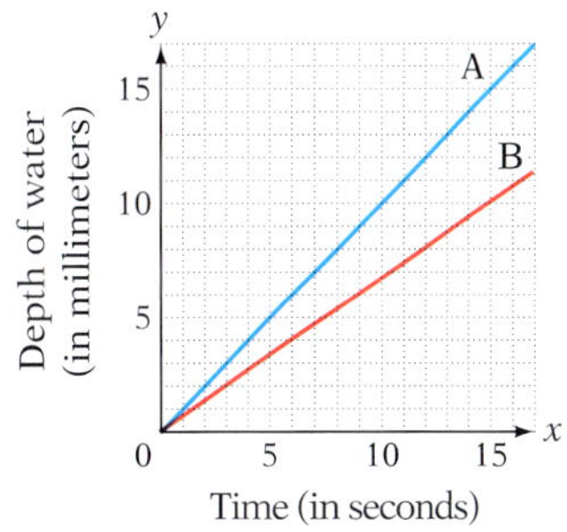

27. **Health Science** The American National Standards Institute (ANSI) states that the slope for a wheelchair ramp must not exceed $\frac{1}{12}$.

a. Does a ramp that is 6 in. high and 5 ft long meet the requirements of ANSI?

b. Does a ramp that is 12 in. high and 170 in. long meet the requirements of ANSI?

28. **Health Science** A ramp for a wheelchair must be 14 in. high. What minimum length must this ramp be so that it meets the ANSI requirements given in Exercise 27?

2 Graph by using the slope and the y-intercept.

29. $y = \frac{1}{2}x + 2$

30. $y = \frac{2}{3}x - 3$

31. $y = -\frac{3}{2}x$

32. $y = \frac{3}{4}x$

33. $y = -\frac{1}{2}x + 2$

34. $y = \frac{2}{3}x - 1$

35. $x - 3y = 3$

36. $3x + 2y = 8$

37. $4x + y = 2$

38. Graph the line that passes through the point $(-1, -3)$ and has slope $\frac{4}{3}$.

39. Graph the line that passes through the point $(-2, -3)$ and has slope $\frac{5}{4}$.

40. Graph the line that passes through the point $(-3, 0)$ and has slope -3.

41. Graph the line that passes through the point $(2, 0)$ and has slope -1.

3.4 Applying Concepts

42. Let $f(x)$ be the digit in the nth decimal place of the repeating digit is $0.\overline{387}$. For example, $f(3) = 7$ because 7 is the digit in the third decimal place. Find $f(14)$.

Complete the sentences.

43. If a line has a slope of 2, then the value of y increases/decreases by _______ as the value of x increases by 1.

44. If a line has a slope of -3, then the value of y increases/decreases by _______ as the value of x increases by 1.

45. If a line has a slope of $\frac{1}{2}$, then the value of y increases/decreases by _______ as the value of x increases by 1.

46. If a line has a slope of $-\frac{2}{3}$, then the value of y increases/decreases by _______ as the value of x increases by 1.

Determine the value of k such that the points whose coordinates are given below lie on the same line.

47. $(3, 2), (4, 6), (5, k)$

48. $(k, 1), (0, -1), (2, -2)$

49. Match each equation with its graph.

i. $y = -2x + 4$

ii. $y = 2x - 4$

iii. $y = 2$

iv. $2x + 4y = 0$

v. $y = \frac{1}{2}x + 4$

vi. $y = -\frac{1}{4}x - 2$

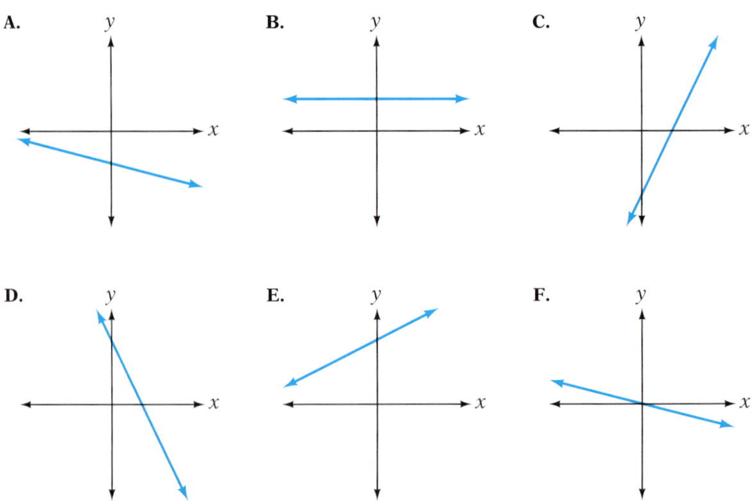

50. Graph $y = 2x + 3$ and $y = 2x - 1$ on the same coordinate system. Explain how the graphs are different and how they are the same. If b is any real number, how is the graph of $y = 2x + b$ related to the two graphs you have drawn?

51. Explain how you can use the slope of a line to determine whether three given points lie on the same line. Then use your procedure to determine whether all of the following points lie on the same line.
a. (2, 5), (−1, −1), (3, 7) **b.** (−1, 5), (0, 3), (−3, 4)

3.5 Finding Equations of Lines

1 Find the equation of a line given a point and the slope

VIDEO & DVD CD TUTOR WEB SSM

When the slope of a line and a point on the line are known, the equation of the line can be determined. If the particular point is the y-intercept, use the slope-intercept form, $y = mx + b$, to find the equation.

Find the equation of the line that contains the point (0, 3) and has slope $\frac{1}{2}$.

The known point is the y-intercept, (0, 3).

Use the slope-intercept form. $y = mx + b$

Replace m with $\frac{1}{2}$, the given slope, and replace b with 3, the y-coordinate of the y-intercept. $y = \frac{1}{2}x + 3$

The equation of the line is $y = \frac{1}{2}x + 3$.

One method of finding the equation of a line when the slope and any point on the line are known involves using the *point-slope formula.* This formula is derived from the formula for the slope of a line.

Let (x_1, y_1) be the given point on the line, and let (x, y) be another point on the line.

Use the formula for the slope of a line. $\frac{y - y_1}{x - x_1} = m$

Multiply both sides of the equation by $(x - x_1)$. $\frac{y - y_1}{x - x_1}(x - x_1) = m(x - x_1)$

Then simplify. $y - y_1 = m(x - x_1)$

Point-Slope Formula

Let m be the slope of a line, and let (x_1, y_1) be the coordinates of a point on the line. The equation of the line can be found using the **point-slope formula:**

$$y - y_1 = m(x - x_1)$$

Find the equation of the line that contains the point (4, −1) and has slope $-\frac{3}{4}$.

Use the point-slope formula.

$$y - y_1 = m(x - x_1)$$

Substitute the slope, $-\frac{3}{4}$, and the coordinates of the given point, (4, −1), into the point-slope formula. Then simplify.

$$y - (-1) = -\frac{3}{4}(x - 4)$$
$$y + 1 = -\frac{3}{4}x + 3$$
$$y = -\frac{3}{4}x + 2$$

The equation of the line is $y = -\frac{3}{4}x + 2$.

Find the equation of the line that passes through the point whose coordinates are (4, 3) and whose slope is undefined.

Because the slope is undefined, the point-slope formula cannot be used to find the equation. Instead, recall that when the slope is undefined, the line is vertical and that the equation of a vertical line is $x = a$, where a is the x-coordinate of the x-intercept. Because the line is vertical and passes through (4, 3), the x-intercept is (4, 0).

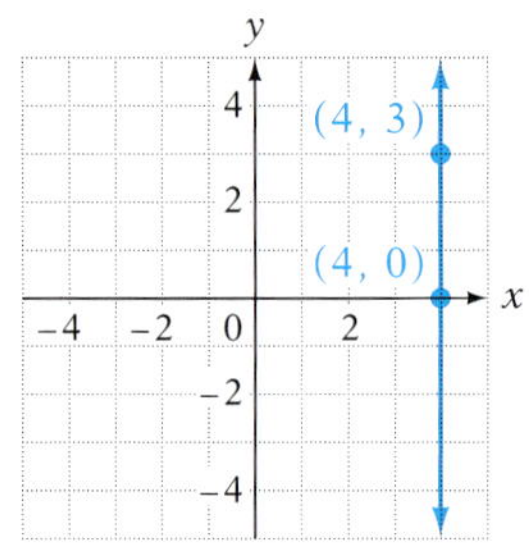

The equation of the line is $x = 4$.

Example 1 Find the equation of the line that contains the point (−2, 4) and has slope 2.

Solution

$$y - y_1 = m(x - x_1)$$
$$y - 4 = 2[x - (-2)]$$
$$y - 4 = 2(x + 2)$$
$$y - 4 = 2x + 4$$
$$y = 2x + 8$$

- **Use the point-slope formula.**
- **Substitute the slope, 2, and the coordinates of the given point, (−2, 4), into the point-slope formula.**

The equation of the line is $y = 2x + 8$.

Problem 1 Find the equation of the line that contains the point (4, −3) and has slope −3.

Solution See page S12.

2 Find the equation of a line given two points

VIDEO & DVD CD TUTOR WEB SSM

The point-slope formula and the formula for slope are used to find the equation of a line when two points are known.

Find the equation of the line containing the points (3, 2) and (−5, 6).

To use the point-slope formula, we must know the slope. Use the formula for slope to determine the slope of the line between the two given points.

Let $(x_1, y_1) = (3, 2)$ and $(x_2, y_2) = (-5, 6)$.

$$m = \frac{y_2 - y_1}{x_2 - x_1} = \frac{6 - 2}{-5 - 3} = \frac{4}{-8} = -\frac{1}{2}$$

Now use the point-slope formula with $m = -\frac{1}{2}$ and $(x_1, y_1) = (3, 2)$.

$$y - y_1 = m(x - x_1)$$
$$y - 2 = -\frac{1}{2}(x - 3)$$
$$y - 2 = -\frac{1}{2}x + \frac{3}{2}$$
$$y = -\frac{1}{2}x + \frac{7}{2}$$

The equation of the line is $y = -\frac{1}{2}x + \frac{7}{2}$.

Example 2 Find the equation of the line containing the points (2, 3) and (4, 1).

Solution

$$m = \frac{y_2 - y_1}{x_2 - x_1} = \frac{1 - 3}{4 - 2} = \frac{-2}{2} = -1$$

- **Find the slope. Let $(x_1, y_1) = (2, 3)$ and $(x_2, y_2) = (4, 1)$.**

$$y - y_1 = m(x - x_1)$$
$$y - 3 = -1(x - 2)$$
$$y - 3 = -x + 2$$
$$y = -x + 5$$

- **Substitute the slope and the coordinates of either one of the known points into the point-slope formula.**

The equation of the line is $y = -x + 5$.

Problem 2 Find the equation of the line containing the points (2, 0) and (5, 3).

Solution See page S12.

3 Application problems

VIDEO & DVD CD TUTOR WEB SSM

Linear functions can be used to model a variety of applications in science and business. For each application, data are collected and the independent and dependent variables are selected. Then a linear function is determined that models the data.

Example 3 In 1950, there were 13 million adults 65 years old or older in the United States. Data from the Census Bureau shows that the population of these adults has been increasing at a constant rate of approximately 0.5 million per year. This rate of increase is expected to continue through the year 2010. Find the equation of the line that approximates the population of adults 65 years old or older in terms of the year. Use your equation to approximate the population of these adults in 2005.

Strategy Select the independent and dependent variables. Because we want to determine the population of adults 65 years or older, that quantity is the *dependent* variable, y. The year is the *independent* variable.

From the data, the ordered pair (1950, 13), where the y-coordinate is in millions of people, is a point on the line. The slope of the line is the *rate* of increase, 0.5 million per year.

Solution Use the point-slope formula.

$$y - y_1 = m(x - x_1)$$

$$y - 13 = 0.5(x - 1950) \quad \text{• } m = 0.5, (x_1, y_1) = (1950, 13)$$
$$y - 13 = 0.5x - 975$$
$$y = 0.5x - 962$$

The linear function is $f(x) = 0.5x - 962$.

$$f(x) = 0.5x - 962$$
$$f(2005) = 0.5(2005) - 962 \quad \text{• Evaluate the function at 2005 to predict the population in 2005.}$$
$$= 1002.5 - 962$$
$$= 40.5$$

There will be approximately 40.5 million people 65 years or older in 2005.

Problem 3 Gabriel Daniel Fahrenheit invented the mercury thermometer in 1717. In terms of readings on this thermometer, water freezes at 32°F and boils at 212°F. In 1742 Anders Celsius invented the Celsius temperature scale. On this scale, water freezes at 0°C and boils at 100°C. Find a linear function that can be used to predict the Celsius temperature when the Fahrenheit temperature is known.

Solution See page S12.

3.5 Concept Review

Identify the following statements as always true, sometimes true, or never true.

1. Increasing the value of m in the equation $y = mx + b$ increases the slope of the line that is the graph of the equation.

2. Decreasing the value of b in the equation $y = mx + b$ decreases the slope of the line that is the graph of the equation.

3. The point-slope formula for the equation of a line is $y = m(x - x_1)$.

4. $x = a$ is the equation for a vertical line.

5. The line represented by the equation $y = 2x - \frac{1}{2}$ has slope $-\frac{1}{2}$ and x-intercept 2.

6. If $y = mx + b$, then m represents the rate of change of y with respect to x.

7. A horizontal line has no slope.

3.5 Exercises

1 **1.** What is the point-slope formula and how is it used?

2. Explain one way to find the equation of a line given its slope and its y-intercept.

Find the equation of the line that contains the given point and has the given slope.

3. Point (0, 5), $m = 2$

4. Point (0, 3), $m = 1$

5. Point (2, 3), $m = \frac{1}{2}$

6. Point (5, 1), $m = \frac{2}{3}$

7. Point (3, 0), $m = -\frac{5}{3}$

8. Point (−2, 0), $m = \frac{3}{2}$

9. Point (−1, 7), $m = -3$

10. Point (−2, 4), $m = -4$

11. Point (0, 0), $m = \frac{1}{2}$

12. Point (0, 0), $m = \frac{3}{4}$

13. Point (2, −3), $m = 3$

14. Point (4, −5), $m = 2$

15. Point (3, 5), $m = -\frac{2}{3}$

16. Point (5, 1), $m = -\frac{4}{5}$

17. Point (0, −3), $m = -1$

18. Point (2, 0), $m = \frac{5}{6}$

19. Point (3, −4), slope is undefined

20. Point (−2, 5), slope is undefined

21. Point (−2, −3), $m = 0$

22. Point (−3, −2), $m = 0$

23. Point (4, −5), $m = -2$

24. Point (−3, 5), $m = 3$

25. Point (−5, −1), slope is undefined

26. Point (0, 4), slope is undefined

2 Find the equation of the line that contains the given points.

27. $P_1(0, 2)$, $P_2(3, 5)$

28. $P_1(0, 4)$, $P_2(1, 5)$

29. $P_1(0, -3)$, $P_2(-4, 5)$

30. $P_1(0, -2)$, $P_2(-3, 4)$

31. $P_1(-1, 3)$, $P_2(2, 4)$

32. $P_1(-1, 1)$, $P_2(4, 4)$

33. $P_1(0, 3)$, $P_2(2, 0)$

34. $P_1(0, 4)$, $P_2(2, 0)$

35. $P_1(-2, -3)$, $P_2(-1, -2)$

36. $P_1(4, 1)$, $P_2(3, -2)$

37. $P_1(2, 3)$, $P_2(5, 5)$

38. $P_1(7, 2)$, $P_2(4, 4)$

39. $P_1(2, 0)$, $P_2(0, -1)$

40. $P_1(0, 4)$, $P_2(-2, 0)$

41. $P_1(3, -4)$, $P_2(-2, -4)$

42. $P_1(-3, 3)$, $P_2(-2, 3)$

43. $P_1(0, 0)$, $P_2(4, 3)$

44. $P_1(2, -5)$, $P_2(0, 0)$

45. $P_1(-2, 5)$, $P_2(-2, -5)$

46. $P_1(3, 2)$, $P_2(3, -4)$

47. $P_1(2, 1)$, $P_2(-2, -3)$

48. $P_1(-3, -2)$, $P_2(1, -4)$

49. $P_1(0, 3)$, $P_2(3, 0)$

50. $P_1(1, -3)$, $P_2(-2, 4)$

3 **51.** **Aeronautics** The pilot of a Boeing 757 jet takes off from Boston's Logan Airport, which is at sea level, and climbs to a cruising altitude of 32,000 ft at a constant rate of 1200 ft/min. Write a linear equation for the height of the plane in terms of the time after take-off. Use your equation to find the height of the plane 11 min after take-off.

52. **Physical Fitness** A jogger running at 9 mph burns approximately 14 calories per minute. Write a linear equation for the number of calories burned by the jogger in terms of the number of minutes run. Use your equation to find the number of calories burned after jogging for 32 min.

53. **Telecommunications** A cellular phone company offers several different options for using a cellular telephone. One option, for people who plan on using the phone only in emergencies, costs the user $4.95 per month plus $.59 per minute for each minute the phone is used. Write a linear equation for the monthly cost of the phone in terms of the number of minutes the phone is used. Use your equation to find the monthly cost of using the cellular phone for 13 min in one month.

54. **Aeronautics** An Airbus 320 plane takes off from Denver International Airport, which is 5200 ft above sea level, and climbs to 30,000 ft at a constant rate of 1000 ft/min. Write a linear equation for the height of the plane in terms of the time after take-off. Use your equation to find the height of the plane 8 min after take-off.

55. **Nutrition** There are approximately 126 calories in a 2-ounce serving of lean hamburger and approximately 189 calories in a 3-ounce serving. Write a linear equation for the number of calories in lean hamburger in terms of the size of the serving. Use your equation to estimate the number of calories in a 5-ounce serving of lean hamburger.

56. **Chemistry** At sea level, the boiling point of water is 100°C. At an altitude of 2 km, the boiling point of water is 93°C. Write a linear equation for the boiling point of water in terms of the altitude above sea level. Use your equation to predict the boiling point of water on the top of Mount Everest, which is approximately 8.85 km above sea level. Round to the nearest degree.

57. **History** In 1927, Charles Lindbergh made history by making the first transatlantic flight from New York to Paris. It took Lindbergh approximately 33.5 h to make the trip. In 1997, the Concorde could make the trip in approximately 3.3 h. Write a linear equation of the time, in hours, it takes to cross the Atlantic in terms of the year. Use your equation to predict how long a flight between the two cities would have taken in 1967. Round your answer to the nearest tenth. On the basis of your answer, do you think a linear model accurately predicts how the flying time between New York and Paris has changed?

58. **Computers** On the basis of data from the U.S. Department of Commerce, there were 24 million homes with computers in 1991. The average rate of growth in computers in homes is expected to increase by 2.4 million homes per year through 2005. Write a linear equation for the number of computers in homes in terms of the year. Use your equation to predict the number of computers in homes in 2001.

59. **Automotive Technology** The gas tank of a certain car contains 16 gal when the driver of the car begins a trip. Each mile driven by the driver decreases the amount of gas in the tank by 0.032 gal. Write a linear equation for the number of gallons of gas in the tank in terms of the number of miles driven. Use your equation to find the number of gallons in the tank after driving 150 mi.

60. **Oceanography** Whales, dolphins, and porpoises communicate using high-pitched sounds that travel through the water. The speed at which the sound travels depends on many factors, one of which is the depth of the water. At approximately 1000 m below sea level, the speed of sound is 1480 m/s. Below 1000 m, the speed of sound increases at a constant rate of 0.017 m/s for each additional meter below 1000 m. Write a linear equation for the speed of sound in terms of the number of meters below sea level. Use your equation to approximate the speed of sound 2500 m below sea level. Round to the nearest meter per second.

3.5 Applying Concepts

61. What is the x-intercept of the graph of $y = mx + b$?

62. A line contains the points $(4, -1)$ and $(2, 1)$. Find the coordinates of three other points that are on this line.

63. Given that f is a linear function for which $f(1) = 3$ and $f(-1) = 5$, determine $f(4)$.

64. Find the equation of the line that passes through the midpoint of the line segment between $P_1(2, 5)$ and $P_2(-4, 1)$ and has slope -2.

65. If $y = mx + b$, where m is a given constant, how does the graph of this equation change as the value of b changes?

66. Explain the similarities and differences between the point-slope formula and the slope-intercept form of a straight line.

67. Explain why the point-slope formula cannot be used to find the equation of a line that is parallel to the y-axis.

3.6 Parallel and Perpendicular Lines

1 Find equations of parallel and perpendicular lines

VIDEO & DVD CD TUTOR WEB SSM

Two lines that have the same slope do not intersect and are called **parallel lines.**

The slope of each of the lines at the right is $\frac{2}{3}$.

The lines are parallel.

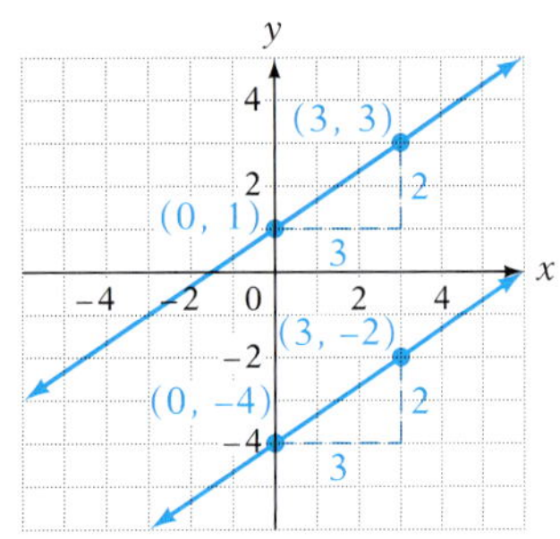

Slopes of Parallel Lines

Two nonvertical lines with slopes of m_1 and m_2 are parallel if and only if $m_1 = m_2$. Any two vertical lines are parallel.

Is the line that contains the points $(-2, 1)$ and $(-5, -1)$ parallel to the line that contains the points $(1, 0)$ and $(4, 2)$?

Find the slope of the line through $(-2, 1)$ and $(-5, -1)$.

$$m_1 = \frac{-1 - 1}{-5 - (-2)} = \frac{-2}{-3} = \frac{2}{3}$$

Find the slope of the line through $(1, 0)$ and $(4, 2)$.

$$m_2 = \frac{2 - 0}{4 - 1} = \frac{2}{3}$$

Because $m_1 = m_2$, the lines are parallel.

Find the equation of the line that contains the point (2, 3) and is parallel to the line $y = \frac{1}{2}x - 4$.

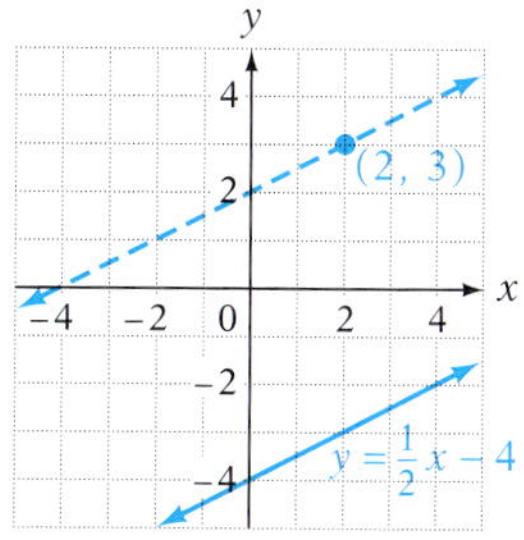

The slope of the given line is $\frac{1}{2}$. Because parallel lines have the same slope, the slope of the unknown line is also $\frac{1}{2}$.

Use the point-slope formula. $\quad y - y_1 = m(x - x_1)$

Substitute $m = \frac{1}{2}$ and $(x_1, y_1) = (2, 3)$. $\quad y - 3 = \frac{1}{2}(x - 2)$

Solve for y.

$$y - 3 = \frac{1}{2}x - 1$$

$$y = \frac{1}{2}x + 2$$

The equation of the line is $y = \frac{1}{2}x + 2$.

Find the equation of the line that contains the point $(-1, 4)$ and is parallel to the line $2x - 3y = 5$.

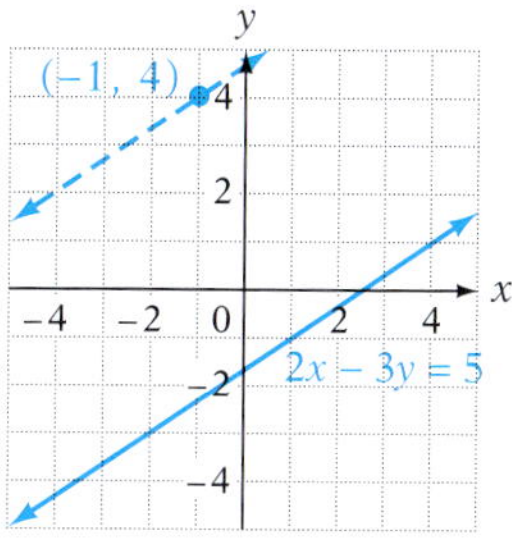

Because the lines are parallel, the slope of the unknown line is the same as the slope of the given line. Solve $2x - 3y = 5$ for y and determine its slope.

$$2x - 3y = 5$$

$$-3y = -2x + 5$$

$$y = \frac{2}{3}x - \frac{5}{3}$$

The slope of the given line is $\frac{2}{3}$. Because the lines are parallel, the slope of the unknown line is also $\frac{2}{3}$.

Use the point-slope formula. $\quad y - y_1 = m(x - x_1)$

Substitute $m = \frac{2}{3}$ and $(x_1, y_1) = (-1, 4)$. $\quad y - 4 = \frac{2}{3}[x - (-1)]$

Solve for y.

$$y - 4 = \frac{2}{3}(x + 1)$$

$$y - 4 = \frac{2}{3}x + \frac{2}{3}$$

$$y = \frac{2}{3}x + \frac{14}{3}$$

The equation of the line is $y = \frac{2}{3}x + \frac{14}{3}$.

Two lines that intersect at right angles are **perpendicular lines.**

Any horizontal line is perpendicular to any vertical line. For example, $x = 3$ is perpendicular to $y = -2$ as shown in the graph at the right.

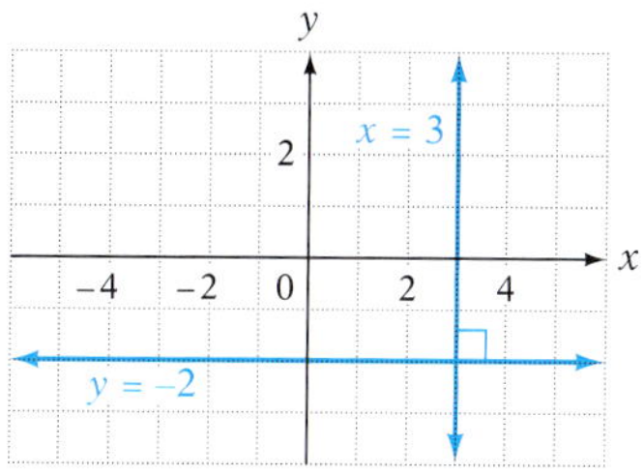

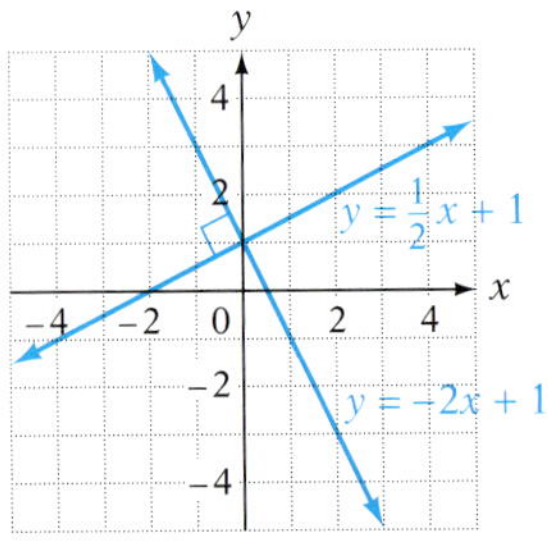

Slopes of Perpendicular Lines

If m_1 and m_2 are the slopes of two lines, neither of which is vertical, then the lines are perpendicular if and only if $m_1 \cdot m_2 = -1$.

A vertical line is perpendicular to a horizontal line.

Solving $m_1 \cdot m_2 = -1$ for m_1 gives $m_1 = -\frac{1}{m_2}$. This last equation states that the slopes of perpendicular lines are **negative reciprocals** of each other.

Is the line that contains the points (4, 2) and (−2, 5) perpendicular to the line that contains the points (−4, 3) and (−3, 5)?

Find the slope of the line that passes through the points (4, 2) and (−2, 5).
$$m_1 = \frac{5-2}{-2-4} = \frac{3}{-6} = -\frac{1}{2}$$

Find the slope of the line that passes through the points (−4, 3) and (−3, 5).
$$m_2 = \frac{5-3}{-3-(-4)} = \frac{2}{1} = 2$$

Find the product of the two slopes.
$$m_1 \cdot m_2 = -\frac{1}{2}(2) = -1$$

Because $m_1 \cdot m_2 = -1$, the lines are perpendicular.

Are the graphs of the lines whose equations are $3x + 4y = 8$ and $8x + 6y = 5$ perpendicular?

To determine whether the lines are perpendicular, solve each equation for y and find the slope of each line.

$$3x + 4y = 8$$
$$4y = -3x + 8$$
$$y = -\frac{3}{4}x + 2$$
$$m_1 = -\frac{3}{4}$$

$$8x + 6y = 5$$
$$6y = -8x + 5$$
$$y = -\frac{4}{3}x + \frac{5}{6}$$
$$m_2 = -\frac{4}{3}$$

Then check whether the equation $m_1 \cdot m_2 = -1$ is true.
$$m_1 \cdot m_2 = \left(-\frac{3}{4}\right)\left(-\frac{4}{3}\right) = 1$$

Because $m_1 \cdot m_2 = 1 \neq -1$, the lines are not perpendicular.

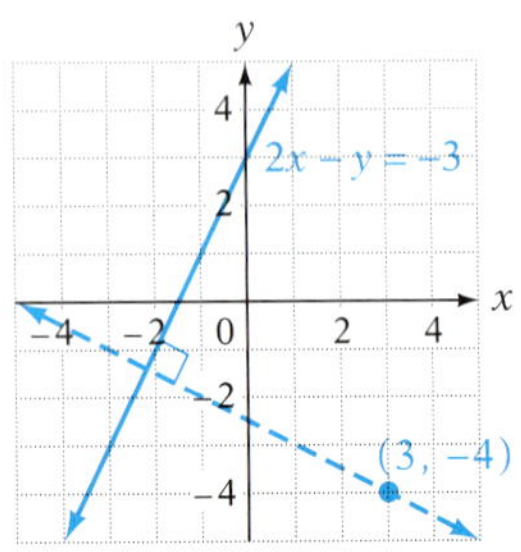

Find the equation of the line that contains the point (3, −4) and is perpendicular to the line $2x - y = -3$.

Determine the slope of the given line by solving the equation for y.
$$2x - y = -3$$
$$-y = -2x - 3$$
$$y = 2x + 3$$

The slope is 2.

The slope of the line perpendicular to the given line is $-\frac{1}{2}$, the negative reciprocal of 2.

Use the point-slope formula. $y - y_1 = m(x - x_1)$

Substitute $m = -\frac{1}{2}$ and $(x_1, y_1) = (3, -4)$. $y - (-4) = -\frac{1}{2}(x - 3)$

Solve for y.

$$y + 4 = -\frac{1}{2}x + \frac{3}{2}$$
$$y = -\frac{1}{2}x - \frac{5}{2}$$

The equation of the perpendicular line is $y = -\frac{1}{2}x - \frac{5}{2}$.

Example 1 Is the line that contains the points (−4, 2) and (1, 6) parallel to the line that contains the points (2, −4) and (7, 0)?

Solution $(x_1, y_1) = (-4, 2), (x_2, y_2) = (1, 6)$

$$m_1 = \frac{6 - 2}{1 - (-4)} = \frac{4}{5}$$ • Find the slope of each line.

$(x_1, y_1) = (2, -4), (x_2, y_2) = (7, 0)$

$$m_2 = \frac{0 - (-4)}{7 - 2} = \frac{4}{5}$$

$$m_1 = m_2 = \frac{4}{5}$$ • Check whether the slopes are equal.

Yes, the lines are parallel.

Problem 1 Is the line that contains the points (−2, −3) and (7, 1) perpendicular to the line that contains the points (4, 1) and (6, −5)?

Solution See page S13.

Example 2 Are the lines $4x - y = -2$ and $x + 4y = -12$ perpendicular?

Solution

$$4x - y = -2$$
$$-y = -4x - 2$$
$$y = 4x + 2 \qquad m_1 = 4$$

• Solve each equation for y to find the slopes.

$$x + 4y = -12$$
$$4y = -x - 12$$
$$y = -\frac{1}{4}x - 3 \qquad m_2 = -\frac{1}{4}$$

$$m_1 \cdot m_2 = 4\left(-\frac{1}{4}\right) = -1$$ • Find the product of the slopes.

Yes, the lines are perpendicular.

Problem 2 Are the lines $5x + 2y = 2$ and $5x + 2y = -6$ parallel?

Solution See page S13.

Example 3 Find the equation of the line that contains the point (3, −1) and is parallel to the line $3x - 2y = 4$.

Solution

$$3x - 2y = 4$$
$$-2y = -3x + 4$$
$$y = \frac{3}{2}x - 2$$
• Solve the equation for y to find the slope.

$$m = \frac{3}{2}$$

$$y - y_1 = m(x - x_1)$$
• Use the point-slope formula.

$$y - (-1) = \frac{3}{2}(x - 3)$$
• $m = \frac{3}{2}$, $(x_1, y_1) = (3, -1)$

$$y + 1 = \frac{3}{2}x - \frac{9}{2}$$
$$y = \frac{3}{2}x - \frac{11}{2}$$

Problem 3 Find the equation of the line that contains the point (−2, 3) and is perpendicular to the line $x - 4y = 3$.

Solution See page S13.

3.6 Concept Review

Determine whether the following statements are always true, sometimes true, or never true.

1. Perpendicular lines have the same y-intercept.

2. Parallel lines have the same slope.

3. The lines given by the equations $y = \frac{3}{2}x + 2$ and $y = -\frac{3}{2}x + 2$ are perpendicular.

4. Two lines are perpendicular if $m_1 \cdot m_2 = 1$.

5. A vertical line is perpendicular to a horizontal line.

6. A line parallel to the y-axis has zero slope.

7. The lines $y = \frac{1}{2}x - 2$ and $y = \frac{1}{2}x - 4$ are parallel.

3.6 Exercises

1

1. Given the slopes of two lines, explain how to determine whether the two lines are parallel.

2. Given the slopes of two lines, how can you to determine whether the two lines are perpendicular?

3. The slope of a line is -5. What is the slope of any line parallel to this line?

4. The slope of a line is $\frac{3}{2}$. What is the slope of any line parallel to this line?

5. The slope of a line is 4. What is the slope of any line perpendicular to this line?

6. The slope of a line is $-\frac{4}{5}$. What is the slope of any line perpendicular to this line?

7. Is the line $x = -2$ perpendicular to the line $y = 3$?

8. Is the line $y = \frac{1}{2}$ perpendicular to the line $y = -4$?

9. Is the line $x = -3$ parallel to the line $y = \frac{1}{3}$?

10. Is the line $x = 4$ parallel to the line $y = -4$?

11. Is the line $y = \frac{2}{3}x - 4$ parallel to the line $y = -\frac{3}{2}x - 4$?

12. Is the line $y = -2x + \frac{2}{3}$ parallel to the line $y = -2x + 3$?

13. Is the line $y = \frac{4}{3}x - 2$ perpendicular to the line $y = -\frac{3}{4}x + 2$?

14. Is the line $y = \frac{1}{2}x + \frac{3}{2}$ perpendicular to the line $y = -\frac{1}{2}x + \frac{3}{2}$?

15. Are the lines $2x + 3y = 2$ and $2x + 3y = -4$ parallel?

16. Are the lines $2x - 4y = 3$ and $2x + 4y = -3$ parallel?

17. Are the lines $x - 4y = 2$ and $4x + y = 8$ perpendicular?

18. Are the lines $4x - 3y = 2$ and $4x + 3y = -7$ perpendicular?

19. Is the line that contains the points $(3, 2)$ and $(1, 6)$ parallel to the line that contains the points $(-1, 3)$ and $(-1, -1)$?

20. Is the line that contains the points $(4, -3)$ and $(2, 5)$ parallel to the line that contains the points $(-2, -3)$ and $(-4, 1)$?

21. Is the line that contains the points $(-3, 2)$ and $(4, -1)$ perpendicular to the line that contains the points $(1, 3)$ and $(-2, -4)$?

22. Is the line that contains the points $(-1, 2)$ and $(3, 4)$ perpendicular to the line that contains the points $(-1, 3)$ and $(-4, 1)$?

23. Is the line that contains the points $(-5, 0)$ and $(0, 2)$ parallel to the line that contains the points $(5, 1)$ and $(0, -1)$?

24. Is the line that contains the points $(3, 5)$ and $(-3, 3)$ perpendicular to the line that contains the points $(2, -5)$ and $(-4, 4)$?

25. Find the equation of the line that contains the point $(-2, -4)$ and is parallel to the line $2x - 3y = 2$.

26. Find the equation of the line that contains the point $(3, 2)$ and is parallel to the line $3x + y = -3$.

27. Find the equation of the line that contains the point $(4, 1)$ and is perpendicular to the line $y = -3x + 4$.

28. Find the equation of the line that contains the point $(2, -5)$ and is perpendicular to the line $y = \frac{5}{2}x - 4$.

29. Find the equation of the line that contains the point $(-1, -3)$ and is perpendicular to the line $3x - 5y = 2$.

30. Find the equation of the line that contains the point $(-1, 3)$ and is perpendicular to the line $2x + 4y = -1$.

3.6 Applying Concepts

31. If the graphs of $A_1x + B_1y = C_1$ and $A_2x + B_2y = C_2$ are parallel, express $\frac{A_1}{B_1}$ in terms of A_2 and B_2.

32. If the graphs of $A_1x + B_1y = C_1$ and $A_2x + B_2y = C_2$ are perpendicular, express $\frac{A_1}{B_1}$ in terms of A_2 and B_2.

For Exercises 33 and 34, suppose a ball is being twirled at the end of a string and the center of rotation is the origin of a coordinate system. If the string breaks, the initial path of the ball is on a line that is perpendicular to the radius of the circle.

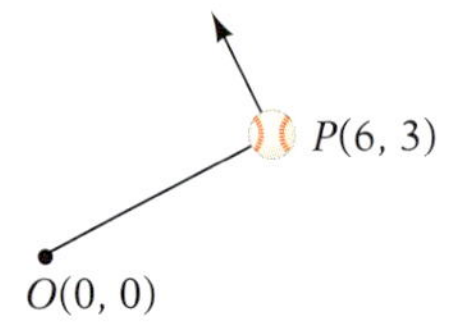

33. Suppose the string breaks when the ball is at the point whose coordinates are $P(6, 3)$. Find the equation of the line on which the initial path lies.

34. Suppose the string breaks when the ball is at the point whose coordinates are $P(2, 8)$. Find the equation of the line on which the initial path lies.

35. The graphs of $y = -\frac{1}{2}x + 2$ and $y = \frac{2}{3}x - 5$ intersect at the point whose coordinates are $(6, -1)$. Find the equation of a line whose graph intersects the graphs of the given lines to form a right triangle. (*Hint:* There is more than one answer to this question.)

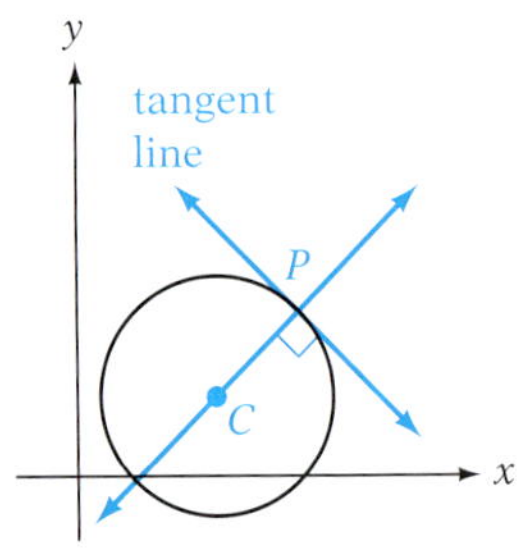

36. **Geometry** A theorem from geometry states that a line passing through the center of a circle and through a point P on the circle is perpendicular to the tangent line at P. (See the figure at the right.)

a. If the coordinates of P are $(5, 4)$ and the coordinates of C are $(3, 2)$, what is the equation of the tangent line?

b. What are the x- and y-intercepts of the tangent line?

3.7 Inequalities in Two Variables

1 Graph the solution set of an inequality in two variables

VIDEO & DVD CD TUTOR WEB SSM

The graph of the linear equation $y = x - 1$ separates the plane into three sets: the set of points on the line, the set of points above the line, and the set of points below the line.

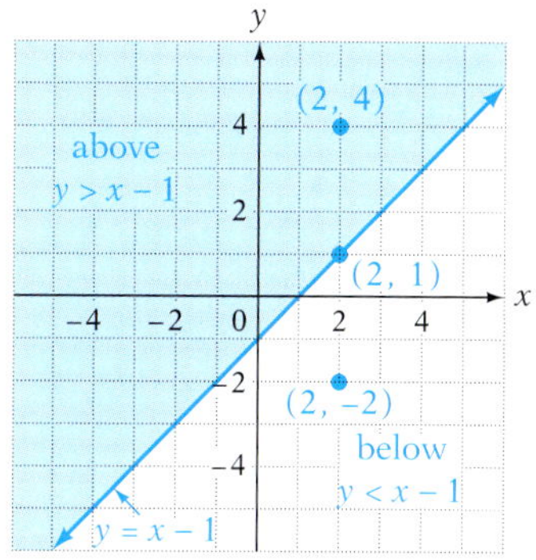

The point whose coordinates are (2, 1) is a solution of $y = x - 1$ and is a point on the line.

The point whose coordinates are (2, 4) is a solution of $y > x - 1$ and is a point above the line.

The point whose coordinates are (2, −2) is a solution of $y < x - 1$ and is a point below the line.

The set of points on the line are the solutions of the equation $y = x - 1$. The set of points above the line are the solutions of the inequality $y > x - 1$. These points form a **half-plane.** The set of points below the line are solutions of the inequality $y < x - 1$. These points also form a half-plane.

An inequality of the form $y > mx + b$ or $Ax + By > C$ is a **linear inequality in two variables.** (The inequality symbol could be replaced by $\geq$, $<$, or $\leq$.) The solution set of a linear inequality in two variables is a half-plane.

The following illustrates the procedure for graphing the solution set of a linear inequality in two variables.

Take Note

When solving the inequality at the right for *y*, we divide both sides of the inequality by −4, so the inequality symbol must be reversed.

Graph the solution set of $3x - 4y < 12$.

Solve the inequality for *y*.

$$\begin{aligned} 3x - 4y &< 12 \\ -4y &< -3x + 12 \\ y &> \frac{3}{4}x - 3 \end{aligned}$$

Change the inequality $y > \frac{3}{4}x - 3$ to the equality $y = \frac{3}{4}x - 3$, and graph the line.

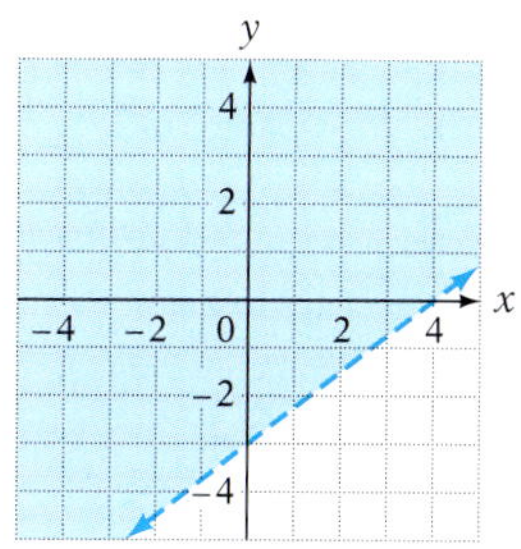

If the inequality contains $\leq$ or $\geq$, the line belongs to the solution set and is shown by a *solid line*. If the inequality contains $<$ or $>$, the line is not part of the solution set and is shown by a *dashed line*.

If the inequality contains $>$ or $\geq$, shade the upper half-plane. If the inequality contains $<$ or $\leq$, shade the lower half-plane.

As a check, use the ordered pair (0, 0) to determine whether the correct region of the plane has been shaded. If (0, 0) is a solution of the inequality, then (0, 0) should be in the shaded region. If (0, 0) is not a solution of the inequality, then (0, 0) should not be in the shaded region.

As shown below, (0, 0) is a solution of the inequality in the example above.

$$y > \frac{3}{4}x - 3$$
$$0 > \frac{3}{4}(0) - 3$$
$$0 > 0 - 3$$
$$0 > -3 \quad \text{True}$$

Because (0, 0) is a solution of the inequality, (0, 0) should be in the shaded region. The solution set as graphed above is correct.

If the line passes through the point (0, 0), another point, such as (0, 1), must be used as a check.

From the graph of $y > \frac{3}{4}x - 3$, note that for a given value of x, more than one value of y can be paired with the value of x. For instance, (4, 1), (4, 3), (5, 1), and $\left(5, \frac{9}{4}\right)$ are all ordered pairs that belong to the graph. Because there are ordered pairs with the same first component and different second components, the inequality does not represent a function. The inequality is a relation but not a function.

Example 1 Graph the solution set of $x + 2y \le 4$.

Solution

$$x + 2y \le 4$$
$$2y \le -x + 4$$
$$y \le -\frac{1}{2}x + 2$$

• Solve the inequality for y.

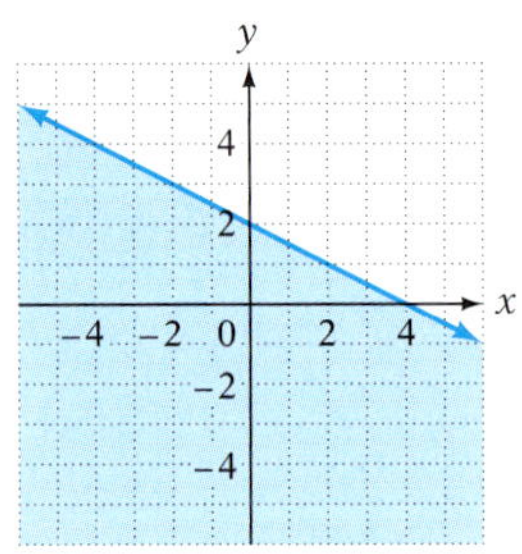

• Graph $y = -\frac{1}{2}x + 2$ as a solid line. Shade the lower half-plane.

Problem 1 Graph the solution set of $x + 3y > 6$.

Solution See page S13.

Example 2 Graph the solution set of $x \ge -1$.

Solution

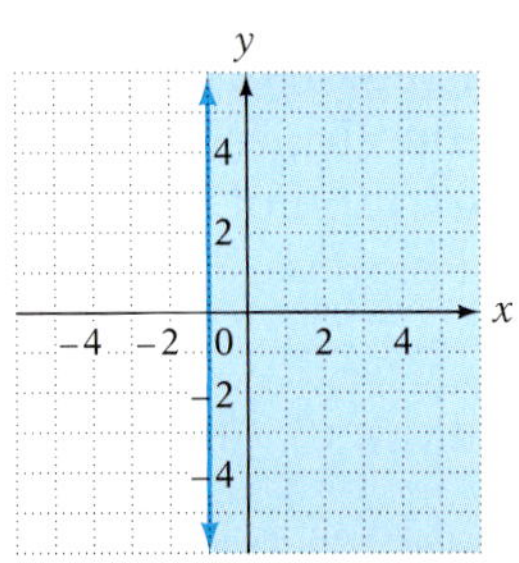

• Graph $x = -1$ as a solid line.
• The point (0, 0) satisfies the inequality. Shade the half-plane to the right of the line.

Problem 2 Graph the solution set of $y < 2$.

Solution See page S13.

3.7 Concept Review

Determine whether the following statements are always true, sometimes true, or never true.

1. It is possible to write a linear inequality in two variables that has no solutions.
2. The exponents on the variables in a linear inequality in two variables are 1.
3. The graph of a linear inequality is a half-plane.
4. The graph of a linear inequality in two variables represents a function.
5. The solution of a linear inequality in two variables containing $\leq$ or $\geq$ includes the line separating the half-planes.
6. The solution of the inequality $y > x + 2$ is all the points above the line $y = x + 2$.

3.7 Exercises

1 **1.** What is a half-plane?

2. Explain a method that you can use to check that the graph of a linear inequality in two variables has been shaded correctly.

3. Is (0, 0) a solution of $y > 2x - 7$?

4. Is (0, 0) a solution of $y < 5x + 3$?

5. Is (0, 0) a solution of $y \leq -\frac{2}{3}x - 8$?

6. Is (0, 0) a solution of $y \geq -\frac{3}{4}x + 9$?

Graph the solution set.

7. $y \leq \frac{3}{2}x - 3$

8. $y \geq \frac{4}{3}x - 4$

9. $y < \frac{4}{5}x - 2$

10. $y < \frac{3}{5}x - 3$

11. $y < -\frac{1}{3}x + 2$

12. $y < -\frac{4}{3}x + 3$

13. $x + 3y < 4$

14. $2x - 5y \leq 10$

15. $2x + 3y \geq 6$

16. $3x + 2y < 4$

17. $-x + 2y > -8$

18. $-3x + 2y > 2$

19. $y - 4 < 0$

20. $x + 2 \geq 0$

21. $6x + 5y < 15$

22. $3x - 5y < 10$

23. $-5x + 3y \geq -12$

24. $3x + 4y \geq 12$

3.7 Applying Concepts

25. Does the inequality $y < 3x - 1$ represent a function? Explain your answer.

26. Are there any points whose coordinates satisfy both $y \leq x + 3$ and $y \geq -\frac{1}{2}x + 1$? If so, give the coordinates of three such points. If not, explain why not.

27. Are there any points whose coordinates satisfy both $y \leq x - 1$ and $y \geq x + 2$? If so, give the coordinates of three such points. If not, explain why not.

Focus on Problem Solving

Find a Pattern

Polya's four recommended problem-solving steps are stated below.

1. Understand the problem.
2. Devise a plan.
3. Carry out the plan.
4. Review your solution

Point of Interest

Carl Friedrich Gauss (1777–1855) has been called the "Prince of Mathematicians" by some historians. He applied his genius to many areas of mathematics and science. A unit of magnetism, the gauss, *is named in his honor. Some electronic equipment (televisions, for instance) contain a* degausser *that controls magnetic fields.*

One of the several ways of devising a plan is first to try to find a pattern. Karl Friedrich Gauss supposedly used this method to solve a problem that was given to his math class when he was in elementary school. As the story goes, his teacher wanted to grade some papers while the class worked on a math problem. The problem given to the class was to find the sum

$$1 + 2 + 3 + 4 + \cdots + 100$$

Gauss quickly solved the problem by seeing a pattern. Here is what he saw.

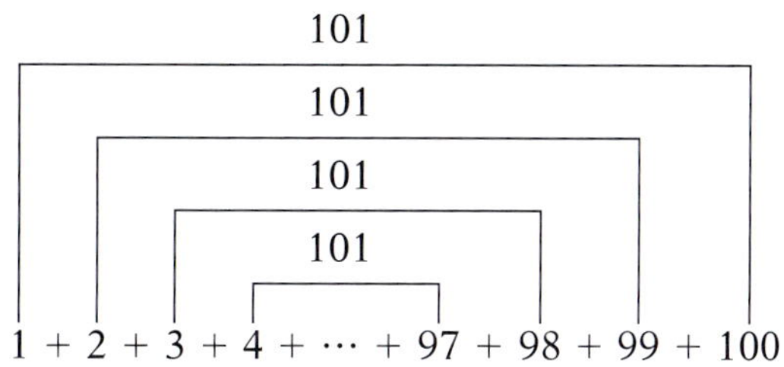

Note that

$$1 + 100 = 101$$
$$2 + 99 = 101$$
$$3 + 98 = 101$$
$$4 + 97 = 101$$

Gauss noted that there were 50 sums of 101. Therefore, the sum of the first 100 natural numbers is

$$1 + 2 + 3 + 4 + \cdots + 97 + 98 + 99 + 100 = 50(101) = 5050$$

Try to solve the following problems by finding a pattern.

1. Find the sum $2 + 4 + 6 + \cdots + 96 + 98 + 100$.
2. Find the sum $1 + 3 + 5 + \cdots + 97 + 99 + 101$.
3. Find another method of finding the sum $1 + 3 + 5 + \cdots + 97 + 99 + 101$ given in the previous exercise.
4. Find the sum $\frac{1}{1 \cdot 2} + \frac{1}{2 \cdot 3} + \frac{1}{3 \cdot 4} + \cdots + \frac{1}{49 \cdot 50}$.

   Hint: $\frac{1}{1 \cdot 2} = \frac{1}{2}$, $\frac{1}{1 \cdot 2} + \frac{1}{2 \cdot 3} = \frac{2}{3}$, $\frac{1}{1 \cdot 2} + \frac{1}{2 \cdot 3} + \frac{1}{3 \cdot 4} = \frac{3}{4}$

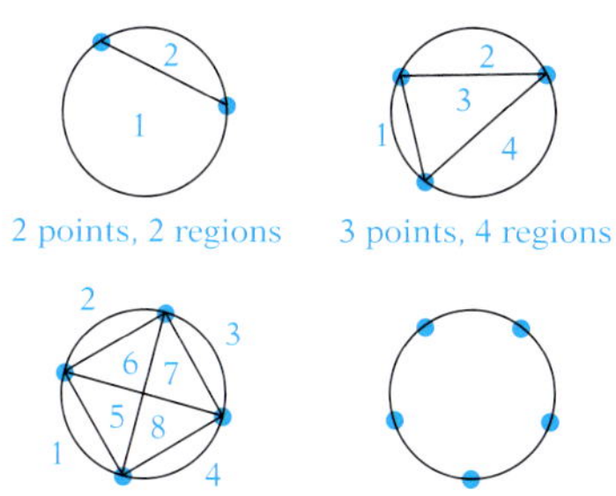

2 points, 2 regions
3 points, 4 regions
4 points, 8 regions
5 points, ? regions

5. The following problem shows that checking a few cases does not always result in a conjecture that is true for *all* cases. Select any two points on a circle and draw a *chord,* the line connecting the points (see the drawing in the left margin). The chord divides the circle into two regions. Now select three different points and draw chords connecting each of the three points with every other point. The chords divide the circle into four regions. Now select four points and connect each of the points with every other point. Make a conjecture about the relationship between the number of regions and the number of points on the circle. Does your conjecture work for five points? six points?

6. A *polygonal number* is a number that can be represented by arranging that number of dots in rows to form a geometric figure such as a triangle, square, pentagon, or hexagon. For instance, the first four *triangular* numbers, 3, 6, 10, and 15, are given (and shown) below. What are the next two triangular numbers?

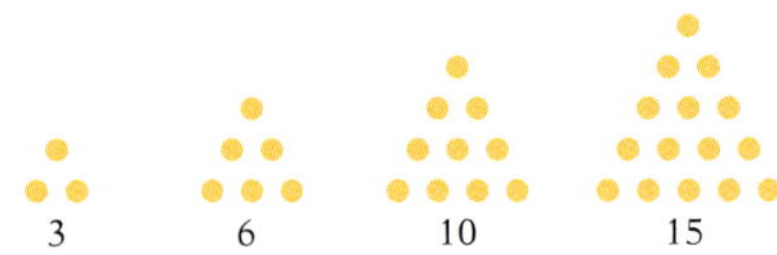

Projects & Group Activities

Evaluating a Function with a Graphing Calculator

You can use a graphing calculator to evaluate some functions. Shown below are the keystrokes needed to evaluate $f(x) = 3x^2 + 2x - 1$ for $x = -2$ on a TI-83. (There are other methods of evaluating functions. This is just one of them.) Try these keystrokes. The calculator should display 7 as the value of the function.

[Y=] [CLEAR] 3 [X,T,θ,n] [X²] [+]
2 [X,T,θ,n] [−] 1 [2nd] QUIT [(−)]
2 [STO] [X,T,θ,n] [ENTER]
[VARS] [▶] 1 1 [ENTER]

Introduction to Graphing Calculators

There are a variety of computer programs and calculators that can graph an equation. A computer or graphing calculator screen is divided into pixels. Depending on the computer or calculator, there are approximately 6000 to 790,000 pixels available on the screen. The greater the number of pixels, the smoother the graph will appear. A portion of a screen is shown at the left. Each little rectangle represents one pixel.

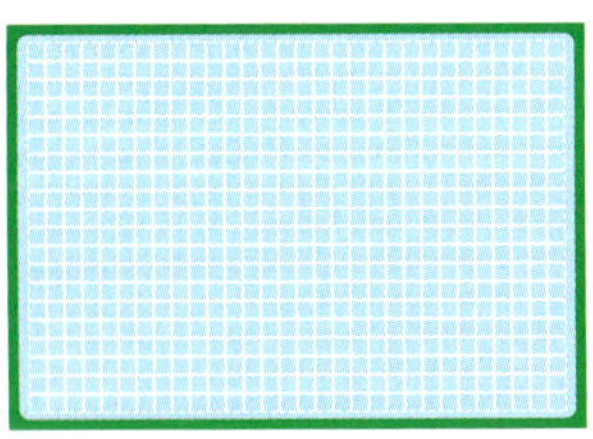
Calculator Screen

A graphing calculator draws a graph in a manner similar to the way we have shown in this chapter. Values of x are chosen and ordered pairs calculated. Then a graph is drawn through those points by illuminating pixels (an abbreviation for "picture element") on the screen.

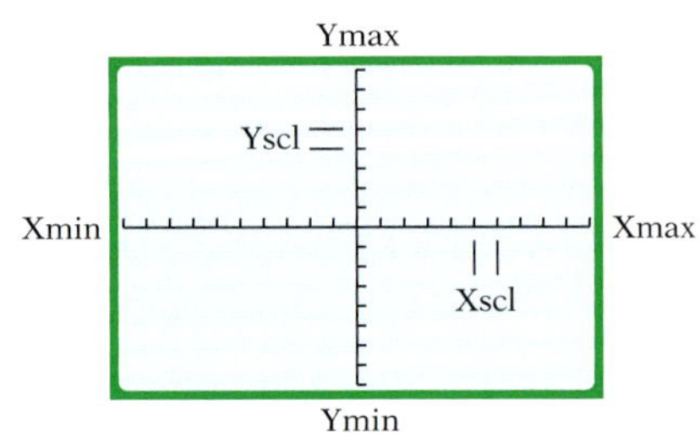

Graphing utilities can display only a portion of the xy-plane, called a window. The window [Xmin, Xmax] by [Ymin, Ymax] consists of those points (x, y) that satisfy both of the following:

$$\text{Xmin} \le x \le \text{Xmax} \quad \text{and} \quad \text{Ymin} \le y \le \text{Ymax}$$

The user sets these values before a graph is drawn.

The numbers Xscl and Yscl are the distances between the tick marks that are drawn on the x- and y-axes. If you do not want tick marks on the axes, set Xscl = 0 and Yscl = 0.

The graph at the right is a portion of the graph of $y = \frac{1}{2}x + 1$, as it was drawn with a graphing calculator. The window is

Xmin = −4.7, Xmax = 4.7
Ymin = −3.1, Ymax = 3.1
Xscl = 1 and Yscl = 1

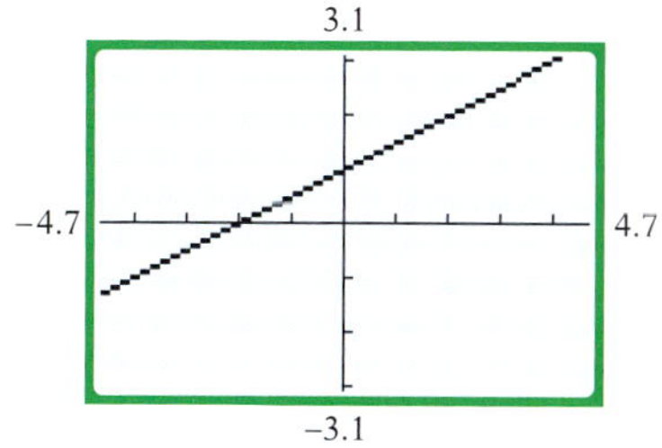

Using interval notation, this is shown as [−4.7, 4.7] by [−3.1, 3.1].

The window [−4.7, 4.7] by [−3.1, 3.1] gives "nice" coordinates in the sense that each time the ▶ or the ◀ is pressed, the change in x is 0.1. The reason for this is that the horizontal distance from the middle of the first pixel to the middle of the last pixel is 94 units. By using Xmin = −4.7 and Xmax = 4.7, we have[1]

$$\text{Change in } x = \frac{\text{Xmax} - \text{Xmin}}{94} = \frac{4.7 - (-4.7)}{94} = \frac{9.4}{94} = 0.1$$

Similarly, the vertical distance from the middle of the first pixel to the middle of the last pixel is 62 units. Therefore, using Ymin = −3.1 and Ymax = 3.1 will give nice coordinates in the vertical direction.

Take Note

The Appendix "Guidelines for Using Graphing Calculators" contains keystroking suggestions to use with this project.

Graph each of the following by using a graphing calculator.

1. $y = 2x + 1$ — For $2x$, you may enter $2 \times x$ or just $2x$. The times sign × is not necessary on many graphing calculators.
2. $y = -x + 2$ — Many calculators use the (−) key to enter a negative sign.
3. $3x + 2y = 6$ — Solve for y. Then enter the equation.
4. $y = 50x$ — You must adjust the viewing window. Try the window given by [−4.7, 4.7] by [−250, 250] with Yscl = 50.
5. $y = \frac{2}{3}x - 3$ — You may enter $\frac{2}{3}x$ as $2x/3$ or $(2/3)x$. Although entering $2/3x$ works on some calculators, it is not recommended.
6. $4x + 3y = 75$ — You must adjust the viewing window.

Wind-Chill Index

The wind-chill index is the temperature of still air that would have the same effect on exposed human skin as a given combination of wind speed and air temperature. For example, given a wind speed of 10 mph and a temperature reading of 20°F, the wind-chill index is 9°F.

In the fall of 2001, the U.S. National Weather Service began using a new wind-chill formula. The following ordered pairs are derived from the new formula.

[1]Some calculators have screen widths of 126 pixels. For those calculators, use Xmin = −6.3 and Xmax = 6.3 to obtain nice coordinates.

In each ordered pair, the abscissa is the air temperature in degrees Fahrenheit, and the ordinate is the wind-chill index when the wind speed is 10 mph.

{(35, 27), (30, 21), (25, 15), (20, 9), (15, 3), (10, −4), (5, −10), (0, −16), (−5, −22), (−10, −28), (−15, −35), (−20, −41), (−25, −47), (−30, −53), (−35, −59)}

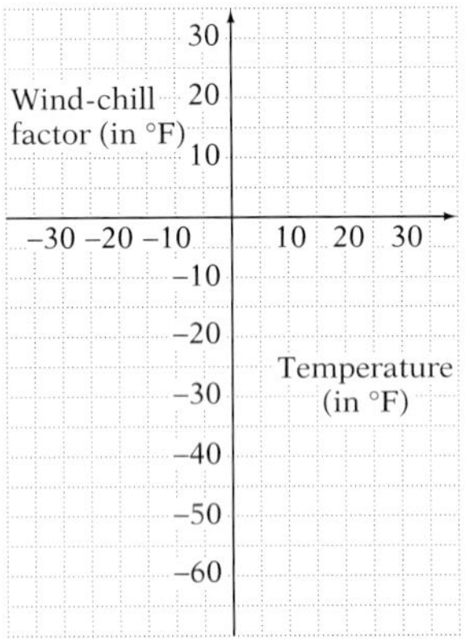

1. Use the coordinate axes at the left to graph the ordered pairs.
2. List the domain of the relation. List the range of the relation.
3. Is the set of ordered pairs a function?
4. The equation of the line that approximately models the graph of the ordered pairs above is $y = 1.2357x - 16$, where x is the air temperature in degrees Fahrenheit and y is the wind-chill index. Evaluate the function $f(x) = 1.2357x - 16$ to determine whether any of the ordered pairs listed above do not satisfy this function. Round to the nearest integer.
5. What does the model predict the wind-chill index is when the air temperature is −40°F and the wind speed is 10 mph?
6. **a.** What is the x-intercept of the graph of $f(x) = 1.2357x - 16$? Round to the nearest integer. What does the x-intercept represent?
   b. What is the y-intercept of the graph of $f(x) = 1.2357x - 16$? What does the y-intercept represent?

In the set of ordered pairs that follows, the abscissa is the air temperature in degrees Fahrenheit, and the ordinate is the wind-chill index when the wind speed is 20 mph. Note that the abscissa in each case is the same as in the ordered pairs above. However, the wind speed has increased to 20 mph.

{(35, 24), (30, 17), (25, 11), (20, 4), (15, −2), (10, −9), (5, −15), (0, −22), (−5, −29), (−10, −35), (−15, −42), (−20, −48), (−25, −55), (−30, −61), (−35, −68)}

7. Use the same coordinate axes to graph these ordered pairs.
8. Where do the points representing this relation lie in relation to the points drawn for Exercise 1? What does this mean with respect to the temperature remaining constant and the wind speed increasing?
9. The equation of the line that approximately models the graph of the ordered pairs in this second relation is $y = 1.3114x - 22$, where x is the air temperature in degrees Fahrenheit and y is the wind-chill index. Compare this with the model for the first relation. Are the lines parallel? Explain why that might be the case.
10. Using the old wind-chill formula and a wind speed of 10 mph gave a wind-chill index of −10°F when the air temperature was 20°F and a wind-chill index of −67°F when the air temperature was −20°F.
    a. Find the linear equation that models these data. [*Hint:* Use the ordered pairs (20, −10) and (−20, −67).]
    b. Compare the y-intercept of your equation to the y-intercept of the equation $y = 1.2357x - 16$, which models the new wind-chill index for a wind speed of 10 mph. Use the difference to determine whether the new formula yields a wind-chill index that is warmer or colder, given the same air temperature and wind speed.

Chapter Summary

Key Words / Examples

A **rectangular coordinate system** is formed by two number lines, one horizontal and one vertical, that intersect at the zero point of each line. The number lines that make up a coordinate system are called the **coordinate axes** or simply the **axes**. The **origin** is the point of intersection of the two coordinate axes. Generally, the horizontal axis is labeled the ***x*-axis**, and the vertical axis is labeled the ***y*-axis**. A rectangular coordinate system divides the **plane** determined by the axes into four regions called **quadrants**. [3.1.1, p. 196]

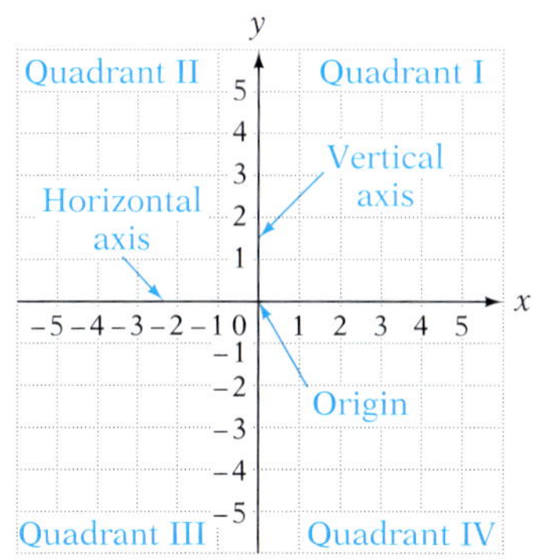

An **ordered pair (*x*, *y*)** is used to locate a point in a plane. The first number in an ordered pair is called the **abscissa**. The second number in an ordered pair is called the **ordinate**. [3.1.1, pp. 196–197]

(3, 4) is an ordered pair.

3 is the abscissa.

4 is the ordinate.

A **scatter diagram** is a graph of ordered-pair data. [3.1.3, p. 201]

A **relation** is any set of ordered pairs. [3.2.1, p. 211]

$\{(2, 3), (2, 4), (3, 4), (5, 7)\}$

A **function** is a set of ordered pairs in which no two ordered pairs have the same first coordinate and different second coordinates. [3.2.1, p. 212]

$\{(2, 3), (3, 5), (5, 7), (6, 9)\}$

Functional notation is used for those equations that represent functions. For the equation at the right, ***x*** is the **independent variable** and ***y*** is the **dependent variable**. [3.2.1, pp. 213–214]

In functional notation, $y = 3x + 7$ is written as $f(x) = 3x + 7$.

The **domain of a function** is the set of first coordinates of all the ordered pairs of the function. The **range of a function** is the set of second coordinates of all the ordered pairs of the function. [3.2.1, p. 212]

$\{(1, 2), (2, 4), (5, 7), (8, 3)\}$

Domain: $\{1, 2, 5, 8\}$

Range: $\{2, 3, 4, 7\}$

A **linear function** is one that can be expressed in the form $f(x) = mx + b$. The graph of a linear function is a straight line. The point at which the graph crosses the x-axis is called the ***x*-intercept**, and the point at which a graph crosses the y-axis is called the ***y*-intercept**. [3.3.1 and 3.3.2, pp. 223 and 228]

$f(x) = -\dfrac{3}{2}x + 3$

The **slope** of a line is a measure of the slant, or tilt, of a line. The symbol for slope is ***m***. A line that slants upward to the right has a **positive slope**, and a line that slants downward to the right has a **negative slope**. [3.4.1, pp. 235–236]

The line $y = 2x - 3$ slants upward to the right.

The line $y = -5x + 2$ slants downward to the right.

A horizontal line has **zero slope**. The slope of a vertical line is **undefined**. [3.4.1, p. 237]

$y = 3$ is an equation of a line with zero slope.

The slope of the line $x = -3$ is undefined.

An inequality of the form $\boldsymbol{y > mx + b}$ or of the form $\boldsymbol{Ax + By > C}$ is a **linear inequality in two variables**. (The symbol $>$ could be replaced by $\geq$, $<$, or $\leq$.) The solution set of an inequality in two variables is a **half-plane**. [3.7.1, p. 259]

Essential Rules and Procedures

The **distance formula** $d = \sqrt{(x_2 - x_1)^2 + (y_2 - y_1)^2}$ can be used to find the distance between any two points in a plane. [3.1.2, p. 199]

$(x_1, y_1) = (2, 4), (x_2, y_2) = (-3, 5)$

$$d = \sqrt{(-3 - 2)^2 + (5 - 4)^2} = \sqrt{25 + 1} = \sqrt{26}$$

The **midpoint formula** $(x_m, y_m) = \left(\frac{x_1 + x_2}{2}, \frac{y_1 + y_2}{2}\right)$ is used to find the midpoint of a line segment. [3.1.2, p. 200]

$(x_1, y_1) = (2, 3), (x_2, y_2) = (-6, -1)$

$$x_m = \frac{2 + (-6)}{2} = -2, \quad y_m = \frac{3 + (-1)}{2} = 1$$

$(x_m, y_m) = (-2, 1)$

The **average rate of change** of y with respect to x is $\frac{\text{change in } y}{\text{change in } x}$. [3.1.4, p. 204]

In 1998, 1.7 million households used electronic bill payment. This number increased to 9.9 million in 2002. (*Source:* Yankee Group)

$$\frac{9.9 - 1.7}{2002 - 1998} = 2.05$$

The average annual rate of change in the number of households using electronic bill payment from 1998 to 2002 was 2.05 million.

To find the *x*-intercept of $Ax + By = C$, let $y = 0$. **To find the *y*-intercept**, let $x = 0$. [3.3.2, pp. 228–229]

$3x + 4y = 12$

Let $y = 0$:	Let $x = 0$:
$3x + 4(0) = 12$	$3(0) + 4y = 12$
$x = 4$	$y = 3$
The x-intercept is (4, 0).	The y-intercept is (0, 3).

The **slope formula** $m = \frac{y_2 - y_1}{x_2 - x_1}$, $x_1 \neq x_2$, is used to find the slope of a line when two points on the line are known. [3.4.1, p. 236]

$(x_1, y_1) = (-3, 2), (x_2, y_2) = (1, 4)$

$$m = \frac{4 - 2}{1 - (-3)} = \frac{1}{2}$$

The equation $y = mx + b$ is called the **slope-intercept form of a straight line**. [3.4.2, p. 239]

$y = -3x + 2$
The slope is -3; the y-intercept is $(0, 2)$.

The equation $y - y_1 = m(x - x_1)$ is called the **point-slope formula**. [3.5.1, p. 246]

The equation of the line that passes through the point $(4, 2)$ and has slope -3 is

$$\begin{aligned} y - y_1 &= m(x - x_1) \\ y - 2 &= -3(x - 4) \\ y - 2 &= -3x + 12 \\ y &= -3x + 14 \end{aligned}$$

Two lines that have the same slope do not intersect and are called **parallel lines**. [3.6.1, p. 252]

$y = 3x - 4, \quad m_1 = 3$
$y = 3x + 2, \quad m_2 = 3$

Because $m_1 = m_2$, the lines are parallel.

Two lines that intersect at right angles are called **perpendicular lines**. The product of the slopes of two nonvertical perpendicular lines is -1. [3.6.1, pp. 253–254]

$y = \frac{1}{2}x - 1, \quad m_1 = \frac{1}{2}$
$y = -2x + 2, \quad m_2 = -2$

Because $m_1 \cdot m_2 = -1$, the lines are perpendicular.

Chapter Review Exercises

1. Determine the ordered-pair solution of $y = \frac{x}{x - 2}$ that corresponds to $x = 4$.

2. Find the midpoint and length of the line segment with endpoints $(-2, 4)$ and $(3, 5)$.

3. Graph the ordered-pair solutions of $y = x^2 - 2$ when $x = -2, -1, 0, 1,$ and 2.

4. Draw a line through all points with an ordinate of -2.

5. Given $P(x) = 3x + 4$, evaluate $P(-2)$ and $P(a)$.

6. Find the domain and range of the function $\{(-1, 0), (0, 2), (1, 2), (2, 0), (5, 3)\}$.

7. Find the range of $f(x) = x^2 - 2$ if the domain is $\{-2, -1, 0, 1, 2\}$.

8. What value of x is excluded from the domain of $f(x) = \frac{x}{x + 4}$?

9. Find the x- and y-intercepts and graph $4x - 6y = 12$.

10. Graph: $y = -2x + 2$

11. Graph: $2x - 3y = -6$

12. Find the slope of the line that contains the points $(3, -2)$ and $(-1, 2)$.

13. Graph $3x + 2y = -4$ by using the x- and y-intercepts.

14. Graph the line that passes through the point $(-1, 4)$ and has slope $-\frac{1}{3}$.

15. Find the equation of the line that contains the point $(-3, 4)$ and has slope $\frac{5}{2}$.

16. Find the equation of the line that contains the points $(-2, 4)$ and $(4, -3)$.

17. Find the equation of the line that contains the point $(3, -2)$ and is parallel to the line $y = -3x + 4$.

18. Find the equation of the line that contains the point $(-2, -4)$ and is parallel to the line $2x - 3y = 4$.

19. Find the equation of the line that contains the point $(2, 5)$ and is perpendicular to the line $y = -\frac{2}{3}x + 6$.

20. Find the equation of the line that contains the point $(-3, -1)$ and is perpendicular to the line $4x - 2y = 7$.

21. Graph the solution set of $y \geq 2x - 3$.

22. Graph the solution set of $3x - 2y < 6$.

23. The table below shows the average number of years that women and men were estimated to live beyond age 65 in 1950, 1960, 1970, 1980, 1990, and 2000. (*Source:* 2000 Social Security Trustees Report) Figures are rounded to the nearest whole number. Graph the scatter diagram for these data.

Women, x	15	16	17	18	19	19
Men, y	13	13	13	14	15	16

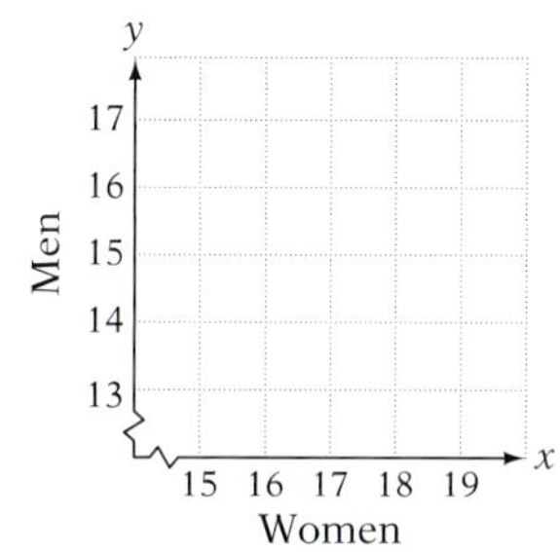

24. The table below shows the number of foreign-born people living in the United States each decade from 1960 to 2000. (*Source:* U.S. Census Bureau)

a. What was the average annual rate of change in the number of foreign-born people living in the United States from 1980 to 2000? Round to the nearest thousand.

b. Find the average annual rate of change in the number of foreign-born people living in the United States from 1970 to 1980. Was this more than or less than the average annual rate of change in the number of foreign-born people living in the United States from 1980 to 1990?

Year	1960	1970	1980	1990	2000
Number of foreign born (in millions)	9.7	9.6	14.1	19.8	28.4

25. A car is traveling at 55 mph. The equation that describes the distance, d (in miles), traveled is $d = 55t$, where t is the number of hours driven. Use the coordinate axes below to graph this equation for $0 \leq t \leq 6$. The point (4, 220) is on the graph. Write a sentence that describes the meaning of this ordered pair.

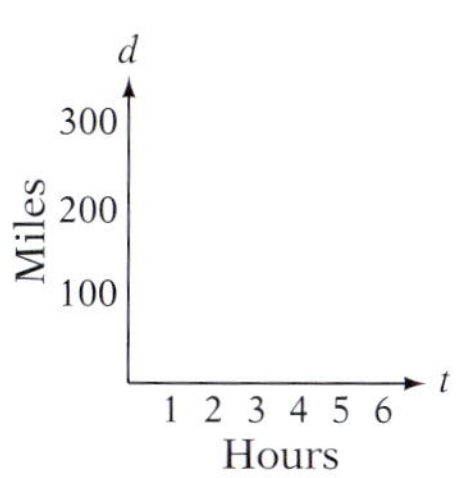

26. The graph below shows the relationship between the cost of manufacturing calculators and the number of calculators manufactured. Find the slope of the line between the two points shown on the graph. Write a sentence that states the meaning of the slope.

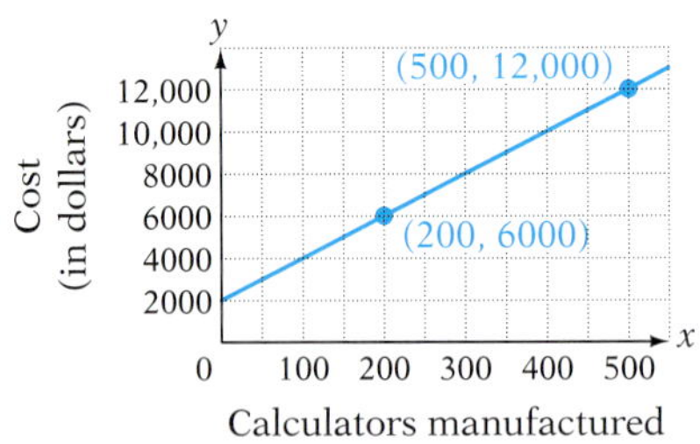

27. A building contractor estimates that the cost to build a new home is \$25,000 plus \$80 for each square foot of floor space in the house. Determine a linear function that will give the cost to build a house that contains a given number of square feet. Use this model to determine the cost of building a house that contains 2000 ft^2.

Chapter Test

1. Graph the ordered-pair solutions of $P(x) = 2 - x^2$ when $x = -2, -1, 0, 1,$ and 2.

2. Find the ordered-pair solution of $y = 2x + 6$ that corresponds to $x = -3$.

3. Graph: $y = \frac{2}{3}x - 4$

4. Graph: $2x + 3y = -3$

5. Find the equation of the vertical line that contains the point $(-2, 3)$.

6. Find the midpoint and length of the line segment with endpoints $(4, 2)$ and $(-5, 8)$.

7. Find the slope of the line that contains the points $(-2, 3)$ and $(4, 2)$.

8. Given $P(x) = 3x^2 - 2x + 1$, evaluate $P(2)$.

9. Graph $2x - 3y = 6$ by using the x- and y-intercepts.

10. Graph the line that passes through the point $(-2, 3)$ and has slope $-\frac{3}{2}$.

11. Find the equation of the line that contains the point $(-5, 2)$ and has slope $\frac{2}{5}$.

12. What value of x is excluded from the domain of $f(x) = \frac{2x + 1}{x}$?

13. Find the equation of the line that contains the points $(3, -4)$ and $(-2, 3)$.

14. Find the equation of the horizontal line that contains the point $(4, -3)$.

15. Find the domain and range of the function $\{(-4, 2), (-2, 2), (0, 0), (3, 5)\}$.

16. Find the equation of the line that contains the point $(1, 2)$ and is parallel to the line $y = -\frac{3}{2}x - 6$.

17. Find the equation of the line that contains the point $(-2, -3)$ and is perpendicular to the line $y = -\frac{1}{2}x - 3$.

18. Graph the solution set of $3x - 4y > 8$.

19. The director of a baseball camp estimates that 100 students will enroll if the tuition is \$250. For each \$20 increase in tuition, 6 fewer students will enroll. Determine a linear function that will predict the number of students who will enroll at a given tuition. Use this model to predict enrollment when the tuition is \$300.

20. The graph at the right shows the relationship between the value of a rental house and the depreciation allowed for income tax purposes. Find the slope between the two points shown on the graph. Write a sentence that states the meaning of the slope.

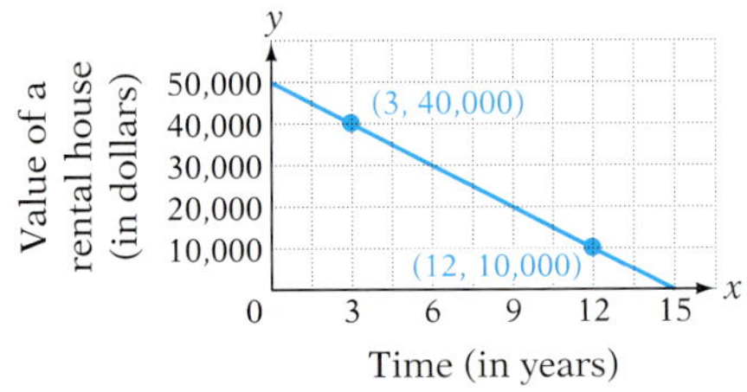

21. The table below shows the Consumer Confidence Index, as reported by the Conference Board, a New York-based research group, each month from August 2000 to January 2001.

a. Find the average rate of change per month in the Consumer Confidence Index from August 2000 to January 2001.

b. The Consumer Confidence Index in October 2001 was 85.5. Find the average rate of change per month in the Consumer Confidence Index from January 2001 to October 2001.

Month	Aug.	Sept.	Oct.	Nov.	Dec.	Jan.
Consumer Confidence Index	140.8	142.5	135.8	132.6	128.3	114.4

Cumulative Review Exercises

1. Identify the property that justifies the statement. $(x + y) \cdot 2 = 2 \cdot (x + y)$

2. Solve: $3 - \frac{x}{2} = \frac{3}{4}$

3. Solve: $2[y - 2(3 - y) + 4] = 4 - 3y$

4. Solve: $\frac{1 - 3x}{2} + \frac{7x - 2}{6} = \frac{4x + 2}{9}$

5. Solve: $x - 3 < -4$ or $2x + 2 > 3$

6. Solve: $8 - |2x - 1| = 4$

7. Solve: $|3x - 5| < 5$

8. Simplify: $4 - 2(4 - 5)^3 + 2$

9. Evaluate $(a - b)^2 \div ab$ when $a = 4$ and $b = -2$.

10. Graph: $\{x \mid x < -2\} \cup \{x \mid x > 0\}$

11. Solve: $3x - 1 < 4$ and $x - 2 > 2$

12. Given $P(x) = x^2 + 5$, evaluate $P(-3)$.

13. Find the ordered-pair solution of $y = -\frac{5}{4}x + 3$ that corresponds to $x = -8$.

14. Find the slope of the line that contains the points $(-1, 3)$ and $(3, -4)$.

15. Find the equation of the line that contains the point $(-1, 5)$ and has slope $\frac{3}{2}$.

16. Find the equation of the line that contains the points $(4, -2)$ and $(0, 3)$.

17. Find the equation of the line that contains the point $(2, 4)$ and is parallel to the line $y = -\frac{3}{2}x + 2$.

18. Find the equation of the line that contains the point $(4, 0)$ and is perpendicular to the line $3x - 2y = 5$.

19. Graph $3x - 5y = 15$ by using the x- and y-intercepts.

20. Graph the line that passes through the point $(-3, 1)$ and has slope $-\frac{3}{2}$.

21. Graph the solution set of $3x - 2y \geq 6$.

22. A furniture store uses a markup rate of 60%. The store sells a solid oak curio cabinet for \$1074. Find the cost of the curio cabinet.

23. Two planes are 1800 mi apart and are traveling toward each other. One plane is traveling twice as fast as the other plane. The planes will meet in 3 h. Find the speed of each plane.

24. A grocer combines coffee that costs \$8.00 per pound with coffee that costs \$3.00 per pound. How many pounds of each should be used to make 80 lb of a blend that costs \$5.00 per pound?

25. The relationship between the depreciated value of a truck for income tax purposes and its age in years is shown in the graph at the right. Write the equation for the line that represents the depreciated value of the truck. Write a sentence that explains the meaning of the slope in the context of this exercise.

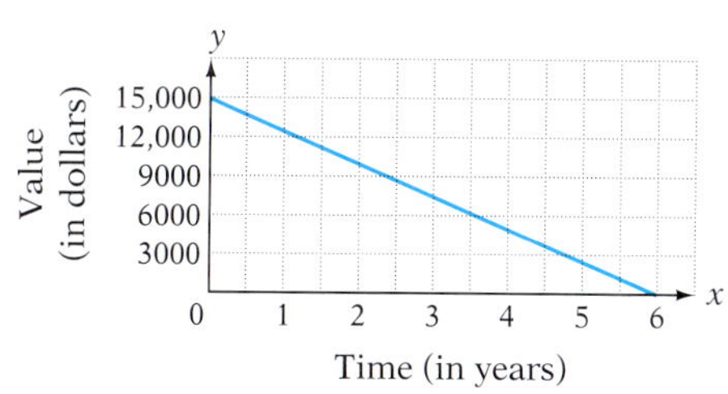

Suppose you are the manager of a soccer league that consists of 12 teams. One of your responsibilities is to create a schedule in which each team plays every other team once. How many games must be scheduled? One method of solving this problem is to solve an easier problem first. This approach is explained in the **Focus on Problem Solving on page 326.**

Applications

4

Systems of Equations and Inequalities

Objectives

Section 4.1

1. Solve a system of linear equations by graphing
2. Solve a system of linear equations by the substitution method

Section 4.2

1. Solve a system of two linear equations in two variables by the addition method
2. Solve a system of three linear equations in three variables by the addition method

Section 4.3

1. Solve systems of linear equations by using matrices
2. Evaluate determinants
3. Solve systems of linear equations by using Cramer's Rule

Section 4.4

1. Rate-of-wind and water-current problems
2. Application problems

Section 4.5

1. Graph the solution set of a system of linear inequalities

WEB **Need help? For online student resources, such as section quizzes, visit this web site: math.college.hmco.com**

Prep Test

1. Simplify: $10\left(\frac{3}{5}x + \frac{1}{2}y\right)$
2. Evaluate $3x + 2y - z$ for $x = -1$, $y = 4$, and $z = -2$.
3. Given $3x - 2z = 4$, find the value of x when $z = -2$.
4. Solve: $3x + 4(-2x - 5) = -5$
5. Solve: $0.45x + 0.06(-x + 4000) = 630$
6. Graph: $y = \frac{1}{2}x - 4$
7. Graph: $3x - 2y = 6$
8. Graph: $y > -\frac{3}{5}x + 1$

Go Figure

Chris can beat Pat by 100 m in a 1000-meter race. Pat can beat Leslie by 10 m in a 100-meter race. If both run at this given rate, by how many meters will Chris beat Leslie in a 1000-meter race?

4.1 Solving Systems of Linear Equations by Graphing and by the Substitution Method

1 Solve a system of linear equations by graphing

VIDEO & DVD CD TUTOR WWW WEB SSM

A **system of equations** is two or more equations considered together. The system at the right is a system of two linear equations in two variables. The graphs of the equations are straight lines.

$$3x + 4y = 7$$
$$2x - 3y = 6$$

A **solution of a system of equations in two variables** is an ordered pair that is a solution of each equation of the system.

Is $(3, -2)$ a solution of the system of equations shown at the right?

$$2x - 3y = 12$$
$$5x + 2y = 11$$

Replace x by 3 and y by -2.

$$\begin{array}{r|l} 2x - 3y = & 12 \\ \hline 2(3) - 3(-2) & 12 \\ 6 - (-6) & \\ 12 = 12 & \end{array} \qquad \begin{array}{r|l} 5x + 2y = & 11 \\ \hline 5(3) + 2(-2) & 11 \\ 15 + (-4) & \\ 11 = 11 & \end{array}$$

Yes, because $(3, -2)$ is a solution of each equation, it is a solution of the system of equations.

A solution of a system of linear equations can be found by graphing the lines that represent the equations of the system on the same set of coordinate axes. Three examples of systems of linear equations in two variables are shown below, along with the graphs of the equations of the systems.

System I	System II	System III
$x + 2y = 4$	$2x + 3y = 6$	$x - 2y = 4$
$2x + y = -1$	$4x + 6y = -12$	$2x - 4y = 8$

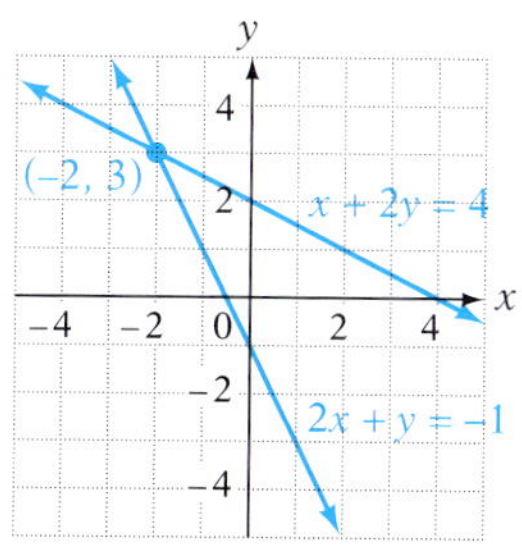

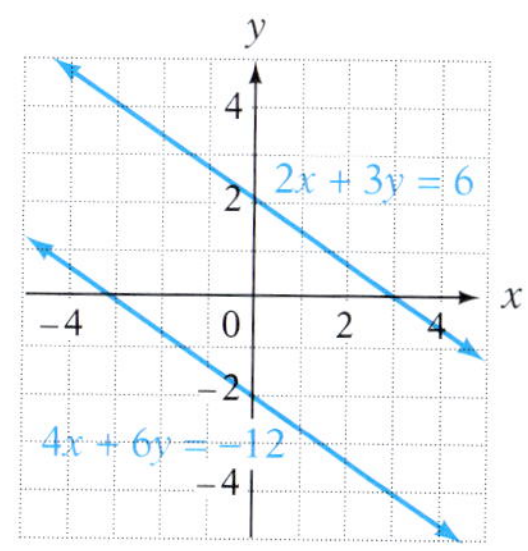

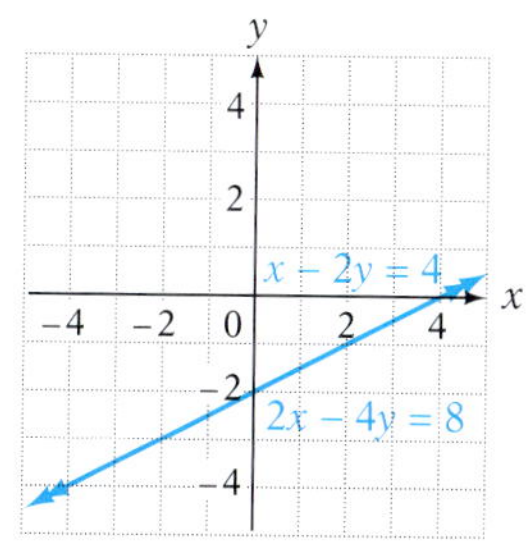

For System I, the two lines intersect at a single point whose coordinates are $(-2, 3)$. Because this point lies on both lines, it is a solution of each equation of the system of equations. We can check this by replacing x by -2 and y by 3. The check is shown below.

Check:

$$\begin{array}{r|l} x + 2y = & 4 \\ \hline -2 + 2(3) & 4 \\ -2 + 6 & 4 \\ 4 = 4 & \end{array} \qquad \begin{array}{r|l} 2x + y = & -1 \\ \hline 2(-2) + 3 & -1 \\ -4 + 3 & -1 \\ -1 = -1 & \end{array}$$

• **Replace x by -2 and y by 3.**

The ordered pair $(-2, 3)$ is a solution of System I.

When the graphs of a system of equations intersect at only one point, the system is called an **independent system of equations.** System I is an independent system of equations.

System II above and the graph of the equations of that system are shown again at the right. Note in this case that the graphs of the lines are parallel and do not intersect. Because the graphs do not intersect, there is no point that is on both lines. Therefore, the system of equations has no solution.

$2x + 3y = 6$
$4x + 6y = -12$

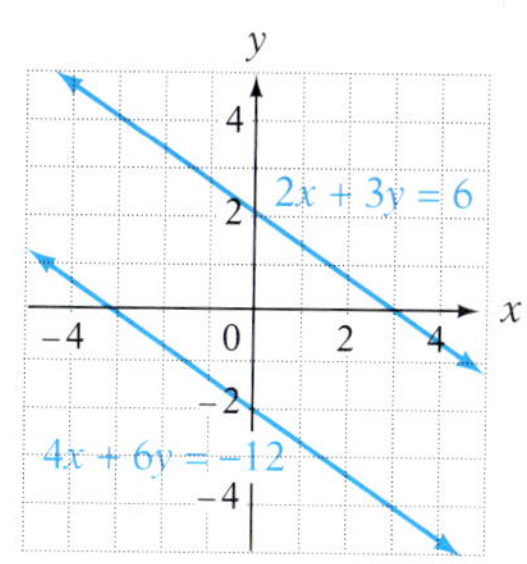

When a system of equations has no solution, it is called an **inconsistent system of equations.** System II is an inconsistent system of equations.

$$x - 2y = 4$$
$$2x - 4y = 8$$

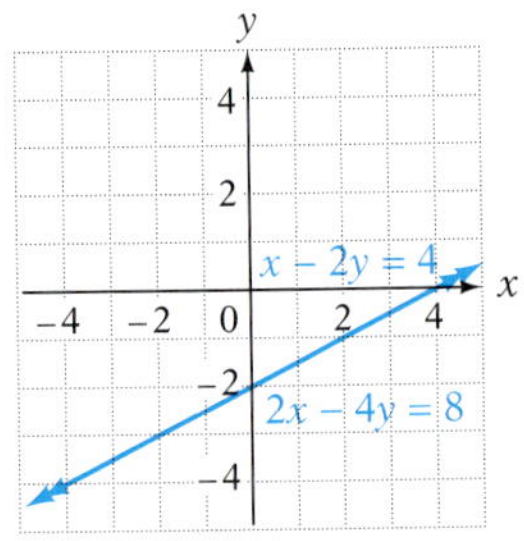

System III from the preceding page and the graph of the equations of that system are shown again at the right. Note that the graph of $x - 2y = 4$ lies directly on top of the graph of $2x - 4y = 8$. Thus the two lines intersect at an infinite number of points. Because the graphs intersect at an infinite number of points, there are an infinite number of solutions of this system of equations. Because each equation represents the same set of points, the solutions of the system of equations can be stated by using the ordered pairs of either one of the equations. Therefore, we can say, "The solutions are the ordered pairs that satisfy $x - 2y = 4$," or we can solve the equation for y and say, "The solutions are the ordered pairs that satisfy $y = \frac{1}{2}x - 2$." We normally state this solution using ordered pairs: "The solutions are the ordered pairs $\left(x, \frac{1}{2}x - 2\right)$."

When the two equations in a system of equations represent the same line, the system is called a **dependent system of equations.** System III is a dependent system of equations.

These systems illustrate the three possibilities for a system of linear equations in two variables.

1. The graphs intersect at one point.
   The solution of the system of equations is the ordered pair (x, y) whose coordinates are the point of intersection.
   The system of equations is *independent*.

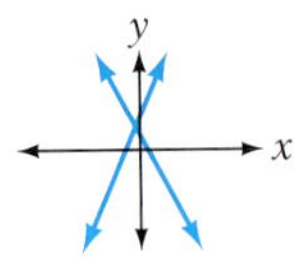

2. The lines are parallel and never intersect.
   There is no solution of the system of equations.
   The system of equations is *inconsistent*.

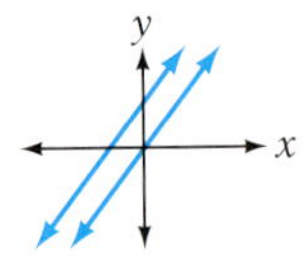

3. The graphs are the same line, and they intersect at infinitely many points.
   There are an infinite number of solutions of the system of equations.
   The system of equations is *dependent*.

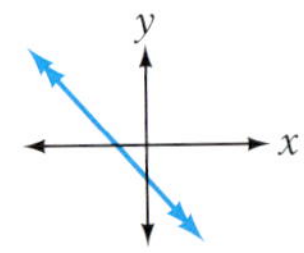

> **Take Note**
>
> Keep in mind the distinction between independent, dependent, and inconsistent systems of equations. You should be able to express your understanding of these terms by using graphs.

> **Take Note**
>
> The Projects & Group Activities feature at the end of this chapter shows how to use a graphing calculator to solve a system of equations.

Solve by graphing: $2x - y = 3$
$4x - 2y = 6$

Graph each line.
The system of equations is dependent.
Solve one of the equations for y.

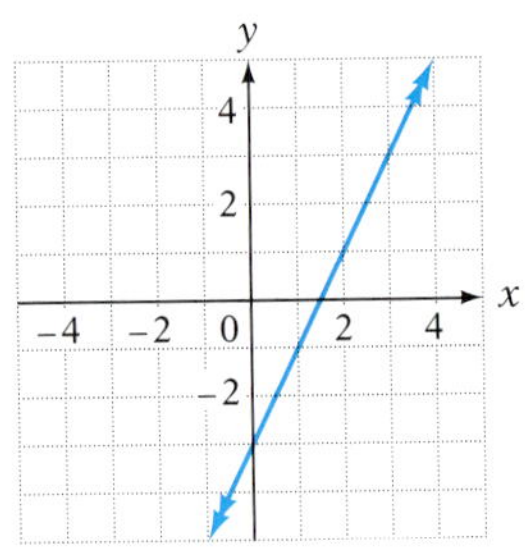

$$2x - y = 3$$
$$-y = -2x + 3$$
$$y = 2x - 3$$

The solutions are the ordered pairs $(x, 2x - 3)$.

Example 1 Solve by graphing: $2x - y = 3$
$3x + y = 2$

Solution

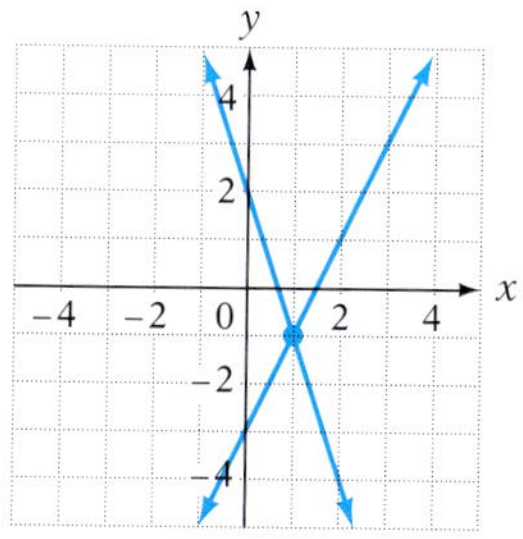

The solution is (1, −1).

Problem 1 Solve by graphing: $x + y = 1$
$2x + y = 0$

Solution See page S13.

Example 2 Solve by graphing.

A. $2x + 3y = 6$
$y = -\frac{2}{3}x + 1$

B. $x - 2y = 6$
$y = \frac{1}{2}x - 3$

Solution

A.

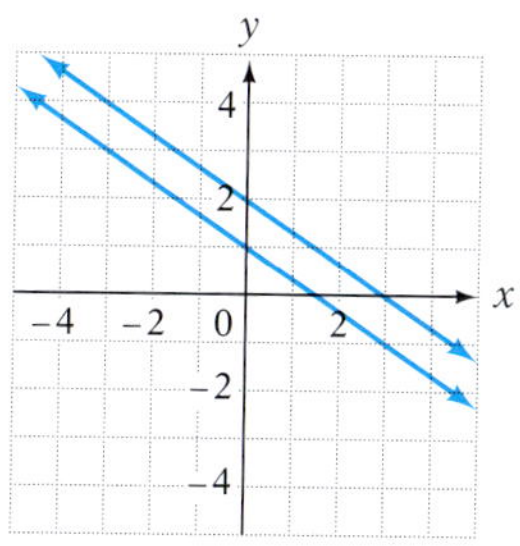

The lines are parallel and therefore do not intersect. The system of equations has no solution. The system of equations is inconsistent.

B.

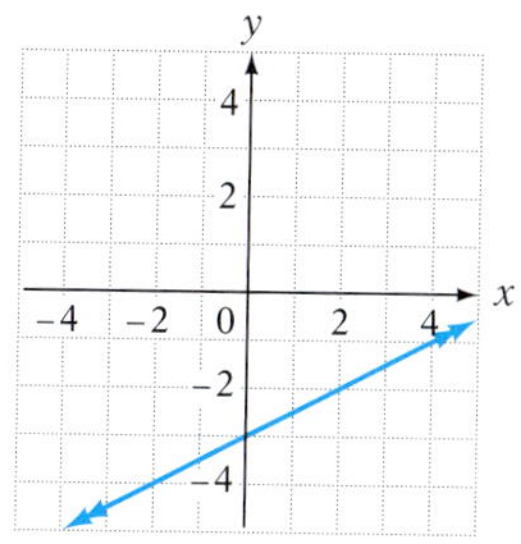

The system of equations is dependent. The solutions are the ordered pairs $\left(x, \frac{1}{2}x - 3\right)$.

Problem 2 Solve by graphing.

A. $2x + 5y = 10$
$y = -\frac{2}{5}x - 2$

B. $3x - 4y = 12$
$y = \frac{3}{4}x - 3$

Solution See page S13.

2 Solve a system of linear equations by the substitution method

VIDEO & DVD   CD TUTOR   WEB   SSM

A graphical solution of a system of equations is based on approximating the coordinates of a point of intersection. An algebraic method called the **substitution method** can be used to find an exact solution of a system of equations.

To use the substitution method, we must write one of the equations of the system in terms of x or in terms of y.

Solve by the substitution method: $3x + y = 5$
$4x + 5y = 3$

(1) $3x + y = 5$
(2) $4x + 5y = 3$

Solve equation (1) for y. The result is equation (3).

$3x + y = 5$
(3) $y = -3x + 5$

Use equation (2). Substitute $-3x + 5$ for y and solve for x.

$$\begin{aligned} 4x + 5y &= 3 \\ 4x + 5(-3x + 5) &= 3 \\ 4x - 15x + 25 &= 3 \\ -11x &= -22 \\ x &= 2 \end{aligned}$$

Now substitute the value of x into equation (3) and solve for y.

$$\begin{aligned} y &= -3x + 5 \\ y &= -3(2) + 5 \\ y &= -6 + 5 \\ y &= -1 \end{aligned}$$

The solution is $(2, -1)$.

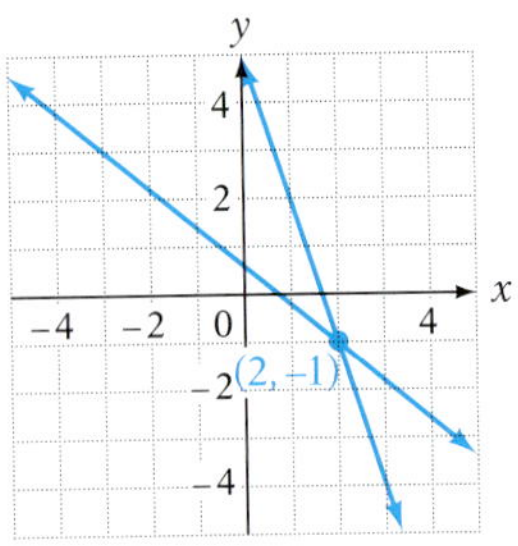

The graph of the system is shown at the left. Note that the graphs intersect at the point whose coordinates are $(2, -1)$, the solution of the system of equations.

Example 3 Solve by substitution.

A. $3x - 3y = 2$
$y = x + 2$

B. $9x + 3y = 12$
$y = -3x + 4$

Solution

A. (1) $3x - 3y = 2$
(2) $y = x + 2$

$$\begin{aligned} 3x - 3(x + 2) &= 2 \\ 3x - 3x - 6 &= 2 \\ -6 &= 2 \end{aligned}$$

- **Equation (2) states that $y = x + 2$.**
- **Substitute $x + 2$ for y in equation (1).**

$-6 = 2$ is not a true equation. The system of equations is inconsistent. The system does not have a solution.

B. (1) $9x + 3y = 12$
(2) $y = -3x + 4$

$$\begin{aligned} 9x + 3(-3x + 4) &= 12 \\ 9x - 9x + 12 &= 12 \\ 12 &= 12 \end{aligned}$$

- **Equation (2) states that $y = -3x + 4$.**
- **Substitute $-3x + 4$ for y in equation (1).**

$12 = 12$ is a true equation. The system of equations is dependent. The solutions are the ordered pairs $(x, -3x + 4)$.

Problem 3 Solve by substitution.

A. $3x - y = 3$
$6x + 3y = -4$

B. $6x - 3y = 6$
$2x - y = 2$

Solution See pages S13–S14.

4.1 Concept Review

Determine whether the following statements are always true, sometimes true, or never true.

1. A solution of a system of equations in two variables is an ordered pair that is a solution of every equation in the system.
2. An inconsistent system of equations has no solution.
3. A system of two equations with two unknowns has an infinite number of solutions.
4. A system of two equations with the same slope and different y-intercepts has an infinite number of solutions.
5. When the graphs of a system of equations intersect at only one point, the system of equations is called an independent system of equations.
6. The system $y = \frac{3}{2}x + 2$ and $y = -4x + 2$ has exactly one solution.

4.1 Exercises

1 Determine whether the ordered pair is a solution of the system of equations.

1. $(0, -1)$
$3x - 2y = 2$
$x + 2y = 6$

2. $(2, 1)$
$x + y = 3$
$2x - 3y = 1$

3. $(-3, -5)$
$x + y = -8$
$2x + 5y = -31$

4. $(1, -1)$
$3x - y = 4$
$7x + 2y = -5$

State whether the system of equations is independent, inconsistent, or dependent.

5.
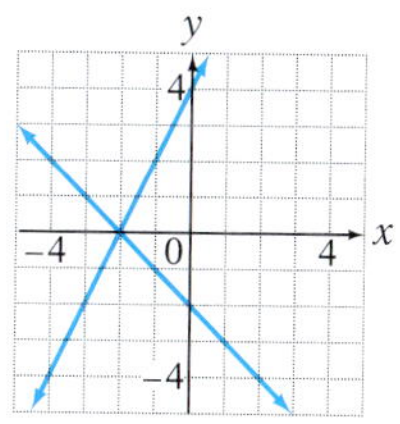

6.
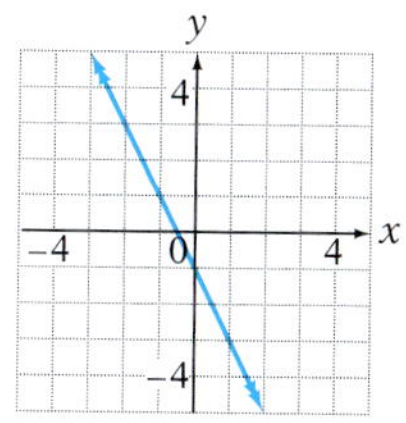

7.
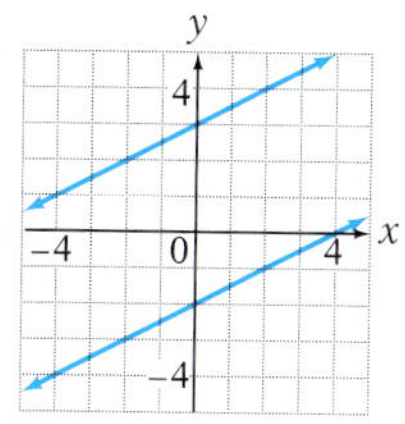

8.
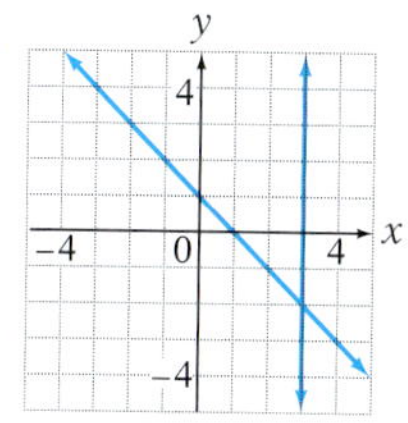

9. Describe the solution of an independent system of linear equations in two variables.

10. How does a dependent system of equations differ from an independent system of equations?

Solve by graphing.

11. $\begin{aligned} x + y &= 2 \\ x - y &= 4 \end{aligned}$

12. $\begin{aligned} x + y &= 1 \\ 3x - y &= -5 \end{aligned}$

13. $\begin{aligned} x - y &= -2 \\ x + 2y &= 10 \end{aligned}$

14. $\begin{aligned} 2x - y &= 5 \\ 3x + y &= 5 \end{aligned}$

15. $\begin{aligned} 3x - 2y &= 6 \\ y &= 3 \end{aligned}$

16. $\begin{aligned} x &= 4 \\ 3x - 2y &= 4 \end{aligned}$

17. $\begin{aligned} x &= 4 \\ y &= -1 \end{aligned}$

18. $\begin{aligned} x + 2 &= 0 \\ y - 1 &= 0 \end{aligned}$

19. $\begin{aligned} y &= x - 5 \\ 2x + y &= 4 \end{aligned}$

20. $\begin{aligned} 2x - 5y &= 4 \\ y &= x + 1 \end{aligned}$

21. $\begin{aligned} y &= \frac{1}{2}x - 2 \\ x - 2y &= 8 \end{aligned}$

22. $\begin{aligned} 2x + 3y &= 6 \\ y &= -\frac{2}{3}x + 1 \end{aligned}$

23. $\begin{aligned} 2x - 5y &= 10 \\ y &= \frac{2}{5}x - 2 \end{aligned}$

24. $\begin{aligned} 3x - 2y &= 6 \\ y &= \frac{3}{2}x - 3 \end{aligned}$

25. $\begin{aligned} 3x - 4y &= 12 \\ 5x + 4y &= -12 \end{aligned}$

26. $\begin{aligned} 2x - 3y &= 6 \\ 2x - 5y &= 10 \end{aligned}$

2 **27.** When you solve a system of equations using the substitution method, how can you tell whether the system of equations is inconsistent?

28. When you solve a system of equations using the substitution method, how can you tell whether the system of equations is dependent?

Solve by substitution.

29. $\begin{aligned} 3x - 2y &= 4 \\ x &= 2 \end{aligned}$

30. $\begin{aligned} y &= -2 \\ 2x + 3y &= 4 \end{aligned}$

31. $\begin{aligned} y &= 2x - 1 \\ x + 2y &= 3 \end{aligned}$

32. $\begin{aligned} y &= -x + 1 \\ 2x - y &= 5 \end{aligned}$

33. $\begin{aligned} 4x - 3y &= 5 \\ y &= 2x - 3 \end{aligned}$

34. $\begin{aligned} 3x + 5y &= -1 \\ y &= 2x - 8 \end{aligned}$

35. $\begin{aligned} x &= 2y + 4 \\ 4x + 3y &= -17 \end{aligned}$

36. $\begin{aligned} 3x - 2y &= -11 \\ x &= 2y - 9 \end{aligned}$

37. $\begin{aligned} 5x + 4y &= -1 \\ y &= 2 - 2x \end{aligned}$

38. $\begin{aligned} 3x + 2y &= 4 \\ y &= 1 - 2x \end{aligned}$

39. $\begin{aligned} 7x - 3y &= 3 \\ x &= 2y + 2 \end{aligned}$

40. $\begin{aligned} 3x - 4y &= 6 \\ x &= 3y + 2 \end{aligned}$

41. $\begin{aligned} 2x + 2y &= 7 \\ y &= 4x + 1 \end{aligned}$

42. $\begin{aligned} 3x + 7y &= -5 \\ y &= 6x - 5 \end{aligned}$

43. $\begin{aligned} 3x + y &= 5 \\ 2x + 3y &= 8 \end{aligned}$

44. $\begin{aligned} 4x + y &= 9 \\ 3x - 4y &= 2 \end{aligned}$

45. $\begin{aligned} x + 3y &= 5 \\ 2x + 3y &= 4 \end{aligned}$

46. $\begin{aligned} x - 4y &= 2 \\ 2x - 5y &= 1 \end{aligned}$

47. $\begin{aligned} 3x + 4y &= 14 \\ 2x + y &= 1 \end{aligned}$

48. $\begin{aligned} 5x + 3y &= 8 \\ 3x + y &= 8 \end{aligned}$

49. $\begin{aligned} 3x + 5y &= 0 \\ x - 4y &= 0 \end{aligned}$

50. $\begin{aligned} 2x - 7y &= 0 \\ 3x + y &= 0 \end{aligned}$

51. $\begin{aligned} 5x - 3y &= -2 \\ -x + 2y &= -8 \end{aligned}$

52. $\begin{aligned} 2x + 7y &= 1 \\ -x + 4y &= 7 \end{aligned}$

53. $x + 3y = 4$
$x = 5 - 3y$

54. $6x + 2y = 7$
$y = -3x + 2$

55. $2x - 4y = 16$
$-x + 2y = -8$

56. $3x - 12y = -24$
$-x + 4y = 8$

57. $3x - y = 10$
$6x - 2y = 5$

58. $6x - 4y = 3$
$3x - 2y = 9$

59. $y = 3x + 2$
$y = 2x + 3$

60. $y = 3x - 7$
$y = 2x - 5$

61. $x = 2y + 1$
$x = 3y - 1$

62. $x = 4y + 1$
$x = -2y - 5$

63. $y = 5x - 1$
$y = 5 - x$

64. $y = 3 - 2x$
$y = 2 - 3x$

4.1 Applying Concepts

For what values of k will the system of equations be inconsistent?

65. $2x - 2y = 5$
$kx - 2y = 3$

66. $6x - 3y = 4$
$3x - ky = 1$

67. $x = 6y + 6$
$kx - 3y = 6$

68. $x = 2y + 2$
$kx - 8y = 2$

Solve using two variables.

69. **Number Problem** The sum of two numbers is 44. One number is 8 less than the other number. Find the numbers.

70. **Number Problem** The sum of two numbers is 76. One number is 12 less than the other number. Find the numbers.

71. **Number Problem** The sum of two numbers is 19. Five less than twice the first number equals the second number. Find the numbers.

72. **Number Problem** The sum of two numbers is 22. Three times the first plus 2 equals the second number. Find the numbers.

Solve. (*Hint:* These equations are not linear equations. First rewrite the equations as linear equations by substituting x for $\frac{1}{a}$ and y for $\frac{1}{b}$.)

73. $\frac{2}{a} + \frac{3}{b} = 4$
$\frac{4}{a} + \frac{1}{b} = 3$

74. $\frac{2}{a} + \frac{1}{b} = 1$
$\frac{8}{a} - \frac{2}{b} = 0$

75. $\frac{1}{a} + \frac{3}{b} = 2$
$\frac{4}{a} - \frac{1}{b} = 3$

76. $\frac{3}{a} + \frac{4}{b} = -1$
$\frac{1}{a} + \frac{6}{b} = 2$

Use a graphing utility to solve each of the following systems of equations. Round answers to the nearest hundredth.

77. $y = -\frac{1}{2}x + 2$
$y = 2x - 1$

78. $y = 1.2x + 2$
$y = -1.3x - 3$

79. $y = \sqrt{2}x - 1$
$y = -\sqrt{3}x + 1$

80. $y = \pi x - \frac{2}{3}$
$y = -x + \frac{\pi}{2}$

81. **Demography** Data from the 2000 Census support the following model of the population decline since 1990 of Pittsburgh, Pennsylvania: $y_1 = -0.0036x + 9.559$, where y_1 is the population, in millions, of Pittsburgh in year x. According to data from the same census, Orlando, Florida, has experienced an increase in population that can be modeled by $y_2 = 0.0419x - 82.156$, where y_2 is the population, in millions, of Orlando in year x.

   a. Assuming that this model is accurate for future years, will the population of Orlando exceed the population of Pittsburgh in the year 2010?

   b. In what year will the population of Orlando first exceed the population of Pittsburgh? Round to the nearest whole number.

   c. What does the slope of the line that models Pittsburgh's population mean in the context of this problem?

   d. What does the slope of the line that models Orlando's population mean in the context of this problem?

82. Suppose $\begin{aligned} a_1x + b_1y &= c_1 \\ a_2x + b_2y &= c_2 \end{aligned}$ is a system of equations and $\dfrac{a_1}{b_1} = \dfrac{a_2}{b_2}$. Can the system of equations be independent? Explain your answer.

83. The Substitution Property of Equality states that if $a = b$, then a may be replaced by b in any expression that involves a. Explain how this property applies to solving a system of equations by using the substitution method.

4.2 Solving Systems of Linear Equations by the Addition Method

1 Solve a system of two linear equations in two variables by the addition method

The **addition method** is an alternative method for solving a system of equations. This method is based on the Addition Property of Equations. Use the addition method when it is not convenient to solve one equation for one variable in terms of another variable.

Note, for the system of equations at the right, the effect of adding equation (2) to equation (1). Because $-3y$ and $3y$ are additive inverses, adding the equations results in an equation with only one variable.

$$\begin{aligned} (1)\quad 5x - 3y &= 14 \\ (2)\quad 2x + 3y &= -7 \\ 7x + 0y &= 7 \\ 7x &= 7 \end{aligned}$$

The solution of the resulting equation is the first component of the ordered-pair solution of the system.

$$\begin{aligned} 7x &= 7 \\ x &= 1 \end{aligned}$$

The second component is found by substituting the value of x into equation (1) or (2) and then solving for y. Equation (1) is used here.

$$\begin{aligned} (1)\quad 5x - 3y &= 14 \\ 5(1) - 3y &= 14 \\ 5 - 3y &= 14 \\ -3y &= 9 \\ y &= -3 \end{aligned}$$

The solution is $(1, -3)$.

Point of Interest

There are records of Babylonian mathematicians solving systems of equations 3600 years ago. Here is a system of equations from that time (in our modern notation):

$$\frac{2}{3}x = \frac{1}{2}y + 500$$
$$x + y = 1800$$

We say modern notation *for many reasons. Foremost is the fact that using variables did not become widespread until the 17th century. There are many other reasons, however. The equals sign had not been invented, 2 and 3 did not look like they do today, and zero had not even been considered as a possible number.*

Sometimes adding the two equations does not eliminate one of the variables. In this case, use the Multiplication Property of Equations to rewrite one or both of the equations so that when the equations are added, one of the variables is eliminated. To do this, first choose which variable to eliminate. The coefficients of that variable must be additive inverses. Multiply each equation by a constant that will produce coefficients that are additive inverses.

Solve by the addition method:

$3x + 4y = 2$
$2x + 5y = -1$

$$\begin{aligned} (1)\quad 3x + 4y &= 2 \\ (2)\quad 2x + 5y &= -1 \end{aligned}$$

Eliminate x. Multiply equation (1) by 2 and equation (2) by -3. Note how the constants are selected. The negative sign is used so that the coefficients will be additive inverses.

$$\begin{aligned} 2(3x + 4y) &= 2(2) \\ -3(2x + 5y) &= -3(-1) \end{aligned}$$

The coefficients of the x-terms are additive inverses.

$$\begin{aligned} 6x + 8y &= 4 \\ -6x - 15y &= 3 \end{aligned}$$

Add the equations.
Solve for y.

$$\begin{aligned} -7y &= 7 \\ y &= -1 \end{aligned}$$

Substitute the value of y into one of the equations, and solve for x. Equation (1) is used here.

$$\begin{aligned} (1)\quad 3x + 4y &= 2 \\ 3x + 4(-1) &= 2 \\ 3x - 4 &= 2 \\ 3x &= 6 \\ x &= 2 \end{aligned}$$

The solution is $(2, -1)$.

Solve by the addition method:

$\frac{2}{3}x + \frac{1}{2}y = 4$
$\frac{1}{4}x - \frac{3}{8}y = -\frac{3}{4}$

$$\begin{aligned} (1)\quad \frac{2}{3}x + \frac{1}{2}y &= 4 \\ (2)\quad \frac{1}{4}x - \frac{3}{8}y &= -\frac{3}{4} \end{aligned}$$

Clear the fractions. Multiply each equation by the LCM of the denominators.

$$\begin{aligned} 6\left(\frac{2}{3}x + \frac{1}{2}y\right) &= 6(4) \\ 8\left(\frac{1}{4}x - \frac{3}{8}y\right) &= 8\left(-\frac{3}{4}\right) \\ 4x + 3y &= 24 \\ 2x - 3y &= -6 \end{aligned}$$

Eliminate y. Add the equations.
Solve for x.

$$\begin{aligned} 6x &= 18 \\ x &= 3 \end{aligned}$$

Substitute the value of x into equation (1), and solve for y.

$$\frac{2}{3}x + \frac{1}{2}y = 4$$
$$\frac{2}{3}(3) + \frac{1}{2}y = 4$$
$$2 + \frac{1}{2}y = 4$$
$$\frac{1}{2}y = 2$$
$$y = 4$$

The solution is (3, 4).

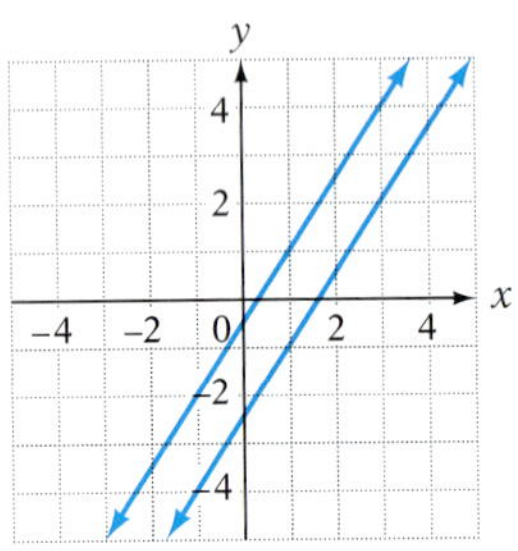

To solve the system of equations at the right, eliminate x. Multiply equation (1) by -2 and add to equation (2).

$$(1)\quad 3x - 2y = 5$$
$$(2)\quad 6x - 4y = 1$$

$$-6x + 4y = -10$$
$$6x - 4y = 1$$
$$0 = -9$$

$0 = -9$ is not a true equation. The system of equations is inconsistent. The system does not have a solution.

The graph of the equations of the system is shown at the left. Note that the lines are parallel and therefore do not intersect.

Example 1 Solve by the addition method.

A. $3x - 2y = 2x + 5$
$2x + 3y = -4$

B. $4x - 8y = 36$
$3x - 6y = 27$

Solution

A. (1) $3x - 2y = 2x + 5$
(2) $2x + 3y = -4$

$$x - 2y = 5$$
$$2x + 3y = -4$$
• Write equation (1) in the form $Ax + By = C$.

$$-2(x - 2y) = -2(5)$$
$$2x + 3y = -4$$
• To eliminate x, multiply each side of equation (1) by -2.

$$-2x + 4y = -10$$
$$2x + 3y = -4$$
$$7y = -14$$
• Add the equations.
$$y = -2$$

$$2x + 3y = -4$$
$$2x + 3(-2) = -4$$
• Replace y in equation (2) by its value.
$$2x - 6 = -4$$
$$2x = 2$$
$$x = 1$$

The solution is $(1, -2)$.

B. (1) $4x - 8y = 36$
(2) $3x - 6y = 27$

$$3(4x - 8y) = 3(36)$$
$$-4(3x - 6y) = -4(27)$$
• To eliminate x, multiply each side of equation (1) by 3 and each side of equation (2) by -4.

$$12x - 24y = 108$$
$$-12x + 24y = -108$$
$$0 = 0$$
• Add the equations.

Take Note

The result of adding the equations is $0 = 0$. It is not $x = 0$. It is not $y = 0$. There is no variable in the equation $0 = 0$. It does not indicate that the solution is $(0, 0)$. Rather, it indicates a dependent system of equations.

$0 = 0$ is a true equation. The system of equations is dependent. The solutions are the ordered pairs $\left(x, \frac{1}{2}x - \frac{9}{2}\right)$.

Problem 1 Solve by the addition method.

A. $2x + 5y = 6$
$3x - 2y = 6x + 2$

B. $2x + y = 5$
$4x + 2y = 6$

Solution See page S14.

2 Solve a system of three linear equations in three variables by the addition method

VIDEO & DVD   CD TUTOR   WEB   SSM

An equation of the form $Ax + By + Cz = D$, where A, B, and C are coefficients and D is a constant, is a **linear equation in three variables.** Examples of these equations are shown at the right. The graph of a linear equation in three variables is a plane.

$$2x + 4y - 3z = 7$$
$$x - 6y + z = -3$$

Graphing an equation in three variables requires a third coordinate axis perpendicular to the xy-plane. The third axis is commonly called the z-axis. The result is a three-dimensional coordinate system called the ***xyz*-coordinate system.** To help visualize a three-dimensional coordinate system, think of a corner of a room: the floor is the xy-plane, one wall is the yz-plane, and the other wall is the xz-plane. A three-dimensional coordinate system is shown at the right.

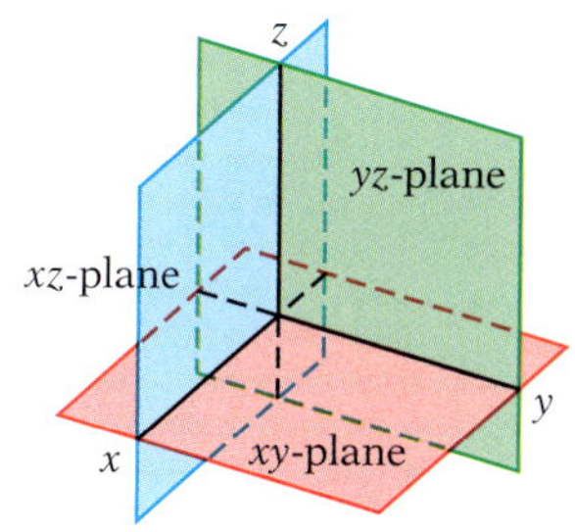

The graph of a point in an xyz-coordinate system is an **ordered triple** (x, y, z). Graphing an ordered triple requires three moves, the first along the x-axis, the second parallel to the y-axis, and the third parallel to the z-axis. The graphs of the points $(-4, 2, 3)$ and $(3, 4, -2)$ are shown at the right.

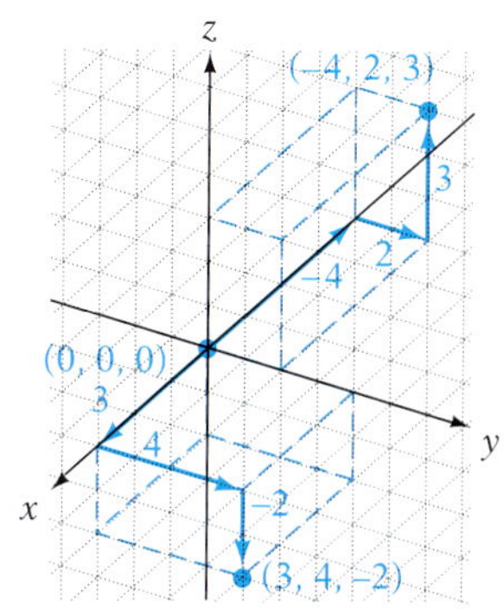

The graph of a linear equation in three variables is a plane. That is, if all the solutions of a linear equation in three variables were plotted in an xyz-coordinate system, the graph would look like a large piece of paper extending infinitely. The graph of $x + y + z = 3$ is shown at the right.

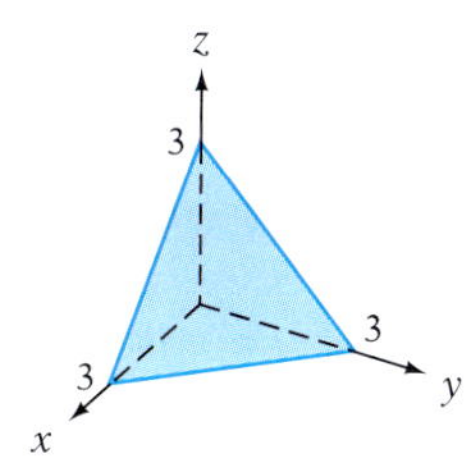

There are different ways in which three planes can be oriented in an xyz-coordinate system. The systems of equations represented by the planes below are inconsistent.

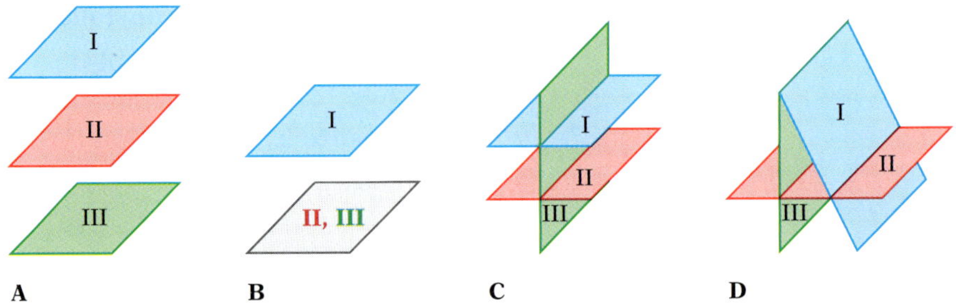

Graphs of Inconsistent Systems of Equations

For a system of three equations in three variables to have a solution, the graphs of the planes must intersect at a single point, they must intersect along a common line, or all equations must have a graph that is the same plane. These situations are shown in the figures that follow.

The three planes shown in Figure E below intersect at a point. A system of equations represented by planes that intersect at a point is independent.

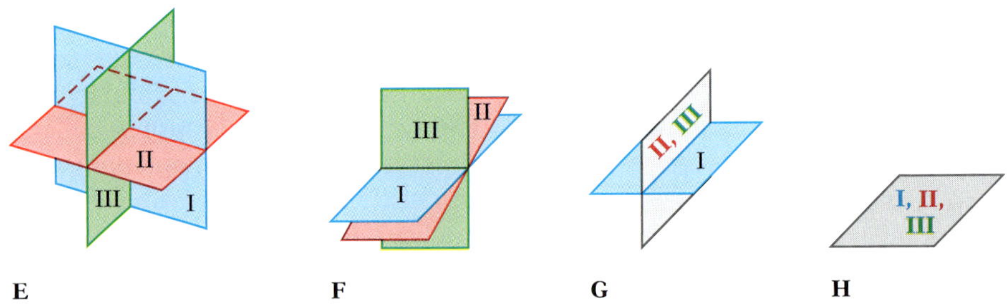

An Independent System of Equations

Dependent Systems of Equations

The planes shown in Figures F and G above intersect along a common line. The system of equations represented by the planes in Figure H has a graph that is the same plane. The systems of equations represented by these three graphs are dependent.

Just as a solution of an equation in two variables is an ordered pair (x, y), a **solution of an equation in three variables** is an ordered triple (x, y, z). For example, $(2, 1, -3)$ is a solution of the equation $2x - y - 2z = 9$. The ordered triple $(1, 3, 2)$ is not a solution.

A **system of linear equations in three variables** is shown at the right. A **solution of a system of equations in three variables** is an ordered triple that is a solution of each equation of the system.

$$\begin{aligned} x - 2y + z &= 6 \\ 3x + y - 2z &= 2 \\ 2x - 3y + 5z &= 1 \end{aligned}$$

Point of Interest

In the early 1980s, Stephen Hoppe became interested in winning Monopoly strategies. Finding these strategies required solving a system that contained 123 equations with 123 variables!

A system of linear equations in three variables can be solved by using the addition method. First, eliminate one variable from any two of the given equations. Then eliminate the same variable from any other two equations. The result will be a system of two equations in two variables. Solve this system by the addition method.

Solve: $x + 4y - z = 10$
$3x + 2y + z = 4$
$2x - 3y + 2z = -7$

$$\begin{aligned} (1)\quad x + 4y - z &= 10 \\ (2)\quad 3x + 2y + z &= 4 \\ (3)\quad 2x - 3y + 2z &= -7 \end{aligned}$$

Eliminate z from equations (1) and (2) by adding the two equations.

$$\begin{aligned} x + 4y - z &= 10 \\ 3x + 2y + z &= 4 \\ (4)\quad 4x + 6y &= 14 \end{aligned}$$

Eliminate z from equations (1) and (3). Multiply equation (1) by 2 and add it to equation (3).

$$\begin{aligned} 2x + 8y - 2z &= 20 \\ 2x - 3y + 2z &= -7 \\ (5)\quad 4x + 5y &= 13 \end{aligned}$$

Solve the system of two equations in two variables.

$$\begin{aligned} (4)\quad 4x + 6y &= 14 \\ (5)\quad 4x + 5y &= 13 \end{aligned}$$

Eliminate x. Multiply equation (5) by -1 and add it to equation (4).

$$\begin{aligned} 4x + 6y &= 14 \\ -4x - 5y &= -13 \\ y &= 1 \end{aligned}$$

Substitute the value of y into equation (4) or (5), and solve for x. Equation (4) is used here.

$$\begin{aligned} 4x + 6y &= 14 \\ 4x + 6(1) &= 14 \\ 4x + 6 &= 14 \\ 4x &= 8 \\ x &= 2 \end{aligned}$$

Substitute the value of y and the value of x into one of the equations in the original system. Equation (2) is used here.

$$\begin{aligned} 3x + 2y + z &= 4 \\ 3(2) + 2(1) + z &= 4 \\ 6 + 2 + z &= 4 \\ 8 + z &= 4 \\ z &= -4 \end{aligned}$$

The solution is $(2, 1, -4)$.

Solve: $2x - 3y - z = 1$
$x + 4y + 3z = 2$
$4x - 6y - 2z = 5$

$$\begin{aligned} (1)\quad 2x - 3y - z &= 1 \\ (2)\quad x + 4y + 3z &= 2 \\ (3)\quad 4x - 6y - 2z &= 5 \end{aligned}$$

Eliminate x from equations (1) and (2). Multiply equation (2) by -2 and add to equation (1).

$$\begin{aligned} 2x - 3y - z &= 1 \\ -2x - 8y - 6z &= -4 \\ -11y - 7z &= -3 \end{aligned}$$

Eliminate x from equations (1) and (3). Multiply equation (1) by -2 and add to equation (3). This is not a true equation.

$$\begin{aligned} -4x + 6y + 2z &= -2 \\ 4x - 6y - 2z &= 5 \\ 0 &= 3 \end{aligned}$$

The system is inconsistent.

Example 2 Solve: $3x - y + 2z = 1$
$2x + 3y + 3z = 4$
$x + y - 4z = -9$

Solution

$$\begin{aligned} (1)\quad 3x - y + 2z &= 1 \\ (2)\quad 2x + 3y + 3z &= 4 \\ (3)\quad x + y - 4z &= -9 \end{aligned}$$

$$\begin{aligned} 3x - y + 2z &= 1 \\ x + y - 4z &= -9 \\ 4x - 2z &= -8 \end{aligned}$$

• **Eliminate y. Add equations (1) and (3).**

$$(4) \qquad 2x - z = -4$$

• **Multiply each side of the equation by $\frac{1}{2}$.**

$$\begin{aligned} 9x - 3y + 6z &= 3 \\ 2x + 3y + 3z &= 4 \\ (5) \qquad 11x + 9z &= 7 \end{aligned}$$

• **Multiply equation (1) by 3 and add to equation (2).**

$$\begin{aligned} (4) \qquad 2x - z &= -4 \\ (5) \qquad 11x + 9z &= 7 \end{aligned}$$

• **Solve the system of two equations.**

$$\begin{aligned} 18x - 9z &= -36 \\ 11x + 9z &= 7 \\ 29x &= -29 \\ x &= -1 \end{aligned}$$

• **Multiply equation (4) by 9 and add to equation (5).**

$$\begin{aligned} 2x - z &= -4 \\ 2(-1) - z &= -4 \\ -2 - z &= -4 \\ -z &= -2 \\ z &= 2 \end{aligned}$$

• **Replace x by -1 in equation (4).**

$$\begin{aligned} x + y - 4z &= -9 \\ -1 + y - 4(2) &= -9 \\ -1 + y - 8 &= -9 \\ -9 + y &= -9 \\ y &= 0 \end{aligned}$$

• **Replace x by -1 and z by 2 in equation (3).**

The solution is $(-1, 0, 2)$.

Problem 2 Solve:

$$\begin{aligned} x - y + z &= 6 \\ 2x + 3y - z &= 1 \\ x + 2y + 2z &= 5 \end{aligned}$$

Solution See page S14.

4.2 Concept Review

Determine whether the following statements are always true, sometimes true, or never true.

1. The graph of a linear equation in three variables is a plane.

2. A solution of an equation in three variables is an ordered triple (x, y, z).

3. The solution of a system of linear equations in three variables is a straight line.

4. A system of linear equations in three variables that has an infinite number of solutions is an independent system.

5. When you are solving a system of equations by the addition method and the final result is $0 = 5$, the system is inconsistent and has an infinite number of solutions.

4.2 Exercises

1 1. If you begin to solve the system of equations $\begin{aligned} x + y &= 6 \\ x - y &= 4 \end{aligned}$ by adding the two equations, the result is $2x = 10$, which means $x = 5$. Does this mean that 5 is a solution of the system of equations? Why or why not?

2. Suppose you must solve the system of equations $\begin{aligned} 2x - 5y &= 13 \\ 3x + 2y &= 10 \end{aligned}$.

a. Explain how you would eliminate x from the system of equations.
b. Explain how you would eliminate y from the system of equations.

Solve by the addition method.

3. $\begin{aligned} x - y &= 5 \\ x + y &= 7 \end{aligned}$

4. $\begin{aligned} x + y &= 1 \\ 2x - y &= 5 \end{aligned}$

5. $\begin{aligned} 3x + y &= 4 \\ x + y &= 2 \end{aligned}$

6. $\begin{aligned} x - 3y &= 4 \\ x + 5y &= -4 \end{aligned}$

7. $\begin{aligned} 3x + y &= 7 \\ x + 2y &= 4 \end{aligned}$

8. $\begin{aligned} x - 2y &= 7 \\ 3x - 2y &= 9 \end{aligned}$

9. $\begin{aligned} 3x - y &= 4 \\ 6x - 2y &= 8 \end{aligned}$

10. $\begin{aligned} x - 2y &= -3 \\ -2x + 4y &= 6 \end{aligned}$

11. $\begin{aligned} 2x + 5y &= 9 \\ 4x - 7y &= -16 \end{aligned}$

12. $\begin{aligned} 8x - 3y &= 21 \\ 4x + 5y &= -9 \end{aligned}$

13. $\begin{aligned} 4x - 6y &= 5 \\ 2x - 3y &= 7 \end{aligned}$

14. $\begin{aligned} 3x + 6y &= 7 \\ 2x + 4y &= 5 \end{aligned}$

15. $\begin{aligned} 3x - 5y &= 7 \\ x - 2y &= 3 \end{aligned}$

16. $\begin{aligned} 3x + 4y &= 25 \\ 2x + y &= 10 \end{aligned}$

17. $\begin{aligned} 3x + 2y &= 16 \\ 2x - 3y &= -11 \end{aligned}$

18. $\begin{aligned} 2x - 5y &= 13 \\ 5x + 3y &= 17 \end{aligned}$

19. $\begin{aligned} 4x + 4y &= 5 \\ 2x - 8y &= -5 \end{aligned}$

20. $\begin{aligned} 3x + 7y &= 16 \\ 4x - 3y &= 9 \end{aligned}$

21. $\begin{aligned} 5x + 4y &= 0 \\ 3x + 7y &= 0 \end{aligned}$

22. $\begin{aligned} 3x - 4y &= 0 \\ 4x - 7y &= 0 \end{aligned}$

23. $\begin{aligned} 3x - 6y &= 6 \\ 9x - 3y &= 8 \end{aligned}$

24. $\begin{aligned} 4x - 8y &= 5 \\ 8x + 2y &= 1 \end{aligned}$

25. $\begin{aligned} 5x + 2y &= 2x + 1 \\ 2x - 3y &= 3x + 2 \end{aligned}$

26. $\begin{aligned} 3x + 3y &= y + 1 \\ x + 3y &= 9 - x \end{aligned}$

27. $\begin{aligned} \frac{2}{3}x - \frac{1}{2}y &= 3 \\ \frac{1}{3}x - \frac{1}{4}y &= \frac{3}{2} \end{aligned}$

28. $\begin{aligned} \frac{3}{4}x + \frac{1}{3}y &= -\frac{1}{2} \\ \frac{1}{2}x - \frac{5}{6}y &= -\frac{7}{2} \end{aligned}$

29. $\begin{aligned} \frac{2}{5}x - \frac{1}{3}y &= 1 \\ \frac{3}{5}x + \frac{2}{3}y &= 5 \end{aligned}$

30. $\frac{5}{6}x + \frac{1}{3}y = \frac{4}{3}$
$\frac{2}{3}x - \frac{1}{2}y = \frac{11}{6}$

31. $\frac{3}{4}x + \frac{2}{5}y = -\frac{3}{20}$
$\frac{3}{2}x - \frac{1}{4}y = \frac{3}{4}$

32. $\frac{2}{5}x - \frac{1}{2}y = \frac{13}{2}$
$\frac{3}{4}x - \frac{1}{5}y = \frac{17}{2}$

33. $4x - 5y = 3y + 4$
$2x + 3y = 2x + 1$

34. $5x - 2y = 8x - 1$
$2x + 7y = 4y + 9$

35. $2x + 5y = 5x + 1$
$3x - 2y = 3y + 3$

36. When you solve a system of equations using the addition method, how can you tell whether the system of equations is inconsistent?

37. When you solve a system of equations using the addition method, how can you tell whether the system of equations is dependent?

2 **38.** Describe the graph of a linear equation in three variables.

39. Describe the solution of an independent system of linear equations in three variables.

Solve by the addition method.

40. $x + 2y - z = 1$
$2x - y + z = 6$
$x + 3y - z = 2$

41. $x + 3y + z = 6$
$3x + y - z = -2$
$2x + 2y - z = 1$

42. $2x - y + 2z = 7$
$x + y + z = 2$
$3x - y + z = 6$

43. $x - 2y + z = 6$
$x + 3y + z = 16$
$3x - y - z = 12$

44. $3x + y = 5$
$3y - z = 2$
$x + z = 5$

45. $2y + z = 7$
$2x - z = 3$
$x - y = 3$

46. $x - y + z = 1$
$2x + 3y - z = 3$
$-x + 2y - 4z = 4$

47. $2x + y - 3z = 7$
$x - 2y + 3z = 1$
$3x + 4y - 3z = 13$

48. $2x + 3z = 5$
$3y + 2z = 3$
$3x + 4y = -10$

49. $3x + 4z = 5$
$2y + 3z = 2$
$2x - 5y = 8$

50. $2x + 4y - 2z = 3$
$x + 3y + 4z = 1$
$x + 2y - z = 4$

51. $x - 3y + 2z = 1$
$x - 2y + 3z = 5$
$2x - 6y + 4z = 3$

52. $2x + y - z = 5$
$x + 3y + z = 14$
$3x - y + 2z = 1$

53. $3x - y - 2z = 11$
$2x + y - 2z = 11$
$x + 3y - z = 8$

54. $3x + y - 2z = 2$
$x + 2y + 3z = 13$
$2x - 2y + 5z = 6$

55. $4x + 5y + z = 6$
$2x - y + 2z = 11$
$x + 2y + 2z = 6$

56. $2x - y + z = 6$
$3x + 2y + z = 4$
$x - 2y + 3z = 12$

57. $3x + 2y - 3z = 8$
$2x + 3y + 2z = 10$
$x + y - z = 2$

58. $3x - 2y + 3z = -4$
$2x + y - 3z = 2$
$3x + 4y + 5z = 8$

59. $3x - 3y + 4z = 6$
$4x - 5y + 2z = 10$
$x - 2y + 3z = 4$

60. $3x - y + 2z = 2$
$4x + 2y - 7z = 0$
$2x + 3y - 5z = 7$

61. $2x + 2y + 3z = 13$
$-3x + 4y - z = 5$
$5x - 3y + z = 2$

62. $2x - 3y + 7z = 0$
$x + 4y - 4z = -2$
$3x + 2y + 5z = 1$

63. $5x + 3y - z = 5$
$3x - 2y + 4z = 13$
$4x + 3y + 5z = 22$

4.2 Applying Concepts

Solve. (*Hint:* First multiply both sides of each equation in the system by a multiple of 10 so that the coefficients and constants are integers.)

64. $0.2x - 0.3y = 0.5$
$0.3x - 0.2y = 0.5$

65. $0.4x - 0.9y = -0.1$
$0.3x + 0.2y = 0.8$

66. $1.25x - 0.25y = -1.5$
$1.5x + 2.5y = 1$

67. $2.25x + 1.5y = 3$
$1.75x + 2.25y = 1.25$

68. $1.5x + 2.5y + 1.5z = 8$
$0.5x - 2y - 1.5z = -1$
$2.5x - 1.5y + 2z = 2.5$

69. $1.6x - 0.9y + 0.3z = 2.9$
$1.6x + 0.5y - 0.1z = 3.3$
$0.8x - 0.7y + 0.1z = 1.5$

Solve.

70. The point of intersection of the graphs of the equations $Ax + 3y = 6$ and $2x + By = -4$ is $(3, -2)$. Find A and B.

71. The point of intersection of the graphs of the equations $Ax + 3y + 2z = 8$, $2x + By - 3z = -12$, and $3x - 2y + Cz = 1$ is $(3, -2, 4)$. Find A, B, and C.

72. The distance between a point and a line is the perpendicular distance from the point to the line. Find the distance between the point $(3, 1)$ and the line $y = x$.

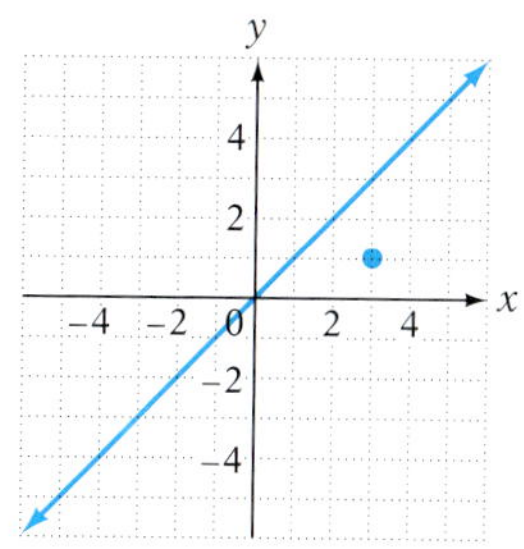

73. **Coin Problem** A coin bank contains only nickels, dimes, and quarters. There is a total of 30 coins in the bank. The value of all the coins is \$3.25. Find the number of nickels, dimes, and quarters in the bank. (*Hint:* There is more than one solution.)

The following are not linear systems of equations. However, they can be solved by using a modification of the addition method. Solve each system of equations.

74. $\frac{1}{x} - \frac{2}{y} = 3$
$\frac{2}{x} + \frac{3}{y} = -1$

75. $\frac{1}{x} + \frac{2}{y} = 3$
$\frac{1}{x} - \frac{3}{y} = -2$

76. $\frac{3}{x} + \frac{2}{y} = 1$
$\frac{2}{x} + \frac{4}{y} = -2$

77. $\frac{3}{x} - \frac{5}{y} = -\frac{3}{2}$
$\frac{1}{x} - \frac{2}{y} = -\frac{2}{3}$

78. For Exercise 74, solve each equation of the system for y. Then use a graphing utility to verify the solution you determined algebraically. Note that the graphs are not straight lines.

79. Describe the graph of each of the following equations in an xyz-coordinate system.

a. $x = 3$ **b.** $y = 4$ **c.** $z = 2$ **d.** $y = x$

80. Explain, graphically, the following situations when they are related to a system of three linear equations in three variables.
  a. The system of equations has no solution.
  b. The system of equations has exactly one solution.
  c. The system of equations has infinitely many solutions.

4.3 Solving Systems of Equations by Using Matrices and by Using Determinants

1 Solve systems of linear equations by using matrices

VIDEO & DVD CD TUTOR WEB SSM

Point of Interest

The word matrix *was first used (in a mathematical context) in 1850. The root of this word is the Latin* mater. *A matrix was thought of as being an object from which something else originates. The idea was that a determinant, discussed below, originated (was born) from a matrix. Today, matrices are one of the most widely used tools of applied mathematics.*

A **matrix** is a rectangular array of numbers. Each number in the matrix is called an **element** of the matrix. The matrix at the right, with three rows and four columns, is called a 3×4 (read "3 by 4") matrix.

$$A = \begin{bmatrix} 1 & -3 & 2 & 4 \\ 0 & 4 & -3 & 2 \\ 6 & -5 & 4 & -1 \end{bmatrix}$$

A matrix of m rows and n columns is said to be of **order $m \times n$.** Matrix A above has order 3×4. The notation a_{ij} refers to the element of a matrix in the ith row and the jth column. For matrix A, $a_{23} = -3$, $a_{31} = 6$, and $a_{13} = 2$.

The elements $a_{11}, a_{22}, a_{33}, \ldots, a_{nn}$ form the **main diagonal** of a matrix. The elements 1, 5, and 7 form the main diagonal of the matrix above.

By considering only the coefficients and constants for the following system of equations, we can form the corresponding 3×4 **augmented matrix.**

Take Note

For augmented matrices, it is common practice to show a vertical line between the coefficients of the variables and the constant terms.

System of Equations | **Augmented Matrix**

$$\begin{aligned} 3x - 2y + z &= 2 \\ x - 3z &= -2 \\ 2x - y + 4z &= 5 \end{aligned} \qquad \left[\begin{array}{rrr|r} 3 & -2 & 1 & 2 \\ 1 & 0 & -3 & -2 \\ 2 & -1 & 4 & 5 \end{array}\right]$$

Note that when a term is missing from one of the equations of the system, the coefficient of that term is 0, and a 0 is entered in the matrix.

A system of equations can be written from an augmented matrix.

$$\left[\begin{array}{rrr|r} 2 & -1 & 4 & 1 \\ 1 & 1 & 0 & 3 \\ 3 & -2 & -1 & 5 \end{array}\right] \qquad \begin{aligned} 2x - y + 4z &= 1 \\ x + y &= 3 \\ 3x - 2y - z &= 5 \end{aligned}$$

Example 1 Write the augmented matrix for $\begin{aligned} 2x - 3y &= 4 \\ x + 5y &= 0 \end{aligned}$.

Solution The augmented matrix is $\left[\begin{array}{rr|r} 2 & -3 & 4 \\ 1 & 5 & 0 \end{array}\right]$.

Problem 1 Write the system of equations that corresponds to the augmented matrix $\left[\begin{array}{ccc|c} 2 & -3 & 1 & 4 \\ 1 & 0 & -2 & 3 \\ 0 & 1 & 2 & -3 \end{array}\right]$.

Solution See page S14.

A system of equations can be solved by writing the system in matrix form and then performing operations on the matrix similar to those performed on the equations of the system. These operations are called **elementary row operations.**

Elementary Row Operations

1. Interchange two rows.
2. Multiply all the elements in a row by the same nonzero number.
3. Replace a row by the sum of that row and a multiple of any other row.

Each of these elementary row operations has as its basis the operations that can be performed on a system of equations. These operations do not change the solution of the system of equations. Here are some examples of each row operation.

> **Take Note**
>
> For each example of a row operation at the right, we are using the same system of equations. The solution of this system of equations is (2, 1). You should verify that (2, 1) is also a solution of the new system of equations. This means that these row operations have not changed the solution.

1. Interchange two rows.

Original System		This notation means to interchange rows 1 and 2.		**New System**
$2x - 3y = 1$ $4x + 5y = 13$	$\left[\begin{array}{cc|c} 2 & -3 & 1 \\ 4 & 5 & 13 \end{array}\right]$	$R_1 \leftrightarrow R_2$	$\left[\begin{array}{cc|c} 4 & 5 & 13 \\ 2 & -3 & 1 \end{array}\right]$	$4x + 5y = 13$ $2x - 3y = 1$
The solution is (2, 1).				The solution is (2, 1).

2. Multiply all the elements in a row by the same nonzero number.

Original System		This notation means to multiply row 2 by 3.		**New System**
$2x - 3y = 1$ $4x + 5y = 13$	$\left[\begin{array}{cc|c} 2 & -3 & 1 \\ 4 & 5 & 13 \end{array}\right]$	$3R_2 \rightarrow$	$\left[\begin{array}{cc|c} 2 & -3 & 1 \\ 12 & 15 & 39 \end{array}\right]$	$2x - 3y = 1$ $12x + 15y = 39$
The solution is (2, 1).				The solution is (2, 1).

> **Take Note**
>
> Note that for the third elementary row operation, the row being multiplied is *not* changed. For the example at the right, row 1 is not changed. The row being added to is replaced by the result of the operation. It is not absolutely necessary to follow this convention, but for consistency, we will always use this method.

3. Replace a row by the sum of that row and a multiple of any other row.

Original System

$$\begin{aligned} 2x - 3y &= 1 \\ 4x + 5y &= 13 \end{aligned} \qquad \left[\begin{array}{cc|c} 2 & -3 & 1 \\ 4 & 5 & 13 \end{array}\right] \quad -2R_1 + R_2 \rightarrow \quad \left[\begin{array}{cc|c} 2 & -3 & 1 \\ 0 & 11 & 11 \end{array}\right]$$

This notation means to replace row 2 by the sum of that row and −2 times row 1.

The solution is (2, 1).

New System

$$\begin{aligned} 2x - 3y &= 1 \\ 11y &= 11 \end{aligned}$$

The solution is (2, 1).

Note that we replace the row that follows the addition. See the Take Note at the left.

Example 2 Let $A = \left[\begin{array}{ccc|c} 1 & 3 & -4 & 6 \\ 3 & 2 & 0 & -1 \\ -2 & -5 & 3 & 4 \end{array}\right]$. Perform the following elementary row operations on A.

A. $R_1 \leftrightarrow R_3$ **B.** $-2R_3$ **C.** $2R_3 + R_1$

Solution **A.** $R_1 \leftrightarrow R_3$ means to interchange row 1 and row 3.

$$\left[\begin{array}{ccc|c} 1 & 3 & -4 & 6 \\ 3 & 2 & 0 & -1 \\ -2 & -5 & 3 & 4 \end{array}\right] \quad R_1 \leftrightarrow R_3 \quad \left[\begin{array}{ccc|c} -2 & -5 & 3 & 4 \\ 3 & 2 & 0 & -1 \\ 1 & 3 & -4 & 6 \end{array}\right]$$

B. $-2R_3$ means to multiply row 3 by −2.

$$\left[\begin{array}{ccc|c} 1 & 3 & -4 & 6 \\ 3 & 2 & 0 & -1 \\ -2 & -5 & 3 & 4 \end{array}\right] \quad -2R_3 \rightarrow \quad \left[\begin{array}{ccc|c} 1 & 3 & -4 & 6 \\ 3 & 2 & 0 & -1 \\ 4 & 10 & -6 & -8 \end{array}\right]$$

C. $2R_3 + R_1$ means to multiply row 3 by 2 and then add the result to row 1.

$$\left[\begin{array}{ccc|c} 1 & 3 & -4 & 6 \\ 3 & 2 & 0 & -1 \\ -2 & -5 & 3 & 4 \end{array}\right] \quad 2R_3 + R_1 \rightarrow \quad \left[\begin{array}{ccc|c} -3 & -7 & 2 & 14 \\ 3 & 2 & 0 & -1 \\ -2 & -5 & 3 & 4 \end{array}\right]$$

Problem 2 Let $B = \left[\begin{array}{ccc|c} 1 & 8 & -2 & 3 \\ 2 & -3 & 4 & 1 \\ 3 & 5 & -7 & 3 \end{array}\right]$. Perform the following elementary row operations on B.

A. $R_2 \leftrightarrow R_3$ **B.** $3R_2$ **C.** $-3R_1 + R_3$

Solution See page S15.

Elementary row operations are used to solve a system of equations. The goal is to use the elementary row operations to rewrite the augmented matrix with 1's down the main diagonal and 0's to the left of the 1's in all rows except the first. This is called the **row echelon form** of the matrix. Here are some examples of the row echelon form.

$$\left[\begin{array}{cc|c} 1 & 3 & -2 \\ 0 & 1 & 3 \end{array}\right] \quad \left[\begin{array}{ccc|c} 1 & -2 & 3 & 1 \\ 0 & 1 & 2.5 & -4 \\ 0 & 0 & 1 & 2 \end{array}\right] \quad \left[\begin{array}{ccc|c} 1 & 4 & \frac{1}{2} & -3 \\ 0 & 1 & 3 & 0 \\ 0 & 0 & 1 & -\frac{2}{3} \end{array}\right]$$

We will follow a very definite procedure to rewrite an augmented matrix in row echelon form. For a 2×3 augmented matrix, use elementary row operations to

Step 1: Change a_{11} to a 1.
Step 2: Change a_{21} to a 0.
Step 3: Change a_{22} to a 1.

$$\left[\begin{array}{cc|c} a_{11} & a_{12} & a_{13} \\ a_{21} & a_{22} & a_{23} \end{array}\right]$$

Write the matrix $\left[\begin{array}{cc|c} 3 & -6 & 12 \\ 2 & 1 & -3 \end{array}\right]$ in row echelon form.

Step 1: Change a_{11} to 1. One way to do this is to multiply row 1 by the reciprocal of a_{11}.

$$\left[\begin{array}{cc|c} 3 & -6 & 12 \\ 2 & 1 & -3 \end{array}\right] \; \tfrac{1}{3}R_1 \rightarrow \left[\begin{array}{cc|c} 1 & -2 & 4 \\ 2 & 1 & -3 \end{array}\right]$$

Step 2: Change a_{21} to 0 by multiplying row 1 by the opposite of a_{21} and then adding to row 2.

$$\left[\begin{array}{cc|c} 1 & -2 & 4 \\ 2 & 1 & -3 \end{array}\right] \; -2R_1 + R_2 \rightarrow \left[\begin{array}{cc|c} 1 & -2 & 4 \\ 0 & 5 & -11 \end{array}\right]$$

Step 3: Change a_{22} to 1 by multiplying by the reciprocal of a_{22}.

$$\left[\begin{array}{cc|c} 1 & -2 & 4 \\ 0 & 5 & -11 \end{array}\right] \; \tfrac{1}{5}R_2 \rightarrow \left[\begin{array}{cc|c} 1 & -2 & 4 \\ 0 & 1 & -2.2 \end{array}\right]$$

A row echelon form of the matrix is $\left[\begin{array}{cc|c} 1 & -2 & 4 \\ 0 & 1 & -2.2 \end{array}\right]$.

The row echelon form of a matrix is not unique and depends on the elementary row operations that are used. For instance, suppose we again start with $\left[\begin{array}{cc|c} 3 & -6 & 12 \\ 2 & 1 & -3 \end{array}\right]$ and follow the elementary row operations below.

$$\left[\begin{array}{cc|c} 3 & -6 & 12 \\ 2 & 1 & -3 \end{array}\right] \; -1R_2 + R_1 \rightarrow \left[\begin{array}{cc|c} 1 & -7 & 15 \\ 2 & 1 & -3 \end{array}\right] \; -2R_1 + R_2 \rightarrow \left[\begin{array}{cc|c} 1 & -7 & 15 \\ 0 & 15 & -33 \end{array}\right] \; \tfrac{1}{15}R_2 \rightarrow \left[\begin{array}{cc|c} 1 & -7 & 15 \\ 0 & 1 & -\frac{11}{5} \end{array}\right]$$

In this case, we come to $\left[\begin{array}{cc|c} 1 & -7 & 15 \\ 0 & 1 & -\frac{11}{5} \end{array}\right]$ as the row echelon form, rather than $\left[\begin{array}{cc|c} 1 & -2 & 4 \\ 0 & 1 & -\frac{11}{5} \end{array}\right]$ as in the first case. Row echelon form is not unique.

The order in which the elements in an augmented 3×4 matrix are changed is as follows:

$$\left[\begin{array}{ccc|c} a_{11} & a_{12} & a_{13} & a_{14} \\ a_{21} & a_{22} & a_{23} & a_{24} \\ a_{31} & a_{32} & a_{33} & a_{34} \end{array}\right]$$

Step 1: Change a_{11} to a 1.
Step 2: Change a_{21} and a_{31} to 0's.
Step 3: Change a_{22} to a 1.
Step 4: Change a_{32} to a 0.
Step 5: Change a_{33} to a 1.

Example 3 Write $\left[\begin{array}{ccc|c} 2 & 1 & 3 & -1 \\ 1 & 3 & 5 & -1 \\ -3 & -1 & 1 & 2 \end{array}\right]$ in row echelon form.

Solution **Step 1:** Change a_{11} to 1 by interchanging row 1 and row 2.

Note: We could have chosen to multiply row 1 by $\frac{1}{2}$. The sequence of steps to get to row echelon form is not unique.

$$\left[\begin{array}{ccc|c} 2 & 1 & 3 & -1 \\ 1 & 3 & 5 & -1 \\ -3 & -1 & 1 & 2 \end{array}\right] \quad R_1 \leftrightarrow R_2 \quad \left[\begin{array}{ccc|c} 1 & 3 & 5 & -1 \\ 2 & 1 & 3 & -1 \\ -3 & -1 & 1 & 2 \end{array}\right]$$

Step 2: Change a_{21} to 0 by multiplying row 1 by the opposite of a_{21} and then adding to row 2.

$$\left[\begin{array}{ccc|c} 1 & 3 & 5 & -1 \\ 2 & 1 & 3 & -1 \\ -3 & -1 & 1 & 2 \end{array}\right] \quad -2R_1 + R_2 \rightarrow \quad \left[\begin{array}{ccc|c} 1 & 3 & 5 & -1 \\ 0 & -5 & -7 & 1 \\ -3 & -1 & 1 & 2 \end{array}\right]$$

Change a_{31} to 0 by multiplying row 1 by the opposite of a_{31} and then adding to row 3.

$$\left[\begin{array}{ccc|c} 1 & 3 & 5 & -1 \\ 0 & -5 & -7 & 1 \\ -3 & -1 & 1 & 2 \end{array}\right] \quad 3R_1 + R_3 \rightarrow \quad \left[\begin{array}{ccc|c} 1 & 3 & 5 & -1 \\ 0 & -5 & -7 & 1 \\ 0 & 8 & 16 & -1 \end{array}\right]$$

Step 3: Change a_{22} to 1 by multiplying by the reciprocal of a_{22}.

$$\left[\begin{array}{ccc|c} 1 & 3 & 5 & -1 \\ 0 & -5 & -7 & 1 \\ 0 & 8 & 16 & -1 \end{array}\right] \quad -\frac{1}{5}R_2 \rightarrow \quad \left[\begin{array}{ccc|c} 1 & 3 & 5 & -1 \\ 0 & 1 & \frac{7}{5} & -\frac{1}{5} \\ 0 & 8 & 16 & -1 \end{array}\right]$$

Step 4: Change a_{32} to 0 by multiplying row 2 by the opposite of a_{32} and then adding to row 3.

$$\left[\begin{array}{ccc|c} 1 & 3 & 5 & -1 \\ 0 & 1 & \frac{7}{5} & -\frac{1}{5} \\ 0 & 8 & 16 & -1 \end{array}\right] \quad -8R_2 + R_3 \rightarrow \quad \left[\begin{array}{ccc|c} 1 & 3 & 5 & -1 \\ 0 & 1 & \frac{7}{5} & -\frac{1}{5} \\ 0 & 0 & \frac{24}{5} & \frac{3}{5} \end{array}\right]$$

Step 5: Change a_{33} to 1 by multiplying by the reciprocal of a_{33}.

$$\left[\begin{array}{ccc|c} 1 & 3 & 5 & -1 \\ 0 & 1 & \frac{7}{5} & -\frac{1}{5} \\ 0 & 0 & \frac{24}{5} & \frac{3}{5} \end{array}\right] \quad \frac{5}{24}R_3 \rightarrow \quad \left[\begin{array}{ccc|c} 1 & 3 & 5 & -1 \\ 0 & 1 & \frac{7}{5} & -\frac{1}{5} \\ 0 & 0 & 1 & \frac{1}{8} \end{array}\right]$$

A row echelon form of the matrix is $\left[\begin{array}{ccc|c} 1 & 3 & 5 & -1 \\ 0 & 1 & \frac{7}{5} & -\frac{1}{5} \\ 0 & 0 & 1 & \frac{1}{8} \end{array}\right]$.

Problem 3 Write $\left[\begin{array}{ccc|c} 1 & -3 & 2 & 1 \\ -4 & 14 & 0 & -2 \\ 2 & -5 & -3 & 16 \end{array}\right]$ in row echelon form.

Solution See page S15.

If an augmented matrix is in row echelon form, the corresponding system of equations can be solved by substitution. For instance, consider the following matrix in row echelon form and the corresponding system of equations.

$$\left[\begin{array}{ccc|c} 1 & -3 & 4 & 7 \\ 0 & 1 & 3 & -6 \\ 0 & 0 & 1 & -1 \end{array}\right] \qquad \begin{aligned} x - 3y + 4z &= 7 \\ y + 3z &= -6 \\ z &= -1 \end{aligned}$$

From the last equation of the system above, we have $z = -1$. Substitute this value into the second equation and solve for y. Thus $y = -3$.

$$\begin{aligned} y + 3z &= -6 \\ y + 3(-1) &= -6 \\ y - 3 &= -6 \\ y &= -3 \end{aligned}$$

Substitute $y = -3$ and $z = -1$ in the first equation of the system and solve for x. Thus $x = 2$.

$$\begin{aligned} x - 3y + 4z &= 7 \\ x - 3(-3) + 4(-1) &= 7 \\ x + 9 - 4 &= 7 \\ x &= 2 \end{aligned}$$

The solution of the system of equations is $(2, -3, -1)$.

The process of solving a system of equations by using elementary row operations is called the **Gaussian elimination method.**

Example 4 Solve by using the Gaussian elimination method:

$$\begin{aligned} 2x - 5y &= 19 \\ 3x + 4y &= -6 \end{aligned}$$

Solution Write the augmented matrix and then use elementary row operations to rewrite the matrix in row echelon form.

$$\left[\begin{array}{cc|c} 2 & -5 & 19 \\ 3 & 4 & -6 \end{array}\right] \xrightarrow[\frac{1}{2}R_1]{\text{Change } a_{11} \text{ to } 1.} \left[\begin{array}{cc|c} 1 & -\frac{5}{2} & \frac{19}{2} \\ 3 & 4 & -6 \end{array}\right]$$

• **Note that we multiplied R_1 by the reciprocal of a_{11}.**

Point of Interest

Johann Carl Friedrich Gauss (1777–1855) is considered one of the greatest mathematicians of all time. He contributed not only to mathematics but to astronomy and physics as well. A unit of magnetism called a gauss was named in his honor. The image above is that of Gauss on a German deutsche mark note.

$$\left[\begin{array}{cc|c} 1 & -\frac{5}{2} & \frac{19}{2} \\ 3 & 4 & -6 \end{array}\right] \xrightarrow[-3R_1 + R_2]{\text{Change } a_{21} \text{ to } 0.} \left[\begin{array}{cc|c} 1 & -\frac{5}{2} & \frac{19}{2} \\ 0 & \frac{23}{2} & -\frac{69}{2} \end{array}\right]$$

• Note that we multiplied R_1 by the opposite of a_{21}.

$$\left[\begin{array}{cc|c} 1 & -\frac{5}{2} & \frac{19}{2} \\ 0 & \frac{23}{2} & -\frac{69}{2} \end{array}\right] \xrightarrow[\frac{2}{23}R_2]{\text{Change } a_{22} \text{ to } 1.} \left[\begin{array}{cc|c} 1 & -\frac{5}{2} & \frac{19}{2} \\ 0 & 1 & -3 \end{array}\right]$$

• This is row echelon form.

$$\begin{aligned} (1) \qquad x - \frac{5}{2}y &= \frac{19}{2} \\ (2) \qquad y &= -3 \end{aligned}$$

• Write the system of equations corresponding to the matrix that is in row echelon form.

$$\begin{aligned} x - \frac{5}{2}(-3) &= \frac{19}{2} \\ x + \frac{15}{2} &= \frac{19}{2} \\ x &= 2 \end{aligned}$$

• Substitute −3 for *y* in equation (1) and solve for *x*.

The solution is $(2, -3)$.

Problem 4 Solve by using the Gaussian elimination method:

$$\begin{aligned} 4x - 5y &= 17 \\ 3x + 2y &= 7 \end{aligned}$$

Solution See page S15.

The Gaussian elimination method can be extended to systems of equations with more than two variables.

Example 5 Solve by using the Gaussian elimination method:

$$\begin{aligned} x + 2y - z &= 9 \\ 2x - y + 2z &= -1 \\ -2x + 3y - 2z &= 7 \end{aligned}$$

Solution

$$\left[\begin{array}{ccc|c} 1 & 2 & -1 & 9 \\ 2 & -1 & 2 & -1 \\ -2 & 3 & -2 & 7 \end{array}\right] \xrightarrow[-2R_1 + R_2]{a_{11} \text{ is 1. Change } a_{21} \text{ to } 0.} \left[\begin{array}{ccc|c} 1 & 2 & -1 & 9 \\ 0 & -5 & 4 & -19 \\ -2 & 3 & -2 & 7 \end{array}\right]$$

$$\left[\begin{array}{ccc|c} 1 & 2 & -1 & 9 \\ 0 & -5 & 4 & -19 \\ -2 & 3 & -2 & 7 \end{array}\right] \xrightarrow[2R_1 + R_3]{\text{Change } a_{31} \text{ to } 0.} \left[\begin{array}{ccc|c} 1 & 2 & -1 & 9 \\ 0 & -5 & 4 & -19 \\ 0 & 7 & -4 & 25 \end{array}\right]$$

$$\left[\begin{array}{ccc|c} 1 & 2 & -1 & 9 \\ 0 & -5 & 4 & -19 \\ 0 & 7 & -4 & 25 \end{array}\right] \xrightarrow[-\frac{1}{5}R_2]{\text{Change } a_{22} \text{ to } 1.} \left[\begin{array}{ccc|c} 1 & 2 & -1 & 9 \\ 0 & 1 & -\frac{4}{5} & \frac{19}{5} \\ 0 & 7 & -4 & 25 \end{array}\right]$$

$$\left[\begin{array}{ccc|c} 1 & 2 & -1 & 9 \\ 0 & 1 & -\frac{4}{5} & \frac{19}{5} \\ 0 & 7 & -4 & 25 \end{array}\right] \xrightarrow[-7R_2 + R_3]{\text{Change } a_{32} \text{ to } 0.} \left[\begin{array}{ccc|c} 1 & 2 & -1 & 9 \\ 0 & 1 & -\frac{4}{5} & \frac{19}{5} \\ 0 & 0 & \frac{8}{5} & -\frac{8}{5} \end{array}\right]$$

$$\left[\begin{array}{ccc|c} 1 & 2 & -1 & 9 \\ 0 & 1 & -\frac{4}{5} & \frac{19}{5} \\ 0 & 0 & \frac{8}{5} & -\frac{8}{5} \end{array}\right] \xrightarrow[\frac{5}{8}R_2]{\text{Change } a_{33} \text{ to } 1.} \left[\begin{array}{ccc|c} 1 & 2 & -1 & 9 \\ 0 & 1 & -\frac{4}{5} & \frac{19}{5} \\ 0 & 0 & 1 & -1 \end{array}\right]$$

• This is row echelon form.

(1) $x + 2y - z = 9$

(2) $y - \frac{4}{5}z = \frac{19}{5}$

(3) $z = -1$

• Write the system of equations corresponding to the matrix that is in row echelon form.

$$y - \frac{4}{5}(-1) = \frac{19}{5}$$
$$y + \frac{4}{5} = \frac{19}{5}$$
$$y = 3$$

• Substitute -1 for z in equation (2) and solve for y.

$$x + 2y - z = 9$$
$$x + 2(3) - (-1) = 9$$
$$x + 7 = 9$$
$$x = 2$$

• Substitute -1 for z and 3 for y in equation (1) and solve for x.

The solution is $(2, 3, -1)$.

Problem 5 Solve by using the Gaussian elimination method:

$$2x + 3y + 3z = -2$$
$$x + 2y - 3z = 9$$
$$3x - 2y - 4z = 1$$

Solution See page S15.

2 Evaluate deteminants

As previously stated in this section, a **matrix** is a rectangular array of numbers. Each number in the matrix is called an **element** of the matrix. The matrix shown below, with 3 rows and 4 columns, is a 3×4 matrix.

$$\begin{bmatrix} 1 & 4 & -3 & 6 \\ -2 & 5 & 2 & 0 \\ -1 & 3 & 7 & -4 \end{bmatrix}$$

A **square matrix** is one that has the same number of rows as columns. A 2×2 matrix and a 3×3 matrix are shown at the right.

$$\begin{bmatrix} -1 & 3 \\ 5 & 2 \end{bmatrix} \qquad \begin{bmatrix} 4 & 0 & 1 \\ 5 & -3 & 7 \\ 2 & 1 & 4 \end{bmatrix}$$

Associated with every square matrix is a number called its **determinant.**

Determinant of a 2 × 2 Matrix

The determinant of a 2 × 2 matrix $\begin{bmatrix} a_{11} & a_{12} \\ a_{21} & a_{22} \end{bmatrix}$ is written $\begin{vmatrix} a_{11} & a_{12} \\ a_{21} & a_{22} \end{vmatrix}$. The value of this determinant is given by the formula

$$\begin{vmatrix} a_{11} & a_{12} \\ a_{21} & a_{22} \end{vmatrix} = a_{11}a_{22} - a_{21}a_{12}$$

Note that vertical bars are used to represent the determinant and that brackets are used to represent the matrix.

Find the value of the determinant $\begin{vmatrix} 3 & 4 \\ -1 & 2 \end{vmatrix}$.

$$\begin{vmatrix} 3 & 4 \\ -1 & 2 \end{vmatrix} = 3 \cdot 2 - (-1)(4) = 6 - (-4) = 10$$

The value of the determinant is 10.

For a square matrix whose order is 3 × 3 or greater, the value of the determinant of that matrix is found by using 2 × 2 determinants.

The **minor of an element** in a 3 × 3 determinant is the 2 × 2 determinant that is obtained by eliminating the row and column that contain that element.

Find the minor of −3 for the determinant $\begin{vmatrix} 2 & -3 & 4 \\ 0 & 4 & 8 \\ -1 & 3 & 6 \end{vmatrix}$.

The minor of −3 is the 2 × 2 determinant created by eliminating the row and column that contain −3.

Eliminate the row and column as shown: $\begin{vmatrix} 2 & -3 & 4 \\ 0 & 4 & 8 \\ -1 & 3 & 6 \end{vmatrix}$

The minor of −3 is $\begin{vmatrix} 0 & 8 \\ -1 & 6 \end{vmatrix}$.

Take Note

The only difference between the cofactor and the minor is one of sign. The definition at the right can be stated symbolically as follows: If C_{ij} is the cofactor and M_{ij} is the minor of the determinant element a_{ij}, then $C_{ij} = (-1)^{i+j}M_{ij}$. If $i + j$ is an even number, then $(-1)^{i+j} = 1$ and $C_{ij} = M_{ij}$. If $i + j$ is an odd number, then $(-1)^{i+j} = -1$ and $C_{ij} = -M_{ij}$.

Definition of a Cofactor

The **cofactor** of an element of a determinant is $(-1)^{i+j}$ times the minor of that element, where i is the row number of the element and j is the column number of the element.

For the determinant $\begin{vmatrix} 3 & -2 & 1 \\ 2 & -5 & -4 \\ 0 & 3 & 1 \end{vmatrix}$, find the cofactor of −2 and of −5.

Because -2 is in the first row and the second column, $i = 1$ and $j = 2$. Therefore, $i + j = 1 + 2 = 3$, and $(-1)^{i+j} = (-1)^3 = -1$. The cofactor of -2 is $(-1)\begin{vmatrix} 2 & -4 \\ 0 & 1 \end{vmatrix}$.

Because -5 is in the second row and the second column, $i = 2$ and $j = 2$. Therefore, $i + j = 2 + 2 = 4$, and $(-1)^{i+j} = (-1)^4 = 1$. The cofactor of -5 is $1 \cdot \begin{vmatrix} 3 & 1 \\ 0 & 1 \end{vmatrix}$.

Note from this example that the cofactor of an element is -1 times the minor of that element or 1 times the minor of that element, depending on whether the sum $i + j$ is an odd or an even integer.

The value of a 3×3 or larger determinant can be found by **expanding by cofactors** of *any* row or *any* column. The result of expanding by cofactors using the first row of a 3×3 determinant is shown here.

$$\begin{vmatrix} a_{11} & a_{12} & a_{13} \\ a_{21} & a_{22} & a_{23} \\ a_{31} & a_{32} & a_{33} \end{vmatrix} = a_{11}(-1)^{1+1}\begin{vmatrix} a_{22} & a_{23} \\ a_{32} & a_{33} \end{vmatrix} + a_{12}(-1)^{1+2}\begin{vmatrix} a_{21} & a_{23} \\ a_{31} & a_{33} \end{vmatrix} + a_{13}(-1)^{1+3}\begin{vmatrix} a_{21} & a_{22} \\ a_{31} & a_{32} \end{vmatrix}$$

$$= a_{11}\begin{vmatrix} a_{22} & a_{23} \\ a_{32} & a_{33} \end{vmatrix} - a_{12}\begin{vmatrix} a_{21} & a_{23} \\ a_{31} & a_{33} \end{vmatrix} + a_{13}\begin{vmatrix} a_{21} & a_{22} \\ a_{31} & a_{32} \end{vmatrix}$$

Find the value of the determinant $\begin{vmatrix} 2 & -3 & 2 \\ 1 & 3 & -1 \\ 0 & -2 & 2 \end{vmatrix}$.

Expand by cofactors of the first row.

$$\begin{vmatrix} 2 & -3 & 2 \\ 1 & 3 & -1 \\ 0 & -2 & 2 \end{vmatrix} = 2\begin{vmatrix} 3 & -1 \\ -2 & 2 \end{vmatrix} - (-3)\begin{vmatrix} 1 & -1 \\ 0 & 2 \end{vmatrix} + 2\begin{vmatrix} 1 & 3 \\ 0 & -2 \end{vmatrix}$$

$$= 2(6 - 2) - (-3)(2 - 0) + 2(-2 - 0)$$

$$= 2(4) - (-3)(2) + 2(-2) = 8 - (-6) + (-4)$$

$$= 10$$

To illustrate a statement made earlier, the value of this determinant will now be found by expanding by cofactors of the second column.

$$\begin{vmatrix} 2 & -3 & 2 \\ 1 & 3 & -1 \\ 0 & -2 & 2 \end{vmatrix} = -3(-1)^{1+2}\begin{vmatrix} 1 & -1 \\ 0 & 2 \end{vmatrix} + 3(-1)^{2+2}\begin{vmatrix} 2 & 2 \\ 0 & 2 \end{vmatrix} + (-2)(-1)^{3+2}\begin{vmatrix} 2 & 2 \\ 1 & -1 \end{vmatrix}$$

$$= -3(-1)\begin{vmatrix} 1 & -1 \\ 0 & 2 \end{vmatrix} + 3 \cdot 1\begin{vmatrix} 2 & 2 \\ 0 & 2 \end{vmatrix} + (-2)(-1)\begin{vmatrix} 2 & 2 \\ 1 & -1 \end{vmatrix}$$

$$= 3(2 - 0) + 3(4 - 0) + 2(-2 - 2)$$

$$= 3(2) + 3(4) + 2(-4) = 6 + 12 + (-8)$$

$$= 10$$

Note that the value of the determinant is the same whether the first row or the second column is used to expand by cofactors. *Any row or column* can be used to evaluate a determinant by expanding by cofactors.

Example 6 Evaluate the determinant. **A.** $\begin{vmatrix} 3 & -2 \\ 6 & -4 \end{vmatrix}$ **B.** $\begin{vmatrix} -2 & 3 & 1 \\ 4 & -2 & 0 \\ 1 & -2 & 3 \end{vmatrix}$

Solution **A.** $\begin{vmatrix} 3 & -2 \\ 6 & -4 \end{vmatrix} = 3(-4) - (6)(-2) = -12 + 12 = 0$

The value of the determinant is 0.

B.
$$\begin{vmatrix} -2 & 3 & 1 \\ 4 & -2 & 0 \\ 1 & -2 & 3 \end{vmatrix} = -2\begin{vmatrix} -2 & 0 \\ -2 & 3 \end{vmatrix} - 3\begin{vmatrix} 4 & 0 \\ 1 & 3 \end{vmatrix} + 1\begin{vmatrix} 4 & -2 \\ 1 & -2 \end{vmatrix}$$
$$= -2(-6 + 0) - 3(12 - 0) + 1(-8 + 2)$$
$$= -2(-6) - 3(12) + 1(-6)$$
$$= 12 - 36 - 6$$
$$= -30$$

The value of the determinant is -30.

Problem 6 Evaluate the determinant.

A. $\begin{vmatrix} -1 & -4 \\ 3 & -5 \end{vmatrix}$ **B.** $\begin{vmatrix} 1 & 4 & -2 \\ 3 & 1 & 1 \\ 0 & -2 & 2 \end{vmatrix}$

Solution See page S16.

3 Solve systems of linear equations by using Cramer's Rule

The connection between determinants and systems of equations can be understood by solving a general system of linear equations.

Solve: $a_{11}x + a_{12}y = b_1$
$a_{21}x + a_{22}y = b_2$

$$(1) \quad a_{11}x + a_{12}y = b_1$$
$$(2) \quad a_{21}x + a_{22}y = b_2$$

Eliminate y. Multiply equation (1) by a_{22} and equation (2) by $-a_{12}$.

$$a_{11}a_{22}x + a_{12}a_{22}y = b_1a_{22}$$
$$-a_{21}a_{12}x - a_{12}a_{22}y = -b_2a_{12}$$

Add the equations.

$$(a_{11}a_{22} - a_{21}a_{12})x = b_1a_{22} - b_2a_{12}$$

Assuming $a_{11}a_{22} - a_{21}a_{12} \neq 0$, solve for x.

$$x = \frac{b_1a_{22} - b_2a_{12}}{a_{11}a_{22} - a_{21}a_{12}}$$

The denominator $a_{11}a_{22} - a_{21}a_{12}$ is the determinant of the coefficients of x and y. This is called the **coefficient determinant.**

$$a_{11}a_{22} - a_{21}a_{12} = \begin{vmatrix} a_{11} & a_{12} \\ a_{21} & a_{22} \end{vmatrix}$$

Coefficients of x (first column)

Coefficients of y (second column)

The numerator for x, $b_1a_{22} - b_2a_{12}$, is the determinant obtained by replacing the first column in the coefficient determinant by the constants b_1 and b_2. This is called the **numerator determinant.**

$$b_1a_{22} - b_2a_{12} = \begin{vmatrix} b_1 & a_{12} \\ b_2 & a_{22} \end{vmatrix}$$

Constants of the equations (first column)

Following a similar procedure and eliminating x, we can also express the y-component of the solution in determinant form. These results are summarized in Cramer's Rule.

Cramer's Rule

The solution of the system of equations $\begin{aligned} a_{11}x + a_{12}y &= b_1 \\ a_{21}x + a_{22}y &= b_2 \end{aligned}$ is given by $x = \dfrac{D_x}{D}$ and $y = \dfrac{D_y}{D}$, where

$$D = \begin{vmatrix} a_{11} & a_{12} \\ a_{21} & a_{22} \end{vmatrix},\ D_x = \begin{vmatrix} b_1 & a_{12} \\ b_2 & a_{22} \end{vmatrix},\ D_y = \begin{vmatrix} a_{11} & b_1 \\ a_{21} & b_2 \end{vmatrix}, \text{ and } D \neq 0.$$

Point of Interest

Cramer's Rule is named after Gabriel Cramer, who used it in a book he published in 1750. However, this rule was published in 1683 by the Japanese mathematician Seki Kown. That publication occurred seven years before Cramer's birth.

Example 7 Solve by using Cramer's Rule: $\begin{aligned} 2x - 3y &= 8 \\ 5x + 6y &= 11 \end{aligned}$

Solution

$$D = \begin{vmatrix} 2 & -3 \\ 5 & 6 \end{vmatrix} = 27$$

• Find the value of the coefficient determinant.

$$D_x = \begin{vmatrix} 8 & -3 \\ 11 & 6 \end{vmatrix} = 81$$

$$D_y = \begin{vmatrix} 2 & 8 \\ 5 & 11 \end{vmatrix} = -18$$

• Find the value of each of the numerator determinants.

$$x = \frac{D_x}{D} = \frac{81}{27} = 3$$

$$y = \frac{D_y}{D} = \frac{-18}{27} = -\frac{2}{3}$$

• Use Cramer's Rule to write the solutions.

The solution is $\left(3, -\dfrac{2}{3}\right)$.

Problem 7 Solve by using Cramer's Rule: $\begin{aligned} 6x - 6y &= 5 \\ 2x - 10y &= -1 \end{aligned}$

Solution See page S16.

For the system shown at the right, $D = 0$. Therefore, $\dfrac{D_x}{D}$ and $\dfrac{D_y}{D}$ are undefined.

$$\begin{aligned} 6x - 9y &= 5 \\ 4x - 6y &= 4 \end{aligned}$$

When $D = 0$, the system of equations is dependent if both D_x and D_y are zero. The system of equations is inconsistent if $D = 0$ and either D_x or D_y is not zero.

$$D = \begin{vmatrix} 6 & -9 \\ 4 & -6 \end{vmatrix} = 0$$

A procedure similar to that followed for two equations in two variables can be used to extend Cramer's Rule to three equations in three variables.

Cramer's Rule for Three Equations in Three Variables

The solution of the system of equations
$$\begin{aligned} a_{11}x + a_{12}y + a_{13}z &= b_1 \\ a_{21}x + a_{22}y + a_{23}z &= b_2 \\ a_{31}x + a_{32}y + a_{33}z &= b_3 \end{aligned}$$
is given by $x = \frac{D_x}{D}$, $y = \frac{D_y}{D}$, and $z = \frac{D_z}{D}$, where

$$D = \begin{vmatrix} a_{11} & a_{12} & a_{13} \\ a_{21} & a_{22} & a_{23} \\ a_{31} & a_{32} & a_{33} \end{vmatrix}, D_x = \begin{vmatrix} b_1 & a_{12} & a_{13} \\ b_2 & a_{22} & a_{23} \\ b_3 & a_{32} & a_{33} \end{vmatrix}, D_y = \begin{vmatrix} a_{11} & b_1 & a_{13} \\ a_{21} & b_2 & a_{23} \\ a_{31} & b_3 & a_{33} \end{vmatrix},$$

$$D_z = \begin{vmatrix} a_{11} & a_{12} & b_1 \\ a_{21} & a_{22} & b_2 \\ a_{31} & a_{32} & b_3 \end{vmatrix}, \text{ and } D \neq 0.$$

Example 8 Solve by using Cramer's Rule:
$$\begin{aligned} 3x - y + z &= 5 \\ x + 2y - 2z &= -3 \\ 2x + 3y + z &= 4 \end{aligned}$$

Solution

$$D = \begin{vmatrix} 3 & -1 & 1 \\ 1 & 2 & -2 \\ 2 & 3 & 1 \end{vmatrix} = 28$$

• **Find the value of the coefficient determinant.**

$$D_x = \begin{vmatrix} 5 & -1 & 1 \\ -3 & 2 & -2 \\ 4 & 3 & 1 \end{vmatrix} = 28$$

• **Find the value of each of the numerator determinants.**

$$D_y = \begin{vmatrix} 3 & 5 & 1 \\ 1 & -3 & -2 \\ 2 & 4 & 1 \end{vmatrix} = 0$$

$$D_z = \begin{vmatrix} 3 & -1 & 5 \\ 1 & 2 & -3 \\ 2 & 3 & 4 \end{vmatrix} = 56$$

$$x = \frac{D_x}{D} = \frac{28}{28} = 1$$

• **Use Cramer's Rule to write the solution.**

$$y = \frac{D_y}{D} = \frac{0}{28} = 0$$

$$z = \frac{D_z}{D} = \frac{56}{28} = 2$$

The solution is (1, 0, 2).

Problem 8 Solve by using Cramer's Rule:
$$\begin{aligned} 2x - y + z &= -1 \\ 3x + 2y - z &= 3 \\ x + 3y + z &= -2 \end{aligned}$$

Solution See page S16.

4.3 Concept Review

Determine whether the following statements are always true, sometimes true, or never true.

1. A 3×3 determinant can be evaluated by expanding about any row or column of the matrix.

2. Cramer's Rule can be used to solve a system of linear equations in three variables.

3. A matrix has the same number of rows and columns.

4. Associated with a matrix is a number called its determinant.

5. A system of two linear equations in two variables has no solution, exactly one solution, or infinitely many solutions.

4.3 Exercises

1

1. Explain how to write the augmented matrix of a system of equations.

2. Is the matrix $\begin{bmatrix} 1 & -2 & 3 & 1 \\ 0 & 1 & 2 & 8 \\ 0 & 1 & 0 & 1 \end{bmatrix}$ in row echelon form? Why or why not?

Write each matrix in row echelon form.

3. $\left[\begin{array}{cc|c} 2 & -4 & 1 \\ 3 & -7 & -1 \end{array}\right]$

4. $\left[\begin{array}{cc|c} 4 & 2 & -2 \\ 7 & 4 & -1 \end{array}\right]$

5. $\left[\begin{array}{cc|c} 5 & -2 & 3 \\ -7 & 3 & 1 \end{array}\right]$

6. $\left[\begin{array}{cc|c} 2 & 5 & -4 \\ 3 & 1 & 2 \end{array}\right]$

7. $\left[\begin{array}{ccc|c} 1 & 4 & 1 & -2 \\ 3 & 11 & -1 & 2 \\ 2 & 3 & 1 & 4 \end{array}\right]$

8. $\left[\begin{array}{ccc|c} 1 & 2 & 2 & -1 \\ -4 & -10 & -1 & 3 \\ 3 & 4 & 2 & -2 \end{array}\right]$

9. $\left[\begin{array}{ccc|c} -2 & 6 & -1 & 3 \\ 1 & -2 & 2 & 1 \\ 3 & -6 & 7 & 6 \end{array}\right]$

10. $\left[\begin{array}{ccc|c} 2 & 6 & 10 & 3 \\ 3 & 8 & 15 & 0 \\ 1 & 2 & 3 & -1 \end{array}\right]$

11. What is the solution of the system of equations that has $\left[\begin{array}{ccc|c} 1 & -1 & 3 & -2 \\ 0 & 1 & -1 & 1 \\ 0 & 0 & 1 & 3 \end{array}\right]$ as the row echelon form of the augmented matrix for the system of equations?

12. What is the solution of the system of equations that has $\left[\begin{array}{ccc|c} 1 & -3 & 2 & 4 \\ 0 & 1 & -2 & 3 \\ 0 & 0 & 1 & -1 \end{array}\right]$ as the row echelon form of the augmented matrix for the system of equations?

Solve by using the Gaussian elimination method.

13. $\begin{aligned} 3x + y &= 6 \\ 2x - y &= -1 \end{aligned}$

14. $\begin{aligned} 2x + y &= 3 \\ x - 4y &= 6 \end{aligned}$

15. $\begin{aligned} x - 3y &= 8 \\ 3x - y &= 0 \end{aligned}$

16. $\begin{aligned} 2x + 3y &= 16 \\ x - 4y &= -14 \end{aligned}$

17. $\begin{aligned} y &= 4x - 10 \\ 2y &= 5x - 11 \end{aligned}$

18. $\begin{aligned} 2y &= 4 - 3x \\ y &= 1 - 2x \end{aligned}$

19. $\begin{aligned} 2x - y &= -4 \\ y &= 2x - 8 \end{aligned}$

20. $\begin{aligned} 3x - 2y &= -8 \\ y &= \frac{3}{2}x - 2 \end{aligned}$

21. $\begin{aligned} 4x - 3y &= -14 \\ 3x + 4y &= 2 \end{aligned}$

22. $\begin{aligned} 5x + 2y &= 3 \\ 3x + 4y &= 13 \end{aligned}$

23. $\begin{aligned} 5x + 4y + 3z &= -9 \\ x - 2y + 2z &= -6 \\ x - y - z &= 3 \end{aligned}$

24. $\begin{aligned} x - y - z &= 0 \\ 3x - y + 5z &= -10 \\ x + y - 4z &= 12 \end{aligned}$

25. $\begin{aligned} 5x - 5y + 2z &= 8 \\ 2x + 3y - z &= 0 \\ x + 2y - z &= 0 \end{aligned}$

26. $\begin{aligned} 2x + y - 5z &= 3 \\ 3x + 2y + z &= 15 \\ 5x - y - z &= 5 \end{aligned}$

27. $\begin{aligned} 2x + 3y + z &= 5 \\ 3x + 3y + 3z &= 10 \\ 4x + 6y + 2z &= 5 \end{aligned}$

28. $\begin{aligned} x - 2y + 3z &= 2 \\ 2x + y + 2z &= 5 \\ 2x - 4y + 6z &= -4 \end{aligned}$

29. $\begin{aligned} 3x + 2y + 3z &= 2 \\ 6x - 2y + z &= 1 \\ 3x + 4y + 2z &= 3 \end{aligned}$

30. $\begin{aligned} 2x + 3y - 3z &= -1 \\ 2x + 3y + 3z &= 3 \\ 4x - 4y + 3z &= 4 \end{aligned}$

31. $\begin{aligned} 5x - 5y - 5z &= 2 \\ 5x + 5y - 5z &= 6 \\ 10x + 10y + 5z &= 3 \end{aligned}$

32. $\begin{aligned} 3x - 2y + 2z &= 5 \\ 6x + 3y - 4z &= -1 \\ 3x - y + 2z &= 4 \end{aligned}$

33. $\begin{aligned} 4x + 4y - 3z &= 3 \\ 8x + 2y + 3z &= 0 \\ 4x - 4y + 6z &= -3 \end{aligned}$

2 **34.** How do you find the value of the determinant associated with a 2×2 matrix?

Evaluate the determinant.

35. $\begin{vmatrix} 2 & -1 \\ 3 & 4 \end{vmatrix}$

36. $\begin{vmatrix} 5 & 1 \\ -1 & 2 \end{vmatrix}$

37. $\begin{vmatrix} 6 & -2 \\ -3 & 4 \end{vmatrix}$

38. $\begin{vmatrix} -3 & 5 \\ 1 & 7 \end{vmatrix}$

39. $\begin{vmatrix} 3 & 6 \\ 2 & 4 \end{vmatrix}$

40. $\begin{vmatrix} 5 & -10 \\ 1 & -2 \end{vmatrix}$

41. $\begin{vmatrix} 1 & -1 & 2 \\ 3 & 2 & 1 \\ 1 & 0 & 4 \end{vmatrix}$

42. $\begin{vmatrix} 4 & 1 & 3 \\ 2 & -2 & 1 \\ 3 & 1 & 2 \end{vmatrix}$

43. $\begin{vmatrix} 3 & -1 & 2 \\ 0 & 1 & 2 \\ 3 & 2 & -2 \end{vmatrix}$

44. $\begin{vmatrix} 4 & 5 & -2 \\ 3 & -1 & 5 \\ 2 & 1 & 4 \end{vmatrix}$

45. $\begin{vmatrix} 4 & 2 & 6 \\ -2 & 1 & 1 \\ 2 & 1 & 3 \end{vmatrix}$

46. $\begin{vmatrix} 3 & 6 & -3 \\ 4 & -1 & 6 \\ -1 & -2 & 3 \end{vmatrix}$

3 **47.** Explain how Cramer's Rule is used to solve a system of two equations in two unknowns.

48. Explain why Cramer's Rule cannot be used to solve a system of equations for which the determinant of the coefficient matrix is 0.

Solve by using Cramer's Rule, if possible.

49. $\begin{aligned} 2x - 5y &= 26 \\ 5x + 3y &= 3 \end{aligned}$

50. $\begin{aligned} 3x + 7y &= 15 \\ 2x + 5y &= 11 \end{aligned}$

51. $\begin{aligned} x - 4y &= 8 \\ 3x + 7y &= 5 \end{aligned}$

52. $\begin{aligned} 5x + 2y &= -5 \\ 3x + 4y &= 11 \end{aligned}$

53. $\begin{aligned} 2x + 3y &= 4 \\ 6x - 12y &= -5 \end{aligned}$

54. $\begin{aligned} 5x + 4y &= 3 \\ 15x - 8y &= -21 \end{aligned}$

55. $\begin{aligned} 2x + 5y &= 6 \\ 6x - 2y &= 1 \end{aligned}$

56. $\begin{aligned} 7x + 3y &= 4 \\ 5x - 4y &= 9 \end{aligned}$

57. $\begin{aligned} -2x + 3y &= 7 \\ 4x - 6y &= 9 \end{aligned}$

58. $\begin{aligned} 9x + 6y &= 7 \\ 3x + 2y &= 4 \end{aligned}$

59. $\begin{aligned} 2x - 5y &= -2 \\ 3x - 7y &= -3 \end{aligned}$

60. $\begin{aligned} 8x + 7y &= -3 \\ 2x + 2y &= 5 \end{aligned}$

61. $\begin{aligned} 2x - y + 3z &= 9 \\ x + 4y + 4z &= 5 \\ 3x + 2y + 2z &= 5 \end{aligned}$

62. $\begin{aligned} 3x - 2y + z &= 2 \\ 2x + 3y + 2z &= -6 \\ 3x - y + z &= 0 \end{aligned}$

63. $\begin{aligned} 3x - y + z &= 11 \\ x + 4y - 2z &= -12 \\ 2x + 2y - z &= -3 \end{aligned}$

64. $\begin{aligned} x + 2y + 3z &= 8 \\ 2x - 3y + z &= 5 \\ 3x - 4y + 2z &= 9 \end{aligned}$

65. $\begin{aligned} 4x - 2y + 6z &= 1 \\ 3x + 4y + 2z &= 1 \\ 2x - y + 3z &= 2 \end{aligned}$

66. $\begin{aligned} x - 3y + 2z &= 1 \\ 2x + y - 2z &= 3 \\ 3x - 9y + 6z &= -3 \end{aligned}$

67. $\begin{aligned} 5x - 4y + 2z &= 4 \\ 3x - 5y + 3z &= -4 \\ 3x + y - 5z &= 12 \end{aligned}$

68. $\begin{aligned} 2x + 4y + z &= 7 \\ x + 3y - z &= 1 \\ 3x + 2y - 2z &= 5 \end{aligned}$

69. $\begin{aligned} 3x + 5y - 2z &= 2 \\ 6x - 5y + 4z &= 2 \\ 9x - 8z &= -1 \end{aligned}$

4.3 Applying Concepts

Solve for x.

70. $\begin{vmatrix} 3 & 2 \\ 4 & x \end{vmatrix} = -11$

71. $\begin{vmatrix} 1 & 0 & 2 \\ 4 & 3 & -1 \\ 0 & 2 & x \end{vmatrix} = -24$

72. $\begin{vmatrix} -2 & 1 & 3 \\ 0 & x & 4 \\ -1 & 2 & -3 \end{vmatrix} = -24$

Complete.

73. If all the elements in one row or one column of a 2×2 matrix are zeros, the value of the determinant of the matrix is ______.

74. If all the elements in one row or one column of a 3×3 matrix are zeros, the value of the determinant of the matrix is ______.

75. a. The value of the determinant $\begin{vmatrix} x & x & a \\ y & y & b \\ z & z & c \end{vmatrix}$ is ______.

b. If two columns of a 3×3 matrix contain identical elements, the value of the determinant is ______.

76. Show that $\begin{vmatrix} a & b \\ c & d \end{vmatrix} = -\begin{vmatrix} c & d \\ a & b \end{vmatrix}$.

77. **Surveying** Surveyors use a formula to find the area of a plot of land. The *surveyor's area formula* states that if the vertices (x_1, y_1), (x_2, y_2), (x_3, y_3), ..., (x_n, y_n) of a simple polygon are listed counterclockwise around the perimeter, then the area of the polygon is

$$A = \frac{1}{2}\left\{\begin{vmatrix} x_1 & x_2 \\ y_1 & y_2 \end{vmatrix} + \begin{vmatrix} x_2 & x_3 \\ y_2 & y_3 \end{vmatrix} + \begin{vmatrix} x_3 & x_4 \\ y_3 & y_4 \end{vmatrix} + \cdots + \begin{vmatrix} x_n & x_1 \\ y_n & y_1 \end{vmatrix}\right\}$$

Use the surveyor's area formula to find the area of the polygon with vertices (9, −3), (26, 6), (18, 21), (16, 10), and (1, 11). Measurements given are in feet.

78. Consider the system of equations $\begin{aligned} 3x - 2y &= 12 \\ 2x + 5y &= -11 \end{aligned}$. Graph the equations. Now multiply the second equation by 3, add it to the first equation, and replace the second equation by the new equation. What is the new system of equations? Graph the equations of this system on the same coordinate system. The lines intersect at the same point, so the new system of equations is equivalent to the original system. Explain how this is related to the third elementary row operation.

79. Suppose that applying elementary row operations to an augmented matrix produces the matrix $\left[\begin{array}{ccc|c} 1 & 2 & -3 & 5 \\ 0 & 1 & -2 & -2 \\ 0 & 0 & 0 & -3 \end{array}\right]$. Explain why the original system of equations has no solution.

4.4 Application Problems

1 Rate-of-wind and water-current problems

VIDEO & DVD CD TUTOR WEB SSM

Motion problems that involve an object moving with or against a wind or current normally require two variables to solve.

Solve: A motorboat traveling with the current can go 24 mi in 2 h. Against the current, it takes 3 h to go the same distance. Find the rate of the motorboat in calm water and the rate of the current.

Strategy *for solving rate-of-wind and water-current problems*

▶ Choose one variable to represent the rate of the object in calm conditions and a second variable to represent the rate of the wind or current. Using these variables, express the rate of the object with and against the wind or current. Use the equation $rt = d$ to write expressions for the distance traveled by the object. The results can be recorded in a table.

Rate of the boat in calm water: x
Rate of the current: y

	Rate	·	Time	=	Distance
With current	$x + y$	·	2	=	$2(x + y)$
Against current	$x - y$	·	3	=	$3(x - y)$

▶ Determine how the expressions for the distance are related.

The distance traveled with the current is 24 mi. $2(x + y) = 24$
The distance traveled against the current is 24 mi. $3(x - y) = 24$

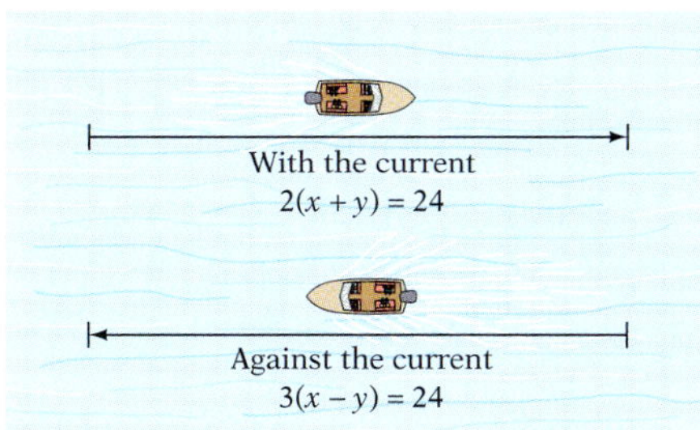

Solve the system of equations.

$$2(x + y) = 24 \quad \rightarrow \quad \frac{1}{2} \cdot 2(x + y) = \frac{1}{2} \cdot 24 \quad \rightarrow \quad x + y = 12$$

$$3(x - y) = 24 \quad \rightarrow \quad \frac{1}{3} \cdot 3(x - y) = \frac{1}{3} \cdot 24 \quad \rightarrow \quad x - y = 8$$

$$2x = 20$$
$$x = 10$$

Replace x by 10 in the equation $x + y = 12$. Solve for y.

$$x + y = 12$$
$$10 + y = 12$$
$$y = 2$$

The rate of the boat in calm water is 10 mph.
The rate of the current is 2 mph.

Example 1 Flying with the wind, a plane flew 1000 mi in 5 h. Flying against the wind, the plane could fly only 500 mi in the same amount of time. Find the rate of the plane in calm air and the rate of the wind.

Strategy

▶ Rate of the plane in still air: p
Rate of the wind: w

	Rate	Time	Distance
With wind	$p + w$	5	$5(p + w)$
Against wind	$p - w$	5	$5(p - w)$

▶ The distance traveled with the wind is 1000 mi.
The distance traveled against the wind is 500 mi.

Solution

$$5(p + w) = 1000$$
$$5(p - w) = 500$$

$$p + w = 200 \quad \text{• Multiply each side of the equations by } \frac{1}{5}.$$
$$p - w = 100$$
$$2p = 300 \quad \text{• Add the equations.}$$
$$p = 150$$

$$p + w = 200 \quad \text{• Substitute the value of } p \text{ into one of the equations.}$$
$$150 + w = 200$$
$$w = 50$$

The rate of the plane in calm air is 150 mph.
The rate of the wind is 50 mph.

Problem 1 A rowing team rowing with the current traveled 18 mi in 2 h. Against the current, the team rowed 10 mi in 2 h. Find the rate of the current.

Solution See page S16.

2 Application problems

VIDEO & DVD | CD TUTOR | WEB | SSM

The application problems in this section are varieties of those problems solved earlier in the text. Each of the strategies for the problems in this section will result in a system of equations.

Solve: A store owner purchased twenty 60-watt light bulbs and thirty fluorescent lights for a total cost of \$40. A second purchase, at the same prices, included thirty 60-watt bulbs and ten fluorescent lights for a total cost of \$25. Find the cost of a 60-watt bulb and of a fluorescent light.

Strategy *for solving an application problem in two variables*

▶ Choose one variable to represent one of the unknown quantities and a second variable to represent the other unknown quantity. Write numerical or variable expressions for all the remaining quantities. These results can be recorded in two tables, one for each of the conditions.

Cost of 60-watt bulb: b
Cost of a fluorescent light: f

First purchase:

	Amount	·	Unit cost	=	Value
60-watt	20	·	b	=	$20b$
Fluorescent	30	·	f	=	$30f$

Second purchase:

	Amount	·	Unit cost	=	Value
60-watt	30	·	b	=	$30b$
Fluorescent	10	·	f	=	$10f$

▶ Determine a system of equations. The strategies presented in Chapter 2 can be used to determine the relationships between the expressions in the tables. Each table will give one equation of the system.

The total of the first purchase was \$40. $20b + 30f = 40$
The total of the second purchase was \$25. $30b + 10f = 25$

Solve the system of equations.

$$\begin{aligned} 20b + 30f &= 40 \\ 30b + 10f &= 25 \end{aligned} \quad \rightarrow \quad \begin{aligned} 3(20b + 30f) &= 3 \cdot 40 \\ -2(30b + 10f) &= -2 \cdot 25 \end{aligned} \quad \rightarrow \quad \begin{aligned} 60b + 90f &= 120 \\ -60b - 20f &= -50 \\ 70f &= 70 \\ f &= 1 \end{aligned}$$

Replace f by 1 in the equation $20b + 30f = 40$. Solve for b.

$$\begin{aligned} 20b + 30f &= 40 \\ 20b + 30(1) &= 40 \\ 20b + 30 &= 40 \\ 20b &= 10 \\ b &= 0.5 \end{aligned}$$

The cost of a 60-watt bulb was \$.50.
The cost of a fluorescent light was \$1.00.

Example 2 The total value of the nickels and dimes in a coin bank is \$2.50. If the nickels were dimes and the dimes were nickels, the total value of the coins would be \$3.05. Find the number of dimes and the number of nickels in the bank.

Strategy ▶ Number of nickels in the bank: n
Number of dimes in the bank: d

Coins in the bank now:

Coin	Number	Value	Total value
Nickels	n	5	$5n$
Dimes	d	10	$10d$

Coins in the bank if the nickels were dimes and the dimes were nickels:

Coin	Number	Value	Total value
Nickels	d	5	$5d$
Dimes	n	10	$10n$

▶ The value of the nickels and dimes in the bank is \$2.50. The value of the nickels and dimes in the bank would be \$3.05.

Solution

(1) $5n + 10d = 250$
(2) $10n + 5d = 305$

$$10n + 20d = 500$$
$$-10n - 5d = -305$$

• Multiply equation (1) by 2 and equation (2) by −1.

$$15d = 195$$
$$d = 13$$

• Add the equations.

$$5n + 10d = 250$$
$$5n + 10(13) = 250$$
$$5n + 130 = 250$$
$$5n = 120$$
$$n = 24$$

• Substitute the value of *d* into one of the equations.

There are 24 nickels and 13 dimes in the bank.

Problem 2 A citrus fruit grower purchased 25 orange trees and 20 grapefruit trees for \$290. The next week, at the same prices, the grower bought 20 orange trees and 30 grapefruit trees for \$330. Find the cost of an orange tree and the cost of a grapefruit tree.

Solution See pages S16–S17.

Example 3 is an application problem that requires more than two variables. The solution of this problem also illustrates how the substitution method can be used to solve a system of linear equations in three variables.

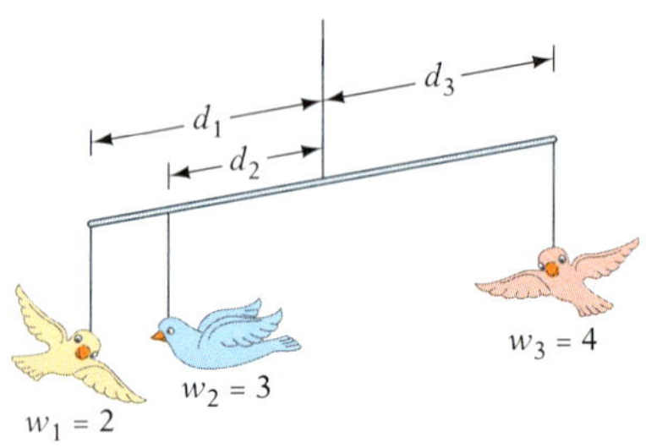

Example 3 An artist is creating a mobile from which three objects will be suspended from a light rod that is 18 in. long as shown at the left. The weight, in ounces, of each object is shown in the diagram. For the mobile to balance, the objects must be positioned so that $w_1d_1 + w_2d_2 = w_3d_3$. The artist wants d_1 to be 1.5 times d_2. Find the distances d_1, d_2, and d_3 so that the mobile will balance.

Strategy There are three unknowns for this problem. Use the information in the problem to write three equations with d_1, d_2, and d_3 as the variables.

The length of the rod is 18 in. Therefore, $d_1 + d_3 = 18$. Using $w_1d_1 + w_2d_2 = w_3d_3$, we have $2d_1 + 3d_2 = 4d_3$. The artist wants d_1 to be 1.5 times d_2. Thus $d_1 = 1.5d_2$.

Solution

(1) $d_1 + d_3 = 18$
(2) $2d_1 + 3d_2 = 4d_3$
(3) $d_1 = 1.5d_2$

We will solve the system of equations by substitution. Use equation (3) to replace d_1 in equation (1) and in equation (2).

$$1.5d_2 + d_3 = 18$$
$$2(1.5d_2) + 3d_2 = 4d_3$$

$$
\begin{aligned}
&(4) \qquad 1.5d_2 + d_3 = 18 \\
&(5) \qquad 6d_2 = 4d_3 \\
&(6) \qquad 1.5d_2 = d_3
\end{aligned}
$$
• Divide each side of equation (5) by 4.

Replace d_3 by $1.5d_2$ in equation (4) and solve for d_2.

$$
\begin{aligned}
1.5d_2 + 1.5d_2 &= 18 \\
3d_2 &= 18 \\
d_2 &= 6
\end{aligned}
$$

For equation (6), $d_3 = 1.5d_2 = 1.5(6) = 9$.

Substituting the value of d_3 into equation (1), we have $d_1 = 9$.

The values are $d_1 = 9$ in., $d_2 = 6$ in., and $d_3 = 9$ in.

Problem 3 A science museum charges \$10 for an admission ticket, but members receive a discount of \$3, and students are admitted for half price. Last Saturday, 750 tickets were sold for a total of \$5400. If 20 more student tickets than full-price tickets were sold, how many of each type of ticket were sold?

Solution See page S17.

4.4 Concept Review

Determine whether the following statements are always true, sometimes true, or never true.

1. The rate of a current in a river is x mph, and the rate of a boat in still water is y mph. The rate of a boat going down the river is $x + y$, and the rate going up the river is $x - y$.

2. The speed of a plane is 500 mph. There is a headwind of 50 mph. The speed of the plane relative to an observer on the ground is 550 mph.

3. A contractor bought 50 yd of nylon carpet for x dollars per yard and bought 100 yd of wool carpet for y dollars per yard. The total cost of the carpet is $50x + 100y$.

4. A coin bank contains nickels and dimes. The total value of the coins is \$.70. The bank had 5 dimes and 4 nickels.

4.4 Exercises

1 Rate-of-wind and water-current problems

1. Flying with the wind, a small plane flew 320 mi in 2 h. Against the wind, the plane could fly only 280 mi in the same amount of time. Find the rate of the plane in calm air and the rate of the wind.

2. A jet plane flying with the wind went 2100 mi in 4 h. Against the wind, the plane could fly only 1760 mi in the same amount of time. Find the rate of the plane in calm air and the rate of the wind.

3. A cabin cruiser traveling with the current went 48 mi in 3 h. Against the current, it took 4 h to travel the same distance. Find the rate of the cabin cruiser in calm water and the rate of the current.

4. A motorboat traveling with the current went 48 mi in 2 h. Against the current, it took 3 h to travel the same distance. Find the rate of the boat in calm water and the rate of the current.

5. Flying with the wind, a pilot flew 450 mi between two cities in 2.5 h. The return trip against the wind took 3 h. Find the rate of the plane in calm air and the rate of the wind.

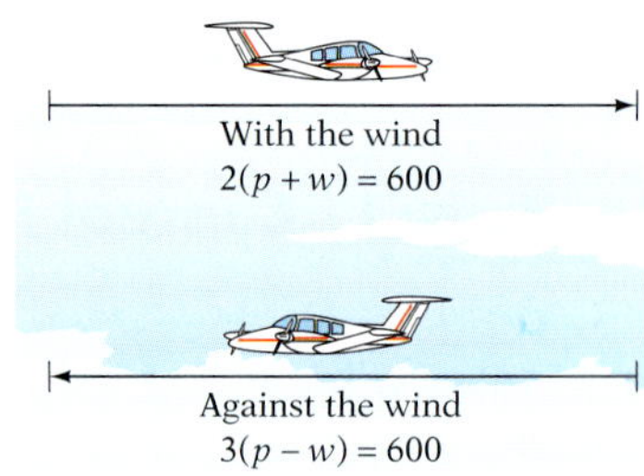

6. A turbo-prop plane flying with the wind flew 600 mi in 2 h. Flying against the wind, the plane required 3 h to travel the same distance. Find the rate of the wind and the rate of the plane in calm air.

7. A motorboat traveling with the current went 88 km in 4 h. Against the current, the boat could go only 64 km in the same amount of time. Find the rate of the boat in calm water and the rate of the current.

8. A rowing team rowing with the current traveled 18 km in 2 h. Rowing against the current, the team rowed 12 km in the same amount of time. Find the rate of the rowing team in calm water and the rate of the current.

9. A plane flying with a tailwind flew 360 mi in 3 h. Against the wind, the plane required 4 h to fly the same distance. Find the rate of the plane in calm air and the rate of the wind.

10. Flying with the wind, a plane flew 1000 mi in 4 h. Against the wind, the plane required 5 h to fly the same distance. Find the rate of the plane in calm air and the rate of the wind.

11. A motorboat traveling with the current went 54 mi in 3 h. Against the current, it took 3.6 h to travel the same distance. Find the rate of the boat in calm water and the rate of the current.

12. A plane traveling with the wind flew 3625 mi in 6.25 h. Against the wind, the plane required 7.25 h to fly the same distance. Find the rate of the plane in calm air and the rate of the wind.

2 Solve.

13. **Construction** A carpenter purchased 50 ft of redwood and 90 ft of pine for a total cost of \$31.20. A second purchase, at the same prices, included 200 ft of redwood and 100 ft of pine for a total cost of \$78. Find the cost per foot of redwood and of pine.

14. **Business** A merchant mixed 10 lb of a cinnamon tea with 5 lb of spice tea. The 15-pound mixture cost \$40. A second mixture included 12 lb of the cinnamon tea and 8 lb of the spice tea. The 20-pound mixture cost \$54. Find the cost per pound of the cinnamon tea and of the spice tea.

15. **Home Economics** During one month, a homeowner used 400 units of electricity and 120 units of gas for a total cost of \$147.20. The next month, 350 units of electricity and 200 units of gas were used for a total cost of \$144. Find the cost per unit of gas.

16. **Construction** A contractor buys 20 yd^2 of nylon carpet and 28 yd^2 of wool carpet for \$1360. A second purchase, at the same prices, includes 15 yd^2 of nylon carpet and 20 yd^2 of wool carpet for \$990. Find the cost per square yard of the wool carpet.

17. **Coin Problem** The total value of the quarters and dimes in a coin bank is \$6.90. If the quarters were dimes and the dimes were quarters, the total value of the coins would be \$7.80. Find the number of quarters in the bank.

18. **Coin Problem** A coin bank contains only nickels and dimes. The total value of the coins in the bank is \$2.50. If the nickels were dimes and the dimes were nickels, the total value of the coins would be \$3.20. Find the number of nickels in the bank.

19. **Manufacturing** A company manufactures both mountain bikes and trail bikes. The cost of materials for a mountain bike is \$70, and the cost of materials for a trail bike is \$50. The cost of labor to manufacture a mountain bike is \$80, and the cost of labor to manufacture a trail bike is \$40. During a week in which the company has budgeted \$2500 for materials and \$2600 for labor, how many mountain bikes does the company plan to manufacture?

20. **Manufacturing** A company manufactures both color and black-and-white monitors. The cost of materials for a black-and-white monitor is \$40, and the cost of materials for a color monitor is \$160. The cost of labor to manufacture a black-and-white monitor is \$60, and the cost of labor to manufacture a color monitor is \$100. During a week in which the company has budgeted \$8400 for materials and \$5600 for labor, how many color monitors does the company plan to manufacture?

21. **Health Science** A pharmacist has two vitamin-supplement powders. The first powder is 25% vitamin B_1 and 15% vitamin B_2. The second is 15% vitamin B_1 and 20% vitamin B_2. How many milligrams of each of the two powders should the pharmacist use to make a mixture that contains 117.5 mg of vitamin B_1 and 120 mg of vitamin B_2?

22. Chemistry A chemist has two alloys, one of which is 10% gold and 15% lead and the other of which is 30% gold and 40% lead. How many grams of each of the two alloys should be used to make an alloy that contains 60 g of gold and 88 g of lead?

23. Business On Monday, a computer manufacturing company sent out three shipments. The first order, which contained a bill for \$114,000, was for 4 Model II, 6 Model VI, and 10 Model IX computers. The second shipment, which contained a bill for \$72,000, was for 8 Model II, 3 Model VI, and 5 Model IX computers. The third shipment, which contained a bill for \$81,000, was for 2 Model II, 9 Model VI, and 5 Model IX computers. What does the manufacturer charge for a Model VI computer?

24. Health Care A relief organization supplies blankets, cots, and lanterns to victims of fires, floods, and other natural disasters. One week the organization purchased 15 blankets, 5 cots, and 10 lanterns for a total cost of \$1250. The next week, at the same prices, the organization purchased 20 blankets, 10 cots, and 15 lanterns for a total cost of \$2000. The next week, at the same prices, the organization purchased 10 blankets, 15 cots, and 5 lanterns for a total cost of \$1625. Find the cost of one blanket, the cost of one cot, and the cost of one lantern.

25. Investments An investor has a total of \$18,000 deposited in three different accounts, which earn annual interest of 9%, 7%, and 5%. The amount deposited in the 9% account is twice the amount in the 5% account. If the three accounts earn total annual interest of \$1340, how much money is deposited in each account?

26. Investments An investor has a total of \$15,000 deposited in three different accounts, which earn annual interest of 9%, 6%, and 4%. The amount deposited in the 6% account is \$2000 more than the amount in the 4% account. If the three accounts earn total annual interest of \$980, how much money is deposited in each account?

27. Art A sculptor is creating a mobile from which three objects will be suspended from a light rod that is 15 in. long as shown at the right. The weight, in ounces, of each object is shown in the diagram. For the mobile to balance, the objects must be positioned so that $w_1d_1 = w_2d_2 + w_3d_3$. The artist wants d_3 to be three times d_2. Find the distances d_1, d_2, and d_3 so that the mobile will balance.

28. Art A mobile is made by suspending three objects from a light rod that is 20 in. long as shown at the right. The weight, in ounces, of each object is shown in the diagram. For the mobile to balance, the objects must be positioned so that $w_1d_1 + w_2d_2 = w_3d_3$. The artist wants d_3 to be twice d_2. Find the distances d_1, d_2, and d_3 so that the mobile will balance.

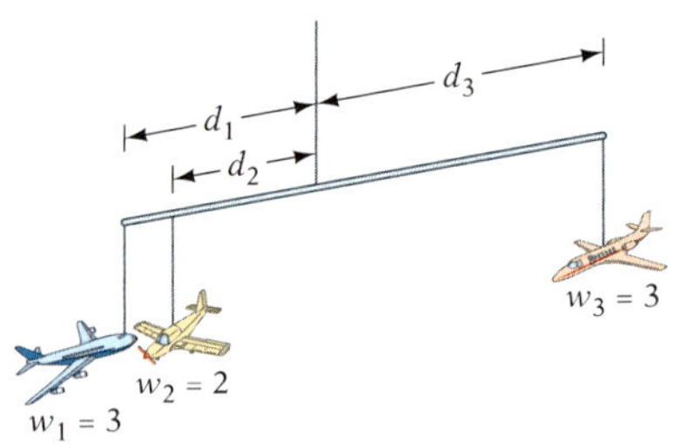

29. Coin Problem A coin bank contains only nickels, dimes, and quarters. The value of the 19 coins in the bank is \$2. There are twice as many nickels as dimes. Find the number of each type of coin in the bank.

30. **Coin Problem** The total value of the nickels, dimes, and quarters in a coin bank is \$4. If the nickels were dimes and the dimes were nickels, the total value of the coins would be \$3.75. If the quarters were dimes and the dimes were quarters, the total value of the coins would be \$6.25. Find the number of nickels, dimes, and quarters in the bank.

31. **Investments** A financial planner invested \$33,000 of a client's money, part at 9%, part at 12%, and the remainder at 8%. The total annual income from these three investments was \$3290. The amount invested at 12% was \$5000 less than the combined amount invested at 9% and 8%. Find the amount invested at each rate.

32. **Education** The total number of elementary school students, secondary school students, and students enrolled in colleges in the United States is 69 million. The number enrolled in colleges is 1 million more than the number in secondary school. The number enrolled in elementary school is 7 million more than the total number of students in secondary schools and colleges (*Source:* **www.infoplease.com**). Find the number of students in elementary school, the number in secondary school, and the number enrolled in colleges in the United States.

According to Smith Travel Research, there are 112,000 hotel rooms in Las Vegas, 95,100 hotel rooms in Orlando, and 80,800 hotel rooms in Los Angeles–Long Beach.

33. **Travel Industry** Smith Travel Research listed the three largest U.S. hotel markets as Las Vegas, Nevada; Orlando, Florida; and Los Angeles–Long Beach, California. These three markets have a total of 1268 hotels. Los Angeles–Long Beach has 38 more hotels than the total number in Las Vegas and Orlando. Las Vegas has 85 fewer hotels than Orlando. Find the number of hotels in each of the three markets.

34. **Labor Market** The U.S. Bureau of Labor Statistics has projected that the top three occupations with the largest numerical job decline between 1994 and 2005 are farmers, typists/word processors, and clerks (bookkeeping, accounting, and auditing). These three occupations are expected to decline by a total of 663,000 jobs. The decline in the number of jobs in farming will be 117,000 less than the combined loss in jobs for typists/word processors and clerks. The job decline in clerk positions will be 95,000 less than that for farmers. Find the job decline for each of the three occupations.

4.4 Applying Concepts

35. **Geometry** Two angles are complementary. The larger angle is 9° more than eight times the measure of the smaller angle. Find the measure of the two angles. (Complementary angles are two angles whose sum is 90°.)

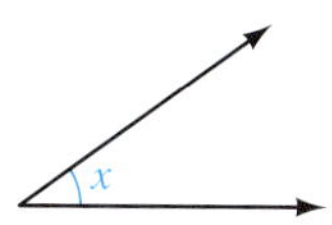

36. **Geometry** Two angles are supplementary. The larger angle is 40° more than three times the measure of the smaller angle. Find the measure of the two angles. (Supplementary angles are two angles whose sum is 180°.)

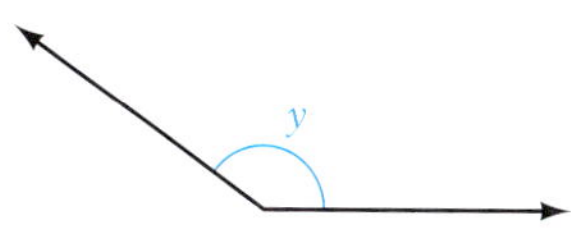

$x + y = 180°$

37. **Coin Problem** The total value of the nickels, dimes, and quarters in a coin bank is \$3.50. If the nickels were dimes and the dimes were nickels, the total value of the coins would be \$4.25. If the quarters were dimes and the dimes were quarters, the total value of the coins would be \$4.25. Find the number of nickels, dimes, and quarters in the bank.

38. **Health Care** The following table shows the per capita, out-of-pocket health care payments, in U.S. dollars, for ten countries in the 1990s. (*Source:* The Organization for Economic Cooperation and Development, Health Data 2001) In the right-hand column is the linear equation that approximately models the data for each country. In each equation, $x = 90$ for 1990, $x = 91$ for 1991, and so on; y is the per capita, out-of-pocket payments for health care.

Country	*1990*	*1991*	*1992*	*1993*	*1994*	*1995*	*1996*	*1997*	*1998*	*Model Equation*
Australia	216	236	245	258	274	281	306	311	338	$y = 14.3x - 1070.3$
Canada	242	258	270	291	308	338	348	373	392	$y = 19.1x - 1485.2$
Denmark	232	232	247	281	304	307	325	332	354	$y = 16.2x - 1229.2$
Finland	201	220	229	268	264	291	301	297	296	$y = 13.0x - 955.9$
Iceland	184	190	223	252	252	283	308	313	340	$y = 19.9x - 1610.0$
Ireland	141	155	171	182	200	206	193	196	175	$y = 5.5x - 332.4$
Korea	196	214	252	276	289	324	355	356	308	$y = 18.8x - 1481.6$
New Zealand	135	138	170	195	187	201	207	210	234	$y = 11.5x - 897.8$
United Kingdom	103	114	130	128	140	143	156	154	168	$y = 7.45x - 563.0$
United States	552	566	578	578	562	568	584	618	652	$y = 9.3x - 290.0$

a. Use the model equations to determine whether payments are decreasing in any of the countries listed.

b. Use the model equations to determine in which country payments are increasing at the most rapid rate and in which country payments are increasing least rapidly.

c. The graph of the model equation for Korea contains the point $(87, 154)$. Write a sentence that gives an interpretation of this ordered pair.

d. For each pair of countries given below, determine during what year the model equations predict that the per capita, out-of-pocket health care payments were the same. What was the payment during that year? Round to the nearest whole number.

(1) Australia and Denmark

(2) Canada and Finland

(3) New Zealand and the United Kingdom

39. Read the article "Matrix Mathematics: How to Win at Monopoly" by Dr. Crypton in the September 1985 issue of *Science Digest*, and prepare a report. Include in your report the relationship between systems of equations and Monopoly.

4.5 Solving Systems of Linear Inequalities

1 Graph the solution set of a system of linear inequalities

VIDEO & DVD CD TUTOR WEB SSM

Two or more inequalities considered together are called a **system of inequalities.** The **solution set of a system of inequalities** is the intersection of the solution sets of the individual inequalities. To graph the solution set of a system of inequalities, first graph the solution set of each inequality. The solution set of the system of inequalities is the region of the plane represented by the intersection of the two shaded areas.

Take Note

You can use a test point to check that the correct region has been denoted as the solution set. We can see from the graph that the point (2, 4) is in the solution set, and as shown below, it is a solution of each inequality in the system. This indicates that the solution set as graphed is correct.

$$2x - y \le 3$$
$$2(2) - (4) \le 3$$
$$0 \le 3 \quad \text{True}$$

$$3x + 2y > 8$$
$$3(2) + 2(4) > 8$$
$$14 > 8 \quad \text{True}$$

Graph the solution set: $2x - y \le 3$
$3x + 2y > 8$

Solve each equation for y.

$$2x - y \le 3 \qquad 3x + 2y > 8$$
$$-y \le -2x + 3 \qquad 2y > -3x + 8$$
$$y \ge 2x - 3 \qquad y > -\frac{3}{2}x + 4$$

Graph $y = 2x - 3$ as a solid line. Because the inequality is $\ge$, shade above the line.

Graph $y = -\frac{3}{2}x + 4$ as a dashed line. Because the inequality is $>$, shade above the line.

The solution set of the system is the region of the plane represented by the intersection of the solution sets of the individual inequalities.

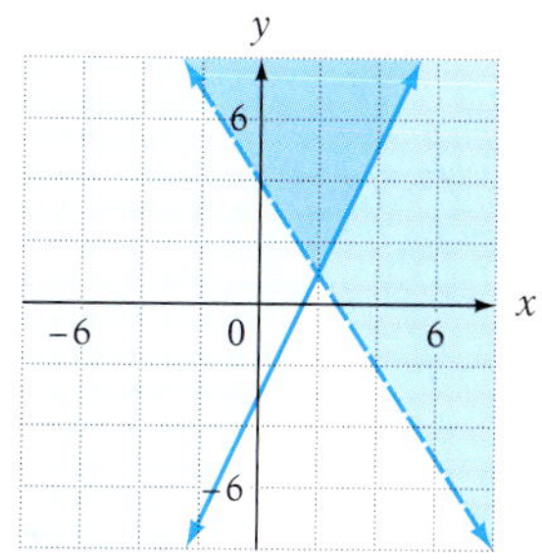

Point of Interest

Large systems of linear inequalities containing over 100 inequalities have been used to solve application problems in such diverse areas as providing health care and hardening a nuclear missile silo.

Graph the solution set: $-x + 2y \ge 4$
$x - 2y \ge 6$

Solve each equation for y.

$$-x + 2y \ge 4 \qquad x - 2y \ge 6$$
$$2y \ge x + 4 \qquad -2y \ge -x + 6$$
$$y \ge \frac{1}{2}x + 2 \qquad y \le \frac{1}{2}x - 3$$

Shade above the solid line $y = \frac{1}{2}x + 2$.

Shade below the solid line $y = \frac{1}{2}x - 3$.

Because the solution sets of the two inequalities do not intersect, the solution set of the system is the empty set.

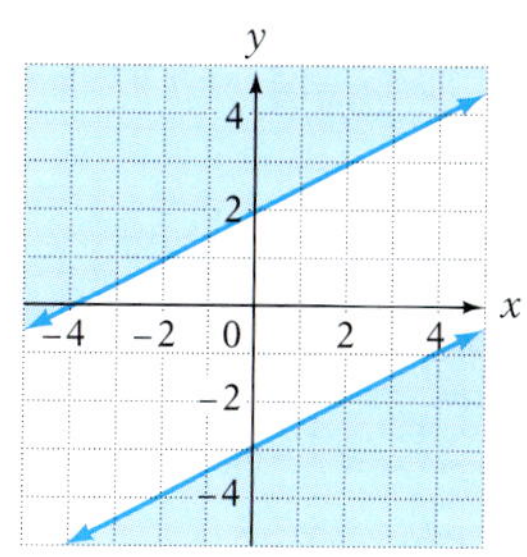

Example 1 Graph the solution set.

A. $y \geq x - 1$
$y < -2x$

B. $2x + 3y > 9$
$y < -\frac{2}{3}x + 1$

Solution **A.** Shade above the solid line $y = x - 1$.

Shade below the dashed line $y = -2x$.

The solution set of the system is the intersection of the solution sets of the individual inequalities.

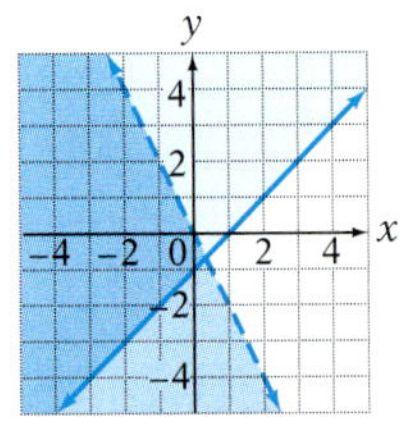

B. Solve $2x + 3y > 9$ for y.

$$2x + 3y > 9$$
$$3y > -2x + 9$$
$$y > -\frac{2}{3}x + 3$$

Shade above the dashed line $y = -\frac{2}{3}x + 3$.

Shade below the dashed line $y = -\frac{2}{3}x + 1$.

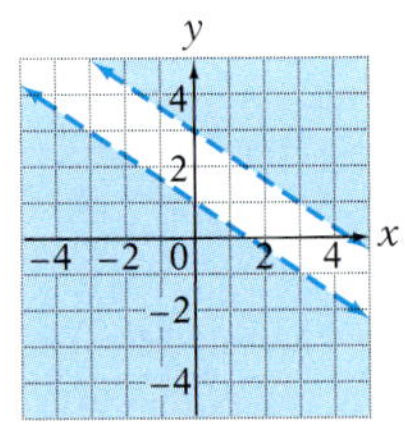

The intersection of the system is the empty set because the solution sets of the two inequalities do not intersect.

Problem 1 Graph the solution set.

A. $y \geq 2x - 3$
$y > -3x$

B. $3x + 4y > 12$
$y < \frac{3}{4}x - 1$

Solution See page S17.

4.5 Concept Review

Determine whether the following statements are always true, sometimes true, or never true.

1. The solution set of a system of linear inequalities is an infinite set.

2. The solution set of a system of linear inequalities is the empty set.

3. The solution set of a system of linear inequalities is the intersection of the solution sets of the individual inequalities.

4.5 Exercises

1 **1.** Which ordered pair is a solution of the system of inequalities?

$2x - y < 4$
$x - 3y \geq 6$

a. $(5, 1)$ **b.** $(-3, -5)$

2. Which ordered pair is a solution of the system of inequalities?

$3x - 2y \geq 6$
$x + y < 5$

a. $(-2, 3)$ **b.** $(3, -2)$

3. What is the solution set of a system of linear inequalities in two variables?

4. If the solution set of a system of linear inequalities in two variables is the empty set, what can be said about the slopes of the lines of the linear equations?

Graph the solution set.

5. $y \leq x - 3$
$y \leq -x + 5$

6. $y > 2x - 4$
$y < -x + 5$

7. $y > 3x - 3$
$y \geq -2x + 2$

8. $y \leq -\dfrac{1}{2}x + 3$
$y \geq x - 3$

9. $2x + y \geq -2$
$6x + 3y \leq 6$

10. $x + y \geq 5$
$3x + 3y \leq 6$

11. $3x - 2y < 6$
$y \leq 3$

12. $x \leq 2$
$3x + 2y > 4$

13. $y > 2x - 6$
$x + y < 0$

14. $x < 3$
$y < -2$

15. $x + 1 \geq 0$
$y - 3 \leq 0$

16. $5x - 2y \geq 10$
$3x + 2y \geq 6$

17. $2x + y \geq 4$
$3x - 2y < 6$

18. $3x - 4y < 12$
$x + 2y < 6$

19. $x - 2y \leq 6$
$2x + 3y \leq 6$

20. $x - 3y > 6$
$2x + y > 5$

21. $x - 2y \leq 4$
$3x + 2y \leq 8$

22. $3x - 2y < 0$
$5x + 3y > 9$

4.5 Applying Concepts

Graph the solution set.

23. $\begin{aligned} 2x + 3y &\le 15 \\ 3x - y &\le 6 \\ y &\ge 0 \end{aligned}$

24. $\begin{aligned} x + y &\le 6 \\ x - y &\le 2 \\ x &\ge 0 \end{aligned}$

25. $\begin{aligned} x - y &\le 5 \\ 2x - y &\ge 6 \\ y &\ge 0 \end{aligned}$

26. $\begin{aligned} x - 3y &\le 6 \\ 5x - 2y &\ge 4 \\ y &\ge 0 \end{aligned}$

27. $\begin{aligned} 2x - y &\le 4 \\ 3x + y &< 1 \\ y &\le 0 \end{aligned}$

28. $\begin{aligned} x - y &\le 4 \\ 2x + 3y &> 6 \\ x &\ge 0 \end{aligned}$

Focus on Problem Solving

Solve an Easier Problem

We begin this problem-solving feature with a restatement of the four steps of Polya's recommended problem-solving model.

1. Understand the problem.
2. Devise a plan.
3. Carry out the plan.
4. Review your solution.

One of the several methods of devising a plan is to try to solve an easier problem. Suppose you are in charge of your softball league, which consists of 15 teams. You must devise a schedule in which each team plays every other team once. How many games must be scheduled?

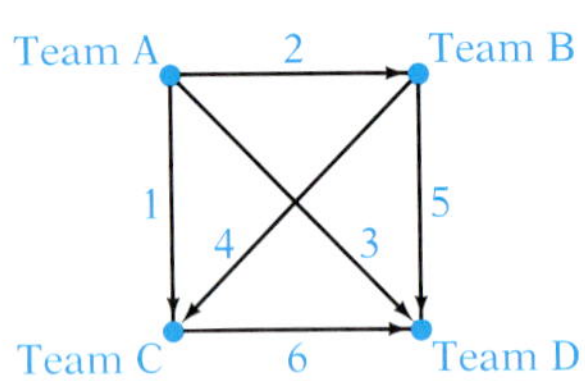

To solve this problem, we will attempt an easier problem first. Suppose that your league contains only a small number of teams. For instance, if there were only 1 team, you would schedule 0 games. If there were 2 teams, you would schedule 1 game. If there were 3 teams, you would schedule 3 games. The diagram at the left shows that 6 games must be scheduled when there are 4 teams in the league.

Here is a table of our results so far. (Remember that making a table is another strategy to be used in problem solving.)

Number of Teams	*Number of Games*	*Possible Pattern*
1	0	0
2	1	1
3	3	1 + 2
4	6	1 + 2 + 3

1. Draw a diagram with 5 dots to represent the teams. Draw lines from each dot to a second dot, and determine the number of games required.
2. What is the apparent pattern for the number of games required?
3. Assuming that the pattern continues, how many games must be scheduled for the 15 teams of the original problem?

After solving a problem, good problem solvers ask whether it is possible to solve the problem in a different manner. Here is a possible alternative method of solving the scheduling problem.

Begin with one of the 15 teams (say team A) and ask, "How many games must this team play?" Because there are 14 teams left to play, you must schedule 14 games. Now move to team B. It is already scheduled to play team A, and it does not play itself, so there are 13 teams left for it to play. Consequently, you must schedule 14 + 13 games.

4. Continue this reasoning for the remaining teams and determine the number of games that must be scheduled. Does this answer correspond to the answer you obtained with the first method?
5. WEB Visit **www.allsports.com** to get a listing of the professional sports teams and how they are separated into leagues and divisions. To appreciate how complicated the game schedules can be, try to create a partial schedule where each team in one of the conferences in the National Football League—for instance, the NFC East—plays each other team in that conference twice. An actual schedule would require scheduling an additional eight games for each team in the conference with selected teams in the other conferences.
6. Making connections to other problems you have solved is an important step toward becoming an excellent problem solver. How does the answer to the scheduling of teams relate to the triangular numbers discussed in the Focus on Problem Solving in the chapter titled Linear Functions and Inequalities in Two Variables?

Projects & Group Activities

Using a Graphing Calculator to Solve a System of Equations

A graphing calculator can be used to solve a system of equations. For this procedure to work on most calculators, it is necessary that the point of intersection be on the screen. This means that you may have to experiment with Xmin, Xmax, Ymin, and Ymax values until the graphs intersect on the screen.

To solve a system of equations graphically, solve each equation for y. Then graph the equations of the system. Their point of intersection is the solution.

For instance, to solve the system of equations

$$\begin{aligned} 4x - 3y &= 7 \\ 5x + 4y &= 2 \end{aligned}$$

first solve each equation for y.

$$4x - 3y = 7 \quad \Rightarrow \quad y = \frac{4}{3}x - \frac{7}{3}$$

$$5x + 4y = 2 \quad \Rightarrow \quad y = -\frac{5}{4}x + \frac{1}{2}$$

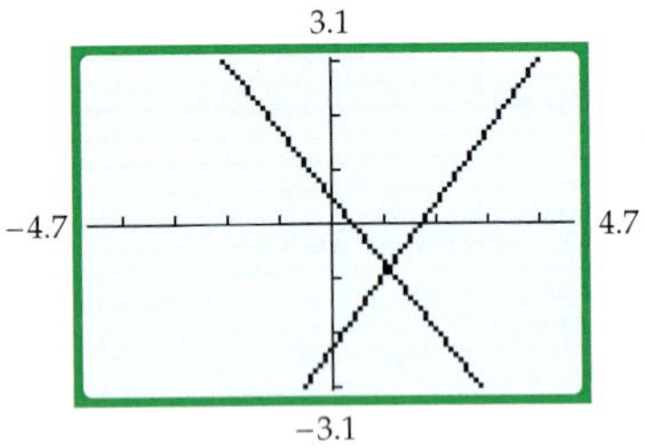

The keystrokes needed to solve this system using a TI-83 follow. We are using a viewing window of $[-4.7, 4.7]$ by $[-3.1, 3.1]$. The approximate solution is $(1.096774, -0.870968)$.

[Y=] [CLEAR] 4 [X, T, θ, n] [÷] 3 [−]
7 [÷] 3 [ENTER] [CLEAR]
[(−)] 5 [X, T, θ, n] [÷] 4 [+] 1 [÷]
2 [GRAPH]

Once the calculator has drawn the graphs, use the TRACE feature and move the cursor to the approximate point of intersection. This will give you an approximate solution of the system of equations. A more accurate solution can be found by using the following keystrokes.

[2nd] [CALC] 5 [ENTER] [ENTER] [ENTER]

Some of the exercises in the first section of this chapter asked you to solve a system of equations by graphing. Try those exercises again, this time using your graphing calculator.

Here is an example of using a graphing calculator to solve an investment problem.

Solve: A marketing manager deposited \$8000 in two simple interest accounts, one with an interest rate of 3.5% and the other with an interest rate of 4.5%. How much is deposited in each account if both accounts earn the same annual interest?

Amount invested at 3.5%: x
Amount invested at 4.5%: $8000 - x$

Interest earned on the 3.5% account: $0.035x$
Interest earned on the 4.5% account: $0.045(8000 - x)$

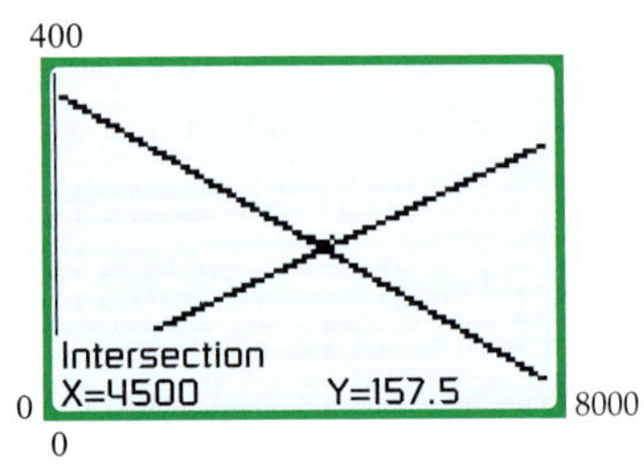

Enter $0.035x$ into Y1. Enter $0.045(8000 - x)$ into Y2. Graph the equations. (We used a window of Xmin = 0, Xmax = 8000, Ymin = 0, Ymax = 400.) Use the intersect feature to to find the point of intersection. At the point of intersection, $x = 4500$. This is the amount in the 3.5% account. The amount in the 4.5% account is $8000 - x = 8000 - 4500 = 3500$.

\$4500 is invested at 3.5%. \$3500 is invested at 4.5%.

Solve each of the following problems using a graphing calculator.

1. Suppose that a breadmaker costs \$190 and that the ingredients and electricity to make one loaf of bread cost \$.95. If a comparable loaf of bread at a grocery store costs \$1.98, how many loaves of bread must you make before the breadmaker pays for itself?

2. Suppose a natural gas clothes washer costs \$295 and uses \$.25 of gas and water to wash a large load of clothes. If a laundromat charges \$1.75 to do a load of laundry, how many loads of clothes must you wash before having purchased the washing machine becomes more economical?

3. Suppose a natural gas clothes dryer costs \$260 and uses \$.40 of gas to dry a load of clothes for 1 h. The laundromat charges \$1.75 to use a dryer for 1 h.
   a. How many loads of clothes must you dry before having purchased the gas dryer becomes more economical?
   b. What is the y-coordinate at the point of intersection? What does this represent in the context of the problem?

4. When Mitch Deerfield changed jobs, he rolled over the \$7500 in his retirement account into two simple interest accounts. On one account, the annual simple interest rate is 6.25%; on the second account, the annual simple interest rate is 5.75%.
   a. How much is invested in each account if the accounts earn the same amount of annual interest?
   b. What is the y-coordinate at the point of intersection? What does this represent in the context of the problem?

Solving a First-Degree Equation with a Graphing Calculator

A first degree equation in one variable can be solved graphically[1] by using a graphing calculator. The idea is to rewrite the equation as a system of equations and then solve the system of equations as we illustrated on the previous page. For instance, to solve the equation

$$2x + 1 = 5x - 5$$

write the equation as the following system of equations.

$$\begin{aligned} y &= 2x + 1 \\ y &= 5x - 5 \end{aligned}$$

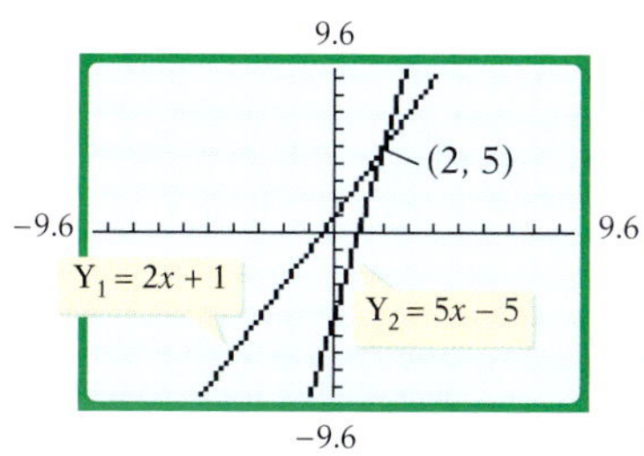

Note that we have used the left and right sides of the original equation to form the system of equations.

Now graph the equations and find the point of intersection. We have shown this at the left, using a viewing window of [−9.6, 9.6] by [−9.6, 9.6]. The solution of the original equation, $2x + 1 = 5x - 5$, is the x-coordinate of the point of intersection. Thus the solution of the equation is 2.

Recall that not all equations have a solution. Consider the equation

$$x - 3(x + 2) = 2x - 2(2x - 2)$$

[1]Some graphing calculators offer nongraphical techniques for solving an equation.

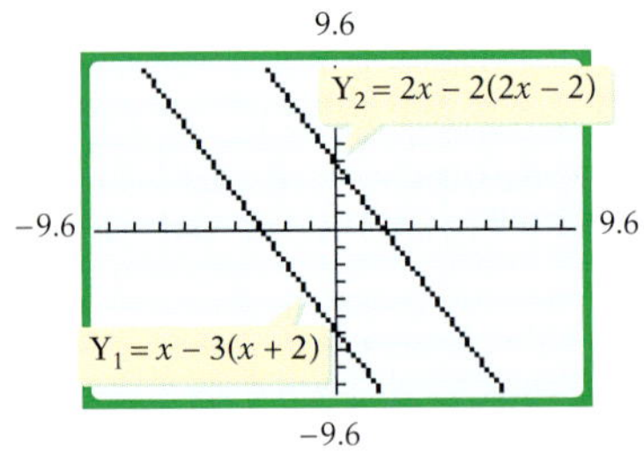

The graphs of the left and right sides of the equation are shown at the left. Note that the lines appear to be parallel and therefore do not intersect. Because the lines do not intersect, there is no solution of the system of equations and thus no solution of the original equation. We algebraically verify this result below.

$$
\begin{aligned}
x - 3(x + 2) &= 2x - 2(2x - 2) \\
x - 3x - 6 &= 2x - 4x + 4 && \text{• Use the Distributive Property.} \\
-2x - 6 &= -2x + 4 && \text{• Simplify.} \\
-6 &= 4 && \text{• Add } 2x \text{ to each side of the equation.}
\end{aligned}
$$

Because $-6 = 4$ is not a true equation, there is no solution.

Note that in the third line of the solution, we obtained the expressions $-2x - 6$ and $-2x + 4$. Consider the graphs of the equations $y = -2x - 6$ and $y = -2x + 4$; the slopes of the lines are equal (both are -2), so therefore the graphs of the lines are parallel.

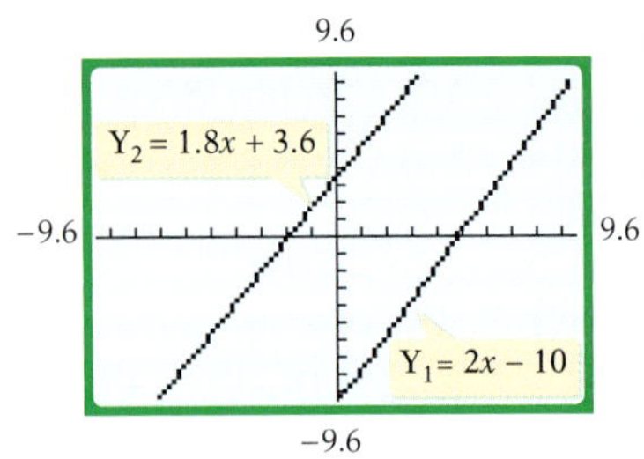

Before we leave graphical solutions, a few words of caution. Graphs can be deceiving! They can appear not to intersect when they do, and they can appear to intersect when they do not. For instance, an attempt to solve

$$2x - 10 = 1.8x + 3.6$$

graphically is shown at the left.

If you use the viewing window $[-9.6, 9.6]$ by $[-9.6, 9.6]$, the graphs will appear to be parallel. However, the graphs of $y = 2x - 10$ and $y = 1.8x + 3.6$ do not have the same slope and therefore must intersect. For this equation, you need a larger viewing window to see the point of intersection.

Solve the following equations by using a graphing calculator.

1. $3x - 1 = 5x + 1$

2. $3x + 2 = 4$

3. $3 + 2(2x - 4) = 5(x - 3)$

4. $2x - 4 = 5x - 3(x + 2) + 2$

5. Explain how Problem 4 is related to an *identity* as explained in Section 1 in the chapter titled Solving Equations and Inequalities.

6. Find an appropriate viewing window so that you can determine the solution of the equation $2x - 10 = 1.8x + 3.6$ given above. What is the solution?

Chapter Summary

Key Words

A **system of equations** is two or more equations considered together. A **solution of a system of equations in two variables** is an ordered pair that is a solution of each equation of the system. [4.1.1, p. 278]

Examples

The solution of the system

$$\begin{aligned} x + y &= 2 \\ x - y &= 4 \end{aligned}$$

is the ordered pair (3, −1). (3, −1) is the only ordered pair that is a solution of both equations.

When the graphs of a system of equations intersect at only one point, the system is called an **independent system of equations**. [4.1.1, p. 279]

$$\begin{aligned} 3x - 2y &= 4 \\ 5x + 2y &= 6 \\ 8x \qquad &= 10 \end{aligned}$$

• Add the equations.

When the graphs of a system of equations coincide, the system is called a **dependent system of equations**. [4.1.1, p. 280]

$$\begin{aligned} -3x + 2y &= 4 \\ 3x - 2y &= -4 \\ 0 &= 0 \end{aligned}$$

• Add the equations.

When the graphs of a system of equations do not intersect, the system has no solution and is called an **inconsistent system of equations**. [4.1.1, p. 279]

$$\begin{aligned} 2x - 3y &= -2 \\ -2x + 3y &= 5 \\ 0 &= 3 \end{aligned}$$

• Add the equations.

An equation of the form $Ax + By + Cz = D$ is called a **linear equation in three variables**. [4.2.2, p. 289]

$3x + 2y - 5z = 12$ is a linear equation in three variables.

A **solution of a system of equations in three variables** is an ordered triple that is a solution of each equation of the system. [4.2.2, p. 290]

The ordered triple (1, 2, 1) is a solution of the system of equations

$$\begin{aligned} 3x + y - 3z &= 2 \\ -x + 2y + 3z &= 6 \\ 2x + 2y - 2z &= 4 \end{aligned}$$

A **matrix** is a rectangular array of numbers. [4.3.1, p. 296]

$\begin{bmatrix} 2 & 3 & 6 \\ -1 & 2 & 4 \end{bmatrix}$ is a 2×3 matrix.

An **augmented matrix** is one consisting of the coefficients and constants of a system of equations. [4.3.1, p. 296]

The augmented matrix associated with the system

$$\begin{aligned} 4x - 3y + z &= 7 \\ x + y - 5z &= 14 \\ 2x + 3z &= -4 \end{aligned}$$

is

$$\left[\begin{array}{ccc|c} 4 & -3 & 1 & 7 \\ 1 & 1 & -5 & 14 \\ 2 & 0 & 3 & -4 \end{array}\right]$$

The process of solving a system of equations by using elementary row operations is called the **Gaussian elimination method**. [4.3.1, p. 301]

A **square matrix** has the same number of rows as columns. [4.3.2, p. 303]

$\begin{bmatrix} 2 & 3 \\ 4 & -1 \end{bmatrix}$ is a square matrix.

A **determinant** is a number associated with a square matrix. [4.3.2, p. 303]

$$\begin{vmatrix} 2 & -3 \\ -4 & 5 \end{vmatrix} = 2(5) - (-4)(-3) = 10 - 12 = -2$$

The **minor of an element** in a 3×3 determinant is the 2×2 determinant obtained by eliminating the row and column that contain the element. [4.3.2, p. 304]

$\begin{vmatrix} 2 & 1 & 3 \\ 4 & 6 & 2 \\ 1 & 8 & 3 \end{vmatrix}$ The minor of 4 is $\begin{vmatrix} 1 & 3 \\ 8 & 3 \end{vmatrix}$.

The **cofactor of an element of a determinant** is $(-1)^{i+j}$ times the minor of the element, where i is the row number of the element and j is its column number. [4.3.2, p. 304]

In the determinant above, 4 is from the second row and the first column. The cofactor of 4 is

$$(-1)^{2+1}\begin{vmatrix} 1 & 3 \\ 8 & 3 \end{vmatrix} = -\begin{vmatrix} 1 & 3 \\ 8 & 3 \end{vmatrix}$$

The evaluation of the determinant of a 3×3 or larger matrix is accomplished by **expanding by cofactors**. [4.3.2, p. 305]

Expand by cofactors of the first column.

$$\begin{vmatrix} 2 & 1 & 3 \\ 4 & 6 & 2 \\ 1 & 8 & 3 \end{vmatrix} = 2\begin{vmatrix} 6 & 2 \\ 8 & 3 \end{vmatrix} - 4\begin{vmatrix} 1 & 3 \\ 8 & 3 \end{vmatrix} + 1\begin{vmatrix} 1 & 3 \\ 6 & 2 \end{vmatrix}$$

Inequalities considered together are called a **system of inequalities**. The **solution set of a system of inequalities** is the intersection of the solution sets of the individual inequalities. [4.5.1, p. 323]

$x + y > 3$
$x - y > -2$

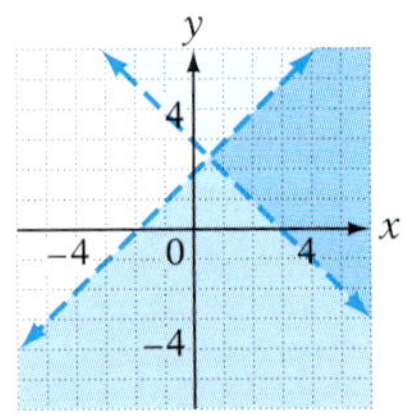

Essential Rules and Procedures

A system of equations can be solved by:

a. Graphing [4.1.1, pp. 279–280]

$$y = \frac{1}{2}x + 2$$
$$y = \frac{5}{2}x - 2$$

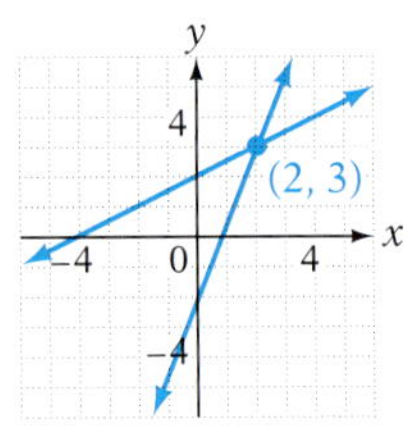

b. The substitution method [4.1.2, pp. 281–282]

(1) $2x - 3y = 4$
(2) $y = -x + 2$

Substitute the value of y into equation (1).

$$2x - 3(-x + 2) = 4$$

c. The addition method [4.2.1 and 4.2.2, pp. 286–288, 290–292]

$$\begin{aligned} -2x + 3y &= 7 \\ 2x - 5y &= 2 \\ -2y &= 9 \end{aligned}$$

• Add the equations.

Elementary row operations to rewrite a matrix in **row echelon form**:

1. Interchange two rows.
2. Multiply all the elements in a row by the same nonzero number.
3. Replace a row by the sum of that row and a multiple of any other row. [4.3.1, p. 297]

$$\begin{bmatrix} 1 & 4 & -2 & 3 \\ 0 & 1 & 3 & -1 \\ 0 & 0 & 1 & 6 \end{bmatrix}$$ Row echelon form

Cramer's Rule [4.3.3, pp. 307–308]

For two variables: $a_{11}x + a_{12}y = b_1$
$a_{21}x + a_{22}y = b_2$

$x = \dfrac{D_x}{D}$, $y = \dfrac{D_y}{D}$, where $D = \begin{vmatrix} a_{11} & a_{12} \\ a_{21} & a_{22} \end{vmatrix}$, $D_x = \begin{vmatrix} b_1 & a_{12} \\ b_2 & a_{22} \end{vmatrix}$

$D_y = \begin{vmatrix} a_{11} & b_1 \\ a_{21} & b_2 \end{vmatrix}$, and $D \neq 0$.

$2x - y = 6$
$x + 3y = 4$

$D = \begin{vmatrix} 2 & -1 \\ 1 & 3 \end{vmatrix} = 7$, $D_x = \begin{vmatrix} 6 & -1 \\ 4 & 3 \end{vmatrix} = 22$,

$D_y = \begin{vmatrix} 2 & 6 \\ 1 & 4 \end{vmatrix} = 2$

$x = \dfrac{D_x}{D} = \dfrac{22}{7}$, $y = \dfrac{D_y}{D} = \dfrac{2}{7}$

For three variables: $a_{11}x + a_{12}y + a_{13}z = b_1$
$a_{21}x + a_{22}y + a_{23}z = b_2$
$a_{31}x + a_{32}y + a_{33}z = b_3$

$x = \dfrac{D_x}{D}$, $y = \dfrac{D_y}{D}$, and $z = \dfrac{D_z}{D}$, where $D = \begin{vmatrix} a_{11} & a_{12} & a_{13} \\ a_{21} & a_{22} & a_{23} \\ a_{31} & a_{32} & a_{33} \end{vmatrix}$,

$D_x = \begin{vmatrix} b_1 & a_{12} & a_{13} \\ b_2 & a_{22} & a_{23} \\ b_3 & a_{32} & a_{33} \end{vmatrix}$, $D_y = \begin{vmatrix} a_{11} & b_1 & a_{13} \\ a_{21} & b_2 & a_{23} \\ a_{31} & b_3 & a_{33} \end{vmatrix}$, $D_z = \begin{vmatrix} a_{11} & a_{12} & b_1 \\ a_{21} & a_{22} & b_2 \\ a_{31} & a_{32} & b_3 \end{vmatrix}$, and

$D \neq 0$

$x - y + z = 2$
$2x + y - 2z = -2$
$x - 2y + 3z = 6$

$D = \begin{vmatrix} 1 & -1 & 1 \\ 2 & 1 & -2 \\ 1 & -2 & 3 \end{vmatrix} = 2$,

$D_x = \begin{vmatrix} 2 & -1 & 1 \\ -2 & 1 & -2 \\ 6 & -2 & 3 \end{vmatrix} = 2$,

$D_y = \begin{vmatrix} 1 & 2 & 1 \\ 2 & -2 & -2 \\ 1 & 6 & 3 \end{vmatrix} = 4$,

$D_z = \begin{vmatrix} 1 & -1 & 2 \\ 2 & 1 & -2 \\ 1 & -2 & 6 \end{vmatrix} = 6$

$$x = \frac{D_x}{D} = \frac{2}{2} = 1,\ y = \frac{D_y}{D} = \frac{4}{2} = 2,$$

$$z = \frac{D_z}{D} = \frac{6}{2} = 3$$

Chapter Review Exercises

1. Solve by substitution: $\begin{aligned} 2x - 6y &= 15 \\ x &= 3y + 8 \end{aligned}$

2. Solve by substitution: $\begin{aligned} 3x + 12y &= 18 \\ x + 4y &= 6 \end{aligned}$

3. Solve by the addition method: $\begin{aligned} 3x + 2y &= 2 \\ x + y &= 3 \end{aligned}$

4. Solve by the addition method: $\begin{aligned} 5x - 15y &= 30 \\ x - 3y &= 6 \end{aligned}$

5. Solve by the addition method: $\begin{aligned} 3x + y &= 13 \\ 2y + 3z &= 5 \\ x + 2z &= 11 \end{aligned}$

6. Solve by the addition method:
$\begin{aligned} 3x - 4y - 2z &= 17 \\ 4x - 3y + 5z &= 5 \\ 5x - 5y + 3z &= 14 \end{aligned}$

7. Evaluate the determinant: $\begin{vmatrix} 6 & 1 \\ 2 & 5 \end{vmatrix}$

8. Evaluate the determinant: $\begin{vmatrix} 1 & 5 & -2 \\ -2 & 1 & 4 \\ 4 & 3 & -8 \end{vmatrix}$

9. Solve by using Cramer's Rule: $\begin{aligned} 2x - y &= 7 \\ 3x + 2y &= 7 \end{aligned}$

10. Solve by using Cramer's Rule: $\begin{aligned} 3x - 4y &= 10 \\ 2x + 5y &= 15 \end{aligned}$

11. Solve by using Cramer's Rule: $\begin{aligned} x + y + z &= 0 \\ x + 2y + 3z &= 5 \\ 2x + y + 2z &= 3 \end{aligned}$

12. Solve by using Cramer's Rule: $\begin{aligned} x + 3y + z &= 6 \\ 2x + y - z &= 12 \\ x + 2y - z &= 13 \end{aligned}$

13. Solve by the addition method: $\begin{aligned} x - 2y + z &= 7 \\ 3x - z &= -1 \\ 3y + z &= 1 \end{aligned}$

14. Solve by using Cramer's Rule: $\begin{aligned} 3x - 2y &= 2 \\ -2x + 3y &= 1 \end{aligned}$

15. Solve by using the Gaussian elimination method:
$\begin{aligned} 2x - 2y - 6z &= 1 \\ 4x + 2y + 3z &= 1 \\ 2x - 3y - 3z &= 3 \end{aligned}$

16. Evaluate the determinant: $\begin{vmatrix} 3 & -2 & 5 \\ 4 & 6 & 3 \\ 1 & 2 & 1 \end{vmatrix}$

17. Solve by using Cramer's Rule: $\begin{aligned} 4x - 3y &= 17 \\ 3x - 2y &= 12 \end{aligned}$

18. Solve by using the Gaussian elimination method:
$\begin{aligned} 3x + 2y - z &= -1 \\ x + 2y + 3z &= -1 \\ 3x + 4y + 6z &= 0 \end{aligned}$

19. Solve by graphing: $\begin{aligned} x + y &= 3 \\ 3x - 2y &= -6 \end{aligned}$

20. Solve by graphing: $\begin{aligned} 2x - y &= 4 \\ y &= 2x - 4 \end{aligned}$

21. Graph the solution set: $\begin{aligned} x + 3y &\le 6 \\ 2x - y &\ge 4 \end{aligned}$

22. Graph the solution set: $\begin{aligned} 2x + 4y &\ge 8 \\ x + y &\le 3 \end{aligned}$

23. A cabin cruiser traveling with the current went 60 mi in 3 h. Against the current, it took 5 h to travel the same distance. Find the rate of the cabin cruiser in calm water and the rate of the current.

24. A plane flying with the wind flew 600 mi in 3 h. Flying against the wind, the plane required 4 h to travel the same distance. Find the rate of the plane in calm air and the rate of the wind.

25. At a movie theater, admission tickets are \$5 for children and \$8 for adults. The receipts for one Friday evening were \$2500. The next day there were three times as many children as the preceding evening and only half as many adults as the night before, yet the receipts were still \$2500. Find the number of children who attended the movie Friday evening.

26. A chef wants to prepare a low-fat, low-sodium meal using lean meat, roasted potatoes, and green beans. A 1-ounce serving of meat contains 50 Cal, 20 g of protein, and 16 mg of sodium. A 1-ounce serving of potatoes contains 9 Cal, 1 g of protein, and 3 mg of sodium. A 1-ounce serving of green beans contains 12 Cal, 2 g of protein, and 17 mg of sodium. If the chef wants to prepare a meal that contains 243 Cal, 73 g of protein, and 131 mg of sodium, how many ounces of each ingredient should be prepared?

Chapter Test

1. Solve by substitution: $\begin{aligned} 3x + 2y &= 4 \\ x &= 2y - 1 \end{aligned}$

2. Solve by substitution: $\begin{aligned} 5x + 2y &= -23 \\ 2x + y &= -10 \end{aligned}$

3. Solve by substitution: $\begin{aligned} y &= 3x - 7 \\ y &= -2x + 3 \end{aligned}$

4. Solve by using the Gaussian elimination method:
$\begin{aligned} 3x + 4y &= -2 \\ 2x + 5y &= 1 \end{aligned}$

5. Solve by the addition method: $\begin{aligned} 4x - 6y &= 5 \\ 6x - 9y &= 4 \end{aligned}$

6. Solve by the addition method:
$\begin{aligned} 3x - y &= 2x + y - 1 \\ 5x + 2y &= y + 6 \end{aligned}$

7. Solve by the addition method:

$$\begin{aligned} 2x + 4y - z &= 3 \\ x + 2y + z &= 5 \\ 4x + 8y - 2z &= 7 \end{aligned}$$

8. Solve by using the Gaussian elimination method:

$$\begin{aligned} x - y - z &= 5 \\ 2x + z &= 2 \\ 3y - 2z &= 1 \end{aligned}$$

9. Evaluate the determinant: $\begin{vmatrix} 3 & -1 \\ -2 & 4 \end{vmatrix}$

10. Evaluate the determinant: $\begin{vmatrix} 1 & -2 & 3 \\ 3 & 1 & 1 \\ 2 & -1 & -2 \end{vmatrix}$

11. Solve by using Cramer's Rule:

$$\begin{aligned} x - y &= 3 \\ 2x + y &= -4 \end{aligned}$$

12. Solve by using Cramer's Rule:

$$\begin{aligned} x - y + z &= 2 \\ 2x - y - z &= 1 \\ x + 2y - 3z &= -4 \end{aligned}$$

13. Solve by using Cramer's Rule:

$$\begin{aligned} 3x + 2y + 2z &= 2 \\ x - 2y - z &= 1 \\ 2x - 3y - 3z &= -3 \end{aligned}$$

14. Solve by graphing:

$$\begin{aligned} 2x - 3y &= -6 \\ 2x - y &= 2 \end{aligned}$$

15. Solve by graphing:

$$\begin{aligned} x - 2y &= -5 \\ 3x + 4y &= -15 \end{aligned}$$

16. Graph the solution set:

$$\begin{aligned} 2x - y &< 3 \\ 4x + 3y &< 11 \end{aligned}$$

17. Graph the solution set:

$$\begin{aligned} x + y &> 2 \\ 2x - y &< -1 \end{aligned}$$

18. Use a graphing utility to solve the system. Round to the nearest hundredth.

$$\begin{aligned} 2x + 3y &= 7 \\ x - 2y &= -5 \end{aligned}$$

19. A plane flying with the wind went 350 mi in 2 h. The return trip, flying against the wind, took 2.8 h. Find the rate of the plane in calm air and the rate of the wind.

20. A clothing manufacturer purchased 60 yd of cotton and 90 yd of wool for a total cost of $1800. Another purchase, at the same prices, included 80 yd of cotton and 20 yd of wool for a total cost of $1000. Find the cost per yard of the cotton and of the wool.

Cumulative Review Exercises

1. Solve: $\frac{3}{2}x - \frac{3}{8} + \frac{1}{4}x = \frac{7}{12}x - \frac{5}{6}$

2. Find the equation of the line that contains the points $(2, -1)$ and $(3, 4)$.

3. Simplify: $3[x - 2(5 - 2x) - 4x] + 6$

4. Evaluate $a + bc \div 2$ when $a = 4$, $b = 8$, and $c = -2$.

5. Solve: $2x - 3 < 9$ or $5x - 1 < 4$

6. Solve: $|x - 2| - 4 < 2$

7. Solve: $|2x - 3| > 5$

8. Given $f(x) = 3x^3 - 2x^2 + 1$, evaluate $f(-3)$.

9. Find the range of $f(x) = 3x^2 - 2x$ if the domain is $\{-2, -1, 0, 1, 2\}$.

10. Given $F(x) = x^2 - 3$, find $F(2)$.

11. Given $f(x) = 3x - 4$, write $f(2 + h) - f(2)$ in simplest form.

12. Graph: $\{x \mid x \le 2\} \cap \{x \mid x > -3\}$

13. Find the equation of the line that contains the point $(-2, 3)$ and has slope $-\frac{2}{3}$.

14. Find the equation of the line that contains the point $(-1, 2)$ and is perpendicular to the line $2x - 3y = 7$.

15. Find the distance between the points $(-4, 1)$ and $(2, 0)$.

16. Find the midpoint of the line connecting the points $(-4, 3)$ and $(3, 5)$.

17. Graph $2x - 5y = 10$ by using the slope and y-intercept.

18. Graph the solution set of the inequality $3x - 4y \ge 8$.

19. Solve by substitution: $\begin{aligned} 3x - 2y &= 7 \\ y &= 2x - 1 \end{aligned}$

20. Solve by the addition method: $\begin{aligned} 3x + 2z &= 1 \\ 2y - z &= 1 \\ x + 2y &= 1 \end{aligned}$

21. Evaluate the determinant: $\begin{vmatrix} 2 & -5 & 1 \\ 3 & 1 & 2 \\ 6 & -1 & 4 \end{vmatrix}$

22. Solve by graphing: $\begin{aligned} 5x - 2y &= 10 \\ 3x + 2y &= 6 \end{aligned}$

23. Solve by using Cramer's Rule: $\begin{aligned} 4x - 3y &= 17 \\ 3x - 2y &= 12 \end{aligned}$

24. Graph the solution set: $\begin{aligned} 3x - 2y &\geq 4 \\ x + y &< 3 \end{aligned}$

25. A student's grades on five sociology exams were 68, 82, 90, 73, and 95. What is the lowest score this student can receive on the sixth exam and still have earned a total of at least 480 points?

26. How many milliliters of pure water must be added to 100 ml of a 4% salt solution to make a 2.5% salt solution?

27. Flying with the wind, a small plane required 2 h to fly 150 mi. Against the wind, it took 3 h to fly the same distance. Find the rate of the wind.

28. A restaurant manager buys 100 lb of hamburger and 50 lb of steak for a total cost of \$490. A second purchase, at the same prices, includes 150 lb of hamburger and 100 lb of steak. The total cost is \$860. Find the price of one pound of steak.

29. Find the lower and upper limits of a 12,000-ohm resistor with a 15% tolerance.

30. The graph below shows the relationship between the monthly income, in dollars, and sales, in thousands of dollars, of an account executive. Find the slope of the line between the two points shown on the graph. Write a sentence that states the meaning of the slope.

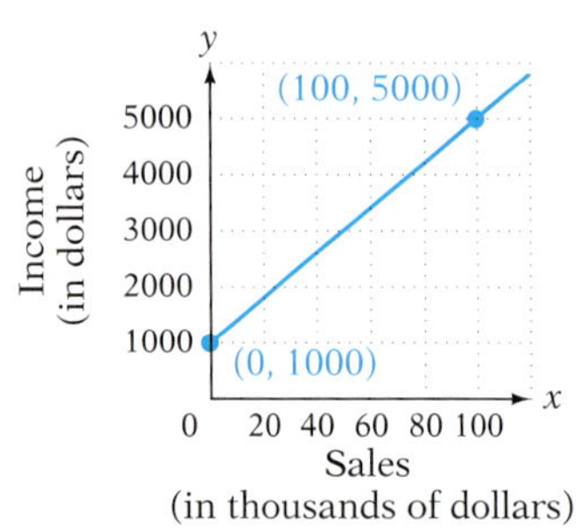

On May 17, 2002, University of Hawaii astronomers discovered 11 new moons orbiting Jupiter. Our knowledge of our solar system is expanding as telescopes become more powerful and as the spacecraft NASA has sent into space send back more and more photos of the outer planets.

Jupiter and its moons are approximately 629,000,000 miles from the sun. Distances this great, as well as very small measurements, are generally expressed in scientific notation, as illustrated in the **Project on page 389.**

Applications

5 Polynomials

Objectives

Section 5.1

1 Evaluate polynomial functions

2 Add and subtract polynomials

Section 5.2

1 Multiply monomials

2 Simplify powers of monomials

Section 5.3

1 Multiply a polynomial by a monomial

2 Multiply two polynomials

3 Multiply two binomials using the FOIL method

4 Multiply binomials that have special products

5 Application problems

Section 5.4

1 Integer exponents

2 Scientific notation

3 Application problems

Section 5.5

1 Divide a polynomial by a monomial

2 Divide polynomials

3 Synthetic division

4 Evaluate a polynomial using synthetic division

Need help? For online student resources, such as section quizzes, visit this web site: math.college.hmco.com

Prep Test

1. Subtract: $-2 - (-3)$
2. Multiply: $-3(6)$
3. Simplify: $-\frac{24}{-36}$
4. Evaluate $3n^4$ when $n = -2$.
5. If $\frac{a}{b}$ is a fraction in simplest form, what number is not a possible value of b?
6. Are $2x^2$ and $2x$ like terms?
7. Simplify: $3x^2 - 4x + 1 + 2x^2 - 5x - 7$
8. Simplify: $-4y + 4y$
9. Simplify: $-3(2x - 8)$
10. Simplify: $3xy - 4y - 2(5xy - 7y)$

Go Figure

I have two more sisters than brothers. Each of my sisters has two more sisters than brothers. How many more sisters than brothers does my youngest brother have?

5.1 Introduction to Polynomials

1 Evaluate polynomial functions

VIDEO & DVD CD TUTOR WEB SSM

A **monomial** is a number, a variable, or a product of numbers and variables. For instance,

7	b	$\frac{2}{3}a$	$12xy^2$
A number	A variable	A product of a number and a variable	A product of a number and variables

The expression $3\sqrt{x}$ is not a monomial because $\sqrt{x}$ cannot be written as a product of variables. The expression $\frac{2x}{y^2}$ is not a monomial because it is a *quotient* of variables.

The **degree of a monomial** is the sum of the exponents of the variables.

x	degree 1 $(x = x^1)$
$3x^2$	degree 2
$4x^2y$	degree 3
$6x^3y^4z^2$	degree 9
x^n	degree n

The degree of a nonzero constant term is zero.

6	degree 0

A **polynomial** is a variable expression in which the terms are monomials.

A polynomial of one term is a **monomial.** $5x$

A polynomial of two terms is a **binomial.** $5x^2y + 6x$

A polynomial of three terms is a **trinomial.** $3x^2 + 9xy - 5y$

Polynomials with more than three terms do not have special names.

The **degree of a polynomial** is the greatest of the degrees of any of its terms.

$3x + 2$ degree 1
$3x^2 + 2x - 4$ degree 2
$4x^3y^2 + 6x^4$ degree 5
$3x^{2n} - 5x^n - 2$ degree $2n$

The terms of a polynomial in one variable are usually arranged so that the exponents of the variable decrease from left to right. This is called **descending order.**

$2x^2 - x + 8$

$3y^3 - 3y^2 + y - 12$

For a polynomial in more than one variable, descending order may refer to any one of the variables.

The polynomial at the right is shown first in descending order of the x variable and then in descending order of the y variable.

$2x^2 + 3xy + 5y^2$

$5y^2 + 3xy + 2x^2$

The **linear function** given by $f(x) = mx + b$ is an example of a polynomial function. It is a polynomial function of degree one. A second-degree polynomial function, called a **quadratic function,** is given by the equation $f(x) = ax^2 + bx + c,\ a \neq 0$. A third-degree polynomial function is called a **cubic function.** In general, a **polynomial function** is an expression whose terms are monomials.

To evaluate a polynomial function, replace the variable by its value and simplify.

Given $P(x) = x^3 - 3x^2 + 4$, evaluate $P(-3)$.

Substitute -3 for x and simplify.

$$P(x) = x^3 - 3x^2 + 4$$
$$P(-3) = (-3)^3 - 3(-3)^2 + 4$$
$$P(-3) = -27 - 27 + 4$$
$$P(-3) = -50$$

Example 1 To overcome the resistance of the tires on the road and the wind, the horsepower (hp), P, required by a cyclist to keep a certain bicycle moving at v mph is given by $P(v) = 0.00003v^3 + 0.00211v$. How much horsepower must the cyclist supply to keep this bicycle moving at 20 mph?

Strategy To find the horsepower, evaluate the function when $v = 20$.

Solution

$$P(v) = 0.00003v^3 + 0.00211v$$
$$P(20) = 0.00003(20)^3 + 0.00211(20)$$ • Replace v by 20.
$$= 0.2822$$ • Simplify.

The cyclist must supply 0.2822 hp.

Problem 1 The velocity of air that is expelled during a cough can be modeled by $v(r) = 600r^2 - 1000r^3$, where v is the velocity of the air in centimeters per second and r is the radius of the trachea in centimeters. What is the velocity of expelled air during a cough when the radius of the trachea is 0.4 cm?

Solution See page S17.

The **leading coefficient** of a polynomial function is the coefficient of the variable with the largest exponent. The **constant term** is the term without a variable.

Find the leading coefficient, the constant term, and the degree of the polynomial function $P(x) = 7x^4 - 3x^2 + 2x - 4$.

The leading coefficient is 7, the constant term is -4, and the degree is 4.

Example 2 Find the leading coefficient, the constant term, and the degree of the polynomial $P(x) = 5x^6 - 4x^5 - 3x^2 + 7$.

Solution The leading coefficient is 5, the constant term is 7, and the degree is 6.

Problem 2 Find the leading coefficient, the constant term, and the degree of the polynomial $R(x) = -3x^4 + 3x^3 + 3x^2 - 2x - 12$.

Solution See page S17.

The following three functions do not represent polynomial functions.

$f(x) = 3x^2 + 2x^{-1}$ A polynomial function does not have a variable raised to a negative power.

$g(x) = 2\sqrt{x} - 3$ A polynomial function does not have a variable expression within a radical.

$h(x) = \dfrac{x}{x - 1}$ A polynomial function does not have a variable in the denominator of a fraction.

Example 3 State whether the function is a polynomial function.

A. $P(x) = 3x^{1/2} + 2x^2 - 3$ **B.** $T(x) = 3\sqrt{x} - 2x^2 - 3x + 2$

C. $R(x) = 14x^3 - \pi x^2 + 3x + 2$

Solution

A. This is not a polynomial function. A polynomial function does not have a variable raised to a fractional power.

B. This is not a polynomial function. A polynomial function does not have a variable expression within a radical.

C. This is a polynomial function.

Problem 3 State whether the function is a polynomial function.

A. $R(x) = 5x^{14} - 5$ **B.** $V(x) = -x^{-1} + 2x - 7$

C. $P(x) = 2x - 2\sqrt{x} - 3$

Solution See page S17.

The graph of a linear function is a straight line and can be found by plotting just two points. The graph of a polynomial function of degree greater than one is a curve. Consequently, many points may have to be found before an accurate graph can be drawn.

Evaluating the quadratic function given by the equation $f(x) = x^2 - x - 6$ when $x = -3, -2, -1, 0, 1, 2, 3$, and 4 gives the points shown in Figure 1. For instance, $f(-3) = 6$, so $(-3, 6)$ is graphed; $f(2) = -4$, so $(2, -4)$ is graphed; and $f(4) = 6$, so $(4, 6)$ is graphed. Evaluating the function when x is not an integer, such as $x = -\frac{3}{2}$ and $x = \frac{5}{2}$, produces more points to graph, as shown in Figure 2. Connecting the points with a smooth curve results in Figure 3, which is the graph of f.

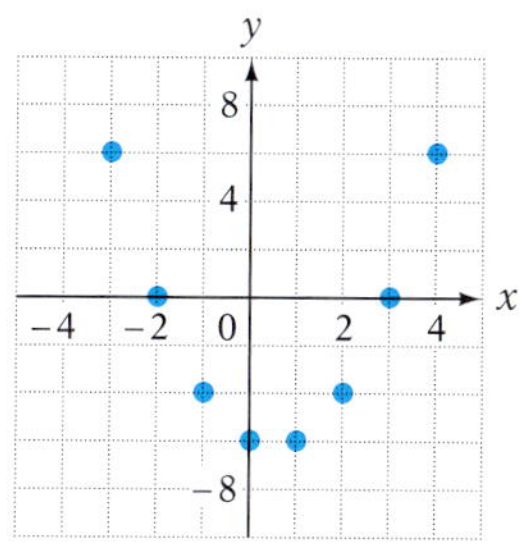

Figure 1

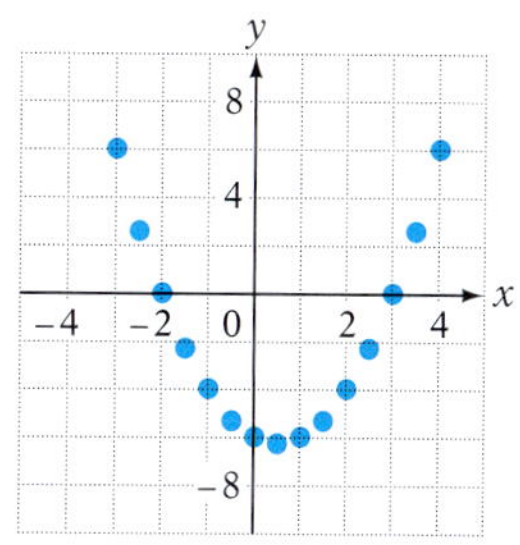

Figure 2

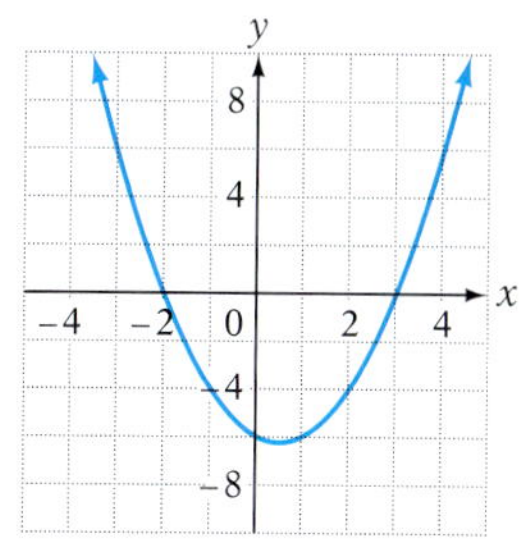

Figure 3

Here is an example of graphing a cubic function, $P(x) = x^3 - 2x^2 - 5x + 6$. Evaluating the function when $x = -2, -1, 0, 1, 2, 3$, and 4 gives the graph in Figure 4. Evaluating at some noninteger values gives the graph in Figure 5. Finally, connecting the dots with a smooth curve gives the graph in Figure 6.

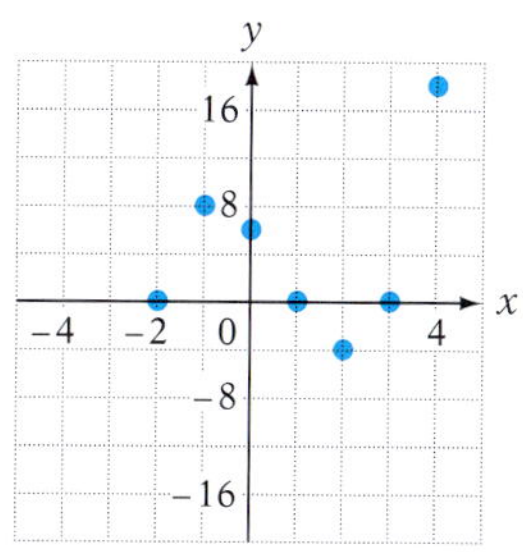

Figure 4

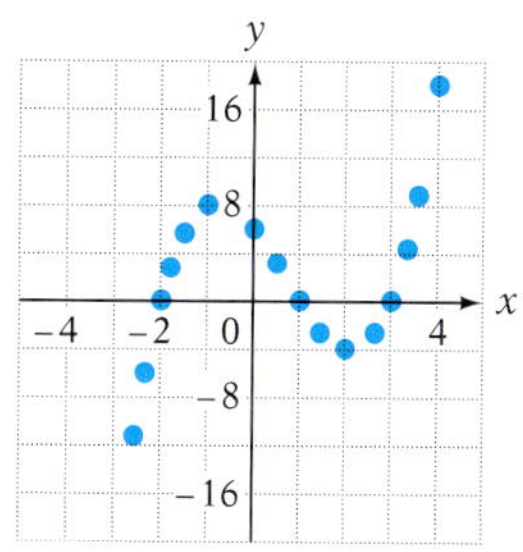

Figure 5

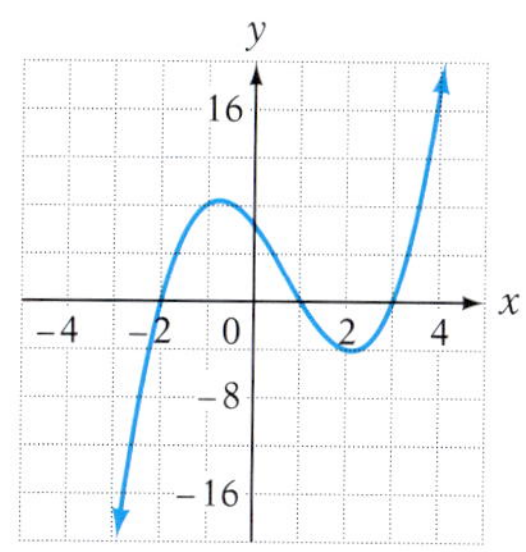

Figure 6

Example 4 Graph: $f(x) = x^2 - 2$

Solution

x	$y = f(x)$
−3	7
−2	2
−1	−1
0	−2
1	−1
2	2
3	7

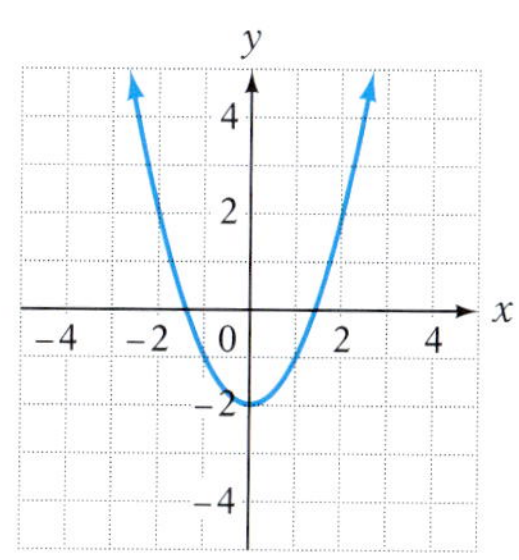

Problem 4 Graph: $f(x) = x^2 - 2x$

Solution See page S18.

Example 5 Graph: $f(x) = x^3 - 1$

Solution

x	$y = f(x)$
-2	-9
-1	-2
0	-1
1	0
2	7

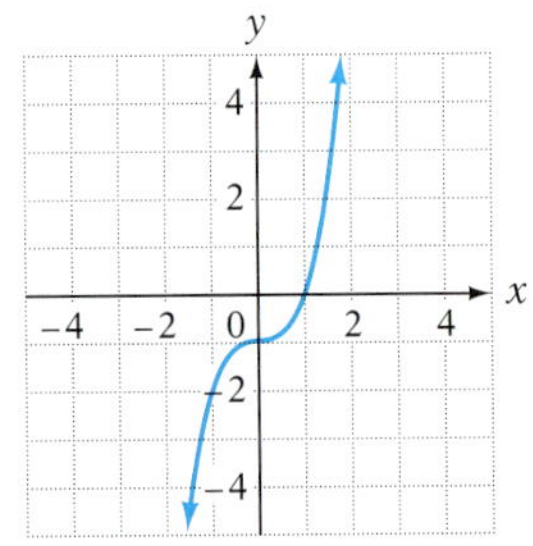

Problem 5 Graph: $F(x) = -x^3 + 1$

Solution See page S18.

It may be necessary to plot a large number of points before drawing the graph in Example 5. Graphing utilities also create graphs by plotting a large number of points and then connecting the points to form a curve. Using a graphing utility, enter the equation $y = x^3 - 1$ and verify the graph drawn in Example 5. Now trace along the graph and verify that $(-2, -9)$, $(-1, -2)$, $(0, -1)$, $(1, 0)$, and $(2, 7)$ are the coordinates of points on the graph. Follow the same procedure for Problem 5.

Example 6 Graph $f(x) = x^2 - 2x - 3$. Find two values of x, to the nearest tenth, for which $f(x) = 2$.

Solution To find the values of x for which $f(x) = 2$, use a graphing utility to graph f.

Recall that $f(x)$ is the y-coordinate of a point on the graph. Thus the values of x for which $f(x) = 2$ are those for which y is 2. The x-coordinates where the line $y = 2$ crosses the graph of f are the desired values.

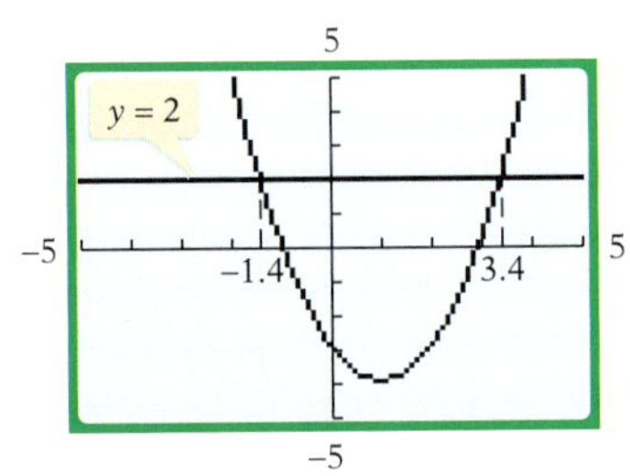

Using the TRACE feature of the graphing utility, move the cursor along the graph of f until $y = 2$. The values of x to the nearest tenth are -1.4 and 3.4.

Problem 6 Graph $S(x) = x^3 - x - 2$. Find a value of x, to the nearest tenth, for which $S(x) = 2$.

Solution See page S18.

2 Add or subtract polynomials

VIDEO & DVD CD TUTOR WEB SSM

Polynomials can be added by combining like terms. Either a vertical or a horizontal format can be used.

Add $(3x^2 + 2x - 7) + (7x^3 - 3 + 4x^2)$. Use a horizontal format.

Use the Commutative and Associative Properties of Addition to rearrange and group like terms.

$$(3x^2 + 2x - 7) + (7x^3 - 3 + 4x^2)$$
$$= 7x^3 + (3x^2 + 4x^2) + 2x + (-7 - 3)$$

Combine like terms.

$$= 7x^3 + 7x^2 + 2x - 10$$

Add $(4x^2 + 5x - 3) + (7x^3 - 7x + 1) + (2x - 3x^2 + 4x^3 + 1)$. Use a vertical format.

Arrange the terms of each polynomial in descending order with like terms in the same column.

$$\begin{array}{rrrr} & 4x^2 & + 5x & - 3 \\ 7x^3 & & - 7x & + 1 \\ 4x^3 & - 3x^2 & + 2x & + 1 \\ \hline 11x^3 & + x^2 & & - 1 \end{array}$$

Add the terms in each column.

The additive inverse of the polynomial $x^2 + 5x - 4$ is $-(x^2 + 5x - 4)$.

To simplify the additive inverse of a polynomial, change the sign of every term inside the parentheses.

$$-(x^2 + 5x - 4) = -x^2 - 5x + 4$$

Take Note

This is the same definition used for subtraction of integers: Subtraction is addition of the opposite.

To subtract two polynomials, add the additive inverse of the second polynomial to the first.

Subtract $(3x^2 - 7xy + y^2) - (-4x^2 + 7xy - 3y^2)$. Use a horizontal format.

Rewrite the subtraction as the addition of the additive inverse.

$$(3x^2 - 7xy + y^2) - (-4x^2 + 7xy - 3y^2)$$
$$= (3x^2 - 7xy + y^2) + (4x^2 - 7xy + 3y^2)$$

Combine like terms.

$$= 7x^2 - 14xy + 4y^2$$

Subtract $(6x^3 - 3x + 7) - (3x^2 - 5x + 12)$. Use a vertical format.

Rewrite the subtraction as the addition of the additive inverse.

$$\begin{aligned}&(6x^3 - 3x + 7) - (3x^2 - 5x + 12)\\ &= (6x^3 - 3x + 7) + (-3x^2 + 5x - 12)\end{aligned}$$

Arrange the terms of each polynomial in descending order with like terms in the same column.

$$\begin{array}{rrrr} 6x^3 & & -3x & +7 \\ & -3x^2 & +5x & -12 \\ \hline 6x^3 & -3x^2 & +2x & -5 \end{array}$$

Add the terms in each column.

Example 7 Add: $(3x^2 - 2x + 5) + (2 - 3x + 4x^3) + (x - 3x^2 - x^3 - 4)$
Use a vertical format.

Solution

$$\begin{array}{rrrr} & 3x^2 & -2x & +5 \\ 4x^3 & & -3x & +2 \\ -x^3 & -3x^2 & +x & -4 \\ \hline 3x^3 & & -4x & +3 \end{array}$$

- Arrange the terms of each polynomial in descending order with like terms in the same column.
- Combine like terms in each column.

Problem 7 Add: $(5x^2 + 3x - 1) + (2x^2 + 4x - 6) + (8 - 7x + x^2)$
Use a vertical format.

Solution See page S18.

Example 8 Subtract: $(3x^2 - 2x + 4) - (7x^2 + 3x - 12)$
Use a vertical format.

Solution $(3x^2 - 2x + 4) + (-7x^2 - 3x + 12)$

- Rewrite the subtraction as the addition of the additive inverse.

$$\begin{array}{rrr} 3x^2 & -2x & +4 \\ -7x^2 & -3x & +12 \\ \hline -4x^2 & -5x & +16 \end{array}$$

- Arrange the polynomials in descending order in a vertical format.
- Combine like terms in each column.

Problem 8 Subtract: $(-5x^2 + 2x - 3) - (6x^2 + 3x - 7)$
Use a vertical format.

Solution See page S18.

Functional notation may be used when adding or subtracting polynomials. For instance, if $P(x) = 3x^2 - 2x + 4$ and $R(x) = -5x^3 + 4x + 7$, then

$$\begin{aligned}P(x) + R(x) &= (3x^2 - 2x + 4) + (-5x^3 + 4x + 7)\\ &= -5x^3 + 3x^2 + 2x + 11\end{aligned}$$

$$\begin{aligned}P(x) - R(x) &= (3x^2 - 2x + 4) - (-5x^3 + 4x + 7)\\ &= (3x^2 - 2x + 4) + (5x^3 - 4x - 7)\\ &= 5x^3 + 3x^2 - 6x - 3\end{aligned}$$

Example 9 Given $P(x) = -3x^2 + 2x - 6$ and $R(x) = 4x^3 - 3x + 4$, find $S(x) = P(x) + R(x)$.

Solution
$$S(x) = (-3x^2 + 2x - 6) + (4x^3 - 3x + 4)$$
$$= 4x^3 - 3x^2 - x - 2$$

Problem 9 Given $P(x) = 4x^3 - 3x^2 + 2$ and $R(x) = -2x^2 + 2x - 3$, find $D(x) = P(x) - R(x)$.

Solution See page S18.

5.1 Concept Review

Determine whether the following statements are always true, sometimes true, or never true.

1. The degree of a polynomial is the greatest of the degrees of its individual terms.
2. The leading coefficient of $4 - 2x - 3x^2$ is 4.
3. Terms with the same variable parts can be added.
4. The additive inverse of a polynomial is a polynomial with every sign changed to the opposite sign.
5. The sum of two trinomials is a trinomial.
6. $f(x) = 3x^2 - 2x + x^{-1}$ is a polynomial function.

5.1 Exercises

1

1. Explain how to determine the degree of a polynomial in one variable. Give two examples of a third-degree polynomial.
2. Explain how to evaluate a polynomial.
3. Given $P(x) = 3x^2 - 2x - 8$, evaluate $P(3)$.
4. Given $P(x) = -3x^2 - 5x + 8$, evaluate $P(-5)$.
5. Given $R(x) = 2x^3 - 3x^2 + 4x - 2$, evaluate $R(2)$.
6. Given $R(x) = -x^3 + 2x^2 - 3x + 4$, evaluate $R(-1)$.
7. Given $f(x) = x^4 - 2x^2 - 10$, evaluate $f(-1)$.
8. Given $f(x) = x^5 - 2x^3 + 4x$, evaluate $f(2)$.

9. **Oceanography** The length of a water wave close to land depends on several factors, one of which is the depth of the water. However, the length, L (in meters), of deep-water waves (those far away from land) can be approximated by $L(s) = 0.641s^2$, where s is the speed of the wave in meters per second. Find the length of a deep-water wave that has a speed of 6 m/s. Round to the nearest tenth of a meter.

10. **Physics** The distance, s (in feet), that an object on the moon will fall in t seconds is given by $s(t) = 2.735t^2$. How far will an object fall on the moon in 3 s?

11. **Sports** The total number of softball games, T, that must be scheduled in a league that has n teams such that each team plays every other team twice is given by $T(n) = n^2 - n$. What is the total number of games that must be scheduled for a league that has 8 teams?

12. **Geometry** A diagonal of a polygon is a line from a vertex to a nonadjacent vertex. Seven of the possible diagonals for a decagon (10-sided figure) are shown at the right. The total number, T, of diagonals for an n-sided polygon is given by $T(n) = \frac{1}{2}n^2 - \frac{3}{2}n$. Find the total number of diagonals for a decagon.

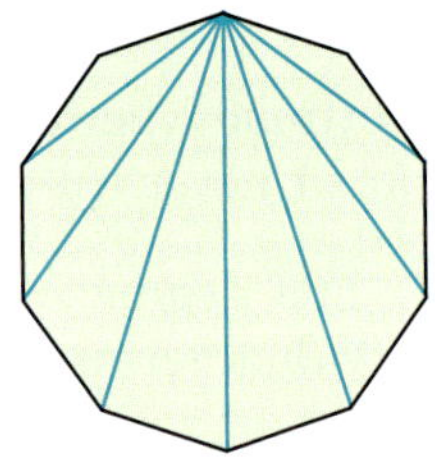

13. **Food Science** Baked Alaska is a dessert that is made by putting a 1-inch meringue coating around a hemisphere of ice cream. The amount of meringue that is needed depends on the radius of the hemisphere of ice cream and is given by $M(r) = 6.14r^2 + 6.14r + 2.094$, where $M(r)$ is the volume of meringue needed (in cubic inches) and r is the radius of the ice cream in inches. Find the amount of meringue that is needed for a baked Alaska that has a mound of ice cream with a 6-inch radius. Round to the nearest whole number.

14. Suppose $f(x) = x^2 + 1$ and $g(x) = x^3 + x$. Is $f(x) > g(x)$ or is $f(x) < g(x)$ when $0 < x < 1$? Explain your answer.

State whether the function is a polynomial function. For those that are polynomial functions, identify **a.** the leading coefficient, **b.** the constant term, and **c.** the degree.

15. $P(x) = -x^2 + 3x + 8$

16. $P(x) = 3x^4 - 3x - 7$

17. $R(x) = \dfrac{x}{x + 1}$

18. $R(x) = \dfrac{3x^2 - 2x + 1}{x}$

19. $f(x) = \sqrt{x} - x^2 + 2$

20. $f(x) = x^2 - \sqrt{x + 2} - 8$

21. $g(x) = 3x^5 - 2x^2 + \pi$

22. $g(x) = -4x^5 - 3x^2 + x - \sqrt{7}$

23. $P(x) = 3x^2 - 5x^3 + 2$

24. $P(x) = x^2 - 5x^4 - x^6$

25. $R(x) = 14$

26. $R(x) = \dfrac{1}{x} + 2$

Graph.

27. $P(x) = x^2 - 1$

28. $P(x) = 2x^2 + 3$

29. $R(x) = x^3 + 2$

30. $R(x) = x^4 + 1$

31. $f(x) = x^3 - 2x$

32. $f(x) = x^2 - x - 2$

2 33. What is the additive inverse of a polynomial?

34. Explain **a.** how to add and **b.** how to subtract two polynomials.

Simplify. Use a vertical format.

35. $(5x^2 + 2x - 7) + (x^2 - 8x + 12)$

36. $(3x^2 - 2x + 7) + (-3x^2 + 2x - 12)$

37. $(x^2 - 3x + 8) - (2x^2 - 3x + 7)$

38. $(2x^2 + 3x - 7) - (5x^2 - 8x - 1)$

For Exercises 39–42, simplify using a horizontal format.

39. $(3y^2 - 7y) + (2y^2 - 8y + 2)$

40. $(-2y^2 - 4y - 12) + (5y^2 - 5y)$

41. $(2a^2 - 3a - 7) - (-5a^2 - 2a - 9)$

42. $(3a^2 - 9a) - (-5a^2 + 7a - 6)$

43. Given $P(x) = 3x^3 - 4x^2 - x + 1$ and $R(x) = 2x^3 + 5x - 8$, find $P(x) + R(x)$.

44. Given $P(x) = 5x^3 - 3x - 7$ and $R(x) = 2x^3 - 3x^2 + 8$, find $P(x) - R(x)$.

45. Given $P(x) = x^{2n} + 7x^n - 3$ and $R(x) = -x^{2n} + 2x^n + 8$, find $P(x) + R(x)$.

46. Given $P(x) = 2x^{2n} - x^n - 1$ and $R(x) = 5x^{2n} + 7x^n + 1$, find $P(x) - R(x)$.

47. Given $P(x) = 3x^4 - 3x^3 - x^2$ and $R(x) = 3x^3 - 7x^2 + 2x$, find $S(x) = P(x) + R(x)$.

48. Given $P(x) = 3x^4 - 2x + 1$ and $R(x) = 3x^5 - 5x - 8$, find $S(x) = P(x) + R(x)$.

49. Given $P(x) = x^2 + 2x + 1$ and $R(x) = 2x^3 - 3x^2 + 2x - 7$, find $D(x) = P(x) - R(x)$.

50. Given $P(x) = 2x^4 - 2x^2 + 1$ and $R(x) = 3x^3 - 2x^2 + 3x + 8$, find $D(x) = P(x) - R(x)$.

5.1 Applying Concepts

The value of the x-coordinates of the x-intercepts of a function are called the **zeros** of the function. Find, to the nearest tenth, the zeros of the following functions.

51. $f(x) = 2x - 3$

52. $f(x) = 1.4x - 2.6$

53. $f(x) = x^2 - 4x + 2$

54. $f(x) = 2x^2 + 3x - 1$

55. $f(x) = x^3 - 3x + 1$

56. $f(x) = x^3 + x^2 - 1$

57. Given $f(x) = x^2 - x + 3$, find, to the nearest tenth, the values of x for which $f(x) = 4$.

58. Given $f(x) = x^3 - 2x^2 + x - 1$, find, to the nearest tenth, the value of x for which $f(x) = -2$.

Two polynomials are equal if the coefficients of like powers are equal. For example,

$$6x^3 - 7x^2 + 2x - 8 = 2x - 7x^2 - 8 + 6x^3$$

In Exercises 59 and 60, use the definition of equality of polynomials to find the value of k that makes the equation an identity.

59. $(2x^3 + 3x^2 + kx + 5) - (x^3 + x^2 - 5x - 2) = x^3 + 2x^2 + 3x + 7$

60. $(6x^3 + kx^2 - 2x - 1) - (4x^3 - 3x^2 + 1) = 2x^3 - x^2 - 2x - 2$

61. If $P(x)$ is a third-degree polynomial and $Q(x)$ is a fourth-degree polynomial, what can be said about the degree of $P(x) + Q(x)$? Give some examples of polynomials that support your answer.

62. If $P(x)$ is a fifth-degree polynomial and $Q(x)$ is a fourth-degree polynomial, what can be said about the degree of $P(x) - Q(x)$? Give some examples of polynomials that support your answer.

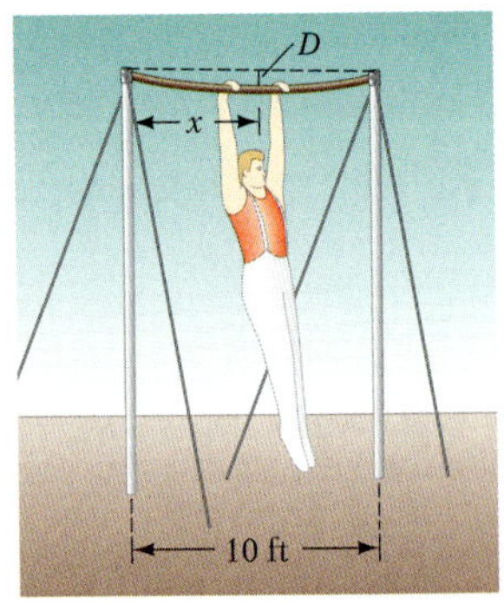

63. **Sports** The deflection, D (in inches), of a beam that is uniformly loaded is given by the polynomial function $D(x) = 0.005x^4 - 0.1x^3 + 0.5x^2$, where x is the distance in feet from one end of the beam. See the figure at the right. The maximum deflection occurs when x is the midpoint of the beam. Determine the maximum deflection for the beam in the diagram.

64. If $P(2) = 3$ and $P(x) = 2x^3 - 4x^2 - 2x + c$, find the value of c.

65. If $P(-1) = -3$ and $P(x) = 4x^4 - 3x^2 + 6x + c$, find the value of c.

66. Graph $f(x) = x^2$, $g(x) = (x - 3)^2$, and $h(x) = (x + 4)^2$ on the same coordinate grid. From the graphs, make a conjecture about the shape and location of $k(x) = (x - 2)^2$. Test your conjecture by graphing k.

67. Graph $f(x) = x^2$, $g(x) = x^2 - 3$, and $h(x) = x^2 + 4$ on the same coordinate grid. From the graphs, make a conjecture about the shape and location of $k(x) = x^2 - 2$. Test your conjecture by graphing k.

68. **Engineering** Construction of the Golden Gate Bridge, which is a suspension bridge, was completed in 1937. The length of the main span of the bridge is 4200 ft. The height, in feet, of the suspension cables above the roadway varies from 0 ft at the center of the bridge to 525 ft at the towers that support the cables.

Distance from the center of the bridge	*Height of the cables above the roadway*
0 ft	0 ft
1050 ft	150 ft
2100 ft	525 ft

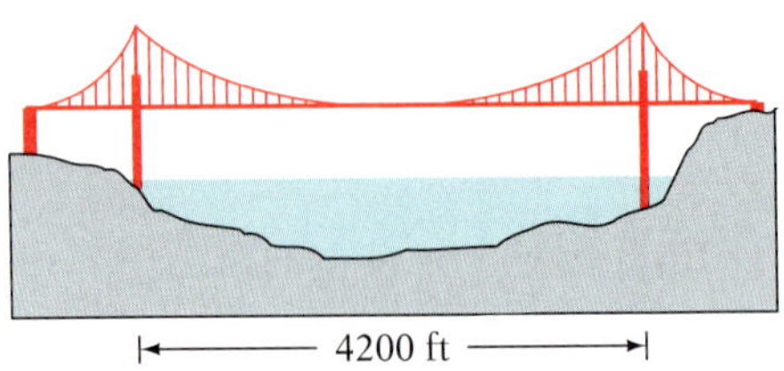

The function that approximately models the data is $f(x) = \frac{1}{8820}x^2 + 25$, where x is the distance from the center of the bridge and $f(x)$ is the height of the cables above the roadway. Use this model to approximate the height of the cables at a distance of **a.** 1000 ft from the center of the bridge and **b.** 1500 ft from the center of the bridge. Round to the nearest tenth.

69. Write a report on *polygonal numbers*.

70. Explain the similarities and differences among the graphs of $f(x) = x^2$, $g(x) = (x - 3)^2$, and $h(x) = x^2 - 3$.

5.2 Multiplication of Monomials

1 Multiply monomials

VIDEO & DVD CD TUTOR WEB SSM

Recall that in the exponential expression x^5, x is the base and 5 is the exponent. The exponent indicates the number of times the base occurs as a factor.

The product of exponential expressions with the *same* base can be simplified by writing each expression in factored form and writing the result with an exponent.

$$x^3 \cdot x^2 = \underbrace{\overbrace{(x \cdot x \cdot x)}^{3 \text{ factors}} \cdot \overbrace{(x \cdot x)}^{2 \text{ factors}}}_{5 \text{ factors}}$$
$$= x \cdot x \cdot x \cdot x \cdot x$$
$$= x^5$$

Adding the exponents results in the same product.

$$x^3 \cdot x^2 = x^{3+2} = x^5$$

Rule for Multiplying Exponential Expressions

If m and n are integers, then $\boldsymbol{x^m \cdot x^n = x^{m+n}}$.

For example, in the expression $a^2 \cdot a^6 \cdot a$, the bases are the same. The expression can be simplified by adding the exponents. Recall that $a = a^1$.

$$a^2 \cdot a^6 \cdot a = a^{2+6+1} = a^9$$

> **Take Note**
> The Rule for Multiplying Exponential Expressions requires that the bases be the same. The expression x^3y^2 cannot be simplified.

Example 1 Multiply: $(2xy)(3x^2y)$

Solution
$(2xy)(3x^2y)$
$= (2 \cdot 3)(x \cdot x^2)(y \cdot y)$ • Use the Commutative and Associative Properties of Multiplication to rearrange and group factors.
$= 6x^{1+2}y^{1+1}$ • Multiply variables with the same base by adding the exponents.
$= 6x^3y^2$

Problem 1 Multiply: $(3x^2)(6x^3)$

Solution See page S18.

Example 2 Multiply: $(2x^2y)(-5xy^4)$

Solution

$(2x^2y)(-5xy^4)$
$= [2(-5)](x^2 \cdot x)(y \cdot y^4)$ • Use the Properties of Multiplication to rearrange and group factors.

$= -10x^3y^5$ • Multiply variables with the same base by adding the exponents.

Problem 2 Multiply: $(-3xy^2)(-4x^2y^3)$

Solution See page S18.

2 Simplify powers of monomials

VIDEO & DVD CD TUTOR WEB SSM

Point of Interest

One of the first symbolic representations of power was given by Diophantus (ca. A.D. 250) in his book Arithmetica. *He used* Δ^Y *for* x^2 *and* k^Y *for* x^3*. The symbol* Δ^Y *was the first two letters of the Greek words* dunamis, *meaning "power";* k^Y *was from the Greek word* kubos, *meaning "cube." He also combined these symbols to denote higher powers. For instance,* Δk^Y *was the symbol for* x^5.

A power of a monomial can be simplified by rewriting the expression in factored form and then using the Rule for Multiplying Exponential Expressions.

$$(x^2)^3 = x^2 \cdot x^2 \cdot x^2 = x^{2+2+2} = x^6$$

$$(x^4y^3)^2 = (x^4y^3)(x^4y^3) = x^4 \cdot y^3 \cdot x^4 \cdot y^3 = (x^4 \cdot x^4)(y^3 \cdot y^3) = x^{4+4}y^{3+3} = x^8y^6$$

Note that multiplying each exponent inside the parentheses by the exponent outside the parentheses gives the same result.

$$(x^2)^3 = x^{2 \cdot 3} = x^6 \qquad (x^4y^3)^2 = x^{4 \cdot 2}y^{3 \cdot 2} = x^8y^6$$

Rule for Simplifying Powers of Exponential Expressions

If m and n are integers, then $(x^m)^n = x^{mn}$.

Simplify: $(x^5)^2$

Multiply the exponents. $(x^5)^2 = x^{5 \cdot 2} = x^{10}$

Rule for Simplifying Powers of Products

If m, n, and p are integers, then $(x^my^n)^p = x^{mp}y^{np}$.

Simplify: $(c^5d^3)^6$

Multiply each exponent inside the parentheses by the exponent outside the parentheses.

$$(c^5d^3)^6 = c^{5 \cdot 6}d^{3 \cdot 6} = c^{30}d^{18}$$

Simplify: $(3a^2b)^3$

Multiply each exponent inside the parentheses by the exponent outside the parentheses.

$$(3a^2b)^3 = 3^{1 \cdot 3}a^{2 \cdot 3}b^{1 \cdot 3} = 3^3a^6b^3 = 27a^6b^3$$

Example 3 Simplify: $(-2x)(-3xy^2)^3$

Solution $(-2x)(-3xy^2)^3$

$= (-2x)(-3)^3x^3y^6$ • Multiply each exponent in $-3xy^2$ by the exponent outside the parentheses.

$= (-2x)(-27)x^3y^6$ • Simplify $(-3)^3$.

$= [-2(-27)](x \cdot x^3)y^6$ • Use the Properties of Multiplication to rearrange and group factors.

$= 54x^4y^6$ • Multiply variables with the same base by adding the exponents.

Problem 3 Simplify: $(3x)(2x^2y)^3$

Solution See page S18.

5.2 Concept Review

Determine whether the statement is always true, sometimes true, or never true.

1. In the expression $(6x)^4$, x is the base and 4 is the exponent.

2. To multiply $x^m \cdot x^n$, multiply the exponents.

3. The expression "a power of a monomial" means the monomial is the base of an exponential expression.

4. $x^6 \cdot x \cdot x^8 = x^{14}$

5. $(4x^3y^5)^2 = 4x^6y^{10}$

6. $a^2 \cdot b^7 = ab^9$

5.2 Exercises

1

1. Explain how to multiply two exponential expressions with the same base. Provide an example.

2. Explain how to simplify the power of an exponential expression. Provide an example.

State whether the expression is the product of two exponential expressions or the power of an exponential expression.

3. a. $b^4 \cdot b^8$ b. $(b^4)^8$

4. a. $(2z)^2$ b. $2z \cdot z$

5. a. $(3a^4)^5$ b. $(3a^4)(5a)$

6. a. $x(-xy^4)$ b. $(-xy)^4$

Multiply.

7. $(x)(2x)$

8. $(-3y)(y)$

9. $(3x)(4x)$

10. $(7y^3)(7y^2)$

11. $(-2a^3)(-3a^4)$

12. $(5a^6)(-2a^5)$

13. $(x^2y)(xy^4)$
14. $(x^2y^4)(xy^7)$
15. $(-2x^4)(5x^5y)$
16. $(-3a^3)(2a^2b^4)$
17. $(x^2y^4)(x^5y^4)$
18. $(a^2b^4)(ab^3)$
19. $(2xy)(-3x^2y^4)$
20. $(-3a^2b)(-2ab^3)$
21. $(x^2yz)(x^2y^4)$
22. $(-ab^2c)(a^2b^5)$
23. $(a^2b^3)(ab^2c^4)$
24. $(x^2y^3z)(x^3y^4)$
25. $(-a^2b^2)(a^3b^6)$
26. $(xy^4)(-xy^3)$
27. $(-6a^3)(a^2b)$
28. $(2a^2b^3)(-4ab^2)$
29. $(-5y^4z)(-8y^6z^5)$
30. $(3x^2y)(-4xy^2)$
31. $(10ab^2)(-2ab)$
32. $(x^2y)(yz)(xyz)$
33. $(xy^2z)(x^2y)(z^2y^2)$
34. $(-2x^2y^3)(3xy)(-5x^3y^4)$
35. $(4a^2b)(-3a^3b^4)(a^5b^2)$
36. $(3ab^2)(-2abc)(4ac^2)$

2 Simplify.

37. $(2^2)^3$
38. $(3^2)^2$
39. $(-2)^2$
40. $(-3)^3$
41. $(-2^2)^3$
42. $(-2^3)^3$
43. $(x^3)^3$
44. $(y^4)^2$
45. $(x^7)^2$
46. $(y^5)^3$
47. $(-x^2)^2$
48. $(-x^2)^3$
49. $(2x)^2$
50. $(3y)^3$
51. $(-2x^2)^3$
52. $(-3y^3)^2$
53. $(x^2y^3)^2$
54. $(x^3y^4)^5$
55. $(3x^2y)^2$
56. $(-2ab^3)^4$
57. $(a^2)(3a^2)^3$
58. $(b^2)(2a^3)^4$
59. $(-2x)(2x^3)^2$
60. $(2y)(-3y^4)^3$
61. $(x^2y)(x^2y)^3$
62. $(a^3b)(ab)^3$
63. $(ab^2)^2(ab)^2$
64. $(x^2y)^2(x^3y)^3$
65. $(-2x)(-2x^3y)^3$
66. $(-3y)(-4x^2y^3)^3$
67. $(-2x)(-3xy^2)^2$
68. $(-3y)(-2x^2y)^3$
69. $(ab^2)(-2a^2b)^3$
70. $(a^2b^2)(-3ab^4)^2$
71. $(-2a^3)(3a^2b)^3$
72. $(-3b^2)(2ab^2)^3$
73. $(-3ab)^2(-2ab)^3$
74. $(-3a^2b)^3(-3ab)^3$

5.2 Applying Concepts

Simplify.

75. $(6x)(2x^2) + (4x^2)(5x)$
76. $(2a^7)(7a^2) - (6a^3)(5a^6)$
77. $(3a^2b^2)(2ab) - (9ab^2)(a^2b)$
78. $(3x^2y^2)^2 - (2xy)^4$
79. $(5xy^3)(3x^4y^2) - (2x^3y)(x^2y^4)$
80. $a^2(ab^2)^3 - a^3(ab^3)^2$
81. $4a^2(2ab)^3 - 5b^2(a^5b)$
82. $9x^3(3x^2y)^2 - x(x^3y)^2$
83. $-2xy(x^2y)^3 - 3x^5(xy^2)^2$
84. $5a^2b(ab^2)^2 + b^3(2a^2b)^2$

85. $a^n \cdot a^n$ **86.** $(a^n)^2$ **87.** $(a^2)^n$ **88.** $a^2 \cdot a^n$

For Exercises 89–92, answer true or false. If the answer is false, correct the right side of the equation.

89. $(-a)^5 = -a^5$ **90.** $(-b)^8 = b^8$

91. $(x^2)^5 = x^{2+5} = x^7$ **92.** $x^3 + x^3 = 2x^{3+3} = 2x^6$

93. Evaluate $(2^3)^2$ and $2^{(3^2)}$. Are the results the same? If not, which expression has the larger value?

94. What is the Order of Operations for the expression x^{m^n}?

95. If n is a positive integer and $x^n = y^n$, when is $x = y$?

96. **Geometry** The length of a rectangle is $4ab$. The width is $2ab$. Find the perimeter of the rectangle in terms of ab.

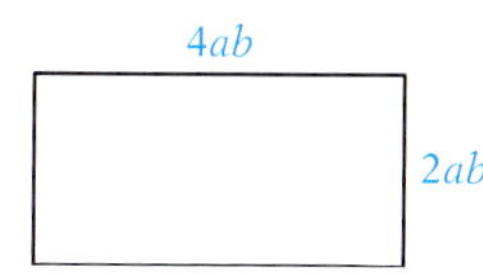

97. The distance a rock will fall in t seconds is $16t^2$ ft (neglecting air resistance). Find other examples of quantities that can be expressed in terms of an exponential expression, and explain where the expression is used.

5.3 Multiplication of Polynomials

1 Multiply a polynomial by a monomial

VIDEO & DVD CD TUTOR WEB SSM

To multiply a polynomial by a monomial, use the Distributive Property and the Rule for Multiplying Exponential Expressions.

Simplify: $-2x(x^2 - 4x - 3)$

$-2x(x^2 - 4x - 3)$
$= -2x(x^2) - (-2x)(4x) - (-2x)(3)$ • Use the Distributive Property.
$= -2x^3 + 8x^2 + 6x$ • Use the Rule for Multiplying Exponential Expressions.

Example 1 Multiply. **A.** $(5x + 4)(-2x)$ **B.** $x^3(2x^2 - 3x + 2)$

Solution **A.** $(5x + 4)(-2x)$
$= 5x(-2x) + 4(-2x)$ • Use the Distributive Property.
$= -10x^2 - 8x$ • Use the Rule for Multiplying Exponential Expressions.

B. $x^3(2x^2 - 3x + 2)$
$= 2x^5 - 3x^4 + 2x^3$

Problem 1 Multiply. **A.** $(-2y + 3)(-4y)$ **B.** $-a^2(3a^2 + 2a - 7)$

Solution See page S18.

2 Multiply two polynomials

VIDEO & DVD CD TUTOR WEB SSM

Multiplication of two polynomials requires the repeated application of the Distributive Property.

$$\begin{aligned}(y-2)(y^2+3y+1) &= (y-2)(y^2)+(y-2)(3y)+(y-2)(1)\\ &= y^3-2y^2+3y^2-6y+y-2\\ &= y^3+y^2-5y-2\end{aligned}$$

A convenient method of multiplying two polynomials is to use a vertical format similar to that used for multiplication of whole numbers.

Multiply each term in the trinomial by −2.
Multiply each term in the trinomial by y.
Like terms must be in the same column.
Add the terms in each column.

$$\begin{array}{rrrr} & y^2 & +\,3y & +\,1\\ & & y & -\,2\\ \hline & -\,2y^2 & -\,6y & -\,2\\ y^3 & +\,3y^2 & +\,y & \\ \hline y^3 & +\,y^2 & -\,5y & -\,2\end{array}$$

Example 2 Multiply: $(2b^3 - b + 1)(2b + 3)$

Solution

$$\begin{array}{rrrrr} & 2b^3 & & -\,b & +\,1\\ & & & 2b & +\,3\\ \hline & 6b^3 & & -\,3b & +\,3\\ 4b^4 & & -\,2b^2 & +\,2b & \\ \hline 4b^4 & +\,6b^3 & -\,2b^2 & -\,b & +\,3\end{array}$$

- **Multiply $2b^3 - b + 1$ by 3.**
- **Multiply $2b^3 - b + 1$ by $2b$. Arrange the terms in descending order.**
- **Add the terms in each column.**

Problem 2 Multiply: $(2y^3 + 2y^2 - 3)(3y - 1)$

Solution See page S18.

Example 3 Multiply: $(4a^3 - 5a - 2)(3a - 2)$

Solution

$$\begin{array}{rrrrr} & 4a^3 & & -\,5a & -\,2\\ & & & 3a & -\,2\\ \hline & -\,8a^3 & & +\,10a & +\,4\\ 12a^4 & & -\,15a^2 & -\,6a & \\ \hline 12a^4 & -\,8a^3 & -\,15a^2 & +\,4a & +\,4\end{array}$$

- **Multiply $4a^3 - 5a - 2$ by −2.**
- **Multiply $4a^3 - 5a - 2$ by $3a$.**
- **Add the terms in each column.**

Problem 3 Multiply: $(3x^3 - 2x^2 + x - 3)(2x + 5)$

Solution See page S18.

3 Multiply two binomials using the FOIL Method

VIDEO & DVD CD TUTOR WEB SSM

It is often necessary to find the product of two binomials. The product can be found using a method called **FOIL**, which is based on the Distributive Property. The letters of FOIL stand for **F**irst, **O**uter, **I**nner, and **L**ast.

Multiply: $(2x + 3)(x + 5)$

Multiply the **F**irst terms. $(2x + 3)(x + 5)$ $\quad 2x \cdot x = 2x^2$

Multiply the **O**uter terms. $(2x + 3)(x + 5)$ $\quad 2x \cdot 5 = 10x$

Multiply the **I**nner terms. $(2x + 3)(x + 5)$ $\quad 3 \cdot x = 3x$

> **Take Note**
>
> FOIL is not really a different way of multiplying. It is based on the Distributive Property.
>
> $(2x + 3)(x + 5)$
> $= 2x(x + 5) + 3(x + 5)$
> $\quad$ F $\quad$ O $\quad$ I $\quad$ L
> $= 2x^2 + 10x + 3x + 15$
> $= 2x^2 + 13x + 15$

Multiply the Last terms. $\quad (2x + 3)(x + 5) \quad 3 \cdot 5 = 15$

Add the products. $\quad (2x + 3)(x + 5) \quad$ F O I L $= 2x^2 + 10x + 3x + 15$

Combine like terms. $\quad = 2x^2 + 13x + 15$

Example 4 Multiply: $(4x - 3)(3x - 2)$

Solution

$$(4x - 3)(3x - 2)$$
$$= 4x(3x) + 4x(-2) + (-3)(3x) + (-3)(-2)$$
$$= 12x^2 - 8x - 9x + 6$$
$$= 12x^2 - 17x + 6$$

- **Use the FOIL method.**
- **Combine like terms.**

Problem 4 Multiply: $(4y - 5)(3y - 3)$

Solution See page S18.

Example 5 Multiply: $(3x - 2y)(x + 4y)$

Solution

$$(3x - 2y)(x + 4y)$$
$$= 3x(x) + 3x(4y) + (-2y)(x) + (-2y)(4y)$$
$$= 3x^2 + 12xy - 2xy - 8y^2$$
$$= 3x^2 + 10xy - 8y^2$$

- **Use the FOIL method.**
- **Combine like terms.**

Problem 5 Multiply: $(3a + 2b)(3a - 5b)$

Solution See page S18.

4 Multiply binomials that have special products

The expression $(a + b)(a - b)$ is the product of the sum and difference of two terms. The first binomial in the expression is a sum; the second is a difference. The two terms are a and b. The first term in each binomial is a. The second term in each binomial is b.

The expression $(a + b)^2$ is the square of a binomial. The first term in the binomial is a. The second term in the binomial is b.

Using FOIL, it is possible to find a pattern for the product of the sum and difference of two terms and for the square of a binomial.

The Sum and Difference of Two Terms

$$(\mathbf{a} + \mathbf{b})(\mathbf{a} - \mathbf{b}) = a^2 - ab + ab - b^2$$
$$= \mathbf{a^2 - b^2}$$

Square of first term (a^2)

Square of second term (b^2)

The Square of a Binomial

$$(\mathbf{a} + \mathbf{b})^2 = (a + b)(a + b) = a^2 + ab + ab + b^2$$
$$= \mathbf{a^2 + 2ab + b^2}$$

- Square of first term → a^2
- Twice the product of the two terms → $2ab$
- Square of last term → b^2

$$(\mathbf{a} - \mathbf{b})^2 = (a - b)(a - b) = a^2 - ab - ab + b^2$$
$$= \mathbf{a^2 - 2ab + b^2}$$

- Square of first term → a^2
- Twice the product of the two terms → $-2ab$
- Square of last term → b^2

Example 6 Multiply: $(2x + 3)(2x - 3)$

Solution $(2x + 3)(2x - 3)$ • $(2x + 3)(2x - 3)$ is the product of the sum and difference of two terms.

$= (2x)^2 - 3^2$ • Square the first term. Square the second term.

$= 4x^2 - 9$ • Simplify.

Problem 6 Multiply: $(2a + 5c)(2a - 5c)$

Solution See page S18.

Example 7 Multiply: $(4c + 5d)^2$

Solution $(4c + 5d)^2$ • $(4c + 5d)^2$ is the square of a binomial.

$= (4c)^2 + 2(4c)(5d) + (5d)^2$ • Square the first term. Find twice the product of the two terms. Square the second term.

$= 16c^2 + 40cd + 25d^2$ • Simplify.

Problem 7 Multiply: $(3x + 2y)^2$

Solution See page S18.

Example 8 Multiply: $(3x - 2)^2$

Solution $(3x - 2)^2$ • $(3x - 2)^2$ is the square of a binomial.

$= (3x)^2 + 2(3x)(-2) + (-2)^2$ • Square the first term. Find twice the product of the two terms. Square the last term.

$= 9x^2 - 12x + 4$ • Simplify.

Problem 8 Multiply: $(6x - y)^2$

Solution See page S18.

5 Application problems

Example 9 The radius of a circle is $(x - 4)$ ft. Find the area of the circle in terms of the variable x. Leave the answer in terms of π.

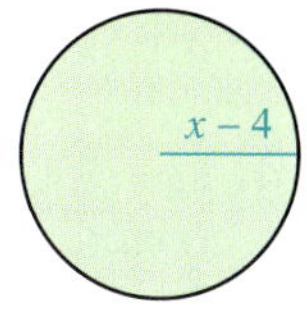

Strategy To find the area, replace the variable r in the formula $A = \pi r^2$ with the given value, and solve for A.

Solution

$A = \pi r^2$

$A = \pi(x - 4)^2$ • **This is the square of a binomial.**

$A = \pi(x^2 - 8x + 16)$

$A = \pi x^2 - 8\pi x + 16\pi$

The area is $(\pi x^2 - 8\pi x + 16\pi)$ ft^2.

Problem 9 The length of a rectangle is $(x + 7)$ m. The width is $(x - 4)$ m. Find the area of the rectangle in terms of the variable x.

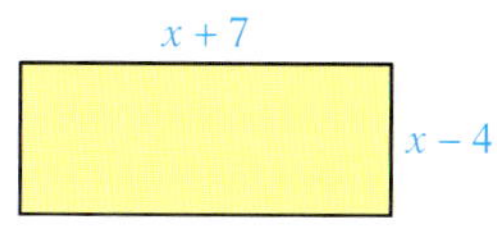

Solution See page S18.

Example 10 The length of a side of a square is $(3x + 5)$ in. Find the area of the square in terms of the variable x.

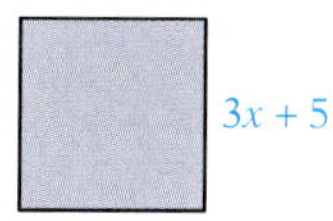

Strategy To find the area of the square, replace the variable s in the formula $A = s^2$ with the given value and simplify.

Solution

$A = s^2$

$A = (3x + 5)^2$ • **This is the square of a binomial.**

$A = 9x^2 + 30x + 25$

The area is $(9x^2 + 30x + 25)$ in^2.

Problem 10 The base of a triangle is $(x + 3)$ cm and the height is $(4x - 6)$ cm. Find the area of the triangle in terms of the variable x.

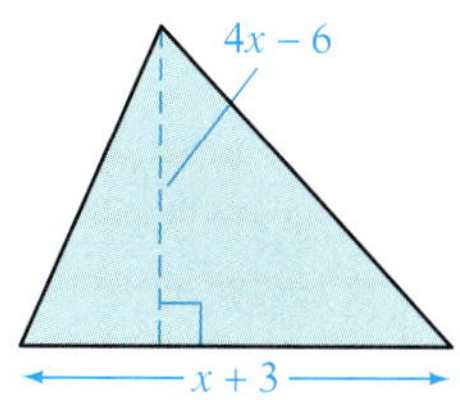

Solution See page S19.

5.3 Concept Review

Determine whether the statement is always true, sometimes true, or never true.

1. To multiply a monomial times a polynomial, use the Distributive Property to multiply each term of the polynomial by the monomial.

2. To multiply two polynomials, multiply each term of one polynomial by the other polynomial.

3. A binomial is a polynomial of degree 2.

4. $(x + 7)(x - 7)$ is the product of the sum and difference of the same two terms.

5. To square a binomial means to multiply it times itself.

6. The square of a binomial is a trinomial.

7. The FOIL method is used to multiply two polynomials.

8. Using the FOIL method, the terms $3x$ and 5 are the "First" terms in $(3x + 5)(2x + 7)$.

5.3 Exercises

1 Multiply.

1. $x(x - 2)$

2. $y(3 - y)$

3. $-x(x + 7)$

4. $-y(7 - y)$

5. $3a^2(a - 2)$

6. $4b^2(b + 8)$

7. $-5x^2(x^2 - x)$

8. $-6y^2(y + 2y^2)$

9. $-x^3(3x^2 - 7)$

10. $-y^4(2y^2 - y^6)$

11. $2x(6x^2 - 3x)$

12. $3y(4y - y^2)$

13. $(2x - 4)3x$

14. $(2x + 1)2x$

15. $-xy(x^2 - y^2)$

16. $-x^2y(2xy - y^2)$

17. $x(2x^3 - 3x + 2)$

18. $y(-3y^2 - 2y + 6)$

19. $-a(-2a^2 - 3a - 2)$

20. $-b(5b^2 + 7b - 35)$

21. $x^2(3x^4 - 3x^2 - 2)$

22. $y^3(-4y^3 - 6y + 7)$

23. $2y^2(-3y^2 - 6y + 7)$

24. $4x^2(3x^2 - 2x + 6)$

25. $(a^2 + 3a - 4)(-2a)$

26. $(b^3 - 2b + 2)(-5b)$

27. $-3y^2(-2y^2 + y - 2)$

28. $-5x^2(3x^2 - 3x - 7)$

29. $xy(x^2 - 3xy + y^2)$

30. $ab(2a^2 - 4ab - 6b^2)$

2 Multiply.

31. $(x^2 + 3x + 2)(x + 1)$

32. $(x^2 - 2x + 7)(x - 2)$

33. $(a - 3)(a^2 - 3a + 4)$

34. $(2x - 3)(x^2 - 3x + 5)$

35. $(-2b^2 - 3b + 4)(b - 5)$

36. $(-a^2 + 3a - 2)(2a - 1)$

37. $(3x - 5)(-2x^2 + 7x - 2)$

38. $(2a - 1)(-a^2 - 2a + 3)$

39. $(x^3 - 3x + 2)(x - 4)$

40. $(y^3 + 4y^2 - 8)(2y - 1)$

41. $(3y - 8)(5y^2 + 8y - 2)$

42. $(4y - 3)(3y^2 + 3y - 5)$

43. $(5a^3 - 15a + 2)(a - 4)$

44. $(3b^3 - 5b^2 + 7)(6b - 1)$

45. $(y + 2)(y^3 + 2y^2 - 3y + 1)$

46. $(2a - 3)(2a^3 - 3a^2 + 2a - 1)$

3 For the product of the binomials, name **a.** the first terms, **b.** the outer terms, **c.** the inner terms, and **d.** the last terms.

47. $(y - 8)(2y + 3)$

48. $(3d + 4)(d - 1)$

Multiply.

49. $(x + 1)(x + 3)$

50. $(y + 2)(y + 5)$

51. $(a - 3)(a + 4)$

52. $(b - 6)(b + 3)$

53. $(y + 3)(y - 8)$

54. $(x + 10)(x - 5)$

55. $(y - 7)(y - 3)$

56. $(a - 8)(a - 9)$

57. $(2x + 1)(x + 7)$

58. $(y + 2)(5y + 1)$

59. $(3x - 1)(x + 4)$

60. $(7x - 2)(x + 4)$

61. $(4x - 3)(x - 7)$

62. $(2x - 3)(4x - 7)$

63. $(3y - 8)(y + 2)$

64. $(5y - 9)(y + 5)$

65. $(3x + 7)(3x + 11)$

66. $(5a + 6)(6a + 5)$

67. $(7a - 16)(3a - 5)$

68. $(5a - 12)(3a - 7)$

69. $(3b + 13)(5b - 6)$

70. $(x + y)(2x + y)$

71. $(2a + b)(a + 3b)$

72. $(3x - 4y)(x - 2y)$

73. $(2a - b)(3a + 2b)$

74. $(5a - 3b)(2a + 4b)$

75. $(2x + y)(x - 2y)$

76. $(3x - 7y)(3x + 5y)$

77. $(2x + 3y)(5x + 7y)$

78. $(5x + 3y)(7x + 2y)$

79. $(3a - 2b)(2a - 7b)$

80. $(5a - b)(7a - b)$

81. $(a - 9b)(2a + 7b)$

82. $(2a + 5b)(7a - 2b)$

83. $(10a - 3b)(10a - 7b)$

84. $(12a - 5b)(3a - 4b)$

85. $(5x + 12y)(3x + 4y)$

86. $(11x + 2y)(3x + 7y)$

87. $(2x - 15y)(7x + 4y)$

88. $(5x + 2y)(2x - 5y)$

89. $(8x - 3y)(7x - 5y)$

90. $(2x - 9y)(8x - 3y)$

4 Multiply.

91. $(y - 5)(y + 5)$

92. $(y + 6)(y - 6)$

93. $(2x + 3)(2x - 3)$

94. $(4x - 7)(4x + 7)$

95. $(x + 1)^2$

96. $(y - 3)^2$

97. $(3a - 5)^2$

98. $(6x - 5)^2$

99. $(3x - 7)(3x + 7)$

100. $(9x - 2)(9x + 2)$

101. $(2a + b)^2$

102. $(x + 3y)^2$

103. $(x - 2y)^2$

104. $(2x - 3y)^2$

105. $(4 - 3y)(4 + 3y)$

106. $(4x - 9y)(4x + 9y)$

107. $(5x + 2y)^2$

108. $(2a - 9b)^2$

5 Geometry Problems

109. The length of a rectangle is $5x$ ft. The width is $(2x - 7)$ ft. Find the area of the rectangle in terms of the variable x.

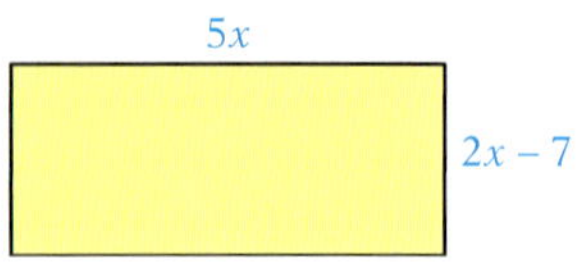

110. The width of a rectangle is $(x - 6)$ m. The length is $(2x + 3)$ m. Find the area of the rectangle in terms of the variable x.

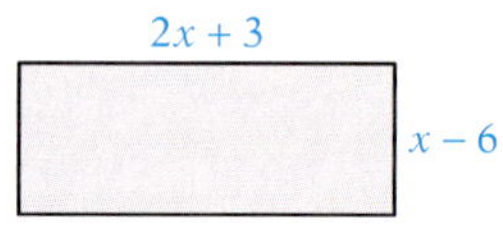

111. The width of a rectangle is $(3x + 1)$ in. The length of the rectangle is twice the width. Find the area of the rectangle in terms of the variable x.

112. The width of a rectangle is $(4x - 3)$ cm. The length of the rectangle is twice the width. Find the area of the rectangle in terms of the variable x.

113. The length of a side of a square is $(2x + 1)$ km. Find the area of the square in terms of the variable x.

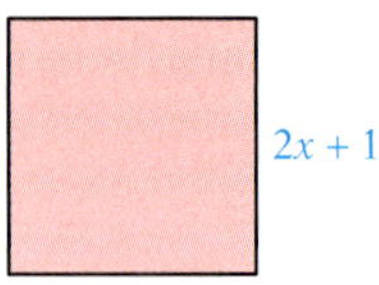

114. The length of a side of a square is $(2x - 3)$ yd. Find the area of the square in terms of the variable x.

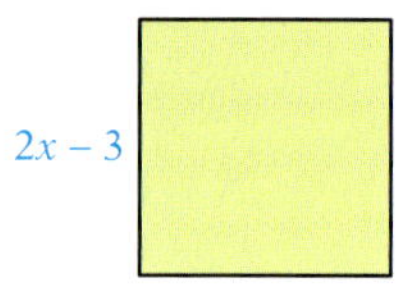

115. The base of a triangle is $4x$ m and the height is $(2x + 5)$ m. Find the area of the triangle in terms of the variable x.

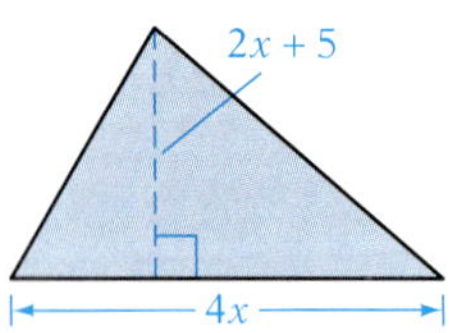

116. The base of a triangle is $(2x + 6)$ in. and the height is $(x - 8)$ in. Find the area of the triangle in terms of the variable x.

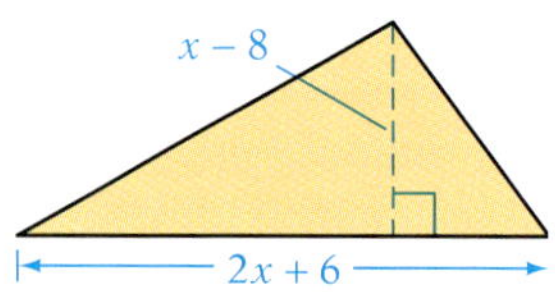

117. The radius of a circle is $(x + 4)$ cm. Find the area of the circle in terms of the variable x. Leave the answer in terms of π.

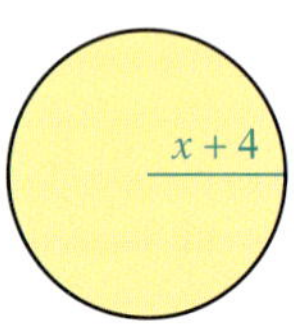

118. The radius of a circle is $(x - 3)$ ft. Find the area of the circle in terms of the variable x. Leave the answer in terms of π.

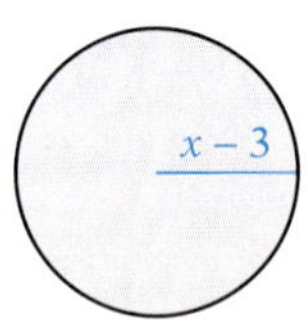

119. **Sports** A softball diamond has dimensions 45 ft by 45 ft. A base path border x feet wide lies on both the first-base side and the third-base side of the diamond. Express the total area of the softball diamond and the base path in terms of the variable x.

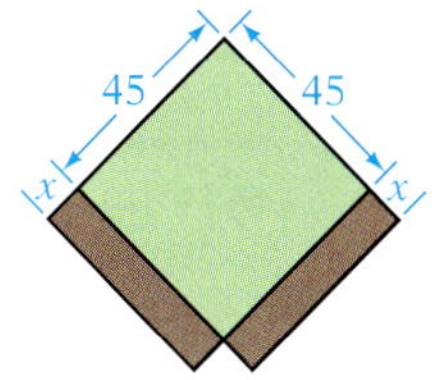

120. **Sports** An athletic field has dimensions 30 yd by 100 yd. An end zone that is w yards wide borders each end of the field. Express the total area of the field and the end zones in terms of the variable w.

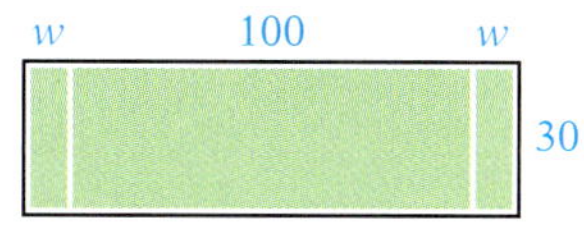

5.3 Applying Concepts

Simplify.

121. $(a + b)^2 - (a - b)^2$

122. $(x + 3y)^2 + (x + 3y)(x - 3y)$

123. $(3a^2 - 4a + 2)^2$

124. $(x + 4)^3$

125. $3x^2(2x^3 + 4x - 1) - 6x^3(x^2 - 2)$

126. $(3b + 2)(b - 6) + (4 + 2b)(3 - b)$

127. $x^n(x^n + 1)$

128. $(x^n + 1)(x^n - 1)$

129. $(x^n + 1)(x^n + 1)$

130. $(x^n - 1)^2$

131. $(x + 1)(x - 1)$

132. $(x + 1)(-x^2 + x - 1)$

133. $(x + 1)(x^3 - x^2 + x - 1)$

134. $(x + 1)(-x^4 + x^3 - x^2 + x - 1)$

Use the pattern of the answers to Exericses 131–134 to write the product.

135. $(x + 1)(x^5 - x^4 + x^3 - x^2 + x - 1)$

136. $(x + 1)(-x^6 + x^5 - x^4 + x^3 - x^2 + x - 1)$

Solve.

137. Find $(4n^3)^2$ if $2n - 3 = 4n - 7$.

138. What polynomial has quotient $x^2 + 2x - 1$ when divided by $x + 3$?

139. What polynomial has quotient $3x - 4$ when divided by $4x + 5$?

140. Subtract $4x^2 - x - 5$ from the product of $x^2 + x + 3$ and $x - 4$.

141. Add $x^2 + 2x - 3$ to the product of $2x - 5$ and $3x + 1$.

142. If a polynomial of degree 3 is multiplied by a polynomial of degree 2, what is the degree of the resulting polynomial?

143. Is it possible to multiply a polynomial of degree 2 by a polynomial of degree 2 and have the product be a polynomial of degree 3? If so, give an example. If not, explain why not.

5.4 Integer Exponents and Scientific Notation

1 Integer exponents

VIDEO & DVD CD TUTOR WEB SSM

The quotient of two exponential expressions with the *same* base can be simplified by writing each expression in factored form, dividing by the common factors, and then writing the result with an exponent.

$$\frac{x^5}{x^2} = \frac{\overset{1}{\cancel{x}} \cdot \overset{1}{\cancel{x}} \cdot x \cdot x \cdot x}{\underset{1}{\cancel{x}} \cdot \underset{1}{\cancel{x}}} = x^3$$

Note that subtracting the exponents results in the same quotient.

$$\frac{x^5}{x^2} = x^{5-2} = x^3$$

To divide two monomials with the same base, subtract the exponents of the like bases.

Simplify: $\dfrac{a^7}{a^3}$

The bases are the same. Subtract the exponent in the denominator from the exponent in the numerator.

$$\frac{a^7}{a^3} = a^{7-3} = a^4$$

Simplify: $\dfrac{r^8s^6}{r^7s}$

Subtract the exponents of the like bases.

$$\frac{r^8s^6}{r^7s} = r^{8-7}s^{6-1} = rs^5$$

Recall that for any number a, $a \neq 0$, $\dfrac{a}{a} = 1$. This property is true for exponential expressions as well. For example, for $x \neq 0$, $\dfrac{x^4}{x^4} = 1$.

This expression also can be simplified using the rule for dividing exponential expressions with the same base.

$$\frac{x^4}{x^4} = x^{4-4} = x^0$$

Because $\dfrac{x^4}{x^4} = 1$ and $\dfrac{x^4}{x^4} = x^{4-4} = x^0$, the following definition of zero as an exponent is used.

Zero as an Exponent

If $\mathbf{x \neq 0}$, then $\mathbf{x^0 = 1}$. The expression $\mathbf{0^0}$ is not defined.

Take Note

For the example at the right, we indicated that $a \neq 0$. If we try to evaluate $(12a^3)^0$ when $a = 0$, we get $[12(0)^3]^0 = [12(0)]^0 = 0^0$. However, 0^0 is not defined. Therefore we must assume that $a \neq 0$. To avoid stating this for every example or exercise, we will assume that variables do not take on values that result in the expression 0^0.

Simplify: $(12a^3)^0,\ a \neq 0$

Any nonzero expression to the zero power is 1. $\qquad (12a^3)^0 = 1$

Simplify: $-(xy^4)^0,\ x \neq 0,\ y \neq 0$

Any nonzero expression to the zero power is 1. Because the negative sign is outside the parentheses, the answer is -1. $\qquad -(xy^4)^0 = -(1) = -1$

The meaning of a negative exponent can be developed by examining the quotient $\dfrac{x^4}{x^6}$.

Point of Interest

In the 15th century, the expression $\mathbf{12^{2\bar{m}}}$ *was used to mean* $\mathbf{12x^{-2}}$*. The use of* $\bar{\mathbf{m}}$ *reflected an Italian influence. In Italy, m was used for minus and p was used for plus. It was understood that* $\mathbf{2\bar{m}}$ *referred to an unnamed variable. Isaac Newton, in the 17th century, advocated the current use of a negative exponent.*

The expression can be simplified by writing the numerator and denominator in factored form, dividing by the common factors, and then writing the result with an exponent.

$$\frac{x^4}{x^6} = \frac{\overset{1}{\cancel{x}} \cdot \overset{1}{\cancel{x}} \cdot \overset{1}{\cancel{x}} \cdot \overset{1}{\cancel{x}}}{\underset{1}{\cancel{x}} \cdot \underset{1}{\cancel{x}} \cdot \underset{1}{\cancel{x}} \cdot \underset{1}{\cancel{x}} \cdot x \cdot x} = \frac{1}{x^2}$$

Now simplify the same expression by subtracting the exponents of the like bases.

$$\frac{x^4}{x^6} = x^{4-6} = x^{-2}$$

Because $\dfrac{x^4}{x^6} = \dfrac{1}{x^2}$ and $\dfrac{x^4}{x^6} = x^{-2}$, the expressions $\dfrac{1}{x^2}$ and x^{-2} must be equal. The following definition of a negative exponent is used.

Definition of Negative Exponents

If n is a positive integer and $\boldsymbol{x \neq 0}$, then $\boldsymbol{x^{-n} = \dfrac{1}{x^n}}$ and $\boldsymbol{\dfrac{1}{x^{-n}} = x^n}$.

Take Note

Note from the example at the right that 2^{-4} is a *positive* number. A negative exponent does not indicate a negative number.

Evaluate 2^{-4}.

Write the expression with a positive exponent. Then simplify. $\qquad 2^{-4} = \dfrac{1}{2^4} = \dfrac{1}{16}$

Now that negative exponents have been defined, the Rule for Dividing Exponential Expressions can be stated.

Rule for Dividing Exponential Expressions

If m and n are integers and $\boldsymbol{x \neq 0}$, then $\boldsymbol{\dfrac{x^m}{x^n} = x^{m-n}}$.

Example 1 Write $\frac{3^{-3}}{3^2}$ with a positive exponent. Then evaluate.

Solution

$$\frac{3^{-3}}{3^2} = 3^{-3-2}$$ • 3^{-3} and 3^2 have the same base. Subtract the exponents.

$$= 3^{-5}$$

$$= \frac{1}{3^5}$$ • Use the Definition of Negative Exponents to write the expression with a positive exponent.

$$= \frac{1}{243}$$ • Evaluate.

Problem 1 Write $\frac{2^{-2}}{2^3}$ with a positive exponent. Then evaluate.

Solution See page S19.

The rules for simplifying exponential expressions and powers of exponential expressions are true for all integers. These rules are restated here.

Rules for Exponents

If m, n, and p are integers, then

$$x^m \cdot x^n = x^{m+n} \qquad (x^m)^n = x^{mn} \qquad (x^m y^n)^p = x^{mp}y^{np}$$

$$\frac{x^m}{x^n} = x^{m-n}, x \neq 0 \qquad x^{-n} = \frac{1}{x^n}, x \neq 0 \qquad x^0 = 1, x \neq 0$$

An exponential expression is in simplest form when it is written with only positive exponents.

Simplify: $a^{-7}b^3$

Rewrite a^{-7} with a positive exponent. $a^{-7}b^3 = \frac{b^3}{a^7}$

Example 2 Simplify: $\frac{x^{-4}y^6}{xy^2}$

Solution

$$\frac{x^{-4}y^6}{xy^2} = x^{-4-1}y^{6-2}$$ • Divide variables with the same base by subtracting the exponents.

$$= x^{-5}y^4$$

$$= \frac{y^4}{x^5}$$ • Write the expression with only positive exponents.

Problem 2 Simplify: $\frac{b^8}{a^{-5}b^6}$

Solution See page S19.

Simplify: $6d^{-4}$, $d \neq 0$

$$6d^{-4} = 6 \cdot \frac{1}{d^4} = \frac{6}{d^4}$$ • Use the Definition of Negative Exponents to rewrite the expression with a positive exponent.

Take Note

In the example at the bottom of this page, the exponent on d is -4 (negative 4). The d^{-4} is written in the denominator as d^4. The exponent on 6 is 1 (positive 1). The 6 remains in the numerator.

Note that we indicated $d \neq 0$. This is necessary because division by zero is not defined. In this textbook, we will assume that values of the variables are chosen so that division by zero does not occur.

Example 3 Simplify: **A.** $\dfrac{-35a^6b^{-2}}{25a^{-2}b^5}$ **B.** $(-2x)(3x^{-2})^{-3}$

Solution **A.** $\dfrac{-35a^6b^{-2}}{25a^{-2}b^5} = -\dfrac{35a^6b^{-2}}{25a^{-2}b^5}$ • A negative sign is placed in front of a fraction.

$= -\dfrac{\overset{1}{\cancel{5}} \cdot 7a^{6-(-2)}b^{-2-5}}{\underset{1}{\cancel{5}} \cdot 5}$ • Factor the coefficients. Divide by the common factors. Divide variables with the same base by subtracting the exponents.

$= -\dfrac{7a^8b^{-7}}{5}$

$= -\dfrac{7a^8}{5b^7}$ • Write the expression with only positive exponents.

B. $(-2x)(3x^{-2})^{-3} = (-2x)(3^{-3}x^6)$ • Use the Rule for Simplifying Powers of Products.

$= \dfrac{-2x \cdot x^6}{3^3}$ • Write the expression with positive exponents.

$= -\dfrac{2x^7}{27}$ • Use the Rule for Multiplying Exponential Expressions, and simplify the numerical exponential expression.

Problem 3 Simplify: **A.** $\dfrac{12x^{-8}y^4}{-16xy^{-3}}$ **B.** $(-3ab)(2a^3b^{-2})^{-3}$

Solution See page S19.

2 Scientific notation

Very large and very small numbers are encountered in the fields of science and engineering. For example, the charge of an electron is 0.000000000000000000160 coulomb. These numbers can be written more easily in scientific notation. In **scientific notation,** a number is expressed as a product of two factors, one a number between 1 and 10 and the other a power of 10.

To change a number written in decimal notation to scientific notation, write it in the form $a \times 10^n$, where a is a number between 1 and 10 and n is an integer.

For numbers greater than 10, move the decimal point to the right of the first digit. The exponent n is positive and equal to the number of places the decimal point has been moved.

$240{,}000 = 2.4 \times 10^5$

$93{,}000{,}000 = 9.3 \times 10^7$

For numbers less than 1, move the decimal point to the right of the first nonzero digit. The exponent n is negative. The absolute value of the exponent is equal to the number of places the decimal point has been moved.

$0.00030 = 3.0 \times 10^{-4}$

$0.0000832 = 8.32 \times 10^{-5}$

Take Note

There are two steps to writing a number in scientific notation: (1) determine the number between 1 and 10, and (2) determine the exponent on 10.

Look at the last example above: $0.0000832 = 8.32 \times 10^{-5}$. Using the Definition of Negative Exponents,

$$10^{-5} = \frac{1}{10^5} = \frac{1}{100{,}000} = 0.00001$$

Because $10^{-5} = 0.00001$, we can write

$$8.32 \times 10^{-5} = 8.32 \times 0.00001 = 0.0000832$$

which is the number we started with. We have not changed the value of the number; we have just written it in another form.

Example 4 Write the number in scientific notation.

A. 824,300,000,000 **B.** 0.000000961

Solution **A.** $824{,}300{,}000{,}000 = 8.243 \times 10^{11}$

- **Move the decimal point 11 places to the left. The exponent on 10 is 11.**

B. $0.000000961 = 9.61 \times 10^{-7}$

- **Move the decimal point 7 places to the right. The exponent on 10 is −7.**

Problem 4 Write the number in scientific notation.

A. 57,000,000,000 **B.** 0.000000017

Solution See page S19.

Changing a number written in scientific notation to decimal notation also requires moving the decimal point.

When the exponent on 10 is positive, move the decimal point to the right the same number of places as the exponent.

$$3.45 \times 10^9 = 3{,}450{,}000{,}000$$

$$2.3 \times 10^8 = 230{,}000{,}000$$

When the exponent on 10 is negative, move the decimal point to the left the same number of places as the absolute value of the exponent.

$$8.1 \times 10^{-3} = 0.0081$$

$$6.34 \times 10^{-6} = 0.00000634$$

Example 5 Write the number in decimal notation.

A. 7.329×10^6 **B.** 6.8×10^{-10}

Solution **A.** $7.329 \times 10^6 = 7{,}329{,}000$

- **The exponent on 10 is positive. Move the decimal point 6 places to the right.**

B. $6.8 \times 10^{-10} = 0.00000000068$

- **The exponent on 10 is negative. Move the decimal point 10 places to the left.**

Problem 5 Write the number in decimal notation.

A. 5×10^{12} **B.** 4.0162×10^{-9}

Solution See page S19.

Numerical calculations involving numbers that have more digits than a handheld calculator is able to handle can be performed using scientific notation.

Example 6 Simplify: $\dfrac{2{,}400{,}000{,}000 \times 0.0000063}{0.00009 \times 480}$

Solution

$$\frac{2{,}400{,}000{,}000 \times 0.0000063}{0.00009 \times 480}$$

$$= \frac{2.4 \times 10^{9} \times 6.3 \times 10^{-6}}{9 \times 10^{-5} \times 4.8 \times 10^{2}}$$ • Write the numbers in scientific notation.

$$= \frac{(2.4)(6.3) \times 10^{9+(-6)-(-5)-2}}{(9)(4.8)}$$ • Simplify.

$$= 0.35 \times 10^{6}$$

$$= 3.5 \times 10^{5}$$ • Write in scientific notation.

Problem 6 Simplify: $\dfrac{5{,}600{,}000 \times 0.000000081}{900 \times 0.000000028}$

Solution See page S19.

3 Application problems

Example 7 How many miles does light travel in 1 day? The speed of light is 186,000 mi/s. Write the answer in scientific notation.

Strategy To find the distance traveled:

- Write the speed of light in scientific notation.
- Write the number of seconds in 1 day in scientific notation.
- Use the equation $d = rt$, where r is the speed of light and t is the number of seconds in 1 day.

Solution $186{,}000 = 1.86 \times 10^{5}$

$24 \cdot 60 \cdot 60 = 86{,}400 = 8.64 \times 10^{4}$

$$d = rt$$
$$d = (1.86 \times 10^{5})(8.64 \times 10^{4})$$
$$d = 1.86 \times 8.64 \times 10^{9}$$
$$d = 16.0704 \times 10^{9}$$
$$d = 1.60704 \times 10^{10}$$

Light travels 1.60704×10^{10} mi in one day.

Problem 7 A computer can do an arithmetic operation in 1×10^{-7} s. How many arithmetic operations can the computer perform in 1 min? Write the answer in scientific notation.

Solution See page S19.

5.4 Concept Review

Determine whether the statement is always true, sometimes true, or never true.

1. The expression $\dfrac{x^5}{y^3}$ can be simplified by subtracting the exponents.

2. The rules of exponents can be applied to expressions that contain an exponent of zero or contain negative exponents.

3. The expression 3^{-2} represents the reciprocal of 3^2.

4. $5x^0 = 0$

5. The expression 4^{-3} represents a negative number.

6. To be in simplest form, an exponential expression cannot contain any negative exponents.

7. $2x^{-5} = \frac{1}{2x^5}$

8. $x^4y^{-6} = \frac{1}{x^4y^6}$

5.4 Exercises

1 Write with a positive or zero exponent. Then evaluate.

1. 5^{-2}
2. 3^{-3}
3. $\frac{1}{8^{-2}}$
4. $\frac{1}{12^{-1}}$
5. $\frac{3^{-2}}{3}$
6. $\frac{5^{-3}}{5}$
7. $\frac{2^3}{2^3}$
8. $\frac{3^{-2}}{3^{-2}}$
9. Explain how to rewrite a variable that has a negative exponent as an expression with a positive exponent.
10. Explain how to divide two nonzero exponential expressions with the same base. Provide an example.

Simplify.

11. $\frac{y^7}{y^3}$
12. $\frac{z^9}{z^2}$
13. $\frac{a^8}{a^5}$
14. $\frac{c^{12}}{c^5}$
15. $\frac{p^5}{p}$
16. $\frac{w^9}{w}$
17. $\frac{4x^8}{2x^5}$
18. $\frac{12z^7}{4z^3}$
19. $\frac{22k^5}{11k^4}$
20. $\frac{14m^{11}}{7m^{10}}$
21. $\frac{m^9n^7}{m^4n^5}$
22. $\frac{y^5z^6}{yz^3}$
23. $\frac{6r^4}{4r^2}$
24. $\frac{8x^9}{12x^6}$
25. $\frac{-16a^7}{24a^6}$
26. $\frac{-18b^5}{27b^4}$
27. x^{-2}
28. y^{-10}
29. $\frac{1}{a^{-6}}$
30. $\frac{1}{b^{-4}}$

31. $4x^{-7}$ **32.** $-6y^{-1}$ **33.** $\frac{5}{b^{-8}}$ **34.** $\frac{-3}{v^{-3}}$

35. $\frac{1}{3x^{-2}}$ **36.** $\frac{2}{5c^{-6}}$ **37.** $(ab^5)^0$ **38.** $(32x^3y^4)^0$

39. $\frac{y^3}{y^8}$ **40.** $\frac{z^4}{z^6}$ **41.** $\frac{a^5}{a^{11}}$ **42.** $\frac{m}{m^7}$

43. $\frac{4x^2}{12x^5}$ **44.** $\frac{6y^8}{8y^9}$ **45.** $\frac{-12x}{-18x^6}$ **46.** $\frac{-24c^2}{-36c^{11}}$

47. $\frac{x^6y^5}{x^8y}$ **48.** $\frac{a^3b^2}{a^2b^3}$ **49.** $\frac{2m^6n^2}{5m^9n^{10}}$ **50.** $\frac{5r^3t^7}{6r^5t^7}$

51. $\frac{pq^3}{p^4q^4}$ **52.** $\frac{a^4b^5}{a^5b^6}$ **53.** $\frac{3x^4y^5}{6x^4y^8}$ **54.** $\frac{14a^3b^6}{21a^5b^6}$

55. $\frac{14x^4y^6z^2}{16x^3y^9z}$ **56.** $\frac{24a^2b^7c^9}{36a^7b^5c}$ **57.** $\frac{15mn^9p^3}{30m^4n^9p}$

58. $\frac{25x^4y^7z^2}{20x^5y^9z^{11}}$ **59.** $(-2xy^{-2})^3$ **60.** $(-3x^{-1}y^2)^2$

61. $(3x^{-1}y^{-2})^2$ **62.** $(5xy^{-3})^{-2}$ **63.** $(2x^{-1})(x^{-3})$

64. $(-2x^{-5})x^7$ **65.** $(-5a^2)(a^{-5})^2$ **66.** $(2a^{-3})(a^7b^{-1})^3$

67. $(-2ab^{-2})(4a^{-2}b)^{-2}$ **68.** $(3ab^{-2})(2a^{-1}b)^{-3}$ **69.** $(-5x^{-2}y)(-2x^{-2}y^2)$

70. $\frac{a^{-3}b^{-4}}{a^2b^2}$ **71.** $\frac{3x^{-2}y^2}{6xy^2}$ **72.** $\frac{2x^{-2}y}{8xy}$

73. $\frac{3x^{-2}y}{xy}$ **74.** $\frac{2x^{-1}y^4}{x^2y^3}$ **75.** $\frac{2x^{-1}y^{-4}}{4xy^2}$

76. $\frac{12a^2b^3}{-27a^2b^2}$ **77.** $\frac{-16xy^4}{96x^4y^4}$ **78.** $\frac{-8x^2y^4}{44y^2z^5}$

2 **79.** Why might a number be written in scientific notation instead of decimal notation?

80. a. Explain how to write 0.00000076 in scientific notation.
b. Explain how to write 4.3×10^8 in decimal notation.

Determine whether the number is written in scientific notation. If not, explain why not.

81. 39.4×10^3
82. 0.8×10^{-6}
83. $7.1 \times 10^{2.4}$
84. 5.8×10^{-132}

Write the number in scientific notation.

85. 2,370,000
86. 75,000
87. 0.00045
88. 0.000076
89. 309,000
90. 819,000,000
91. 0.000000601
92. 0.00000000096

Write the number in decimal notation.

93. 7.1×10^5
94. 2.3×10^7
95. 4.3×10^{-5}
96. 9.21×10^{-7}
97. 6.71×10^8
98. 5.75×10^9
99. 7.13×10^{-6}
100. 3.54×10^{-8}

101. **Physics** Light travels approximately 16,000,000,000 mi in one day. Write this number in scientific notation.

102. **Chemistry** Avogadro's number is used in chemistry. Its value is approximately 602,300,000,000,000,000,000,000. Write this number in scientific notation.

103. **Geology** The mass of Earth is 5,980,000,000,000,000,000,000,000 kg. Write this number in scientific notation.

104. **Astronomy** A parsec is a distance measurement that is used by astronomers. One parsec is 3,086,000,000,000,000,000 cm. Write this number in scientific notation.

105. **Electricity** The electric charge on an electron is approximately 0.00000000000000000016 coulomb. Write this number in scientific notation.

106. **Light** The length of an infrared light wave is approximately 0.0000037 m. Write this number in scientific notation.

107. **Computers** One unit used to measure the speed of a computer is the picosecond. One picosecond is 0.000000000001 second. Write this number in scientific notation.

108. **Astronomy** One light-year is the distance traveled by light in one year. One light-year is 5,880,000,000,000 mi. Write this number in scientific notation.

Multiply or divide.

109. $(1.9 \times 10^{12})(3.5 \times 10^7)$
110. $(4.2 \times 10^7)(1.8 \times 10^{-5})$

111. $(2.3 \times 10^{-8})(1.4 \times 10^{-6})$

112. $(3 \times 10^{-20})(2.4 \times 10^{9})$

113. $\dfrac{6.12 \times 10^{14}}{1.7 \times 10^{9}}$

114. $\dfrac{6 \times 10^{-8}}{2.5 \times 10^{-2}}$

115. $\dfrac{5.58 \times 10^{-7}}{3.1 \times 10^{11}}$

116. $\dfrac{9.03 \times 10^{6}}{4.3 \times 10^{-5}}$

3 Solve. Write the answer in scientific notation.

117. **Computer Science** A computer can do an arithmetic operation in 5×10^{-7} s. How many arithmetic operations can the computer perform in 1 h?

118. **Computer Science** A computer can do an arithmetic operation in 2×10^{-9} s. How many arithmetic operations can the computer perform in 1 min?

119. **Physics** How many meters does light travel in 8 h? The speed of light is 3×10^{8} m/s.

120. **Physics** How many meters does light travel in 1 day? The speed of light is 3×10^{8} m/s.

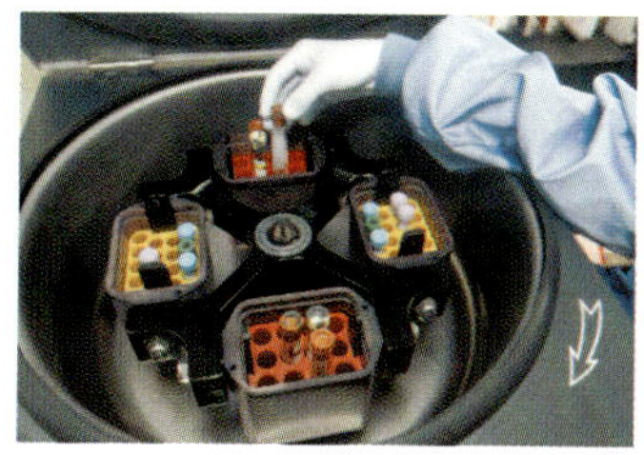

121. **Physics** A high-speed centrifuge makes 4×10^{8} revolutions each minute. Find the time in seconds for the centrifuge to make one revolution.

122. **Physics** The mass of an electron is 9.109×10^{-31} kg. The mass of a proton is 1.673×10^{-27} kg. How many times heavier is a proton than an electron?

123. **Geology** The mass of Earth is 5.9×10^{24} kg. The mass of the sun is 2×10^{30} kg. How many times heavier is the sun than Earth?

124. **Astronomy** It took 11 min for the commands from a computer on Earth to travel to the rover Sojourner on Mars, a distance of 119 million miles. How fast did the signals from Earth to Mars travel? Write the answer in scientific notation.

125. **Astronomy** How long does it take light to travel to Earth from the sun? The sun is 9.3×10^{7} mi from Earth, and light travels 1.86×10^{5} mi/s.

126. **Measurement** The weight of 31 million orchid seeds is 1 oz. Find the weight of one orchid seed.

127. **The Federal Government** In 2000, the gross national debt was approximately 6.6×10^{12} dollars. How much would each American have to pay in order to pay off the debt? Use 3×10^{8} as the number of citizens.

128. **Astronomy** The distance to Saturn is 8.86×10^{8} mi. A satellite leaves Earth traveling at a constant rate of 1×10^{5} mph. How long does it take for the satellite to reach Saturn?

129. **Astronomy** The diameter of Neptune is 3×10^4 mi. Use the formula $SA = 4\pi r^2$ to find the surface area of Neptune in square miles.

130. **Biology** The radius of a cell is 1.5×10^{-4} mm. Use the formula $V = \frac{4}{3}\pi r^3$ to find the volume of the cell.

131. **Chemistry** One gram of hydrogen contains 6.023×10^{23} atoms. Find the weight of one atom of hydrogen.

132. **Astronomy** Our galaxy is estimated to be 5.6×10^{19} mi across. How long (in hours) would it take a space ship to cross the galaxy traveling at 25,000 mph?

5.4 Applying Concepts

Evaluate.

133. $8^{-2} + 2^{-5}$

134. $9^{-2} + 3^{-3}$

135. Evaluate 2^x and 2^{-x} when $x = -2, -1, 0, 1,$ and 2.

136. Evaluate 3^x and 3^{-x} when $x = -2, -1, 0, 1,$ and 2.

Write in decimal notation.

137. 2^{-4}

138. 25^{-2}

Simplify.

139. $\left(\frac{9x^2y^4}{3xy^2}\right) - \left(\frac{12x^5y^6}{6x^4y^4}\right)$

140. $\left(\frac{6x^2 + 9x}{3x}\right) + \left(\frac{8xy^2 + 4y^2}{4y^2}\right)$

141. $\left(\frac{6x^4yz^3}{2x^2y^3}\right)\left(\frac{2x^2z^3}{4y^2z}\right) \div \left(\frac{6x^2y^3}{x^4y^2z}\right)$

142. $\left(\frac{5x^2yz^3}{3x^4yz^3}\right) \div \left(\frac{10x^2y^5z^4}{2y^3z}\right) \div \left(\frac{5y^4z^2}{x^2y^6z}\right)$

Complete.

143. If $m = n$ and $a \neq 0$, then $\frac{a^m}{a^n} =$ _____.

144. If $m = n + 1$ and $a \neq 0$, then $\frac{a^m}{a^n} =$ _____.

Solve.

145. $(-4.8)^x = 1$

146. $-6.3^x = -1$

Determine whether each equation is true or false. If the equation is false, change the right side of the equation to make a true equation.

147. $(2a)^{-3} = \frac{2}{a^3}$

148. $((a^{-1})^{-1})^{-1} = \frac{1}{a}$

149. $(2 + 3)^{-1} = 2^{-1} + 3^{-1}$

150. If $x \neq \frac{1}{3}$, then $(3x - 1)^0 = (1 - 3x)^0$.

151. Why is the condition $x \neq \frac{1}{3}$ given in Exercise 150?

152. If x is a nonzero real number, is x^{-2} always positive, always negative, or positive or negative depending on whether x is positive or negative? Explain your answer.

153. If x is a nonzero real number, is x^{-3} always positive, always negative, or positive or negative depending on whether x is positive or negative? Explain your answer.

5.5 Division of Polynomials

1 Divide a polynomial by a monomial

VIDEO & DVD CD TUTOR WEB SSM

Note that $\frac{8 + 4}{2}$ can be simplified by first adding the terms in the numerator and then dividing the result. It can also be simplified by first dividing each term in the numerator by the denominator and then adding the result.

$$\frac{8 + 4}{2} = \frac{12}{2} = 6$$

$$\frac{8 + 4}{2} = \frac{8}{2} + \frac{4}{2} = 4 + 2 = 6$$

To divide a polynomial by a monomial, divide each term in the numerator by the denominator, and write the sum of the quotients.

$$\frac{a + b}{c} = \frac{a}{c} + \frac{b}{c}$$

Divide: $\frac{6x^2 + 4x}{2x}$

Divide each term of the polynomial $6x^2 + 4x$ by the monomial $2x$. Simplify.

$$\frac{6x^2 + 4x}{2x} = \frac{6x^2}{2x} + \frac{4x}{2x}$$
$$= 3x + 2$$

Example 1 Divide: $\frac{6x^3 - 3x^2 + 9x}{3x}$

Solution

$$\frac{6x^3 - 3x^2 + 9x}{3x} = \frac{6x^3}{3x} - \frac{3x^2}{3x} + \frac{9x}{3x}$$
$$= 2x^2 - x + 3$$

- **Divide each term of the polynomial by the monomial 3*x*.**
- **Simplify each expression.**

Problem 1 Divide: $\frac{4x^3y + 8x^2y^2 - 4xy^3}{2xy}$

Solution See page S19.

Example 2 Divide: $\dfrac{12x^2y - 6xy + 4x^2}{2xy}$

Solution

$$\frac{12x^2y - 6xy + 4x^2}{2xy}$$

$$= \frac{12x^2y}{2xy} - \frac{6xy}{2xy} + \frac{4x^2}{2xy}$$ • Divide each term of the polynomial by the monomial $2xy$.

$$= 6x - 3 + \frac{2x}{y}$$ • Simplify each expression.

Problem 2 Divide: $\dfrac{24x^2y^2 - 18xy + 6y}{6xy}$

Solution See page S19.

2 Divide polynomials

To divide polynomials, use a method similar to that used for division of whole numbers. The same equation used to check division of whole numbers is used to check polynomial division.

(Quotient × Divisor) + Remainder = Dividend

For example:

$$\begin{array}{r} 3 \\ 5\overline{)17} \\ \underline{-15} \\ 2 \end{array}$$

(Quotient × Divisor)	+	Remainder	=	Dividend
(3 × 5)	+	2	=	17

Divide: $(x^2 - 5x + 8) \div (x - 3)$

Step 1

$$\begin{array}{r} x \phantom{{}-5x+8} \\ x - 3\overline{)x^2 - 5x + 8} \\ \underline{x^2 - 3x} \phantom{{}+8} \\ -2x + 8 \end{array}$$

Think: $x\overline{)x^2} = \dfrac{x^2}{x} = x$

Multiply: $x(x - 3) = x^2 - 3x$

Subtract: $(x^2 - 5x) - (x^2 - 3x) = -2x$

Bring down $+ 8$.

Step 2

$$\begin{array}{r} x - 2 \\ x - 3\overline{)x^2 - 5x + 8} \\ \underline{x^2 - 3x} \phantom{{}+8} \\ -2x + 8 \\ \underline{-2x + 6} \\ 2 \end{array}$$

Think: $x\overline{)-2x} = \dfrac{-2x}{x} = -2$

Multiply: $-2(x - 3) = -2x + 6$

Subtract: $(-2x + 8) - (-2x + 6) = 2$

The remainder is 2.

Check:

Quotient × Divisor + Remainder = Dividend

$(x - 2)(x - 3) + 2 = x^2 - 3x - 2x + 6 + 2 = x^2 - 5x + 8$

$$(x^2 - 5x + 8) \div (x - 3) = x - 2 + \frac{2}{x - 3}$$

If a term is missing in the dividend, insert the term with zero as its coefficient. This helps keep like terms in the same column. This is illustrated below.

Divide: $\dfrac{6 - 6x^2 + 4x^3}{2x + 3}$

Arrange the terms in descending order. Note that there is no term of x in $4x^3 - 6x^2 + 6$. Insert a zero as $0x$ for the missing term so that like terms will be in the same columns.

$$
\begin{array}{r}
2x^2 - 6x + 9 \\
2x + 3\overline{)4x^3 - 6x^2 + 0x + 6} \\
\underline{4x^3 + 6x^2} \qquad\qquad \\
-12x^2 + 0x \qquad \\
\underline{-12x^2 - 18x} \qquad \\
18x + 6 \\
\underline{18x + 27} \\
-21
\end{array}
$$

$$\frac{4x^3 - 6x^2 + 6}{2x + 3} = 2x^2 - 6x + 9 - \frac{21}{2x + 3}$$

Example 3 Divide. **A.** $\dfrac{12x^2 - 11x + 10}{4x - 5}$ **B.** $\dfrac{x^3 + 1}{x + 1}$

Solution **A.**

$$
\begin{array}{r}
3x + 1 \\
4x - 5\overline{)12x^2 - 11x + 10} \\
\underline{12x^2 - 15x} \qquad \\
4x + 10 \\
\underline{4x - 5} \\
15
\end{array}
$$

$$\frac{12x^2 - 11x + 10}{4x - 5} = 3x + 1 + \frac{15}{4x - 5}$$

B.

$$
\begin{array}{r}
x^2 - x + 1 \\
x + 1\overline{)x^3 + 0x^2 + 0x + 1} \\
\underline{x^3 + x^2} \qquad\qquad \\
-x^2 + 0x \qquad \\
\underline{-x^2 - x} \qquad \\
x + 1 \\
\underline{x + 1} \\
0
\end{array}
$$

$$\frac{x^3 + 1}{x + 1} = x^2 - x + 1$$

Problem 3 Divide. **A.** $\dfrac{15x^2 + 17x - 20}{3x + 4}$ **B.** $\dfrac{3x^3 + 8x^2 - 6x + 2}{3x - 1}$

Solution See page S19.

3 Synthetic division

VIDEO & DVD CD TUTOR WEB SSM

Synthetic division is a shorter method of dividing a polynomial by a binomial of the form $x - a$. This method of dividing uses only the coefficients of the variable terms.

Both long division and synthetic division are used below to simplify the expression $(3x^2 - 4x + 6) \div (x - 2)$.

LONG DIVISION

Compare the coefficients in this problem worked by long division with the coefficients in the same problem worked by synthetic division.

$$
\begin{array}{r}
3x + 2 \\
x - 2\overline{)3x^2 - 4x + 6} \\
\underline{3x^2 - 6x} \phantom{{}+6} \\
2x + 6 \\
\underline{2x - 4} \\
10
\end{array}
$$

$$(3x^2 - 4x + 6) \div (x - 2) = 3x + 2 + \frac{10}{x - 2}$$

SYNTHETIC DIVISION

$x - a = x - 2;\ a = 2$

Value of a Coefficients of the dividend

$$
\begin{array}{r|rrr}
2 & 3 & -4 & 6 \\
 & & & \\
\hline
 & 3 & &
\end{array}
$$

Bring down the 3.

Multiply $2 \cdot 3$ and add the product (6) to -4.

$$
\begin{array}{r|rrr}
2 & 3 & -4 & 6 \\
 & & 6 & \\
\hline
 & 3 & 2 &
\end{array}
$$

Multiply $2 \cdot 2$ and add the product (4) to 6.

$$
\begin{array}{r|rrr}
2 & 3 & -4 & 6 \\
 & & 6 & 4 \\
\hline
 & 3 & 2 & 10
\end{array}
$$

Coefficients of the quotient   Remainder

The degree of the first term of the quotient is one degree less than the degree of the first term of the dividend.

$$(3x^2 - 4x + 6) \div (x - 2) = 3x + 2 + \frac{10}{x - 2}$$

Check: $(3x + 2)(x - 2) + 10$
$= 3x^2 - 6x + 2x - 4 + 10$
$= 3x^2 - 4x + 6$

Divide: $(2x^3 + 3x^2 - 4x + 8) \div (x + 3)$

Write down the value of a and the coefficients of the dividend.

$x - a = x + 3 = x - (-3);$
$a = -3$

Bring down the 2. Multiply $-3 \cdot 2$ and add the product (-6) to 3. Continue until all the coefficients have been used.

$$
\begin{array}{r|rrrr}
-3 & 2 & 3 & -4 & 8 \\
 & & -6 & 9 & -15 \\
\hline
 & 2 & -3 & 5 & -7
\end{array}
$$

Coefficients of the quotient   Remainder

Write the quotient. The degree of the quotient is one degree less than the degree of the dividend.

$$(2x^3 + 3x^2 - 4x + 8) \div (x + 3) = 2x^2 - 3x + 5 - \frac{7}{x + 3}$$

Example 4 Divide: **A.** $(5x^2 - 3x + 7) \div (x - 1)$
B. $(3x^4 - 8x^2 + 2x + 1) \div (x + 2)$

Solution **A.**

$$\begin{array}{r|rrr} 1 & 5 & -3 & 7 \\ & & 5 & 2 \\ \hline & 5 & 2 & 9 \end{array}$$

• $x - a = x - 1$; $a = 1$

$$(5x^2 - 3x + 7) \div (x - 1) = 5x + 2 + \frac{9}{x - 1}$$

B.

$$\begin{array}{r|rrrrr} -2 & 3 & 0 & -8 & 2 & 1 \\ & & -6 & 12 & -8 & 12 \\ \hline & 3 & -6 & 4 & -6 & 13 \end{array}$$

• Insert a zero for the missing term.
$x - a = x + 2$; $a = -2$

$$(3x^4 - 8x^2 + 2x + 1) \div (x + 2) = 3x^3 - 6x^2 + 4x - 6 + \frac{13}{x + 2}$$

Problem 4 Divide: **A.** $(6x^2 + 8x - 5) \div (x + 2)$
B. $(2x^4 - 3x^3 - 8x^2 - 2) \div (x - 3)$

Solution See page S20.

4 Evaluate a polynomial using synthetic division

VIDEO & DVD CD TUTOR WEB SSM

A polynomial can be evaluated by using synthetic division. Consider the polynomial $P(x) = 2x^4 - 3x^3 + 4x^2 - 5x + 1$. One way to evaluate the polynomial when $x = 2$ is to replace x by 2 and then simplify the numerical expression.

$$\begin{aligned} P(x) &= 2x^4 - 3x^3 + 4x^2 - 5x + 1 \\ P(2) &= 2(2)^4 - 3(2)^3 + 4(2)^2 - 5(2) + 1 \\ P(2) &= 2(16) - 3(8) + 4(4) - 5(2) + 1 \\ P(2) &= 32 - 24 + 16 - 10 + 1 = 15 \end{aligned}$$

Now use synthetic division to divide $(2x^4 - 3x^3 + 4x^2 - 5x + 1) \div (x - 2)$.

$$\begin{array}{r|rrrrr} 2 & 2 & -3 & 4 & -5 & 1 \\ & & 4 & 2 & 12 & 14 \\ \hline & 2 & 1 & 6 & 7 & 15 \end{array}$$

$\underbrace{2 \quad 1 \quad 6 \quad 7}_{\text{Coefficients of the quotient}} \quad \underbrace{15}_{\text{Remainder}}$

Note that the remainder is 15, which is the same value as $P(2)$. This is not a coincidence. The following theorem states that this situation is always true.

Remainder Theorem

If the polynomial $P(x)$ is divided by $x - a$, the remainder is $P(a)$.

Evaluate $P(x) = x^4 - 3x^2 + 4x - 5$ when $x = -2$ by using the Remainder Theorem.

The value at which the polynomial is evaluated

-2	1	0	-3	4	-5	A 0 is inserted for the x^3 term.
		-2	4	-2	-4	
	1	-2	1	2	-9	← The remainder

$P(-2) = -9$

Example 5 Use the Remainder Theorem to evaluate $P(-2)$ when $P(x) = x^3 - 3x^2 + x + 3$.

Solution Use synthetic division with $a = -2$.

-2	1	-3	1	3
		-2	10	-22
	1	-5	11	-19

By the Remainder Theorem, $P(-2) = -19$.

Problem 5 Use the Remainder Theorem to evaluate $P(3)$ when $P(x) = 2x^3 - 4x - 5$.

Solution See page S20.

5.5 Concept Review

Determine whether the statement is always true, sometimes true, or never true.

1. For $c \neq 0$, $\dfrac{a-b}{c} = \dfrac{a}{c} - \dfrac{b}{c}$.

2. For $x \neq 0$, $\dfrac{9x^2 + 6x}{3x} = 3x + 6x$.

3. Given $(6x^2 + x - 6) \div (3x + 2) = 2x - 1 - \dfrac{4}{3x+2}$, then $(3x + 2)(2x - 1) - \dfrac{4}{3x+2} = 6x^2 + x - 6$.

5.5 Exercises

1 Divide.

1. $\dfrac{2x + 2}{2}$

2. $\dfrac{5y + 5}{5}$

3. $\dfrac{10a - 25}{5}$

4. $\dfrac{16b - 40}{8}$

5. $\dfrac{3a^2 + 2a}{a}$

6. $\dfrac{6y^2 + 4y}{y}$

7. $\dfrac{4b^3 - 3b}{b}$

8. $\dfrac{12x^2 - 7x}{x}$

9. $\dfrac{3x^2 - 6x}{3x}$

10. $\dfrac{10y^2 - 6y}{2y}$

11. $\dfrac{5x^2 - 10x}{-5x}$

12. $\dfrac{3y^2 - 27y}{-3y}$

13. $\dfrac{x^3 + 3x^2 - 5x}{x}$

14. $\dfrac{a^3 - 5a^2 + 7a}{a}$

15. $\dfrac{x^6 - 3x^4 - x^2}{x^2}$

16. $\dfrac{a^8 - 5a^5 - 3a^3}{a^2}$

17. $\dfrac{5x^2y^2 + 10xy}{5xy}$

18. $\dfrac{8x^2y^2 - 24xy}{8xy}$

19. $\dfrac{9y^6 - 15y^3}{-3y^3}$

20. $\dfrac{4x^4 - 6x^2}{-2x^2}$

21. $\dfrac{3x^2 - 2x + 1}{x}$

22. $\dfrac{8y^2 + 2y - 3}{y}$

23. $\dfrac{-3x^2 + 7x - 6}{x}$

24. $\dfrac{2y^2 - 6y + 9}{y}$

25. $\dfrac{16a^2b - 20ab + 24ab^2}{4ab}$

26. $\dfrac{22a^2b + 11ab - 33ab^2}{11ab}$

27. $\dfrac{9x^2y + 6xy - 3xy^2}{xy}$

2 28. Explain how you can check the result of dividing two polynomials.

Divide by using long division.

29. $(x^2 + 3x - 40) \div (x - 5)$

30. $(x^2 - 14x + 24) \div (x - 2)$

31. $(x^3 - 3x^2 + 2) \div (x - 3)$

32. $(x^3 + 4x^2 - 8) \div (x + 4)$

33. $(6x^2 + 13x + 8) \div (2x + 1)$

34. $(12x^2 + 13x - 14) \div (3x - 2)$

35. $(10x^2 + 9x - 5) \div (2x - 1)$

36. $(18x^2 - 3x + 2) \div (3x + 2)$

37. $(8x^3 - 9) \div (2x - 3)$

38. $(64x^3 + 4) \div (4x + 2)$

39. $(6x^4 - 13x^2 - 4) \div (2x^2 - 5)$

40. $(12x^4 - 11x^2 + 10) \div (3x^2 + 1)$

41. $\dfrac{3x^3 - 8x^2 - 33x - 10}{3x + 1}$

42. $\dfrac{8x^3 - 38x^2 + 49x - 10}{4x - 1}$

43. $\dfrac{4 - 7x + 5x^2 - x^3}{x - 3}$

44. $\dfrac{4 + 6x - 3x^2 + 2x^3}{2x + 1}$

45. $\dfrac{16x^2 - 13x^3 + 2x^4 - 9x + 20}{x - 5}$

46. $\dfrac{x + 3x^4 - x^2 + 5x^3 - 2}{x + 2}$

47. $\dfrac{x^3 - 4x^2 + 2x - 1}{x^2 + 1}$

48. $\dfrac{3x^3 - 2x^2 - 8}{x^2 + 5}$

49. $\dfrac{2x^3 - x + 4 - 3x^2}{x^2 - 1}$

50. $\dfrac{2 - 3x^2 + 5x^3}{x^2 + 3}$

51. $\dfrac{6x^3 + 2x^2 + x + 4}{2x^2 - 3}$

52. $\dfrac{9x^3 + 6x^2 + 2x + 1}{3x^2 + 2}$

3 Divide by using synthetic division.

53. $(2x^2 - 6x - 8) \div (x + 1)$

54. $(3x^2 + 19x + 20) \div (x + 5)$

55. $(3x^2 - 4) \div (x - 1)$

56. $(4x^2 - 8) \div (x - 2)$

57. $(x^2 - 9) \div (x + 4)$

58. $(x^2 - 49) \div (x + 5)$

59. $(2x^2 + 24) \div (2x + 4)$

60. $(3x^2 - 15) \div (x + 3)$

61. $(2x^3 - x^2 + 6x + 9) \div (x + 1)$

62. $(3x^3 + 10x^2 + 6x - 4) \div (x + 2)$

63. $(x^3 - 6x^2 + 11x - 6) \div (x - 3)$

64. $(x^3 - 4x^2 + x + 6) \div (x + 1)$

65. $(6x - 3x^2 + x^3 - 9) \div (x + 2)$

66. $(5 - 5x + 4x^2 + x^3) \div (x - 3)$

67. $(x^3 + x - 2) \div (x + 1)$

68. $(x^3 + 2x + 5) \div (x - 2)$

69. $(18 + x - 4x^3) \div (2 - x)$

70. $(12 - 3x^2 + x^3) \div (3 + x)$

71. $\dfrac{16x^2 - 13x^3 + 2x^4 - 9x + 20}{x - 5}$

72. $\dfrac{2x^3 - x^2 - 10x + 15 + x^4}{x - 2}$

73. $\dfrac{5 + 5x - 8x^2 + 4x^3 - 3x^4}{2 - x}$

74. $\dfrac{3 - 13x - 5x^2 + 9x^3 - 2x^4}{3 - x}$

75. $\dfrac{3x^4 + 3x^3 - x^2 + 3x + 2}{x + 1}$

76. $\dfrac{4x^4 + 12x^3 - x^2 - x + 2}{x + 3}$

77. $\dfrac{2x^4 - x^2 + 2}{x - 3}$

78. $\dfrac{x^4 - 3x^3 - 30}{x + 2}$

79. $\dfrac{x^3 + 125}{x + 5}$

80. $\dfrac{x^3 + 343}{x + 7}$

4 Use the Remainder Theorem to evaluate the polynomial.

81. $P(x) = 2x^2 - 3x - 1$; $P(3)$

82. $Q(x) = 3x^2 - 5x - 1$; $Q(2)$

83. $R(x) = x^3 - 2x^2 + 3x - 1$; $R(4)$

84. $F(x) = x^3 + 4x^2 - 3x + 2$; $F(3)$

85. $P(z) = 2z^3 - 4z^2 + 3z - 1$; $P(-2)$

86. $R(t) = 3t^3 + t^2 - 4t + 2$; $R(-3)$

87. $Q(x) = x^4 + 3x^3 - 2x^2 + 4x - 9$; $Q(2)$

88. $Y(z) = z^4 - 2z^3 - 3z^2 - z + 7$; $Y(3)$

89. $F(x) = 2x^4 - x^3 - 2x - 5$; $F(-3)$

90. $Q(x) = x^4 - 2x^3 + 4x - 2$; $Q(-2)$

91. $P(x) = x^3 - 3$; $P(5)$

92. $S(t) = 4t^3 + 5$; $S(-4)$

93. $R(t) = 4t^4 - 3t^2 + 5$; $R(-3)$

94. $P(z) = 2z^4 + z^2 - 3$; $P(-4)$

95. $Q(x) = x^5 - 4x^3 - 2x^2 + 5x - 2$; $Q(2)$

96. $T(x) = 2x^5 + 4x^4 - x^2 + 4$; $T(3)$

5.5 Applying Concepts

For what value of k will the remainder be zero?

97. $(x^3 - 3x^2 - x + k) \div (x - 3)$

98. $(x^3 - 2x^2 + x + k) \div (x - 2)$

99. $(x^2 + kx - 6) \div (x - 3)$

100. $(x^3 + kx + k - 1) \div (x - 1)$

In Exercises 101 and 102, replace the question marks with expressions to make a true statement.

101. If $\dfrac{x^2 - x - 6}{x - 3} = x + 2$, then $x^2 - x - 6 = (?)(?)$.

102. If $\dfrac{x^2 + 2x - 3}{x - 2} = x + 4 + \dfrac{5}{x - 2}$, then $x^2 + 2x - 3 = (?)(?) + ?$.

Solve.

103. The product of a monomial and $4b$ is $12a^2b$. Find the monomial.

104. The product of a monomial and $6x$ is $24xy^2$. Find the monomial.

105. The quotient of a polynomial and $2x + 1$ is $2x - 4 + \dfrac{7}{2x + 1}$. Find the polynomial.

106. The quotient of a polynomial and $x - 3$ is $x^2 - x + 8 + \dfrac{22}{x - 3}$. Find the polynomial.

Focus on Problem Solving

Dimensional Analysis

In solving application problems, it may be useful to include units in order to organize the problem so that the answer is in the proper units. Using units to organize and check the correctness of an application is called **dimensional analysis**. We use the operations of multiplying units and dividing units in applying dimensional analysis to application problems.

The Rule for Multiplying Exponential Expressions states that we multiply two expressions with the same base by adding the exponents.

$$x^4 \cdot x^6 = x^{4+6} = x^{10}$$

In calculations that involve quantities, the units are operated on algebraically.

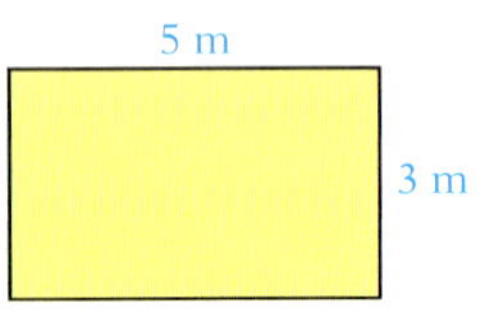

A rectangle measures 3 m by 5 m. Find the area of the rectangle.

$$A = LW = (3 \text{ m})(5 \text{ m}) = (3 \cdot 5)(\text{m} \cdot \text{m}) = 15 \text{ m}^2$$

The area of the rectangle is 15 m^2 (square meters).

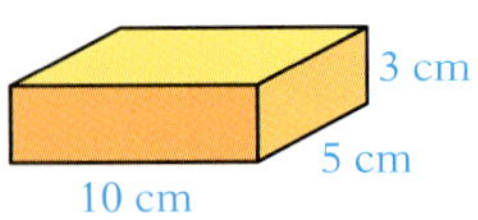

A box measures 10 cm by 5 cm by 3 cm. Find the volume of the box.

$$V = LWH = (10 \text{ cm})(5 \text{ cm})(3 \text{ cm}) = (10 \cdot 5 \cdot 3)(\text{cm} \cdot \text{cm} \cdot \text{cm}) = 150 \text{ cm}^3$$

The volume of the box is 150 cm^3 (cubic centimeters).

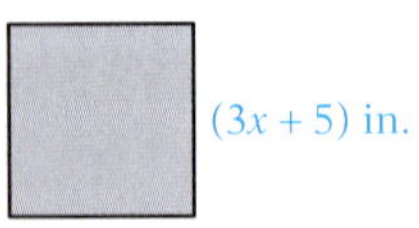

Find the area of a square whose side measures $(3x + 5)$ in.

$$A = s^2 = [(3x + 5) \text{ in.}]^2 = (3x + 5)^2 \text{ in}^2 = (9x^2 + 30x + 25) \text{ in}^2$$

The area of the square is $(9x^2 + 30x + 25)$ in^2 (square inches).

Dimensional analysis is used in the conversion of units.

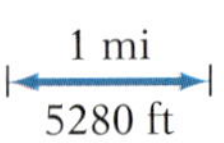

The following example converts the unit miles to feet. The equivalent measures 1 mi and 5280 ft are used to form the following rates, which are called conversion factors: $\dfrac{1 \text{ mi}}{5280 \text{ ft}}$ and $\dfrac{5280 \text{ ft}}{1 \text{ mi}}$. Because 1 mi = 5280 ft, both of the conversion factors $\dfrac{1 \text{ mi}}{5280 \text{ ft}}$ and $\dfrac{5280 \text{ ft}}{1 \text{ mi}}$ are equal to 1.

To convert 3 mi to feet, multiply 3 mi by the conversion factor $\frac{5280 \text{ ft}}{1 \text{ mi}}$.

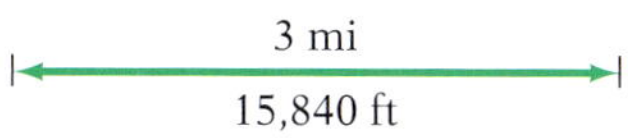

$$3 \text{ mi} = 3 \text{ mi} \cdot 1 = \frac{3 \text{ mi}}{1} \cdot \frac{5280 \text{ ft}}{1 \text{ mi}} = \frac{3 \cancel{\text{mi}} \cdot 5280 \text{ ft}}{1 \cancel{\text{mi}}} = 3 \cdot 5280 \text{ ft} = 15{,}840 \text{ ft}$$

There are two important points in the above illustration. First, you can think of dividing the numerator and denominator by the common unit "mile" just as you would divide the numerator and denominator of a fraction by a common factor. Second, the conversion factor $\frac{5280 \text{ ft}}{1 \text{ mi}}$ is equal to 1, and multiplying an expression by 1 does not change the value of the expression.

In the application problem that follows, the units are kept in the problem while the problem is worked.

In 2000, a horse named Fusaichi Pegasus ran a 1.25-mile race in 2.02 min. Find Fusaichi Pegasus's average speed for the race in miles per hour. Round to the nearest tenth.

Strategy To find the average speed, use the formula $r = \frac{d}{t}$, where r is the speed, d is the distance, and t is the time. Use the conversion factor $\frac{60 \text{ min}}{1 \text{ h}}$.

Solution

$$r = \frac{d}{t} = \frac{1.25 \text{ mi}}{2.02 \text{ min}} = \frac{1.25 \text{ mi}}{2.02 \text{ min}} \cdot \frac{60 \text{ min}}{1 \text{ h}}$$

$$= \frac{75 \text{ mi}}{2.02 \text{ h}} \approx 37.1 \text{ mph}$$

Fusaichi Pegasus's average speed was 37.1 mph.

Try each of the following problems. Round to the nearest tenth or to the nearest cent.

1. Convert 88 ft/s to miles per hour.

2. Convert 8 m/s to kilometers per hour (1 km = 1000 m).

3. A carpet is to be placed in a meeting hall that is 36 ft wide and 80 ft long. At $21.50 per square yard, how much will it cost to carpet the meeting hall?

4. A carpet is to be placed in a room that is 20 ft wide and 30 ft long. At $22.25 per square yard, how much will it cost to carpet the area?

5. Find the number of gallons of water in a fish tank that is 36 in. long and 24 in. wide and is filled to a depth of 16 in. (1 gal = 231 in^3).

6. Find the number of gallons of water in a fish tank that is 24 in. long and 18 in. wide and is filled to a depth of 12 in. (1 gal = 231 in^3).

7. A $\frac{1}{4}$-acre commercial lot is on sale for $2.15 per square foot. Find the sale price of the commercial lot (1 acre = 43,560 ft^2).

8. A 0.75-acre industrial parcel was sold for $98,010. Find the parcel's price per square foot (1 acre = 43,560 ft^2).

9. A new driveway will require 800 ft^3 of concrete. Concrete is ordered by the cubic yard. How much concrete should be ordered?

10. A piston-engined dragster traveled 1320 ft in 4.477 s at Joliet, Illinois, on June 2, 2001. Find the average speed of the dragster in miles per hour.

11. The Marianas Trench in the Pacific Ocean is the deepest part of the ocean. Its depth is 6.85 mi. The speed of sound under water is 4700 ft/s. Find the time it takes sound to travel from the surface to the bottom of the Marianas Trench and back.

Projects & Group Activities

Population and Land Allocation*

Rhode Island

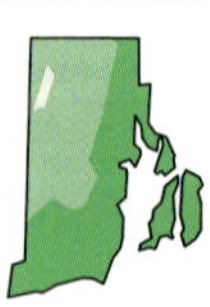

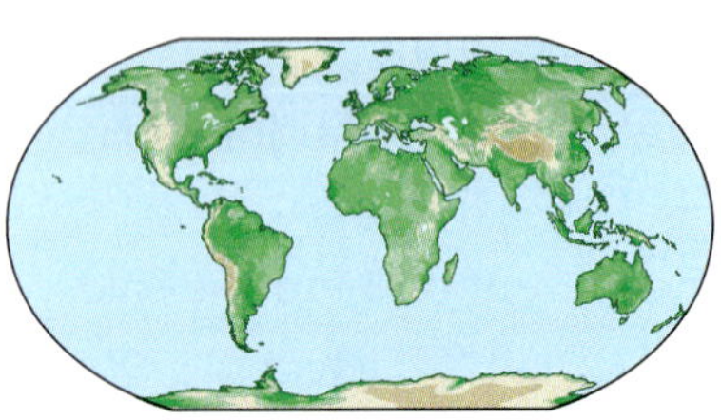

In this project, you are asked to determine hypothetical land allocation for the world's population today. Use the figure 6×10^9 for the current world population and the figure 2.9×10^8 for the current U.S. population. One square mile is approximately 2.8×10^7 ft^2.

1. If every person in the world moved to Texas and each person were given an equal amount of land, how many square feet of land would each person have? The area of Texas is 2.619×10^5 mi^2.

2. If every person in the United States moved to Rhode Island and each person were given an equal amount of land, how many square feet of land would each person have? The area of Rhode Island is 1.0×10^3 mi^2. Round to the nearest whole number.

3. Suppose every person in the world were given a plot of land the size of a two-car garage (22 ft $\times$ 22 ft).
   a. How many people would fit in a square mile? Round to the nearest hundred.
   b. How many square miles would be required to accommodate the entire world population? Round to the nearest hundred.

4. If the total land area of Earth were divided equally, how many acres of land would each person be allocated? Use a figure of 5.7×10^7 mi^2 for the land area of Earth. One acre is 43,560 ft^2. Round to the nearest tenth.

5. If every person on Earth were given a plot of land the size of a two-car garage, what would be the carrying capacity of Earth? Round to the nearest hundred billion.

Source:* **www.infoplease.com

Astronomical Distances and Scientific Notation

Astronomers have units of measurement that are useful for measuring vast distances in space. Two of these units are the astronomical unit and the light-year. An **astronomical unit** is the average distance between Earth and the sun. A **light-year** is the distance a ray of light travels in 1 year.

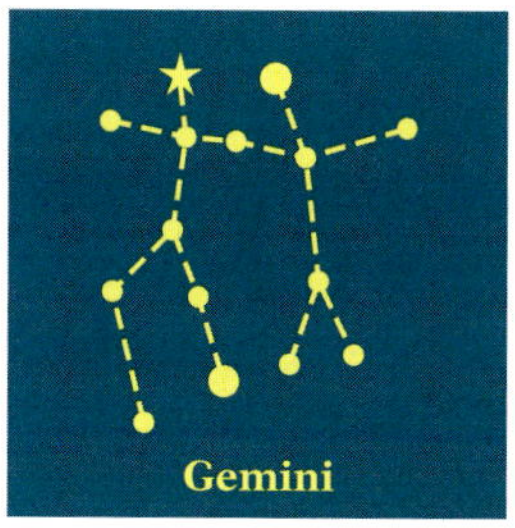

1. Light travels at a speed of 1.86×10^5 mi/s. Find the measure of 1 light-year in miles. Use a 365-day year.
2. The distance between Earth and the star Alpha Centauri is approximately 25 trillion miles. Find the distance between Earth and Alpha Centauri in light-years. Round to the nearest hundredth.
3. The Coma cluster of galaxies is approximately 2.8×10^8 light-years from Earth. Find the distance, in miles, from the Coma Cluster to Earth. Write the answer in scientific notation.
4. One astronomical unit (A.U.) is 9.3×10^7 mi. The star Pollux in the constellation Gemini is 1.8228×10^{12} mi from Earth. Find the distance from Pollux to Earth in astronomical units.
5. One light-year is equal to approximately how many astronomical units? Round to the nearest thousand.

Shown below are data on the planets in our solar system. The planets are listed in alphabetical order.

Point of Interest

In November 2001, the Hubble Space Telescope took photos of the atmosphere of a planet orbiting a star 150 light-years from Earth in the constellation Pegasus. The planet is about the size of Jupiter and orbits close to the star HD209458. It was the first discovery of an atmosphere around a planet outside our solar system.

Planet	*Distance from the Sun (in kilometers)*	*Mass (in kilograms)*
Earth	1.50×10^8	5.97×10^{24}
Jupiter	7.79×10^8	1.90×10^{27}
Mars	2.28×10^8	6.42×10^{23}
Mercury	5.79×10^7	3.30×10^{23}
Neptune	4.50×10^9	1.02×10^{26}
Pluto	5.87×10^9	1.25×10^{22}
Saturn	1.43×10^9	5.68×10^{26}
Uranus	2.87×10^9	8.68×10^{25}
Venus	1.08×10^8	4.87×10^{24}

6. Arrange the planets in order from closest to the sun to farthest from the sun.
7. Arrange the planets in order from the one with the least mass to the one with the greatest mass.
8. Write a rule for ordering numbers written in scientific notation.

Chapter Summary

Key Words

A **monomial** is a number, a variable, or a product of numbers and variables. [5.1.1, p. 342]

Examples: 5 is a number. y is a variable. $8a^2b^2$ is a product of numbers and variables. 5, y, and $8a^2b^2$ are monomials.

The **degree of a monomial** is the sum of the exponents of the variables. [5.1.1, p. 342]

The degree of $3x^4y^2 - 5x + 1$ is 7.

A **polynomial** is a variable expression in which the terms are monomials. A polynomial of one term is a **monomial**. A polynomial of two terms is a **binomial**. A polynomial of three terms is a **trinomial**. [5.1.1, p. 343]

5, y, and $8a^2b^2$ are monomials. $x + 9$, $y^2 - 3$, and $6a + 7b$ are binomials. $x^2 + 2x - 1$ is a trinomial.

The **degree of a polynomial** is the greatest of the degrees of any of its terms. [5.1.1, p. 343]

The degree of the polynomials $x^3 + 3x^2y^2 - 4xy - 3$ is 4.

The terms of a polynomial in one variable are usually arranged so that the exponents of the variable decrease from left to right. This is called **descending order**. [5.1.1, p. 343]

$7y^4 + 5y^3 - y^2 + 6y - 8$ is written in descending order.

A **polynomial** function is an expression whose terms are monomials. The **leading coefficient** of a polynomial function is the coefficient of the variable with the largest exponent. The **constant term** is the term without a variable. [5.1.1, pp. 343, 344]

For the polynomial function $P(x) = 6x^3 - 5x^2 + 4x - 7$, the leading coefficient is 6 and the constant term is -7.

Synthetic division is a shorter method of dividing a polynomial by a binomial of the form $x - a$. This method uses only the coefficients of the variable terms. [5.5.3, p. 379]

Essential Rules and Procedures

To add polynomials, combine like terms. [5.1.2, p. 347]

$$(8x^2 + 2x - 9) + (-3x^2 + 5x - 7)$$
$$= (8x^2 - 3x^2) + (2x + 5x) + (-9 - 7)$$
$$= 5x^2 + 7x - 16$$

To subtract two polynomials, add the additive inverse of the second polynomial to the first polynomial. [5.1.2, p. 347]

$$(3y^2 - 8y + 6) - (-y^2 + 4y - 5)$$
$$= (3y^2 - 8y + 6) + (y^2 - 4y + 5)$$
$$= 4y^2 - 12y + 11$$

Rule for Multiplying Exponential Expressions [5.2.1, p. 353]
$x^m \cdot x^n = x^{m+n}$

$b^5 \cdot b^4 = b^{5+4} = b^9$

Rule for Simplifying Powers of Exponential Expressions [5.2.2, p. 354]
$(x^m)^n = x^{mn}$

$(y^3)^7 = y^{3(7)} = y^{21}$

Rule for Simplifying Powers of Products [5.2.2, p. 354]
$(x^m y^n)^p = x^{mp} y^{np}$

$(x^6 y^4 z^5)^2 = x^{6(2)} y^{4(2)} z^{5(2)} = x^{12} y^8 z^{10}$

The FOIL Method [5.3.3, p. 358]
To find the product of two binomials, add the products of the **F**irst terms, the **O**uter terms, the **I**nner terms, and the **L**ast terms.

$$\begin{aligned}(4x + 3)(2x - 5)\\ &= (4x)(2x) + (4x)(-5) + (3)(2x) + (3)(-5)\\ &= 8x^2 - 20x + 6x - 15\\ &= 8x^2 - 14x - 15\end{aligned}$$

The Sum and Difference of Two Terms [5.3.4, p. 359]
$(a + b)(a - b) = a^2 - b^2$

$$\begin{aligned}(5x + 3)(5x - 3) &= (5x)^2 - 3^2\\ &= 25x^2 - 9\end{aligned}$$

The Square of a Binomial [5.3.4, p. 360]
$(a + b)^2 = a^2 + 2ab + b^2$
$(a - b)^2 = a^2 - 2ab + b^2$

$$\begin{aligned}(6x - 5)^2 &= (6x)^2 + 2(6x)(-5) + (-5)^2\\ &= 36x^2 - 60x + 25\end{aligned}$$

Zero as an Exponent [5.4.1, p. 366]
For $x \neq 0$, $x^0 = 1$.

$17^0 = 1$ $\quad (5y)^0 = 1, y \neq 0$

Definition of Negative Exponents [5.4.1, p. 367]
For $x \neq 0$, $x^{-n} = \dfrac{1}{x^n}$ and $\dfrac{1}{x^{-n}} = x^n$.

$x^{-6} = \dfrac{1}{x^6}$ and $\dfrac{1}{x^{-6}} = x^6$

Rule for Dividing Exponential Expressions [5.4.1, p. 367]
For $x \neq 0$, $\dfrac{x^m}{x^n} = x^{m-n}$.

$\dfrac{y^8}{y^3} = y^{8-3} = y^5$

Scientific Notation [5.4.2, p. 369]
To express a number in scientific notation, write it in the form $a \times 10^n$, where a is a number between 1 and 10 and n is an integer. If the number is greater than 10, the exponent on 10 will be positive. If the number is less than 1, the exponent on 10 will be negative.

$367{,}000{,}000 = 3.67 \times 10^8$

$0.0000059 = 5.9 \times 10^{-6}$

To change a number written in scientific notation to decimal notation, move the decimal point to the right if the exponent on 10 is positive and to the left if the exponent on 10 is negative. Move the decimal point the same number of places as the absolute value of the exponent on 10.

$2.418 \times 10^7 = 24{,}180{,}000$

$9.06 \times 10^{-5} = 0.0000906$

Chapter Review Exercises

1. Add: $(3x^2 - 2x - 6) + (-x^2 - 3x + 4)$

2. Subtract: $(5x^2 - 8xy + 2y^2) - (x^2 - 3y^2)$

3. Multiply: $(5x^2yz^4)(2xy^3z^{-1})(7x^{-2}y^{-2}z^3)$

4. Multiply: $(2x^{-1}y^2z^5)^4(-3x^3yz^{-3})$

5. Simplify: $\dfrac{3x^4yz^{-1}}{-12xy^3z^2}$

6. Simplify: $\dfrac{(2a^4b^{-3}c^2)^3}{(2a^3b^2c^{-1})^4}$

7. Write 93,000,000 in scientific notation.

8. Write 2.54×10^{-3} in decimal notation.

9. Simplify: $\dfrac{3 \times 10^{-3}}{15 \times 10^2}$

10. Given $P(x) = 2x^3 - x + 7$, evaluate $P(-2)$.

11. Graph: $y = x^2 + 1$

12. Identify **a.** the leading coefficient, **b.** the constant term, and **c.** the degree of the polynomial $P(x) = 3x^5 - 3x^2 + 7x + 8$.

13. Use the Remainder Theorem to evaluate $P(x) = -2x^3 + 2x^2 - 4$ when $x = -3$.

14. Divide: $\dfrac{15x^2 + 2x - 2}{3x - 2}$

15. Divide: $\dfrac{12x^2 - 16x - 7}{6x + 1}$

16. Divide: $\dfrac{4x^3 + 27x^2 + 10x + 2}{x + 6}$

17. Divide: $\dfrac{x^4 - 4}{x - 4}$

18. Multiply: $4x^2y(3x^3y^2 + 2xy - 7y^3)$

19. Multiply: $a^{2n+3}(a^n - 5a + 2)$

20. Simplify: $5x^2 - 4x[x - (3x + 2) + x]$

21. Multiply: $(x^{2n} - x)(x^{n+1} - 3)$

22. Multiply: $(x + 6)(x^3 - 3x^2 - 5x + 1)$

23. Multiply: $(x - 4)(3x + 2)(2x - 3)$

24. Multiply: $(5a + 2b)(5a - 2b)$

25. Simplify: $(4x - 3y)^2$

26. The most distant object visible from Earth without the aid of a telescope is the Great Galaxy of Andromeda. It takes light from the Great Galaxy of Andromeda 2.2×10^6 years to travel to Earth. Light travels about 5.9×10^{12} mph. How far from Earth is the Great Galaxy of Andromeda?

27. Light from the sun supplies Earth with 2.4×10^{14} horsepower. Earth receives only 2.2×10^{-7} of the power generated by the sun. How much power is generated by the sun?

28. The length of a rectangle is $(5x + 3)$ cm. The width is $(2x - 7)$ cm. Find the area of the rectangle in terms of the variable x.

29. The length of the side of a cube is $(3x - 1)$ ft. Find the volume of the cube in terms of the variable x.

30. Find the area of the figure shown at the right. All dimensions given are in inches.

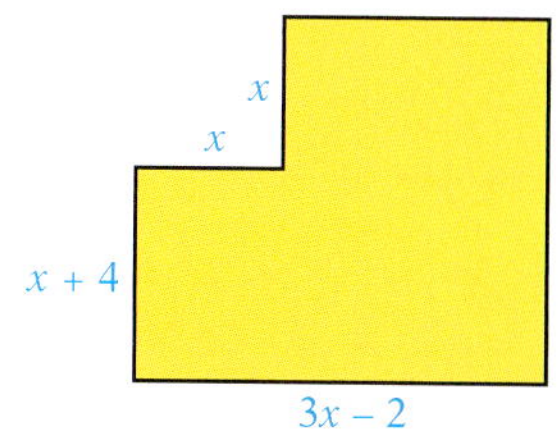

Chapter Test

1. Subtract: $(6x^3 - 7x^2 + 6x - 7) - (4x^3 - 3x^2 + 7)$

2. Simplify: $(-4a^2b)^3(-ab^4)$

3. Simplify: $\dfrac{(2a^{-4}b^2)^3}{4a^{-2}b^{-1}}$

4. Write the number 0.000000501 in scientific notation.

5. Write the number of seconds in 1 week in scientific notation.

6. Simplify: $(2x^{-3}y)^{-4}$

7. Simplify: $-5x[3 - 2(2x - 4) - 3x]$

8. Multiply: $(3a + 4b)(2a - 7b)$

9. Multiply: $(3t^3 - 4t^2 + 1)(2t^2 - 5)$

10. Simplify: $(3z - 5)^2$

11. Divide: $(4x^3 + x - 15) \div (2x - 3)$

12. Divide: $(x^3 - 5x^2 + 5x + 5) \div (x - 3)$

13. Given $P(x) = 3x^2 - 8x + 1$, evaluate $P(2)$.

14. Use the Remainder Theorem to evaluate $P(x) = -x^3 + 4x - 8$ when $x = -2$.

15. Add: $(3x^3 - 2x^2 - 4) + (8x^2 - 8x + 7)$

16. Multiply: $2x(2x^2 - 3x)$

17. Simplify: $\dfrac{12x^2}{-3x^{-4}}$

18. Subtract: $(3a^2 - 2a - 7) - (5a^3 + 2a - 10)$

19. Divide: $\dfrac{16x^5 - 8x^3 + 20x}{4x}$

20. Multiply: $(-2xy^2)(3x^2y^4)$

21. Simplify: $\dfrac{27xy^3}{3x^4y^3}$

22. Multiply: $(2x - 7y)(5x - 4y)$

23. Simplify: $(a^2b^{-3})^2$

24. The length of a rectangle is $(5x + 1)$ ft. The width is $(2x - 1)$ ft. Find the area of the rectangle in terms of the variable x.

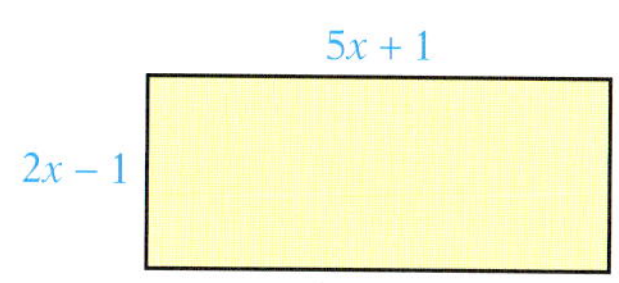

25. A space vehicle travels 2.4×10^5 mi from Earth to the moon at an average velocity of 2×10^4 mph. How long does it take the space vehicle to reach the moon?

Cumulative Review Exercises

1. Simplify: $8 - 2[-3 - (-1)]^2 + 4$

2. Evaluate $\frac{2a - b}{b - c}$ when $a = 4$, $b = -2$, and $c = 6$.

3. Identify the property that justifies the statement $2x + (-2x) = 0$.

4. Simplify: $2x - 4[x - 2(3 - 2x) + 4]$

5. Solve: $\frac{2}{3} - y = \frac{5}{6}$

6. Solve: $8x - 3 - x = -6 + 3x - 8$

7. Simplify: $\frac{x^3 - 3}{x - 3}$

8. Solve: $3 - |2 - 3x| = -2$

9. Given $P(x) = 3x^2 - 2x + 2$, evaluate $P(-2)$.

10. What values of x are excluded from the domain of the function $f(x) = \frac{x + 1}{x + 2}$?

11. Find the range of the function given by $F(x) = 3x^2 - 4$ if the domain is $\{-2, -1, 0, 1, 2\}$.

12. Find the slope of the line that contains the points $(-2, 3)$ and $(4, 2)$.

13. Find the equation of the line that contains the point $(-1, 2)$ and has slope $-\frac{3}{2}$.

14. Find the equation of the line that contains the point $(-2, 4)$ and is perpendicular to the line $3x + 2y = 4$.

15. Solve by using Cramer's Rule:
$$\begin{aligned} 2x - 3y &= 2 \\ x + y &= -3 \end{aligned}$$

16. Solve by the addition method:
$$\begin{aligned} x - y + z &= 0 \\ 2x + y - 3z &= -7 \\ -x + 2y + 2z &= 5 \end{aligned}$$

17. Graph $3x - 4y = 12$ by using the x- and y-intercepts.

18. Graph the solution set of $-3x + 2y < 6$

19. Solve by graphing: $x - 2y = 3$
$-2x + y = -3$

20. Graph the solution set: $2x + y < 3$
$-6x + 3y \geq 4$

21. Simplify: $(4a^{-2}b^3)(2a^3b^{-1})^{-2}$

22. Simplify: $\dfrac{(5x^3y^{-3}z)^{-2}}{y^4z^{-2}}$

23. Simplify: $3 - (3 - 3^{-1})^{-1}$

24. Multiply: $(2x + 3)(2x^2 - 3x + 1)$

25. Divide: $(8x^2 + 4x - 3) \div (2x - 3)$

26. Multiply: $(a^2b^5)(ab^2)$

27. Multiply: $(a - 3b)(a + 4b)$

28. Write 3.5×10^{-8} in decimal notation.

29. The sum of two integers is twenty-four. The difference between four times the smaller integer and nine is three less than twice the larger integer. Find the integers.

30. How many ounces of pure gold that costs \$360 per ounce must be mixed with 80 oz of an alloy that costs \$120 per ounce to make a mixture that costs \$200 per ounce?

31. Two bicycles are 25 mi apart and are traveling toward each other. One cyclist is traveling at $\frac{2}{3}$ the rate of the other cyclist. They pass in 2 h. Find the rate of each cyclist.

32. If \$3000 is invested at an annual simple interest rate of 7.5%, how much additional money must be invested at an annual simple interest rate of 10% so that the total interest earned is 9% of the total investment?

33. The graph below shows the relationship between the distance traveled in miles and the time of travel in hours. Find the slope of the line between the two points on the graph. Write a sentence that states the meaning of the slope.

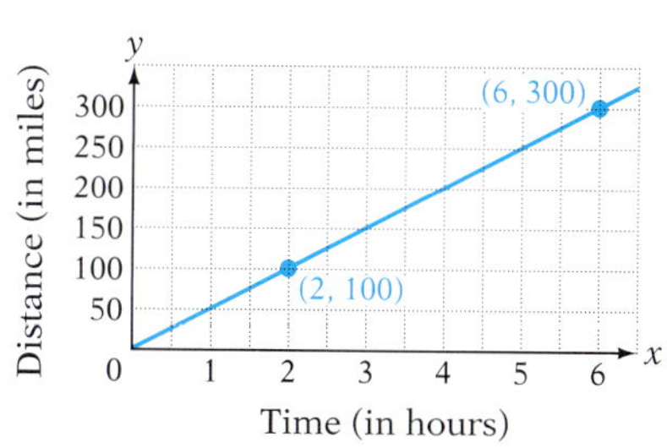

Leonardo DaVinci designed the first parachute in 1485. People have been jumping out of planes ever since. The highest parachute jump ever made was one by U.S. Air Force Captain Joseph W. Kittinger, Jr. in 1960. Kittinger, who was performing high-altitude escape experiments for the Air Force, jumped from 102,800 feet above sea level—over 19 miles up! His free fall lasted more then 4.5 minutes and he reached a velocity of 614 mph. Factors that affect velocity include the initial altitude of the jump, the size of the parachute, and the weight and body position of the diver. The velocity of a falling object can be modeled by a quadratic equation, as shown in the **Project on pages 436–438.**

Applications

6 Factoring

Objectives

Section 6.1

1. Factor a monomial from a polynomial
2. Factor by grouping

Section 6.2

1. Factor trinomials of the form $x^2 + bx + c$
2. Factor completely

Section 6.3

1. Factor trinomials of the form $ax^2 + bx + c$ using trial factors
2. Factor trinomials of the form $ax^2 + bx + c$ by grouping

Section 6.4

1. Factor the difference of two perfect squares and factor perfect-square trinomials
2. Factor the sum or the difference of two cubes
3. Factor trinomials that are quadratic in form
4. Factor completely

Section 6.5

1. Solve equations by factoring
2. Application problems

WWW WEB **Need help? For online student resources, such as section quizzes, visit this web site: math.college.hmco.com**

Prep Test

1. Write 30 as a product of prime numbers.
2. Simplify: $-3(4y - 5)$
3. Simplify: $-(a - b)$
4. Simplify: $2(a - b) - 5(a - b)$
5. Solve: $4x = 0$
6. Solve: $2x + 1 = 0$
7. Multiply: $(x + 4)(x - 6)$
8. Multiply: $(2x - 5)(3x + 2)$
9. Simplify: $\dfrac{x^5}{x^2}$
10. Simplify: $\dfrac{6x^4y^3}{2xy^2}$

Go Figure

Write the expressions in order from greatest to least. Do not use a calculator

$$2^{150}, 3^{100}, 5^{50}$$

6.1 Common Factors

1 Factor a monomial from a polynomial

VIDEO & DVD CD TUTOR WEB SSM

Take Note

12 is the GCF of 24 and 60 because 12 is the largest integer that divides evenly into both 24 and 60.

The **greatest common factor (GCF)** of two or more integers is the greatest integer that is a factor of all the integers.

$$24 = 2 \cdot 2 \cdot 2 \cdot 3$$
$$60 = 2 \cdot 2 \cdot 3 \cdot 5$$
$$\text{GCF} = 2 \cdot 2 \cdot 3 = 12$$

The GCF of two or more monomials is the product of the GCF of the coefficients and the common variable factors.

$$6x^3y = 2 \cdot 3 \cdot x \cdot x \cdot x \cdot y$$
$$8x^2y^2 = 2 \cdot 2 \cdot 2 \cdot x \cdot x \cdot y \cdot y$$
$$\text{GCF} = 2 \cdot x \cdot x \cdot y = 2x^2y$$

Note that the exponent of each variable in the GCF is the same as the *smallest* exponent of that variable in either of the monomials.

The GCF of $6x^3y$ and $8x^2y^2$ is $2x^2y$.

Example 1 Find the GCF of $12a^4b$ and $18a^2b^2c$.

Solution

$12a^4b = 2 \cdot 2 \cdot 3 \cdot a^4 \cdot b$ • Factor each monomial.
$18a^2b^2c = 2 \cdot 3 \cdot 3 \cdot a^2 \cdot b^2 \cdot c$

$\text{GCF} = 2 \cdot 3 \cdot a^2 \cdot b = 6a^2b$ • The common variable factors are a^2 and b. c is not a common factor.

Problem 1 Find the GCF of $4x^6y$ and $18x^2y^6$.

Solution See page S20.

The Distributive Property is used to multiply factors of a polynomial. To **factor a polynomial** means to write the polynomial as a product of other polynomials.

Factors		**Polynomial**
$2x(x + 5)$	$=$	$2x^2 + 10x$

Multiply → (Factors to Polynomial); Factor ← (Polynomial to Factors)

In the example above, $2x$ is the GCF of the terms $2x^2$ and $10x$. It is a **common monomial factor** of the terms. $x + 5$ is a **binomial factor** of $2x^2 + 10x$.

Example 2 Factor. **A.** $5x^3 - 35x^2 + 10x$ **B.** $16x^2y + 8x^4y^2 - 12x^4y^5$

Solution **A.** The GCF is $5x$.

• Find the GCF of the terms of the polynomial.

$$\frac{5x^3}{5x} = x^2, \quad \frac{-35x^2}{5x} = -7x, \quad \frac{10x}{5x} = 2$$

• Divide each term of the polynomial by the GCF.

$$5x^3 - 35x^2 + 10x$$
$$= 5x(x^2) + 5x(-7x) + 5x(2)$$

• Use the quotients to rewrite the polynomial, expressing each term as a product with the GCF as one of the factors.

$$= 5x(x^2 - 7x + 2)$$

• Use the Distributive Property to write the polynomial as a product of factors.

B. $16x^2y = 2 \cdot 2 \cdot 2 \cdot 2 \cdot x^2 \cdot y$
$8x^4y^2 = 2 \cdot 2 \cdot 2 \cdot x^4 \cdot y^2$
$12x^4y^5 = 2 \cdot 2 \cdot 3 \cdot x^4 \cdot y^5$
The GCF is $4x^2y$.

$$16x^2y + 8x^4y^2 - 12x^4y^5$$
$$= 4x^2y(4) + 4x^2y(2x^2y) + 4x^2y(-3x^2y^4)$$
$$= 4x^2y(4 + 2x^2y - 3x^2y^4)$$

Problem 2 Factor. **A.** $14a^2 - 21a^4b$ **B.** $6x^4y^2 - 9x^3y^2 + 12x^2y^4$

Solution See page S20.

2 Factor by grouping

In the examples below, the binomials in parentheses are called **binomial factors.**

$$2a(a + b)$$
$$3xy(x - y)$$

The Distributive Property is used to factor a common binomial factor from an expression.

In the expression at the right, the common binomial factor is $y - 3$. The Distributive Property is used to write the expression as a product of factors.

$$x(y - 3) + 4(y - 3)$$
$$= (y - 3)(x + 4)$$

Example 3 Factor: $y(x + 2) + 3(x + 2)$

Solution

$$y(x + 2) + 3(x + 2)$$
$$= (x + 2)(y + 3)$$

• The common binomial factor is $x + 2$.

Problem 3 Factor: $a(b - 7) + b(b - 7)$

Solution See page S20.

Sometimes a binomial factor must be rewritten before a common binomial factor can be found.

Factor: $a(a - b) + 5(b - a)$

$a - b$ and $b - a$ are different binomials.

Note that $(b - a) = (-a + b) = -(a - b)$.

Rewrite $(b - a)$ as $-(a - b)$ so that there is a common factor.

$$\begin{aligned} a(a - b) + 5(b - a) &= a(a - b) + 5[-(a - b)] \\ &= a(a - b) - 5(a - b) \\ &= (a - b)(a - 5) \end{aligned}$$

Example 4 Factor: $2x(x - 5) + y(5 - x)$

Solution

$2x(x - 5) + y(5 - x)$

$= 2x(x - 5) - y(x - 5)$ • Rewrite $5 - x$ as $-(x - 5)$ so that there is a common factor.

$= (x - 5)(2x - y)$ • Write the expression as a product of factors.

Problem 4 Factor: $3y(5x - 2) - 4(2 - 5x)$

Solution See page S20.

Some polynomials can be factored by grouping the terms so that a common binomial factor is found.

Factor: $2x^3 - 3x^2 + 4x - 6$

$2x^3 - 3x^2 + 4x - 6$

Group the first two terms and the last two terms. $= (2x^3 - 3x^2) + (4x - 6)$

Factor out the GCF from each group. $= x^2(2x - 3) + 2(2x - 3)$

Write the expression as a product of factors. $= (2x - 3)(x^2 + 2)$

Example 5 Factor: $3y^3 - 4y^2 - 6y + 8$

Solution

$3y^3 - 4y^2 - 6y + 8$

$= (3y^3 - 4y^2) - (6y - 8)$ • Group the first two terms and the last two terms. Note that $-6y + 8 = -(6y - 8)$.

$= y^2(3y - 4) - 2(3y - 4)$ • Factor out the GCF from each group.

$= (3y - 4)(y^2 - 2)$ • Write the expression as a product of factors.

Problem 5 Factor: $y^5 - 5y^3 + 4y^2 - 20$

Solution See page S20.

6.1 Concept Review

Determine whether the statement is always true, sometimes true, or never true.

1. To factor a polynomial means to rewrite it as multiplication.

2. The expression $3x(x + 5)$ is a sum, and the expression $3x^2 + 15x$ is a product.

3. The greatest common factor of two numbers is the largest number that divides evenly into both numbers.

4. x is a monomial factor of $x(2x - 1)$. $2x - 1$ is a binomial factor of $x(2x - 1)$.

5. $y(y + 4) + 6$ is in factored form.

6. A common monomial factor is a factor of every term of a polynomial.

7. $a^2 - 8a - ab + 8b = a(a - 8) - b(a + 8)$

8. The binomial $d + 9$ is a common binomial factor of $2c(d + 9) - 7(d + 9)$.

6.1 Exercises

1 1. Explain the meaning of "a factor of a polynomial" and the meaning of "to factor a polynomial."

2. Explain why the statement is true.

   a. The terms of the binomial $3x - 6$ have a common factor.
   b. The expression $3x^2 + 15$ is not in factored form.
   c. $5y - 7$ is a factor of $y(5y - 7)$.

Find the greatest common factor.

3. x^7, x^3
4. y^6, y^{12}
5. x^2y^4, xy^6
6. a^5b^3, a^3b^8
7. $x^2y^4z^6, xy^8z^2$
8. ab^2c^3, a^3b^2c
9. $14a^3, 49a^7$
10. $12y^2, 27y^4$
11. $3x^2y^2, 5ab^2$
12. $8x^2y^3, 7ab^4$
13. $9a^2b^4, 24a^4b^2$
14. $15a^4b^2, 9ab^5$
15. $ab^3, 4a^2b, 12a^2b^3$
16. $12x^2y, x^4y, 16x$
17. $2x^2y, 4xy, 8x$
18. $16x^2, 8x^4y^2, 12xy$
19. $3x^2y^2, 6x, 9x^3y^3$
20. $4a^2b^3, 8a^3, 12ab^4$

Factor.

21. $5a + 5$
22. $7b - 7$
23. $16 - 8a^2$
24. $12 + 12y^2$
25. $8x + 12$
26. $16a - 24$
27. $30a - 6$
28. $20b + 5$
29. $7x^2 - 3x$
30. $12y^2 - 5y$
31. $3a^2 + 5a^5$
32. $9x - 5x^2$
33. $14y^2 + 11y$
34. $6b^3 - 5b^2$
35. $2x^4 - 4x$
36. $3y^4 - 9y$
37. $10x^4 - 12x^2$
38. $12a^5 - 32a^2$
39. $x^2y - xy^3$
40. $a^2b + a^4b^2$
41. $2a^5b + 3xy^3$
42. $5x^2y - 7ab^3$
43. $6a^2b^3 - 12b^2$
44. $8x^2y^3 - 4x^2$
45. $6a^2bc + 4ab^2c$
46. $10x^2yz^2 + 15xy^3z$
47. $18x^2y^2 - 9a^2b^2$
48. $9a^2x - 27a^3x^3$
49. $6x^3y^3 - 12x^6y^6$
50. $3a^2b^2 - 12a^5b^5$
51. $x^3 - 3x^2 - x$
52. $a^3 + 4a^2 + 8a$
53. $2x^2 + 8x - 12$
54. $a^3 - 3a^2 + 5a$
55. $b^3 - 5b^2 - 7b$
56. $5x^2 - 15x + 35$
57. $8y^2 - 12y + 32$
58. $3x^3 + 6x^2 + 9x$
59. $5y^3 - 20y^2 + 10y$
60. $2x^4 - 4x^3 + 6x^2$
61. $3y^4 - 9y^3 - 6y^2$
62. $2x^3 + 6x^2 - 14x$
63. $3y^3 - 9y^2 + 24y$
64. $2y^5 - 3y^4 + 7y^3$
65. $6a^5 - 3a^3 - 2a^2$
66. $x^3y - 3x^2y^2 + 7xy^3$
67. $8x^2y^2 - 4x^2y + x^2$
68. $x^4y^4 - 3x^3y^3 + 6x^2y^2$
69. $4x^5y^5 - 8x^4y^4 + x^3y^3$
70. $16x^2y - 8x^3y^4 - 48x^2y^2$

2 Factor.

71. $x(a + b) + 2(a + b)$
72. $a(x + y) + 4(x + y)$
73. $x(b + 2) - y(b + 2)$
74. $d(z - 8) + 5(z - 8)$
75. $a(y - 4) - b(y - 4)$
76. $c(x - 6) - 7(x - 6)$

Rewrite the expression so that there is a common binomial factor.

77. $a(x - 2) - b(2 - x)$
78. $a(x - 2) + b(2 - x)$
79. $b(a - 7) + 3(7 - a)$
80. $b(a - 7) - 3(7 - a)$
81. $x(a - 2b) - y(2b - a)$
82. $x(a - 2b) + y(2b - a)$

Factor.

83. $a(x - 2) + 5(2 - x)$
84. $a(x - 7) + b(7 - x)$
85. $b(y - 3) + 3(3 - y)$
86. $c(a - 2) - b(2 - a)$
87. $a(x - y) - 2(y - x)$
88. $3(a - b) - x(b - a)$
89. $z(c + 5) - 8(5 + c)$
90. $y(ab + 4) - 7(4 + ab)$
91. $w(3x - 4) - (4 - 3x)$
92. $d(6y - 1) - (1 - 6y)$
93. $x^3 + 4x^2 + 3x + 12$
94. $x^3 - 4x^2 - 3x + 12$
95. $2y^3 + 4y^2 + 3y + 6$
96. $3y^3 - 12y^2 + y - 4$
97. $ab + 3b - 2a - 6$

98. $yz + 6z - 3y - 18$

99. $x^2a - 2x^2 - 3a + 6$

100. $x^2y + 4x^2 + 3y + 12$

101. $3ax - 3bx - 2ay + 2by$

102. $8 + 2c + 4a^2 + a^2c$

103. $x^2 - 3x + 4ax - 12a$

104. $t^2 + 4t - st - 4s$

105. $xy - 5y - 2x + 10$

106. $2y^2 - 10y + 7xy - 35x$

107. $21x^2 + 6xy - 49x - 14y$

108. $4a^2 + 5ab - 10b - 8a$

109. $2ra + a^2 - 2r - a$

110. $2ab - 3b^2 - 3b + 2a$

111. $4x^2 + 3xy - 12y - 16x$

112. $8s + 12r - 6s^2 - 9rs$

113. $10xy^2 - 15xy + 6y - 9$

114. $10a^2b - 15ab - 4a + 6$

6.1 Applying Concepts

Factor by grouping.

115. **a.** $2x^2 + 6x + 5x + 15$ **b.** $2x^2 + 5x + 6x + 15$

116. **a.** $3x^2 + 3xy - xy - y^2$ **b.** $3x^2 - xy + 3xy - y^2$

117. **a.** $2a^2 - 2ab - 3ab + 3b^2$ **b.** $2a^2 - 3ab - 2ab + 3b^2$

Compare your answers to parts (a) and (b) of Exercises 115–117 in order to answer Exercise 118.

118. Do different groupings of the terms in a polynomial affect the binomial factoring?

A whole number is a perfect number if it is the sum of all of its factors that are less than itself. For example, 6 is a perfect number because all the factors of 6 that are less than 6 are 1, 2, and 3, and $1 + 2 + 3 = 6$.

119. Find a perfect number between 20 and 30.

120. Find a perfect number between 490 and 500.

Solve.

121. **Geometry** In the equation $P = 2L + 2W$, what is the effect on P when the quantity $L + W$ doubles?

122. **Geometry** Write an expression in factored form for the shaded portion in the diagram.

a.

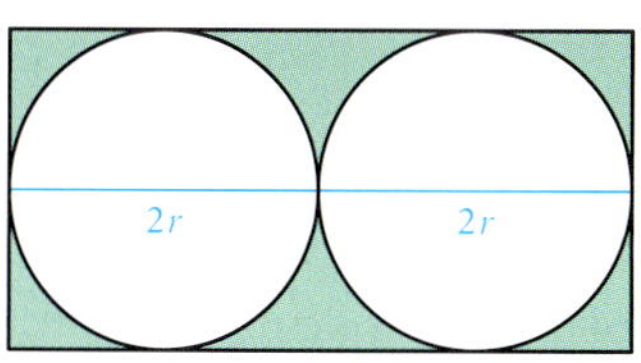

b.

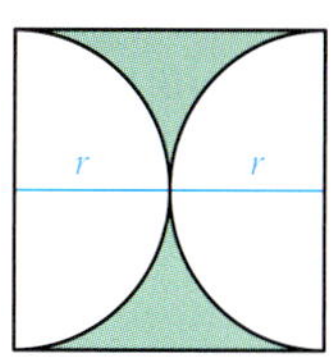

6.2 Factoring Polynomials of the Form $x^2 + bx + c$

1 Factor trinomials of the form $x^2 + bx + c$

VIDEO & DVD CD TUTOR WEB SSM

Trinomials of the form $x^2 + bx + c$, where b and c are integers, are shown at the right.

$x^2 + 9x + 14, \quad b = 9, \quad c = 14$
$x^2 - x - 12, \quad b = -1, \quad c = -12$
$x^2 - 2x - 15, \quad b = -2, \quad c = -15$

To factor a trinomial of this form means to express the trinomial as the product of two binomials. Some trinomials expressed as the product of binomials (factored form) are shown at the right.

Trinomial		Factored Form
$x^2 + 9x + 14$	=	$(x + 2)(x + 7)$
$x^2 - x - 12$	=	$(x + 3)(x - 4)$
$x^2 - 2x - 15$	=	$(x + 3)(x - 5)$

The method by which the factors of a trinomial are found is based on FOIL. Consider the following binomial products, noting the relationship between the constant terms of the binomials and the terms of the trinomials.

Signs in the binomials are the same

$(x + 6)(x + 2) = x^2 + 2x + 6x + (6)(2) = x^2 + 8x + 12$

Sum of 6 and 2 → $8x$
Product of 6 and 2 → 12

$(x - 3)(x - 4) = x^2 - 4x - 3x + (-3)(-4) = x^2 - 7x + 12$

Sum of −3 and −4 → $-7x$
Product of −3 and −4 → 12

Signs in the binomials are opposite

$(x + 3)(x - 5) = x^2 - 5x + 3x + (3)(-5) = x^2 - 2x - 15$

Sum of 3 and −5 → $-2x$
Product of 3 and −5 → -15

$(x - 4)(x + 6) = x^2 + 6x - 4x + (-4)(6) = x^2 + 2x - 24$

Sum of −4 and 6 → $2x$
Product of −4 and 6 → -24

Points to Remember in Factoring $x^2 + bx + c$

1. In the trinomial, the coefficient of x is the sum of the constant terms of the binomials.
2. In the trinomial, the constant term is the product of the constant terms of the binomials.
3. When the constant term of the trinomial is positive, the constant terms of the binomials have the same sign as the coefficient of x in the trinomial.
4. When the constant term of the trinomial is negative, the constant terms of the binomials have opposite signs.

Success at factoring a trinomial depends on remembering these four points. For example, to factor

$$x^2 - 2x - 24$$

find two numbers whose sum is −2 and whose product is −24 [Points 1 and 2]. Because the constant term of the trinomial is negative (−24), the numbers will have opposite signs [Point 4].

A systematic method of finding these numbers involves listing the factors of the constant term of the trinomial and the sum of those factors.

Factors of −24	Sum of the Factors
1, −24	1 + (−24) = −23
−1, 24	−1 + 24 = 23
2, −12	2 + (−12) = −10
−2, 12	−2 + 12 = 10
3, −8	3 + (−8) = −5
−3, 8	−3 + 8 = 5
4, −6	**4 + (−6) = −2**
−4, 6	−4 + 6 = 2

Take Note

Always check your proposed factorization to ensure accuracy.

4 and −6 are two numbers whose sum is −2 and whose product is −24. Write the binomial factors of the trinomial.

$$x^2 - 2x - 24 = (x + 4)(x - 6)$$

Check: $(x + 4)(x - 6) = x^2 - 6x + 4x - 24 = x^2 - 2x - 24$

By the Commutative Property of Multiplication, the binomial factors can also be written as

$$x^2 - 2x - 24 = (x - 6)(x + 4)$$

Example 1 Factor: $x^2 + 18x + 32$

Solution

Factors of 32	Sum
1, 32	33
2, 16	**18**
4, 8	12

- Try only positive factors of 32 [Point 3].
- Once the correct pair is found, the other factors need not be tried.

$$x^2 + 18x + 32 = (x + 2)(x + 16)$$

- Write the factors of the trinomial.

Check $(x + 2)(x + 16) = x^2 + 16x + 2x + 32$
$= x^2 + 18x + 32$

Problem 1 Factor: $x^2 - 8x + 15$

Solution See page S20.

Example 2 Factor: $x^2 - 6x - 16$

Solution

Factors of −16	Sum
1, −16	−15
−1, 16	15
2, −8	**−6**
−2, 8	6
4, −4	0

- The factors must be of opposite signs [Point 4].

$$x^2 - 6x - 16 = (x + 2)(x - 8)$$

- Write the factors of the trinomial.

Check $(x + 2)(x - 8) = x^2 - 8x + 2x - 16$
$= x^2 - 6x - 16$

Problem 2 Factor: $x^2 + 3x - 18$

Solution See page S20.

Not all trinomials can be factored when using only integers. Consider the trinomial $x^2 - 6x - 8$.

Factors of −8	Sum
1, −8	−7
−1, 8	7
2, −4	−2
−2, 4	2

Because none of the pairs of factors of −8 have a sum of −6, the trinomial is not factorable using integers. The trinomial is said to be **nonfactorable over the integers.**

2 Factor completely

VIDEO & DVD CD TUTOR WEB SSM

A polynomial is **factored completely** when it is written as a product of factors that are nonfactorable over the integers.

> **Take Note**
> The first step in *any* factoring problem is to determine whether the terms of the polynomial have a *common factor*. If they do, factor it out first.

Example 3 Factor: $3x^3 + 15x^2 + 18x$

Solution The GCF of $3x^3$, $15x^2$, and $18x$ is $3x$. • Find the GCF of the terms of the polynomial.

$3x^3 + 15x^2 + 18x$
$= 3x(x^2) + 3x(5x) + 3x(6)$ • Factor out the GCF.
$= 3x(x^2 + 5x + 6)$ • Write the polynomial as a product of factors.

• Factor the trinomial $x^2 + 5x + 6$. Try only positive factors of 6.

Factors of 6	Sum
1, 6	7
2, 3	5

$3x^3 + 15x^2 + 18x = 3x(x + 2)(x + 3)$

Check $3x(x + 2)(x + 3) = 3x(x^2 + 3x + 2x + 6)$
$= 3x(x^2 + 5x + 6)$
$= 3x^3 + 15x^2 + 18x$

Problem 3 Factor: $3a^2b - 18ab - 81b$

Solution See page S20.

Example 4 Factor: $x^2 + 9xy + 20y^2$

Solution There is no common factor.

Factors of 20	Sum
1, 20	21
2, 10	12
4, 5	9

• Try only positive factors of 20.

$$x^2 + 9xy + 20y^2 = (x + 4y)(x + 5y)$$

Check $(x + 4y)(x + 5y) = x^2 + 5xy + 4xy + 20y^2$
$= x^2 + 9xy + 20y^2$

Take Note

$4y$ and $5y$ are placed in the binomials. This is necessary so that the middle term contains xy and the last term contains y^2.

Problem 4 Factor: $4x^2 - 40xy + 84y^2$

Solution See page S20.

6.2 Concept Review

Determine whether the statement is always true, sometimes true, or never true.

1. The value of b in the trinomial $x^2 - 8x + 7$ is 8.

2. The factored form of $x^2 - 5x - 14$ is $(x - 7)(x + 2)$.

3. To factor a trinomial of the form $x^2 + bx + c$ means to rewrite the polynomial as a product of two binomials.

4. In the factoring of a trinomial, if the constant term is positive, then the signs in both binomial factors will be the same.

5. In the factoring of a trinomial, if the constant term is negative, then the signs in both binomial factors will be negative.

6. The first step in factoring a trinomial is to determine whether the terms of the trinomial have a common factor.

6.2 Exercises

1 Complete the table by listing the pairs of factors of the number and the sum of each pair.

1.

Factors of 18	Sum

2.

Factors of 12	Sum

3.

Factors of −21	Sum

4.

Factors of −10	Sum

5.

Factors of −28	Sum

6.

Factors of −32	Sum

Factor.

7. $x^2 + 3x + 2$ **8.** $x^2 + 5x + 6$ **9.** $x^2 - x - 2$ **10.** $x^2 + x - 6$

11. $a^2 + a - 12$ **12.** $a^2 - 2a - 35$ **13.** $a^2 - 3a + 2$ **14.** $a^2 - 5a + 4$

15. $a^2 + a - 2$ **16.** $a^2 - 2a - 3$ **17.** $b^2 - 6b + 9$ **18.** $b^2 + 8b + 16$

19. $b^2 + 7b - 8$ **20.** $y^2 - y - 6$ **21.** $y^2 + 6y - 55$ **22.** $z^2 - 4z - 45$

23. $y^2 - 5y + 6$ **24.** $y^2 - 8y + 15$ **25.** $z^2 - 14z + 45$ **26.** $z^2 - 14z + 49$

27. $z^2 - 12z - 160$ **28.** $p^2 + 2p - 35$ **29.** $p^2 + 12p + 27$ **30.** $p^2 - 6p + 8$

31. $x^2 + 20x + 100$ **32.** $x^2 + 18x + 81$ **33.** $b^2 + 9b + 20$ **34.** $b^2 + 13b + 40$

35. $x^2 - 11x - 42$ **36.** $x^2 + 9x - 70$ **37.** $b^2 - b - 20$ **38.** $b^2 + 3b - 40$

39. $y^2 - 14y - 51$ **40.** $y^2 - y - 72$ **41.** $p^2 - 4p - 21$ **42.** $p^2 + 16p + 39$

43. $y^2 - 8y + 32$ **44.** $y^2 - 9y + 81$ **45.** $x^2 - 20x + 75$ **46.** $p^2 + 24p + 63$

47. $x^2 - 15x + 56$ **48.** $x^2 + 21x + 38$ **49.** $x^2 + x - 56$ **50.** $x^2 + 5x - 36$

51. $a^2 - 21a - 72$ **52.** $a^2 - 7a - 44$ **53.** $a^2 - 15a + 36$ **54.** $a^2 - 21a + 54$

55. $z^2 - 9z - 136$ **56.** $z^2 + 14z - 147$ **57.** $c^2 - c - 90$ **58.** $c^2 - 3c - 180$

2 Factor.

59. $2x^2 + 6x + 4$
60. $3x^2 + 15x + 18$
61. $3a^2 + 3a - 18$
62. $4x^2 - 4x - 8$
63. $ab^2 + 2ab - 15a$
64. $ab^2 + 7ab - 8a$
65. $xy^2 - 5xy + 6x$
66. $xy^2 + 8xy + 15x$
67. $z^3 - 7z^2 + 12z$
68. $2a^3 + 6a^2 + 4a$
69. $3y^3 - 15y^2 + 18y$
70. $4y^3 + 12y^2 - 72y$
71. $3x^2 + 3x - 36$
72. $2x^3 - 2x^2 - 4x$
73. $5z^2 - 15z - 140$
74. $6z^2 + 12z - 90$
75. $2a^3 + 8a^2 - 64a$
76. $3a^3 - 9a^2 - 54a$
77. $x^2 - 5xy + 6y^2$
78. $x^2 + 4xy - 21y^2$
79. $a^2 - 9ab + 20b^2$
80. $a^2 - 15ab + 50b^2$
81. $x^2 - 3xy - 28y^2$
82. $s^2 + 2st - 48t^2$
83. $y^2 - 15yz - 41z^2$
84. $y^2 + 85yz + 36z^2$
85. $z^4 - 12z^3 + 35z^2$
86. $z^4 + 2z^3 - 80z^2$
87. $b^4 - 22b^3 + 120b^2$
88. $b^4 - 3b^3 - 10b^2$
89. $2y^4 - 26y^3 - 96y^2$
90. $3y^4 + 54y^3 + 135y^2$
91. $x^4 + 7x^3 - 8x^2$
92. $x^4 - 11x^3 - 12x^2$
93. $4x^2y + 20xy - 56y$
94. $3x^2y - 6xy - 45y$
95. $8y^2 - 32y + 24$
96. $10y^2 - 100y + 90$
97. $c^3 + 13c^2 + 30c$
98. $c^3 + 18c^2 - 40c$
99. $3x^3 - 36x^2 + 81x$
100. $4x^3 + 4x^2 - 24x$
101. $x^2 - 8xy + 15y^2$
102. $y^2 - 7xy - 8x^2$
103. $a^2 - 13ab + 42b^2$
104. $y^2 + 4yz - 21z^2$
105. $y^2 + 8yz + 7z^2$
106. $y^2 - 16yz + 15z^2$
107. $3x^2y + 60xy - 63y$
108. $4x^2y - 68xy - 72y$
109. $3x^3 + 3x^2 - 36x$
110. $4x^3 + 12x^2 - 160x$
111. $4z^3 + 32z^2 - 132z$
112. $5z^3 - 50z^2 - 120z$

6.2 Applying Concepts

Factor.

113. $20 + c^2 + 9c$
114. $x^2y - 54y - 3xy$
115. $45a^2 + a^2b^2 - 14a^2b$
116. $12p^2 - 96p + 3p^3$

Find all integers k such that the trinomial can be factored over the integers.

117. $x^2 + kx + 35$
118. $x^2 + kx + 18$
119. $x^2 - kx + 21$
120. $x^2 - kx + 14$

Determine the positive integer values of k for which the following polynomials are factorable over the integers.

121. $y^2 + 4y + k$

122. $z^2 + 7z + k$

123. $a^2 - 6a + k$

124. $c^2 - 7c + k$

125. $x^2 - 3x + k$

126. $y^2 + 5y + k$

127. Exercises 121–126 included the requirement that $k > 0$. If k is allowed to be any integer, how many different values of k are possible for each polynomial? Explain your answer.

6.3 Factoring Polynomials of the Form $ax^2 + bx + c$

1 Factor trinomials of the form $ax^2 + bx + c$ using trial factors

VIDEO & DVD CD TUTOR WEB SSM

Trinomials of the form $ax^2 + bx + c$, where a, b, and c are integers and $a \neq 0$, are shown at the right.

$3x^2 - x + 4$, $a = 3$, $b = -1$, $c = 4$
$4x^2 + 5x - 8$, $a = 4$, $b = 5$, $c = -8$

These trinomials differ from those in the previous section in that the coefficient of x^2 is not 1. There are various methods of factoring these trinomials. The method described in this objective is factoring trinomials using trial factors.

To **factor a trinomial of the form $ax^2 + bx + c$** means to express the polynomial as the product of two binomials. Factoring such polynomials by trial and error may require testing many trial factors. To reduce the number of trial factors, remember the following points.

Points to Remember in Factoring $ax^2 + bx + c$

1. If the terms of the trinomial do not have a common factor, then the terms of a binomial factor cannot have a common factor.
2. When the constant term of the trinomial is positive, the constant terms of the binomials have the same sign as the coefficient of x in the trinomial.
3. When the constant term of the trinomial is negative, the constant terms of the binomials have opposite signs.

Factor: $10x^2 - x - 3$

The terms of the trinomial do not have a common factor; therefore, a binomial factor will not have a common factor.

Because the constant term, c, of the trinomial is negative (−3), the constant terms of the binomial factors will have opposite signs [Point 3].

Find the factors of a (10) and the factors of c (−3).

Factors of 10	Factors of −3
1, 10	1, −3
2, 5	−1, 3

Using these factors, write trial factors, and use the Outer and Inner products of FOIL to check the middle term.

Trial Factors	Middle Term
$(x + 1)(10x - 3)$	$-3x + 10x = 7x$
$(x - 1)(10x + 3)$	$3x - 10x = -7x$
$\mathbf{(2x + 1)(5x - 3)}$	$\mathbf{-6x + 5x = -x}$
$(2x - 1)(5x + 3)$	$6x - 5x = x$
$(10x + 1)(x - 3)$	$-30x + x = -29x$
$(10x - 1)(x + 3)$	$30x - x = 29x$
$(5x + 1)(2x - 3)$	$-15x + 2x = -13x$
$(5x - 1)(2x + 3)$	$15x - 2x = 13x$

From the list of trial factors, $10x^2 - x - 3 = (2x + 1)(5x - 3)$.

Check: $(2x + 1)(5x - 3) = 10x^2 - 6x + 5x - 3$
$= 10x^2 - x - 3$

All the trial factors for this trinomial were listed in this example. However, once the correct binomial factors are found, it is not necessary to continue checking the remaining trial factors.

Factor: $4x^2 - 27x + 18$

The terms of the trinomial do not have a common factor; therefore, a binomial factor will not have a common factor.

Because the constant term, c, of the trinomial is positive (18), the constant terms of the binomial factors will have the same sign as the coefficient of x. Because the coefficient of x is −27, both signs will be negative [Point 2].

Find the factors of a (4) and the negative factors of c (18).

Factors of 4	Factors of 18
1, 4	−1, −18
2, 2	−2, −9
	−3, −6

Using these factors, write trial factors, and use the Outer and Inner products of FOIL to check the middle term.

Trial Factors	Middle Term
$(x - 1)(4x - 18)$	Common factor
$(x - 2)(4x - 9)$	$-9x - 8x = -17x$
$(x - 3)(4x - 6)$	Common factor
$(2x - 1)(2x - 18)$	Common factor
$(2x - 2)(2x - 9)$	Common factor
$(2x - 3)(2x - 6)$	Common factor
$(4x - 1)(x - 18)$	$-72x - x = -73x$
$(4x - 2)(x - 9)$	Common factor
$\mathbf{(4x - 3)(x - 6)}$	$\mathbf{-24x - 3x = -27x}$

The correct factors have been found.

$$4x^2 - 27x + 18 = (4x - 3)(x - 6)$$

This last example illustrates that many of the trial factors may have common factors and thus need not be tried. For the remainder of this chapter, the trial factors with a common factor will not be listed.

Example 1 Factor: $3x^2 + 20x + 12$

Solution

Factors of 3	Factors of 12
1, 3	1, 12
	2, 6
	3, 4

• Because 20 is positive, only the positive factors of 12 need be tried.

Trial Factors	Middle Term
$(x + 3)(3x + 4)$	$4x + 9x = 13x$
$(3x + 1)(x + 12)$	$36x + x = 37x$
$(3x + 2)(x + 6)$	$18x + 2x = 20x$

• Write the trial factors. Use FOIL to check the middle term.

$$3x^2 + 20x + 12 = (3x + 2)(x + 6)$$

Check $(3x + 2)(x + 6) = 3x^2 + 18x + 2x + 12$
$= 3x^2 + 20x + 12$

Problem 1 Factor: $6x^2 - 11x + 5$

Solution See page S21.

Example 2 Factor: $6x^2 - 5x - 6$

Solution

Factors of 6	Factors of −6
1, 6	1, −6
2, 3	−1, 6
	2, −3
	−2, 3

• Find the factors of *a* (6) and the factors of *c* (−6).

Trial Factors	Middle Term
$(x - 6)(6x + 1)$	$x - 36x = -35x$
$(x + 6)(6x - 1)$	$-x + 36x = 35x$
$(2x - 3)(3x + 2)$	$4x - 9x = -5x$

• Write the trial factors. Use FOIL to check the middle term.

$$6x^2 - 5x - 6 = (2x - 3)(3x + 2)$$

Check $(2x - 3)(3x + 2) = 6x^2 + 4x - 9x - 6$
$= 6x^2 - 5x - 6$

Problem 2 Factor: $8x^2 + 14x - 15$

Solution See page S21.

Example 3 Factor: $15 - 2x - x^2$

Solution

Factors of 15	Factors of −1
1, 15 3, 5	1, −1

• The terms have no common factors. The coefficient of x^2 is −1.

Trial Factors	Middle Term
$(1 + x)(15 - x)$	$-x + 15x = 14x$
$(1 - x)(15 + x)$	$x - 15x = -14x$
$(3 + x)(5 - x)$	$-3x + 5x = 2x$
$(3 - x)(5 + x)$	$3x - 5x = -2x$

• Write the trial factors. Use FOIL to check the middle term.

$$15 - 2x - x^2 = (3 - x)(5 + x)$$

Check

$$\begin{aligned}(3 - x)(5 + x) &= 15 + 3x - 5x - x^2\\ &= 15 - 2x - x^2\end{aligned}$$

Problem 3 Factor: $24 - 2y - y^2$

Solution See page S21.

The first step in factoring a trinomial is to determine whether there is a common factor. If there is a common factor, factor out the GCF of the terms.

Example 4 Factor: $3x^3 - 23x^2 + 14x$

Solution The GCF of $3x^3$, $23x^2$, and $14x$ is x.

• Find the GCF of the terms of the polynomial.

$$3x^3 - 23x^2 + 14x = x(3x^2 - 23x + 14)$$

• Factor out the GCF.

Factors of 3	Factors of 14
1, 3	−1, −14 −2, −7

• Factor the trinomial $3x^2 - 23x + 14$.

Trial Factors	Middle Term
$(x - 1)(3x - 14)$	$-14x - 3x = -17x$
$(x - 14)(3x - 1)$	$-x - 42x = -43x$
$(x - 2)(3x - 7)$	$-7x - 6x = -13x$
$(x - 7)(3x - 2)$	$-2x - 21x = -23x$

$$3x^3 - 23x^2 + 14x = x(x - 7)(3x - 2)$$

Check

$$\begin{aligned}x(x - 7)(3x - 2) &= x(3x^2 - 2x - 21x + 14)\\ &= x(3x^2 - 23x + 14)\\ &= 3x^3 - 23x^2 + 14x\end{aligned}$$

Problem 4 Factor: $4a^2b^2 - 30a^2b + 14a^2$

Solution See page S21.

2 Factor trinomials of the form $ax^2 + bx + c$ by grouping

VIDEO & DVD CD TUTOR WEB SSM

> **Take Note**
> In this objective, we are using the skills taught in Objective 2 of Section 1 of this chapter. You may want to review that material before studying this objective.

In the previous objective, trinomials of the form $ax^2 + bx + c$ were factored using trial factors. In this objective, factoring by grouping is used.

To factor $ax^2 + bx + c$, first find two factors of $a \cdot c$ whose sum is b. Use the two factors to rewrite the middle term of the trinomial as the sum of two terms. Then use factoring by grouping to write the factorization of the trinomial.

Factor: $2x^2 + 13x + 15$

$a = 2, c = 15, a \cdot c = 2 \cdot 15 = 30$

Find two positive factors of 30 whose sum is 13.

Positive Factors of 30	Sum
1, 30	31
2, 15	17
3, 10	13
5, 6	11

The factors are 3 and 10.

Use the factors 3 and 10 to rewrite $13x$ as $3x + 10x$.

$$2x^2 + 13x + 15 = 2x^2 + 3x + 10x + 15$$

Factor by grouping.

$$= (2x^2 + 3x) + (10x + 15)$$
$$= x(2x + 3) + 5(2x + 3)$$
$$= (2x + 3)(x + 5)$$

Check: $(2x + 3)(x + 5) = 2x^2 + 10x + 3x + 15$
$= 2x^2 + 13x + 15$

Factor: $6x^2 - 11x - 10$

$a = 6, c = -10, a \cdot c = 6(-10) = -60$

Find two factors of -60 whose sum is -11.

Factors of −60	Sum
1, −60	−59
−1, 60	59
2, −30	−28
−2, 30	28
3, −20	−17
−3, 20	17
4, −15	−11

The required sum has been found. The remaining factors need not be checked. The factors are 4 and -15.

Use the factors 4 and -15 to rewrite $-11x$ as $4x - 15x$.

$$6x^2 - 11x - 10 = 6x^2 + 4x - 15x - 10$$

Factor by grouping.

Note: $-15x - 10 = -(15x + 10)$

$$= (6x^2 + 4x) - (15x + 10)$$
$$= 2x(3x + 2) - 5(3x + 2)$$
$$= (3x + 2)(2x - 5)$$

Check: $(3x + 2)(2x - 5) = 6x^2 - 15x + 4x - 10$
$= 6x^2 - 11x - 10$

Factor: $3x^2 - 2x - 4$

$a = 3, c = -4, a \cdot c = 3(-4) = -12$

Find two factors of -12 whose sum is -2.

Factors of -12	Sum
$1, -12$	-11
$-1, 12$	11
$2, -6$	-4
$-2, 6$	4
$3, -4$	-1
$-3, 4$	1

No integer factors of -12 have a sum of -2. Therefore, $3x^2 - 2x - 4$ is nonfactorable over the integers.

Example 5 Factor: $2x^2 + 19x - 10$

Solution $a \cdot c = 2(-10) = -20$ • Find $a \cdot c$.

$-1(20) = -20$
$-1 + 20 = 19$ • Find two numbers whose product is -20 and whose sum is 19.

$2x^2 + 19x - 10$
$= 2x^2 - x + 20x - 10$ • Rewrite $19x$ as $-x + 20x$.
$= (2x^2 - x) + (20x - 10)$ • Factor by grouping.
$= x(2x - 1) + 10(2x - 1)$
$= (2x - 1)(x + 10)$

Problem 5 Factor: $2a^2 + 13a - 7$

Solution See page S21.

Example 6 Factor: $8y^2 - 10y - 3$

Solution $a \cdot c = 8(-3) = -24$ • Find $a \cdot c$.

$2(-12) = -24$
$2 + (-12) = -10$ • Find two numbers whose product is -24 and whose sum is -10.

$8y^2 - 10y - 3$
$= 8y^2 + 2y - 12y - 3$ • Rewrite $-10y$ as $2y - 12y$.
$= (8y^2 + 2y) - (12y + 3)$ • Factor by grouping.
$= 2y(4y + 1) - 3(4y + 1)$
$= (4y + 1)(2y - 3)$

Problem 6 Factor: $4a^2 - 11a - 3$

Solution See page S21.

Remember that the first step in factoring a trinomial is to determine whether there is a common factor. If there is a common factor, factor out the GCF of the terms.

Example 7 Factor: $24x^2y - 76xy + 40y$

Solution $24x^2y - 76xy + 40y$
$= 4y(6x^2 - 19x + 10)$
• The terms of the polynomial have a common factor, $4y$. Factor out the GCF.

$a \cdot c = 6(10) = 60$
• To factor $6x^2 - 19x + 10$, first find $a \cdot c$.

$-4(-15) = 60$
$-4 + (-15) = -19$
• Find two numbers whose product is 60 and whose sum is -19.

$6x^2 - 19x + 10$
$= 6x^2 - 4x - 15x + 10$
• Rewrite $-19x$ as $-4x - 15x$.
$= (6x^2 - 4x) - (15x - 10)$
• Factor by grouping.
$= 2x(3x - 2) - 5(3x - 2)$
$= (3x - 2)(2x - 5)$

$24x^2y - 76xy + 40y$
$= 4y(6x^2 - 19x + 10)$
$= 4y(3x - 2)(2x - 5)$
• Write the complete factorization of the given polynomial.

Problem 7 Factor: $15x^3 + 40x^2 - 80x$

Solution See page S21.

6.3 Concept Review

Determine whether the statement is always true, sometimes true, or never true.

1. The value of a in the trinomial $3x^2 + 8x - 7$ is -7.
2. The factored form of $2x^2 - 7x - 15$ is $(2x + 3)(x - 5)$.
3. To check the factorization of a trinomial of the form $ax^2 + bx + c$, multiply the two binomial factors.
4. The terms of the binomial $3x - 9$ have a common factor.
5. If the terms of a trinomial of the form $ax^2 + bx + c$ do not have a common factor, then one of its binomial factors cannot have a common factor.
6. To factor a trinomial of the form $ax^2 + bx + c$ by grouping, first find two numbers whose product is ac and whose sum is b.

6.3 Exercises

1 Factor by the method of using trial factors.

1. $2x^2 + 3x + 1$
2. $5x^2 + 6x + 1$
3. $2y^2 + 7y + 3$
4. $3y^2 + 7y + 2$
5. $2a^2 - 3a + 1$
6. $3a^2 - 4a + 1$
7. $2b^2 - 11b + 5$
8. $3b^2 - 13b + 4$

9. $2x^2 + x - 1$
10. $4x^2 - 3x - 1$
11. $2x^2 - 5x - 3$
12. $3x^2 + 5x - 2$
13. $6z^2 - 7z + 3$
14. $9z^2 + 3z + 2$
15. $6t^2 - 11t + 4$
16. $10t^2 + 11t + 3$
17. $8x^2 + 33x + 4$
18. $7x^2 + 50x + 7$
19. $3b^2 - 16b + 16$
20. $6b^2 - 19b + 15$
21. $2z^2 - 27z - 14$
22. $4z^2 + 5z - 6$
23. $3p^2 + 22p - 16$
24. $7p^2 + 19p + 10$
25. $6x^2 - 17x + 12$
26. $15x^2 - 19x + 6$
27. $5b^2 + 33b - 14$
28. $8x^2 - 30x + 25$
29. $6a^2 + 7a - 24$
30. $14a^2 + 15a - 9$
31. $18t^2 - 9t - 5$
32. $12t^2 + 28t - 5$
33. $15a^2 + 26a - 21$
34. $6a^2 + 23a + 21$
35. $8y^2 - 26y + 15$
36. $18y^2 - 27y + 4$
37. $8z^2 + 2z - 15$
38. $10z^2 + 3z - 4$
39. $3x^2 + 14x - 5$
40. $15a^2 - 22a + 8$
41. $12x^2 + 25x + 12$
42. $10b^2 + 43b - 9$
43. $3z^2 + 95z + 10$
44. $8z^2 - 36z + 1$
45. $3x^2 + xy - 2y^2$
46. $6x^2 + 10xy + 4y^2$
47. $28 + 3z - z^2$
48. $15 - 2z - z^2$
49. $8 - 7x - x^2$
50. $12 + 11x - x^2$
51. $9x^2 + 33x - 60$
52. $16x^2 - 16x - 12$
53. $24x^2 - 52x + 24$
54. $60x^2 + 95x + 20$
55. $35a^4 + 9a^3 - 2a^2$
56. $15a^4 + 26a^3 + 7a^2$
57. $15b^2 - 115b + 70$
58. $25b^2 + 35b - 30$
59. $10x^3 + 12x^2 + 2x$
60. $9x^3 - 39x^2 + 12x$
61. $10y^3 - 44y^2 + 16y$
62. $14y^3 + 94y^2 - 28y$
63. $4yz^3 + 5yz^2 - 6yz$
64. $2yz^3 - 17yz^2 + 8yz$
65. $20b^4 + 41b^3 + 20b^2$
66. $6b^4 - 13b^3 + 6b^2$
67. $9x^3y + 12x^2y + 4xy$
68. $9a^3b - 9a^2b^2 - 10ab^3$

Factor by grouping.

2 **69.** $2t^2 - t - 10$
70. $2t^2 + 5t - 12$
71. $3p^2 - 16p + 5$
72. $6p^2 + 5p + 1$
73. $12y^2 - 7y + 1$
74. $6y^2 - 5y + 1$
75. $5x^2 - 62x - 7$
76. $9x^2 - 13x - 4$
77. $12y^2 + 19y + 5$
78. $5y^2 - 22y + 8$
79. $7a^2 + 47a - 14$
80. $11a^2 - 54a - 5$
81. $4z^2 + 11z + 6$
82. $6z^2 - 25z + 14$
83. $22p^2 + 51p - 10$
84. $14p^2 - 41p + 15$
85. $8y^2 + 17y + 9$
86. $12y^2 - 145y + 12$
87. $6b^2 - 13b + 6$
88. $20b^2 + 37b + 15$
89. $33b^2 + 34b - 35$
90. $15b^2 - 43b + 22$
91. $18y^2 - 39y + 20$
92. $24y^2 + 41y + 12$
93. $15x^2 - 82x + 24$
94. $13z^2 + 49z - 8$
95. $10z^2 - 29z + 10$
96. $15z^2 - 44z + 32$
97. $36z^2 + 72z + 35$
98. $16z^2 + 8z - 35$
99. $14y^2 - 29y + 12$
100. $8y^2 + 30y + 25$

101. $6x^2 + 35x - 6$

102. $4x^2 + 6x + 2$

103. $12x^2 + 33x - 9$

104. $15y^2 - 50y + 35$

105. $30y^2 + 10y - 20$

106. $2x^3 - 11x^2 + 5x$

107. $2x^3 - 3x^2 - 5x$

108. $3a^2 + 5ab - 2b^2$

109. $2a^2 - 9ab + 9b^2$

110. $4y^2 - 11yz + 6z^2$

111. $2y^2 + 7yz + 5z^2$

112. $12 - x - x^2$

113. $18 + 17x - x^2$

114. $21 - 20x - x^2$

115. $360y^2 + 4y - 4$

116. $10t^2 - 5t - 50$

117. $16t^2 + 40t - 96$

118. $3p^3 - 16p^2 + 5p$

119. $6p^3 + 5p^2 + p$

120. $26z^2 + 98z - 24$

121. $30z^2 - 87z + 30$

122. $12a^3 + 14a^2 - 48a$

123. $42a^3 + 45a^2 - 27a$

124. $36p^2 - 9p^3 - p^4$

125. $9x^2y - 30xy^2 + 25y^3$

126. $8x^2y - 38xy^2 + 35y^3$

127. $9x^3y - 24x^2y^2 + 16xy^3$

128. $45a^3b - 78a^2b^2 + 24ab^3$

6.3 Applying Concepts

Factor.

129. $6y + 8y^3 - 26y^2$

130. $22p^2 - 3p^3 + 16p$

131. $a^3b - 24ab - 2a^2b$

132. $3xy^2 - 14xy + 2xy^3$

133. $25t^2 + 60t - 10t^3$

134. $3xy^3 + 2x^3y - 7x^2y^2$

Factor.

135. $2(y + 2)^2 - (y + 2) - 3$

136. $3(a + 2)^2 - (a + 2) - 4$

137. $10(x + 1)^2 - 11(x + 1) - 6$

138. $4(y - 1)^2 - 7(y - 1) - 2$

Find all integers k such that the trinomial can be factored over the integers.

139. $2x^2 + kx + 3$

140. $2x^2 + kx - 3$

141. $3x^2 + kx + 2$

142. $3x^2 + kx - 2$

143. $2x^2 + kx + 5$

144. $2x^2 + kx - 5$

145. Given that $x + 2$ is a factor of $x^3 - 2x^2 - 5x + 6$, factor $x^3 - 2x^2 - 5x + 6$ completely.

146. In your own words, explain how the signs of the last terms of the two binomial factors of a trinomial are determined.

147. **Geometry** The area of a rectangle is $(3x^2 + x - 2)$ ft^2. Find the dimensions of the rectangle in terms of the variable x. Given that $x > 0$, specify the dimension that is the length and the dimension that is the width. Can x be negative? Can $x = 0$? Explain your answers.

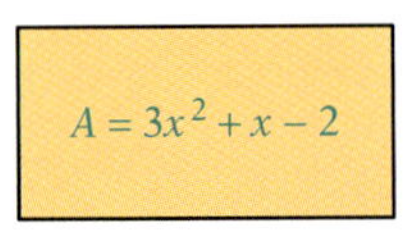

148. **Geometry** Write the area of the shaded region in factored form.

a.

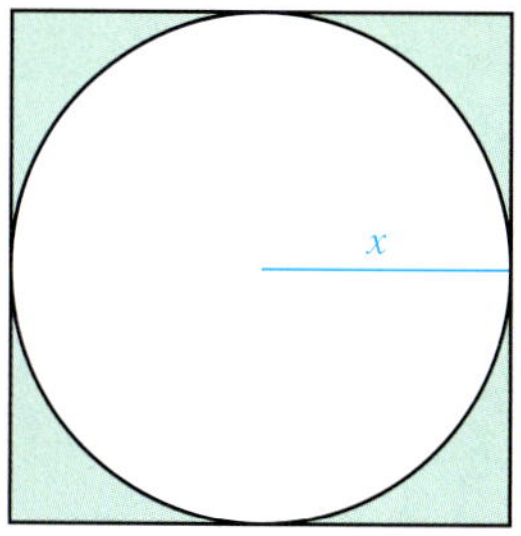

b.

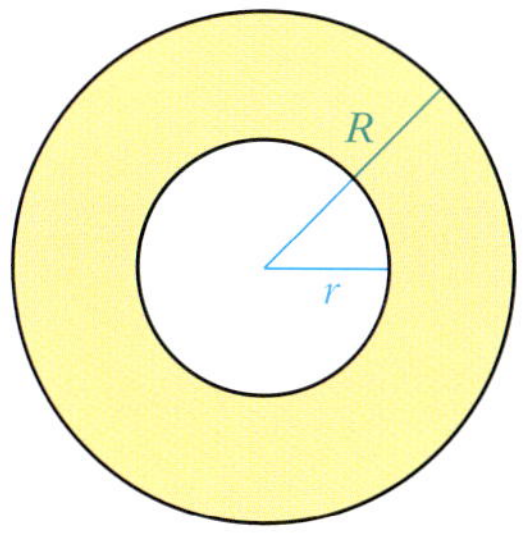

c.

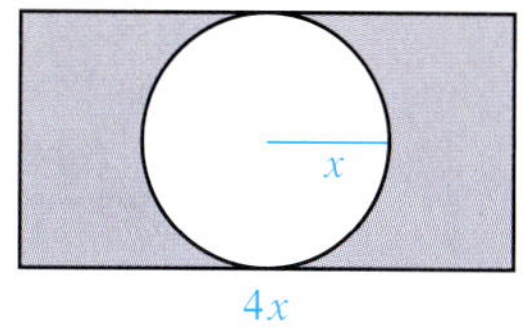

6.4 Special Factoring

1 Factor the difference of two perfect squares and factor perfect-square trinomials

The product of a term and itself is called a **perfect square.** The exponents on variables of perfect squares are always even numbers.

Term		*Perfect Square*
5	$5 \cdot 5 =$	25
x	$x \cdot x =$	x^2
$3y^4$	$3y^4 \cdot 3y^4 =$	$9y^8$
x^n	$x^n \cdot x^n =$	x^{2n}

The **square root of a perfect square** is one of the two equal factors of the perfect square. $\sqrt{\ }$ is the symbol for square root. To find the exponent of the square root of a variable term, divide the exponent by 2. For the examples at the right, assume that x and y represent positive numbers.

$\sqrt{25} = 5$

$\sqrt{x^2} = x$

$\sqrt{9y^8} = 3y^4$

$\sqrt{x^{2n}} = x^n$

Take Note

$(a + b)$ is the sum of the two terms a and b. $(a - b)$ is the difference of the two terms a and b. a and b are the square roots of a^2 and b^2.

The factors of the **difference of two perfect squares** are the sum and difference of the square roots of the perfect squares.

Factors of the Difference of Two Perfect Squares

$$a^2 - b^2 = (a + b)(a - b)$$

The *sum* of two perfect squares, $a^2 + b^2$, is nonfactorable over the integers.

Example 1 Factor: $25x^2 - 1$

Solution $25x^2 - 1 = (5x)^2 - 1^2$ • Write the binomial as the difference of two perfect squares.

$= (5x + 1)(5x - 1)$ • The factors are the sum and difference of the square roots of the perfect squares.

Problem 1 Factor: $x^2 - 36y^4$

Solution See page S21.

A perfect-square trinomial is the square of a binomial.

Factors of a Perfect-Square Trinomial

$$a^2 + 2ab + b^2 = (a + b)^2$$
$$a^2 - 2ab + b^2 = (a - b)^2$$

In factoring a **perfect-square trinomial,** remember that the terms of the binomial are the square roots of the perfect squares of the trinomial. The sign in the binomial is the sign of the middle term of the trinomial.

Factor: $4x^2 - 12x + 9$

Because $4x^2$ is a perfect square and 9 is a perfect square, try factoring $4x^2 - 12x + 9$ as the square of a binomial.

$$4x^2 - 12x + 9 \stackrel{?}{=} (2x - 3)^2$$

Check:
$(2x - 3)^2 = (2x - 3)(2x - 3) = 4x^2 - 6x - 6x + 9 = 4x^2 - 12x + 9$

The factorization checks. Thus $4x^2 - 12x + 9 = (2x - 3)^2$.

It is important to check a proposed factorization as we did above. The next example illustrates the importance of this check.

Factor: $x^2 + 13x + 36$

Because x^2 is a perfect square and 36 is a perfect square, try factoring $x^2 + 13x + 36$ as the square of a binomial.

$$x^2 + 13x + 36 \stackrel{?}{=} (x + 6)^2$$

Check: $(x + 6)^2 = (x + 6)(x + 6) = x^2 + 6x + 6x + 36 = x^2 + 12x + 36$

In this case, the proposed factorization of $x^2 + 13x + 36$ does *not* check. Try another factorization. The numbers 4 and 9 are factors of 36 whose sum is 13.

$$x^2 + 13x + 36 = (x + 4)(x + 9)$$

Example 2 Factor: $4x^2 - 20x + 25$

Solution $4x^2 - 20x + 25 = (2x - 5)^2$

Problem 2 Factor: $9x^2 + 12x + 4$

Solution See page S22.

2 Factor the sum or the difference of two cubes

VIDEO & DVD CD TUTOR WEB SSM

The product of the same three factors is called a **perfect cube.** The exponents on variables of perfect cubes are always divisible by 3.

Term		*Perfect Cube*
2	$2 \cdot 2 \cdot 2 =$	8
$3y$	$3y \cdot 3y \cdot 3y =$	$27y^3$
x^2	$x^2 \cdot x^2 \cdot x^2 =$	x^6
x^n	$x^n \cdot x^n \cdot x^n =$	x^{3n}

The **cube root** of a perfect cube is one of the three equal factors of the perfect cube. $\sqrt[3]{}$ is the symbol for cube root. To find the exponent of the cube root of a perfect-cube variable term, divide the exponent by 3.

$$\sqrt[3]{8} = 2$$
$$\sqrt[3]{27y^3} = 3y$$
$$\sqrt[3]{x^6} = x^2$$
$$\sqrt[3]{x^{3n}} = x^n$$

The following rules are used to factor the sum or difference of two perfect cubes.

Factoring the Sum or Difference of Two Cubes

$$a^3 + b^3 = (a + b)(a^2 - ab + b^2)$$
$$a^3 - b^3 = (a - b)(a^2 + ab + b^2)$$

To factor $8x^3 - 27$:

Write the binomial as the difference of two perfect cubes.

$$8x^3 - 27 = (2x)^3 - 3^3$$

The terms of the binomial factor are the cube roots of the perfect cubes. The sign of the binomial factor is the same sign as in the given binomial. The trinomial factor is obtained from the binomial factor.

$$= (2x - 3)(4x^2 + 6x + 9)$$

Square of the first term ($4x^2$)

Opposite of the product of the two terms ($6x$)

Square of the last term (9)

Example 3 Factor. **A.** $x^3y^3 - 1$ **B.** $(x + y)^3 - x^3$

Solution **A.** $x^3y^3 - 1$

$= (xy)^3 - 1^3$ • Write the binomial as the difference of two perfect cubes.

$= (xy - 1)(x^2y^2 + xy + 1)$

B. $(x + y)^3 - x^3$
$= [(x + y) - x][(x + y)^2 + x(x + y) + x^2]$ • This is a difference of two perfect cubes.
$= y(x^2 + 2xy + y^2 + x^2 + xy + x^2)$ • Simplify.
$= y(3x^2 + 3xy + y^2)$

Problem 3 Factor. **A.** $8x^3 + y^3z^3$ **B.** $(x - y)^3 + (x + y)^3$

Solution See page S22.

3 Factor trinomials that are quadratic in form

VIDEO & DVD CD TUTOR WEB SSM

Certain trinomials can be expressed as quadratic trinomials by making suitable variable substitutions. A trinomial is **quadratic in form** if it can be written as

$$au^2 + bu + c$$

If we let $x^2 = u$, the trinomial $x^4 + 5x^2 + 6$ can be written as shown at the right.

$$x^4 + 5x^2 + 6 = (x^2)^2 + 5(x^2) + 6$$

The trinomial is quadratic in form.

$$= u^2 + 5u + 6$$

If we let $xy = u$, the trinomial $2x^2y^2 + 3xy - 9$ can be written as shown at the right.

$$2x^2y^2 + 3xy - 9 = 2(xy)^2 + 3(xy) - 9$$

The trinomial is quadratic in form.

$$= 2u^2 + 3u - 9$$

When a trinomial that is quadratic in form is factored, the variable part of the first term in each binomial factor will be u. For example, because $x^4 + 5x^2 + 6$ is quadratic in form when $x^2 = u$, the first term in each binomial factor will be x^2.

$$x^4 + 5x^2 + 6 = (x^2)^2 + 5(x^2) + 6$$
$$= (x^2 + 2)(x^2 + 3)$$

The trinomial $x^2y^2 - 2xy - 15$ is quadratic in form when $xy = u$. The first term in each binomial factor will be xy.

$$x^2y^2 - 2xy - 15 = (xy)^2 - 2(xy) - 15$$
$$= (xy + 3)(xy - 5)$$

Example 4 Factor. **A.** $6x^2y^2 - xy - 12$ **B.** $2x^4 + 5x^2 - 12$

Solution **A.** $6x^2y^2 - xy - 12$
$= (3xy + 4)(2xy - 3)$ • The trinomial is quadratic in form when $xy = u$.

B. $2x^4 + 5x^2 - 12$
$= (x^2 + 4)(2x^2 - 3)$ • The trinomial is quadratic in form when $x^2 = u$.

Problem 4 Factor. **A.** $6x^2y^2 - 19xy + 10$ **B.** $3x^4 + 4x^2 - 4$

Solution See page S22.

4 Factor completely

VIDEO & DVD CD TUTOR WEB SSM

When factoring a polynomial completely, ask the following questions about the polynomial.

1. Is there a common factor? If so, factor out the GCF.
2. If the polynomial is a binomial, is it the difference of two perfect squares, the sum of two perfect cubes, or the difference of two perfect cubes? If so, factor.
3. If the polynomial is a trinomial, is it a perfect-square trinomial or the product of two binomials? If so, factor.
4. If the polynomial has four terms, can it be factored by grouping? If so, factor.
5. Is each factor nonfactorable over the integers? If not, factor.

Take Note

Remember that you may have to factor more than once in order to write the polynomial as a product of *prime* factors.

Example 5 Factor. **A.** $x^2y + 2x^2 - y - 2$ **B.** $x^{4n} - y^{4n}$

Solution **A.** $x^2y + 2x^2 - y - 2$

$= (x^2y + 2x^2) - (y + 2)$ • Factor by grouping.

$= x^2(y + 2) - (y + 2)$

$= (y + 2)(x^2 - 1)$

$= (y + 2)(x + 1)(x - 1)$ • Factor the difference of two perfect squares.

B. $x^{4n} - y^{4n}$

$= (x^{2n})^2 - (y^{2n})^2$ • Factor the difference of two perfect squares.

$= (x^{2n} + y^{2n})(x^{2n} - y^{2n})$

$= (x^{2n} + y^{2n})[(x^n)^2 - (y^n)^2]$ • Factor the difference of two perfect squares.

$= (x^{2n} + y^{2n})(x^n + y^n)(x^n - y^n)$

Problem 5 Factor. **A.** $4x - 4y - x^3 + x^2y$ **B.** $x^{4n} - x^{2n}y^{2n}$

Solution See page S22.

6.4 Concept Review

Determine whether the following statements are always true, sometimes true, or never true.

1. $a^2b^2c^2$ is an example of a perfect-square trinomial.

2. The factors of $a^3 + b^3$ are $(a + b)(a + b)(a + b)$.

3. The difference of two perfect cubes is not factorable.

4. Factoring out the GCF is the first thing to try in factoring polynomials.

5. The polynomial $b^3 + 1$ is not factorable.

6.4 Exercises

1 Which of the expressions are perfect squares?

1. 4; 8; $25x^6$; $12y^{10}$; $100x^4y^4$

2. 9; 18; $15a^8$; $49b^{12}$; $64a^{16}b^2$

Name the square root of the expression.

3. $16z^8$

4. $36d^{10}$

5. $81a^4b^6$

6. $25m^2n^{12}$

7. Explain how to factor the difference of two perfect squares.

8. What is a perfect-square trinomial?

Factor.

9. $x^2 - 16$

10. $y^2 - 49$

11. $4x^2 - 1$

12. $81x^2 - 4$

13. $b^2 - 2b + 1$

14. $a^2 + 14a + 49$

15. $16x^2 - 40x + 25$

16. $49x^2 + 28x + 4$

17. $x^2y^2 - 100$

18. $a^2b^2 - 25$

19. $x^2 + 4$

20. $a^2 + 16$

21. $x^2 + 6xy + 9y^2$

22. $4x^2y^2 + 12xy + 9$

23. $4x^2 - y^2$

24. $49a^2 - 16b^4$

25. $a^{2n} - 1$

26. $b^{2n} - 16$

27. $a^2 + 4a + 4$

28. $b^2 - 18b + 81$

29. $x^2 - 12x + 36$

30. $y^2 - 6y + 9$

31. $16x^2 - 121$

32. $49y^2 - 36$

33. $1 - 9a^2$

34. $16 - 81y^2$

35. $4a^2 + 4a - 1$

36. $9x^2 + 12x - 4$

37. $b^2 + 7b + 14$

38. $y^2 - 5y + 25$

39. $25 - a^2b^2$

40. $64 - x^2y^2$

41. $25a^2 - 40ab + 16b^2$

42. $4a^2 - 36ab + 81b^2$

43. $x^{2n} + 6x^n + 9$

44. $y^{2n} - 16y^n + 64$

2 Which of the expressions are perfect cubes?

45. 4; 8; x^9; a^8b^8; $27c^{15}d^{18}$

46. 9; 27; y^{12}; m^3n^6; $64mn^9$

Name the cube root of the expression.

47. $8x^9$ **48.** $27y^{15}$ **49.** $64a^6b^{18}$ **50.** $125c^{12}d^3$

Factor.

51. $x^3 - 27$ **52.** $y^3 + 125$ **53.** $8x^3 - 1$

54. $64a^3 + 27$ **55.** $x^3 - y^3$ **56.** $x^3 - 8y^3$

57. $m^3 + n^3$ **58.** $27a^3 + b^3$ **59.** $64x^3 + 1$

60. $1 - 125b^3$ **61.** $27x^3 - 8y^3$ **62.** $64x^3 + 27y^3$

63. $x^3y^3 + 64$ **64.** $8x^3y^3 + 27$ **65.** $16x^3 - y^3$

66. $27x^3 - 8y^3$ **67.** $8x^3 - 9y^3$ **68.** $27a^3 - 16$

69. $(a - b)^3 - b^3$ **70.** $a^3 + (a + b)^3$

71. $x^{6n} + y^{3n}$ **72.** $x^{3n} + y^{3n}$

73. What does it mean for a polynomial to be quadratic in form?

74. Do all polynomials that are quadratic in form factor? If not, give an example of one that does not factor.

3 Factor.

75. $x^2y^2 - 8xy + 15$ **76.** $x^2y^2 - 8xy - 33$ **77.** $x^2y^2 - 17xy + 60$

78. $a^2b^2 + 10ab + 24$ **79.** $x^4 - 9x^2 + 18$ **80.** $y^4 - 6y^2 - 16$

81. $b^4 - 13b^2 - 90$ **82.** $a^4 + 14a^2 + 45$ **83.** $x^4y^4 - 8x^2y^2 + 12$

84. $a^4b^4 + 11a^2b^2 - 26$ **85.** $x^{2n} + 3x^n + 2$ **86.** $a^{2n} - a^n - 12$

87. $3x^2y^2 - 14xy + 15$ **88.** $5x^2y^2 - 59xy + 44$ **89.** $6a^2b^2 - 23ab + 21$

90. $10a^2b^2 + 3ab - 7$ **91.** $2x^4 - 13x^2 - 15$ **92.** $3x^4 + 20x^2 + 32$

93. $2x^{2n} - 7x^n + 3$ **94.** $4x^{2n} + 8x^n - 5$ **95.** $6a^{2n} + 19a^n + 10$

4 Factor.

96. $5x^2 + 10x + 5$ **97.** $12x^2 - 36x + 27$ **98.** $3x^4 - 81x$

99. $27a^4 - a$ **100.** $7x^2 - 28$ **101.** $20x^2 - 5$

102. $y^4 - 10y^3 + 21y^2$ **103.** $y^5 + 6y^4 - 55y^3$ **104.** $x^4 - 16$

105. $16x^4 - 81$

106. $8x^5 - 98x^3$

107. $16a - 2a^4$

108. $x^3y^3 - x^3$

109. $x^3 + 2x^2 - x - 2$

110. $2x^3 - 3x^2 - 8x + 12$

111. $2x^3 + 4x^2 - 3x - 6$

112. $3x^3 - 3x^2 + 4x - 4$

113. $x^3 + x^2 - 16x - 16$

114. $4x^3 + 8x^2 - 9x - 18$

115. $a^3b^6 - b^3$

116. $x^6y^6 - x^3y^3$

117. $x^4 - 2x^3 - 35x^2$

118. $x^4 + 15x^3 - 56x^2$

119. $4x^2 + 4x - 1$

120. $8x^4 - 40x^3 + 50x^2$

121. $6x^5 + 74x^4 + 24x^3$

122. $x^4 - y^4$

123. $16a^4 - b^4$

124. $x^6 + y^6$

125. $x^4 - 5x^2 - 4$

126. $a^4 - 25a^2 - 144$

127. $3b^5 - 24b^2$

128. $16a^4 - 2a$

129. $x^4y^2 - 5x^3y^3 + 6x^2y^4$

130. $a^4b^2 - 8a^3b^3 - 48a^2b^4$

131. $16x^3y + 4x^2y^2 - 42xy^3$

132. $24a^2b^2 - 14ab^3 - 90b^4$

133. $x^3 - 2x^2 - x + 2$

134. $x^3 - 2x^2 - 4x + 8$

135. $8xb - 8x - 4b + 4$

136. $4xy + 8x + 4y + 8$

137. $4x^2y^2 - 4x^2 - 9y^2 + 9$

138. $4x^4 - x^2 - 4x^2y^2 + y^2$

139. $x^5 - 4x^3 - 8x^2 + 32$

140. $x^6y^3 + x^3 - x^3y^3 - 1$

141. $a^{2n+2} - 6a^{n+2} + 9a^2$

142. $x^{2n+1} + 2x^{n+1} + x$

143. $2x^{n+2} - 7x^{n+1} + 3x^n$

144. $3b^{n+2} + 4b^{n+1} - 4b^n$

6.4 Applying Concepts

Find all integers k such that the trinomial is a perfect-square trinomial.

145. $4x^2 - kx + 25$

146. $9x^2 - kx + 1$

147. $16x^2 + kxy + y^2$

148. $49x^2 + kxy + 64y^2$

Factor.

149. $ax^3 + b - bx^3 - a$

150. $xy^2 - 2b - x + 2by^2$

151. $y^{8n} - 2y^{4n} + 1$

152. $x^{6n} - 1$

Solve.

153. **Number Problem** The product of two numbers is 63. One of the two numbers is a perfect square. The other is a prime number. Find the sum of the two numbers.

154. **Number Problem** What is the smallest whole number by which 250 can be multiplied so that the product will be a perfect square?

155. **Number Problem** Palindromic numbers are natural numbers that remain unchanged when their digits are written in reverse order. Find all perfect squares less than 500 that are palindromic numbers.

156. **Food Science** A large circular cookie is cut from a square piece of dough. The diameter of the cookie is x cm. The piece of dough is x cm on a side and 1 cm deep. In terms of x, how many cubic centimeters of dough are left over? Use the approximation 3.14 for π.

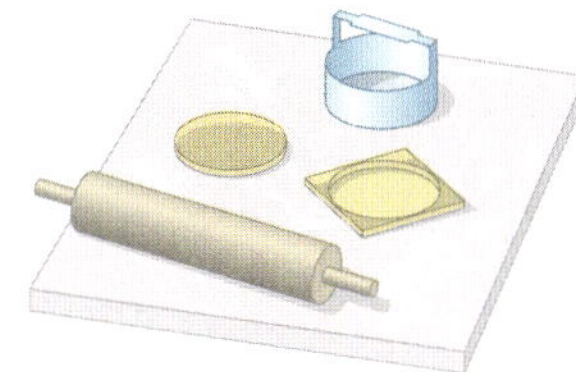

157. Factor $x^4 + 64$. (*Suggestion:* Add and subtract $16x^2$ so that the expression becomes $(x^4 + 16x^2 + 64) - 16x^2$. Now factor the difference of two squares.)

158. Using the strategy of Exercise 157, factor $x^4 + 4$. (*Suggestion:* Add and subtract $4x^2$.)

159. Can a third-degree polynomial have factors $x - 1$, $x + 1$, $x - 3$, and $x + 4$? Why or why not?

6.5 Solving Equations by Factoring

1 Solve equations by factoring

VIDEO & DVD CD TUTOR WEB SSM

Consider the equation $ab = 0$. If a is not zero, then b must be zero. Conversely, if b is not zero, then a must be zero. This is summarized in the Principle of Zero Products.

Principle of Zero Products

If the product of two factors is zero, then at least one of the factors must be zero.

$$\text{If } ab = 0, \text{ then } a = 0 \text{ or } b = 0.$$

The Principle of Zero Products is used to solve equations.

Solve: $(x - 4)(x + 2) = 0$

By the Principle of Zero Products, if $(x - 4)(x + 2) = 0$, then $x - 4 = 0$ or $x + 2 = 0$.

$$\begin{aligned} x - 4 &= 0 \\ x &= 4 \end{aligned} \qquad \begin{aligned} x + 2 &= 0 \\ x &= -2 \end{aligned}$$

Check:

$$\begin{array}{c|c} (x - 4)(x + 2) & 0 \\ \hline (4 - 4)(4 + 2) & 0 \\ 0 \cdot 6 & 0 \\ 0 = 0 & \end{array} \qquad \begin{array}{c|c} (x - 4)(x + 2) & 0 \\ \hline (-2 - 4)(-2 + 2) & 0 \\ -6 \cdot 0 & 0 \\ 0 = 0 & \end{array}$$

-2 and 4 check as solutions. The solutions are -2 and 4.

An equation of the form $ax^2 + bx + c = 0$, $a \neq 0$, is a **quadratic equation.** A quadratic equation is in **standard form** when the polynomial is written in descending order and is equal to zero.

Some quadratic equations can be solved by factoring and then using the Principle of Zero Products.

> **Take Note**
>
> Note the steps involved in solving a quadratic equation by factoring:
>
> 1. Write in standard form.
> 2. Factor.
> 3. Set each factor equal to 0.
> 4. Solve each equation.
> 5. Check the solutions.

Solve: $2x^2 - x = 1$

$$2x^2 - x = 1$$

Write the equation in standard form. $\quad 2x - x - 1 = 0$

Factor the trinomial. $\quad (2x + 1)(x - 1) = 0$

Use the Principle of Zero Products. $\quad 2x + 1 = 0 \qquad x - 1 = 0$

Solve each equation for x. $\quad 2x = -1 \qquad x = 1$

$$x = -\frac{1}{2}$$

The solutions check. $\quad$ The solutions are $-\frac{1}{2}$ and 1.

The Principle of Zero Products can be extended to more than two factors. For example, if $abc = 0$, then $a = 0$, $b = 0$, or $c = 0$.

Solve: $x^3 - x^2 - 4x + 4 = 0$

Factor by grouping.

$$\begin{aligned} x^3 - x^2 - 4x + 4 &= 0 \\ (x^3 - x^2) - (4x - 4) &= 0 \\ x^2(x - 1) - 4(x - 1) &= 0 \\ (x - 1)(x^2 - 4) &= 0 \\ (x - 1)(x + 2)(x - 2) &= 0 \end{aligned}$$

Use the Principle of Zero Products. $\quad x - 1 = 0 \qquad x + 2 = 0 \qquad x - 2 = 0$

Solve each equation for x. $\quad x = 1 \qquad x = -2 \qquad x = 2$

The solutions are -2, 1, and 2.

Example 1 Solve.

A. $3x^2 + 5x = 2$ **B.** $(x + 4)(x - 3) = 8$

C. $x^3 - x^2 - 25x + 25 = 0$

Solution **A.**

$$3x^2 + 5x = 2$$

$$3x^2 + 5x - 2 = 0$$

• **Write the equation in standard form.**

$$(3x - 1)(x + 2) = 0$$

• **Factor the trinomial.**

$$3x - 1 = 0 \qquad x + 2 = 0$$

• **Let each factor equal zero (the Principle of Zero Products).**

$$3x = 1 \qquad x = -2$$

$$x = \frac{1}{3}$$

• **Solve each equation for x.**

The solutions are $\frac{1}{3}$ and -2. • **Write the solutions.**

B. $(x + 4)(x - 3) = 8$

$x^2 + x - 12 = 8$ • Write the equation in standard form by first multiplying the binomials.

$x^2 + x - 20 = 0$ • Subtract 8 from each side of the equation.

$(x + 5)(x - 4) = 0$ • Factor the trinomial.

$x + 5 = 0 \quad x - 4 = 0$ • Let each factor equal zero.

$x = -5 \quad x = 4$ • Solve each equation for *x*.

The solutions are −5 and 4. • Write the solutions.

C. $x^3 - x^2 - 25x + 25 = 0$

$(x^3 - x^2) - (25x - 25) = 0$ • Factor by grouping.

$x^2(x - 1) - 25(x - 1) = 0$

$(x^2 - 25)(x - 1) = 0$

$(x + 5)(x - 5)(x - 1) = 0$

$x + 5 = 0 \quad x - 5 = 0 \quad x - 1 = 0$ • Let each factor equal zero.

$x = -5 \quad x = 5 \quad x = 1$ • Solve each equation for *x*.

The solutions are −5, 1, and 5. • Write the solutions.

Problem 1 Solve.

A. $4x^2 + 11x = 3$ **B.** $(x - 2)(x + 5) = 8$

C. $x^3 + 4x^2 - 9x - 36 = 0$

Solution See page S22.

The Principle of Zero Products is used to find elements in the domain of a quadratic function that correspond to a given element in the range.

Take Note

You can check the values of *c* by evaluating the function.

$f(x) = x^2 - 3x - 5$
$f(4) = 4^2 - 3(4) - 5$
$= -1 \ \checkmark$

$f(-1) = (-1)^2 - 3(-1) - 5$
$= -1 \ \checkmark$

Example 2 Given that −1 is in the range of the function defined by $f(x) = x^2 - 3x - 5$, find two values of *c* for which $f(c) = -1$.

Solution

$f(c) = -1$

$c^2 - 3c - 5 = -1$ • $f(c) = c^2 - 3c - 5$

$c^2 - 3c - 4 = 0$ • Solve for *c*.

$(c - 4)(c + 1) = 0$

$c - 4 = 0 \quad c + 1 = 0$

$c = 4 \quad c = -1$

The values of *c* are −1 and 4.

Problem 2 Given that 4 is in the range of the function defined by $s(t) = t^2 - t - 2$, find two values of *c* for which $s(c) = 4$.

Solution See page S22.

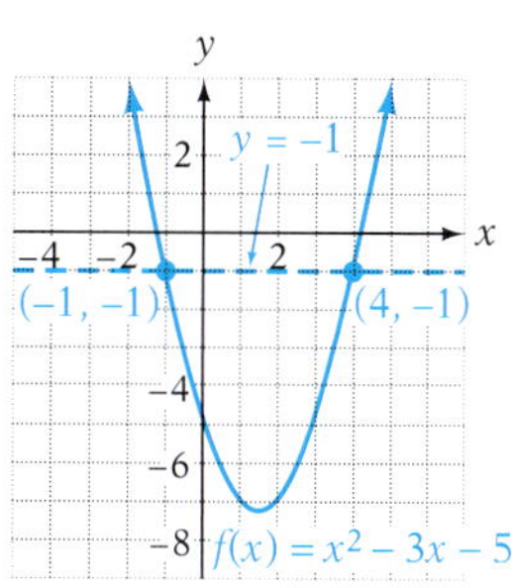

In Example 2, there are two values in the domain that can be paired with the range element −1. The two values are −1 and 4. Two ordered pairs that belong to the function are (−1, −1) and (4, −1). *Remember:* A function can have different first elements paired with the same second element. A function cannot have the same first element paired with different second elements.

The graph of $f(x) = x^2 - 3x - 5$ is shown at the left.

2 Application problems

VIDEO & DVD · CD TUTOR · WEB · SSM

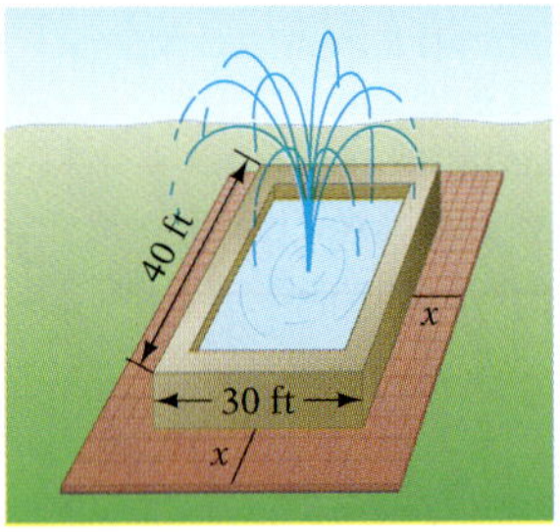

Example 3 An architect wants to design a fountain to be placed on the front lawn of a new art museum. The base of the fountain is to be 30 ft by 40 ft with a uniform brick walkway around the fountain. The total area of the fountain and the walkway is 2576 ft^2. How wide is the walkway?

Strategy Let x represent the width of the brick walkway. Then the width of the fountain and walkway is $30 + 2x$, and the length of the fountain and walkway is $40 + 2x$. Use the formula $A = LW$, where $A = 2576$, to write an equation.

Solution

$$A = LW$$
$$2576 = (40 + 2x)(30 + 2x)$$ • Replace A, L, and W by their values.
$$2576 = 1200 + 140x + 4x^2$$ • Multiply.
$$0 = 4x^2 + 140x - 1376$$
$$0 = x^2 + 35x - 344$$ • Divide each side by 4.
$$0 = (x - 8)(x + 43)$$ • Factor and use the Principle of Zero Products.

$$x - 8 = 0 \qquad x + 43 = 0$$
$$x = 8 \qquad x = -43$$

A walkway −43 ft wide would not make sense.

The walkway must be 8 ft wide.

Problem 3 A diagonal of a polygon is a line from one vertex of the polygon to a nonadjacent vertex. The number of diagonals, D, of a polygon with n sides is given by $D = \frac{n(n-3)}{2}$. Find the number of sides for a polygon that has 20 diagonals.

Solution See page S22.

6.5 Concept Review

Determine whether the following statements are always true, sometimes true, or never true.

1. The Principle of Zero Products can be stated as follows: If $ab = 0$, then $a = 0$ or $b = 0$.

2. One side of an equation must be zero in order for us to solve the equation by factoring.

3. Two solutions of an equation are found when solving an equation by the Principle of Zero Products.

4. The Principle of Zero Products can be used to solve a second-degree equation.

5. If n is an integer, then $n + 1$ and $n + 3$ represent odd integers.

6. By the Principle of Zero Products, if $(x + 2)(x + 3) = 4$, then $x + 2 = 4$ or $x + 3 = 4$.

6.5 Exercises

1 1. What is a quadratic equation? How does it differ from a linear equation? Give an example of each type of equation.

2. How is the Principle of Zero Products used to solve some quadratic equations?

Solve.

3. $(y + 4)(y + 6) = 0$
4. $(a - 5)(a - 2) = 0$
5. $x(x - 7) = 0$
6. $b(b + 8) = 0$
7. $3z(2z + 5) = 0$
8. $4y(3y - 2) = 0$
9. $(2x + 3)(x - 7) = 0$
10. $(4a - 1)(a + 9) = 0$
11. $b^2 - 49 = 0$
12. $4z^2 - 1 = 0$
13. $9t^2 - 16 = 0$
14. $x^2 + x - 6 = 0$
15. $y^2 + 4y - 5 = 0$
16. $a^2 - 8a + 16 = 0$
17. $2b^2 - 5b - 12 = 0$
18. $t^2 - 8t = 0$
19. $x^2 - 9x = 0$
20. $2y^2 - 10y = 0$
21. $3a^2 - 12a = 0$
22. $b^2 - 4b = 32$
23. $z^2 - 3z = 28$
24. $2x^2 - 5x = 12$
25. $3t^2 + 13t = 10$
26. $4y^2 - 19y = 5$
27. $5b^2 - 17b = -6$
28. $6a^2 + a = 2$
29. $8x^2 - 10x = 3$
30. $z(z - 1) = 20$
31. $y(y - 2) = 35$
32. $t(t + 1) = 42$
33. $x(x - 12) = -27$
34. $x(2x - 5) = 12$
35. $y(3y - 2) = 8$
36. $2b^2 - 6b = b - 3$
37. $3a^2 - 4a = 20 - 15a$
38. $2t^2 + 5t = 6t + 15$
39. $(y + 5)(y - 7) = -20$
40. $(x + 2)(x - 6) = 20$
41. $(b + 5)(b + 10) = 6$
42. $(a - 9)(a - 1) = -7$
43. $(t - 3)^2 = 1$
44. $(y - 4)^2 = 4$
45. $(3 - x)^2 + x^2 = 5$
46. $(2 - b)^2 + b^2 = 10$
47. $(a - 1)^2 = 3a - 5$
48. $2x^3 + x^2 - 8x - 4 = 0$
49. $x^3 + 4x^2 - x - 4 = 0$
50. $12x^3 - 8x^2 - 3x + 2 = 0$

Find the values c in the domain of f for which $f(c)$ is the indicated value.

51. $f(x) = x^2 - 3x + 3; f(c) = 1$

52. $f(x) = x^2 + 4x - 2; f(c) = 3$

53. $f(x) = 2x^2 - x - 5; f(c) = -4$

54. $f(x) = 6x^2 - 5x - 9; f(c) = -3$

55. $f(x) = 4x^2 - 4x + 3; f(c) = 2$

56. $f(x) = x^2 - 6x + 12; f(c) = 3$

57. $f(x) = x^3 + 9x^2 - x - 14; f(c) = -5$

58. $f(x) = x^3 + 3x^2 - 4x - 11; f(c) = 1$

2 **59.** **Number Problem** The sum of a number and its square is 90. Find the number.

60. **Number Problem** The sum of a number and its square is 132. Find the number.

61. **Number Problem** The sum of the squares of two consecutive positive integers is equal to 145. Find the two integers.

62. **Number Problem** The sum of the squares of two consecutive odd integers is equal to 290. Find the two integers.

63. **Number Problem** The sum of the cube of a number and the product of the number and twelve is equal to seven times the square of the number. Find the number.

64. **Number Problem** The sum of the cube of a number and the product of the number and seven is equal to eight times the square of the number. Find the number.

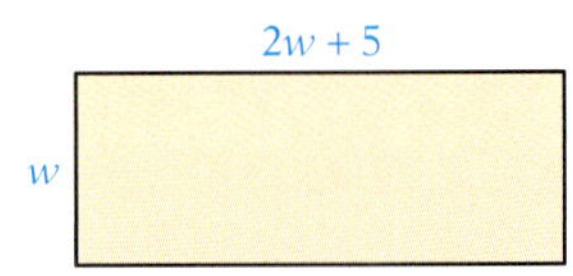

65. **Geometry** The length of a rectangle is 5 in. more than twice the width. The area is 168 in^2. Find the width and length of the rectangle.

66. **Geometry** The width of a rectangle is 5 ft less than the length. The area of the rectangle is 300 ft^2. Find the length and width of the rectangle.

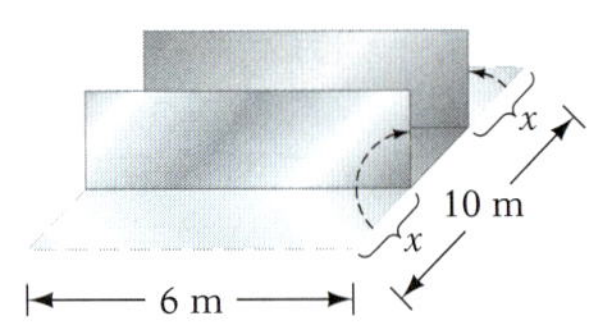

67. **Construction** A trough is made from a rectangular piece of metal by folding up the sides as shown in the figure at the right. What should the value of x be so that the volume is 72 m^3?

68. **Geometry** The number of possible diagonals, D, in a polygon with n sides is given by $D = \dfrac{n(n-3)}{2}$. Find the number of sides for a polygon with 54 diagonals.

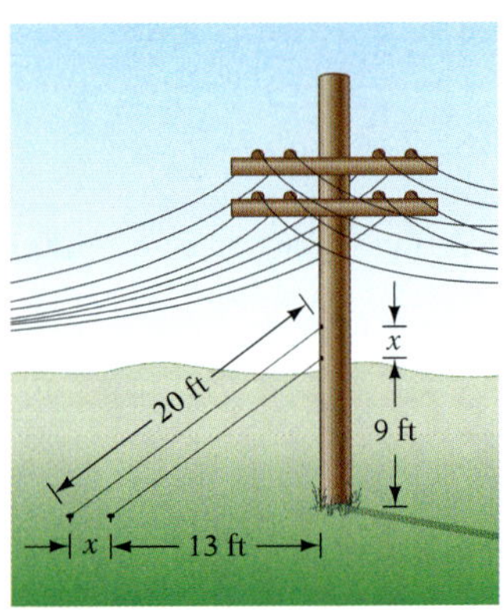

69. **Space Science** A small rocket is launched with an acceleration of 10 m/s^2. The velocity, v, of the rocket after it has traveled s meters is given by $v^2 = 20s$. Find the velocity of this rocket after it has traveled 500 m.

70. **Construction** A pole is being secured with two guy wires as shown in the figure at the right. What is the value of x so that the 20-foot guy wire will attach as shown?

71. **Geometry** The length of the base of a triangle is three times the height. The area of the triangle is 24 cm^2. Find the base and height of the triangle.

72. **Geometry** The height of a triangle is 4 cm more than twice the length of the base. The area of the triangle is 35 cm^2. Find the height of the triangle.

73. **Physics** An object is thrown downward, with an initial speed of 16 ft/s, from the top of a building 480 ft high. How many seconds later will the object hit the ground? Use the equation $d = vt + 16t^2$, where d is the distance in feet, v is the initial speed in feet per second, and t is the time in seconds.

74. **Physics** A stone is thrown into an empty well with an initial speed of 8 ft/s. The well is 624 ft deep. How many seconds later will the stone hit the bottom of the well? Use the equation $d = vt + 16t^2$, where d is the distance in feet, v is the initial speed in feet per second, and t is the time in seconds.

75. **Geometry** The length of a rectangle is 6 cm, and the width is 3 cm. If both the length and the width are increased by equal amounts, the area of the rectangle is increased by 70 cm^2. Find the length and width of the larger rectangle.

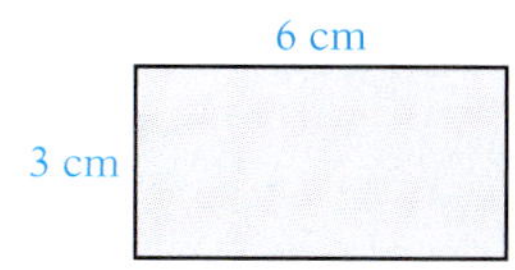

76. **Geometry** The width of a rectangle is 4 cm, and the length is 8 cm. If both the width and the length are increased by equal amounts, the area of the rectangle is increased by 64 cm^2. Find the length and width of the larger rectangle.

6.5 Applying Concepts

77. Find $3n^2 + 2n - 1$ if $n(n + 6) = 16$.

78. Find $4y^2 - y + 3$ if $(y - 1)(y + 2) = 4$.

79. Find $-a^2 + 5a - 2$ if $6a(a - 1) = 36$.

80. **Geometry** The perimeter of a rectangular garden is 44 m. The area of the garden is 120 m^2. Find the length and width of the garden.

81. **Construction** A rectangular piece of cardboard is 10 in. longer than it is wide. Squares 2 in. on a side are to be cut from each corner and then the sides folded up to make an open box with a volume of 112 in^3. Find the length and width of the piece of cardboard.

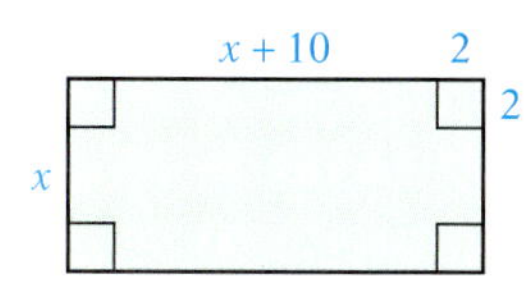

82. **Geometry** The sides of a rectangular box have areas of 16 cm^2, 20 cm^2, and 80 cm^2. Find the volume of the box.

83. Write an equation whose solutions are 3, 2, and -1.

84. The following seems to show that $1 = 2$. Explain the error.

$$\begin{aligned} a &= b \\ a^2 &= ab && \text{• Multiply each side of the equation by } a. \\ a^2 - b^2 &= ab - b^2 && \text{• Subtract } b^2 \text{ from each side of the equation.} \\ (a - b)(a + b) &= b(a - b) && \text{• Factor.} \\ a + b &= b && \text{• Divide each side by } a - b. \\ b + b &= b && \text{• Because } a = b\text{, substitute } b \text{ for } a. \\ 2b &= b \\ 2 &= 1 && \text{• Divide both sides by } b. \end{aligned}$$

Focus on Problem Solving

The Trial-and-Error Method

A topic in Section 3 of this chapter is factoring trinomials using trial factors. This method involves writing trial factors and then using the FOIL method to determine which pair of factors is the correct one. The **trial-and-error method** of arriving at a solution to a problem involves performing repeated tests or experiments until a satisfactory conclusion is reached. However, not all problems solved using the trial-and-error method have a strategy by which to determine the answer. Here is an example:

Explain how you could cut through a cube so that the face of the resulting solid is **a.** a square, **b.** an equilateral triangle, **c.** a trapezoid, and **d.** a hexagon.

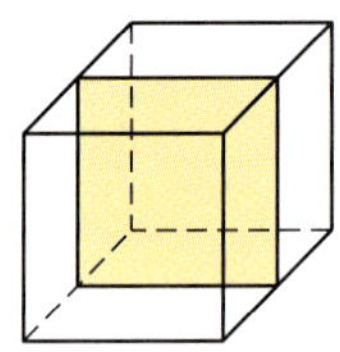

There is no formula to apply to this problem; there is no computation to perform. This problem requires picturing a cube and the results after cutting through it at different places on its surface and at different angles. For part (a), cutting perpendicular to the top and bottom of the cube and parallel to two of its sides will result in a square. The other shapes may prove more difficult.

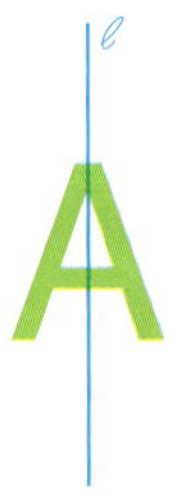

When solving problems of this type, keep an open mind. Sometimes when using the trial-and-error method, we are hampered by narrowness of vision; we cannot expand our thinking to include other possibilities. Then when we see someone else's solution, it appears so obvious! For example, for the question above, it is necessary to conceive of cutting through the cube at places other than the top surface; we need to be open to the idea of beginning the cut at one of the corner points of the cube.

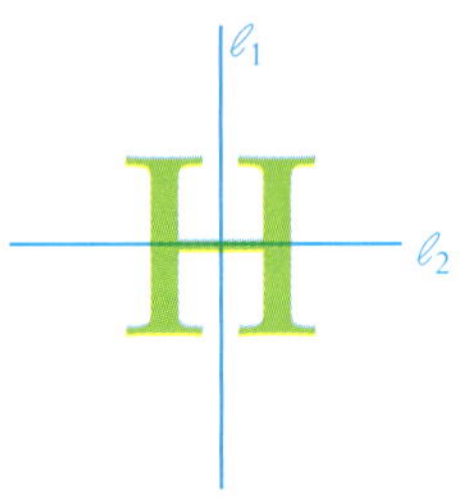

Look at the letter A printed at the left. If the letter were folded along line ℓ, the two sides of the letter would match exactly. This letter has **symmetry** with respect to line ℓ. Line ℓ is called the **axis of symmetry.** Now consider the letter H printed below at the left. Both lines ℓ_1 and ℓ_2 are axes of symmetry for this letter; the letter could be folded along either line and the two sides would match exactly. Does the letter A have more than one axis of symmetry? Find axes of symmetry for other capital letters of the alphabet. Which lower-case letters have one axis of symmetry? Do any of the lower-case letters have more than one axis of symmetry?

How many axes of symmetry does a square have? In determining lines of symmetry for a square, begin by drawing a square as shown at the right. The horizontal line of symmetry and the vertical line of symmetry may be immediately obvious to you. But there are two others. Do you see that a line drawn through opposite corners of the square is also a line of symmetry?

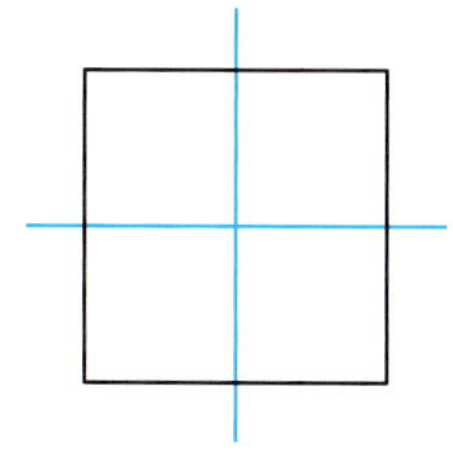

Find the number of axes of symmetry for each of the following figures: equilateral triangle, rectangle, regular pentagon, regular hexagon, and circle.

Many of the questions in this text that require an answer of "always true, sometimes true, or never true" are best solved by the trial-and-error method. For example, consider the following statement:

If two rectangles have the same area, then they have the same perimeter.

Try some numbers. Each of two rectangles, one measuring 6 units by 2 units and another measuring 4 units by 3 units, has an area of 12 square units, but the perimeter of the first is 16 units and the perimeter of the second is 14 units. Thus the answer "always true" has been eliminated. We still need to determine whether there is a case when the statement is true. After experimenting with a lot of numbers, you may come to realize that we are trying to determine whether it is possible for two different pairs of factors of a number to have the same sum. Is it?

Don't be afraid to make many experiments, and remember that *errors*, or tests that "don't work," are a part of the trial-and-*error* process.

Projects & Group Activities

Prime and Composite Numbers

A **prime number** is a natural number greater than 1 whose only natural number factors are itself and 1. The number 11 is a prime number because the only natural number factors of 11 are 11 and 1.

Eratosthenes, a Greek philosopher and astronomer who lived from 270 to 190 B.C., devised a method of identifying prime numbers. It is called the **Sieve of Eratosthenes.** The procedure is illustrated in the following paragraphs.

~~1~~	(2)	(3)	~~4~~	(5)	~~6~~	(7)	~~8~~	~~9~~	~~10~~
(11)	~~12~~	(13)	~~14~~	~~15~~	~~16~~	(17)	~~18~~	(19)	~~20~~
~~21~~	~~22~~	(23)	~~24~~	~~25~~	~~26~~	~~27~~	~~28~~	(29)	~~30~~
(31)	~~32~~	~~33~~	~~34~~	~~35~~	~~36~~	(37)	~~38~~	~~39~~	~~40~~
(41)	~~42~~	(43)	~~44~~	~~45~~	~~46~~	(47)	~~48~~	~~49~~	~~50~~
~~51~~	~~52~~	(53)	~~54~~	~~55~~	~~56~~	~~57~~	~~58~~	(59)	~~60~~
(61)	~~62~~	~~63~~	~~64~~	~~65~~	~~66~~	(67)	~~68~~	~~69~~	~~70~~
(71)	~~72~~	(73)	~~74~~	~~75~~	~~76~~	~~77~~	~~78~~	(79)	~~80~~
~~81~~	~~82~~	(83)	~~84~~	~~85~~	~~86~~	~~87~~	~~88~~	(89)	~~90~~
~~91~~	~~92~~	~~93~~	~~94~~	~~95~~	~~96~~	(97)	~~98~~	~~99~~	~~100~~

List all the natural numbers from 1 to 100. Cross out the number 1, because it is not a prime number. The number 2 is prime; circle it. Cross out all the other multiples of 2 (4, 6, 8, . . .), because they are not prime. The number 3 is prime; circle it. Cross out all the other multiples of 3 (6, 9, 12, . . .) that are not already crossed out. The number 4, the next consecutive number in the list, has already been crossed out. The number 5 is prime; circle it. Cross out all the other multiples of 5 that are not already crossed out. Continue in this manner until all the prime numbers less than 100 are circled.

A **composite number** is a natural number greater than 1 that has a natural-number factor other than itself and 1. The number 21 is a composite number because it has factors of 3 and 7. All the numbers crossed out in the table above, except the number 1, are composite numbers.

Solve.

1. Use the Sieve of Eratosthenes to find the prime numbers between 100 and 200.

2. How many prime numbers are even numbers?

3. Find the "twin primes" between 1 and 200. Twin primes are two prime numbers whose difference is 2. For instance, 3 and 5 are twin primes; 5 and 7 are also twin primes.

4. a. List two prime numbers that are consecutive natural numbers.

b. Can there be any other pairs of prime numbers that are consecutive natural numbers?

5. Some primes are the sum of a square and 1. For example, $5 = 2^2 + 1$. Find another prime p such that $p = n^2 + 1$, where n is a natural number.

6. Find a prime p such that $p = n^2 - 1$, where n is a natural number.

7. a. $4!$ (which is read "4 factorial") is equal to $4 \cdot 3 \cdot 2 \cdot 1$. Show that $4! + 2$, $4! + 3$, and $4! + 4$ are all composite numbers.

b. $5!$ (which is read "5 factorial") is equal to $5 \cdot 4 \cdot 3 \cdot 2 \cdot 1$. Will $5! + 2$, $5! + 3$, $5! + 4$, and $5! + 5$ generate four consecutive composite numbers?

c. Use the notation $6!$ to represent a list of five consecutive composite numbers.

Problems Involving Falling Objects

When an object is dropped, the object accelerates due to Earth's gravitational pull. This means that the speed of the object continually increases. If we assume that there is little or no air resistance, then the height h, in feet, of an object as it drops is given by the equation

$$h = -16t^2 + h_0$$

where t is the number of seconds the object has been falling and h_0 is the initial height, in feet, from which the object was dropped.

Note that the equation above applies to an object that is *dropped*. If an object is *thrown* either up or down, then the equation must reflect the effect of the initial vertical velocity. Again assuming that there is little or no air resistance, the equation for the height h, in feet, of an object that is thrown is

$$h = -16t^2 + v_0t + h_0$$

where t is the number of seconds since the object was thrown, v_0 is the initial vertical velocity in feet per second, and h_0 is the initial height, in feet, from which the object was thrown. In this equation, **v_0 is positive for an object moving upward, negative for an object moving downward, and zero for an object that is stationary.**

Equations for Problems Involving Falling Objects

(neglecting air resistance)

For an object that is dropped: $h = -16t^2 + h_0$

For an object that is projected: $h = -16t^2 + v_0t + h_0$

h = height, in feet
t = time in motion, in seconds
v_0 = initial vertical velocity, in feet per second
h_0 = initial height, in feet

Suppose you drop a basketball from the top of the gymnasium bleachers, a height of 16 ft, to the floor below. How long after you drop the basketball does it hit the ground?

$h = -16t^2 + h_0$ • Use the formula for an object that is dropped.

$0 = -16t^2 + 16$ • $h = 0$, the height of the basketball when it hits the ground. $h_0 = 16$, the initial height.

$0 = -16(t^2 - 1)$ • Factor -16 from $-16t^2 + 16$.

$0 = -16(t + 1)(t - 1)$ • Factor the difference of two squares.

$t + 1 = 0 \qquad t - 1 = 0$ • Use the Principle of Zero Products.

$t = -1 \qquad t = 1$ • Solve each equation for t.

The time cannot be a negative number. The solution $t = -1$ is not possible.

The basketball hits the ground 1 s after you drop it.

An arrow is shot upward at an initial velocity of 144 ft/s. Find the time for the arrow to return to Earth.

$h = -16t^2 + v_0t + h_0$ • Use the formula for an object that is projected.

$0 = -16t^2 + 144t + 0$ • $h = 0$, the height of the arrow when it returns to Earth. $v_0 = 144$, the initial velocity. (The initial velocity is positive because the object is moving upward.) $h_0 = 0$, the initial height.

$0 = -16t^2 + 144t$

$0 = -16t(t - 9)$ • Factor $-16t$ from $-16t^2 + 144t$.

$-16t = 0 \qquad t - 9 = 0$ • Use the Principle of Zero Products.

$t = 0 \qquad t = 9$ • Solve each equation for t.

The solution $t = 0$ is the time when the arrow is shot upward.

The arrow returns to Earth after 9 s.

Solve.

1. A bundle of leaflets is dropped from a plane at an altitude of 1600 ft. How many seconds later will the bundle hit the ground?

2. You drop your glasses from the top of a cliff that is 256 ft above the ocean. How many seconds later will your glasses hit the water?

3. A roofer tosses an old shingle from the roof of a building 320 ft high. The initial velocity is 16 ft/s. How many seconds later will the shingle hit the ground?

4. To test its bounce, a tennis ball is thrown onto a tennis court. It rebounds straight up with an initial velocity of 32 ft/s. How many seconds later will the tennis ball return to the ground?

5. In a slow-pitch softball game, the pitcher releases the ball at a height of 4 ft with an initial vertical velocity of 24 ft/s. The batter hits the ball when it is 4 ft off the ground. For how many seconds is the ball in the air before it is hit?

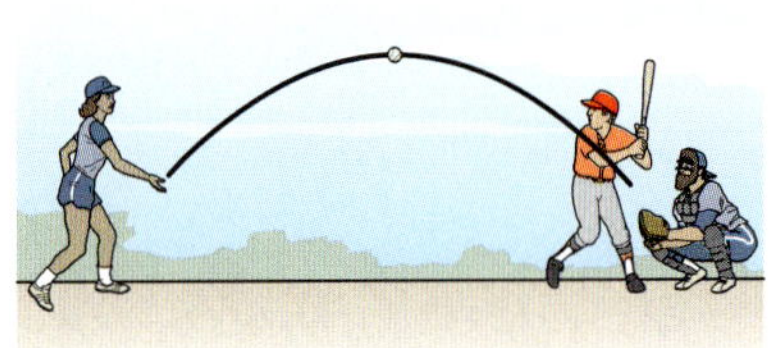

6. A basketball player shoots at a basket 25 ft away. The ball is released at a height of 8 ft with an initial vertical velocity of 33 ft/s. The height of the basket is 10 ft. How many seconds after the ball is released does it hit the basket? Explain what the solution you discarded represents.

7. An arrow is shot upward with an initial velocity of 128 ft/s. The equation $16t(t - a) = 0$, where t is the time in seconds and a is a constant, describes the times at which the arrow is on the ground. Suppose you know that the arrow is on the ground at 0 seconds and at 8 seconds. What is the value of a?

8. Suppose you double the height from which an object is dropped. Is the falling time of the object doubled? Why or why not?

9. Make up a problem involving an object that is dropped that can be solved by factoring a quadratic equation.

10. Make up a problem involving a projected object that can be solved by factoring a quadratic equation.

Chapter Summary

Key Words	*Examples*
The **greatest common factor** (GCF) of two or more integers is the greatest integer that is a factor of all the integers. [6.1.1, p. 398]	The greatest common factor of 12 and 18 is 6 because 6 is the greatest integer that divides evenly into both 12 and 18.
To **factor a polynomial** means to write the polynomial as a product of other polynomials. [6.1.1, p. 399]	To factor $8x + 12$ means to write it as the product $4(2x + 3)$. The expression $8x + 12$ is a sum. The expression $4(2x + 3)$ is a product (the polynomial $2x + 3$ is multiplied by 4).
A polynomial that does not factor using only integers is **nonfactorable over the integers.** [6.2.1, p. 406]	The trinomial $x^2 + x + 2$ is nonfactorable over the integers. There are no two integers whose product is 2 and whose sum is 1.
To **factor a trinomial of the form $ax^2 + bx + c$** means to express the trinomial as the product of two binomials. [6.3.1, p. 410]	To factor $3x^2 + 7x + 2$ means to write it as the product $(3x + 1)(x + 2)$.
The product of a term and itself is a **perfect square**. [6.4.2, p. 419]	$(5x)(5x) = 25x^2$; $25x^2$ is a perfect square.
The **square root of a perfect square** is one of the two equal factors of the perfect square. [6.4.1, p. 419]	$\sqrt{25x^2} = 5x$
The product of three factors is called a **perfect cube**. [6.4.2, p. 421]	$(2x)(2x)(2x) = 8x^3$; $8x^3$ is a perfect cube.
The **cube root** of a perfect cube is one of the three equal factors of the perfect cube. [6.4.2, p. 421]	$\sqrt[3]{8x^3} = 2x$
A trinomial is **quadratic in form** if it can be written as $au^2 + bu + c$ [6.4.3, p. 422]	Let $x^2 = u$. $x^4 + 7x^2 + 12 = u^2 + 7u + 12$ $x^4 + 7x^2 + 12$ is quadratic in form.
An equation of the form $ax^2 + bx + c = 0$, $a \neq 0$, is a **quadratic equation.** A quadratic equation is in **standard form** when the polynomial is in descending order and equal to zero. [6.5.1, p. 428]	The equation $2x^2 + 7x + 3 = 0$ is a quadratic equation in standard form.

Essential Rules and Procedures

Factoring the Difference of Two Squares [6.4.1, p. 419]
The difference of two squares is the product of the sum and difference of two terms.
$a^2 - b^2 = (a + b)(a - b)$

$y^2 - 81 = (y + 9)(y - 9)$

Factoring a Perfect-Square Trinomial [6.4.1, p. 420]
A perfect-square trinomial is the square of a binomial.
$a^2 + 2ab + b^2 = (a + b)^2$
$a^2 - 2ab + b^2 = (a - b)^2$

$x^2 + 10x + 25 = (x + 5)^2$
$x^2 - 10x + 25 = (x - 5)^2$

Factoring the Sum or Difference of Two Cubes [6.4.2, p. 421]
$a^3 + b^3 = (a + b)(a^2 - ab + b^2)$
$a^3 - b^3 = (a - b)(a^2 + ab + b^2)$

$x^3 + 64 = (x + 4)(x^2 - 4x + 16)$
$8b^3 - 1 = (2b - 1)(4b^2 + 2b + 1)$

General Factoring Strategy [6.4.4, p. 423]
1. Is there a common factor? If so, factor out the GCF.
2. If the polynomial is a binomial, is it the difference of two perfect squares, the sum of two perfect cubes, or the difference of two perfect cubes? If so, factor.
3. If the polynomial is a trinomial, is it a perfect-square trinomial or the product of two binomials? If so, factor.
4. If the polynomial has four terms, can it be factored by grouping? If so, factor.
5. Is each factor nonfactorable over the integers? If not, factor.

$24x + 6 = 6(4x + 1)$
$2x^2y + 6xy + 8y = 2y(x^2 + 3x + 4)$
$4x^2 - 49 = (2x + 7)(2x - 7)$
$9x^2 + 6x + 1 = (3x + 1)^2$
$2x^2 + 7x + 5 = (2x + 5)(x + 1)$
$x^3 - 3x^2 + 2x - 6 = (x^3 - 3x^2) + (2x - 6)$
$= x^2(x - 3) + 2(x - 3)$
$= (x - 3)(x^2 + 2)$
$x^4 - 16 = (x^2 + 4)(x^2 - 4)$
$= (x^2 + 4)(x + 2)(x - 2)$

The Principle of Zero Products [6.5.1, p. 427]
If the product of two factors is zero, then at least one of the factors must be zero.

If $a \cdot b = 0$, then $a = 0$ or $b = 0$.

The Principle of Zero Products is used to solve a quadratic equation by factoring.

$x^2 + x = 12$
$x^2 + x - 12 = 0$
$(x + 4)(x - 3) = 0$

$x + 4 = 0 \qquad x - 3 = 0$
$x = -4 \qquad x = 3$

Chapter Review Exercises

1. Factor: $14y^9 - 49y^6 + 7y^3$
2. Factor: $3a^2 - 12a + ab - 4b$
3. Factor: $c^2 + 8c + 12$
4. Factor: $a^3 - 5a^2 + 6a$
5. Factor: $6x^2 - 29x + 28$
6. Factor: $3y^2 + 16y - 12$
7. Factor: $18a^2 - 3a - 10$
8. Factor: $a^2b^2 - 1$

9. Factor: $4y^2 - 16y + 16$

10. Solve: $a(5a + 1) = 0$

11. Factor: $12a^2b + 3ab^2$

12. Factor: $b^2 - 13b + 30$

13. Factor: $10x^2 + 25x + 4xy + 10y$

14. Factor: $3a^2 - 15a - 42$

15. Factor: $n^4 - 2n^3 - 3n^2$

16. Factor: $2x^2 - 5x + 6$

17. Factor: $6x^2 - 7x + 2$

18. Factor: $16x^2 + 49$

19. Solve: $(x - 2)(2x - 3) = 0$

20. Factor: $7x^2 - 7$

21. Factor: $3x^5 - 9x^4 - 4x^3$

22. Factor: $4x(x - 3) - 5(3 - x)$

23. Factor: $a^2 + 5a - 14$

24. Factor: $y^2 + 5y - 36$

25. Factor: $5x^2 - 50x - 120$

26. Solve: $(x + 1)(x - 5) = 16$

27. Factor: $5x^{n+5} + x^{n+3} + 4x^2$

28. Factor: $x(y - 3) + 4(3 - y)$

29. Factor: $2ax + 4bx - 3ay - 6by$

30. Factor: $x^2 + 12x + 35$

31. Factor: $12 + x - x^2$

32. Factor: $x^2 - 16x + 63$

33. Factor: $6x^2 - 31x + 18$

34. Factor: $24x^2 + 61x - 8$

35. Factor: $x^2y^2 - 9$

36. Factor: $4x^2 + 12xy + 9y^2$

37. Factor: $x^{2n} - 12x^n + 36$

38. Factor: $36 - a^{2n}$

39. Factor: $64a^3 - 27b^3$

40. Factor: $8 - y^{3n}$

41. Factor: $15x^4 + x^2 - 6$

42. Factor: $36x^8 - 36x^4 + 5$

43. Factor: $21x^4y^4 + 23x^2y^2 + 6$

44. Factor: $3a^6 - 15a^4 - 18a^2$

45. Factor: $x^{4n} - 8x^{2n} + 16$

46. Factor: $3a^4b - 3ab^4$

47. Solve: $x^3 - x^2 - 6x = 0$

48. Solve: $6x^2 + 60 = 39x$

49. Solve: $x^3 - 16x = 0$

50. Solve: $y^3 + y^2 - 36y - 36 = 0$

51. Factor: $16x^2 - 94x + 33$

52. Factor: $9x^2 - 30x + 25$

53. Factor: $12y^2 + 16y - 3$

54. Factor: $3x^2 + 36x + 108$

55. The length of a playing field is twice the width. The area is 5000 yd^2. Find the length and width of the playing field.

56. The length of a hockey field is 20 yd less than twice the width. The area of the field is 6000 yd^2. Find the length and width of the hockey field.

57. The sum of the squares of two consecutive positive integers is forty-one. Find the two integers.

58. The size of a motion picture on the screen is given by the equation $S = d^2$, where d is the distance between the projector and the screen. Find the distance between the projector and the screen when the size of the picture is 400 ft^2.

59. A rectangular garden plot has dimensions 15 ft by 12 ft. A uniform path around the garden increases the total area to 270 ft^2. What is the width of the larger rectangle?

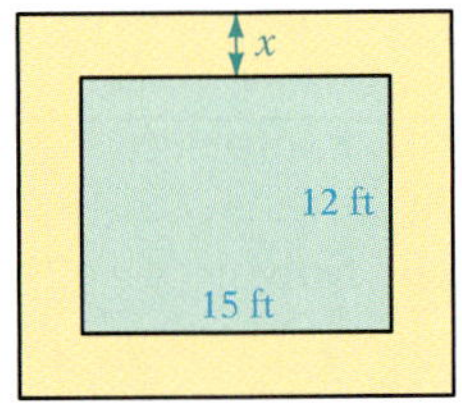

60. The length of each side of a square is extended 4 ft. The area of the resulting square is 576 ft^2. Find the length of a side of the original square.

Chapter Test

1. Factor: $6x^2y^2 + 9xy^2 + 12y^2$

2. Factor: $6x^3 - 8x^2 + 10x$

3. Factor: $p^2 + 5p + 6$

4. Factor: $a(x - 2) + b(2 - x)$

5. Solve: $(2a - 3)(a + 7) = 0$

6. Factor: $a^2 - 19a + 48$

7. Factor: $x^3 + 2x^2 - 15x$

8. Factor: $8x^2 + 20x - 48$

9. Factor: $ab + 6a - 3b - 18$

10. Solve: $4x^2 - 1 = 0$

11. Factor: $6x^2 + 19x + 8$

12. Factor: $6a^4 - 13a^2 - 5$

13. Factor: $2b^2 - 32$

14. Factor: $4a^2 - 12ab + 9b^2$

15. Factor: $px + x - p - 1$

16. Factor: $27x^3 - 8$

17. Factor: $2x^2 + 4x - 5$

18. Factor: $4x^2 - 49y^2$

19. Solve: $x(x - 8) = -15$

20. Factor: $p^2 + 12p + 36$

21. Factor: $18x^2 - 48xy + 32y^2$

22. Factor: $2y^4 - 14y^3 - 16y^2$

23. The length of a rectangle is 3 cm more than twice the width. The area of the rectangle is 90 cm^2. Find the length and width of the rectangle.

24. The length of the base of a triangle is three times the height. The area of the triangle is 24 in^2. Find the length of the base of the triangle.

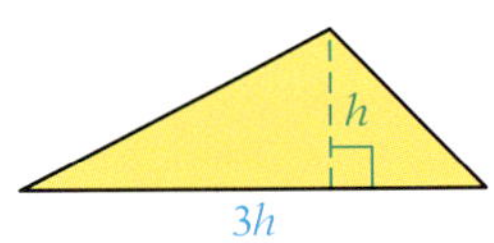

25. The product of two consecutive negative integers is one hundred fifty-six. Find the two integers.

Cumulative Review Exercises

1. Subtract: $4 - (-5) - 6 - 11$

2. Divide: $0.372 \div (-0.046)$
Round to the nearest tenth.

3. Simplify: $(3 - 7)^2 \div (-2) - 3 \cdot (-4)$

4. Evaluate $-2a^2 \div (2b) - c$ when $a = -4$, $b = 2$, and $c = -1$.

5. Identify the property that justifies the statement.
$(3 + 8) + 7 = 3 + (8 + 7)$

6. Multiply: $-\frac{3}{4}(-24x^2)$

7. Simplify: $-2[3x - 4(3 - 2x) - 8x]$

8. Solve: $-\frac{5}{7}x = -\frac{10}{21}$

9. Solve: $4 + 3(x - 2) = 13$

10. Solve: $3x - 2 = 12 - 5x$

11. Solve: $-2 + 4[3x - 2(4 - x) - 3] = 4x + 2$

12. 120% of what number is 42?

13. Solve: $-4x - 2 \geq 10$

14. Solve: $9 - 2(4x - 5) < 3(7 - 6x)$

15. Graph: $y = \frac{3}{4}x - 2$

16. Graph: $f(x) = -3x - 3$

17. Find the domain and range of the relation $\{(-5, -4), (-3, -2), (-1, 0), (1, 2), (3, 4)\}$. Is the relation a function?

18. Evaluate $f(x) = 6x - 5$ at $x = 11$.

19. Graph the solution set of $x + 3y > 2$.

20. Solve by substitution: $\begin{aligned} 6x + y &= 7 \\ x - 3y &= 17 \end{aligned}$

21. Solve by the addition method: $2x - 3y = -4$
$5x + y = 7$

22. Add: $(3y^3 - 5y^2 - 6) + (2y^2 - 8y + 1)$

23. Simplify: $(-3a^4b^2)^3$

24. Multiply: $(x + 2)(x^2 - 5x + 4)$

25. Divide: $(8x^2 + 4x - 3) \div (2x - 3)$

26. Simplify: $(x^{-4}y^2)^3$

27. Factor: $3a - 3b - ax + bx$

28. Factor: $x^2 + 3xy - 10y^2$

29. Factor: $6a^4 + 22a^3 + 12a^2$

30. Factor: $25a^2 - 36b^2$

31. Factor: $12x^2 - 36xy + 27y^2$

32. Solve: $3x^2 + 11x - 20 = 0$

33. Find the range of the function given by the equation $f(x) = \frac{4}{5}x - 3$ if the domain is $\{-10, -5, 0, 5, 10\}$.

34. The daily high temperatures, in degrees Celsius, during one week were recorded as follows: $-4°$, $-7°$, $2°$, $0°$, $-1°$, $-6°$, $-5°$. Find the average daily high temperature for the week.

35. The width of a rectangle is 40% of the length. The perimeter of the rectangle is 42 cm. Find the length and width of the rectangle.

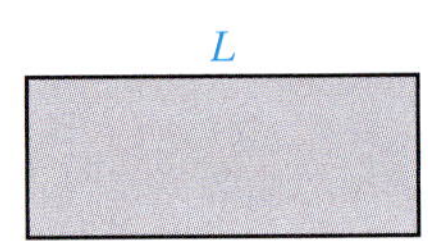

36. A board 10 ft long is cut into two pieces. Four times the length of the shorter piece is 2 ft less than three times the length of the longer piece. Find the length of each piece.

37. Company A rents cars for $6 a day and 25¢ for every mile driven. Company B rents cars for $15 a day and 10¢ per mile driven. You want to rent a car for 6 days. What is the maximum number of miles you can drive a Company A car if it is to cost you less than a Company B car?

38. An investment of $4000 is made at an annual simple interest rate of 8%. How much additional money must be invested at an annual simple interest rate of 11% so that the total interest earned is $1035?

39. A stereo that regularly sells for $165 is on sale for $99. Find the discount rate.

40. Find three consecutive even integers such that five times the middle integer is twelve more than twice the sum of the first and third.

Voltage is the measure of the strength of electrons flowing through a wire. A transformer is an electrical device that takes electricity of one voltage and changes that voltage by increasing or decreasing it. You see transformers at the top of utility poles, such as the one in the photo above. Transformers are the topic of the **Project on pages 498–499.** *The equations used in this application contain rational expressions, the topic of this chapter.*

Applications

7 Rational Expressions

Objectives

Section 7.1

1. Find the domain of a rational function
2. Simplify rational expressions

Section 7.2

1. Multiply and divide rational expressions
2. Add and subtract rational expressions

Section 7.3

1. Simplify complex fractions

Section 7.4

1. Solve fractional equations
2. Work problems
3. Uniform motion problems

Section 7.5

1. Proportions
2. Variation problems

Section 7.6

1. Solve literal equations

Need help? For online student resources, such as section quizzes, visit this web site: math.college.hmco.com

Prep Test

1. Find the LCM of 10 and 25.

For Exercises 2 to 5, add, subtract, multiply, or divide.

2. $-\frac{3}{8} \cdot \frac{4}{9}$

3. $-\frac{4}{5} \div \frac{8}{15}$

4. $-\frac{5}{6} + \frac{7}{8}$

5. $-\frac{3}{8} - \left(-\frac{7}{12}\right)$

6. Simplify: $\dfrac{\frac{2}{3} - \frac{1}{4}}{\frac{1}{8} - 2}$

7. Evaluate $\dfrac{2x - 3}{x^2 - x + 1}$ for $x = 2$.

8. Solve: $4(2x + 1) = 3(x - 2)$

9. Solve: $10\left(\frac{t}{2} + \frac{t}{5}\right) = 10(1)$

10. Two planes start from the same point and fly in opposite directions. The first plane is flying 20 mph slower than the second plane. In 2 h, the planes are 480 mi apart. Find the rate of each plane.

Go Figure

If 6 machines can fill 12 boxes of cereal in 7 minutes, how many boxes of cereal can 14 machines fill in 12 minutes?

7.1 Introduction to Rational Functions

1 Find the domain of a rational function

VIDEO & DVD CD TUTOR WEB SSM

An expression in which the numerator and denominator are polynomials is called a **rational expression.** Examples of rational expressions are shown at the right.

$$\frac{9}{z} \qquad \frac{3x + 4}{2x^2 + 1} \qquad \frac{x^3 - x + 1}{x^2 - 3x - 5}$$

The expression $\frac{\sqrt{x} + 3}{x}$ is not a rational expression because $\sqrt{x} + 3$ is not a polynomial.

A function that is written in terms of a rational expression is a **rational function.** Each of the following equations represents a rational function.

$$f(x) = \frac{x^2 + 3}{2x - 1} \qquad g(t) = \frac{3}{t^2 - 4} \qquad R(z) = \frac{z^2 + 3z - 1}{z^2 + z - 12}$$

To evaluate a rational function, replace the variable by its value. Then simplify.

Example 1 Given $f(x) = \dfrac{3x - 4}{x^2 - 2x + 1}$, find $f(-3)$.

Solution

$$f(x) = \frac{3x - 4}{x^2 - 2x + 1}$$

$$f(-3) = \frac{3(-3) - 4}{(-3)^2 - 2(-3) + 1} \qquad \text{• Substitute } -3 \text{ for } x.$$

$$f(-3) = \frac{-9 - 4}{9 + 6 + 1}$$

$$f(-3) = \frac{-13}{16}$$

$$f(-3) = -\frac{13}{16}$$

Problem 1 Given $f(x) = \dfrac{3 - 5x}{x^2 + 5x + 6}$, find $f(2)$.

Solution See page S22.

Because division by zero is not defined, the domain of a rational function must exclude those numbers for which the value of the polynomial in the denominator is zero.

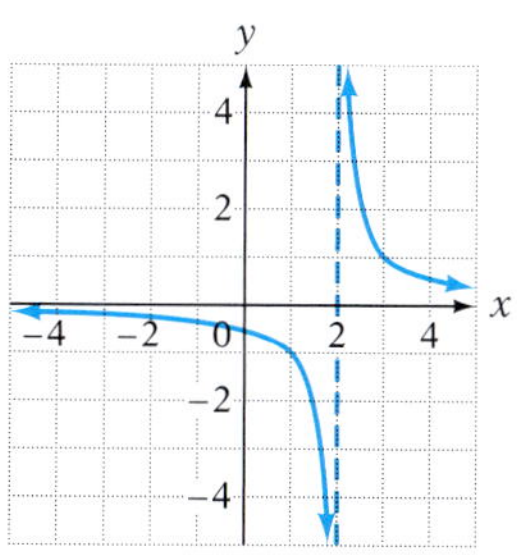

The graph of $f(x) = \dfrac{1}{x - 2}$ is shown at the left. Note that the graph never intersects the line $x = 2$ (shown as a dashed line). This value of x is excluded from the domain of $f(x) = \dfrac{1}{x - 2}$.

Determine the domain of $g(x) = \dfrac{x^2 + 4}{3x - 6}$.

The domain of g must exclude values of x for which the denominator is zero.

Set the denominator equal to zero. $\quad 3x - 6 = 0$

Solve for x. $\quad 3x = 6$

This value must be *excluded* from the domain. $\quad x = 2$

The domain of g is $\{x \mid x \neq 2\}$.

Example 2 Find the domain of $f(x) = \dfrac{2x - 6}{x^2 - 3x - 4}$.

Solution The domain must exclude values of x for which $x^2 - 3x - 4 = 0$. Solve this equation for x. Because $x^2 - 3x - 4$ factors over the integers, solve the equation by using the Principle of Zero Products.

$$x^2 - 3x - 4 = 0$$
$$(x + 1)(x - 4) = 0$$
$$x + 1 = 0 \qquad x - 4 = 0$$
$$x = -1 \qquad x = 4$$

When $x = -1$ and $x = 4$, the value of the denominator is zero. Therefore, these values must be excluded from the domain of f.

The domain is $\{x \mid x \neq -1, 4\}$.

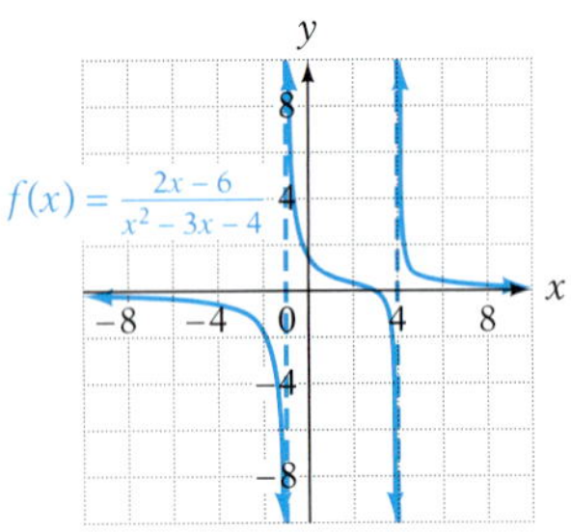

Problem 2 Find the domain of $g(x) = \dfrac{5 - x}{x^2 - 4}$.

Solution See page S22.

The graph of the function in Example 2 is shown at the left. Note that the graph never intersects the lines $x = -1$ and $x = 4$ (shown as dashed lines). These are the two values of x excluded from the domain of f.

Example 3 Find the domain of $f(x) = \dfrac{3x + 2}{x^2 + 1}$.

Solution The domain must exclude values of x for which $x^2 + 1 = 0$. It is not possible for $x^2 + 1 = 0$, because $x^2 \geq 0$, and a positive number added to a number equal to or greater than zero cannot equal zero. Therefore, there are no real numbers that must be excluded from the domain of f.

The domain is $\{x \mid x \in \text{real numbers}\}$.

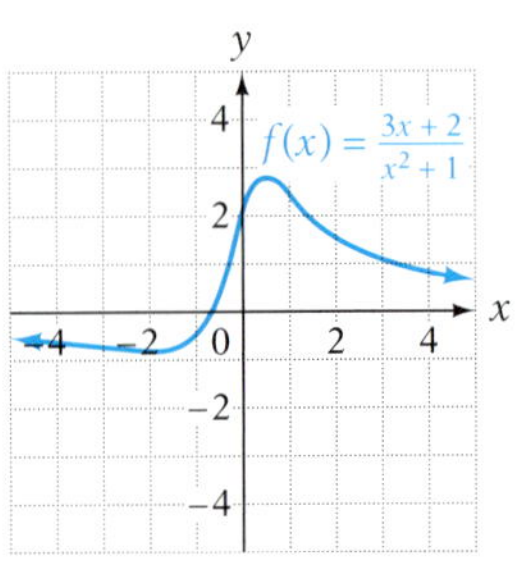

Problem 3 Find the domain of $p(x) = \dfrac{6x}{x^2 + 4}$.

Solution See page S23.

The graph of the function in Example 3 is shown at the left. Because the domain of f is all real numbers, there are points on the graph for all real numbers.

2 Simplify rational expressions

VIDEO & DVD · CD TUTOR · WEB · SSM

A rational expression is in **simplest form** when the numerator and denominator have no common factors other than 1. The Multiplication Property of One is used to write a rational expression in simplest form.

Simplify: $\dfrac{x^2 - 25}{x^2 + 13x + 40}$

$$\frac{x^2 - 25}{x^2 + 13x + 40} = \frac{(x-5)(x+5)}{(x+8)(x+5)} = \frac{(x-5)}{(x+8)} \cdot \frac{(x+5)}{(x+5)} = \frac{x-5}{x+8} \cdot 1 = \frac{x-5}{x+8},$$

$x \neq -8, -5$

The requirement $x \neq -8, -5$ must be included because division by 0 is not allowed.

The simplification above is usually shown with slashes to indicate that a common factor has been removed:

$$\frac{x^2 - 25}{x^2 + 13x + 40} = \frac{(x - 5)\overset{1}{\cancel{(x + 5)}}}{(x + 8)\underset{1}{\cancel{(x + 5)}}} = \frac{x - 5}{x + 8}, x \neq -8, -5$$

We will show a simplification with slashes. We will also omit the restrictions that prevent division by zero. Nonetheless, those restrictions *always* are implied.

Example 4 Simplify.

A. $\frac{x^2 - 16}{x^2 + 11x + 28}$ **B.** $\frac{12 + 5x - 2x^2}{2x^2 - 3x - 20}$ **C.** $\frac{x^{2n} + x^n - 2}{x^{2n} - 1}$

Solution

A. $\frac{x^2 - 16}{x^2 + 11x + 28} = \frac{(x + 4)(x - 4)}{(x + 4)(x + 7)}$ • Factor the numerator and the denominator.

$= \frac{\overset{1}{\cancel{(x + 4)}}(x - 4)}{\underset{1}{\cancel{(x + 4)}}(x + 7)}$ • Divide by the common factors.

$= \frac{x - 4}{x + 7}$ • Write the answer in simplest form.

B. $\frac{12 + 5x - 2x^2}{2x^2 - 3x - 20} = \frac{(4 - x)(3 + 2x)}{(x - 4)(2x + 5)}$ • Factor the numerator and the denominator.

$= \frac{\overset{-1}{\cancel{(4 - x)}}(3 + 2x)}{\underset{1}{\cancel{(x - 4)}}(2x + 5)}$ • Divide by the common factors. Remember that $4 - x = -(x - 4)$. Therefore, $\frac{4 - x}{x - 4} = \frac{-(x - 4)}{x - 4} = \frac{-1}{1} = -1$.

$= -\frac{2x + 3}{2x + 5}$ • Write the answer in simplest form.

C. $\frac{x^{2n} + x^n - 2}{x^{2n} - 1} = \frac{(x^n - 1)(x^n + 2)}{(x^n - 1)(x^n + 1)}$ • Factor the numerator and the denominator.

$= \frac{\overset{1}{\cancel{(x^n - 1)}}(x^n + 2)}{\underset{1}{\cancel{(x^n - 1)}}(x^n + 1)}$ • Divide by the common factors.

$= \frac{x^n + 2}{x^n + 1}$

Problem 4 Simplify.

A. $\frac{6x^4 - 24x^3}{12x^3 - 48x^2}$ **B.** $\frac{20x - 15x^2}{15x^3 - 5x^2 - 20x}$

C. $\frac{x^{2n} + x^n - 12}{x^{2n} - 3x^n}$

Solution See page S23.

Take Note

Recall that

$$(b - a) = -(a - b)$$

Therefore,

$$(4 - x) = -(x - 4)$$

In general,

$$\frac{b - a}{a - b} = \frac{-\overset{1}{\cancel{(a - b)}}}{\underset{1}{\cancel{a - b}}} = -1$$

7.1 Concept Review

Determine whether the following statements are always true, sometimes true, or never true.

1. $\dfrac{x^2 + 2x + 1}{x^{\frac{1}{2}} - 2x + 4}$ is a rational expression.

2. A rational expression with a variable in the denominator will have restrictions on the value of x.

3. When simplifying rational expressions, divide the numerator and denominator by the GCF of the numerator and denominator.

4. If $x \neq y$, the rational expression $\dfrac{x - y}{y - x}$ can be reduced to -1.

5. $\dfrac{a + 4}{a^2 + 6a + 8} = \dfrac{a + 4}{(a + 4)(a + 2)} = \dfrac{\cancel{a + 4}}{\cancel{(a + 4)}(a + 2)} = \dfrac{0}{a + 2} = 0$

6. For all values of a, $\dfrac{a(a + 4)}{a} = a + 4$.

7.1 Exercises

1 1. What is a rational function? Give an example of a rational function.

2. What values are excluded from the domain of a rational function?

3. Given $f(x) = \dfrac{2}{x - 3}$, find $f(4)$.

4. Given $f(x) = \dfrac{-7}{5 - x}$, find $f(-2)$.

5. Given $f(x) = \dfrac{x - 2}{x + 4}$, find $f(-2)$.

6. Given $f(x) = \dfrac{x - 3}{2x - 1}$, find $f(3)$.

7. Given $f(x) = \dfrac{1}{x^2 - 2x + 1}$, find $f(-2)$.

8. Given $f(x) = \dfrac{-3}{x^2 - 4x + 2}$, find $f(-1)$.

9. Given $f(x) = \dfrac{x - 2}{2x^2 + 3x + 8}$, find $f(3)$.

10. Given $f(x) = \dfrac{x^2}{3x^2 - 3x + 5}$, find $f(4)$.

11. Given $f(x) = \dfrac{x^2 - 2x}{x^3 - x + 4}$, find $f(-1)$.

12. Given $f(x) = \dfrac{8 - x^2}{x^3 - x^2 + 4}$, find $f(-3)$.

Find the domain of the function.

13. $H(x) = \dfrac{4}{x - 3}$

14. $G(x) = \dfrac{-2}{x + 2}$

15. $f(x) = \dfrac{x}{x + 4}$

16. $g(x) = \dfrac{3x}{x - 5}$

17. $R(x) = \dfrac{5x}{3x + 9}$

18. $p(x) = \dfrac{-2x}{6 - 2x}$

19. $q(x) = \dfrac{4 - x}{(x - 4)(x + 2)}$

20. $h(x) = \dfrac{2x + 1}{(x + 1)(x + 5)}$

21. $V(x) = \dfrac{x^2}{(2x + 5)(3x - 6)}$

22. $F(x) = \dfrac{x^2 - 1}{(4x + 8)(3x - 1)}$

23. $f(x) = \dfrac{x^2 + 1}{x}$

24. $g(x) = \dfrac{2x^3 - x - 1}{x^2}$

25. $k(x) = \dfrac{x + 1}{x^2 + 1}$

26. $P(x) = \dfrac{2x + 3}{2x^2 + 3}$

27. $f(x) = \dfrac{2x - 1}{x^2 + x - 6}$

28. $G(x) = \dfrac{3 - 4x}{x^2 + 4x - 5}$

29. $A(x) = \dfrac{5x + 2}{x^2 + 2x - 24}$

30. $h(x) = \dfrac{3x}{x^2 - 4}$

31. $f(x) = \dfrac{4x - 7}{3x^2 + 12}$

32. $g(x) = \dfrac{x^2 + x + 1}{5x^2 + 1}$

33. $G(x) = \dfrac{x^2 + 1}{6x^2 - 13x + 6}$

34. $A(x) = \dfrac{5x - 7}{x(x - 2)(x - 3)}$

35. $f(x) = \dfrac{x^2 + 8x + 4}{2x^3 + 9x^2 - 5x}$

36. $H(x) = \dfrac{x^4 - 1}{2x^3 + 2x^2 - 24x}$

2 **37.** When is a rational expression in simplest form?

38. Are the rational expressions $\dfrac{x(x - 2)}{2(x - 2)}$ and $\dfrac{x}{2}$ equal for all values of x? Why or why not?

Simplify.

39. $\dfrac{4 - 8x}{4}$

40. $\dfrac{8y + 2}{2}$

41. $\dfrac{6x^2 - 2x}{2x}$

42. $\dfrac{3y - 12y^2}{3y}$

43. $\dfrac{8x^2(x - 3)}{4x(x - 3)}$

44. $\dfrac{16y^4(y + 8)}{12y^3(y + 8)}$

45. $\dfrac{2x - 6}{3x - x^2}$

46. $\dfrac{3a^2 - 6a}{12 - 6a}$

47. $\dfrac{6x^3 - 15x^2}{12x^2 - 30x}$

48. $\dfrac{-36a^2 - 48a}{18a^3 + 24a^2}$

49. $\dfrac{a^2 + 4a}{4a - 16}$

50. $\dfrac{3x - 6}{x^2 + 2x}$

51. $\dfrac{16x^3 - 8x^2 + 12x}{4x}$

52. $\dfrac{3x^3y^3 - 12x^2y^2 + 15xy}{3xy}$

53. $\dfrac{-10a^4 - 20a^3 + 30a^2}{-10a^2}$

54. $\dfrac{-7a^5 - 14a^4 + 21a^3}{-7a^3}$

55. $\dfrac{3x^{3n} - 9x^{2n}}{12x^{2n}}$

56. $\dfrac{8a^n}{4a^{2n} - 8a^n}$

57. $\dfrac{x^2 - 7x + 12}{x^2 - 9x + 20}$

58. $\dfrac{x^2 - x - 20}{x^2 - 2x - 15}$

59. $\dfrac{x^2 - xy - 2y^2}{x^2 - 3xy + 2y^2}$

60. $\dfrac{2x^2 + 7xy - 4y^2}{4x^2 - 4xy + y^2}$

61. $\dfrac{6 - x - x^2}{3x^2 - 10x + 8}$

62. $\dfrac{3x^2 + 10x - 8}{8 - 14x + 3x^2}$

63. $\dfrac{14 - 19x - 3x^2}{3x^2 - 23x + 14}$

64. $\dfrac{x^2 + x - 12}{x^2 - x - 12}$

65. $\dfrac{a^2 - 7a + 10}{a^2 + 9a + 14}$

66. $\dfrac{x^2 - 2x}{x^2 + 2x}$

67. $\dfrac{a^2 - b^2}{a^3 + b^3}$

68. $\dfrac{x^4 - y^4}{x^2 + y^2}$

69. $\dfrac{x^3 + y^3}{3x^3 - 3x^2y + 3xy^2}$

70. $\dfrac{3x^3 + 3x^2 + 3x}{9x^3 - 9}$

71. $\dfrac{x^3 - 4xy^2}{3x^3 - 2x^2y - 8xy^2}$

72. $\dfrac{4a^2 - 8ab + 4b^2}{4a^2 - 4b^2}$

73. $\dfrac{4x^3 - 14x^2 + 12x}{24x + 4x^2 - 8x^3}$

74. $\dfrac{6x^3 - 15x^2 - 75x}{150x + 30x^2 - 12x^3}$

75. $\dfrac{x^2 - 4}{a(x + 2) - b(x + 2)}$

76. $\dfrac{x^2(a - 2) - a + 2}{ax^2 - ax}$

77. $\dfrac{x^4 + 3x^2 + 2}{x^4 - 1}$

78. $\dfrac{x^4 - 2x^2 - 3}{x^4 + 2x^2 + 1}$

79. $\dfrac{x^2y^2 + 4xy - 21}{x^2y^2 - 10xy + 21}$

80. $\dfrac{6x^2y^2 + 11xy + 4}{9x^2y^2 + 9xy - 4}$

81. $\dfrac{a^{2n} - a^n - 2}{a^{2n} + 3a^n + 2}$

82. $\dfrac{a^{2n} + a^n - 12}{a^{2n} - 2a^n - 3}$

7.1 Applying Concepts

83. Evaluate $h(x) = \frac{x + 2}{x - 3}$ when $x = 2.9, 2.99, 2.999$, and 2.9999. On the basis of your evaluations, complete the following sentence. As x becomes closer to 3, the values of $h(x)$ ______.

84. Evaluate $h(x) = \frac{x + 2}{x - 3}$ when $x = 3.1, 3.01, 3.001$, and 3.0001. On the basis of your evaluations, complete the following sentence. As x becomes closer to 3, the values of $h(x)$ ______.

85. **Photography** The focal length of a camera lens is the distance from the lens to the point where parallel rays of light come to a focus.

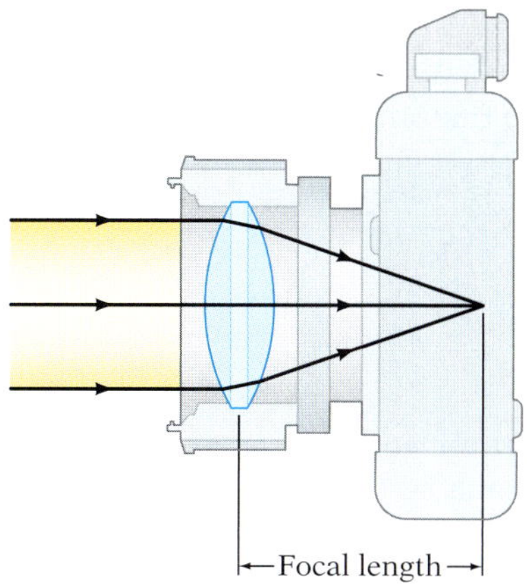

The relationship among the focal length (F), the distance between the object and the lens (x), and the distance between the lens and the film (y) is given by $\frac{1}{F} = \frac{1}{x} + \frac{1}{y}$. A camera used by a professional photographer has a dial that allows the focal length to be set at a constant value. Suppose a photographer chooses a focal length of 50 mm. Substituting this value into the equation, solving for y, and using $y = f(x)$ notation yield $f(x) = \frac{50x}{x - 50}$.

a. Graph this equation for $50 < x \le 6000$.
b. The point whose coordinates are (2000, 51), to the nearest integer, is on the graph of the function. Give an interpretation of this ordered pair.
c. Give a reason for choosing the domain so that $x > 50$.
d. Photographers refer to depth of field as a range of distances in which an object remains in focus. Use the graph to explain why the depth of field is larger for objects that are far from the lens than for objects that are close to the lens.

86. Suppose that $F(x) = \frac{g(x)}{h(x)}$ and that, for some real number a, $g(a) = 0$ and $h(a) = 0$. Is $F(x)$ in simplest form? Explain your answer.

87. Why can the numerator and denominator of a rational expression be divided by their common factors? What conditions must be placed on the value of the variables when a rational expression is simplified?

7.2 Operations on Rational Expressions

1 Multiply and divide rational expressions

VIDEO & DVD CD TUTOR WEB SSM

The product of two fractions is a fraction whose numerator is the product of the numerators of the two fractions and whose denominator is the product of the denominators of the two fractions.

$$\frac{a}{b} \cdot \frac{c}{d} = \frac{ac}{bd}$$

$$\frac{5}{a+2} \cdot \frac{b-3}{3} = \frac{5(b-3)}{(a+2)3} = \frac{5b-15}{3a+6}$$

The product of two rational expressions can often be simplified by factoring the numerator and the denominator.

Simplify: $\dfrac{x^2 - 2x}{2x^2 + x - 15} \cdot \dfrac{2x^2 - x - 10}{x^2 - 4}$

$$\frac{x^2 - 2x}{2x^2 + x - 15} \cdot \frac{2x^2 - x - 10}{x^2 - 4}$$

Factor the numerator and denominator of each fraction.

$$= \frac{x(x-2)}{(x+3)(2x-5)} \cdot \frac{(x+2)(2x-5)}{(x+2)(x-2)}$$

Multiply.

$$= \frac{x(x-2)(x+2)(2x-5)}{(x+3)(2x-5)(x+2)(x-2)}$$

Divide by the common factors.

$$= \frac{x\overset{1}{\cancel{(x-2)}}\overset{1}{\cancel{(x+2)}}\overset{1}{\cancel{(2x-5)}}}{(x+3)\underset{1}{\cancel{(2x-5)}}\underset{1}{\cancel{(x+2)}}\underset{1}{\cancel{(x-2)}}}$$

Write the answer in simplest form.

$$= \frac{x}{x+3}$$

Example 1 Multiply.

A. $\dfrac{2x^2 - 6x}{3x - 6} \cdot \dfrac{6x - 12}{8x^3 - 12x^2}$ **B.** $\dfrac{6x^2 + x - 2}{6x^2 + 7x + 2} \cdot \dfrac{2x^2 + 9x + 4}{4 - 7x - 2x^2}$

Solution

A. $$\frac{2x^2 - 6x}{3x - 6} \cdot \frac{6x - 12}{8x^3 - 12x^2} = \frac{2x(x-3)}{3(x-2)} \cdot \frac{6(x-2)}{4x^2(2x-3)}$$

$$= \frac{12x(x-3)\overset{1}{\cancel{(x-2)}}}{12x^2\underset{1}{\cancel{(x-2)}}(2x-3)} = \frac{x-3}{x(2x-3)}$$

B. $$\frac{6x^2 + x - 2}{6x^2 + 7x + 2} \cdot \frac{2x^2 + 9x + 4}{4 - 7x - 2x^2}$$

$$= \frac{(2x-1)(3x+2)}{(3x+2)(2x+1)} \cdot \frac{(2x+1)(x+4)}{(1-2x)(4+x)}$$

$$= \frac{\overset{-1}{\cancel{(2x-1)}}\overset{1}{\cancel{(3x+2)}}\overset{1}{\cancel{(2x+1)}}\overset{1}{\cancel{(x+4)}}}{\underset{1}{\cancel{(3x+2)}}\underset{1}{\cancel{(2x+1)}}\underset{1}{\cancel{(1-2x)}}\underset{1}{\cancel{(x+4)}}} = -1$$

Problem 1 Multiply.

A. $\dfrac{12 + 5x - 3x^2}{x^2 + 2x - 15} \cdot \dfrac{2x^2 + x - 45}{3x^2 + 4x}$

B. $\dfrac{2x^2 - 13x + 20}{x^2 - 16} \cdot \dfrac{2x^2 + 9x + 4}{6x^2 - 7x - 5}$

Solution See page S23.

The **reciprocal of a rational expression** is the rational expression with the numerator and denominator interchanged.

Rational expression		Reciprocal
$\dfrac{a}{b}$	$\dfrac{b}{a}$	
$\dfrac{a^2 - 2y}{4}$	$\dfrac{4}{a^2 - 2y}$	

To divide two rational expressions, multiply by the reciprocal of the divisor.

$$\frac{a}{b} \div \frac{c}{d} = \frac{a}{b} \cdot \frac{d}{c} = \frac{ad}{bc}$$

$$\frac{2}{a} \div \frac{5}{b} = \frac{2}{a} \cdot \frac{b}{5} = \frac{2b}{5a}$$

$$\frac{x+y}{2} \div \frac{x-y}{5} = \frac{x+y}{2} \cdot \frac{5}{x-y} = \frac{(x+y)5}{2(x-y)} = \frac{5x+5y}{2x-2y}$$

Example 2 Divide.

A. $\dfrac{12x^2y^2 - 24xy^2}{5z^2} \div \dfrac{4x^3y - 8x^2y}{3z^4}$

B. $\dfrac{3y^2 - 10y + 8}{3y^2 + 8y - 16} \div \dfrac{2y^2 - 7y + 6}{2y^2 + 5y - 12}$

Solution **A.** $\dfrac{12x^2y^2 - 24xy^2}{5z^2} \div \dfrac{4x^3y - 8x^2y}{3z^4}$

$$= \frac{12x^2y^2 - 24xy^2}{5z^2} \cdot \frac{3z^4}{4x^3y - 8x^2y}$$

$$= \frac{12xy^2(x-2)}{5z^2} \cdot \frac{3z^4}{4x^2y(x-2)}$$

$$= \frac{36xy^2z^4\overset{1}{\cancel{(x-2)}}}{20x^2yz^2\underset{1}{\cancel{(x-2)}}} = \frac{9yz^2}{5x}$$

B. $\dfrac{3y^2 - 10y + 8}{3y^2 + 8y - 16} \div \dfrac{2y^2 - 7y + 6}{2y^2 + 5y - 12}$

$$= \frac{3y^2 - 10y + 8}{3y^2 + 8y - 16} \cdot \frac{2y^2 + 5y - 12}{2y^2 - 7y + 6}$$

$$= \frac{(y-2)(3y-4)}{(3y-4)(y+4)} \cdot \frac{(y+4)(2y-3)}{(y-2)(2y-3)}$$

$$= \frac{\overset{1}{\cancel{(y-2)}}\overset{1}{\cancel{(3y-4)}}\overset{1}{\cancel{(y+4)}}\overset{1}{\cancel{(2y-3)}}}{\underset{1}{\cancel{(3y-4)}}\underset{1}{\cancel{(y+4)}}\underset{1}{\cancel{(y-2)}}\underset{1}{\cancel{(2y-3)}}} = 1$$

Problem 2 Divide.

A. $\dfrac{6x^2 - 3xy}{10ab^4} \div \dfrac{16x^2y^2 - 8xy^3}{15a^2b^2}$

B. $\dfrac{6x^2 - 7x + 2}{3x^2 + x - 2} \div \dfrac{4x^2 - 8x + 3}{5x^2 + x - 4}$

Solution See page S23.

2 Add and subtract rational expressions

When adding rational expressions in which the denominators are the same, add the numerators. The denominator of the sum is the common denominator. Write the answer in simplest form.

$$\frac{a}{c} + \frac{b}{c} = \frac{a+b}{c}$$

$$\frac{4x}{15} + \frac{8x}{15} = \frac{4x + 8x}{15} = \frac{12x}{15} = \frac{4x}{5}$$

$$\frac{a}{a^2 - b^2} + \frac{b}{a^2 - b^2} = \frac{a+b}{a^2 - b^2} = \frac{a+b}{(a-b)(a+b)} = \frac{\overset{1}{\cancel{(a+b)}}}{(a-b)\underset{1}{\cancel{(a+b)}}} = \frac{1}{a-b}$$

When subtracting rational expressions in which the denominators are the same, subtract the numerators. The denominator of the difference is the common denominator. Write the answer in simplest form.

$$\frac{y}{y-3} - \frac{3}{y-3} = \frac{y-3}{y-3} = \frac{\overset{1}{\cancel{(y-3)}}}{\underset{1}{\cancel{(y-3)}}} = 1$$

$$\frac{7x - 12}{2x^2 + 5x - 12} - \frac{3x - 6}{2x^2 + 5x - 12} = \frac{(7x - 12) - (3x - 6)}{2x^2 + 5x - 12}$$

$$= \frac{7x - 12 - 3x + 6}{2x^2 + 5x - 12}$$

$$= \frac{4x - 6}{2x^2 + 5x - 12}$$

$$= \frac{2(2x - 3)}{(2x - 3)(x + 4)}$$

$$= \frac{2\overset{1}{\cancel{(2x - 3)}}}{\underset{1}{\cancel{(2x - 3)}}(x + 4)} = \frac{2}{x + 4}$$

Take Note

Be careful with signs when subtracting algebraic fractions. Note that we must subtract the entire numerator.

$(7x - 12) - (3x - 6) =$
$7x - 12 - 3x + 6$

Before two rational expressions with different denominators can be added or subtracted, both rational expressions must be expressed in terms of a common denominator. This common denominator is the LCM of the denominators of the rational expressions.

The LCM of two or more polynomials is the simplest polynomial that contains the factors of each polynomial. To find the LCM, first factor each polynomial completely. The LCM is the product of each factor the greatest number of times it occurs in any one factorization.

To find the LCM of $3x^2 + 15x$ and $6x^4 + 24x^3 - 30x^2$, factor each polynomial.

$$3x^2 + 15x = 3x(x + 5)$$
$$6x^4 + 24x^3 - 30x^2 = 6x^2(x^2 + 4x - 5) = 6x^2(x - 1)(x + 5)$$

The LCM is the product of the LCM of the numerical coefficients and each variable factor the greatest number of times it occurs in any one factorization.

$$\text{LCM} = 6x^2(x - 1)(x + 5)$$

Take Note

$x^2 - 2x = x(x - 2)$

$3x - 6 = 3(x - 2)$

The LCM of $x(x - 2)$ and $3(x - 2)$ is $3x(x - 2)$.

Write the fractions $\frac{x + 2}{x^2 - 2x}$ and $\frac{5x}{3x - 6}$ in terms of the LCM of the denominators.

Find the LCM of the denominators. — The LCM is $3x(x - 2)$.

For each fraction, multiply the numerator and denominator by the factor whose product with the denominator is the LCM.

$$\frac{x + 2}{x^2 - 2x} = \frac{x + 2}{x(x - 2)} \cdot \frac{3}{3} = \frac{3x + 6}{3x(x - 2)}$$
$$\frac{5x}{3x - 6} = \frac{5x}{3(x - 2)} \cdot \frac{x}{x} = \frac{5x^2}{3x(x - 2)}$$

Add: $\frac{3x}{2x - 3} + \frac{3x + 6}{2x^2 + x - 6}$

The LCM of $2x - 3$ and $2x^2 + x - 6$ is $(2x - 3)(x + 2)$.

$$\frac{3x}{2x - 3} + \frac{3x + 6}{2x^2 + x - 6}$$

Rewrite each fraction in terms of the LCM of the denominators.

$$= \frac{3x}{2x - 3} \cdot \frac{x + 2}{x + 2} + \frac{3x + 6}{(2x - 3)(x + 2)}$$
$$= \frac{3x^2 + 6x}{(2x - 3)(x + 2)} + \frac{3x + 6}{(2x - 3)(x + 2)}$$

Add the fractions.

$$= \frac{(3x^2 + 6x) + (3x + 6)}{(2x - 3)(x + 2)}$$
$$= \frac{3x^2 + 9x + 6}{(2x - 3)(x + 2)}$$

Factor the numerator to determine whether there are common factors in the numerator and denominator.

$$= \frac{3(x^2 + 3x + 2)}{(2x - 3)(x + 2)}$$
$$= \frac{3(x + 2)(x + 1)}{(2x - 3)(x + 2)}$$
$$= \frac{3\overset{1}{\cancel{(x + 2)}}(x + 1)}{(2x - 3)\underset{1}{\cancel{(x + 2)}}} = \frac{3(x + 1)}{2x - 3}$$

Take Note

Note the steps involved in adding or subtracting rational expressions:

1. Find the LCM of the denominators.
2. Rewrite each fraction in terms of the common denominator.
3. Add or subtract the rational expressions.
4. Simplify the resulting sum or difference.

Example 3 Subtract: $\dfrac{x}{2x-4} - \dfrac{4-x}{x^2-2x}$

Solution

$$\frac{x}{2x-4} - \frac{4-x}{x^2-2x}$$

$$= \frac{x}{2(x-2)} \cdot \frac{x}{x} - \frac{4-x}{x(x-2)} \cdot \frac{2}{2}$$

• Write each fraction in terms of the LCM. The LCM is $2x(x-2)$.

$$= \frac{x^2}{2x(x-2)} - \frac{8-2x}{2x(x-2)}$$

• Subtract the fractions.

$$= \frac{x^2-(8-2x)}{2x(x-2)}$$

$$= \frac{x^2+2x-8}{2x(x-2)}$$

$$= \frac{(x+4)(x-2)}{2x(x-2)}$$

$$= \frac{(x+4)\overset{1}{\cancel{(x-2)}}}{2x\underset{1}{\cancel{(x-2)}}} = \frac{x+4}{2x}$$

• Divide by the common factors.

Problem 3 Add: $\dfrac{a-3}{a^2-5a} + \dfrac{a-9}{a^2-25}$

Solution See page S23.

Example 4 Simplify: $\dfrac{6x-23}{2x^2+x-6} + \dfrac{3x}{2x-3} - \dfrac{5}{x+2}$

Solution

$$\frac{6x-23}{2x^2+x-6} + \frac{3x}{2x-3} - \frac{5}{x+2}$$

$$= \frac{6x-23}{(2x-3)(x+2)} + \frac{3x}{2x-3} \cdot \frac{x+2}{x+2} - \frac{5}{x+2} \cdot \frac{2x-3}{2x-3}$$

$$= \frac{6x-23}{(2x-3)(x+2)} + \frac{3x^2+6x}{(2x-3)(x+2)} - \frac{10x-15}{(2x-3)(x+2)}$$

$$= \frac{(6x-23)+(3x^2+6x)-(10x-15)}{(2x-3)(x+2)}$$

$$= \frac{6x-23+3x^2+6x-10x+15}{(2x-3)(x+2)}$$

$$= \frac{3x^2+2x-8}{(2x-3)(x+2)} = \frac{(3x-4)(x+2)}{(2x-3)(x+2)} = \frac{3x-4}{2x-3}$$

Problem 4 Simplify: $\dfrac{x-1}{x-2} - \dfrac{7-6x}{2x^2-7x+6} + \dfrac{4}{2x-3}$

Solution See page S23.

7.2 Concept Review

Determine whether the following statements are always true, sometimes true, or never true.

1. To divide an expression by a nonzero rational number, multiply the expression by the reciprocal of the divisor.

2. To multiply rational expressions, multiply the numerators and multiply the denominators.

3. $\dfrac{3}{x+y} + \dfrac{2}{x+y} = \dfrac{3+2}{(x+y)+(x+y)} = \dfrac{5}{2x+2y}$

4. $\dfrac{2x}{x(x+4)} - \dfrac{2}{x+4} = \dfrac{2\cancel{x}}{\cancel{x}(x+4)} - \dfrac{2}{x+4} = \dfrac{2}{x+4} - \dfrac{2}{x+4} = 0$

7.2 Exercises

1 1. Explain how to multiply two rational expressions.

2. Explain how to divide two rational expressions.

Multiply or divide.

3. $\dfrac{27a^2b^5}{16xy^2} \cdot \dfrac{20x^2y^3}{9a^2b}$

4. $\dfrac{15x^2y^4}{24ab^3} \cdot \dfrac{28a^2b^4}{35xy^4}$

5. $\dfrac{3x-15}{4x^2-2x} \cdot \dfrac{20x^2-10x}{15x-75}$

6. $\dfrac{2x^2+4x}{8x^2-40x} \cdot \dfrac{6x^3-30x^2}{3x^2+6x}$

7. $\dfrac{x^2y^3}{x^2-4x-5} \cdot \dfrac{2x^2-13x+15}{x^4y^3}$

8. $\dfrac{2x^2-5x+3}{x^6y^3} \cdot \dfrac{x^4y^4}{2x^2-x-3}$

9. $\dfrac{x^2-3x+2}{x^2-8x+15} \cdot \dfrac{x^2+x-12}{8-2x-x^2}$

10. $\dfrac{x^2+x-6}{12+x-x^2} \cdot \dfrac{x^2+x-20}{x^2-4x+4}$

11. $\dfrac{x^{n+1}+2x^n}{4x^2-6x} \cdot \dfrac{8x^2-12x}{x^{n+1}-x^n}$

12. $\dfrac{x^{2n}+2x^n}{x^{n+1}+2x} \cdot \dfrac{x^2-3x}{x^{n+1}-3x^n}$

13. $\dfrac{12+x-6x^2}{6x^2+29x+28} \cdot \dfrac{2x^2+x-21}{4x^2-9}$

14. $\dfrac{x^2+5x+4}{4+x-3x^2} \cdot \dfrac{3x^2+2x-8}{x^2+4x}$

15. $\dfrac{x^{2n} - x^n - 6}{x^{2n} + x^n - 2} \cdot \dfrac{x^{2n} - 5x^n - 6}{x^{2n} - 2x^n - 3}$

16. $\dfrac{x^{2n} + 3x^n + 2}{x^{2n} - x^n - 6} \cdot \dfrac{x^{2n} + x^n - 12}{x^{2n} - 1}$

17. $\dfrac{x^3 - y^3}{2x^2 + xy - 3y^2} \cdot \dfrac{2x^2 + 5xy + 3y^2}{x^2 + xy + y^2}$

18. $\dfrac{x^4 - 5x^2 + 4}{3x^2 - 4x - 4} \cdot \dfrac{3x^2 - 10x - 8}{x^2 - 4}$

19. $\dfrac{6x^2y^4}{35a^2b^5} \div \dfrac{12x^3y^3}{7a^4b^5}$

20. $\dfrac{12a^4b^7}{13x^2y^2} \div \dfrac{18a^5b^6}{26xy^3}$

21. $\dfrac{2x - 6}{6x^2 - 15x} \div \dfrac{4x^2 - 12x}{18x^3 - 45x^2}$

22. $\dfrac{4x^2 - 4y^2}{6x^2y^2} \div \dfrac{3x^2 + 3xy}{2x^2y - 2xy^2}$

23. $\dfrac{2x^2 - 2y^2}{14x^2y^4} \div \dfrac{x^2 + 2xy + y^2}{35xy^3}$

24. $\dfrac{8x^3 + 12x^2y}{4x^2 - 9y^2} \div \dfrac{16x^2y^2}{4x^2 - 12xy + 9y^2}$

25. $\dfrac{2x^2 - 5x - 3}{2x^2 + 7x + 3} \div \dfrac{2x^2 - 3x - 20}{2x^2 - x - 15}$

26. $\dfrac{3x^2 - 10x - 8}{6x^2 + 13x + 6} \div \dfrac{2x^2 - 9x + 10}{4x^2 - 4x - 15}$

27. $\dfrac{x^2 - 8x + 15}{x^2 + 2x - 35} \div \dfrac{15 - 2x - x^2}{x^2 + 9x + 14}$

28. $\dfrac{2x^2 + 13x + 20}{8 - 10x - 3x^2} \div \dfrac{6x^2 - 13x - 5}{9x^2 - 3x - 2}$

29. $\dfrac{x^{2n} + x^n}{2x - 2} \div \dfrac{4x^n + 4}{x^{n+1} - x^n}$

30. $\dfrac{x^{2n} - 4}{4x^n + 8} \div \dfrac{x^{n+1} - 2x}{4x^3 - 12x^2}$

31. $\dfrac{2x^2 - 13x + 21}{2x^2 + 11x + 15} \div \dfrac{2x^2 + x - 28}{3x^2 + 4x - 15}$

32. $\dfrac{2x^2 - 13x + 15}{2x^2 - 3x - 35} \div \dfrac{6x^2 + x - 12}{6x^2 + 13x - 28}$

33. $\dfrac{14 + 17x - 6x^2}{3x^2 + 14x + 8} \div \dfrac{4x^2 - 49}{2x^2 + 15x + 28}$

34. $\dfrac{16x^2 - 9}{6 - 5x - 4x^2} \div \dfrac{16x^2 + 24x + 9}{4x^2 + 11x + 6}$

35. $\dfrac{2x^{2n} - x^n - 6}{x^{2n} - x^n - 2} \div \dfrac{2x^{2n} + x^n - 3}{x^{2n} - 1}$

36. $\dfrac{x^{4n} - 1}{x^{2n} + x^n - 2} \div \dfrac{x^{2n} + 1}{x^{2n} + 3x^n + 2}$

37. $\dfrac{6x^2 + 6x}{3x + 6x^2 + 3x^3} \div \dfrac{x^2 - 1}{1 - x^3}$

38. $\dfrac{x^3 + y^3}{2x^3 + 2x^2y} \div \dfrac{3x^3 - 3x^2y + 3xy^2}{6x^2 - 6y^2}$

2 Write each fraction in terms of the LCM of the denominators.

39. $\dfrac{3}{4x^2y}, \dfrac{17}{12xy^4}$

40. $\dfrac{5}{16a^3b^3}, \dfrac{7}{30a^5b}$

41. $\dfrac{3x}{2x - 3}, \dfrac{5x}{2x + 3}$

42. $\dfrac{2}{7y - 3}, \dfrac{-3}{7y + 3}$

43. $\dfrac{2x}{x^2 - 9}, \dfrac{x + 1}{x - 3}$

44. $\dfrac{3x}{x^2 - 16}, \dfrac{2x}{4x - 16}$

45. $\dfrac{3x}{x^2 - 1}, \dfrac{5x}{x^2 - 2x + 1}$

46. $\dfrac{2x}{x^2 + x - 6}, \dfrac{-4x}{x^2 + 5x + 6}$

Add or subtract.

47. $\dfrac{3}{2xy} - \dfrac{7}{2xy} - \dfrac{9}{2xy}$

48. $-\dfrac{3}{4x^2} + \dfrac{8}{4x^2} - \dfrac{3}{4x^2}$

49. $\dfrac{x}{x^2 - 3x + 2} - \dfrac{2}{x^2 - 3x + 2}$

50. $\dfrac{3x}{3x^2 + x - 10} - \dfrac{5}{3x^2 + x - 10}$

51. $\dfrac{3}{2x^2y} - \dfrac{8}{5x} - \dfrac{9}{10xy}$

52. $\dfrac{2}{5ab} - \dfrac{3}{10a^2b} + \dfrac{4}{15ab^2}$

53. $\dfrac{2}{3x} - \dfrac{3}{2xy} + \dfrac{4}{5xy} - \dfrac{5}{6x}$

54. $\dfrac{3}{4ab} - \dfrac{2}{5a} + \dfrac{3}{10b} - \dfrac{5}{8ab}$

55. $\dfrac{2x - 1}{12x} - \dfrac{3x + 4}{9x}$

56. $\dfrac{3x - 4}{6x} - \dfrac{2x - 5}{4x}$

57. $\dfrac{3x + 2}{4x^2y} - \dfrac{y - 5}{6xy^2}$

58. $\dfrac{2y - 4}{5xy^2} + \dfrac{3 - 2x}{10x^2y}$

59. $\dfrac{2x}{x - 3} - \dfrac{3x}{x - 5}$

60. $\dfrac{3a}{a - 2} - \dfrac{5a}{a + 1}$

61. $\dfrac{3}{2a - 3} + \dfrac{2a}{3 - 2a}$

62. $\dfrac{x}{2x - 5} - \dfrac{2}{5x - 2}$

63. $\dfrac{3}{x + 5} + \dfrac{2x + 7}{x^2 - 25}$

64. $\dfrac{x}{4 - x} - \dfrac{4}{x^2 - 16}$

65. $\dfrac{2}{x} - 3 - \dfrac{10}{x - 4}$

66. $\dfrac{6a}{a - 3} - 5 + \dfrac{3}{a}$

67. $\dfrac{1}{2x - 3} - \dfrac{5}{2x} + 1$

68. $\dfrac{5}{x} - \dfrac{5x}{5 - 6x} + 2$

69. $\dfrac{3}{x^2 - 1} + \dfrac{2x}{x^2 + 2x + 1}$

70. $\dfrac{1}{x^2 - 6x + 9} - \dfrac{1}{x^2 - 9}$

71. $\dfrac{x}{x + 3} - \dfrac{3 - x}{x^2 - 9}$

72. $\dfrac{1}{x + 2} - \dfrac{3x}{x^2 + 4x + 4}$

73. $\dfrac{2x - 3}{x + 5} - \dfrac{x^2 - 4x - 19}{x^2 + 8x + 15}$

74. $\dfrac{-3x^2 + 8x + 2}{x^2 + 2x - 8} - \dfrac{2x - 5}{x + 4}$

75. $\dfrac{x^n}{x^{2n} - 1} - \dfrac{2}{x^n + 1}$

76. $\dfrac{2}{x^n - 1} + \dfrac{x^n}{x^{2n} - 1}$

77. $\dfrac{2x-2}{4x^2-9} - \dfrac{5}{3-2x}$

78. $\dfrac{x^2+4}{4x^2-36} - \dfrac{13}{x+3}$

79. $\dfrac{x-2}{x+1} - \dfrac{3-12x}{2x^2-x-3}$

80. $\dfrac{3x-4}{4x+1} + \dfrac{3x+6}{4x^2+9x+2}$

81. $\dfrac{x+1}{x^2+x-6} - \dfrac{x+2}{x^2+4x+3}$

82. $\dfrac{x+1}{x^2+x-12} - \dfrac{x-3}{x^2+7x+12}$

83. $\dfrac{x-1}{2x^2+11x+12} + \dfrac{2x}{2x^2-3x-9}$

84. $\dfrac{x-2}{4x^2+4x-3} + \dfrac{3-2x}{6x^2+x-2}$

85. $\dfrac{x}{x-3} - \dfrac{2}{x+4} - \dfrac{14}{x^2+x-12}$

86. $\dfrac{x^2}{x^2+x-2} + \dfrac{3}{x-1} - \dfrac{4}{x+2}$

87. $\dfrac{x^2+6x}{x^2+3x-18} - \dfrac{2x-1}{x+6} + \dfrac{x-2}{3-x}$

88. $\dfrac{2x^2-2x}{x^2-2x-15} - \dfrac{2}{x+3} + \dfrac{x}{5-x}$

89. $\dfrac{4-20x}{6x^2+11x-10} - \dfrac{4}{2-3x} + \dfrac{x}{2x+5}$

90. $\dfrac{x}{4x-1} + \dfrac{2}{2x+1} + \dfrac{6}{8x^2+2x-1}$

91. $\dfrac{2x^2}{x^4-1} - \dfrac{1}{x^2-1} + \dfrac{1}{x^2+1}$

92. $\dfrac{x^2-12}{x^4-16} + \dfrac{1}{x^2-4} - \dfrac{1}{x^2+4}$

7.2 Applying Concepts

Simplify.

93. $\dfrac{(x+1)^2}{1-2x} \cdot \dfrac{2x-1}{x+1}$

94. $\left(\dfrac{2m}{3}\right)^2 \div \left(\dfrac{m^2}{6} + \dfrac{m}{2}\right)$

95. $\left(\dfrac{y-2}{x^2}\right)^3 \cdot \left(\dfrac{x}{2-y}\right)^2$

96. $\dfrac{b+3}{b-1} \div \dfrac{b+3}{b-2} \cdot \dfrac{b-1}{b+4}$

97. $\left(\dfrac{y+1}{y-1}\right)^2 - 1$

98. $\left(\dfrac{1}{3} - \dfrac{2}{a}\right) \div \left(\dfrac{3}{a} - 2 + \dfrac{a}{4}\right)$

99. $\dfrac{3x^2+6x}{4x^2-16} \cdot \dfrac{2x+8}{x^2+2x} \div \dfrac{3x-9}{5x-20}$

100. $\dfrac{5y^2-20}{3y^2-12y} \cdot \dfrac{9y^3+6y^2}{2y^2-4y} \div \dfrac{y^3+2y^2}{2y^2-8y}$

101. $\dfrac{a^2 + a - 6}{4 + 11a - 3a^2} \cdot \dfrac{15a^2 - a - 2}{4a^2 + 7a - 2} \div \dfrac{6a^2 - 7a - 3}{4 - 17a + 4a^2}$

102. $\dfrac{25x - x^3}{x^4 - 1} \cdot \dfrac{3 - x - 4x^2}{2x^2 + 7x - 15} \div \dfrac{4x^3 - 23x^2 + 15x}{3 - 5x + 2x^2}$

103. $\left(\dfrac{x + 1}{2x - 1} - \dfrac{x - 1}{2x + 1}\right) \cdot \left(\dfrac{2x - 1}{x} - \dfrac{2x - 1}{x^2}\right)$

104. $\left(\dfrac{y - 2}{3y + 1} - \dfrac{y + 2}{3y - 1}\right) \cdot \left(\dfrac{3y + 1}{y} - \dfrac{3y - 1}{y^2}\right)$

Solve.

105. Use $x = 3$ and $y = 5$ to show that $\dfrac{1}{x} + \dfrac{1}{y} \neq \dfrac{1}{x + y}$.

106. Use $x = 3$ and $y = 5$ to show that $\dfrac{1}{x} - \dfrac{1}{y} \neq \dfrac{1}{x - y}$.

Rewrite the expression as the sum of two fractions in simplest form.

107. $\dfrac{3x + 6y}{xy}$

108. $\dfrac{5a + 8b}{ab}$

109. $\dfrac{4a^2 + 3ab}{a^2b^2}$

110. **Manufacturing** Manufacturers who package their product in cans would like to design the can so that the minimum amount of aluminum is needed. If a soft drink can contains 12 oz (355 cm^3), the function that relates the surface area of the can (the amount of aluminum needed) to the radius of the bottom of the can is given by the equation $f(r) = 2\pi r^2 + \dfrac{710}{r}$, where r is measured in centimeters.

a. Express the right side of this equation with a common denominator.
b. Graph the equation for $0 < r \leq 7.15$.
c. The point whose coordinates are (7, 409), to the nearest integer, is on the graph of f. Write a sentence that gives an interpretation of this ordered pair.
d. Use a graphing utility to determine the radius of the can that has a minimum surface area. Round to the nearest tenth.
e. The height of the can is determined from $h = \dfrac{355}{\pi r^2}$. Use the answer to part (d) to determine the height of the can that has a minimum surface area. Round to the nearest tenth.
f. Determine the minimum surface area. Round to the nearest tenth.

111. **Manufacturing** A manufacturer wants to make square tissues and package them in a box. The manufacturer has determined that to be competitive, the box needs to hold 175 tissues, which means that the volume of the box will be 132 in^3. The amount of cardboard (surface area) that will be necessary to build this box is given by $f(x) = 2x^2 + \frac{528}{x}$, where x is the height of the box in inches.

a. Express the right side of this equation with a common denominator.

b. Graph the equation for $1.8 < x \leq 10$.

c. The point whose coordinates are (4, 164) is on the graph. Write a sentence that explains the meaning of this point.

d. With a graphing utility, determine, to the nearest tenth, the height of the box that uses the minimum amount of cardboard.

e. Determine the minimum amount of cardboard. Round to the nearest tenth.

112. For adding or subtracting fractions, any common denominator will do. Explain the advantages and disadvantages of using the LCM of the denominators.

113. A student incorrectly tried to add the fractions $\frac{1}{5}$ and $\frac{2}{3}$ by adding the the numerators and the denominators. The procedure was shown as

$$\frac{1}{5} + \frac{2}{3} = \frac{1+2}{5+3} = \frac{3}{8}$$

Write the fractions $\frac{1}{5}, \frac{2}{3}$, and $\frac{3}{8}$ in order from smallest to largest. Now take any two other fractions, add the numerators and the denominators, and then write the fractions in order from smallest to largest. Do you see a pattern? If so, explain it. If not, try a few more examples until you find a pattern.

7.3 Complex Fractions

1 Simplify complex fractions

A **complex fraction** is a fraction whose numerator or denominator contains one or more fractions. Examples of complex fractions are shown below.

$$\frac{5}{2+\frac{1}{2}} \qquad \frac{5+\frac{1}{y}}{5-\frac{1}{y}} \qquad \frac{x+4+\frac{1}{x+2}}{x-2+\frac{1}{x+2}}$$

Simplify: $\dfrac{\frac{1}{x}+\frac{1}{y}}{\frac{1}{x}-\frac{1}{y}}$

Multiply the numerator and denominator of the complex fraction by the LCM of the denominators. The LCM of x and y is xy.

$$\frac{\frac{1}{x}+\frac{1}{y}}{\frac{1}{x}-\frac{1}{y}} = \frac{\frac{1}{x}+\frac{1}{y}}{\frac{1}{x}-\frac{1}{y}} \cdot \frac{xy}{xy} = \frac{\frac{1}{x}\cdot xy + \frac{1}{y}\cdot xy}{\frac{1}{x}\cdot xy - \frac{1}{y}\cdot xy} = \frac{y+x}{y-x}$$

Note that after the numerator and denominator of the complex fraction have been multiplied by the LCM of the denominators, no fraction remains in the numerator or denominator.

The method shown above of simplifying a complex fraction by multiplying the numerator and denominator by the LCM of the denominators is used in Example 1. However, a different approach is to rewrite the numerator and denominator of the complex fraction as single fractions and then divide the numerator by the denominator. Here the example shown above is simplified by using this alternative method.

Take Note

Recall that the fraction bar can be read "divided by."

Rewrite the numerator and denominator of the complex fraction as single fractions.

$$\frac{\frac{1}{x}+\frac{1}{y}}{\frac{1}{x}-\frac{1}{y}} = \frac{\frac{1}{x}\cdot\frac{y}{y}+\frac{1}{y}\cdot\frac{x}{x}}{\frac{1}{x}\cdot\frac{y}{y}-\frac{1}{y}\cdot\frac{x}{x}} = \frac{\frac{y}{xy}+\frac{x}{xy}}{\frac{y}{xy}-\frac{x}{xy}} = \frac{\frac{y+x}{xy}}{\frac{y-x}{xy}}$$

Divide the numerator of the complex fraction by the denominator.

$$= \frac{y+x}{xy} \div \frac{y-x}{xy} = \frac{y+x}{xy}\cdot\frac{xy}{y-x}$$

Multiply the fractions. Simplify.

$$= \frac{(y+x)xy}{xy(y-x)} = \frac{y+x}{y-x}$$

Note that this is the same result we got before.

Example 1 Simplify. **A.** $\dfrac{2-\frac{11}{x}+\frac{15}{x^2}}{3-\frac{5}{x}-\frac{12}{x^2}}$ **B.** $\dfrac{2x-1+\frac{7}{x+4}}{3x-8+\frac{17}{x+4}}$

Solution **A.**

$$\frac{2-\frac{11}{x}+\frac{15}{x^2}}{3-\frac{5}{x}-\frac{12}{x^2}} = \frac{2-\frac{11}{x}+\frac{15}{x^2}}{3-\frac{5}{x}-\frac{12}{x^2}}\cdot\frac{x^2}{x^2}$$

• The LCM of the denominators is x^2.

$$= \frac{2\cdot x^2-\frac{11}{x}\cdot x^2+\frac{15}{x^2}\cdot x^2}{3\cdot x^2-\frac{5}{x}\cdot x^2-\frac{12}{x^2}\cdot x^2}$$

$$= \frac{2x^2-11x+15}{3x^2-5x-12}$$

$$= \frac{(2x-5)(x-3)}{(3x+4)(x-3)} = \frac{2x-5}{3x+4}$$

B. $$\frac{2x - 1 + \frac{7}{x+4}}{3x - 8 + \frac{17}{x+4}} = \frac{2x - 1 + \frac{7}{x+4}}{3x - 8 + \frac{17}{x+4}} \cdot \frac{x+4}{x+4}$$

• The LCM of the denominators is $x + 4$.

$$= \frac{(2x-1)(x+4) + \frac{7}{x+4}(x+4)}{(3x-8)(x+4) + \frac{17}{x+4}(x+4)}$$

$$= \frac{2x^2 + 7x - 4 + 7}{3x^2 + 4x - 32 + 17}$$

$$= \frac{2x^2 + 7x + 3}{3x^2 + 4x - 15}$$

$$= \frac{(2x+1)(x+3)}{(3x-5)(x+3)} = \frac{2x+1}{3x-5}$$

Problem 1 Simplify. **A.** $\dfrac{3 + \frac{16}{x} + \frac{16}{x^2}}{6 + \frac{5}{x} - \frac{4}{x^2}}$ **B.** $\dfrac{2x + 5 + \frac{14}{x-3}}{4x + 16 + \frac{49}{x-3}}$

Solution See page S24.

7.3 Concept Review

Determine whether the following statements are always true, sometimes true, or never true.

1. The numerator of a complex fraction contains one or more fractions.

2. We can simplify a complex fraction by multiplying the numerator and denominator of the complex fraction by the LCM of the denominators.

3. The LCM of the denominators in the complex fraction $\dfrac{1 - \frac{3}{x-3}}{\frac{2}{x} + 1}$ is $x(x - 3)$.

4. $\dfrac{c^{-1}}{a^{-1} + b^{-1}} = \dfrac{a+b}{c}$

7.3 Exercises

1

1. What is a complex fraction?

2. What is the general goal of simplifying a complex fraction?

Simplify.

3. $\dfrac{2 - \dfrac{1}{3}}{4 + \dfrac{11}{3}}$

4. $\dfrac{3 + \dfrac{5}{2}}{8 - \dfrac{7}{2}}$

5. $\dfrac{3 - \dfrac{2}{3}}{5 + \dfrac{5}{6}}$

6. $\dfrac{1 + \dfrac{1}{x}}{1 - \dfrac{1}{x^2}}$

7. $\dfrac{\dfrac{1}{y^2} - 1}{1 + \dfrac{1}{y}}$

8. $\dfrac{a - 2}{\dfrac{4}{a} - a}$

9. $\dfrac{\dfrac{25}{a} - a}{5 + a}$

10. $\dfrac{\dfrac{1}{a^2} - \dfrac{1}{a}}{\dfrac{1}{a^2} + \dfrac{1}{a}}$

11. $\dfrac{\dfrac{1}{b} + \dfrac{1}{2}}{\dfrac{4}{b^2} - 1}$

12. $\dfrac{2 - \dfrac{4}{x + 2}}{5 - \dfrac{10}{x + 2}}$

13. $\dfrac{4 + \dfrac{12}{2x - 3}}{5 + \dfrac{15}{2x - 3}}$

14. $\dfrac{\dfrac{3}{2a - 3} + 2}{\dfrac{-6}{2a - 3} - 4}$

15. $\dfrac{\dfrac{-5}{b - 5} - 3}{\dfrac{10}{b - 5} + 6}$

16. $\dfrac{\dfrac{x}{x + 1} - \dfrac{1}{x}}{\dfrac{x}{x + 1} + \dfrac{1}{x}}$

17. $\dfrac{\dfrac{2a}{a - 1} - \dfrac{3}{a}}{\dfrac{1}{a - 1} + \dfrac{2}{a}}$

18. $\dfrac{\dfrac{3}{x}}{\dfrac{9}{x^2}}$

19. $\dfrac{\dfrac{x}{3x - 2}}{\dfrac{x}{9x^2 - 4}}$

20. $\dfrac{\dfrac{a^2 - b^2}{4a^2b}}{\dfrac{a + b}{16ab^2}}$

21. $\dfrac{1 - \dfrac{1}{x} - \dfrac{6}{x^2}}{1 - \dfrac{4}{x} + \dfrac{3}{x^2}}$

22. $\dfrac{1 - \dfrac{3}{x} - \dfrac{10}{x^2}}{1 + \dfrac{11}{x} + \dfrac{18}{x^2}}$

23. $\dfrac{\dfrac{15}{x^2} - \dfrac{2}{x} - 1}{\dfrac{4}{x^2} - \dfrac{5}{x} + 4}$

24. $\dfrac{1 - \dfrac{2x}{3x - 4}}{x - \dfrac{32}{3x - 4}}$

25. $\dfrac{1 - \dfrac{12}{3x + 10}}{x - \dfrac{8}{3x + 10}}$

26. $\dfrac{x - 1 + \dfrac{2}{x - 4}}{x + 3 + \dfrac{6}{x - 4}}$

27. $\dfrac{x-5-\dfrac{18}{x+2}}{x+7+\dfrac{6}{x+2}}$

28. $\dfrac{x-4+\dfrac{9}{2x+3}}{x+3-\dfrac{5}{2x+3}}$

29. $\dfrac{\dfrac{1}{a}-\dfrac{3}{a-2}}{\dfrac{2}{a}+\dfrac{5}{a-2}}$

30. $\dfrac{\dfrac{2}{b}-\dfrac{5}{b+3}}{\dfrac{3}{b}+\dfrac{3}{b+3}}$

31. $\dfrac{\dfrac{1}{y^2}-\dfrac{1}{xy}-\dfrac{2}{x^2}}{\dfrac{1}{y^2}-\dfrac{3}{xy}+\dfrac{2}{x^2}}$

32. $\dfrac{\dfrac{2}{b^2}-\dfrac{5}{ab}-\dfrac{3}{a^2}}{\dfrac{2}{b^2}+\dfrac{7}{ab}+\dfrac{3}{a^2}}$

33. $\dfrac{\dfrac{x-1}{x+1}-\dfrac{x+1}{x-1}}{\dfrac{x-1}{x+1}+\dfrac{x+1}{x-1}}$

34. $\dfrac{\dfrac{y}{y+2}-\dfrac{y}{y-2}}{\dfrac{y}{y+2}+\dfrac{y}{y-2}}$

35. $4-\dfrac{2}{2-\dfrac{3}{x}}$

7.3 Applying Concepts

Simplify.

36. $\dfrac{x^{-1}+y^{-1}}{x^{-1}-y^{-1}}$

37. $\dfrac{x^{-1}}{y^{-1}}+\dfrac{y}{x}$

38. $\dfrac{x^{-1}+y}{x^{-1}-y}$

39. $\dfrac{x-\dfrac{1}{x}}{1+\dfrac{1}{x}}$

40. $1-\dfrac{1}{1-\dfrac{1}{b-2}}$

41. $2-\dfrac{2}{2-\dfrac{2}{c-1}}$

42. $3-\dfrac{2}{1-\dfrac{2}{3-\dfrac{2}{x}}}$

43. $a+\dfrac{a}{2+\dfrac{1}{1-\dfrac{2}{a}}}$

44. $a-\dfrac{1}{2-\dfrac{2}{2-\dfrac{2}{a}}}$

45. $\dfrac{\dfrac{1}{x+h}-\dfrac{1}{x}}{h}$

46. $\dfrac{\dfrac{1}{(x+h)^2}-\dfrac{1}{x^2}}{h}$

47. Find the sum of the reciprocals of three consecutive even integers.

48. The total resistance, R, of three resistors in parallel is given by the formula $R=\dfrac{1}{\dfrac{1}{R_1}+\dfrac{1}{R_2}+\dfrac{1}{R_3}}$. Find the total parallel resistance when $R_1=2$ ohms, $R_2=4$ ohms, and $R_3=8$ ohms.

49. **Car Loans** The interest rate on a loan to purchase a car affects the monthly payment. The function that relates the monthly payment for a 5-year loan (60-month loan) to the monthly interest rate is given by

$$P(x) = \frac{Cx}{\left[1 - \frac{1}{(x+1)^{60}}\right]}$$

where x is the monthly interest rate (as a decimal), C is the loan amount for the car, and $P(x)$ is the monthly payment.

a. Simplify the complex fraction.

b. Graph this equation for $0 < x \leq 0.025$ and $C = \$20{,}000$.

c. What is the interval of *annual* interest rates for the domain in part (b)?

d. The point whose coordinates are approximately (0.005, 386.66) is on the graph of this equation. Write a sentence that gives an interpretation of this ordered pair.

e. Use a graphing utility to determine the monthly payment for a car with a loan amount of \$20,000 and an annual interest rate of 8%. Round to the nearest dollar.

50. Write a report on continued fractions. Give an example of a continued fraction that can be used to approximate $\sqrt{2}$.

7.4 Rational Equations

1 Solve fractional equations

VIDEO & DVD CD TUTOR WEB SSM

To solve an equation containing fractions, *clear denominators* by multiplying each side of the equation by the *LCM of the denominators*. Then solve for the variable.

Solve: $\frac{3x}{x-5} = 5 - \frac{5}{x-5}$

$$\frac{3x}{x-5} = 5 - \frac{5}{x-5}$$

Multiply each side of the equation by the LCM of the denominators.

$$(x-5)\left(\frac{3x}{x-5}\right) = (x-5)\left(5 - \frac{5}{x-5}\right)$$

$$3x = (x-5)5 - (x-5)\left(\frac{5}{x-5}\right)$$

Simplify.

$$3x = 5x - 25 - 5$$
$$3x = 5x - 30$$

Solve the equation for x.

$$-2x = -30$$
$$x = 15$$

15 checks as a solution.

The solution is 15.

Occasionally, a value of the variable that appears to be a solution will make one of the denominators zero. In this case, the equation has no solution for that value of the variable.

Solve: $\dfrac{3x}{x-3} = 2 + \dfrac{9}{x-3}$

$$\frac{3x}{x-3} = 2 + \frac{9}{x-3}$$

Multiply each side of the equation by the LCM of the denominators.

$$(x-3)\left(\frac{3x}{x-3}\right) = (x-3)\left(2 + \frac{9}{x-3}\right)$$

Use the Distributive Property.

$$3x = (x-3)2 + (x-3)\left(\frac{9}{x-3}\right)$$
$$3x = 2x - 6 + 9$$
$$3x = 2x + 3$$
$$x = 3$$

Substituting 3 into the equation results in division by zero. Because division by zero is not defined, the equation has no solution.

$$\frac{3x}{x-3} = 2 + \frac{9}{x-3}$$
$$\frac{3(3)}{3-3} = 2 + \frac{9}{3-3}$$
$$\frac{9}{0} = 2 + \frac{9}{0}$$

Take Note

If each side of an equation is multiplied by a variable expression, it is essential that the solutions be checked. As this example shows, a proposed solution to an equation may not check when it is substituted into the original equation.

Multiplying each side of an equation by a variable expression may produce an equation with different solutions from the original equation. Thus, any time you multiply each side of an equation by a variable expression, you must check the resulting solution.

Example 1 Solve. **A.** $\dfrac{1}{4} = \dfrac{5}{x+5}$ **B.** $\dfrac{2x}{x-2} = \dfrac{1}{3x-4} + 2$

Solution **A.**

$$\frac{1}{4} = \frac{5}{x+5}$$

$$4(x+5)\frac{1}{4} = 4(x+5)\frac{5}{x+5}$$

• **Multiply each side of the equation by the LCM of the denominators.**

$$x + 5 = 4(5)$$
$$x + 5 = 20$$
$$x = 15$$

15 checks as a solution.
The solution is 15.

B.

$$\frac{2x}{x-2} = \frac{1}{3x-4} + 2$$

$$(x-2)(3x-4)\frac{2x}{x-2} = (x-2)(3x-4)\left(\frac{1}{3x-4} + 2\right)$$

$$(3x-4)2x = (x-2)(3x-4)\left(\frac{1}{3x-4}\right) + (x-2)(3x-4)2$$

$$6x^2 - 8x = x - 2 + 6x^2 - 20x + 16$$
$$6x^2 - 8x = 6x^2 - 19x + 14$$
$$11x = 14$$

$$x = \frac{14}{11}$$

$\frac{14}{11}$ checks as a solution.

The solution is $\frac{14}{11}$.

Problem 1 Solve. **A.** $\frac{5}{2x - 3} = \frac{-2}{x + 1}$ **B.** $\frac{4x + 1}{2x - 1} = 2 + \frac{3}{x - 3}$

Solution See page S24.

2 Work problems

VIDEO & DVD CD TUTOR WEB SSM

Point of Interest

The following problem was recorded in the Jiuzhang, a Chinese text that dates to the Han dynasty (about 200 B.C. to A.D. 200). "A reservoir has 5 channels bringing water to it. The first channel can fill the reservoir in $\frac{1}{3}$ day, the second in 1 day, the third in $2\frac{1}{2}$ days, the fourth in 3 days, and the fifth in 5 days. If all channels are open, how long does it take to fill the reservoir?" This problem is the earliest known work problem.

Rate of work is that part of a task that is completed in one unit of time. If a mason can build a retaining wall in 12 h, then in 1 h the mason can build $\frac{1}{12}$ of the wall. The mason's rate of work is $\frac{1}{12}$ of the wall each hour. If an apprentice can build the wall in x hours, the rate of work for the apprentice is $\frac{1}{x}$ of the wall each hour.

In solving a work problem, the goal is to determine the time it takes to complete a task. The basic equation that is used to solve work problems is

Rate of work × Time worked = Part of task completed

For example, if a pipe can fill a tank in 5 h, then in 2 h the pipe will fill $\frac{1}{5} \times 2 = \frac{2}{5}$ of the tank. In t hours, the pipe will fill $\frac{1}{5} \times t = \frac{t}{5}$ of the tank.

Solve: A mason can build a wall in 10 h. An apprentice can build a wall in 15 h. How long would it take them to build the wall if they worked together?

Strategy for solving a work problem

▶ For each person or machine, write a numerical or variable expression for the rate of work, the time worked, and the part of the task completed. The results can be recorded in a table.

Unknown time to build the wall working together: t

	Rate of work	·	Time worked	=	Part of task completed
Mason	$\frac{1}{10}$	·	t	=	$\frac{t}{10}$
Apprentice	$\frac{1}{15}$	·	t	=	$\frac{t}{15}$

▶ Determine how the parts of the task completed are related. Use the fact that the sum of the parts of the task completed must equal 1, the complete task.

The sum of the part of the task completed by the mason and the part of the task completed by the apprentice is 1.

$$\frac{t}{10} + \frac{t}{15} = 1$$
$$30\left(\frac{t}{10} + \frac{t}{15}\right) = 30(1)$$
$$3t + 2t = 30$$
$$5t = 30$$
$$t = 6$$

Working together, they would build the wall in 6 h.

Example 2 An electrician requires 12 h to wire a house. The electrician's apprentice can wire a house in 16 h. After working alone on one job for 4 h, the electrician quits, and the apprentice completes the task. How long does it take the apprentice to finish wiring the house?

Strategy

▶ Time required for the apprentice to finish wiring the house: t

	Rate	Time	Part
Electrician	$\frac{1}{12}$	4	$\frac{4}{12}$
Apprentice	$\frac{1}{16}$	t	$\frac{t}{16}$

▶ The sum of the part of the task completed by the electrician and the part of the task completed by the apprentice is 1.

Solution

$$\frac{4}{12} + \frac{t}{16} = 1$$
$$\frac{1}{3} + \frac{t}{16} = 1$$
$$48\left(\frac{1}{3} + \frac{t}{16}\right) = 48(1)$$
$$16 + 3t = 48$$
$$3t = 32$$
$$t = \frac{32}{3}$$

It takes the apprentice $10\frac{2}{3}$ h to finish wiring the house.

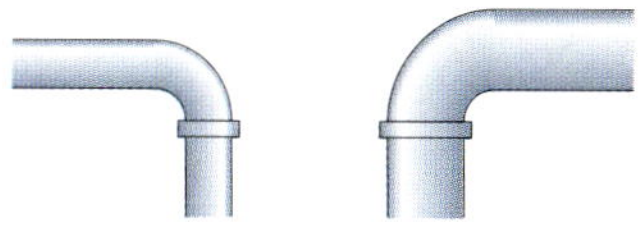

Problem 2 Two water pipes can fill a tank with water in 6 h. The larger pipe working alone can fill the tank in 9 h. How long would it take the smaller pipe, working alone, to fill the tank?

Solution See page S24.

3 Uniform motion problems

A car that travels constantly in a straight line at 55 mph is in uniform motion. **Uniform motion** means that the speed of an object does not change.

The basic equation used to solve uniform motion problems is

Distance = Rate × Time

An alternative form of this equation can be written by solving the equation for time. This form of the equation is used to solve the following problem.

$$\frac{\textbf{Distance}}{\textbf{Rate}} = \textbf{Time}$$

Solve: A motorist drove 150 mi on country roads before driving 50 mi on mountain roads. The rate of speed on the country roads was three times the rate on the mountain roads. The time spent traveling the 200 mi was 5 h. Find the rate of the motorist on the country roads.

***Strategy** for solving a uniform motion problem*

▶ For each object, write a numerical or variable expression for the distance, rate, and time. The results can be recorded in a table.

The unknown rate of speed on the mountain roads: r
Rate of speed on the country roads: $3r$

	Distance	÷	Rate	=	Time
Country roads	150	÷	$3r$	=	$\frac{150}{3r}$
Mountain roads	50	÷	r	=	$\frac{50}{r}$

▶ Determine how the times traveled by each object are related. For example, it may be known that the times are equal, or the total time may be known.

The total time for the trip is 5 h.

$$\frac{150}{3r} + \frac{50}{r} = 5$$
$$\frac{50}{r} + \frac{50}{r} = 5$$
$$r\left(\frac{50}{r} + \frac{50}{r}\right) = r(5)$$
$$50 + 50 = 5r$$
$$100 = 5r$$
$$20 = r$$

The rate of speed on the country roads was $3r$. Replace r with 20 and evaluate.

$$3r = 3(20) = 60$$

The rate of speed on the country roads was 60 mph.

Example 3 A marketing executive traveled 810 mi on a corporate jet in the same amount of time that it took to travel an additional 162 mi by helicopter. The rate of the jet was 360 mph greater than the rate of the helicopter. Find the rate of the jet.

Strategy

▶ Rate of the helicopter: r
Rate of the jet: $r + 360$

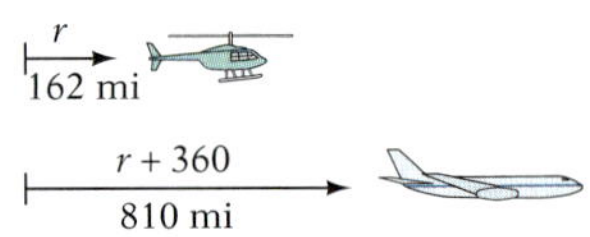

	Distance	Rate	Time
Jet	810	$r + 360$	$\frac{810}{r + 360}$
Helicopter	162	r	$\frac{162}{r}$

▶ The time traveled by jet is equal to the time traveled by helicopter.

Solution

$$\frac{810}{r + 360} = \frac{162}{r}$$
$$r(r + 360)\left(\frac{810}{r + 360}\right) = r(r + 360)\left(\frac{162}{r}\right)$$
$$810r = (r + 360)162$$
$$810r = 162r + 58{,}320$$
$$648r = 58{,}320$$
$$r = 90$$

• The rate of the helicopter was 90 mph.

$$r + 360 = 90 + 360 = 450$$

• Substitute the value of r into the variable expression for the rate of the jet.

The rate of the jet was 450 mph.

Problem 3 A plane can fly at a rate of 150 mph in calm air. Traveling with the wind, the plane flew 700 mi in the same amount of time it took to fly 500 mi against the wind. Find the rate of the wind.

Solution See page S25.

7.4 Concept Review

Determine whether the following statements are always true, sometimes true, or never true.

1. $\frac{3x}{x-2} + \frac{4}{x} - 3$ is a rational equation.

2. Multiplying both sides of an equation by the same variable expression produces an equivalent equation.

3. -2 and 5 cannot be solutions of the equation $\frac{1}{x+2} - 3 = \frac{x-2}{x-5}$.

4. An equation that contains a variable in a denominator can be simplified to an equivalent equation with no variables in the denominator.

5. If it takes 3 h to fill a tub with water, then $\frac{1}{3}$ of the tub can be filled in 1 h.

7.4 Exercises

1

1. Explain the difference between dealing with the denominator in simplifying a rational expression and dealing with the denominator in solving a rational equation.

2. Explain why it is necessary to check the solution of a rational equation.

Solve.

3. $\frac{x}{2} + \frac{5}{6} = \frac{x}{3}$

4. $\frac{x}{5} - \frac{2}{9} = \frac{x}{15}$

5. $1 - \frac{3}{y} = 4$

6. $7 + \frac{6}{y} = 5$

7. $\frac{8}{2x-1} = 2$

8. $3 = \frac{18}{3x-4}$

9. $\frac{4}{x-4} = \frac{2}{x-2}$

10. $\frac{x}{3} = \frac{x+1}{7}$

11. $\frac{x-2}{5} = \frac{1}{x+2}$

12. $\frac{x+4}{10} = \frac{6}{x-3}$

13. $\frac{3}{x-2} = \frac{4}{x}$

14. $\frac{5}{x} = \frac{2}{x+3}$

15. $\dfrac{3}{x-4}+2=\dfrac{5}{x-4}$

16. $\dfrac{5}{y+3}-2=\dfrac{7}{y+3}$

17. $\dfrac{8}{x-5}=\dfrac{3}{x}$

18. $\dfrac{16}{2-x}=\dfrac{4}{x}$

19. $5+\dfrac{8}{a-2}=\dfrac{4a}{a-2}$

20. $\dfrac{-4}{a-4}=3-\dfrac{a}{a-4}$

21. $\dfrac{x}{2}+\dfrac{20}{x}=7$

22. $3x=\dfrac{4}{x}-\dfrac{13}{2}$

23. $\dfrac{6}{x-5}=\dfrac{1}{x}$

24. $\dfrac{8}{x-2}=\dfrac{4}{x+1}$

25. $\dfrac{x}{x+2}=\dfrac{6}{x+5}$

26. $\dfrac{x}{x-2}=\dfrac{3}{x-4}$

27. $-\dfrac{5}{x+7}+1=\dfrac{4}{x+7}$

28. $5-\dfrac{2}{2x-5}=\dfrac{3}{2x-5}$

29. $\dfrac{2}{4y^2-9}+\dfrac{1}{2y-3}=\dfrac{3}{2y+3}$

30. $\dfrac{5}{x-2}-\dfrac{2}{x+2}=\dfrac{3}{x^2-4}$

31. $\dfrac{5}{x^2-7x+12}=\dfrac{2}{x-3}+\dfrac{5}{x-4}$

32. $\dfrac{9}{x^2+7x+10}=\dfrac{5}{x+2}-\dfrac{3}{x+5}$

2 33. If a gardener can mow a lawn in 20 min, what portion of the lawn can the gardener mow in 1 min?

34. If one person can complete a task in 2 h and another person can complete the same task in 3 h, will it take more or less than 2 h to complete the task when both people are working? Explain your answer.

Work problems

35. A large biotech firm uses two computers to process the daily results of its research studies. One computer can process the data in 2 h; the other computer takes 3 h to do the same job. How long would it take to process the data if both computers were used?

36. Two college students have started their own business building computers from kits. Working alone, one student can build a computer in 20 h. When the second student helps, they can build a computer in 7.5 h. How long would it take the second student, working alone, to build the computer?

37. One solar heating panel can raise the temperature of water 1 degree in 30 min. A second solar heating panel can raise the temperature 1 degree in 45 min. How long would it take to raise the temperature of the water 1 degree with both solar panels operating?

38. One member of a gardening team can landscape a new lawn in 36 h. The other member of the team can do the job in 45 h. How long would it take to landscape a lawn if both gardeners worked together?

39. One member of a telephone crew can wire new telephone lines in 5 h. It takes 7.5 h for the other member of the crew to do the job. How long would it take to wire new telephone lines if both members of the crew worked together?

40. As the spring flood waters began to recede in Texas, a young family was faced with pumping the water from their basement. One pump they were using could dispose of 9000 gal in 3 h. A second pump could dispose of the same number of gallons in 4.5 h. How many hours would it take to dispose of 9000 gal if both pumps were working?

41. A new machine can package transistors four times as fast as an older machine. Working together, the machines can package the transistors in 8 h. How long would it take the new machine, working alone, to package the transistors?

42. A ski resort can "manufacture" enough snow to open its steepest run in 12 h, whereas naturally falling snow would have to last for 36 h to provide enough snow. If the resort makes snow at the same time as it is snowing naturally, how long will it be before the run can be opened?

43. The larger of two printers being used to print the payroll for a major corporation requires 40 min to print the payroll. After both printers have been operating for 10 min, the larger printer malfunctions. The smaller printer requires 50 more minutes to complete the payroll. How long would it take the smaller printer, working alone, to print the payroll?

44. An experienced bricklayer can work twice as fast as an apprentice bricklayer. After they worked together on a job for 8 h, the experienced bricklayer quit. The apprentice required 12 more hours to finish the job. How long would it take the experienced bricklayer, working alone, to do the job?

45. A roofer requires 12 h to shingle a roof. After the roofer and an apprentice work on a roof for 3 h, the roofer moves on to another job. The apprentice requires 12 more hours to finish the job. How long would it take the apprentice, working alone, to do the job?

46. A welder requires 25 h to do a job. After the welder and an apprentice work on a job for 10 h, the welder quits. The apprentice finishes the job in 17 h. How long would it take the apprentice, working alone, to do the job?

47. Two clerks are addressing envelopes for a mayoral candidate's political campaign. One clerk can address one envelope every 30 s, whereas it takes 40 s for the second clerk to address one envelope. How many minutes will it take them, working together, to address 140 envelopes?

48. Three machines fill soda bottles. The machines can fill the daily quota of soda bottles in 12 h, 15 h, and 20 h, respectively. How long would it take to fill the daily quota of soda bottles with all three machines working?

49. With both hot and cold water running, a bathtub can be filled in 10 min. The drain will empty the tub in 15 min. A child turns both faucets on and leaves the drain open. How long will it be before the bathtub starts to overflow?

50. The inlet pipe can fill a water tank in 30 min. The outlet pipe can empty the tank in 20 min. How long would it take to empty a full tank with both pipes open?

51. An oil tank has two inlet pipes and one outlet pipe. One inlet pipe can fill the tank in 12 h, and the other inlet pipe can fill the tank in 20 h. The outlet pipe can empty the tank in 10 h. How long would it take to fill the tank with all three pipes open?

52. Water from a tank is being used for irrigation at the same time as the tank is being filled. The two inlet pipes can fill the tank in 6 h and 12 h, respectively. The outlet pipe can empty the tank in 24 h. How long would it take to fill the tank with all three pipes open?

3 Uniform motion problems

53. Two skaters take off for an afternoon of rollerblading in Central Park. The first skater can cover 15 mi in the same time it takes the second skater, traveling 3 mph slower than the first skater, to cover 12 mi. Find the rate of each rollerblader.

54. A commercial jet travels 1620 mi in the same amount of time it takes a corporate jet to travel 1260 mi. The rate of the commercial jet is 120 mph greater than the rate of the corporate jet. Find the rate of each jet.

55. A passenger train travels 295 mi in the same amount of time it takes a freight train to travel 225 mi. The rate of the passenger train is 14 mph greater than the rate of the freight train. Find the rate of each train.

56. The rate of a bicyclist is 7 mph more than the rate of a long-distance runner. The bicyclist travels 30 mi in the same amount of time it takes the runner to travel 16 mi. Find the rate of the runner.

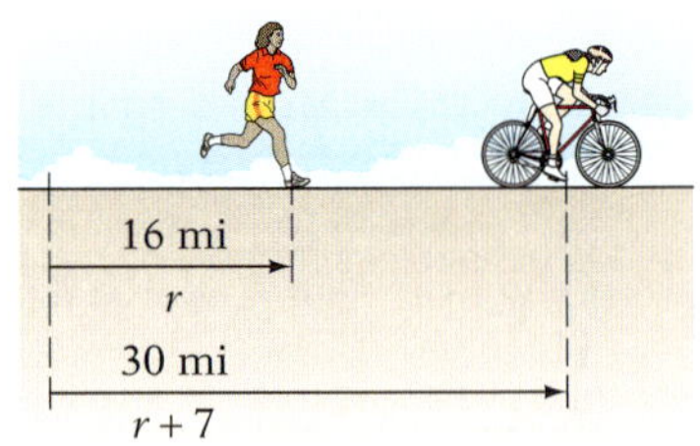

57. A cyclist rode 40 mi before having a flat tire and then walking 5 mi to a service station. The cycling rate was four times the walking rate. The time spent cycling and walking was 5 h. Find the rate at which the cyclist was riding.

58. A sales executive traveled 32 mi by car and then an additional 576 mi by plane. The rate of the plane was nine times the rate of the car. The total time of the trip was 3 h. Find the rate of the plane.

59. A motorist drove 72 mi before running out of gas and then walking 4 mi to a gas station. The driving rate of the motorist was twelve times the walking rate. The time spent driving and walking was 2.5 h. Find the rate at which the motorist walks.

60. An insurance representative traveled 735 mi by commercial jet and then an additional 105 mi by helicopter. The rate of the jet was four times the rate of the helicopter. The entire trip took 2.2 h. Find the rate of the jet.

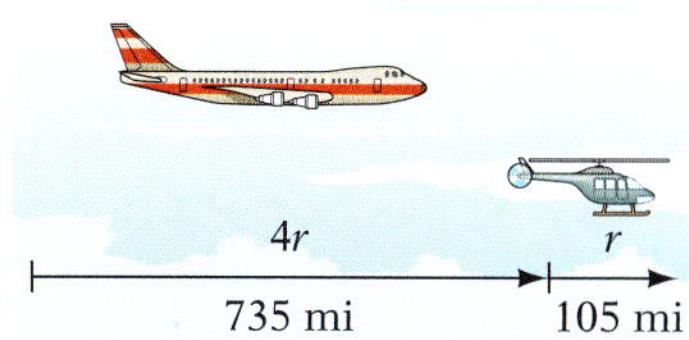

61. A business executive can travel the 480 ft between two terminals of an airport by walking on a moving sidewalk in the same time required to walk 360 ft without using the moving sidewalk. If the rate of the moving sidewalk is 2 ft/s, find the rate at which the executive can walk.

62. A cyclist and a jogger start from a town at the same time and head for a destination 18 mi away. The rate of the cyclist is twice the rate of the jogger. The cyclist arrives 1.5 h before the jogger. Find the rate of the cyclist.

63. A single-engine plane and a commercial jet leave an airport at 10 A.M. and head for an airport 960 mi away. The rate of the jet is four times the rate of the single-engine plane. The single-engine plane arrives 4 h after the jet. Find the rate of each plane.

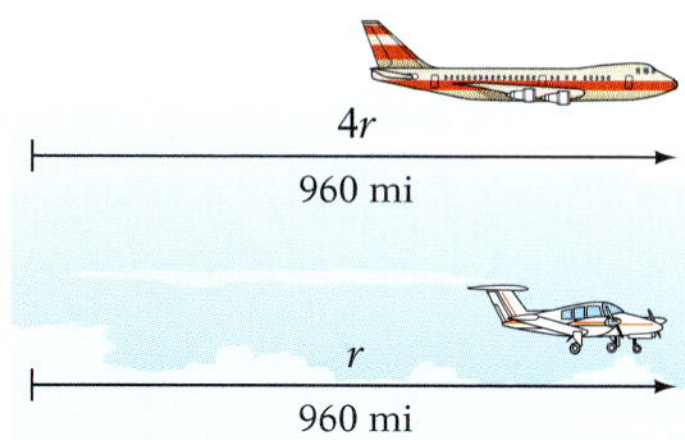

64. Marlys can row a boat 3 mph faster than she can swim. She is able to row 10 mi in the same time it takes her to swim 4 mi. Find Marlys's rate swimming.

65. A cruise ship can sail 28 mph in calm water. Sailing with the Gulf Stream, the ship can sail 170 mi in the same amount of time as it takes to sail 110 mi against the Gulf Stream. Find the rate of the Gulf Stream.

66. A commercial jet can fly 500 mph in calm air. Traveling with the jet stream, the plane flew 2420 mi in the same amount of time as it takes to fly 1580 mi against the jet stream. Find the rate of the jet stream.

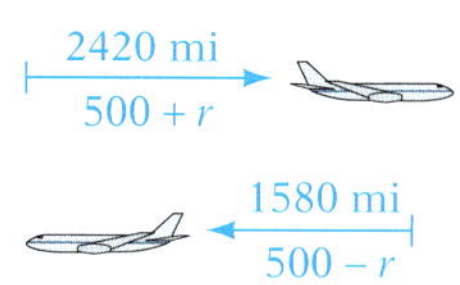

67. A tour boat used for river excursions can travel 7 mph in calm water. The amount of time it takes to travel 20 mi with the current is the same amount of time it takes to travel 8 mi against the current. Find the rate of the current.

68. A canoe can travel 8 mph in still water. Traveling with the current of a river, the canoe can travel 15 mi in the same amount of time it takes to travel 9 mi against the current. Find the rate of the current.

7.4 Applying Concepts

69. **Number Problem** The denominator of a fraction is 4 more than the numerator. If both the numerator and the denominator of the fraction are increased by 3, the new fraction is $\frac{5}{6}$. Find the original fraction.

70. **Work Problem** One pipe can fill a tank in 3 h, a second pipe can fill the tank in 4 h, and a third pipe can fill the tank in 6 h. How long would it take to fill the tank with all three pipes operating?

71. **Work Problem** One printer can print a company's paychecks in 24 min, a second printer can print the checks in 16 min, and a third printer can do the job in 12 min. How long would it take to print the checks with all three printers operating?

72. **Uniform Motion Problem** By increasing your speed by 10 mph, you can drive the 200-mile trip to your hometown in 40 min less time than the trip usually takes you. How fast do you usually drive?

73. **Uniform Motion Problem** Because of weather conditions, a bus driver reduced the usual speed along a 165-mile bus route by 5 mph. The bus arrived only 15 min later than its usual arrival time. How fast does the bus usually travel?

74. **Work Problem** If a pump can fill a pool in A hours and a second pump can fill the pool in B hours, find a formula, in terms of A and B, for the time it takes both pumps, working together, to fill the pool.

75. **Uniform Motion Problem** If a parade is 1 mi long and is proceeding at 3 mph, how long will it take a runner, jogging at 5 mph, to run from the beginning of the parade to the end and then back to the beginning?

76. Write a report on the Rhind Papyrus.

77. What is a unit fraction? What role did unit fractions play in early Egyptian mathematics?

7.5 Proportions and Variation

1 Proportions

VIDEO & DVD CD TUTOR WEB SSM

Quantities such as 3 feet, 5 liters, and 2 miles are number quantities written with units. In these examples, the units are feet, liters, and miles.

A **ratio** is the quotient of two quantities that have the same unit.

The weekly wages of a painter are \$800. The painter spends \$150 a week for food. The ratio of wages spent for food to the total weekly wages is written as shown below.

$\frac{\$150}{\$800} = \frac{150}{800} = \frac{3}{16}$ A ratio is in simplest form when the two numbers do not have a common factor. The units are not written.

A **rate** is the quotient of two quantities that have different units.

A car travels 120 mi on 3 gal of gas. The miles-to-gallon rate is written as shown below.

$\frac{120 \text{ mi}}{3 \text{ gal}} = \frac{40 \text{ mi}}{1 \text{ gal}}$ A rate is in simplest form when the two numbers do not have a common factor. The units are written as part of the rate.

A **proportion** is an equation that states that two ratios or rates are equal. For example, $\frac{90\text{ km}}{4\text{ L}} = \frac{45\text{ km}}{2\text{ L}}$ and $\frac{3}{4} = \frac{x+2}{16}$ are proportions.

Note that a proportion is a special kind of fractional equation. Many application problems can be solved by using proportions.

The sales tax on a car that costs \$24,000 is \$1320. Find the sales tax on a car that costs \$29,000.

Write a proportion using x to represent the sales tax.

$$\frac{1320}{24{,}000} = \frac{x}{29{,}000}$$

To solve the proportion, multiply each side of the equation by the denominators.

$$\frac{66}{1200} = \frac{x}{29{,}000}$$

$$(1200)(29{,}000)\left(\frac{66}{1200}\right) = (1200)(29{,}000)\left(\frac{x}{29{,}000}\right)$$

$$(29{,}000)(66) = 1200x$$

$$1{,}914{,}000 = 1200x$$

$$1595 = x$$

The sales tax on the \$29,000 car is \$1595.

Example 1 A stock investment of 50 shares pays a dividend of \$106. At this rate, how many additional shares are needed to earn a dividend of \$424?

Strategy To find the additional number of shares that are required, write and solve a proportion using x to represent the additional number of shares. Then $50 + x$ is the total number of shares of stock.

Solution

$$\frac{106}{50} = \frac{424}{50 + x}$$

$$\frac{53}{25} = \frac{424}{50 + x}$$

$$25(50 + x)\frac{53}{25} = 25(50 + x)\frac{424}{50 + x}$$

$$(50 + x)53 = (25)424$$

$$2650 + 53x = 10{,}600$$

$$53x = 7950$$

$$x = 150$$

An additional 150 shares of stock are required.

Problem 1 Two pounds of cashews cost \$5.80. At this rate, how much would 15 lb of cashews cost?

Solution See page S25.

2 Variation problems

VIDEO & DVD CD TUTOR WEB SSM

Direct variation is a special function that can be expressed as the equation $y = kx$, where k is a constant. The equation $y = kx$ is read "y varies directly as x" or "y is proportional to x." The constant k is called the **constant of variation** or the **constant of proportionality.**

The circumference (C) of a circle varies directly as the diameter (d). The direct variation equation is written $C = \pi d$. The constant of variation is π.

A nurse earns \$28 per hour. The nurse's total wage (w) is directly proportional to the number of hours (h) worked. The equation of variation is $w = 28h$. The constant of proportionality is 28.

In general, a direct variation equation can be written in the form $y = kx^n$, where n is a positive number. For example, the equation $y = kx^2$ is read "y varies directly as the square of x."

The area (A) of a circle varies directly as the square of the radius (r) of the circle. The direct variation equation is $A = \pi r^2$.

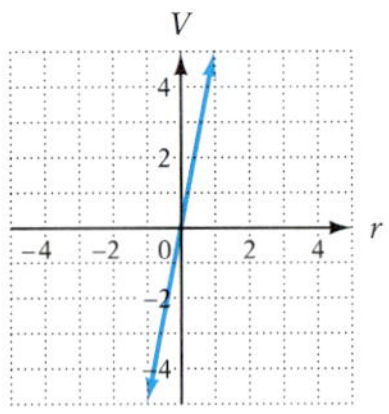

The graph of $V = 5r$

Given that V varies directly as r and that $V = 20$ when $r = 4$, the constant of variation can be found by writing the basic direct variation equation, replacing V and r by the given values, and solving for the constant of variation.

$$V = kr$$
$$20 = k \cdot 4$$
$$5 = k$$

The direct variation equation can then be written by substituting the value of k into the basic direct variation equation.

$$V = 5r$$

The tension (T) in a spring varies directly as the distance (x) it is stretched. If $T = 8$ lb when $x = 2$ in., find T when $x = 4$ in.

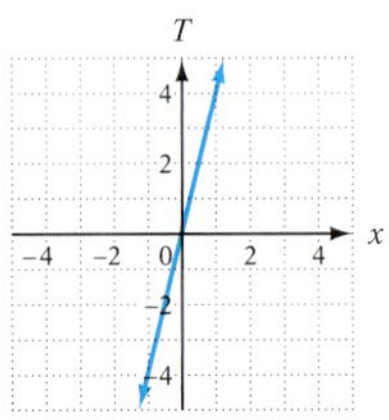

The graph of $T = 4x$

Write the basic direct variation equation. $T = kx$

Replace T and x by the given values. $8 = k \cdot 2$

Solve for the constant of variation. $4 = k$

Write the direct variation equation. $T = 4x$

To find T when $x = 4$ in., substitute 4 for x in the basic variation equation and solve for T.

$$T = 4x$$
$$T = 4 \cdot 4$$
$$T = 16$$

When $x = 4$ in., the tension is 16 lb.

Example 2 The amount (A) of medication prescribed for a person varies directly with the person's weight (W). For a person who weighs 50 kg, 2 ml of medication are prescribed. How many milliliters of medication are required for a person who weighs 75 kg?

Strategy To find the required amount of medication:

- ▶ Write the basic direct variation equation, replace the variables by the given values, and solve for k.

▶ Write the direct variation equation, replacing k by its value. Substitute 75 for W, and solve for A.

Solution

$$A = kW$$
$$2 = k \cdot 50$$
$$\frac{1}{25} = k$$
$$A = \frac{1}{25}W \quad \text{• This is the direct variation equation.}$$
$$A = \frac{1}{25} \cdot 75 = 3 \quad \text{• Replace } W \text{ by 75.}$$

The required amount of medication is 3 ml.

Problem 2 The distance (s) a body falls from rest varies directly as the square of the time (t) of the fall. An object falls 64 ft in 2 s. How far will it fall in 5 s?

Solution See page S25.

A **joint variation** is a variation in which a variable varies directly as the product of two or more other variables. A joint variation can be expressed as the equation $z = kxy$, where k is a constant. The equation $z = kxy$ is read "z varies jointly as x and y."

The area (A) of a triangle varies jointly as the base (b) and the height (h). The joint variation equation is written $A = \frac{1}{2}bh$. The constant of variation is $\frac{1}{2}$.

An **inverse variation** is a function that can be expressed as the equation $y = \frac{k}{x}$, where k is a constant. The equation $y = \frac{k}{x}$ is read "y varies inversely as x" or "y is inversely proportional to x."

In general, an inverse variation equation can be written $y = \frac{k}{x^n}$, where n is a positive number. For example, the equation $y = \frac{k}{x^2}$ is read "y varies inversely as the square of x."

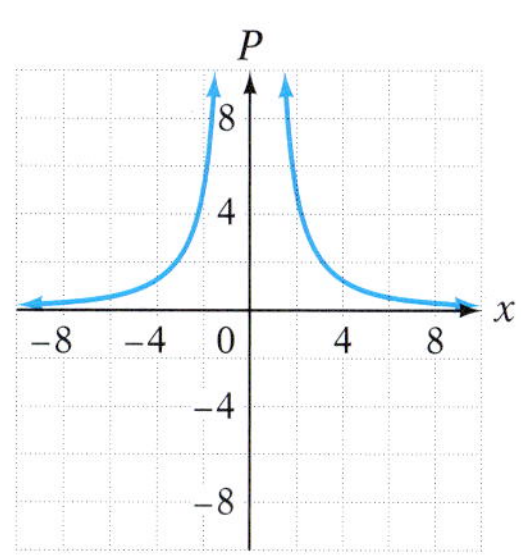

The graph of $P = \frac{20}{x^2}$

Given that P varies inversely as the square of x and that $P = 5$ when $x = 2$, the variation constant can be found by writing the basic inverse variation equation, replacing P and x by the given values, and solving for the constant variation.

$$P = \frac{k}{x^2}$$
$$5 = \frac{k}{2^2}$$
$$5 = \frac{k}{4}$$
$$20 = k$$

The inverse variation equation can then be found by substituting the value of k into the basic inverse variation equation.

$$P = \frac{20}{x^2}$$

The length (L) of a rectangle with fixed area is inversely proportional to the width (w). If $L = 6$ ft when $w = 2$ ft, find L when $w = 3$ ft.

Write the basic inverse variation equation. $\quad L = \frac{k}{w}$

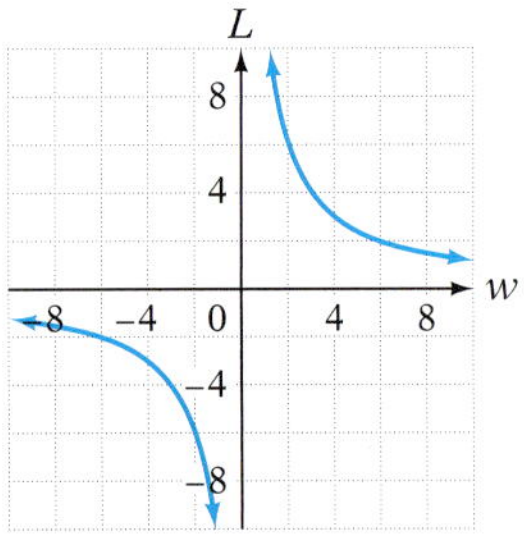

The graph of $L = \frac{12}{w}$

Replace L and w by the given values. $6 = \frac{k}{2}$

Solve for the constant of variation. $12 = k$

Write the inverse variation equation. $L = \frac{12}{w}$

To find L when $w = 3$ ft, substitute 3 for w in the inverse variation equation and solve for L.

$$L = \frac{12}{w}$$

$$L = \frac{12}{3}$$

$$L = 4$$

When $w = 3$ ft, the length is 4 ft.

Example 3 A company that produces personal computers has determined that the number of computers it can sell (s) is inversely proportional to the price (P) of the computer. Two thousand computers can be sold when the price is $2500. How many computers can be sold when the price of a computer is $2000?

Strategy To find the number of computers:

▶ Write the basic inverse variation equation, replace the variables by the given values, and solve for k.

▶ Write the inverse variation equation, replacing k by its value. Substitute 2000 for P, and solve for s.

Solution

$$s = \frac{k}{P}$$

$$2000 = \frac{k}{2500}$$

$$5{,}000{,}000 = k$$

$$s = \frac{5{,}000{,}000}{P}$$ • This is the inverse variation equation.

$$s = \frac{5{,}000{,}000}{2000} = 2500$$ • Replace P by 2000.

At a price of $2000, 2500 computers can be sold.

Problem 3 The resistance (R) to the flow of electric current in a wire of fixed length is inversely proportional to the square of the diameter (d) of the wire. If a wire of diameter 0.01 cm has a resistance of 0.5 ohm, what is the resistance in a wire that is 0.02 cm in diameter?

Solution See page S25.

A **combined variation** is a variation in which two or more types of variation occur at the same time. For example, in physics, the volume (V) of a gas varies directly as the temperature (T) and inversely as the pressure (P). This combined variation is written $V = \frac{kT}{P}$.

Example 4 The pressure (P) of a gas varies directly as the temperature (T) and inversely as the volume (V). When $T = 50°$ and $V = 275\text{ in}^3$, $P = 20\text{ lb/in}^2$. Find the pressure of a gas when $T = 60°$ and $V = 250\text{ in}^3$.

Strategy To find the pressure:

- ▶ Write the basic combined variation equation, replace the variables by the given values, and solve for k.
- ▶ Write the combined variation equation, replacing k by its value. Substitute 60 for T and 250 for V, and solve for P.

Solution

$$P = \frac{kT}{V}$$

$$20 = \frac{k(50)}{275}$$

$$110 = k$$

$$P = \frac{110T}{V}$$ • This is the combined variation equation.

$$P = \frac{110(60)}{250} = 26.4$$ • Replace T by 60 and V by 250.

The pressure is 26.4 lb/in^2.

Problem 4 The strength (s) of a rectangular beam varies jointly as its width (W) and as the square of its depth (d) and inversely as its length (L). If the strength of a beam 2 in. wide, 12 in. deep, and 12 ft long is 1200 lb, find the strength of a beam 4 in. wide, 8 in. deep, and 16 ft long.

Solution See page S25.

7.5 Concept Review

Determine whether the following statements are always true, sometimes true, or never true.

1. If x varies inversely as y, then when x is doubled, y is doubled.

2. If a varies inversely as b, then ab is a constant.

3. If a varies jointly as b and c, then $a = \frac{kb}{c}$.

4. If the length of a rectangle is held constant, then the area of the rectangle varies directly as the width.

5. If the area of a rectangle is held constant, then the length varies directly as the width.

6. The circumference of a circle varies directly as the diameter. If the diameter of a circle is doubled, then the circumference of the circle will be doubled.

7.5 Exercises

1 1. How does a ratio differ from a rate?

2. What is a proportion?

Solve the proportion.

3. $\frac{x+1}{10} = \frac{2}{5}$

4. $\frac{4}{x+2} = \frac{3}{4}$

5. $\frac{x}{4} = \frac{x-2}{8}$

6. $\frac{8}{x-5} = \frac{3}{x}$

7. $\frac{8}{x-2} = \frac{4}{x+1}$

8. $\frac{4}{x-4} = \frac{2}{x-2}$

9. $\frac{8}{3x-2} = \frac{2}{2x+1}$

10. $\frac{x-2}{x-5} = \frac{2x}{2x+5}$

11. **Environmental Science** In a wildlife preserve, 60 ducks are captured, tagged, and then released. Later, 200 ducks are examined, and three of the 200 ducks are found to have tags. Estimate the number of ducks in the preserve.

12. **Political Science** A pre-election survey showed that 7 out of every 12 voters would vote in an election. At this rate, how many people would be expected to vote in a city of 210,000?

13. **Food Service** A caterer estimates that 2 gal of fruit punch will serve 30 people. How much additional punch is necessary to serve 75 people?

14. **Nutrition** If a 56-gram serving of pasta contains 7 g of protein, how many grams of protein are in a 454-gram box of the pasta?

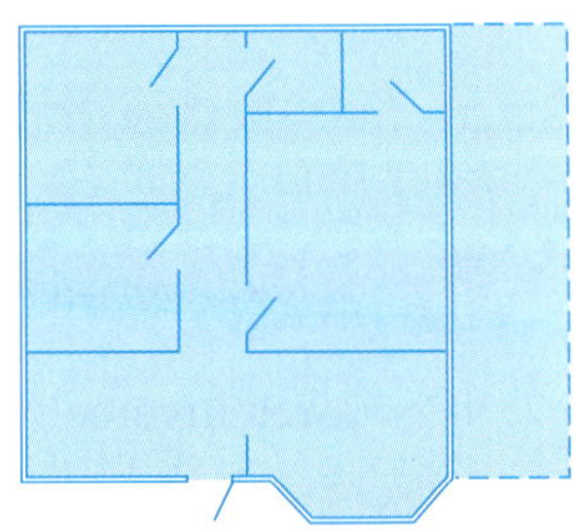

15. **Architecture** On an architectural drawing, $\frac{1}{4}$ in. represents 1 ft. Find the dimensions of a room that measures $4\frac{1}{4}$ in. by $5\frac{1}{2}$ in. on the drawing.

16. **Construction** A contractor estimated that 15 ft^2 of window space will be allowed for every 160 ft^2 of floor space. Using this estimate, how much window space will be allowed for 3200 ft^2 of floor space?

17. **Energy** Walking 2 miles per hour will use 325 calories per hour. Walking at the same rate, how many miles would a person need to walk to lose 1 pound? (The burning of 3500 calories is equivalent to the loss of 1 pound.) Round to the nearest hundredth.

18. **Construction** To tile 24 ft^2 of area, 120 ceramic tiles are needed. At this rate, how many tiles are needed to tile 300 ft^2?

19. **Health Science** Three-fourths of an ounce of a medication are required for a 120-pound adult. At the same rate, how many additional ounces of medication are required for a 200-pound adult?

20. **Physics** An object that weighs 100 lb on Earth would weigh 90.5 lb on Venus. At this rate, what is the weight on Venus of an object that weighs 150 lb on Earth?

21. **Environmental Science** Six ounces of an insecticide are mixed with 15 gal of water to make a spray for spraying an orange grove. At the same rate, how much additional insecticide is required to be mixed with 100 gal of water?

22. **Ecology** In an attempt to estimate the number of Siberian tigers in a preserve, a wildlife management team captured, tagged, and released 50 Siberian tigers. A few months later, they captured 150 Siberian tigers of which 30 had tags. Estimate the number of Siberian tigers in the region.

23. **Shipping** The United States Postal Service delivers 200 billion pieces of mail a year. How many pieces of mail does the Postal Service deliver per week? Assume that the Postal Service delivers mail 300 days during the year and 6 days per week.

24. **Construction** A contractor estimated that 30 ft^3 of cement is required to make a concrete floor that measures 90 ft^2. Using this estimate, find how many additional cubic feet of cement would be required to make a concrete floor that measures 120 ft^2.

25. **Graphic Arts** The accompanying picture of a whale uses a scale of 1 in. represents 48 ft. Estimate the actual length of the whale.

26. **Graphic Arts** The accompanying picture of an elephant uses a scale of 1 in. represents 12 ft. Estimate the actual height of the elephant.

2 **27.** What is a direct variation?

28. What is an inverse variation?

29. **Scuba Diving** The pressure (p) on a diver in the water varies directly as the depth (d). If the pressure is 3.6 lb/in^2 when the depth is 8 ft, what is the pressure when the depth is 30 ft?

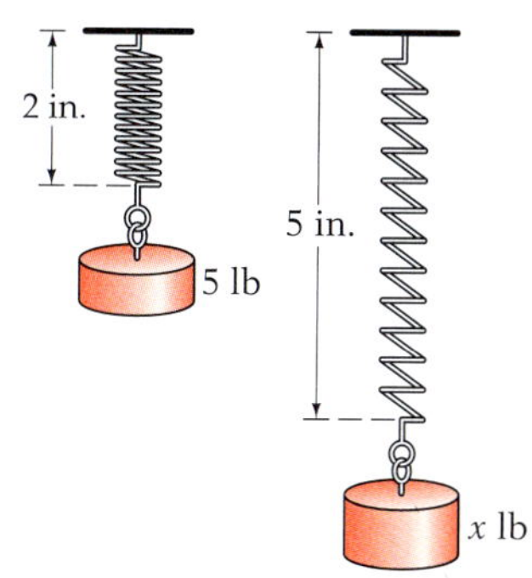

30. **Physics** The distance (d) a spring will stretch varies directly as the force (f) applied to the spring. If a force of 5 lb is required to stretch a spring 2 in., what force is required to stretch the spring 5 in.?

31. **Business** The profit (P) realized by a company varies directly as the number of products it sells (s). If a company makes a profit of \$2500 on the sale of 20 products, what is the profit when the company sells 300 products?

32. **Farming** The number of bushels of wheat (b) produced by a farm is directly proportional to the number of acres (A) planted in wheat. If a 25-acre farm yields 1125 bushels of wheat, what is the yield of a farm that has 220 acres of wheat?

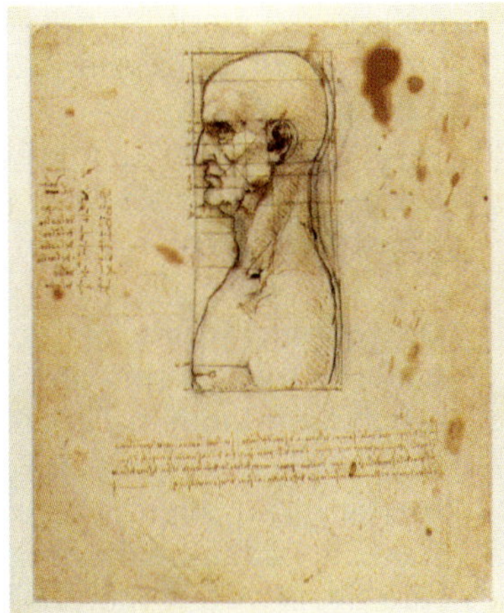

33. **Art** Leonardo da Vinci observed that the length of a person's face varies directly as the length of the person's chin. If a person whose face length is 9 in. has a chin length of 1.5 in., what is the length of a person's face whose chin length is 1.7 in.?

34. **Art** Leonardo da Vinci observed that a person's height varies directly as the width of the person's shoulders. If a person 70 in. tall has a shoulder width of 17.5 in., what is the height of a person whose shoulder width is 16 in.?

35. **Optics** The distance (d) a person can see to the horizon from a point above the surface of Earth varies directly as the square root of the height (H). If, for a height of 500 ft, the horizon is 19 mi away, how far is the horizon from a point that is 800 ft high? Round to the nearest hundredth.

36. **Physics** The period (p) of a pendulum, or the time it takes for the pendulum to make one complete swing, varies directly as the square root of the length (L) of the pendulum. If the period of a pendulum is 1.5 s when the length is 2 ft, find the period when the length is 4.5 ft. Round to the nearest hundredth.

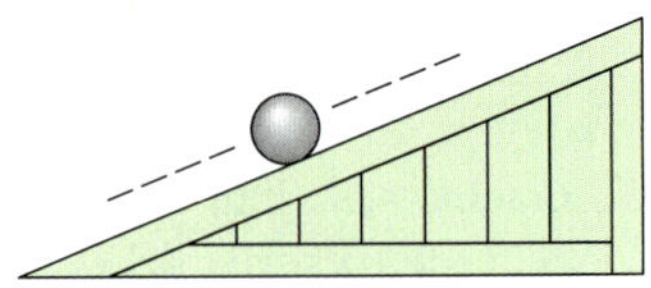

37. **Physics** The distance (s) a ball will roll down an inclined plane is directly proportional to the square of the time (t). If the ball rolls 5 ft in 1 s, how far will it roll in 4 s?

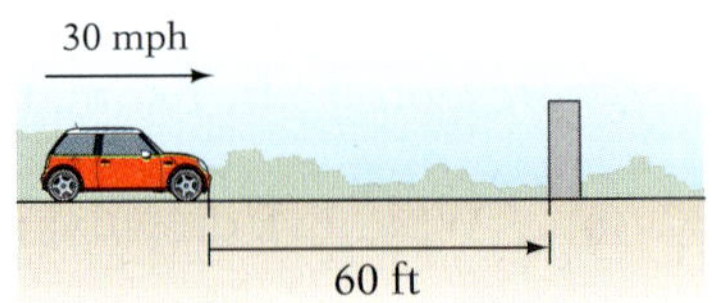

38. **Automotive Technology** The stopping distance (s) of a car varies directly as the square of its speed (v). If a car traveling 30 mph requires 60 ft to stop, find the stopping distance for a car traveling 55 mph. Round to the nearest tenth.

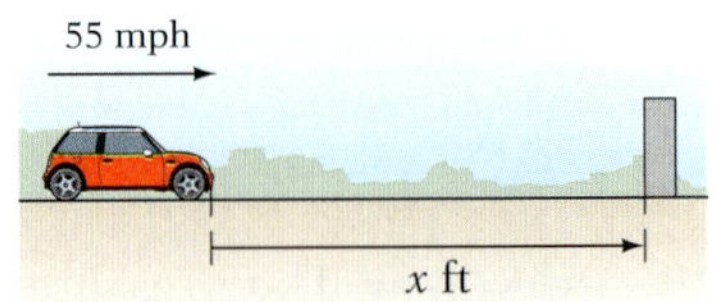

39. **Physics** For a constant temperature, the pressure (P) of a gas varies inversely as the volume (V). If the pressure is 25 lb/in^2 when the volume is 400 ft^3, find the pressure when the volume is 150 ft^3.

40. **Physics** The pressure (p) in a liquid varies directly as the product of the depth (d) and the density (D) of the liquid. If the pressure is 37.5 lb/in^2 when the depth is 100 in. and the density is 1.2 lb/in^2, find the pressure when the density remains the same and the depth is 60 in.

41. **Mechanical Engineering** The speed (v) of a gear varies inversely as the number of teeth (t). If a gear that has 48 teeth makes 20 revolutions per minute, how many revolutions per minute will a gear that has 30 teeth make?

42. **Electricity** The current (I) in a wire varies directly as the voltage (v) and inversely as the resistance (r). If the current is 27.5 amps when the voltage is 110 volts and the resistance is 4 ohms, find the current when the voltage is 195 volts and the resistance is 12 ohms.

43. **Magnetism** The repulsive force (f) between the north poles of two magnets is inversely proportional to the square of the distance (d) between them. If the repulsive force is 18 lb when the distance is 3 in., find the repulsive force when the distance is 1.2 in.

44. **Physics** The intensity (I) of a light source is inversely proportional to the square of the distance (d) from the source. If the intensity is 8 lumens at a distance of 6 m, what is the intensity when the distance is 4 m?

45. **Electricity** The resistance (R) of a wire varies directly as the length (L) of the wire and inversely as the square of the diameter (d). If the resistance is 9 ohms in 50 ft of wire that has a diameter of 0.05 in., find the resistance in 50 ft of a similar wire that has a diameter of 0.02 in.

46. **Music** The frequency of vibration (f) of a guitar string varies directly as the square root of the tension (T) and inversely as the length (L) of the string. If the frequency is 40 vibrations per second when the tension is 25 lb and the length of the string is 3 ft, find the frequency when the tension is 36 lb and the string is 4 ft long.

47. **Sailing** The wind force (w) on a vertical surface varies directly as the product of the area (A) of the surface and the square of the wind velocity (v). When the wind is blowing at 30 mph, the force on an area of 10 ft^2 is 45 lb. Find the force on this area when the wind is blowing at 60 mph.

48. **Electricity** The power (P) in an electric circuit is directly proportional to the product of the current (I) and the square of the resistance (R). If the power is 100 watts when the current is 4 amps and the resistance is 5 ohms, find the power when the current is 2 amps and the resistance is 10 ohms.

7.5 Applying Concepts

49. **a.** Graph $y = kx$ when $k = 2$.
b. What kind of function does the graph represent?

50. **a.** Graph $y = kx$ when $k = \frac{1}{2}$.
b. What kind of function does the graph represent?

51. **a.** Graph $y = \frac{k}{x}$ when $k = 2$ and $x > 0$.
b. Is this the graph of a function?

52. **a.** Graph $y = \dfrac{k}{x^2}$ when $k = 2$ and $x > 0$.
b. Is this the graph of a function?

53. In the inverse variation equation $y = \dfrac{k}{x}$, what is the effect on x when y doubles?

54. In the direct variation equation $y = kx$, what is the effect on y when x doubles?

Complete using the word *directly* or *inversely*.

55. If a varies directly as b and inversely as c, then c varies ______ as b and ______ as a.

56. If a varies ______ as b and c, then abc is constant.

57. If the width of a rectangle is held constant, the area of the rectangle varies ______ as the length.

58. If the area of a rectangle is held constant, the length of the rectangle varies ______ as the width.

59. A number h is the harmonic mean of the numbers a and b if the reciprocal of h is equal to the average of the reciprocals of a and b.
a. Write an expression for the harmonic mean of a and b.
b. Find the harmonic mean of 10 and 15.

60. Write a report on financial ratios used by stock market analysts. Include the price-earnings ratio, current ratio, quick ratio, percent yield, and ratio of stockholders' equity to liabilities.

7.6 Literal Equations

1 Solve literal equations

VIDEO & DVD CD TUTOR WEB SSM

A *literal equation* is an equation that contains more than one variable. Examples of literal equations are shown at the right.

$$3x - 2y = 4$$
$$v^2 = v_0^2 + 2as$$

Formulas are used to express relationships among physical quantities. A **formula** is a literal equation that states a rule about measurement. Examples of formulas are shown below.

$$s = vt - 16t^2 \quad \text{(Physics)}$$
$$c^2 = a^2 + b^2 \quad \text{(Geometry)}$$
$$A = P(1 + r)^t \quad \text{(Business)}$$

The Addition and Multiplication Properties of Equations can be used to solve a literal equation for one of the variables. The goal is to rewrite the equation so that the variable being solved for is alone on one side of the equation, and all the other numbers and variables are on the other side.

Solve $C = \frac{5}{9}(F - 32)$ for F.

$$C = \frac{5}{9}(F - 32)$$

Use the Distributive Property to remove parentheses.

$$C = \frac{5}{9}F - \frac{160}{9}$$

Multiply each side of the equation by the LCM of the denominators.

$$9C = 5F - 160$$

Add 160 to each side of the equation.

$$9C + 160 = 5F$$

Divide each side of the equation by the coefficient 5.

$$\frac{9C + 160}{5} = F$$

Example 1 **A.** Solve $A = P + Prt$ for P. **B.** Solve $\frac{S}{S - C} = R$ for C.

Solution **A.**

$$A = P + Prt$$

$$A = (1 + rt)P$$

• Factor P from $P + Prt$.

$$\frac{A}{1 + rt} = \frac{(1 + rt)P}{1 + rt}$$

• Divide each side of the equation by $1 + rt$.

$$\frac{A}{1 + rt} = P$$

B.

$$\frac{S}{S - C} = R$$

• Multiply each side of the equation by $S - C$.

$$(S - C)\frac{S}{S - C} = (S - C)R$$

$$S = SR - CR$$

$$CR + S = SR$$

• Add CR to each side of the equation.

$$CR = SR - S$$

• Subtract S from each side of the equation.

$$C = \frac{SR - S}{R}$$

• Divide each side of the equation by R.

Problem 1 **A.** Solve $\frac{1}{R_1} + \frac{1}{R_2} = \frac{1}{R}$ for R. **B.** Solve $t = \frac{r}{r + 1}$ for r.

Solution See page S26.

7.6 Concept Review

Determine whether the following statements are always true, sometimes true, or never true.

1. A formula is a literal equation that states a rule about measurement.

2. A literal equation contains two variables.

3. The formula $\frac{1}{R_1} + \frac{1}{R_2} = \frac{1}{R}$ is taken from physics. When we solve for R_2, we get $R_2 = R - R_1$.

4. Consider the formula $P = 2L + 2W$. When we double the perimeter, the length and width will both be doubled.

7.6 Exercises

1 Solve the formula for the variable given.

1. $P = 2L + 2W$; W (Geometry)

2. $F = \frac{9}{5}C + 32$; C (Temperature conversion)

3. $S = C - rC$; C (Business)

4. $A = P + Prt$; t (Business)

5. $PV = nRT$; R (Chemistry)

6. $A = \frac{1}{2}bh$; h (Geometry)

7. $F = \frac{Gm_1m_2}{r^2}$; m_2 (Physics)

8. $\frac{P_1V_1}{T_1} = \frac{P_2V_2}{T_2}$; P_2 (Chemistry)

9. $I = \frac{E}{R + r}$; R (Physics)

10. $S = V_0t - 16t^2$; V_0 (Physics)

11. $A = \frac{1}{2}h(b_1 + b_2)$; b_2 (Geometry)

12. $V = \frac{1}{3}\pi r^2 h$; h (Geometry)

13. $\frac{1}{R} = \frac{1}{R_1} + \frac{1}{R_2}$; R_2 (Physics)

14. $\frac{1}{f} = \frac{1}{a} + \frac{1}{b}$; b (Physics)

15. $a_n = a_1 + (n - 1)d$; d (Mathematics)

16. $P = \frac{R - C}{n}$; R (Business)

17. $S = 2WH + 2WL + 2LH$; H (Geometry)

18. $S = 2\pi r^2 + 2\pi rH$; H (Geometry)

Solve for x.

19. $ax + by + c = 0$

20. $x = ax + b$

21. $ax + b = cx + d$

22. $y - y_1 = m(x - x_1)$

23. $\frac{a}{x} = \frac{b}{c}$

24. $\frac{1}{x} + \frac{1}{a} = b$

25. $\frac{1}{a} + \frac{1}{b} = \frac{1}{x}$

26. $a(a - x) = b(b - x)$

7.6 Applying Concepts

Solve for x.

27. $\frac{x - y}{y} = \frac{x + 5}{2y}$

28. $\frac{x - 2}{y} = \frac{x + 2}{5y}$

29. $\frac{x}{x + y} = \frac{2x}{4y}$

30. $\frac{2x}{x - 2y} = \frac{x}{2y}$

31. $\frac{x - y}{2x} = \frac{x - 3y}{5y}$

32. $\frac{x - y}{x} = \frac{2x}{9y}$

Solve for the given variable.

33. $\frac{w_1}{w_2} = \frac{f_2 - f}{f - f_1}; f$

34. $v = \frac{v_1 + v_2}{1 + \frac{v_1 v_2}{c^2}}; v_1$

Focus on Problem Solving

Implication

Sentences that are constructed using "If..., then..." occur frequently in problem-solving situations. These sentences are called **implications.** The sentence "If it rains, then I will stay home" is an implication. The phrase *it rains* is called the **antecedent** of the implication. The phrase *I will stay home* is the **consequent** of the implication.

The sentence "If $x = 4$, then $x^2 = 16$" is a true sentence. The **contrapositive** of an implication is formed by switching the antecedent and the consequent and then negating each one. The contrapositive of "If $x = 4$, then $x^2 = 16$" is "If $x^2 \neq 16$, then $x \neq 4$." This sentence is also true. It is a principle of logic that an implication and its contrapositive are either both true or both false.

The **converse** of an implication is formed by switching the antecedent and the consequent. The converse of "If $x = 4$, then $x^2 = 16$" is "If $x^2 = 16$, then $x = 4$." Note that the converse is not a true statement, because if $x^2 = 16$, then x could equal -4. The converse of an implication may or may not be true.

Those statements for which the implication and the converse are both true are very important. They can be stated in the form "x if and only if y." For instance, a number is divisible by 5 if and only if the last digit of the number is 0 or 5. The implication "If a number is divisible by 5, then it ends in 0 or 5" and the converse "If a number ends in 0 or 5, then it is divisible by 5" are both true.

For each of the following, state the contrapositive and the converse of the implication. If the converse and the implication are both true statements, write a sentence using the phrase "if and only if."

1. If I live in Chicago, then I live in Illinois.
2. If today is June 1, then yesterday was May 31.
3. If today is not Thursday, then tomorrow is not Friday.
4. If a number is divisible by 8, then it is divisible by 4.
5. If a number is an even number, then it is divisible by 2.
6. If a number is a multiple of 6, then it is a multiple of 3.
7. If $4z = 20$, then $z = 5$.
8. If an angle measures 90°, then it is a right angle.
9. If p is a prime number greater than 2, then p is an odd number.
10. If the equation of a graph is $y = mx + b$, then the graph of the equation is a straight line.
11. If $a = 0$ or $b = 0$, then $ab = 0$.
12. If the coordinates of a point are (5, 0), then the point is on the x-axis.
13. If a quadrilateral is a square, then the quadrilateral has four sides of equal length.
14. If $x = y$, then $x^2 = y^2$.

Projects & Group Activities

Continued Fractions

The following complex fraction is called a **continued fraction.**

$$1 + \cfrac{1}{1 + \cfrac{1}{1 + \cfrac{1}{1 + \cfrac{1}{1 + \cdots}}}}$$

The dots indicate that the pattern continues to repeat forever.

A **convergent** for a continued fraction is an approximation of the repeated pattern. For instance,

$$c_2 = 1 + \cfrac{1}{1 + \cfrac{1}{1 + 1}} \qquad c_3 = 1 + \cfrac{1}{1 + \cfrac{1}{1 + \cfrac{1}{1 + 1}}} \qquad c_4 = 1 + \cfrac{1}{1 + \cfrac{1}{1 + \cfrac{1}{1 + \cfrac{1}{1 + 1}}}}$$

1. Calculate c_5 for the continued fraction at the bottom of the previous page.

The fraction above is related to the golden rectangle, which has been used in architectural designs as diverse as the Parthenon in Athens, built around 440 B.C., and the United Nations building. A golden rectangle is one for which

$$\frac{\text{length}}{\text{width}} = \frac{\text{length} + \text{width}}{\text{length}}$$

An example of a golden rectangle is shown below.

Here is another continued fraction, which was discovered by Leonhard Euler (1707–1793). Calculating the convergents of this continued fraction yields approximations that are closer and closer to π.

$$\pi = 3 + \cfrac{1^2}{6 + \cfrac{3^2}{6 + \cfrac{5^2}{6 + \cfrac{7^2}{6 + \cdots}}}}$$

2. Calculate $c_5 = 3 + \cfrac{1^2}{6 + \cfrac{3^2}{6 + \cfrac{5^2}{6 + \cfrac{7^2}{6 + \cfrac{9^2}{6 + 11^2}}}}}$

Graphing Rational Functions

The domain of a function is the set of the first coordinates of all the ordered pairs of the function. When a function is given by an equation, the domain of the function is all real numbers for which the function evaluates to a real number. For rational functions, we must exclude from the domain all those values of the variable for which the denominator of the rational function is zero. The graphing calculator is a useful tool to show the graphs of functions with excluded values in the domain of the function.

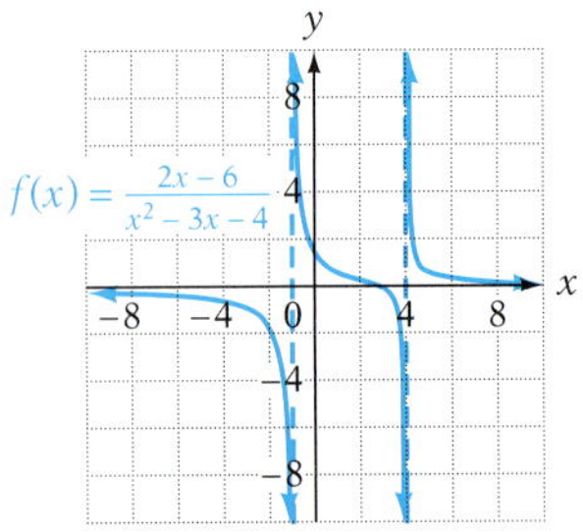

The domain is $\{x \mid x \neq -1, x \neq 4\}$.

The graph of the function $f(x) = \frac{2x-6}{x^2-3x-4}$ is shown at the left. Note that the graph never intersects the lines $x = -1$ and $x = 4$ (shown as dashed lines). These are the two values excluded from the domain of f. The graph of the function f gets closer to the dashed line $x = 4$ as x gets closer to 4. The graph of the function f also gets closer to $x = -1$ as x gets closer to -1. The lines that are "approached" by the function are called **asymptotes.**

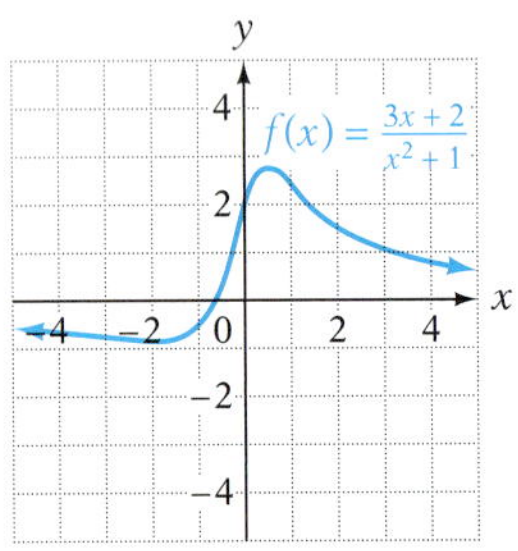

The domain is $\{x \mid x \in \text{real numbers}\}$.

The graph of the function $f(x) = \frac{3x+2}{x^2+1}$ is shown at the left. The domain must exclude values of x for which $x^2 + 1 = 0$. It is not possible that $x^2 + 1 = 0$, because $x^2 \geq 0$, and a positive number added to a number equal to or greater than zero cannot equal zero. Therefore, there are no real numbers that must be excluded from the domain of f.

Exercises 13–36 in Section 7.1 ask you to find the domains of rational functions by algebraic means. Try these exercises again, this time using a graphing calculator.

Transformers

Transformers are devices used to increase or decrease the voltage in an electrical system. For example, if the voltage in a wire is 120 volts, it can pass through a transformer and "come out the other side" with a voltage of 60 volts. Transformers have extensive applications; for instance, they are commonly used to decrease the voltage from a mainframe to individual houses (look for the big gray boxes on telephone poles). Transformers are also used in wall adapters to decrease wall voltage, for example, for an answering machine. A transformer must have voltage and current entering and leaving; therefore, the voltage and current of both sides can be determined by an equation.

$$V_1 = \frac{V_2 I_2}{I_1}$$

where V_1 and I_1 are the voltage and current on one side of a transformer, and V_2 and I_2 are the voltage and current on the other side of the transformer.

A transformer takes 1000 volts and reduces it to 100 volts. If the current is 30 amperes when the voltage is 1000 volts, find the current after the voltage is reduced.

$$V_1 = \frac{V_2 I_2}{I_1}$$

$$1000 = \frac{100 I_2}{30}$$

• The volts are 1000 when the amperes are 30, so $V_1 = 1000$ and $I_1 = 30$. $V_2 = 100$. Find I_2.

$$30{,}000 = 100 I_2$$

$$300 = I_2$$

After the voltage is reduced, the current is 300 amperes.

On one side of a transformer, the current is 40 amperes and the voltage is 500 volts. If the current on the other side is 400 amperes, what is the voltage on that side?

$$V_1 = \frac{V_2 I_2}{I_1}$$

$$500 = \frac{V_2 \cdot 400}{40}$$

• The volts are 500 when the amperes are 40, so $V_1 = 500$ and $I_1 = 40$. $I_2 = 400$. Find V_2.

$$20{,}000 = 400V_2$$

$$50 = V_2$$

After the current is increased, the voltage is 50 volts.

Solve.

1. A transformer takes 2000 volts and reduces it to 1000 volts. If the current is 40 amperes when the voltage is 2000 volts, find the current after the voltage is reduced.

2. A wall adapter for a cassette tape player reduces 180 volts to 9 volts. If the current was 15 amperes at 180 volts, find the current at 9 volts.

3. A wall adapter for a portable CD player reduces 120 volts to 9 volts. If the current was 12 amperes at 120 volts, find the current at 9 volts.

4. The current on one side of a transformer is 12 amperes and the voltage is 120 volts. If the current on the other side is 40 amperes, what is the voltage on that side?

5. The voltage on one side of a transformer is 12 volts and the current is 4.5 amperes. Find the voltage on the other side if the current on that side is 60 amperes.

6. The current on one side of a transformer is 500 amperes and the voltage is 8 volts. If the current on the other side is 400 amperes, what is the voltage on that side?

Chapter Summary

Key Words

A fraction in which the numerator and denominator are polynomials is called a **rational expression**. [7.1.1, p. 448]

Examples

$\frac{x^2 + 2x + 1}{x - 3}$ is a rational expression.

A rational expression is in **simplest form** when the numerator and denominator have no common factors. [7.1.2, p. 450]	$\frac{x^2+3x+2}{(x+1)^2} = \frac{(x+1)(x+2)}{(x+1)(x+1)} = \frac{x+2}{x+1}$
The **reciprocal of a rational expression** is the rational expression with the numerator and denominator interchanged. [7.2.1, p. 457]	The reciprocal of $\frac{a^2-b^2}{x-y}$ is $\frac{x-y}{a^2-b^2}$.
The **least common multiple (LCM) of two or more polynomials** is the simplest polynomial that contains the factors of each polynomial. [7.2.2, p. 459]	$4x^2-12x = 4x(x-3)$ $3x^3-21x^2+36x = 3x(x-3)(x-4)$ $\text{LCM} = 12x(x-3)(x-4)$
A **complex fraction** is a fraction whose numerator or denominator contains one or more fractions. [7.3.1, p. 466]	$\frac{\frac{1}{x}+\frac{1}{y}}{\frac{1}{x}}$ is a complex fraction.
A **ratio** is the quotient of two quantities that have the same unit. When a ratio is in simplest form, the units are not written. [7.5.1, p. 482]	$\frac{\$35}{\$100}$ written as a ratio in simplest form is $\frac{7}{20}$.
A **rate** is the quotient of two quantities that have different units. [7.5.1, p. 482]	$\frac{65 \text{ mi}}{2 \text{ gal}}$ is a rate.
A **proportion** is an equation that states the equality of two ratios or rates. [7.5.1, p. 483]	$\frac{75}{3} = \frac{x}{42}$ is a proportion.
Direct variation is a special function that can be expressed as the equation $y = kx^n$, where k is a constant called the **constant of variation** or the **constant of proportionality.** [7.5.2, p. 484]	$E = mc^2$ is Einstein's formula relating energy and mass. c^2 is the constant of proportionality.
Inverse variation is a function that can be expressed as the equation $y = \frac{k}{x^n}$, where k is a constant. [7.5.2, p. 485]	$I = \frac{k}{d^2}$ gives the intensity of a light source at a distance d from the source.
Joint variation is a variation in which a variable varies directly as the product of two or more variables. A joint variation can be expressed as the equation $z = kxy$, where k is a constant. [7.5.2, p. 485]	$C = kAT$ is a formula for the cost of insulation, where A is the area to be insulated and T is the thickness of the insulation.
Combined variation is a variation in which two or more types of variation occur at the same time. [7.5.2, p. 486]	$V = \frac{kT}{P}$ is a formula that states that the volume of a gas is directly proportional to the temperature and inversely proportional to the pressure.
A **formula** is a literal equation that states rules about measurements. [7.6.1, p. 492]	$P = 2L + 2W$ is the formula for the perimeter of a rectangle.

Essential Rules and Procedures

To multiply fractions

$\frac{a}{b} \cdot \frac{c}{d} = \frac{ac}{bd}$ [7.2.1, p. 456]

$$\frac{x^2 + x}{3} \cdot \frac{2}{5x} = \frac{(x^2 + x) \cdot 2}{3(5x)} = \frac{2x(x + 1)}{15x} = \frac{2(x + 1)}{15} = \frac{2x + 2}{15}$$

To divide fractions

$\frac{a}{b} \div \frac{c}{d} = \frac{a}{b} \cdot \frac{d}{c}$ [7.2.1, p. 457]

$$\frac{3x}{x - 5} \div \frac{3}{x + 4} = \frac{3x}{x - 5} \cdot \frac{x + 4}{3} = \frac{3x(x + 4)}{3(x - 5)} = \frac{x(x + 4)}{x - 5} = \frac{x^2 + 4x}{x - 5}$$

To add fractions

$\frac{a}{c} + \frac{b}{c} = \frac{a + b}{c}$ [7.2.2, p. 458]

$$\frac{2x - 7}{x^2 + 4} + \frac{x + 2}{x^2 + 4} = \frac{(2x - 7) + (x + 2)}{x^2 + 4} = \frac{3x - 5}{x^2 + 4}$$

To subtract fractions

$\frac{a}{c} - \frac{b}{c} = \frac{a - b}{c}$ [7.2.2, p. 458]

$$\frac{5x - 6}{x + 2} - \frac{2x + 4}{x + 2} = \frac{(5x - 6) - (2x + 4)}{x + 2} = \frac{3x - 10}{x + 2}$$

Equation for Work Problems

Rate of work × Time worked = Part of task completed [7.4.2, p. 473]

A roofer requires 24 h to shingle a roof. An apprentice can shingle the roof in 36 h. How long would it take to shingle the roof if both roofers worked together?

$$\frac{t}{24} + \frac{t}{36} = 1$$

Equation for Uniform Motion Problems

Distance = Rate × Time or $\frac{\text{Distance}}{\text{Rate}} = \text{Time}$ [7.4.3, p. 475]

A motorcycle travels 195 mi in the same amount of time it takes a car to travel 159 mi. The rate of the motorcycle is 12 mph greater than the rate of the car. Find the rate of the car.

$$\frac{195}{r + 12} = \frac{159}{r}$$

Chapter Review Exercises

1. Given $P(x) = \frac{x}{x-3}$, find $P(4)$.

2. Given $P(x) = \frac{x^2 - 2}{3x^2 - 2x + 5}$, find $P(-2)$.

3. Find the domain of $g(x) = \frac{2x}{x-3}$.

4. Find the domain of $f(x) = \frac{2x - 7}{3x^2 + 3x - 18}$.

5. Find the domain of $F(x) = \frac{x^2 - x}{3x^2 + 4}$.

6. Simplify: $\frac{6a^{5n} + 4a^{4n} - 2a^{3n}}{2a^{3n}}$

7. Simplify: $\frac{16 - x^2}{x^3 - 2x^2 - 8x}$

8. Simplify: $\frac{x^3 - 27}{x^2 - 9}$

9. Multiply: $\frac{a^6b^4 + a^4b^6}{a^5b^4 - a^4b^4} \cdot \frac{a^2 - b^2}{a^4 - b^4}$

10. Multiply: $\frac{x^3 - 8}{x^3 + 2x^2 + 4x} \cdot \frac{x^3 + 2x^2}{x^2 - 4}$

11. Multiply: $\frac{16 - x^2}{6x - 6} \cdot \frac{x^2 + 5x + 6}{x^2 - 8x + 16}$

12. Divide: $\frac{x^{2n} - 5x^n + 4}{x^{2n} - 2x^n - 8} \div \frac{x^{2n} - 4x^n + 3}{x^{2n} + 8x^n + 12}$

13. Divide: $\frac{27x^3 - 8}{9x^3 + 6x^2 + 4x} \div \frac{9x^2 - 12x + 4}{9x^2 - 4}$

14. Divide: $\frac{3 - x}{x^2 + 3x + 9} \div \frac{x^2 - 9}{x^3 - 27}$

15. Add: $\frac{5}{3a^2b^3} + \frac{7}{8ab^4}$

16. Subtract: $\frac{3x^2 + 2}{x^2 - 4} - \frac{9x - x^2}{x^2 - 4}$

17. Simplify: $\frac{8}{9x^2 - 4} + \frac{5}{3x - 2} - \frac{4}{3x + 2}$

18. Simplify: $\frac{6x}{3x^2 - 7x + 2} - \frac{2}{3x - 1} + \frac{3x}{x - 2}$

19. Simplify: $\frac{x}{x-3} - 4 - \frac{2x - 5}{x + 2}$

20. Simplify: $\dfrac{x - 6 + \frac{6}{x-1}}{x + 3 - \frac{12}{x-1}}$

21. Simplify: $\dfrac{x + \frac{3}{x-4}}{3 + \frac{x}{x-4}}$

22. Solve: $\frac{5x}{2x - 3} + 4 = \frac{3}{2x - 3}$

23. Solve: $\frac{x}{x-3} = \frac{2x+5}{x+1}$

24. Solve: $\frac{6}{x-3} - \frac{1}{x+3} = \frac{51}{x^2-9}$

25. Solve: $\frac{30}{x^2+5x+4} + \frac{10}{x+4} = \frac{4}{x+1}$

26. Solve $I = \frac{1}{R}V$ for R.

27. Solve $Q = \frac{N-S}{N}$ for N.

28. Solve $S = \frac{a}{1-r}$ for r.

29. A car uses 4 tanks of fuel to travel 1800 mi. At this rate, how many tanks of fuel would be required for a trip of 3000 mi?

30. On a certain map, 2.5 in. represents 10 mi. How many miles would be represented by 12 in.?

31. An electrician requires 65 min to install a ceiling fan. The electrician and an apprentice, working together, take 40 min to install the fan. How long would it take the apprentice, working alone, to install the ceiling fan?

32. The inlet pipe can fill a tub in 24 min. The drain pipe can empty the tub in 15 min. How long would it take to empty a full tub with both pipes open?

33. Three students can paint a dormitory room in 8 h, 16 h, and 16 h, respectively. How long would it take to do the job if all three students worked together?

34. A canoeist can travel 10 mph in calm water. The amount of time it takes to travel 60 mi with the current is the same amount of time it takes to travel 40 mi against the current. Find the rate of the current.

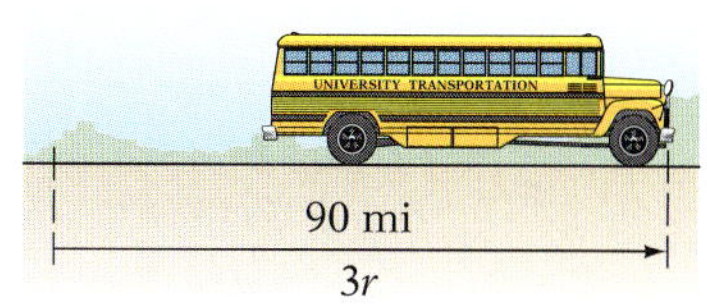

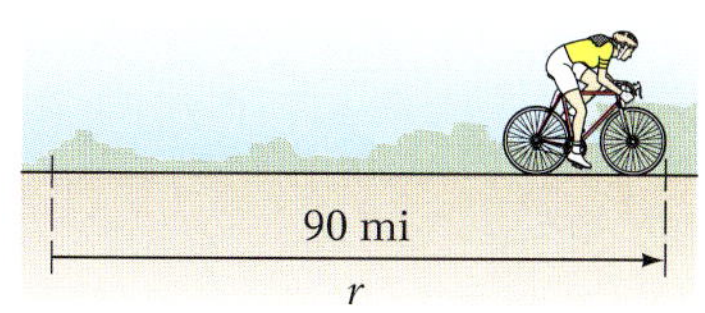

35. A bus and a cyclist leave a school at 8 A.M. and head for a stadium 90 mi away. The rate of the bus is three times the rate of the cyclist. The cyclist arrives 4 h after the bus. Find the rate of the bus.

36. A tractor travels 10 mi in the same amount of time it takes a car to travel 15 mi. The rate of the tractor is 15 mph less than the rate of the car. Find the rate of the tractor.

37. The pressure (p) of wind on a flat surface varies jointly as the area (A) of the surface and the square of the wind's velocity (v). If the pressure on 22 ft^2 is 10 lb when the wind's velocity is 10 mph, find the pressure on the same surface when the wind's velocity is 20 mph.

38. The illumination (I) produced by a light varies inversely as the square of the distance (d) from the light. If the illumination produced 10 ft from a light is 12 lumens, find the illumination 2 ft from the light.

39. The electrical resistance (r) of a cable varies directly as its length (l) and inversely as the square of its diameter (d). If a cable 16,000 ft long and $\frac{1}{4}$ in. in diameter has a resistance of 3.2 ohms, what is the resistance of a cable that is 8000 ft long and $\frac{1}{2}$ in. in diameter?

Chapter Test

1. Simplify: $\dfrac{v^3 - 4v}{2v^2 - 5v + 2}$

2. Simplify: $\dfrac{2a^2 - 8a + 8}{4 + 4a - 3a^2}$

3. Multiply: $\dfrac{3x^2 - 12}{5x - 15} \cdot \dfrac{2x^2 - 18}{x^2 + 5x + 6}$

4. Given $P(x) = \dfrac{3 - x^2}{x^3 - 2x^2 + 4}$, find $P(-1)$.

5. Divide: $\dfrac{2x^2 - x - 3}{2x^2 - 5x + 3} \div \dfrac{3x^2 - x - 4}{x^2 - 1}$

6. Multiply: $\dfrac{x^{2n} - x^n - 2}{x^{2n} + x^n} \cdot \dfrac{x^{2n} - x^n}{x^{2n} - 4}$

7. Simplify: $\dfrac{2}{x^2} + \dfrac{3}{y^2} - \dfrac{5}{2xy}$

8. Simplify: $\dfrac{3x}{x - 2} - 3 + \dfrac{4}{x + 2}$

9. Find the domain of $f(x) = \dfrac{3x^2 - x + 1}{x^2 - 9}$.

10. Subtract: $\dfrac{x + 2}{x^2 + 3x - 4} - \dfrac{2x}{x^2 - 1}$

11. Simplify: $\dfrac{1 - \dfrac{1}{x} - \dfrac{12}{x^2}}{1 + \dfrac{6}{x} + \dfrac{9}{x^2}}$

12. Simplify: $\dfrac{1 - \dfrac{1}{x + 2}}{1 - \dfrac{3}{x + 4}}$

13. Solve: $\dfrac{3}{x + 1} = \dfrac{2}{x}$

14. Solve: $\dfrac{4x}{2x - 1} = 2 - \dfrac{1}{2x - 1}$

15. Solve $ax = bx + c$ for x.

16. The inlet pipe can fill a water tank in 48 min. The outlet pipe can empty the tank in 30 min. How long would it take to empty a full tank with both pipes open?

17. An interior designer uses two rolls of wallpaper for every 45 ft^2 of wall space in an office. At this rate, how many rolls of wallpaper are needed for an office that has 315 ft^2 of wall space?

18. One landscaper can till the soil for a lawn in 30 min, whereas it takes a second landscaper 15 min to do the same job. How long would it take to till the soil for the lawn if both landscapers worked together?

19. A cyclist travels 20 mi in the same amount of time it takes a hiker to walk 6 mi. The rate of the cyclist is 7 mph greater than the rate of the hiker. Find the rate of the cyclist.

20. The stopping distance (s) of a car varies directly as the square of the speed (v) of the car. For a car traveling at 50 mph, the stopping distance is 170 ft. Find the stopping distance of a car that is traveling at 30 mph.

Cumulative Review Exercises

1. Simplify: $8 - 4[-3 - (-2)]^2 \div 5$

2. Solve: $\dfrac{2x - 3}{6} - \dfrac{x}{9} = \dfrac{x - 4}{3}$

3. Solve: $5 - |x - 4| = 2$

4. Find the domain of $f(x) = \dfrac{x}{x - 3}$.

5. Given $P(x) = \dfrac{x - 1}{2x - 3}$, find $P(-2)$.

6. Write 0.000000035 in scientific notation.

7. Solve: $\dfrac{x}{x + 1} = 1$

8. Solve: $(9x - 1)(x - 4) = 0$

9. Simplify: $\dfrac{(2a^{-2}b^3)}{(4a)^{-1}}$

10. Solve: $x - 3(1 - 2x) \geq 1 - 4(2 - 2x)$

11. Multiply: $(2a^2 - 3a + 1)(-2a^2)$

12. Factor: $2x^{2n} + 3x^n - 2$

13. Factor: $x^3y^3 - 27$

14. Simplify: $\dfrac{x^4 + x^3y - 6x^2y^2}{x^3 - 2x^2y}$

15. Find the equation of the line that contains the point $(-2, -1)$ and is parallel to the line $3x - 2y = 6$.

16. Simplify: $(x - x^{-1})^{-1}$

17. Graph $-3x + 5y = -15$ by using the x- and y-intercepts.

18. Graph the solution set: $\begin{aligned} x + y &\leq 3 \\ -2x + y &> 4 \end{aligned}$

19. Divide: $\dfrac{4x^3 + 2x^2 - 10x + 1}{x - 2}$

20. Divide: $\dfrac{16x^2 - 9y^2}{16x^2y - 12xy^2} \div \dfrac{4x^2 - xy - 3y^2}{12x^2y^2}$

21. Find the domain of $f(x) = \dfrac{x^2 + 6x + 7}{3x^2 + 5}$.

22. Subtract: $\dfrac{5x}{3x^2 - x - 2} - \dfrac{2x}{x^2 - 1}$

23. Evaluate the determinant: $\begin{vmatrix} 6 & 5 \\ 2 & -3 \end{vmatrix}$

24. Simplify: $\dfrac{x - 4 + \dfrac{5}{x + 2}}{x + 2 - \dfrac{1}{x + 2}}$

25. Solve: $\begin{aligned} x + y + z &= 3 \\ -2x + y + 3z &= 2 \\ 2x - 4y + z &= -1 \end{aligned}$

26. Given $f(x) = x^2 - 3x + 3$ and $f(c) = 1$, find c.

27. Solve the proportion: $\dfrac{2}{x - 3} = \dfrac{5}{2x - 3}$

28. Solve: $\dfrac{3}{x^2 - 36} = \dfrac{2}{x - 6} - \dfrac{5}{x + 6}$

29. Multiply: $(a + 5)(a^3 - 3a + 4)$

30. Solve $I = \dfrac{E}{R + r}$ for r.

31. Are the graphs of $4x + 3y = 12$ and $3x + 4y = 16$ perpendicular?

32. Graph: $f(x) = 1 - x + x^3$

33. The sum of two integers is 15. Five times the smaller integer is five more than twice the larger integer. Find the integers.

34. How many pounds of almonds that cost \$5.40 per pound must be mixed with 50 lb of peanuts that cost \$2.60 per pound to make a mixture that costs \$4.00 per pound?

35. A pre-election survey showed that three out of five voters would vote in an election. At this rate, how many people would be expected to vote in a city of 125,000?

36. A new computer can work six times as fast as an older computer. Working together, the computers can complete a job in 12 min. How long would it take the new computer, working alone, to do the job?

37. A plane can fly at a rate of 300 mph in calm air. Traveling with the wind, the plane flew 900 mi in the same amount of time it took to fly 600 mi against the wind. Find the rate of the wind.

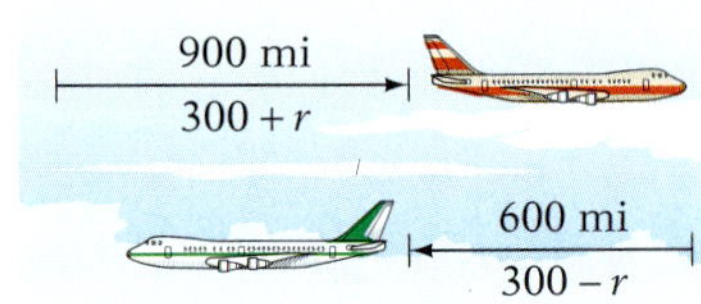

38. The frequency of vibration (f) in an open pipe organ varies inversely as the length (L) of the pipe. If the air in a pipe 2 m long vibrates 60 times per minute, find the frequency in a pipe that is 1.5 m long.

You may have heard discussions about the mysterious effect of Mona Lisa's expression and of her eyes. But perhaps you have not looked at da Vinci's painting from a geometric viewpoint. You will be doing just that in the **Project** *on the golden rectangle* **on page 553.** *The ratio of the length to the width of the golden rectangle involves an expression containing a radical, a topic of this chapter.*

Applications

8 Rational Exponents and Radicals

Objectives

Section 8.1

1. Simplify expressions with rational exponents
2. Write exponential expressions as radical expressions and radical expressions as exponential expressions
3. Simplify radical expressions that are roots of perfect powers

Section 8.2

1. Simplify radical expressions
2. Add and subtract radical expressions
3. Multiply radical expressions
4. Divide radical expressions

Section 8.3

1. Find the domain of a radical function
2. Graph a radical function

Section 8.4

1. Solve equations containing one or more radical expressions
2. Application problems

Section 8.5

1. Simplify complex numbers
2. Add and subtract complex numbers
3. Multiply complex numbers
4. Divide complex numbers

WEB

Need help? For online student resources, such as section quizzes, visit this web site: math.college.hmco.com

Prep Test

1. Complete: $48 = ? \cdot 3$

For Exercises 2 to 7, simplify.

2. 2^5

3. $6\left(\frac{3}{2}\right)$

4. $\frac{1}{2} - \frac{2}{3} + \frac{1}{4}$

5. $(3 - 7x) - (4 - 2x)$

6. $\frac{3x^5y^6}{12x^4y}$

7. $(3x - 2)^2$

For Exercises 8 and 9, multiply.

8. $(2 + 4x)(5 - 3x)$

9. $(6x - 1)(6x + 1)$

10. Solve: $x^2 - 14x - 5 = 10$

Go Figure

You are planning a large dinner party. If you seat 5 people at each table, you end up with only 2 people at the last table. If you seat 3 people at each table, you have 9 people left over with no place to sit. There are fewer than 10 tables. How many guests are coming to the dinner party?

8.1 Rational Exponents and Radical Expressions

1 Simplify expressions with rational exponents

VIDEO & DVD CD TUTOR WEB SSM

Point of Interest

Nicolas Chuquet (ca. 1475), a French physician, wrote an algebra text in which he used a notation for expressions with fractional exponents. He wrote R^26 to mean $6^{1/2}$ and R^315 to mean $15^{1/3}$. This was an improvement over earlier notations that used words for these expressions.

In this section, the definition of an exponent is extended beyond integers so that any rational number can be used as an exponent. The definition is expressed in such a way that the Rules of Exponents hold true for rational exponents.

Consider the expression $\left(a^{\frac{1}{n}}\right)^n$ for $a > 0$ and n a positive integer. Now simplify, assuming that the Rule for Simplifying a Power of an Exponential Expression is true.

$$\left(a^{\frac{1}{n}}\right)^n = a^{\frac{1}{n}\cdot n} = a^1 = a$$

Because $\left(a^{\frac{1}{n}}\right)^n = a$, the number $a^{\frac{1}{n}}$ is the number whose nth power is a.

$$25^{\frac{1}{2}} = 5 \text{ because } (5)^2 = 25.$$

$$8^{\frac{1}{3}} = 2 \text{ because } (2)^3 = 8.$$

In the expression $a^{\frac{1}{n}}$, if a is a negative number and n is a positive even integer, then $a^{\frac{1}{n}}$ is not a real number.

$(-4)^{\frac{1}{2}}$ is not a real number, because there is no real number whose second power is -4.

When n is a positive odd integer, a can be a positive or a negative number.

$$(-27)^{\frac{1}{3}} = -3 \text{ because } (-3)^3 = -27.$$

Using the definition of $a^{\frac{1}{n}}$ and the Rules of Exponents, it is possible to define any exponential expression that contains a rational exponent.

Definition of $a^{\frac{m}{n}}$

If m and n are positive integers and $a^{\frac{1}{n}}$ is a real number, then

$$a^{\frac{m}{n}} = \left(a^{\frac{1}{n}}\right)^m$$

Take Note

Although we can simplify an expression by rewriting $a^{\frac{m}{n}}$ in the form $(a^m)^{\frac{1}{n}}$, it is usually easier to simplify $\left(a^{\frac{1}{n}}\right)^m$. For instance, simplifying $\left(27^{\frac{1}{3}}\right)^2$ is easier than simplifying $(27^2)^{\frac{1}{3}}$.

The expression $a^{\frac{m}{n}}$ can also be written $a^{\frac{m}{n}} = a^{m \cdot \frac{1}{n}} = (a^m)^{\frac{1}{n}}$. However, rewriting $a^{\frac{m}{n}}$ as $(a^m)^{\frac{1}{n}}$ is not as useful as rewriting it as $\left(a^{\frac{1}{n}}\right)^m$. See the Take Note at the left.

As shown above, expressions that contain rational exponents do not always represent real numbers when the base of the exponential expression is a negative number. For this reason, all variables in this chapter represent positive numbers unless otherwise stated.

Example 1 Simplify. **A.** $27^{\frac{2}{3}}$ **B.** $32^{-\frac{2}{5}}$ **C.** $(-49)^{\frac{3}{2}}$

Solution

A. $27^{\frac{2}{3}} = (3^3)^{\frac{2}{3}}$ • Rewrite 27 as 3^3.

$= 3^{3\left(\frac{2}{3}\right)}$ • Use the Rule for Simplifying a Power of an Exponential Expression.

$= 3^2$

$= 9$ • Simplify.

B. $32^{-\frac{2}{5}} = (2^5)^{-\frac{2}{5}}$ • Rewrite 32 as 2^5.

$= 2^{-2}$ • Use the Rule for Simplifying a Power of an Exponential Expression.

$= \dfrac{1}{2^2}$ • Use the Rule of Negative Exponents.

$= \dfrac{1}{4}$ • Simplify.

C. $(-49)^{\frac{3}{2}}$ • The base of the exponential expression is a negative number, while the denominator of the exponent is a positive even number.

$(-49)^{\frac{3}{2}}$ is not a real number.

Take Note

Note that $32^{-\frac{2}{5}} = \frac{1}{4}$, a positive number. A negative exponent does not affect the sign of a number.

Problem 1 Simplify. **A.** $64^{\frac{2}{3}}$ **B.** $16^{-\frac{3}{4}}$ **C.** $(-81)^{\frac{3}{4}}$

Solution See page S26.

Example 2 Simplify. **A.** $b^{\frac{1}{2}} \cdot b^{\frac{2}{3}} \cdot b^{-\frac{1}{4}}$ **B.** $(x^6y^4)^{\frac{3}{2}}$ **C.** $\left(\dfrac{8a^3b^{-4}}{64a^{-9}b^2}\right)^{\frac{2}{3}}$

Solution

A. $b^{\frac{1}{2}} \cdot b^{\frac{2}{3}} \cdot b^{-\frac{1}{4}} = b^{\frac{1}{2}+\frac{2}{3}-\frac{1}{4}}$ • Use the Rule for Multiplying Exponential Expressions.

$= b^{\frac{6}{12}+\frac{8}{12}-\frac{3}{12}}$

$= b^{\frac{11}{12}}$

B. $(x^6y^4)^{\frac{3}{2}} = x^{6\left(\frac{3}{2}\right)}y^{4\left(\frac{3}{2}\right)}$ • Use the Rule for Simplifying Powers of Products.

$= x^9y^6$

C. $\left(\dfrac{8a^3b^{-4}}{64a^{-9}b^2}\right)^{\frac{2}{3}} = \left(\dfrac{a^{12}}{8b^6}\right)^{\frac{2}{3}}$ • Use the Rule for Dividing Exponential Expressions.

$= \left(\dfrac{a^{12}}{2^3b^6}\right)^{\frac{2}{3}}$ • Rewrite 8 as 2^3.

$= \dfrac{a^8}{2^2b^4}$ • Use the Rule for Simplifying Powers of Quotients.

$= \dfrac{a^8}{4b^4}$

Problem 2 Simplify. **A.** $\dfrac{x^{\frac{1}{2}}y^{-\frac{5}{4}}}{x^{-\frac{4}{3}}y^{\frac{1}{3}}}$ **B.** $\left(x^{\frac{3}{4}}y^{\frac{1}{2}}z^{-\frac{2}{3}}\right)^{-\frac{4}{3}}$ **C.** $\left(\dfrac{16a^{-2}b^{\frac{4}{3}}}{9a^4b^{-\frac{2}{3}}}\right)^{-\frac{1}{2}}$

Solution See page S26.

2 Write exponential expressions as radical expressions and radical expressions as exponential expressions

Point of Interest

The radical sign was introduced in 1525 in a book by Christoff Rudolff called Coss. He modified the symbol to indicate square roots, cube roots, and fourth roots. The idea of using an index, as we do in our modern notation, did not occur until some years later.

As stated in the previous objective, $a^{\frac{1}{n}}$ is the number whose nth power is a. This can be stated as $a^{\frac{1}{n}}$ is the **nth root** of a. The expression $\sqrt[n]{a}$ is another symbol for the nth root of a.

Alternative Notation for the nth Root of a

If a is a real number and n is a positive integer, then $\sqrt[n]{a} = a^{\frac{1}{n}}$.

In the expression $\sqrt[n]{a}$, the symbol $\sqrt{\ }$ is called a **radical,** n is the **index** of the radical, and a is the **radicand.** When $n = 2$, the radical expression represents a square root and the index 2 is usually not written.

An exponential expression with a rational exponent can be written as a radical expression.

Definition of the nth Root of a^m

If $a^{\frac{1}{n}}$ is a real number, then $a^{\frac{m}{n}} = a^{m \cdot \frac{1}{n}} = (a^m)^{\frac{1}{n}} = \sqrt[n]{a^m}$.

The expression $a^{\frac{m}{n}}$ can also be written $a^{\frac{m}{n}} = a^{\frac{1}{n} \cdot m} = \left(\sqrt[n]{a}\right)^m$.

The exponential expression at the right has been written as a radical expression.

$$y^{\frac{2}{3}} = (y^2)^{\frac{1}{3}} = \sqrt[3]{y^2}$$

The radical expressions at the right have been written as exponential expressions.

$$\sqrt[5]{x^6} = (x^6)^{\frac{1}{5}} = x^{\frac{6}{5}}$$
$$\sqrt{17} = (17)^{\frac{1}{2}} = 17^{\frac{1}{2}}$$

Write $(5x)^{\frac{2}{5}}$ as a radical expression.

The denominator of the rational exponent is the index of the radical. The numerator is the power of the radicand.

$$(5x)^{\frac{2}{5}} = \sqrt[5]{(5x)^2}$$

Simplify.

$$= \sqrt[5]{25x^2}$$

Write $\sqrt[3]{x^4}$ as an exponential expression with a rational exponent.

The index of the radical is the denominator of the rational exponent. The power of the radicand is the numerator of the rational exponent.

$$\sqrt[3]{x^4} = (x^4)^{\frac{1}{3}} = x^{\frac{4}{3}}$$

Example 3 Rewrite the exponential expression as a radical expression.

A. $(3x)^{\frac{2}{3}}$ **B.** $-2x^{\frac{2}{3}}$

Solution

A. $(3x)^{\frac{2}{3}} = \sqrt[3]{(3x)^2}$
$= \sqrt[3]{9x^2}$

• **The denominator of the rational exponent is the index of the radical. The numerator is the power of the radicand.**

B. $-2x^{\frac{2}{3}} = -2(x^2)^{\frac{1}{3}}$
$= -2\sqrt[3]{x^2}$

• **The -2 is not raised to the power.**

Problem 3 Rewrite the exponential expression as a radical expression.

A. $(2x^3)^{\frac{3}{4}}$ **B.** $-5a^{\frac{5}{6}}$

Solution See page S26.

Example 4 Rewrite the radical expression as an exponential expression.

A. $\sqrt[7]{x^5}$ **B.** $\sqrt[3]{a^3 + b^3}$

Solution

A. $\sqrt[7]{x^5} = (x^5)^{\frac{1}{7}} = x^{\frac{5}{7}}$

• **The index of the radical is the denominator of the rational exponent. The power of the radicand is the numerator of the rational exponent.**

B. $\sqrt[3]{a^3 + b^3} = (a^3 + b^3)^{\frac{1}{3}}$

• **Note that $(a^3 + b^3)^{\frac{1}{3}} \neq a + b$.**

Problem 4 Rewrite the radical expression as an exponential expression.

A. $\sqrt[3]{3ab}$ **B.** $\sqrt[4]{x^4 + y^4}$

Solution See page S26.

3 Simplify radical expressions that are roots of perfect powers

VIDEO & DVD CD TUTOR WEB SSM

Every positive number has two square roots, one a positive and one a negative number. For example, because $(5)^2 = 25$ and $(-5)^2 = 25$, there are two square roots of 25: 5 and -5.

The symbol $\sqrt{}$ is used to indicate the positive or **principal square root.** To indicate the negative square root of a number, a negative sign is placed in front of the radical.

$\sqrt{25} = 5$

$-\sqrt{25} = -5$

The square root of zero is zero. $\sqrt{0} = 0$

The square root of a negative number is not a real number, because the square of a real number must be positive. $\sqrt{-25}$ is not a real number.

The square root of a squared negative number is a positive number. $\sqrt{(-5)^2} = \sqrt{25} = 5$

For any real number a, $\sqrt{a^2} = |a|$ and $-\sqrt{a^2} = -|a|$. If a is a positive real number, then $\sqrt{a^2} = a$ and $\left(\sqrt{a}\right)^2 = a$.

The cube root of a positive number is positive. $\sqrt[3]{8} = 2$, because $2^3 = 8$.

The cube root of a negative number is negative. $\sqrt[3]{-8} = -2$, because $(-2)^3 = -8$.

For any real number a, $\sqrt[3]{a^3} = a$.

The following holds true for finding the nth root of a real number.

If n is an even integer, then $\sqrt[n]{a^n} = |a|$ and $-\sqrt[n]{a^n} = -|a|$. If n is an odd integer, then $\sqrt[n]{a^n} = a$.

For example,

$$\sqrt[6]{y^6} = |y| \qquad -\sqrt[12]{x^{12}} = -|x| \qquad \sqrt[5]{b^5} = b$$

For the remainder of this chapter, we will assume that variable expressions inside a radical represent positive numbers. Therefore, it is not necessary to use the absolute value signs.

> **Take Note**
>
> Note that when the index is an even natural number, the nth root requires absolute value symbols.
>
> $\sqrt[6]{y^6} = |y|$ but $\sqrt[5]{y^5} = y$
>
> Because we stated that variables within radicals represent *positive* numbers, we will omit the absolute value symbols when writing an answer.

Simplify: $\sqrt[4]{x^4y^8}$

The radicand is a perfect fourth power because the exponents on the variables are divisible by 4. Write the radical expression as an exponential expression.

$$\sqrt[4]{x^4y^8} = (x^4y^8)^{\frac{1}{4}}$$

Use the Rule for Simplifying Powers of Products.

$$= xy^2$$

Simplify: $\sqrt[3]{125c^9d^6}$

The radicand is a perfect cube because 125 is a perfect cube ($125 = 5^3$) and all the exponents on the variables are divisible by 3.

$\sqrt[3]{125c^9d^6} = (5^3c^9d^6)^{\frac{1}{3}}$

Use the Rule for Simplifying Powers of Products.

$= 5c^3d^2$

Note that **a variable expression is a perfect power if the exponents on the factors are evenly divisible by the index of the radical.**

The chart below shows roots of perfect powers. Knowledge of these roots is very helpful in simplifying radical expressions.

Square Roots		*Cube Roots*	*Fourth Roots*	*Fifth Roots*
$\sqrt{1} = 1$	$\sqrt{36} = 6$	$\sqrt[3]{1} = 1$	$\sqrt[4]{1} = 1$	$\sqrt[5]{1} = 1$
$\sqrt{4} = 2$	$\sqrt{49} = 7$	$\sqrt[3]{8} = 2$	$\sqrt[4]{16} = 2$	$\sqrt[5]{32} = 2$
$\sqrt{9} = 3$	$\sqrt{64} = 8$	$\sqrt[3]{27} = 3$	$\sqrt[4]{81} = 3$	$\sqrt[5]{243} = 3$
$\sqrt{16} = 4$	$\sqrt{81} = 9$	$\sqrt[3]{64} = 4$	$\sqrt[4]{256} = 4$	
$\sqrt{25} = 5$	$\sqrt{100} = 10$	$\sqrt[3]{125} = 5$	$\sqrt[4]{625} = 5$	

Take Note

From the chart, $\sqrt[5]{243} = 3$, which means that $3^5 = 243$. From this we know that $(-3)^5 = -243$, which means $\sqrt[5]{-243} = -3$.

Simplify: $\sqrt[5]{-243x^5y^{15}}$

From the chart, 243 is a perfect fifth power, and each exponent is divisible by 5. Therefore, the radicand is a perfect fifth power.

$\sqrt[5]{-243x^5y^{15}} = -3xy^3$

Example 5 Simplify.

A. $\sqrt{49x^2y^{12}}$ **B.** $\sqrt[3]{-125a^6b^9}$ **C.** $-\sqrt[4]{16a^4b^8}$

Solution

A. $\sqrt{49x^2y^{12}} = 7xy^6$ • The radicand is a perfect square. Divide each exponent by 2.

B. $\sqrt[3]{-125a^6b^9} = -5a^2b^3$ • The radicand is a perfect cube. Divide each exponent by 3.

C. $-\sqrt[4]{16a^4b^8} = -2ab^2$ • The radicand is a perfect fourth power. Divide each exponent by 4.

Problem 5 Simplify.

A. $\sqrt{121x^{10}y^4}$ **B.** $\sqrt[3]{-8x^{12}y^3}$ **C.** $-\sqrt[4]{81x^{12}y^8}$

Solution See page S26.

8.1 Concept Review

Determine whether the following statements are always true, sometimes true, or never true.

1. For real numbers, $\sqrt{x^2} = x$.
2. $\sqrt{(-2)^2} = -2$
3. $\sqrt[3]{(-3)^3} = -3$
4. For positive integers a and n, $\sqrt[n]{a} = a^{\frac{1}{n}}$.
5. For positive integers a, m, and n, $(a^m)^{\frac{1}{n}} = a^{\frac{m}{n}}$.
6. The nth root of a negative number is negative.

8.1 Exercises

1

1. What is the nth root of a number?
2. What is the meaning of the exponential expression $a^{\frac{m}{n}}$?

Simplify.

3. $8^{\frac{1}{3}}$
4. $16^{\frac{1}{2}}$
5. $9^{\frac{3}{2}}$
6. $25^{\frac{3}{2}}$
7. $27^{-\frac{2}{3}}$
8. $64^{-\frac{1}{3}}$
9. $32^{\frac{2}{5}}$
10. $16^{\frac{3}{4}}$
11. $(-25)^{\frac{5}{2}}$
12. $(-36)^{\frac{1}{4}}$
13. $\left(\frac{25}{49}\right)^{-\frac{3}{2}}$
14. $\left(\frac{8}{27}\right)^{-\frac{2}{3}}$
15. $x^{\frac{1}{2}}x^{\frac{1}{2}}$
16. $a^{\frac{1}{3}}a^{\frac{5}{3}}$
17. $y^{-\frac{1}{4}}y^{\frac{3}{4}}$
18. $x^{\frac{2}{5}} \cdot x^{-\frac{4}{5}}$
19. $x^{-\frac{2}{3}} \cdot x^{\frac{3}{4}}$
20. $x \cdot x^{-\frac{1}{2}}$
21. $a^{\frac{1}{3}} \cdot a^{\frac{3}{4}} \cdot a^{-\frac{1}{2}}$
22. $y^{-\frac{1}{6}} \cdot y^{\frac{2}{3}} \cdot y^{\frac{1}{2}}$
23. $\dfrac{a^{\frac{1}{2}}}{a^{\frac{3}{2}}}$
24. $\dfrac{b^{\frac{1}{3}}}{b^{\frac{4}{3}}}$
25. $\dfrac{y^{-\frac{3}{4}}}{y^{\frac{1}{4}}}$
26. $\dfrac{x^{-\frac{3}{5}}}{x^{\frac{1}{5}}}$

27. $\dfrac{y^{\frac{2}{3}}}{y^{-\frac{5}{6}}}$

28. $\dfrac{b^{\frac{3}{4}}}{b^{-\frac{3}{2}}}$

29. $(x^2)^{-\frac{1}{2}}$

30. $(a^8)^{-\frac{3}{4}}$

31. $\left(x^{-\frac{2}{3}}\right)^6$

32. $\left(y^{-\frac{5}{6}}\right)^{12}$

33. $\left(a^{-\frac{1}{2}}\right)^{-2}$

34. $\left(b^{-\frac{2}{3}}\right)^{-6}$

35. $\left(x^{-\frac{3}{8}}\right)^{-\frac{4}{5}}$

36. $\left(y^{-\frac{3}{2}}\right)^{-\frac{2}{9}}$

37. $\left(a^{\frac{1}{2}} \cdot a\right)^2$

38. $\left(b^{\frac{2}{3}} \cdot b^{\frac{1}{6}}\right)^6$

39. $\left(x^{-\frac{1}{2}} \cdot x^{\frac{3}{4}}\right)^{-2}$

40. $\left(a^{\frac{1}{2}} \cdot a^{-2}\right)^3$

41. $\left(y^{-\frac{1}{2}} \cdot y^{\frac{3}{2}}\right)^{\frac{2}{3}}$

42. $\left(b^{-\frac{2}{3}} \cdot b^{\frac{1}{4}}\right)^{-\frac{4}{3}}$

43. $(x^8y^2)^{\frac{1}{2}}$

44. $(a^3b^9)^{\frac{2}{3}}$

45. $(x^4y^2z^6)^{\frac{3}{2}}$

46. $(a^8b^4c^4)^{\frac{3}{4}}$

47. $(x^{-3}y^6)^{-\frac{1}{3}}$

48. $(a^2b^{-6})^{-\frac{1}{2}}$

49. $\left(x^{-2}y^{\frac{1}{3}}\right)^{-\frac{3}{4}}$

50. $\left(a^{-\frac{2}{3}}b^{\frac{2}{3}}\right)^{\frac{3}{2}}$

51. $\left(\dfrac{x^{\frac{1}{2}}}{y^{-2}}\right)^4$

52. $\left(\dfrac{b^{-\frac{3}{4}}}{a^{-\frac{1}{2}}}\right)^8$

53. $\dfrac{x^{\frac{1}{4}} \cdot x^{-\frac{1}{2}}}{x^{\frac{2}{3}}}$

54. $\dfrac{b^{\frac{1}{2}} \cdot b^{-\frac{3}{4}}}{b^{\frac{1}{4}}}$

55. $\left(\dfrac{y^{\frac{2}{3}} \cdot y^{-\frac{5}{6}}}{y^{\frac{1}{9}}}\right)^9$

56. $\left(\dfrac{a^{\frac{1}{3}} \cdot a^{-\frac{2}{3}}}{a^{\frac{1}{2}}}\right)^4$

57. $\left(\dfrac{b^2 \cdot b^{-\frac{3}{4}}}{b^{-\frac{1}{2}}}\right)^{-\frac{1}{2}}$

58. $\dfrac{\left(x^{-\frac{5}{6}} \cdot x^3\right)^{-\frac{2}{3}}}{x^{\frac{4}{3}}}$

59. $\left(a^{\frac{2}{3}}b^2\right)^6(a^3b^3)^{\frac{1}{3}}$

60. $\left(x^3y^{-\frac{1}{2}}\right)^{-2}(x^{-3}y^2)^{\frac{1}{6}}$

61. $(16m^{-2}n^4)^{-\frac{1}{2}}\left(mn^{\frac{1}{2}}\right)$

62. $(27m^3n^{-6})^{\frac{1}{3}}\left(m^{-\frac{1}{3}}n^{\frac{5}{6}}\right)^6$

63. $\left(\dfrac{x^{\frac{1}{2}}y^{-\frac{3}{4}}}{y^{\frac{2}{3}}}\right)^{-6}$

64. $\left(\dfrac{x^{\frac{1}{2}}y^{-\frac{5}{4}}}{y^{-\frac{3}{4}}}\right)^{-4}$

65. $\left(\dfrac{2^{-6}b^{-3}}{a^{-\frac{1}{2}}}\right)^{-\frac{2}{3}}$

66. $\left(\dfrac{49c^{\frac{5}{3}}}{a^{-\frac{1}{4}}b^{\frac{5}{6}}}\right)^{-\frac{3}{2}}$

67. $\dfrac{(x^{-2}y^4)^{\frac{1}{2}}}{\left(x^{\frac{1}{2}}\right)^4}$

68. $\dfrac{(x^{-3})^{\frac{1}{3}}}{(x^9y^6)^{\frac{1}{6}}}$

69. $a^{-\frac{1}{4}}\left(a^{\frac{5}{4}} - a^{\frac{9}{4}}\right)$

70. $x^{\frac{4}{3}}\left(x^{\frac{2}{3}} + x^{-\frac{1}{3}}\right)$

71. $y^{\frac{2}{3}}\left(y^{\frac{1}{3}} + y^{-\frac{2}{3}}\right)$

72. $b^{-\frac{2}{5}}\left(b^{-\frac{3}{5}} - b^{\frac{7}{5}}\right)$

73. $a^{\frac{1}{6}}\left(a^{\frac{5}{6}} - a^{-\frac{7}{6}}\right)$

74. $\left(x^{\frac{n}{3}}\right)^{3n}$

75. $\left(a^{\frac{2}{n}}\right)^{-5n}$

76. $x^n \cdot x^{\frac{n}{2}}$

77. $a^{\frac{n}{2}} \cdot a^{-\frac{n}{3}}$

78. $\dfrac{y^{\frac{n}{2}}}{y^{-n}}$

79. $\dfrac{b^{\frac{m}{3}}}{b^m}$

80. $\left(x^{\frac{2}{n}}\right)^n$

81. $(x^{5n})^{2n}$

82. $\left(x^{\frac{n}{4}}y^{\frac{n}{8}}\right)^8$

83. $\left(x^{\frac{n}{2}}y^{\frac{n}{3}}\right)^6$

2 Rewrite the exponential expression as a radical expression.

84. $3^{\frac{1}{4}}$

85. $5^{\frac{1}{2}}$

86. $a^{\frac{3}{2}}$

87. $b^{\frac{4}{3}}$
88. $(2t)^{\frac{5}{2}}$
89. $(3x)^{\frac{2}{3}}$
90. $-2x^{\frac{2}{3}}$
91. $-3a^{\frac{2}{5}}$
92. $(a^2b)^{\frac{2}{3}}$
93. $(x^2y^3)^{\frac{3}{4}}$
94. $(a^2b^4)^{\frac{3}{5}}$
95. $(a^3b^7)^{\frac{3}{2}}$
96. $(4x + 3)^{\frac{3}{4}}$
97. $(3x - 2)^{\frac{1}{3}}$
98. $x^{-\frac{2}{3}}$

Rewrite the radical expression as an exponential expression.

99. $\sqrt{14}$
100. $\sqrt{7}$
101. $\sqrt[3]{x}$
102. $\sqrt[4]{y}$
103. $\sqrt[3]{x^4}$
104. $\sqrt[4]{a^3}$
105. $\sqrt[5]{b^3}$
106. $\sqrt[4]{b^5}$
107. $\sqrt[3]{2x^2}$
108. $\sqrt[5]{4y^7}$
109. $-\sqrt{3x^5}$
110. $-\sqrt[4]{4x^5}$
111. $3x\sqrt[3]{y^2}$
112. $2y\sqrt{x^3}$
113. $\sqrt{a^2 + 2}$

3 Simplify.

114. $\sqrt{x^{16}}$
115. $\sqrt{y^{14}}$
116. $-\sqrt{x^8}$
117. $-\sqrt{a^6}$
118. $\sqrt{x^2y^{10}}$
119. $\sqrt{a^{14}b^6}$
120. $\sqrt{25x^6}$
121. $\sqrt{121y^{12}}$
122. $\sqrt[3]{x^3y^9}$
123. $\sqrt[3]{a^6b^{12}}$
124. $-\sqrt[3]{x^{15}y^3}$
125. $-\sqrt[3]{a^9b^9}$
126. $\sqrt[3]{27a^9}$
127. $\sqrt[3]{125b^{15}}$
128. $\sqrt[3]{-8x^3}$
129. $\sqrt[3]{-a^6b^9}$
130. $\sqrt{16a^4b^{12}}$
131. $\sqrt{25x^8y^2}$
132. $\sqrt{-16x^4y^2}$
133. $\sqrt{-9a^6b^8}$
134. $\sqrt[3]{27x^9}$
135. $\sqrt[3]{8a^{21}b^6}$
136. $\sqrt[3]{-64x^9y^{12}}$
137. $\sqrt[3]{-27a^3b^{15}}$
138. $\sqrt[4]{x^{16}}$
139. $\sqrt[4]{y^{12}}$
140. $\sqrt[4]{16x^{12}}$
141. $\sqrt[4]{81a^{20}}$
142. $-\sqrt[4]{x^8y^{12}}$
143. $-\sqrt[4]{a^{16}b^4}$
144. $\sqrt[5]{x^{20}y^{10}}$
145. $\sqrt[5]{a^5b^{25}}$
146. $\sqrt[4]{81x^4y^{20}}$
147. $\sqrt[4]{16a^8b^{20}}$
148. $\sqrt[5]{32a^5b^{10}}$
149. $\sqrt[5]{-32x^{15}y^{20}}$
150. $\sqrt[5]{243x^{10}y^{40}}$
151. $\sqrt{\dfrac{16x^2}{y^{14}}}$
152. $\sqrt{\dfrac{49a^4}{b^{24}}}$
153. $\sqrt[3]{\dfrac{27b^3}{a^9}}$
154. $\sqrt[3]{\dfrac{64x^{15}}{y^6}}$
155. $\sqrt{(2x + 3)^2}$
156. $\sqrt{(4x + 1)^2}$
157. $\sqrt{x^2 + 2x + 1}$
158. $\sqrt{x^2 + 4x + 4}$

8.1 Applying Concepts

Simplify.

159. $\sqrt[3]{\sqrt{x^6}}$

160. $\sqrt{\sqrt[3]{y^6}}$

161. $\sqrt[5]{\sqrt[3]{b^{15}}}$

162. $\sqrt{\sqrt{16x^{12}}}$

163. $\sqrt[5]{\sqrt{a^{10}b^{20}}}$

164. $\sqrt[3]{\sqrt{64x^{36}y^{30}}}$

For what value of p is the given equation true?

165. $y^p y^{\frac{2}{5}} = y$

166. $\dfrac{y^p}{y^{\frac{3}{4}}} = y^{\frac{1}{2}}$

167. $x^p x^{-\frac{1}{2}} = x^{\frac{1}{4}}$

8.2 Operations on Radical Expressions

1 Simplify radical expressions

Point of Interest

The Latin expression for irrational numbers was numerus surdus, *which means "inaudible number." A prominent 16th-century mathematician wrote of irrational numbers, "just as an infinite number is not a number, so an irrational number is not a true number, but lies hidden in some sort of cloud of infinity." In 1872, Richard Dedekind wrote a paper that established the first logical treatment of irrational numbers.*

If a number is not a perfect power, its root can only be approximated; examples include $\sqrt{5}$ and $\sqrt[3]{3}$. These numbers are *irrational numbers*. Their decimal representations never terminate or repeat.

$$\sqrt{5} = 2.2360679\ldots \qquad \sqrt[3]{3} = 1.4422495\ldots$$

A radical expression is not in simplest form when the radicand contains a factor greater than 1 that is a perfect power. The Product Property of Radicals is used to simplify radical expressions whose radicands are not perfect powers.

Product Property of Radicals

If $\sqrt[n]{a}$ and $\sqrt[n]{b}$ are positive real numbers, then

$$\sqrt[n]{ab} = \sqrt[n]{a} \cdot \sqrt[n]{b} \quad \text{and} \quad \sqrt[n]{a} \cdot \sqrt[n]{b} = \sqrt[n]{ab}$$

Simplify: $\sqrt{48}$

Write the radicand as the product of a perfect square and a factor that does not contain a perfect square. $\qquad \sqrt{48} = \sqrt{16 \cdot 3}$

Use the Product Property of Radicals to write the expression as a product. $\qquad = \sqrt{16}\sqrt{3}$

Simplify $\sqrt{16}$. $\qquad = 4\sqrt{3}$

Note that 48 must be written as the product of a perfect square and *a factor that does not contain a perfect square.* Therefore, it would not be correct to rewrite $\sqrt{48}$ as $\sqrt{4 \cdot 12}$ and simplify the expression as shown at the right. Although 4 is a perfect-square factor of 48, 12 contains a perfect square $(12 = 4 \cdot 3)$ and $\sqrt{12}$ can be simplified. Remember to find the largest perfect power that is a factor of the radicand.

$$\begin{aligned}\sqrt{48} &= \sqrt{4 \cdot 12}\\ &= \sqrt{4}\sqrt{12}\\ &= 2\sqrt{12}\end{aligned}$$

Not in simplest form

Simplify: $\sqrt[3]{x^7}$

Write the radicand as the product of a perfect cube and a factor that does not contain a perfect cube. $\sqrt[3]{x^7} = \sqrt[3]{x^6 \cdot x}$

Use the Product Property of Radicals to write the expression as a product. $= \sqrt[3]{x^6}\sqrt[3]{x}$

Simplify. $= x^2\sqrt[3]{x}$

Example 1 Simplify: $\sqrt[4]{32x^7}$

Solution $\sqrt[4]{32x^7} = \sqrt[4]{16x^4(2x^3)}$

- Write the radicand as the product of a perfect fourth power and factors that do not contain a perfect fourth power.

$= \sqrt[4]{16x^4}\sqrt[4]{2x^3}$

- Use the Product Property of Radicals to write the expression as a product.

$= 2x\sqrt[4]{2x^3}$

- Simplify.

Problem 1 Simplify: $\sqrt[5]{128x^7}$

Solution See page S26.

2 Add and subtract radical expressions

VIDEO & DVD CD TUTOR WEB SSM

The Distributive Property is used to simplify the sum or difference of radical expressions that have the same radicand and the same index. For example,

$$3\sqrt{5} + 8\sqrt{5} = (3 + 8)\sqrt{5} = 11\sqrt{5}$$
$$2\sqrt[3]{3x} - 9\sqrt[3]{3x} = (2 - 9)\sqrt[3]{3x} = -7\sqrt[3]{3x}$$

Radical expressions that are in simplest form and have unlike radicands or different indices cannot be simplified by the Distributive Property. The following expressions cannot be simplified by the Distributive Property.

$3\sqrt[4]{2} - 6\sqrt[4]{3}$ • The radicands are different.

$2\sqrt[4]{4x} + 3\sqrt[3]{4x}$ • The indices are different.

Simplify: $3\sqrt{32x^2} - 2x\sqrt{2} + \sqrt{128x^2}$

First simplify each term. Then combine like terms by using the Distributive Property.

$$\begin{aligned} 3\sqrt{32x^2} - 2x\sqrt{2} + \sqrt{128x^2} &= 3\sqrt{16x^2}\sqrt{2} - 2x\sqrt{2} + \sqrt{64x^2}\sqrt{2} \\ &= 3 \cdot 4x\sqrt{2} - 2x\sqrt{2} + 8x\sqrt{2} \\ &= 12x\sqrt{2} - 2x\sqrt{2} + 8x\sqrt{2} \\ &= 18x\sqrt{2} \end{aligned}$$

Example 2 Subtract.

A. $5b\sqrt[4]{32a^7b^5} - 2a\sqrt[4]{162a^3b^9}$ **B.** $5\sqrt[5]{2x^7y^{11}} - y\sqrt[5]{64^7y^6}$

Solution **A.** $5b\sqrt[4]{32a^7b^5} - 2a\sqrt[4]{162a^3b^9}$

$$\begin{aligned} &= 5b\sqrt[4]{16a^4b^4}\sqrt[4]{2a^3b} - 2a\sqrt[4]{81b^8}\sqrt[4]{2a^3b} \\ &= 5b \cdot 2ab\sqrt[4]{2a^3b} - 2a \cdot 3b^2\sqrt[4]{2a^3b} \\ &= 10ab^2\sqrt[4]{2a^3b} - 6ab^2\sqrt[4]{2a^3b} \\ &= 4ab^2\sqrt[4]{2a^3b} \end{aligned}$$

B. $5\sqrt[5]{2x^7y^{11}} - y\sqrt[5]{64x^7y^6} = 5\sqrt[5]{x^5y^{10}}\sqrt[5]{2x^2y} - y\sqrt[5]{32x^5y^5}\sqrt[5]{2x^2y}$
$= 5 \cdot xy^2\sqrt[5]{2x^2y} - y \cdot 2xy\sqrt[5]{2x^2y}$
$= 5xy^2\sqrt[5]{2x^2y} - 2xy^2\sqrt[5]{2x^2y}$
$= 3xy^2\sqrt[5]{2x^2y}$

Problem 2 Add or subtract.

A. $3xy\sqrt[3]{81x^5y} - \sqrt[3]{192x^8y^4}$ **B.** $4a\sqrt[3]{54a^7b^9} + a^2b\sqrt[3]{128a^4b^6}$

Solution See page S26.

3 Multiply radical expressions

VIDEO & DVD CD TUTOR WEB SSM

The Product Property of Radicals is used to multiply radical expressions with the same index.

$$\sqrt{3x} \cdot \sqrt{5y} = \sqrt{3x \cdot 5y} = \sqrt{15xy}$$

Multiply: $\sqrt[3]{2a^5b}\sqrt[3]{16a^2b^2}$

Use the Product Property of Radicals to multiply the radicands. Then simplify.

$$\sqrt[3]{2a^5b}\sqrt[3]{16a^2b^2} = \sqrt[3]{32a^7b^3} = \sqrt[3]{8a^6b^3}\sqrt[3]{4a} = 2a^2b\sqrt[3]{4a}$$

Multiply: $\sqrt{2x}(\sqrt{8x} - \sqrt{3})$

Use the Distributive Property to remove parentheses. Then simplify.

$$\sqrt{2x}(\sqrt{8x} - \sqrt{3}) = \sqrt{16x^2} - \sqrt{6x} = 4x - \sqrt{6x}$$

Example 3 Multiply: $\sqrt{3x}(\sqrt{27x^2} - \sqrt{3x})$

Solution $\sqrt{3x}(\sqrt{27x^2} - \sqrt{3x}) = \sqrt{81x^3} - \sqrt{9x^2}$
$= \sqrt{81x^2}\sqrt{x} - \sqrt{9x^2}$
$= 9x\sqrt{x} - 3x$

Problem 3 Multiply: $\sqrt{5b}(\sqrt{3b} - \sqrt{10})$

Solution See page S26.

To multiply $(\sqrt[3]{x} - 1)(\sqrt[3]{x} + 7)$, use the FOIL method. Then combine like terms.

$$(\sqrt[3]{x} - 1)(\sqrt[3]{x} + 7) = \sqrt[3]{x^2} + 7\sqrt[3]{x} - \sqrt[3]{x} - 7 = \sqrt[3]{x^2} + 6\sqrt[3]{x} - 7$$

> **Take Note**
>
> The concept of conjugate is used in a number of different instances. Make sure you understand this idea.
>
> The conjugate of $\sqrt{3} - 4$ is $\sqrt{3} + 4$.
>
> The conjugate of $\sqrt{3} + 4$ is $\sqrt{3} - 4$.
>
> The conjugate of $\sqrt{5a} + \sqrt{b}$ is $\sqrt{5a} - \sqrt{b}$.

The expressions $a + b$ and $a - b$, which are the sum and difference of two terms, are called **conjugates** of each other. The product of conjugates of the form $(a + b)(a - b)$ is $a^2 - b^2$.

$$(\sqrt{x} - 3)(\sqrt{x} + 3) = (\sqrt{x})^2 - 3^2 = x - 9$$

Example 4 Multiply.

A. $(2\sqrt[3]{x} - 3)(3\sqrt[3]{x} - 4)$ **B.** $(\sqrt{xy} - 2)(\sqrt{xy} + 2)$

Solution **A.** $(2\sqrt[3]{x} - 3)(3\sqrt[3]{x} - 4)$ • **Use the FOIL method.**
$= 6\sqrt[3]{x^2} - 8\sqrt[3]{x} - 9\sqrt[3]{x} + 12$
$= 6\sqrt[3]{x^2} - 17\sqrt[3]{x} + 12$

B. $(\sqrt{xy} - 2)(\sqrt{xy} + 2)$ • Use $(a - b)(a + b) = a^2 - b^2$.
$= (\sqrt{xy})^2 - 2^2$
$= xy - 4$

Problem 4 Multiply.
A. $(2\sqrt[3]{2x} - 3)(\sqrt[3]{2x} - 5)$ **B.** $(2\sqrt{x} - 3)(2\sqrt{x} + 3)$

Solution See page S26.

4 Divide radical expressions

VIDEO & DVD CD TUTOR WEB SSM

The Quotient Property of Radicals is used to divide radical expressions with the same index.

Quotient Property of Radicals

If $\sqrt[n]{a}$ and $\sqrt[n]{b}$ are real numbers, and $b \neq 0$, then

$$\sqrt[n]{\frac{a}{b}} = \frac{\sqrt[n]{a}}{\sqrt[n]{b}} \quad \text{and} \quad \frac{\sqrt[n]{a}}{\sqrt[n]{b}} = \sqrt[n]{\frac{a}{b}}$$

Simplify: $\sqrt[3]{\frac{81x^5}{y^6}}$

Use the Quotient Property of Radicals. Then simplify each radical expression.

$$\sqrt[3]{\frac{81x^5}{y^6}} = \frac{\sqrt[3]{81x^5}}{\sqrt[3]{y^6}} = \frac{\sqrt[3]{27x^3}\sqrt[3]{3x^2}}{\sqrt[3]{y^6}} = \frac{3x\sqrt[3]{3x^2}}{y^2}$$

Simplify: $\frac{\sqrt{5a^4b^7c^2}}{\sqrt{ab^3c}}$

Use the Quotient Property of Radicals. Then simplify the radicand.

$$\frac{\sqrt{5a^4b^7c^2}}{\sqrt{ab^3c}} = \sqrt{\frac{5a^4b^7c^2}{ab^3c}} = \sqrt{5a^3b^4c} = \sqrt{a^2b^4}\sqrt{5ac} = ab^2\sqrt{5ac}$$

Simplest Form of a Radical Expression

A radical expression is in simplest form when *all* of the following conditions are met.

1. The radicand contains no factor greater than 1 that is a perfect power of the index.
2. There is no fraction under the radical sign.
3. No radical remains in the denominator of the radical expression.

The procedure used to remove a radical from a denominator is called **rationalizing the denominator.** This is illustrated in the next example.

Simplify: $\frac{2}{\sqrt{x}}$

Multiply the expression by 1 in the form $\frac{\sqrt{x}}{\sqrt{x}}$.

$$\frac{2}{\sqrt{x}} = \frac{2}{\sqrt{x}} \cdot \frac{\sqrt{x}}{\sqrt{x}} = \frac{2\sqrt{x}}{\sqrt{x^2}} = \frac{2\sqrt{x}}{x}$$

Then simplify.

Example 5 Simplify. **A.** $\frac{5}{\sqrt{5x}}$ **B.** $\frac{3x}{\sqrt[3]{4x}}$

Solution

A.
$$\frac{5}{\sqrt{5x}} = \frac{5}{\sqrt{5x}} \cdot \frac{\sqrt{5x}}{\sqrt{5x}}$$
• Multiply the expression by $\frac{\sqrt{5x}}{\sqrt{5x}}$.
$$= \frac{5\sqrt{5x}}{\sqrt{25x^2}} = \frac{5\sqrt{5x}}{5x} = \frac{\sqrt{5x}}{x}$$

B.
$$\frac{3x}{\sqrt[3]{4x}} = \frac{3x}{\sqrt[3]{4x}} \cdot \frac{\sqrt[3]{2x^2}}{\sqrt[3]{2x^2}}$$
• Multiply the expression by $\frac{\sqrt[3]{2x^2}}{\sqrt[3]{2x^2}}$. $\sqrt[3]{4x} \cdot \sqrt[3]{2x^2} = \sqrt[3]{8x^3}$. $8x^3$ is a perfect cube.
$$= \frac{3x\sqrt[3]{2x^2}}{\sqrt[3]{8x^3}} = \frac{3x\sqrt[3]{2x^2}}{2x} = \frac{3\sqrt[3]{2x^2}}{2}$$

Take Note

Note that multiplying by $\frac{\sqrt[3]{4x}}{\sqrt[3]{4x}}$ will not rationalize the denominator of $\frac{3x}{\sqrt[3]{4x}}$.

$$\frac{3x}{\sqrt[3]{4x}} \cdot \frac{\sqrt[3]{4x}}{\sqrt[3]{4x}} = \frac{3x\sqrt[3]{4x}}{\sqrt[3]{16x^2}}$$

Because $16x^2$ is not a perfect cube, the denominator still contains a radical expression.

Problem 5 Simplify. **A.** $\frac{y}{\sqrt{3y}}$ **B.** $\frac{3}{\sqrt[3]{3x^2}}$

Solution See page S26.

To simplify a fraction that has a square-root radical expression with two terms in the denominator, multiply the numerator and denominator by the conjugate of the denominator. Then simplify. This is illustrated in the two examples that follow.

Take Note

Here are two examples of using a conjugate to simplify a radical expression.

$$\frac{3}{5-\sqrt{7}} = \frac{3}{5-\sqrt{7}} \cdot \frac{5+\sqrt{7}}{5+\sqrt{7}} = \frac{15+3\sqrt{7}}{25-7} = \frac{3(5+\sqrt{7})}{18} = \frac{5+\sqrt{7}}{6}$$

$$\frac{\sqrt{x}-\sqrt{y}}{\sqrt{x}+\sqrt{y}} = \frac{\sqrt{x}-\sqrt{y}}{\sqrt{x}+\sqrt{y}} \cdot \frac{\sqrt{x}-\sqrt{y}}{\sqrt{x}-\sqrt{y}} = \frac{\sqrt{x^2}-\sqrt{xy}-\sqrt{xy}+\sqrt{y^2}}{(\sqrt{x})^2-(\sqrt{y})^2} = \frac{x-2\sqrt{xy}+y}{x-y}$$

Example 6 Simplify. **A.** $\dfrac{2-\sqrt{5}}{3+\sqrt{2}}$ **B.** $\dfrac{\sqrt{3}+\sqrt{y}}{\sqrt{3}-\sqrt{y}}$

Solution **A.** $\dfrac{2-\sqrt{5}}{3+\sqrt{2}} = \dfrac{2-\sqrt{5}}{3+\sqrt{2}} \cdot \dfrac{3-\sqrt{2}}{3-\sqrt{2}} = \dfrac{6-2\sqrt{2}-3\sqrt{5}+\sqrt{10}}{9-2}$

$= \dfrac{6-2\sqrt{2}-3\sqrt{5}+\sqrt{10}}{7}$

B. $\dfrac{\sqrt{3}+\sqrt{y}}{\sqrt{3}-\sqrt{y}} = \dfrac{\sqrt{3}+\sqrt{y}}{\sqrt{3}-\sqrt{y}} \cdot \dfrac{\sqrt{3}+\sqrt{y}}{\sqrt{3}+\sqrt{y}}$

$= \dfrac{\sqrt{9}+\sqrt{3y}+\sqrt{3y}+\sqrt{y^2}}{(\sqrt{3})^2-(\sqrt{y})^2} = \dfrac{3+2\sqrt{3y}+y}{3-y}$

Problem 6 Simplify. **A.** $\dfrac{4+\sqrt{2}}{3-\sqrt{3}}$ **B.** $\dfrac{\sqrt{2}+\sqrt{x}}{\sqrt{2}-\sqrt{x}}$

Solution See page S27.

8.2 Concept Review

Determine whether the following statements are always true, sometimes true, or never true.

1. If a is a real number, then $\sqrt{a}$ represents a real number.
2. If b is a real number, then $\sqrt[3]{b}$ represents a real number.
3. For positive integers x and y, $\sqrt[a]{x} \cdot \sqrt[b]{y} = \sqrt[ab]{xy}$.
4. The conjugate of $\sqrt{a} + \sqrt{b}$ is found by replacing b with its opposite.
5. If $x < 0$, then $\sqrt{x^2} = -x$.
6. For positive integers x and y, $\sqrt[a]{x} \cdot \sqrt[a]{y} = \sqrt[a]{xy}$.

8.2 Exercises

1 Explain why the simplification is incorrect.

1. $\sqrt{32} = \sqrt{4 \cdot 8} = \sqrt{4}\sqrt{8} = 2\sqrt{8}$

2. $\sqrt[3]{128} = \sqrt[3]{8 \cdot 16} = \sqrt[3]{8}\sqrt[3]{16} = 2\sqrt[3]{16}$

Simplify.

3. $\sqrt{18}$
4. $\sqrt{40}$
5. $\sqrt{98}$
6. $\sqrt{128}$
7. $\sqrt[3]{72}$
8. $\sqrt[3]{54}$
9. $\sqrt[3]{16}$
10. $\sqrt[3]{128}$

11. $\sqrt{x^4y^3z^5}$

12. $\sqrt{x^3y^6z^9}$

13. $\sqrt{8a^3b^8}$

14. $\sqrt{24a^9b^6}$

15. $\sqrt{45x^2y^3z^5}$

16. $\sqrt{60xy^7z^{12}}$

17. $\sqrt[3]{-125x^2y^4}$

18. $\sqrt[4]{16x^9y^5}$

19. $\sqrt[3]{-216x^5y^9}$

20. $\sqrt[3]{a^8b^{11}c^{15}}$

21. $\sqrt[3]{a^5b^8}$

22. $\sqrt[4]{64x^8y^{10}}$

2 Simplify.

23. $\sqrt{2} + \sqrt{2}$

24. $\sqrt{5} - \sqrt{5}$

25. $4\sqrt[3]{7} - \sqrt[3]{7}$

26. $3\sqrt[3]{11} - 8\sqrt[3]{11}$

27. $2\sqrt{x} - 8\sqrt{x}$

28. $3\sqrt{y} + 12\sqrt{y}$

29. $\sqrt{8} - \sqrt{32}$

30. $\sqrt{27} - \sqrt{75}$

31. $\sqrt{128x} - \sqrt{98x}$

32. $\sqrt{48x} + \sqrt{147x}$

33. $\sqrt{27a} - \sqrt{8a}$

34. $\sqrt{18b} + \sqrt{75b}$

35. $2\sqrt{2x^3} + 4x\sqrt{8x}$

36. $5y\sqrt{8y} + 2\sqrt{50y^3}$

37. $x\sqrt{75xy} - \sqrt{27x^3y}$

38. $3\sqrt{8x^2y^3} - 2x\sqrt{32y^3}$

39. $2\sqrt{32x^2y^3} - xy\sqrt{98y}$

40. $6y\sqrt{x^3y} - 2\sqrt{x^3y^3}$

41. $7b\sqrt{a^5b^3} - 2ab\sqrt{a^3b^3}$

42. $2a\sqrt{27ab^5} + 3b\sqrt{3a^3b}$

43. $\sqrt[3]{128} + \sqrt[3]{250}$

44. $\sqrt[3]{16} - \sqrt[3]{54}$

45. $2\sqrt[3]{3a^4} - 3a\sqrt[3]{81a}$

46. $2b\sqrt[3]{16b^2} + \sqrt[3]{128b^5}$

47. $3\sqrt[3]{x^5y^7} - 8xy\sqrt[3]{x^2y^4}$

48. $3\sqrt[4]{32a^5} - a\sqrt[4]{162a}$

49. $2a\sqrt[4]{16ab^5} + 3b\sqrt[4]{256a^5b}$

50. $2\sqrt{50} - 3\sqrt{125} + \sqrt{98}$

51. $3\sqrt{108} - 2\sqrt{18} - 3\sqrt{48}$

52. $\sqrt{9b^3} - \sqrt{25b^3} + \sqrt{49b^3}$

53. $\sqrt{4x^7y^5} + 9x^2\sqrt{x^3y^5} - 5xy\sqrt{x^5y^3}$

54. $2x\sqrt{8xy^2} - 3y\sqrt{32x^3} + \sqrt{8x^3y^2}$

55. $5a\sqrt{3a^3b} + 2a^2\sqrt{27ab} - 4\sqrt{75a^5b}$

56. $\sqrt[3]{54xy^3} - 5\sqrt[3]{2xy^3} + \sqrt[3]{128xy^3}$

3 Multiply.

57. $\sqrt{8}\sqrt{32}$

58. $\sqrt{14}\sqrt{35}$

59. $\sqrt[3]{4}\sqrt[3]{8}$

60. $\sqrt[3]{6}\sqrt[3]{36}$

61. $\sqrt{x^2y^5}\sqrt{xy}$

62. $\sqrt{a^3b}\sqrt{ab^4}$

63. $\sqrt{2x^2y}\sqrt{32xy}$

64. $\sqrt{5x^3y}\sqrt{10x^3y^4}$

65. $\sqrt[3]{x^2y}\sqrt[3]{16x^4y^2}$

66. $\sqrt[3]{4a^2b^3}\sqrt[3]{8ab^5}$

67. $\sqrt[4]{12ab^3}\sqrt[4]{4a^5b^2}$

68. $\sqrt[4]{36a^2b^4}\sqrt[4]{12a^5b^3}$

69. $\sqrt{3}(\sqrt{27} - \sqrt{3})$

70. $\sqrt{10}(\sqrt{10} - \sqrt{5})$

71. $\sqrt{x}(\sqrt{x} - \sqrt{2})$

72. $\sqrt{y}(\sqrt{y} - \sqrt{5})$

73. $\sqrt{2x}(\sqrt{8x} - \sqrt{32})$

74. $\sqrt{3a}(\sqrt{27a^2} - \sqrt{a})$

75. $(\sqrt{x} - 3)^2$

76. $(\sqrt{2x} + 4)^2$

77. $(4\sqrt{5} + 2)^2$

78. $2\sqrt{3x^2} \cdot 3\sqrt{12xy^3} \cdot \sqrt{6x^3y}$

79. $2\sqrt{14xy} \cdot 4\sqrt{7x^2y} \cdot 3\sqrt{8xy^2}$

80. $\sqrt[3]{8ab}\sqrt[3]{4a^2b^3}\sqrt[3]{9ab^4}$

81. $\sqrt[3]{2a^2b}\sqrt[3]{4a^3b^2}\sqrt[3]{8a^5b^6}$

82. $(\sqrt{2} - 3)(\sqrt{2} + 4)$

83. $(\sqrt{5} - 5)(2\sqrt{5} + 2)$

84. $(\sqrt{y} - 2)(\sqrt{y} + 2)$

85. $(\sqrt{x} - y)(\sqrt{x} + y)$

86. $(\sqrt{2x} - 3\sqrt{y})(\sqrt{2x} + 3\sqrt{y})$

87. $(2\sqrt{3x} - \sqrt{y})(2\sqrt{3x} + \sqrt{y})$

88. $(\sqrt{a} - 2)(\sqrt{a} - 3)$

89. $(\sqrt{x} + 4)(\sqrt{x} - 7)$

90. $(\sqrt[3]{a} + 2)(\sqrt[3]{a} + 3)$

91. $(\sqrt[3]{x} - 4)(\sqrt[3]{x} + 5)$

92. $(2\sqrt{x} - \sqrt{y})(3\sqrt{x} + \sqrt{y})$

4 **93.** When is a radical expression in simplest form?

94. Explain what it means to rationalize the denominator of a radical expression and how to do so.

Simplify.

95. $\dfrac{\sqrt{32x^2}}{\sqrt{2x}}$

96. $\dfrac{\sqrt{60y^4}}{\sqrt{12y}}$

97. $\dfrac{\sqrt{42a^3b^5}}{\sqrt{14a^2b}}$

98. $\dfrac{\sqrt{65ab^4}}{\sqrt{5ab}}$

99. $\dfrac{1}{\sqrt{5}}$

100. $\dfrac{1}{\sqrt{2}}$

101. $\dfrac{1}{\sqrt{2x}}$

102. $\dfrac{2}{\sqrt{3y}}$

103. $\dfrac{5}{\sqrt{5x}}$

104. $\dfrac{9}{\sqrt{3a}}$

105. $\sqrt{\dfrac{x}{5}}$

106. $\sqrt{\dfrac{y}{2}}$

107. $\dfrac{3}{\sqrt[3]{2}}$

108. $\dfrac{5}{\sqrt[3]{9}}$

109. $\dfrac{3}{\sqrt[3]{4x^2}}$

110. $\dfrac{5}{\sqrt[3]{3y}}$

111. $\dfrac{\sqrt{40x^3y^2}}{\sqrt{80x^2y^3}}$

112. $\dfrac{\sqrt{15a^2b^5}}{\sqrt{30a^5b^3}}$

113. $\dfrac{\sqrt{24a^2b}}{\sqrt{18ab^4}}$

114. $\dfrac{\sqrt{12x^3y}}{\sqrt{20x^4y}}$

115. $\dfrac{2}{\sqrt{5}+2}$

116. $\dfrac{5}{2-\sqrt{7}}$

117. $\dfrac{3}{\sqrt{y}-2}$

118. $\dfrac{-7}{\sqrt{x}-3}$

119. $\dfrac{\sqrt{2}-\sqrt{3}}{\sqrt{2}+\sqrt{3}}$

120. $\dfrac{\sqrt{3}+\sqrt{4}}{\sqrt{2}+\sqrt{3}}$

121. $\dfrac{4-\sqrt{2}}{2-\sqrt{3}}$

122. $\dfrac{3-\sqrt{x}}{3+\sqrt{x}}$

123. $\dfrac{\sqrt{3}-\sqrt{5}}{\sqrt{2}+\sqrt{5}}$

124. $\dfrac{\sqrt{2}+\sqrt{3}}{\sqrt{3}-\sqrt{2}}$

125. $\dfrac{3}{\sqrt[4]{8x^3}}$

126. $\dfrac{-3}{\sqrt[4]{27y^2}}$

127. $\dfrac{4}{\sqrt[5]{16a^2}}$

128. $\dfrac{a}{\sqrt[5]{81a^4}}$

129. $\dfrac{2x}{\sqrt[5]{64x^3}}$

130. $\dfrac{3y}{\sqrt[4]{32y^2}}$

131. $\dfrac{\sqrt{a}+a\sqrt{b}}{\sqrt{a}-a\sqrt{b}}$

132. $\dfrac{\sqrt{3}-3\sqrt{y}}{\sqrt{3}+3\sqrt{y}}$

133. $\dfrac{3\sqrt{xy}+2\sqrt{xy}}{\sqrt{x}-\sqrt{y}}$

134. $\dfrac{2\sqrt{x}+3\sqrt{y}}{\sqrt{x}-4\sqrt{y}}$

8.2 Applying Concepts

Simplify.

135. $(\sqrt{8}-\sqrt{2})^3$

136. $(\sqrt{27}-\sqrt{3})^3$

137. $(\sqrt{2}-3)^3$

138. $(\sqrt{5}+2)^3$

139. $\dfrac{3}{\sqrt{y+1}+1}$

140. $\dfrac{2}{\sqrt{x+4}+2}$

Rewrite as an expression with a single radical.

141. $\dfrac{\sqrt[3]{(x+y)^2}}{\sqrt{x+y}}$

142. $\dfrac{\sqrt[4]{(a+b)^3}}{\sqrt{a+b}}$

143. $\sqrt[4]{2y}\sqrt{x+3}$

144. $\sqrt[4]{2x}\sqrt{y-2}$

145. $\sqrt{a}\sqrt[3]{a+3}$

146. $\sqrt{b}\sqrt[3]{b-1}$

Write in exponential form and then simplify.

147. $\sqrt{16^{\frac{1}{2}}}$ **148.** $\sqrt[3]{4^{\frac{3}{2}}}$ **149.** $\sqrt[4]{32^{-\frac{4}{5}}}$ **150.** $\sqrt{243^{-\frac{4}{5}}}$

151. **National Debt** The approximate national debt, in trillions of dollars, for the even years from 1992 to 2000 is given in the table at the right (*Source:* **www.publicdebt.treas.gov/**). A function that models this data is $D(x) = 3.50x^{0.2}$, where $x = 2$ corresponds to the year 1992 and $D(x)$ is the national debt in trillions of dollars.

a. Write this function in radical form.
b. Find the value of the national debt predicted by the model for the year 1995.
c. Assuming that the model is accurate past 2000, what will the national debt be in the year 2010?

Year	*National Debt*
1992	4.06
1994	4.69
1996	5.22
1998	5.53
2000	5.67

152. By what factor must you multiply a number in order to double its square root? to triple its square root? to double its cube root? to triple its cube root? Explain.

8.3 Radical Functions

1 Find the domain of a radical function

VIDEO & DVD CD TUTOR WEB SSM

A **radical function** is one that contains a variable underneath a radical or a fractional exponent. Examples of radical functions are shown at the right.

$$f(x) = 3\sqrt[4]{x^5} - 7$$
$$g(x) = 3x - 2x^{\frac{1}{2}} + 5$$

Note that these are *not* polynomial functions because polynomial functions do not contain variable radical expressions or variables raised to a fractional power.

The domain of a radical function is a set of real numbers for which the radical expression is a real number. For example, −9 is one number that would be excluded from the domain of $f(x) = \sqrt{x + 5}$ because

$$f(-9) = \sqrt{-9 + 5} = \sqrt{-4}, \text{ which is not a real number.}$$

State the domain of $f(x) = \sqrt{x + 5}$ in set-builder notation.

The value of $\sqrt{x + 5}$ is a real number when $x + 5$ is greater than or equal to zero: $x + 5 \geq 0$. Solving this inequality for x results in $x \geq -5$.

The domain of f is $\{x \mid x \geq -5\}$.

State the domain of $F(x) = \sqrt[3]{2x - 6}$.

Because the cube root of a real number is a real number, $\sqrt[3]{2x - 6}$ is a real number for all values of x. [For instance, $F(-1) = \sqrt[3]{2(-1) - 6} = \sqrt[3]{-8} = -2$.] Therefore, the domain of F is all real numbers.

These last two examples suggest the following:

If the index of a radical expression is an even number, the radicand must be greater than or equal to zero to ensure that the value of the expression will be a real number. If the index of a radical expression is an odd number, the radicand may be a positive or a negative number.

Example 1 State the domain of each function in set-builder notation.

A. $V(x) = \sqrt[4]{6 - 4x}$ **B.** $R(x) = \sqrt[5]{x + 4}$

Solution

A.
$$6 - 4x \geq 0$$
$$-4x \geq -6$$
$$x \leq \frac{3}{2}$$

• *V* contains an even root. Therefore, the radicand must be greater than or equal to zero.

The domain is $\left\{x \mid x \leq \frac{3}{2}\right\}$.

B. $\{x \mid x \in \text{real numbers}\}$

• Because *R* contains an odd root, the radicand may be positive or negative.

Problem 1 State the domain of each function in interval notation.

A. $Q(x) = \sqrt[3]{6x + 12}$ **B.** $T(x) = (3x + 9)^{\frac{1}{2}}$

Solution See page S27.

2 Graph a radical function

The graph of a radical function is produced in the same manner as the graph of any other function. The function is evaluated at several values in the domain of the function, and the resulting ordered pairs are graphed. Ordered pairs must be graphed until an accurate graph can be made.

Take Note

By first determining the domain of the function, you discover the possible values of *x* to use when evaluating the function. Any value of *x* in the domain can be used. We used "nice" values—values of *x* that result in a perfect-square radicand. However, any value of *x* can be used. For instance, if $x = 3$, then

$$f(3) = \sqrt{5} \approx 2.24$$

and the ordered pair (3, 2.24) belongs to the graph. This ordered pair is just a little more difficult to graph.

Graph: $f(x) = \sqrt{x + 2}$

Because *f* contains an even root, the radicand must be positive. To determine the domain of *f*, solve the inequality $x + 2 \geq 0$. The solution is $x \geq -2$, so the domain is $\{x \mid x \geq -2\}$. Now determine ordered pairs of the function by choosing values of *x* from the domain. Some possible choices are shown in the following table.

x	$f(x) = \sqrt{x + 2}$	y
-2	$f(-2) = \sqrt{-2 + 2} = \sqrt{0} = 0$	0
-1	$f(-1) = \sqrt{-1 + 2} = \sqrt{1} = 1$	1
2	$f(2) = \sqrt{2 + 2} = \sqrt{4} = 2$	2
7	$f(7) = \sqrt{7 + 2} = \sqrt{9} = 3$	3

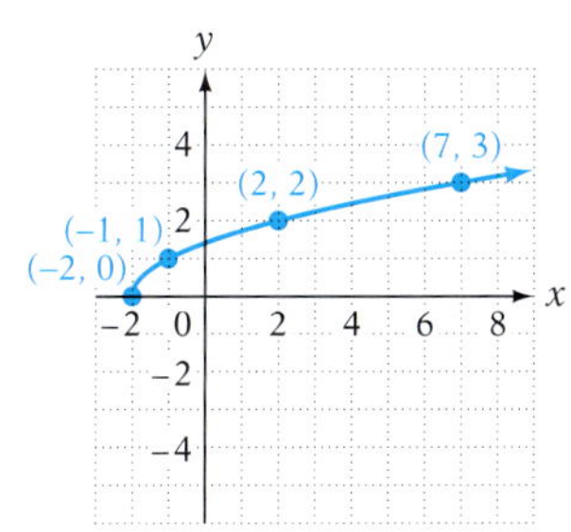

Example 2 Graph: $H(x) = \sqrt[3]{x}$

Solution Because H contains only an odd root, the domain of H is all real numbers. Choose some values of x in the domain of H, and evaluate the function for those values. Some possible choices are given in the following table.

x	$H(x) = \sqrt[3]{x}$	y
-8	$H(-8) = \sqrt[3]{-8} = -2$	-2
-1	$H(-1) = \sqrt[3]{-1} = -1$	-1
0	$H(0) = \sqrt[3]{0} = 0$	0
1	$H(1) = \sqrt[3]{1} = 1$	1
8	$H(8) = \sqrt[3]{8} = 2$	2

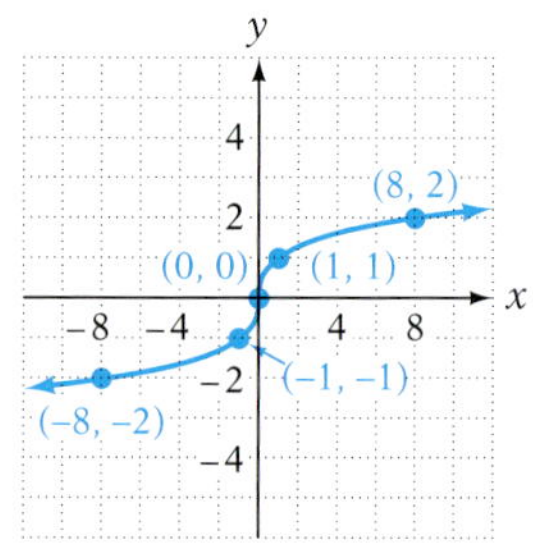

Problem 2 Graph: $F(x) = \sqrt{x - 2}$

Solution See page S27.

A graphing utility can be used to graph radical functions. See the Appendix for instructions on how to enter a radical function on a graphing utility.

Example 3 Graph: $Z(x) = 4 + (8 - 6x)^{\frac{1}{2}}$

Solution Because Z involves an even root, the radicand must be positive. The domain is $\left\{x \mid x \le \frac{4}{3}\right\}$.

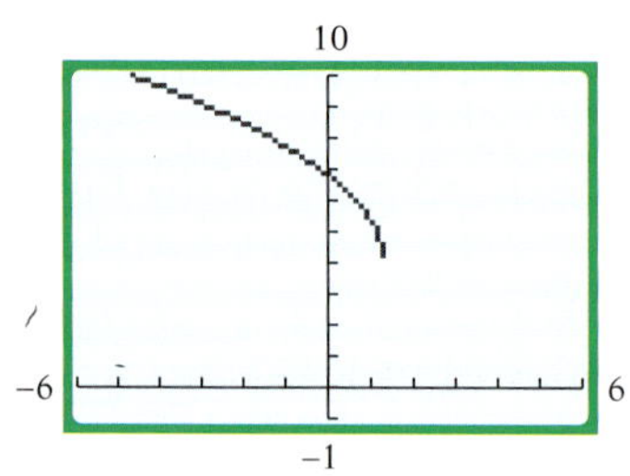

Problem 3 Graph: $y(x) = 2 - \sqrt[3]{x - 1}$

Solution See page S27.

8.3 Concept Review

Determine whether the following statements are always true, sometimes true, or never true.

1. The number 5 is not in the domain of $f(x) = \sqrt{4 - x}$.

2. The domain of $f(x) = \sqrt[3]{x - 2}$ is the set of all real numbers.

3. The domain of $f(x) = \dfrac{3}{x^2 + 4}$ is all the real numbers except -2 and 2.

4. If x is a real number, the expression $f(x) = \sqrt{-x^2}$ represents a radical function.

8.3 Exercises

1 1. What is a radical function?

2. What is the difference between a radical function and a polynomial function?

State the domain of each function in set-builder notation.

3. $f(x) = 2x^{\frac{1}{3}}$

4. $r(x) = -3\sqrt[5]{2x}$

5. $g(x) = -2\sqrt{x + 1}$

6. $h(x) = 3x^{\frac{1}{4}} - 2$

7. $f(x) = 2x\sqrt{x} - 3$

8. $y(x) = -3\sqrt[3]{1 + x}$

9. $C(x) = -3x^{\frac{3}{4}} + 1$

10. $G(x) = 6x^{\frac{2}{5}} + 5$

11. $F(x) = 4(3x - 6)^{\frac{1}{2}}$

State the domain of each function in interval notation.

12. $f(x) = -2(4x - 12)^{\frac{1}{2}}$

13. $g(x) = 2(2x - 10)^{\frac{2}{3}}$

14. $J(x) = 4 - (3x - 3)^{\frac{2}{5}}$

15. $V(x) = x - \sqrt{12 - 4x}$

16. $Y(x) = -6 + \sqrt{6 - x}$

17. $h(x) = 3\sqrt[4]{(x - 2)^3}$

18. $g(x) = \dfrac{2}{3}\sqrt[4]{(4 - x)^3}$

19. $f(x) = x - (4 - 6x)^{\frac{1}{2}}$

20. $F(x) = (9 + 12x)^{\frac{1}{2}} - 4$

2 Graph.

21. $F(x) = \sqrt{x}$

22. $G(x) = -\sqrt{x}$

23. $h(x) = -\sqrt[3]{x}$

24. $K(x) = \sqrt[3]{x + 1}$

25. $f(x) = -\sqrt{x + 2}$

26. $g(x) = \sqrt{x - 1}$

27. $S(x) = -\sqrt[4]{x}$

28. $C(x) = (x + 2)^{\frac{1}{4}}$

29. $F(x) = (x - 2)^{\frac{1}{2}}$

30. $f(x) = -(x - 1)^{\frac{1}{2}}$

31. $Q(x) = (x - 3)^{\frac{1}{3}}$

32. $H(x) = (-x)^{\frac{1}{3}}$

Graph.

33. $f(x) = 2x^{\frac{2}{5}} - 1$

34. $h(x) = 3x^{\frac{2}{5}} + 2$

35. $g(x) = 3 - (5 - 2x)^{\frac{1}{2}}$

36. $y(x) = 1 + (4 - 8x)^{\frac{1}{2}}$

37. $V(x) = \sqrt[5]{(x - 2)^2}$

38. $A(x) = x\sqrt{3x - 9}$

39. $F(x) = 2x - 3\sqrt[3]{x} - 1$

40. $f(x) = 2x - \sqrt{x} - 1$

41. $f(x) = 3x\sqrt{4x + 8}$

8.3 Applying Concepts

42. **Sports** Many major league baseball parks have a symmetric design as shown in the figure at the right. One decision that the designer must make is the shape of the outfield. One possible design uses the function

$$f(x) = k + (400 - k)\sqrt{1 - \frac{x^2}{a^2}}$$

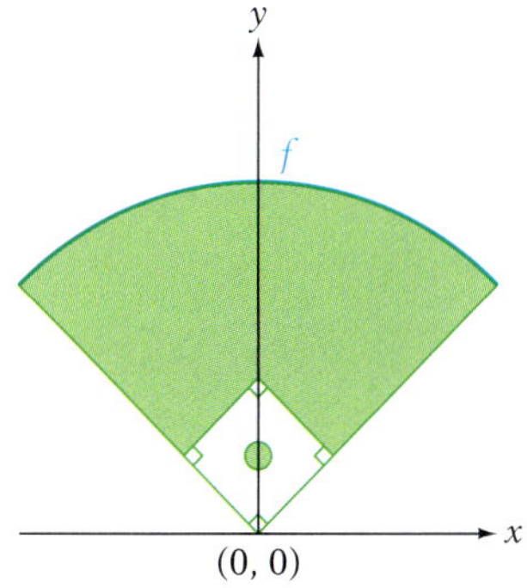

to determine the shape of the outfield.

a. Graph this equation for $k = 0$, $a = 287$, and $-240 \le x \le 240$.

b. What is the maximum value of this function for the given interval?

c. The equation of the right-field foul line is $y = x$. Where does the foul line intersect the graph of f? That is, find the point on the graph of f for which $y = x$.

d. If the units on the axes are feet, what is the distance from home plate to the base of the right-field wall?

43. **Currency** According to the Bureau of Engraving and Printing, the average life span of different denominations of currency is as shown in the table at the right. The function that approximately models these data is $f(x) = 1.38x^{\frac{2}{5}}$, where x is the denomination of the bill and $f(x)$ is the average life span in years.

a. In the context of the problem, what is the domain of the function $f(x) = 1.38x^{\frac{2}{5}}$? Why is 0 not in the domain? Why are negative numbers not in the domain?

b. If 2 were in the domain of the function, what life span would the model predict for the \$2 bill? Round to the nearest tenth. Do you think this estimate is reasonable?

Currency	*Average Life Span*
\$1 bill	1.8 years
\$5 bill	2 years
\$10 bill	3 years
\$20 bill	4 years
\$50 bill	9 years
\$100 bill	9 years

44. Writing the fraction $\frac{2}{4}$ in lowest terms as $\frac{1}{2}$, it appears that $(x^2)^{\frac{1}{4}} = x^{\frac{1}{2}}$. Using $-10 \le x \le 10$, graph $f(x) = (x^2)^{\frac{1}{4}}$ and then graph $g(x) = x^{\frac{1}{2}}$. Are the graphs the same? If they are, try graphing $g(x) = x^{\frac{1}{2}}$ first and then $f(x) = (x^2)^{\frac{1}{4}}$. Explain why the graphs are not the same. In your explanation, include why $x^{\frac{2}{4}} = x^{\frac{1}{2}}$ is not always a true statement.

8.4 Solving Equations Containing Radical Expressions

1 Solve equations containing one or more radical expressions

VIDEO & DVD · CD TUTOR · WEB · SSM

An equation that contains a variable expression in a radicand is a **radical equation.**

$$\left.\begin{aligned} \sqrt[3]{2x-5} + x &= 7 \\ \sqrt{x+1} - \sqrt{x} &= 4 \end{aligned}\right\} \text{Radical equations}$$

Take Note

You might assume that the converse of this is also true. That is, if $a^2 = b^2$, then $a = b$. But this is not the case. For example, if $a = -4$ and $b = 4$, then $(-4)^2 = 4^2$ but $-4 \ne 4$.

The following property is used to solve a radical equation.

Property of Raising Each Side of an Equation to a Power

If two numbers are equal, then the same powers of the numbers are equal.

If $a = b$, then $a^n = b^n$.

Solve: $\sqrt{x-2} - 6 = 0$

Rewrite the equation with the radical on one side of the equation and the constant on the other side.

$$\sqrt{x-2} - 6 = 0$$
$$\sqrt{x-2} = 6$$

Square each side of the equation.

$$\left(\sqrt{x-2}\right)^2 = 6^2$$

Solve the resulting equation.

$$x - 2 = 36$$
$$x = 38$$

Check the solution.

Check: $\sqrt{x-2} - 6 = 0$

$\sqrt{38-2} - 6$	0
$\sqrt{36} - 6$	0
$6 - 6$	0

$$0 = 0$$

38 checks as a solution.
The solution is 38.

Example 1 Solve. **A.** $\sqrt{3x-2} - 8 = -3$ **B.** $\sqrt[3]{3x-1} = -4$

Solution **A.** $\sqrt{3x-2} - 8 = -3$

$$\sqrt{3x-2} = 5$$

• Rewrite the equation so that the radical is alone on one side of the equation.

$$\left(\sqrt{3x-2}\right)^2 = 5^2$$

• Square each side of the equation.

$$3x - 2 = 25$$
$$3x = 27$$
$$x = 9$$

• Solve the resulting equation.

Check: $\sqrt{3x-2} - 8 = -3$

• Check the solution.

$\sqrt{3 \cdot 9 - 2} - 8$	-3
$\sqrt{27-2} - 8$	-3
$\sqrt{25} - 8$	-3
$5 - 8$	-3

$$-3 = -3$$

The solution is 9.

B. $\sqrt[3]{3x-1} = -4$

$$\left(\sqrt[3]{3x-1}\right)^3 = (-4)^3$$

• Cube each side of the equation.

$$3x - 1 = -64$$
$$3x = -63$$
$$x = -21$$

• Solve the resulting equation.

Check: $\sqrt[3]{3x-1} = -4$

• Check the solution.

$\sqrt[3]{3(-21) - 1}$	-4
$\sqrt[3]{-63-1}$	-4
$\sqrt[3]{-64}$	-4

$$-4 = -4$$

The solution is -21.

Problem 1 Solve. **A.** $\sqrt{4x + 5} - 12 = -5$ **B.** $\sqrt[4]{x - 8} = 3$

Solution See page S27.

> **Take Note**
> Reread the previous Take Note. The fact that the converse of the Property of Raising Each Side of an Equation to a Power is *not* true is what creates the potential for extraneous solutions.

When you raise both sides of an equation to an even power, the resulting equation may have a solution that is not a solution of the original equation. Therefore, it is necessary to check the solution of a radical equation.

Example 2 Solve. **A.** $x + 2\sqrt{x - 1} = 9$ **B.** $\sqrt{x + 7} = \sqrt{x} + 1$

Solution **A.** $x + 2\sqrt{x - 1} = 9$

$2\sqrt{x - 1} = 9 - x$ • Rewrite the equation with the radical on one side of the equation.

$(2\sqrt{x - 1})^2 = (9 - x)^2$ • Square each side of the equation.

$4(x - 1) = 81 - 18x + x^2$

$4x - 4 = 81 - 18x + x^2$

$0 = x^2 - 22x + 85$ • Write the quadratic equation in standard form.

$0 = (x - 5)(x - 17)$ • Factor.

$x - 5 = 0 \qquad x - 17 = 0$ • Use the Principle of Zero Products.

$x = 5 \qquad x = 17$

Check:

$x + 2\sqrt{x - 1} = 9$	
$5 + 2\sqrt{5 - 1}$	9
$5 + 2\sqrt{4}$	9
$5 + 2 \cdot 2$	9
$5 + 4$	9
$9 = 9$	

$x + 2\sqrt{x - 1} = 9$	
$17 + 2\sqrt{17 - 1}$	9
$17 + 2\sqrt{16}$	9
$17 + 2 \cdot 4$	9
$17 + 8$	9
$25 \neq 9$	

The solution is 5.

> **Take Note**
> You must check the proposed solutions to radical equations. The proposed solutions of the equation on the right were 5 and 17. However, 17 did not check as a solution.

17 does not check as a solution. It is called an **extraneous** solution of the equation.

B. $\sqrt{x + 7} = \sqrt{x} + 1$ • A radical appears on each side of the equation.

$(\sqrt{x + 7})^2 = (\sqrt{x} + 1)^2$ • Square each side of the equation.

$x + 7 = x + 2\sqrt{x} + 1$

$6 = 2\sqrt{x}$ • Simplify the resulting equation.

$3 = \sqrt{x}$ • The equation contains a radical.

$3^2 = (\sqrt{x})^2$ • Square each side of the equation.

$9 = x$

Check: • Check the solution.

$\sqrt{x + 7} = \sqrt{x} + 1$	
$\sqrt{9 + 7}$	$\sqrt{9} + 1$
$\sqrt{16}$	$3 + 1$
$4 = 4$	

The solution is 9.

> **Take Note**
> Note that
> $(\sqrt{x} + 1)^2$
> $= (\sqrt{x} + 1)(\sqrt{x} + 1)$
> $= x + 2\sqrt{x} + 1$

Problem 2 Solve. **A.** $x + 3\sqrt{x + 2} = 8$ **B.** $\sqrt{x + 5} = 5 - \sqrt{x}$

Solution See page S27.

2 Application problems

A right triangle contains one 90° angle. The side opposite the 90° angle is called the **hypotenuse.** The other two sides are called **legs.**

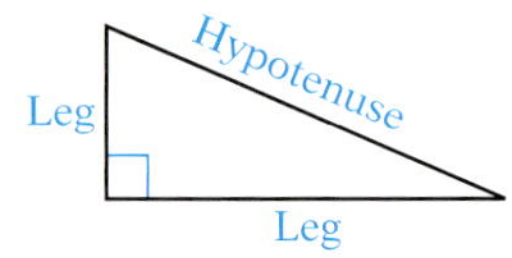

Pythagoras, a Greek mathematician, is credited with the discovery that the square of the hypotenuse of a right triangle is equal to the sum of the squares of the two legs. This is called the Pythagorean Theorem.

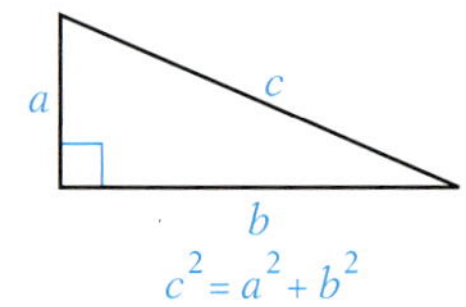

Pythagorean Theorem

The square of the hypotenuse, c, of a right triangle is equal to the sum of the squares of the two legs, a and b.

$$c^2 = a^2 + b^2$$

Example 3 A ladder 20 ft long is leaning against a building. How high on the building will the ladder reach when the bottom of the ladder is 8 ft from the building? Round to the nearest tenth.

Strategy To find the distance, use the Pythagorean Theorem. The hypotenuse is the length of the ladder. One leg is the distance from the bottom of the ladder to the base of the building. The distance along the building from the ground to the top of the ladder is the unknown leg.

Solution

$$\begin{aligned} c^2 &= a^2 + b^2 \\ 20^2 &= 8^2 + b^2 \\ 400 &= 64 + b^2 \\ 336 &= b^2 \\ (336)^{\frac{1}{2}} &= (b^2)^{\frac{1}{2}} \\ \sqrt{336} &= b \\ 18.3 &\approx b \end{aligned}$$

The distance is 18.3 ft.

Problem 3 Find the diagonal of a rectangle that is 6 cm in length and 3 cm in width. Round to the nearest tenth.

Solution See page S28.

8.4 Concept Review

Determine whether the following statements are always true, sometimes true, or never true.

1. If $a^n = b^n$, then $a = b$.

2. When you raise both sides of an equation to an even power, the resulting equation has a solution that is not a solution of the original equation.

3. If two sides of a triangle are known, the third side can be found by using the Pythagorean Theorem.

4. The solution of $\sqrt[3]{-x} = 8$ is -512.

8.4 Exercises

1 **1.** What is the first step when solving $\sqrt{x} + 3 = 9$?

2. Why is it necessary to check the proposed solutions of a radical equation?

Solve.

3. $\sqrt{x} = 5$

4. $\sqrt{y} = 2$

5. $\sqrt[3]{a} = 3$

6. $\sqrt[3]{y} = 5$

7. $\sqrt{3x} = 12$

8. $\sqrt{5x} = 10$

9. $\sqrt[3]{4x} = -2$

10. $\sqrt[3]{6x} = -3$

11. $\sqrt{2x} = -4$

12. $\sqrt{5x} = -5$

13. $\sqrt{3x - 2} = 5$

14. $\sqrt{5x - 4} = 9$

15. $\sqrt{3 - 2x} = 7$

16. $\sqrt{9 - 4x} = 4$

17. $7 = \sqrt{1 - 3x}$

18. $6 = \sqrt{8 - 7x}$

19. $\sqrt[3]{4x - 1} = 2$

20. $\sqrt[3]{5x + 2} = 3$

21. $\sqrt[3]{1 - 2x} = -3$

22. $\sqrt[3]{3 - 2x} = -2$

23. $\sqrt[3]{9x + 1} = 4$

24. $\sqrt{3x + 9} - 12 = 0$

25. $\sqrt{4x - 3} - 5 = 0$

26. $\sqrt{x - 2} = 4$

27. $\sqrt[3]{x - 3} + 5 = 0$

28. $\sqrt[3]{x - 2} = 3$

29. $\sqrt[3]{2x - 6} = 4$

30. $\sqrt[4]{4x + 1} = 2$

31. $\sqrt[4]{2x - 9} = 3$

32. $\sqrt{2x - 3} - 2 = 1$

33. $\sqrt{3x - 5} - 5 = 3$

34. $\sqrt[3]{2x - 3} + 5 = 2$

35. $\sqrt[3]{x - 4} + 7 = 5$

36. $\sqrt{5x - 16} + 1 = 4$

37. $\sqrt{3x - 5} - 2 = 3$

38. $\sqrt{2x - 1} - 8 = -5$

39. $\sqrt{7x + 2} - 10 = -7$

40. $\sqrt[3]{4x - 3} - 2 = 3$

41. $\sqrt[3]{1 - 3x} + 5 = 3$

42. $1 - \sqrt{4x + 3} = -5$

43. $7 - \sqrt{3x + 1} = -1$

44. $\sqrt{x + 1} = 2 - \sqrt{x}$

45. $\sqrt{2x + 4} = 3 - \sqrt{2x}$

46. $\sqrt{x^2 + 3x - 2} - x = 1$

47. $\sqrt{x^2 - 4x - 1} + 3 = x$

48. $\sqrt{x^2 - 3x - 1} = 3$

49. $\sqrt{x^2 - 2x + 1} = 3$

50. $\sqrt{2x + 5} - \sqrt{3x - 2} = 1$

51. $\sqrt{4x + 1} - \sqrt{2x + 4} = 1$

52. $\sqrt{5x - 1} - \sqrt{3x - 2} = 1$

53. $\sqrt{5x + 4} - \sqrt{3x + 1} = 1$

54. $\sqrt[4]{x^2 + 2x + 8} - 2 = 0$

55. $\sqrt[4]{x^2 + x - 1} - 1 = 0$

56. $4\sqrt{x + 1} - x = 1$

57. $3\sqrt{x - 2} + 2 = x$

58. $x + 3\sqrt{x - 2} = 12$

59. $x + 2\sqrt{x + 1} = 7$

2 **60.** **Sports** The equation $s = 16.97\sqrt[9]{n}$ can be used to predict the maximum speed, s (in feet per second), of n rowers on a scull. Find, to the nearest whole number, how many rowers are needed to travel at 20 ft/s. Does doubling the number of rowers double the maximum speed of the scull?

61. **Meteorology** The sustained wind velocity, v (in meters per second), in a hurricane is given by $v = 6.3\sqrt{1013 - p}$, where p is the air pressure in millibars (mb). If the velocity of the wind in a hurricane is 64 m/s, what is the air pressure? Round to the nearest tenth. What happens to wind speed in a hurricane as air pressure decreases?

62. **Astronomy** The time, T (in days), that it takes a planet to revolve around the sun can be approximated by the equation $T = 0.407\sqrt{d^3}$, where d is the mean distance of the planet from the sun in millions of miles. It takes Venus approximately 226 days to complete one revolution of the sun. To the nearest million miles, what is the mean distance of Venus from the sun?

63. **Astronomy** The time, T (in days), that it takes a moon of Saturn to revolve around Saturn can be approximated by the equation $T = 0.373\sqrt{d^3}$, where d is the mean distance of the moon from Saturn in units of 100,000 km. It takes the moon Tethys approximately 1.89 days to complete one revolution of Saturn. To the nearest 1000 km, what is the mean distance of Tethys from Saturn?

64. **Construction** The maximum velocity of a roller coaster depends on the vertical drop from the top of the highest hill to the bottom of that hill. The formula $v = 8\sqrt{h}$ gives the relationship between maximum velocity, v (in feet per second), and height, h (in feet). The maximum velocity of the Magnum XL-200 roller coaster in Sandusky, Ohio, is approximately 114 ft/s. How tall, to the nearest foot, is the highest hill for this roller coaster?

65. **Health Science** The number of calories an animal uses per day (called the metabolic rate of the animal) can be approximated by $M = 126.4\sqrt[4]{W^3}$, where M is the metabolic rate and W is the weight of the animal in pounds. Find, to the nearest hundred pounds, the weight of an elephant whose metabolic rate is 60,000 calories per day.

66. **Construction** A 12-foot ladder is leaning against a building. How high on the building will the ladder reach when the bottom of the ladder is 4 ft from the building? Round to the nearest tenth.

67. **Construction** A 26-foot ladder is leaning against a building. How far is the bottom of the ladder from the wall when the ladder reaches a height of 24 ft on the building?

68. **Uniform Motion** Two joggers start from the same point and travel in the directions shown in the figure at the right. One jogger is traveling 8 ft/s, and the second jogger is traveling 9 ft/s. What is the distance between the joggers after 1 min? Round to the nearest tenth.

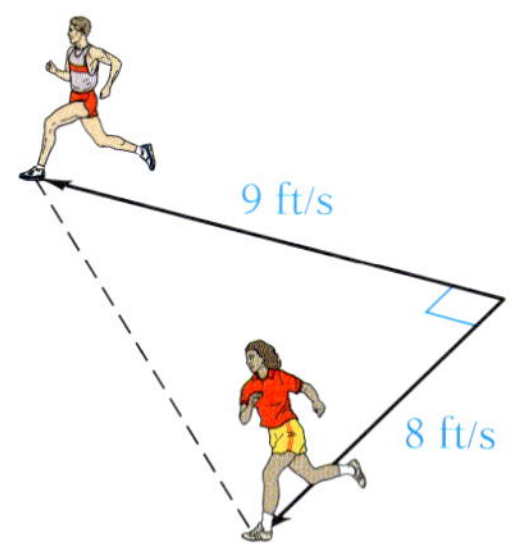

69. **Uniform Motion** Two joggers leave the same point at the same time, one jogging east at 3 m/s and the other jogging south at 3.5 m/s. What is the distance between the joggers after 3 min? Round to the nearest hundredth.

70. **Geometry** The diagonal of a rectangle is a line drawn from one vertex to the opposite vertex. Find the length of the diagonal in the rectangle shown at the right. Round to the nearest tenth.

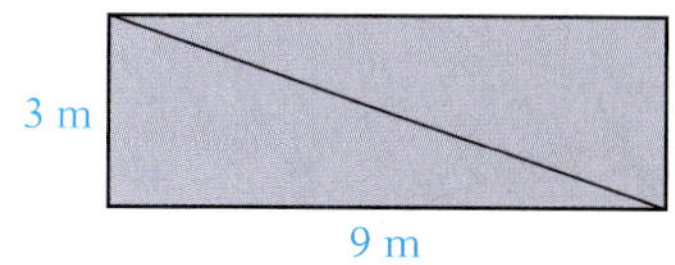

71. **Geometry** The diagonal of a rectangle is a line drawn from one vertex to the opposite vertex. Find the length of the diagonal in the rectangle shown at the right. Round to the nearest tenth.

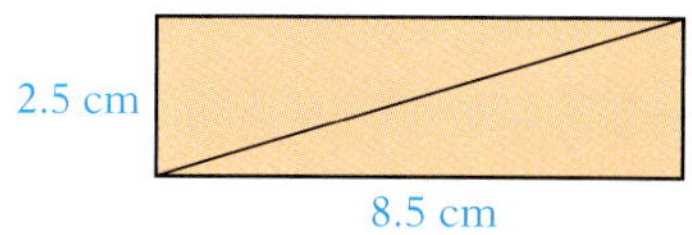

72. **Oceanography** How far would a submarine periscope have to be above the water to locate a ship 3.2 mi away? The equation for the distance in miles that the lookout can see is $d = \sqrt{1.5h}$, where h is the height in feet above the surface of the water. Round to the nearest hundredth.

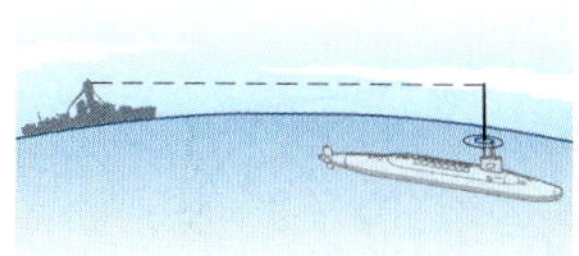

73. **Oceanography** How far would a submarine periscope have to be above the water to locate a ship 3.5 mi away? The equation for the distance in miles that the lookout can see is $d = \sqrt{1.5h}$, where h is in height in feet above the surface of the water. Round to the nearest hundredth.

74. **Physics** An object is dropped from a bridge. Find the distance the object has fallen when the speed reaches 80 ft/s. Use the equation $v = 8\sqrt{d}$, where v is the speed of the object and d is the distance.

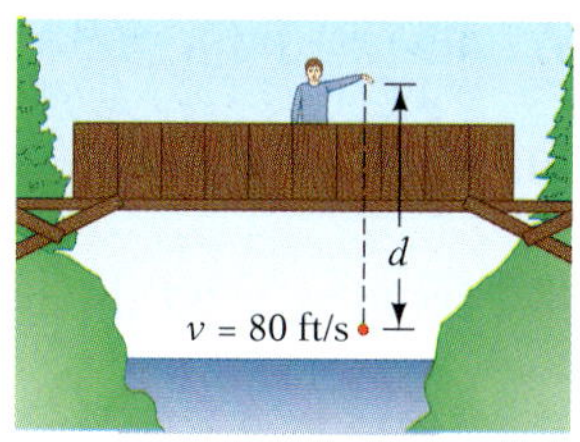

75. **Physics** An object is dropped from a high building. Find the distance the object has fallen when the speed reaches 120 ft/s. Use the equation $v = 8\sqrt{d}$, where v is the speed of the object and d is the distance.

76. **Automotive Technology** Find the distance required for a car to reach a velocity of 60 m/s when the acceleration is 10 m/s^2. Use the equation $v = \sqrt{2as}$, where v is the velocity, a is the acceleration, and s is the distance.

77. **Automotive Technology** Find the distance required for a car to reach a velocity of 48 ft/s when the acceleration is 12 ft/s^2. Use the equation $v = \sqrt{2as}$, where v is the velocity, a is the acceleration, and s is the distance.

78. **Clocks** Find the length of a pendulum that makes one swing in 3 s. The equation for the time of one swing of a pendulum is $T = 2\pi\sqrt{\frac{L}{32}}$, where T is the time in seconds and L is the length in feet. Round to the nearest hundredth.

79. **Clocks** Find the length of a pendulum that makes one swing in 2.4 s. The equation for the time of one swing of a pendulum is $T = 2\pi\sqrt{\frac{L}{32}}$, where T is the time in seconds and L is the length in feet. Round to the nearest hundredth.

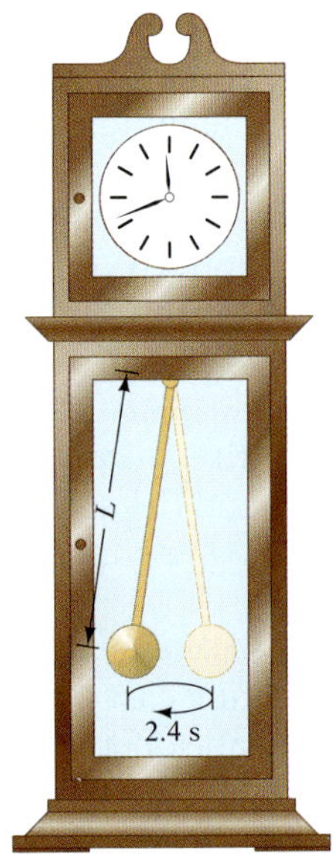

8.4 Applying Concepts

Solve.

80. $x^{\frac{3}{4}} = 8$ **81.** $x^{\frac{2}{3}} = 9$ **82.** $x^{\frac{5}{4}} = 32$

Solve the formula for the given variable.

83. $v = \sqrt{64d}$; d **84.** $a^2 + b^2 = c^2$; a

85. $V = \pi r^2 h$; r **86.** $V = \frac{4}{3}\pi r^3$; r

Solve.

87. $\sqrt{3x - 2} = \sqrt{2x - 3} + \sqrt{x - 1}$ **88.** $\sqrt{2x + 3} + \sqrt{x + 2} = \sqrt{x + 5}$

89. **Geometry** A box has a base that measures 4 in. by 6 in. The height of the box is 3 in. Find the greatest distance between two corners. Round to the nearest hundredth.

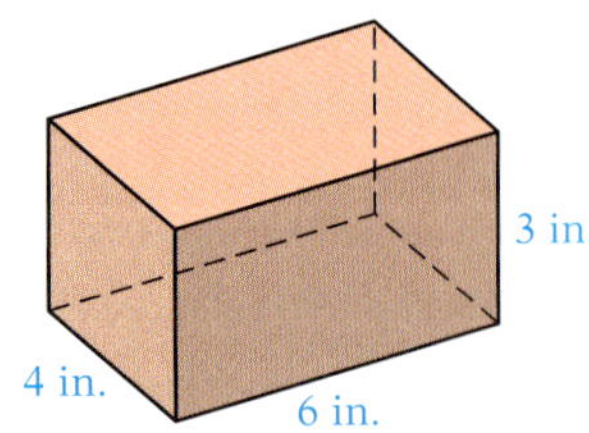

90. **Number Problem** Find three odd integers a, b, and c such that $a^2 + b^2 = c^2$.

91. **Geometry** Find the length of the side labeled x.

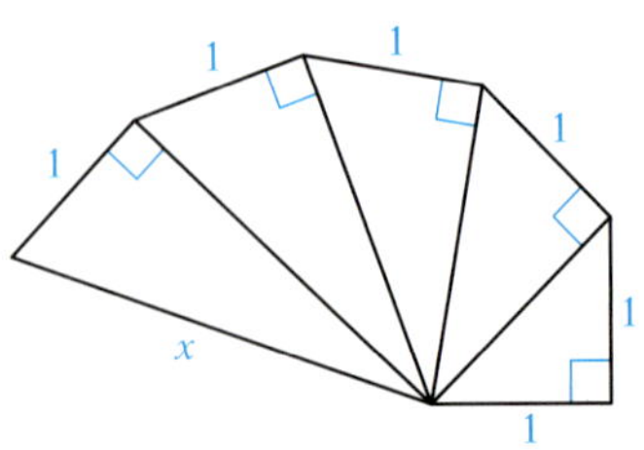

92. If a and b are nonnegative real numbers, is the equation always true, sometimes true, or never true?

$$\sqrt{a^2 + b^2} = a + b$$

Write a paragraph that supports your answer.

93. If a and b are both positive real numbers and $a > b$, is a^b less than, equal to, or greater than b^a? Does your answer depend on a and b?

8.5 Complex Numbers

1 Simplify complex numbers

VIDEO & DVD CD TUTOR WEB SSM

The radical expression $\sqrt{-4}$ is not a real number, because there is no real number whose square is -4. However, the solution of an algebraic equation is sometimes the square root of a negative number.

For example, the equation $x^2 + 1 = 0$ does not have a real number solution, because there is no real number whose square is a negative number.

$$x^2 + 1 = 0$$
$$x^2 = -1$$

During the late 17th century, a new number, called an **imaginary number,** was defined so that a negative number would have a square root. The letter i was chosen to represent the number whose square is -1.

$$\boldsymbol{i^2 = -1}$$

An imaginary number is defined in terms of i.

Point of Interest

The first written occurrence of an imaginary number was in a book published in 1545 by Hieronimo Cardano, where he wrote (in our modern notation) $5 + \sqrt{-15}$. He went on to say that the number "is as refined as it is useless." It was not until the 20th century that applications of complex numbers were found.

Definition of $\sqrt{-a}$

If a is a positive real number, then the principal square root of negative a is the imaginary number $i\sqrt{a}$.

$$\sqrt{-a} = i\sqrt{a}$$

Here are some examples.

$$\sqrt{-16} = i\sqrt{16} = 4i$$
$$\sqrt{-12} = i\sqrt{12} = 2i\sqrt{3}$$
$$\sqrt{-21} = i\sqrt{21}$$
$$\sqrt{-1} = i\sqrt{1} = i$$

It is customary to write i in front of a radical to avoid confusing $\sqrt{a}\,i$ with $\sqrt{ai}$.

The real numbers and imaginary numbers make up the complex numbers.

Definition of a Complex Number

A **complex number** is a number of the form $\boldsymbol{a + bi}$, where a and b are real numbers and $i = \sqrt{-1}$. The number a is the **real part** of $a + bi$, and b is the **imaginary part.**

Take Note

The *imaginary part* of a complex number is a real number. As another example, the imaginary part of $6 - 8i$ is -8.

Examples of complex numbers are shown at the right.

Real part — Imaginary part

$a + bi$
$3 + 2i$
$8 - 10i$

Complex numbers $a + bi$:
- Real numbers $a + 0i$
- Imaginary numbers $0 + bi$

A **real number** is a complex number in which $b = 0$.

An **imaginary number** is a complex number in which $a = 0$.

Example 1 Simplify: $\sqrt{-80}$

Solution $\sqrt{-80} = i\sqrt{80} = i\sqrt{16 \cdot 5} = 4i\sqrt{5}$

Problem 1 Simplify: $\sqrt{-45}$

Solution See page S28.

Simplify: $\sqrt{20} - \sqrt{-50}$

Write the complex number in the form $a + bi$.

$\sqrt{20} - \sqrt{-50} = \sqrt{20} - i\sqrt{50}$

Use the Product Property of Radicals to simplify each radical.

$= \sqrt{4 \cdot 5} - i\sqrt{25 \cdot 2}$
$= 2\sqrt{5} - 5i\sqrt{2}$

Example 2 Simplify: $\sqrt{25} + \sqrt{-40}$

Solution $\sqrt{25} + \sqrt{-40} = \sqrt{25} + i\sqrt{40} = \sqrt{25} + i\sqrt{4 \cdot 10}$
$= 5 + 2i\sqrt{10}$

Problem 2 Simplify: $\sqrt{98} - \sqrt{-60}$

Solution See page S28.

2 Add and subtract complex numbers

VIDEO & DVD CD TUTOR WEB SSM

Addition and Subtraction of Complex Numbers

To add two complex numbers, add the real parts and add the imaginary parts. To subtract two complex numbers, subtract the real parts and subtract the imaginary parts.

$$(a + bi) + (c + di) = (a + c) + (b + d)i$$

$$(a + bi) - (c + di) = (a - c) + (b - d)i$$

Example 3 Add: $(3 + 2i) + (6 - 5i)$

Solution $(3 + 2i) + (6 - 5i)$
$= (3 + 6) + (2 - 5)i$ • Add the real parts and add the imaginary parts.
$= 9 - 3i$

Problem 3 Subtract: $(-4 + 2i) - (6 - 8i)$

Solution See page S28.

Add: $(3 + \sqrt{-12}) + (7 - \sqrt{-27})$

Write each complex number in the form $a + bi$. $(3 + \sqrt{-12}) + (7 - \sqrt{-27})$
$= (3 + i\sqrt{12}) + (7 - i\sqrt{27})$

Use the Product Property of Radicals to simplify each radical. $= (3 + i\sqrt{4 \cdot 3}) + (7 - i\sqrt{9 \cdot 3})$
$= (3 + 2i\sqrt{3}) + (7 - 3i\sqrt{3})$

Add the complex numbers. $= 10 - i\sqrt{3}$

Example 4 Subtract: $(9 - \sqrt{-8}) - (5 + \sqrt{-32})$

Solution $(9 - \sqrt{-8}) - (5 + \sqrt{-32})$
$= (9 - i\sqrt{8}) - (5 + i\sqrt{32})$ • Write each complex number in the form $a + bi$.
$= (9 - i\sqrt{4 \cdot 2}) - (5 + i\sqrt{16 \cdot 2})$ • Simplify each radical.
$= (9 - 2i\sqrt{2}) - (5 + 4i\sqrt{2})$
$= 4 - 6i\sqrt{2}$

Problem 4 Subtract: $(16 - \sqrt{-45}) - (3 + \sqrt{-20})$

Solution See page S28.

3 Multiply complex numbers

VIDEO & DVD CD TUTOR WEB SSM

When multiplying complex numbers, the term i^2 is frequently a part of the product. Recall that $i^2 = -1$.

Multiply: $2i \cdot 3i$

Multiply the imaginary numbers. $2i \cdot 3i = 6i^2$

Replace i^2 by -1. Then simplify. $= 6(-1) = -6$

When multiplying square roots of negative numbers, first rewrite the radical expressions using i.

Multiply: $\sqrt{-6} \cdot \sqrt{-24}$

Write each radical as the product of a real number and i. $\sqrt{-6} \cdot \sqrt{-24} = i\sqrt{6} \cdot i\sqrt{24}$

Multiply the imaginary numbers. $= i^2\sqrt{144}$

Replace i^2 by -1. $= -\sqrt{144}$

Simplify. $= -12$

Take Note

This example illustrates an important point. When working with a number that has a square root of a negative number, always rewrite the number with i before continuing.

Note from this example that it would have been incorrect to multiply the radicands of the two radical expressions. To illustrate,

$$\sqrt{-6} \cdot \sqrt{-24} = \sqrt{(-6)(-24)} = \sqrt{144} = 12, \textit{ not } -12$$

Multiply: $4i(3 - 2i)$

Use the Distributive Property to remove parentheses.	$4i(3 - 2i) = 12i - 8i^2$
Replace i^2 by -1.	$= 12i - 8(-1)$
Write the answer in the form $a + bi$.	$= 8 + 12i$

Example 5 Multiply: $\sqrt{-8}\left(\sqrt{6} - \sqrt{-2}\right)$

Solution

$\sqrt{-8}\left(\sqrt{6} - \sqrt{-2}\right)$

$= i\sqrt{8}\left(\sqrt{6} - i\sqrt{2}\right)$ • **Write each complex number in the form $a + bi$.**

$= i\sqrt{48} - i^2\sqrt{16}$ • **Use the Distributive Property.**

$= i\sqrt{16 \cdot 3} - (-1)\sqrt{16}$ • **Simplify each radical. Replace i^2 by -1.**

$= 4i\sqrt{3} + 4$

$= 4 + 4i\sqrt{3}$ • **Write the answer in the form $a + bi$.**

Problem 5 Multiply: $\sqrt{-3}\left(\sqrt{27} - \sqrt{-6}\right)$

Solution See page S28.

The product of two complex numbers can be found by using the FOIL Method. For example,

$$\begin{aligned}(2 + 4i)(3 - 5i) &= 6 - 10i + 12i - 20i^2 \\ &= 6 + 2i - 20i^2 \\ &= 6 + 2i - 20(-1) \\ &= 26 + 2i\end{aligned}$$

The conjugate of $a + bi$ is $a - bi$.

The product of conjugates of the form $(a + bi)(a - bi)$ is $a^2 + b^2$.

$$\mathbf{(a + bi)(a - bi)} = a^2 - b^2i^2 = a^2 - b^2(-1) = a^2 + b^2$$

For example, $(2 + 3i)(2 - 3i) = 2^2 + 3^2 = 4 + 9 = 13$.

Note that the product of a complex number and its conjugate is a real number.

Example 6 Multiply.

A. $(3 - 4i)(2 + 5i)$ **B.** $\left(\frac{9}{10} + \frac{3}{10}i\right)\left(1 - \frac{1}{3}i\right)$

C. $(4 + 5i)(4 - 5i)$ **D.** $(6 + i)^2$

Solution

A. $(3 - 4i)(2 + 5i)$

$= 6 + 15i - 8i - 20i^2$ • **Use the FOIL method.**

$= 6 + 7i - 20i^2$ • **Combine like terms.**

$= 6 + 7i - 20(-1)$ • **Replace i^2 by -1.**

$= 26 + 7i$ • **Write the answer in the form $a + bi$.**

B. $\left(\frac{9}{10}+\frac{3}{10}i\right)\left(1-\frac{1}{3}i\right)$

$= \frac{9}{10} - \frac{3}{10}i + \frac{3}{10}i - \frac{1}{10}i^2$ • Use the FOIL method.

$= \frac{9}{10} - \frac{1}{10}i^2$ • Combine like terms.

$= \frac{9}{10} - \frac{1}{10}(-1)$ • Replace i^2 by -1.

$= \frac{9}{10} + \frac{1}{10} = 1$ • Simplify.

C. $(4+5i)(4-5i) = 4^2 + 5^2 = 16 + 25 = 41$ • The product of conjugates of the form $(a+bi)(a-bi)$ is a^2+b^2.

D. $(6+i)^2 = 36 + 12i + i^2 = 36 + 12i + (-1) = 35 + 12i$ • $(6+i)^2 = (6+i)(6+i)$

Problem 6 Multiply.

A. $(4-3i)(2-i)$ **B.** $(3-i)\left(\frac{3}{10}+\frac{1}{10}i\right)$

C. $(3+6i)(3-6i)$ **D.** $(1+5i)^2$

Solution See page S28.

4 Divide complex numbers

VIDEO & DVD CD TUTOR WEB SSM

A fraction containing one or more complex numbers is in simplest form when no imaginary number remains in the denominator.

Simplify: $\frac{4-5i}{2i}$

Multiply the expression by 1 in the form $\frac{i}{i}$. $\frac{4-5i}{2i} = \frac{4-5i}{2i}\cdot\frac{i}{i}$

$= \frac{4i-5i^2}{2i^2}$

Replace i^2 by -1. $= \frac{4i-5(-1)}{2(-1)}$

Simplify. $= \frac{5+4i}{-2}$

Write the answer in the form $a+bi$. $= -\frac{5}{2} - 2i$

Example 7 Simplify: $\frac{5+4i}{3i}$

Solution $\frac{5+4i}{3i} = \frac{5+4i}{3i}\cdot\frac{i}{i} = \frac{5i+4i^2}{3i^2} = \frac{5i+4(-1)}{3(-1)} = \frac{-4+5i}{-3}$

$= \frac{4}{3} - \frac{5}{3}i$

Problem 7 Simplify: $\dfrac{2 - 3i}{4i}$

Solution See page S28.

Simplify: $\dfrac{3 + 2i}{1 + i}$

To simplify a fraction that has a complex number in the denominator, multiply the numerator and denominator by the conjugate of the complex number. The conjugate of $1 + i$ is $1 - i$.

$$\frac{3 + 2i}{1 + i} = \frac{(3 + 2i)}{(1 + i)} \cdot \frac{(1 - i)}{(1 - i)}$$
$$= \frac{3 - 3i + 2i - 2i^2}{1^2 + 1^2}$$
$$= \frac{3 - i - 2(-1)}{2}$$
$$= \frac{5 - i}{2} = \frac{5}{2} - \frac{1}{2}i$$

Example 8 Simplify: $\dfrac{5 - 3i}{4 + 2i}$

Solution

$$\frac{5 - 3i}{4 + 2i} = \frac{(5 - 3i)}{(4 + 2i)} \cdot \frac{(4 - 2i)}{(4 - 2i)} = \frac{20 - 10i - 12i + 6i^2}{4^2 + 2^2}$$
$$= \frac{20 - 22i + 6(-1)}{20}$$
$$= \frac{14 - 22i}{20}$$
$$= \frac{7 - 11i}{10}$$
$$= \frac{7}{10} - \frac{11}{10}i$$

Problem 8 Simplify: $\dfrac{2 + 5i}{3 - 2i}$

Solution See page S28.

8.5 Concept Review

Determine whether the following statements are always true, sometimes true, or never true.

1. The product of two imaginary numbers is a real number.

2. The sum of two complex numbers is a real number.

3. A complex number is a number of the form $a + bi$, where a and b are real numbers and $\sqrt{-1} = i$.

4. The product of two complex numbers is a real number.

8.5 Exercises

1 1. What is an imaginary number? What is a complex number?

2. Are all real numbers also complex numbers? Are all complex numbers also real numbers?

Simplify.

3. $\sqrt{-4}$
4. $\sqrt{-64}$
5. $\sqrt{-98}$
6. $\sqrt{-72}$
7. $\sqrt{-27}$
8. $\sqrt{-75}$
9. $\sqrt{16} + \sqrt{-4}$
10. $\sqrt{25} + \sqrt{-9}$
11. $\sqrt{12} - \sqrt{-18}$
12. $\sqrt{60} - \sqrt{-48}$
13. $\sqrt{160} - \sqrt{-147}$
14. $\sqrt{96} - \sqrt{-125}$
15. $\sqrt{-4a^2}$
16. $\sqrt{-16b^6}$
17. $\sqrt{-49x^{12}}$
18. $\sqrt{-32x^3y^2}$
19. $\sqrt{-144a^3b^5}$
20. $\sqrt{-18a^{10}b^9}$
21. $\sqrt{4a} + \sqrt{-12a^2}$
22. $\sqrt{25b} - \sqrt{-48b^2}$
23. $\sqrt{18b^5} - \sqrt{-27b^3}$
24. $\sqrt{a^5b^2} - \sqrt{-a^5b^2}$
25. $\sqrt{-50x^3y^3} + x\sqrt{25x^4y^3}$
26. $\sqrt{-121xy} + \sqrt{60x^2y^2}$
27. $\sqrt{-49a^5b^2} - ab\sqrt{-25a^3}$
28. $\sqrt{-16x^2y} - x\sqrt{-49y}$
29. $\sqrt{12a^3} + \sqrt{-27b^3}$

2 Add or subtract.

30. $(2 + 4i) + (6 - 5i)$
31. $(6 - 9i) + (4 + 2i)$
32. $(-2 - 4i) - (6 - 8i)$
33. $(3 - 5i) + (8 - 2i)$
34. $(8 - \sqrt{-4}) - (2 + \sqrt{-16})$
35. $(5 - \sqrt{-25}) - (11 - \sqrt{-36})$
36. $(12 - \sqrt{-50}) + (7 - \sqrt{-8})$
37. $(5 - \sqrt{-12}) - (9 + \sqrt{-108})$
38. $(\sqrt{8} + \sqrt{-18}) + (\sqrt{32} - \sqrt{-72})$
39. $(\sqrt{40} - \sqrt{-98}) - (\sqrt{90} + \sqrt{-32})$
40. $(5 - 3i) + 2i$
41. $(6 - 8i) + 4i$
42. $(7 + 2i) + (-7 - 2i)$
43. $(8 - 3i) + (-8 + 3i)$
44. $(9 + 4i) + 6$
45. $(4 + 6i) + 7$

3 Multiply.

46. $(7i)(-9i)$
47. $(-6i)(-4i)$

48. $\sqrt{-2}\,\sqrt{-8}$

49. $\sqrt{-5}\,\sqrt{-45}$

50. $\sqrt{-3}\,\sqrt{-6}$

51. $\sqrt{-5}\,\sqrt{-10}$

52. $2i(6 + 2i)$

53. $-3i(4 - 5i)$

54. $\sqrt{-2}\left(\sqrt{8} + \sqrt{-2}\right)$

55. $\sqrt{-3}\left(\sqrt{12} - \sqrt{-6}\right)$

56. $(5 - 2i)(3 + i)$

57. $(2 - 4i)(2 - i)$

58. $(6 + 5i)(3 + 2i)$

59. $(4 - 7i)(2 + 3i)$

60. $(1 - i)\left(\frac{1}{2} + \frac{1}{2}i\right)$

61. $\left(\frac{4}{5} - \frac{2}{5}i\right)\left(1 + \frac{1}{2}i\right)$

62. $\left(\frac{6}{5} + \frac{3}{5}i\right)\left(\frac{2}{3} - \frac{1}{3}i\right)$

63. $(2 - i)\left(\frac{2}{5} + \frac{1}{5}i\right)$

64. $(4 - 3i)(4 + 3i)$

65. $(8 - 5i)(8 + 5i)$

66. $(3 - i)(3 + i)$

67. $(7 - i)(7 + i)$

68. $(6 + i)^2$

69. $(4 - 3i)^2$

70. $(5 - 2i)^2$

71. $(-1 + i)^2$

4 Simplify.

72. $\frac{3}{i}$

73. $\frac{4}{5i}$

74. $\frac{2 - 3i}{-4i}$

75. $\frac{16 + 5i}{-3i}$

76. $\frac{4}{5 + i}$

77. $\frac{6}{5 + 2i}$

78. $\frac{2}{2 - i}$

79. $\frac{5}{4 - i}$

80. $\frac{1 - 3i}{3 + i}$

81. $\frac{2 + 12i}{5 + i}$

82. $\frac{\sqrt{-10}}{\sqrt{8} - \sqrt{-2}}$

83. $\frac{\sqrt{-2}}{\sqrt{12} - \sqrt{-8}}$

84. $\frac{2 - 3i}{3 + i}$

85. $\frac{3 + 5i}{1 - i}$

8.5 Applying Concepts

Note the pattern when successive powers of i are simplified.

$i^1 = i$ | $i^5 = i \cdot i^4 = i(1) = i$

$i^2 = -1$ | $i^6 = i^2 \cdot i^4 = -1$

$i^3 = i^2 \cdot i = -i$ | $i^7 = i^3 \cdot i^4 = -i$

$i^4 = i^2 \cdot i^2 = (-1)(-1) = 1$ | $i^8 = i^4 \cdot i^4 = 1$

86. When the exponent on i is a multiple of 4, the power equals ______.

Use the pattern above to simplify the power of i.

87. i^6

88. i^9

89. i^{57}

90. i^{65}

91. i^{-6}

92. i^{-34}

93. i^{-58}

94. i^{-180}

95. a. Is $3i$ a solution of $2x^2 + 18 = 0$?
b. Is $-3i$ a solution of $2x^2 + 18 = 0$?

96. a. Is $1 + 3i$ a solution of $x^2 - 2x - 10 = 0$?
b. Is $1 - 3i$ a solution of $x^2 - 2x - 10 = 0$?

97. Solve the following equations. Describe the solution by using the following terms: integer, rational number, irrational number, real number, and imaginary number. Note that more than one term may be used to describe the answer.

a. $x^2 + 3 = 7$

b. $x^2 + 1 = 0$

c. $\frac{5}{8}x = \frac{2}{3}$

d. $x^2 + 1 = 9$

e. $x^{\frac{3}{4}} = 8$

f. $\sqrt[3]{x} = -27$

Focus on Problem Solving

Another Look at Polya's Four-Step Process

Polya's four general steps to follow when attempting to solve a problem are to understand the problem, devise a plan, carry out the plan, and review the solution. In the process of devising a plan (Step 2), it may be appropriate to write a mathematical expression or an equation. We will illustrate this with the following problem.

Number the letters of the alphabet in sequence from 1 to 26. (See the list on the next page.) Find a word for which the product of the numerical values of the letters of the word equals 1,000,000. We will agree that a "word" is any sequence of letters that contains at least one vowel but is not necessarily in the dictionary.

Understand the Problem

Consider REZB. The product of the letters is $18 \cdot 5 \cdot 26 \cdot 2 = 4680$. This "word" is a sequence of letters with at least one vowel. However, the product of the numerical values of the letters is not 1,000,000. Thus this word does not solve our problem.

Devise a Plan

A = 1	N = 14
B = 2	O = 15
C = 3	P = 16
D = 4	Q = 17
E = 5	R = 18
F = 6	S = 19
G = 7	T = 20
H = 8	U = 21
I = 9	V = 22
J = 10	W = 23
K = 11	X = 24
L = 12	Y = 25
M = 13	Z = 26

Actually, we should have known that the product of the letters in REZB could not equal 1,000,000. The letter R has a factor of 9, and the letter Z has a factor of 13. Neither of these two numbers is a factor of 1,000,000. Consequently, R and Z cannot be letters in the word we are trying to find. This observation leads to an important observation: each of the letters that make up our word must be factors of 1,000,000. To find these letters, consider the prime factorization of 1,000,000.

$$1{,}000{,}000 = 2^6 \cdot 5^6$$

Looking at the prime factorization, we note that only letters that contain only 2 or 5 as factors are possible candidates. These letters are B, D, E, H, J, P, T, and Y. One additional point: Because 1 times any number is the number, A can be part of any word we construct.

Our task is now to construct a word from these letters such that the product is 1,000,000. From the prime factorization above, we must have 2 as a factor 6 times and 5 as a factor 6 times.

Carry Out the Plan

We must construct a word with the characteristics described in our plan. Here is a possibility:

THEBEYE

Review the Solution

You should multiply the values of all of the letters and verify that the product is 1,000,000. To ensure that you have an understanding of the problem, try to find other "words" that satisfy the condition of the problem.

Projects & Group Activities

Graphing Complex Numbers

That a negative number might be the solution of an equation was not readily accepted by mathematicians until the middle of the 16th century. Hieronimo Cardano was the first to take notice of these numbers in his book *Ars Magna* (*The Great Art*), which was published in 1545. However, he called these solutions fictitious. In that same book, in one case, he also considered a solution of an equation of the form $5 + \sqrt{-15}$, a number we now call a complex number.

Complex numbers were thought to have no real or geometric interpretation and therefore to be "imaginary." One mathematician called them "ghosts of real numbers." Then, in 1806, Jean-Robert Argand published a book that included a geometric description of complex numbers. (Actually, Caspar Wessel first proposed this idea in 1797. However, his work went unnoticed.)

Take Note

4 is the complex number $4 + 0i$, and $2i$ is the complex number $0 + 2i$.

An Argand diagram consists of a horizontal and a vertical axis. The horizontal axis is the *real axis;* the vertical axis is the *imaginary axis.* A complex number of the form $a + bi$ is shown as the point in the plane whose coordinates are (a, b). The graphs of $2 + 4i$, $-4 - 3i$, 4, which is $4 + 0i$, and $2i$, which is $0 + 2i$, are shown at the right.

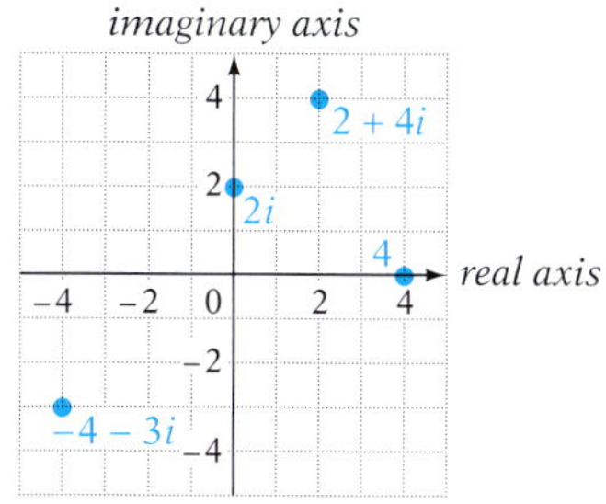

By using the Argand diagram, it is possible to give various interpretations of complex numbers and the operations on them. We will focus on only one small aspect: a geometric explanation of multiplying by i, the imaginary unit.

Before doing this, recall that real number multiplication is repeated addition. For instance, $2 \cdot 3 = 2 + 2 + 2$. Geometrically, this can be shown on a number line.

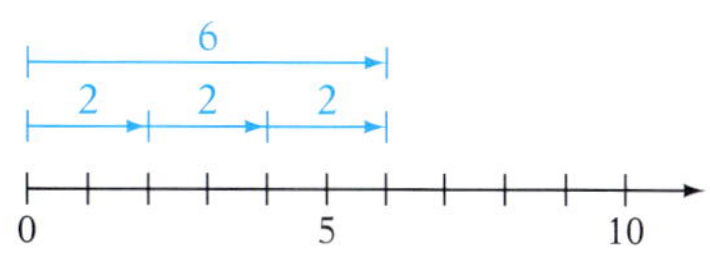

Thus, multiplying by 3 triples the size of the arrow that represents 2. In general, multiplying a real number by a number greater than 1 increases the magnitude of the number.

Consider the complex number $3 + 4i$. Multiply this number by i.

$$i(3 + 4i) = -4 + 3i$$

The two complex numbers are shown in the graph at the right.

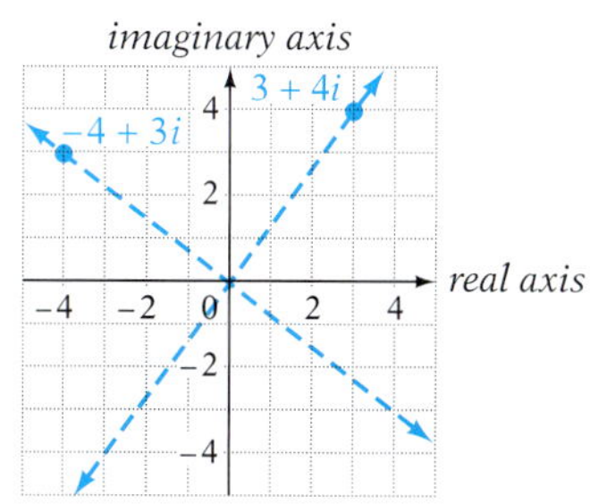

1. Find the slope of the line that passes through the origin and the point (3, 4).
2. Find the slope of the line that passes through the origin and the point (−4, 3).
3. Find the product of the slopes of the two lines. What does this product tell you?
4. Find the distance of the points (3, 4) and (−4, 3) from the origin.

A conclusion that can be drawn from the answers to the foregoing questions is that multiplying a number by i rotates the line on which the number lies by 90°. This is shown in the figure at the right.

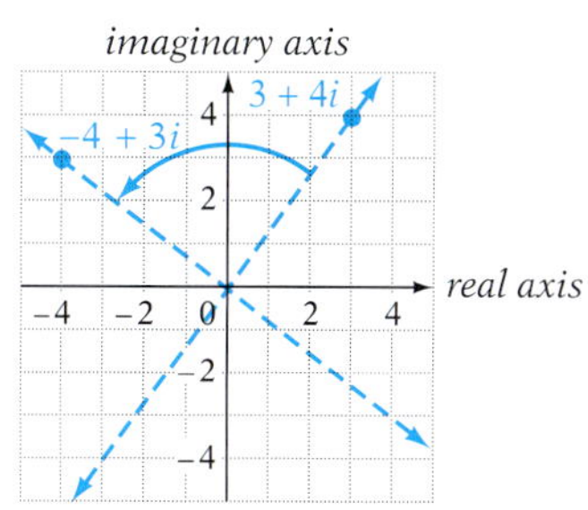

This property of complex numbers is important in the study of electricity and magnetism. Complex numbers are used extensively by electrical engineers to study electrical circuits.

Solving Radical Equations with a Graphing Calculator

The radical equation $\sqrt{x-2} - 6 = 0$ was solved algebraically at the beginning of Section 8.4. To solve $\sqrt{x-2} - 6 = 0$ with a graphing calculator, use the left side of the equation and write an equation in two variables.

$$y = \sqrt{x-2} - 6$$

The graph of $y = \sqrt{x-2} - 6$ is shown below. The solution set of the equation $y = \sqrt{x-2} - 6$ is the set of ordered pairs (x, y) whose coordinates make the equation a true statement. The x-coordinate where $y = 0$ is the solution of the equation $\sqrt{x-2} - 6 = 0$. The solution is the x-intercept of the curve given by the equation $y = \sqrt{x-2} - 6$. The solution is 38.

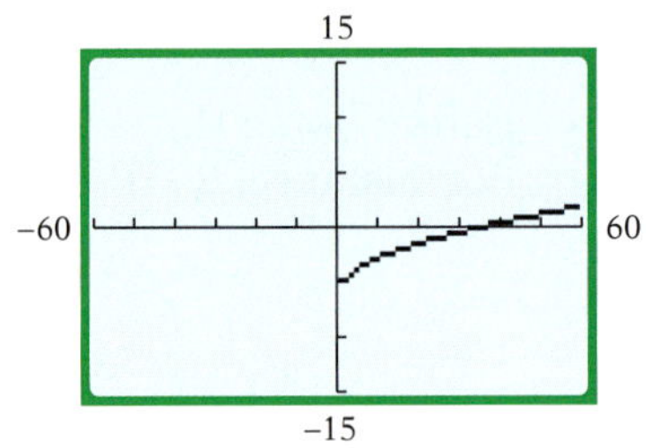

The solution of the radical equation $\sqrt[3]{x-2} = -0.8$ is 1.488. A graphing calculator can be used to find any rational or irrational solution to a predetermined degree of accuracy. The graph of $y = \sqrt[3]{x-2} + 0.8$ is shown below. The graph intersects the x-axis at $x = 1.488$.

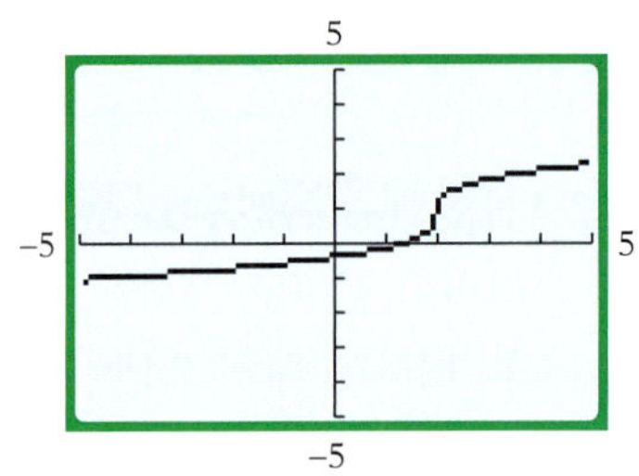

To graph the equation $y = \sqrt[3]{x-2} + 0.8$ on a TI-83, press

[Y=] [MATH] 4 [X, T, θ, n] [−] 2 [)] [+] .8 [ZOOM] 6

Press [2nd] [CALC] 2 to find the x-intercept of the equation.

You were to solve Exercises 3–59 in Section 8.4 by algebraic means. Try some of those exercises again, this time using a graphing calculator.

Use a graphing calculator to find the solution of the following equations. Round to the nearest thousandth.

1. $\sqrt{x + 0.3} = 1.3$

2. $\sqrt[3]{x + 1.2} = -1.1$

3. $\sqrt[4]{3x - 1.5} = 1.4$

The Golden Rectangle

Rectangle A

Rectangle B

Rectangle C

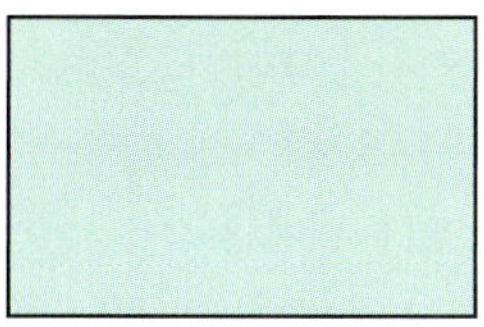

Rectangle D

Take the time to look closely at each of the four rectangles shown at the left. Which rectangle seems to be the most pleasant to look at? Record the responses of all the students in your class. Which rectangle was named most frequently as being the most appealing?

The length of rectangle A is three times the width. The ratio of the length to the width is 3 to 1.

Rectangle B is a square. Its length is equal to its width. The ratio of the length to the width is 1 to 1.

The ratio of the length of rectangle C to its width is 1.2 to 1.

The shape of rectangle D is referred to as the golden rectangle. In a golden rectangle, the ratio of the length to the width is approximately 1.6 to 1.

Did most students in the class select the golden rectangle as the one most pleasant to look at? The early Greeks considered a golden rectangle to be the most pleasing to the eye. Therefore, they used it extensively in their art and architecture. The Parthenon in Athens, Greece, is an example of the use of the golden rectangle in Greek architecture. It was built about 435 B.C. as a temple to the goddess Athena.

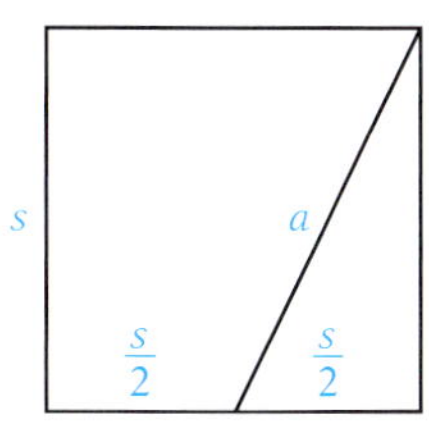

1. Construct a golden rectangle. Beginning with a square with each side of length s, draw a line segment from the midpoint of the base of the square to a vertex, as shown. Call this distance a. Make a rectangle whose width is that of the square and whose length is $\frac{s}{2} + a$.
2. Find the area of the rectangle in terms of s.
3. Find the exact ratio of the length to the width.
4. Find some examples of the golden rectangle in art and architecture. (*Note:* The World Wide Web is a wonderful source of information on the golden rectangle. Simply enter "golden rectangle" in a search engine. It will respond with hundreds of web sites where you can learn about this rectangle's appearance in art, architecture, music, nature, and mathematics.)

Chapter Summary

Key Words

The ***nth root of a*** is $a^{\frac{1}{n}}$. The expression $\sqrt[n]{a}$ is another symbol for the nth root of a. In the expression $\sqrt[n]{a}$, the symbol $\sqrt{}$ is called a **radical**, n is the **index** of the radical, and a is the **radicand**. [8.1.2, p. 512]

Examples: $125^{\frac{1}{3}} = \sqrt[3]{125} = 5$

The index is 3 and the radicand is 125.

If $a^{\frac{1}{n}}$ is a real number, then $a^{\frac{m}{n}} = \sqrt[n]{a^m} = (\sqrt[n]{a})^m$. [8.1.2, p. 512]

$8^{\frac{2}{3}} = (\sqrt[3]{8})^2 = 2^2 = 4$

The symbol $\sqrt{}$ is used to indicate the positive square root, or **principal square root**, of a number. [8.1.3, p. 514]

$\sqrt{16} = 4$
$-\sqrt{16} = -4$

The expressions $a + b$ and $a - b$ are called **conjugates** of each other. The product of conjugates of the form $(a + b)(a - b)$ is $a^2 - b^2$. [8.2.3, p. 521]

$(x - 3)(x + 3) = x^2 - 3^2 = x^2 - 9$

The procedure used to remove a radical from the denominator of a radical expression is called **rationalizing the denominator**. [8.2.4, p. 523]

$$\frac{2}{1 - \sqrt{3}} = \frac{2}{1 - \sqrt{3}} \cdot \frac{1 + \sqrt{3}}{1 + \sqrt{3}} = \frac{2(1 + \sqrt{3})}{(1 - \sqrt{3})(1 + \sqrt{3})} = \frac{2 + 2\sqrt{3}}{1 - 3} = -1 - \sqrt{3}$$

A **radical equation** is an equation that contains a variable expression in a radicand. [8.4.1, p. 533]

$\sqrt{x - 2} - 3 = 6$ is a radical equation.

A **complex number** is a number of the form $a + bi$, where a and b are real numbers and $i = \sqrt{-1}$. For the complex number $a + bi$, a is the **real part** of the complex number, and b is the **imaginary part** of the complex number. [8.5.1, p. 541]

$3 + 2i$ is a complex number.

Essential Rules and Procedures

Definition of $a^{\frac{m}{n}}$

If m and n are positive integers and $a^{\frac{1}{n}}$ is a real number, then $a^{\frac{m}{n}} = \left(a^{\frac{1}{n}}\right)^m$. [8.1.1, p. 511]

$8^{\frac{2}{3}} = \left(8^{\frac{1}{3}}\right)^2 = 2^2 = 4$

Definition of the nth root of a

If a is a real number, then $a^{\frac{1}{n}} = \sqrt[n]{a}$. [8.1.2, p. 512]

$x^{\frac{1}{3}} = \sqrt[3]{x}$

Definition of the nth root of a^m

If $a^{\frac{1}{n}}$ is a real number, then $a^{\frac{m}{n}} = \sqrt[n]{a^m}$. [8.1.2, p. 512]

$b^{\frac{3}{4}} = \sqrt[4]{b^3}$

Product Property of Radicals
If $\sqrt[n]{a}$ and $\sqrt[n]{b}$ are positive real numbers, then $\sqrt[n]{ab} = \sqrt[n]{a}\sqrt[n]{b}$. [8.2.1, p. 519]

$\sqrt{9 \cdot 7} = \sqrt{9}\sqrt{7} = 3\sqrt{7}$

Quotient Property of Radicals
If $\sqrt[n]{a}$ and $\sqrt[n]{b}$ are positive real numbers, and $b \neq 0$, then $\sqrt[n]{\frac{a}{b}} = \frac{\sqrt[n]{a}}{\sqrt[n]{b}}$. [8.2.4, p. 522]

$\sqrt[3]{\frac{5}{27}} = \frac{\sqrt[3]{5}}{\sqrt[3]{27}} = \frac{\sqrt[3]{5}}{3}$

Property of Raising Both Sides of an Equation to a Power
If a and b are real numbers and $a = b$, then $a^n = b^n$. [8.4.1, p. 533]

If $x = 4$, then $x^2 = 16$.

Pythagorean Theorem
$c^2 = a^2 + b^2$ [8.4.2, p. 536]

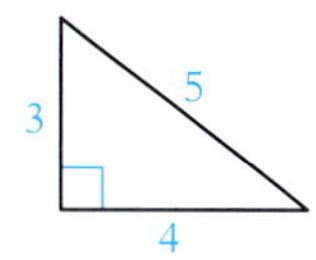

$3^2 + 4^2 = 5^2$

Principal Square Root of a Negative Number
If a is a positive real number, then $\sqrt{-a} = i\sqrt{a}$. [8.5.1, p. 541]

$\sqrt{-8} = i\sqrt{8} = 2i\sqrt{2}$

Addition Property of Complex Numbers
If $a + bi$ and $c + di$ are complex numbers, then $(a + bi) + (c + di) = (a + c) + (b + d)i$. [8.5.2, p. 542]

$$(2 + 4i) + (3 + 6i) = (2 + 3) + (4 + 6)i = 5 + 10i$$

Subtraction Property of Complex Numbers
If $a + bi$ and $c + di$ are complex numbers, then $(a + bi) - (c + di) = (a - c) + (b - d)i$. [8.5.2, p. 542]

$$(4 + 3i) - (7 + 4i) = (4 - 7) + (3 - 4)i = -3 - i$$

Chapter Review Exercises

1. Simplify: $81^{-\frac{1}{4}}$

2. Simplify: $\dfrac{x^{-\frac{3}{2}}}{x^{\frac{7}{2}}}$

3. Simplify: $(a^{16})^{-\frac{5}{8}}$

4. Simplify: $(16x^{-4}y^{12})(100x^{6}y^{-2})^{\frac{1}{2}}$

5. Rewrite $3x^{\frac{3}{4}}$ as a radical expression.

6. Rewrite $7y\sqrt[3]{x^2}$ as an exponential expression.

7. Simplify: $\sqrt[4]{81a^8b^{12}}$

8. Simplify: $-\sqrt{49x^6y^{16}}$

9. Simplify: $\sqrt[3]{-8a^6b^{12}}$

10. Simplify: $\sqrt{18a^3b^6}$

11. Simplify: $\sqrt[5]{-64a^8b^{12}}$

12. Simplify: $\sqrt[4]{x^6y^8z^{10}}$

13. Add: $\sqrt{54} + \sqrt{24}$

14. Subtract: $\sqrt{48x^5y} - x\sqrt{80x^3y}$

15. Subtract: $\sqrt{50a^4b^3} - ab\sqrt{18a^2b}$

16. Simplify: $4x\sqrt{12x^2y} + \sqrt{3x^4y} - x^2\sqrt{27y}$

17. Multiply: $\sqrt{32}\,\sqrt{50}$

18. Multiply: $\sqrt[3]{16x^4y}\,\sqrt[3]{4xy^5}$

19. Multiply: $\sqrt{3x}(3 + \sqrt{3x})$

20. Multiply: $(5 - \sqrt{6})^2$

21. Multiply: $(\sqrt{3} + 8)(\sqrt{3} - 2)$

22. Simplify: $\dfrac{\sqrt{125x^6}}{\sqrt{5x^3}}$

23. Simplify: $\dfrac{8}{\sqrt{3y}}$

24. Simplify: $\dfrac{x + 2}{\sqrt{x} + \sqrt{2}}$

25. Simplify: $\dfrac{\sqrt{x} + \sqrt{y}}{\sqrt{x} - \sqrt{y}}$

26. Simplify: $\sqrt{-36}$

27. Simplify: $\sqrt{-50}$

28. Simplify: $\sqrt{49} - \sqrt{-16}$

29. Simplify: $\sqrt{200} + \sqrt{-12}$

30. Add: $(5 + 2i) + (4 - 3i)$

31. Subtract: $(-8 + 3i) - (4 - 7i)$

32. Add: $(9 - \sqrt{-16}) + (5 + \sqrt{-36})$

33. Subtract: $(\sqrt{50} + \sqrt{-72}) - (\sqrt{162} - \sqrt{-8})$

34. Add: $(3 - 9i) + 7$

35. Multiply: $(8i)(2i)$

36. Multiply: $i(3 - 7i)$

37. Multiply: $\sqrt{-12}\,\sqrt{-6}$

38. Multiply: $(6 - 5i)(4 + 3i)$

39. Simplify: $\dfrac{-6}{i}$

40. Simplify: $\dfrac{5 + 2i}{3i}$

41. Simplify: $\dfrac{7}{2 - i}$

42. Simplify: $\dfrac{\sqrt{16}}{\sqrt{4} - \sqrt{-4}}$

43. Simplify: $\dfrac{5 + 9i}{1 - i}$

44. Solve: $\sqrt[3]{9x} = -6$

45. State the domain of $f(x) = \sqrt[4]{3x - 2}$ in set-builder notation.

46. State the domain of $f(x) = \sqrt[3]{x - 5}$ in set-builder notation.

47. Graph: $g(x) = 3x^{\frac{1}{3}}$

48. Graph: $f(x) = \sqrt{x - 3}$

49. Solve: $\sqrt[3]{3x - 5} = 2$

50. Solve: $\sqrt{4x + 9} + 10 = 11$

51. Find the width of a rectangle that has a diagonal of 13 in. and a length of 12 in.

52. The velocity of the wind determines the amount of power generated by a windmill. A typical equation for this relationship is $v = 4.05\sqrt[3]{P}$, where v is the velocity in miles per hour and P is the power in watts. Find the amount of power generated by a wind of 20 mph. Round to the nearest whole number.

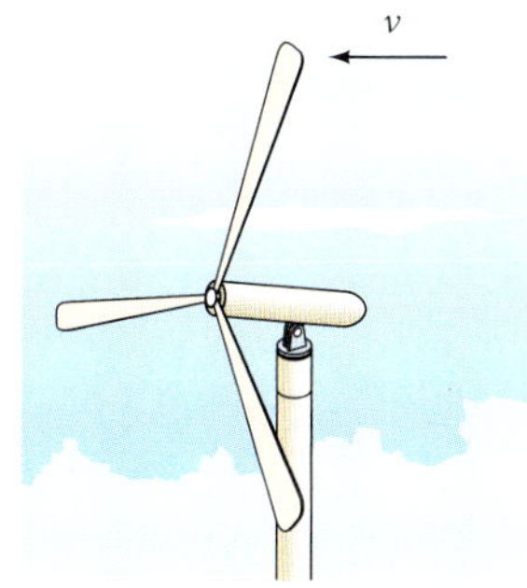

53. Find the distance required for a car to reach a velocity of 88 ft/s when the acceleration is 16 ft/s^2. Use the equation $v = \sqrt{2as}$, where v is the velocity, a is the acceleration, and s is the distance.

54. A 12-ft ladder is leaning against a building. How far from the building is the bottom of the ladder when the top of the ladder touches the building 10 ft above the ground? Round to the nearest hundredth.

Chapter Test

1. Simplify: $\dfrac{r^{\frac{2}{3}}r^{-1}}{r^{-\frac{1}{2}}}$

2. Simplify: $\dfrac{\left(2x^{\frac{1}{3}}y^{-\frac{2}{3}}\right)^6}{(x^{-4}y^8)^{\frac{1}{4}}}$

3. Simplify: $\left(\dfrac{4a^4}{b^2}\right)^{-\frac{3}{2}}$

4. Rewrite $3y^{\frac{2}{5}}$ as a radical expression.

5. Rewrite $\frac{1}{2}\sqrt[4]{x^3}$ as an exponential expression.

6. State the domain of $f(x) = \sqrt{4 - x}$ in set-builder notation.

7. State the domain of $f(x) = (2x - 3)^{\frac{1}{3}}$ in interval notation.

8. Simplify: $\sqrt[3]{27a^4b^3c^7}$

9. Add: $\sqrt{18a^3} + a\sqrt{50a}$

10. Subtract: $\sqrt[3]{54x^7y^3} - x\sqrt[3]{128x^4y^3} - x^2\sqrt[3]{2xy^3}$

11. Multiply: $\sqrt{3x}\left(\sqrt{x} - \sqrt{25x}\right)$

12. Multiply: $\left(2\sqrt{3} + 4\right)\left(3\sqrt{3} - 1\right)$

13. Multiply: $\left(\sqrt{a} - 3\sqrt{b}\right)\left(2\sqrt{a} + 5\sqrt{b}\right)$

14. Multiply: $\left(2\sqrt{x} + \sqrt{y}\right)^2$

15. Simplify: $\dfrac{\sqrt{32x^5y}}{\sqrt{2xy^3}}$

16. Simplify: $\dfrac{4 - 2\sqrt{5}}{2 - \sqrt{5}}$

17. Simplify: $\dfrac{\sqrt{x}}{\sqrt{x} - \sqrt{y}}$

18. Multiply: $\left(\sqrt{-8}\right)\left(\sqrt{-2}\right)$

19. Subtract: $(5 - 2i) - (8 - 4i)$

20. Multiply: $(2 + 5i)(4 - 2i)$

21. Simplify: $\dfrac{2 + 3i}{1 - 2i}$

22. Add: $(2 + i) + (2 - i)$

23. Solve: $\sqrt{x + 12} - \sqrt{x} = 2$

24. Solve: $\sqrt[3]{2x - 2} + 4 = 2$

25. A guy wire is attached to a point 30 ft above the ground on a lamppost. The wire is anchored to the ground at a point 6 ft from the base of the post. Find the length of the guy wire. Round to the nearest tenth.

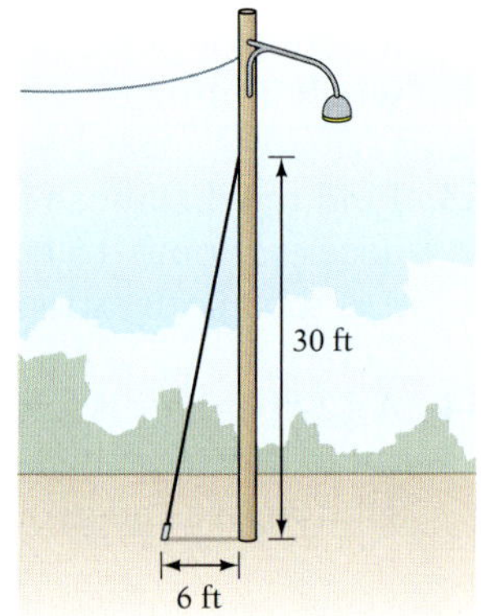

Cumulative Review Exercises

1. Identify the property that justifies the statement.
$(a + 2)b = ab + 2b$

2. Simplify: $2x - 3[x - 2(x - 4) + 2x]$

3. Find $A \cap B$ given $A = \{2, 4, 6\}$ and $B = \{1, 3, 5\}$.

4. Solve: $\sqrt[3]{2x - 5} + 3 = 6$

5. Solve: $5 - \frac{2}{3}x = 4$

6. Solve: $2[4 - 2(3 - 2x)] = 4(1 - x)$

7. Solve: $3x - 4 \leq 8x + 1$
Write the solution set in set-builder notation.

8. Solve: $5 < 2x - 3 < 7$
Write the solution set in set-builder notation.

9. Solve: $|7 - 3x| > 1$

10. Factor: $64a^2 - b^2$

11. Factor: $x^5 + 2x^3 - 3x$

12. Solve: $3x^2 + 13x - 10 = 0$

13. Graph: $g(x) = \sqrt{1 - x}$

14. Is the line $x - 2y = 4$ perpendicular to the line $2x + y = 4$?

15. Simplify: $(3^{-1}x^3y^{-5})(3^{-1}y^{-2})^{-2}$

16. Simplify: $\left(\dfrac{x^{-\frac{1}{2}}y^{\frac{3}{4}}}{y^{-\frac{5}{4}}}\right)^4$

17. Subtract: $\sqrt{20x^3} - x\sqrt{45x}$

18. Multiply: $(\sqrt{5} - 3)(\sqrt{5} - 2)$

19. Simplify: $\dfrac{\sqrt[3]{4x^5y^4}}{\sqrt[3]{8x^2y^5}}$

20. Simplify: $\dfrac{3i}{2 - i}$

21. Graph: $\{x \mid x > -1\} \cap \{x \mid x \leq 3\}$

22. State the domain of $g(x) = \sqrt[5]{2x + 5}$ in set-builder notation.

23. Given $f(x) = 3x^2 - 2x + 1$, evaluate $f(-3)$.

24. Find an equation of the line that passes through the points $(2, 3)$ and $(-1, 2)$.

25. Evaluate the determinant: $\begin{vmatrix} 1 & 2 & -3 \\ 0 & -1 & 2 \\ 3 & 1 & -2 \end{vmatrix}$

26. Solve by using Cramer's Rule: $\begin{aligned} 2x - y &= 4 \\ -2x + 3y &= 5 \end{aligned}$

27. Find the slope and y-intercept of the equation $3x - 2y = -6$ and then graph the line.

28. Graph the solution set of $3x + 2y \leq 4$.

29. An investment of \$2500 is made at an annual simple interest rate of 7.2%. How much additional money must be invested at an annual simple interest rate of 8.4% so that the total interest earned is \$516?

30. The width of a rectangle is 6 ft less than the length. The area of the rectangle is 72 ft^2. Find the length and width of the rectangle.

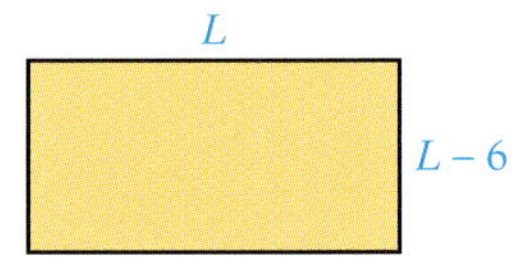

31. A sales executive traveled 25 mi by car and then an additional 625 mi by plane. The rate by plane was five times the rate by car. The total time of the trip was 3 h. Find the rate of the plane.

32. How long does it take light to travel to Earth from the moon when the moon is 232,500 mi from Earth? Light travels 1.86×10^5 mi/s.

33. The graph shows the amount invested and the annual interest income from an investment. Find the slope of the line between the two points shown on the graph. Then write a sentence that states the meaning of the slope.

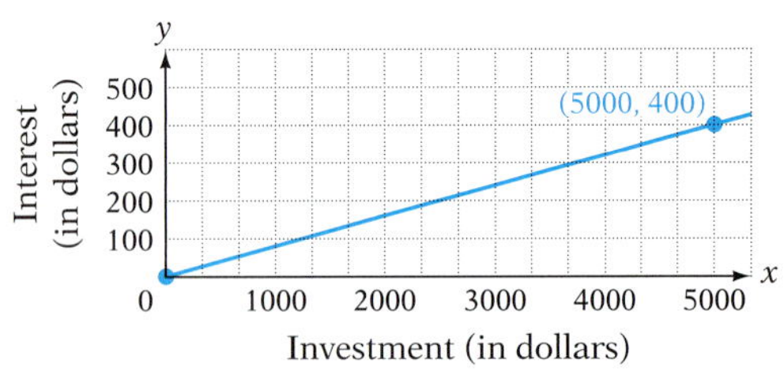

FedEx is actually a family of companies that includes, among others, FedEx Express, FedEx Ground, and FedEx Freight. The objective of FedEx, as for any business, is to earn a profit. Profit is the difference between a company's revenue (the total amount of money the company earns by selling its products or services) and its costs (the total amount of money the company spends in doing business). Quadratic equations are used in the **Project on pages 625–626** *to determine the maximum profit companies can earn.*

Applications

9 Quadratic Equations and Inequalities

Objectives

Section 9.1

1. Solve quadratic equations by factoring
2. Write a quadratic equation given its solutions
3. Solve quadratic equations by taking square roots

Section 9.2

1. Solve quadratic equations by completing the square
2. Solve quadratic equations by using the quadratic formula

Section 9.3

1. Equations that are quadratic in form
2. Radical equations
3. Fractional equations

Section 9.4

1. Application problems

Section 9.5

1. Solve nonlinear inequalities

Section 9.6

1. Graph a quadratic function
2. Find the x-intercepts of a parabola

Section 9.7

1. Minimum and maximum problems
2. Applications of minimum and maximum

WEB **Need help? For online student resources, such as section quizzes, visit this web site: math.college.hmco.com**

Prep Test

1. Simplify: $\sqrt{18}$
2. Simplify: $\sqrt{-9}$
3. Simplify: $\dfrac{3x - 2}{x - 1} - 1$
4. Evaluate $b^2 - 4ac$ when $a = 2$, $b = -4$, and $c = 1$.
5. Is $4x^2 + 28x + 49$ a perfect-square trinomial?
6. Factor: $4x^2 - 4x + 1$
7. Factor: $9x^2 - 4$
8. Graph $\{x \mid x < -1\} \cap \{x \mid x < 4\}$
9. Solve: $x(x - 1) = x + 15$
10. Solve: $\dfrac{4}{x - 3} = \dfrac{16}{x}$

Go Figure

The numeral $0.AAA...$ is a repeating decimal. Given that $A \neq B$ and $\sqrt{0.AAA...} = \sqrt{B}$, find the value of A.

9.1 Solving Quadratic Equations by Factoring or by Taking Square Roots

1 Solve quadratic equations by factoring

VIDEO & DVD CD TUTOR WEB SSM

A **quadratic equation** is an equation of the form $ax^2 + bx + c = 0$, where a and b are coefficients, c is a constant, and $a \neq 0$.

$$
\begin{array}{llll}
3x^2 - x + 2 = 0, & a = 3, & b = -1, & c = 2 \\
-x^2 + 4 = 0, & a = -1, & b = 0, & c = 4 \\
6x^2 - 5x = 0, & a = 6, & b = -5, & c = 0
\end{array}
$$

A quadratic equation is in **standard form** when the polynomial is in descending order and equal to zero.

Because the degree of the polynomial $ax^2 + bx + c$ is 2, a quadratic equation is also called a **second-degree equation.**

In the chapter entitled "Factoring," the Principle of Zero Products was used to solve some quadratic equations. That procedure is reviewed here.

Principle of Zero Products

If the product of two factors is zero, then at least one of the factors must be zero.

If $ab = 0$, then $a = 0$ or $b = 0$.

The Principle of Zero Products states that if the product of two factors is zero, then at least one of the factors must be zero.

Take Note

Recall that the steps involved in solving a quadratic equation by factoring are

1. Write the equation in standard form.
2. Factor.
3. Use the Principle of Zero Products to set each factor equal to 0.
4. Solve each equation.
5. Check the solutions.

Solve by factoring: $x^2 - 6x = -9$

	$x^2 - 6x = -9$
Write the equation in standard form.	$x^2 - 6x + 9 = 0$
Factor.	$(x - 3)(x - 3) = 0$
Use the Principle of Zero Products.	$x - 3 = 0 \qquad x - 3 = 0$
Solve each equation.	$x = 3 \qquad x = 3$
	3 checks as a solution.
Write the solutions.	The solution is 3.

When a quadratic equation has two solutions that are the same number, the solution is called a **double root** of the equation. The solution 3 is a double root of the equation $x^2 - 6x = -9$.

Example 1 Solve for x by factoring: $x^2 - 4ax - 5a^2 = 0$

Solution

$x^2 - 4ax - 5a^2 = 0$ • This is a literal equation. Solve for x in terms of a.

$(x + a)(x - 5a) = 0$ • Factor.

$x + a = 0 \qquad x - 5a = 0$

$x = -a \qquad x = 5a$

The solutions are $-a$ and $5a$.

Problem 1 Solve for x by factoring: $x^2 - 3ax - 4a^2 = 0$

Solution See page S28.

2 Write a quadratic equation given its solutions

VIDEO & DVD CD TUTOR WEB SSM

As shown below, the solutions of the equation $(x - r_1)(x - r_2) = 0$ are r_1 and r_2.

$(x - r_1)(x - r_2) = 0$

$x - r_1 = 0 \qquad x - r_2 = 0$

$x = r_1 \qquad x = r_2$

Check:

$$\begin{array}{c|c} (x - r_1)(x - r_2) = 0 & \\ \hline (r_1 - r_1)(r_1 - r_2) & 0 \\ 0 \cdot (r_1 - r_2) & \\ 0 = 0 & \end{array} \qquad \begin{array}{c|c} (x - r_1)(x - r_2) = 0 & \\ \hline (r_2 - r_1)(r_2 - r_2) & 0 \\ (r_2 - r_1) \cdot 0 & \\ 0 = 0 & \end{array}$$

Using the equation $(x - r_1)(x - r_2) = 0$ and the fact that r_1 and r_2 are solutions of this equation, it is possible to write a quadratic equation given its solutions.

Write a quadratic equation that has solutions 4 and −5.

	$(x - r_1)(x - r_2) = 0$
Replace r_1 by 4 and r_2 by −5.	$(x - 4)[x - (-5)] = 0$
Simplify.	$(x - 4)(x + 5) = 0$
Multiply.	$x^2 + x - 20 = 0$

Example 2 Write a quadratic equation that has integer coefficients and has solutions $\frac{2}{3}$ and $\frac{1}{2}$.

Solution

$$(x - r_1)(x - r_2) = 0$$

$$\left(x - \frac{2}{3}\right)\left(x - \frac{1}{2}\right) = 0$$ • Replace r_1 by $\frac{2}{3}$ and r_2 by $\frac{1}{2}$.

$$x^2 - \frac{7}{6}x + \frac{1}{3} = 0$$ • Multiply.

$$6\left(x^2 - \frac{7}{6}x + \frac{1}{3}\right) = 6 \cdot 0$$ • Multiply each side of the equation by the LCM of the denominators.

$$6x^2 - 7x + 2 = 0$$

Problem 2 Write a quadratic equation that has integer coefficients and has solutions $-\frac{2}{3}$ and $\frac{1}{6}$.

Solution See page S28.

3 Solve quadratic equations by taking square roots

VIDEO & DVD CD TUTOR WWW WEB SSM

The solution of the quadratic equation $x^2 = 16$ is shown at the right.

$$x^2 = 16$$
$$x^2 - 16 = 0$$
$$(x + 4)(x - 4) = 0$$

Note that the solution is the positive or the negative square root of 16, 4 or −4.

$$x + 4 = 0 \qquad x - 4 = 0$$
$$x = -4 \qquad x = 4$$

The solution can also be found by taking the square root of each side of the equation and writing the positive and the negative square roots of the number. The notation $x = \pm 4$ means $x = 4$ or $x = -4$.

$$x^2 = 16$$
$$\sqrt{x^2} = \sqrt{16}$$
$$x = \pm\sqrt{16}$$
$$x = \pm 4$$

The solutions are 4 and −4.

Solve by taking square roots: $3x^2 = 54$

$$3x^2 = 54$$

Solve for x^2.

$$x^2 = 18$$

Take the square root of each side of the equation.

$$\sqrt{x^2} = \sqrt{18}$$
$$x = \pm\sqrt{18}$$

Simplify.

$$x = \pm 3\sqrt{2}$$

$3\sqrt{2}$ and $-3\sqrt{2}$ check as solutions.

Write the solutions. The solutions are $3\sqrt{2}$ and $-3\sqrt{2}$.

Solving a quadratic equation by taking the square root of each side of the equation can lead to solutions that are complex numbers.

Solve by taking square roots: $2x^2 + 18 = 0$

Solve for x^2.

$$2x^2 + 18 = 0$$
$$2x^2 = -18$$
$$x^2 = -9$$

Take the square root of each side of the equation.

$$\sqrt{x^2} = \sqrt{-9}$$
$$x = \pm\sqrt{-9}$$

Simplify.

$$x = \pm 3i$$

$3i$ and $-3i$ check as solutions.

Write the solutions. The solutions are $3i$ and $-3i$.

An equation containing the square of a binomial can be solved by taking square roots.

Example 3 Solve by taking square roots: $3(x - 2)^2 + 12 = 0$

Solution

$$3(x - 2)^2 + 12 = 0$$
$$3(x - 2)^2 = -12$$ • Solve for $(x - 2)^2$.
$$(x - 2)^2 = -4$$
$$\sqrt{(x - 2)^2} = \sqrt{-4}$$ • Take the square root of each side of the equation. Then simplify.
$$x - 2 = \pm\sqrt{-4}$$
$$x - 2 = \pm 2i$$
$$x - 2 = 2i \qquad x - 2 = -2i$$ • Solve for x.
$$x = 2 + 2i \qquad x = 2 - 2i$$

The solutions are $2 + 2i$ and $2 - 2i$.

Problem 3 Solve by taking square roots: $2(x + 1)^2 + 24 = 0$

Solution See page S28.

9.1 Concept Review

Determine whether the following statements are always true, sometimes true, or never true.

1. A quadratic equation has two real roots.
2. A quadratic equation is also called a second-degree equation.
3. If $(x - 3)(x + 5) = 8$, then $x - 3 = 8$ or $x + 5 = 8$.
4. A quadratic equation has one root.

9.1 Exercises

1 **1.** Explain why the restriction that $a \neq 0$ is given in the definition of a quadratic equation.

2. What does the Principle of Zero Products state? How is it used to solve a quadratic equation?

Write the quadratic equation in standard form and with the coefficient of x^2 positive. Name the values of a, b, and c.

3. $2x^2 - 4x = 5$
4. $x^2 = 3x + 1$
5. $5x = 4x^2 - 6$
6. $3x^2 = 7$

Solve by factoring.

7. $x^2 - 4x = 0$
8. $y^2 + 6y = 0$
9. $t^2 - 25 = 0$
10. $p^2 - 81 = 0$
11. $s^2 - s - 6 = 0$
12. $v^2 + 4v - 5 = 0$
13. $y^2 - 6y + 9 = 0$
14. $x^2 + 10x + 25 = 0$
15. $9z^2 - 18z = 0$
16. $4y^2 + 20y = 0$
17. $r^2 - 3r = 10$
18. $p^2 + 5p = 6$
19. $v^2 + 10 = 7v$
20. $t^2 - 16 = 15t$
21. $2x^2 - 9x - 18 = 0$
22. $3y^2 - 4y - 4 = 0$
23. $4z^2 - 9z + 2 = 0$
24. $2s^2 - 9s + 9 = 0$
25. $3w^2 + 11w = 4$
26. $2r^2 + r = 6$
27. $6x^2 = 23x + 18$
28. $6x^2 = 7x - 2$
29. $4 - 15u - 4u^2 = 0$
30. $3 - 2y - 8y^2 = 0$
31. $x + 18 = x(x - 6)$
32. $t + 24 = t(t + 6)$
33. $4s(s + 3) = s - 6$
34. $3v(v - 2) = 11v + 6$
35. $u^2 - 2u + 4 = (2u - 3)(u + 2)$
36. $(3v - 2)(2v + 1) = 3v^2 - 11v - 10$
37. $(3x - 4)(x + 4) = x^2 - 3x - 28$

Solve for x by factoring.

38. $x^2 + 14ax + 48a^2 = 0$
39. $x^2 - 9bx + 14b^2 = 0$
40. $x^2 + 9xy - 36y^2 = 0$
41. $x^2 - 6cx - 7c^2 = 0$
42. $x^2 - ax - 20a^2 = 0$
43. $2x^2 + 3bx + b^2 = 0$
44. $3x^2 - 4cx + c^2 = 0$
45. $3x^2 - 14ax + 8a^2 = 0$
46. $3x^2 - 11xy + 6y^2 = 0$

2 Write a quadratic equation that has integer coefficients and has as solutions the given pair of numbers.

47. 2 and 5 **48.** 3 and 1 **49.** -2 and -4

50. -1 and -3 **51.** 6 and -1 **52.** -2 and 5

53. 3 and -3 **54.** 5 and -5 **55.** 4 and 4

56. 2 and 2 **57.** 0 and 5 **58.** 0 and -2

59. 0 and 3 **60.** 0 and -1 **61.** 3 and $\frac{1}{2}$

62. 2 and $\frac{2}{3}$ **63.** $-\frac{3}{4}$ and 2 **64.** $-\frac{1}{2}$ and 5

65. $-\frac{5}{3}$ and -2 **66.** $-\frac{3}{2}$ and -1

67. $-\frac{2}{3}$ and $\frac{2}{3}$ **68.** $-\frac{1}{2}$ and $\frac{1}{2}$

69. $\frac{1}{2}$ and $\frac{1}{3}$ **70.** $\frac{3}{4}$ and $\frac{2}{3}$

71. $\frac{6}{5}$ and $-\frac{1}{2}$ **72.** $\frac{3}{4}$ and $-\frac{3}{2}$

73. $-\frac{1}{4}$ and $-\frac{1}{2}$ **74.** $-\frac{5}{6}$ and $-\frac{2}{3}$

75. $\frac{3}{5}$ and $-\frac{1}{10}$ **76.** $\frac{7}{2}$ and $-\frac{1}{4}$

3 Solve by taking square roots.

77. $y^2 = 49$ **78.** $x^2 = 64$ **79.** $z^2 = -4$ **80.** $v^2 = -16$

81. $s^2 - 4 = 0$ **82.** $r^2 - 36 = 0$ **83.** $4x^2 - 81 = 0$ **84.** $9x^2 - 16 = 0$

85. $y^2 + 49 = 0$ **86.** $z^2 + 16 = 0$ **87.** $v^2 - 48 = 0$ **88.** $s^2 - 32 = 0$

89. $r^2 - 75 = 0$ **90.** $u^2 - 54 = 0$ **91.** $z^2 + 18 = 0$

92. $t^2 + 27 = 0$ **93.** $(x - 1)^2 = 36$ **94.** $(x + 2)^2 = 25$

95. $3(y + 3)^2 = 27$ **96.** $4(s - 2)^2 = 36$ **97.** $5(z + 2)^2 = 125$

98. $(x - 2)^2 = -4$ **99.** $(x + 5)^2 = -25$ **100.** $(x - 8)^2 = -64$

101. $3(x-4)^2=-12$

102. $5(x+2)^2=-125$

103. $3(x-9)^2=-27$

104. $2(y-3)^2=18$

105. $\left(v-\frac{1}{2}\right)^2=\frac{1}{4}$

106. $\left(r+\frac{2}{3}\right)^2=\frac{1}{9}$

107. $\left(x-\frac{2}{5}\right)^2=\frac{9}{25}$

108. $\left(y+\frac{1}{3}\right)^2=\frac{4}{9}$

109. $\left(a+\frac{3}{4}\right)^2=\frac{9}{16}$

110. $4\left(x-\frac{1}{2}\right)^2=1$

111. $3\left(x-\frac{5}{3}\right)^2=\frac{4}{3}$

112. $2\left(x+\frac{3}{5}\right)^2=\frac{8}{25}$

113. $(x+5)^2-6=0$

114. $(t-1)^2-15=0$

115. $(s-2)^2-24=0$

116. $(y+3)^2-18=0$

117. $(z+1)^2+12=0$

118. $(r-2)^2+28=0$

119. $(v-3)^2+45=0$

120. $(x+5)^2+32=0$

121. $\left(u+\frac{2}{3}\right)^2-18=0$

122. $\left(z-\frac{1}{2}\right)^2-20=0$

123. $\left(x+\frac{1}{2}\right)^2+40=0$

124. $\left(r-\frac{3}{2}\right)^2+48=0$

9.1 Applying Concepts

Write a quadratic equation that has as solutions the given pair of numbers.

125. $\sqrt{2}$ and $-\sqrt{2}$

126. $\sqrt{5}$ and $-\sqrt{5}$

127. i and $-i$

128. $2i$ and $-2i$

129. $2\sqrt{2}$ and $-2\sqrt{2}$

130. $3\sqrt{2}$ and $-3\sqrt{2}$

131. $2\sqrt{3}$ and $-2\sqrt{3}$

132. $i\sqrt{2}$ and $-i\sqrt{2}$

133. $2i\sqrt{3}$ and $-2i\sqrt{3}$

Solve for x.

134. $2a^2x^2=32b^2$

135. $5y^2x^2=125z^2$

136. $(x+a)^2-4=0$

137. $2(x-y)^2-8=0$

138. $(2x-1)^2=(2x+3)^2$

139. $(x-4)^2=(x+2)^2$

Solve.

140. Show that the solutions of the equation $ax^2+bx=0$ are 0 and $-\frac{b}{a}$.

141. Show that the solutions of the equation $ax^2+c=0$, $a>0$, $c>0$, are $\frac{\sqrt{ca}}{a}i$ and $-\frac{\sqrt{ca}}{a}i$.

142. **Chemistry** A chemical reaction between carbon monoxide and water vapor is used to increase the ratio of hydrogen gas in certain gas mixtures. In the process, carbon dioxide is also formed. For a certain reaction, the concentration of carbon dioxide, x, in moles per liter, is given by $0.58 = \dfrac{x^2}{(0.02 - x)^2}$. Solve this equation for x. Round to the nearest ten-thousandth.

143. Show that if the complex number $z_1 + z_2 i$ is a solution of the equation $ax^2 + bx + c = 0$, where a, b, and c are real numbers and $a \neq 0$. Then $z_1 - z_2 i$, the complex conjugate, is also a solution of the equation.

144. Write a paper on the history of solving quadratic equations. Include a discussion of methods used by the Babylonians, the Egyptians, the Greeks, and the Hindus.

9.2 Solving Quadratic Equations by Completing the Square and by Using the Quadratic Formula

1 Solve quadratic equations by completing the square

VIDEO & DVD · CD TUTOR · WEB · SSM

Recall that a perfect-square trinomial is the square of a binomial. Some examples of perfect-square trinomials follow.

Perfect-square trinomial		Square of a binomial
$x^2 + 8x + 16$	=	$(x + 4)^2$
$x^2 - 10x + 25$	=	$(x - 5)^2$
$x^2 + 2ax + a^2$	=	$(x + a)^2$

For each perfect-square trinomial, the square of $\frac{1}{2}$ the coefficient of x equals the constant term.

$$\left(\frac{1}{2}\ \textbf{coefficient of } x\right)^2 = \textbf{Constant term}$$

$$x^2 + 8x + 16, \quad \left(\frac{1}{2}\cdot 8\right)^2 = 16$$

$$x^2 - 10x + 25, \quad \left[\frac{1}{2}(-10)\right]^2 = 25$$

$$x^2 + 2ax + a^2, \quad \left(\frac{1}{2}\cdot 2a\right)^2 = a^2$$

To complete the square on $x^2 + bx$, add $\left(\frac{1}{2}b\right)^2$ to $x^2 + bx$.

Point of Interest

Early attempts to solve quadratic equations were primarily geometric. The Persian mathematician al-Khwarizmi (c. A.D. 800) essentially completed a square of $x^2 + 12x$ as shown below.

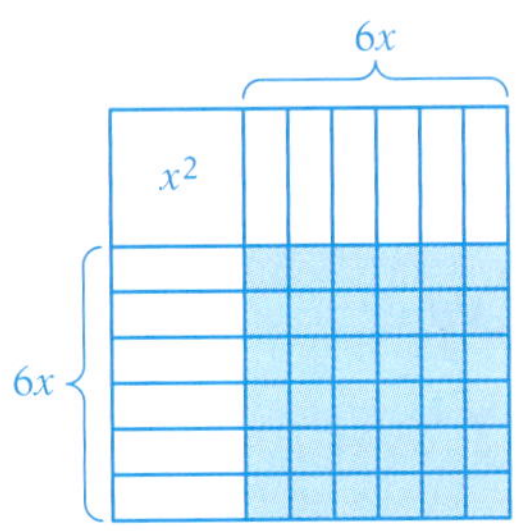

Complete the square on $x^2 - 12x$. Write the resulting perfect-square trinomial as the square of a binomial.

Find the constant term. $\left[\frac{1}{2}(-12)\right]^2 = (-6)^2 = 36$

Complete the square on $x^2 - 12x$ by adding the constant term. $x^2 - 12x + 36$

Write the resulting perfect-square trinomial as the square of a binomial. $x^2 - 12x + 36 = (x - 6)^2$

Complete the square on $z^2 + 3z$. Write the resulting perfect-square trinomial as the square of a binomial.

Find the constant term. $\left(\frac{1}{2} \cdot 3\right)^2 = \left(\frac{3}{2}\right)^2 = \frac{9}{4}$

Complete the square on $z^2 + 3z$ by adding the constant term. $z^2 + 3z + \frac{9}{4}$

Write the resulting perfect-square trinomial as the square of a binomial. $z^2 + 3z + \frac{9}{4} = \left(z + \frac{3}{2}\right)^2$

Not all quadratic equations can be solved by factoring, but any quadratic equation can be solved by completing the square. Add to each side of the equation the term that completes the square. Rewrite the equation in the form $(x + a)^2 = b$. Then take the square root of each side of the equation.

Solve by completing the square: $x^2 - 4x - 14 = 0$

$$x^2 - 4x - 14 = 0$$

Add 14 to each side of the equation.

$$x^2 - 4x = 14$$

Add the constant term that completes the square on $x^2 - 4x$ to each side of the equation.
$\left[\frac{1}{2}(-4)\right]^2 = 4$

$$x^2 - 4x + 4 = 14 + 4$$

Factor the perfect-square trinomial.

$$(x - 2)^2 = 18$$

Take the square root of each side of the equation.

$$\sqrt{(x - 2)^2} = \sqrt{18}$$
$$x - 2 = \pm\sqrt{18}$$

Simplify.

$$x - 2 = \pm 3\sqrt{2}$$

Solve for x.

$$x - 2 = 3\sqrt{2} \qquad x - 2 = -3\sqrt{2}$$
$$x = 2 + 3\sqrt{2} \qquad x = 2 - 3\sqrt{2}$$

Check:

$$\begin{array}{c|c} x^2 - 4x - 14 & 0 \\ \hline (2 + 3\sqrt{2})^2 - 4(2 + 3\sqrt{2}) - 14 & 0 \\ 4 + 12\sqrt{2} + 18 - 8 - 12\sqrt{2} - 14 & 0 \\ 0 & 0 \end{array}$$

$$\begin{array}{c|c} x^2 - 4x - 14 & 0 \\ \hline (2 - 3\sqrt{2})^2 - 4(2 - 3\sqrt{2}) - 14 & 0 \\ 4 - 12\sqrt{2} + 18 - 8 + 12\sqrt{2} - 14 & 0 \\ 0 & 0 \end{array}$$

Write the solutions. The solutions are $2 + 3\sqrt{2}$ and $2 - 3\sqrt{2}$.

Take Note

The exact solutions of the equation $x^2 - 4x - 14 = 0$ are $2 + 3\sqrt{2}$ and $2 - 3\sqrt{2}$. 6.243 and −2.243 are approximate solutions of the equation.

As shown above, the solutions of the equation $x^2 - 4x - 14 = 0$ are $2 + 3\sqrt{2}$ and $2 - 3\sqrt{2}$. These are the exact solutions. However, in some situations it may be preferable to have decimal approximations of the solutions to a quadratic equation. Approximate solutions can be found by using a calculator and then rounding to the desired degree of accuracy.

$$2 + 3\sqrt{2} \approx 6.243 \text{ and } 2 - 3\sqrt{2} \approx -2.243$$

To the nearest thousandth, approximate solutions of the equation $x^2 - 4x - 14 = 0$ are 6.243 and −2.243.

When a, the coefficient of the x^2 term, is not 1, divide each side of the equation by a before completing the square.

Solve $2x^2 - x - 2 = 0$. Find the exact solutions and approximate the solutions to the nearest thousandth.

Take Note

This example illustrates all the steps required in solving a quadratic equation by completing the square.

1. Write the equation in the form $ax^2 + bx = -c$.
2. Multiply both sides of the equation by $\frac{1}{a}$.
3. Complete the square on $x^2 + \frac{b}{a}x$. Add to both sides of the equation the number that completes the square.
4. Factor the perfect-square trinomial.
5. Take the square root of each side of the equation.
6. Solve the resulting equation for x.
7. Check the solutions.

$$2x^2 - x - 2 = 0$$

$$2x^2 - x = 2$$

Divide each side of the equation by the coefficient of x^2.

$$\frac{2x^2 - x}{2} = \frac{2}{2}$$

The coefficient of the x^2 term is now 1.

$$x^2 - \frac{1}{2}x = 1$$

Add the term that completes the square on $x^2 - \frac{1}{2}x$ to each side of the equation. Factor the perfect-square trinomial.

$$x^2 - \frac{1}{2}x + \frac{1}{16} = 1 + \frac{1}{16}$$

$$\left(x - \frac{1}{4}\right)^2 = \frac{17}{16}$$

Take the square root of each side of the equation.

$$\sqrt{\left(x - \frac{1}{4}\right)^2} = \sqrt{\frac{17}{16}}$$

$$x - \frac{1}{4} = \pm\sqrt{\frac{17}{16}}$$

Simplify.

$$x - \frac{1}{4} = \pm\frac{\sqrt{17}}{4}$$

Solve for x.

$$x - \frac{1}{4} = \frac{\sqrt{17}}{4} \qquad x - \frac{1}{4} = -\frac{\sqrt{17}}{4}$$

$$x = \frac{1}{4} + \frac{\sqrt{17}}{4} \qquad x = \frac{1}{4} - \frac{\sqrt{17}}{4}$$

Check the solutions. $\frac{1+\sqrt{17}}{4}$ and $\frac{1-\sqrt{17}}{4}$ check as solutions.

$$\frac{1+\sqrt{17}}{4} \approx 1.281 \qquad \frac{1-\sqrt{17}}{4} \approx -0.781$$

Write the solutions. The exact solutions are $\frac{1+\sqrt{17}}{4}$ and $\frac{1-\sqrt{17}}{4}$. To the nearest thousandth, the solutions are 1.281 and -0.781.

Point of Interest

Mathematicians have studied quadratic equations for centuries. Many of the initial equations were a result of trying to solve a geometry problem. One of the most famous, which dates from around 500 B.C., is "squaring the circle." The question was "Is it possible to construct a square whose area is that of a given circle?" For these early mathematicians, to construct *meant to draw with only a straightedge and a compass. It was approximately 2300 years later that mathematicians were able to prove that such a construction is impossible.*

Example 1 Solve by completing the square.

A. $4x^2 - 8x + 1 = 0$ **B.** $x^2 + 4x + 5 = 0$

Solution

A. $4x^2 - 8x + 1 = 0$

$$4x^2 - 8x = -1$$

• Subtract 1 from each side of the equation.

$$\frac{4x^2 - 8x}{4} = \frac{-1}{4}$$

$$x^2 - 2x = -\frac{1}{4}$$

• The coefficient of the x^2 term must be 1. Divide each side of the equation by 4.

$$x^2 - 2x + 1 = -\frac{1}{4} + 1$$

• Complete the square. $\left[\frac{1}{2}(-2)\right]^2 = 1$

$$(x-1)^2 = \frac{3}{4}$$

• Factor the perfect-square trinomial.

$$\sqrt{(x-1)^2} = \sqrt{\frac{3}{4}}$$

$$x - 1 = \pm\sqrt{\frac{3}{4}}$$

• Take the square root of each side of the equation.

$$x - 1 = \pm\frac{\sqrt{3}}{2}$$

• Simplify.

$$x - 1 = \frac{\sqrt{3}}{2} \qquad x - 1 = -\frac{\sqrt{3}}{2}$$

• Solve for x.

$$x = 1 + \frac{\sqrt{3}}{2} \qquad x = 1 - \frac{\sqrt{3}}{2}$$

$$x = \frac{2+\sqrt{3}}{2} \qquad x = \frac{2-\sqrt{3}}{2}$$

The solutions are $\frac{2+\sqrt{3}}{2}$ and $\frac{2-\sqrt{3}}{2}$.

B. $x^2 + 4x + 5 = 0$

$$x^2 + 4x = -5$$

- Subtract 5 from each side of the equation.

$$x^2 + 4x + 4 = -5 + 4$$

- Complete the square.

$$(x + 2)^2 = -1$$

- Factor the perfect-square trinomial.

$$\sqrt{(x + 2)^2} = \sqrt{-1}$$
$$x + 2 = \pm\sqrt{-1}$$
$$x + 2 = \pm i$$

- Take the square root of each side of the equation.
- Simplify.

$$x + 2 = i \qquad\qquad x + 2 = -i$$
$$x = -2 + i \qquad\qquad x = -2 - i$$

- Solve for x.

The solutions are $-2 + i$ and $-2 - i$.

Problem 1 Solve by completing the square.

A. $4x^2 - 4x - 1 = 0$ **B.** $2x^2 + x - 5 = 0$

Solution See page S29.

2 Solve quadratic equations by using the quadratic formula

VIDEO & DVD CD TUTOR WEB SSM

A general formula known as the **quadratic formula** can be derived by applying the method of completing the square to the standard form of a quadratic equation. This formula can be used to solve any quadratic equation.

The solution of the equation $ax^2 + bx + c = 0$ by completing the square is shown below.

$$ax^2 + bx + c = 0$$

Subtract the constant term from each side of the equation.

$$ax^2 + bx + c - c = 0 - c$$
$$ax^2 + bx = -c$$

Divide each side of the equation by a, the coefficient of x^2.

$$\frac{ax^2 + bx}{a} = \frac{-c}{a}$$
$$x^2 + \frac{b}{a}x = -\frac{c}{a}$$

Complete the square by adding $\left(\frac{1}{2} \cdot \frac{b}{a}\right)^2$ to each side of the equation.

$$x^2 + \frac{b}{a}x + \left(\frac{1}{2} \cdot \frac{b}{a}\right)^2 = \left(\frac{1}{2} \cdot \frac{b}{a}\right)^2 - \frac{c}{a}$$
$$x^2 + \frac{b}{a}x + \frac{b^2}{4a^2} = \frac{b^2}{4a^2} - \frac{c}{a}$$

Simplify the right side of the equation.

$$x^2 + \frac{b}{a}x + \frac{b^2}{4a^2} = \frac{b^2}{4a^2} - \left(\frac{c}{a} \cdot \frac{4a}{4a}\right)$$
$$x^2 + \frac{b}{a}x + \frac{b^2}{4a^2} = \frac{b^2}{4a^2} - \frac{4ac}{4a^2}$$
$$x^2 + \frac{b}{a}x + \frac{b^2}{4a^2} = \frac{b^2 - 4ac}{4a^2}$$

Factor the perfect-square trinomial on the left side of the equation.

$$\left(x + \frac{b}{2a}\right)^2 = \frac{b^2 - 4ac}{4a^2}$$

Take the square root of each side of the equation.

$$\sqrt{\left(x + \frac{b}{2a}\right)^2} = \sqrt{\frac{b^2 - 4ac}{4a^2}}$$

$$x + \frac{b}{2a} = \pm\sqrt{\frac{b^2 - 4ac}{4a^2}}$$

$$x + \frac{b}{2a} = \pm\frac{\sqrt{b^2 - 4ac}}{2a}$$

Solve for x.

$$x + \frac{b}{2a} = \frac{\sqrt{b^2 - 4ac}}{2a} \qquad x + \frac{b}{2a} = -\frac{\sqrt{b^2 - 4ac}}{2a}$$

$$x = -\frac{b}{2a} + \frac{\sqrt{b^2 - 4ac}}{2a} \qquad x = -\frac{b}{2a} - \frac{\sqrt{b^2 - 4ac}}{2a}$$

$$= \frac{-b + \sqrt{b^2 - 4ac}}{2a} \qquad = \frac{-b - \sqrt{b^2 - 4ac}}{2a}$$

Point of Interest

Although mathematicians have studied quadratic equations since around 500 B.C., it was not until the 18th century that the formula was written as it is today. Of further note, the word quadratic *has the same Latin root as does the word* square.

Quadratic Formula

The solutions of $ax^2 + bx + c = 0$, $a \neq 0$, are

$$\frac{-b + \sqrt{b^2 - 4ac}}{2a} \quad \text{and} \quad \frac{-b - \sqrt{b^2 - 4ac}}{2a}$$

The quadratic formula is frequently written in the form

$$x = \frac{-b \pm \sqrt{b^2 - 4ac}}{2a}$$

Solve by using the quadratic formula: $2x^2 + 5x + 3 = 0$

The equation $2x^2 + 5x + 3 = 0$ is in standard form. $a = 2$, $b = 5$, $c = 3$ Replace a, b, and c in the quadratic formula with these values.

$$x = \frac{-b \pm \sqrt{b^2 - 4ac}}{2a}$$

$$= \frac{-(5) \pm \sqrt{(5)^2 - 4(2)(3)}}{2(2)}$$

$$= \frac{-5 \pm \sqrt{25 - 24}}{4}$$

$$= \frac{-5 \pm \sqrt{1}}{4} = \frac{-5 \pm 1}{4}$$

$$x = \frac{-5 + 1}{4} = \frac{-4}{4} = -1 \qquad x = \frac{-5 - 1}{4} = \frac{-6}{4} = -\frac{3}{2}$$

The solutions are -1 and $-\frac{3}{2}$.

Take Note

The solutions of this quadratic equation are rational numbers. When this happens, the equation could have been solved by factoring and using the Principle of Zero Products. This may be easier than using the quadratic formula.

Solve $3x^2 = 4x + 6$ by using the quadratic formula. Find the exact solutions and approximate the solutions to the nearest thousandth.

Write the equation in standard form by subtracting $4x$ and 6 from each side of the equation.

$$3x^2 = 4x + 6$$

$$3x^2 - 4x - 6 = 0$$

$$x = \frac{-b \pm \sqrt{b^2 - 4ac}}{2a}$$

$a = 3$, $b = -4$, $c = -6$

Replace a, b, and c in the quadratic formula with these values.

$$= \frac{-(-4) \pm \sqrt{(-4)^2 - 4(3)(-6)}}{2(3)}$$
$$= \frac{4 \pm \sqrt{16 - (-72)}}{6}$$
$$= \frac{4 \pm \sqrt{88}}{6} = \frac{4 \pm 2\sqrt{22}}{6}$$
$$= \frac{\not{2}(2 \pm \sqrt{22})}{\not{2} \cdot 3} = \frac{2 \pm \sqrt{22}}{3}$$

Check:

$3x^2 = 4x + 6$	
$3\left(\frac{2 + \sqrt{22}}{3}\right)^2$	$4\left(\frac{2 + \sqrt{22}}{3}\right) + 6$
$3\left(\frac{4 + 4\sqrt{22} + 22}{9}\right)$	$\frac{8}{3} + \frac{4\sqrt{22}}{3} + \frac{18}{3}$
$3\left(\frac{26 + 4\sqrt{22}}{9}\right)$	$\frac{26}{3} + \frac{4\sqrt{22}}{3}$
$\frac{26 + 4\sqrt{22}}{3} =$	$\frac{26 + 4\sqrt{22}}{3}$

$3x^2 = 4x + 6$	
$3\left(\frac{2 - \sqrt{22}}{3}\right)^2$	$4\left(\frac{2 - \sqrt{22}}{3}\right) + 6$
$3\left(\frac{4 - 4\sqrt{22} + 22}{9}\right)$	$\frac{8}{3} - \frac{4\sqrt{22}}{3} + \frac{18}{3}$
$3\left(\frac{26 - 4\sqrt{22}}{9}\right)$	$\frac{26}{3} - \frac{4\sqrt{22}}{3}$
$\frac{26 - 4\sqrt{22}}{3} =$	$\frac{26 - 4\sqrt{22}}{3}$

The exact solutions are $\frac{2 + \sqrt{22}}{3}$ and $\frac{2 - \sqrt{22}}{3}$.

$$\frac{2 + \sqrt{22}}{3} \approx 2.230 \qquad \frac{2 - \sqrt{22}}{3} \approx -0.897$$

To the nearest thousandth, the solutions are 2.230 and −0.897.

Solve by using the quadratic formula: $4x^2 = 8x - 13$

$$4x^2 = 8x - 13$$

Write the equation in standard form.

$$4x^2 - 8x + 13 = 0$$
$$a = 4, b = -8, c = 13$$

Replace a, b, and c in the quadratic formula by their values.

$$x = \frac{-b \pm \sqrt{b^2 - 4ac}}{2a}$$
$$= \frac{-(-8) \pm \sqrt{(-8)^2 - 4 \cdot 4 \cdot 13}}{2 \cdot 4}$$

Simplify.

$$= \frac{8 \pm \sqrt{64 - 208}}{8}$$
$$= \frac{8 \pm \sqrt{-144}}{8}$$

Write the answer in the form $a + bi$.

$$= \frac{8 \pm 12i}{8} = \frac{2 \pm 3i}{2} = 1 \pm \frac{3}{2}i$$

Check:

$4x^2 = 8x - 13$	
$4\left(1 + \frac{3}{2}i\right)^2$	$8\left(1 + \frac{3}{2}i\right) - 13$
$4\left(1 + 3i - \frac{9}{4}\right)$	$8 + 12i - 13$
$4\left(-\frac{5}{4} + 3i\right)$	$-5 + 12i$
$-5 + 12i = -5 + 12i$	

$4x^2 = 8x - 13$	
$4\left(1 - \frac{3}{2}i\right)^2$	$8\left(1 - \frac{3}{2}i\right) - 13$
$4\left(1 - 3i - \frac{9}{4}\right)$	$8 - 12i - 13$
$4\left(-\frac{5}{4} - 3i\right)$	$-5 - 12i$
$-5 - 12i = -5 - 12i$	

The solutions are $1 + \frac{3}{2}i$ and $1 - \frac{3}{2}i$.

Example 2 Solve by using the quadratic formula.

A. $4x^2 + 12x + 9 = 0$ **B.** $2x^2 - x + 5 = 0$

Solution **A.** $4x^2 + 12x + 9 = 0$

• $a = 4, b = 12, c = 9$

$$x = \frac{-b \pm \sqrt{b^2 - 4ac}}{2a}$$

• Replace a, b, and c in the quadratic formula by their values. Then simplify.

$$= \frac{-12 \pm \sqrt{12^2 - 4 \cdot 4 \cdot 9}}{2 \cdot 4}$$

$$= \frac{-12 \pm \sqrt{144 - 144}}{8}$$

$$= \frac{-12 \pm \sqrt{0}}{8} = \frac{-12}{8} = -\frac{3}{2}$$

• The equation has a double root.

The solution is $-\frac{3}{2}$.

B. $2x^2 - x + 5 = 0$

• $a = 2, b = -1, c = 5$

$$x = \frac{-b \pm \sqrt{b^2 - 4ac}}{2a}$$

• Replace a, b, and c in the quadratic formula by their values. Then simplify.

$$= \frac{-(-1) \pm \sqrt{(-1)^2 - 4 \cdot 2 \cdot 5}}{2 \cdot 2}$$

$$= \frac{1 \pm \sqrt{1 - 40}}{4}$$

$$= \frac{1 \pm \sqrt{-39}}{4} = \frac{1 \pm i\sqrt{39}}{4}$$

The solutions are $\frac{1}{4} + \frac{\sqrt{39}}{4}i$ and $\frac{1}{4} - \frac{\sqrt{39}}{4}i$.

Problem 2 Solve by using the quadratic formula.

A. $x^2 + 6x - 9 = 0$ **B.** $4x^2 = 4x - 1$

Solution See page S29.

In Example 2A, the solution of the equation is a double root, and in Example 2B, the solutions are complex numbers.

In the quadratic formula, the quantity $b^2 - 4ac$ is called the **discriminant.** When a, b, and c are real numbers, the discriminant determines whether a quadratic equation will have a double root, two real number solutions that are not equal, or two complex number solutions.

Effect of the Discriminant on the Solutions of a Quadratic Equation

1. If $b^2 - 4ac = 0$, the equation has one real number solution, a double root.
2. If $b^2 - 4ac > 0$, the equation has two real number solutions that are not equal.
3. If $b^2 - 4ac < 0$, the equation has two complex number solutions.

The equation $x^2 - 4x - 5 = 0$ has two real number solutions because the discriminant is greater than zero.

$a = 1, b = -4, c = -5$

$$\begin{aligned} b^2 - 4ac &= (-4)^2 - 4(1)(-5) \\ &= 16 + 20 \\ &= 36 \end{aligned}$$

$36 > 0$

Example 3 Use the discriminant to determine whether $4x^2 - 2x + 5 = 0$ has one real number solution, two real number solutions, or two complex number solutions.

Solution

$$\begin{aligned} b^2 - 4ac &= (-2)^2 - 4(4)(5) \\ &= 4 - 80 \\ &= -76 \end{aligned}$$

• $a = 4, b = -2, c = 5$

$-76 < 0$ • The discriminant is less than 0.

The equation has two complex number solutions.

Problem 3 Use the discriminant to determine whether $3x^2 - x - 1 = 0$ has one real number solution, two real number solutions, or two complex number solutions.

Solution See page S29.

9.2 Concept Review

Determine whether the following statements are always true, sometimes true, or never true.

1. A quadratic equation can be solved by completing the square.

2. The expression $b^2 - 4ac$ is called the discriminant.

3. If $b^2 > 4ac$, the quadratic equation has two complex number solutions.

4. The quadratic formula can be used to solve quadratic equations that have irrational roots.

5. $x^2 + 5x + 25$ is a perfect-square trinomial.

6. To complete the square on $3x^2 + 6x$, add 9.

9.2 Exercises

1 **1.** What is the next step when using completing the square to solve $x^2 + 6x = 4$?

2. Are there some quadratic equations that cannot be solved by completing the square? Why or why not?

Solve by completing the square.

3. $x^2 - 4x - 5 = 0$

4. $y^2 + 6y + 5 = 0$

5. $v^2 + 8v - 9 = 0$

6. $w^2 - 2w - 24 = 0$

7. $z^2 - 6z + 9 = 0$

8. $u^2 + 10u + 25 = 0$

9. $r^2 + 4r - 7 = 0$

10. $s^2 + 6s - 1 = 0$

11. $x^2 - 6x + 7 = 0$

12. $y^2 + 8y + 13 = 0$

13. $z^2 - 2z + 2 = 0$

14. $t^2 - 4t + 8 = 0$

First try to solve by factoring. If you are unable to solve the equation by factoring, solve the equation by completing the square.

15. $t^2 - t - 1 = 0$

16. $u^2 - u - 7 = 0$

17. $y^2 - 6y = 4$

18. $w^2 + 4w = 2$

19. $x^2 = 8x - 15$

20. $z^2 = 4z - 3$

21. $v^2 = 4v - 13$

22. $x^2 = 2x - 17$

23. $p^2 + 6p = -13$

24. $x^2 + 4x = -20$

25. $y^2 - 2y = 17$

26. $x^2 + 10x = 7$

27. $z^2 = z + 4$

28. $r^2 = 3r - 1$

29. $x^2 + 13 = 2x$

30. $x^2 + 27 = 6x$

31. $2y^2 + 3y + 1 = 0$

32. $2t^2 + 5t - 3 = 0$

33. $4r^2 - 8r = -3$

34. $4u^2 - 20u = -9$

35. $6y^2 - 5y = 4$

36. $6v^2 - 7v = 3$

37. $4x^2 - 4x + 5 = 0$

38. $4t^2 - 4t + 17 = 0$

39. $9x^2 - 6x + 2 = 0$

40. $9y^2 - 12y + 13 = 0$

41. $2s^2 = 4s + 5$

42. $3u^2 = 6u + 1$

43. $2r^2 = 3 - r$

44. $2x^2 = 12 - 5x$

45. $y - 2 = (y - 3)(y + 2)$

46. $8s - 11 = (s - 4)(s - 2)$

47. $6t - 2 = (2t - 3)(t - 1)$

48. $2z + 9 = (2z + 3)(z + 2)$

49. $(x - 4)(x + 1) = x - 3$

50. $(y - 3)^2 = 2y + 10$

Solve by completing the square. Approximate the solutions to the nearest thousandth.

51. $z^2 + 2z = 4$

52. $t^2 - 4t = 7$

53. $2x^2 = 4x - 1$

54. $3y^2 = 5y - 1$

55. $4z^2 + 2z - 1 = 0$

56. $4w^2 - 8w = 3$

2

57. Write the quadratic formula. What does each variable in the formula represent?

58. Write the expression that appears under the radical symbol in the quadratic formula. What is this quantity called? What can it be used to determine?

59. Suppose you must solve the quadratic equation $x^2 = 3x + 5$. Does it matter whether you rewrite the equation as $x^2 - 3x - 5 = 0$ or as $0 = -x^2 + 3x + 5$ before you begin?

60. Which method of solving a quadratic equation do you prefer: completing the square or using the quadratic formula? Why?

Solve by using the quadratic formula.

61. $x^2 - 3x - 10 = 0$

62. $z^2 - 4z - 8 = 0$

63. $y^2 + 5y - 36 = 0$

64. $z^2 - 3z - 40 = 0$

65. $w^2 = 8w + 72$

66. $t^2 = 2t + 35$

67. $v^2 = 24 - 5v$

68. $x^2 = 18 - 7x$

69. $2y^2 + 5y - 3 = 0$

70. $4p^2 - 7p + 3 = 0$

71. $8s^2 = 10s + 3$

72. $12t^2 = 5t + 2$

First try to solve by factoring. If you are unable to solve the equation by factoring, solve the equation by using the quadratic formula.

73. $v^2 - 2v - 7 = 0$

74. $t^2 - 2t - 11 = 0$

75. $y^2 - 8y - 20 = 0$

76. $x^2 = 14x - 24$

77. $v^2 = 12v - 24$

78. $2z^2 - 2z - 1 = 0$

79. $4x^2 - 4x - 7 = 0$

80. $2p^2 - 8p + 5 = 0$

81. $2s^2 - 3s + 1 = 0$

82. $4w^2 - 4w - 1 = 0$

83. $3x^2 + 10x + 6 = 0$

84. $3v^2 = 6v - 2$

85. $6w^2 = 19w - 10$

86. $z^2 + 2z + 2 = 0$

87. $p^2 - 4p + 5 = 0$

88. $y^2 - 2y + 5 = 0$

89. $x^2 + 6x + 13 = 0$

90. $s^2 - 4s + 13 = 0$

91. $t^2 - 6t + 10 = 0$

92. $2w^2 - 2w + 5 = 0$

93. $4v^2 + 8v + 3 = 0$

94. $2x^2 + 6x + 5 = 0$

95. $2y^2 + 2y + 13 = 0$

96. $4t^2 - 6t + 9 = 0$

97. $3v^2 + 6v + 1 = 0$

98. $2r^2 = 4r - 11$

99. $3y^2 = 6y - 5$

100. $2x(x - 2) = x + 12$

101. $10y(y + 4) = 15y - 15$

102. $(3s - 2)(s + 1) = 2$

103. $(2t + 1)(t - 3) = 9$

Solve by using the quadratic formula. Approximate the solutions to the nearest thousandth.

104. $x^2 - 6x - 6 = 0$

105. $p^2 - 8p + 3 = 0$

106. $r^2 - 2r = 4$

107. $w^2 + 4w = 1$

108. $3t^2 = 7t + 1$

109. $2y^2 = y + 5$

Use the discriminant to determine whether the quadratic equation has one real number solution, two real number solutions, or two complex number solutions.

110. $2z^2 - z + 5 = 0$

111. $3y^2 + y + 1 = 0$

112. $9x^2 - 12x + 4 = 0$

113. $4x^2 + 20x + 25 = 0$

114. $2v^2 - 3v - 1 = 0$

115. $3w^2 + 3w - 2 = 0$

116. $2p^2 + 5p + 1 = 0$

117. $2t^2 + 9t + 3 = 0$

118. $5z^2 + 2 = 0$

Classify each equation as linear or quadratic.

119. $2x^2 + 3x = 1$

120. $2x = 3x - 1$

121. $4x - 2 = 5$

122. $4x^2 - 2x = 5$

123. $6x(x - 2) = 7$

124. $6x - 2 = 7$

9.2 Applying Concepts

Solve.

125. $\sqrt{2}y^2 + 3y - 2\sqrt{2} = 0$

126. $\sqrt{3}z^2 + 10z - 3\sqrt{3} = 0$

127. $\sqrt{2}x^2 + 5x - 3\sqrt{2} = 0$

128. $\sqrt{3}w^2 + w - 2\sqrt{3} = 0$

129. $t^2 - t\sqrt{3} + 1 = 0$

130. $y^2 + y\sqrt{7} + 2 = 0$

Solve for x.

131. $x^2 - ax - 2a^2 = 0$

132. $x^2 - ax - 6a^2 = 0$

133. $2x^2 + 3ax - 2a^2 = 0$

134. $2x^2 - 7ax + 3a^2 = 0$

135. $x^2 - 2x - y = 0$

136. $x^2 - 4xy - 4 = 0$

For what values of p does the quadratic equation have two real number solutions that are not equal? Write the answer in set-builder notation.

137. $x^2 - 6x + p = 0$

138. $x^2 + 10x + p = 0$

For what values of p does the quadratic equation have two complex number solutions? Write the answer in set-builder notation.

139. $x^2 - 2x + p = 0$ **140.** $x^2 + 4x + p = 0$

Solve.

141. Find all values of x that satisfy the equation $x^2 + ix + 2 = 0$.

142. Show that the equation $2x^2 + bx - 2 = 0$ always has real number solutions regardless of the value of b.

143. **Sports** The height, h (in feet), of a baseball above the ground t seconds after it is hit can be approximated by the equation $h = -16t^2 + 70t + 4$. Using this equation, determine when the ball will hit the ground. Round to the nearest hundredth. (*Hint:* The ball hits the ground when $h = 0$.)

144. **Sports** After a baseball is hit, there are two equations that can be considered. One gives the height, h (in feet), the ball is above the ground t seconds after it is hit. The second is the distance, s (in feet), the ball is from home plate t seconds after it is hit. A model of this situation is given by $h = -16t^2 + 70t + 4$ and $s = 44.5t$. Using this model, determine whether the ball will clear a fence 325 ft from home plate. Round to the nearest tenth.

145. **Conservation** The National Forest Management Act of 1976 specifies that harvesting timber in national forests must be accomplished in conjunction with environmental considerations. One such consideration is providing a habitat for the spotted owl. One model of the survival of the spotted owl requires the solution of the equation $x^2 - s_a x - s_j s_s f = 0$ for x. Different values of s_a, s_j, s_s, and f are given in the table at the right. The values are particularly important because they relate to the survival of the owl. If $x > 1$, then the model predicts a growth in the population; if $x = 1$, the population remains steady; if $x < 1$, the population decreases. The important solution of the equation is the larger of the two roots of the equation.

a. Determine the larger root of this equation for values provided by the U.S. Forest Service. Round to the nearest hundredth. Does it predict that the population will increase, remain steady, or decrease?

b. Determine the larger root of this equation for the values provided by R. Lande in *Oecologia* (Vol. 75, 1988). Round to the nearest hundredth. Does it predict that the population will increase, remain steady, or decrease?

	U.S. Forest Service	*Lande*
s_j	0.34	0.11
s_s	0.97	0.71
s_a	0.97	0.94
f	0.24	0.24

146. **Social Security** There are many models of the solvency of the current Social Security system. Based on data from a recent Social Security trustees' report, one possible model of the amount of taxes received is given by the function $A(x) = 6.145x^2 - 73.954x + 422.744$. A model for the amount of benefits paid out is given by $B(x) = 10.287x^2 - 156.632x + 486.345$. For each of these models, x is the number of years after 1994 and the amounts are in billions of dollars. According to these models, during what year after 2000 will the amount paid in benefits first exceed the amount received in taxes?

147. Here is an outline for an alternative method for deriving the quadratic formula. Fill in the details of this derivation. Begin with the equation $ax^2 + bx + c = 0$. Subtract c from each side of the equation. Multiply each side of the equation by $4a$. Now add b^2 to each side. Supply the remaining steps until you have reached the quadratic formula.

148. Explain how to complete the square of $x^2 + bx$.

9.3 Equations That Are Reducible to Quadratic Equations

1 Equations that are quadratic in form

VIDEO & DVD CD TUTOR WEB SSM

Certain equations that are not quadratic equations can be expressed in quadratic form by making suitable substitutions. An equation is **quadratic in form** if it can be written as $au^2 + bu + c = 0$.

To see that the equation at the right is quadratic in form, let $x^2 = u$. Replace x^2 by u. The equation is quadratic in form.

$$x^4 - 4x^2 - 5 = 0$$
$$(x^2)^2 - 4(x^2) - 5 = 0$$
$$u^2 - 4u - 5 = 0$$

To see that the equation at the right is quadratic in form, let $y^{\frac{1}{2}} = u$. Replace $y^{\frac{1}{2}}$ by u. The equation is quadratic in form.

$$y - y^{\frac{1}{2}} - 6 = 0$$
$$\left(y^{\frac{1}{2}}\right)^2 - \left(y^{\frac{1}{2}}\right) - 6 = 0$$
$$u^2 - u - 6 = 0$$

The key to recognizing equations that are quadratic in form is that when the equation is written in standard form, the exponent on one variable term is $\frac{1}{2}$ the exponent on the other variable term.

Solve: $z + 7z^{\frac{1}{2}} - 18 = 0$

The equation $z + 7z^{\frac{1}{2}} - 18 = 0$ is quadratic in form.

$$z + 7z^{\frac{1}{2}} - 18 = 0$$
$$\left(z^{\frac{1}{2}}\right)^2 + 7\left(z^{\frac{1}{2}}\right) - 18 = 0$$

To solve this equation, let $z^{\frac{1}{2}} = u$.

$$u^2 + 7u - 18 = 0$$

Solve for u by factoring.

$$(u - 2)(u + 9) = 0$$

$$u - 2 = 0 \qquad u + 9 = 0$$
$$u = 2 \qquad u = -9$$

Replace u by $z^{\frac{1}{2}}$.

$$z^{\frac{1}{2}} = 2 \qquad z^{\frac{1}{2}} = -9$$

Solve for z by squaring each side of the equation.

$$\left(z^{\frac{1}{2}}\right)^2 = 2^2 \qquad \left(z^{\frac{1}{2}}\right)^2 = (-9)^2$$
$$z = 4 \qquad z = 81$$

Check the solution. When each side of the equation has been squared, the resulting equation may have a solution that is not a solution of the original equation.

Check:

$z + 7z^{\frac{1}{2}} - 18$	$= 0$
$4 + 7(4)^{\frac{1}{2}} - 18$	0
$4 + 7 \cdot 2 - 18$	
$4 + 14 - 18$	

$$0 = 0$$

$$\begin{array}{r|l} z + 7z^{\frac{1}{2}} - 18 & = 0 \\ \hline 81 + 7(81)^{\frac{1}{2}} - 18 & 0 \\ 81 + 7 \cdot 9 - 18 & \\ 81 + 63 - 18 & \\ 126 & \neq 0 \end{array}$$

4 checks as a solution, but 81 does not.

Write the solution. The solution is 4.

Example 1 Solve: **A.** $x^4 + x^2 - 12 = 0$ **B.** $x^{\frac{2}{3}} - 2x^{\frac{1}{3}} - 3 = 0$

Solution **A.**

$$\begin{aligned} x^4 + x^2 - 12 &= 0 \\ (x^2)^2 + (x^2) - 12 &= 0 \\ u^2 + u - 12 &= 0 \\ (u - 3)(u + 4) &= 0 \end{aligned}$$

- The equation is quadratic in form.
- Let $x^2 = u$.
- Solve for u by factoring.

$$\begin{aligned} u - 3 &= 0 & u + 4 &= 0 \\ u &= 3 & u &= -4 \end{aligned}$$

$$\begin{aligned} x^2 &= 3 & x^2 &= -4 \\ \sqrt{x^2} &= \sqrt{3} & \sqrt{x^2} &= \sqrt{-4} \\ x &= \pm\sqrt{3} & x &= \pm 2i \end{aligned}$$

- Replace u by x^2.
- Solve for x by taking square roots.

The solutions are $\sqrt{3}$, $-\sqrt{3}$, $2i$, and $-2i$.

B.

$$\begin{aligned} x^{\frac{2}{3}} - 2x^{\frac{1}{3}} - 3 &= 0 \\ \left(x^{\frac{1}{3}}\right)^2 - 2\left(x^{\frac{1}{3}}\right) - 3 &= 0 \\ u^2 - 2u - 3 &= 0 \\ (u - 3)(u + 1) &= 0 \end{aligned}$$

- The equation is quadratic in form.
- Let $x^{\frac{1}{3}} = u$.
- Solve for u by factoring.

$$\begin{aligned} u - 3 &= 0 & u + 1 &= 0 \\ u &= 3 & u &= -1 \\ x^{\frac{1}{3}} &= 3 & x^{\frac{1}{3}} &= -1 \\ \left(x^{\frac{1}{3}}\right)^3 &= 3^3 & \left(x^{\frac{1}{3}}\right)^3 &= (-1)^3 \\ x &= 27 & x &= -1 \end{aligned}$$

- Replace u by $x^{\frac{1}{3}}$.
- Solve for x by cubing both sides of the equation.

The solutions are 27 and -1.

Problem 1 Solve. **A.** $x - 5x^{\frac{1}{2}} + 6 = 0$ **B.** $4x^4 + 35x^2 - 9 = 0$

Solution See page S29.

2 Radical equations

VIDEO & DVD CD TUTOR WEB SSM

Certain equations containing a radical can be solved by first solving the equation for the radical expression and then squaring each side of the equation.

Remember that when each side of an equation has been squared, the resulting equation may have a solution that is not a solution of the original equation. Therefore, the solutions of a radical equation must be checked.

Solve: $\sqrt{x + 2} + 4 = x$

	$\sqrt{x+2}+4=x$
Solve for the radical expression.	$\sqrt{x+2}=x-4$
Square each side of the equation.	$(\sqrt{x+2})^2=(x-4)^2$
Simplify.	$x+2=x^2-8x+16$
Write the equation in standard form.	$0=x^2-9x+14$
Solve for x by factoring.	$0=(x-7)(x-2)$
	$x-7=0 \qquad x-2=0$
	$x=7 \qquad\quad x=2$

Check the solution.

Check:

$$\begin{array}{r|l} \sqrt{x+2}+4 & x \\ \hline \sqrt{7+2}+4 & 7 \\ \sqrt{9}+4 & \\ 3+4 & \\ 7=7 & \end{array} \qquad \begin{array}{r|l} \sqrt{x+2}+4 & x \\ \hline \sqrt{2+2}+4 & 2 \\ \sqrt{4}+4 & \\ 2+4 & \\ 6\neq 2 & \end{array}$$

7 checks as a solution, but 2 does not.

Write the solution.

Example 2 Solve: $\sqrt{7y-3}+3=2y$

Solution

$\sqrt{7y-3}+3=2y$

$\sqrt{7y-3}=2y-3$ • Solve for the radical expression.

$(\sqrt{7y-3})^2=(2y-3)^2$ • Square each side of the equation.

$7y-3=4y^2-12y+9$

$0=4y^2-19y+12$ • Write the equation in standard form.

$0=(4y-3)(y-4)$ • Solve for y by factoring.

$4y-3=0 \qquad y-4=0$

$4y=3 \qquad y=4$ • 4 checks as a solution.

$y=\frac{3}{4}$ $\frac{3}{4}$ does not check as a solution.

The solution is 4.

Problem 2 Solve: $\sqrt{2x+1}+x=7$

Solution See page S30.

If an equation contains more than one radical, the procedure of solving for the radical expression and squaring each side of the equation may have to be repeated.

Example 3 Solve: $\sqrt{2y+1}-\sqrt{y}=1$

Solution

$\sqrt{2y+1}-\sqrt{y}=1$

$\sqrt{2y+1}=\sqrt{y}+1$ • Solve for one of the radical expressions.

$(\sqrt{2y+1})^2=(\sqrt{y}+1)^2$ • Square each side of the equation.

$$2y + 1 = y + 2\sqrt{y} + 1$$
$$y = 2\sqrt{y}$$ • Solve for the radical expression.
$$y^2 = (2\sqrt{y})^2$$ • Square each side of the equation.
$$y^2 = 4y$$
$$y^2 - 4y = 0$$
$$y(y - 4) = 0$$
$$y = 0 \qquad y - 4 = 0$$
$$y = 4$$ • 0 and 4 check as solutions.

The solutions are 0 and 4.

Problem 3 Solve: $\sqrt{2x - 1} + \sqrt{x} = 2$

Solution See page S30.

3 Fractional equations

VIDEO & DVD CD TUTOR WEB SSM

After each side of a fractional equation has been multiplied by the LCM of the denominators, the resulting equation is sometimes a quadratic equation. The solutions to the resulting equation must be checked, because multiplying each side of an equation by a variable expression may produce an equation that has a solution that is not a solution of the original equation.

Solve: $\frac{1}{r} + \frac{1}{r + 1} = \frac{3}{2}$

$$\frac{1}{r} + \frac{1}{r + 1} = \frac{3}{2}$$

Multiply each side of the equation by the LCM of the denominators.

$$2r(r + 1)\left(\frac{1}{r} + \frac{1}{r + 1}\right) = 2r(r + 1) \cdot \frac{3}{2}$$
$$2(r + 1) + 2r = r(r + 1) \cdot 3$$
$$2r + 2 + 2r = (r^2 + r) \cdot 3$$
$$4r + 2 = 3r^2 + 3r$$

Write the equation in standard form.

$$0 = 3r^2 - r - 2$$
$$0 = (3r + 2)(r - 1)$$

Solve for r by factoring.

$$3r + 2 = 0 \qquad r - 1 = 0$$
$$3r = -2 \qquad r = 1$$
$$r = -\frac{2}{3}$$

$-\frac{2}{3}$ and 1 check as solutions.

Write the solutions. The solutions are $-\frac{2}{3}$ and 1.

Example 4 Solve: $\frac{18}{2a - 1} + 3a = 17$

Solution

$$\frac{18}{2a - 1} + 3a = 17$$ • The LCM of the denominators is $2a - 1$.
$$(2a - 1)\left(\frac{18}{2a - 1} + 3a\right) = (2a - 1)17$$

$$(2a - 1)\frac{18}{2a - 1} + (2a - 1)(3a) = (2a - 1)17$$

$$18 + 6a^2 - 3a = 34a - 17$$

$$6a^2 - 37a + 35 = 0$$

• Write the equation in standard form.

$$(6a - 7)(a - 5) = 0$$

• Solve for x by factoring.

$$6a - 7 = 0 \qquad a - 5 = 0$$

$$6a = 7 \qquad a = 5$$

$$a = \frac{7}{6}$$

The solutions are $\frac{7}{6}$ and 5.

Problem 4 Solve: $3y + \dfrac{25}{3y - 2} = -8$

Solution See page S30.

9.3 Concept Review

Determine whether the following statements are always true, sometimes true, or never true.

1. An equation that is quadratic in form can be solved by the quadratic formula.
2. Squaring both sides on a radical equation produces an extraneous root.
3. $x + 3\sqrt{x} - 8 = 0$ is quadratic in form.
4. $\sqrt[4]{x} + 2\sqrt{3x} - 8 = 0$ is quadratic in form.

9.3 Exercises

1

1. What does it mean for an equation to be quadratic in form?

2. Explain how to show that $x^4 - 2x^2 - 3 = 0$ is quadratic in form.

Solve.

3. $x^4 - 13x^2 + 36 = 0$
4. $y^4 - 5y^2 + 4 = 0$
5. $z^4 - 6z^2 + 8 = 0$
6. $t^4 - 12t^2 + 27 = 0$
7. $p - 3p^{\frac{1}{2}} + 2 = 0$
8. $v - 7v^{\frac{1}{2}} + 12 = 0$

9. $x - x^{\frac{1}{2}} - 12 = 0$

10. $w - 2w^{\frac{1}{2}} - 15 = 0$

11. $z^4 + 3z^2 - 4 = 0$

12. $y^4 + 5y^2 - 36 = 0$

13. $x^4 + 12x^2 - 64 = 0$

14. $x^4 - 81 = 0$

15. $p + 2p^{\frac{1}{2}} - 24 = 0$

16. $v + 3v^{\frac{1}{2}} - 4 = 0$

17. $y^{\frac{2}{3}} - 9y^{\frac{1}{3}} + 8 = 0$

18. $z^{\frac{2}{3}} - z^{\frac{1}{3}} - 6 = 0$

19. $x^6 - 9x^3 + 8 = 0$

20. $y^6 + 9y^3 + 8 = 0$

21. $z^8 - 17z^4 + 16 = 0$

22. $v^4 - 15v^2 - 16 = 0$

23. $p^{\frac{2}{3}} + 2p^{\frac{1}{3}} - 8 = 0$

24. $w^{\frac{2}{3}} + 3w^{\frac{1}{3}} - 10 = 0$

25. $2x - 3x^{\frac{1}{2}} + 1 = 0$

26. $3y - 5y^{\frac{1}{2}} - 2 = 0$

2 Solve.

27. $\sqrt{x + 1} + x = 5$

28. $\sqrt{x - 4} + x = 6$

29. $x = \sqrt{x} + 6$

30. $\sqrt{2y - 1} = y - 2$

31. $\sqrt{3w + 3} = w + 1$

32. $\sqrt{2s + 1} = s - 1$

33. $\sqrt{4y + 1} - y = 1$

34. $\sqrt{3s + 4} + 2s = 12$

35. $\sqrt{10x + 5} - 2x = 1$

36. $\sqrt{t + 8} = 2t + 1$

37. $\sqrt{p + 11} = 1 - p$

38. $x - 7 = \sqrt{x - 5}$

39. $\sqrt{x - 1} - \sqrt{x} = -1$

40. $\sqrt{y} + 1 = \sqrt{y + 5}$

41. $\sqrt{2x - 1} = 1 - \sqrt{x - 1}$

42. $\sqrt{x + 6} + \sqrt{x + 2} = 2$

43. $\sqrt{t + 3} + \sqrt{2t + 7} = 1$

44. $\sqrt{5 - 2x} = \sqrt{2 - x} + 1$

3 Solve.

45. $x = \dfrac{10}{x - 9}$

46. $z = \dfrac{5}{z - 4}$

47. $\dfrac{t}{t + 1} = \dfrac{-2}{t - 1}$

48. $\dfrac{2v}{v - 1} = \dfrac{5}{v + 2}$

49. $\dfrac{y - 1}{y + 2} + y = 1$

50. $\dfrac{2p - 1}{p - 2} + p = 8$

51. $\dfrac{3r + 2}{r + 2} - 2r = 1$

52. $\dfrac{2v + 3}{v + 4} + 3v = 4$

53. $\dfrac{2}{2x + 1} + \dfrac{1}{x} = 3$

54. $\dfrac{3}{s} - \dfrac{2}{2s - 1} = 1$

55. $\dfrac{16}{z - 2} + \dfrac{16}{z + 2} = 6$

56. $\dfrac{2}{y + 1} + \dfrac{1}{y - 1} = 1$

57. $\dfrac{t}{t - 2} + \dfrac{2}{t - 1} = 4$

58. $\dfrac{4t + 1}{t + 4} + \dfrac{3t - 1}{t + 1} = 2$

59. $\dfrac{5}{2p - 1} + \dfrac{4}{p + 1} = 2$

60. $\dfrac{3w}{2w + 3} + \dfrac{2}{w + 2} = 1$

61. $\dfrac{2v}{v + 2} + \dfrac{3}{v + 4} = 1$

62. $\dfrac{x + 3}{x + 1} - \dfrac{x - 2}{x + 3} = 5$

9.3 Applying Concepts

63. $\frac{x^2}{4} + \frac{x}{2} = 6$

64. $3\left(\frac{x+1}{2}\right)^2 = 54$

65. $\frac{x+2}{3} + \frac{2}{x-2} = 3$

66. $\frac{x^4}{4} + 1 = \frac{5x^2}{4}$

67. $\frac{x^4}{3} - \frac{8x^2}{3} = 3$

68. $\frac{x^2}{4} + \frac{x}{2} + \frac{1}{8} = 0$

69. $\frac{x^4}{8} + \frac{x^2}{4} = 3$

70. $\sqrt{x^4 - 2} = x$

71. $\sqrt{x^4 + 4} = 2x$

72. $(\sqrt{x} - 2)^2 - 5\sqrt{x} + 14 = 0$ (*Hint:* Let $u = \sqrt{x} - 2$.)

73. $(\sqrt{x} + 3)^2 - 4\sqrt{x} - 17 = 0$ (*Hint:* $u = \sqrt{x} + 3$.)

74. **Sports** According to the Compton's Interactive Encyclopedia, the minimum dimensions of a football used in the National Football Association games are 10.875 in. long and 20.75 in. in circumference at the center. A possible model for the cross section of a football is given by $y = \pm 3.3041\sqrt{1 - \frac{x^2}{29.7366}}$, where x is the distance from the center of the football and y is the radius of the football at x.

a. What is the domain of the equation?

b. Graph $y = 3.3041\sqrt{1 - \frac{x^2}{29.7366}}$ and $y = -3.3041\sqrt{1 - \frac{x^2}{29.7366}}$ on the same coordinate axes. Explain why the ± symbol occurs in the equation.

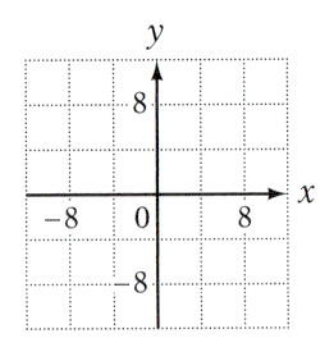

c. Determine the radius of the football when x is 3 in. Round to the nearest ten-thousandth.

9.4 Applications of Quadratic Equations

1 Application problems

VIDEO & DVD CD TUTOR WEB SSM

The application problems in this section are similar to those problems that were solved earlier in the text. Each of the strategies for the problems in this section will result in a quadratic equation.

Solve: A small pipe takes 16 min longer to empty a tank than does a larger pipe. Working together, the pipes can empty the tank in 6 min. How long would it take each pipe, working alone, to empty the tank?

Strategy ***for solving an application problem***

▶ Determine the type of problem. Is it a uniform motion problem, a geometry problem, an integer problem, or a work problem?

The problem is a work problem.

▶ Choose a variable to represent the unknown quantity. Write numerical or variable expressions for all the remaining quantities. These results can be recorded in a table.

The unknown time of the larger pipe: t
The unknown time of the smaller pipe: $t + 16$

	Rate of work	·	Time worked	=	Part of task completed
Larger pipe	$\frac{1}{t}$	·	6	=	$\frac{6}{t}$
Smaller pipe	$\frac{1}{t+16}$	·	6	=	$\frac{6}{t+16}$

▶ Determine how the quantities are related. If necessary, review the strategies presented in earlier chapters.

The sum of the parts of the task completed must equal 1.

$$\frac{6}{t} + \frac{6}{t+16} = 1$$
$$t(t+16)\left(\frac{6}{t} + \frac{6}{t+16}\right) = t(t+16) \cdot 1$$
$$(t+16)6 + 6t = t^2 + 16t$$
$$6t + 96 + 6t = t^2 + 16t$$
$$0 = t^2 + 4t - 96$$
$$0 = (t+12)(t-8)$$

$$t + 12 = 0 \qquad t - 8 = 0$$
$$t = -12 \qquad t = 8$$

The solution $t = -12$ is not possible because time cannot be a negative number.

The time for the smaller pipe is $t + 16$. Replace t by 8 and evaluate.

$$t + 16 = 8 + 16 = 24$$

The larger pipe requires 8 min to empty the tank.
The smaller pipe requires 24 min to empty the tank.

Example 1 In 8 h, two campers rowed 15 mi down a river and then rowed back to their campsite. The rate of the river's current was 1 mph. Find the rate at which the campers row in calm water.

Strategy

▶ This is a uniform motion problem.

▶ Unknown rowing rate of the campers: r

	Distance	÷	Rate	=	Time
Down river	15	÷	$r + 1$	=	$\frac{15}{r+1}$
Up river	15	÷	$r - 1$	=	$\frac{15}{r-1}$

▶ The total time of the trip was 8 h.

Solution

$$\frac{15}{r+1} + \frac{15}{r-1} = 8$$
$$(r+1)(r-1)\left(\frac{15}{r+1} + \frac{15}{r-1}\right) = (r+1)(r-1)8$$
$$(r-1)15 + (r+1)15 = (r^2 - 1)8$$
$$15r - 15 + 15r + 15 = 8r^2 - 8$$
$$30r = 8r^2 - 8$$
$$0 = 8r^2 - 30r - 8$$
$$0 = 2(4r^2 - 15r - 4)$$
$$0 = 2(4r + 1)(r - 4)$$

$$4r + 1 = 0 \qquad r - 4 = 0$$
$$4r = -1 \qquad r = 4$$
$$r = -\frac{1}{4}$$

• The solution $r = -\frac{1}{4}$ is not possible because the rate cannot be a negative number.

The rowing rate is 4 mph.

Problem 1 The length of a rectangle is 3 m more than the width. The area is 54 m^2. Find the length of the rectangle.

Solution See page S30.

9.4 Concept Review

Determine whether the following statements are always true, sometimes true, or never true.

1. Let x be an integer. Then x^2 and $(x + 2)^2$ represent the squares of two consecutive odd integers.

2. Let x be an integer. Then x^2, $(x + 1)^2$, and $(x + 2)^2$ represent the squares of three consecutive integers.

3. If t is the time to complete an amount of work, then $\frac{1}{t}$ is the rate of doing work.

4. Let r be the rate of rowing a boat in calm water. If the river has a current of 2 mph, then the rate of rowing down the river is $r + 2$, and the rate of rowing up the river is $r - 2$.

9.4 Exercises

1 Solve.

1. **Farming** A rectangular corral is constructed in a pasture. The length of the rectangle is 8 ft more than twice the width. The area of the rectangle is to be 640 ft^2. What are the length and width of the rectangle?

2. **Manufacturing** The surface area of the ice cream cone shown at the right is given by $A = \pi r^2 + \pi rs$, where r is the radius of the circular top of the ice cream cone and s is the slant height of the cone. If the area of the cone is 11.25π in^2 and the slant height of the cone is 6 in., find the radius of the cone.

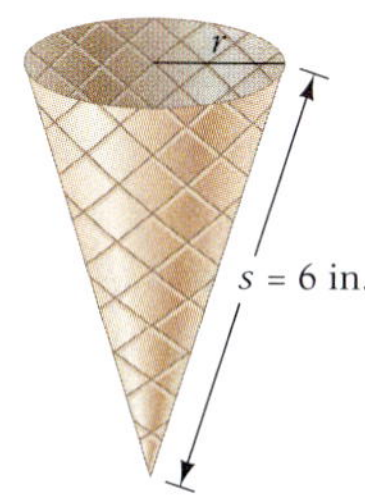

3. **Geometry** The length of a rectangle is 2 ft less than three times the width of the rectangle. The area of the rectangle is 65 ft^2. Find the length and width of the rectangle.

4. **Geometry** The length of a rectangle is 2 cm less than twice the width. The area of the rectangle is 180 cm^2. Find the length and width of the rectangle.

5. **Geometry** A square piece of cardboard is formed into a box by cutting 10-centimeter squares from each of the four corners and then folding up the sides, as shown in the figure. If the volume of the box is to be 49,000 cm^3, what size square piece of cardboard is needed?

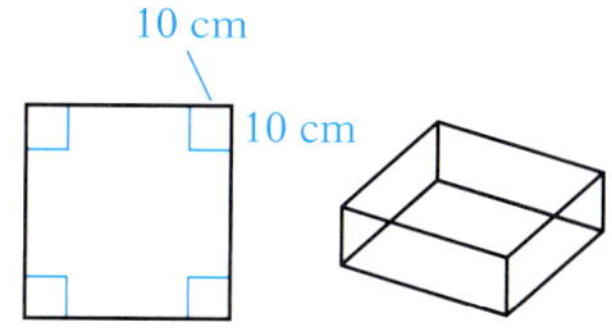

6. **Geography** The state of Colorado is almost perfectly rectangular, with its north border 111 mi longer than its west border. The state encompasses 104,000 mi^2. Estimate the dimensions of Colorado. Round to the nearest mile.

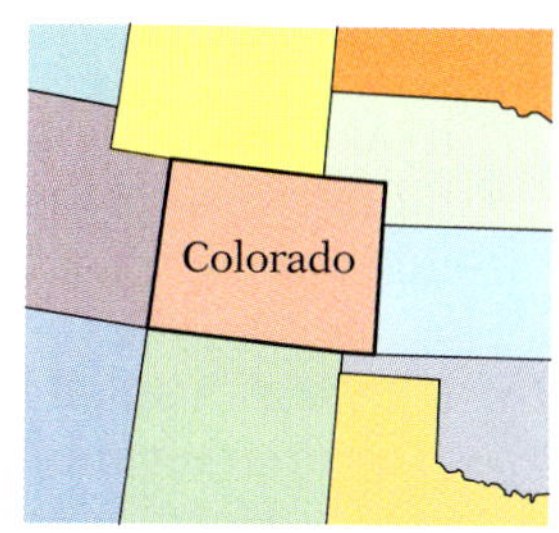

7. **Transportation** An 18-wheeler left a grain depot with a load of wheat and traveled 550 mi to deliver the wheat before returning to the depot. Because of the lighter load on the return trip, the average speed of the truck returning was 5 mph faster than its average speed going. Find the rate returning if the entire trip, not counting unloading time or rest stops, was 21 h.

8. **Racing** The *Tour de France* is a bicycle race that begins in Dublin, Ireland, but is primarily a race through France. On each day, racers cover a certain distance depending on the steepness of the terrain. Suppose that on one particular day, the racers must complete 210 mi. One cyclist, traveling 10 mph faster than a second cyclist, covers this distance in 2.4 h less time than the second cyclist. Find the rate of the first cyclist.

9. **Physics** The height of a projectile fired upward is given by the formula $s = v_0 t - 16t^2$, where s is the height, v_0 is the initial velocity, and t is the time. Find the time for a projectile to return to Earth if it has an initial velocity of 200 ft/s.

10. **Physics** The height of a projectile fired upward is given by the formula $s = v_0t - 16t^2$, where s is the height, v_0 is the initial velocity, and t is the time. Find the times at which a projectile with an initial velocity of 128 ft/s will be 64 ft above the ground. Round to the nearest hundredth of a second.

German Autobahn System

11. **Automotive Technology** In Germany, there are no speed limits on some portions of the autobahn (highway). Other portions have a speed limit of 180 km/h (approximately 112 mph). The distance, d (in meters), required to stop a car traveling at v kilometers per hour is $d = 0.019v^2 + 0.69v$. Approximate, to the nearest tenth, the maximum speed a driver can be going and still be able to stop within 150 m.

12. **Sports** A penalty kick in soccer is made from a penalty mark that is 36 ft from a goal that is 8 ft high. A possible equation for the flight of a penalty kick is $h = -0.002x^2 + 0.35x$, where h is the height (in feet) of the ball x feet from the penalty mark. Assuming that the flight of the kick is toward the goal and that it is not touched by the goalie, will the ball land in the net?

13. **Aeronautics** A model rocket is launched with an initial velocity of 200 ft/s. The height, h, of the rocket t seconds after the launch is given by $h = -16t^2 + 200t$. How many seconds after launch will the rocket be 300 ft above the ground? Round to the nearest hundredth of a second.

14. **Sports** A diver jumps from a platform that is 10 m high. The height of the diver t seconds after the beginning of the dive is given by $h = -4.9t^2 + 3.2t + 10.5$. To the nearest hundredth of a second, how long after the beginning of the dive will the diver enter the water?

15. **Petroleum Engineering** A small pipe can fill an oil tank in 6 min more time than it takes a larger pipe to fill the same tank. Working together, both pipes can fill the tank in 4 min. How long would it take each pipe, working alone, to fill the tank?

16. **Metallurgy** A small heating unit takes 8 h longer to melt a piece of iron than does a larger unit. Working together, the heating units can melt the iron in 3 h. How long would it take each heating unit, working alone, to melt the iron?

17. **Recreation** A cruise ship made a trip of 100 mi in 8 h. The ship traveled the first 40 mi at a constant rate before increasing its speed by 5 mph. Another 60 mi was traveled at the increased speed. Find the rate of the cruise ship for the first 40 mi.

18. **Sports** A cyclist traveled 60 mi at a constant rate before reducing the speed by 2 mph. Another 40 mi was traveled at the reduced speed. The total time for the 100-mi trip was 9 h. Find the rate during the first 60 mi.

19. **Aeronautics** The rate of a single-engine plane in calm air is 100 mph. Flying with the wind, the plane can fly 240 mi in 1 h less time than is required to make the return trip of 240 mi. Find the rate of the wind.

20. **Transportation** A car travels 120 mi. A second car, traveling 10 mph faster than the first car, makes the same trip in 1 h less time. Find the speed of each car.

21. **Sports** To prepare for an upcoming race, a sculling crew rowed 16 mi down a river and back in 6 h. If the rate of the river's current is 2 mph, find the sculling crew's rate of rowing in calm water.

22. **Sports** A fishing boat traveled 30 mi down a river and then returned. The total time for the round trip was 4 h, and the rate of the river's current was 4 mph. Find the rate of the boat in still water.

9.4 Applying Concepts

Solve.

23. **Number Problem** The numerator of a fraction is 3 less than the denominator. The sum of the fraction and four times its reciprocal is $\frac{17}{2}$. Find the fraction.

24. **Integer Problem** Find two consecutive even integers whose cubes differ by 488.

25. **Geometry** An open box is formed from a rectangular piece of cardboard whose length is 8 cm more than its width by cutting squares whose sides are 2 cm in length from each corner and then folding up the sides. Find the dimensions of the box if its volume is 256 cm^3.

26. **Construction** The height of an arch is given by the equation

$$h(x) = -\frac{3}{64}x^2 + 27, \ -24 \le x \le 24$$

where $|x|$ is the distance in feet from the center of the arch.

a. What is the maximum height of the arch?

b. What is the height of the arch 8 ft to the right of center?

c. How far from the center is the arch 8 ft tall? Round to the nearest hundredth.

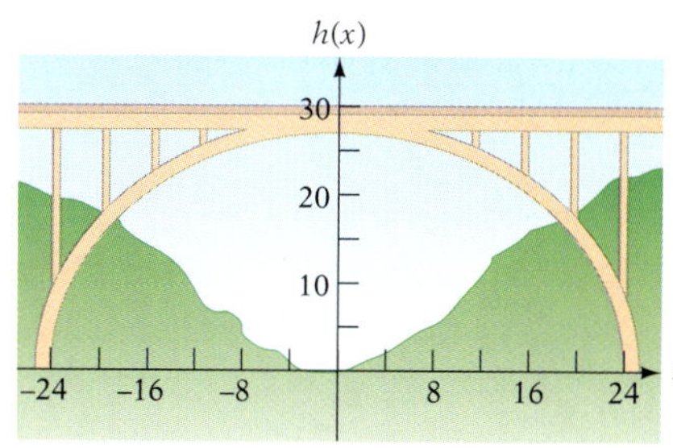

27. **Physics** Using Torricelli's Principle, it can be shown that the depth d, in centimeters, of a liquid in a bottle with a hole of area 0.5 cm^2 in its side can be approximated by $d = 0.0034t^2 - 0.52518t + 20$, where t is the time, in seconds, since a stopper was removed from the hole. When will the depth be 10 cm? Round to the nearest tenth.

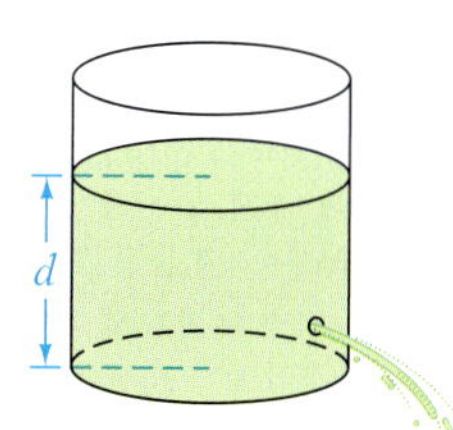

28. **Geometry** The volumes of two spheres differ by 372π cm^3. The radius of the larger sphere is 3 cm more than the radius of the smaller sphere. Find the radius of the larger sphere. $\left(\textit{Hint:} \text{ The formula for the volume of a sphere is } V = \frac{4}{3}\pi r^3.\right)$

29. **Geometry** A perfectly spherical scoop of mint chocolate chip ice cream is placed in a cone as shown at the right. How far is the bottom of the scoop of ice cream from the bottom of the cone? Round to the nearest tenth. (*Hint:* A line segment from the center of the scoop of ice cream to the point at which the ice cream touches the cone is perpendicular to the edge of the cone.)

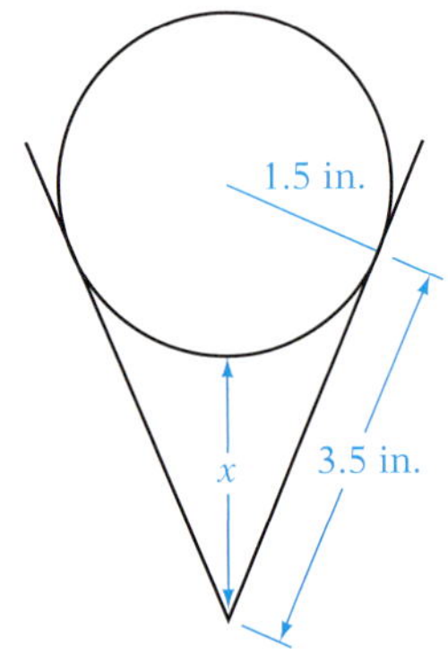

9.5 Nonlinear Inequalities

1 Solve nonlinear inequalities

VIDEO & DVD   CD TUTOR   WEB   SSM

A **quadratic inequality in one variable** is one that can be written in the form $ax^2 + bx + c < 0$ or $ax^2 + bx + c > 0$, where $a \neq 0$. The symbols $\leq$ and $\geq$ can also be used.

Quadratic inequalities can be solved by algebraic means. However, it is often easier to use a graphical method to solve these inequalities. The graphical method is used in the example that follows.

Solve and graph the solution set of $x^2 - x - 6 < 0$.

Factor the trinomial.

$$x^2 - x - 6 < 0$$
$$(x - 3)(x + 2) < 0$$

On a number line, draw vertical lines at the numbers that make each factor equal to zero.

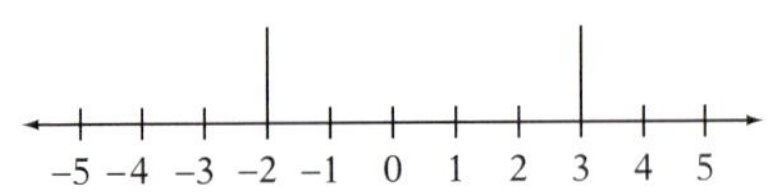

$$x - 3 = 0 \qquad x + 2 = 0$$
$$x = 3 \qquad x = -2$$

For each factor, place plus signs above the number line for those regions where the factor is positive and negative signs where the factor is negative. $x - 3$ is positive for $x > 3$ and $x + 2$ is positive for $x > -2$.

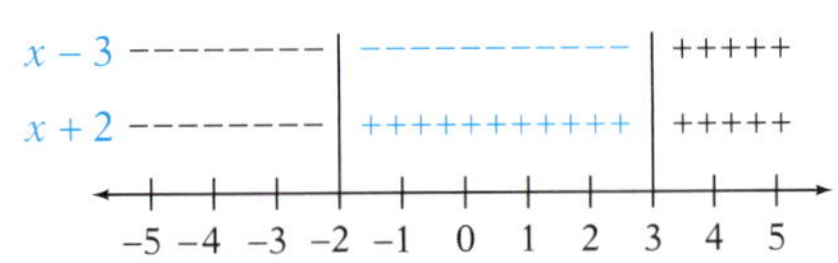

Because $x^2 - x - 6 < 0$, the solution set will be the regions where one factor is positive and the other factor is negative.

Take Note

For each factor, choose a number in each region. For example, when $x = -4$, $x - 3$ is negative; when $x = 1$, $x - 3$ is negative; and when $x = 4$, $x - 3$ is positive. When $x = -4$, $x + 2$ is negative; when $x = 1$, $x + 2$ is positive; and when $x = 4$, $x + 2$ is positive.

Write the solution set. $\{x \mid -2 < x < 3\}$

The graph of the solution set of the inequality $x^2 - x - 6 < 0$ is shown at the right.

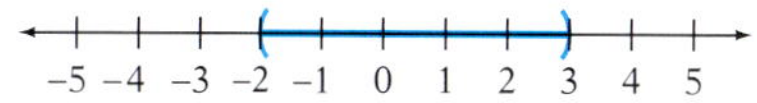

This method of solving quadratic inequalities can be used on any polynomial that can be factored into linear factors.

Solve and graph the solution set of $x^3 - 4x^2 - 4x + 16 > 0$.

Factor the polynomial by grouping.

$$x^3 - 4x^2 - 4x + 16 > 0$$
$$x^2(x - 4) - 4(x - 4) > 0$$
$$(x^2 - 4)(x - 4) > 0$$
$$(x - 2)(x + 2)(x - 4) > 0$$

On a number line, identify for each factor the regions where the factor is positive and where the factor is negative.

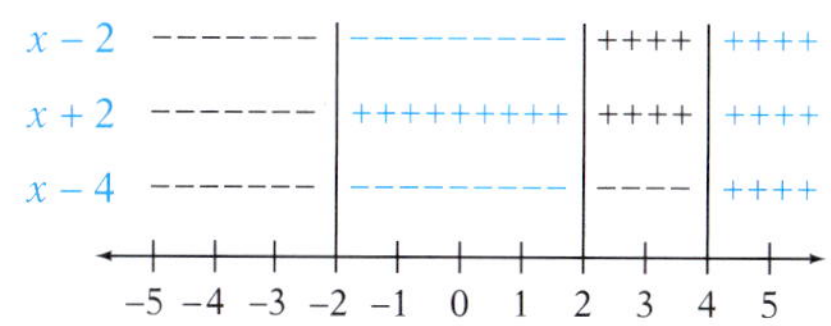

There are two regions where the product of the three factors is positive.

Write the solution set. $\{x \mid -2 < x < 2 \text{ or } x > 4\}$

The graph of the solution set of the inequality $x^3 - 4x^2 - 4x + 16 > 0$ is shown at the right.

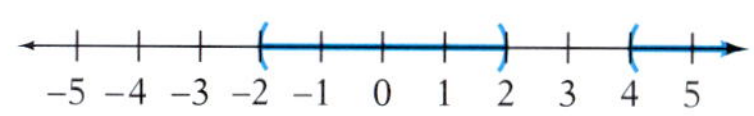

Example 1 Solve and graph the solution set of $2x^2 - x - 3 \geq 0$.

Solution

$$2x^2 - x - 3 \geq 0$$
$$(2x - 3)(x + 1) \geq 0$$

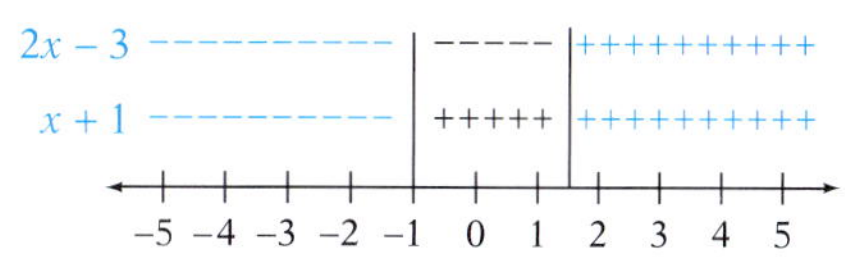

$$\left\{x \,\middle|\, x \leq -1 \text{ or } x \geq \frac{3}{2}\right\}$$

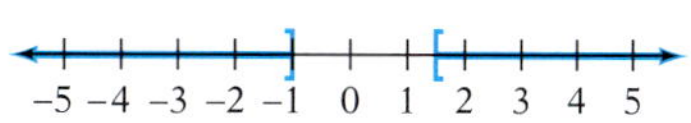

Problem 1 Solve and graph the solution set of $2x^2 - x - 10 \leq 0$.

Solution See page S30.

The graphical method can be used to solve rational inequalities.

Solve: $\dfrac{2x - 5}{x - 4} \le 1$

$$\frac{2x - 5}{x - 4} \le 1$$

Rewrite the inequality so that 0 appears on the right side of the inequality.

$$\frac{2x - 5}{x - 4} - 1 \le 0$$

Then simplify.

$$\frac{2x - 5}{x - 4} - \frac{x - 4}{x - 4} \le 0$$

$$\frac{x - 1}{x - 4} \le 0$$

On a number line, identify for each factor of the numerator and each factor of the denominator the regions where the factor is positive and where the factor is negative.

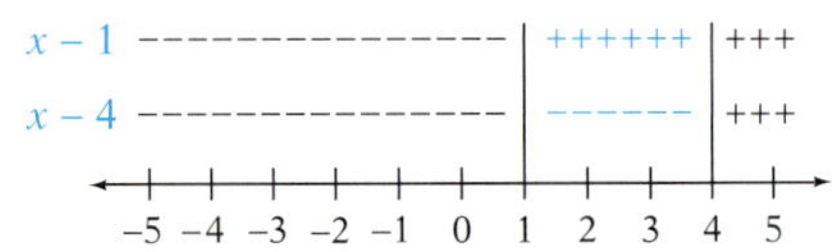

The region where the quotient of the two factors is negative is between 1 and 4.

Write the solution set. $\{x \mid 1 \le x < 4\}$

Note that 1 is part of the solution set, but 4 is not part of the solution set because the denominator of the rational expression is zero when $x = 4$.

Example 2 Solve and graph the solution set of $\dfrac{x + 4}{x - 3} \ge 0$.

Solution $\dfrac{x + 4}{x - 3} \ge 0$

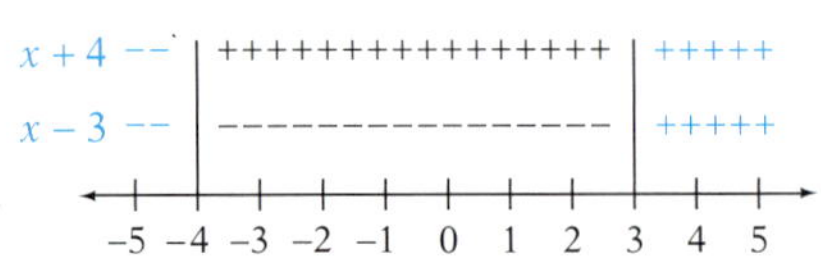

$\{x \mid x > 3 \text{ or } x \le -4\}$

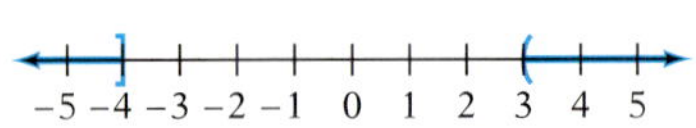

Problem 2 Solve and graph the solution set of $\dfrac{x}{x - 2} \le 0$.

Solution See page S31.

9.5 Concept Review

Determine whether the following statements are always true, sometimes true, or never true.

1. The end points of a solution set of a quadratic inequality are not included in the solution set.

2. The solution set of $(x - 3)(x - 2)(x + 2) > 0$ is 3, 2, and -2.

3. The solution set of a quadratic inequality is an infinite set.

4. The solution set of $\frac{1}{x - 1} > 0$ is $\{x \mid x > 1\}$.

9.5 Exercises

1 **1.** If $(x - 3)(x - 5) > 0$, what must be true of the values of $x - 3$ and $x - 5$?

2. For the nonlinear inequality $\frac{x - 2}{x - 3} \geq 1$, is 3 a possible element of the solution set? Why or why not?

Solve and graph the solution set.

3. $(x - 4)(x + 2) > 0$

4. $(x + 1)(x - 3) > 0$

5. $x^2 - 3x + 2 \geq 0$

6. $x^2 + 5x + 6 > 0$

7. $x^2 - x - 12 < 0$

8. $x^2 + x - 20 < 0$

9. $(x - 1)(x + 2)(x - 3) < 0$

10. $(x + 4)(x - 2)(x - 1) \geq 0$

11. $\frac{x - 4}{x + 2} > 0$

12. $\frac{x + 2}{x - 3} > 0$

13. $\frac{x - 3}{x + 1} \leq 0$

14. $\frac{x - 1}{x} > 0$

15. $\frac{(x - 1)(x + 2)}{x - 3} \leq 0$

16. $\frac{(x + 3)(x - 1)}{x - 2} \geq 0$

Solve.

17. $x^2 - 16 > 0$

18. $x^2 - 4 \geq 0$

19. $x^2 - 4x + 4 > 0$

20. $x^2 + 6x + 9 > 0$

21. $x^2 - 9x \leq 36$

22. $x^2 + 4x > 21$

23. $2x^2 - 5x + 2 \geq 0$

24. $4x^2 - 9x + 2 < 0$

25. $4x^2 - 8x + 3 < 0$

26. $2x^2 + 11x + 12 \geq 0$

27. $(x - 6)(x + 3)(x - 2) \leq 0$

28. $(x + 5)(x - 2)(x - 3) > 0$

29. $(2x - 1)(x - 4)(2x + 3) > 0$

30. $(x - 2)(3x - 1)(x + 2) \leq 0$

31. $x^3 + 3x^2 - x - 3 \leq 0$

32. $x^3 + x^2 - 9x - 9 < 0$

33. $x^3 - x^2 - 4x + 4 \geq 0$

34. $2x^3 + 3x^2 - 8x - 12 \geq 0$

35. $\dfrac{3x}{x - 2} > 1$

36. $\dfrac{2x}{x + 1} < 1$

37. $\dfrac{2}{x + 1} \geq 2$

38. $\dfrac{3}{x - 1} < 2$

39. $\dfrac{x}{(x - 1)(x + 2)} \geq 0$

40. $\dfrac{x - 2}{(x + 1)(x - 1)} \leq 0$

41. $\dfrac{1}{x} < 2$

42. $\dfrac{x}{2x - 1} \geq 1$

9.5 Applying Concepts

Graph the solution set.

43. $(x + 2)(x - 3)(x + 1)(x + 4) > 0$

44. $(x - 1)(x + 3)(x - 2)(x - 4) \geq 0$

45. $(x^2 + 2x - 8)(x^2 - 2x - 3) < 0$

46. $(x^2 + 2x - 3)(x^2 + 3x + 2) \geq 0$

47. $(x^2 + 1)(x^2 - 3x + 2) > 0$

48. $(x^2 - 9)(x^2 + 5x + 6) \leq 0$

49. $\dfrac{x^2(3 - x)(2x + 1)}{(x + 4)(x + 2)} \geq 0$

50. $\dfrac{1}{x} + x > 2$

51. $3x - \dfrac{1}{x} \leq 2$

52. $x^2 - x < \dfrac{1 - x}{x}$

9.6 Properties of Quadratic Functions

1 Graph a quadratic function

VIDEO & DVD CD TUTOR WEB SSM

Recall that a linear function is one that can be expressed by the equation $f(x) = mx + b$. The graph of a linear function has certain characteristics. It is a straight line with slope m and y-intercept $(0, b)$. A **quadratic function** is one that can be expressed by the equation $f(x) = ax^2 + bx + c, a \neq 0$. The graph of this function, which is called a **parabola,** also has certain characteristics. The graph of a quadratic function can be drawn by finding ordered pairs that belong to the function.

Take Note

Sometimes the value of the independent variable is called the **input** because it is *put in* place of the independent variable. The result of evaluating the function is called the **output.**

An **input/output table** shows the results of evaluating a function for various values of the independent variable. An input/output table for $f(x) = x^2 - 2x - 3$ is shown at the right.

Graph: $f(x) = x^2 - 2x - 3$

By evaluating the function for various values of x, find enough ordered pairs to determine the shape of the graph.

x	$f(x) = x^2 - 2x - 3$	$f(x)$	(x, y)
-2	$f(-2) = (-2)^2 - 2(-2) - 3$	5	$(-2, 5)$
-1	$f(-1) = (-1)^2 - 2(-1) - 3$	0	$(-1, 0)$
0	$f(0) = (0)^2 - 2(0) - 3$	-3	$(0, -3)$
1	$f(1) = (1)^2 - 2(1) - 3$	-4	$(1, -4)$
2	$f(2) = (2)^2 - 2(2) - 3$	-3	$(2, -3)$
3	$f(3) = (3)^2 - 2(3) - 3$	0	$(3, 0)$
4	$f(4) = (4)^2 - 2(4) - 3$	5	$(4, 5)$

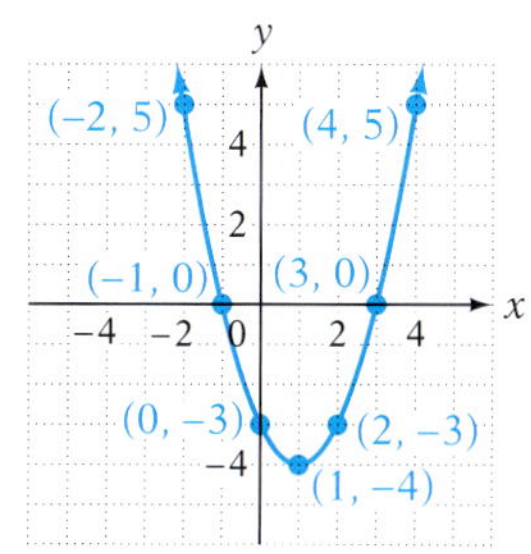

$f(x) = x^2 - 2x - 3$

Because the value of $f(x) = x^2 - 2x - 3$ is a real number for all values of x, the domain of f is all real numbers. From the graph, it appears that no value of y is less than -4. Thus the range is $\{y \mid y \geq -4\}$. The range can also be determined algebraically, as shown below, by completing the square.

Take Note

In completing the square, 1 is both added and subtracted. Because $1 - 1 = 0$, the expression $x^2 - 2x - 3$ is not changed. Note that

$$(x - 1)^2 - 4$$
$$= (x^2 - 2x + 1) - 4$$
$$= x^2 - 2x - 3$$

which is the original expression.

$$f(x) = x^2 - 2x - 3$$

Group the variable terms. $\quad f(x) = (x^2 - 2x) - 3$

Complete the square of $x^2 - 2x$. Add and subtract $\left[\frac{1}{2}(-2)\right]^2 = 1$ to $x^2 - 2x$. $\quad f(x) = (x^2 - 2x + 1) - 1 - 3$

Factor and combine like terms. $\quad f(x) = (x - 1)^2 - 4$

The square of a number is always positive. $\quad (x - 1)^2 \geq 0$

Subtract 4 from each side of the inequality. $\quad (x - 1)^2 - 4 \geq -4$

Replace $(x - 1)^2 - 4$ with $f(x)$. $\quad f(x) \geq -4$

$$y \geq -4$$

From the last inequality, the range is $\{y \mid y \geq -4\}$.

In general, the graph of $f(x) = ax^2 + bx + c, a \neq 0$, resembles a "cup" shape as shown below. The parabola opens up when $a > 0$ and opens down when

> **Take Note**
>
> The suspension cables for some bridges, such as the Golden Gate bridge, have the shape of a parabola. Parabolic shapes are also used for the mirrors in telescopes and in the design of certain antennas.

$a < 0$. The **vertex** of the parabola is the point with the smallest y-coordinate when $a > 0$ and is the point with the largest y-coordinate when $a < 0$.

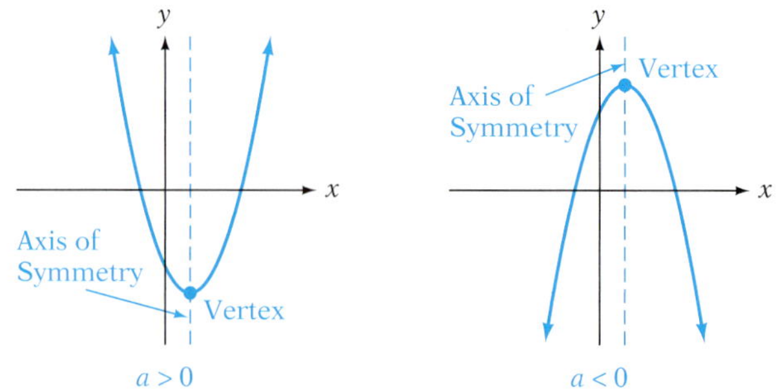

The **axis of symmetry** is a line that passes through the vertex of the parabola and is parallel to the y-axis. To understand the axis of symmetry, think of folding the graph along that line. The two portions of the graph will match up.

The vertex and axis of symmetry of a parabola can be found by completing the square.

Find the vertex and the axis of symmetry of the graph of $F(x) = x^2 + 4x + 3$.

To find the coordinates of the vertex, complete the square. $\quad F(x) = x^2 + 4x + 3$

Group the variable terms. $\quad F(x) = (x^2 + 4x) + 3$

Complete the square of $x^2 + 4x$. $\quad F(x) = (x^2 + 4x + 4) - 4 + 3$

Add and subtract $\left[\frac{1}{2}(4)\right]^2 = 4$ to and from $x^2 + 4x$.

Factor and combine like terms. $\quad F(x) = (x + 2)^2 - 1$

Because a, the coefficient of x^2, is positive ($a = 1$), the parabola opens up and the vertex is the point with the least y-coordinate. Because $(x + 2)^2 \geq 0$ for all values of x, the least y-coordinate occurs when $(x + 2)^2 = 0$. When $x = -2$, this expression equals zero.

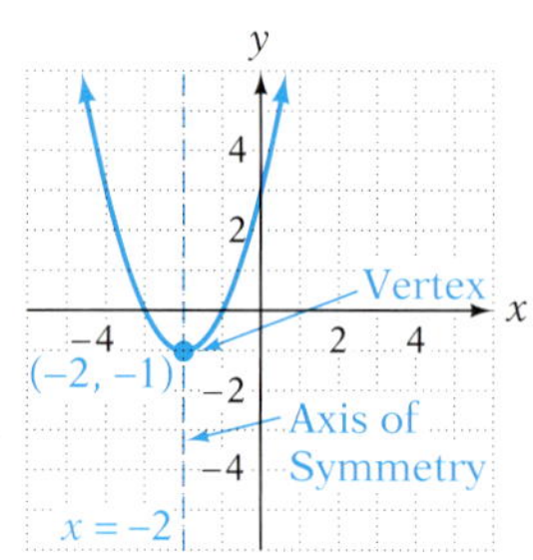

The x-coordinate of the vertex is -2.

To find the y-coordinate of the vertex, evaluate the function at $x = -2$.

$$F(x) = (x + 2)^2 - 1$$
$$F(-2) = (-2 + 2)^2 - 1 = -1$$

The y-coordinate of the vertex is -1.

From these results, the coordinates of the vertex are $(-2, -1)$. The axis of symmetry is the line whose equation is $x = -2$.

By following the process illustrated in the last example and completing the square of $f(x) = ax^2 + bx + c$, we can find a formula for the coordinates of the vertex of a parabola.

Vertex and Axis of Symmetry of a Parabola

Let $f(x) = ax^2 + bx + c$ be the equation of a parabola. The coordinates of the vertex are $\left(-\frac{b}{2a}, f\left(-\frac{b}{2a}\right)\right)$. The equation of the axis of symmetry is $x = -\frac{b}{2a}$.

The coordinates of the vertex and the axis of symmetry can be used to graph a parabola.

Graph $y = x^2 + 2x - 3$ using the vertex and the axis of symmetry.

Find the x-coordinate of the vertex. $a = 1$ and $b = 2$.

$$x = -\frac{b}{2a} = -\frac{2}{2(1)} = -1$$

Find the y-coordinate of the vertex by replacing x with -1 and solving for y.

$$y = x^2 + 2x - 3$$
$$y = (-1)^2 + 2(-1) - 3$$
$$y = 1 - 2 - 3$$
$$y = -4$$

The vertex is $(-1, -4)$.

The axis of symmetry is the line $x = -1$.

Find some ordered-pair solutions of the equation and record these in a table. Because the graph is symmetric to the line $x = -1$, choose values of x greater than -1.

x	y
0	-3
1	0
2	5

Graph the ordered-pair solutions on a rectangular coordinate system. Use symmetry to locate points of the graph on the other side of the axis of symmetry. Remember that corresponding points on the graph are the same distance from the axis of symmetry.

Draw a parabola through the points.

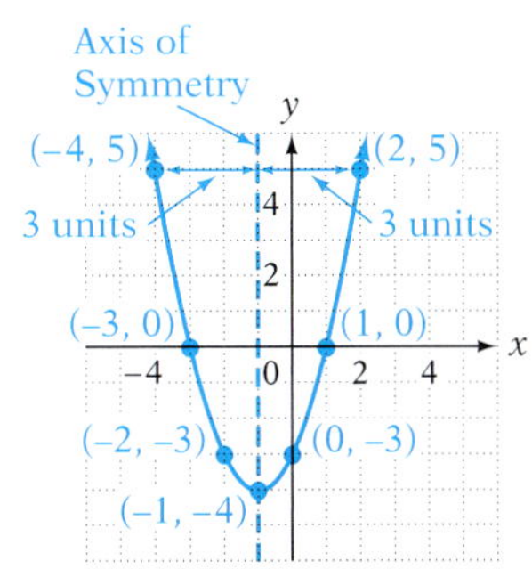

Using a graphing utility, enter the equation $y = x^2 + 2x - 3$ and verify the graph shown above. Now trace along the graph and verify that $(-1, -4)$ are the coordinates of the vertex.

Example 1 Find the vertex and the axis of symmetry of the parabola whose equation is $y = -3x^2 + 6x + 1$. Then sketch its graph.

Solution

$-\frac{b}{2a} = -\frac{6}{2(-3)} = 1$ • **Find the x-coordinate of the vertex. $a = -3$ and $b = 6$.**

$y = -3x^2 + 6x + 1$ • **Find the y-coordinate of the vertex by replacing x by 1 and solving for y.**
$y = -3(1)^2 + 6(1) + 1$
$y = 4$

The vertex is (1, 4).

The axis of symmetry is the line $x = 1$. • **The axis of symmetry is the line $x = -\frac{b}{2a}$.**

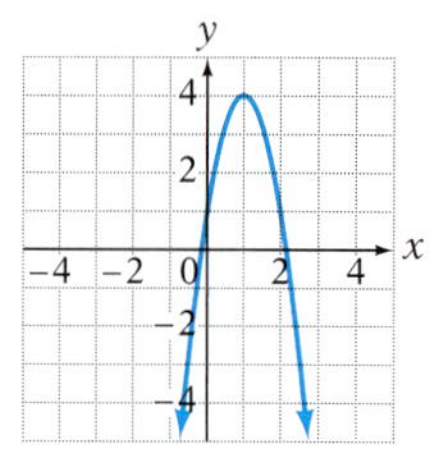

• **Because a is negative, the parabola opens down. Find a few ordered pairs, and use symmetry to sketch the graph.**

Problem 1 Find the vertex and the axis of symmetry of the parabola whose equation is $y = x^2 - 2$. Then sketch its graph.

Solution See page S31.

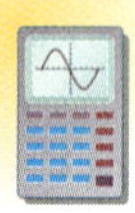

Using a graphing utility, enter the equation $y = -3x^2 + 6x + 1$ and verify the graph drawn in Example 1. Now trace along the graph and verify that the vertex is (1, 4). Follow the same procedure for Problem 1.

Example 2 Graph $f(x) = -2x^2 - 4x + 3$. State the domain and range of the function.

Solution Because a is negative $(a = -2)$, the graph of f will open down. The x-coordinate of the vertex is

$$x = -\frac{b}{2a} = -\frac{-4}{2(-2)} = -1.$$

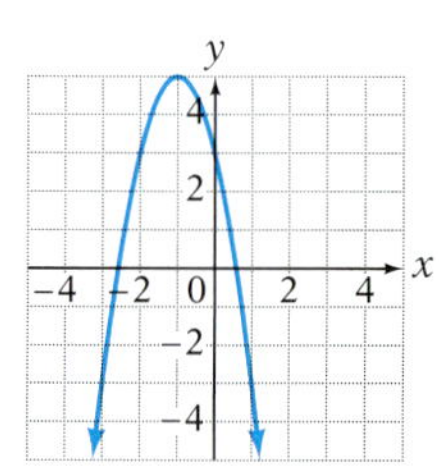

The y-coordinate of the vertex is $f(-1) = -2(-1)^2 - 4(-1) + 3 = 5$. The vertex is $(-1, 5)$.

Evaluate $f(x)$ for various values of x, and use symmetry to draw the graph.

Because $f(x) = -2x^2 - 4x + 3$ is a real number for all values of x, the domain of the function is $\{x \mid x \in \text{real numbers}\}$. The

Take Note

Once the coordinates of the vertex are found, the range of the quadratic function can be determined.

vertex of the parabola is the highest point on the graph. Because the y-coordinate at that point is 5, the range is $\{y \mid y \leq 5\}$.

Problem 2 Graph $g(x) = x^2 + 4x - 2$. State the domain and range of the function.

Solution See page S31.

2 Find the *x*-intercepts of a parabola

VIDEO & DVD CD TUTOR WEB SSM

Recall that a point at which a graph crosses the x- or y-axis is called an *intercept* of the graph. The x-intercepts of the graph of an equation occur when $y = 0$; the y-intercepts occur when $x = 0$.

The graph of $y = x^2 + 3x - 4$ is shown at the right. The points whose coordinates are $(-4, 0)$ and $(1, 0)$ are x-intercepts of the graph.

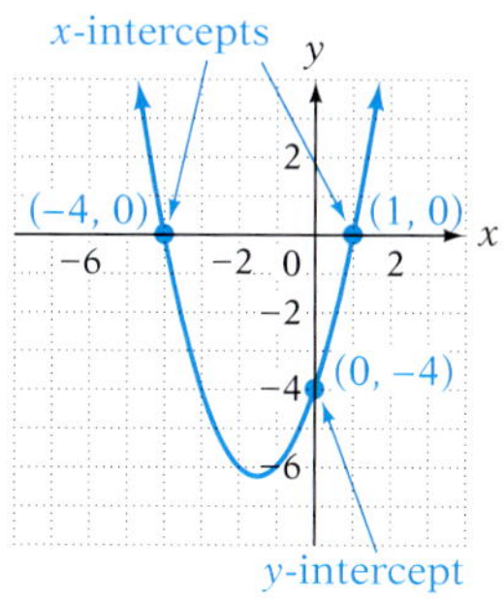

Find the x-intercepts for the parabola whose equation is $y = 4x^2 - 4x + 1$.

$$y = 4x^2 - 4x + 1$$

To find x-intercepts, let $y = 0$.

$$0 = 4x^2 - 4x + 1$$

Solve for x by factoring and using the Principle of Zero Products.

$$0 = (2x - 1)(2x - 1)$$

$$2x - 1 = 0 \qquad 2x - 1 = 0$$
$$2x = 1 \qquad 2x = 1$$
$$x = \frac{1}{2} \qquad x = \frac{1}{2}$$

The x-intercept is $\left(\frac{1}{2}, 0\right)$.

In this last example, the parabola has only one x-intercept. In this case, the parabola is said to be *tangent* to the x-axis at $x = \frac{1}{2}$.

Find the x-intercepts of $y = 2x^2 - x - 6$.

$$y = 2x^2 - x - 6$$

To find the x-intercepts, let $y = 0$.

$$0 = 2x^2 - x - 6$$

Solve for x by factoring and using the Principle of Zero Products.

$$0 = (2x + 3)(x - 2)$$

$$2x + 3 = 0 \qquad x - 2 = 0$$
$$x = -\frac{3}{2} \qquad x = 2$$

The x-intercepts are $\left(-\frac{3}{2}, 0\right)$ and $(2, 0)$.

Take Note

A **real zero** of a function is the x-coordinate of the x-intercept of the graph of the function. Because the x-intercepts of the graph of $f(x) = 2x^2 - x - 6$ are $\left(-\frac{3}{2}, 0\right)$ and $(2, 0)$, the zeros are $-\frac{3}{2}$ and 2.

If the equation above, $y = 2x^2 - x - 6$, were written in functional notation as $f(x) = 2x^2 - x - 6$, then to find the x-intercepts you would let $f(x) = 0$ and solve for x. A value of x for which $f(x) = 0$ is a *zero* of the function. Thus, $-\frac{3}{2}$ and 2 are *zeros* of $f(x) = 2x^2 - x - 6$.

Find the zeros of $f(x) = x^2 - 2x - 1$.

To find the zeros, let $f(x) = 0$ and solve for x.

$$f(x) = x^2 - 2x - 1$$
$$0 = x^2 - 2x - 1$$

Because $x^2 - 2x - 1$ does not easily factor, use the quadratic formula to solve for x.

$a = 1$, $b = -2$, and $c = -1$.

$$x = \frac{-b \pm \sqrt{b^2 - 4ac}}{2a}$$
$$= \frac{-(-2) \pm \sqrt{(-2)^2 - 4(1)(-1)}}{2(1)}$$
$$= \frac{2 \pm \sqrt{4 + 4}}{2}$$
$$= \frac{2 \pm \sqrt{8}}{2}$$
$$= \frac{2 \pm 2\sqrt{2}}{2}$$
$$= 1 \pm \sqrt{2}$$

The zeros of the function are $1 - \sqrt{2}$ and $1 + \sqrt{2}$.

The graph of $f(x) = x^2 - 2x - 1$ is shown at the right. Note that the zeros are the x-intercepts of the graph of f.

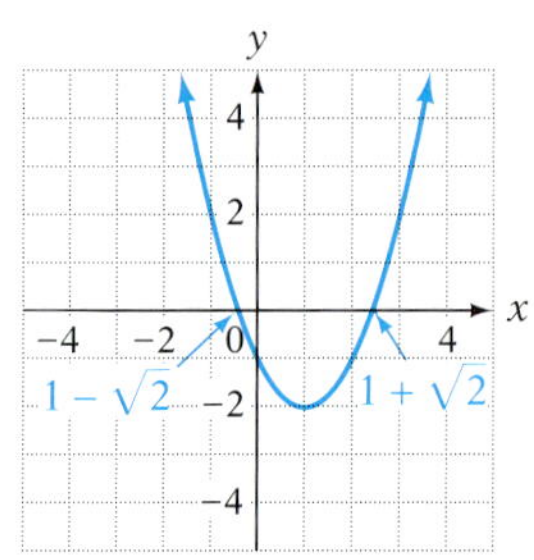

The preceding examples suggest that there is a relationship among the x-intercepts of the graph of a function, the zeros of a function, and the solutions of an equation. In fact, those three concepts are different ways of discussing the same number. The choice depends on the focus of the discussion. If we are discussing graphing, then intercept is our focus; if we are discussing functions, then the zero of the function is our focus; and if we are discussing equations, then the solution of the equation is our focus.

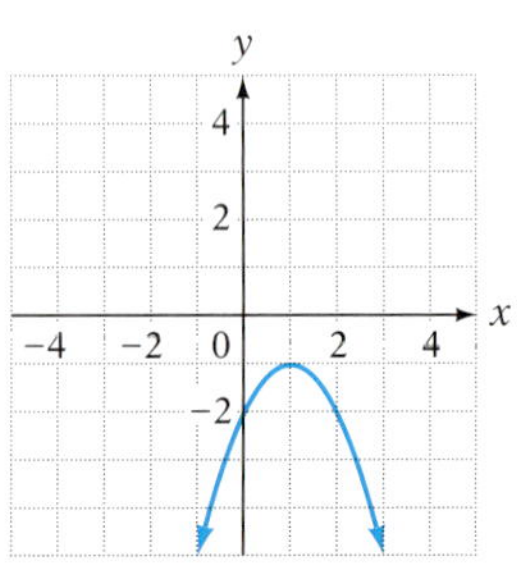

The graph of a parabola may not have x-intercepts. The graph of $y = -x^2 + 2x - 2$ is shown at the left. Note that the graph does not pass through the x-axis and thus there are no x-intercepts. This means that there are no real number zeros of $f(x) = -x^2 + 2x - 2$ and no real number solutions of $-x^2 + 2x - 2 = 0$.

Using the quadratic formula, we find that the solutions of the equation $-x^2 + 2x - 2 = 0$ are $1 - i$ and $1 + i$. Thus the zeros of $f(x) = -x^2 + 2x - 2$ are the complex numbers $1 - i$ and $1 + i$.

Example 3 Find the x-intercepts of the parabola given by each equation.

A. $y = x^2 + 2x - 2$ **B.** $y = 4x^2 + 4x + 1$

Solution **A.** $y = x^2 + 2x - 2$

$0 = x^2 + 2x - 2$ • Let $y = 0$.

$$x = \frac{-b \pm \sqrt{b^2 - 4ac}}{2a}$$

• The equation is nonfactorable over the integers. Use the quadratic formula to solve for x.

$$= \frac{-(2) \pm \sqrt{(2)^2 - 4(1)(-2)}}{2 \cdot 1}$$

$$= \frac{-2 \pm \sqrt{4 + 8}}{2}$$

$$= \frac{-2 \pm \sqrt{12}}{2}$$

$$= \frac{-2 \pm 2\sqrt{3}}{2} = -1 \pm \sqrt{3}$$

The x-intercepts are $(-1 + \sqrt{3}, 0)$ and $(-1 - \sqrt{3}, 0)$

B. $y = 4x^2 + 4x + 1$

$0 = 4x^2 + 4x + 1$ • Let $y = 0$.

$0 = (2x + 1)(2x + 1)$ • Solve for x by factoring.

$2x + 1 = 0$ $\quad$ $2x + 1 = 0$

$2x = -1$ $\quad$ $2x = -1$

$x = -\frac{1}{2}$ $\quad$ $x = -\frac{1}{2}$ • The equation has a double root.

The x-intercept is $\left(-\frac{1}{2}, 0\right)$.

Problem 3 Find the x-intercepts of the parabola given by each equation.

A. $y = 2x^2 - 5x + 2$ **B.** $y = x^2 + 4x + 4$

Solution See page S31.

The zeros of a quadratic function can always be determined exactly by solving a quadratic equation. However, it is possible to approximate the real zeros by graphing the function and determining the x-intercepts.

Example 4 Graph $f(x) = 2x^2 - x - 5$. Estimate the real zeros of the function to the nearest tenth.

Solution Graph the function. Then use the features of a graphing utility to estimate the zeros.

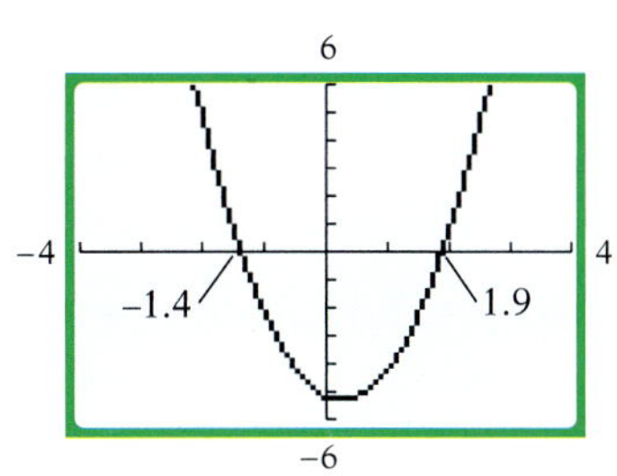

The approximate values of the zeros are -1.4 and 1.9.

Problem 4 Graph $f(x) = -x^2 - 4x + 6$. Estimate the real zeros of the function to the nearest tenth.

Solution See page S31.

Recall that the discriminant of $ax^2 + bx + c$ is the expression $b^2 - 4ac$ and that this expression can be used to determine whether $ax^2 + bx + c = 0$ has zero, one, or two real number solutions. Because there is a connection between the solutions of $ax^2 + bx + c = 0$ and the x-intercepts of the graph of $y = ax^2 + bx + c$, the discriminant can be used to determine the number of x-intercepts of a parabola.

Effect of the Discriminant on the Number of x-Intercepts of a Parabola

1. If $b^2 - 4ac = 0$, the parabola has one x-intercept.
2. If $b^2 - 4ac > 0$, the parabola has two x-intercepts.
3. If $b^2 - 4ac < 0$, the parabola has no x-intercepts.

Use the discriminant to determine the number of x-intercepts of the parabola whose equation is $y = 2x^2 - x + 2$.

$a = 2, b = -1, c = 2$

Evaluate the discriminant.

$$b^2 - 4ac = (-1)^2 - 4(2)(2) = 1 - 16 = -15$$

$$-15 < 0$$

The discriminant is less than zero, so the parabola has no x-intercepts.

Example 5 Use the discriminant to determine the number of x-intercepts of the parabola whose equation is $y = x^2 - 6x + 9$.

Solution

$$b^2 - 4ac = (-6)^2 - 4(1)(9)$$ • $a = 1, b = -6, c = 9$

$$= 36 - 36$$

$$= 0$$ • The discriminant is equal to zero.

The parabola has one x-intercept.

Problem 5 Use the discriminant to determine the number of x-intercepts of the parabola whose equation is $y = x^2 - x - 6$.

Solution See page S31.

9.6 Concept Review

Determine whether the following statements are always true, sometimes true, or never true.

1. The axis of symmetry of a parabola goes through the origin.
2. The vertex is the lowest point on the graph of a quadratic function.
3. A parabola has two x-intercepts.
4. If the discriminant for a quadratic function is zero, the graph of the function has one x-intercept.
5. The graph of a quadratic function has one y-intercept.
6. The domain of a quadratic function is the set of real numbers.

9.6 Exercises

1 1. What is a quadratic function?

2. What are the vertex and axis of symmetry of the graph of a parabola?

3. The axis of symmetry of a parabola is the line $x = -5$. What is the x-coordinate of the vertex of the parabola?

4. The axis of symmetry of a parabola is the line $x = 8$. What is the x-coordinate of the vertex of the parabola?

5. The vertex of a parabola is $(7, -9)$. What is the equation of the axis of symmetry of the parabola?

6. The vertex of a parabola is $(-4, 10)$. What is the equation of the axis of symmetry of the parabola?

Find the vertex and axis of symmetry of the parabola given by each equation. Then sketch its graph.

7. $y = x^2$

8. $y = -x^2$

9. $y = x^2 - 2$

10. $y = x^2 + 2$

11. $y = -x^2 + 3$

12. $y = -x^2 - 1$

13. $y = \frac{1}{2}x^2$

14. $y = 2x^2$

15. $y = 2x^2 - 1$

16. $y = -\frac{1}{2}x^2 + 2$

17. $y = x^2 - 2x$

18. $y = x^2 + 2x$

19. $y = -2x^2 + 4x$

20. $y = \frac{1}{2}x^2 - x$

21. $y = x^2 - x - 2$

22. $y = x^2 - 3x + 2$

23. $y = 2x^2 - x - 5$

24. $y = 2x^2 - x - 3$

Graph the function. State the domain and range of the function.

25. $f(x) = 2x^2 - 4x - 5$

26. $f(x) = 2x^2 + 8x + 3$

27. $f(x) = -2x^2 - 3x + 2$

28. $f(x) = 2x^2 - 7x + 3$

29. $f(x) = x^2 - 4x + 4$

30. $f(x) = -x^2 + 6x - 9$

31. $f(x) = x^2 + 4x - 3$

32. $f(x) = x^2 - 2x - 2$

33. $f(x) = -x^2 - 4x - 5$

34. $f(x) = -x^2 + 4x + 1$

2 **35.** How can you find the x-intercepts for the graph of a quadratic function?

36. How can the discriminant be used to determine the number of x-intercepts for the graph of a quadratic function?

Find the x-intercepts of the parabola given by each equation.

37. $y = x^2 - 4$

38. $y = x^2 - 9$

39. $y = 2x^2 - 4x$

40. $y = 3x^2 + 6x$

41. $y = x^2 - x - 2$

42. $y = x^2 - 2x - 8$

43. $y = 2x^2 - 5x - 3$

44. $y = 4x^2 + 11x + 6$

45. $y = 3x^2 - 19x - 14$

46. $y = 6x^2 + 7x + 2$

47. $y = 3x^2 - 19x + 20$

48. $y = 3x^2 + 19x + 28$

49. $y = 9x^2 - 12x + 4$

50. $y = x^2 - 2$

51. $y = 9x^2 - 2$

52. $y = 2x^2 - x - 1$

53. $y = 4x^2 - 4x - 15$

54. $y = x^2 + 2x - 1$

55. $y = x^2 + 4x - 3$

56. $y = x^2 + 6x + 10$

57. $y = -x^2 - 4x - 5$

58. $y = x^2 - 2x - 2$

59. $y = -x^2 - 2x + 1$

60. $y = -x^2 + 4x + 1$

Find the zeros of the function.

61. $f(x) = x^2 + 3x + 2$

62. $f(x) = x^2 - 6x + 9$

63. $f(x) = -x^2 + 4x - 5$

64. $f(x) = -x^2 + 3x + 8$

65. $f(x) = 2x^2 - 3x$

66. $f(x) = -3x^2 + 4x$

67. $f(x) = 2x^2 - 4$

68. $f(x) = 3x^2 + 6$

69. $f(x) = 2x^2 + 3x + 2$

70. $f(x) = 3x^2 - x + 4$

71. $f(x) = -3x^2 + 4x - 1$

72. $f(x) = -2x^2 + x + 5$

Graph the function. Estimate the real zeros of the function to the nearest tenth.

73. $f(x) = x^2 + 3x - 1$

74. $f(x) = x^2 - 2x - 4$

75. $f(x) = 2x^2 - 3x - 7$

76. $f(x) = -2x^2 - x + 2$

77. $f(x) = x^2 + 6x + 12$

78. $f(x) = x^2 - 3x + 9$

Use the discriminant to determine the number of x-intercepts of the graph of the parabola.

79. $y = 2x^2 + x + 1$

80. $y = 2x^2 + 2x - 1$

81. $y = -x^2 - x + 3$

82. $y = -2x^2 + x + 1$

83. $y = x^2 - 8x + 16$

84. $y = x^2 - 10x + 25$

85. $y = -3x^2 - x - 2$

86. $y = -2x^2 + x - 1$

87. $y = 4x^2 - x - 2$

88. $y = 2x^2 + x + 4$

89. $y = -2x^2 - x - 5$

90. $y = -3x^2 + 4x - 5$

91. $y = x^2 + 8x + 16$

92. $y = x^2 - 12x + 36$

93. $y = x^2 + x - 3$

94. The zeros of the function $f(x) = x^2 - 2x - 3$ are -1 and 3. What are the x-intercepts of the graph of the equation $y = x^2 - 2x - 3$?

95. The zeros of the function $f(x) = x^2 - x - 20$ are -4 and 5. What are the x-intercepts of the graph of the equation $y = x^2 - x - 20$?

9.6 Applying Concepts

Find the value of k such that the graph of the equation contains the given point.

96. $y = x^2 - 3x + k$; $(2, 5)$

97. $y = x^2 + 2x + k$; $(-3, 1)$

98. $y = 2x^2 + kx - 3$; $(4, -3)$

99. $y = 3x^2 + kx - 6$; $(-2, 4)$

Solve.

100. One root of the quadratic equation $2x^2 - 5x + k = 0$ is 4. What is the other root?

101. What is the value of k if the vertex of the parabola $y = x^2 - 8x + k$ is a point on the x-axis?

102. The roots of the function $f(x) = mx^2 + nx + 6$ are -2 and 3. What are the roots of the function $g(x) = nx^2 + mx - 6$?

103. The point (x_1, y_1) lies in Quadrant I and is a solution of the equation $y = 3x^2 - 2x - 1$. Given $y_1 = 5$, find x_1.

104. The point (x_1, y_1) lies in Quadrant II and is a solution of the equation $y = 2x^2 + 5x - 3$. Given $y_1 = 9$, find x_1.

105. Match each equation with its graph.

i. $f(x) = x^2$
ii. $f(x) = x^2 - 1$
iii. $f(x) = -x^2$
iv. $f(x) = -x^2 + 3$
v. $f(x) = x^2 + 3x - 2$
vi. $f(x) = -x^2 - 3x + 2$

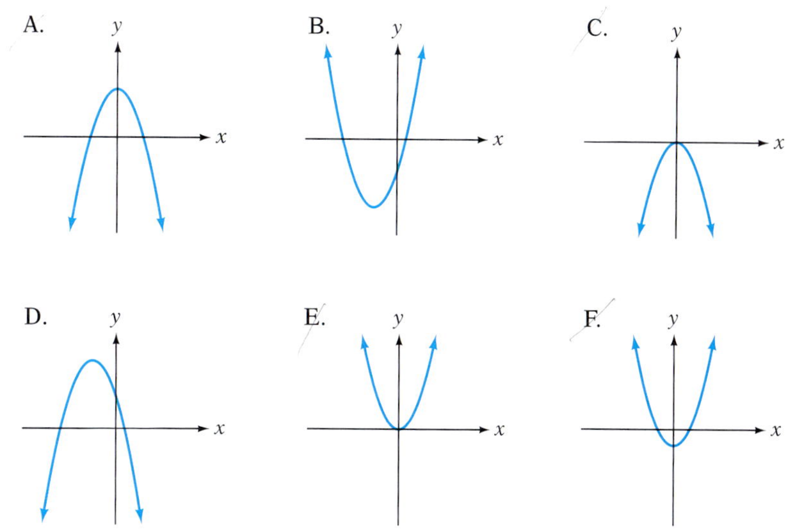

Physics The drag force F, in pounds, on a certain bicyclist is given by $F(v) = 0.151v^2$, where v is the velocity of the bicyclist in miles per hour.

106. Find the vertex of the parabola given by $F(v) = 0.151v^2$, and state what the vertex means in the context of this problem.

107. Find the drag force on the bicyclist when she or he is traveling 15 mph.

108. Graph the equation for the interval [0, 30].

An equation of the form $y = ax^2 + bx + c$ can be written in the form $y = a(x - h)^2 + k$, where (h, k) are the coordinates of the vertex of the parabola. Use the process of completing the square to rewrite the equation in the form $y = a(x - h)^2 + k$. Find the vertex. (*Hint:* Review the example at the bottom of page 601.)

109. $y = x^2 - 4x + 7$

110. $y = x^2 - 2x - 2$

111. $y = x^2 + x + 2$

112. $y = x^2 - x - 3$

Using $y = a(x - h)^2 + k$ as the equation of a parabola with the vertex at (h, k), find the equation of the parabola that has the given characteristics. Write the final equation in the form $y = ax^2 + bx + c$.

113. Vertex (1, 2); the graph passes through $P(2, 5)$

114. Vertex (0, −3); the graph passes through $P(3, -2)$

9.7 Minimum or Maximum Value of a Quadratic Function

1 Minimum and maximum problems

VIDEO & DVD CD TUTOR WEB SSM

The graph of $f(x) = x^2 - 2x + 3$ is shown at the right. Because a is positive, the parabola opens up. The vertex of the parabola is the lowest point on the parabola. It is the point that has the minimum y-coordinate. Therefore, the value of the function at this point is a **minimum.**

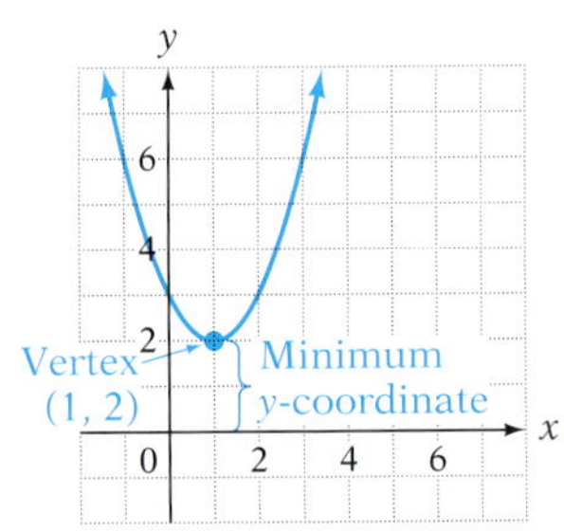

The graph of $f(x) = -x^2 + 2x + 1$ is shown at the right. Because a is negative, the parabola opens down. The vertex of the parabola is the highest point on the parabola. It is the point that has the maximum y-coordinate. Therefore, the value of the function at this point is a **maximum.**

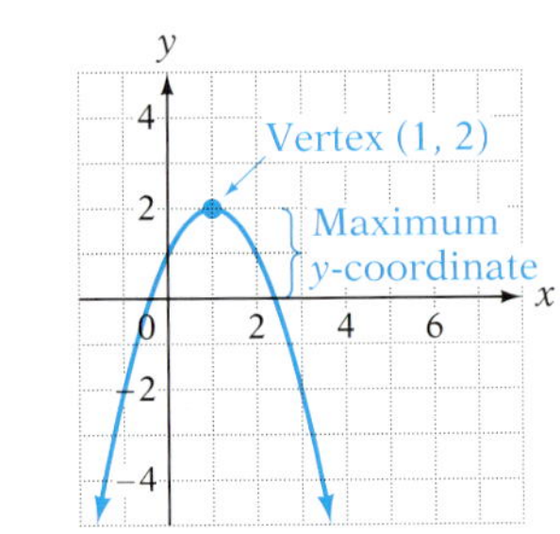

To find the minimum or maximum value of a quadratic function, first find the x-coordinate of the vertex. Then evaluate the function at that value.

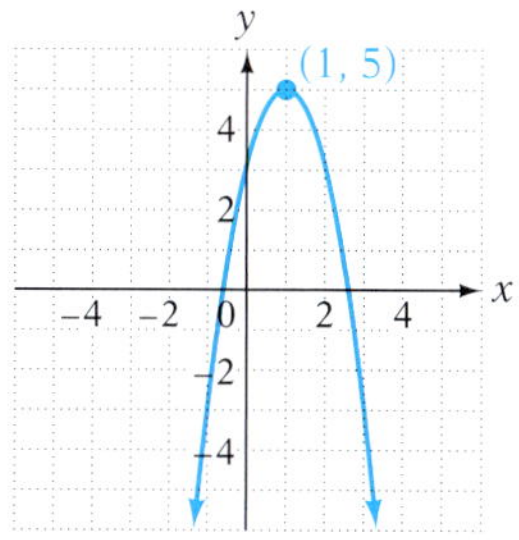

$f(x) = -2x^2 + 4x + 3$

Example 1 Find the maximum or minimum value of the function $f(x) = -2x^2 + 4x + 3$.

Solution

$$x = -\frac{b}{2a} = -\frac{4}{2(-2)} = 1$$

• **Find the x-coordinate of the vertex. $a = -2$, $b = 4$.**

$$f(x) = -2x^2 + 4x + 3$$
$$f(1) = -2(1)^2 + 4(1) + 3$$
$$f(1) = 5$$

• **Evaluate the function at $x = 1$.**

Because $a < 0$, the graph of f opens down. Therefore, the function has a maximum value.

The maximum value of the function is 5.

Problem 1 Find the maximum or minimum value of $f(x) = 2x^2 - 3x + 1$.

Solution See page S31.

2 Applications of minimum and maximum

VIDEO & DVD · CD TUTOR · WEB · SSM

Point of Interest

Calculus is a branch of mathematics that demonstrates, among other things, how to find the maximum or minimum of functions other than quadratic functions. These are very important problems in applied mathematics. For instance, an automotive engineer wants to design a car whose shape will minimize the effect of air flow. The same engineer tries to maximize the efficiency of a car's engine. Similarly, an economist may try to determine what business practices will minimize cost and maximize profit.

A mason is forming a rectangular floor for a storage shed. The perimeter of the rectangle is 44 ft. What dimensions of the rectangle would give the floor a maximum area? What is the maximum area?

We are given the perimeter of the rectangle, and we want to find the dimensions of the rectangle that will yield the maximum area for the floor. Use the equation for the perimeter of a rectangle.

$$P = 2L + 2W$$

$P = 44$ $\qquad 44 = 2L + 2W$

Divide both sides of the equation by 2. $\qquad 22 = L + W$

Solve the equation for W. $\qquad 22 - L = W$

Now use the equation for the area of a rectangle. Use substitution to express the area in terms of L.

$$A = LW$$

From the equation above, $W = 22 - L$.

Substitute $22 - L$ for W. $\qquad A = L(22 - L)$

$$A = 22L - L^2$$

The area of the rectangle is $22L - L^2$. To find the length of the rectangle, find the L-coordinate of the vertex of the function $f(L) = -L^2 + 22L$.

For the equation $f(L) = -L^2 + 22L$, $a = -1$ and $b = 22$.

$$L = -\frac{b}{2a} = -\frac{22}{2(-1)} = 11$$

The length of the rectangle is 11 ft.

To find the width, replace L in $22 - L$ by the L-coordinate of the vertex and evaluate.

$$W = 22 - L$$
$$W = 22 - 11 = 11$$

The width of the rectangle is 11 ft.

The dimensions of the rectangle that would give the floor a maximum area are 11 ft by 11 ft.

To find the maximum area of the floor, evaluate the function $f(L) = -L^2 + 22L$ at the L-coordinate of the vertex.

$$f(L) = -L^2 + 22L$$
$$f(11) = -(11)^2 + 22(11) = -121 + 242 = 121$$

The maximum area of the floor is 121 ft^2.

The graph of the function $f(L) = -L^2 + 22L$ is shown at the right. Note that the vertex of the parabola is (11, 121). For any value of L less than 11, the area of the floor will be less than 121 ft^2. For any value of L greater than 11, the area of the floor will be less than 121 ft^2. 121 is the maximum value of the function, and the maximum value occurs when $L = 11$.

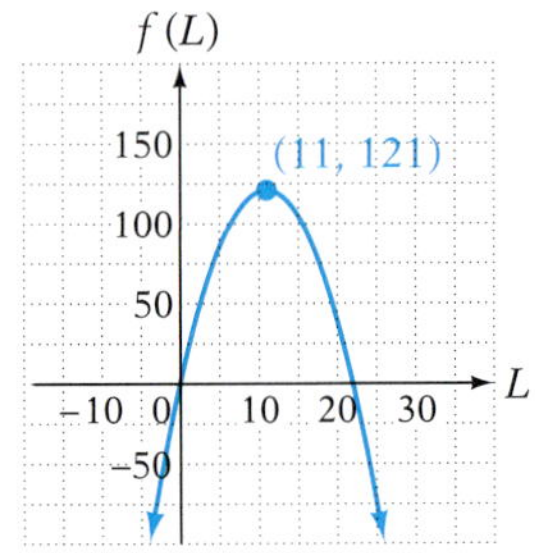

Example 2 A mining company has determined that the cost in dollars (c) per ton of mining a mineral is given by

$$c(x) = 0.2x^2 - 2x + 12$$

where x is the number of tons of the mineral that is mined. Find the number of tons of the mineral that should be mined to minimize the cost. What is the minimum cost?

Strategy

- To find the number of tons that will minimize the cost, find the x-coordinate of the vertex.
- To find the minimum cost, evaluate the function at the x-coordinate of the vertex.

Solution

$$x = -\frac{b}{2a} = -\frac{-2}{2(0.2)} = 5$$

To minimize the cost, 5 tons should be mined.

$$c(x) = 0.2x^2 - 2x + 12$$
$$c(5) = 0.2(5)^2 - 2(5) + 12 = 5 - 10 + 12 = 7$$

The minimum cost per ton is \$7.

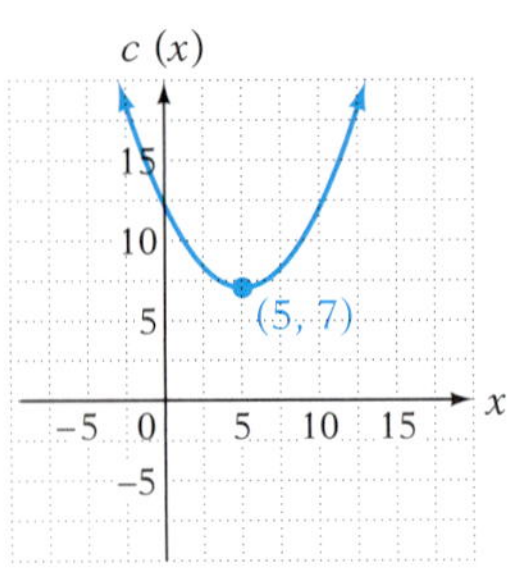

Note: The graph of the function $c(x) = 0.2x^2 - 2x + 12$ is shown at the left. The vertex of the parabola is (5, 7). For any value of x less than 5, the cost per ton is greater than \$7. For any value of x greater than 5, the cost per ton is greater than \$7. 7 is the minimum value of the function, and the minimum value occurs when $x = 5$.

Problem 2 The height in feet (s) of a ball thrown straight up is given by $s(t) = -16t^2 + 64t$, where t is the time in seconds. Find the time it takes the ball to reach its maximum height. What is the maximum height?

Solution See page S31.

Example 3 Find two numbers whose difference is 10 and whose product is a minimum.

Strategy Let x represent one number. Because the difference between the two numbers is 10, $x + 10$ represents the other number. Then their product is represented by $x(x + 10) = x^2 + 10x$.

- ▶ To find one of the two numbers, find the x-coordinate of the vertex of $f(x) = x^2 + 10x$.
- ▶ To find the other number, replace x in $x + 10$ by the x-coordinate of the vertex and evaluate.

Solution

$$x = -\frac{b}{2a} = -\frac{10}{2(1)} = -5$$

$$x + 10 = -5 + 10 = 5$$

The numbers are −5 and 5.

Problem 3 A rectangular fence is being constructed along a stream to enclose a picnic area. If 100 ft of fence is available, what dimensions of the rectangle will produce the maximum area for picnicking?

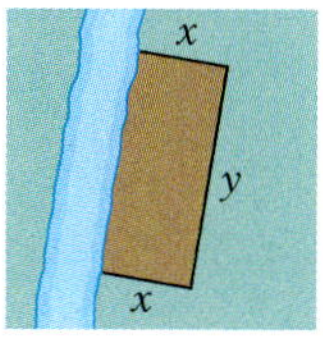

Solution See page S32.

9.7 Concept Review

Determine whether the following statements are always true, sometimes true, or never true.

1. A quadratic function has a minimum value.

2. The y-coordinate of the vertex of a quadratic function is the minimum value of the function.

3. The maximum or minimum of a quadratic function is the value of the function evaluated at the x-coordinate of the vertex.

4. The x-value of the vertex of $y = ax^2 + bx + c$, $a \neq 0$, is given by $-\frac{b}{2a}$.

5. The maximum value of a quadratic function is a positive number.

6. The function $f(x) = -x^2 + 3x - 2$ will have a maximum value.

9.7 Exercises

1 **1.** What is the minimum value or the maximum value of a quadratic function?

2. Describe how to find the minimum or maximum value of a quadratic function.

3. Does the function have a minimum or a maximum value?
a. $f(x) = -x^2 + 6x - 1$ **b.** $f(x) = 2x^2 - 4$ **c.** $f(x) = -5x^2 + x$

4. Does the function have a minimum or a maximum value?
a. $f(x) = 3x^2 - 2x + 4$ **b.** $f(x) = -x^2 + 9$ **c.** $f(x) = 6x^2 - 3x$

Find the minimum or maximum value of the quadratic function.

5. $f(x) = x^2 - 2x + 3$

6. $f(x) = 2x^2 + 4x$

7. $f(x) = -2x^2 + 4x - 3$

8. $f(x) = -2x^2 + 4x - 5$

9. $f(x) = -2x^2 - 3x + 4$

10. $f(x) = -2x^2 - 3x$

11. $f(x) = 2x^2 + 3x - 8$

12. $f(x) = 3x^2 + 3x - 2$

13. $f(x) = -3x^2 + x - 6$

14. $f(x) = -x^2 - x + 2$

15. $f(x) = x^2 - 5x + 3$

16. $f(x) = 3x^2 + 5x + 2$

17. Which of the following parabolas has the highest minimum value?
a. $y = x^2 - 2x - 3$ **b.** $y = x^2 - 10x + 20$ **c.** $y = 3x^2 - 6$

18. Which of the following parabolas has the highest maximum value?
a. $y = -2x^2 + 2x - 1$ **b.** $y = -x^2 + 8x - 2$ **c.** $y = -4x^2 + 3$

19. The vertex of a parabola that opens up is $(-4, 7)$. Does the function have a maximum or a minimum value? What is the maximum or minimum value of the function?

20. The vertex of a parabola that opens down is $(3, -5)$. Does the function have a maximum or a minimum value? What is the maximum or minimum value of the function?

2 Solve.

21. **Physics** The height s, in feet, of a rock thrown upward at an initial speed of 64 ft/s from a cliff 50 ft above an ocean beach is given by the function

$s(t) = -16t^2 + 64t + 50$, where t is the time in seconds. Find the maximum height above the beach that the rock will attain.

22. **Sports** An event in the Summer Olympics is 10-meter springboard diving. In this event, the height s, in meters, of a diver above the water t seconds after jumping is given by $s(t) = -4.9t^2 + 7.8t + 10$. What is the maximum height that the diver will be above the water? Round to the nearest tenth.

23. **Business** A tour operator believes that the profit P, in dollars, from selling x tickets is given by $P(x) = 40x - 0.25x^2$. Using this model, find the maximum profit that the tour operator can expect.

24. **Manufacturing** A manufacturer of microwave ovens believes that the revenue R, in dollars, that the company receives is related to the price P, in dollars, of an oven by the function $R(P) = 125P - 0.25P^2$. What price will give the maximum revenue?

25. **Chemistry** A pool is treated with a chemical to reduce the amount of algae. The amount of algae in the pool t days after the treatment can be approximated by the function $A(t) = 40t^2 - 400t + 500$. How many days after treatment will the pool have the least amount of algae?

26. **Structural Engineering** The suspension cable that supports a small footbridge hangs in the shape of a parabola. The height h, in feet, of the cable above the bridge is given by the function $h(x) = 0.25x^2 - 0.8x + 25$, where x is the distance in feet from one end of the bridge. What is the minimum height of the cable above the bridge?

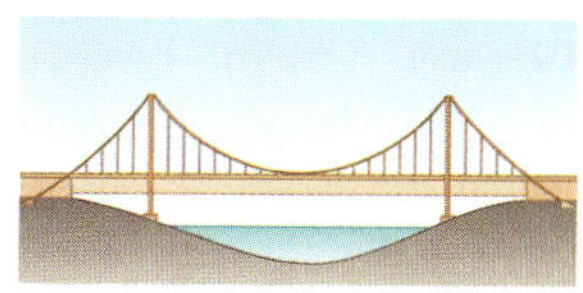

27. **Telescopes** An equation of the thickness h, in inches, of the mirror at the Palomar Mountain Observatory is given by $h(x) = 0.000379x^2 - 0.0758x + 24$, where x is measured from the edge of the mirror. Find the minimum thickness of the mirror.

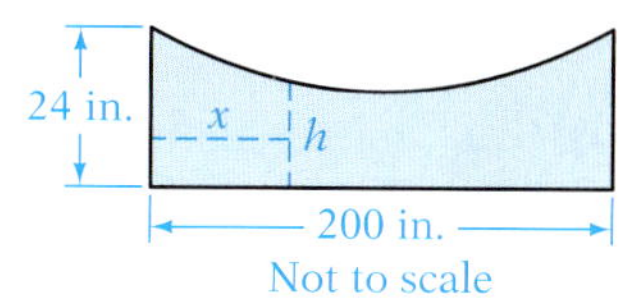

28. **Art** The Buckingham Fountain in Chicago shoots water from a nozzle at the base of the fountain. The height h, in feet, of the water above the ground t seconds after it leaves the nozzle is given by the function $h(t) = -16t^2 + 90t + 15$. What is the maximum height of the water? Round to the nearest tenth.

29. **Fire Science** The height s, in feet, of water squirting from a fire hose nozzle is given by the equation $s(x) = -\frac{1}{30}x^2 + 2x + 5$, where x is the horizontal distance from the nozzle. How high on a building 40 ft from the fire hose will the water land?

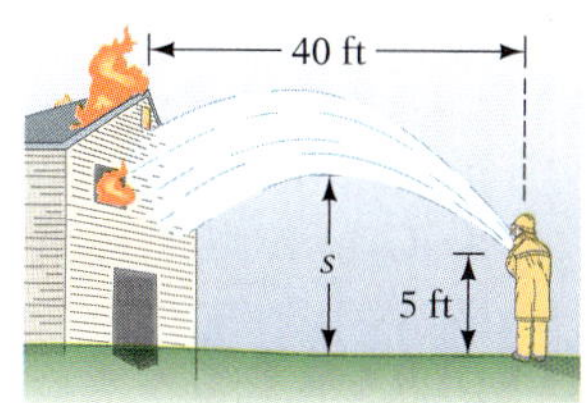

30. **Automotive Technology** On wet concrete, the stopping distance s, in feet, of a car traveling v miles per hour is given by $s(v) = 0.055v^2 + 1.1v$. At what speed could a car be traveling and still stop at a stop sign 44 ft away?

31. **Number Problem** Find two numbers whose sum is 20 and whose product is a maximum.

32. **Number Problem** Find two numbers whose difference is 14 and whose product is a minimum.

33. **Ranching** A rancher has 200 ft of fencing to build a rectangular corral alongside an existing fence. Determine the dimensions of the corral that will maximize the enclosed area.

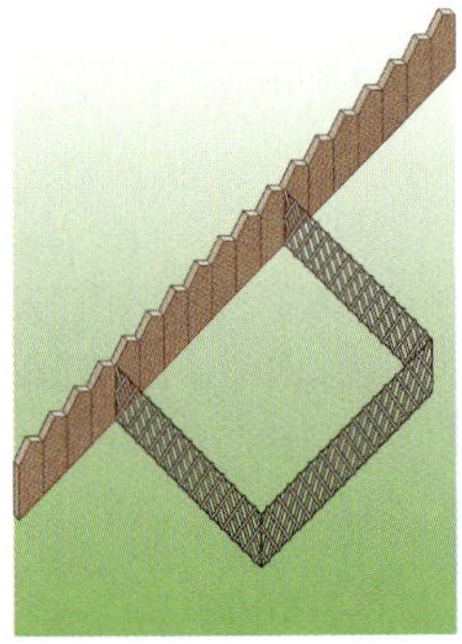

9.7 Applying Concepts

34. Show that the minimum value of

$$S(x) = (2 - x)^2 + (5 - x)^2 + (4 - x)^2 + (7 - x)^2$$

occurs when x is the average of the numbers 2, 5, 4, and 7.

Use a graphing utility to find the minimum or maximum value of the function. Round to the nearest tenth. See the Graphing Calculator Appendix for assistance.

35. $f(x) = x^4 - 2x^2 + 4$

36. $f(x) = x^4 + 2x^3 + 1$

37. $f(x) = -x^6 + x^4 - x^3 + x$

38. $f(x) = -x^8 + x^6 - x^4 + 5x^2 + 7$

39. On the basis of Exercises 35–38, make a conjecture about the relationship between the sign of the leading coefficient of a polynomial function of even degree and whether that function will have a minimum or a maximum value.

Solve.

40. **Traffic Control** Traffic engineers try to determine the effect a traffic light has at an intersection. By gathering data about the intersection, engineers can determine the approximate number of cars that enter the intersection in the horizontal direction and those that enter in the vertical direction. The engineers would also collect information on the time it takes a stopped car to regain the normal posted speed limit. One model of this situation is $T = \left(\frac{H + V}{2}\right)R^2 + (0.08H - 1.08V)R + 0.58V$, where H is the number of cars arriving at the intersection from the horizontal direction, V is the number of cars arriving at the intersection from the vertical direction, and R is the percent of time the light is red in the horizontal direction. T is the total delay time for all cars and is measured as the number of times the traffic light changes from red to green and back to red.

a. Graph this equation for $H = 100$, $V = 150$, and $0 \le R \le 1$.
b. Write a sentence that explains why the graph is drawn only for $0 \le R \le 1$.
c. What percent of the time should the traffic light remain red in the horizontal direction to minimize T? Round to the nearest whole percent.

41. Write a paper on the history of higher-degree equations. Include discoveries made by Del Farro, Ferrari, and Ruffini.

Focus on Problem Solving

Inductive and Deductive Reasoning

Consider the following sums of odd positive integers.

$1 + 3 = 4 = 2^2$ Sum of the first two odd numbers is 2^2.

$1 + 3 + 5 = 9 = 3^2$ Sum of the first three odd numbers is 3^2.

$1 + 3 + 5 + 7 = 16 = 4^2$ Sum of the first four odd numbers is 4^2.

$1 + 3 + 5 + 7 + 9 = 25 = 5^2$ Sum of the first five odd numbers is 5^2.

1. Make a conjecture about the value of the sum

$$1 + 3 + 5 + 7 + 9 + 11 + 13 + 15$$

without adding the numbers.

If the pattern continues, a possible conjecture is that the sum of the first eight odd positive integers is $8^2 = 64$. By adding the numbers you can verify that this is correct. Inferring that the pattern established by a few cases is valid for all cases is an example of **inductive reasoning.**

2. Use inductive reasoning to find the next figure in the sequence below.

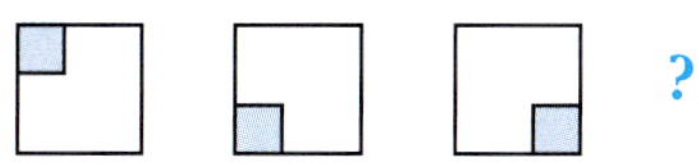

The fact that a pattern appears to be true does not *prove* that it is true for all cases. For example, consider the polynomial $n^2 - n + 41$. If we begin substituting positive integer values for n and evaluating the expression, we produce the following table.

n	$n^2 - n + 41$	
1	$1^2 - 1 + 41 = 41$	41 is a prime number.
2	$2^2 - 2 + 41 = 43$	43 is a prime number.
3	$3^2 - 3 + 41 = 47$	47 is a prime number.
4	$4^2 - 4 + 41 = 53$	53 is a prime number.

Even if we try a number like 30, we have $30^2 - 30 + 41 = 911$ and 911 is a prime number. Thus it appears that the value of this polynomial is a prime number for any value of n. However, the conjecture is not true. For instance, if $n = 41$, we have $41^2 - 41 + 41 = 41^2 = 1681$, which is not a prime number because it is divisible by 41. This illustrates that inductive reasoning may lead to incorrect conclusions and that an inductive proof must be available to prove conjectures. There is such a proof, called *mathematical induction,* that you may study in a future math course.

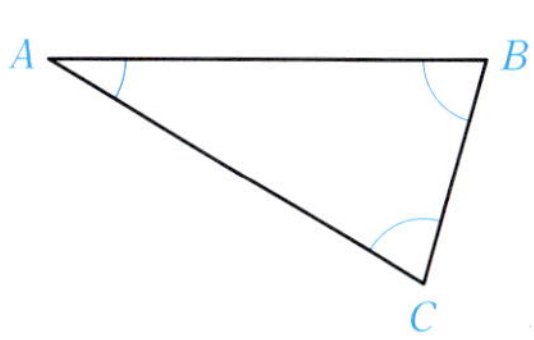

$\angle A + \angle B + \angle C = 180°$

Now consider the true statement that the sum of the measures of the interior angles of a triangle is 180°. The figure on the left is a triangle; therefore, the sum of the measures of the interior angles must be 180°. This is an example of *deductive reasoning.* **Deductive reasoning** uses a rule or statement of fact to reach a conclusion.

3. Use deductive reasoning to complete the following: "All even numbers are divisible by 2. Because 41,386 is an even number, . . ."

4. a. Use the pattern given below to find the three missing terms in __ __ __ 64.

   A 2 3 4 B 6 7 8 C 10 11 12 D 14 15 16 . . .

   b. Is this inductive or deductive reasoning?

5. a. Draw the next figure in the sequence:

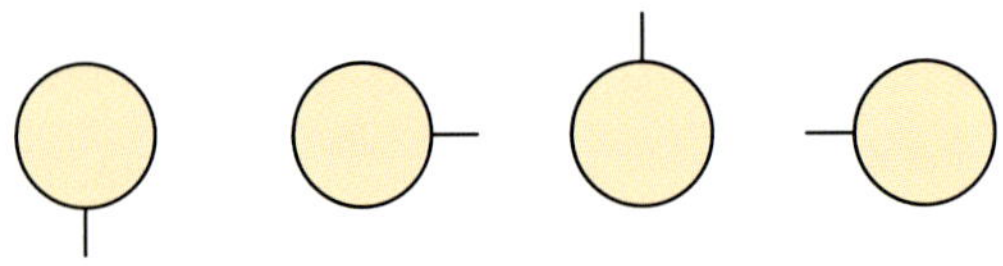

   b. Is this inductive or deductive reasoning?

6. a. If ◊◊◊◊◊ = ‡‡‡ and ‡‡‡ = ∇∇∇∇, how many ◊'s equal ∇∇∇∇∇∇∇∇?

   b. Is this inductive or deductive reasoning?

7. a. If ♠♠ = ♦♦♦♦♦♦, and ♦♦♦ = ♣♣, and ♣♣♣♣ = ♥, how many ♥'s equal ♠♠♠♠♠♠?

   b. Is this inductive or deductive reasoning?

For Exercises 8 to 11, determine whether inductive or deductive reasoning is being used.

8. The Atlanta Braves have won eight games in a row. Therefore, the Atlanta Braves will win their next game.

9. All quadrilaterals have four sides. A square is a quadrilateral. Therefore, a square has four sides.

10. Every English setter likes to hunt. Duke is an English setter, so Duke likes to hunt.

11. All kitchen remodeling jobs cost more than the contractor's estimate. The contractor estimated the cost of remodeling my kitchen at $35,000. Therefore, it will cost more than $35,000 to have my kitchen remodeled.

Projects & Group Activities

Completing the Square

Essentially all of the investigations into mathematics before the Renaissance were geometric. The solutions of quadratic equations were calculated from a construction of a certain area. Proofs of theorems, even theorems about numbers, were based entirely on geometry. In this project, we will examine the solution of a quadratic equation.

Solve: $x^2 + 6x = 7$

Begin with a line of unknown length, x, and one of length 6, the coefficient of x. Using these lines, construct a rectangle as shown in Figure 1.

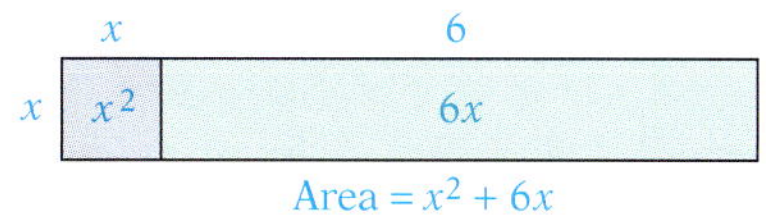

Figure 1

Now draw another area that has exactly the same area as Figure 1 by cutting one-half of the rectangle off and placing it on the bottom of the square. See Figure 2.

The shaded area in Figure 2 has exactly the same area as Figure 1. However, when the unshaded area is added to Figure 2 to make a square, the area of the square is 9 square units larger than that of Figure 1. In equation form,

$$(\text{Area of Figure 1}) + 9 = \text{Area of Figure 2}$$

or

$$x^2 + 6x + 9 = (x + 3)^2$$

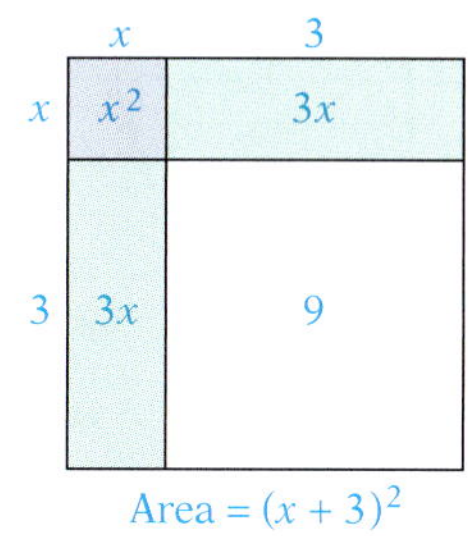

Figure 2

From the original equation, $x^2 + 6x = 7$. Thus,

$$\begin{aligned} x^2 + 6x + 9 &= (x + 3)^2 \\ 7 + 9 &= (x + 3)^2 && \bullet\ x^2 + 6x = 7 \\ 16 &= (x + 3)^2 \\ 4 &= x + 3 && \bullet\ \text{See note below.} \\ 1 &= x \end{aligned}$$

Note: Although early mathematicians knew that a quadratic equation may have two solutions, both solutions were allowed only if they were positive. After all, a geometric construction could not have a negative length. Therefore, the solution of this equation was 1; the solution -7 would have been dismissed as *fictitious,* the actual word that was frequently used through the 15th century for negative-number solutions of an equation.

Try to solve the quadratic equation $x^2 + 4x = 12$ by geometrically completing the square.

Using a Graphing Calculator to Solve a Quadratic Equation

Recall that an x-intercept of the graph of an equation is a point at which the graph crosses the x-axis. For the graph in Figure 3 on page 524, the x-intercepts are $(-2, 0)$ and $(3, 0)$.

Recall also that to find the x-intercept of a graph, set $y = 0$ and then solve for x. For the equation in Figure 3, if we set $y = 0$, the resulting equation is $0 = x^2 - x - 6$, which is a quadratic equation. Solving this equation by factoring, we have

$$\begin{aligned} 0 &= x^2 - x - 6 \\ 0 &= (x + 2)(x - 3) \end{aligned}$$

$$\begin{aligned} x + 2 &= 0 \qquad & x - 3 &= 0 \\ x &= -2 & x &= 3 \end{aligned}$$

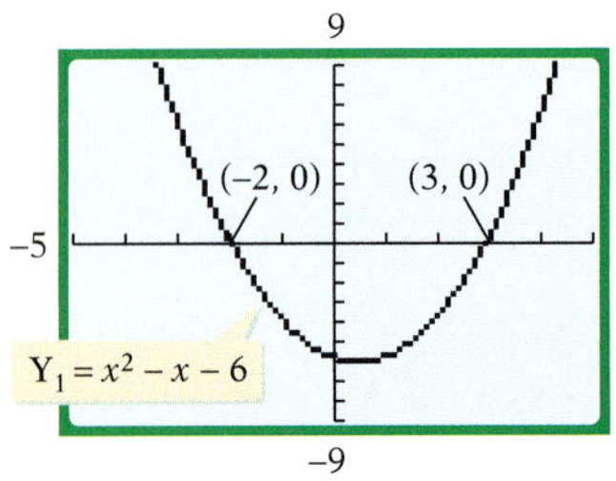

Figure 3

Thus the solutions of the equation are the x-coordinates of the x-intercepts of the graph.

This connection between the solutions of an equation and the x-intercepts of a graph allows us to find approximations of the real number solutions of an equation graphically. For example, to use a TI-83 to approximate the solutions of $2x^2 + 5x - 1 = 0$ graphically, use the following keystrokes to graph $y = 2x^2 + 5x - 1$ and find the decimal approximations of the x-coordinates of the x-intercepts. Use the window shown in the graph in Figure 4.

[Y=] [CLEAR] 2 [X, T, θ, n] [x^2] [+] 5 [X, T, θ, n] [−] 1 [2nd] [CALC] 2

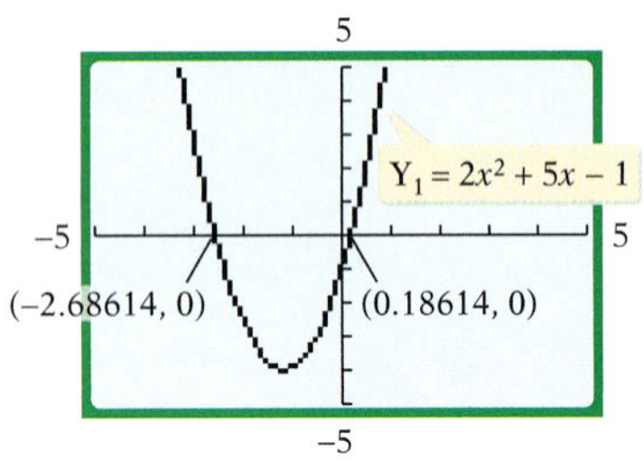

Figure 4

Use the arrow keys to move to the left of the leftmost x-intercept. Press ENTER. Use the arrow keys to move to a point just to the right of the leftmost x-intercept. Press ENTER twice. The x-coordinate at the bottom of the screen is the approximation of one solution of the equation. To find the other solution, use the same procedure but move the cursor first to the left and then to the right of the rightmost x-intercept.

Attempting to find the solutions of an equation graphically will not necessarily yield all the solutions. Because the x-coordinates of the x-intercepts of a graph are *real* numbers, only real number solutions can be found. For instance, consider the equation $x^2 + 4x + 5 = 0$. The graph of $y = x^2 + 4x + 5$ is shown in Figure 5. Note that the graph has no x-intercepts and that consequently it has no real number solutions. However, $x^2 + 4x + 5 = 0$ does have complex number solutions that can be obtained by using the quadratic formula. They are $-2 + i$ and $-2 - i$.

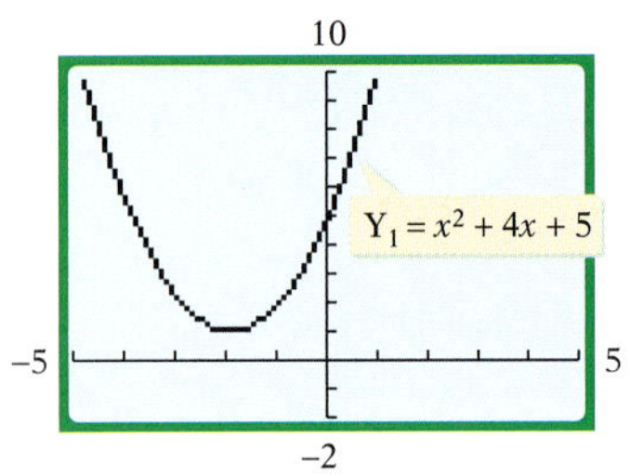

Figure 5

Use a graphing calculator to approximate the solutions of the equation.

1. $x^2 + 6x + 1 = 0$
2. $-x^2 + x + 3 = 0$
3. $2x^2 + 4x - 1 = 0$
4. $x^2 + 4x + 5 = 0$

Finding the Maximum or Minimum of a Function Using a Graphing Calculator

Recall that for a quadratic function, the maximum or minimum value of the function is the y-coordinate of the vertex of the graph of the function. We determined algebraically that the coordinates of the vertex are $\left(-\frac{b}{2a}, f\left(-\frac{b}{2a}\right)\right)$. It is also possible to use a graphing calculator to approximate a maximum or minimum value. The method used to find the value depends on the calculator model. However, it is generally true for most calculators that the point at which the function has a maximum or minimum must be shown on the screen. If it is not, you must select a different viewing window.

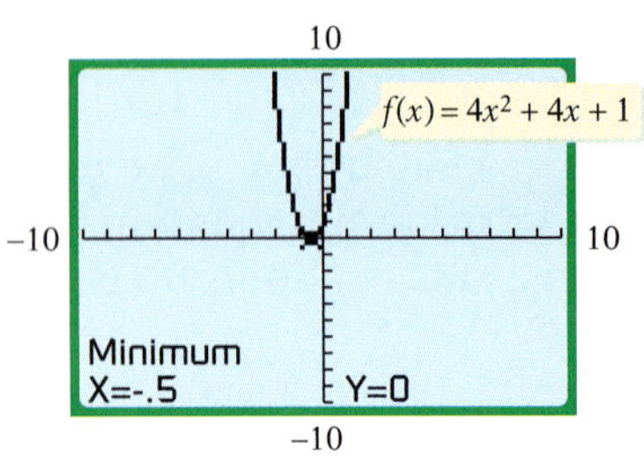

For a TI-83 graphing calculator, press [2nd] CALC 3 to determine a minimum value or [2nd] CALC 4 to determine a maximum value. Move the cursor to a point on the curve that is to the left of the minimum (or maximum). Press [ENTER]. Move the cursor to a point on the curve that is to the right of the minimum (or maximum). Press [ENTER] twice. The minimum (or maximum) value is shown as the y-coordinate at the bottom of the screen.

1. Find the maximum or minimum value of $f(x) = -x^2 - 3x + 2$ by using a graphing calculator.
2. Find the maximum or minimum value of $f(x) = x^2 - 4x - 2$ by using a graphing calculator.
3. Find the maximum or minimum value of $f(x) = -\sqrt{2}x^2 - \pi x + 2$ by using a graphing calculator.
4. Find the maximum or minimum value of $f(x) = \frac{x^2}{\sqrt{5}} - \sqrt{7}x - \sqrt{11}$ by using a graphing calculator.
5. A manufacturer of camera lenses estimated that the average monthly cost C of production is given by the function $C(x) = 0.1x^2 - 20x + 2000$, where x is the number of lenses produced each month. Find the number of lenses the company should produce in order to minimize the average cost.
6. The net annual income I, in dollars, of a family physician can be modeled by the equation $I(x) = -290(x - 48)^2 + 148{,}000$, where x is the age of the physician and $27 \le x \le 70$. Find (a) the age at which the physician's income will be a maximum and (b) the maximum income.

Business Applications of Maximum and Minimum Values of Quadratic Functions

A company's **revenue,** R, is the total amount of money the company earned by selling its products. The **cost,** C, is the total amount of money the company spent to manufacture and sell its products. A company's **profit,** P, is the difference between the revenue and the cost.

$$P = R - C$$

A company's revenue and cost may be represented by equations.

The owners of a travel agency are selling tickets for a tour. The travel agency must pay a cost per person of \$10. A maximum of 180 people can be accommodated on the tour. For every \$.25 increase in the price of a ticket over the \$10 cost, the owners estimate that they will sell 1 ticket less. Therefore, the price per ticket can be represented by

$$\begin{aligned} p &= 10 + 0.25(180 - x) \\ &= 55 - 0.25x \end{aligned}$$

where x represents the number of tickets sold. Determine the maximum profit and the cost per ticket that yields the maximum profit.

Solution

The cost is represented by $10x$ and the revenue is represented by $x(55 - 0.25x)$, where x represents the number of tickets sold.

Use the equation $P = R - C$ to represent the profit.

$$\begin{aligned} P &= R - C \\ P &= x(55 - 0.25x) - 10x \\ P &= -0.25x^2 + 45x \end{aligned}$$

The graph of $P = -0.25x^2 + 45x$ is a parabola that opens down. Find the x-coordinate of the vertex.

$$x = -\frac{b}{2a} = -\frac{45}{2(-0.25)} = 90$$

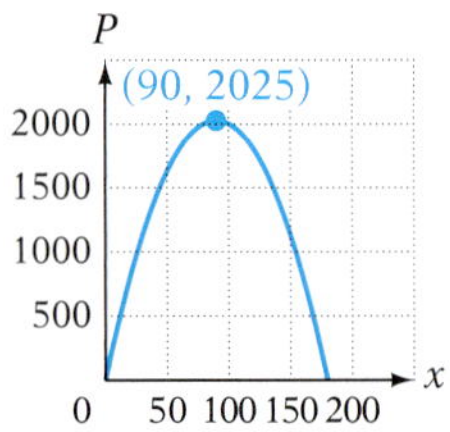

$P = -0.25x^2 + 45x$

To find the maximum profit, evaluate $P = -0.25x^2 + 45x$ at 90.

$$P = -0.25x^2 + 45x$$
$$P = -0.25(90^2) + 45(90)$$
$$P = 2025$$

To find the price per ticket, evaluate $p = 55 - 0.25x$ at 90.

$$p = 55 - 0.25x$$
$$p = 55 - 0.25(90)$$
$$p = 32.50$$

The travel agency can expect a maximum profit of $2025 when 90 people take the tour at a ticket price of $32.50 per person.

The graph of the profit function is shown at the left.

1. A company's total monthly cost, in dollars, for producing and selling n DVDs per month is $C = 25n + 4000$. The company's revenue, in dollars, from selling all n DVDs is $R = 275n - 0.2n^2$.
   a. How many DVDs must the company produce and sell each month in order to make the maximum monthly profit?
   b. What is the maximum monthly profit?

2. A shipping company has determined that its cost to deliver x packages per week is $1000 + 6x$. The price per package that the company charges when it sends x packages per week is $24 - 0.02x$. Determine the company's maximum weekly profit and the price per package that yields the maximum profit.

Chapter Summary

Key Words	Examples
A **quadratic equation** is an equation of the form $ax^2 + bx + c = 0$, where $a \neq 0$. A quadratic equation is also called a **second-degree equation**. [9.1.1, p. 564]	$3x^2 + 4x - 7 = 0$ and $x^2 - 1 = 0$ are quadratic equations.
A quadratic equation is in **standard form** when the polynomial is in descending order and equal to zero. [9.1.1, p. 564]	$x^2 - 5x + 6 = 0$ is a quadratic equation in standard form.
When a quadratic equation has two solutions that are the same number, the solution is called a **double root** of the equation. [9.1.1, p. 565]	$x^2 - 4x + 4 = 0$ $(x - 2)(x - 2) = 0$ $x - 2 = 0 \quad x - 2 = 0$ $x = 2 \quad x = 2$ 2 is a double root.
For an equation of the form $ax^2 + bx + c = 0$, the quantity $b^2 - 4ac$ is called the **discriminant**. [9.2.2, p. 578]	$2x^2 - 3x + 2 = 0$ $a = 2, b = -3, c = 2$ $b^2 - 4ac = (-3)^2 - 4(2)(2)$ $= 9 - 16 = -7$

A **quadratic inequality in one variable** is one that can be written in the form $ax^2 + bx + c > 0$ or $ax^2 + bx + c < 0$, where $a \neq 0$. The symbols $\leq$ and $\geq$ can also be used. [9.5.1, p. 596]

$3x^2 + 5x - 8 \leq 0$ is a quadratic inequality.

The ***x*-coordinate of the vertex of a parabola** is given by $-\frac{b}{2a}$. [9.6.1, p. 603]

$y = 2x^2 - 8x + 5 \quad a = 2, b = -8$
The x-coordinate of the vertex is $-\frac{-8}{2(2)} = 2$.

The ***x*-intercepts** of the graph of a parabola occur when $y = 0$. [9.6.2, p. 605]

$$x^2 - 5x + 6 = 0$$
$$(x - 3)(x - 2) = 0$$
$$x - 3 = 0 \qquad x - 2 = 0$$
$$x = 3 \qquad x = 2$$

The x-intercepts are (3, 0) and (2, 0).

The graph of a **quadratic function** $f(x) = ax^2 + bx = c$ is a **parabola** that opens up when $a > 0$ and down when $a < 0$. The **axis of symmetry** is a line that passes through the vertex of the parabola. [9.6.1, pp. 601–602]

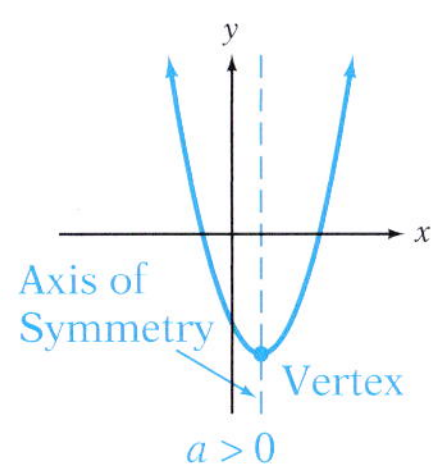

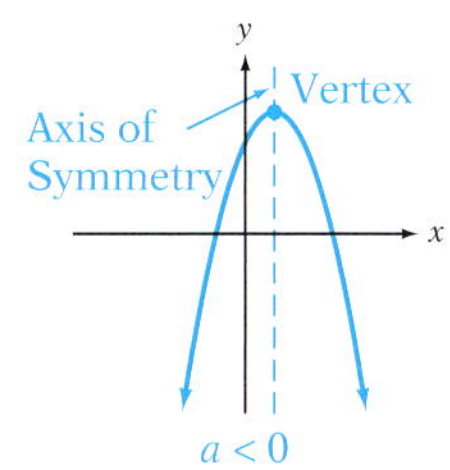

The graph of $f(x) = ax^2 + bx + x$ has a **minimum** value if $a > 0$ and a **maximum** value if $a < 0$. [9.7.1, p. 614]

Essential Rules and Procedures

Principle of Zero Products

If $ab = 0$, then $a = 0$ or $b = 0$. [9.1.1, p. 564]

$$(x - 3)(x + 4) = 0$$
$$x - 3 = 0 \qquad x + 4 = 0$$
$$x = 3 \qquad x = -4$$

A quadratic equation can be solved by **taking square roots of each side of the equation**. [9.1.3, p. 566]

$$(x + 2)^2 - 9 = 0$$
$$(x + 2)^2 = 9$$
$$\sqrt{(x + 2)^2} = \sqrt{9}$$
$$x + 2 = \pm\sqrt{9}$$
$$x + 2 = \pm 3$$
$$x + 2 = 3 \qquad x + 2 = -3$$
$$x = 1 \qquad x = -5$$

A quadratic equation can be solved by **completing the square**. [9.2.1, p. 572]

$$x^2 + 4x - 1 = 0$$
$$x^2 + 4x = 1$$
$$x^2 + 4x + 4 = 1 + 4$$
$$(x + 2)^2 = 5$$
$$\sqrt{(x + 2)^2} = \sqrt{5}$$
$$x + 2 = \pm\sqrt{5}$$

$$x + 2 = \sqrt{5} \qquad x + 2 = -\sqrt{5}$$
$$x = -2 + \sqrt{5} \qquad x = -2 - \sqrt{5}$$

Quadratic Formula

$$x = \frac{-b \pm \sqrt{b^2 - 4ac}}{2a}$$ [9.2.2, p. 576]

$$2x^2 - 3x + 4 = 0$$
$$a = 2, b = -3, c = 4$$
$$x = \frac{-(-3) \pm \sqrt{(-3)^2 - 4(2)(4)}}{2(2)}$$
$$= \frac{3 \pm \sqrt{9 - 32}}{4} = \frac{3 \pm \sqrt{-23}}{4}$$
$$= \frac{3 \pm i\sqrt{23}}{4} = \frac{3}{4} \pm \frac{\sqrt{23}}{4}i$$

Vertex of a parabola

The coordinates of the vertex of a parabola are $\left(-\frac{b}{2a}, f\left(-\frac{b}{2a}\right)\right)$. [9.6.1, p. 603]

$$f(x) = x^2 - 2x - 4 \quad a = 1, b = -2$$
$$-\frac{b}{2a} = -\frac{-2}{2(1)} = 1$$
$$f(1) = 1^2 - 2(1) - 4 = -5$$

The coordinates of the vertex are $(1, -5)$.

Discriminant [9.6.2, p. 579]

If $b^2 - 4ac = 0$, the equation has a double root.

$x^2 + 8x + 16 = 0$ has a double root because

$$b^2 - 4ac = 8^2 - 4(1)(16) = 0$$

If $b^2 - 4ac > 0$, the equation has two real number solutions that are not equal.

$2x^2 + 3x - 5 = 0$ has two unequal real number solutions because

$$b^2 - 4ac = 3^2 - 4(2)(-5) = 49$$

If $b^2 - 4ac < 0$, the equation has two complex number solutions.

$3x^2 + 2x + 4 = 0$ has two complex number solutions because

$$b^2 - 4ac = 2^2 - 4(3)(4) = 4 - 48$$
$$= -44$$

Chapter Review Exercises

1. Solve: $2x^2 - 3x = 0$

2. Solve for x: $6x^2 + 9xc = 6c^2$

3. Solve: $x^2 = 48$

4. Solve: $\left(x + \frac{1}{2}\right)^2 + 4 = 0$

5. Find the minimum value of the function $f(x) = x^2 - 7x + 8$.

6. Find the maximum value of the function $f(x) = -2x^2 + 4x + 1$.

7. Write a quadratic equation that has integer coefficients and has solutions $\frac{1}{3}$ and -3.

8. Solve: $2x^2 + 9x = 5$

9. Solve: $2(x + 1)^2 - 36 = 0$

10. Solve: $x^2 + 6x + 10 = 0$

11. Solve: $\frac{2}{x - 4} + 3 = \frac{x}{2x - 3}$

12. Solve: $x^4 - 6x^2 + 8 = 0$

13. Solve: $\sqrt{2x - 1} + \sqrt{2x} = 3$

14. Solve: $2x^{\frac{2}{3}} + 3x^{\frac{1}{3}} - 2 = 0$

15. Solve: $\sqrt{3x - 2} + 4 = 3x$

16. Solve: $x^2 - 6x - 2 = 0$

17. Solve: $\frac{2x}{x - 4} + \frac{6}{x + 1} = 11$

18. Solve: $2x^2 - 2x = 1$

19. Solve: $2x = 4 - 3\sqrt{x - 1}$

20. Solve: $3x = \frac{9}{x - 2}$

21. Solve: $\frac{3x + 7}{x + 2} + x = 3$

22. Solve: $\frac{x - 2}{2x + 3} - \frac{x - 4}{x} = 2$

23. Solve: $1 - \frac{x + 4}{2 - x} = \frac{x - 3}{x + 2}$

24. Find the axis of symmetry of the parabola whose equation is $y = -x^2 + 6x - 5$.

25. Find the vertex of the parabola whose equation is $y = -x^2 + 3x - 2$.

26. Use the discriminant to determine the number of x-intercepts of the parabola whose equation is $y = -2x^2 + 2x - 3$.

27. Use the discriminant to determine the number of x-intercepts of the parabola whose equation is $y = 3x^2 - 2x - 4$.

28. Find the x-intercepts of the parabola whose equation is $y = 4x^2 + 12x + 4$.

29. Find the x-intercepts of the parabola whose equation is $y = -2x^2 - 3x + 2$.

30. Find the zeros of $f(x) = 3x^2 + 2x + 2$.

31. Solve: $(x + 3)(2x - 5) < 0$

32. Solve: $(x - 2)(x + 4)(2x + 3) \le 0$

33. Solve and graph the solution set of $\frac{x - 2}{2x - 3} \ge 0$.

34. Solve and graph the solution set of $\frac{(2x - 1)(x + 3)}{x - 4} \le 0$.

35. Graph $f(x) = x^2 + 2x - 4$. State the domain and range.

36. Find the vertex and axis of symmetry of the parabola whose equation is $y = x^2 - 2x + 3$. Then graph the equation.

37. The sum of the squares of three consecutive even integers is fifty-six. Find the three integers.

38. The length of a rectangle is 2 cm more than twice the width. The area of the rectangle is 60 cm^2. Find the length and width of the rectangle.

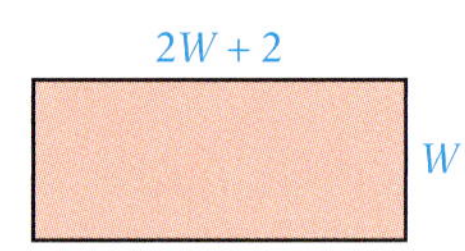

39. An older computer requires 12 min longer to print the payroll than does a newer computer. Together the computers can print the payroll in 8 min. Find the time for the newer computer, working alone, to complete the payroll.

40. A car travels 200 mi. A second car, traveling 10 mph faster than the first car, makes the same trip in 1 h less time. Find the speed of each car.

Chapter Test

1. Solve: $2x^2 + x = 6$

2. Solve: $12x^2 + 7x - 12 = 0$

3. Find the maximum value of the function $f(x) = -x^2 + 8x - 7$.

4. Write the quadratic equation that has integer coefficients and has solutions $-\frac{1}{3}$ and 3.

5. Solve: $2(x + 3)^2 - 36 = 0$

6. Solve: $x^2 + 4x - 1 = 0$

7. Find the zeros of $g(x) = x^2 + 3x - 8$.

8. Solve: $3x^2 - x + 8 = 0$

9. Solve: $\frac{2x}{x-1} + \frac{3}{x+2} = 1$

10. Solve: $2x + 7x^{\frac{1}{2}} - 4 = 0$

11. Solve: $x^4 - 11x^2 + 18 = 0$

12. Solve: $\sqrt{2x+1} + 5 = 2x$

13. Solve: $\sqrt{x-2} = \sqrt{x} - 2$

14. Use the discriminant to determine the number of x-intercepts of the parabola $y = 3x^2 + 2x - 4$.

15. Find the x-intercepts of the parabola whose equation is $y = 2x^2 + 5x - 12$.

16. Find the axis of symmetry of the parabola whose equation is $y = 2x^2 + 6x + 3$.

17. Graph $y = \frac{1}{2}x^2 + x - 4$. State the domain and range.

18. Solve and graph the solution set of $\frac{2x-3}{x+4} \le 0$.

19. The base of a triangle is 3 ft more than three times the height. The area of the triangle is 30 ft^2. Find the base and height of the triangle.

20. The rate of a river's current is 2 mph. A canoe was rowed 6 mi down the river and back in 4 h. Find the rowing rate in calm water.

Cumulative Review Exercises

1. Evaluate $2a^2 - b^2 \div c^2$ when $a = 3$, $b = -4$, and $c = -2$.

2. Solve: $\frac{2x-3}{4} - \frac{x+4}{6} = \frac{3x-2}{8}$

3. Find the slope of the line that contains the points $(3, -4)$ and $(-1, 2)$.

4. Find the equation of the line that contains the point $(1, 2)$ and is parallel to the line $x - y = 1$.

5. Factor: $-3x^3y + 6x^2y^2 - 9xy^3$

6. Factor: $6x^2 - 7x - 20$

7. Factor: $a^nx + a^ny - 2x - 2y$

8. Divide: $(3x^3 - 13x^2 + 10) \div (3x - 4)$

9. Multiply: $\dfrac{x^2 + 2x + 1}{8x^2 + 8x} \cdot \dfrac{4x^3 - 4x^2}{x^2 - 1}$

10. Find the distance between the points $(-2, 3)$ and $(2, 5)$.

11. Solve $S = \dfrac{n}{2}(a + b)$ for b.

12. Multiply: $-2i(7 - 4i)$

13. Simplify: $a^{-\frac{1}{2}}\left(a^{\frac{1}{2}} - a^{\frac{3}{2}}\right)$

14. Simplify: $\dfrac{\sqrt[3]{8x^4y^5}}{\sqrt[3]{16xy^6}}$

15. Solve: $\dfrac{x}{x + 2} - \dfrac{4x}{x + 3} = 1$

16. Solve: $\dfrac{x}{2x + 3} - \dfrac{3}{4x^2 - 9} = \dfrac{x}{2x - 3}$

17. Solve: $x^4 - 6x^2 + 8 = 0$

18. Solve: $\sqrt{3x + 1} - 1 = x$

19. Solve: $|3x - 2| < 8$

20. Find the x- and y-intercepts of the graph of $6x - 5y = 15$.

21. Graph the solution set: $\begin{aligned} x + y &\le 3 \\ 2x - y &< 4 \end{aligned}$

22. Solve the system of equations:
$$\begin{aligned} x + y + z &= 2 \\ -x + 2y - 3z &= -9 \\ x - 2y - 2z &= -1 \end{aligned}$$

23. Given $f(x) = \dfrac{2x - 3}{x^2 - 1}$ find $f(-2)$.

24. Find the domain of the function $f(x) = \dfrac{x - 2}{x^2 - 2x - 15}$.

25. Solve and graph the solution set of $x^3 + x^2 - 6x < 0$.

26. Solve and graph the solution set of $\dfrac{(x - 1)(x - 5)}{x + 3} \ge 0$.

27. Fifty ounces of pure orange juice are added to 200 oz of a fruit punch that is 10% orange juice. What is the percent concentration of orange juice in the resulting mixture?

28. A piston rod for an automobile is $9\frac{3}{8}$ in. with a tolerance of $\frac{1}{64}$ in. Find the lower and upper limits of the length of the piston rod.

29. The base of a triangle is $(x + 8)$ ft. The height is $(2x - 4)$ ft. Find the area of the triangle in terms of the variable x.

30. The graph shows the relationship between the cost of a building and the depreciation allowed for income tax purposes. Find the slope of the line between the two points shown on the graph. Write a sentence that states the meaning of the slope.

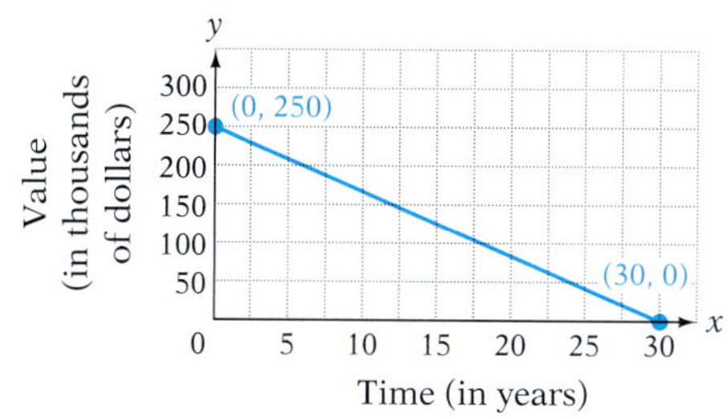

The Toyota dealership pictured above is offering rebates of $3500 and discounts off the price of a new car. Composite functions, a topic in this chapter, can be used to determine which is the better deal: take the rebate first and then the discount, or take the discount first and then the rebate. This problem is a focus of the **Project on pages 684–685.**

Applications

10 Functions and Relations

Objectives

Section 10.1

1. Graph functions

Section 10.2

1. Graph by using translations

Section 10.3

1. Perform operations on functions
2. Find the composition of two functions

Section 10.4

1. Determine whether a function is one-to-one
2. Find the inverse of a function

Section 10.5

1. Graph parabolas
2. Find the equation of a circle and then graph the circle
3. Graph an ellipse with center at the origin
4. Graph a hyperbola with center at the origin

WWW WEB **Need help? For online student resources, such as section quizzes, visit this web site: math.college.hmco.com**

Prep Test

1. Evaluate $-\frac{b}{2a}$ for $b = -4$ and $a = 2$.

2. Given $y = -x^2 + 2x + 1$, find the value of y when $x = -2$.

3. Given $f(x) = x^2 - 3x + 2$, find $f(-4)$.

4. Evaluate $p(r) = r^2 - 5$ when $r = 2 + h$.

5. Solve: $0 = 3x^2 - 7x - 6$

6. Solve: $0 = x^2 - 4x + 1$

7. Solve $x = 2y + 4$ for y.

8. Find the domain and range of the relation $\{(-2, 4), (3, 5), (4, 6), (6, 5)\}$. Is the relation a function?

9. What value is excluded from the domain of $f(x) = \frac{3}{x - 8}$?

10. Graph: $x = -2$

11. Find the distance between the points whose coordinates are $(-2, 3)$ and $(4, -1)$. Round to the nearest hundredth.

Go Figure

Each time the two hands of a certain standard, analog, 12-hour clock form a 180° angle, a bell chimes once. From noon today until noon tomorrow, how many chimes will be heard?

10.1 Graphs of Functions

1 Graph functions

VIDEO & DVD CD TUTOR WEB SSM

The graphs of the polynomial functions $L(x) = mx + b$ (a straight line) and $Q(x) = ax^2 + bx + c$, $a \neq 0$ (a parabola) have been discussed earlier. The graphs of other functions can also be drawn by finding ordered pairs that belong to the function, plotting the points that correspond to the ordered pairs, and then drawing a curve through the points.

Graph: $F(x) = x^3$

Select several values of x and evaluate the function.

> **Take Note**
>
> The units along the x-axis are different from those along the y-axis. If the units along the x-axis were the same as those along the y-axis, the graph would appear narrower than the one shown.

x	$F(x) = x^3$	$F(x)$	(x, y)
-2	$(-2)^3$	-8	$(-2, -8)$
-1	$(-1)^3$	-1	$(-1, -1)$
0	0^3	0	$(0, 0)$
1	1^3	1	$(1, 1)$
2	2^3	8	$(2, 8)$

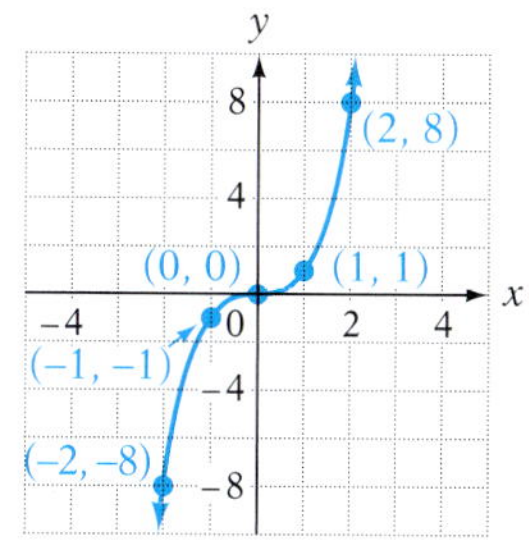

Plot the ordered pairs and draw a graph through the points.

Graph: $g(x) = x^3 - 4x + 5$

Select several values of x and evaluate the function.

x	$g(x) = x^3 - 4x + 5$	$g(x)$	(x, y)
-3	$(-3)^3 - 4(-3) + 5$	-10	$(-3, -10)$
-2	$(-2)^3 - 4(-2) + 5$	5	$(-2, 5)$
-1	$(-1)^3 - 4(-1) + 5$	8	$(-1, 8)$
0	$(0)^3 - 4(0) + 5$	5	$(0, 5)$
1	$(1)^3 - 4(1) + 5$	2	$(1, 2)$
2	$(2)^3 - 4(2) + 5$	5	$(2, 5)$

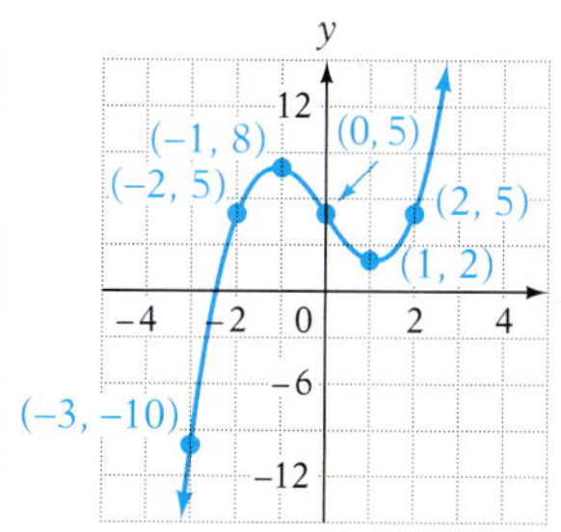

Plot the ordered pairs and draw a graph through the points.

Note from the graphs of the two different cubic functions that the shapes of the graphs can be quite different. The following graphs of typical cubic polynomial functions show the general shape of the graph of a cubic polynomial.

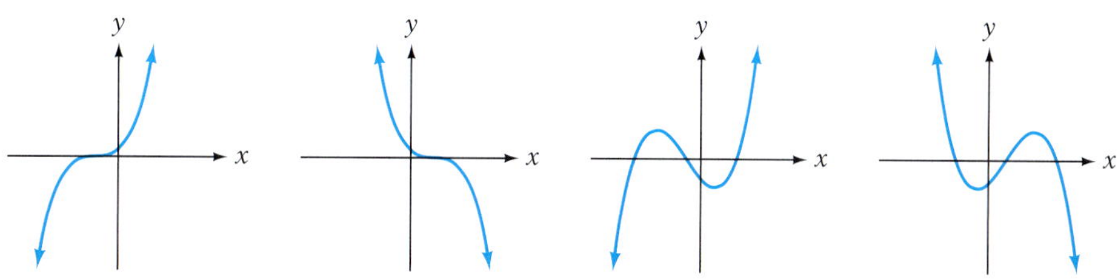

Graph: $f(x) = |x + 2|$

This is an **absolute value function.**

x	$f(x) = \|x + 2\|$	$f(x)$	(x, y)
-3	$\|-3 + 2\|$	1	$(-3, 1)$
-2	$\|-2 + 2\|$	0	$(-2, 0)$
-1	$\|-1 + 2\|$	1	$(-1, 1)$
0	$\|0 + 2\|$	2	$(0, 2)$
1	$\|1 + 2\|$	3	$(1, 3)$
2	$\|2 + 2\|$	4	$(2, 4)$

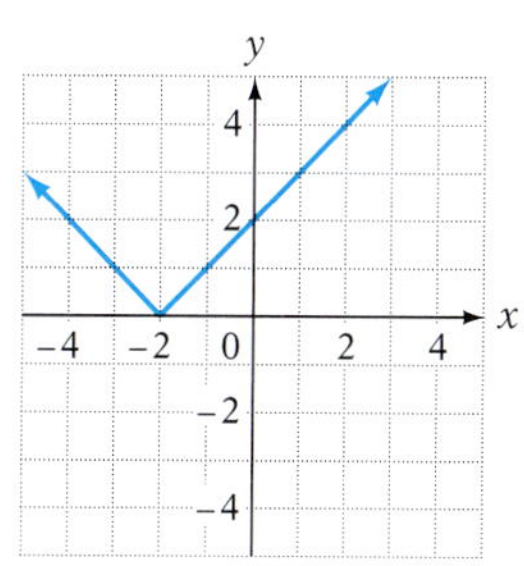

In general, the graph of the absolute value of a linear polynomial is V-shaped.

Graph: $R(x) = \sqrt{2x - 4}$

This is a **radical function.** Because the square root of a negative number is not a real number, the domain of this function requires that $2x - 4 \geq 0$. Solve this inequality for x.

$$2x - 4 \geq 0$$
$$2x \geq 4$$
$$x \geq 2$$

The domain is $\{x \mid x \geq 2\}$. This means that only values of x that are greater than or equal to 2 can be chosen as values at which to evaluate the function. In this case, some of the y-coordinates must be approximated.

x	$f(x) = \sqrt{2x - 4}$	$f(x)$	(x, y)
2	$\sqrt{2(2) - 4}$	0	(2, 0)
3	$\sqrt{2(3) - 4}$	1.41	(3, 1.41)
4	$\sqrt{2(4) - 4}$	2	(4, 2)
5	$\sqrt{2(5) - 4}$	2.45	(5, 2.45)
6	$\sqrt{2(6) - 4}$	2.83	(6, 2.83)

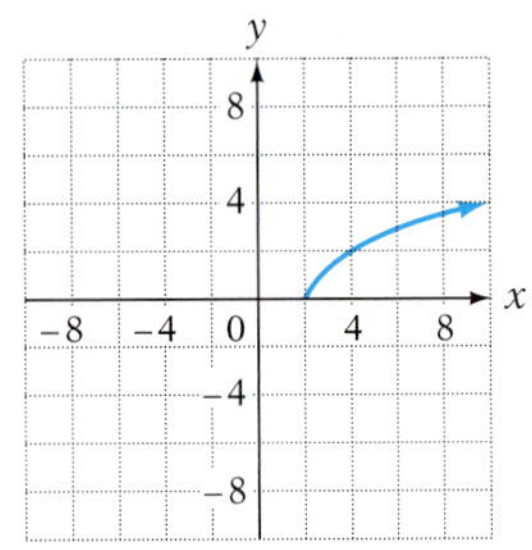

Recall that a function is a special type of relation, one for which no two ordered pairs have the same first coordinate. Graphically, this means that the graph of a function cannot pass through two points that have the same x-coordinate and different y-coordinates. For instance, the graph at the right is not the graph of a function because there are ordered pairs with the same x-coordinate and different y-coordinates.

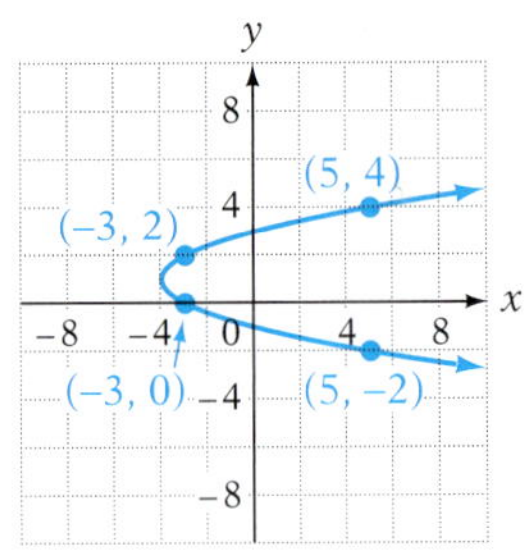

The last graph illustrates a general statement that can be made about whether a graph defines a function. It is called the **vertical line test.**

Vertical Line Test

A graph defines a function if any vertical line intersects the graph at no more than one point.

For example, the graph of a nonvertical straight line is the graph of a function. Any vertical line intersects the graph no more than once. The graph of a circle, however, is not the graph of a function. There are vertical lines that intersect the graph at more than one point.

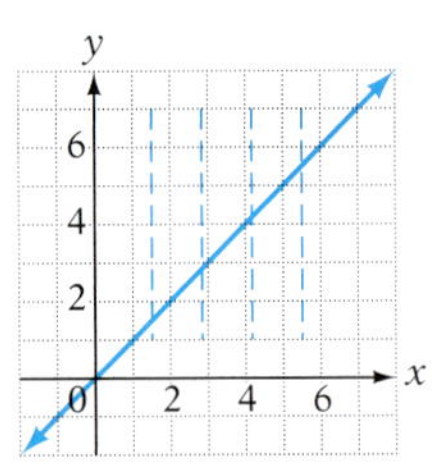

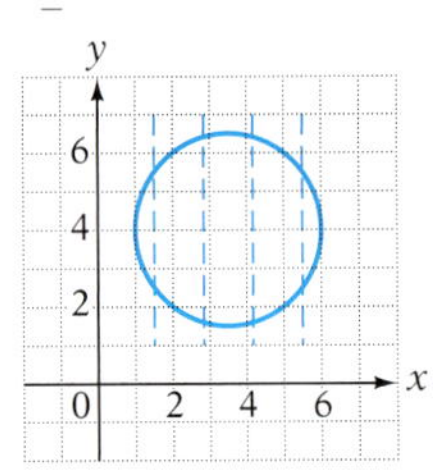

There may be practical situations in which a graph is not the graph of a function. The problem below is an example.

One of the causes of smog is an inversion layer where temperatures at higher altitudes are warmer than those at lower altitudes. The graph at the right shows the altitudes at which various temperatures were recorded. As shown by the dashed lines in the graph, there are two altitudes at which the temperature was 25°C. This means that there are two ordered pairs (shown in the graph) with the same first coordinate but different second coordinates. The graph does not define a function.

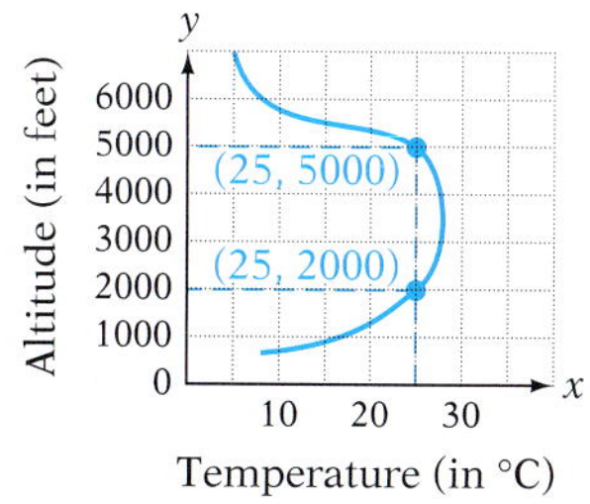

When a graph does define a function, the domain and range can be estimated from the graph.

Take Note

To determine the domain, think of collapsing the graph onto the x-axis and determining what interval on the x-axis the graph covers. To determine the range, think of collapsing the graph onto the y-axis and determining what interval on the y-axis the graph covers.

Determine the domain and range of the function given by the graph at the right. Write the answer in set-builder notation.

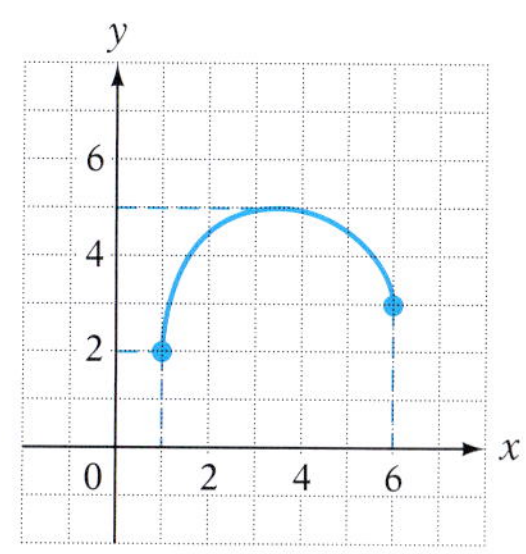

The solid dots on the graph indicate its beginning and ending points.

The domain is the set of x-coordinates. The domain is $\{x \mid 1 \le x \le 6\}$.

The range is the set of y-coordinates. The range is $\{y \mid 2 \le y \le 5\}$.

Take Note

In this first example, the domain and range are given in set-builder notation. For the second example, the domain and range are given in interval notation. Either method may be used.

Determine the domain and range of the function given by the graph at the right. Write the answer in interval notation.

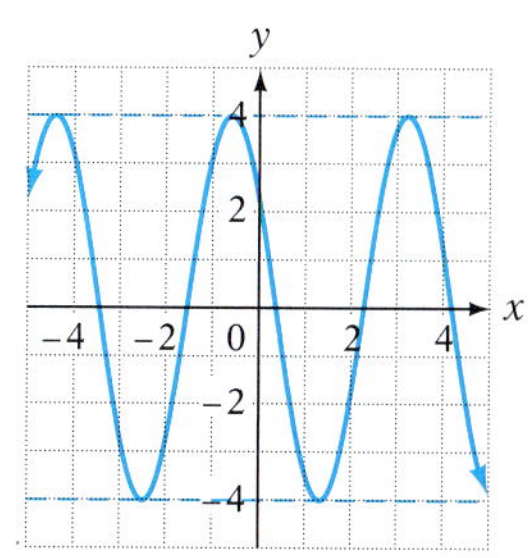

The arrows on the graph indicate that the graph continues in the same manner.

The domain is the set of x-coordinates. The domain is $(-\infty, \infty)$.

The range is the set of y-coordinates. The range is $[-4, 4]$.

Example 1 Use the vertical line test to determine whether the graph shown at the right is the graph of a function.

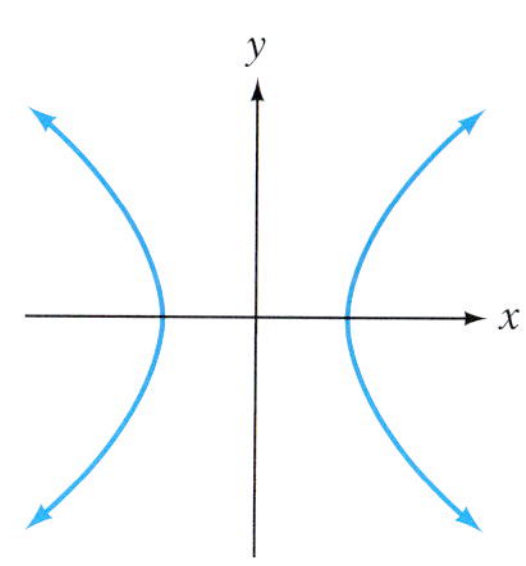

Solution

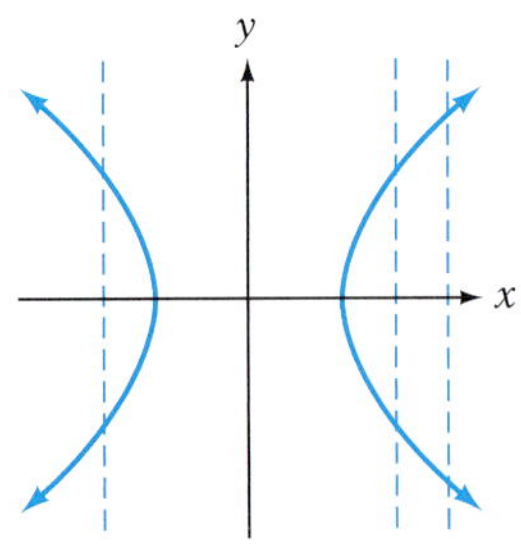

• A vertical line can intersect the graph at more than one point. Therefore, the relation includes ordered pairs with the same first coordinate and different second coordinates.

The graph is not the graph of a function.

Problem 1 Use the vertical line test to determine whether the graph shown at the right is the graph of a function.

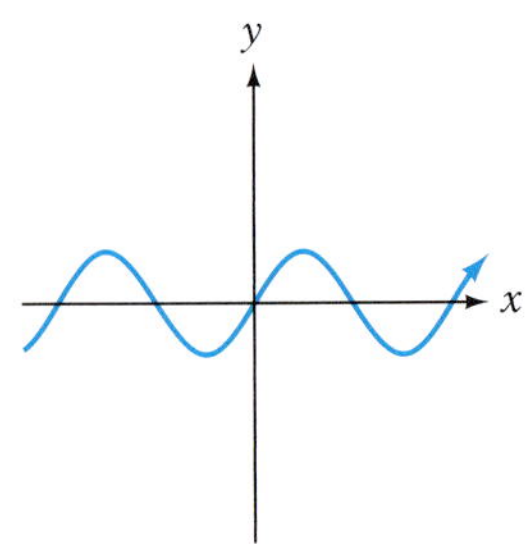

Solution See page S32.

Example 2 Graph. Use set-builder notation to state the domain and range of the function defined by each equation.

A. $f(x) = |x| + 2$ **B.** $f(x) = -x^3 - 2x^2 + 2$

C. $f(x) = \sqrt{2 - x}$

Solution **A.**

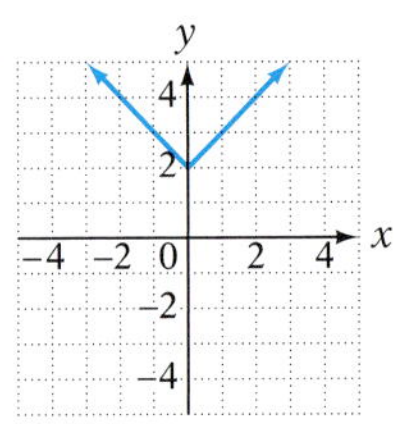

• This is an absolute value function. The graph is V-shaped.

The domain is $\{x \mid x \in \text{real numbers}\}$.
The range is $\{y \mid y \geq 2\}$.

B.

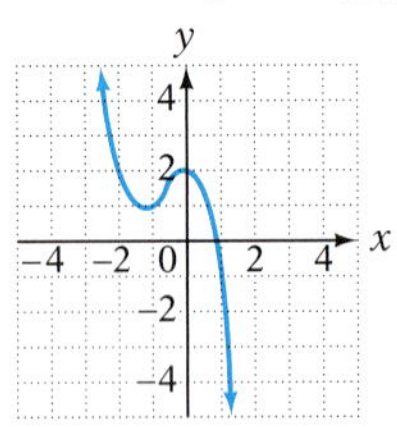

• This is a cubic function. Some ordered pairs are (−2, 2), (−1, 1), (0, 2), and (1, −1).

The domain is $\{x \mid x \in \text{real numbers}\}$.
The range is $\{y \mid y \in \text{real numbers}\}$.

C. 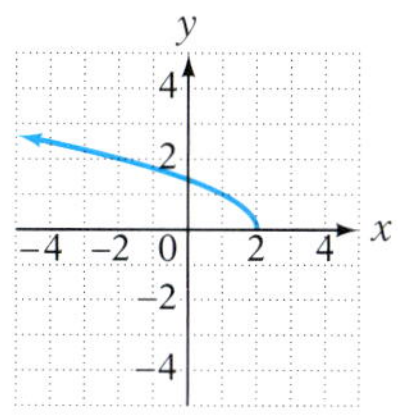

- **This is a radical function. The numbers under the radicand must be positive.**

$2 - x \geq 0$

$-x \geq -2$

$x \leq 2$

The domain is $\{x \mid x \leq 2\}$.
The range is $\{y \mid y \geq 0\}$.

Problem 2 Graph. Use set-builder notation to state the domain and range of the function defined by each equation.

A. $f(x) = |x + 2|$ **B.** $f(x) = x^3 + 2x + 1$

C. $f(x) = -\sqrt{x - 1}$

Solution See page S32.

10.1 Concept Review

Determine whether the following statements are always true, sometimes true, or never true.

1. The value of an absolute value function is positive.
2. A graph defines a function if any vertical line intersects the graph at no more than one point.
3. The range of a polynomial function is the real numbers.
4. A relation is a function.
5. The symbol $f(x)$ represents a value in the range of the function.
6. The domain of a cubic polynomial is the real numbers.

10.1 Exercises

1

1. What is the vertical line test?

2. When we describe the domain or range of a function in interval notation, what is the difference between using a bracket and using a parenthesis?

Use the vertical line test in order to determine whether the graph is the graph of a function.

3.

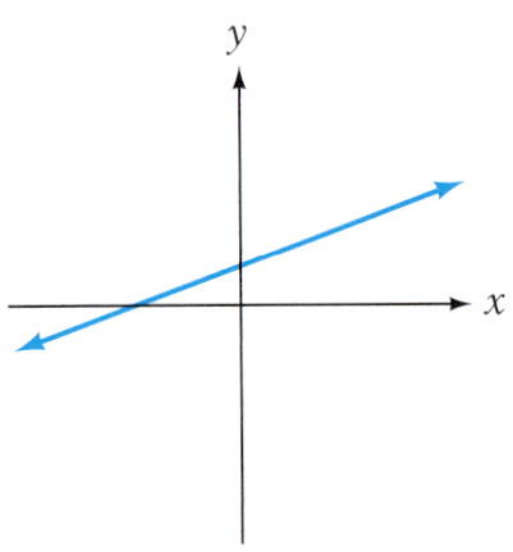

4.

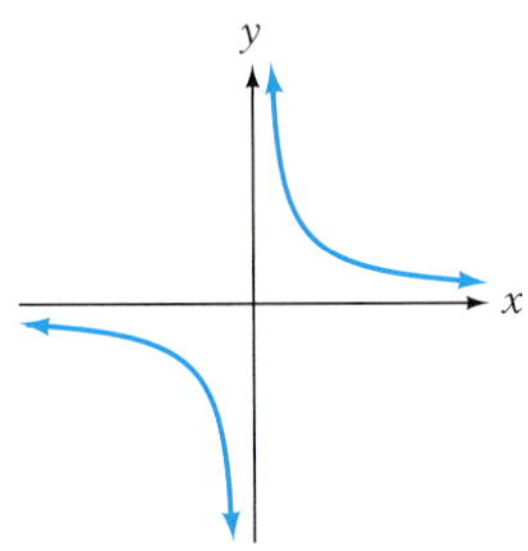

5.

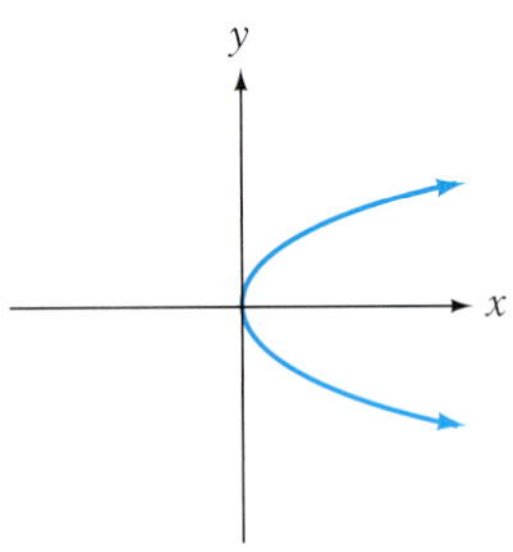

6.

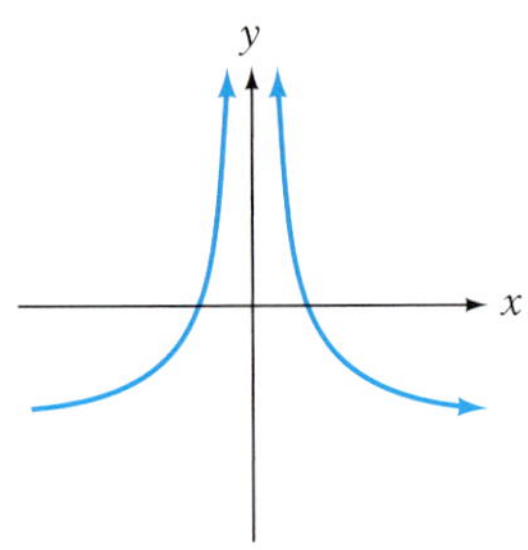

7.

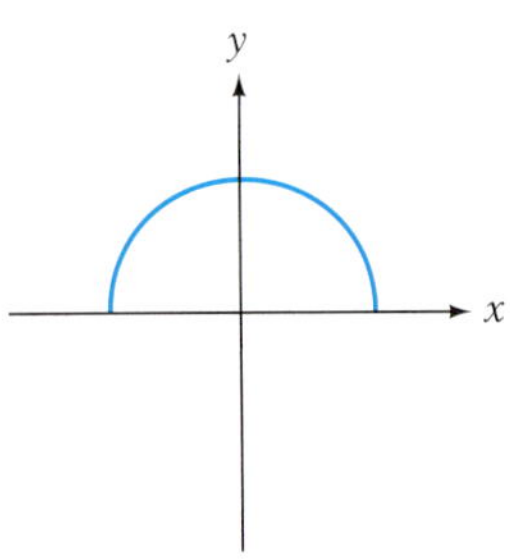

8.

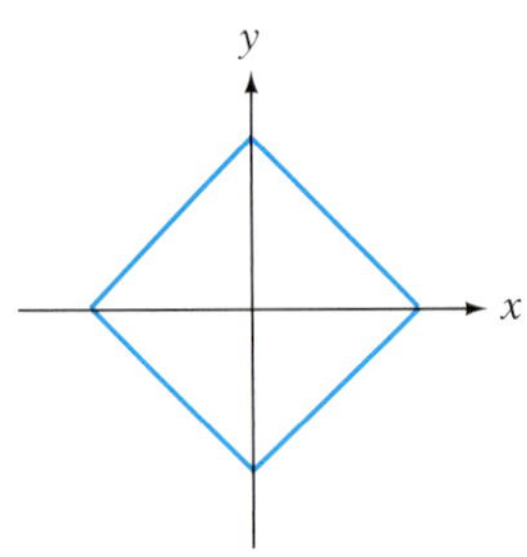

Graph the function defined by each equation. Use set-builder notation to state the domain and range of the function.

9. $f(x) = 2x - 1$

10. $f(x) = 4$

11. $f(x) = 2x^2 - 1$

12. $f(x) = 2x^2 - 3$

13. $f(x) = |x - 1|$

14. $f(x) = |x - 3|$

15. $f(x) = x^3 - 1$

16. $f(x) = -x^3 + 1$

17. $f(x) = \sqrt{1 + x}$

18. $f(x) = \sqrt{4 - x}$

19. $f(x) = \frac{1}{2}x^2$

20. $f(x) = \frac{1}{3}x^2$

21. $f(x) = |x| + 1$

22. $f(x) = |x| - 1$

23. $f(x) = x^3 + 2x^2$

24. $f(x) = x^3 - 3x^2$

25. $f(x) = -\sqrt{x + 2}$

26. $f(x) = -\sqrt{x - 3}$

27. $f(x) = 2x^2 - 4x - 3$

28. $f(x) = -1$

29. $f(x) = 2|x| - 1$

30. $f(x) = 2|x| + 2$

31. $f(x) = 2x^3 + 3x$

32. $f(x) = -2x^3 + 4x$

33. $f(x) = -\frac{1}{2}x - 2$

34. $f(x) = -3$

35. $f(x) = x^2 + 2x - 4$

36. $f(x) = -x^2 + 2x - 1$

37. $f(x) = -|x|$

38. $f(x) = -|x + 2|$

39. $f(x) = x^3 - x^2 + x - 1$

40. $f(x) = x^3 + x^2 - x + 1$

10.1 Applying Concepts

Which of the following relations are not functions?

41. **a.** $f(x) = x$ **b.** $f(x) = \left|\frac{x}{2}\right|$ **c.** $\{(3, 1), (1, 3), (3, 0), (0, 3)\}$

42. **a.** $f(x) = -x$ **b.** $f(x) = \dfrac{2}{\sqrt{x}}$ **c.** $\{(1, 4), (4, 1), (1, -4), (-4, 1)\}$

43. If $f(x) = \sqrt{x - 2}$ and $f(a) = 4$. find a.

44. If $f(x) = \sqrt{x + 5}$ and $f(a) = 3$, find a.

45. The graph of the function f is shown at the right. For this function, which of the following are true?
a. $f(4) = 1$ **b.** $f(0) = 3$ **c.** $f(-3) = 2$

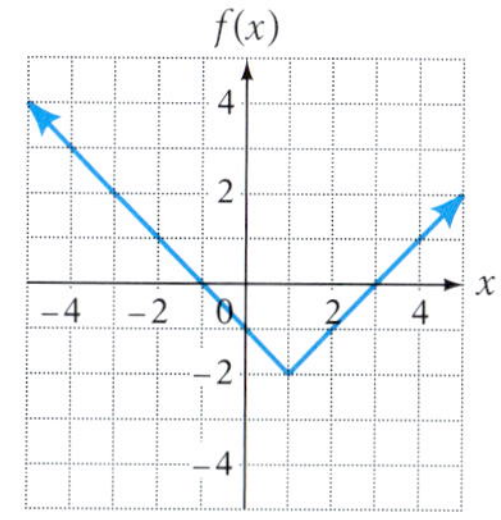

46. Let $f(a, b) =$ the sum of a and b.
Let $g(a, b) =$ the product of a and b.
Find $f(2, 5) + g(2, 5)$.

47. Let $f(a, b) =$ the greatest common divisor of a and b.
Let $g(a, b) =$ the least common multiple of a and b.
Find $f(14, 35) + g(14, 35)$.

48. Given $f(x) = (x + 2)(x - 2)$, for what values of x is $f(x)$ negative? Write your answer in set-builder notation.

49. Given $f(x) = (x + 1)(x - 1)$, for what values of x is $f(x)$ negative? Write your answer in set-builder notation.

50. Let $f(x)$ be the digit in the xth decimal place of the repeating decimal $0.\overline{387}$. For example, $f(3) = 7$ because 7 is the digit in the third decimal place. Find $f(14)$.

51. Given $f(x) = -|x + 3|$, for what value of x is $f(x)$ greatest?

52. Given $f(x) = |2x - 2|$, for what value of x is $f(x)$ smallest?

53. **a.** For what real value of x does $x^3 = 0$?
b. What is the x-intercept of the graph of $f(x) = x^3$?

54. **a.** For what real value of x does $x^3 - 1 = 0$?
b. What is the x-intercept of the graph of $f(x) = x^3 - 1$?

A *turning point* in the graph of a function is a point where the graph changes direction. For instance, the graph of $f(x) = x^3 - 4x + 5$ shown at the right has two turning points. In general, the number of turning points that the graph of a polynomial has depends on the degree of the polynomial. However, not all graphs of cubic functions have two turning points. For instance, the graph of $f(x) = x^3$ has no turning points.

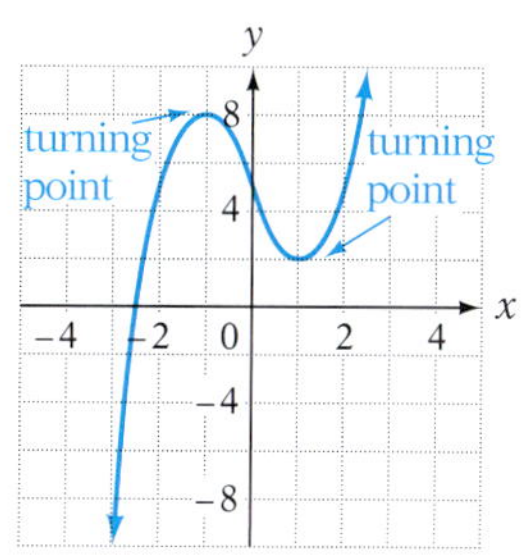

By using a graphing utility, graph the following and answer the questions. You may need to set Ymin and Ymax to fairly large numbers to accommodate these graphs.

55. Graph $f(x) = x^4 - 5$. How many turning points does the graph have?

56. Graph $f(x) = x^4 - 3x^3 - 7x^2 + x + 6$. How many turning points does the graph have?

57. Make up six more fourth-degree polynomial functions and graph each one. Record the number of turning points for each graph. Make a conjecture about the number of turning points the graph of a fourth-degree polynomial will have. Graph a few more fourth-degree polynomials and determine whether your conjecture is correct for those graphs. If not, modify your conjecture and try a few more graphs.

10.2 Translations of Graphs

1 Graph by using translations

VIDEO & DVD   CD TUTOR   WEB   SSM

The graphs of $f(x) = |x|$ and $g(x) = |x| + 2$ are shown in Figure 1. Note that for a given x-coordinate, the y-coordinate on the graph of g is 2 units higher than that on the graph of f. The graph of g is said to be a **vertical translation,** or **vertical shift,** of the graph of f.

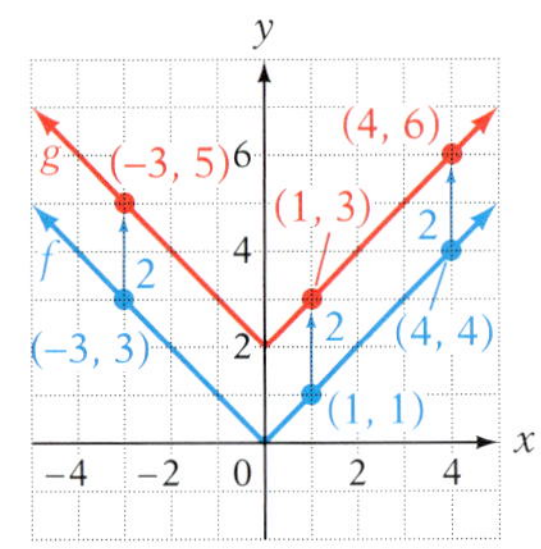

Figure 1

Take Note

Remember that $g(x)$ and $f(x)$ are y-coordinates of an ordered pair. The fact that $g(x)$ is 2 units more than $f(x)$ means that, for a given value of x, the y values for g are 2 units higher than the y values for f.

Note that because $f(x) = |x|$,

$$\begin{aligned} g(x) &= |x| + 2 \\ &= f(x) + 2 \end{aligned}$$

Thus $g(x)$ is 2 units more than $f(x)$.

Now consider the graphs of the functions $f(x) = |x|$ and $h(x) = |x| - 3$, shown in Figure 2. Note that for a given x-coordinate, the y-coordinate on the graph of h is 3 units lower than that on the graph of f. The graph of h is a vertical translation of the graph of f.

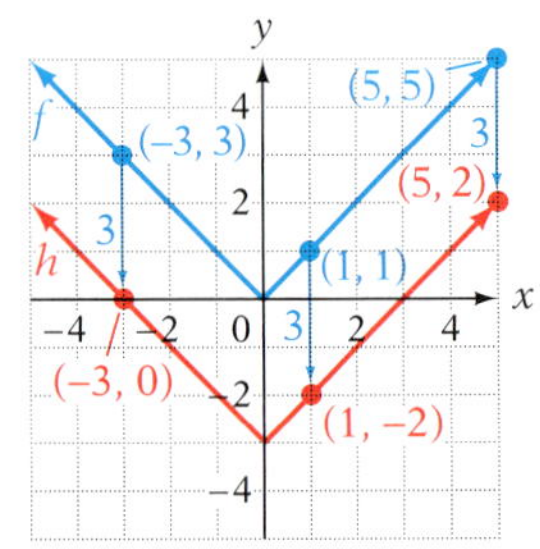

Figure 2

In Figure 1, the graph of $g(x) = |x| + 2$ is the graph of f shifted *up* 2 units, whereas in Figure 2, the graph of $h(x) = |x| - 3$ is the graph of f shifted *down* 3 units.

Vertical Translation

If f is a function and c is a positive constant, then

$y = f(x) + c$ is the graph of $y = f(x)$ shifted up c units.

$y = f(x) - c$ is the graph of $y = f(x)$ shifted down c units.

Example 1 Given the graph of the function $f(x) = \sqrt[3]{x}$ shown at the right in blue, graph $g(x) = \sqrt[3]{x} + 3$ by using a vertical translation.

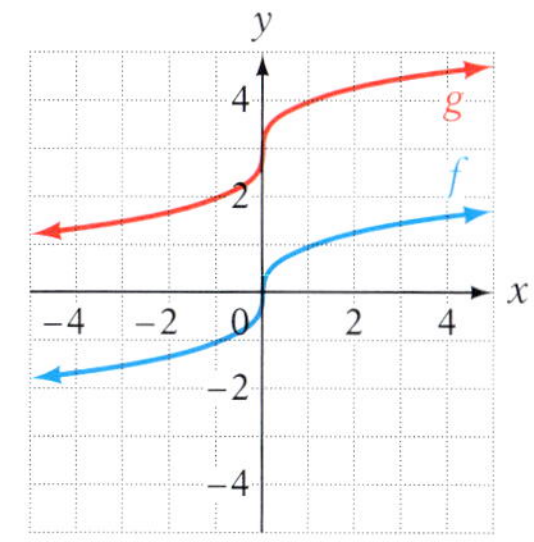

Solution The graph of $g(x) = \sqrt[3]{x} + 3$ is the graph of $f(x) = \sqrt[3]{x}$ shifted 3 units up. This graph is shown in red at the right.

Problem 1 Given the graph of the function $f(x) = x^2 - 2x$ shown at the right, graph $g(x) = x^2 - 2x - 3$ by using a vertical translation.

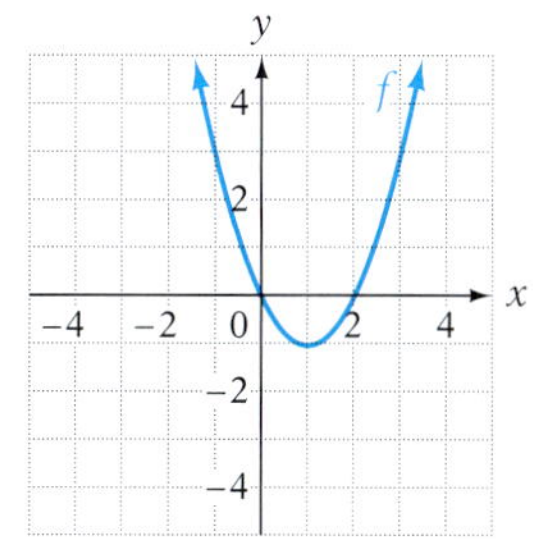

Solution See page S32.

The graphs of $f(x) = |x|$ and $g(x) = |x - 2|$ are shown in Figure 3. Note that the graph of g is the graph of f shifted 2 units to the right. The graph of g is a **horizontal translation,** or **horizontal shift,** of the graph of f. In this situation, each x-coordinate is moved to the right 2 units, but the y-coordinates are unchanged.

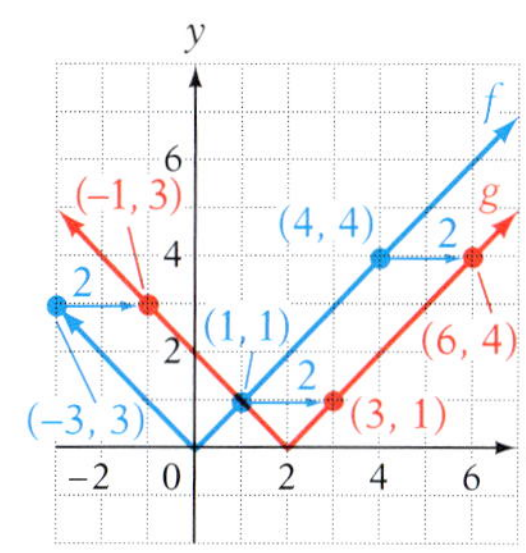

Figure 3

The graphs of $f(x) = |x|$ and $h(x) = |x + 3|$ are shown in Figure 4. In this case, the graph of f is translated 3 units to the left.

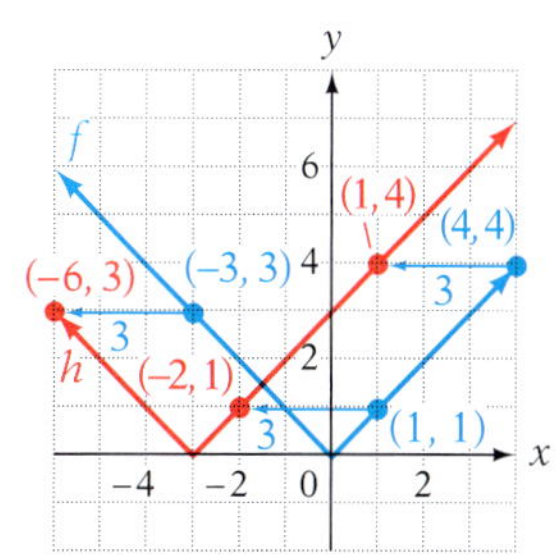

Figure 4

> **Take Note**
>
> Recall that evaluation of a function is accomplished by replacing a variable by another quantity.
>
> $f(x) = |x|$
>
> $f(-3) = |-3|$
>
> $f(x - 2) = |x - 2| = g(x)$
>
> $f(x + 3) = |x + 3| = h(x)$

For horizontal translations, note that when $f(x) = |x|$,

$$f(x - 2) = |x - 2| = g(x)$$

is a translation to the right and

$$f(x + 3) = |x + 3| = h(x)$$

is a translation to the left.

One way to determine the direction of the horizontal shift is to ask, "For what value of x is $y = 0$?" For instance, the horizontal shift of g in Figure 3 is 2 units to the right because $y = 0$ when $x = 2$.

$$g(x) = |x - 2|$$

$$g(2) = |2 - 2| = 0$$

Horizontal Translation

If f is a function and c is a positive constant, then

$y = f(x - c)$ is the graph of $y = f(x)$ shifted to the right c units.

$y = f(x + c)$ is the graph of $y = f(x)$ shifted to the left c units.

Point of Interest

Translations of a figure or pattern may be applied in the design of wallpaper, in the background of a web page, and in the use of shadows to create an illusion of a letter raised off a page.

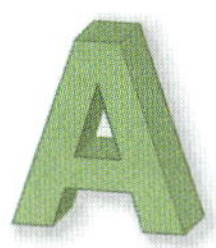

Example 2 Given the graph of the function $F(x) = x^3$ shown at the right in blue, graph $G(x) = (x + 2)^3$ by using a horizontal translation.

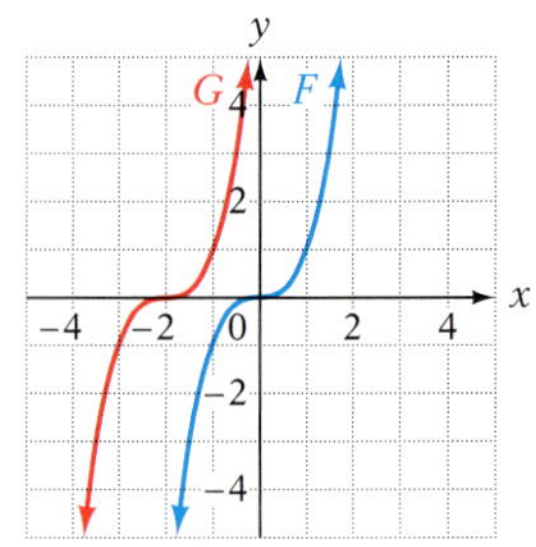

Solution The graph of G (shown in red) is the graph of F shifted horizontally 2 units to the left. Note that if

$$(x + 2)^3 = 0, \text{ then } x = -2.$$

The result, -2, is one confirmation that the graph should be shifted 2 units to the left.

Problem 2 Given the graph of the function $f(x) = \sqrt{x}$ shown at the right, graph $g(x) = \sqrt{x - 3}$ by using a horizontal translation.

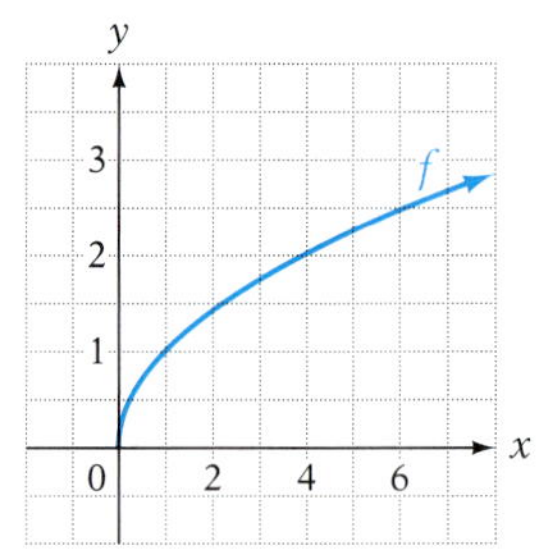

Solution See page S32.

It is possible for a graph to involve both a horizontal and a vertical translation.

Example 3 Given the graph of the function $f(x) = |x|$ shown at the right in blue, graph $A(x) = |x + 1| - 3$ by using both a horizontal and a vertical translation.

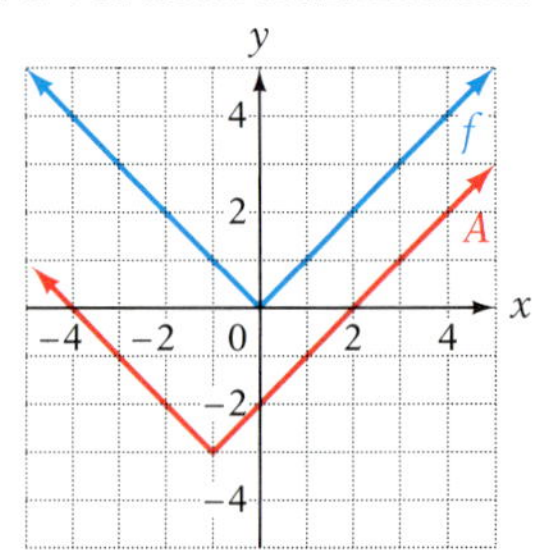

Solution The graph of A includes a horizontal shift of 1 unit to the left ($|x + 1| = 0$ means that $x = -1$) and a vertical shift of 3 units down. The graph of A is shown in red at the right.

Problem 3 Given the graph of the function $g(x) = x^3$ shown at the right, graph $V(x) = (x - 2)^3 + 1$ by using both a horizontal and a vertical translation.

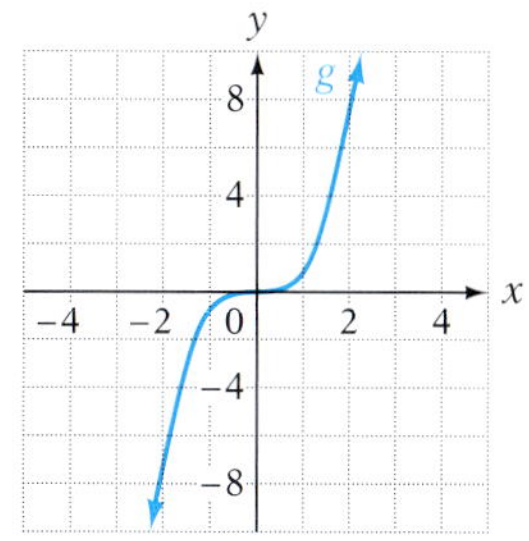

Solution See page S32.

10.2 Concept Review

Determine whether the following statements are always true, sometimes true, or never true.

1. If c is a positive constant, the graph of $y = f(x) + c$ is the graph of $y = f(x)$ shifted upward c units.

2. If c is a positive constant, the graph of $y = f(x + c)$ is the graph of $y = f(x)$ translated c units to the right.

3. The graph of $y = f(x - 2)$ is the graph of $y = f(x)$ translated 2 units to the right.

4. The graph of $y = |x| + 2$ is the graph of $y = |x|$ translated 2 units to the left.

5. If c_1 and c_2 are positive constants, the graph of $y = f(x + c_1) + c_2$ is the graph of $y = f(x)$ translated c_1 units to the right and shifted up c_2 units.

6. The graph of $y = (x - 2)^3 + 3$ is the graph of $y = x^3$ translated 2 units to the right and shifted 3 units up.

10.2 Exercises

1 1. Explain the meaning of a vertical or horizontal translation of a graph.

2. How does the graph of $y = f(x - 2) + 1$ differ from the graph of $y = f(x)$?

In Exercises 3–28, use translations to draw the graphs. Many of these exercises refer to Figures 1–3 below.

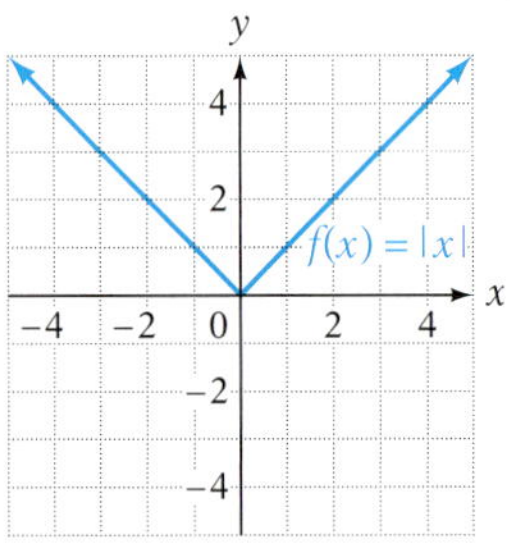

Figure 1

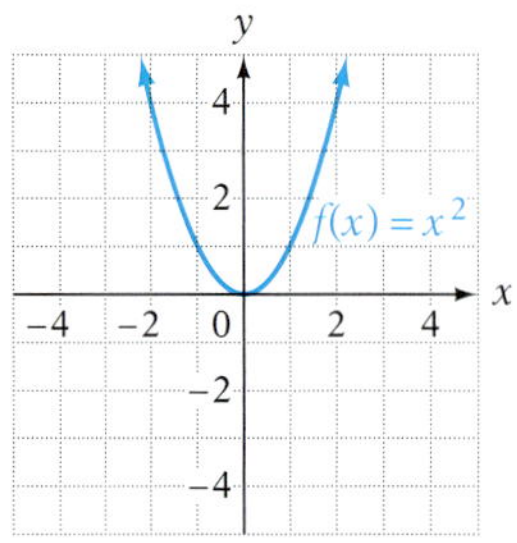

Figure 2

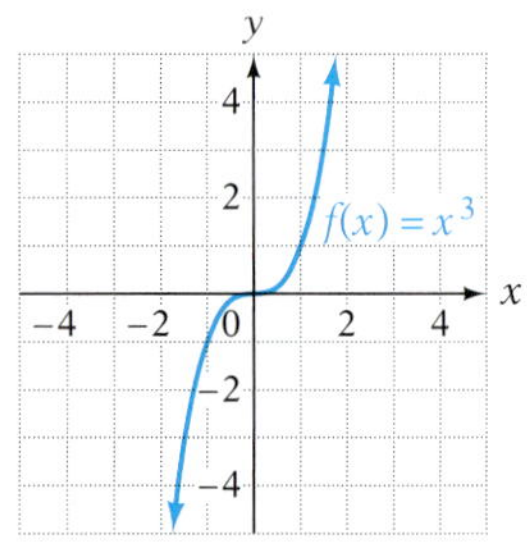

Figure 3

3. Given the graph of $f(x) = |x|$ shown in Figure 1 above, graph $g(x) = |x| + 4$.

4. Given the graph of $f(x) = |x|$ shown in Figure 1 above, graph $g(x) = |x| - 3$.

5. Given the graph of $f(x) = x^2$ shown in Figure 2 above, graph $g(x) = x^2 - 2$.

6. Given the graph of $f(x) = x^2$ shown in Figure 2 above, graph $g(x) = x^2 + 1$.

7. Given the graph of $f(x) = x^3$ shown in Figure 3 above, graph $g(x) = x^3 + 1$.

8. Given the graph of $f(x) = x^3$ shown in Figure 3 above, graph $g(x) = x^3 - 2$.

9. Given the graph of $f(x) = x^2$ shown in Figure 2 above, graph $g(x) = (x - 3)^2$.

10. Given the graph of $f(x) = x^2$ shown in Figure 2 above, graph $g(x) = (x + 2)^2$.

11. Given the graph of $f(x) = |x|$ shown in Figure 1 above, graph $g(x) = |x - 1|$.

12. Given the graph of $f(x) = |x|$ shown in Figure 1 above, graph $g(x) = |x + 4|$.

13. Given the graph of $f(x) = x^3$ shown in Figure 3 above, graph $g(x) = (x - 4)^3$.

14. Given the graph of $f(x) = x^3$ shown in Figure 3 above, graph $g(x) = (x + 3)^3$.

15. Given the graph of $f(x) = -x^2$ shown below, graph $g(x) = -x^2 + 2$.

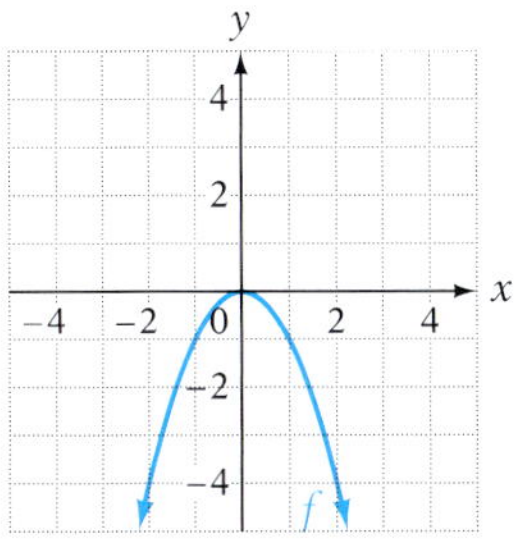

16. Given the graph of $f(x) = -|x|$ shown below, graph $g(x) = -|x| - 2$.

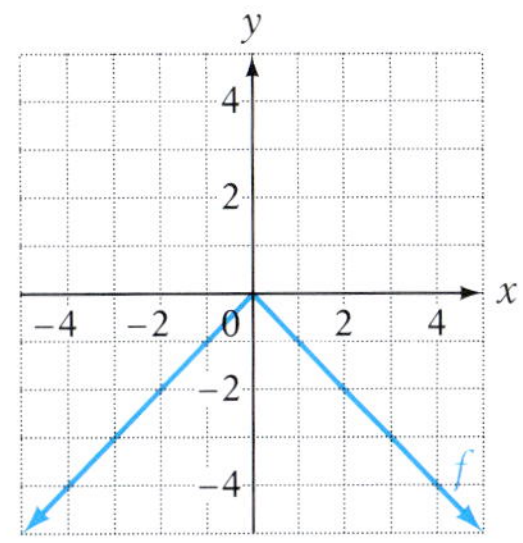

17. Given the graph of $f(x) = x^3$ shown in Figure 3 on page 650, graph $g(x) = (x + 1)^3 - 2$.

18. Given the graph of $f(x) = x^3$ shown in Figure 3 on page 650, graph $g(x) = (x - 2)^3 + 1$.

19. Given the graph of $f(x) = x^2$ shown in Figure 2 on page 650, graph $g(x) = (x - 3)^2 - 4$.

20. Given the graph of $f(x) = x^2$ shown in Figure 2 on page 650, graph $g(x) = (x + 2)^2 + 4$.

21. Given the graph of $f(x) = \sqrt{x}$ shown below, graph $g(x) = \sqrt{x - 1} + 2$.

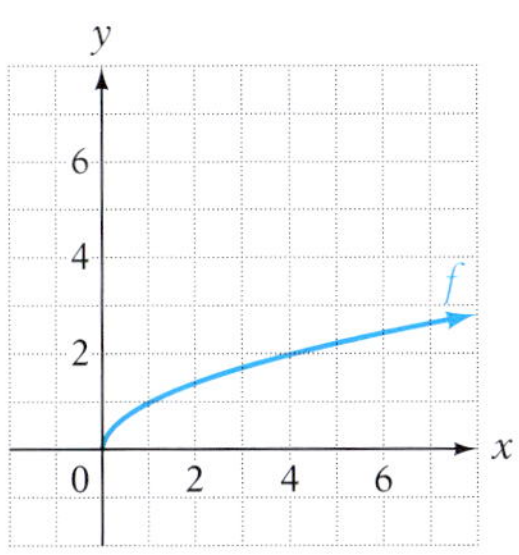

22. Given the graph of $f(x) = -\sqrt{x}$ shown below, graph $g(x) = -\sqrt{x + 1} - 3$.

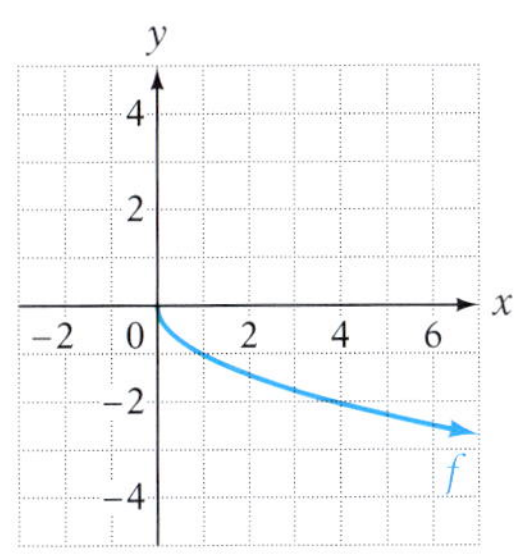

23. Given the graph of $f(x) = x^2 - 2x$ shown below, graph $g(x) = x^2 - 2x - 1$.

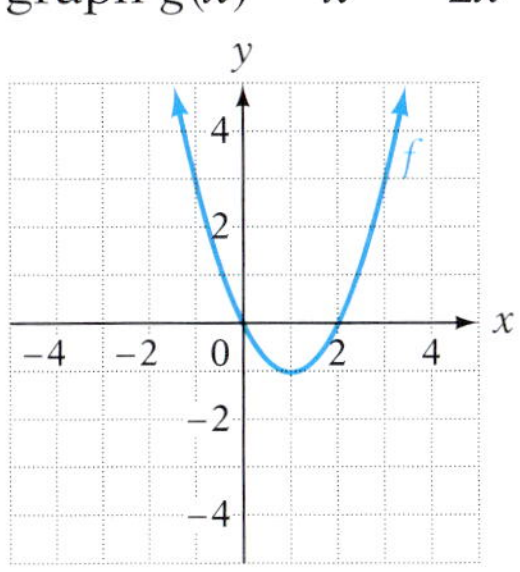

24. Given the graph of $f(x) = x^3 - 4x$ shown below, graph $g(x) = x^3 - 4x + 2$.

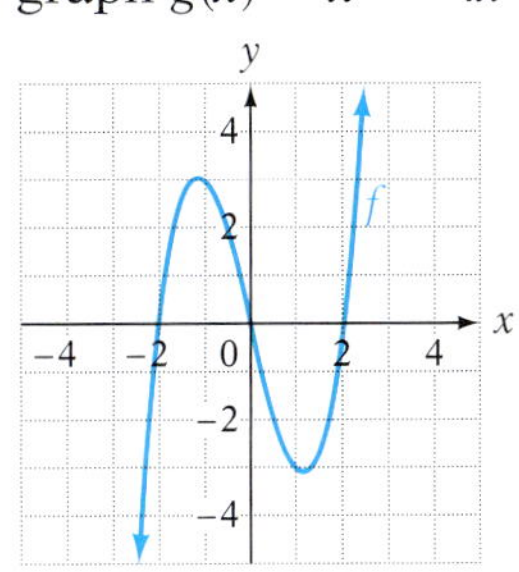

25. Given the graph of $f(x) = -x^2$ shown in Exercise 15, graph $g(x) = -(x + 1)^2 - 2$.

26. Given the graph of $f(x) = -x^2$ shown in Exercise 15, graph $g(x) = -(x - 1)^2 + 1$.

27. Given the graph of $f(x) = -\sqrt{x}$ shown in Exercise 22, graph $g(x) = -\sqrt{x - 2} + 1$.

28. Given the graph of $f(x) = \sqrt{x}$ shown in Exercise 21, graph $g(x) = \sqrt{x + 2} - 3$.

10.2 Applying Concepts

In Exercises 29–36, use translations to draw the graph. These exercises refer to Figures 4–6 below. No equations are given for the graphs.

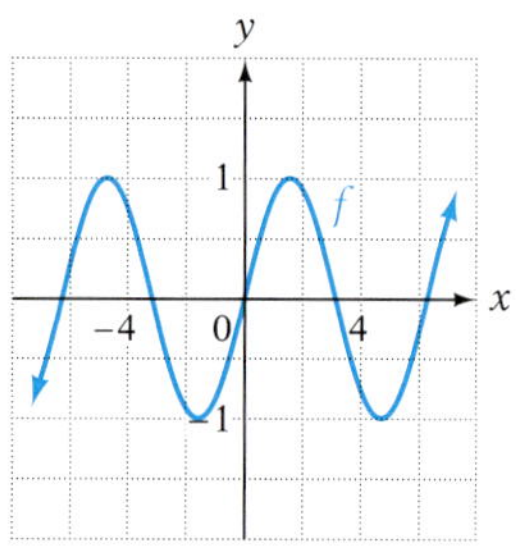

Figure 4

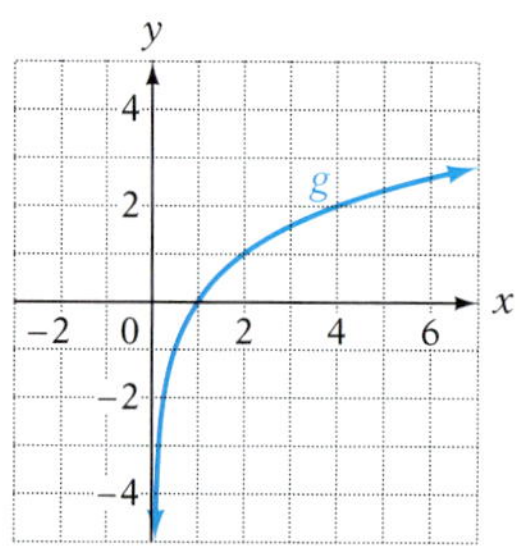

Figure 5

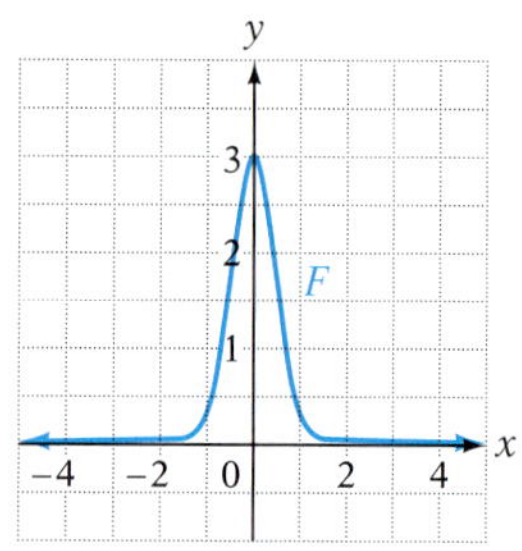

Figure 6

29. Given the graph of $y = f(x)$ shown in Figure 4 above, graph $y = f(x) + 1$.

30. Given the graph of $y = f(x)$ shown in Figure 4 above, graph $y = f(x) - 2$.

31. Given the graph of $y = g(x)$ shown in Figure 5 above, graph $y = g(x) - 3$.

32. Given the graph of $y = g(x)$ shown in Figure 5 above, graph $y = g(x) + 1$.

33. Given the graph of $y = f(x)$ shown in Figure 4 above, graph $y = f(x - 2)$.

34. Given the graph of $y = f(x)$ shown in Figure 4 above, graph $y = f(x + 1)$.

35. Given the graph of $y = F(x)$ shown in Figure 6 above, graph $y = F(x + 3)$.

36. Given the graph of $y = F(x)$ shown in Figure 6 above, graph $y = F(x - 1)$.

37. Given the triangle shown below, draw the triangle for which every value of x is translated 3 units to the right and every value of y is translated 2 units up.

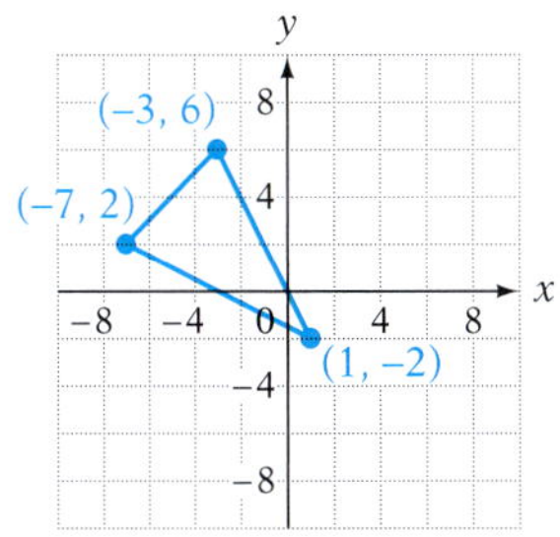

38. Given the rectangle shown below, draw the rectangle for which every value of x is translated 2 units to the right and every value of y is translated 1 unit up.

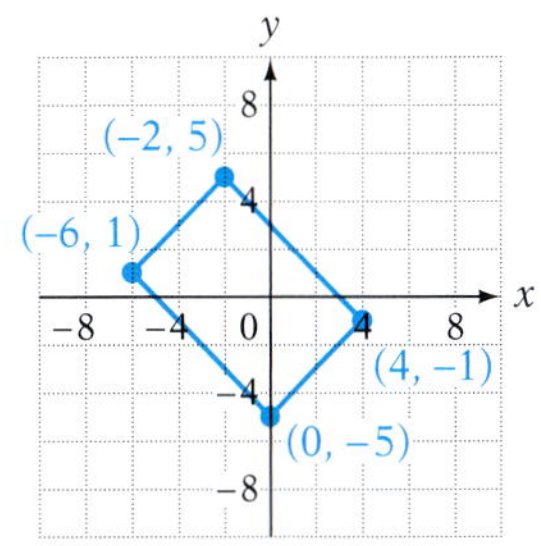

10.3 Algebra of Functions

1 Perform operations on functions

VIDEO & DVD · CD TUTOR · WEB · SSM

The operations of addition, subtraction, multiplication, and division of functions are defined as follows.

Operations on Functions

If f and g are functions and x is an element of the domain of each function, then

$$(f+g)(x) = f(x) + g(x)$$
$$(f-g)(x) = f(x) - g(x)$$
$$(f \cdot g)(x) = f(x) \cdot g(x)$$
$$\left(\frac{f}{g}\right)(x) = \frac{f(x)}{g(x)},\ g(x) \neq 0$$

Given $f(x) = x^2 + 1$ and $g(x) = 3x - 2$, find $(f+g)(3)$ and $(f \cdot g)(-1)$.

$$\begin{aligned}(f+g)(3) &= f(3) + g(3)\\ &= [(3)^2 + 1] + [3(3) - 2]\\ &= 10 + 7 = 17\end{aligned}$$

$$\begin{aligned}(f \cdot g)(-1) &= f(-1) \cdot g(-1)\\ &= [(-1)^2 + 1] \cdot [3(-1) - 2]\\ &= 2 \cdot (-5) = -10\end{aligned}$$

Consider the functions f and g above. Find $S(x)$, the sum of the two functions.

Use the definition of addition. $\quad S(x) = (f+g)(x)$
$\quad = f(x) + g(x)$

Substitute $x^2 + 1$ for $f(x)$ and $3x - 2$ for $g(x)$. $\quad = (x^2 + 1) + (3x - 2)$

$$S(x) = x^2 + 3x - 1$$

Now evaluate $S(3)$.

$$\begin{aligned} S(3) &= (3)^2 + 3(3) - 1 \\ &= 9 + 9 - 1 \\ &= 17 = (f + g)(3) \end{aligned}$$

Note that $S(3) = 17$ and $(f + g)(3) = 17$. This shows that adding $f(x) + g(x)$ and then evaluating is the same as evaluating $f(x)$ and $g(x)$ and then adding. The same is true for the other operations on functions. For instance, let $P(x)$ be the product of the functions f and g. Then

$$\begin{aligned} P(x) &= (f \cdot g)(x) \\ &= f(x) \cdot g(x) \\ &= (x^2 + 1)(3x - 2) \\ &= 3x^3 - 2x^2 + 3x - 2 \end{aligned}$$

$$\begin{aligned} P(-1) &= 3(-1)^3 - 2(-1)^2 + 3(-1) - 2 \\ &= -3 - 2 - 3 - 2 \\ &= -10 = (f \cdot g)(-1) \end{aligned}$$

Given $f(x) = 2x^2 - 5x + 3$ and $g(x) = x^2 - 1$, find $\left(\frac{f}{g}\right)(1)$.

$$\begin{aligned} \left(\frac{f}{g}\right)(1) &= \frac{f(1)}{g(1)} \\ &= \frac{2(1)^2 - 5(1) + 3}{(1)^2 - 1} \\ &= \frac{0}{0} \leftarrow \text{This is not a real number.} \end{aligned}$$

Because $\frac{0}{0}$ is not defined, the expression $\left(\frac{f}{g}\right)(1)$ cannot be evaluated.

Example 1 Given $f(x) = x^2 - x + 1$ and $g(x) = x^3 - 4$, find $(f - g)(3)$.

Solution

$$\begin{aligned} (f - g)(3) &= f(3) - g(3) \\ &= (3^2 - 3 + 1) - (3^3 - 4) \\ &= 7 - 23 \\ &= -16 \end{aligned}$$

$$(f - g)(3) = -16$$

Problem 1 Given $f(x) = x^2 + 2x$ and $g(x) = 5x - 2$, find $(f + g)(-2)$.

Solution See page S32.

Example 2 Given $f(x) = x^2 + 2$ and $g(x) = 2x + 3$, find $(f \cdot g)(-2)$.

Solution

$$\begin{aligned} (f \cdot g)(-2) &= f(-2) \cdot g(-2) \\ &= [(-2)^2 + 2] \cdot [2(-2) + 3] \\ &= 6 \cdot [-1] \\ &= -6 \end{aligned}$$

$$(f \cdot g)(-2) = -6$$

Problem 2 Given $f(x) = 4 - x^2$ and $g(x) = 3x - 4$, find $(f \cdot g)(3)$.

Solution See page S32.

Example 3 Given $f(x) = x^2 + 4x + 4$ and $g(x) = x^3 - 2$, find $\left(\frac{f}{g}\right)(3)$.

Solution

$$\left(\frac{f}{g}\right)(3) = \frac{f(3)}{g(3)} = \frac{3^2 + 4(3) + 4}{3^3 - 2} = \frac{25}{25} = 1$$

$$\left(\frac{f}{g}\right)(3) = 1$$

Problem 3 Given $f(x) = x^2 - 4$ and $g(x) = x^2 + 2x + 1$, find $\left(\frac{f}{g}\right)(4)$.

Solution See page S32.

2 Find the composition of two functions

VIDEO & DVD CD TUTOR WEB SSM

A function can be evaluated at the value of another function. Consider

$$f(x) = 2x + 7 \quad \text{and} \quad g(x) = x^2 + 1$$

The expression $f[g(-2)]$ means to evaluate the function f at $g(-2)$.

$$g(-2) = (-2)^2 + 1 = 4 + 1 = 5$$
$$f[g(-2)] = f(5) = 2(5) + 7 = 10 + 7 = 17$$

Definition of the Composition of Two Functions

Let f and g be two functions such that $g(x)$ is in the domain of f for all x in the domain of g. Then the **composition** of the two functions, denoted by $f \circ g$, is the function whose value at x is given by $(f \circ g)(x) = f[g(x)]$.

The function defined by $f[g(x)]$ is called the **composite** of f and g.

The function machine at the right illustrates the composition of the two functions $g(x) = x^2$ and $f(x) = 2x$. Note that a composite function combines two functions. First one function pairs an input with an output. Then that output is used as the input for a second function, which in turn produces a final output.

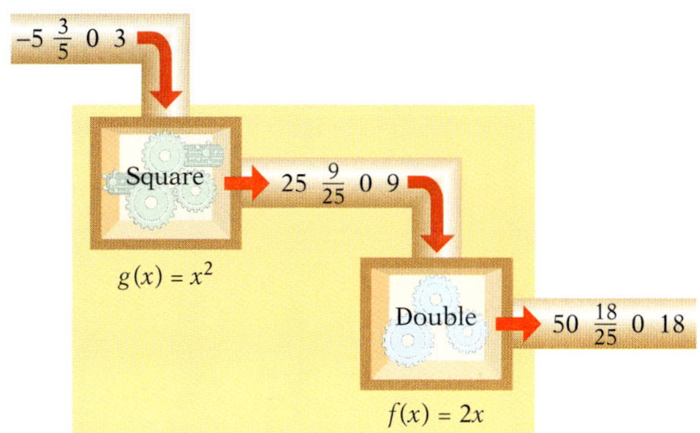

$(f \circ g)(x) = f[g(x)]$

The requirement in the definition of the composition of two functions that $g(x)$ be in the domain of f for all x in the domain of g is important. For instance, let

$$f(x) = \frac{1}{x-1} \qquad \text{and} \qquad g(x) = 3x - 5$$

When $x = 2$,

$$g(2) = 3(2) - 5 = 1$$

$$f[g(2)] = f(1) = \frac{1}{1-1} = \frac{1}{0} \leftarrow \text{This is not a real number.}$$

In this case, $g(2)$ is not in the domain of f. Thus the composition is not defined at 2.

Given $f(x) = x^3 - x + 1$ and $g(x) = 2x^2 - 10$, evaluate $(g \circ f)(2)$.

$$\begin{aligned} f(2) &= (2)^3 - (2) + 1 \\ &= 7 \\ (g \circ f)(2) &= g[f(2)] \\ &= g(7) \\ &= 2(7)^2 - 10 = 88 \\ (g \circ f)(2) &= 88 \end{aligned}$$

Given $f(x) = 3x - 2$ and $g(x) = x^2 - 2x$, evaluate $(f \circ g)(x)$.

$$\begin{aligned} (f \circ g)(x) &= f[g(x)] = 3(x^2 - 2x) - 2 \\ &= 3x^2 - 6x - 2 \end{aligned}$$

When we evaluate compositions of functions, the order in which the functions are applied is important. In the two diagrams that follow, the *square function*, $g(x) = x^2$, and the *double function*, $f(x) = 2x$, are interchanged. Note that the final outputs are different. Therefore, $\mathbf{(f \circ g)(x) \neq (g \circ f)(x)}$.

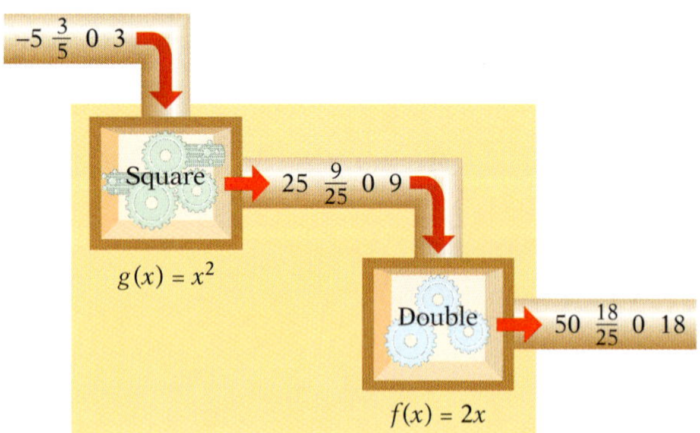

$(f \circ g)(x) = f[g(x)]$

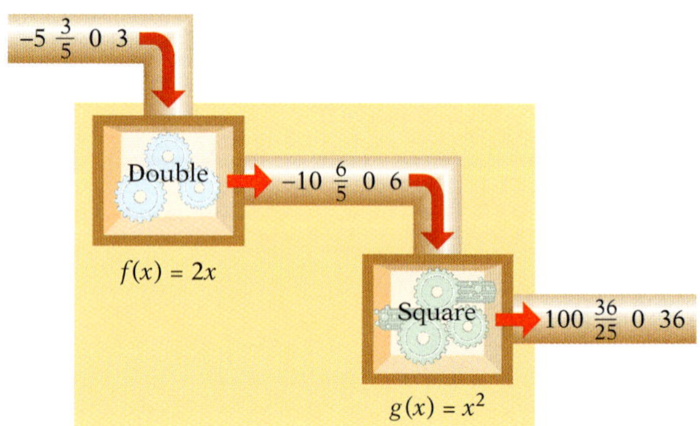

$(g \circ f)(x) = g[f(x)]$

Example 4 Given $f(x) = x^2 - 1$ and $g(x) = 3x + 4$, evaluate each composite function.

A. $f[g(0)]$ **B.** $g[f(x)]$

Solution **A.** $g(x) = 3x + 4$

$g(0) = 3(0) + 4 = 4$ • To evaluate $f[g(0)]$, first evaluate $g(0)$.

$f(x) = x^2 - 1$

$f[g(0)] = f(4) = 4^2 - 1 = 15$ • Substitute the value of $g(0)$ for x in $f(x)$; $g(0) = 4$.

$f[g(0)] = 15$

B. $g[f(x)] = g(x^2 - 1)$ • $f(x) = x^2 - 1$

$= 3(x^2 - 1) + 4$ • Substitute $x^2 - 1$ for x in the function $g(x)$; $g(x) = 3x + 4$.

$= 3x^2 - 3 + 4$

$= 3x^2 + 1$

Problem 4 Given $g(x) = 3x - 2$ and $h(x) = x^2 + 1$, evaluate each composite function.

A. $g[h(0)]$ **B.** $h[g(x)]$

Solution See page S33.

10.3 Concept Review

Determine whether the following statements are always true, sometimes true, or never true.

1. $(f \circ g)(x) = f[g(x)]$
2. If $f(x) = 2x$ and $g(x) = \frac{1}{2}x$, then $(f \circ g)(x) = (g \circ f)(x)$.
3. If $f(x) = 2x + 1$ and $g(x) = x + 4$, then $(f \circ g)(x) = (2x + 1)(x + 4)$.
4. If $f(x) = x^2 - 4$ and $g(x) = x^3$, then $(f \circ g)(x) = (x^2 - 4)^3$.

10.3 Exercises

1 Given $f(x) = 2x^2 - 3$ and $g(x) = -2x + 4$, find:

1. $(f - g)(2)$
2. $(f - g)(3)$
3. $(f + g)(0)$
4. $(f + g)(1)$
5. $(f \cdot g)(2)$
6. $(f \cdot g)(-1)$
7. $\left(\frac{f}{g}\right)(4)$
8. $\left(\frac{f}{g}\right)(-1)$
9. $\left(\frac{g}{f}\right)(-3)$

Given $f(x) = 2x^2 + 3x - 1$ and $g(x) = 2x - 4$, find:

10. $(f + g)(-3)$
11. $(f + g)(1)$
12. $(f + g)(-2)$

13. $(f - g)(4)$
14. $(f \cdot g)(-2)$
15. $(f \cdot g)(1)$

16. $\left(\dfrac{f}{g}\right)(2)$
17. $\left(\dfrac{f}{g}\right)(-3)$
18. $(f \cdot g)\left(\dfrac{1}{2}\right)$

Given $f(x) = x^2 + 3x - 5$ and $g(x) = x^3 - 2x + 3$, find:

19. $(f - g)(2)$
20. $(f \cdot g)(-3)$
21. $\left(\dfrac{f}{g}\right)(-2)$

2 22. Explain the meaning of the notation $f[g(2)]$.

23. What is the meaning of the notation $(f \circ g)(x)$?

Given $f(x) = 2x - 3$ and $g(x) = 4x - 1$, evaluate the composite function.

24. $f[g(0)]$
25. $g[f(0)]$
26. $f[g(2)]$
27. $g[f(-2)]$
28. $f[g(x)]$
29. $g[f(x)]$

Given $g(x) = x^2 + 3$ and $h(x) = x - 2$, evaluate the composite function.

30. $g[h(0)]$
31. $h[g(0)]$
32. $g[h(4)]$
33. $h[g(-2)]$
34. $g[h(x)]$
35. $h[g(x)]$

Given $f(x) = x^2 + x + 1$ and $h(x) = 3x + 2$, evaluate the composite function.

36. $f[h(0)]$
37. $h[f(0)]$
38. $f[h(-1)]$
39. $h[f(-2)]$
40. $f[h(x)]$
41. $h(f(x)]$

Given $f(x) = x - 2$ and $g(x) = x^3$, evaluate the composite function.

42. $f[g(2)]$
43. $f[g(-1)]$
44. $g[f(2)]$
45. $g[f(-1)]$
46. $f[g(x)]$
47. $g[f(x)]$

10.3 Applying Concepts

Given $g(x) = x^2 - 1$, find:

48. $g(2 + h)$
49. $g(3 + h) - g(3)$
50. $g(-1 + h) - g(-1)$

51. $\dfrac{g(1 + h) - g(1)}{h}$
52. $\dfrac{g(-2 + h) - g(-2)}{h}$
53. $\dfrac{g(a + h) - g(a)}{h}$

Given $f(x) = 2x$, $g(x) = 3x - 1$, and $h(x) = x - 2$, find:

54. $f(g[h(2)])$
55. $g(h[f(1)])$
56. $h(g[f(-1)])$

57. $f(h[g(0)])$
58. $f(g[h(x)])$
59. $g(f[h(x)])$

The graphs of the functions f and g are shown at the right. Use these graphs to determine the values of the following composite functions.

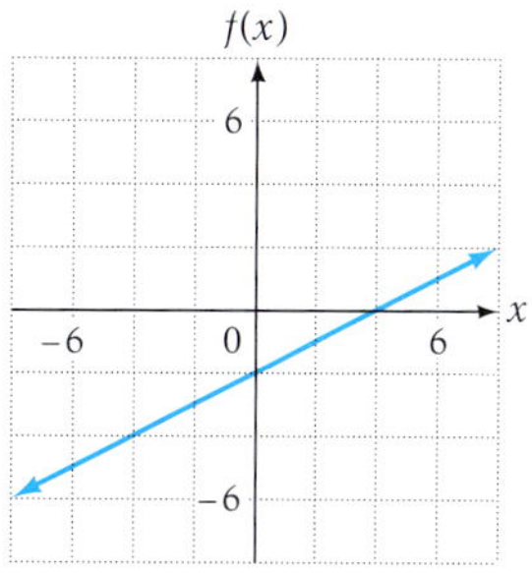

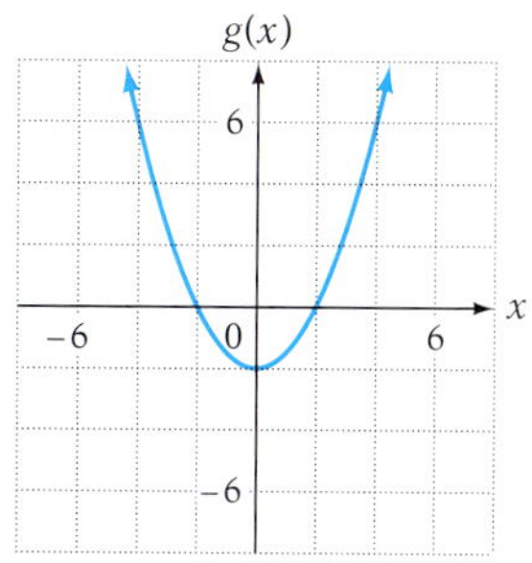

60. $f[g(0)]$

61. $f[g(2)]$

62. $f[g(4)]$

63. $f[g(-2)]$

64. $f[g(-4)]$

65. $g[f(0)]$

66. $g[f(4)]$

67. $g[f(-4)]$

10.4 One-to-One and Inverse Functions

1 Determine whether a function is one-to-one

VIDEO & DVD CD TUTOR WEB SSM

Recall that a function is a set of ordered pairs in which no two ordered pairs that have the same first component have different second components. This means that given any x there is only one y that can be paired with that x. A **one-to-one function** satisfies the additional condition that given any y, there is only one x that can be paired with the given y. One-to-one functions are commonly expressed by writing 1–1.

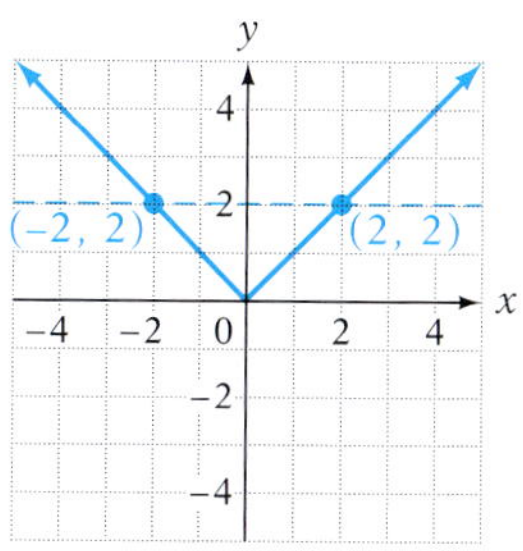

The function given by the equation $y = |x|$ is not a 1–1 function since, given $y = 2$, there are two possible values of x, 2 and -2, which can be paired with the given y-value. The graph at the left illustrates that a horizontal line intersects the graph more than once.

Just as the vertical line test can be used to determine whether a graph represents a function, a **horizontal line test** can be used to determine whether the graph of a function represents a 1–1 function.

Horizontal Line Test

A graph of a function is the graph of a 1–1 function if any horizontal line intersects the graph at no more than one point.

The graph of a quadratic function is shown at the right. Note that a horizontal line can intersect the graph at more than one point. Therefore, this graph is not the graph of a 1–1 function. In general, $f(x) = ax^2 + bx + c$, $a \neq 0$, is not a 1–1 function.

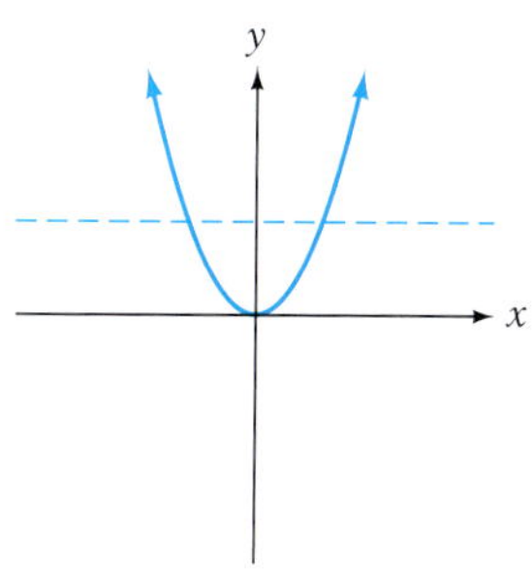

Since any vertical line will intersect the graph at the right at no more than one point, the graph is the graph of a function. Since any horizontal line will intersect the graph at no more than one point, the graph is the graph of a 1–1 function.

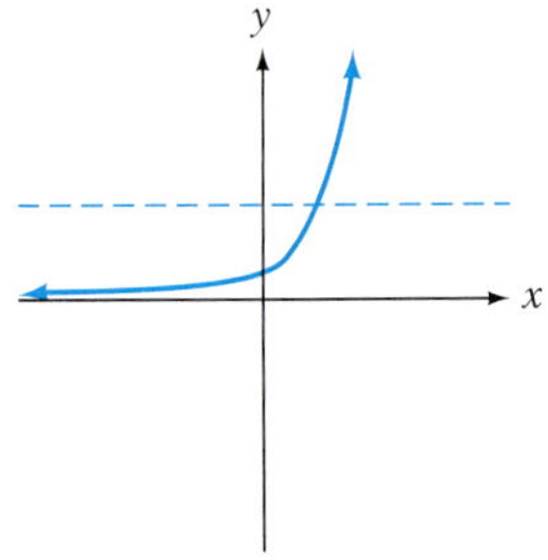

Example 1 Determine whether or not the graph represents the graph of a 1–1 function.

A.

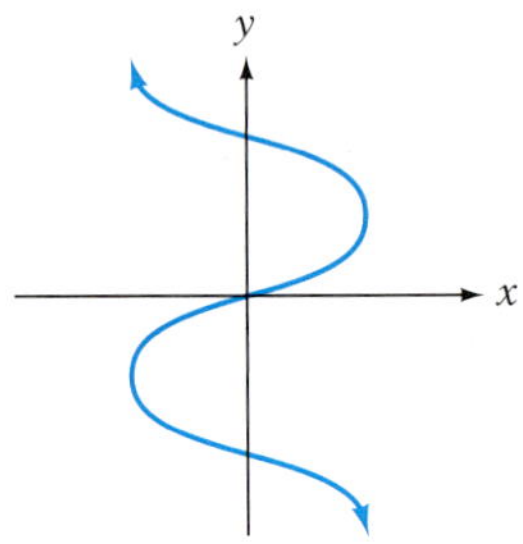

B.

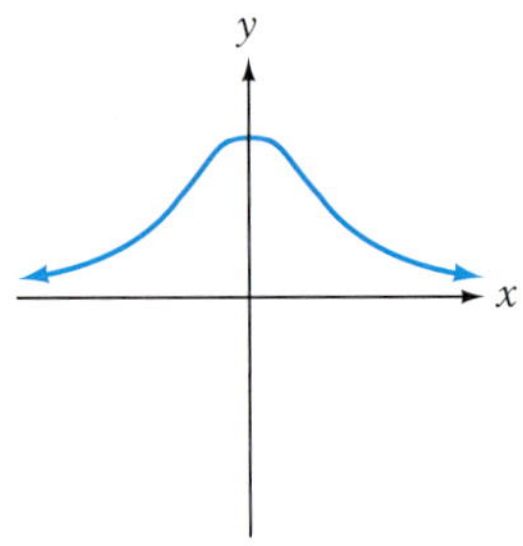

Solution **A.**

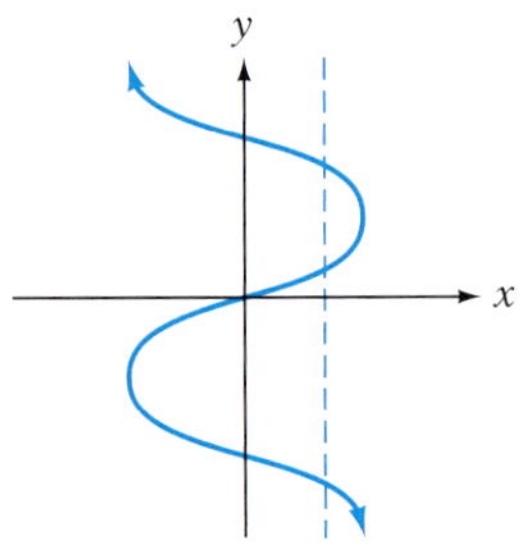

- **A vertical line can intersect the graph at more than one point. The graph does not represent a function.**

This is not the graph of a 1–1 function.

B.

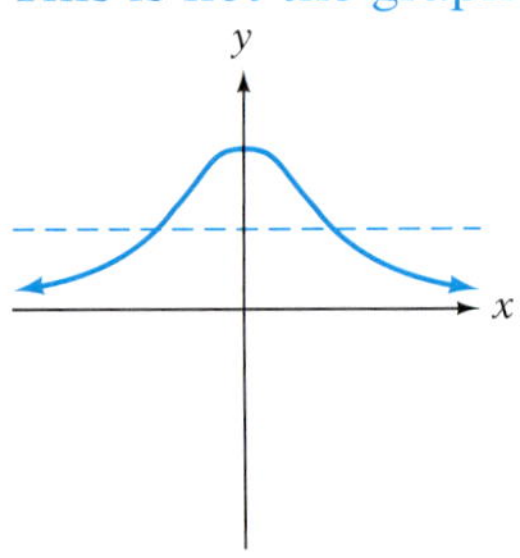

- **A horizontal line can intersect the curve at more than one point.**

This is not the graph of a 1–1 function.

Problem 1 Determine whether or not the graph represents the graph of a 1–1 function.

A.

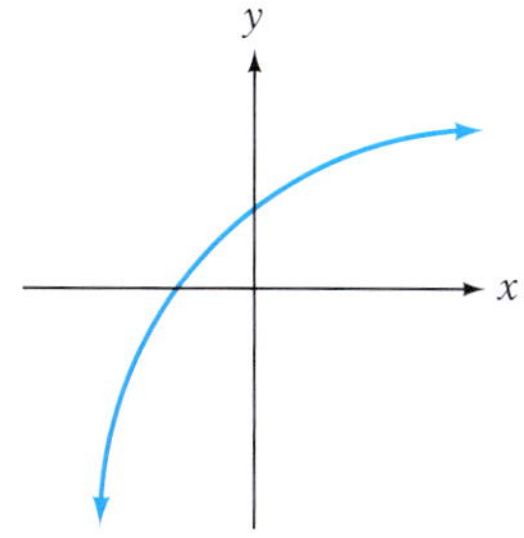

B.

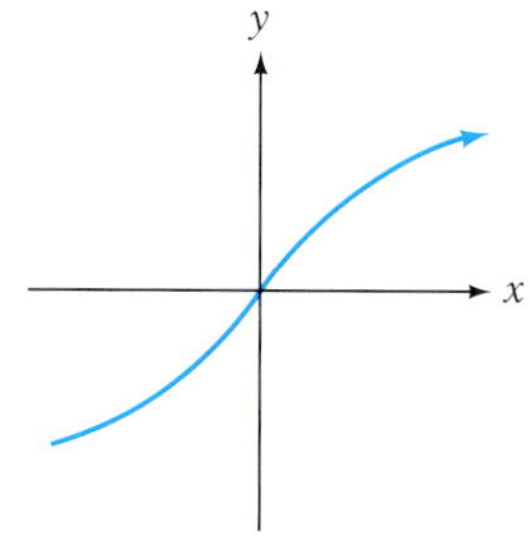

Solution See page S33.

2 Find the inverse of a function

VIDEO & DVD CD TUTOR WEB SSM

The **inverse of a function** is the set of ordered pairs formed by reversing the coordinates of each ordered pair of the function.

For example, the set of ordered pairs of the function defined by $f(x) = 2x$ with domain $\{-2, -1, 0, 1, 2\}$ is $\{(-2, -4), (-1, -2), (0, 0), (1, 2), (2, 4)\}$. The set of ordered pairs of the inverse function is $\{(-4, -2), (-2, -1), (0, 0), (2, 1), (4, 2)\}$.

From the ordered pairs of f, we have

$$\text{Domain} = \{-2, -1, 0, 1, 2\} \quad \text{and} \quad \text{Range} = \{-4, -2, 0, 2, 4\}$$

From the ordered pairs of the inverse function, we have

$$\text{Domain} = \{-4, -2, 0, 2, 4\} \quad \text{and} \quad \text{Range} = \{-2, -1, 0, 1, 2\}$$

Note that the domain of the inverse function is the range of the function, and the range of the inverse function is the domain of the function.

Now consider the function defined by $g(x) = x^2$ with domain $\{-2, -1, 0, 1, 2\}$. The set of ordered pairs of this function is $\{(-2, 4), (-1, 1), (0, 0), (1, 1), (2, 4)\}$. Reversing the ordered pairs gives $\{(4, -2), (1, -1), (0, 0), (1, 1), (4, 2)\}$. These ordered pairs do not satisfy the condition of a function, because there are ordered pairs with the same first coordinate and different second coordinates. This example illustrates that not all functions have an inverse function.

The graphs of $f(x) = 2x$ and $g(x) = x^2$ with the set of real numbers as the domain are shown below.

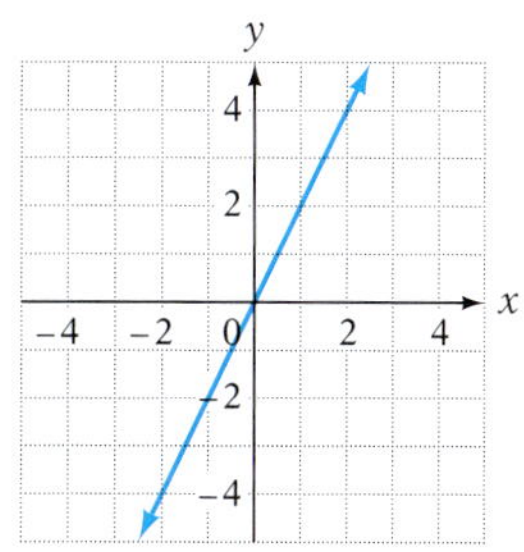

$f(x) = 2x$

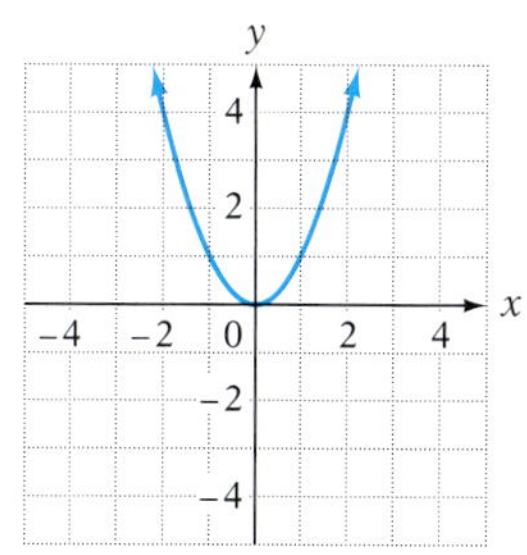

$g(x) = x^2$

By the horizontal line test, f is a 1–1 function but g is not.

Condition for an Inverse Function

A function f has an inverse function if and only if f is a 1–1 function.

> **Take Note**
> It is important to note that f^{-1} is the symbol for the *inverse* function and does not mean reciprocal.
> $$f^{-1}(x) \neq \frac{1}{f(x)}$$

The symbol f^{-1} is used to denote the inverse of the function f. The symbol $f^{-1}(x)$ is read "f inverse of x."

$f^{-1}(x)$ is *not* the reciprocal of $f(x)$ but is the notation for the inverse of a 1–1 function.

To find the inverse of a function, interchange x and y. Then solve for y.

> **Take Note**
> If the ordered pairs of f are given by (x, y), then the ordered pairs of f^{-1} are given by (y, x). That is, x and y are interchanged. This is the reason for Step 2 at the right.

Find the inverse of the function defined by $f(x) = 3x + 6$.

$$f(x) = 3x + 6$$

Replace $f(x)$ by y. $\quad y = 3x + 6$

Interchange x and y. $\quad x = 3y + 6$

Solve for y. $\quad x - 6 = 3y$

$$\frac{1}{3}x - 2 = y$$

Replace y by $f^{-1}(x)$. $\quad f^{-1}(x) = \frac{1}{3}x - 2$

The inverse of the function is given by $f^{-1}(x) = \frac{1}{3}x - 2$.

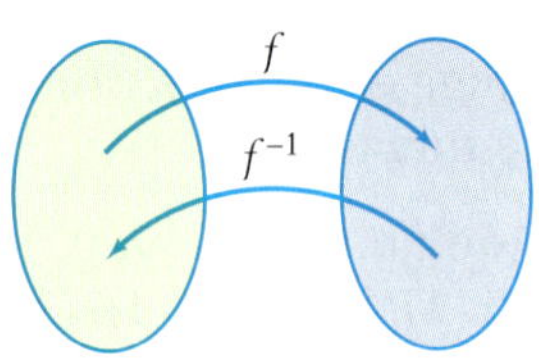

Domain of f
Range of f^{-1}

Range of f
Domain of f^{-1}

The fact that the ordered pairs of the inverse of a function are the reverse of those of the function has a graphical interpretation. The function shown at the left below includes the points (−2, 0), (−1, 2), (1, 4), and (5, 6). In the graph in the middle, the points with these coordinates reversed are plotted. The inverse function is graphed by drawing a smooth curve through those points, as shown in the figure on the right.

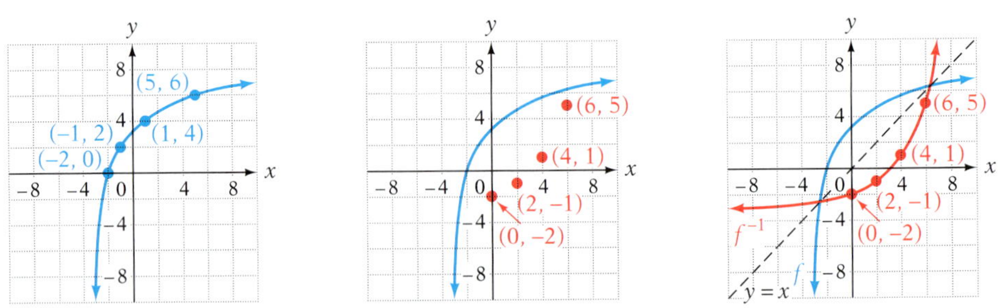

Note that the dashed graph of $y = x$ is shown in the figure on the right. If two functions are inverses of each other, then their graphs are mirror images with respect to the graph of the line $y = x$.

The composition of a function and its inverse have a special property.

Composition of Inverse Functions Property

$$f^{-1}[f(x)] = x \quad \text{and} \quad f[f^{-1}(x)] = x$$

This property can be used to determine whether two functions are inverses of each other.

Are $f(x) = 2x - 4$ and $g(x) = \frac{1}{2}x + 2$ inverses of each other?

To determine whether the functions are inverses, use the Composition of Inverse Functions Property.

$$\begin{aligned} f[g(x)] &= 2\left(\frac{1}{2}x + 2\right) - 4 \\ &= x + 4 - 4 \\ &= x \end{aligned} \qquad \begin{aligned} g[f(x)] &= \frac{1}{2}(2x - 4) + 2 \\ &= x - 2 + 2 \\ &= x \end{aligned}$$

Because $f[g(x)] = x$ and $g[f(x)] = x$, the functions are inverses of each other.

Example 2 Find the inverse of the function defined by the equation $f(x) = 2x - 4$.

Solution

$$\begin{aligned} f(x) &= 2x - 4 \\ y &= 2x - 4 && \bullet \text{ Think of the function as the equation } y = 2x - 4. \\ x &= 2y - 4 && \bullet \text{ Interchange } x \text{ and } y. \\ 2y &= x + 4 && \bullet \text{ Solve for } y. \\ y &= \frac{1}{2}x + 2 \end{aligned}$$

$$f^{-1}(x) = \frac{1}{2}x + 2$$

Problem 2 Find the inverse of the function defined by the equation $f(x) = 4x + 2$.

Solution See page S33.

Example 3 Are the functions defined by the equations $f(x) = -2x + 3$ and $g(x) = -\frac{1}{2}x + \frac{3}{2}$ inverses of each other?

Solution

$$\begin{aligned} f[g(x)] &= f\left(-\frac{1}{2}x + \frac{3}{2}\right) && \bullet \text{ Use the property that for inverses, } f[f^{-1}(x)] = f^{-1}[f(x)] = x. \\ &= -2\left(-\frac{1}{2}x + \frac{3}{2}\right) + 3 \\ &= x - 3 + 3 \\ &= x && \bullet\ f[g(x)] = x \end{aligned}$$

$$g[f(x)] = g(-2x + 3)$$
$$= -\frac{1}{2}(-2x + 3) + \frac{3}{2}$$
$$= x - \frac{3}{2} + \frac{3}{2}$$
$$= x \qquad \bullet\ g[f(x)] = x$$

The functions are inverses of each other.

Problem 3 Are the functions defined by the equations $h(x) = 4x + 2$ and $g(x) = \frac{1}{4}x - \frac{1}{2}$ inverses of each other?

Solution See page S33.

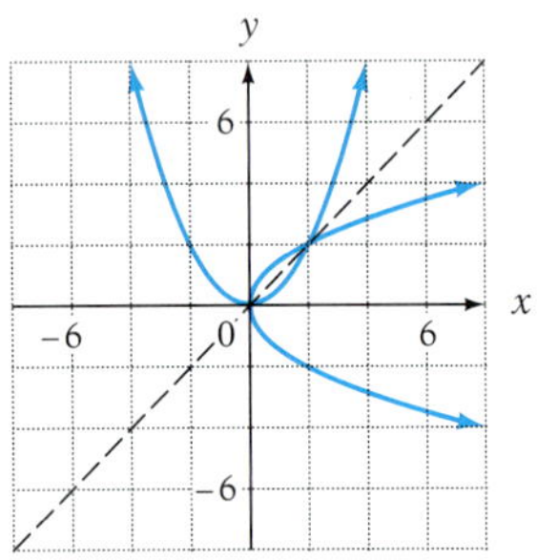

The function given by the equation $f(x) = \frac{1}{2}x^2$ does not have an inverse that is a function. Two of the ordered-pair solutions of this function are (4, 8) and (−4, 8).

The graph of $f(x) = \frac{1}{2}x^2$ is shown at the left. This graph does not pass the horizontal line test for the graph of a 1–1 function. The mirror image of the graph with respect to the line $y = x$ is also shown. This graph does not pass the vertical line test for the graph of a function.

A quadratic function with domain the real numbers does not have an inverse function.

10.4 Concept Review

Determine whether the following statements are always true, sometimes true, or never true.

1. The graph of a function represents the graph of a 1–1 function if any horizontal line intersects the graph at no more than one point.

2. A function has an inverse if and only if it is a 1–1 function.

3. The inverse of the function {(2, 3), (4, 5), (6, 3)} is the function {(3, 2), (5, 4), (3, 6)}.

4. The inverse of a function is a relation.

5. The inverse of a function is a function.

6. The graph of $y = x^2$ is the graph of a 1–1 function.

10.4 Exercises

1 **1.** What is a 1–1 function?

2. What is the horizontal line test?

Determine whether the graph represents the graph of a 1–1 function.

3.

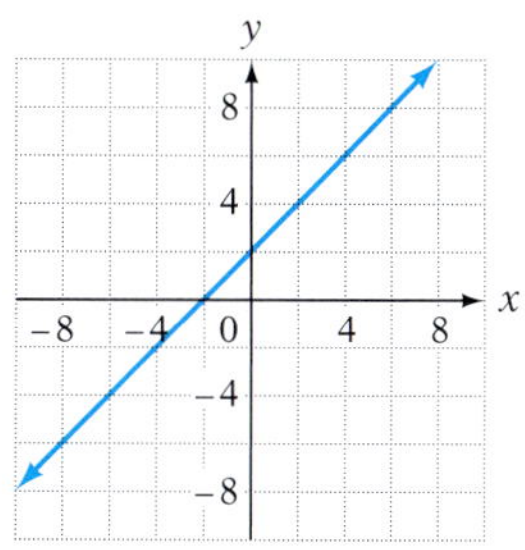

4.

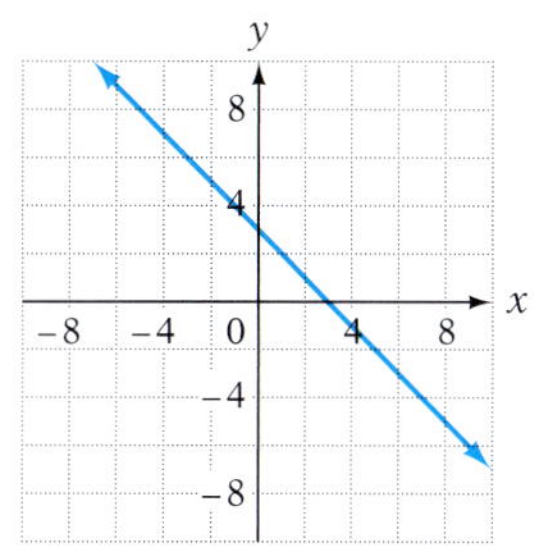

5.

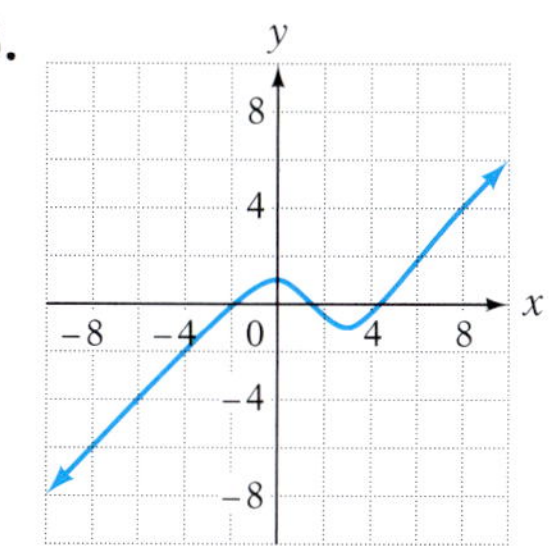

6.

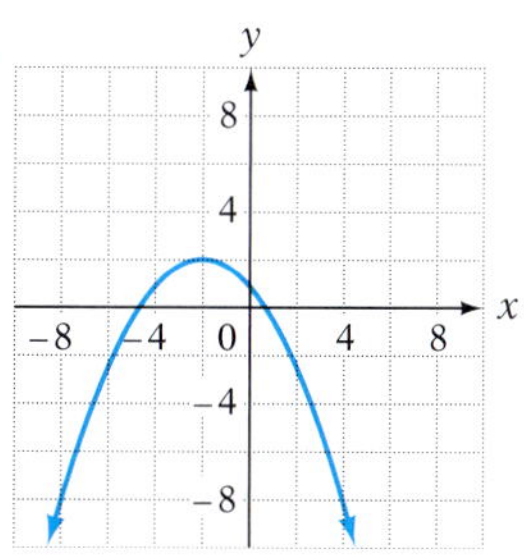

7.

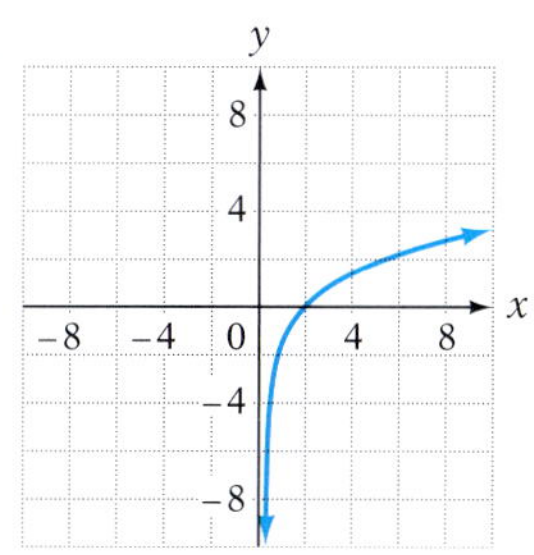

8.

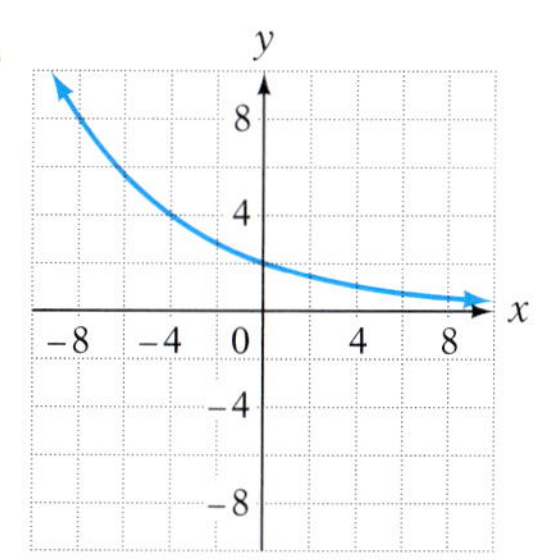

9.

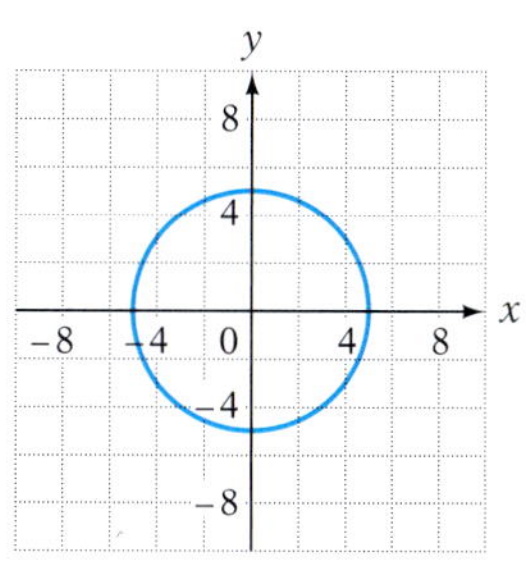

10.

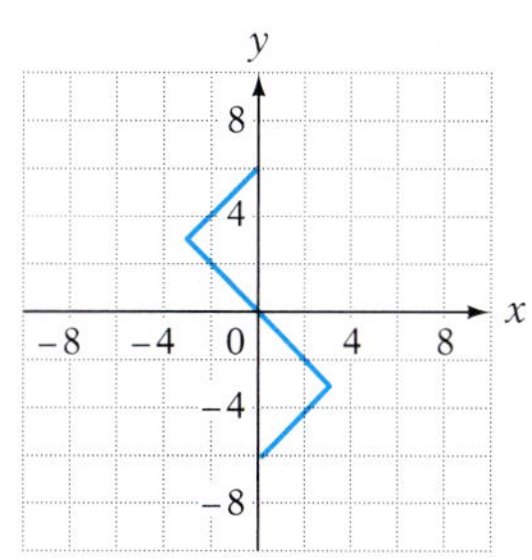

11.

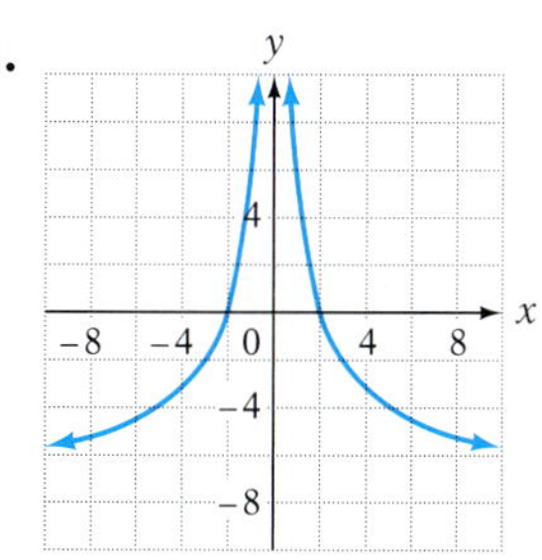

12.

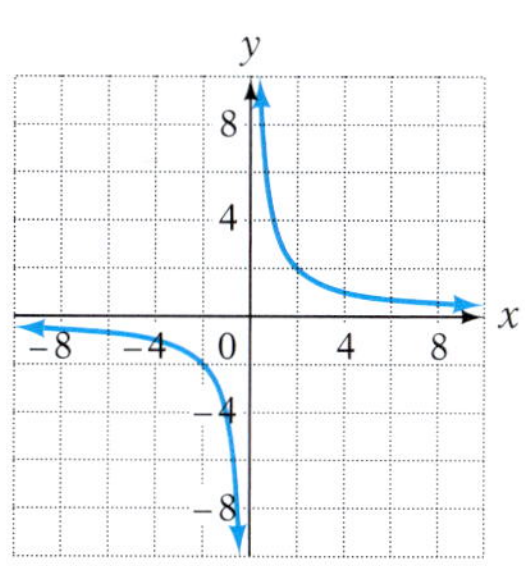

13.

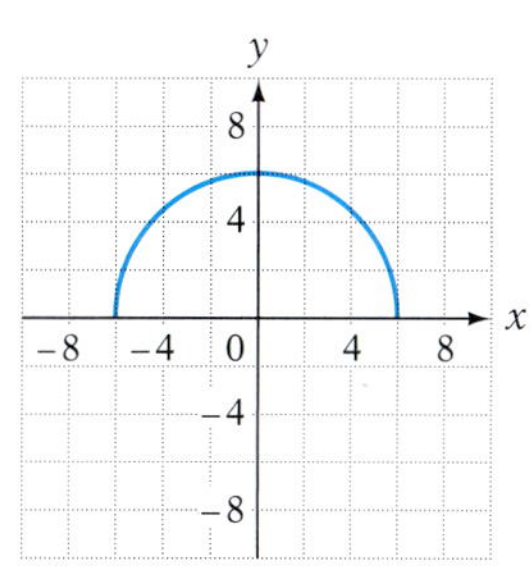

14.

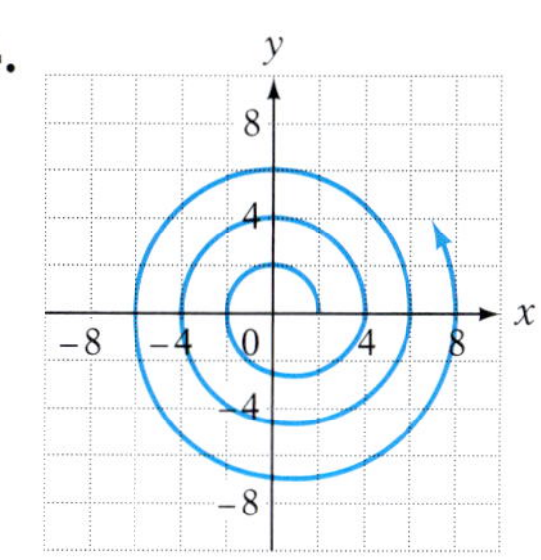

2 **15.** How are the ordered pairs of the inverse of a function related to the function?

16. Why is it that not all functions have an inverse function?

Find the inverse of the function. If the function does not have an inverse function, write "No inverse function."

17. $\{(1, 0), (2, 3), (3, 8), (4, 15)\}$

18. $\{(1, 0), (2, 1), (-1, 0), (-2, 1)\}$

19. $\{(3, 5), (-3, -5), (2, 5), (-2, -5)\}$

20. $\{(-5, -5), (-3, -1), (-1, 3), (1, 7)\}$

21. $f(x) = 4x - 8$

22. $f(x) = 3x + 6$

23. $f(x) = x^2 - 1$

24. $f(x) = 2x + 4$

25. $f(x) = x - 5$

26. $f(x) = \frac{1}{2}x - 1$

27. $f(x) = \frac{1}{3}x + 2$

28. $f(x) = -2x + 2$

29. $f(x) = -3x - 9$

30. $f(x) = 2x^2 + 2$

31. $f(x) = \frac{2}{3}x + 4$

32. $f(x) = \frac{3}{4}x - 4$

33. $f(x) = -\frac{1}{3}x + 1$

34. $f(x) = -\frac{1}{2}x + 2$

35. $f(x) = 2x - 5$

36. $f(x) = 3x + 4$

37. $f(x) = x^2 + 3$

38. $f(x) = 5x - 2$

Given $f(x) = 3x - 5$, find:

39. $f^{-1}(0)$

40. $f^{-1}(2)$

41. $f^{-1}(4)$

State whether the graph is the graph of a function. If it is the graph of a function, does it have an inverse?

42.

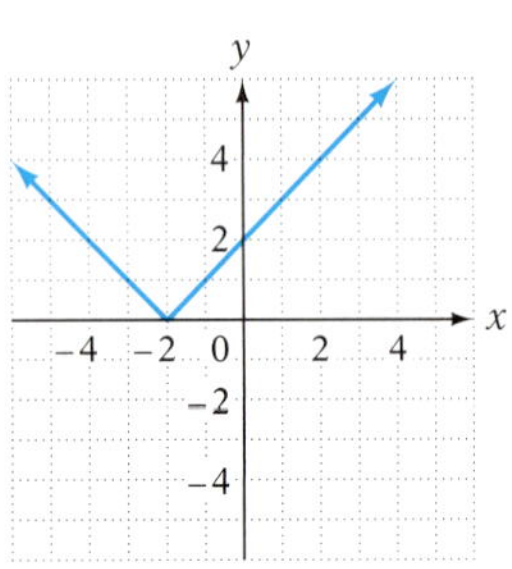

43.

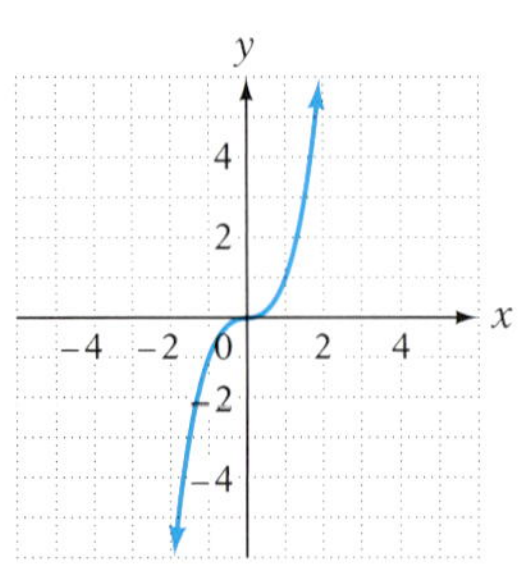

44.

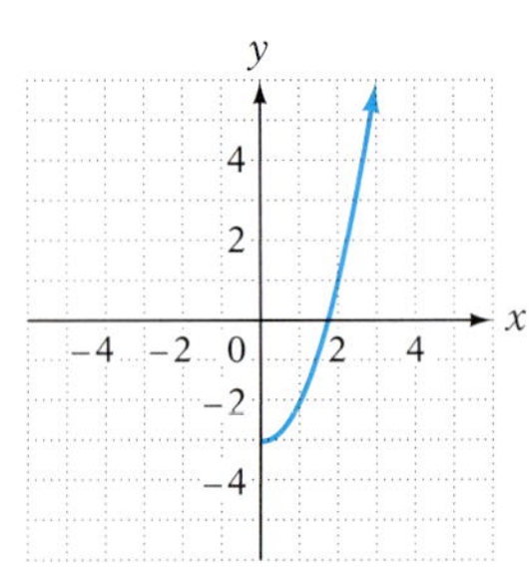

Use the Composition of Inverse Functions Property to determine whether the functions are inverses of each other.

45. $f(x) = 4x; g(x) = \frac{x}{4}$

46. $g(x) = x + 5; h(x) = x - 5$

47. $f(x) = 3x; h(x) = \frac{1}{3x}$

48. $h(x) = x + 2; g(x) = 2 - x$

49. $g(x) = 3x + 2; f(x) = \frac{1}{3}x - \frac{2}{3}$

50. $h(x) = 4x - 1; f(x) = \frac{1}{4}x + \frac{1}{4}$

51. $f(x) = \frac{1}{2}x - \frac{3}{2}; g(x) = 2x + 3$

52. $g(x) = -\frac{1}{2}x - \frac{1}{2}; h(x) = -2x + 1$

Complete.

53. The domain of the inverse function f^{-1} is the _______ of f.

54. The range of the inverse function f^{-1} is the _______ of f.

55. For any function f and its inverse f^{-1}, $f[f^{-1}(3)] =$ _______.

56. For any function f and its inverse f^{-1}, $f^{-1}[f(-4)] =$ _______.

If f is a 1–1 function and $f(0) = 5$, $f(1) = 7$, $f(2) = 9$, and $f(3) = 11$, find:

57. $f^{-1}(5)$

58. $f^{-1}(11)$

59. $f^{-1}(9)$

60. $f^{-1}(7)$

10.4 Applying Concepts

Given the graph of the 1–1 function, draw the graph of the inverse of the function by using the technique shown in this section.

61.

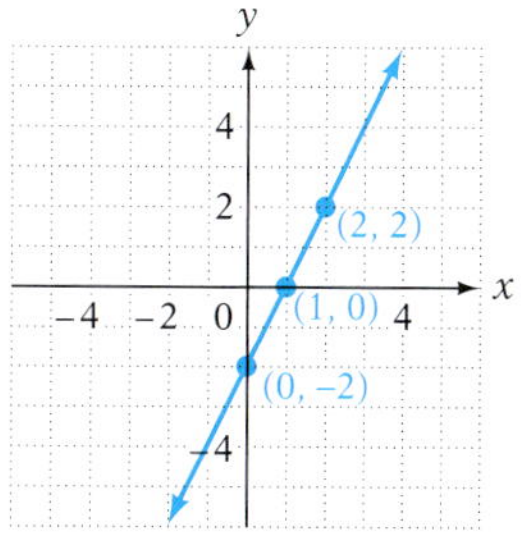

62.

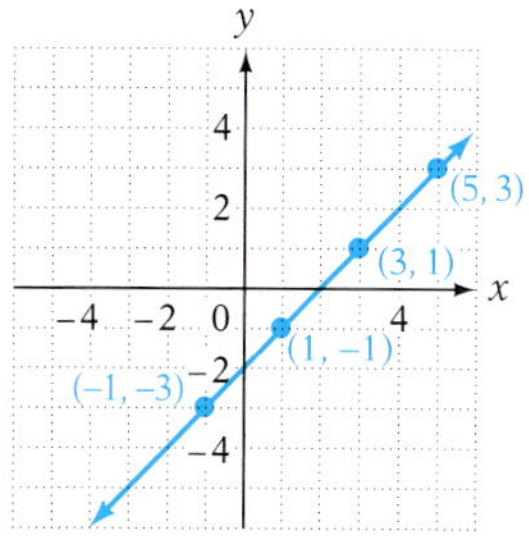

63.

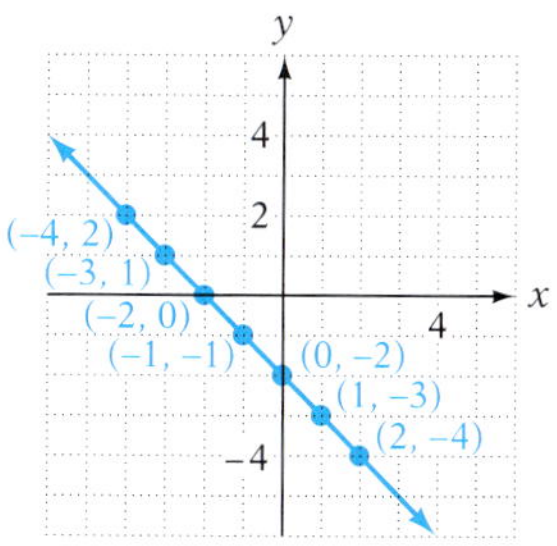

64.

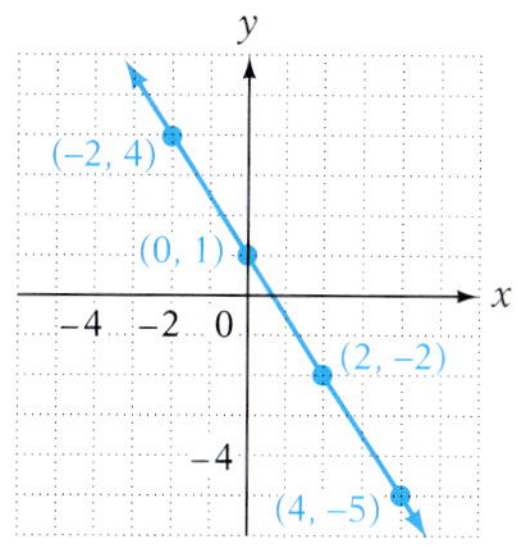

65.

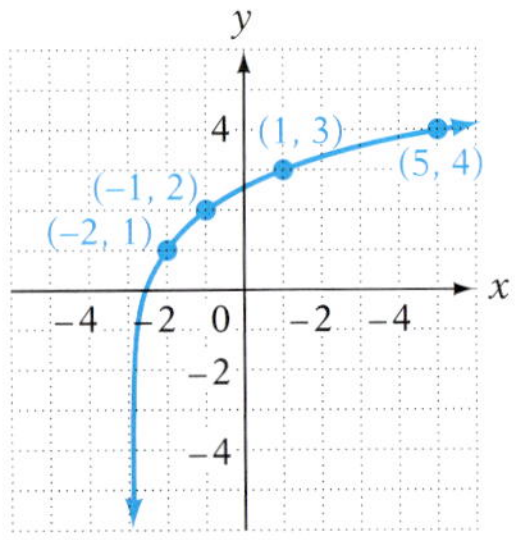

66.

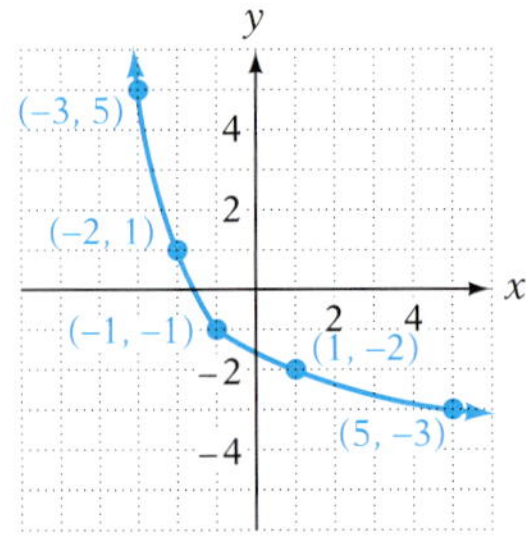

Each of the following tables defines a function. Is the inverse of the function a function? Explain your answer.

67. **Grading Scale Table**

Score	*Grade*
90 – 100	A
80 – 89	B
70 – 79	C
60 – 69	D
0 – 59	F

68. **First-Class Postage**

Weight (in oz)	*Cost*
$0 < w \leq 1$	\$.37
$1 < w \leq 2$	\$.60
$2 < w \leq 3$	\$.83
$3 < w \leq 4$	\$1.06

69. Is the inverse of a constant function a function? Explain your answer.

70. The graphs of all functions given by $f(x) = mx + b, m \neq 0$, are straight lines. Are all of these functions 1–1 functions? If so, explain why. If not, give an example of a linear function that is not 1–1.

10.5 Conic Sections

1 Graph parabolas

VIDEO & DVD CD TUTOR WEB SSM

The **conic sections** are curves that can be constructed from the intersection of a plane and a right circular cone. The four conic sections are the parabola, circle, ellipse, and hyperbola. The **parabola** was introduced earlier. Here we will review some of that previous discussion and look at equations of parabolas that were not discussed before.

Point of Interest

Hypatia (ca. 355–415) is considered the first prominent woman mathematician. She lectured in mathematics and philosophy at the Museum in Alexandria, the most distinguished place of learning in the world at that time. One of the topics on which Hypatia lectured was conic sections. One historian has claimed that with the death (actually the murder) of Hypatia, "the long and glorious history of Greek mathematics was at an end."

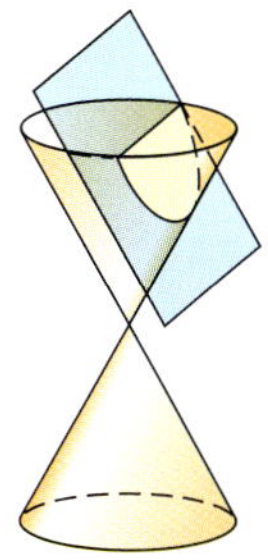

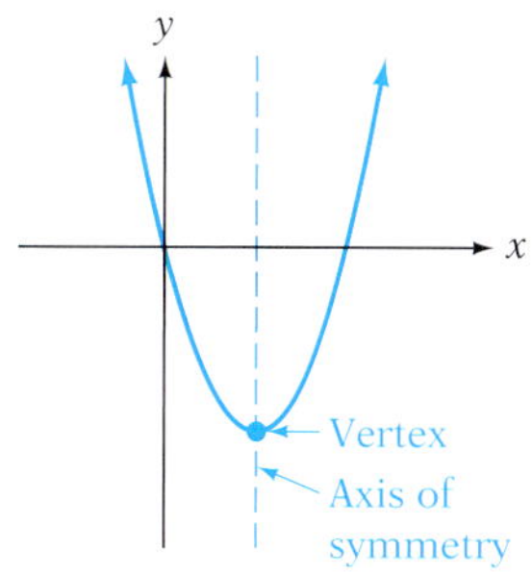

A parabola is a conic section formed by the intersection of a right circular cone and a plane parallel to the side of the cone. Every parabola has an **axis of symmetry** and a **vertex** that is on the axis of symmetry. To understand the axis of symmetry, think of folding the paper along that axis. The two halves of the curve will match up.

The graph of the equation $y = ax^2 + bx + c$, $a \neq 0$, is a parabola with the axis of symmetry parallel to the y-axis. The parabola opens up when $a > 0$ and opens down when $a < 0$. When the parabola opens up, the vertex is the lowest point on the parabola. When the parabola opens down, the vertex is the highest point on the parabola.

The coordinates of the vertex can be found by completing the square.

Find the vertex of the parabola whose equation is $y = x^2 - 4x + 5$.

$y = x^2 - 4x + 5$

Group the terms involving x. $\quad y = (x^2 - 4x) + 5$

Complete the square of $x^2 - 4x$. Note that 4 is added and subtracted. Because $4 - 4 = 0$, the equation is not changed. $\quad y = (x^2 - 4x + 4) - 4 + 5$

Factor the trinomial and combine like terms. $\quad y = (x - 2)^2 + 1$

The coefficient of x^2 is positive, so the parabola opens up. The vertex is the lowest point on the parabola, or the point that has the least y-coordinate.

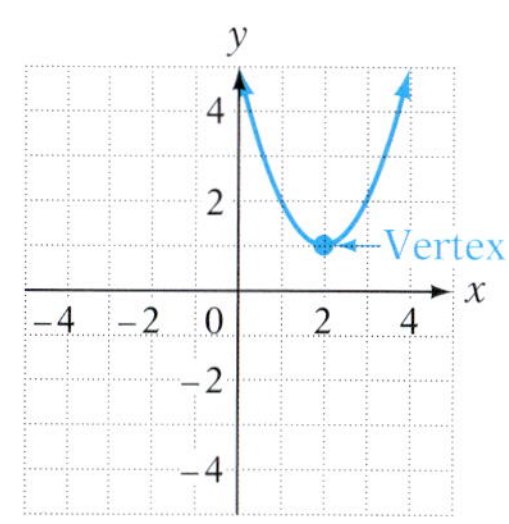

Because $(x - 2)^2 \geq 0$ for all x, the least y-coordinate occurs when $(x - 2)^2 = 0$, which occurs when $x = 2$. This means the x-coordinate of the vertex is 2.

To find the y-coordinate of the vertex, replace x in $y = (x - 2)^2 + 1$ by 2 and solve for y.

$$y = (x - 2)^2 + 1$$
$$= (2 - 2)^2 + 1 = 1$$

The vertex is (2, 1).

By following the procedure of the last example and completing the square on the equation $y = ax^2 + bx + c$, we find that the **x-coordinate of the vertex** is

Point of Interest

The suspension cables for some bridges, such as the Golden Gate bridge, hang in the shape of a parabola. Parabolic shapes are also used for mirrors in telescopes and in certain antenna designs.

$-\frac{b}{2a}$. The y-coordinate of the vertex can then be determined by substituting this value of x into $y = ax^2 + bx + c$ and solving for y.

Because the axis of symmetry is parallel to the y-axis and passes through the vertex, the equation of the **axis of symmetry** is $x = -\frac{b}{2a}$.

Example 1 Find the vertex and the axis of symmetry of the parabola given by the equation $y = x^2 + 2x - 3$. Then sketch the graph of the parabola.

Solution $-\frac{b}{2a} = -\frac{2}{2(1)} = -1$ • The x-coordinate of the vertex is $-\frac{b}{2a}$.

$y = x^2 + 2x - 3$
$y = (-1)^2 + 2(-1) - 3$
$y = -4$ • Find the y-coordinate of the vertex by replacing x by -1 and solving for y.

The vertex is $(-1, -4)$.

The axis of symmetry is the line $x = -1$. • The axis of symmetry is the line $x = -\frac{b}{2a}$.

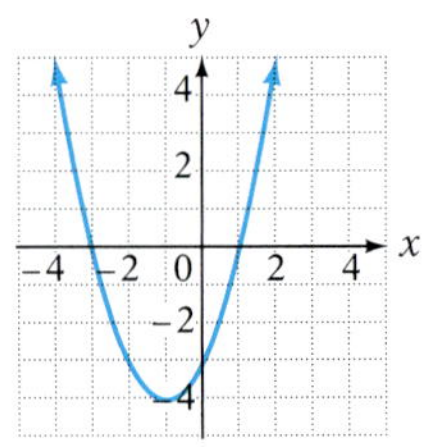

• Because a is positive, the parabola opens up. Use the vertex and axis of symmetry to sketch the graph.

Problem 1 Find the vertex and the axis of symmetry of the parabola given by the equation $y = -x^2 + x + 3$. Then sketch the graph of the parabola.

Solution See page S33.

The graph of an equation of the form $x = ay^2 + by + c$, $a \neq 0$, is also a parabola. In this case, the parabola opens to the right when a is positive and opens to the left when a is negative.

For a parabola of this form, the **y-coordinate of the vertex** is $-\frac{b}{2a}$. The **axis of symmetry** is the line $y = -\frac{b}{2a}$.

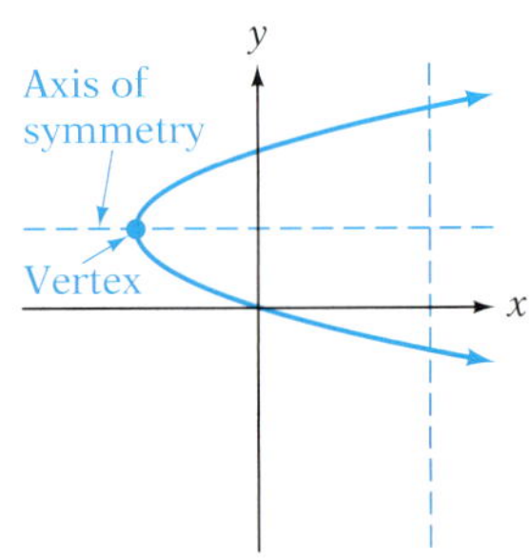

Using the vertical-line test reveals that the graph of a parabola of this form is not the graph of a function. The graph of $x = ay^2 + by + c$ is the graph of a relation.

Example 2 Find the vertex and the axis of symmetry of the parabola whose equation is $x = 2y^2 - 8y + 5$. Then sketch its graph.

Solution

$-\frac{b}{2a} = -\frac{-8}{2(2)} = 2$ • Find the y-coordinate of the vertex. $a = 2, b = -8$

$x = 2y^2 - 8y + 5$
$x = 2(2)^2 - 8(2) + 5$
$x = -3$ • Find the x-coordinate of the vertex by replacing y by 2 and solving for x.

The vertex is $(-3, 2)$.

The axis of symmetry is the line $y = 2$. • The axis of symmetry is the line $y = -\frac{b}{2a}$.

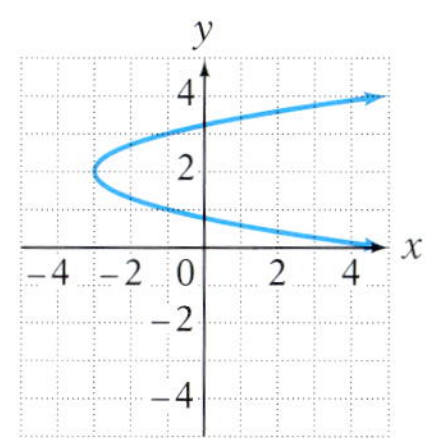

• Because a is positive, the parabola opens to the right. Use the vertex and axis of symmetry to sketch the graph.

Problem 2 Find the vertex and the axis of symmetry of the parabola whose equation is $x = -2y^2 - 4y - 3$. Then sketch its graph.

Solution See page S33.

2 Find the equation of a circle and then graph the circle

VIDEO & DVD | CD TUTOR | WEB | SSM

A **circle** is a conic section formed by the intersection of a cone and a plane that is parallel to the base of the cone.

Take Note

As the angle of the plane that intersects the cone changes, different conic sections are formed. For a parabola, the plane is *parallel to the side* of the cone. For a circle, the plane is *parallel to the base* of the cone.

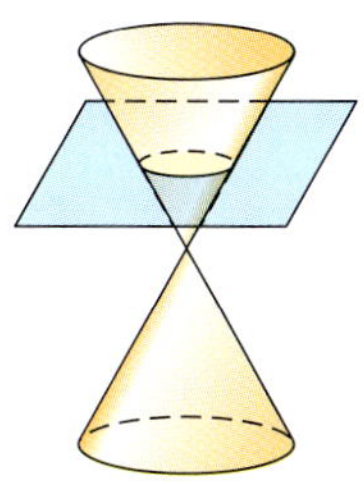

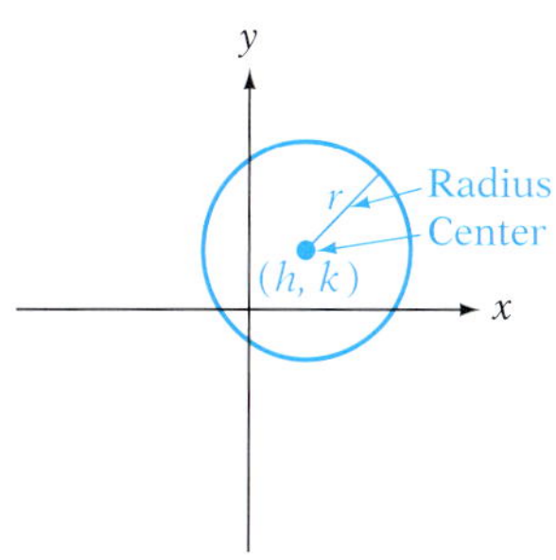

A **circle** can be defined as all the points (x, y) in the plane that are a fixed distance from a given point (h, k) called the **center.** The fixed distance is the **radius** of the circle.

The equation of a circle can be determined by using the **distance formula.**

Let (h, k) be the coordinates of the center of the circle, let r be the radius, and let (x, y) be any point on the circle. Then, by the distance formula,

$$r = \sqrt{(x-h)^2 + (y-k)^2}$$

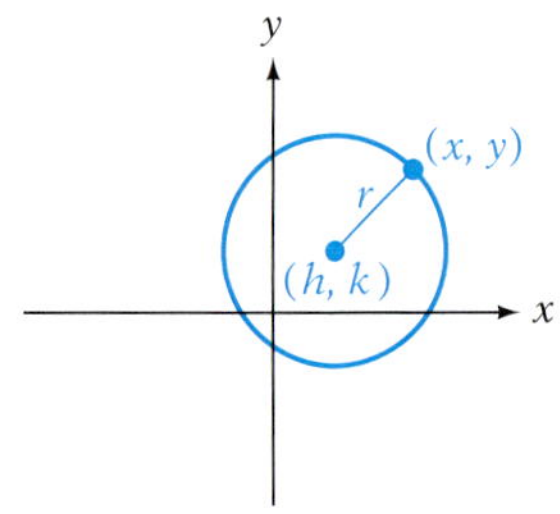

Squaring each side of the equation gives the equation of a circle.

$$r^2 = \left[\sqrt{(x-h)^2 + (y-k)^2}\right]^2$$
$$r^2 = (x-h)^2 + (y-k)^2$$

Standard Form of the Equation of a Circle

Let r be the radius of a circle and let (h, k) be the coordinates of the center of the circle. Then the equation of the circle is given by

$$(x-h)^2 + (y-k)^2 = r^2$$

Find the equation of the circle with radius 4 and center $C(-1, 2)$.

Use the standard form of the equation of a circle. $\quad (x-h)^2 + (y-k)^2 = r^2$

Replace r by 4, h by -1, and k by 2. $\quad [x-(-1)]^2 + (y-2)^2 = 4^2$

$(x+1)^2 + (y-2)^2 = 16$

To sketch the graph of this circle, draw a circle with center $C(-1, 2)$ and radius 4.

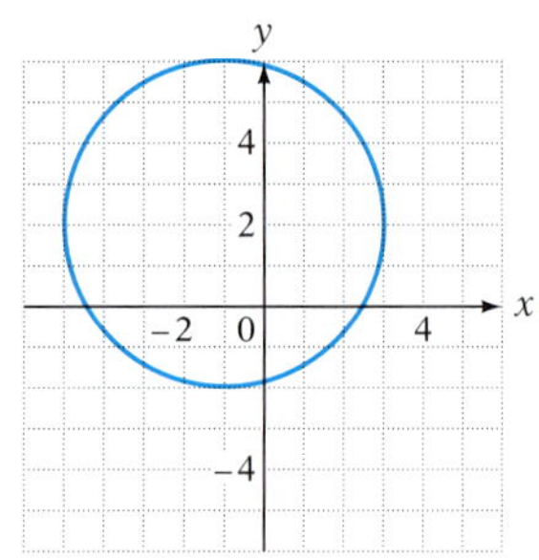

Take Note

Applying the vertical line test reveals that the graph of a circle is not the graph of a function. The graph of a circle is the graph of a relation.

Example 3 Find the equation of the circle with radius 5 and center $C(-1, 3)$. Then sketch its graph.

Solution

$$(x-h)^2 + (y-k)^2 = r^2$$
$$[x-(-1)]^2 + (y-3)^2 = 5^2$$
$$(x+1)^2 + (y-3)^2 = 25$$

Center: $(-1, 3)$

Radius: $\sqrt{25} = 5$

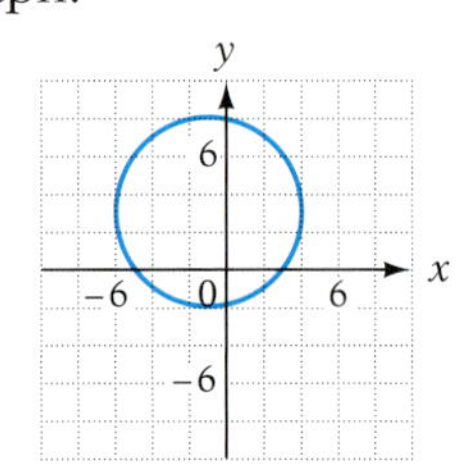

Problem 3 Find the equation of the circle with radius 4 and center $C(2, -3)$. Then sketch its graph.

Solution See page S33.

Find the radius of a circle that passes through the point $P(2, 1)$ and has as its center the point $C(3, -4)$.

Because the center of the circle and a point on the circle are known, use the distance formula to find the radius of the circle.

$$\sqrt{(x_2 - x_1)^2 + (y_2 - y_1)^2} = r$$
$$\sqrt{(3 - 2)^2 + (-4 - 1)^2} = r$$
$$\sqrt{1^2 + (-5)^2} = r$$
$$\sqrt{1 + 25} = r$$
$$\sqrt{26} = r$$

The equation of the circle is $(x - 3)^2 + (y + 4)^2 = 26$. The radius is $\sqrt{26}$.

Example 4 Find the equation of the circle for which a diameter has endpoints $P_1(-4, -1)$ and $P_2(2, 3)$.

Solution

$$x_m = \frac{x_1 + x_2}{2} \qquad y_m = \frac{y_1 + y_2}{2}$$
$$x_m = \frac{-4 + 2}{2} \qquad y_m = \frac{-1 + 3}{2}$$
$$x_m = -1 \qquad y_m = 1$$

• Let $(x_1, y_1) = (-4, -1)$ and $(x_2, y_2) = (2, 3)$. Find the center of the circle by finding the midpoint of the diameter.

$$(x_m, y_m) = (-1, 1)$$

$$r = \sqrt{(x_1 - x_m)^2 + (y - y_m)^2}$$
$$r = \sqrt{[-4 - (-1)]^2 + (-1 - 1)^2}$$
$$r = \sqrt{9 + 4}$$
$$r = \sqrt{13}$$

• Find the radius of the circle. Use either point on the circle and the coordinates of the center of the circle. P_1 is used here.

$$(x + 1)^2 + (y - 1)^2 = 13$$

• Write the equation of the circle with center $C(-1, 1)$ and radius $\sqrt{13}$.

Problem 4 Find the equation of the circle for which a diameter has endpoints $P_1(-2, 1)$ and $P_2(4, -1)$.

Solution See page S33.

3 Graph an ellipse with center at the origin

VIDEO & DVD CD TUTOR WEB SSM

The orbits of the planets around the sun are "oval" shaped. This oval shape can be described as an **ellipse,** which is another of the conic sections.

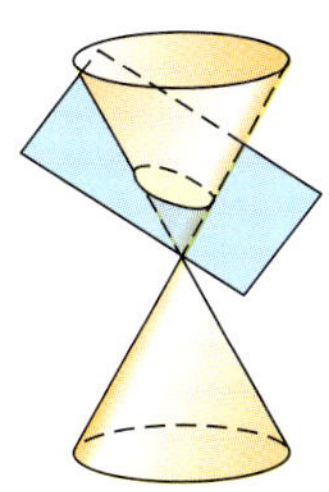

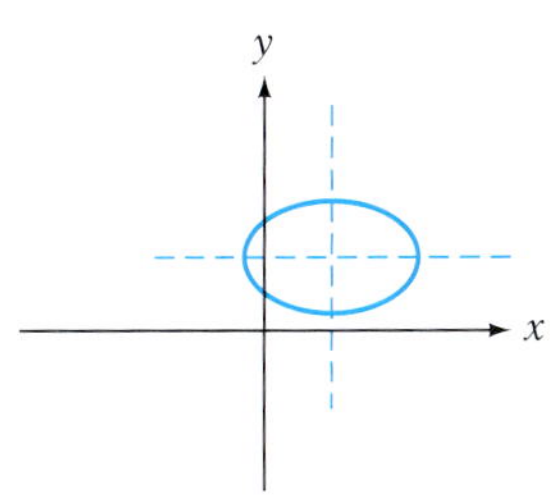

There are two **axes of symmetry** for an ellipse. The intersection of these two axes is the **center** of the ellipse.

An ellipse with center at the origin is shown at the right. Note that there are two x-intercepts and two y-intercepts.

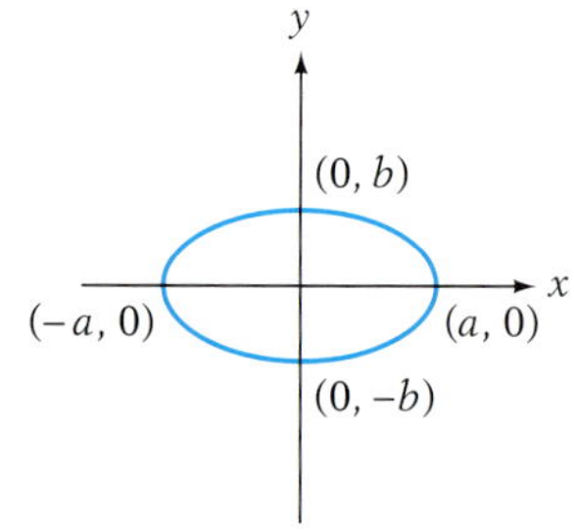

Point of Interest

The word ellipse *comes from the Greek word* ellipsis, *which means "deficient." The method by which the early Greeks analyzed the conics caused a certain area in the construction of the ellipse to be less than another area (deficient). The word* ellipsis *in English, which means "omission," has the same Greek root as the word* ellipse.

Standard Form of the Equation of an Ellipse with Center at the Origin

The equation of an ellipse with center at the origin is $\frac{x^2}{a^2} + \frac{y^2}{b^2} = 1$. The x-intercepts are $(a, 0)$ and $(-a, 0)$. The y-intercepts are $(0, b)$ and $(0, -b)$.

By finding the x- and y-intercepts for an ellipse and using the fact that the ellipse is "oval" shaped, we can sketch a graph of an ellipse.

Example 5 Sketch a graph of the ellipse given by the equation.

A. $\frac{x^2}{9} + \frac{y^2}{4} = 1$ **B.** $\frac{x^2}{16} + \frac{y^2}{12} = 1$

Solution **A.** $\frac{x^2}{9} + \frac{y^2}{4} = 1$ • $a^2 = 9, b^2 = 4$

x-intercepts: $(3, 0)$ and $(-3, 0)$ • The x-intercepts are $(a, 0)$ and $(-a, 0)$.

y-intercepts: $(0, 2)$ and $(0, -2)$ • The y-intercepts are $(0, b)$ and $(0, -b)$.

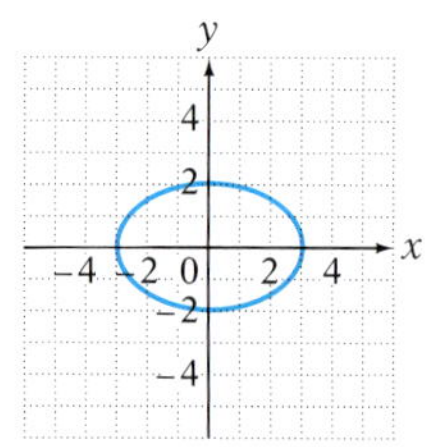

• Use the intercepts and symmetry to sketch the graph of the ellipse.

Take Note

Applying the vertical line test reveals that the graph of an ellipse is not the graph of a function. The graph of an ellipse is the graph of a relation.

B. $\frac{x^2}{16} + \frac{y^2}{12} = 1$ • $a^2 = 16, b^2 = 12$

x-intercepts: $(4, 0)$ and $(-4, 0)$ • The x-intercepts are $(a, 0)$ and $(-a, 0)$.

y-intercepts: $(0, 2\sqrt{3})$ and $(0, -2\sqrt{3})$ • The y-intercepts are $(0, b)$ and $(0, -b)$.

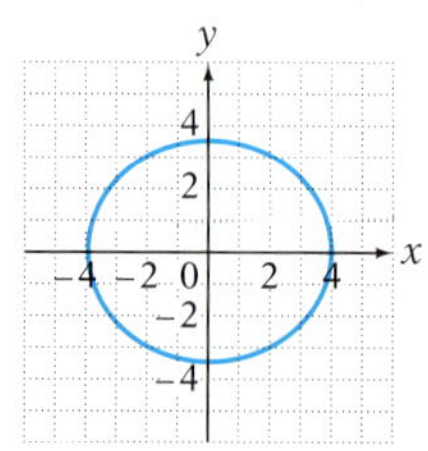

• Use the intercepts and symmetry to sketch the graph of the ellipse. $2\sqrt{3} \approx 3.5$

Point of Interest

Early Greek astronomers thought that each planet had a circular orbit. Today we know that the planets have elliptical orbits. However, in most cases, the ellipse is very nearly a circle. Note that when $a = b$ in the equation $\frac{x^2}{a^2} + \frac{y^2}{b^2} = 1$, the graph is the graph of a circle.

Problem 5 Sketch a graph of the ellipse given by the equation.

A. $\dfrac{x^2}{4} + \dfrac{y^2}{25} = 1$ **B.** $\dfrac{x^2}{18} + \dfrac{y^2}{9} = 1$

Solution See page S33.

4 Graph a hyperbola with center at the origin

VIDEO & DVD CD TUTOR WEB SSM

A **hyperbola** is a conic section that is formed by the intersection of a right circular cone and a plane perpendicular to the base of the cone.

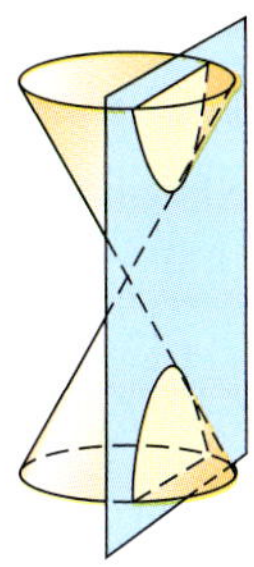

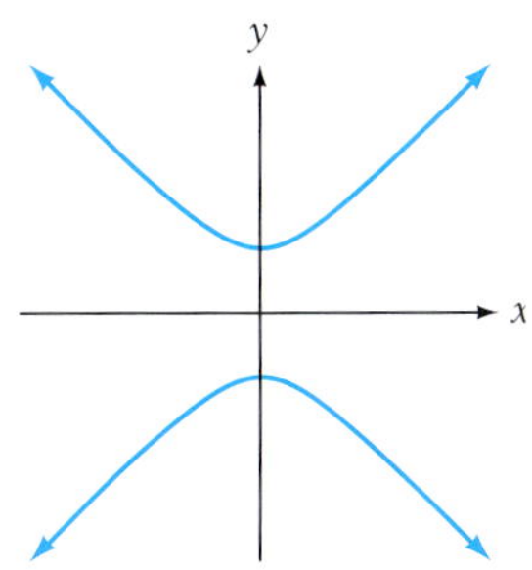

The hyperbola has two **vertices** and an **axis of symmetry** that passes through the vertices. The **center** of a hyperbola is the point halfway between the vertices.

The graphs below show two graphs of a hyperbola with center at the origin.

In the first graph, an axis of symmetry is the x-axis and the vertices are x-intercepts.

In the second graph, an axis of symmetry is the y-axis and the vertices are y-intercepts.

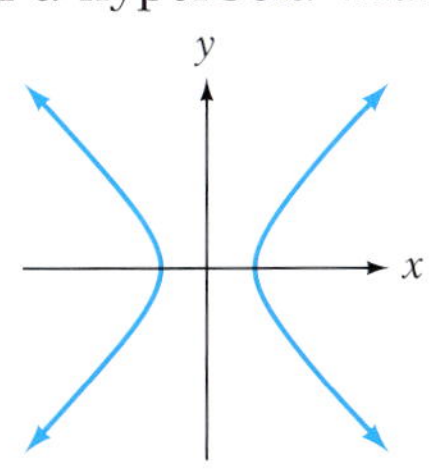

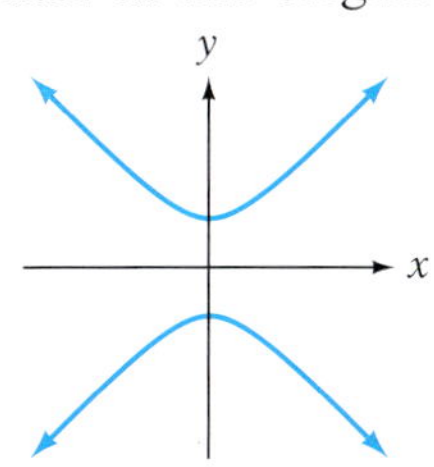

Note that in either case, the graph of a hyperbola is not the graph of a function. The graph of a hyperbola is the graph of a relation.

Standard Form of the Equation of a Hyperbola with Center at the Origin

The equation of a hyperbola for which an axis of symmetry is the x-axis is $\dfrac{x^2}{a^2} - \dfrac{y^2}{b^2} = 1$. The vertices are $(a, 0)$ and $(-a, 0)$.

The equation of a hyperbola for which an axis of symmetry is the y-axis is $\dfrac{y^2}{b^2} - \dfrac{x^2}{a^2} = 1$. The vertices are $(0, b)$ and $(0, -b)$.

Point of Interest

The word hyperbola *comes from the Greek word* yperboli, *which means "exceeding." The method by which the early Greeks analyzed the conics caused a certain area in the construction of the hyperbola to be greater than (to exceed) another area. The word* hyperbole *in English, which means "exaggeration," has the same Greek root as the word* hyperbola.

The word asymptote *comes from the Greek word* asymptotos, *which means "not capable of meeting."*

To sketch a hyperbola, it is helpful to draw two lines that are "approached" by the hyperbola. These two lines are called **asymptotes.** As a point on the hyperbola gets farther from the origin, the hyperbola "gets closer to" the asymptotes.

Because the asymptotes are straight lines, their equations are linear equations. The equations of the asymptotes for a hyperbola with center at the origin are $y = \frac{b}{a}x$ and $y = -\frac{b}{a}x$.

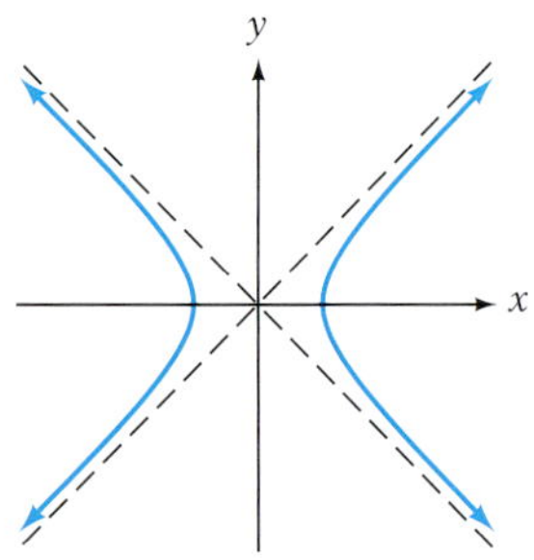

Point of Interest

Hyperbolas are used in "loran" (LOng RAnge Navigation) as a method by which a ship's navigator can determine the position of the ship, as shown in the figure below. They are also used as mirrors in some telescopes to focus incoming light.

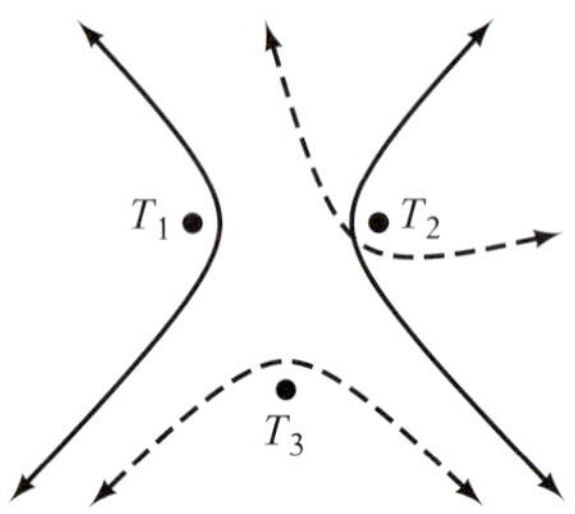

Example 6 Sketch a graph of the hyperbola given by the equation.

A. $\frac{x^2}{16} - \frac{y^2}{4} = 1$ **B.** $\frac{y^2}{16} - \frac{x^2}{25} = 1$

Solution **A.** $\frac{x^2}{16} - \frac{y^2}{4} = 1$ • $a^2 = 16, b^2 = 4$

Axis of symmetry:
x-axis

Vertices:
(4, 0) and (−4, 0) • The vertices are $(a, 0)$ and $(-a, 0)$.

Asymptotes:
$y = \frac{1}{2}x$ and $y = -\frac{1}{2}x$ • The asymptotes are $y = \frac{b}{a}x$ and $y = -\frac{b}{a}x$.

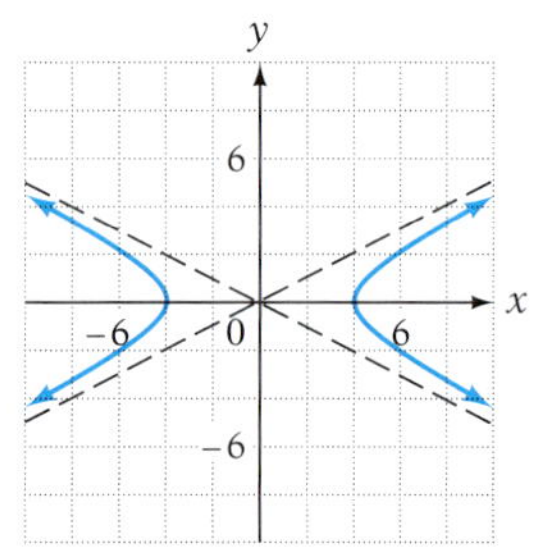

• Sketch the asymptotes. Use symmetry and the fact that the hyperbola will approach the asymptotes to sketch its graph.

B. $\frac{y^2}{16} - \frac{x^2}{25} = 1$ • $a^2 = 25, b^2 = 16$

Axis of symmetry:
y-axis

Vertices:
(0, 4) and (0, −4) • The vertices are $(0, b)$ and $(0, -b)$.

Asymptotes:
$y = \frac{4}{5}x$ and $y = -\frac{4}{5}x$ • The asymptotes are $y = \frac{b}{a}x$ and $y = -\frac{b}{a}x$.

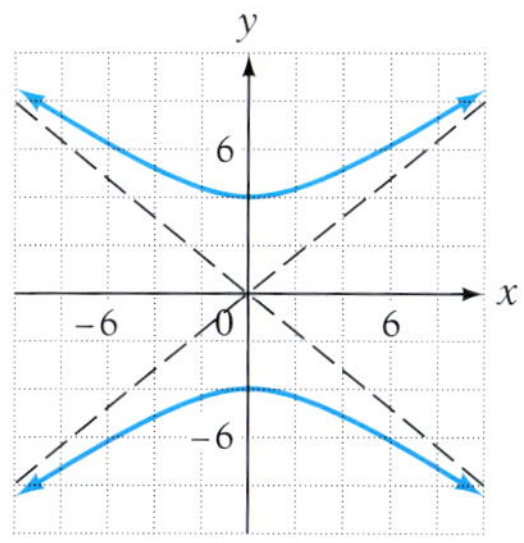

• **Sketch the asymptotes. Use symmetry and the fact that the hyperbola will approach the asymptotes to sketch its graph.**

Problem 6 Sketch a graph of the hyperbola given by the equation.

A. $\frac{x^2}{9} - \frac{y^2}{25} = 1$ **B.** $\frac{y^2}{9} - \frac{x^2}{9} = 1$

Solution See page S33.

10.5 Concept Review

Determine whether the following statements are always true, sometimes true, or never true.

1. The graph of a parabola is the graph of a function.
2. The axis of symmetry of a parabola passes through the vertex.
3. The graph of a parabola has two x-intercepts.
4. The center of the circle given by the equation $(x + 2)^2 + (y + 1)^2 = 16$ is $(2, 1)$.
5. $(x - 2)^2 + (y + 4)^2 = -4$ is the equation of a circle.
6. The graph of an ellipse is the graph of a function.
7. The graph of a hyperbola is the graph of a function.
8. An ellipse has two axes of symmetry.
9. An ellipse with center at the origin has two x-intercepts.
10. A hyperbola with center at the origin has two x-intercepts.

10.5 Exercises

1 State (a) whether the axis of symmetry is a vertical or a horizontal line and (b) the direction in which the parabola opens.

1. $y = 3x^2 - 4x + 7$

2. $y = -x^2 + 5x - 2$

3. $x = y^2 + 2y - 8$

4. $x = -3y^2 - y + 9$

5. $x = -\frac{1}{2}y^2 - 4y - 7$

6. $y = \frac{1}{4}x^2 + 6x - 1$

Find the vertex and the axis of symmetry of the parabola given by the equation. Then sketch its graph.

7. $y = x^2 - 2x - 4$

8. $y = x^2 + 4x - 4$

9. $y = -x^2 + 2x - 3$

10. $y = -x^2 + 4x - 5$

11. $x = y^2 + 6y + 5$

12. $x = y^2 - y - 6$

13. $y = 2x^2 - 4x + 1$

14. $y = 2x^2 + 4x - 5$

15. $y = x^2 - 5x + 4$

16. $y = x^2 + 5x + 6$

17. $x = y^2 - 2y - 5$

18. $x = y^2 - 3y - 4$

19. $y = -3x^2 - 9x$

20. $y = -2x^2 + 6x$

21. $x = -\frac{1}{2}y^2 + 4$

22. $x = -\frac{1}{4}y^2 - 1$

23. $x = \frac{1}{2}y^2 - y + 1$

24. $x = -\frac{1}{2}y^2 + 2y - 3$

25. $y = \frac{1}{2}x^2 + 2x - 6$

26. $y = -\frac{1}{2}x^2 + x - 3$

2 **27.** Describe how the points on the circumference of a circle are related to the center of the circle.

28. What do the values of h, k, and r represent in the equation of a circle in standard form?

Sketch a graph of the circle given by the equation.

29. $(x - 2)^2 + (y + 2)^2 = 9$

30. $(x + 2)^2 + (y - 3)^2 = 16$

31. $(x + 3)^2 + (y - 1)^2 = 25$

32. $(x - 2)^2 + (y + 3)^2 = 4$

33. $(x - 4)^2 + (y + 2)^2 = 1$

34. $(x - 3)^2 + (y - 2)^2 = 16$

35. $(x + 5)^2 + (y + 2)^2 = 4$

36. $(x + 1)^2 + (y - 1)^2 = 9$

37. Find the equation of the circle with radius 2 and center $C(2, -1)$. Then sketch its graph.

38. Find the equation of the circle with radius 3 and center $C(-1, -2)$. Then sketch its graph.

39. Find the equation of the circle that passes through the point $P(1, 2)$ and whose center is the point $C(-1, 1)$. Then sketch its graph.

40. Find the equation of the circle that passes through the point $P(-1, 3)$ and whose center is the point $C(-2, 1)$. Then sketch its graph.

41. Find the equation of the circle for which a diameter has endpoints $P_1(-1, 4)$ and $P_2(-5, 8)$.

42. Find the equation of the circle for which a diameter has endpoints $P_1(2, 3)$ and $P_2(5, -2)$.

43. Find the equation of the circle for which a diameter has endpoints $P_1(-4, 2)$ and $P_2(0, 0)$.

44. Find the equation of the circle for which a diameter has endpoints $P_1(-8, -3)$ and $P_2(0, -4)$.

3 Sketch the graph of the ellipse given by the equation.

45. $\dfrac{x^2}{4} + \dfrac{y^2}{9} = 1$

46. $\dfrac{x^2}{25} + \dfrac{y^2}{16} = 1$

47. $\dfrac{x^2}{25} + \dfrac{y^2}{9} = 1$

48. $\dfrac{x^2}{16} + \dfrac{y^2}{9} = 1$

49. $\dfrac{x^2}{36} + \dfrac{y^2}{16} = 1$

50. $\dfrac{x^2}{49} + \dfrac{y^2}{64} = 1$

51. $\dfrac{x^2}{9} + \dfrac{y^2}{25} = 1$

52. $\dfrac{x^2}{8} + \dfrac{y^2}{25} = 1$

53. $\dfrac{x^2}{12} + \dfrac{y^2}{4} = 1$

54. $\frac{x^2}{16} + \frac{y^2}{36} = 1$

55. $\frac{x^2}{36} + \frac{y^2}{9} = 1$

56. $\frac{x^2}{4} + \frac{y^2}{16} = 1$

4 57. How can you tell from an equation whether the graph will be that of an ellipse or a hyperbola?

58. What are the asymptotes of a hyperbola?

Sketch a graph of the hyperbola given by the equation.

59. $\frac{x^2}{9} - \frac{y^2}{16} = 1$

60. $\frac{x^2}{25} - \frac{y^2}{4} = 1$

61. $\frac{y^2}{16} - \frac{x^2}{9} = 1$

62. $\frac{y^2}{4} - \frac{x^2}{9} = 1$

63. $\frac{x^2}{4} - \frac{y^2}{25} = 1$

64. $\frac{x^2}{9} - \frac{y^2}{49} = 1$

65. $\frac{y^2}{25} - \frac{x^2}{9} = 1$

66. $\frac{y^2}{4} - \frac{x^2}{16} = 1$

67. $\frac{x^2}{25} - \frac{y^2}{16} = 1$

68. $\frac{x^2}{9} - \frac{y^2}{9} = 1$

69. $\frac{y^2}{16} - \frac{x^2}{4} = 1$

70. $\frac{y^2}{9} - \frac{x^2}{36} = 1$

71. $\frac{x^2}{25} - \frac{y^2}{9} = 1$

72. $\frac{x^2}{16} - \frac{y^2}{25} = 1$

10.5 Applying Concepts

Use the vertex and the direction in which the parabola opens to determine the domain and range of the relation.

73. $y = x^2 - 4x - 2$

74. $y = -x^2 + 2x - 3$

75. $x = y^2 + 6y - 5$

76. $x = -y^2 - 2y + 6$

Parabolas have a unique property that is important in the design of telescopes and antennas. If a parabola has a mirrored surface, then all light rays parallel to the axis of symmetry of the parabola are reflected to a single point called the **focus** of the parabola. The location of this point is p units from the vertex on the axis of symmetry. The value of p is given by $p = \frac{1}{4a}$, where $y = ax^2$ is the equation of a parabola with vertex at the origin. For the graph of $y = \frac{1}{4}x^2$ shown at the right, the coordinates of the focus are (0, 1). For Exercises 77 and 78, find the coordinates of the focus of the parabola given by the equation.

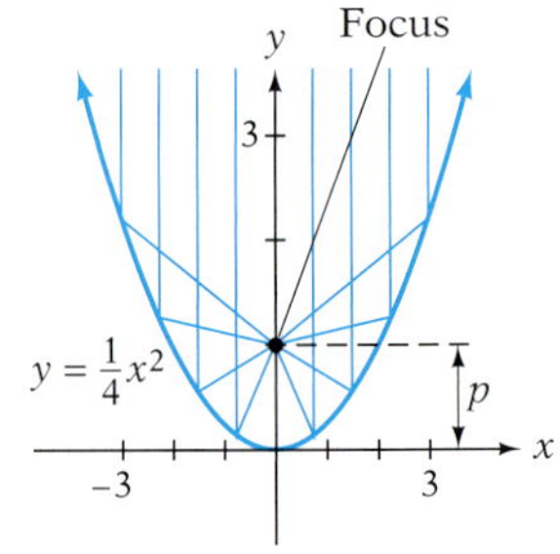

Parallel rays of light are reflected to the focus

$a = \frac{1}{4}$

$p = \frac{1}{4a} = \frac{1}{4(1/4)} = 1$

77. $y = 2x^2$

78. $y = \frac{1}{10}x^2$

In Exercises 79 and 80, write the equation of the circle.

79. The circle has its center at the point $C(3, 0)$ and passes through the origin.

80. The circle has radius 1, is tangent to both the x- and y-axes, and lies in quadrant II.

81. **Geometry** The radius of a sphere is 12 in. What is the radius of the circle that is formed by the intersection of a sphere and a plane that is 6 in. from the center of the sphere?

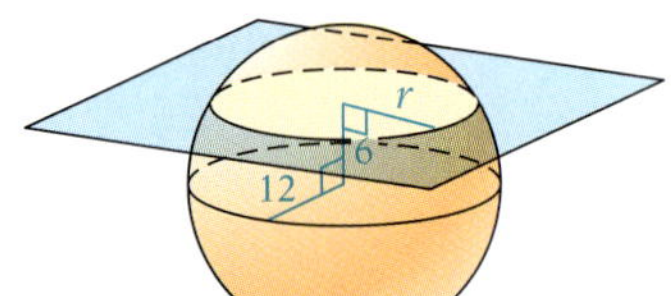

82. Explain why the graph of the equation $\frac{x^2}{9} + \frac{y^2}{9} = 1$ is or is not a circle.

Sketch a graph of the conic section given by the equation. (*Hint:* Divide each term by the number on the right side of the equation.)

83. $4x^2 + y^2 = 16$

84. $x^2 - y^2 = 9$

85. $y^2 - 4x^2 = 16$

86. $9x^2 + 4y^2 = 144$

87. $9x^2 - 25y^2 = 225$

88. $4y^2 - x^2 = 36$

89. **Astronomy** The orbit of Halley's comet is an ellipse with a major axis of approximately 36 AU and a minor axis of approximately 9 AU. (1 AU is 1 astronomical unit and is approximately 92,960,000 mi, the average distance of Earth from the sun.)

a. Determine an equation for the orbit of Halley's comet in terms of astronomical units. See the diagram at the right.

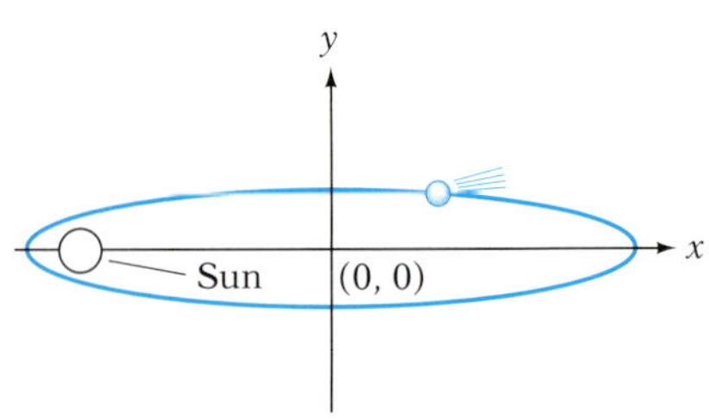

b. The distance of the sun from the center of Halley's comet's elliptical orbit is $\sqrt{a^2 - b^2}$. The aphelion of the orbit (the point at which the comet is farthest from the sun) is a vertex on the major axis. Determine the distance, to the nearest hundred-thousand miles, from the sun to the point at the aphelion of Halley's comet.

c. The perihelion of the orbit (the point at which the comet is closest to the sun) is a vertex on the major axis. Determine the distance, to the nearest hundred-thousand miles, from the sun to the point at the perihelion of Halley's comet.

90. **Astronomy** The orbit of the comet Hale–Bopp is an ellipse as shown at the right. The units are astronomical units (abbreviated AU). 1 AU $\approx$ 92,960,000 mi.

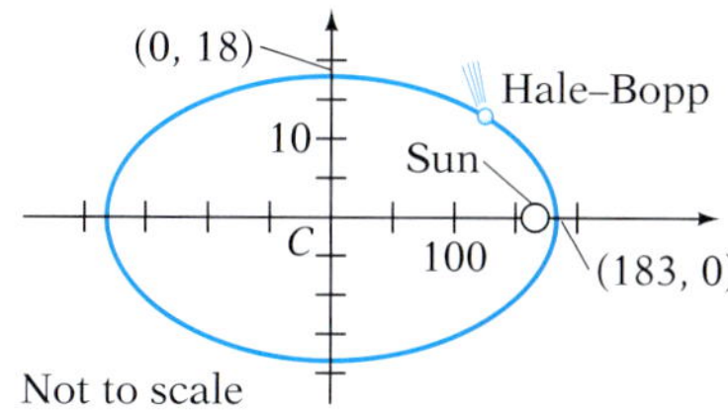

a. Find the equation of the orbit of the comet.

b. The distance from the center, C, of the orbit to the sun is approximately 182.085 AU. Find the aphelion (the point at which the comet is farthest from the sun) in miles. Round to the nearest million miles.

c. Find the perihelion (the point at which the comet is closest to the sun) in miles. Round to the nearest hundred-thousand miles.

91. **Astronomy** As mentioned in a Point of Interest in this section, the orbits of the planets are ellipses. The length of the major axis of Mars' orbit is 3.04 AU (see Exercise 89), and the length of the minor axis is 2.99 AU.

a. Determine an equation for the orbit of Mars.

b. Determine the aphelion to the nearest hundred-thousand miles.

c. Determine the perihelion to the nearest hundred-thousand miles.

92. Prepare a report on the system of navigation called loran (long-range navigation).

Focus on Problem Solving

Use a Variety of Problem-Solving Strategies

We have examined several problem-solving strategies throughout the text. See if you can apply those techniques to the following problems.

1. Eight coins look exactly alike, but one is lighter than the others. Explain how the different coin can be found in two weighings on a balance scale.

2. For the sequence of numbers 1, 1, 2, 3, 5, 8, 13, . . . , identify a possible pattern and then use that pattern to determine the next number in the sequence.

3. Arrange the numbers 1, 2, 3, 4, 5, 6, 7, 8, and 9 in the squares at the right so that the sum of any row or diagonal is 15. (*Suggestion:* Note that 1, 5, 9; 2, 5, 8; 3, 5, 7; and 4, 5, 6 all add to 15. Because 5 is part of each sum, this suggests that 5 be placed in the center of the squares.)

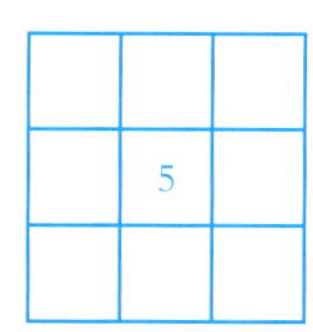

4. A restaurant charges \$10.00 for a pizza that has a diameter of 9 in. Determine the selling price of a pizza with a diameter of 18 in. so that the selling price per square inch is the same as for the 9-inch pizza.

5. You have a balance scale and weights of 1 gram, 4 grams, 8 grams, and 16 grams. Using only these weights, can you weigh something that weighs 3 grams? 7 grams? 12 grams? 19 grams?

6. Can the checkerboard at the right be covered with dominos (which look like 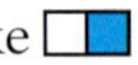) so that every square on the board is covered by a domino? Why or why not? (*Note:* The dominos cannot overlap.)

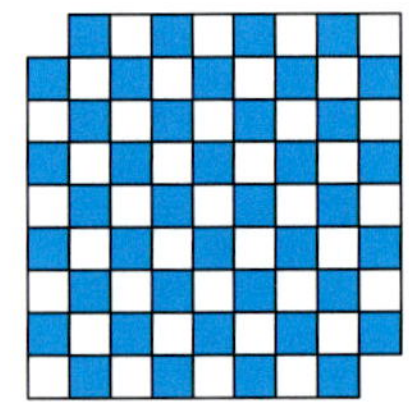

Projects & Group Activities

Applications of the Composition of Functions

Suppose a company's manufacturing cost, in dollars, per compact disc player is given by

$$M(x) = \frac{180x + 2600}{x}$$

where x is the number of compact disc players to be manufactured. An electronics outlet agrees to sell the compact discs by marking up the manufacturing cost per player, $M(x)$, by 30%. Note that the selling price S will be a function of $M(x)$. We can write this as

$$S[M(x)] = 1.30(M(x))$$

Simplifying $S[M(x)]$ yields

$$\begin{aligned} S[M(x)] &= 1.30\left(\frac{180x + 2600}{x}\right) \\ &= 1.30(180) + 1.30\left(\frac{2600}{x}\right) \\ &= 234 + \left(\frac{3380}{x}\right) \end{aligned}$$

The function produced in this manner is referred to as the composition of M by S. The notation $S \circ M$ is used to denote this composition of functions. That is,

$$(S \circ M)(x) = 234 + \left(\frac{3380}{x}\right)$$

Suppose 10,000 compact disc players are manufactured. Then

$$\begin{aligned} (S \circ M)(10{,}000) &= 234 + \frac{3380}{10{,}000} \\ &= 234.338 \end{aligned}$$

If 10,000 compact disc players are manufactured, the selling price per compact disc player is \$234.34.

Solve.

1. Let x be the number of computer monitors to be manufactured. The manufacturing cost, in dollars, per computer monitor is given by the function

$$M(x) = \frac{60x + 34,000}{x}$$

A computer store will sell the monitors by marking up the manufacturing cost per monitor, $M(x)$, by 45%. Thus the selling price S is a function of $M(x)$ given by the equation $S[M(x)] = 1.45[M(x)]$.

a. Express the selling price as a function of the number of monitors to be manufactured. That is, find $S \circ M$.

b. Find $(S \circ M)(24,650)$. What does this represent?

2. The number of bookcases b that a factory can produce per day is a function of the number of hours h it operates.

$$b(h) = 40\,h \quad \text{for } 0 \le h \le 12$$

The daily cost c to manufacture b bookcases is given by the function

$$c(b) = 0.1b^2 + 90b + 800$$

Evaluate each of the following and interpret your answers.

a. $b(5)$ **b.** $c(5)$ **c.** $(c \circ b)(h)$ **d.** $(c \circ b)(10)$

3. A car dealership offers a \$1500 rebate and a 10% discount off the price of a new car. Let p be the sticker price of a new car on the dealer's lot, let r be the rebate, and let d be the discount. Then $r(p) = p - 1500$ and $d(p) = 0.90p$.

a. Write a composite function for the dealer taking the rebate first and then the discount.

b. Write a composite function for the dealer taking the discount first and then the rebate.

c. Which composite function would you prefer the dealer use if you were buying a new car?

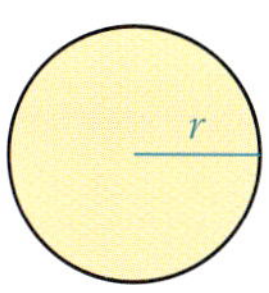

4. a. Write the area of a circle as a function of r.

b. Write the radius of a circle as a function of the circumference.

c. Write the area of a circle as a function of the circumference.

The Eccentricity and Foci of an Ellipse

The graph of an ellipse can be long and thin, or it can have a shape that is very close to a circle. The **eccentricity,** e, of an ellipse is a measure of its "roundness."

The shapes of ellipses with various eccentricities are shown below.

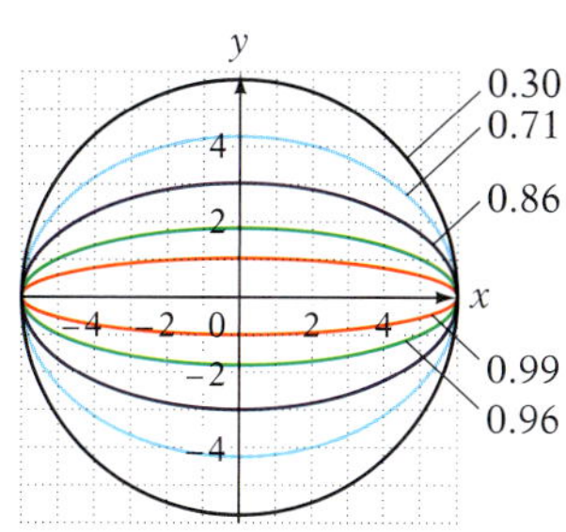

Planet	*Eccentricity*
Mercury	0.206
Venus	0.007
Earth	0.017
Mars	0.093
Jupiter	0.049
Saturn	0.051
Uranus	0.046
Neptune	0.005
Pluto	0.250

1. On the basis of the eccentricities of the ellipses shown above, complete the following sentence. "As the eccentricity of an ellipse gets closer to 1, the ellipses get *flatter/rounder.*"

2. The planets travel around the sun in elliptical orbits. The eccentricities of the orbits of the planets are shown in the table at the left. Which planet has the most nearly circular orbit?

For an ellipse given by the equation $\frac{x^2}{a^2} + \frac{y^2}{b^2} = 1$, $a > b$, a formula for eccentricity is $e = \frac{\sqrt{a^2 - b^2}}{a}$. Use this formula to find the eccentricity of the ellipses in Exercises 3 and 4. If necessary, round to the nearest hundredth.

3. $\frac{x^2}{25} + \frac{y^2}{16} = 1$

4. $\frac{x^2}{9} + \frac{y^2}{4} = 1$

The **foci** of an ellipse are two points on the longer axis, called the **major axis,** of the ellipse. F_1 and F_2 are the foci of the ellipse shown below.

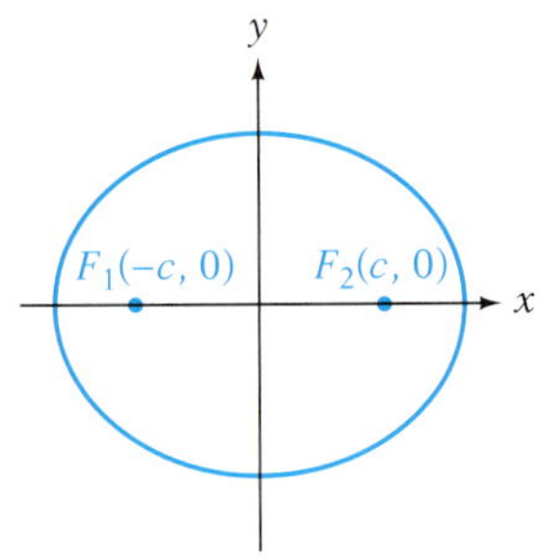

Ellipses have a reflective property that has been used in the design of some buildings. If light or sound emanates from one focus, it is reflected to the other focus. This phenomenon results in what are called "whispering galleries." The rotunda of the Capitol Building in Washington, D.C., is a whispering gallery. A whisper spoken by a person at one focus can be heard clearly by a person at the other focus.

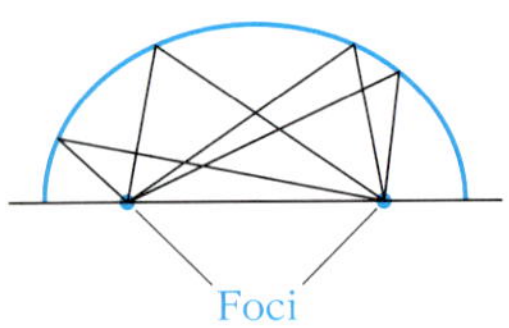

The foci are c units from the center of an ellipse, where $c = \sqrt{a^2 - b^2}$ for an ellipse whose equation is $\frac{x^2}{a^2} + \frac{y^2}{b^2} = 1$, $a > b$.

5. Find the foci for an ellipse whose equation is $\frac{x^2}{169} + \frac{y^2}{144} = 1$.

Chapter Summary

Key Words

The **inverse of a function** is the set of ordered pairs formed by reversing the coordinates of each ordered pair of the function. [10.4.2, p. 661]

Examples

Function: {(2, 4), (3, 5), (6, 7), (8, 9)}
Inverse: {(4, 2), (5, 3), (7, 6), (9, 8)}

Conic sections are curves that can be constructed from the intersection of a plane and a cone. The four conic sections are the **parabola**, **circle**, **ellipse**, and **hyperbola**. [10.5.1, p. 668]

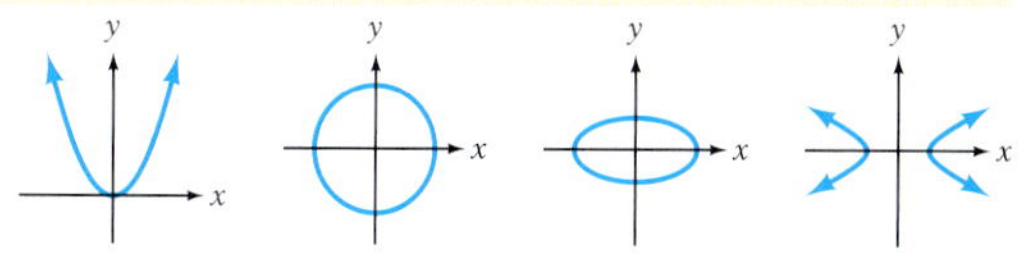

Parabola Circle Ellipse Hyperbola

The **asymptotes** of a hyperbola are the two straight lines that are "approached" by the hyperbola. As the graph of the hyperbola gets farther from the origin, the hyperbola gets "closer to" the asymptotes. [10.5.4, p. 676]

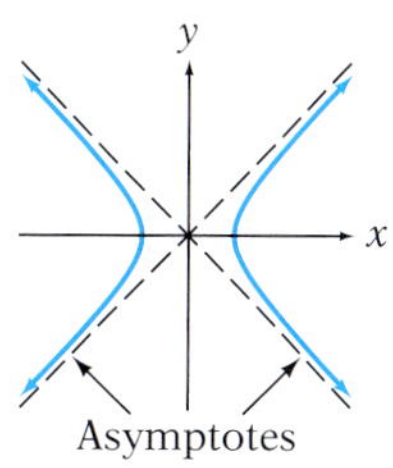

Essential Rules and Procedures

Vertical Line Test

A graph defines a function if any vertical line intersects the graph at no more than one point. [10.1.1, p. 638]

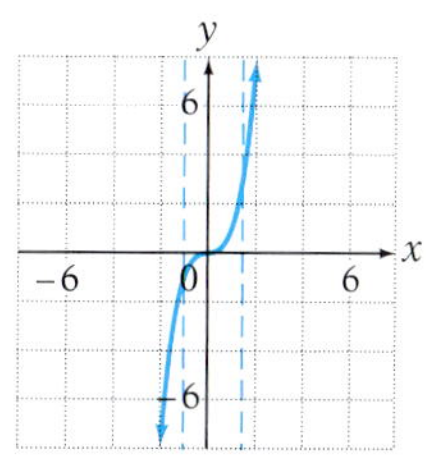

Function

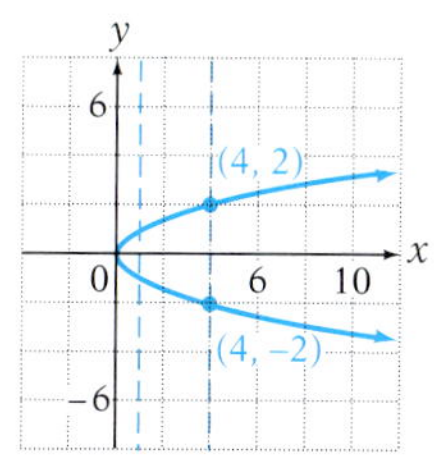

Not a function

Vertical Translation [10.2.1, p. 646]

The graph of $y = f(x) + c$, $c > 0$, is the graph of $y = f(x)$ shifted up c units.

The graph of $y = f(x) - c$, $c > 0$, is the graph of $y = f(x)$ shifted down c units.

The graph $g(x) = x^2 + 1$ is the graph of $f(x) = x^2$ shifted up 1 unit.

The graph of $G(x) = x^3 - 3$ is the graph of $F(x) = x^3$ shifted down 3 units.

Horizontal Translation [10.2.1, p. 647]

The graph of $y = f(x - c)$, $c > 0$, is the graph of $y = f(x)$ shifted horizontally c units to the right.

The graph of $y = f(x + c)$, $c > 0$, is the graph of $y = f(x)$ shifted horizontally c units to the left.

The graph $g(x) = (x - 2)^2$ is the graph of $f(x) = x^2$ shifted 2 units to the right.

The graph of $G(x) = (x + 4)^3$ is the graph of $F(x) = x^3$ shifted 4 units to the left.

Operations on Functions [10.3.1, p. 653]

If f and g are functions and x is an element of the domain of each function, then

$(f + g)(x) = f(x) + g(x)$

$(f - g)(x) = f(x) - g(x)$

$(f \cdot g)(x) = f(x) \cdot g(x)$

$\left(\frac{f}{g}\right)(x) = \frac{f(x)}{g(x)},\ g(x) \neq 0$

Given $f(x) = x + 2$ and $g(x) = 2x$, then

$$\begin{aligned}(f + g)(4) &= f(4) + g(4)\\ &= (4 + 2) + 2(4) = 6 + 8\\ &= 14\end{aligned}$$

$$\begin{aligned}(f - g)(4) &= f(4) - g(4)\\ &= (4 + 2) - 2(4) = 6 - 8\\ &= -2\end{aligned}$$

$$\begin{aligned}(f \cdot g)(4) &= f(4) \cdot g(4)\\ &= (4 + 2) \cdot (2)(4) = 6 \cdot 8\\ &= 48\end{aligned}$$

$$\left(\frac{f}{g}\right)(4) = \frac{f(4)}{g(4)} = \frac{4 + 2}{2(4)} = \frac{6}{8} = \frac{3}{4}$$

Composition of Two Functions

$(f \circ g)(x) = f[g(x)]$ [10.3.2, p. 655]

Given $f(x) = x - 4$ and $g(x) = 4x$, then

$$\begin{aligned}(f \circ g)(2) &= f[g(2)]\\ &= f(8) \qquad \text{because } g(2) = 8\\ &= 8 - 4 = 4\end{aligned}$$

Horizontal Line Test

A function f is a 1–1 function if any horizontal line intersects the graph at no more than one point. [10.4.1, p. 659]

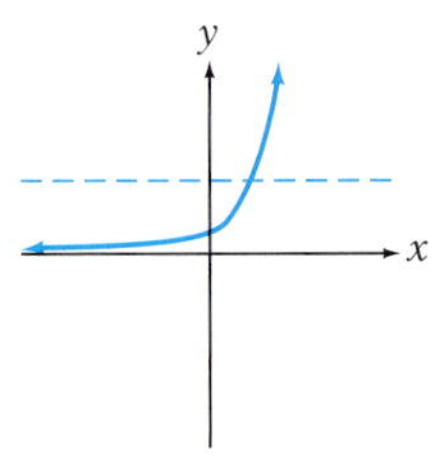

A 1–1 function

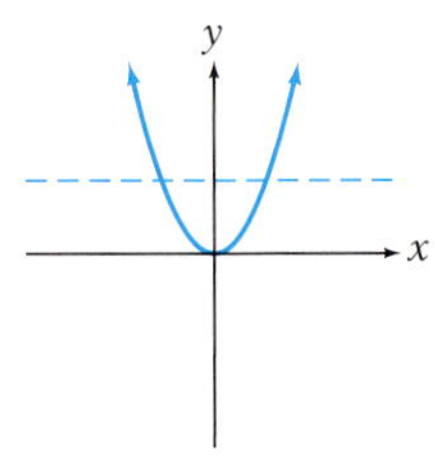

Not a 1–1 function

Condition for an Inverse Function

A function f has an inverse if and only if f is a 1–1 function. [10.4.2, p. 662]

The function $f(x) = x^2$ does not have an inverse function. When $y = 4$, $x = 2$ or -2; therefore, the function $f(x)$ is not a 1–1 function.

Composition of Inverse Functions Property

$f^{-1}[f(x)] = x$ and $f[f^{-1}(x)] = x$ [10.4.2, p. 663]

$$f(x) = 2x - 3 \qquad f^{-1}(x) = \frac{1}{2}x + \frac{3}{2}$$

$$f^{-1}[f(x)] = \frac{1}{2}(2x - 3) + \frac{3}{2} = x$$

$$f[f^{-1}(x)] = 2\left(\frac{1}{2}x + \frac{3}{2}\right) - 3 = x$$

Equation of a parabola [10.5.1, pp. 669–670]

$y = ax^2 + bx + c$

When $a > 0$, the parabola opens up.

When $a < 0$, the parabola opens down.

The x-coordinate of the vertex is $-\frac{b}{2a}$.

The axis of symmetry is the line $x = -\frac{b}{2a}$.

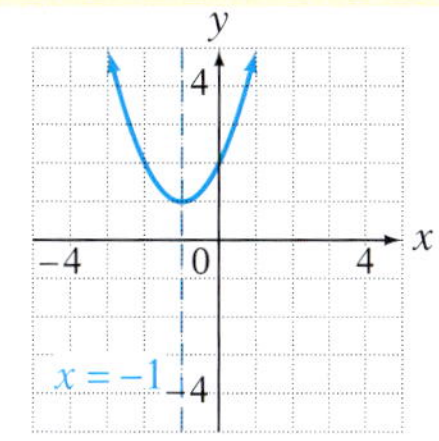

$y = x^2 + 2x + 2$

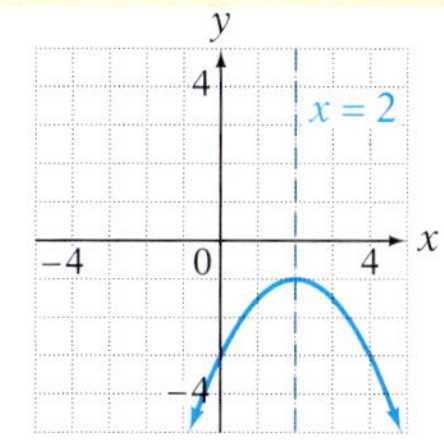

$y = -\frac{1}{2}x^2 + 2x - 3$

$x = ay^2 + by + c$

When $a > 0$, the parabola opens to the right.

When $a < 0$, the parabola opens to the left.

The y-coordinate of the vertex is $-\frac{b}{2a}$.

The axis of symmetry is the line $y = -\frac{b}{2a}$.

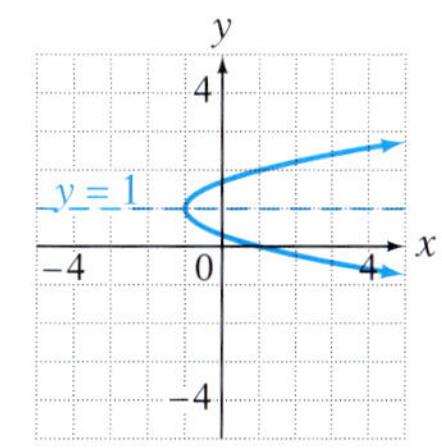

$x = 2y^2 - 4y + 1$

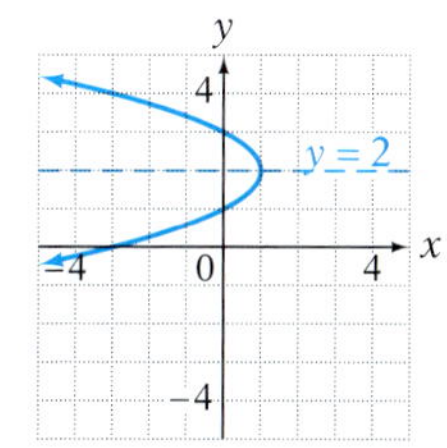

$x = -y^2 + 4y - 3$

Equation of a circle [10.5.2, p. 672]

$(x - h)^2 + (y - k)^2 = r^2$

The center of the circle is (h, k), and the radius is r.

$(x - 2)^2 + (y + 1)^2 = 9$

$(h, k) = (2, -1)$

$r = 3$

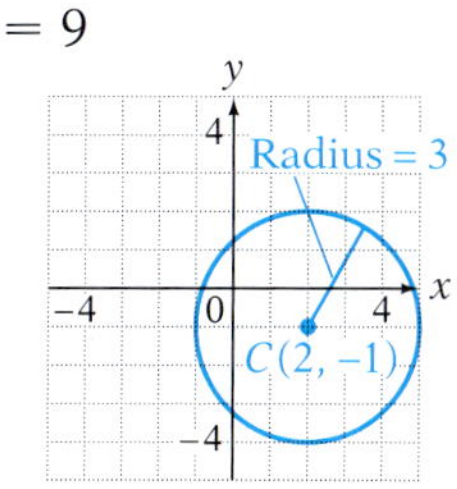

Equation of an ellipse [10.5.3, p. 674]

$\frac{x^2}{a^2} + \frac{y^2}{b^2} = 1$

The x-intercepts are $(a, 0)$ and $(-a, 0)$.

The y-intercepts are $(0, b)$ and $(0, -b)$.

$\frac{x^2}{9} + \frac{y^2}{1} = 1$

x-intercepts:
$(3, 0)$ and $(-3, 0)$

y-intercepts:
$(0, 1)$ and $(0, -1)$

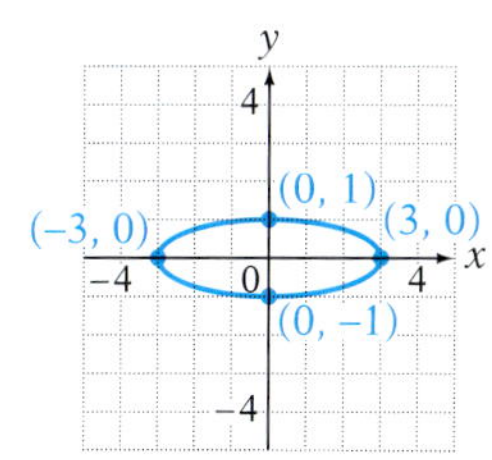

Equation of a hyperbola [10.5.4, pp. 675–676]

$\frac{x^2}{a^2} - \frac{y^2}{b^2} = 1$

An axis of symmetry is the x-axis.

The vertices are $(a, 0)$ and $(-a, 0)$.

The equations of the asymptotes are $y = \frac{b}{a}x$ and $y = -\frac{b}{a}x$.

$\frac{x^2}{9} - \frac{y^2}{4} = 1$

Vertices:
$(3, 0)$ and $(-3, 0)$

Asymptotes:
$y = \frac{2}{3}x$ and $y = -\frac{2}{3}x$

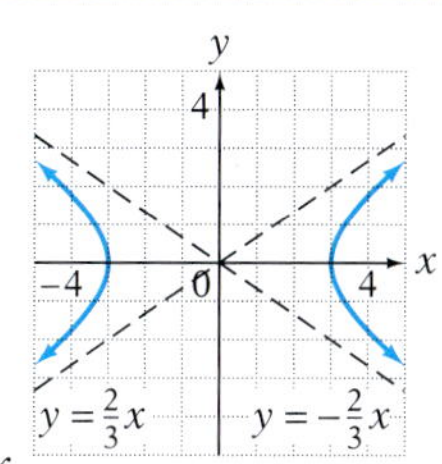

$\frac{y^2}{b^2} - \frac{x^2}{a^2} = 1$

An axis of symmetry is the y-axis.
The vertices are $(0, b)$ and $(0, -b)$.

The equations of the asymptotes are $y = \frac{b}{a}x$ and $y = -\frac{b}{a}x$.

$\frac{y^2}{1} - \frac{x^2}{4} = 1$

Vertices:
$(0, 1)$ and $(0, -1)$

Asymptotes:
$y = \frac{1}{2}x$ and
$y = -\frac{1}{2}x$

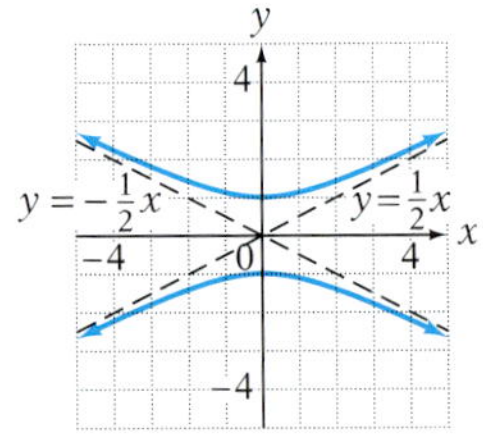

Chapter Review Exercises

1. Determine whether the graph is the graph of a function.

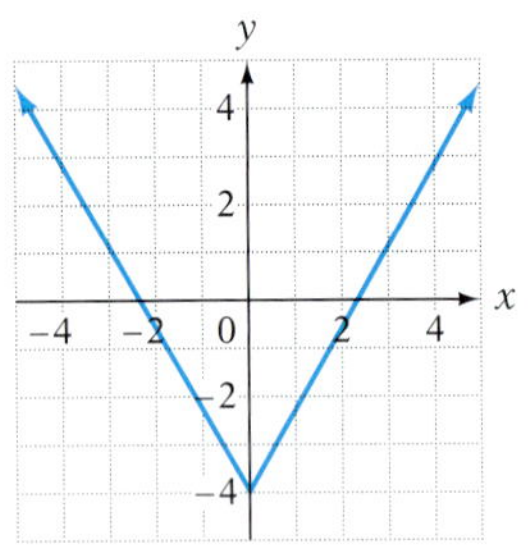

2. Graph $f(x) = |x| - 3$. Use set-builder notation to state the domain and range.

3. Graph $f(x) = 3x^3 - 2$. Use set-builder notation to state the domain and range.

4. Graph $f(x) = \sqrt{x + 4}$. Use interval notation to state the domain and range.

5. Use the graph of $f(x) = x^2$ shown below to graph $g(x) = (x - 1)^2$.

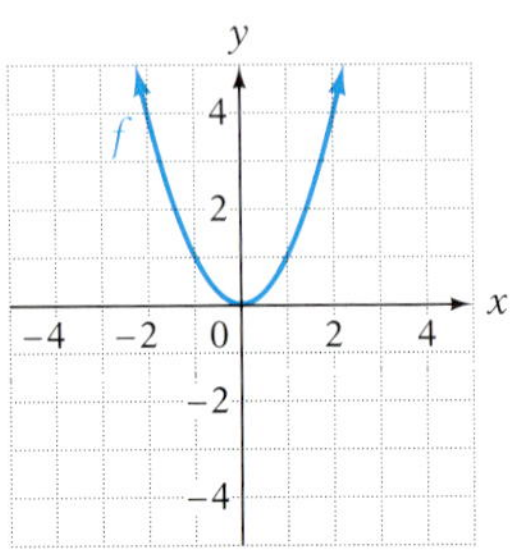

6. Use the graph of $f(x) = x^3$ shown below to graph $g(x) = x^3 + 3$.

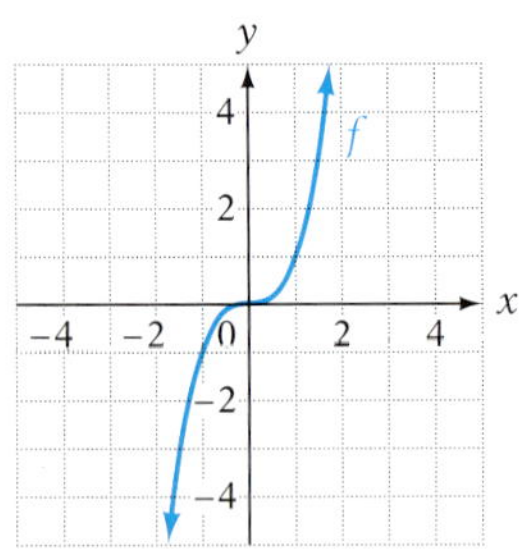

7. Use the graph of $f(x) = |x|$ shown below to graph $g(x) = |x - 1| + 2$.

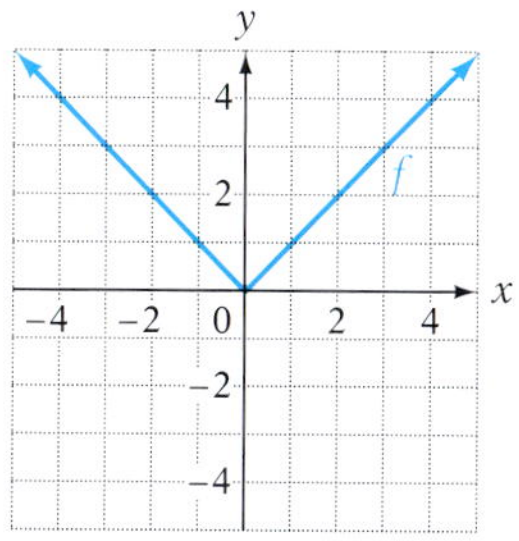

8. Use the graph of $f(x) = \sqrt{x}$ shown below to graph $g(x) = \sqrt{x + 1} - 2$.

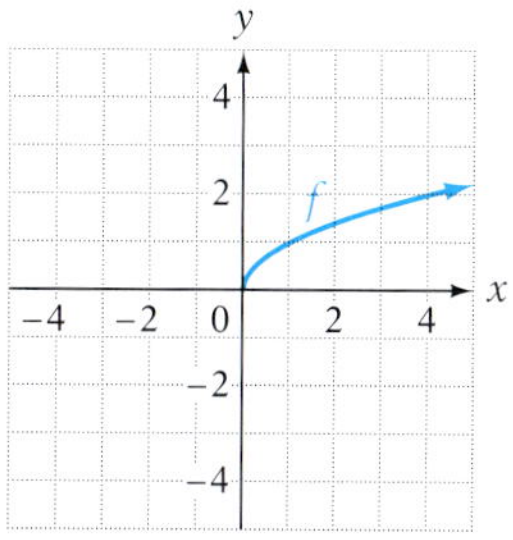

In Exercises 9–12, given $f(x) = x^2 + 2x - 3$ and $g(x) = x^2 - 2$, find:

9. $(f + g)(2)$

10. $(f - g)(-4)$

11. $(f \cdot g)(-4)$

12. $\left(\frac{f}{g}\right)(3)$

13. Given $f(x) = 3x^2 - 4$ and $g(x) = 2x + 1$, find $f[g(x)]$.

14. Given $f(x) = x^2 + 4$ and $g(x) = 4x - 1$, find $f[g(0)]$.

15. Given $f(x) = 6x + 8$ and $g(x) = 4x + 2$, find $g[f(-1)]$.

16. Given $f(x) = 2x^2 + x - 5$ and $g(x) = 3x - 1$, find $g[f(x)]$.

17. Is the set of ordered pairs a 1–1 function?

$\{(-3, 0), (-2, 0), (-1, 1), (0, 1)\}$

18. Find the inverse of the function $\{(-2, 1), (2, 3), (5, -4), (7, 9)\}$.

19. Determine whether the graph is the graph of a 1–1 function.

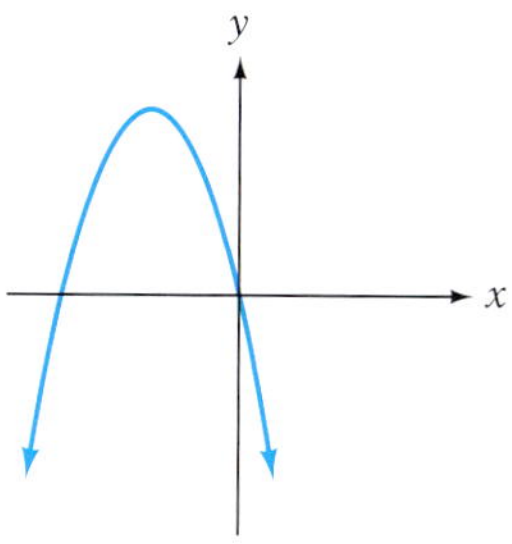

20. Determine whether the graph is the graph of a 1–1 function.

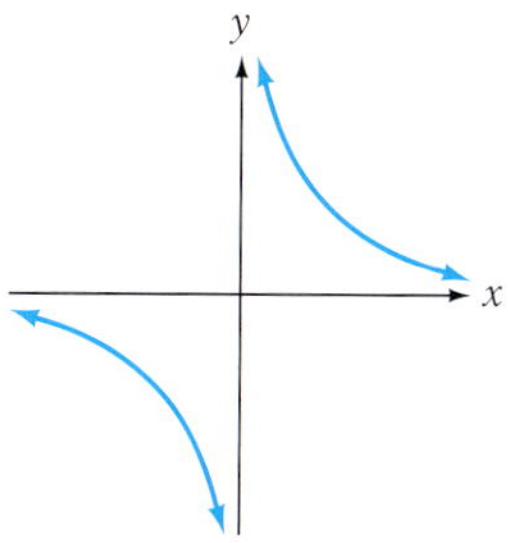

21. Find the inverse of the function $f(x) = \frac{1}{2}x + 8$.

22. Find the inverse of the function $f(x) = -6x + 4$.

23. Find the inverse of the function $f(x) = \frac{2}{3}x - 12$.

24. Are the functions $f(x) = -\frac{1}{4}x + \frac{5}{4}$ and $g(x) = -4x + 5$ inverses of each other?

25. Find the vertex and axis of symmetry of the parabola $y = x^2 - 4x + 8$.

26. Find the vertex and axis of symmetry of the parabola $y = -x^2 + 7x - 8$.

27. Sketch a graph of $y = -2x^2 + x - 2$.

28. Sketch a graph of $x = 2y^2 - 6y + 5$.

29. Find the equation of the circle that passes through the point $P(2, -1)$ and whose center is the point $C(-1, 2)$.

30. Find the equation of the circle with radius 6 and center $C(-1, 5)$.

31. Sketch a graph of $(x + 3)^2 + (y + 1)^2 = 1$.

32. Sketch a graph of $x^2 + (y - 2)^2 = 9$.

33. Find the equation of the circle that passes through the point $P(4, 6)$ and whose center is the point $C(0, -3)$.

34. Sketch a graph of $\dfrac{x^2}{25} + \dfrac{y^2}{9} = 1$.

35. Sketch a graph of $\dfrac{x^2}{25} - \dfrac{y^2}{1} = 1$.

36. Sketch a graph of $\dfrac{y^2}{16} - \dfrac{x^2}{9} = 1$.

Chapter Test

1. Determine whether the graph represents the graph of a function.

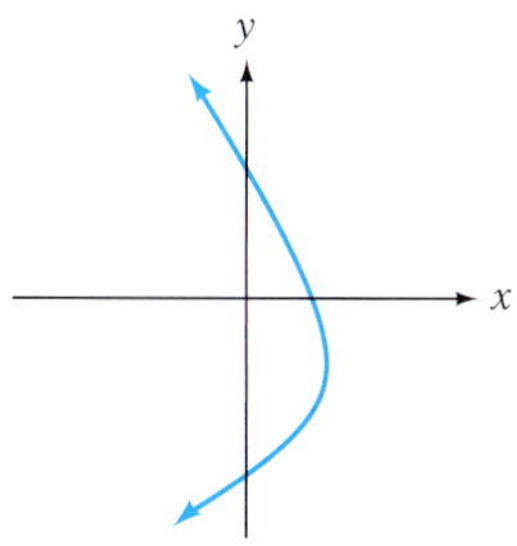

2. Given $f(x) = x^2 + 2x - 3$ and $g(x) = x^3 - 1$, find $(f - g)(2)$.

3. Given $f(x) = x^3 + 1$ and $g(x) = 2x - 3$ find $(f \cdot g)(-3)$.

4. Given $f(x) = 4x - 5$ and $g(x) = x^2 + 3x + 4$, find $\left(\frac{f}{g}\right)(-2)$.

5. Given $f(x) = x^2 + 4$ and $g(x) = 2x^2 + 2x + 1$, find $(f - g)(-4)$.

6. Given $f(x) = 4x + 2$ and $g(x) = \frac{x}{x + 1}$, find $f[g(3)]$.

7. Use the graph of $f(x) = -x^3$ shown below to graph $g(x) = -x^3 - 2$.

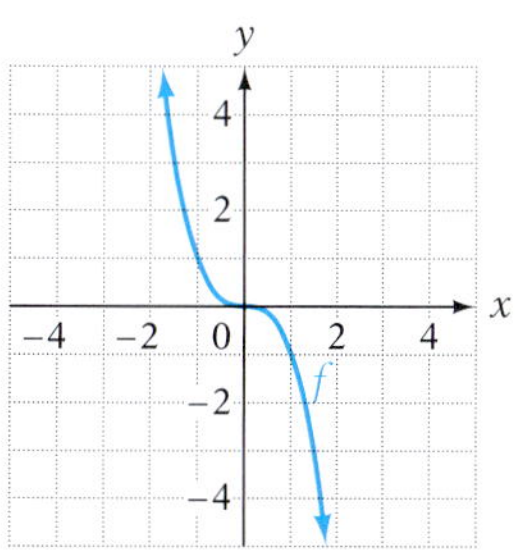

8. Use the graph of $f(x) = -|x|$ shown below to graph $g(x) = -|x| + 3$.

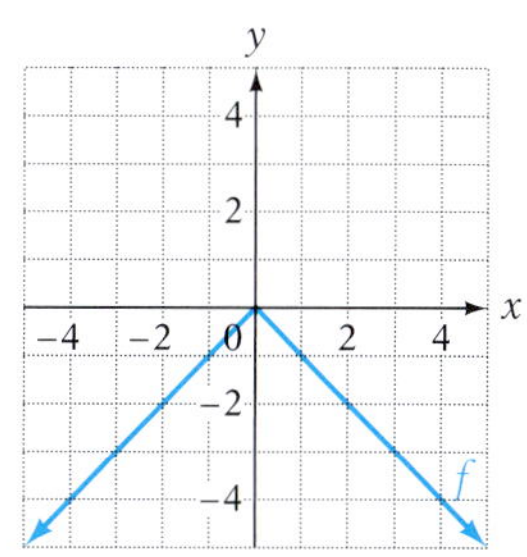

9. Graph $f(x) = -\sqrt{3 - x}$. Use interval notation to state the domain and range.

10. Graph $f(x) = \left|\frac{1}{2}x\right| - 2$. Use set-builder notation to state the domain and range.

11. Use the graph of $f(x) = |x|$ shown below to graph $g(x) = |x + 2| - 3$.

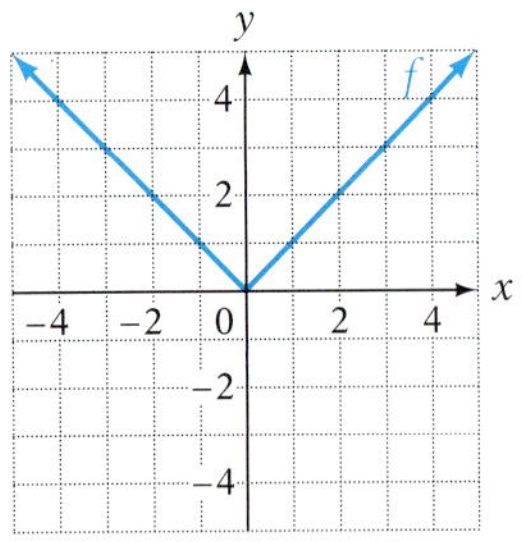

12. Use the graph of $f(x) = -x^2$ shown below to graph $g(x) = -(x - 2)^2 + 2$.

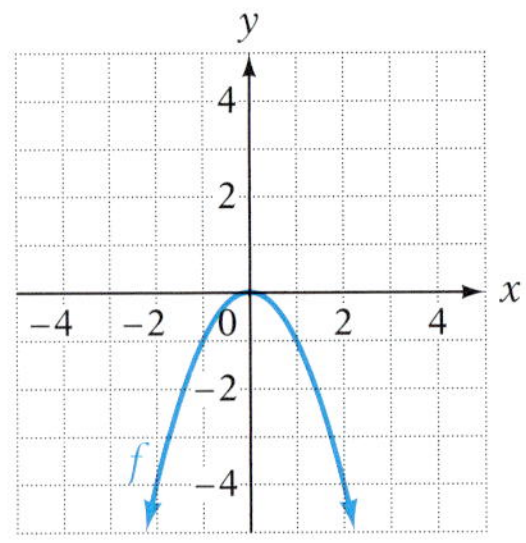

13. Find the inverse of the function $f(x) = \frac{1}{4}x - 4$.

14. Find the inverse of the function $\{(2, 6), (3, 5), (4, 4), (5, 3)\}$.

15. Are the functions $f(x) = \frac{1}{2}x + 2$ and $g(x) = 2x - 4$ inverses of each other?

16. Given $f(x) = 2x^2 - 7$ and $g(x) = x - 1$, find $f[g(x)]$.

17. Find the inverse of the function $f(x) = \frac{1}{2}x - 3$.

18. Are the functions $f(x) = \frac{2}{3}x + 3$ and $g(x) = \frac{3}{2}x - 3$ inverses of each other?

19. Graph $f(x) = x^3 - 3x + 2$. Use set-builder notation to state the domain and range.

20. Determine whether the graph is the graph of a 1–1 function.

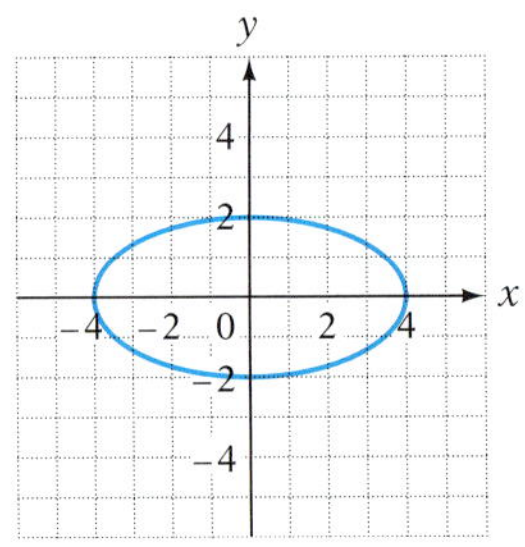

21. Find the axis of symmetry of the parabola $y = -x^2 + 6x - 5$.

22. Find the vertex of the parabola $y = -x^2 + 3x - 2$.

23. Sketch a graph of $y = \frac{1}{2}x^2 + x - 4$.

24. Sketch a graph of $x = y^2 - y - 2$.

25. Find the equation of the circle with radius 4 and center $C(-3, -3)$.

26. Find the equation of the circle that passes through the point $P(2, 4)$ and whose center is the point $C(-1, -3)$.

27. Sketch a graph of $(x - 2)^2 + (y + 1)^2 = 9$.

28. Sketch a graph of $\frac{y^2}{25} - \frac{x^2}{16} = 1$.

29. Sketch a graph of $\frac{x^2}{9} - \frac{y^2}{4} = 1$.

30. Sketch a graph of $\frac{x^2}{16} + \frac{y^2}{4} = 1$.

Cumulative Review Exercises

1. Evaluate $-3a + \left|\frac{3b - ab}{3b - c}\right|$ when $a = 2$, $b = 2$, and $c = -2$.

2. Graph: $\{x \mid x < -3\} \cap \{x \mid x > -4\}$

3. Solve: $\frac{3x - 1}{6} - \frac{5 - x}{4} = \frac{5}{6}$

4. Solve: $4x - 2 < -10$ or $3x - 1 > 8$

5. Graph $f(x) = \frac{1}{4}x^2$. Find the vertex and axis of symmetry.

6. Graph the solution set of $3x - 4y \geq 8$.

7. Solve: $|8 - 2x| \geq 0$

8. Simplify: $\left(\frac{3a^3b}{2a}\right)^2\left(\frac{a^2}{-3b^2}\right)^3$

9. Multiply: $(x - 4)(2x^2 + 4x - 1)$

10. Factor: $a^4 - 2a^2 - 8$

11. Factor: $x^3y + x^2y^2 - 6xy^3$

12. Solve: $(b + 2)(b - 5) = 2b + 14$

13. Solve: $x^2 - 2x > 15$

14. Subtract: $\dfrac{x^2 + 4x - 5}{2x^2 - 3x + 1} - \dfrac{x}{2x - 1}$

15. Solve: $\dfrac{5}{x^2 + 7x + 12} = \dfrac{9}{x + 4} - \dfrac{2}{x + 3}$

16. Simplify: $\dfrac{4 - 6i}{2i}$

17. Find the equation of the line that contains the points $(-3, 4)$ and $(2, -6)$.

18. Find the equation of the line that contains the point $(-3, 1)$ and is perpendicular to the line $2x - 3y = 6$.

19. Solve: $3x^2 = 3x - 1$

20. Solve: $\sqrt{8x + 1} = 2x - 1$

21. Find the minimum value of the function $f(x) = 2x^2 - 3$.

22. Find the range of $f(x) = |3x - 4|$ if the domain is $\{0, 1, 2, 3\}$.

23. Is this set of ordered pairs a function?

$\{(-3, 0), (-2, 0), (-1, 1), (0, 1)\}$

24. Solve: $\sqrt[3]{5x - 2} = 2$

25. Given $g(x) = 3x - 5$ and $h(x) = \frac{1}{2}x + 4$, find $g[h(2)]$.

26. Find the inverse of the function given by $f(x) = -3x + 9$.

27. Find the cost per pound of a tea mixture made from 30 lb of tea costing \$4.50 per pound and 45 lb of tea costing \$3.60 per pound.

28. How many pounds of an 80% copper alloy must be mixed with 50 lb of a 20% copper alloy to make an alloy that is 40% copper?

29. Six ounces of an insecticide are mixed with 16 gal of water to make a spray for spraying an orange grove. How much additional insecticide is required if it is to be mixed with 28 gal of water?

30. A large pipe can fill a tank in 8 min less time than it takes a smaller pipe to fill the same tank. Working together, both pipes can fill the tank in 3 min. How long would it take the larger pipe, working alone, to fill the tank?

31. The distance (d) that a spring stretches varies directly as the force (f) used to stretch the spring. If a force of 50 lb can stretch the spring 30 in., how far can a force of 40 lb stretch the spring?

32. A car traveling at 50 mph overtakes a cyclist who, riding at 10 mph, has had a 2-hour head start. How far from the starting point does the car overtake the cyclist?

33. The frequency of vibration (f) in an open pipe organ varies inversely as the length (L) of the pipe. If the air in a pipe 2 m long vibrates 60 times per minute, find the frequency in a pipe that is 1.5 m long.

How long have you had each of your credit cards? Do you pay the balance on your credit cards every month? Do you send in your payments on time? If you apply for a loan, the lender will want to know the answers to these and other questions. The lender can get the answers from a credit reporting agency. These agencies also provide a credit score based on the information in the report. One of the most widely used credit scores is the FICO® score. The higher the FICO® score, the lower the predicted credit risk for lenders. FICO® scores are a topic of the **Project on pages 748–750.**

Applications

11 Exponential and Logarithmic Functions

Objectives

Section 11.1

1. Evaluate exponential functions
2. Graph exponential functions

Section 11.2

1. Write equivalent exponential and logarithmic equations
2. The properties of logarithms

Section 11.3

1. Graph logarithmic functions

Section 11.4

1. Solve exponential equations
2. Solve logarithmic equations

Section 11.5

1. Application problems

Need help? For online student resources, such as section quizzes, visit this web site: math.college.hmco.com

Prep Test

1. Simplify: 3^{-2}

2. Simplify: $\left(\frac{1}{2}\right)^{-4}$

3. Complete: $\frac{1}{8} = 2^?$

4. Evaluate $f(x) = x^4 + x^3$ for $x = -1$ and $x = 3$.

5. Solve: $3x + 7 = x - 5$

6. Solve: $16 = x^2 - 6x$

7. Evaluate $A(1 + i)^n$ for $A = 5000$, $i = 0.04$, and $n = 6$. Round to the nearest hundredth.

8. Graph: $f(x) = x^2 - 1$

Go Figure

What is the ones digit of $1 + 9 + 9^2 + 9^3 + 9^4 + \cdots + 9^{2000} + 9^{2001}$?

11.1 Exponential Functions

1 Evaluate exponential functions

VIDEO & DVD CD TUTOR WEB SSM

The growth of a \$500 savings account that earns 5% annual interest compounded daily is shown at the right. In 14 years, the savings account contains approximately \$1000, twice the initial amount. The growth of this savings account is an example of an exponential function.

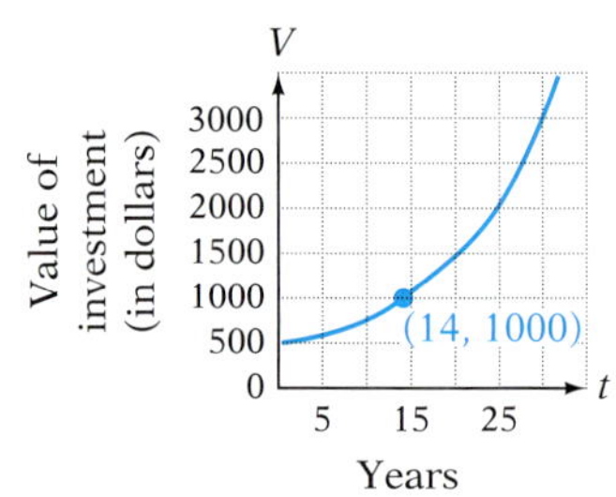

The pressure of the atmosphere at a certain height is shown in the graph at the right. This is another example of an exponential function. From the graph, we read that the air pressure is approximately 6.5 lb/in^2 at an altitude of 20,000 ft.

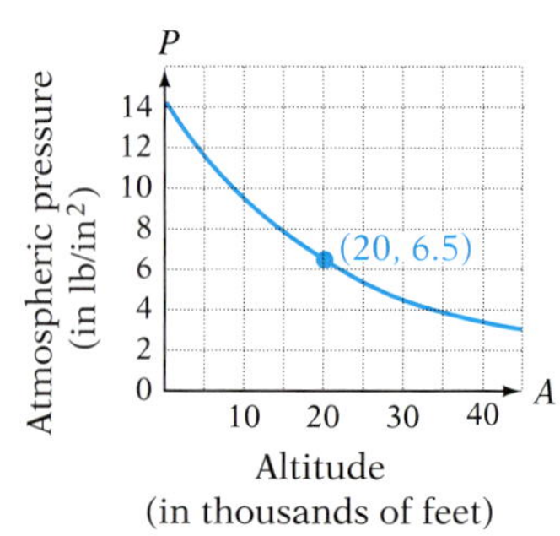

Definition of an Exponential Function

The **exponential function** with base b is defined by

$$f(x) = b^x$$

where $b > 0$, $b \neq 1$, and x is any real number.

In the definition of an exponential function, b, the base, is required to be positive. If the base were a negative number, the value of the function would be a complex number for some values of x. For instance, the value of $f(x) = (-4)^x$ when $x = \frac{1}{2}$ is $f\left(\frac{1}{2}\right) = (-4)^{\frac{1}{2}} = \sqrt{-4} = 2i$. To avoid complex number values of a function, the base of the exponential function is a positive number.

Evaluate $f(x) = 2^x$ at $x = 3$ and $x = -2$.

Substitute 3 for x and simplify. $f(3) = 2^3 = 8$

Substitute -2 for x and simplify. $f(-2) = 2^{-2} = \dfrac{1}{2^2} = \dfrac{1}{4}$

To evaluate an exponential expression for an irrational number such as $\sqrt{2}$, we obtain an approximation to the value of the function by approximating the irrational number. For instance, the value of $f(x) = 4^x$ when $x = \sqrt{2}$ can be approximated by using an approximation of $\sqrt{2}$.

$$f(\sqrt{2}) = 4^{\sqrt{2}} \approx 4^{1.4142} \approx 7.1029$$

Because $f(x) = b^x$ $(b > 0, b \neq 1)$ can be evaluated at both rational and irrational numbers, the domain of f is all real numbers. And because $b^x > 0$ for all values of x, the range of f is the positive real numbers.

Example 1 Evaluate $f(x) = \left(\frac{1}{2}\right)^x$ at $x = 2$ and $x = -3$.

Solution $f(x) = \left(\dfrac{1}{2}\right)^x$

$f(2) = \left(\dfrac{1}{2}\right)^2 = \dfrac{1}{4}$ $\quad f(-3) = \left(\dfrac{1}{2}\right)^{-3} = 2^3 = 8$

Problem 1 Evaluate $f(x) = \left(\frac{2}{3}\right)^x$ at $x = 3$ and $x = -2$.

Solution See page S34.

Example 2 Evaluate $f(x) = 2^{3x-1}$ at $x = 1$ and $x = -1$.

Solution $f(x) = 2^{3x-1}$

$f(1) = 2^{3(1)-1} = 2^2 = 4$ $\quad f(-1) = 2^{3(-1)-1} = 2^{-4} = \dfrac{1}{2^4} = \dfrac{1}{16}$

Problem 2 Evaluate $f(x) = 2^{2x+1}$ at $x = 0$ and $x = -2$.

Solution See page S34.

Example 3 Evaluate $f(x) = (\sqrt{5})^x$ at $x = 4$, $x = -2.1$, and $x = \pi$. Round to the nearest ten-thousandth.

Solution $f(x) = (\sqrt{5})^x$

$f(4) = (\sqrt{5})^4 = 25$ $\qquad f(-2.1) = (\sqrt{5})^{-2.1} \approx 0.1845$ $\qquad f(\pi) = (\sqrt{5})^{\pi} \approx 12.5297$

Problem 3 Evaluate $f(x) = \pi^x$ at $x = 3$, $x = -2$, and $x = \pi$. Round to the nearest ten-thousandth.

Solution See page S34.

A frequently used base in applications of exponential functions is an irrational number designated by e. The number e is approximately 2.71828183. It is an irrational number, so it has a nonterminating, nonrepeating decimal representation.

Natural Exponential Function

The function defined by $f(x) = e^x$ is called the **natural exponential function.**

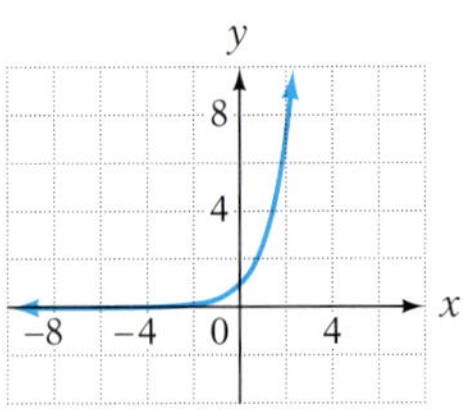

The $\boxed{e^x}$ key on a calculator can be used to evaluate the natural exponential function. The graph of $y = e^x$ is shown at the left.

Example 4 Evaluate $f(x) = e^x$ at $x = 2$, $x = -3$, and $x = \pi$. Round to the nearest ten-thousandth.

Solution $f(x) = e^x$

$f(2) = e^2 \approx 7.3891$ $\qquad f(-3) = e^{-3} \approx 0.0498$ $\qquad f(\pi) = e^{\pi} \approx 23.1407$

Problem 4 Evaluate $f(x) = e^x$ at $x = 1.2$, $x = -2.5$, and $x = e$. Round to the nearest ten-thousandth.

Solution See page S34.

2 Graph exponential functions

VIDEO & DVD CD TUTOR WEB SSM

Some of the properties of an exponential function can be seen by considering its graph.

Point of Interest

The natural exponential function is an extremely important function. It is used extensively in applied problems in virtually all disciplines from archeology to zoology. Leonhard Euler (1707–1783) was the first to use the letter e as the base of the natural exponential function.

To graph $f(x) = 2^x$, think of the function as the equation $y = 2^x$.

Choose values of x, and find the corresponding values of y. The results can be recorded in a table.

Graph the ordered pairs on a rectangular coordinate system.

Connect the points with a smooth curve.

x	$f(x) = y$
-2	$2^{-2} = \frac{1}{4}$
-1	$2^{-1} = \frac{1}{2}$
0	$2^0 = 1$
1	$2^1 = 2$
2	$2^2 = 4$
3	$2^3 = 8$

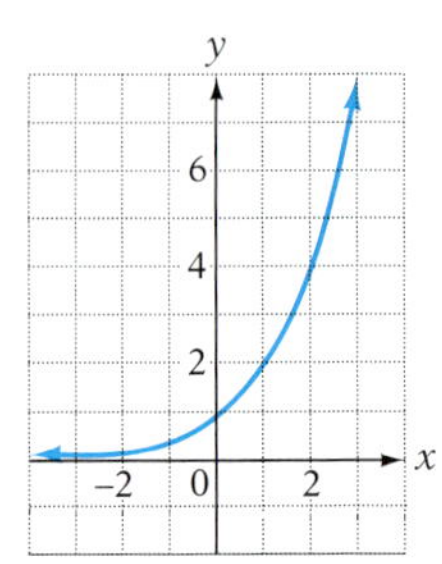

Note that a vertical line would intersect the graph at only one point. Therefore, by the vertical line test, $f(x) = 2^x$ is the graph of a function. Also note that a horizontal line would intersect the graph at only one point. Therefore, $f(x) = 2^x$ is the graph of a 1–1 function.

Graph: $f(x) = \left(\frac{1}{2}\right)^x$

Think of the function as the equation $y = \left(\frac{1}{2}\right)^x$.

Choose values of x, and find the corresponding values of y.

Graph the ordered pairs on a rectangular coordinate system.

Connect the points with a smooth curve.

x	$f(x) = y$
-3	$\left(\frac{1}{2}\right)^{-3} = 8$
-2	$\left(\frac{1}{2}\right)^{-2} = 4$
-1	$\left(\frac{1}{2}\right)^{-1} = 2$
0	$\left(\frac{1}{2}\right)^{0} = 1$
1	$\left(\frac{1}{2}\right)^{1} = \frac{1}{2}$
2	$\left(\frac{1}{2}\right)^{2} = \frac{1}{4}$

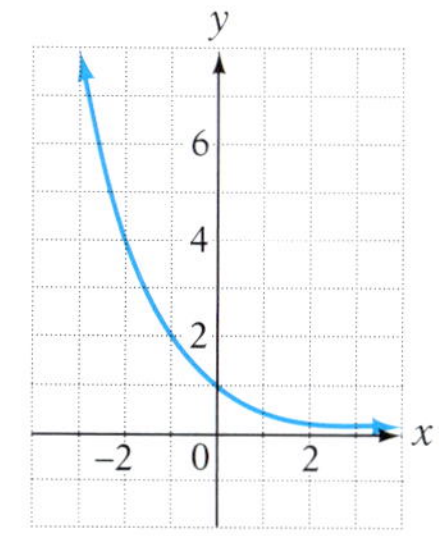

Applying the vertical and horizontal line tests reveals that $f(x) = \left(\frac{1}{2}\right)^x$ is also the graph of a 1–1 function.

Take Note

Note that because $2^{-x} = (2^{-1})^x = \left(\frac{1}{2}\right)^x$, the graphs of $f(x) = 2^{-x}$ and $f(x) = \left(\frac{1}{2}\right)^x$ are the same.

Graph $f(x) = 2^{-x}$.

Think of this as the equation $y = 2^{-x}$.

Choose values of x and find the corresponding values of y.

Graph the ordered pairs on a rectangular coordinate system.

Connect the points with a smooth curve.

x	y
-3	8
-2	4
-1	2
0	1
1	$\frac{1}{2}$
2	$\frac{1}{4}$

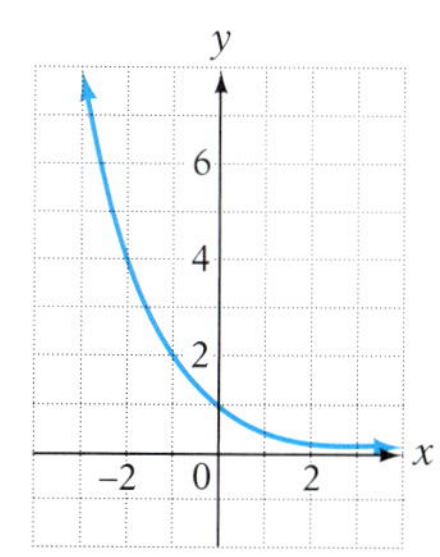

Example 5 Graph. **A.** $f(x) = 3^{\frac{1}{2}x - 1}$ **B.** $f(x) = 2^x - 1$

Solution **A.**

x	y
-2	$\frac{1}{9}$
0	$\frac{1}{3}$
2	1
4	3

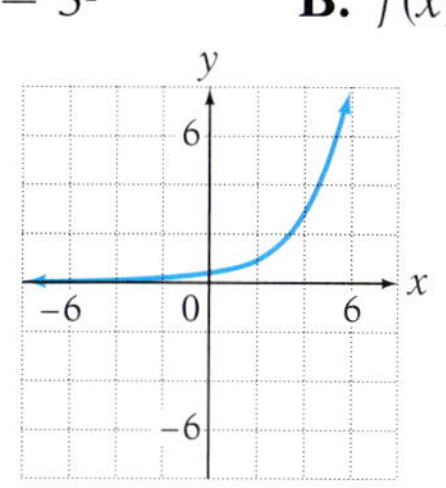

B.

x	y
-2	$-\frac{3}{4}$
-1	$-\frac{1}{2}$
0	0
1	1
2	3
3	7

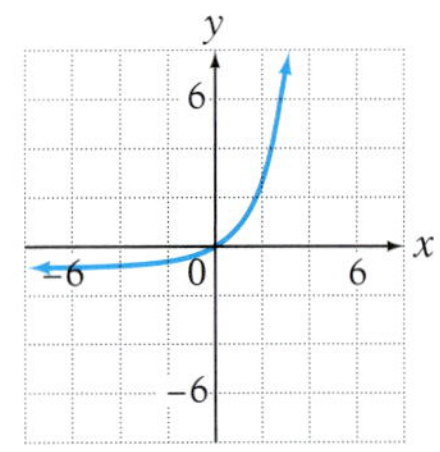

Problem 5 Graph. **A.** $f(x) = 2^{-\frac{1}{2}x}$ **B.** $f(x) = 2^x + 1$

Solution See page S34.

Example 6 Graph. **A.** $f(x) = 3^{x+1}$ **B.** $f(x) = e^x - 1$

Solution **A.**

x	y
-3	$\frac{1}{9}$
-2	$\frac{1}{3}$
-1	1
0	3
1	9

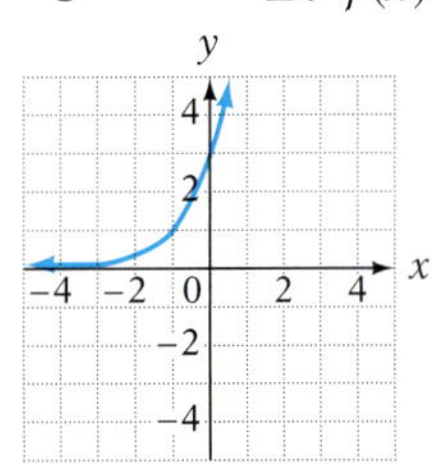

B.

x	y
-2	-0.86
-1	-0.63
0	0
1	1.72
2	6.39

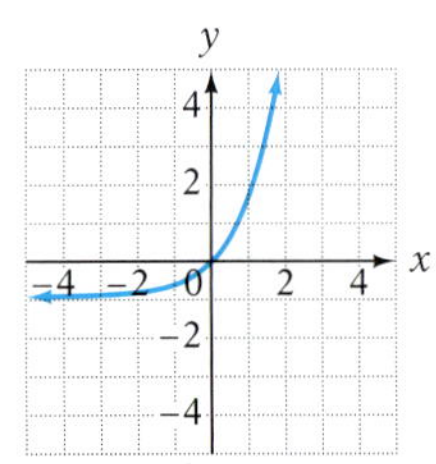

Problem 6 Graph. **A.** $f(x) = 2^{-x} + 2$ **B.** $f(x) = e^{-2x} - 4$

Solution See page S34.

Example 7 Graph $f(x) = -\frac{1}{3}e^{2x} + 2$ and approximate the zero of f to the nearest tenth.

Solution

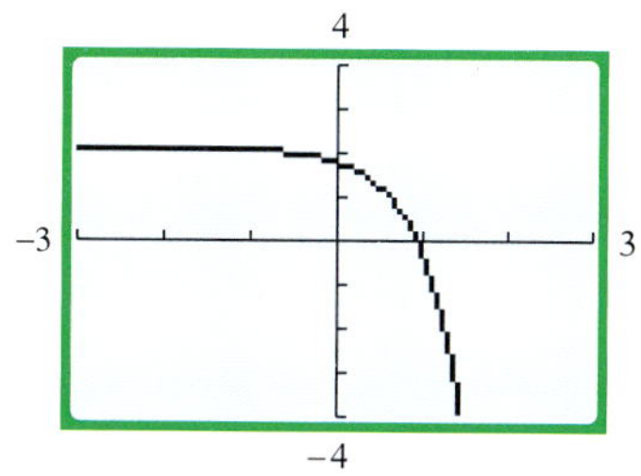

• Recall that a zero of f is a value of x for which $f(x) = 0$. Use the features of a graphing utility to determine the x-intercept of the graph, which is the zero of f.

To the nearest tenth, the zero of f is 0.9.

Problem 7 Graph $f(x) = 2\left(\frac{3}{4}\right)^x - 3$ and approximate, to the nearest tenth, the value of x for which $f(x) = 1$.

Solution See page S34.

11.1 Concept Review

Determine whether the following statements are always true, sometimes true, or never true.

1. The domain of an exponential function $f(x) = b^x$, $b > 0$, $b \neq 1$, is the set of positive numbers.
2. An exponential function $f(x) = b^x$, $b > 0$, $b \neq 1$, is a 1–1 function.
3. The graph of the exponential function $f(x) = b^x$, $b > 0$, $b \neq 1$, passes through the point (0, 0).
4. For the function $f(x) = b^x$, $b > 0$, $b \neq 1$, the base b is a positive integer.
5. An exponential function $f(x) = b^x$, $b > 0$, $b \neq 1$, has two x-intercepts.

11.1 Exercises

1

1. What is an exponential function?
2. What is the natural exponential function?

3. Which of the following cannot be the base of an exponential function?

a. 7 **b.** $\frac{1}{4}$ **c.** -5 **d.** 0.01

4. Which of the following cannot be the base of an exponential function?

a. 0.9 **b.** 476 **c.** 8 **d.** $-\frac{1}{2}$

5. Given $f(x) = 3^x$, evaluate:
a. $f(2)$ **b.** $f(0)$ **c.** $f(-2)$

6. Given $H(x) = 2^x$, evaluate:
a. $H(-3)$ **b.** $H(0)$ **c.** $H(2)$

7. Given $g(x) = 2^{x+1}$, evaluate:
a. $g(3)$ **b.** $g(1)$ **c.** $g(-3)$

8. Given $F(x) = 3^{x-2}$, evaluate:
a. $F(-4)$ **b.** $F(-1)$ **c.** $F(0)$

9. Given $P(x) = \left(\frac{1}{2}\right)^{2x}$, evaluate:

a. $P(0)$ **b.** $P\left(\frac{3}{2}\right)$ **c.** $P(-2)$

10. Given $R(t) = \left(\frac{1}{3}\right)^{3t}$, evaluate:

a. $R\left(-\frac{1}{3}\right)$ **b.** $R(1)$ **c.** $R(-2)$

11. Given $G(x) = e^{\frac{x}{2}}$, evaluate the following. Round to the nearest ten-thousandth.

a. $G(4)$ **b.** $G(-2)$ **c.** $G\left(\frac{1}{2}\right)$

12. Given $f(x) = e^{2x}$, evaluate the following. Round to the nearest ten-thousandth.

a. $f(-2)$ **b.** $f\left(-\frac{2}{3}\right)$ **c.** $f(2)$

13. Given $H(r) = e^{-r+3}$, evaluate the following. Round to the nearest ten-thousandth.
a. $H(-1)$ **b.** $H(3)$ **c.** $H(5)$

14. Given $P(t) = e^{-\frac{1}{2}t}$, evaluate the following. Round to the nearest ten-thousandth.

a. $P(-3)$ **b.** $P(4)$ **c.** $P\left(\frac{1}{2}\right)$

15. Given $F(x) = 2^{x^2}$, evaluate:

a. $F(2)$ **b.** $F(-2)$ **c.** $F\left(\frac{3}{4}\right)$

16. Given $Q(x) = 2^{-x^2}$, evaluate:
a. $Q(3)$ **b.** $Q(-1)$ **c.** $Q(-2)$

17. Given $f(x) = e^{-\frac{x^2}{2}}$, evaluate the following. Round to the nearest ten-thousandth.
a. $f(-2)$ **b.** $f(2)$ **c.** $f(-3)$

18. Given $f(x) = e^{-2x} + 1$, evaluate the following. Round to the nearest ten-thousandth.
a. $f(-1)$ **b.** $f(3)$ **c.** $f(-2)$

2 Graph.

19. $f(x) = 3^x$

20. $f(x) = 3^{-x}$

21. $f(x) = 2^{x+1}$

22. $f(x) = 2^{x-1}$

23. $f(x) = \left(\frac{1}{3}\right)^x$

24. $f(x) = \left(\frac{2}{3}\right)^x$

25. $f(x) = 2^{-x} + 1$

26. $f(x) = 2^x - 3$

27. $f(x) = \left(\frac{1}{3}\right)^{-x}$

28. $f(x) = \left(\frac{3}{2}\right)^{-x}$

29. $f(x) = \left(\frac{1}{2}\right)^{-x} + 2$

30. $f(x) = \left(\frac{1}{2}\right)^{x} - 1$

31. Which of the following functions have the same graph?

a. $f(x) = 3^x$ **b.** $f(x) = \left(\frac{1}{3}\right)^x$ **c.** $f(x) = x^3$ **d.** $f(x) = 3^{-x}$

32. Which of the following functions have the same graph?

a. $f(x) = x^4$ **b.** $f(x) = 4^{-x}$ **c.** $f(x) = 4^x$ **d.** $f(x) = \left(\frac{1}{4}\right)^x$

33. Graph $f(x) = 3^x$ and $f(x) = 3^{-x}$ and find the point of intersection of the two graphs.

34. Graph $f(x) = 2^{x+1}$ and $f(x) = 2^{-x+1}$ and find the point of intersection of the two graphs.

35. Graph $f(x) = \left(\frac{1}{3}\right)^x$. What are the x- and y-intercepts of the graph of the function?

36. Graph $f(x) = \left(\frac{1}{3}\right)^{-x}$. What are the x- and y-intercepts of the graph of the function?

Solve.

37. Graph $f(x) = 2^x - 3$ and approximate the zero of f to the nearest tenth.

38. Graph $f(x) = 5 - 3^x$ and approximate the zero of f to the nearest tenth.

39. Graph $f(x) = e^x$ and approximate, to the nearest tenth, the value of x for which $f(x) = 3$.

40. Graph $f(x) = e^{-2x-3}$ and approximate to the nearest tenth, the value of x for which $f(x) = 2$.

41. **Investments** The exponential function given by $F(n) = 500(1.00021918)^{365n}$ gives the value in n years of a \$500 investment in a certificate of deposit that earns 8% annual interest compounded daily. Graph F and determine in how many years the investment will be worth \$1000.

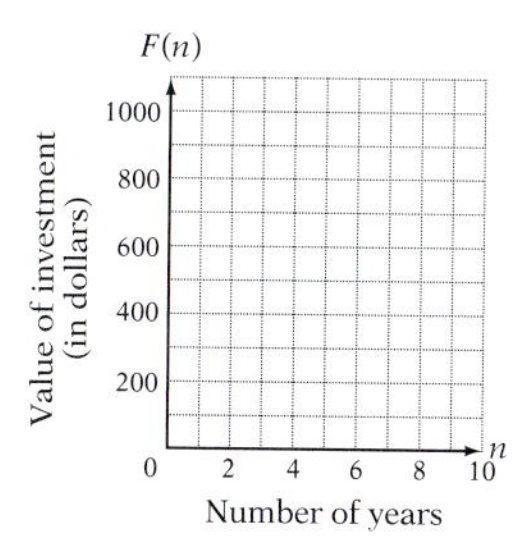

42. **Chemistry** The number of grams of radioactive cesium that remain after t years from an original sample of 30 g is given by $N(t) = 30(2^{-0.0322t})$. Graph N and determine in how many years there will be 20 g of cesium remaining.

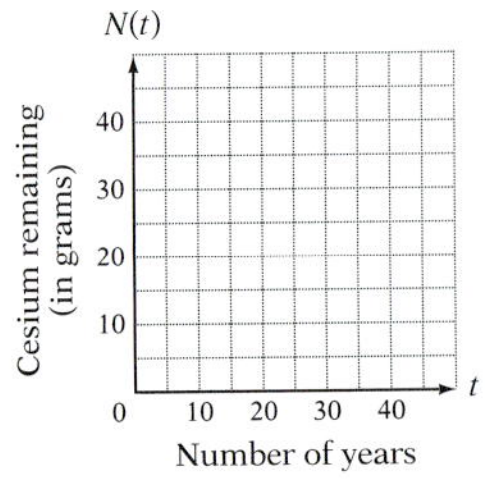

43. **Oceanography** The percent of light that reaches m meters below the surface of the ocean is given by the equation $P(m) = 100e^{-1.38m}$. Graph P and determine the depth to which 50% of the light will reach. Round to the nearest tenth.

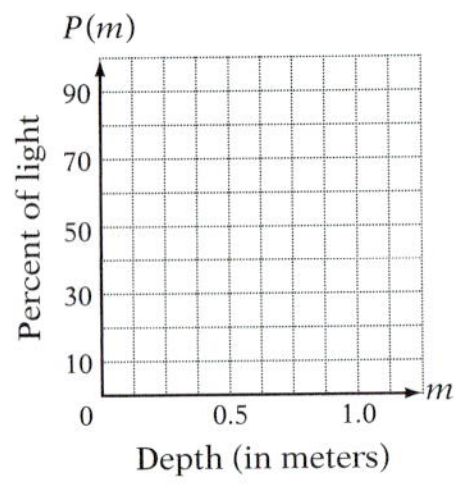

11.1 Applying Concepts

44. Evaluate $\left(1 + \dfrac{1}{n}\right)^n$ for $n = 100, 1000, 10{,}000$, and $100{,}000$ and compare the results with the value of e, the base of the natural exponential function. On the basis of your evaluation, complete the following sentence:

As n increases, $\left(1 + \dfrac{1}{n}\right)^n$ becomes closer to ______.

45. **Physics** If air resistance is ignored, the speed v, in feet per second, of an object t seconds after it has been dropped is given by the equation $v = 32t$. This is true regardless of the mass of the object. However, if air resistance is considered, then the speed depends on the mass (and on other things). For a certain mass, the speed t seconds after it has been dropped is given by $v = 64\left(1 - e^{-\frac{t}{2}}\right)$.

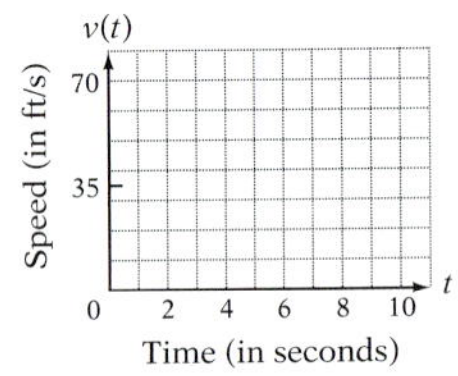

a. Graph this equation. *Suggestion:* Use Xmin = −0.5, Xmax = 10, Xscl = 1, Ymin = −0.5, Ymax = 70, and Yscl = 10.

b. The point whose approximate coordinates are (4, 55.3) is on the graph. Write a sentence that explains the meaning of these coordinates.

46. Graph $f(x) = e^x$, $g(x) = e^{x-2}$, and $h(x) = e^{x+2}$ on the same rectangular coordinate system. Explain how the graphs are similar and how they are different.

47. Graph $f(x) = 2^x$, $g(x) = 2^x - 2$, and $h(x) = 2^x + 2$ on the same rectangular coordinate system. Explain how the graphs are similar and how they are different.

11.2 Introduction to Logarithms

1 Write equivalent exponential and logarithmic equations

VIDEO & DVD CD TUTOR WEB SSM

Because the exponential function is a 1–1 function, it has an inverse function that is called a *logarithm.* A logarithm is used to answer a question similar to the following one: "If $16 = 2^y$, what is the value of y?" Because $16 = 2^4$, the logarithm, base 2, of 16 is 4. This is written $\log_2 16 = 4$. Note that a logarithm is an exponent that solves a certain equation.

Definition of Logarithm

For $x > 0$, $b > 0$, $b \neq 1$, $y = \log_b x$ is equivalent to $x = b^y$.

Read $\log_b x$ as "the logarithm of x, base b" or "log base b of x."

The table below shows equivalent statements written in both exponential and logarithmic form.

Exponential Form		Logarithmic Form
$2^4 = 16$	⟺	$\log_2 16 = 4$
$\left(\frac{2}{3}\right)^2 = \frac{4}{9}$	⟺	$\log_{\frac{2}{3}}\left(\frac{4}{9}\right) = 2$
$10^{-1} = 0.1$	⟺	$\log_{10}(0.1) = -1$

Example 1 Write $4^5 = 1024$ in logarithmic form.

Solution $4^5 = 1024$ is equivalent to $\log_4 1024 = 5$.

Problem 1 Write $3^{-4} = \frac{1}{81}$ in logarithmic form.

Solution See page S34.

Example 2 Write $\log_7 343 = 3$ in exponential form.

Solution $\log_7 343 = 3$ is equivalent to $7^3 = 343$.

Problem 2 Write $\log_{10} 0.0001 = -4$ in exponential form.

Solution See page S34.

Recalling the equations $y = \log_b x$ and $x = b^y$ from the definition of a logarithm, note that because $b^y > 0$ for all values of y, x is always a positive number. Therefore, in the equation $y = \log_b x$, x is a positive number. The logarithm of a negative number is not a real number.

The 1–1 property of exponential functions can be used to evaluate some logarithms.

Equality of Exponents Property

For $b > 0$, $b \neq 1$, if $b^u = b^v$, then $u = v$.

Evaluate: $\log_2 8$

Write an equation.	$\log_2 8 = x$
Write the equation in its equivalent exponential form.	$8 = 2^x$
Write 8 in exponential form using 2 as the base.	$2^3 = 2^x$
Use the Equality of Exponents Property.	$3 = x$
	$\log_2 8 = 3$

Example 3 Evaluate: $\log_3\left(\frac{1}{9}\right)$

Solution

$\log_3\left(\frac{1}{9}\right) = x$ • Write an equation.

$\frac{1}{9} = 3^x$ • Write the equation in its equivalent exponential form.

$3^{-2} = 3^x$ • Write $\frac{1}{9}$ in exponential form using 3 as the base.

$-2 = x$ • Solve for x using the Equality of Exponents Property.

$\log_3\left(\frac{1}{9}\right) = -2$

Problem 3 Evaluate: $\log_4 64$

Solution See page S34.

Solve $\log_4 x = -2$ for x. $\log_4 x = -2$

Write the equation in its equivalent exponential form. $4^{-2} = x$

Solve for x. $\frac{1}{16} = x$

The solution is $\frac{1}{16}$.

In this example, $\frac{1}{16}$ is called the antilogarithm base 4 of negative 2. In general, if $\log_b M = N$, then M is the antilogarithm base b of N. The antilogarithm of a number can be determined by rewriting the logarithmic expression in exponential form. For instance, the exponential form of $\log_5 x = 3$ is $5^3 = x$; x is the antilogarithm base 5 of 3.

Definition of Antilogarithm

If $\log_b M = N$, the **antilogarithm,** base b, of N is M. In exponential form,

$$M = b^N$$

Example 4 Solve $\log_6 x = 2$ for x.

Solution

$\log_6 x = 2$

$6^2 = x$ • Write $\log_6 x = 2$ in its equivalent exponential form.

$36 = x$

The solution is 36.

Problem 4 Solve $\log_2 x = -4$ for x.

Solution See page S34.

> **Take Note**
>
> The logarithms of most numbers are irrational numbers. Therefore, the value in the display of a calculator is an approximation.

Logarithms base 10 are called **common logarithms.** We usually omit the base, 10, when writing the common logarithm of a number. Therefore, $\log_{10} x$ is written $\log x$. To find the common logarithm of most numbers, a calculator is necessary. A calculator was used to find the value of log 384, shown below.

$$\log 384 \approx 2.584331224$$

Mantissa (the decimal part, .584331224)

Characteristic (the integer part, 2)

The decimal part of a common logarithm is called the **mantissa;** the integer part is called the **characteristic.**

> **Take Note**
>
> The integer and decimal parts of a natural logarithm do not have names associated with them as they do in common logarithms.

When e (the base of the natural exponential function) is used as a base of a logarithm, the logarithm is referred to as the **natural logarithm** and is abbreviated $\ln x$. This is read "el en x." Using a calculator, we find that

$$\ln 23 \approx 3.135494216$$

Example 5 Solve $\ln x = -1$ for x. Round to the nearest ten-thousandth.

Solution

$\ln x = -1$ • **In x is the abbreviation for $\log_e x$.**

$e^{-1} = x$ • **Write the equation in its equivalent exponential form.**

$0.3679 \approx x$

The solution is 0.3679.

Problem 5 Solve $\log x = 1.5$ for x. Round to the nearest ten-thousandth.

Solution See page S34.

2 The properties of logarithms

VIDEO & DVD CD TUTOR WEB SSM

Because a logarithm is an exponent, the properties of logarithms are similar to the properties of exponents.

The table at the right shows some powers of 2 and the equivalent logarithmic form.

$2^0 = 1$	$\log_2 1 = 0$
$2^1 = 2$	$\log_2 2 = 1$
$2^2 = 4$	$\log_2 4 = 2$
$2^3 = 8$	$\log_2 8 = 3$
$2^4 = 16$	$\log_2 16 = 4$
$2^5 = 32$	$\log_2 32 = 5$

The table can be used to show that $\log_2 4 + \log_2 8$ equals $\log_2 32$.

$$\log_2 4 + \log_2 8 = 2 + 3 = 5$$
$$\log_2 32 = 5$$
$$\log_2 4 + \log_2 8 = \log_2 32$$

Note that $\log_2 32 = \log_2(4 \times 8) = \log_2 4 + \log_2 8$.

> **Take Note**
>
> Pay close attention to this theorem. Note, for instance, that this theorem states that
>
> $\log_3(4p) = \log_3 4 + \log_3 p$
>
> It also states that
>
> $\log_5 9 + \log_5 z = \log_5(9z)$
>
> It does not state any relationship that involves $\log_b(x + y)$. **This expression cannot be simplified.**

The property of logarithms that states that the logarithm of the product of two numbers equals the sum of the logarithms of the two numbers is similar to the property of exponents that states that to multiply two exponential expressions with the same base, we add the exponents.

Product Property of Logarithms

For any positive real numbers x, y, and b, $b \neq 1$,

$$\log_b(xy) = \log_b x + \log_b y$$

Proof:
Let $\log_b x = m$ and $\log_b y = n$.

Write each equation in its equivalent exponential form. $x = b^m \qquad y = b^n$

Use substitution and the properties of exponents. $xy = b^m b^n$
$xy = b^{m+n}$

Write the equation in its equivalent logarithmic form. $\log_b xy = m + n$

Substitute $\log_b x$ for m and $\log_b y$ for n. $\log_b xy = \log_b x + \log_b y$

The Logarithm Property of Products is used to rewrite logarithmic expressions.

The $\log_b 6z$ is written in **expanded form** as $\log_b 6 + \log_b z$.

$$\log_b 6z = \log_b 6 + \log_b z$$

The $\log_b 12 + \log_b r$ is written as a single logarithm as $\log_b 12r$.

$$\log_b 12 + \log_b r = \log_b 12r$$

The Logarithm Property of Products can be extended to include the logarithm of the product of more than two factors. For example,

$$\log_b xyz = \log_b(xy)z = \log_b xy + \log_b z = \log_b x + \log_b y + \log_b z$$

To write $\log_b 5st$ in expanded form, use the Logarithm Property of Products.

$$\log_b 5st = \log_b 5 + \log_b s + \log_b t$$

A second property of logarithms involves the logarithm of the quotient of two numbers. This property of logarithms is also based on the fact that a logarithm is an exponent and that to divide two exponential expressions with the same base, we substract the exponents.

Take Note

This theorem is used to rewrite expressions such as

$\log_5\left(\frac{m}{8}\right) = \log_5 m - \log_5 8$

It does *not* state any relationship that involves $\frac{\log_b x}{\log_b y}$.

This expression cannot be simplified.

Quotient Property of Logarithms

For any positive real numbers x, y, and b, $b \neq 1$, $\log_b \frac{x}{y} = \log_b x - \log_b y$

Proof:
Let $\log_b x = m$ and $\log_b y = n$.

Write each equation in its equivalent exponential form. $x = b^m \qquad y = b^n$

Use substitution and the properties of exponents. $\frac{x}{y} = \frac{b^m}{b^n}$
$\frac{x}{y} = b^{m-n}$

Write the equation in its equivalent logarithmic form. $\log_b \frac{x}{y} = m - n$

Substitute $\log_b x$ for m and $\log_b y$ for n. $\log_b \frac{x}{y} = \log_b x - \log_b y$

The Logarithm Property of Quotients is used to rewrite logarithmic expressions.

The $\log_b \frac{p}{8}$ is written in expanded form as $\log_b p - \log_b 8$.

$$\log_b \frac{p}{8} = \log_b p - \log_b 8$$

The $\log_b y - \log_b v$ is written as a single logarithm as $\log_b \frac{y}{v}$.

$$\log_b y - \log_b v = \log_b \frac{y}{v}$$

Point of Interest

Logarithms were developed independently by Jobst Burgi (1552–1632) and John Napier (1550–1617) as a means of simplifying the calculations of astronomers. The idea was to devise a method by which two numbers could be multiplied by performing additions. Napier is usually given credit for logarithms because he published his result first.

In Napier's original work, the logarithm of 10,000,000 was 0. After this work was published, Napier, in discussions with Henry Briggs (1561–1631), decided that tables of logarithms would be easier to use if the logarithm of 1 was 0. Napier died before new tables could be determined, and Briggs took on the task. His table consisted of logarithms accurate to 30 decimal places, all accomplished without a calculator!

The logarithms Briggs calculated are the common logarithms mentioned earlier.

A third property of logarithms, useful especially in the computation of a power of a number, is based on the fact that a logarithm is an exponent and that the power of an exponential expression is found by multiplying the exponents.

The table of the powers of 2 shown earlier can be used to show that $\log_2 2^3$ equals $3 \log_2 2$.

$$\log_2 2^3 = \log_2 8 = 3$$
$$3 \log_2 2 = 3 \cdot 1 = 3$$
$$\log_2 2^3 = 3 \log_2 2$$

Power Property of Logarithms

For any positive real numbers x and b, $b \neq 1$, and for any real number r, $\log_b x^r = r \log_b x$.

Proof:
Let $\log_b x = m$.

Write the equation in its equivalent exponential form.	$x = b^m$
Raise each side to the r power.	$x^r = (b^m)^r$ $x^r = b^{mr}$
Write the equation in its equivalent logarithmic form.	$\log_b x^r = mr$
Substitute $\log_b x$ for m.	$\log_b x^r = r \log_b x$

The Power Property of Logarithms is used to rewrite logarithmic expressions.

The $\log_b x^3$ is written in terms of $\log_b x$ as $3 \log_b x$.

$$\log_b x^3 = 3 \log_b x$$

$\frac{2}{3} \log_4 z$ is written with a coefficient of 1 as $\log_4 z^{\frac{2}{3}}$.

$$\frac{2}{3} \log_4 z = \log_4 z^{\frac{2}{3}}$$

The properties of logarithms can be used in combination to simplify expressions that contain logarithms.

Example 6 Write the logarithm in expanded form.

A. $\log_b \frac{xy}{z}$ **B.** $\ln \frac{x^2}{y^3}$ **C.** $\log_8 \sqrt{x^3 y}$

Solution

A. $\log_b \frac{xy}{z} = \log_b(xy) - \log_b z$ • Use the Quotient Property of Logarithms.

$= \log_b x + \log_b y - \log_b z$ • Use the Product Property of Logarithms.

B. $\ln \frac{x^2}{y^3} = \ln x^2 - \ln y^3$ • Use the Quotient Property of Logarithms.

$= 2 \ln x - 3 \ln y$ • Use the Power Property of Logarithms.

C. $\log_8 \sqrt{x^3 y} = \log_8 (x^3 y)^{\frac{1}{2}}$ • Write the radical expression as an exponential expression.

$= \frac{1}{2} \log_8 x^3 y$ • Use the Power Property of Logarithms.

$= \frac{1}{2}(\log_8 x^3 + \log_8 y)$ • Use the Product Property of Logarithms.

$= \frac{1}{2}(3 \log_8 x + \log_8 y)$ • Use the Power Property of Logarithms.

$= \frac{3}{2} \log_8 x + \frac{1}{2} \log_8 y$ • Use the Distributive Property.

Problem 6 Write the logarithm in expanded form.

A. $\log_b \frac{x^2}{y}$ **B.** $\ln y^{\frac{1}{3}} z^3$ **C.** $\log_8 \sqrt[3]{xy^2}$

Solution See page S34.

Example 7 Express as a single logarithm with a coefficient of 1.

A. $3 \log_5 x + \log_5 y - 2 \log_5 z$

B. $\frac{1}{2}(\log_3 x - 3 \log_3 y + \log_3 z)$

C. $\frac{1}{3}(2 \ln x - 4 \ln y)$

Solution

A. $3 \log_5 x + \log_5 y - 2 \log_5 z$

$= \log_5 x^3 + \log_5 y - \log_5 z^2$ • Use the Power Property of Logarithms.

$= \log_5 x^3 y - \log_5 z^2$ • Use the Product Property of Logarithms.

$= \log_5 \frac{x^3 y}{z^2}$ • Use the Quotient Property of Logarithms.

B. $\frac{1}{2}(\log_3 x - 3 \log_3 y + \log_3 z)$

$= \frac{1}{2}(\log_3 x - \log_3 y^3 + \log_3 z)$ • Use the Power Property of Logarithms.

$$= \frac{1}{2}\left(\log_3 \frac{x}{y^3} + \log_3 z\right)$$ • Use the Quotient Property of Logarithms.

$$= \frac{1}{2}\left(\log_3 \frac{xz}{y^3}\right)$$ • Use the Product Property of Logarithms.

$$= \log_3\left(\frac{xz}{y^3}\right)^{\frac{1}{2}} = \log_3 \sqrt{\frac{xz}{y^3}}$$ • Use the Power Property of Logarithms. Write the exponential expression as a radical expression.

C. $\frac{1}{3}(2 \ln x - 4 \ln y)$

$$= \frac{1}{3}(\ln x^2 - \ln y^4)$$ • Use the Power Property of Logarithms.

$$= \frac{1}{3}\left(\ln \frac{x^2}{y^4}\right)$$ • Use the Quotient Property of Logarithms.

$$= \ln\left(\frac{x^2}{y^4}\right)^{\frac{1}{3}} = \ln \sqrt[3]{\frac{x^2}{y^4}}$$ • Use the Power Property of Logarithms. Write the exponential expression as a radical expression.

Problem 7 Express as a single logarithm with a coefficient of 1.

A. $2 \log_b x - 3 \log_b y - \log_b z$

B. $\frac{1}{3}(\log_4 x - 2 \log_4 y + \log_4 z)$

C. $\frac{1}{2}(2 \ln x - 5 \ln y)$

Solution See page S35.

There are three other properties of logarithms that are useful in simplifying logarithmic expressions.

Properties of Logarithms

Logarithmic Property of One

For any positive real number b, $b \neq 1$, $\log_b 1 = 0$.

Inverse Property of Logarithms

For any positive real numbers x and b, $b \neq 1$, $\log_b b^x = x$.

1–1 Property of Logarithms

For any positive real numbers x, y, and b, $b \neq 1$, if $\log_b x = \log_b y$, then $x = y$.

Example 8 Simplify. **A.** $8 \log_4 4$ **B.** $\log_8 1$

Solution **A.** $8 \log_4 4 = \log_4 4^8$ • Use the Power Property of Logarithms.

$= 8$ • Use the Inverse Property of Logarithms.

B. $\log_8 1 = 0$ • Use the Logarithmic Property of One.

Problem 8 Simplify. **A.** $\log_{16} 1$ **B.** $12 \log_3 3$

Solution See page S35.

Although only common logarithms and natural logarithms are programmed into a calculator, the logarithms for other positive bases can be found.

Evaluate $\log_5 22$. Round to the nearest ten-thousandth.

Write an equation.	$\log_5 22 = x$
Write the equation in its equivalent exponential form.	$5^x = 22$
Apply the common logarithm to each side of the equation.	$\log 5^x = \log 22$
Use the Power Property of Logarithms.	$x \log 5 = \log 22$
This is an exact answer.	$x = \dfrac{\log 22}{\log 5}$
This is an approximate answer.	$x \approx 1.9206$
	$\log_5 22 \approx 1.9206$

In the third step above, the natural logarithm, instead of the common logarithm, could have been applied to each side of the equation. The same result would have been obtained.

Using a procedure similar to the one used to evaluate $\log_5 22$, a formula for changing bases can be derived.

Change-of-Base Formula

$$\log_a N = \frac{\log_b N}{\log_b a}$$

Example 9 Evaluate $\log_7 32$. Round to the nearest ten-thousandth.

Solution $\log_7 32 = \dfrac{\ln 32}{\ln 7} \approx 1.7810$

- **Use the Change-of-Base Formula. $N = 32, a = 7, b = e$**

Problem 9 Evaluate $\log_4 2.4$. Round to the nearest ten-thousandth.

Solution See page S35.

Example 10 Rewrite $f(x) = -3 \log_7(2x - 5)$ in terms of natural logarithms.

Solution

$$f(x) = -3 \log_7(2x - 5)$$
$$= -3 \frac{\ln(2x - 5)}{\ln 7}$$
$$= -\frac{3}{\ln 7} \ln(2x - 5)$$

- **Use the Change-of-Base Formula to rewrite $\log_7(2x - 5)$ as $\dfrac{\ln(2x - 5)}{\ln 7}$.**

Problem 10 Rewrite $f(x) = 4 \log_8(3x + 4)$ in terms of common logarithms.

Solution See page S35.

In Example 10, it is important to understand that $-\frac{3}{\ln 7}\ln(2x - 5)$ and $-3 \log_7(2x - 5)$ are *exactly* equal. If common logarithms had been used, the result would have been $f(x) = -\frac{3}{\log 7}\log(2x - 5)$. The expressions $-\frac{3}{\log 7}\log(2x - 5)$ and $-3 \log_7(2x - 5)$ are also *exactly* equal.

If you are working in a base other than base 10 or base e, the Change-of-Base Formula will enable you to calculate the value of the logarithm in that base just as if that base were programmed into the calculator.

11.2 Concept Review

Determine whether the following statements are always true, sometimes true, or never true.

1. For $x > 0$, $b > 0$, $b \neq 1$, $y = \log_b x$ is equivalent to $b^y = x$.

2. $\log_b \frac{x}{y} = \frac{\log_b x}{\log_b y}$

3. $\frac{\log x}{\log y} = \frac{x}{y}$

4. $\log(x + y) = \log x + \log y$

5. $\log_b \sqrt{x} = \frac{1}{2} \log_b x$

6. $\log(x^{-1}) = \frac{1}{\log x}$

7. The inverse of an exponential function is a logarithmic function.

8. If x and y are positive real numbers, $x < y$, and $b > 0$, $b \neq 1$, then $\log_b x < \log_b y$.

11.2 Exercises

1 **1.** **a.** What is a common logarithm?
b. How is the common logarithm of $4z$ written?

2. **a.** What is a natural logarithm?
b. How is the natural logarithm of $3x$ written?

Write the exponential equation in logarithmic form.

3. $5^2 = 25$ **4.** $10^3 = 1000$ **5.** $4^{-2} = \frac{1}{16}$

6. $3^{-3} = \frac{1}{27}$ **7.** $10^y = x$ **8.** $e^y = x$

9. $a^x = w$ **10.** $b^y = c$

Write the logarithmic equation in exponential form.

11. $\log_3 9 = 2$ **12.** $\log_2 32 = 5$ **13.** $\log 0.01 = -2$

14. $\log_5 \frac{1}{5} = -1$ **15.** $\ln x = y$ **16.** $\log x = y$

17. $\log_b u = v$ **18.** $\log_c x = y$

Evaluate.

19. $\log_3 81$ **20.** $\log_7 49$ **21.** $\log_2 128$ **22.** $\log_5 125$

23. $\log 100$ **24.** $\log 0.001$ **25.** $\ln e^3$ **26.** $\ln e^2$

27. $\log_8 1$ **28.** $\log_3 243$ **29.** $\log_5 625$ **30.** $\log_2 64$

Solve for x.

31. $\log_3 x = 2$ **32.** $\log_5 x = 1$ **33.** $\log_4 x = 3$ **34.** $\log_2 x = 6$

35. $\log_7 x = -1$ **36.** $\log_8 x = -2$ **37.** $\log_6 x = 0$ **38.** $\log_4 x = 0$

Solve for x. Round to the nearest hundredth.

39. $\log x = 2.5$ **40.** $\log x = 3.2$ **41.** $\log x = -1.75$ **42.** $\log x = -2.1$

43. $\ln x = 2$ **44.** $\ln x = 4$ **45.** $\ln x = -\frac{1}{2}$ **46.** $\ln x = -1.7$

2 **47.** What is the Product Property of Logarithms?

48. What is the Quotient Property of Logarithms?

Write the logarithm in expanded form.

49. $\log_8(xz)$

50. $\log_7(4y)$

51. $\log_3 x^5$

52. $\log_2 y^7$

53. $\ln\left(\frac{r}{s}\right)$

54. $\ln\left(\frac{z}{4}\right)$

55. $\log_3(x^2y^6)$

56. $\log_4(t^4u^2)$

57. $\log_7\left(\frac{u^3}{v^4}\right)$

58. $\log\left(\frac{s^5}{t^2}\right)$

59. $\log_2(rs)^2$

60. $\log_3(x^2y)^3$

61. $\log_9 x^2yz$

62. $\log_6 xy^2z^3$

63. $\ln\left(\frac{xy^2}{z^4}\right)$

64. $\ln\left(\frac{r^2s}{t^3}\right)$

65. $\log_8\left(\frac{x^2}{yz^2}\right)$

66. $\log_9\left(\frac{x}{y^2z^3}\right)$

67. $\log_7\sqrt{xy}$

68. $\log_8\sqrt[3]{xz}$

69. $\log_2\sqrt{\frac{x}{y}}$

70. $\log_3\sqrt[3]{\frac{r}{s}}$

71. $\ln\sqrt{x^3y}$

72. $\ln\sqrt{x^5y^3}$

73. $\log_7\sqrt{\frac{x^3}{y}}$

74. $\log_b\sqrt[3]{\frac{r^2}{t}}$

Express as a single logarithm with a coefficient of 1.

75. $\log_3 x^3 - \log_3 y$

76. $\log_7 t + \log_7 v^2$

77. $\log_8 x^4 + \log_8 y^2$

78. $\log_2 r^2 + \log_2 s^3$

79. $3 \ln x$

80. $4 \ln y$

81. $3 \log_5 x + 4 \log_5 y$

82. $2 \log_6 x + 5 \log_6 y$

83. $-2 \log_4 x$

84. $-3 \log_2 y$

85. $2 \log_3 x - \log_3 y + 2 \log_3 z$

86. $4 \log_5 r - 3 \log_5 s + \log_5 t$

87. $\log_b x - (2 \log_b y + \log_b z)$

88. $2 \log_2 x - (3 \log_2 y + \log_2 z)$

89. $2(\ln x + \ln y)$

90. $3(\ln r + \ln t)$

91. $\frac{1}{2}(\log_6 x - \log_6 y)$

92. $\frac{1}{3}(\log_8 x - \log_8 y)$

93. $2(\log_4 s - 2\log_4 t + \log_4 r)$

94. $3(\log_9 x + 2\log_9 y - 2\log_9 z)$

95. $\log_5 x - 2(\log_5 y + \log_5 z)$

96. $\log_4 t - 3(\log_4 u + \log_4 v)$

97. $3\ln t - 2(\ln r - \ln v)$

98. $2\ln x - 3(\ln y - \ln z)$

99. $\frac{1}{2}(3\log_4 x - 2\log_4 y + \log_4 z)$

100. $\frac{1}{3}(4\log_5 t - 3\log_5 u - 3\log_5 v)$

Evaluate. Round to the nearest ten-thousandth.

101. $\ln 4$

102. $\ln 6$

103. $\ln\left(\frac{17}{6}\right)$

104. $\ln\left(\frac{13}{17}\right)$

105. $\log_8 6$

106. $\log_4 8$

107. $\log_5 30$

108. $\log_6 28$

109. $\log_3(0.5)$

110. $\log_5(0.6)$

111. $\log_7(1.7)$

112. $\log_6(3.2)$

113. $\log_5 15$

114. $\log_3 25$

115. $\log_{12} 120$

116. $\log_9 90$

Rewrite each function in terms of common logarithms.

117. $f(x) = \log_3(3x - 2)$

118. $f(x) = \log_5(x^2 + 4)$

119. $f(x) = \log_8(4 - 9x)$

120. $f(x) = \log_7(3x^2)$

121. $f(x) = 5\log_9(6x + 7)$

122. $f(x) = 3\log_2(2x^2 - x)$

Rewrite each function in terms of natural logarithms.

123. $f(x) = \log_2(x + 5)$

124. $f(x) = \log_4(3x + 4)$

125. $f(x) = \log_3(x^2 + 9)$

126. $f(x) = \log_7(9 - x^2)$

127. $f(x) = 7\log_8(10x - 7)$

128. $f(x) = 7\log_3(2x^2 - x)$

11.2 Applying Concepts

Use the Properties of Logarithms to solve for x.

129. $\log_8 x = 3\log_8 2$

130. $\log_5 x = 2\log_5 3$

131. $\log_4 x = \log_4 2 + \log_4 3$

132. $\log_3 x = \log_3 4 + \log_3 7$

133. $\log_6 x = 3\log_6 2 - \log_6 4$

134. $\log_9 x = 5\log_9 2 - \log_9 8$

135. $\log x = \frac{1}{3}\log 27$

136. $\log_2 x = \frac{3}{2}\log_2 4$

137. Complete each statement using the equation $\log_a b = c$.

a. $a^c =$ ______ **b.** $\text{antilog}_a(\log_a b) =$ ______

138. Given $f(x) = 3\log_6(2x - 1)$, determine $f(7)$ to the nearest hundredth.

139. Given $S(t) = 8\log_5(6t + 2)$, determine $S(2)$ to the nearest hundredth.

140. Given $P(v) = -3\log_6(4 - 2v)$, determine $P(-4)$ to the nearest hundredth.

141. Given $G(x) = -5\log_7(2x + 19)$, determine $G(-3)$ to the nearest hundredth.

142. Solve for x: $\log_2(\log_2 x) = 3$

143. Solve for x: $\ln(\ln x) = 1$

144. **Biology** To discuss the variety of species that live in a certain environment, a biologist needs a precise definition of *diversity.* Let p_1, $p_2, \ldots, p_n$ be the proportions of n species that live in an environment. The biological diversity, D, of this system is

$$D = -(p_1\log_2 p_1 + p_2\log_2 p_2 + \cdots + p_n\log_2 p_n)$$

The larger the value of D, the greater the diversity of the system. Suppose an ecosystem has exactly five different varieties of grass: rye (R), bermuda (B), blue (L), fescue (F), and St. Augustine (A).

a. Calculate the diversity of this ecosystem if the proportions are as shown in Table 1.

b. Because bermuda and St. Augustine are virulent grasses, after a time the proportions are as shown in Table 2. Does this system have more or less diversity than the one given in Table 1?

c. After an even longer period, the bermuda and St. Augustine completely overrun the environment, and the proportions are as in Table 3. Calculate the diversity of this system. (*Note:* For purposes of the diversity definition, $0\log_2 0 = 0$.) Does it have more or less diversity than the system given in Table 2?

d. Finally, the St. Augustine overruns the bermuda, and the proportions are as in Table 4. Calculate the diversity of this system. Write a sentence that describes your answer.

Table 1

R	B	L	F	A
$\frac{1}{5}$	$\frac{1}{5}$	$\frac{1}{5}$	$\frac{1}{5}$	$\frac{1}{5}$

Table 2

R	B	L	F	A
$\frac{1}{8}$	$\frac{3}{8}$	$\frac{1}{16}$	$\frac{1}{8}$	$\frac{5}{16}$

Table 3

R	B	L	F	A
0	$\frac{1}{4}$	0	0	$\frac{3}{4}$

Table 4

R	B	L	F	A
0	0	0	0	1

145. Explain how common logarithms can be used to determine the number of digits in the expansion of $9^{(9^9)}$.

11.3 Graphs of Logarithmic Functions

1 Graph logarithmic functions

VIDEO & DVD · CD TUTOR · WEB · SSM

The graph of a logarithmic function can be drawn by using the relationship between the exponential and logarithmic functions.

Point of Interest

Although logarithms were originally developed to assist with computations, logarithmic functions have a much broader use today. These functions occur in geology, acoustics, chemistry, and economics, for example.

To graph $f(x) = \log_2 x$, think of the function as the equation $y = \log_2 x$.

$$f(x) = \log_2 x$$
$$y = \log_2 x$$

Write the equivalent exponential equation.

$$x = 2^y$$

Because the equation is solved for x in terms of y, it is easier to choose values of y and find the corresponding values of x. The results can be recorded in a table.

Graph the ordered pairs on a rectangular coordinate system.

Connect the points with a smooth curve.

x	y
$\frac{1}{4}$	-2
$\frac{1}{2}$	-1
1	0
2	1
4	2

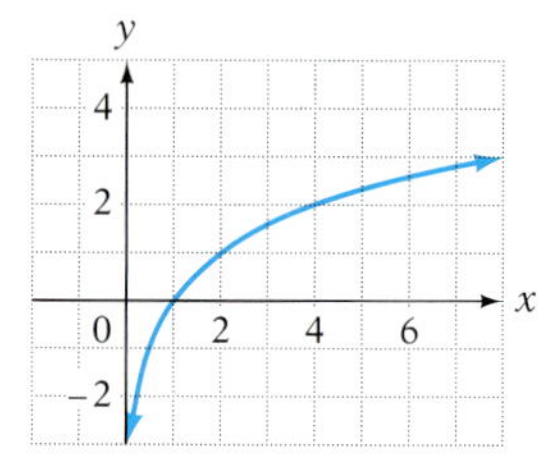

Applying the vertical- and horizontal-line tests reveals that $f(x) = \log_2 x$ is the graph of a 1–1 function.

Recall that the graph of the inverse of a function f is the mirror image of f with respect to the line $y = x$. The graph of $f(x) = 2^x$ was shown earlier. Because $g(x) = \log_2 x$ is the inverse of $f(x) = 2^x$, the graphs of these functions are mirror images of each other with respect to the line $y = x$.

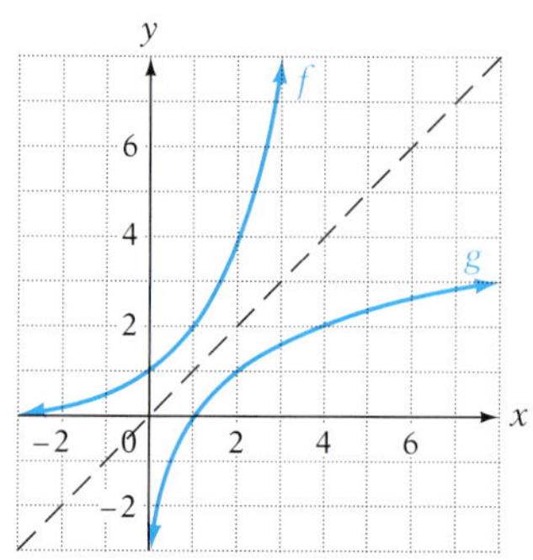

Graph: $f(x) = \log_2 x + 1$

Think of the function as the equation $y = \log_2 x + 1$.

$$f(x) = \log_2 x + 1$$
$$y = \log_2 x + 1$$

Solve the equation for $\log_2 x$.

$$y - 1 = \log_2 x$$

Write the equivalent exponential equation.

$$2^{y-1} = x$$

Choose values of y, and find the corresponding values of x.

Graph the ordered pairs on a rectangular coordinate system.

Connect the points with a smooth curve.

x	y
$\frac{1}{4}$	-1
$\frac{1}{2}$	0
1	1
2	2
4	3

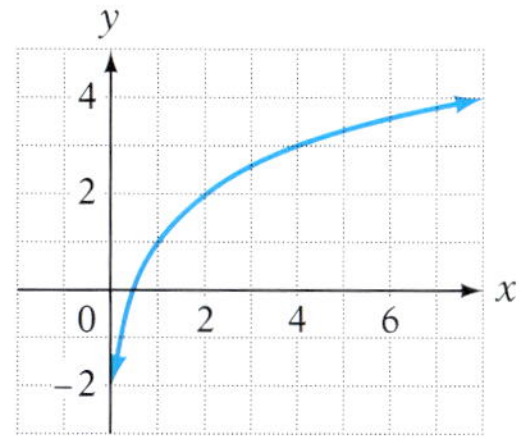

Example 1 Graph. **A.** $f(x) = \log_3 x$ **B.** $f(x) = 2 \log_3 x$

Solution **A.** $f(x) = \log_3 x$

$y = \log_3 x$ • Substitute y for $f(x)$.

$x = 3^y$ • Write the equivalent exponential equation.

x	y
$\frac{1}{9}$	-2
$\frac{1}{3}$	-1
1	0
3	1

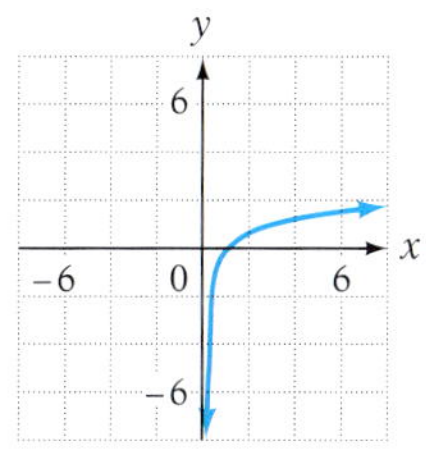

• Choose values of y, and find the corresponding values of x. Graph the ordered pairs on a rectangular coordinate system. Connect the points with a smooth curve.

B. $f(x) = 2 \log_3 x$

$y = 2 \log_3 x$ • Substitute y for $f(x)$.

$\frac{y}{2} = \log_3 x$ • Solve the equation for $\log_3 x$.

$x = 3^{\frac{y}{2}}$ • Write the equivalent exponential equation.

x	y
$\frac{1}{9}$	-4
$\frac{1}{3}$	-2
1	0
3	2

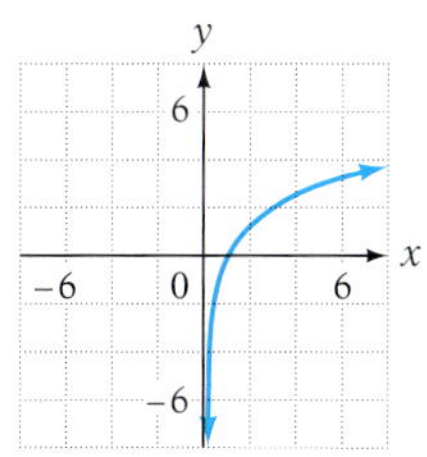

• Choose values of y, and find the corresponding values of x. Graph the ordered pairs on a rectangular coordinate system. Connect the points with a smooth curve.

Problem 1 Graph. **A.** $f(x) = \log_2(x - 1)$ **B.** $f(x) = \log_3 2x$

Solution See page S35.

Example 2 Graph: $f(x) = 2 \ln x + 3$

Solution The graph is shown below. To verify the accuracy of the graph, evaluate $f(x)$ for a few values of x, and compare the results to values found by using the TRACE feature of the graphing utility. For example, $f(1) = 2 \ln(1) + 3 = 3$, so $(1, 3)$ is a point on the graph. When $x = 3$, $f(3) = 2 \ln(3) + 3 \approx 5.2$, so $(3, 5.2)$ is a point on the graph.

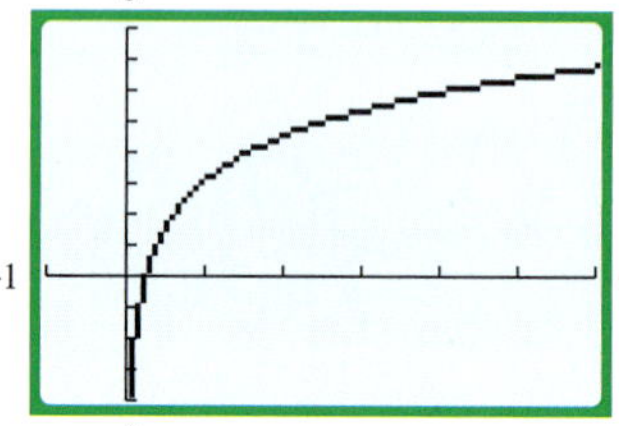

Problem 2 Graph: $f(x) = 10 \log(x - 2)$

Solution See page S35.

The graphs of logarithmic functions to other than base e or base 10 can be drawn with a graphing utility by first using the Change-of-Base Formula $\log_a N = \dfrac{\log_b N}{\log_b a}$ to rewrite the logarithmic function in terms of base e or base 10.

Graph: $f(x) = \log_3 x$

Use the Change-of-Base Formula to rewrite $\log_3 x$ in terms of $\log x$ or $\ln x$. The natural logarithm function $\ln x$ is used here.

$$\log_3 x = \frac{\ln x}{\ln 3}$$

To graph $f(x) = \log_3 x$ using a graphing utility, use the equivalent form $f(x) = \frac{\ln x}{\ln 3}$.

The graph is shown at the right.

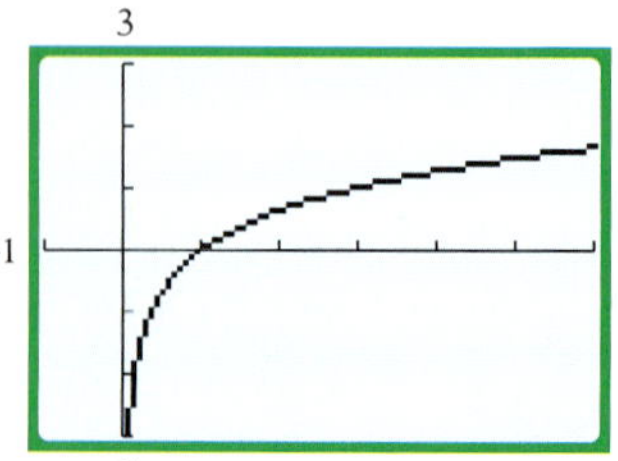

The graph of $f(x) = \log_3 x$ could have been drawn by rewriting $\log_3 x$

in terms of $\log x$, $\log_3 x = \dfrac{\log x}{\log 3}$. The graph of $f(x) = \log_3 x$ is identical to the graph of $f(x) = \dfrac{\log x}{\log 3}$.

The examples that follow were graphed by rewriting the logarithmic function in terms of the natural logarithmic function. The common logarithmic function could also have been used.

Example 3 Graph: $f(x) = -3 \log_2 x$

Solution

$$f(x) = -3 \log_2 x$$
$$= -3\frac{\ln x}{\ln 2} = -\frac{3}{\ln 2} \ln x$$

• Rewrite $\log_2 x$ in terms of $\ln x$. $\log_2 x = \dfrac{\ln x}{\ln 2}$

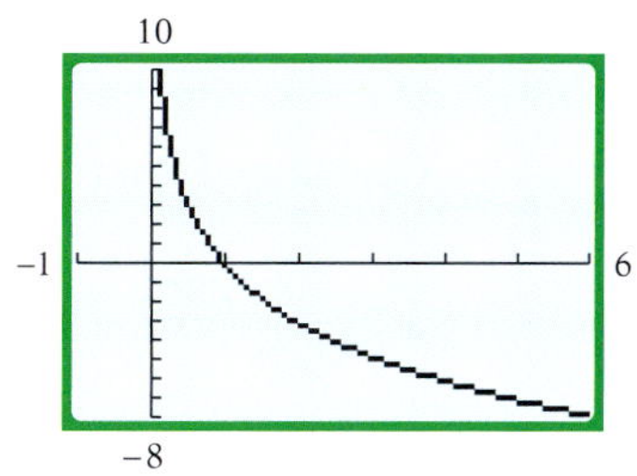

• The graph of $f(x) = -3 \log_2 x$ is the same as the graph of $f(x) = -\dfrac{3}{\ln 2} \ln x$.

Problem 3 Graph: $f(x) = 2 \log_4 x$

Solution See page S35.

11.3 Concept Review

Determine whether the following statements are always true, sometimes true, or never true.

1. The domain of $f(x) = \log_b x$, $b > 0$, $b \neq 1$, is the real numbers.

2. The range of $f(x) = \log_b x$, $b > 0$, $b \neq 1$, is the real numbers.

3. The graphs of $x = 2^y$ and $y = \log_2 x$ are identical graphs.

4. The range of $f(x) = \ln x$ is the real numbers greater than zero.

11.3 Exercises

1 **1.** Is the function $f(x) = \log x$ a 1–1 function? Why or why not?

2. Name two characteristics of the graph of $y = \log_b x$, $b > 1$.

3. What is the relationship between the graph of $x = 3^y$ and that of $y = \log_3 x$?

4. What is the relationship between the graph of $y = 3^x$ and that of $y = \log_3 x$?

Graph.

5. $f(x) = \log_4 x$

6. $f(x) = \log_2(x + 1)$

7. $f(x) = \log_3(2x - 1)$

8. $f(x) = \log_2\left(\frac{1}{2}x\right)$

9. $f(x) = 3 \log_2 x$

10. $f(x) = \frac{1}{2} \log_2 x$

11. $f(x) = -\log_2 x$

12. $f(x) = -\log_3 x$

13. $f(x) = \log_2(x - 1)$

14. $f(x) = \log_3(2 - x)$

15. $f(x) = -\log_2(x - 1)$

16. $f(x) = -\log_2(1 - x)$

Use a graphing utility to graph the following.

17. $f(x) = \log_2 x - 3$

18. $f(x) = \log_3 x + 2$

19. $f(x) = -\log_2 x + 2$

20. $f(x) = -\frac{1}{2}\log_2 x - 1$

21. $f(x) = x - \log_2(1 - x)$

22. $f(x) = x + \log_3(2 - x)$

Graph the functions on the same rectangular coordinate system.

23. $f(x) = 4^x$; $g(x) = \log_4 x$

24. $f(x) = 3^x$; $g(x) = \log_3 x$

25. $f(x) = \left(\frac{1}{2}\right)^x$; $g(x) = \log_{\frac{1}{2}} x$

26. $f(x) = 10^x$; $g(x) = \log_{10} x$

11.3 Applying Concepts

27. Employment The proficiency of a typist decreases (without practice) over time. An equation that approximates this decrease is given by $S = 60 - 7\ln(t + 1)$, where S is the typing speed in words per minute and t is the number of months without typing.

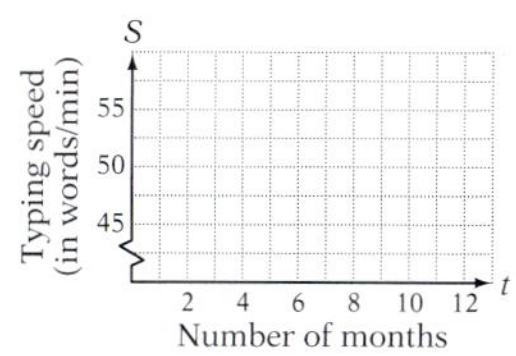

a. Graph the equation.

b. The point whose approximate coordinates are (4, 49) is on the graph. Write a sentence that describes the meaning of this ordered pair.

28. Astronomy Astronomers use the *distance modulus* of a star as a method of determining the star's distance from Earth. The formula is $M = 5 \log s - 5$, where M is the distance modulus and s is the star's distance from Earth in parsecs. (One parsec $\approx 2.1 \times 10^{13}$ miles.)

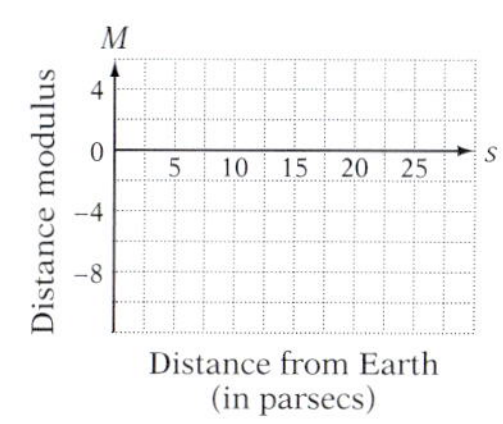

a. Graph the equation.

b. The point whose approximate coordinates are (25.1, 2) is on the graph. Write a sentence that describes the meaning of this ordered pair.

29. **Interest Rates** General interest rate theory suggests that short-term interest rates (less than 2 years) are lower than long-term interest rates (more than 10 years) because short-term securities are less risky than long-term ones. In periods of high inflation, however, the situation is reversed and economists discuss *inverted yield* curves. During the early 1980s, inflation was very high in the United States. The rates for short- and long-term U.S. Treasury securities during 1980 are shown in the table at the right. An equation that models these data is given by $y = 14.33759 - 0.62561 \ln x$, where x is the term of the security in years and y is the interest rate as a percent.

Term (in years)	*Interest Rate*
0.5	15.0%
1	14.0%
5	13.5%
10	12.8%
20	12.5%

a. Graph the equation.

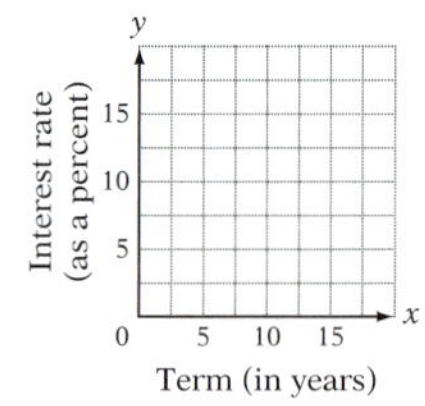

b. According to this model, what is the term, to the nearest tenth of a year, of a security that has a yield of 13%?

c. Determine the interest rate, to the nearest tenth of a percent, that this model predicts for a security that has a 30-year maturity.

Determine whether the ordered pairs in the table belong to a linear function, a quadratic function, an exponential function, or a logarithmic function.

30.

x	y
0	5
1	3
2	1
3	−1
4	−3

31.

x	y
−1	6
0	1
1	2
2	9
3	22

32.

x	y
0	1
1	4
2	16
3	64
4	256

33.

x	y
0	−3
2	−2
4	−1
6	0
8	1

34.

x	y
0	−3
1	−7
2	−9
3	−9
4	−7

35.

x	y
0	160
1	80
2	40
3	20
4	10

36.

x	y
$\frac{1}{2}$	−1
1	0
2	1
4	2
8	3

37.

x	y
$\frac{1}{4}$	−2
1	0
4	2
16	4
256	8

38. Using the Power Property of Logarithms, $\ln x^2 = 2 \ln x$. Graph the equations $f(x) = \ln x^2$ and $g(x) = 2 \ln x$ on the same rectangular coordinate system. Are the graphs the same? Why or why not?

39. Because $f(x) = e^x$ and $g(x) = \ln x$ are inverse functions of each other, $f[g(x)] = x$ and $g[f(x)] = x$. Graph $f[g(x)] = e^{\ln x}$ and $g[f(x)] = \ln e^x$. Explain why the graphs are different even though $f[g(x)] = g[f(x)]$.

11.4 Exponential and Logarithmic Equations

1 Solve exponential equations

VIDEO & DVD CD TUTOR WEB SSM

An **exponential equation** is one in which a variable occurs in the exponent. The examples at the right are exponential equations.

$$6^{2x+1} = 6^{3x-2}$$
$$4^x = 3$$
$$2^{x+1} = 7$$

An exponential equation in which each side of the equation can be expressed in terms of the same base can be solved by using the Equality of Exponents Property. Recall that the Equality of Exponents Property states that

$$\text{If } b^u = b^v, \text{ then } u = v.$$

In Example 1 below, this property is used to solve an exponential equation.

Example 1 Solve and check: $9^{x+1} = 27^{x-1}$

Solution

$9^{x+1} = 27^{x-1}$

$(3^2)^{x+1} = (3^3)^{x-1}$ • Rewrite each side of the equation using the same base.

$3^{2x+2} = 3^{3x-3}$

$2x + 2 = 3x - 3$ • Use the Equality of Exponents Property to equate the exponents.

$2 = x - 3$ • Solve the resulting equation.

$5 = x$

Check:

9^{x+1}	$= 27^{x-1}$
9^{5+1}	27^{5-1}
9^6	27^4
$(3^2)^6$	$(3^3)^4$
3^{12}	$= 3^{12}$

The solution is 5.

Problem 1 Solve and check: $10^{3x+5} = 10^{x-3}$

Solution See page S35.

When both sides of an exponential equation cannot easily be expressed in terms of the same base, logarithms are used to solve the exponential equation.

Example 2 Solve for x. Round to the nearest ten-thousandth.

A. $4^x = 7$ **B.** $3^{2x} = 4$

Solution **A.**

$4^x = 7$

$\log 4^x = \log 7$ • Take the common logarithm of each side of the equation.

$x \log 4 = \log 7$ • Rewrite using the properties of logarithms.

$x = \dfrac{\log 7}{\log 4}$ • Solve for x.

$x \approx 1.4037$

The solution is 1.4037.

Take Note

To evaluate $\dfrac{\log 7}{\log 4}$ on a scientific calculator, use the keystrokes

7 [log] [÷] 4 [log] [=]

The display should read 1.4036775.

B.

$$3^{2x} = 4$$

$$\log 3^{2x} = \log 4$$ • Take the common logarithm of each side of the equation.

$$2x \log 3 = \log 4$$ • Rewrite using the properties of logarithms.

$$2x = \frac{\log 4}{\log 3}$$ • Solve for x.

$$x = \frac{\log 4}{2 \log 3}$$

$$x \approx 0.6309$$

The solution is 0.6309.

Problem 2 Solve for x. Round to the nearest ten-thousandth.

A. $4^{3x} = 25$ **B.** $(1.06)^x = 1.5$

Solution See page S36.

The equations in Example 2 can be solved by graphing. For Example 2A, by subtracting 7 from each side of the equation $4^x = 7$, the equation can be written as $4^x - 7 = 0$. The graph of $f(x) = 4^x - 7$ is shown below. The values of x for which $f(x) = 0$ are the solutions of the equation $4^x = 7$. These values of x are the zeros of the function. By using the features of a graphing utility, it is possible to determine a very accurate solution. The solution to the nearest tenth is shown in the graph.

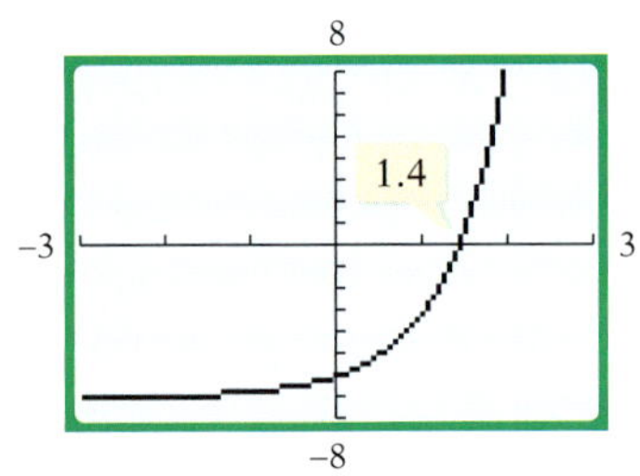

Example 3 Solve $e^x = 2x + 1$ for x. Round to the nearest hundredth.

Solution Rewrite the equation by subtracting $2x + 1$ from each side and writing the equation as $e^x - 2x - 1 = 0$. The zeros of $f(x) = e^x - 2x - 1$ are the solutions of $e^x = 2x + 1$. Graph f and use the features of a graphing utility to estimate the solutions to the nearest hundredth.

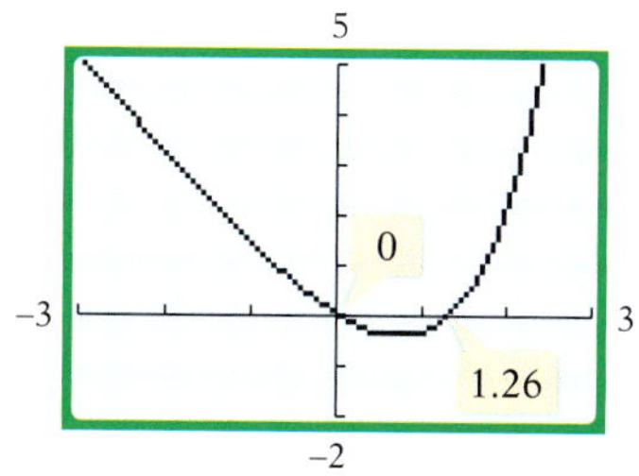

The solutions are 0 and 1.26.

Problem 3 Solve $e^x = x$ for x. Round to the nearest hundredth.

Solution See page S36.

2 Solve logarithmic equations

VIDEO & DVD CD TUTOR WEB SSM

A logarithmic equation can be solved by using the properties of logarithms. Here is an example.

Solve: $\log_9 x + \log_9(x - 8) = 1$

Use the Product Property of Logarithms to rewrite the left side of the equation.

$$\log_9 x + \log_9(x - 8) = 1$$
$$\log_9 x(x - 8) = 1$$

Write the equation in exponential form.

$$9^1 = x(x - 8)$$

Simplify and solve for x.

$$9 = x^2 - 8x$$
$$0 = x^2 - 8x - 9$$
$$0 = (x - 9)(x + 1)$$

$$x - 9 = 0 \qquad x + 1 = 0$$
$$x = 9 \qquad x = -1$$

When x is replaced by 9 in the original equation, 9 checks as a solution. When x is replaced by -1, the original equation contains the expression $\log_9(-1)$. Because the logarithm of a negative number is not a real number, -1 does not check as a solution.

The solution of the equation is 9.

Some logarithmic equations can be solved by using the 1–1 Property of Logarithms, which states that for $x > 0$, $y > 0$, $b > 0$, $b \neq 1$,

$$\text{If } \log_b x = \log_b y, \text{ then } x = y.$$

This property is used in Example 4B.

Example 4 Solve for x.

A. $\log_3(2x - 1) = 2$

B. $\log_2 x - \log_2(x - 1) = \log_2 2$

Solution **A.** $\log_3(2x - 1) = 2$

$$3^2 = 2x - 1 \quad \text{• Rewrite in exponential form.}$$
$$9 = 2x - 1 \quad \text{• Solve for } x.$$
$$10 = 2x$$
$$5 = x$$

The solution is 5.

B. $\log_2 x - \log_2(x - 1) = \log_2 2$

$$\log_2\left(\frac{x}{x-1}\right) = \log_2 2 \quad \text{• Use the Quotient Property of Logarithms.}$$
$$\frac{x}{x-1} = 2 \quad \text{• Use the 1–1 Property of Logarithms.}$$
$$(x-1)\left(\frac{x}{x-1}\right) = (x-1)2 \quad \text{• Solve for } x.$$
$$x = 2x - 2$$
$$-x = -2$$
$$x = 2$$

The solution is 2.

Problem 4 Solve for x.

A. $\log_4(x^2 - 3x) = 1$

B. $\log_3 x + \log_3(x + 3) = \log_3 4$

Solution See page S36.

Some logarithmic equations cannot be solved algebraically. In these cases, a graphical approach may be appropriate.

Example 5 Solve $\ln(2x + 4) = x^2$ for x. Round to the nearest hundredth.

Solution Rewrite the equation by subtracting x^2 from each side and writing the equation as $\ln(2x + 4) - x^2 = 0$. The zeros of the function defined by $f(x) = \ln(2x + 4) - x^2$ are the solutions of the equation. Graph f and then use the features of a graphing utility to estimate the solutions to the nearest hundredth.

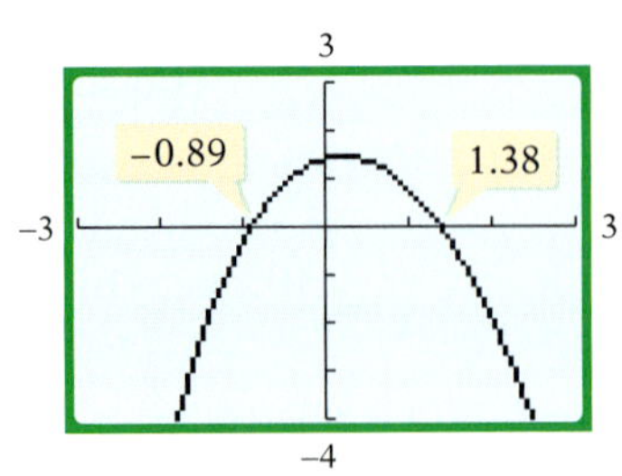

The solutions are −0.89 and 1.38.

Problem 5 Solve $\log(3x - 2) = -2x$ for x. Round to the nearest hundredth.

Solution See page S36.

Recall the equation $\log_9 x + \log_9(x - 8) = 1$ from the beginning of this objective. The algebraic solution showed that although -1 and 9 may be solutions, only 9 satisfies the equation. The extraneous solution was introduced at the second step. The Product Property of Logarithms, $\log_b(xy) = \log_b x + \log_b y$, applies only when both x and y are positive numbers. This occurs when $x > 8$. Therefore, a solution to this equation must be greater than 8. The graphs of $f(x) = \log_9 x + \log_9(x - 8) - 1$ and $g(x) = \log_9 x(x - 8) - 1$ are shown below.

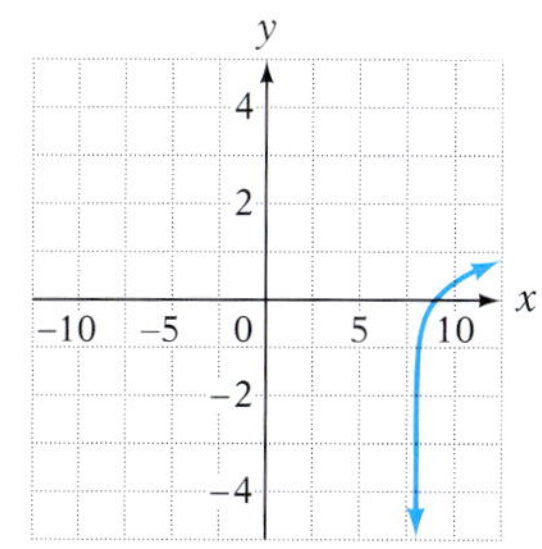

$f(x) = \log_9 x + \log_9(x - 8) - 1$

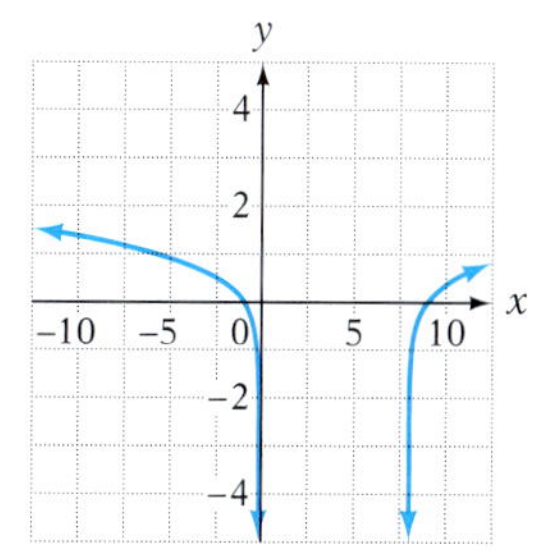

$g(x) = \log_9 x(x - 8) - 1$

Note that the only zero of f is 9, whereas the zeros of g are -1 and 9.

11.4 Concept Review

Determine whether the following are always true, sometimes true, or never true.

1. Suppose $b > 0$, $b \neq 1$. If $b^x = b^y$, then $x = y$.

2. $\log x + \log(x + 2) = \log(2x + 2)$

3. $\log_2(-4) = -2$

4. $\log(2x - 2) - \log x = 4$ is equivalent to $\log \dfrac{2x - 2}{x} = 4$.

5. $\log(2x) + \log 4 = 6$ is equivalent to $\log(8x) = 6$.

6. $\log_{10} 10^4 = 4$

11.4 Exercises

1 **1.** What is an exponential equation?

2. **a.** What does the Equality of Exponents Property state?
b. Provide an example of when you would use this property.

Solve for x. Round to the nearest ten-thousandth.

3. $5^{4x-1} = 5^{x+2}$
4. $7^{4x-3} = 7^{2x+1}$
5. $8^{x-4} = 8^{5x+8}$

6. $10^{4x-5} = 10^{x+4}$
7. $5^x = 6$
8. $7^x = 10$

9. $12^x = 6$
10. $10^x = 5$
11. $\left(\frac{1}{2}\right)^x = 3$

12. $\left(\frac{1}{3}\right)^x = 2$
13. $(1.5)^x = 2$
14. $(2.7)^x = 3$

15. $10^x = 21$
16. $10^x = 37$
17. $2^{-x} = 7$

18. $3^{-x} = 14$
19. $2^{x-1} = 6$
20. $4^{x+1} = 9$

21. $3^{2x-1} = 4$
22. $4^{-x+2} = 12$
23. $9^x = 3^{x+1}$

24. $2^{x-1} = 4^x$
25. $8^{x+2} = 16^x$
26. $9^{3x} = 81^{x-4}$

27. $5^{x^2} = 21$
28. $3^{x^2} = 40$
29. $2^{4x-2} = 20$

30. $4^{3x+8} = 12$
31. $3^{-x+2} = 18$
32. $5^{-x+1} = 15$

33. $4^{2x} = 100$
34. $3^{3x} = 1000$
35. $2.5^{-x} = 4$

36. $3.25^{x+1} = 4.2$
37. $0.25^x = 0.125$
38. $0.1^{5x} = 10^{-2}$

Solve for x by graphing. Round to the nearest hundredth.

39. $3^x = 2$
40. $5^x = 9$
41. $2^x = 2x + 4$

42. $3^x = -x - 1$
43. $e^x = -2x - 2$
44. $e^x = 3x + 4$

2 **45.** What is a logarithmic equation?

46. What does the 1–1 Property of Logarithms state?

Solve for x.

47. $\log_3(x + 1) = 2$
48. $\log_5(x - 1) = 1$
49. $\log_2(2x - 3) = 3$

50. $\log_4(3x + 1) = 2$
51. $\log_2(x^2 + 2x) = 3$
52. $\log_3(x^2 + 6x) = 3$

53. $\log_5\left(\frac{2x}{x-1}\right) = 1$
54. $\log_6 \frac{3x}{x+1} = 1$
55. $\log_7 x = \log_7(1 - x)$

56. $\frac{3}{4} \log x = 3$
57. $\frac{2}{3} \log x = 6$
58. $\log(x - 2) - \log x = 3$

59. $\log_2(x - 3) + \log_2(x + 4) = 3$

60. $\log x - 2 = \log(x - 4)$

61. $\log_3 x + \log_3(x - 1) = \log_3 6$

62. $\log_4 x + \log_4(x - 2) = \log_4 15$

63. $\log_2(8x) - \log_2(x^2 - 1) = \log_2 3$

64. $\log_5(3x) - \log_5(x^2 - 1) = \log_5 2$

65. $\log_9 x + \log_9(2x - 3) = \log_9 2$

66. $\log_6 x + \log_6(3x - 5) = \log_6 2$

67. $\log_8(6x) = \log_8 2 + \log_8(x - 4)$

68. $\log_7(5x) = \log_7 3 + \log_7(2x + 1)$

69. $\log_9(7x) = \log_9 2 + \log_9(x^2 - 2)$

70. $\log_3 x = \log_3 2 + \log_3(x^2 - 3)$

71. $\log(x^2 + 3) - \log(x + 1) = \log 5$

72. $\log(x + 3) + \log(2x - 4) = \log 3$

Solve for x by graphing. Round to the nearest hundredth.

73. $\log x = -x + 2$

74. $\log x = -2x$

75. $\log(2x - 1) = -x + 3$

76. $\log(x + 4) = -2x + 1$

77. $\ln(x + 2) = x^2 - 3$

78. $\ln x = -x^2 + 1$

11.4 Applying Concepts

Solve for x. Round to the nearest ten-thousandth.

79. $8^{\frac{x}{2}} = 6$

80. $4^{\frac{x}{3}} = 2$

81. $5^{\frac{3x}{2}} = 7$

82. $9^{\frac{2x}{3}} = 8$

83. $1.2^{\frac{x}{2} - 1} = 1.4$

84. $5.6^{\frac{x}{3} + 1} = 7.8$

Solve.

85. If $4^x = 7$, find the value of $2^{(6x + 3)}$.

86. **Physics** A model for the distance s (in feet) that an object experiencing air resistance will fall in t seconds is given by $s = 78 \ln\left(\frac{e^{0.8t} + e^{-0.8t}}{2}\right)$.

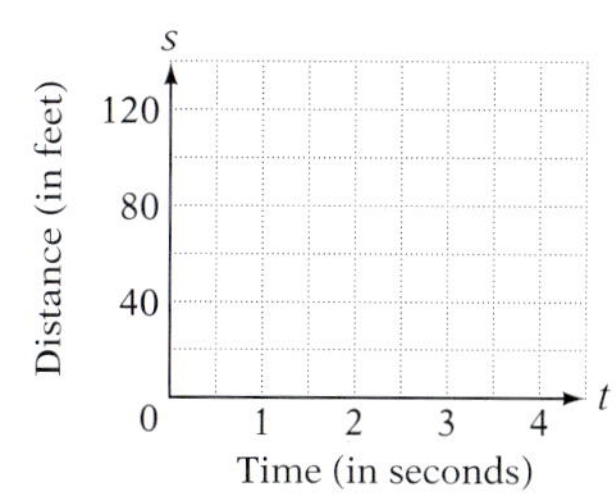

a. Graph this equation. *Suggestion:* Use Xmin = 0, Xmax = 4.5, Ymin = 0, Ymax = 140, and Yscl = 20.

b. Determine, to the nearest hundredth of a second, the time it takes the object to travel 125 ft.

87. **Physics** A model for the distance s (in feet) an object experiencing air resistance will fall in t seconds is given by $s = 312.5 \ln\left(\frac{e^{0.32t} + e^{-0.32t}}{2}\right)$.

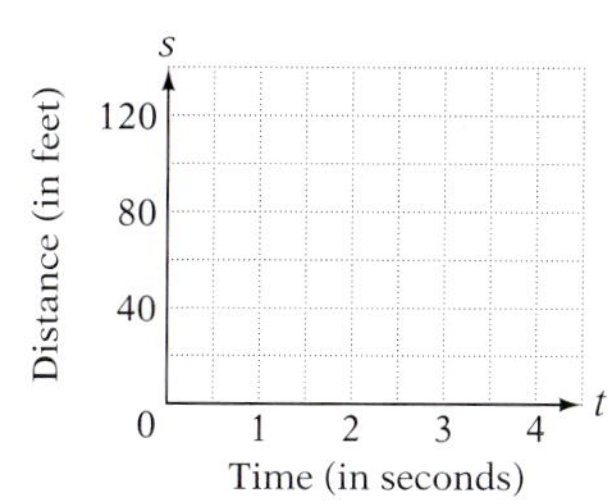

a. Graph this equation. *Suggestion:* Use Xmin = −0.5, Xmax = 5, Xscl = 0.5, Ymin = −0.5, Ymax = 150, and Yscl = 10.

b. Determine, to the nearest hundredth of a second, the time it takes the object to travel 100 ft.

88. The following "proof" appears to show that $0.04 < 0.008$. Explain the error.

$$\begin{aligned} 2 &< 3 \\ 2 \log 0.2 &< 3 \log 0.2 \\ \log(0.2)^2 &< \log(0.2)^3 \\ (0.2)^2 &< (0.2)^3 \\ 0.04 &< 0.008 \end{aligned}$$

11.5 Applications of Exponential and Logarithmic Functions

1 Application problems

VIDEO & DVD CD TUTOR WEB SSM

A biologist places one single-celled bacterium in a culture, and each hour that particular species of bacterium divides into two bacteria. After one hour there will be two bacteria. After two hours, each of the two bacteria will divide and there will be four bacteria. After three hours, each of the four bacteria will divide and there will be eight bacteria.

Time, t	*Number of Bacteria, N*
0	1
1	2
2	4
3	8
4	16

The table at the left shows the number of bacteria in the culture after various intervals of time, t, in hours. Values in this table could also be found by using the exponential equation $N = 2^t$.

The equation $N = 2^t$ is an example of an **exponential growth equation.** In general, any equation that can be written in the form $\boldsymbol{A = A_0 b^{kt}}$, where A is the size at time t, A_0 is the initial size, $b > 1$, and k is a positive real number, is an exponential growth equation. These equations are important not only in population growth studies but also in physics, chemistry, psychology, and economics.

Recall that interest is the amount of money one pays (or receives) when borrowing (or investing) money. **Compound interest** is interest that is computed not only on the original principal but also on the interest already earned. The compound interest formula is an exponential equation.

The **compound interest formula** is $\boldsymbol{P = A(1 + i)^n}$, where A is the original value of an investment, i is the interest rate per compounding period, n is the total number of compounding periods, and P is the value of the investment after n periods.

An investment broker deposits $1000 into an account that earns 12% annual interest compounded quarterly. What is the value of the investment after 2 years? Round to the nearest dollar.

Find i, the interest rate per quarter. The quarterly rate is the annual rate divided by 4, the number of quarters in 1 year.

$$i = \frac{12\%}{4} = \frac{0.12}{4} = 0.03$$

Find n, the number of compounding periods. The investment is compounded quarterly, 4 times a year, for 2 years.

$$n = 4 \cdot 2 = 8$$

Use the compound interest formula.	$P = A(1 + i)^n$
Replace A, i, and n by their values.	$P = 1000(1 + 0.03)^8$
Solve for P.	$P \approx 1267$

The value of the investment after 2 years is \$1267.

Exponential decay is another important example of an exponential equation. One of the most common illustrations of exponential decay is the decay of a radioactive substance.

A radioactive isotope of cobalt has a half-life of approximately 5 years. This means that one-half of any given amount of the cobalt isotope will disintegrate in 5 years.

Time, t	*Amount, A*
0	10
5	5
10	2.5
15	1.25
20	0.625

The table at the left indicates the amount of the initial 10 mg of a cobalt isotope that remains after various intervals of time, t, in years. Values in this table could also be found by using the exponential equation $A = 10\left(\frac{1}{2}\right)^{\frac{t}{5}}$.

The equation $A = 10\left(\frac{1}{2}\right)^{\frac{t}{5}}$ is an example of an **exponential decay equation.**

Compare this equation to the exponential growth equation, and note that for exponential growth, the base of the exponential equation is greater than 1, whereas for exponential decay, the base is between 0 and 1.

A method by which an archeologist can measure the age of a bone is called **carbon dating.** Carbon dating is based on a radioactive isotope of carbon called carbon-14, which has a half-life of approximately 5570 years. The exponential decay equation is given by $A = A_0\left(\frac{1}{2}\right)^{\frac{t}{5570}}$, where A_0 is the original amount of carbon-14 present in the bone, t is the age of the bone, and A is the amount present after t years.

A bone that originally contained 100 mg of carbon-14 now has 70 mg of carbon-14. What is the approximate age of the bone? Round to the nearest year.

Use the exponential decay equation.

$$A = A_0\left(\frac{1}{2}\right)^{\frac{t}{5570}}$$

Replace A_0 and A by their given values, and solve for t.

$$70 = 100\left(\frac{1}{2}\right)^{\frac{t}{5570}}$$

$$70 = 100(0.5)^{\frac{t}{5570}}$$

Divide each side of the equation by 100.

$$0.7 = (0.5)^{\frac{t}{5570}}$$

Take the common logarithm of each side of the equation. Then simplify.

$$\log 0.7 = \log(0.5)^{\frac{t}{5570}}$$

$$\log 0.7 = \frac{t}{5570}\log 0.5$$

$$\frac{5570 \log 0.7}{\log 0.5} = t$$

$$2866 \approx t$$

The age of the bone is approximately 2866 years.

Example 1 The number of words per minute a student can type will increase with practice and can be approximated by the equation $N = 100[1 - (0.9)^t]$, where N is the number of words typed per minute after t days of instruction. In how many days will the student be able to type 60 words per minute?

Strategy To find the number of days, replace N by 60 and solve for t.

Solution

$$N = 100[1 - (0.9)^t]$$
$$60 = 100[1 - (0.9)^t]$$
$$0.6 = 1 - (0.9)^t$$

• **Divide each side of the equation by 100.**

$$-0.4 = -(0.9)^t$$

• **Subtract 1 from each side of the equation.**

$$0.4 = (0.9)^t$$
$$\log 0.4 = \log(0.9)^t$$
$$\log 0.4 = t \log 0.9$$
$$t = \frac{\log 0.4}{\log 0.9} \approx 8.6967184$$

After approximately 9 days the student will type 60 words per minute.

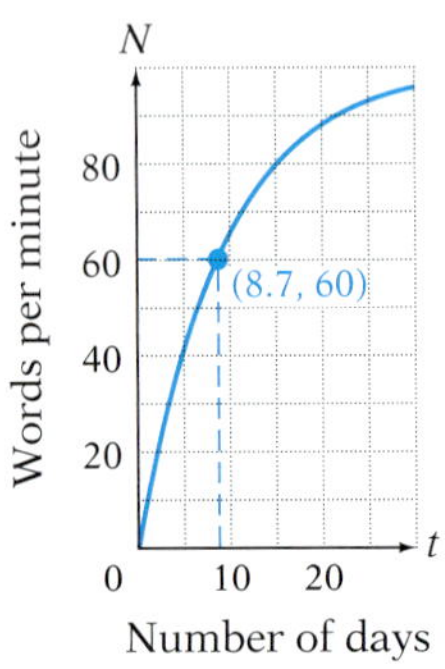

The graph at the left is the graph of $N = 100[1 - (0.9)^t]$. Note that t is approximately 8.7 when $N = 60$.

Problem 1 An investment of \$3000 is placed into an account that earns 12% annual interest compounded monthly. In approximately how many years will the investment be worth twice the original amount? Round to the nearest whole number.

Solution See page S36.

The first applications of logarithms (and the main reason why they were developed) were to reduce computational drudgery. Today, with the widespread use of calculators and computers, the computational uses of logarithms have diminished. However, a number of other applications of logarithms have emerged.

Point of Interest

The pH scale was created by the Danish biochemist Søren Sørensen in 1909 to measure the acidity of water used in the brewing of beer. pH is an abbreviation for pondus hydrogenii, *which translates as "potential hydrogen."*

A chemist measures the acidity or alkalinity of a solution by the formula $\mathbf{pH = -\log(H^+)}$, where H^+ is the concentration of hydrogen ions in the solution. A neutral solution such as distilled water has a pH of 7, acids have a pH less than 7, and alkaline solutions (also called basic solutions) have a pH greater than 7.

Find the pH of vinegar for which $H^+ = 1.26 \times 10^{-3}$. Round to the nearest tenth.

Use the pH equation. $H^+ = 1.26 \times 10^{-3}$.

$$\begin{aligned} \text{pH} &= -\log(H^+) \\ &= -\log(1.26 \times 10^{-3}) \\ &= -(\log 1.26 + \log 10^{-3}) \\ &\approx -[0.1004 + (-3)] = 2.8996 \end{aligned}$$

The pH of vinegar is approximately 2.9.

Logarithmic functions are used to scale very large or very small numbers into numbers that are easier to comprehend. For instance, the *Richter scale* magnitude of an earthquake uses a logarithmic function to convert the intensity of shock waves I into a number M, which for most earthquakes is in the range of 0 to 10. The intensity I of an earthquake is often given in terms of the constant I_0, where I_0 is the intensity of the smallest earthquake, called a **zero-level earthquake,** that can be measured on a seismograph near the earthquake's epicenter. An earthquake with an intensity I_0 has a Richter scale magnitude of $M = \log \frac{I}{I_0}$, where I_0 is the measure of a zero-level earthquake.

Find the Richter scale magnitude of the 1999 Joshua Tree, California, earthquake that had an intensity of I of $12{,}589{,}254I_0$. Round to the nearest tenth.

	$M = \log \frac{I}{I_0}$
$I = 12{,}589{,}254I_0$	$M = \log \frac{12{,}589{,}254I_0}{I_0}$
Divide the numerator and denominator by I_0.	$M = \log 12{,}589{,}254$
Evaluate log 12,589,254.	$M = 7.1$

The 1999 Joshua Tree earthquake had a Richter scale magnitude of 7.1.

Take Note

Note that we do not need to know the value of I_0 in order to determine the Richter scale magnitude of the quake.

If you know the Richter scale magnitude of an earthquake, then you can also determine the intensity of the earthquake.

Find the intensity of the 1999 Taiwan earthquake that measured 7.6 on the Richter scale. Write the answer in terms of I_0.

Replace M in $M = \log \frac{I}{I_0}$ by 7.6.	$7.6 = \log \frac{I}{I_0}$
Write in exponential form.	$10^{7.6} = \frac{I}{I_0}$
Multiply both sides by I_0.	$10^{7.6}I_0 = I$
Evaluate $10^{7.6}$.	$39{,}810{,}717I_0 = I$

The 1999 Taiwan earthquake had an intensity that was approximately 39,811,000 times the intensity of a zero-level earthquake.

Point of Interest

The Richter scale was created by seismologist Charles F. Richter in 1935. Note that a tenfold increase in the intensity level of an earthquake increases the Richter scale magnitude of the earthquake by only 1.

The percent of light that will pass through a substance is given by the equation $\log P = -kd$, where P is the percent of light passing through the substance, k is a constant that depends on the substance, and d is the thickness of the substance in centimeters.

Find the percent of light that will pass through translucent glass for which $k = 0.4$ and d is 0.5 cm.

$$\log P = -kd$$

Replace k and d in the equation by their given values, and solve for P.

$$\log P = -(0.4)(0.5)$$
$$\log P = -0.2$$

Use the relationship between the logarithmic and exponential functions.

$$P = 10^{-0.2}$$
$$P \approx 0.6310$$

Approximately 63.1% of the light will pass through the glass.

Example 2 Find the hydrogen ion concentration, H^+, of orange juice that has a pH of 3.6.

Strategy To find the hydrogen ion concentration, replace pH by 3.6 in the equation $pH = -\log(H^+)$ and solve for H^+.

Solution

$$pH = -\log(H^+)$$
$$3.6 = -\log(H^+)$$
$$-3.6 = \log(H^+)$$
$$10^{-3.6} = H^+$$
$$0.00025 \approx H^+$$

The hydrogen ion concentration is approximately 0.00025.

Problem 2 On September 3, 2000, an earthquake measuring 5.2 struck the Napa Valley, 50 mi north of San Francisco. Find the intensity of the quake in terms of I_0.

Solution See page S36.

11.5 Concept Review

Determine whether the following statements are always true, sometimes true, or never true.

1. In the exponential growth equation $A = A_0 b^{kt}$, A_0 is the initial size of the growth material.

2. An investor deposits $1000 into an account that earns 8% compounded quarterly. The investor receives interest of $20 every quarter.

3. The equation $A = A_0 e^{kt}$ is an exponential growth equation. An equivalent equation to this exponential growth equation is $t = \frac{1}{k} \ln \frac{A}{A_0}$.

4. The hydrogen ion concentration of a solution is given by the equation $pH = -\log(H^+)$. Solving this equation for H^+ gives the equivalent equation $H^+ = 10^{-pH}$.

11.5 Exercises

1 **1.** What is meant by the phrase *exponential decay*?

2. Explain how compound interest differs from simple interest. Is compound interest an example of exponential decay or exponential growth?

For Exercises 3 to 6, use the compound interest formula $P = A(1 + i)^n$, where A is the original value of an investment, i is the interest rate per compounding period, n is the total number of compounding periods, and P is the value of the investment after n periods.

3. **Finance** An investment broker deposits \$1000 into an account that earns 8% annual interest compounded quarterly. What is the value of the investment after 2 years? Round to the nearest dollar.

4. **Finance** A financial advisor recommends that a client deposit \$2500 into a fund that earns 7.5% annual interest compounded monthly. What will be the value of the investment after 3 years? Round to the nearest cent.

5. **Finance** To save for college tuition, the parents of a preschooler invest \$5000 in a bond fund that earns 6% annual interest compounded monthly. In approximately how many years will the investment be worth \$15,000? Round to the nearest whole number.

6. **Finance** A hospital administrator deposits \$10,000 into an account that earns 9% annual interest compounded monthly. In approximately how many years will the investment be worth \$15,000? Round to the nearest whole number.

For Exercises 7 to 10, use the exponential decay equation $A = A_0\left(\frac{1}{2}\right)^{t/k}$, where A is the amount of a radioactive material present after time t, k is the half-life of the radioactive material, and A_0 is the original amount of radioactive substance. Round to the nearest tenth.

7. **Biology** An isotope of technetium is used to prepare images of internal body organs. This isotope has a half-life of approximately 6 h. A patient is injected with 30 mg of this isotope.
a. What will be the amount of technetium in the patient after 3 h?
b. How long (in hours) will it take for the amount of technetium in the patient to reach 20 mg?

8. **Biology** Iodine-131 is an isotope that is used to study the functioning of the thyroid gland. This isotope has a half-life of approximately 8 days. A patient is given an injection that contains 8 micrograms of iodine-131.
a. What will be the amount of iodine in the patient after 5 days?
b. How long (in days) will it take for the amount of iodine in the patient to reach 5 micrograms?

9. **Biology** A sample of promethium-147 (used in some luminous paints) contains 25 mg. One year later, the sample contains 18.95 mg. What is the half-life of promethium-147, in years?

10. **Biology** Francium-223 is a very rare radioactive isotope discovered in 1939 by Marguerite Percy. A 3-microgram sample of francium-223 decays to 2.54 micrograms in 5 min. What is the half-life of francium-223, in minutes?

11. **Geology** Earth's atmospheric pressure changes as you rise above the surface. At an altitude of h kilometers, where $0 < h < 80$, the pressure P in newtons per square centimeter is approximately modeled by the equation $P(h) = 10.13e^{-0.116h}$.

a. What is the approximate pressure at 40 km above Earth?

b. What is the approximate pressure on Earth's surface?

c. Does atmospheric pressure increase or decrease as you rise above Earth's surface?

12. **Demography** The U.S. Bureau of the Census provides information on the various segments of the population in the United States. The following table gives the number of people, in millions, aged 80 and older at the beginning of each decade from 1910 to 2000. An equation that approximately models the data is $y = 0.18808(1.0365)^x$, where $x = 10$ for 1910 and y is the population, in millions, of people aged 80 and over.

Year	*80-year-olds (in millions)*
1910	0.3
1920	0.4
1930	0.5
1940	0.8
1950	1.1
1960	1.6
1970	2.3
1980	2.9
1990	3.9
2000	9.3

a. According to the model, what is the predicted population of this age group in the year 2020? Round to the nearest tenth of a million. (*Hint:* You will need to determine what the x-value is when the year is 2020.)

b. In what year does this model predict that the population of this age group will be 15 million? Round to the nearest year.

For Exercises 13 and 14, use the equation $\text{pH} = -\log(\text{H}^+)$, where H^+ is the hydrogen ion concentration of a solution. Round to the nearest tenth.

13. **Chemistry** Find the pH of milk, for which the hydrogen ion concentration is 3.97×10^{-7}.

14. **Chemistry** Find the pH of a baking soda solution, for which the hydrogen ion concentration is 3.98×10^{-9}.

For Exercises 15 and 16, use the equation $\log P = -kd$, which gives the relationship between the percent P, as a decimal, of light passing through a substance of thickness d.

15. **Optics** The value of k for a swimming pool is approximately 0.05. At what depth, in meters, will the percent of light be 75% of the light at the surface of the pool? Round to the nearest tenth.

16. **Optics** The constant k for a piece of blue stained glass is 20. What percent of light will pass through a piece of this glass that is 0.005 m thick? Round to the nearest tenth of a percent.

The number of decibels, D, of a sound can be given by the equation $D = 10(\log I + 16)$, where I is the power of a sound measured in watts. Use this equation for Exercises 17 and 18. Round to the nearest whole number.

17. **Acoustics** Find the number of decibels of normal conversation. The power of the sound of normal conversation is approximately 3.2×10^{-10} watts.

18. **Acoustics** The loudest sound made by any animal is made by the blue whale and can be heard over 500 mi away. The power of the sound is 630 watts. Find the number of decibels of sound emitted by the blue whale.

19. **Sports** During the 1980s and 1990s, the average time T of a major-league baseball game increased each year. If the year 1981 is represented by $x = 1$, then the function $T(x) = 149.57 + 7.64 \ln x$ approximates the time T in minutes of a major-league baseball game for the years from $x = 1$ to $x = 19$. By how many minutes did the average time of a major-league baseball game increase during the years from 1981 to 1999? Round to the nearest tenth.

20. **Inflation** In 1962, the cost of a first-class stamp was \$.04. In 2001, the cost was \$.34. The increase in cost can be modeled by the equation $C = 0.04e^{0.057t}$, where C is the cost and t is the number of years after 1962. According to this model, in what year did a first-class stamp cost \$.22?

21. **Medicine** The intensity I of an x-ray after it has passed through a material that is x centimeters thick is given by $I = I_0e^{-kx}$, where I_0 is the initial intensity and k is a number that depends on the material. The constant k for copper is 3.2. Find the thickness of copper that is needed so that the intensity of the x-ray after passing through the copper is 25% of the original intensity. Round to the nearest tenth.

22. **Energy** One model for the time it will take for the world's oil supply to be depleted is given by the equation $T = 14.29 \ln (0.00411r + 1)$, where r is the estimated world oil reserves in billions of barrels and T is the time before that amount of oil is depleted. Use this equation to determine how many barrels of oil are necessary to last 20 years. Round to the nearest tenth.

For Exercises 23 to 26, use the Richter scale equation $M = \log \frac{I}{I_0}$, where M is the magnitude of an earthquake, I is the intensity of the shock waves, and I_0 is the measure of a zero-level earthquake.

23. **Geology** On July 14, 2000, an earthquake struck the Kodiak Island Region in Alaska. The earthquake had an intensity of $I = 6{,}309{,}573I_0$. Find the Richter scale magnitude of the earthquake. Round to the nearest tenth.

24. **Geology** The January 26, 2000, earthquake near Gujarat, India, had an intensity of $I = 50{,}118{,}723I_0$. Find the Richter scale magnitude of the earthquake. Round to the nearest tenth.

25. **Geology** An earthquake in Japan on March 2, 1933, measured 8.9 on the Richter scale. Find the intensity of the earthquake in terms of I_0. Round to the nearest whole number.

26. **Geology** An earthquake that occurred in China in 1978 measured 8.2 on the Richter scale. Find the intensity of the earthquake in terms of I_0. Round to the nearest whole number.

Shown at the right is a **seismogram,** which is used to measure the magnitude of an earthquake. The magnitude is determined by the amplitude A of the wave and the time t, in seconds, between the occurrence of two types of waves, called primary waves and secondary waves. As you can see on the graph, a primary wave is abbreviated p-wave, and a secondary wave is abbreviated s-wave. The amplitude A of a wave is one-half the difference between its highest and lowest points. For this graph, A is 23 millimeters. The equation is $M = \log A + 3 \log 8t - 2.92$.

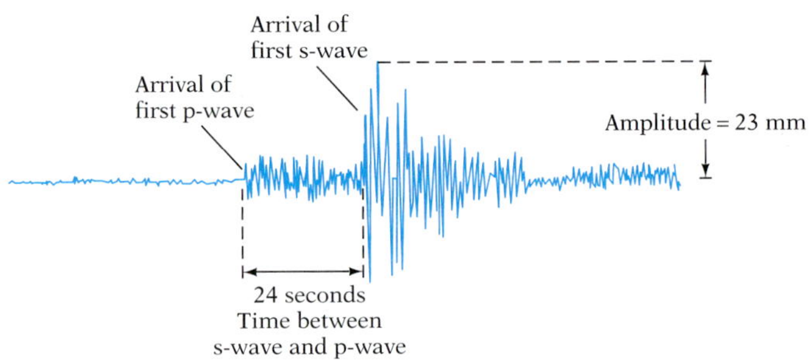

27. **Geology** Determine the magnitude of the earthquake for the seismogram given in the figure. Round to the nearest tenth.

28. **Geology** Find the magnitude of an earthquake that has a seismogram with an amplitude of 30 mm and for which t is 21 s.

29. **Geology** Find the magnitude of an earthquake that has a seismogram with an amplitude of 28 mm and for which t is 28 s.

11.5 Applying Concepts

30. **Biology** At 9 A.M., a culture of bacteria had a population of 1.5×10^6. At noon, the population was 3.0×10^6. If the population is growing exponentially, at what time will the population be 9×10^6? Round to the nearest hour.

31. **Inflation** If the average annual rate of inflation is 5%, in how many years will prices double? Round to the nearest whole number.

32. **Investments** The value of an investment in an account that earns an annual interest rate of 10% compounded daily grows according to the equation $A + A_0\left(1 + \dfrac{0.10}{365}\right)^{365t}$. Find the time for the investment to double in value. Round to the nearest whole number.

33. **Investments** Some banks now use continuous compounding of an amount invested. In this case, the equation that relates the value of an initial investment of A dollars after t years at an annual interest rate of r% is $P = Ae^{rt}$. Using this equation, find the value after 5 years of an investment of \$2500 in an account that earns 5% annual interest.

When a rock is tossed into the air, the mass of the rock remains constant and a reasonable model for the height of the rock can be given by a quadratic function. However, when a rocket is launched straight up from Earth's surface, the rocket is burning fuel, so the mass of the rocket is always changing. The height of the rocket above Earth can be approximated by the equation

$$y(t) = At - 16t^2 + \frac{A}{k}(M + m - kt)\ln\left(1 - \frac{k}{M + m}t\right)$$

where M is the mass of the rocket, m is the mass of the fuel, A is the rate at which fuel is ejected from the engines, k is the rate at which fuel is burned, t is the time in seconds, and $y(t)$ is the height in feet after t seconds.

34. **Rocketry** During the development of the V-2 rocket program in the United States, approximate values for a V-2 rocket were $M = 8000$ lb, $m = 16{,}000$ lb, $A = 8000$ ft/s, and $k = 250$ lb/s.

a. Use a graphing utility to estimate, to the nearest second, the time required for the rocket to reach a height of 1 mi (5280 ft).

b. Use $v(t) = -32t + A\ln\left(\dfrac{M + m}{M + m - kt}\right)$, and the answer to part (a), to determine the velocity of the rocket. Round to the nearest whole number.

c. Determine the domain of the velocity function.

35. **Mortgages** When you purchase a car or home and make monthly payments on the loan, you are amortizing the loan. Part of each monthly payment is interest on the loan, and the remaining part of the payment is a repayment of the loan amount. The amount remaining to be repaid on the loan after x months is given by $y = A(1 + i)^x + B$, where y is the amount of the loan to be repaid. In this equation, $A = \dfrac{Pi - M}{i}$ and $B = \dfrac{M}{i}$, where P is the original loan amount, i is the monthly interest rate $\left(\dfrac{\text{annual interest rate}}{12}\right)$, and M is the monthly payment. For a 30-year home mortgage of \$100,000 with an annual interest rate of 8%, $i = 0.00667$ and $M = 733.76$.

a. How many months are required to reduce the loan amount to \$90,000? Round to the nearest month.

b. How many months are required to reduce the loan amount to one-half the original amount? Round to the nearest month.

c. The total amount of interest that is paid after x months is given by $I = Mx + A(1 + i)^x + B - P$. Determine the month in which the total interest paid exceeds \$100,000. Round to the nearest month.

36. One scientific study suggested that the *carrying capacity* of Earth is around 10 billion people. What is meant by "carrying capacity"? Find the current world population and project when Earth's population would reach 10 billion, assuming population growth rates of 1%, 2%, 3%, 4%, and 5%. Find the current rate of world population growth and use that number to determine when, according to your model, the population will reach 10 billion.

37. Radioactive carbon dating can be used only in certain situations. Other methods of radioactive dating, such as the rubidium-strontium method, are appropriate in other situations. Write a report on radioactive dating. Include in your report when each form of dating is appropriate.

Focus on Problem Solving

Proof by Contradiction

The four-step plan for solving problems that we have used before is restated here.

1. Understand the problem.

2. Devise a plan.

3. Carry out the plan.

4. Review the solution.

One of the techniques that can be used in the second step is a method called *proof by contradiction*. In this method, you assume that the conditions of the problem you are trying to solve can be met and then show that your assumption leads to a condition you already know is not true.

To illustrate this method, suppose we try to prove that $\sqrt{2}$ is a rational number. We begin by recalling that a rational number is one that can be written as the quotient of integers. Therefore, let a and b be two integers with *no common factors*. If $\sqrt{2}$ is a rational number, then

$$\sqrt{2} = \frac{a}{b}$$

$$(\sqrt{2})^2 = \left(\frac{a}{b}\right)^2 \qquad \text{• Square each side.}$$

$$2 = \frac{a^2}{b^2}$$

$$2b^2 = a^2 \qquad \text{• Multiply both sides by } b^2.$$

The last equation states that a^2 is an even number. Because a^2 is an even number, then a is an even number. Now divide each side of the last equation by 2.

$$2b^2 = a^2$$
$$2b^2 = a \cdot a$$
$$b^2 = a \cdot x \qquad \bullet\ x = \frac{a}{2}$$

Because a is an even number, $a \cdot x$ is an even number. Because $a \cdot x$ is an even number, b^2 is an even number, and this in turn means that b is an even number. This result, however, contradicts the assumption that a and b are two integers with *no common factors*. Because this assumption is now known not to be true, we conclude that our original assumption, that $\sqrt{2}$ is a rational number, is false. This proves that $\sqrt{2}$ is an irrational number.

Try a proof by contradiction for the following problem: "Is it possible to write numbers using each of the digits 0, 1, 2, 3, 4, 5, 6, 7, 8, and 9 exactly once so that the sum of the numbers is exactly 100?"[1] Some suggestions for the proof are outlined below.

First note that the sum of the ten digits is 45. This means that some of the digits must be used as tens digits. Let x be the sum of those digits.

a. What is the sum of the remaining units digits?
b. Express "the sum of the units digits and the tens digits equals 100" as an equation. (*Note:* Writing an equation is one of the possible problem-solving strategies.)
c. Solve the equation for x.
d. Explain why this means that it is impossible to satisfy both conditions of the problem.

Projects & Group Activities

Fractals*

Fractals have a wide variety of applications. They have been used to create special effects for movies such as the *Star Wars* and *Star Trek* movies and to explain the behavior of some biological and economic systems. One aspect of fractals that has fascinated mathematicians is that they apparently have *fractional dimension.*

To understand the idea of fractional dimension, one must first understand the terms "scale factor" and "size." Consider a unit square (a square of length 1). By joining four of these squares, we can create another square, the length of which is 2, and the size of which is 4. (Here, size = number of square units.) Four of these larger squares can in turn be put together to make a third square of length 4 and size 16. This process of grouping together four squares can in

[1]G. Polya, *How to Solve It: A New Aspect of Mathematical Method,* copyright © 1957 by G. Polya. Reprinted by permission of Princeton University Press.
*Adapted with permission from "Student Math Notes," by Tami Martin, *News Bulletin,* November 1991.

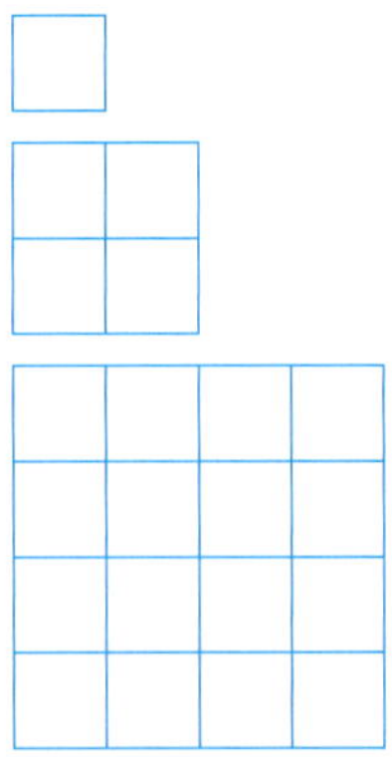

theory be done an infinite number of times; yet at each step, the following quantities will be the same:

$$\text{Scale factor} = \frac{\text{new length}}{\text{old length}} \qquad \text{Size ratio} = \frac{\text{new size}}{\text{old size}}$$

Consider the unit square as Step 1, the 4-unit squares as Step 2, etc.

1. Calculate the scale factor going from **a.** Step 1 to Step 2, **b.** Step 2 to Step 3, and **c.** Step 3 to Step 4.
2. Calculate the size ratio going from **a.** Step 1 to Step 2, **b.** Step 2 to Step 3, and **c.** Step 3 to Step 4.
3. What is **a.** the scale factor and **b.** the size ratio going from Step n to Step $n + 1$?

Mathematicians have defined dimension using the formula $d = \dfrac{\log(\text{size ratio})}{\log(\text{scale factor})}$.

For the squares discussed above, $d = \dfrac{\log(\text{size ratio})}{\log(\text{scale factor})} = \dfrac{\log 4}{\log 2} = 2$.

So by this definition of dimension, squares are two-dimensional figures.

Now consider a unit cube (Step 1). Group eight unit cubes to form a cube that is 2 units on each side (Step 2). Group eight of the cubes from Step 2 to form a cube that is 4 units on each side (Step 3).

4. Calculate **a.** the scale factor and **b.** the size ratio for this process.
5. Show that the cubes are three-dimensional figures.

In each of the above examples, if the process is continued indefinitely, we still have a square or a cube. Consider a process that is more difficult to envision. Let Step 1 be an equilateral triangle whose base has length 1 unit, and let Step 2 be a grouping of three of these equilateral triangles, such that the space between them is another equilateral triangle with a base of length 1 unit. Three shapes from Step 2 are arranged with an equilateral triangle in their center, and so on. It is hard to imagine the result if this is done an infinite number of times, but mathematicians have shown that the result is a single figure of fractional dimension. (Similar processes have been used to create fascinating artistic patterns and to explain scientific phenomena.)

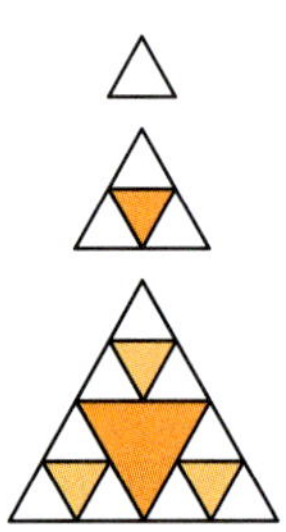

6. Show that for this process **a.** the scale factor is 2 and **b.** the size ratio is 3.
7. Calculate the dimension of the fractal. (Note that it is a *fractional* dimension!)

Credit Reports and FICO® Scores[1]

When a consumer applies for a loan, the lender generally wants to know the consumer's credit history. For this, the lender turns to a credit reporting agency. These agencies maintain files on millions of borrowers. They provide the lender with a credit report, which lists information such as the types of credit the consumer uses, the length of time the consumer's credit accounts have been open, the amounts owed by the consumer, and whether the consumer has paid his or her bills on time. Along with the credit report, the lender can buy a credit score based on the information in the report. The credit score gives the lender a quick measure of the consumer's credit risk, or how likely

[1]Source: **www.myfico.com**

the consumer is to repay a debt. It answers the lender's question: "If I lend this person money (or give this person a credit card), what is the probability that I will be paid back in a timely manner?"

One of the most widely used credit bureau scores is the FICO® score, so named because scores are produced from software developed by Fair, Isaac and Company (FICO). FICO scores range from 300 to 850. The higher the score, the lower the predicted credit risk for lenders.

The following graph shows the delinquency rate, or credit risk, associated with ranges of FICO scores. For example, the delinquency rate of consumers in the 550–599 range of scores is 51%. This means that within the next 2 years, for every 100 borrowers in this range, approximately 51 will default on a loan, file for bankruptcy, or fail to pay a credit card bill within 90 days of the due date.

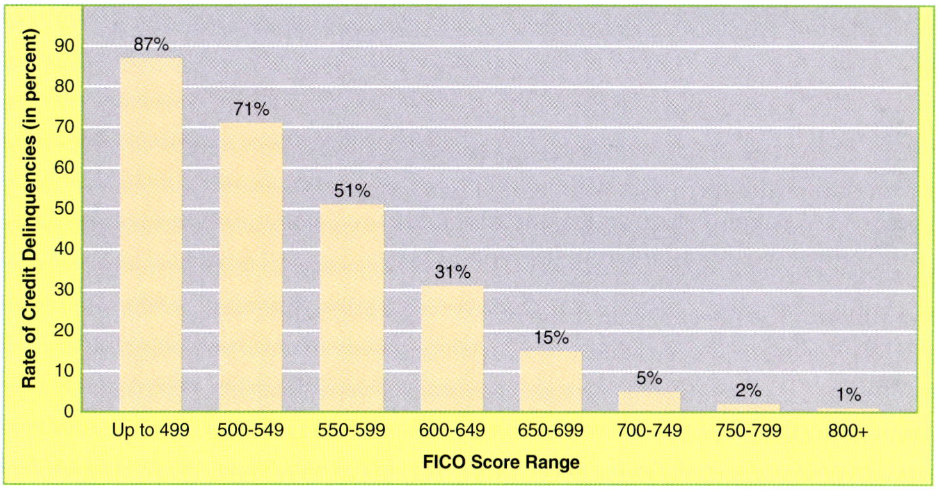

Because the delinquency rate depends on the FICO score, we can let the independent variable x represent the FICO score and the dependent variable y represent the delinquency rate. We will use the middle of each range for the values of x. The resulting ordered pairs are recorded in the table at the right.

x	y
400	87
525	71
575	51
625	31
675	15
725	5
775	2
825	1

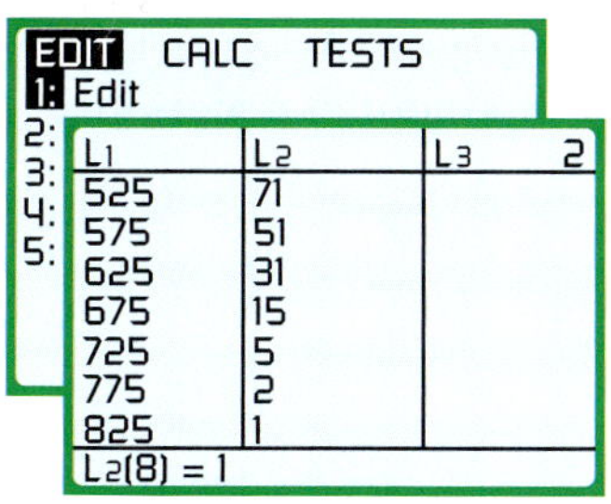

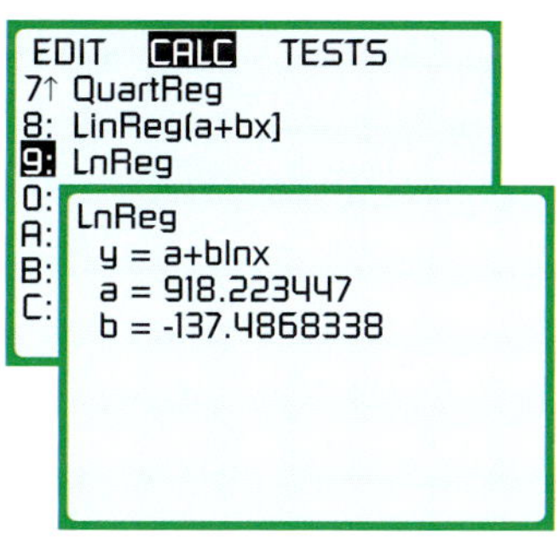

1. Use your graphing calculator to find a logarithmic equation that approximates these data. Here are instructions from the TI-83:

   Press STAT ENTER. This will bring up a table into which you can enter data. Enter the values of x in L1. Enter the values of y in L2.

   Press STAT again, press the right arrow key to highlight CALC, and arrow down to LnReg (ln for the natural logarithm). Press ENTER twice. The values for a and b in the equation $y = a + b \ln x$ will appear on the screen.

2. Use your equation to predict the delinquency rate of a consumer with a score of 500. Round to the nearest whole number.

3. The equation pairs a delinquency rate of 36% with what score?

4. What is the highest FICO score that will result in a delinquency rate of 71% or more?

5. What is the lowest FICO score that will result in a delinquency rate of 5% or lower?

6. WEB Use the Internet to find the name of at least one of the major credit reporting agencies.

Chapter Summary

Key Words	Examples
A function of the form $f(x) = b^x$, where b is a positive real number not equal to 1, is an **exponential function**. The number b is the **base** of the exponential function. [11.1.1, p. 701]	$f(x) = 3^x$; 3 is the base of the function.
The function $f(x) = e^x$ is called the **natural exponential function**. [11.1.1, p. 702]	$f(x) = e^x$; e is an irrational number approximately equal to 2.71828183.
For $x > 0$, $b > 0$, $b \neq 1$, $y = \log_b x$ is equivalent to $x = b^y$. Read $\log_b x$ as "the **logarithm** of x, base b." [11.2.1, p. 709]	$\log_2 8 = 3$ is equivalent to $8 = 2^3$.
Common logarithms are logarithms to the base 10. [11.2.1, p. 711]	$\log_{10} 100 = 2$ is usually written as $\log 100 = 2$.
Natural logarithms are logarithms to the base e. [11.2.1, p. 711]	$\log_e 100 \approx 4.61$ is usually written as $\ln 100 \approx 4.61$.
An **exponential equation** is an equation in which a variable occurs in the exponent. [11.4.1, p. 729]	$2^x = 12$ is an exponential equation.

Essential Rules and Procedures

Equality of Exponents Property For $b > 0$, $b \neq 1$, if $b^u = b^v$, then $u = v$. [11.2.1, p. 709]	If $b^x = b^5$, then $x = 5$.
Product Property of Logarithms For any positive real numbers x, y, and b, $b \neq 1$, $\log_b(xy) = \log_b x + \log_b y$. [11.2.2, p. 711]	$\log_b(3x) = \log_b 3 + \log_b x$
Quotient Property of Logarithms For any positive real numbers x, y, and b, $b \neq 1$, $\log_b \frac{x}{y} = \log_b x - \log_b y$. [11.2.2, p. 712]	$\log_b \frac{x}{20} = \log_b x - \log_b 20$

Power Property of Logarithms
For any positive real numbers x and b, $b \neq 1$, and for any real number r, $\log_b x^r = r \log_b x$. [11.2.2, p. 713]

$\log_b x^5 = 5 \log_b x$

Logarithmic Property of One
For any positive real numbers b, $b \neq 1$, $\log_b 1 = 0$. [11.2.2, p. 715]

$\log_6 1 = 0$

Inverse Property of Logarithms
For any positive real numbers x and b, $b \neq 1$, $\log_b b^x = x$. [11.2.2, p. 715]

$\log_3 3^4 = 4$

1–1 Property of Logarithms
For any positive real numbers x, y, and b, $b \neq 1$, if $\log_b x = \log_b y$, then $x = y$. [11.2.2, p. 715]

If $\log_5(x - 2) = \log_5 3$, then $x - 2 = 3$.

Change-of-Base Formula
$\log_a N = \dfrac{\log_b N}{\log_b a}$ [11.2.2, p. 716]

$\log_3 12 = \dfrac{\log 12}{\log 3}$ $\quad \log_6 16 = \dfrac{\ln 16}{\ln 6}$

Chapter Review Exercises

1. Evaluate: $\log_4 16$

2. Write $\frac{1}{2}(\log_3 x - \log_3 y)$ as a single logarithm with a coefficient of 1.

3. Evaluate the function $f(x) = e^{x-2}$ at $x = 2$.

4. Solve for x: $8^x = 2^{x-6}$

5. Evaluate $f(x) = \left(\dfrac{2}{3}\right)^x$ at $x = 0$.

6. Solve for x: $\log_3 x = -2$

7. Write $2^5 = 32$ in logarithmic form.

8. Solve for x: $\log x + \log(x - 4) = \log 12$

9. Write $\log_6 \sqrt{xy^3}$ in expanded form.

10. Solve for x: $4^{5x-2} = 4^{3x+2}$

11. Solve for x: $3^{7x+1} = 3^{4x-5}$

12. Evaluate $f(x) = 3^{x+1}$ at $x = -2$.

13. Evaluate: $\log_2 16$

14. Solve for x: $\log_6 2x = \log_6 2 + \log_6(3x - 4)$

15. Evaluate $\log_2 5$. Round to the nearest ten-thousandth.

16. Evaluate $\log_6 22$. Round to the nearest ten-thousandth.

17. Solve for x: $4^x = 8^{x-1}$

18. Evaluate $f(x) = \left(\frac{1}{4}\right)^x$ at $x = -1$.

19. Write $\log_5 \sqrt{\frac{x}{y}}$ in expanded form.

20. Solve for x: $\log_5 \frac{7x + 2}{3x} = 1$

21. Solve for x: $\log_5 x = 3$

22. Solve for x: $\log x + \log(2x + 3) = \log 2$

23. Write $3 \log_b x - 5 \log_b y$ as a single logarithm with a coefficient of 1.

24. Evaluate $f(x) = 2^{-x-1}$ at $x = -3$.

25. Evaluate $\log_3 19$. Round to the nearest ten-thousandth.

26. Solve $3^{x+2} = 5$ for x. Round to the nearest ten-thousandth.

27. Graph: $f(x) = 2^x - 3$

28. Graph: $f(x) = \left(\frac{1}{2}\right)^x + 1$

29. Graph: $f(x) = \log_2(2x)$

30. Graph: $f(x) = \log_2 x - 1$

31. Graph $f(x) = \frac{\log x}{x}$ and approximate the zeros of f to the nearest tenth.

32. Use the exponential decay equation $A = A_0\left(\frac{1}{2}\right)^{\frac{t}{k}}$, where A is the amount of a radioactive material present after time t, k is the half-life, and A_0 is the original amount of radioactive material, to find the half-life of a material that decays from 10 mg to 9 mg in 5 h. Round to the nearest whole number.

33. The percent of light that will pass through a translucent material is given by the equation $\log P = -0.5d$, where P is the percent of light that passes through the material, and d is the thickness of the material in centimeters. How thick must this translucent material be so that only 50% of the light that is incident to the material will pass through it? Round to the nearest thousandth.

Chapter Test

1. Evaluate $f(x) = \left(\frac{3}{4}\right)^x$ at $x = 0$.

2. Evaluate $f(x) = 4^{x-1}$ at $x = -2$.

3. Evaluate: $\log_4 64$

4. Solve for x: $\log_4 x = -2$

5. Write $\log_6 \sqrt[3]{x^2y^5}$ in expanded form.

6. Write $\frac{1}{2}(\log_5 x - \log_5 y)$ as a single logarithm with a coefficient of 1.

7. Solve for x: $\log_6 x + \log_6(x - 1) = 1$

8. Evaluate $f(x) = 3^{x+1}$ at $x = -2$.

9. Solve for x: $3^x = 17$

10. Solve for x: $\log_2 x + 3 = \log_2(x^2 - 20)$

11. Solve for x: $5^{6x-2} = 5^{3x+7}$

12. Solve for x: $4^x = 2^{3x+4}$

13. Solve for x: $\log(2x + 1) + \log x = \log 6$

14. Graph: $f(x) = 2^x - 1$

15. Graph: $f(x) = 2^x + 2$

16. Graph: $f(x) = \log_2(3x)$

17. Graph: $f(x) = \log_3(x + 1)$

18. Graph $f(x) = 3 - 2^x$ and approximate the zeros of f to the nearest tenth.

19. Find the value of \$10,000 invested for 6 years at 7.5% compounded monthly. Use the compound interest formula $P = A(1 + i)^n$, where A is the original value of the investment, i is the interest rate per compounding period, and n is the number of compounding periods. Round to the nearest dollar.

20. Use the exponential decay equation $A = A_0\left(\frac{1}{2}\right)^{\frac{t}{k}}$ where A is the amount of a radioactive material present after time t, k is the half-life, and A_0 is the original amount of radioactive material, to find the half-life of a material that decays from 40 mg to 30 mg in 10 h. Round to the nearest whole number.

Cumulative Review Exercises

1. Solve: $4 - 2[x - 3(2 - 3x) - 4x] = 2x$

2. Solve $S = 2WH + 2WL + 2LH$ for L.

3. Solve: $|2x - 5| \leq 3$

4. Factor: $4x^{2n} + 7x^n + 3$

5. Solve: $x^2 + 4x - 5 \leq 0$

6. Simplify: $\dfrac{1 - \frac{5}{x} + \frac{6}{x^2}}{1 + \frac{1}{x} - \frac{6}{x^2}}$

7. Simplify: $\dfrac{\sqrt{xy}}{\sqrt{x} - \sqrt{y}}$

8. Simplify: $y\sqrt{18x^5y^4} - x\sqrt{98x^3y^6}$

9. Simplify: $\dfrac{i}{2 - i}$

10. Find the equation of the line that contains the point $(2, -2)$ and is parallel to the line $2x - y = 5$.

11. Write a quadratic equation that has integer coefficients and has as solutions $\frac{1}{3}$ and -3.

12. Solve: $x^2 - 4x - 6 = 0$

13. Find the range of $f(x) = x^2 - 3x - 4$ if the domain is $\{-1, 0, 1, 2, 3\}$.

14. Given $f(x) = x^2 + 2x + 1$ and $g(x) = 2x - 3$, find $f[g(0)]$.

15. Solve by the addition method:

$$\begin{aligned} 3x - y + z &= 3 \\ x + y + 4z &= 7 \\ 3x - 2y + 3z &= 8 \end{aligned}$$

16. Solve:

$$\begin{aligned} y &= -2x - 3 \\ y &= 2x - 1 \end{aligned}$$

17. Evaluate $f(x) = 3^{-x+1}$ at $x = -4$.

18. Solve for x: $\log_4 x = 3$

19. Solve for x: $2^{3x+2} = 4^{x+5}$

20. Solve for x: $\log x + \log(3x + 2) = \log 5$

21. Graph: $\{x \mid x < 0\} \cap \{x \mid x > -4\}$

22. Graph the solution set of $\dfrac{x + 2}{x - 1} \geq 0$.

23. Graph: $y = -x^2 - 2x + 3$

24. Graph: $f(x) = |x| - 2$

25. Graph: $f(x) = \left(\frac{1}{2}\right)^x - 1$

26. Graph: $f(x) = \log_2 x + 1$

27. An alloy containing 25% tin is mixed with an alloy containing 50% tin. How much of each were used to make 2000 lb of an alloy containing 40% tin?

28. An account executive earns \$500 per month plus 8% commission on the amount of sales. The executive's goal is to earn a minimum of \$3000 per month. What amount of sales will enable the executive to earn \$3000 or more per month?

29. The table below shows the amount of money, in billions of dollars, spent on health care research and development in the United States (*Source: New York Times,* November 11, 2001). Draw a scatter diagram of these data.

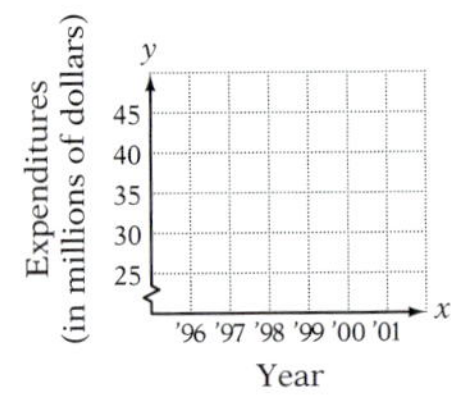

Year, x	*Expenditure, y*
1996	25
1997	28
1998	32
1999	34
2000	37
2001	43

30. A new printer can print checks three times faster than an older printer. The old printer can print the checks in 30 min. How long would it take to print the checks with both printers operating?

31. For a constant temperature, the pressure (P) of a gas varies inversely as the volume (V). If the pressure is 50 lb/in^2 when the volume is 250 ft^3, find the pressure when the volume is 25 ft^3.

32. A contractor buys 45 yd of nylon carpet and 30 yd of wool carpet for \$2340. A second purchase, at the same prices, includes 25 yd of nylon carpet and 80 yd of wool carpet for \$2820. Find the cost per yard of the wool carpet.

33. An investor deposits \$10,000 into an account that earns 9% interest compounded monthly. What is the value of the investment after 5 years? Use the compound interest formula $P = A(1 + i)^n$, where A is the original value of an investment, i is the interest rate per compounding period, n is the total number of compounding periods, and P is the value of the investment after n periods. Round to the nearest dollar.

Final Exam

1. Simplify: $12 - 8[3 - (-2)]^2 \div 5 - 3$

2. Evaluate $\dfrac{a^2 - b^2}{a - b}$ when $a = 3$ and $b = -4$.

3. Simplify: $5 - 2[3x - 7(2 - x) - 5x]$

4. Solve: $\dfrac{3}{4}x - 2 = 4$

5. Solve: $\dfrac{2 - 4x}{3} - \dfrac{x - 6}{12} = \dfrac{5x - 2}{6}$

6. Solve: $8 - |5 - 3x| = 1$

7. Solve: $|2x + 5| < 3$

8. Solve: $2 - 3x < 6$ and $2x + 1 > 4$

9. Find the equation of the line that contains the point $(-2, 1)$ and is perpendicular to the line $3x - 2y = 6$.

10. Simplify: $2a[5 - a(2 - 3a) - 2a] + 3a^2$

11. Simplify: $\dfrac{3}{2 + i}$

12. Write a quadratic equation that has integer coefficients and has solutions $-\frac{1}{2}$ and 2.

13. Factor: $8 - x^3y^3$

14. Factor: $x - y - x^3 + x^2y$

15. Divide: $(2x^3 - 7x^2 + 4) \div (2x - 3)$

16. Divide: $\dfrac{x^2 - 3x}{2x^2 - 3x - 5} \div \dfrac{4x - 12}{4x^2 - 4}$

17. Subtract: $\dfrac{x - 2}{x + 2} - \dfrac{x + 3}{x - 3}$

18. Simplify: $\dfrac{\dfrac{3}{x} + \dfrac{1}{x + 4}}{\dfrac{1}{x} + \dfrac{3}{x + 4}}$

19. Solve: $\dfrac{5}{x - 2} - \dfrac{5}{x^2 - 4} = \dfrac{1}{x + 2}$

20. Solve $a_n = a_1 + (n - 1)d$ for d.

21. Simplify: $\left(\dfrac{4x^2y^{-1}}{3x^{-1}y}\right)^{-2}\left(\dfrac{2x^{-1}y^2}{9x^{-2}y^2}\right)^3$

22. Simplify: $\left(\dfrac{3x^{\frac{2}{3}}y^{\frac{1}{2}}}{6x^2y^{\frac{4}{3}}}\right)^6$

23. Subtract: $x\sqrt{18x^2y^3} - y\sqrt{50x^4y}$

24. Simplify: $\dfrac{\sqrt{16x^5y^4}}{\sqrt{32xy^4}}$

25. Solve by using the quadratic formula: $2x^2 - 3x - 1 = 0$

26. Solve: $x^{\frac{2}{3}} - x^{\frac{1}{3}} - 6 = 0$

27. Find the equation of the line containing the points $(3, -2)$ and $(1, 4)$.

28. Solve: $\dfrac{2}{x} - \dfrac{2}{2x + 3} = 1$

29. Solve by the addition method:
$$\begin{aligned} 3x - 2y &= 1 \\ 5x - 3y &= 3 \end{aligned}$$

30. Evaluate the determinant: $\begin{vmatrix} 3 & 4 \\ -1 & 2 \end{vmatrix}$

31. Solve for x: $\log_3 x - \log_3(x - 3) = \log_3 2$

32. Find the inverse function of $f(x) = \frac{2}{3}x - 4$.

33. Write $2(\log_2 a - \log_2 b)$ as a single logarithm with a coefficient of 1.

34. Graph $2x - 3y = 9$ by using the x- and y-intercepts.

35. Graph the solution set of $3x + 2y > 6$.

36. Graph: $f(x) = -x^2 + 4$

37. Graph: $\dfrac{x^2}{16} + \dfrac{y^2}{4} = 1$

38. Graph: $f(x) = \log_2(x + 1)$

39. Given the graph of $f(x) = \ln x$ shown below, graph $g(x) = \ln(x + 3)$.

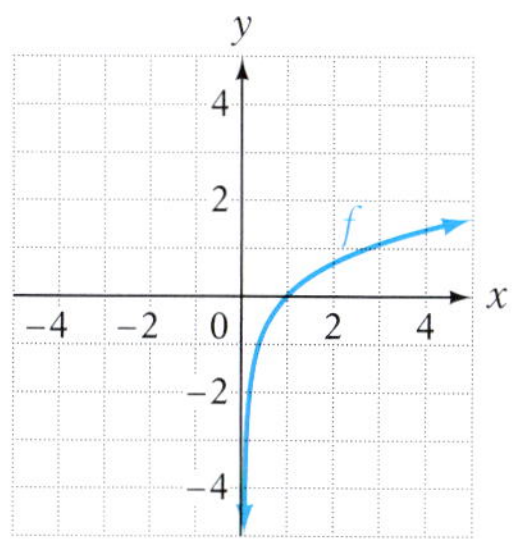

40. An average score of 70–79 in a history class receives a C grade. A student has grades of 64, 58, 82, and 77 on four history tests. Find the range of scores on the fifth test that will give the student a C grade for the course.

41. A jogger and a cyclist set out at 8 A.M. from the same point headed in the same direction. The average speed of the cyclist is two and a half times the average speed of the jogger. In 2 h, the cyclist is 24 mi ahead of the jogger. How far did the cyclist ride?

42. You have a total of \$12,000 invested in two simple interest accounts. On one account, a money market fund, the annual simple interest rate is 8.5%. On the other account, a tax-free bond fund, the annual simple interest rate is 6.4%. The total annual interest earned by the two accounts is \$936. How much do you have invested in each account?

43. The length of a rectangle is 1 ft less than three times the width. The area of the rectangle is 140 ft^2. Find the length and width of the rectangle.

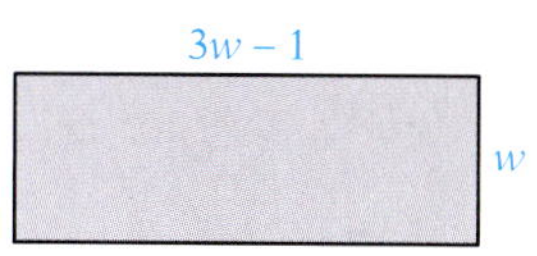

44. Three hundred shares of a utility stock earn a yearly dividend of \$486. How many additional shares of the utility stock would give a total dividend income of \$810?

45. An account executive traveled 45 mi by car and then an additional 1050 mi by plane. The rate of the plane was seven times the rate of the car. The total time of the trip was $3\frac{1}{4}$ h. Find the rate of the plane.

46. An object is dropped from the top of a building. Find the distance the object has fallen when the speed reaches 75 ft/s. Use the equation $v = \sqrt{64d}$, where v is the speed of the object and d is the distance. Round to the nearest whole number.

47. A small plane made a trip of 660 mi in 5 h. The plane traveled the first 360 mi at a constant rate before increasing its speed by 30 mph. Another 300 mi was traveled at the increased speed. Find the rate of the plane for the first 360 mi.

48. The intensity (L) of a light source is inversely proportional to the square of the distance (d) from the source. If the intensity is 8 lumens at a distance of 20 ft, what is the intensity when the distance is 4 ft?

49. A motorboat traveling with the current can go 30 mi in 2 h. Against the current, it takes 3 h to go the same distance. Find the rate of the motorboat in calm water and the rate of the current.

50. An investor deposits \$4000 into an account that earns 9% annual interest compounded monthly. Use the compound interest formula $P = A(1 + i)^n$, where A is the original value of the investment, i is the interest rate per compounding period, n is the total number of compounding periods, and P is the value of the investment after n periods, to find the value of the investment after 2 years. Round to the nearest cent.

R

Review of Beginning Algebra Topics

Objectives

Section R.1

1. Evaluate Variable Expressions
2. Simplify Variable Expressions

Section R.2

1. First-Degree Equations in One Variable
2. Inequalities in One Variable

Section R.3

1. Graph Points in a Rectangular Coordinate System
2. Graph a Linear Equation in Two Variables
3. Evaluate a Function
4. Find the Equation of a Line

Section R.4

1. Multiply and Divide Monomials
2. Add and Subtract Polynomials
3. Multiply Polynomials
4. Divide Polynomials
5. Factor Trinomials of the Form $ax^2 + bx + c$

WEB **Need help? For online student resources, such as section quizzes, visit this web site: math.college.hmco.com**

R.1 Variable Expressions

1 Evaluate Variable Expressions

VIDEO & DVD CD TUTOR WEB SSM

Whenever an expression contains more than one operation, the operations must be performed in a specified order, as listed below in the Order of Operations Agreement.

1.4.2* Order of Operations Agreement

The Order of Operations Agreement

Step 1 Perform operations inside grouping symbols. Grouping symbols include parentheses ( ), brackets [ ], braces { }, the absolute value symbol ||, and fraction bars.

Step 2 Simplify exponential expressions.

Step 3 Do multiplication and division as they occur from left to right.

Step 4 Do addition and subtraction as they occur from left to right.

Example 1 Evaluate: $-2(7 - 3)^2 + 4 - 2(5 - 2)$

Solution $-2(7 - 3)^2 + 4 - 2(5 - 2)$

$= -2(4)^2 + 4 - 2(3)$ • Perform operations inside parentheses.

$= -2(16) + 4 - 2(3)$ • Simplify the exponential expression.

$= -32 + 4 - 2(3)$ • Do the multiplication and division from left to right.

$= -32 + 4 - 6$

$= -28 - 6$ • Do the addition and subtraction from left to right.

$= -28 + (-6) = -34$

Problem 1 Evaluate: $(-4)(6 - 8)^2 - (-12 \div 4)$

Solution See page S37.

1.5.1 Evaluate Variable Expressions

A **variable** is a letter that represents a quantity that is unknown or that can change or vary. An expression that contains one or more variables is a **variable expression.** $3x - 4y + 7z$ is a variable expression. It contains the variables x, y, and z.

Replacing a variable in a variable expression by a number and then simplifying the resulting numerical expression is called **evaluating the variable expression.** The number substituted for the variable is called the **value of the variable.** The result is called the **value of the variable expression.**

*Review this objective for more detailed coverage of this topic.

Example 2 Evaluate $5ab^3 + 2a^2b^2 - 4$ when $a = 3$ and $b = -2$.

Solution $5ab^3 + 2a^2b^2 - 4$

$5(3)(-2)^3 + 2(3)^2(-2)^2 - 4$ • Replace a by 3 and b by -2.

$= 5(3)(-8) + 2(9)(4) - 4$ • Use the Order of Operations Agreement to simplify the numerical expression.

$= -120 + 72 - 4$

$= -48 - 4$

$= -48 + (-4)$

$= -52$

Problem 2 Evaluate $3xy^2 - 3x^2y$ when $x = -2$ and $y = 5$.

Solution See page S37.

2 Simplify Variable Expressions

The Properties of Real Numbers are used to simplify variable expressions.

1.5.3
1.5.4
Simplify variable expressions using the Properties of Addition and Multiplication

Example 3 Simplify: $-\frac{1}{3}(-3y)$

Solution $-\frac{1}{3}(-3y)$

$= \left[-\frac{1}{3}(-3)\right]y$ • Use the Associative Property of Multiplication to regroup factors.

$= 1y$ • Use the Inverse Property of Multiplication.

$= y$ • Use the Multiplication Property of One.

Problem 3 Simplify: $-5(-3a)$

Solution See page S37.

A variable expression is shown at the right. The expression can be rewritten by writing subtraction as addition of the opposite. Note that the expression has four addends. The **terms** of a variable expression are the addends of the expression. The expression has four terms.

$3x^2 - 4xy + 5z - 2$

$3x^2 + (-4xy) + 5z + (-2)$

Four terms

$3x^2 - 4xy + 5z - 2$

Variable terms | Constant term

The terms $3x^2$, $-4xy$, and $5z$ are **variable terms.** The term -2 is a **constant term,** or simply a **constant.**

Like terms of a variable expression are terms that have the same variable parts. The terms $3x$ and $-7x$ are like terms. Constant terms are also like terms. Thus -6 and 9 are like terms.

like terms

$$3x^2 + 3x - 6 - 7x + 9$$

like terms

By using the Commutative Property of Multiplication, we can rewrite the Distributive Property as $ba + ca = (b + c)a$. This form of the Distributive Property is used to **combine like terms** of a variable expression by adding their coefficients. For instance,

$$7x + 9x = (7 + 9)x \quad \text{• Use the Distributive Property: } ba + ca = (b + c)a.$$
$$= 16x$$

Example 4 Simplify: $4x^2 + 5x - 6x^2 - 7x$

Solution

$4x^2 + 5x - 6x^2 - 7x$

$= 4x^2 - 6x^2 + 5x - 7x$

$= (4x^2 - 6x^2) + (5x - 7x)$ • Use the Associative and Commutative Properties of Addition to rearrange and group like terms.

$= -2x^2 + (-2x)$ • Use the Distributive Property to combine like terms.

$= -2x^2 - 2x$

Problem 4 Simplify: $2z^2 - 5z - 3z^2 + 6z$

Solution See page S37.

1.5.5 Simplify variable expressions using the Distributive Property

The Distributive Property is also used to remove parentheses from a variable expression. Here is an example.

$4(2x + 5z) = 4(2x) + 4(5z)$ • Use the Distributive Property: $a(b + c) = ab + ac$

$= (4 \cdot 2)x + (4 \cdot 5)z$ • Use the Associative Property of Multiplication to regroup factors.

$= 8x + 20z$ • Multiply $4 \cdot 2$ and $4 \cdot 5$.

The Distributive Property can be extended to expressions containing more than two terms. For instance,

$$4(2x + 3y + 5z) = 4(2x) + 4(3y) + 4(5z)$$
$$= 8x + 12y + 20z$$

Example 5 Simplify. **a.** $-3(2x + 4)$ **b.** $6(3x - 4y + z)$

Solution

a. $-3(2x + 4) = -3(2x) + (-3)(4)$
$= -6x - 12$

b. $6(3x - 4y + z)$
$= 6(3x) - 6(4y) + 6(z)$
$= 18x - 24y + 6z$

Problem 5 Simplify. **a.** $-3(5y - 2)$ **b.** $-2(4x + 2y - 6z)$

Solution See page S37.

Take Note

Recall that the Distributive Property states that if a, b, and c are real numbers, then

$$a(b + c) = ab + ac$$

To simplify the expression $5 + 3(4x - 2)$, use the Distributive Property to remove the parentheses.

$$5 + 3(4x - 2) = 5 + 3(4x) - 3(2)$$ • Use the Distributive Property.

$$= 5 + 12x - 6$$

$$= 12x - 1$$ • Add the like terms 5 and −6.

Example 6 Simplify.

a. $3(2x - 4) - 5(3x + 2)$

b. $3a - 2[7a - 2(2a + 1)]$

Solution

a. $3(2x - 4) - 5(3x + 2)$

$= 6x - 12 - 15x - 10$

$= -9x - 22$

b. $3a - 2[7a - 2(2a + 1)]$

$= 3a - 2[7a - 4a - 2]$

$= 3a - 2[3a - 2]$

$= 3a - 6a + 4$

$= -3a + 4$

Problem 6 Simplify.

a. $7(-3x - 4y) - 3(3x + y)$

b. $2y - 3[5 - 3(3 + 2y)]$

Solution See page S37.

R.1 Exercises

1 Evaluate the variable expression when $a = 2$, $b = 3$, and $c = -4$.

1. $a - 2c$

2. $-3a + 4b$

3. $3b - 3c$

4. $-3c + 4$

5. $16 \div (2c)$

6. $6b \div (-a)$

7. $3b - (a + c)^2$

8. $(a - b)^2 + 2c$

9. $(b - 3a)^2 + bc$

Evaluate the variable expression when $a = -1$, $b = 3$, $c = -2$, and $d = 4$.

10. $\dfrac{b - a}{d}$

11. $\dfrac{d - b}{a}$

12. $\dfrac{2d + b}{-a}$

13. $\dfrac{b - d}{c - a}$

14. $2(b + c) - 2a$

15. $3(b - a) - bc$

16. $\frac{-4bc}{2a + c}$

17. $\frac{abc}{b - d}$

18. $(d - a)^2 - (b - c)^2$

19. $(-b + d)^2 + (-a + c)^2$

20. $4ab + (2c)^2$

21. $3cd - (4a)^2$

Evaluate the variable expression when $a = 2.7$, $b = -1.6$, and $c = -0.8$.

22. $c^2 - ab$

23. $(a + b)^2 - c$

24. $\frac{b^3}{c} - 4a$

2 Simplify each of the following.

25. $x + 7x$

26. $12y + 9y$

27. $8b - 5b$

28. $4y - 11y$

29. $-12a + 17a$

30. $-15xy + 7xy$

31. $4x + 5x + 2x$

32. $-5x^2 - 10x^2 + x^2$

33. $6x - 2y + 9x$

34. $3x - 7y - 6x + 4x$

35. $5a + 6a - 2a$

36. $2a - 5a + 3a$

37. $12y^2 + 10y^2$

38. $2z^2 - 9z^2$

39. $\frac{3}{4}x - \frac{1}{4}x$

40. $\frac{2}{5}y - \frac{3}{5}y$

41. $-4(5x)$

42. $-2(-8y)$

43. $(6a)(-4)$

44. $-5(7x^2)$

45. $\frac{1}{4}(4x)$

46. $\frac{12x}{5}\left(\frac{5}{12}\right)$

47. $\frac{1}{3}(21x)$

48. $-\frac{5}{8}(24a^2)$

49. $(36y)\left(\frac{1}{12}\right)$

50. $-(z + 4)$

51. $-3(a + 5)$

52. $(4 - 3b)9$

53. $(-2x - 6)8$

54. $3(5x^2 + 2x)$

55. $-5(2y^2 - 1)$

56. $4(x^2 - 3x + 5)$

57. $6(3x^2 - 2xy - y^2)$

58. $5a - (4a + 6)$

59. $3 - (10 + 8y)$

60. $12(y - 2) + 3(7 - 4y)$

61. $-5[2x + 3(5 - x)]$

62. $-3[2x - (x + 7)]$

63. $-5a - 2[2a - 4(a + 7)]$

R.2 Equations and Inequalities

1 First-Degree Equations in One Variable

VIDEO & DVD CD TUTOR WEB SSM

An **equation** expresses the equality of two mathematical expressions. Each of the equations that follows is a **first-degree equation in one variable.** *First degree* means that the variable has an exponent of 1.

$$x + 11 = 14$$
$$3a + 5 = 8a$$
$$2(6y - 1) = 3$$

A **solution** of an equation is a number that, when substituted for the variable, results in a true equation.

3 is a solution of the equation $x + 4 = 7$ because $3 + 4 = 7$.

9 is not a solution of the equation $x + 4 = 7$ because $9 + 4 \neq 7$.

To **solve an equation** means to find a solution of the equation. In solving an equation, the goal is to rewrite the given equation with the variable alone on one side of the equation and a constant term on the other side of the equation; the constant term is the solution of the equation. The following properties of equations are used to rewrite equations in this form.

2.1.2 2.1.3 Solve equations using the Addition and Multiplication Properties of Equations

Properties of Equations

Addition Property of Equations

The same number can be added to each side of an equation without changing the solution of the equation. In symbols, the equation $a = b$ has the same solution as the equation $a + c = b + c$.

Multiplication Property of Equations

Each side of an equation can be multiplied by the same nonzero number without changing the solution of the equation. In symbols, if $c \neq 0$, then the equation $a = b$ has the same solutions as the equation $ac = bc$.

Take Note

Subtraction is defined as addition of the opposite.

$$a - b = a + (-b)$$

The Addition Property of Equations is used to remove a term from one side of the equation by adding the opposite of that term to each side of the equation. Because subtraction is defined in terms of addition, the Addition Property of Equations also makes it possible to subtract the same number from each side of an equation without changing the solution of the equation.

For example, to solve the equation $t + 9 = -4$, subtract the constant term (9) from each side of the equation.

$$t + 9 = -4$$
$$t + 9 - 9 = -4 - 9$$
$$t = -13$$

Now the variable is alone on one side of the equation and a constant term (-13) is on the other side. The solution is the constant. The solution is -13.

Take Note

Division is defined as multiplication by the reciprocal.

$$a \div b = a \cdot \frac{1}{b}$$

The Multiplication Property of Equations is used to remove a coefficient by multiplying each side of the equation by the reciprocal of the coefficient. Because division is defined in terms of multiplication, each side of an equation can be divided by the same nonzero number without changing the solution of the equation.

Take Note

When using the Multiplication Property of Equations, multiply each side of the equation by the reciprocal of the coefficient when the coefficient is a fraction. Divide each side of the equation by the coefficient when the coefficient is an integer or a decimal.

For example, to solve the equation $-5q = 120$, divide each side of the equation by the coefficient -5.

$$-5q = 120$$
$$\frac{-5q}{-5} = \frac{120}{-5}$$
$$q = -24$$

Now the variable is alone on one side of the equation and a constant (-24) is on the other side. The solution is the constant. The solution is -24.

In solving more complicated first-degree equations in one variable, use the following sequence of steps.

2.2.3 Solve general equations

Steps for Solving a First-Degree Equation in One Variable

1. Use the Distributive Property to remove parentheses.
2. Combine any like terms on the right side of the equation and any like terms on the left side of the equation.
3. Use the Addition Property to rewrite the equation with only one variable term.
4. Use the Addition Property to rewrite the equation with only one constant term.
5. Use the Multiplication Property to rewrite the equation with the variable alone on one side of the equation and a constant on the other side of the equation.

If one of the above steps is not needed to solve a given equation, proceed to the next step.

Example 1 Solve.

a. $5x + 9 = 23 - 2x$

b. $8x - 3(4x - 5) = -2x + 6$

Solution

a.

$$5x + 9 = 23 - 2x$$
$$5x + 2x + 9 = 23 - 2x + 2x \qquad \text{• Step 3}$$
$$7x + 9 = 23$$
$$7x + 9 - 9 = 23 - 9 \qquad \text{• Step 4}$$
$$7x = 14$$
$$\frac{7x}{7} = \frac{14}{7} \qquad \text{• Step 5}$$
$$x = 2$$

The solution is 2.

b. $8x - 3(4x - 5) = -2x + 6$

$$
\begin{aligned}
8x - 12x + 15 &= -2x + 6 && \text{• Step 1}\\
-4x + 15 &= -2x + 6 && \text{• Step 2}\\
-4x + 2x + 15 &= -2x + 2x + 6 && \text{• Step 3}\\
-2x + 15 &= 6\\
-2x + 15 - 15 &= 6 - 15 && \text{• Step 4}\\
-2x &= -9\\
\frac{-2x}{-2} &= \frac{-9}{-2} && \text{• Step 5}\\
x &= \frac{9}{2}
\end{aligned}
$$

The solution is $\frac{9}{2}$.

Problem 1 Solve.

a. $4x + 3 = 7x + 9$

b. $4 - (5x - 8) = 4x + 3$

Solution See page S37.

2 Inequalities in One Variable

VIDEO & DVD · CD TUTOR · WEB · SSM

An **inequality** contains the symbol $>$, $<$, $\geq$, or $\leq$. An inequality expresses the relative order of two mathematical expressions. Here are some examples of inequalities in one variable.

$$
\left.\begin{aligned}
&4x \geq 12\\
&2x + 7 \leq 9\\
&x^2 + 1 > 3x
\end{aligned}\right\} \text{ Inequalities in one variable}
$$

2.7.2 Solve an inequality in one variable

A **solution of an inequality in one variable** is a number that, when substituted for the variable, results in a true inequality. For the inequality $x < 4$, 3, 0, and -5 are solutions of the inequality because replacing the variable by these numbers results in a true inequality.

$x < 4$		$x < 4$		$x < 4$	
$3 < 4$	True	$0 < 4$	True	$-5 < 4$	True

The number 7 is not a solution of the inequality $x < 4$ because $7 < 4$ is a false inequality.

Besides the numbers 3, 0, and -5, there are an infinite number of other solutions of the inequality $x < 4$. Any number less than 4 is a solution; for instance, -5.2, $\frac{5}{2}$, π, and 1 are also solutions of the inequality. The set of all the solutions of an inequality is called the **solution set of the inequality.** The solution set of the inequality $x < 4$ is written in set-builder notation as $\{x \mid x < 4\}$. This is read "the set of all x such that x is less than 4." It is written in interval notation as $(-\infty, 4)$.

The graph of the solution set of $x < 4$ is shown at the right.

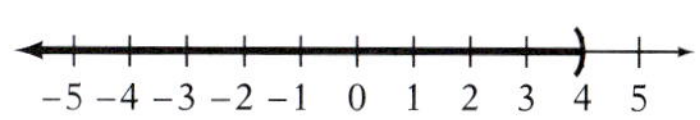

In solving an inequality, the goal is to rewrite the given inequality in the form

variable $<$ *constant* or *variable* $>$ *constant*

The Addition Property of Inequalities is used to rewrite an inequality in this form.

> **Take Note**
> The Addition Property of Inequalities states that the same number can be added to each side of an inequality without changing the solution set of the inequality.

The Addition Property of Inequalities

If $a > b$ and c is a real number, then the inequalities $a > b$ and $a + c > b + c$ have the same solution set.

If $a < b$ and c is a real number, then the inequalities $a < b$ and $a + c < b + c$ have the same solution set.

The Addition Property of Inequalities is also true for the symbols $\leq$ and $\geq$.

The Addition Property of Inequalities is used to remove a term from one side of an inequality by adding the additive inverse of that term to each side of the inequality. Because subtraction is defined in terms of addition, the same number can be subtracted from each side of an inequality without changing the solution set of the inequality.

As shown in the following example, the Addition Property of Inequalities applies to variable terms as well as to constants.

Solve $4x - 5 \leq 3x - 2$. Write the solution set in interval notation.

$$4x - 5 \leq 3x - 2$$

$$4x - 3x - 5 \leq 3x - 3x - 2$$ • Subtract $3x$ from each side of the inequality.

$$x - 5 \leq -2$$ • Simplify.

$$x - 5 + 5 \leq -2 + 5$$ • Add 5 to each side of the inequality.

$$x \leq 3$$ • Simplify.

The solution set is $(-\infty, 3]$.

When we multiply or divide an inequality by a number, the inequality symbol may be reversed, depending on whether the number is positive or negative. Look at the following two examples.

$$3 < 5$$

$$2(3) < 2(5)$$ • Multiply by positive 2. The inequality symbol remains the same. $6 < 10$ is a true statement.

$$6 < 10$$

$$3 < 5$$

$$-2(3) > -2(5)$$ • Multiply by negative 2. The inequality symbol is reversed in order to make the inequality a true statement.

$$-6 > -10$$

This is summarized in the Multiplication Property of Inequalities.

The Multiplication Property of Inequalities

Rule 1
If $a > b$ and $c > 0$, then $ac > bc$.
If $a < b$ and $c > 0$, then $ac < bc$.

Rule 2
If $a > b$ and $c < 0$, then $ac < bc$.
If $a < b$ and $c < 0$, then $ac > bc$.

Rule 1 states that **when each side of an inequality is multiplied by a positive number, the inequality symbol remains the same.** Rule 2 states that **when each side of an inequality is multiplied by a negative number, the inequality symbol must be reversed.**

Here are a few more examples of this property.

Rule 1		Rule 2	
$-4 < -2$	$5 > -3$	$-1 < 7$	$-2 > -6$
$2(-4) < 2(-2)$	$3(5) > 3(-3)$	$-2(-1) > -2(7)$	$-3(-2) < -3(-6)$
$-8 < -4$	$15 > -9$	$2 > -14$	$6 < 18$

Use the Multiplication Property of Inequalities to remove a coefficient other than 1 from one side of an inequality so that the inequality can be rewritten with the variable alone on one side of the inequality and a constant term on the other side. The Multiplication Property of Inequalities is also true for the symbols $\le$ and $\ge$.

Because division is defined in terms of multiplication, when each side of an inequality is divided by a positive number, the inequality symbol remains the same. When each side of an inequality is divided by a negative number, the inequality symbol must be reversed.

Take Note

Solving inequalities in one variable is similar to solving equations in one variable, except that when you multiply or divide by a negative number, you must reverse the inequality symbol.

Solve $-3x < 9$. Write the solution set in set-builder notation.

$$-3x < 9$$

$$\frac{-3x}{-3} > \frac{9}{-3}$$ • Divide each side of the inequality by the coefficient -3 and reverse the inequality symbol.

$$x > -3$$ • Simplify.

$$\{x \mid x > -3\}$$ • Write the answer in set-builder notation.

Example 2 Solve $x + 3 > 4x + 6$. Write the solution set in set-builder notation.

Solution

$$
\begin{aligned}
x + 3 &> 4x + 6 \\
x - 4x + 3 &> 4x - 4x + 6 \\
-3x + 3 &> 6 \\
-3x + 3 - 3 &> 6 - 3 \\
-3x &> 3 \\
\frac{-3x}{-3} &< \frac{3}{-3} \\
x &< -1
\end{aligned}
$$

$\{x \mid x < -1\}$

Problem 2 Solve $3x - 1 \le 5x - 7$. Write the solution set in set-builder notation.

Solution See page S37.

When an inequality contains parentheses, often the first step in solving the inequality is to use the Distributive Property to remove the parentheses.

Example 3 Solve $-2(x - 7) > 3 - 4(2x - 3)$. Write the solution set in interval notation.

Solution

$$
\begin{aligned}
-2(x - 7) &> 3 - 4(2x - 3) \\
-2x + 14 &> 3 - 8x + 12 \\
-2x + 14 &> 15 - 8x \\
-2x + 8x + 14 &> 15 - 8x + 8x \\
6x + 14 &> 15 \\
6x + 14 - 14 &> 15 - 14 \\
6x &> 1 \\
\frac{6x}{6} &> \frac{1}{6} \\
x &> \frac{1}{6}
\end{aligned}
$$

$\left(\frac{1}{6}, \infty\right)$

Problem 3 Solve $3 - 2(3x + 1) \le 7 - 2x$. Write the solution set in interval notation.

Solution See page S37.

R.2 Exercises

1 Solve.

1. $x + 7 = -5$

2. $9 + b = 21$

3. $-9 = z - 8$

4. $b - 11 = 11$

5. $-48 = 6z$

6. $-9a = -108$

7. $-\frac{3}{4}x = 15$

8. $\frac{5}{2}x = -10$

9. $-\frac{x}{4} = -2$

10. $\frac{2x}{5} = -8$

11. $4 - 2b = 2 - 4b$

12. $4y - 10 = 6 + 2y$

13. $5x - 3 = 9x - 7$

14. $3m + 5 = 2 - 6m$

15. $6a - 1 = 2 + 2a$

16. $5x + 7 = 8x + 5$

17. $2 - 6y = 5 - 7y$

18. $4b + 15 = 3 - 2b$

19. $2(x + 1) + 5x = 23$

20. $9n - 15 = 3(2n - 1)$

21. $7a - (3a - 4) = 12$

22. $5(3 - 2y) = 3 - 4y$

23. $9 - 7x = 4(1 - 3x)$

24. $2(3b + 5) - 1 = 10b + 1$

25. $2z - 2 = 5 - (9 - 6z)$

26. $4a + 3 = 7 - (5 - 8a)$

27. $5(6 - 2x) = 2(5 - 3x)$

28. $4(3y + 1) = 2(y - 8)$

29. $2(3b - 5) = 4(6b - 2)$

30. $3(x - 4) = 1 - (2x - 7)$

2 Solve. Write the answer in set-building notation.

31. $x - 5 > -2$

32. $5 + n \geq 4$

33. $-2 + n \geq 0$

34. $x - 3 < 2$

35. $8x \leq -24$

36. $-4x < 8$

37. $3n > 0$

38. $-2n \leq -8$

39. $2x - 1 > 7$

40. $5x - 2 \leq 8$

41. $4 - 3x < 10$

42. $7 - 2x \geq 1$

43. $3x - 1 > 2x + 2$

44. $6x + 4 \leq 8 + 5x$

45. $8x + 1 \geq 2x + 13$

46. $6x + 3 > 4x - 1$

47. $-3 - 4x > -11$

48. $4x - 2 < x - 11$

Solve. Write the answer using interval notation.

49. $4x - 2 > 3x + 1$

50. $7x + 5 \leq 9 + 6x$

51. $9x + 2 \geq 3x + 14$

52. $8x + 1 > 6x + 3$

53. $-5 - 2x > -13$

54. $5x - 3 < x - 11$

55. $4(2x - 1) > 3x - 2(3x - 5)$

56. $2 - 5(x + 1) \geq 3(x - 1) - 8$

57. $3(4x + 3) \leq 7 - 4(x - 2)$

58. $3 + 2(x + 5) \geq x + 5(x + 1) + 1$

59. $3 - 4(x + 2) \leq 6 + 4(2x + 1)$

60. $12 - 2(3x - 2) \geq 5x - 2(5 - x)$

R.3 Linear Equations in Two Variables

1 Graph Points in a Rectangular Coordinate System

VIDEO & DVD · CD TUTOR · WEB · SSM

3.1.1 Points in the plane

A **rectangular coordinate system** is formed by two number lines, one horizontal and one vertical. The point of intersection is called the **origin.** The two axes are called the **coordinate axes** or simply the axes. Generally, the horizontal axis is labeled the x-axis, and the vertical axis is labeled the y-axis.

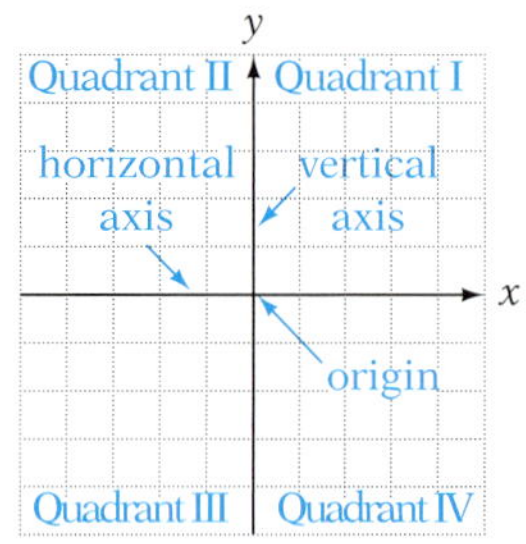

Each point in the plane can be identified by a pair of numbers called an **ordered pair.** The first number of the ordered pair measures a horizontal change from the y-axis and is called the **abscissa,** or ***x*-coordinate.** The second number of the pair measures a vertical change from the x-axis and is called the **ordinate,** or ***y*-coordinate.** The ordered pair (x, y) associated with a point is also called the **coordinates** of the point.

Horizontal change → Vertical change

Ordered pair → (4, 3)

x-coordinate → y-coordinate

To **graph,** or **plot,** a point means to place a dot at the coordinates of the point. For example, to graph the ordered pair (4, 3), start at the origin. Move 4 units to the right and then 3 units up. Draw a dot. To graph (−3, −4), start at the origin. Move 3 units to the left and then 4 units down. Draw a dot.

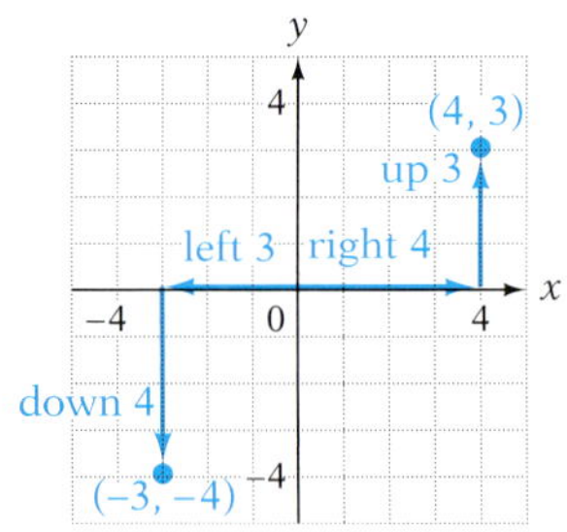

Example 1 Graph the ordered pairs (−4, 2), (4, 3), (0, −1), (2, 0), and (−1, −3).

Solution

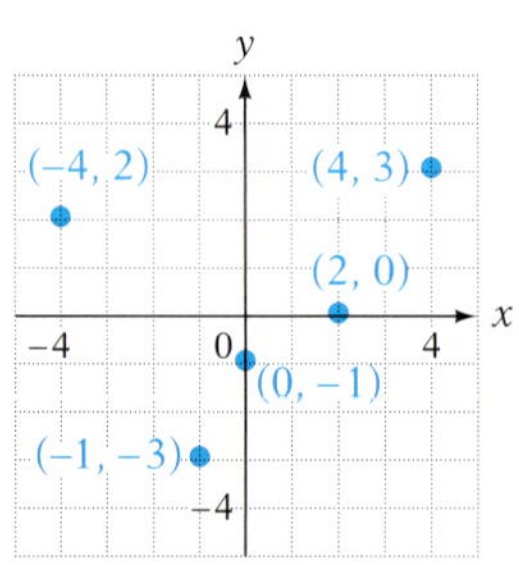

Problem 1 Graph the ordered pairs $(-2, 4)$, $(4, 0)$, $(0, 3)$, $(-3, -4)$ and $(5, -1)$.

Solution See page S38.

2 Graph a Linear Equation in Two Variables

VIDEO & DVD CD TUTOR WEB SSM

The equations that follow are examples of equations in two variables.

$$y = 3x - 4$$
$$2x - y = 7$$
$$y = x^2 + 1$$

A **solution of an equation in two variables** is an ordered pair (x, y) whose coordinates make the equation a true statement.

3.3.1 Solutions of equations in two variables

Is the ordered pair $(-4, 9)$ a solution of the equation $y = -2x + 1$?

$$y = -2x + 1$$

9	$-2(-4) + 1$	• Replace x by -4 and y by 9.
9	$8 + 1$	• Simplify the right side.
$9 = 9$		• Compare the results. If the resulting equation is true, the ordered pair is a solution of the equation. If it is not true, the ordered pair is not a solution of the equation.

Yes, the ordered pair $(-4, 9)$ is a solution of the equation $y = -2x + 1$.

Besides $(-4, 9)$, there are many other ordered pairs that are solutions of the equation $y = -2x + 1$. For example, $(0, 1)$, $(3, -5)$, and $\left(-\frac{3}{2}, 4\right)$ are also solutions. In general, an equation in two variables has an infinite number of solutions. By choosing any value of x and substituting that value into the equation, we can calculate a corresponding value of y.

Find the ordered-pair solution of $y = \frac{2}{5}x - 4$ that corresponds to $x = 5$.

$$y = \frac{2}{5}x - 4$$

$$y = \frac{2}{5}(5) - 4$$ • We are given that $x = 5$. Replace x by 5 in the equation.

$$y = 2 - 4$$ • Simplify the right side.

$$y = -2$$ • When $x = 5$, $y = -2$.

The ordered-pair solution of $y = \frac{2}{5}x - 4$ in which $x = 5$ is $(5, -2)$.

3.3.1 Graph an equation of the form $y = mx + b$

Solutions of an equation in two variables can be graphed in the rectangular coordinate system.

Graph the ordered-pair solutions of $y = -2x + 1$ for $x = -2, -1, 0, 1,$ and 2.

x	$y = -2x + 1$	y	(x, y)
-2	$-2(-2) + 1$	5	$(-2, 5)$
-1	$-2(-1) + 1$	3	$(-1, 3)$
0	$-2(0) + 1$	1	$(0, 1)$
1	$-2(1) + 1$	-1	$(1, -1)$
2	$-2(2) + 1$	-3	$(2, -3)$

- **Use the given values of x to determine ordered-pair solutions of the equation. It is convenient to record these in a table.**

The ordered-pair solutions of $y = -2x + 1$ for $x = -2, -1$ 0, 1, and 2 are $(-2, 5)$, $(-1, 3)$, $(0, 1)$, $(1, -1)$ and $(2, -3)$. These are graphed at the right.

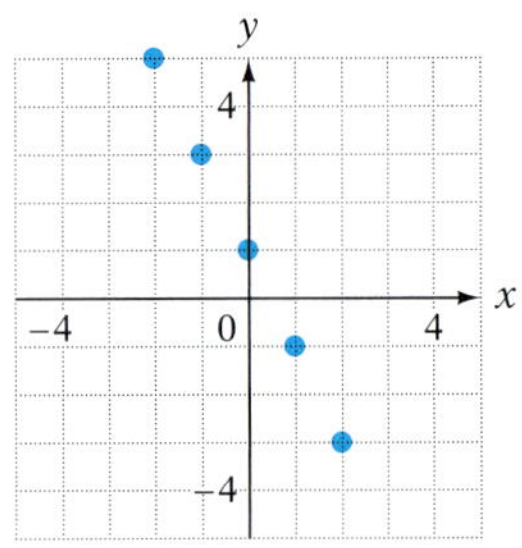

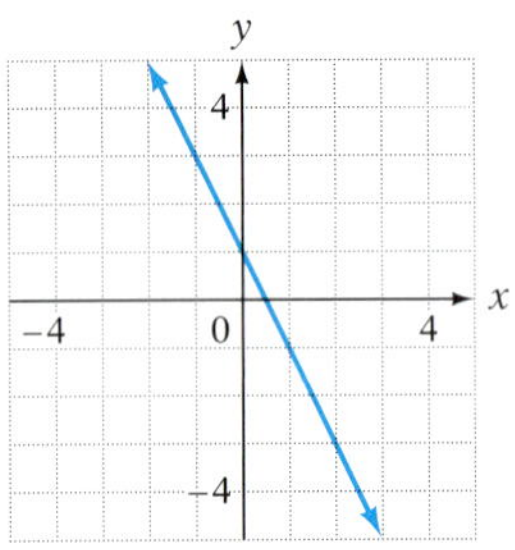

The **graph of an equation in two variables** is a graph of the ordered-pair solutions of the equation. Consider the equation $y = -2x + 1$ above. The ordered-pair solutions $(-2, 5)$, $(-1, 3)$, $(0, 1)$, $(1, -1)$ and $(2, -3)$ are graphed in the figure above. We can choose values of x that are not integers to produce more ordered pairs to graph, such as $\left(\frac{5}{2}, -4\right)$ and $\left(-\frac{1}{2}, 2\right)$. Choosing still other values of x would result in more and more ordered pairs being graphed. The result would be so many dots that the graph would appear as a straight line, as shown at the left. This is the graph of $y = -2x + 1$.

The equation $y = -2x + 1$ is an example of a linear equation because its graph is a straight line. It is also called a first-degree equation in two variables because the exponent on each variable is 1.

Take Note

The equation $y = x^2 + 2x - 4$ is not a linear equation in two variables because there is a term with a variable squared. The equation $y = \frac{5}{x - 3}$ is not a linear equation because a variable occurs in the denominator of a fraction.

Linear Equation in Two Variables

An equation of the form $y = mx + b$, where m is the coefficient of x and b is a constant, is a linear equation in two variables. The graph of a linear equation in two variables is a straight line.

Examples of linear equations are shown at the right.

$$y = 5x + 3 \qquad (m = 5, b = 3)$$
$$y = x - 4 \qquad (m = 1, b = -4)$$
$$y = -\frac{3}{4}x \qquad \left(m = -\frac{3}{4}, b = 0\right)$$

To graph a linear equation, find ordered-pair solutions of the equation. Do this by choosing any value of x and finding the corresponding value of y. Repeat this procedure, choosing different values for x, until you have found the number of solutions desired. Because the graph of a linear equation in two variables is a straight line, and a straight line is determined by two points, it is necessary to find only two solutions. However, it is recommended that at least three points be used to ensure accuracy.

Graph: $y = 2x - 3$

x	$y = 2x - 3$	y
0	$2(0) - 3$	-3
2	$2(2) - 3$	1
-1	$2(-1) - 3$	-5

- **Choose any values of x. Then find the corresponding values of y. The numbers 0, 2, and -1 were chosen arbitrarily for x.**

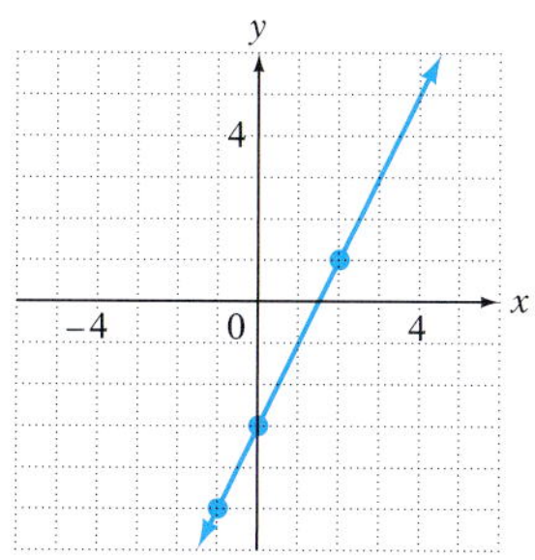

- **Graph the ordered-pair solutions $(0, -3)$, $(2, 1)$, and $(-1, -5)$. Draw a straight line through the points.**

Remember that a graph is a drawing of the ordered-pair solutions of the equation. Therefore, every point on the graph is a solution of the equation, and every solution of the equation is a point on the graph.

When graphing an equation of the form $y = mx + b$, if m is a fraction, choose values of x that will simplify the evaluation. This is illustrated in Example 2. Note that the values of x chosen are multiples of the denominator, 2.

Example 2 Graph $y = -\frac{3}{2}x - 3$.

Solution

x	y
0	-3
-2	0
-4	3

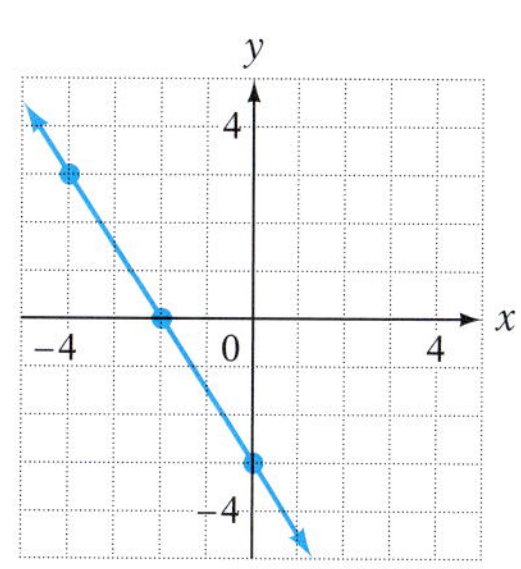

Problem 2 Graph $y = \frac{3}{5}x - 4$.

Solution See page S38.

3.3.2 Graph an equation of the form $Ax + By = C$

An equation of the form $Ax + By = C$ is also a linear equation in two variables. Examples of these equations follow.

$$3x + 4y = 12 \qquad (A = 3, B = 4, C = 12)$$
$$x - 5y = -10 \qquad (A = 1, B = -5, C = -10)$$
$$2x - y = 0 \qquad (A = 2, B = -1, C = 0)$$

One method of graphing an equation of the form $Ax + By = C$ involves first solving the equation for y and then following the same procedure used for graphing an equation of the form $y = mx + b$. To solve the equation for y means to rewrite the equation so that y is alone on one side of the equation and the term containing x and the constant are on the other side of the equation. The Addition and Multiplication Properties of Equations are used to rewrite an equation of the form $Ax + By = C$ in the form $y = mx + b$.

Graph $3x + 2y = 6$.

$3x + 2y = 6$ • **The equation is in the form $Ax + By = C$.**

$2y = -3x + 6$ • **Solve the equation for y. Subtract $3x$ from each side of the equation.**

$y = -\frac{3}{2}x + 3$ • **Divide each side of the equation by 2. Note that each term on the right side is divided by 2.**

• **Find at least three solutions.**

x	y
0	3
2	0
4	−3

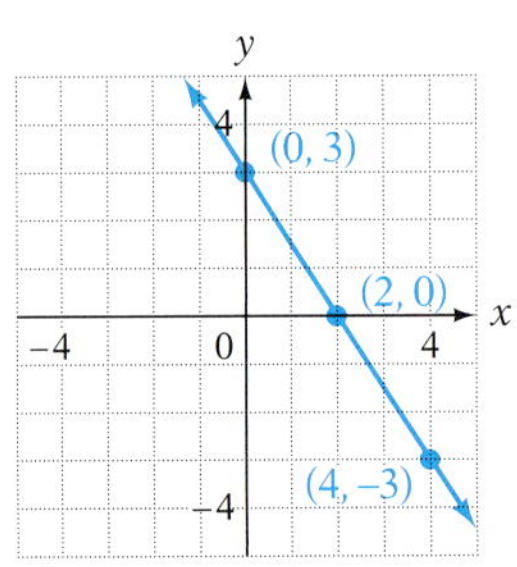

• **Graph the ordered pairs (0, 3), (2, 0), and (4, −3). Draw a straight line through the points.**

Example 3 Graph $2x + 3y = 9$.

Solution

$$2x + 3y = 9$$
$$3y = -2x + 9$$
$$y = -\frac{2}{3}x + 3$$

x	y
−3	5
0	3
3	1

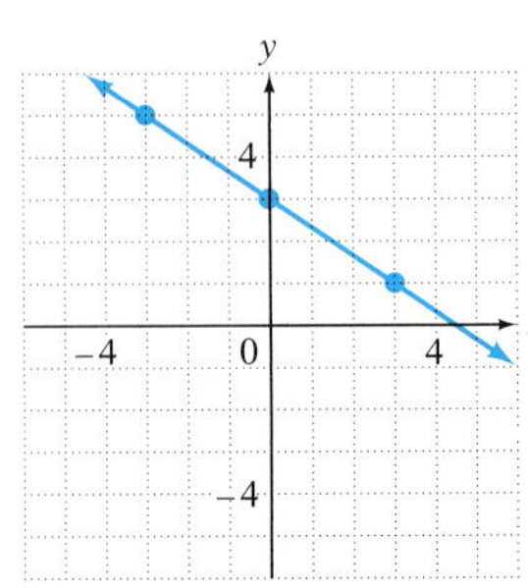

Problem 3 Graph $-3x + 2y = 4$.

Solution See page S38.

3.4.1 Slope

The **slope** of a line is a measure of the slant of the line. The symbol for slope is m.

For an equation of the form $y = mx + b$, m is the slope. Here are a few examples.

The slope of the line $y = -2x + 5$ is -2.

The slope of the line $y = 8x$ is 8.

The slope of the line $y = \frac{3}{4}x - 1$ is $\frac{3}{4}$.

The slope of a line containing two points is the ratio of the change in the y values between the two points to the change in the x values.

Take Note

$$\text{Slope} = m = \frac{\text{change in } y}{\text{change in } x}$$

Slope Formula

The slope of the line containing the two points $P_1(x_1, y_1)$ and $P_2(x_2, y_2)$ is given by

$$m = \frac{y_2 - y_1}{x_2 - x_1}, \qquad x_1 \neq x_2$$

Example 4 Find the slope of the line containing the points $(-4, -3)$ and $(-1, 1)$.

Solution Let $(x_1, y_1) = (-4, -3)$ and $(x_2, y_2) = (-1, 1)$.

$$m = \frac{y_2 - y_1}{x_2 - x_1} = \frac{1 - (-3)}{-1 - (-4)} = \frac{4}{3}$$

The slope is $\frac{4}{3}$.

Problem 4 Find the slope of the line containing the points $(-2, 3)$ and $(1, -3)$.

Solution See page S38.

3.4.2 Slope–intercept form of a straight line

One important characteristic of the graph of a linear equation is its *intercepts*. An ***x*-intercept** is a point at which the graph crosses the x-axis. A ***y*-intercept** is a point at which the graph crosses the y-axis.

The graph of the equation $y = \frac{1}{2}x - 2$ is shown at the right. The x-intercept of the graph is $(4, 0)$. The y-intercept of the graph is $(0, -2)$.

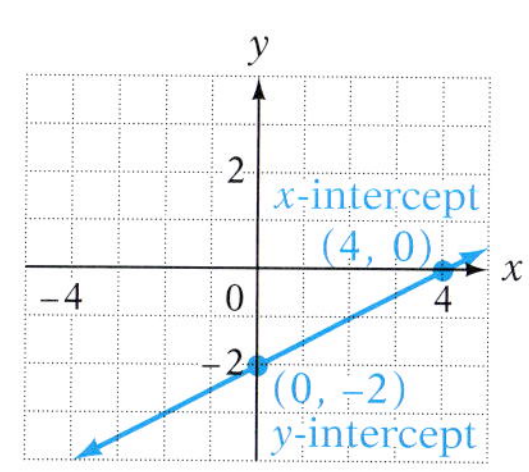

Note that at an x-intercept, the y-coordinate is 0. At a y-intercept, the x-coordinate is 0.

We can find the y-intercept of a linear equation by letting $x = 0$.

Find the y-intercept of the graph of the equation $y = 3x + 4$.

$$\begin{aligned} y &= 3x + 4 \\ y &= 3(0) + 4 && \text{• To find the } y\text{-intercept, let } x = 0. \\ y &= 0 + 4 \\ y &= 4 \end{aligned}$$

The y-intercept is $(0, 4)$.

Note that the constant term of $y = 3x + 4$ is the y-coordinate of the y-intercept.

In general, **for any equation of the form $y = mx + b$, the y-intercept is $(0, b)$.**

Because the slope and the y-intercept can be determined directly from the equation $y = mx + b$, this equation is called the slope–intercept form of a straight line.

Slope–Intercept Form of a Straight Line

The equation $y = mx + b$ is called the **slope–intercept form of a straight line.** The slope of the line is m, the coefficient of x. The y-intercept is $(0, b)$.

The following equations are written in slope–intercept form.

$y = -4x + 3$	Slope $= -4$, y-intercept $= (0, 3)$
$y = \frac{2}{5}x - 1$	Slope $= \frac{2}{5}$, y-intercept $= (0, -1)$
$y = -x$	Slope $= -1$, y-intercept $= (0, 0)$

When an equation is in slope-intercept form, it is possible to draw a graph of the function quickly.

Graph $x + 2y = 4$ by using the slope and y-intercept.

Solve the equation for y.

$$\begin{aligned} x + 2y &= 4 \\ 2y &= -x + 4 \\ y &= -\frac{1}{2}x + 2 \end{aligned}$$

From the equation $y = -\frac{1}{2}x + 2$, the slope is $-\frac{1}{2}$ and the y-intercept is $(0, 2)$.

Take Note

Recall that

slope $= m = \dfrac{\text{change in } y}{\text{change in } x}$.

For the example at the right,

$m = -\dfrac{1}{2} = \dfrac{-1}{2}$

$= \dfrac{\text{change in } y}{\text{change in } x}$

Therefore, the change in y is -1 and the change in x is 2.

Rewrite the slope $-\dfrac{1}{2}$ as $\dfrac{-1}{2}$.

Beginning at the y-intercept, move right 2 units (change in x) and then down 1 unit (change in y).

The point whose coordinates are (2, 1) is a second point on the graph. Draw a straight line through the points (0, 2) and (2, 1).

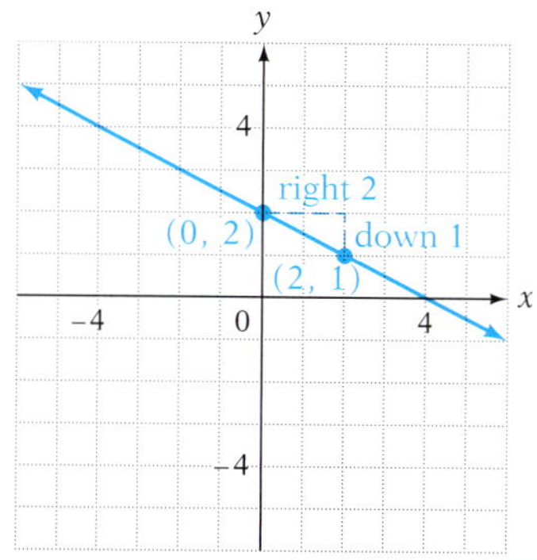

Example 5 Graph $y = -\dfrac{3}{2}x + 4$ by using the slope and the y-intercept.

Solution From the equation, the slope is $-\dfrac{3}{2}$ and the y-intercept is (0, 4).

Rewrite the slope $-\dfrac{3}{2}$ as $\dfrac{-3}{2}$.

Place a dot at the y-intercept.

Starting at the y-intercept, move to the right 2 units (the change in x) and move down 3 units (the change in y). Place a dot at that location. Draw a line through the two points.

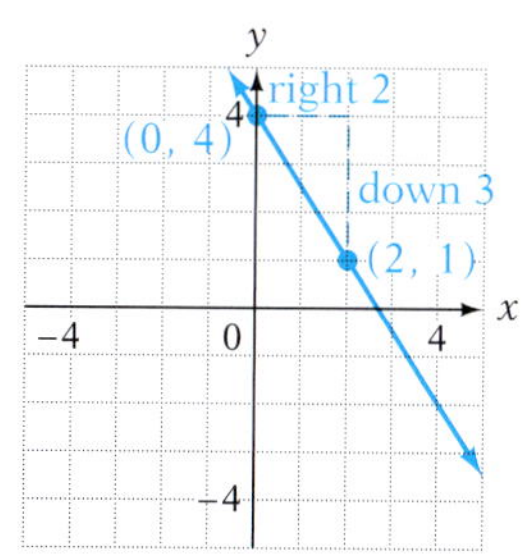

Problem 5 Graph $y = -\dfrac{2}{3}x + 2$ by using the slope and the y-intercept.

Solution See page S38.

3 Evaluate a Function

VIDEO & DVD CD TUTOR WEB SSM

3.2.1 Evaluate a function

A **relation** is a set of ordered pairs. A **function** is a relation in which no two ordered pairs have the same first coordinate and different second coordinates.

The relation {(0, 4), (1, 5), (1, 6), (2, 7)} is not a function because the ordered pairs (1, 5), and (1, 6) have the same first coordinate and different second coordinates.

The relation $\{(0, 4), (1, 5), (2, 6), (3, 7)\}$ is a function.

Take Note

The symbol $f(x)$ is read "the value of f at x" or "f of x."

The phrase "y is a function of x," or a similar phrase with different variables, is used to describe those equations in two variables that define functions. To emphasize that the equation represents a function, functional notation is used. For example, the square function is written in functional notation as follows:

$$f(x) = x^2$$

The process of determining $f(x)$ for a given value of x is called **evaluating the function.** For instance, to evaluate $f(x) = x^2$ when $x = 3$, replace x by 3 and simplify.

$$f(x) = x^2$$
$$f(3) = 3^2 = 9$$

The value of the function is 9 when $x = 3$. An ordered pair of the function is (3, 9).

Example 6 Evaluate $f(x) = 2x - 4$ when $x = 3$. Use your answer to write an ordered pair of the function.

Solution

$$f(x) = 2x - 4$$
$$f(3) = 2(3) - 4$$
$$f(3) = 6 - 4$$
$$f(3) = 2$$

An ordered pair of the function is (3, 2).

Problem 6 Evaluate the function $f(x) = 4 - 2x$ at $x = -3$. Use your answer to write an ordered pair of the function.

Solution See page S38.

4 Find the Equation of a Line

VIDEO & DVD CD TUTOR WEB SSM

3.5.1 Find the equation of a line given a point and the slope

When the slope of a line and a point on the line are known, the equation of the line can be determined by using the point–slope formula.

Point–Slope Formula

Let m be the slope of a line, and let (x_1, y_1) be the coordinates of a point on the line. The equation of the line can be found using the point–slope formula:

$$y - y_1 = m(x - x_1)$$

Example 7 Find the equation of the line that contains the point $(1, -3)$ and has slope -2.

Solution

$$y - y_1 = m(x - x_1)$$
$$y - (-3) = -2(x - 1) \quad \bullet\ (x_1, y_1) = (1, -3),\ m = -2$$
$$y + 3 = -2x + 2$$
$$y = -2x - 1$$

The equation of the line is $y = -2x - 1$.

Problem 7 Find the equation of the line that contains the point $(-2, 2)$ and has slope $-\frac{1}{2}$.

Solution See page S38.

R.3 Exercises

1 Graph the ordered pairs.

1. $(2, 3), (4, 0), (-4, 1), (-2, -2)$
2. $(0, 2), (-4, -1), (2, 0), (1, -3)$
3. $(-2, 5), (3, 4), (0, 0), (-3, -2)$

2 Graph.

4. $y = 2x - 4$
5. $y = 2x + 1$
6. $y = -4x + 1$
7. $y = -3x + 4$
8. $y = 2x$
9. $y = 3x$
10. $y = -\frac{5}{2}x$
11. $y = -\frac{4}{3}x$
12. $y = \frac{2}{3}x + 1$
13. $y = \frac{3}{2}x - 1$
14. $y = -\frac{1}{2}x + 3$
15. $y = -\frac{2}{3}x + 1$
16. $2x - y = 3$
17. $2x + y = -3$
18. $2x + 5y = 10$

19. $x - 4y = 8$

20. $3x - y = -2$

21. $3x - 2y = 8$

Find the slope of the line containing the given points.

22. $P_1(4, 2), P_2(3, 4)$

23. $P_1(2, 1), P_2(3, 4)$

24. $P_1(-1, 3), P_2(2, 4)$

25. $P_1(-2, 1), P_2(2, 2)$

26. $P_1(2, 4), P_2(4, -1)$

27. $P_1(1, 3), P_2(5, -3)$

28. $P_1(3, -4), P_2(3, 5)$

29. $P_1(-1, 2), P_2(-1, 3)$

30. $P_1(4, -2), P_2(3, -2)$

31. $P_1(5, 1), P_2(-2, 1)$

32. $P_1(0, -1), P_2(3, -2)$

33. $P_1(3, 0), P_2(2, -1)$

Give the slope and the y-intercept of the graph of the equation.

34. $y = -3x + 7$

35. $y = \frac{5}{2}x - 4$

36. $y = x$

Graph by using the slope and the y-intercept.

37. $y = \frac{1}{2}x + 2$

38. $y = \frac{2}{3}x - 3$

39. $y = -\frac{3}{2}x$

40. $y = \frac{3}{4}x$

41. $y = -\frac{1}{2}x + 2$

42. $y = \frac{2}{3}x - 1$

43. $x - 3y = 3$

44. $3x + 2y = 8$

45. $4x + y = 2$

3 Evaluate the function for the given value of the variable. Use your answer to write an ordered pair of the function.

46. $f(x) = 2x + 7; x = -2$

47. $g(x) = -3x + 1; x = -4$

48. $h(x) = 4x - 2; x = 3$

49. $p(x) = 6 - 8x; x = -1$

50. $t(x) = 5 - 7x; x = 0$

51. $f(t) = t^2 - t - 3; t = 2$

52. $p(n) = n^2 - 4n - 7; n = -3$

53. $h(x) = -3x^2 + x - 1; x = -2$

54. $g(t) = 4t^3 - 2t; t = -1$

4 Find the equation of the line that contains the given point and has the given slope.

55. Point $(-1, 2)$; $m = -3$

56. Point $(2, -3)$; $m = 3$

57. Point $(4, -5)$; $m = -2$

58. Point $(-3, 5)$; $m = 3$

59. Point $(5, -3)$; $m = -\frac{3}{5}$

60. Point $(3, 1)$; $m = \frac{1}{3}$

61. Point $(-3, -2)$; $m = -\frac{2}{3}$

62. Point $(4, -2)$; $m = \frac{3}{4}$

63. Point $(2, 3)$; $m = -\frac{1}{2}$

R.4 Polynomials

1 Multiply and Divide Monomials

VIDEO & DVD   CD TUTOR   WEB   SSM

A **monomial** is a number, a variable, or a product of a number and variables. The following rules and definitions are used to multiply and divide monomials and to write monomials in simplest form.

5.2.1
5.2.2 Multiply and divide monomials
5.4.1

Rules of Exponents

Rule for Multiplying Exponential Expressions

If m and n are integers, then $x^m \cdot x^n = x^{m+n}$.

Rule for Simplifying the Power of an Exponential Expression

If m and n are integers, then $(x^m)^n = x^{m \cdot n}$.

Rule for Simplifying Powers of Products

If m, n, and p are integers, then $(x^m y^n)^p = x^{m \cdot p} y^{n \cdot p}$.

Rule for Dividing Exponential Expressions

If m and n are integers and $x \neq 0$, then $\frac{x^m}{x^n} = x^{m-n}$.

Rule for Simplifying Powers of Quotients

If m, n, and p are integers and $y \neq 0$, then $\left(\frac{x^m}{y^n}\right)^p = \frac{x^{m \cdot p}}{y^{n \cdot p}}$.

Definition of Zero as an Exponent

If $x \neq 0$, then $x^0 = 1$. The expression 0^0 is undefined.

Definition of Negative Exponents

If n is a positive integer and $x \neq 0$, then $x^{-n} = \frac{1}{x^n}$ and $\frac{1}{x^{-n}} = x^n$.

Simplify: $(3x^4)^2(4x^3)$

$$\begin{aligned}(3x^4)^2(4x^3) &= (3^{1\cdot 2}x^{4\cdot 2})(4x^3) \\ &= (3^2x^8)(4x^3) \\ &= (9x^8)(4x^3) \\ &= (9 \cdot 4)(x^8 \cdot x^3) \\ &= 36x^{8+3} \\ &= 36x^{11}\end{aligned}$$

- Use the Rule for Simplifying Powers of Products to simplify $(3x^4)^2$.
- Use the Rule for Multiplying Exponential Expressions.

An exponential expression is in simplest form when there are no negative exponents in the expression. For example, the expression y^{-7} is not in simplest form; use the Definition of Negative Exponents to rewrite the expression with a positive exponent: $y^{-7} = \frac{1}{y^7}$. The expression $\frac{1}{c^{-4}}$ is not in simplest form; use the Definition of Negative Exponents to rewrite the expression with a positive exponent: $\frac{1}{c^{-4}} = c^4$.

Simplify: $\dfrac{6x^2}{8x^9}$

$$\begin{aligned}\frac{6x^2}{8x^9} &= \frac{3x^2}{4x^9} = \frac{3x^{2-9}}{4} \\ &= \frac{3x^{-7}}{4} = \frac{3}{4} \cdot \frac{x^{-7}}{1} = \frac{3}{4} \cdot \frac{1}{x^7} \\ &= \frac{3}{4x^7}\end{aligned}$$

- Divide the coefficients by their common factors. Use the Rule for Dividing Exponential Expressions.
- Rewrite the expression with only positive exponents.

Simplify: $\left(\dfrac{a^4}{b^3}\right)^{-2}$

$$\begin{aligned}\left(\frac{a^4}{b^3}\right)^{-2} &= \frac{a^{4(-2)}}{b^{3(-2)}} = \frac{a^{-8}}{b^{-6}} \\ &= \frac{b^6}{a^8}\end{aligned}$$

- Use the Rule for Simplifying Powers of Quotients.
- Rewrite the expression with positive exponents.

Example 1 Simplify.

a. $(-2x)(3x^{-2})^{-3}$ **b.** $\left(\dfrac{3a^2b^{-1}}{27a^{-3}b^{-4}}\right)^{-2}$

Solution

a. $(-2x)(3x^{-2})^{-3}$

$$\begin{aligned}&= (-2x)(3^{-3}x^6) \\ &= \frac{-2x \cdot x^6}{3^3} \\ &= -\frac{2x^7}{27}\end{aligned}$$

- Use the Rule for Simplifying Powers of Products.
- Write the expression with positive exponents.
- Use the Rule for Multiplying Exponential Expressions. Simplify 3^3.

b. Use the Rule for Simplifying Powers of Quotients. Then simplify the expression and write it with positive exponents.

$$\left(\frac{3a^2b^{-1}}{27a^{-3}b^{-4}}\right)^{-2} = \left(\frac{a^2b^{-1}}{9a^{-3}b^{-4}}\right)^{-2}$$
$$= \frac{a^{2(-2)}b^{(-1)(-2)}}{9^{1(-2)}a^{(-3)(-2)}b^{(-4)(-2)}}$$
$$= \frac{a^{-4}b^2}{9^{-2}a^6b^8}$$
$$= 9^2a^{-4-6}b^{2-8}$$
$$= 81a^{-10}b^{-6} = \frac{81}{a^{10}b^6}$$

Problem 1 Simplify.

a. $(-2ab)(2a^3b^{-2})^{-3}$ **b.** $\left(\frac{2x^2y^{-4}}{4x^{-2}y^{-5}}\right)^{-3}$

Solution See page S38.

2 Add and Subtract Polynomials

A **polynomial** is a variable expression in which the terms are monomials. The polynomial $15t^2 - 2t + 3$ has three terms: $15t^2$, $-2t$, and 3. Note that each of these three terms is a monomial.

A polynomial of *one* term is a **monomial**. $-7x^2$ is a monomial.
A polynomial of *two* terms is a **binomial**. $4y + 3$ is a binomial.
A polynomial of *three* terms is a **trinomial**. $6b^2 + 5b - 8$ is a trinomial.

5.1.2 Add and subtract polynomials

Polynomials can be added by combining like terms. This is illustrated in Example 2.

Example 2 Add: $(8x^2 - 4x - 9) + (2x^2 + 9x - 9)$

Solution

$$(8x^2 - 4x - 9) + (2x^2 + 9x - 9) = (8x^2 + 2x^2) + (-4x + 9x) + (-9 - 9)$$
$$= 10x^2 + 5x - 18$$

Problem 2 Add: $(-4x^3 + 2x^2 - 8) + (4x^3 + 6x^2 - 7x + 5)$

Solution See page S38.

The additive inverse of the polynomial $(3x^2 - 7x + 8)$ is $-(3x^2 - 7x + 8)$.

To find the additive inverse of a polynomial, change the sign of each term inside the parentheses.

$$-(3x^2 - 7x + 8) = -3x^2 + 7x - 8$$

To subtract two polynomials, add the additive inverse of the second polynomial to the first.

Simplify: $(5a^2 - a + 2) - (-2a^3 + 3a - 3)$

$(5a^2 - a + 2) - (-2a^3 + 3a - 3)$

$= (5a^2 - a + 2) + (2a^3 - 3a + 3)$ • **Rewrite subtraction as addition of the additive inverse.**

$= 2a^3 + 5a^2 - 4a + 5$ • **Combine like terms.**

Example 3 Subtract: $(7c^2 - 9c - 12) - (9c^2 + 5c - 8)$

Solution

$$(7c^2 - 9c - 12) - (9c^2 + 5c - 8) = (7c^2 - 9c - 12) + (-9c^2 - 5c + 8)$$
$$= -2c^2 - 14c - 4$$

Problem 3 Subtract: $(-4w^3 + 8w - 8) - (3w^3 - 4w^2 + 2w - 1)$

Solution See page S38.

3 Multiply Polynomials

5.3.1 Multiply a polynomial by a monomial

The Distributive Property is used to multiply a polynomial by a monomial. Each term of the polynomial is multiplied by the monomial.

Multiply: $3x^3(4x^4 - 2x + 5)$

$3x^3(4x^4 - 2x + 5)$

$= 3x^3(4x^4) - 3x^3(2x) + 3x^3(5)$ • **Use the Distributive Property. Multiply each term of the polynomial by $3x^3$.**

$= 12x^7 - 6x^4 + 15x^3$ • **Use the Rule for Multiplying Exponential Expressions.**

Example 4 Multiply: $2xy(3x^2 - xy + 2y^2)$

Solution $2xy(3x^2 - xy + 2y^2) = 2xy(3x^2) - 2xy(xy) + 2xy(2y^2)$
$$= 6x^3y - 2x^2y^2 + 4xy^3$$

Problem 4 Multiply: $3mn^2(2m^2 - 3mn - 1)$

Solution See page S39.

5.3.2 Multiply two polynomials

A vertical format similar to that used for multiplication of whole numbers is used to multiply two polynomials. The product $(2y - 3)(y^2 + 2y + 5)$ is shown below.

$$\begin{array}{rrrrl}
 & y^2 & + 2y & + 5 & \\
 & & 2y & - 3 & \\
\hline
 & -3y^2 & - 6y & - 15 & \text{This is } -3(y^2 + 2y + 5). \\
2y^3 & + 4y^2 & + 10y & & \text{This is } 2y(y^2 + 2y + 5)\text{. Like terms are placed in the same columns.} \\
\hline
2y^3 & + y^2 & + 4y & - 15 & \text{Add the terms in each column.}
\end{array}$$

Example 5 Multiply: $(2a^2 + 4a - 5)(3a + 5)$

Solution

$$\begin{array}{r} 2a^2 + 4a - 5 \\ 3a + 5 \\ \hline 10a^2 + 20a - 25 \\ 6a^3 + 12a^2 - 15a \quad \\ \hline 6a^3 + 22a^2 + 5a - 25 \end{array}$$

• **Align like terms in the same column.**

Problem 5 Multiply: $(3c^2 - 4c + 5)(2c - 3)$

Solution See page S39.

5.3.3 Multiply two binomials

It is frequently necessary to multiply two binomials. The product is computed by using a method called FOIL. The letters of FOIL stand for **F**irst, **O**uter, **I**nner, and **L**ast. The FOIL method is based on the Distributive Property and involves adding the products of the first terms, the outer terms, the inner terms, and the last terms.

The product $(2x + 3)(3x + 4)$ is shown below using FOIL.

Last; First; Inner; Outer

First terms	Outer terms	Inner terms	Last terms

$$(2x + 3) \cdot (3x + 4) = (2x)(3x) + (2x)(4) + (3)(3x) + (3)(4)$$
$$= 6x^2 + 8x + 9x + 12$$
$$= 6x^2 + 17x + 12$$

Example 6 Multiply: $(4x - 3)(2x + 5)$

Solution

$$\begin{aligned}(4x - 3)(2x + 5) &= (4x)(2x) + (4x)(5) + (-3)(2x) + (-3)(5) \\ &= 8x^2 + 20x + (-6x) + (-15) \\ &= 8x^2 + 14x - 15\end{aligned}$$

Problem 6 Multiply: $(4y - 7)(3y - 5)$

Solution See page S39.

4 Divide Polynomials

5.5.2 Divide Polynomials

To divide two polynomials, use a method similar to that used for division of whole numbers.

To divide $(x^2 - 5x + 8) \div (x - 3)$:

Step 1:

$$\begin{array}{r} x \\ x - 3\overline{)x^2 - 5x + 8} \\ \underline{x^2 - 3x} \downarrow \\ -2x + 8 \end{array}$$

Think: $x\overline{)x^2} = \frac{x^2}{x} = x$

Multiply: $x(x - 3) = x^2 - 3x$

Subtract: $(x^2 - 5x) - (x^2 - 3x) = -2x$
Bring down the $+8$.

Step 2:

$$\begin{array}{r} x - 2 \\ x - 3\overline{)x^2 - 5x + 8} \\ \underline{x^2 - 3x} \\ -2x + 8 \\ \underline{-2x + 6} \\ 2 \end{array}$$

Think: $x\overline{)-2x} = \frac{-2x}{x} = -2$

Multiply: $-2(x - 3) = -2x + 6$

Subtract: $(-2x + 8) - (-2x + 6) = 2$
The remainder is 2.

Check: **(Quotient × Divisor) + Remainder = Dividend**

$(x - 2)(x - 3) + 2 = x^2 - 3x - 2x + 6 + 2 = x^2 - 5x + 8$

$(x^2 - 5x + 8) \div (x - 3) = x - 2 + \dfrac{2}{x - 3}$

Example 7 Divide: $(6x + 2x^3 + 26) \div (x + 2)$

Solution Arrange the terms of the dividend in descending order. There is no x^2 term in $2x^3 + 6x + 26$. Insert $0x^2$ for the missing term so that like terms will be in columns.

$$\begin{array}{r} 2x^2 - 4x + 14 \\ x + 2\overline{)2x^3 + 0x^2 + 6x + 26} \\ \underline{2x^3 + 4x^2} \\ -4x^2 + 6x \\ \underline{-4x^2 - 8x} \\ 14x + 26 \\ \underline{14x + 28} \\ -2 \end{array}$$

Check: $(x + 2)(2x^2 - 4x + 14) - 2$
$= 2x^3 + 6x + 28 - 2 = 2x^3 + 6x + 26$

$(6x + 2x^3 + 26) \div (x + 2) = 2x^2 - 4x + 14 - \dfrac{2}{x + 2}$

Problem 7 Divide: $(x^3 - 7 - 2x) \div (x - 2)$

Solution See page S39.

5 Factor Trinomials of the Form $ax^2 + bx + c$

6.1.1 Factor a monomial from a polynomial

A polynomial is in **factored form** when it is written as a product of other polynomials. Factoring can be thought of as the reverse of multiplication.

Polynomial		Factored Form
$2x^3 + 6x^2 - 10x$	=	$2x(x^2 + 3x - 5)$
$x^2 - 3x - 28$	=	$(x + 4)(x - 7)$

To factor out a common monomial from the terms of a polynomial, first find the greatest common factor (GCF) of the terms.

The GCF of two or more monomials is the product of the GCF of the coefficients and the common variable factors.

$10a^3b = \mathbf{2} \cdot 5 \cdot \boldsymbol{a} \cdot \boldsymbol{a} \cdot a \cdot \boldsymbol{b}$
$4a^2b^2 = \mathbf{2} \cdot 2 \cdot \boldsymbol{a} \cdot \boldsymbol{a} \cdot \boldsymbol{b} \cdot b$
$\text{GCF} = \mathbf{2} \cdot \boldsymbol{a} \cdot \boldsymbol{a} \cdot \boldsymbol{b} = 2a^2b$

Note that the exponent of each variable in the GCF is the same as the smallest exponent of that variable in either of the monomials.

The GCF of $10a^3b$ and $4a^2b^2$ is $2a^2b$.

Factor: $5x^2 - 35x^2 + 10x$

The GCF is $5x$.

$\dfrac{5x^3 - 35x^2 + 10x}{5x} = x^2 - 7x + 2$

- Find the GCF of the terms $5x^3$, $-35x^2$, and $10x$.
- Divide each term of the polynomial by the GCF.

$5x^3 - 35x^2 + 10x = 5x(x^2 - 7x + 2)$

- Write the polynomial as the product of the GCF and the quotient found above.

$5x(x^2 - 7x + 2) = 5x^3 - 35x^2 + 10x$

- Check the factorization by multiplying.

Example 8 Factor: $16x^4y^5 + 8x^4y^2 - 12x^3y$

Solution The GCF is $4x^3y$.

$16x^4y^5 + 8x^4y^2 - 12x^3y = 4x^3y(4xy^4 + 2xy - 3)$

Problem 8 Factor: $6x^4y^2 - 9x^3y^2 + 12x^2y^4$

Solution See page S39.

A **quadratic trinomial** is a trinomial of the form $ax^2 + bx + c$, where a and b are coefficients and c is a constant. Examples of quadratic trinomials follow.

$$x^2 + 9x + 14 \qquad a = 1,\ b = 9,\ c = 14$$
$$x^2 - 2x - 15 \qquad a = 1,\ b = -2,\ c = -15$$
$$3x^2 - x + 4 \qquad a = 3,\ b = -1,\ c = 4$$

To **factor a quadratic trinomial** means to express the trinomial as the product of two binomials. For example,

Trinomial		**Factored Form**
$2x^2 - x - 1$	=	$(2x + 1)(x - 1)$
$y^2 - 3y + 2$	=	$(y - 1)(y - 2)$

We will begin by factoring trinomials of the form $x^2 + bx + c$, where the coefficient of x^2 is 1.

6.2.1 Factor a trinomial of the form $x^2 + bx + c$

The method by which factors of a trinomial are found is based on FOIL. Consider the following binomial products, noting the relationship between the constant term of the binomials and the terms of the trinomial.

Sum of the binomial constants

Product of the binomial constants

$$(x + 4)(x + 5) = x \cdot x + 5x + 4x + 4 \cdot 5 \quad = x^2 + 9x + 20$$
$$(x - 6)(x + 8) = x \cdot x + 8x - 6x + (-6)(8) \quad = x^2 + 2x - 48$$
$$(x - 3)(x - 2) = x \cdot x - 2x - 3x + (-3)(-2) \quad = x^2 - 5x + 6$$

Factor: $x^2 - 7x + 10$

Find two integers whose product is 10 and whose sum is −7.

Negative Factors of 10	Sum	
−1, −10	−11	
−2, −5	**−7**	• These are the correct factors.

$x^2 - 7x + 10 = (x - 2)(x - 5)$ • Write the trinomial as a product of the factors.

Check: $(x - 2)(x - 5) = x^2 - 5x - 2x + 10$
$= x^2 - 7x + 10$

Take Note

Once the correct factors are found, it is not necessary to try the remaining factors.

Example 9 Factor: $x^2 + 6x - 27$

Solution Two factors of −27 whose sum is 6 are −3 and 9.

$x^2 + 6x - 27 = (x - 3)(x + 9)$

Problem 9 Factor: $x^2 - 8x + 15$

Solution See page S39.

6.3.1 Factor a trinomial of the form $ax^2 + bx + c$ by using trial factors

To use the trial factor method to factor a trinomial of the form $ax^2 + bx + c$, where $a \neq 1$, use the factors of a and the factors of c to write all of the possible binomial factors of the trinomial. Then use FOIL to determine the correct factorization. To reduce the number of trial factors that must be considered, remember the following guidelines.

Use the signs of the constant term and the coefficient of x in the trinomial to determine the signs of the binomial factors. If the constant term is positive, the signs of the binomial factors will be the same as the sign of the coefficient of x in the trinomial. If the sign of the constant term is negative, the constant terms in the binomials will have different signs.

Factor: $2x^2 - 7x + 3$

Because the constant term is positive (+3) and the coefficient of x is negative (−7), the binomial constants will be negative.

Positive Factors of 2 ($a = 2$)	Negative Factors of 3 ($c = 3$)
1, 2	−1, −3

Write trial factors. Use the **O**uter and **I**nner products of FOIL to determine the middle term, $-7x$, of the trinomial.

Trial Factors	**Middle Term**	
$(x - 1)(2x - 3)$	$-3x - 2x = -5x$	
$\mathbf{(x - 3)(2x - 1)}$	$\mathbf{-x - 6x = -7x}$	• $-7x$ is the middle term.

$2x^2 - 7x + 3 = (x - 3)(2x - 1)$

Check: $(x - 3)(2x - 1) = 2x^2 - x - 6x + 3 = 2x^2 - 7x + 3$

Factor: $5x^2 + 22x - 15$

The constant term is negative (-15). The binomial constants will have different signs.

Positive Factors of 5 ($a = 5$)	**Factors of -15 ($c = -15$)**
1, 5	$-1, 15$
	$1, -15$
	$-3, 5$
	$3, -5$

Write trial factors. Use the **O**uter and **I**nner products of FOIL to determine the middle term, $22x$, of the trinomial.

Take Note

It is not necessary to test trial factors that have a common factor. If the trinomial does not have a common factor, then its factors cannot have a common factor.

Trial Factors	**Middle Term**	
$(x - 1)(5x + 15)$	common factor	
$(x + 15)(5x - 1)$	$-x + 75x = 74x$	
$(x + 1)(5x - 15)$	common factor	
$(x - 15)(5x + 1)$	$x - 75x = -74x$	
$(x - 3)(5x + 5)$	common factor	
$\mathbf{(x + 5)(5x - 3)}$	$\mathbf{-3x + 25x = 22x}$	• $22x$ is the middle term.
$(x + 3)(5x - 5)$	common factor	
$(x - 5)(5x + 3)$	$3x - 25x = -22x$	

$5x^2 + 22x - 15 = (x + 5)(5x - 3)$

Check: $(x + 5)(5x - 3) = 5x^2 - 3x + 25x - 15 = 5x^2 + 22x - 15$

Example 10 Factor: $2x^2 + 3x - 5$

Solution

Positive factors of 2	**Factors of -5**
1, 2	$1, -5$
	$-1, 5$

Trial Factors	**Middle Term**
$(x + 1)(2x - 5)$	$-5x + 2x = -3x$
$(x - 5)(2x + 1)$	$x - 10x = -9x$
$(x - 1)(2x + 5)$	$5x - 2x = 3x$
$(x + 5)(2x - 1)$	$-x + 10x = 9x$

$2x^2 + 3x - 5 = (x - 1)(2x + 5)$

Check: $(x - 1)(2x + 5) = 2x^2 + 5x - 2x - 5$
$= 2x^2 + 3x - 5$

Problem 10 Factor: $3x^2 - x - 2$

Solution See page S39.

A polynomial is factored completely when it is written as a product of factors that are nonfactorable over the integers.

6.4.4 Factor completely

The first step in any factoring problem is to determine whether the terms of the polynomial have a common factor. If they do, factor it out first.

Example 11 Factor: $5x^2y + 60xy + 100y$

Solution There is a common factor, $5y$.

Factor out the GCF.

$$5x^2y + 60xy + 100y = 5y(x^2 + 12x + 20)$$

Factor $x^2 + 12x + 20$. The two factors of 20 whose sum is 12 are 2 and 10.

$$5y(x^2 + 12x + 20) = 5y(x + 2)(x + 10)$$
$$5x^2y + 60xy + 100y = 5y(x + 2)(x + 10)$$

Check: $5y(x + 2)(x + 10) = (5xy + 10y)(x + 10)$
$= 5x^2y + 50xy + 10xy + 100y$
$= 5x^2y + 60xy + 100y$

Problem 11 Factor: $4a^3 - 4a^2 - 24a$

Solution See page S39.

R.4 Exercises

1 Simplify.

1. $z^3 \cdot z \cdot z^4$
2. $b \cdot b^2 \cdot b^6$
3. $(x^3)^5$
4. $(b^2)^4$
5. $(x^2y^3)^6$
6. $(m^4n^2)^3$
7. $\dfrac{a^8}{a^2}$
8. $\dfrac{c^{12}}{c^5}$
9. $(-m^3n)(m^6n^2)$
10. $(-r^4t^3)(r^2t^9)$
11. $(-2a^3bc^2)^3$
12. $(-4xy^3z^2)^2$
13. $\dfrac{m^4n^7}{m^3n^5}$
14. $\dfrac{a^5b^6}{a^3b^2}$
15. $\dfrac{-16a^7}{24a^6}$
16. $\dfrac{18b^5}{-45b^4}$
17. $(9mn^4p)(-3mp^2)$
18. $(-3v^2wz)(-4vz^4)$
19. $(-2n^2)(-3n^4)^3$
20. $(-3m^3n)(-2m^2n^3)^3$

21. $\dfrac{14x^4y^6z^2}{16x^3y^9z}$
22. $\dfrac{25x^4y^7z^2}{20x^5y^9z^{11}}$
23. $(-2x^3y^2)^3(-xy^2)^4$
24. $(-m^4n^2)^5(-2m^3n^3)^3$
25. $4x^{-7}$
26. $-6y^{-1}$
27. $d^{-4}d^{-6}$
28. $x^{-3}x^{-5}$
29. $\dfrac{x^{-3}}{x^2}$
30. $\dfrac{x^4}{x^{-5}}$
31. $\dfrac{1}{3x^{-2}}$
32. $\dfrac{2}{5c^{-6}}$
33. $(x^2y^{-4})^3$
34. $(x^3y^5)^{-4}$
35. $(3x^{-1}y^{-2})^2$
36. $(5xy^{-3})^{-2}$
37. $(2x^{-1})(x^{-3})$
38. $(-2x^{-5})(x^7)$
39. $\dfrac{3x^{-2}y^2}{6xy^2}$
40. $\dfrac{2x^{-2}y}{8xy}$
41. $\dfrac{2x^{-1}y^{-4}}{4xy^2}$
42. $\dfrac{3a^{-2}b}{ab}$
43. $(x^{-2}y)^2(xy)^{-2}$
44. $(x^{-1}y^2)^{-3}(x^2y^{-4})^{-3}$
45. $\left(\dfrac{x^2y^{-1}}{xy}\right)^{-4}$
46. $\left(\dfrac{x^{-2}y^{-4}}{x^{-2}y}\right)^{-2}$
47. $\left(\dfrac{4a^{-2}b}{8a^3b^{-4}}\right)^2$
48. $\left(\dfrac{6ab^{-2}}{3a^{-2}b}\right)^{-2}$

2 Add or subtract.

49. $(4b^2 - 5b) + (3b^2 + 6b - 4)$
50. $(2c^2 - 4) + (6c^2 - 2c + 4)$
51. $(2a^2 - 7a + 10) + (a^2 + 4a + 7)$
52. $(-6x^2 + 7x + 3) + (3x^2 + x + 3)$
53. $(x^2 - 2x + 1) - (x^2 + 5x + 8)$
54. $(3x^2 + 2x - 2) - (5x^2 - 5x + 6)$
55. $(-2x^3 + x - 1) - (-x^2 + x - 3)$
56. $(2x^2 + 5x - 3) - (3x^3 + 2x - 5)$
57. $(x^3 - 7x + 4) + (2x^2 + x - 10)$
58. $(3y^3 + y^2 + 1) + (-4y^3 - 6y - 3)$
59. $(5x^3 + 7x - 7) + (10x^2 - 8x + 3)$
60. $(3y^3 + 4y + 9) + (2y^2 + 4y - 21)$
61. $(2y^3 + 6y - 2) - (y^3 + y^2 + 4)$
62. $(-2x^2 - x + 4) - (-x^3 + 3x - 2)$
63. $(4y^3 - y - 1) - (2y^2 - 3y + 3)$
64. $(3x^2 - 2x - 3) - (2x^3 - 2x^2 + 4)$

3 Multiply.

65. $4b(3b^3 - 12b^2 - 6)$
66. $-2a^2(3a^2 - 2a + 3)$
67. $3b(3b^4 - 3b^2 + 8)$
68. $-2x^2(2x^2 - 3x - 7)$
69. $-2x^2y(x^2 - 3xy + 2y^2)$
70. $3ab^2(3a^2 - 2ab + 4b^2)$
71. $(x^2 + 3x + 2)(x + 1)$
72. $(x^2 - 2x + 7)(x - 2)$
73. $(a - 3)(a^2 - 3a + 4)$
74. $(2x - 3)(x^2 - 3x + 5)$

75. $(-2b^2 - 3b + 4)(b - 5)$

76. $(-a^2 + 3a - 2)(2a - 1)$

77. $(x^3 - 3x + 2)(x - 4)$

78. $(y^3 + 4y^2 - 8)(2y - 1)$

79. $(y + 2)(y^3 + 2y^2 - 3y + 1)$

80. $(2a - 3)(2a^3 - 3a^2 + 2a - 1)$

81. $(a - 3)(a + 4)$

82. $(b - 6)(b + 3)$

83. $(y - 7)(y - 3)$

84. $(a - 8)(a - 9)$

85. $(2x + 1)(x + 7)$

86. $(y + 2)(5y + 1)$

87. $(3x - 1)(x + 4)$

88. $(7x - 2)(x + 4)$

89. $(4x - 3)(x - 7)$

90. $(2x - 3)(4x - 7)$

91. $(3y - 8)(y + 2)$

92. $(5y - 9)(y + 5)$

93. $(7a - 16)(3a - 5)$

94. $(5a - 12)(3a - 7)$

95. $(x + y)(2x + y)$

96. $(2a + b)(a + 3b)$

97. $(3x - 4y)(x - 2y)$

98. $(2a - b)(3a + 2b)$

99. $(5a - 3b)(2a + 4b)$

100. $(2x + 3)(2x - 3)$

101. $(4x - 7)(4x + 7)$

4 Divide.

102. $(b^2 - 14b + 49) \div (b - 7)$

103. $(x^2 - x - 6) \div (x - 3)$

104. $(2x^2 + 5x + 2) \div (x + 2)$

105. $(2y^2 - 13y + 21) \div (y - 3)$

106. $(x^2 + 1) \div (x - 1)$

107. $(x^2 + 4) \div (x + 2)$

108. $(6x^2 - 7x) \div (3x - 2)$

109. $(6y^2 + 2y) \div (2y + 4)$

110. $(a^2 + 5a + 10) \div (a + 2)$

111. $(b^2 - 8b - 9) \div (b - 3)$

112. $(2y^2 - 9y + 8) \div (2y + 3)$

113. $(3x^2 + 5x - 4) \div (x - 4)$

114. $(8x + 3 + 4x^2) \div (2x - 1)$

115. $(10 + 21y + 10y^2) \div (2y + 3)$

116. $(x^3 + 3x^2 + 5x + 3) \div (x + 1)$

117. $(x^3 - 6x^2 + 7x - 2) \div (x - 1)$

118. $(x^4 - x^2 - 6) \div (x^2 + 2)$

119. $(x^4 + 3x^2 - 10) \div (x^2 - 2)$

5 Factor.

120. $8x + 12$ **121.** $12y^2 - 5y$ **122.** $10x^4 - 12x^2$ **123.** $10x^2yz^2 + 15xy^3z$

124. $x^3 - 3x^2 - x$ **125.** $5x^2 - 15x + 35$ **126.** $3x^3 + 6x^2 + 9x$ **127.** $3y^4 - 9y^3 - 6y^2$

128. $2x^3 + 6x^2 - 14x$ **129.** $x^4y^4 - 3x^3y^3 + 6x^2y^2$

130. $4x^5y^5 - 8x^4y^4 + x^3y^3$ **131.** $16x^2y - 8x^3y^4 - 48x^2y^2$

132. $x^2 + 5x + 6$ **133.** $x^2 + x - 2$ **134.** $x^2 + x - 6$ **135.** $a^2 + a - 12$

136. $a^2 - 2a - 35$ **137.** $a^2 - 3a + 2$ **138.** $a^2 - 5a + 4$ **139.** $b^2 + 7b - 8$

140. $y^2 + 6y - 55$ **141.** $z^2 - 4z - 45$ **142.** $y^2 - 8y + 15$ **143.** $z^2 - 14z + 45$

144. $p^2 + 12p + 27$ **145.** $b^2 + 9b + 20$ **146.** $y^2 - 8y + 32$ **147.** $y^2 - 9y + 81$

148. $p^2 + 24p + 63$ **149.** $x^2 - 15x + 56$ **150.** $5x^2 + 6x + 1$ **151.** $2y^2 + 7y + 3$

152. $2a^2 - 3a + 1$ **153.** $3a^2 - 4a + 1$ **154.** $4x^2 - 3x - 1$ **155.** $2x^2 - 5x - 3$

156. $6t^2 - 11t + 4$ **157.** $10t^2 + 11t + 3$ **158.** $8x^2 + 33x + 4$ **159.** $10z^2 + 3z - 4$

160. $3x^2 + 14x - 5$ **161.** $3z^2 + 95z + 10$ **162.** $8z^2 - 36z + 1$ **163.** $2t^2 - t - 10$

164. $2t^2 + 5t - 12$ **165.** $12y^2 + 19y + 5$ **166.** $5y^2 - 22y + 8$ **167.** $11a^2 - 54a - 5$

168. $4z^2 + 11z + 6$ **169.** $6b^2 - 13b + 6$ **170.** $6x^2 + 35x - 6$ **171.** $3x^2 + 15x + 18$

172. $3a^2 + 3a - 18$ **173.** $ab^2 + 7ab - 8a$ **174.** $3y^3 - 15y^2 + 18y$ **175.** $2y^4 - 26y^3 - 96y^2$

176. $3y^4 + 54y^3 + 135y^2$ **177.** $2x^3 - 11x^2 + 5x$ **178.** $2x^3 + 3x^2 - 5x$ **179.** $10t^2 - 5t - 50$

180. $16t^2 + 40t - 96$ **181.** $6p^3 + 5p^2 + p$ **182.** $12x^2y - 36xy + 27y$

Appendix

Table of Properties

Properties of Real Numbers

The Associative Property of Addition
If a, b, and c are real numbers, then $(a + b) + c = a + (b + c)$.

The Commutative Property of Addition
If a and b are real numbers, then $a + b = b + a$.

The Addition Property of Zero
If a is a real number, then $a + 0 = 0 + a = a$.

The Multiplication Property of Zero
If a is a real number, then $a \cdot 0 = 0 \cdot a = 0$.

The Inverse Property of Addition
If a is a real number, then $a + (-a) = (-a) + a = 0$.

The Associative Property of Multiplication
If a, b, and c are real numbers, then $(a \cdot b) \cdot c = a \cdot (b \cdot c)$.

The Commutative Property of Multiplication
If a and b are real numbers, then $a \cdot b = b \cdot a$.

The Multiplication Property of One
If a is a real number, then $a \cdot 1 = 1 \cdot a = a$.

The Inverse Property of Multiplication
If a is a real number and $a \neq 0$, then $a \cdot \frac{1}{a} = \frac{1}{a} \cdot a = 1$.

Distributive Property
If a, b, and c are real numbers, then $a(b + c) = ab + ac$ or $(b + c)a = ba + ca$.

Properties of Equations

Addition Property of Equations
The same number or variable term can be added to each side of an equation without changing the solution of the equation.

Multiplication Property of Equations
Each side of an equation can be multiplied by the same nonzero number without changing the solution of the equation.

Properties of Inequalities

Addition Property of Inequalities
If $a > b$, then $a + c > b + c$.

If $a < b$, then $a + c < b + c$.

Multiplication Property of Inequalities
If $a > b$ and $c > 0$, then $ac > bc$.

If $a < b$ and $c > 0$, then $ac < bc$.

If $a > b$ and $c < 0$, then $ac < bc$.

If $a < b$ and $c < 0$, then $ac > bc$.

Properties of Exponents

If m and n are integers, then $x^m \cdot x^n = x^{m+n}$.

If m and n are integers, then $(x^m)^n = x^{mn}$.

If $x \neq 0$, then $x^0 = 1$.

If m and n are integers and $x \neq 0$, then $\frac{x^m}{x^n} = x^{m-n}$.

If m, n, and p are integers, then $(x^m \cdot y^n)^p = x^{mp}y^{np}$.

If n is a positive integer and $x \neq 0$, then $x^{-n} = \frac{1}{x^n}$ and $\frac{1}{x^{-n}} = x^n$.

If m, n, and p are integers and $y \neq 0$, then $\left(\frac{x^m}{y^n}\right)^p = \frac{x^{mp}}{y^{np}}$.

Table of Properties (*continued*)

Principle of Zero Products

If $a \cdot b = 0$, then $a = 0$ or $b = 0$.

Properties of Radical Expressions

If a and b are positive real numbers, then $\sqrt[n]{ab} = \sqrt[n]{a}\,\sqrt[n]{b}$.

If a and b are positive real numbers, then $\sqrt[n]{\frac{a}{b}} = \frac{\sqrt[n]{a}}{\sqrt[n]{b}}$.

Property of Squaring Both Sides of an Equation

If a and b are real numbers and $a = b$, then $a^2 = b^2$.

Properties of Logarithms

If x, y, and b are positive real numbers and $b \neq 1$, then $\log_b(xy) = \log_b x + \log_b y$.

If x, y, and b are positive real numbers and $b \neq 1$, then $\log_b \frac{x}{y} = \log_b x - \log_b y$.

If x and b are positive real numbers, $b \neq 1$, and r is any real number, then $\log_b x^r = r \log_b x$.

If x and b are positive real numbers and $b \neq 1$, then $\log_b b^x = x$.

Calculator Guide for the TI-83 and TI-83 Plus

To evaluate an expression

a. Press the [Y=] key. A menu showing \Y1 = through \Y7 = will be displayed vertically with a blinking cursor to the right of \Y1 =. Press [CLEAR], if necessary, to delete an unwanted expression.

b. Input the expression to be evaluated. For example, to input the expression $-3a^2b - 4c$, use the following keystrokes:

[(−)] 3 [ALPHA] A [X²] [ALPHA] B [−] 4 [ALPHA] C [2nd] QUIT

Note the difference between the keys for a negative sign [(−)] and a minus sign [−].

c. Store the value of each variable that will be used in the expression. For example, to evaluate the expression above when $a = 3$, $b = -2$, and $c = -4$, use the following keystrokes:

3 [STO▷] [ALPHA] A [ENTER] [(−)] 2 [STO▷] [ALPHA] B [ENTER] [(−)] 4 [STO▷] [ALPHA] C [ENTER]

These steps store the value of each variable.

d. Press [VARS] [▷] [1] [1] [ENTER]. The value for the expression, Y1, for the given values is displayed; in this case, Y1 = 70.

To graph a function

a. Press the [Y=] key. A menu showing \Y1 = through \Y7 = will be displayed vertically with a blinking cursor to the right of \Y1 =. Press [CLEAR], if necessary, to delete an unwanted expression.

b. Input the expression for each function that is to be graphed. Press [X,T,θ,n] to input x. For example, to input $y = x^3 + 2x^2 - 5x - 6$, use the following keystrokes:

[X,T,θ,n] [∧] 3 [+] 2 [X,T,θ,n] [X²] [−] 5 [X,T,θ,n] [−] 6

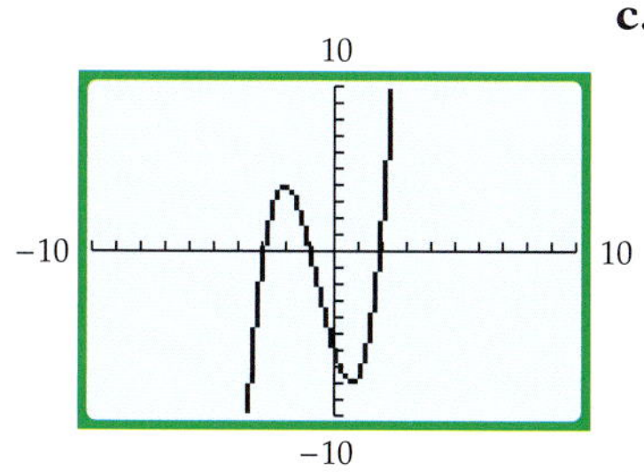

c. Set the viewing window by pressing [WINDOW]. Enter the values for the minimum x-value (Xmin), the maximum x-value (Xmax), the distance between tick marks on the x-axis (Xscl), the minimum y-value (Ymin), the maximum y-value (Ymax), and the distance between tick marks on the y-axis (Yscl). Now press [GRAPH]. For the graph shown at the left, Xmin = −10, Xmax = 10, Xscl = 1, Ymin = −10, Ymax = 10, and Yscl = 1. This is called the standard viewing window. Pressing [ZOOM] [6] is a quick way to set the calculator to the standard viewing window. *Note:* This will also immediately graph the function in that window.

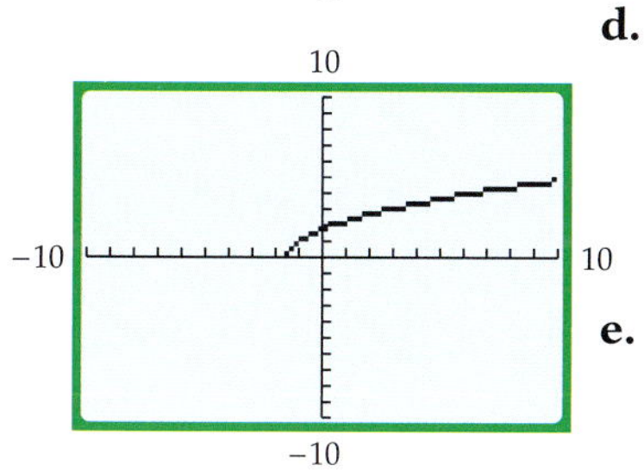

d. Press the [Y=] key. The equal sign has a black rectangle around it. This indicates that the function is active and will be graphed when the [GRAPH] key is pressed. A function is deactivated by using the arrow keys. Move the cursor over the equal sign and press [ENTER]. When the cursor is moved to the right, the black rectangle will not be present and that equation will not be active.

e. Graphing some radical equations requires special care. To graph the equation $y = \sqrt{2x + 3}$ shown at the left, enter the following keystrokes:

[Y=] [CLEAR] [2nd] $\sqrt{\ }$ 2 [X,T,θ,n] [+] 3 [)] [GRAPH]

To display the *x*-coordinates of rectangular coordinates as integers

a. Set the viewing window as follows: Xmin = −47, Xmax = 47, Xscl = 10, Ymin = −31, Ymax = 31, Yscl = 10.

b. Graph the function and use the TRACE feature. Press [TRACE] and then move the cursor with the [◁] and [▷] keys. The values of x and $y = f(x)$ displayed on the bottom of the screen are the coordinates of a point on the graph.

To display the *x*-coordinates of rectangular coordinates in tenths

a. Set the viewing window as follows: [ZOOM] [4]

b. Graph the function. Press [TRACE] and then move the cursor with the [◁] and [▷] keys. The values of x and $y = f(x)$ displayed on the bottom of the screen are the coordinates of a point on the graph.

To evaluate a function for a given value of *x*, or to produce ordered pairs of a function

a. Input the equation; for example, input $Y_1 = 2x^3 - 3x + 2$.

b. Press [2nd] QUIT.

c. To evaluate the function when $x = 3$, press [VARS] [▷] [1] [1] [(] 3 [)] [ENTER]. The value for the function for the given x-value is displayed, in this case, $Y_1(3) = 47$. An ordered pair of the function is (3, 47).

d. Repeat step **c.** to produce as many pairs as desired.

The TABLE feature can also be used to determine ordered pairs.

To use a table

a. Press [2nd] TBLSET to activate the table setup menu.

b. TblStart is the beginning number for the table; ΔTbl is the difference between any two successive x-values in the table.

c. The portion of the table that appears as

```
Indpnt: Auto Ask
Depend: Auto Ask
```

allows you to choose between having the calculator automatically produce the results (Auto) or having the calculator ask you for values of x. You can choose Ask by using the arrow keys.

d. Once a table has been set up, enter an expression for Y_1. Now select TABLE by pressing [2nd] TABLE. A table showing ordered pair solutions of the equation will be displayed on the screen.

Zoom Features

To zoom in or out on a graph

Here are two methods of using ZOOM.

a. The first method uses the built-in features of the calculator. Press [TRACE] and then move the cursor to a point on the graph that is of interest. Press [ZOOM]. The ZOOM menu will appear. Press [2] [ENTER] to zoom in on the graph by the amount shown under the SET FACTORS menu. The center of the new graph is the location at which you placed the cursor. Press [ZOOM] [3] [ENTER] to zoom out on the graph by the amount under the SET FACTORS menu. (The SET FACTORS menu is accessed by pressing [ZOOM] [▷] [4].)

b. The second method uses the ZBOX option under the ZOOM menu. To use this method, press ZOOM 1. A cursor will appear on the graph. Use the arrow keys to move the cursor to a portion of the graph that is of interest. Press ENTER. Now use the arrow keys to draw a box around the portion of the graph you wish to zoom in on. Press ENTER. The portion of the graph defined by the box will be drawn.

c. Pressing ZOOM 6 resets the window to the standard viewing window.

Solving Equations

This discussion is based on the fact that the real number solutions of an equation are related to the x-intercepts of a graph. For instance, the real solutions of the equation $x^2 = x + 1$ are the x-intercepts of the graph of $f(x) = x^2 - x - 1$, which are the zeros of f.

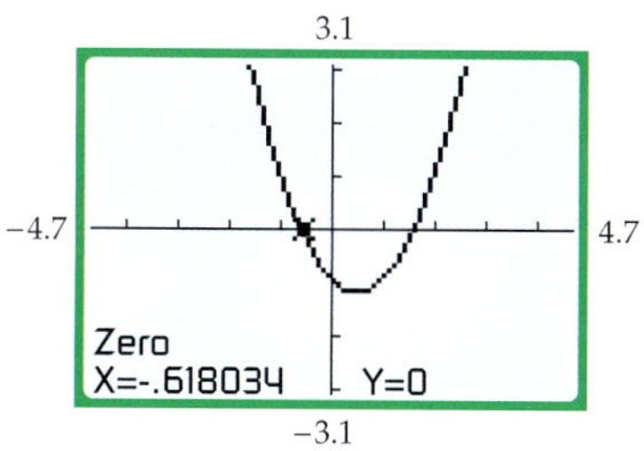

To solve $x^2 = x + 1$, rewrite the equation with all terms on one side: $x^2 - x - 1 = 0$. Think of this equation as $Y_1 = x^2 - x - 1$. The x-intercepts of the graph of Y_1 are the solutions of the equation $x^2 = x + 1$.

a. Enter $x^2 - x - 1$ into Y_1.

b. Graph the equation. You may need to adjust the viewing window so that the x-intercepts are visible.

c. Press 2nd CALC 2.

d. Move the cursor to a point on the curve that is to the left of an x-intercept. Press ENTER.

e. Move the cursor to a point on the curve that is to the right of the same x-intercept. Press ENTER ENTER.

f. The root is shown as the x-coordinate on the bottom of the screen; in this case, the root is approximately -0.618034. To find the next intercept, repeat steps **c.** through **e.** The SOLVER feature under the MATH menu can also be used to find solutions of equations.

Solving Systems of Equations in Two Variables

To solve a system of equations

To solve

$$y = x^2 - 1$$
$$\frac{1}{2}x + y = 1$$

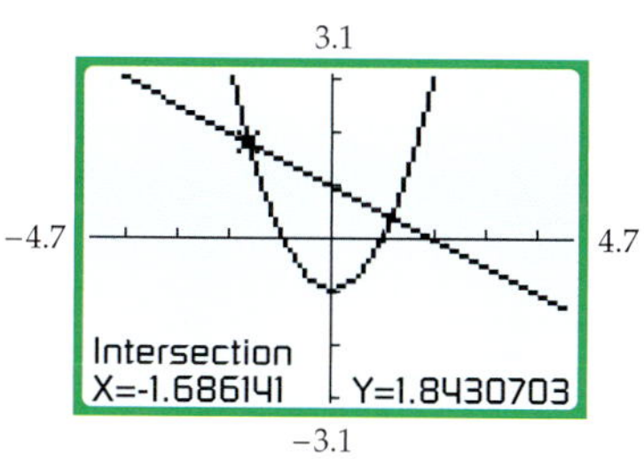

a. If necessary, solve one or both of the equations for y.

b. Enter the first equation as Y_1: $Y_1 = x^2 - 1$.

c. Enter the second equation as Y_2: $Y_2 = 1 - .5x$.

d. Graph both equations. (*Note:* The point of intersection must appear on the screen. It may be necessary to adjust the viewing window so that the points of intersection are displayed.)

e. Press 2nd CALC 5.

f. Move the cursor close to the first point of intersection. Press ENTER ENTER ENTER.

g. The first point of intersection is $(-1.686141, 1.8430703)$.

h. Repeat steps **e.** and **f.** for each point of intersection.

Finding Minimum or Maximum Values of a Function

a. Enter the function into Y_1. The equation $y = x^2 - x - 1$ is used here.

b. Graph the equation. You may need to adjust the viewing window so that the maximum or minimum points are visible.

c. Press [2nd] CALC [3] to determine a minimum value or press [2nd] CALC [4] to determine a maximum value.

d. Move the cursor to a point on the curve that is to the left of the minimum (maximum). Press [ENTER].

e. Move the cursor to a point on the curve that is to the right of the minimum (maximum). Press [ENTER] [ENTER].

f. The minimum (maximum) is shown as the y-coordinate on the bottom of the screen; in this case the minimum value is -1.25.

Statistics

To calculate a linear regression equation

a. Press [STAT] to access the statistics menu. Press 1 to Edit or enter a new list of data. To delete data already in a list, press the up arrow to highlight the list name. For instance, to delete data in L_1, highlight L_1. Then press [CLEAR] and [ENTER]. Now enter each value of the independent variable in L_1. Enter each value of the dependent variable in L_2.

b. When all the data has been entered, press [STAT] [▷] 4 [ENTER]. The values of the slope and y-intercept of the linear regression equation will be displayed on the screen.

c. To graph the linear regression equation, modify step **b.** with these keystrokes: [STAT] [▷] 4 [VARS] [▷] 1 1 [ENTER]. This will store the regression equation in Y_1. Now press [GRAPH]. It may be necessary to adjust the viewing window.

d. To evaluate the regression equation for a value of x, complete step **c.** but do not graph the equation. Now press [VARS] [▷] 1 1 [(] A [)] [ENTER], where A is replaced by the number at which you want to evaluate the expression.

Solutions to Chapter Problems

Solutions to Chapter 1 Problems

SECTION 1.1

Problem 1

$A = \{1, 2, 3, 4\}$

Problem 2

$-5 < -1$
$-1 = -1$
$5 > -1$

The element 5 is greater than -1.

Problem 3

A. 9 **B.** -62

Problem 4

A. $|-5| = 5$ **B.** $-|-9| = -9$

SECTION 1.2

Problem 1

A. $-162 + 98 = -64$

B. $-154 + (-37) = -191$

C. $-36 + 17 + (-21) = -19 + (-21)$
$= -40$

Problem 2

$-8 - 14 = -8 + (-14)$
$= -22$

Problem 3

$4 - (-3) - 12 - (-7) - 20$
$= 4 + 3 + (-12) + 7 + (-20)$
$= 7 + (-12) + 7 + (-20)$
$= -5 + 7 + (-20)$
$= 2 + (-20)$
$= -18$

Problem 4

A. $-38 \cdot 51 = -1938$

B. $-7(-8)(9)(-2) = 56(9)(-2)$
$= 504(-2)$
$= -1008$

Problem 5

A. $(-135) \div (-9) = 15$

B. $\frac{84}{-6} = -14$

C. $-\frac{36}{-12} = -(-3) = 3$

Problem 6

Strategy To find the average daily high temperature:
- ▶ Add the seven temperature readings.
- ▶ Divide the sum by 7.

Solution $-5 + (-6) + 3 + 0 + (-4) + (-7) + (-2)$
$= -11 + 3 + 0 + (-4) + (-7) + (-2)$
$= -8 + 0 + (-4) + (-7) + (-2)$
$= -8 + (-4) + (-7) + (-2)$
$= -12 + (-7) + (-2)$
$= -19 + (-2)$
$= -21$

$-21 \div 7 = -3$

The average daily high temperature was -3°C.

SECTION 1.3

Problem 1

$$\begin{array}{r} 0.16 \\ 25\overline{)4.00} \\ \underline{-2\ 5} \\ 1\ 50 \\ \underline{-1\ 50} \\ 0 \end{array}$$

$\frac{4}{25} = 0.16$

Problem 2

$$\begin{array}{r} 0.444 \\ 9\overline{)4.000} \\ \underline{-3\ 6} \\ 40 \\ \underline{-36} \\ 40 \\ \underline{-36} \\ 4 \end{array}$$

$\frac{4}{9} = 0.\overline{4}$

Problem 3

Prime factorization of 9 and 12:

$9 = 3 \cdot 3 \qquad 12 = 2 \cdot 2 \cdot 3$

$\text{LCM} = 2 \cdot 2 \cdot 3 \cdot 3 = 36$

$\frac{5}{9} - \frac{11}{12} = \frac{20}{36} - \frac{33}{36} = \frac{20}{36} + \frac{-33}{36}$
$= \frac{20 + (-33)}{36} = \frac{-13}{36} = -\frac{13}{36}$

Problem 4

$$-\frac{7}{8}-\frac{5}{6}+\frac{1}{2}=-\frac{21}{24}-\frac{20}{24}+\frac{12}{24}$$
$$=\frac{-21}{24}+\frac{-20}{24}+\frac{12}{24}$$
$$=\frac{-21+(-20)+12}{24}=\frac{-29}{24}=-\frac{29}{24}$$

Problem 5

$$\begin{array}{r} 3.097 \\ 4.9 \\ +\ 3.09 \\ \hline 11.087 \end{array}$$

Problem 6

$$\begin{array}{r} 67.910 \\ -16.127 \\ \hline 51.783 \end{array}$$

$16.127-67.91=-51.783$

Problem 7

$$-\frac{7}{12}\cdot\frac{9}{14}=-\frac{7\cdot 9}{12\cdot 14}=-\frac{\overset{1}{\cancel{7}}\cdot\overset{1}{\cancel{3}}\cdot 3}{2\cdot 2\cdot\underset{1}{\cancel{3}}\cdot 2\cdot\underset{1}{\cancel{7}}}$$
$$=-\frac{3}{8}$$

Problem 8

$$-\frac{3}{8}\div\left(-\frac{5}{12}\right)=\frac{3}{8}\cdot\frac{12}{5}$$
$$=\frac{3\cdot 12}{8\cdot 5}=\frac{3\cdot\overset{1}{\cancel{2}}\cdot\overset{1}{\cancel{2}}\cdot 3}{\underset{1}{\cancel{2}}\cdot\underset{1}{\cancel{2}}\cdot 2\cdot 5}=\frac{9}{10}$$

Problem 9

$$\begin{array}{r} 5.44 \\ \times\ \ 3.8 \\ \hline 4352 \\ 1632\ \\ \hline 20.672 \end{array}$$

$(-5.44)(3.8)=-20.672$

Problem 10

$$\begin{array}{r} 0.231 \\ 1.7.\overline{)0.3.940} \\ -3\,4 \\ \hline 54 \\ -51 \\ \hline 30 \\ -17 \\ \hline 13 \end{array}$$

$-0.394\div 1.7\approx -0.23$

Problem 11

$$125\%=125\left(\frac{1}{100}\right)=\frac{125}{100}=1\frac{1}{4}$$
$$125\%=125(0.01)=1.25$$

Problem 12

$$16\frac{2}{3}\%=16\frac{2}{3}\left(\frac{1}{100}\right)=\frac{50}{3}\left(\frac{1}{100}\right)=\frac{1}{6}$$

Problem 13

$6.08\%=6.08(0.01)=0.0608$

Problem 14

A. $0.043=0.043(100\%)=4.3\%$

B. $2.57=2.57(100\%)=257\%$

Problem 15

$$\frac{5}{9}=\frac{5}{9}(100\%)=\frac{500}{9}\%\approx 55.6\%$$

Problem 16

$$\frac{9}{16}=\frac{9}{16}(100\%)=\frac{900}{16}\%=56\frac{1}{4}\%$$

SECTION 1.4

Problem 1

$(-5)^3=(-5)(-5)(-5)=25(-5)=-125$

$-5^3=-(5\cdot 5\cdot 5)=-(25\cdot 5)=-125$

Problem 2

$(-3)^3=(-3)(-3)(-3)=9(-3)=-27$

$(-3)^4=(-3)(-3)(-3)(-3)$

$\quad=9(-3)(-3)=-27(-3)=81$

Problem 3

$(3^3)(-2)^3=(3)(3)(3)\cdot(-2)(-2)(-2)$

$\quad=27\cdot(-8)=-216$

$$\left(-\frac{2}{5}\right)^2=\left(-\frac{2}{5}\right)\left(-\frac{2}{5}\right)=\frac{2\cdot 2}{5\cdot 5}=\frac{4}{25}$$

Problem 4

$36\div(8-5)^2-(-3)^2\cdot 2$

$\quad=36\div(3)^2-(-3)^2\cdot 2$

$\quad=36\div 9-9\cdot 2$

$\quad=4-9\cdot 2$

$\quad=4-18$

$\quad=-14$

Problem 5

$27\div 3^2+(-3)^2\cdot 4$

$\quad=27\div 9+9\cdot 4$

$\quad=3+9\cdot 4$

$\quad=3+36$

$\quad=39$

Problem 6

$4-3[4-2(6-3)]\div 2$

$\quad=4-3[4-2(3)]\div 2$

$\quad=4-3[4-6]\div 2$

$\quad=4-3(-2)\div 2$

$\quad=4-(-6)\div 2$

$\quad=4-(-3)$

$\quad=7$

SECTION 1.5

Problem 1
-4

Problem 2
$2xy + y^2$

$$\begin{aligned}2(-4)(2) + (2)^2 &= 2(-4)(2) + 4\\ &= -8(2) + 4\\ &= -16 + 4\\ &= -12\end{aligned}$$

Problem 3
$\dfrac{a^2 + b^2}{a + b}$

$$\begin{aligned}\frac{(5)^2 + (-3)^2}{5 + (-3)} &= \frac{25 + 9}{5 + (-3)}\\ &= \frac{34}{2}\\ &= 17\end{aligned}$$

Problem 4
$x^3 - 2(x + y) + z^2$

$$\begin{aligned}&(2)^3 - 2[2 + (-4)] + (-3)^2\\ &\quad = (2)^3 - 2(-2) + (-3)^2\\ &\quad = 8 - 2(-2) + 9\\ &\quad = 8 + 4 + 9\\ &\quad = 12 + 9\\ &\quad = 21\end{aligned}$$

Problem 5
$$V = \frac{1}{3}\pi r^2 h$$

$$V = \frac{1}{3}\pi(4.5)^2(9.5) \qquad \bullet\ r = \frac{1}{2}d = \frac{1}{2}(9) = 4.5$$

$$V = \frac{1}{3}\pi(20.25)(9.5)$$

$$V \approx 201.5$$

The volume is approximately 201.5 cm^3.

Problem 6
$7 + (-7) = 0$

Problem 7
The Associative Property of Addition

Problem 8
A. $9x + 6x = (9 + 6)x = 15x$

B. $-4y - 7y = [-4 + (-7)]y = -11y$

Problem 9
A. $$\begin{aligned}3a - 2b + 5a &= 3a + 5a - 2b\\ &= (3a + 5a) - 2b\\ &= 8a - 2b\end{aligned}$$

B. $$\begin{aligned}&x^2 - 7 + 9x^2 - 14\\ &\quad = x^2 + 9x^2 - 7 - 14\\ &\quad = (x^2 + 9x^2) + (-7 - 14)\\ &\quad = 10x^2 - 21\end{aligned}$$

Problem 10
A. $$\begin{aligned}-7(-2a) &= [-7(-2)]a\\ &= 14a\end{aligned}$$

B. $$\begin{aligned}-\frac{5}{6}(-30y^2) &= \left[-\frac{5}{6}(-30)\right]y^2\\ &= 25y^2\end{aligned}$$

C. $$\begin{aligned}(-5x)(-2) &= (-2)(-5x)\\ &= [-2(-5)]x\\ &= 10x\end{aligned}$$

Problem 11
A. $$\begin{aligned}7(4 + 2y) &= 7(4) + 7(2y)\\ &= 28 + 14y\end{aligned}$$

B. $$\begin{aligned}-(5x - 12) &= -1(5x - 12)\\ &= -1(5x) - (-1)(12)\\ &= -5x + 12\end{aligned}$$

C. $$\begin{aligned}(3a - 1)5 &= (3a)(5) - (1)(5)\\ &= 15a - 5\end{aligned}$$

D. $$\begin{aligned}&-3(6a^2 - 8a + 9)\\ &\quad = -3(6a^2) - (-3)(8a) + (-3)(9)\\ &\quad = -18a^2 + 24a - 27\end{aligned}$$

Problem 12
$$\begin{aligned}&7(x - 2y) - 3(-x - 2y)\\ &\quad = 7x - 14y + 3x + 6y\\ &\quad = 10x - 8y\end{aligned}$$

Problem 13
$$\begin{aligned}&3y - 2[x - 4(2 - 3y)]\\ &\quad = 3y - 2[x - 8 + 12y]\\ &\quad = 3y - 2x + 16 - 24y\\ &\quad = -2x - 21y + 16\end{aligned}$$

SECTION 1.6

Problem 1
A. eighteen less than the cube of x

$x^3 - 18$

B. y decreased by the sum of z and nine

$y - (z + 9)$

C. the difference between the square of q and the sum of r and t

$q^2 - (r + t)$

Problem 2
the unknown number: n
the square of the number: n^2
the product of five and the square of the number: $5n^2$

$5n^2 + n$

Problem 3
the unknown number: n
twice the number: $2n$
the sum of seven and twice the number: $7 + 2n$

$3(7 + 2n)$

Problem 4
the unknown number: n
twice the number: $2n$
the difference between twice the number and 17: $2n - 17$

$$n - (2n - 17) = n - 2n + 17 = -n + 17$$

Problem 5
the unknown number: n

three-fourths of the number: $\frac{3}{4}n$

one-fifth of the number: $\frac{1}{5}n$

$$\frac{3}{4}n + \frac{1}{5}n = \frac{15}{20}n + \frac{4}{20}n = \frac{19}{20}n$$

Problem 6
the time required by the newer model: t
the time required by the older model is twice the time required for the newer model: $2t$

Problem 7
the length of one piece: L
the length of the second piece: $6 - L$

Solutions to Chapter 2 Problems

SECTION 2.1

Problem 1

$$5 - 4x = 8x + 2$$

$$\begin{array}{c|c} 5 - 4\left(\frac{1}{4}\right) & 8\left(\frac{1}{4}\right) + 2 \\ 5 - 1 & 2 + 2 \\ 4 = 4 & \end{array}$$

Yes, $\frac{1}{4}$ is a solution.

Problem 2

$$10x - x^2 = 3x - 10$$

$$\begin{array}{c|c} 10(5) - (5)^2 & 3(5) - 10 \\ 50 - 25 & 15 - 10 \\ 25 \neq 5 & \end{array}$$

No, 5 is not a solution.

Problem 3

$$x - \frac{1}{3} = -\frac{3}{4}$$

$$x - \frac{1}{3} + \frac{1}{3} = -\frac{3}{4} + \frac{1}{3}$$

$$x + 0 = -\frac{9}{12} + \frac{4}{12}$$

$$x = -\frac{5}{12}$$

Check $\quad x - \frac{1}{3} = -\frac{3}{4}$

$$\begin{array}{c|c} -\frac{5}{12} - \frac{1}{3} & -\frac{3}{4} \\ -\frac{3}{4} = -\frac{3}{4} & \end{array}$$

The solution is $-\frac{5}{12}$.

Problem 4

$$-8 = 5 + x$$
$$-8 - 5 = 5 - 5 + x$$
$$-13 = x$$

The solution is -13.

Problem 5

$$-\frac{2x}{5} = 6$$

$$\left(-\frac{5}{2}\right)\left(-\frac{2}{5}x\right) = \left(-\frac{5}{2}\right)(6)$$

$$1x = -15$$
$$x = -15$$

The solution is -15.

Problem 6

$$6x = 10$$
$$\frac{6x}{6} = \frac{10}{6}$$
$$x = \frac{5}{3}$$

The solution is $\frac{5}{3}$.

Problem 7

$$4x - 8x = 16$$
$$-4x = 16$$
$$\frac{-4x}{-4} = \frac{16}{-4}$$
$$x = -4$$

The solution is -4.

Problem 8

$$PB = A$$
$$P(60) = 27$$
$$60P = 27$$
$$\frac{60P}{60} = \frac{27}{60}$$
$$P = 0.45$$

27 is 45% of 60.

Problem 9

Strategy To find the percent of the questions answered correctly, solve the basic percent equation using $B = 80$ and $A = 72$. The percent is unknown.

Solution
$$\begin{aligned} PB &= A \\ P(80) &= 72 \\ 80P &= 72 \\ \frac{80P}{80} &= \frac{72}{80} \\ P &= 0.9 \end{aligned}$$

90% of the questions were answered correctly.

Problem 10

Strategy To determine the sales tax, solve the basic percent equation using $B = 895$ and $P = 6\% = 0.06$. The amount is unknown.

Solution
$$\begin{aligned} PB &= A \\ 895(0.06) &= A \\ 53.7 &= A \end{aligned}$$

The sales tax is \$53.70.

Problem 11

Strategy To determine how much she must deposit into the bank account:

- Find the amount of interest earned on the municipal bond by solving the equation $I = Prt$ for I using $P = 1000$, $r = 6.4\% = 0.064$, and $t = 1$.
- Solve $I = Prt$ for P using the amount of interest earned on the municipal bond for I, $r = 8\% = 0.08$, and $t = 1$.

Solution
$$\begin{aligned} I &= Prt \\ I &= 1000(0.064)(1) \\ I &= 64 \end{aligned}$$

The interest earned on the municipal bond is \$64.

$$\begin{aligned} I &= Prt \\ 64 &= P(0.08)(1) \\ 64 &= 0.08P \\ \frac{64}{0.08} &= \frac{0.08P}{0.08} \\ 800 &= P \end{aligned}$$

Clarissa must deposit \$800 into the bank account.

Problem 12

Strategy To find the number of ounces of cereal in the bowl, solve the equation $Q = Ar$ for A using $Q = 2$ and $r = 25\% = 0.25$.

Solution
$$\begin{aligned} Q &= Ar \\ 2 &= A(0.25) \\ \frac{2}{0.25} &= \frac{A(0.25)}{0.25} \\ 8 &= A \end{aligned}$$

The cereal bowl contains 8 oz of cereal.

SECTION 2.2

Problem 1

$$\begin{aligned} 5x + 7 &= 10 \\ 5x + 7 - 7 &= 10 - 7 \\ 5x &= 3 \\ \frac{5x}{5} &= \frac{3}{5} \\ x &= \frac{3}{5} \end{aligned}$$

The solution is $\frac{3}{5}$.

Problem 2

$$\begin{aligned} 11 &= 11 + 3x \\ 11 - 11 &= 11 - 11 + 3x \\ 0 &= 3x \\ \frac{0}{3} &= \frac{3x}{3} \\ 0 &= x \end{aligned}$$

The solution is 0.

Problem 3

$$\begin{aligned} \frac{5}{2} - \frac{2}{3}x &= \frac{1}{2} \\ 6\left(\frac{5}{2} - \frac{2}{3}x\right) &= 6\left(\frac{1}{2}\right) \\ 6\left(\frac{5}{2}\right) - 6\left(\frac{2}{3}x\right) &= 3 \\ 15 - 4x &= 3 \\ 15 - 15 - 4x &= 3 - 15 \\ -4x &= -12 \\ \frac{-4x}{-4} &= \frac{-12}{-4} \\ x &= 3 \end{aligned}$$

The solution is 3.

Problem 4

$$\begin{aligned} 5x + 4 &= 6 + 10x \\ 5x - 10x + 4 &= 6 + 10x - 10x \\ -5x + 4 &= 6 \\ -5x + 4 - 4 &= 6 - 4 \\ -5x &= 2 \\ \frac{-5x}{-5} &= \frac{2}{-5} \\ x &= -\frac{2}{5} \end{aligned}$$

The solution is $-\frac{2}{5}$.

Problem 5

$$\begin{aligned} 5x - 4(3 - 2x) &= 2(3x - 2) + 6 \\ 5x - 12 + 8x &= 6x - 4 + 6 \\ 13x - 12 &= 6x + 2 \\ 13x - 6x - 12 &= 6x - 6x + 2 \\ 7x - 12 &= 2 \\ 7x - 12 + 12 &= 2 + 12 \\ 7x &= 14 \\ \frac{7x}{7} &= \frac{14}{7} \\ x &= 2 \end{aligned}$$

The solution is 2.

Problem 6

$$\begin{aligned} -2[3x - 5(2x - 3)] &= 3x - 8 \\ -2[3x - 10x + 15] &= 3x - 8 \\ -2[-7x + 15] &= 3x - 8 \\ 14x - 30 &= 3x - 8 \\ 14x - 3x - 30 &= 3x - 3x - 8 \\ 11x - 30 &= -8 \\ 11x - 30 + 30 &= -8 + 30 \\ 11x &= 22 \\ \frac{11x}{11} &= \frac{22}{11} \\ x &= 2 \end{aligned}$$

The solution is 2.

Problem 7

Strategy To find the number of years, replace V with 10,200 in the given equation and solve for t.

Solution

$$\begin{aligned} V &= 450t + 7500 \\ 10{,}200 &= 450t + 7500 \\ 10{,}200 - 7500 &= 450t + 7500 - 7500 \\ 2700 &= 450t \\ \frac{2700}{450} &= \frac{450t}{450} \\ 6 &= t \end{aligned}$$

The value of the investment will be \$10,200 in 6 years.

Problem 8

Strategy The lever is 14 ft long, so $d = 14$. One force is 6 ft from the fulcrum, so $x = 6$. The one force is 40 lb, so $F_1 = 40$. To find the other force when the system balances, replace the variables F_1, x, and d in the lever system equation with the given values, and solve for F_2.

Solution

$$\begin{aligned} F_1x &= F_2(d - x) \\ 40(6) &= F_2(14 - 6) \\ 240 &= F_2(8) \\ 240 &= 8F_2 \\ \frac{240}{8} &= \frac{8F_2}{8} \\ 30 &= F_2 \end{aligned}$$

A 30-pound force must be applied to the other end.

SECTION 2.3

Problem 1

the unknown number: n

nine less than twice a number	is	five times the sum of the number and twelve

$$\begin{aligned} 2n - 9 &= 5(n + 12) \\ 2n - 9 &= 5n + 60 \\ 2n - 5n - 9 &= 5n - 5n + 60 \\ -3n - 9 &= 60 \\ -3n - 9 + 9 &= 60 + 9 \\ -3n &= 69 \\ \frac{-3n}{-3} &= \frac{69}{-3} \\ n &= -23 \end{aligned}$$

The number is -23.

Problem 2

Strategy To find the number of carbon atoms, write and solve an equation using n to represent the number of carbon atoms.

Solution

eight	represents	twice the number of carbon atoms

$$\begin{aligned} 8 &= 2n \\ \frac{8}{2} &= \frac{2n}{2} \\ 4 &= n \end{aligned}$$

There are 4 carbon atoms in a butane molecule.

Problem 3

Strategy To find the number of 10-speed bicycles made each day, write and solve an equation using n to represent the number of 10-speed bicycles and $160 - n$ to represent the number of 3-speed bicycles.

Solution

four times the number of 3-speed bicycles made	equals	30 less than the number of 10-speed bicycles made

$$\begin{aligned} 4(160 - n) &= n - 30 \\ 640 - 4n &= n - 30 \\ 640 - 4n - n &= n - n - 30 \\ 640 - 5n &= -30 \\ 640 - 640 - 5n &= -30 - 640 \\ -5n &= -670 \\ \frac{-5n}{-5} &= \frac{-670}{-5} \\ n &= 134 \end{aligned}$$

There are 134 10-speed bicycles made each day.

Problem 4

Strategy ▶ First consecutive integer: n
Second consecutive integer: $n + 1$
Third consecutive integer: $n + 2$

▶ The sum of the three integers is -12.

Solution

$$\begin{aligned} n + (n + 1) + (n + 2) &= -12 \\ 3n + 3 &= -12 \\ 3n &= -15 \\ n &= -5 \end{aligned}$$

$$\begin{aligned} n + 1 &= -5 + 1 = -4 \\ n + 2 &= -5 + 2 = -3 \end{aligned}$$

The three consecutive integers are -5, -4, and -3.

Problem 5

Strategy ▶ Pounds of \$4.00 hamburger: x
Pounds of \$2.80 hamburger: $75 - x$

	Amount	Cost	Value
\$4.00 hamburger	x	4.00	$4.00x$
\$2.80 hamburger	$75 - x$	2.80	$2.80(75 - x)$
Mixture	75	3.20	$75(3.20)$

▶ The sum of the values before mixing equals the value after mixing.

Solution

$$\begin{aligned} 4.00x + 2.80(75 - x) &= 75(3.20) \\ 4x + 210 - 2.80x &= 240 \\ 1.2x + 210 &= 240 \\ 1.2x &= 30 \\ x &= 25 \end{aligned}$$

$$75 - x = 75 - 25 = 50$$

The mixture must contain 25 lb of the \$4.00 hamburger and 50 lb of the \$2.80 hamburger.

Problem 6

Strategy The distance is 350 mi. Therefore, $d = 350$. The plane is flying into a headwind, so the rate of the plane is the difference between its rate in calm air (175 mph) and the rate of the wind (35 mph): 175 mph $-$ 35 mph $=$ 140 mph. Therefore, $r = 140$. To find the time, solve the equation $d = rt$ for t.

Solution

$$\begin{aligned} d &= rt \\ 350 &= 140t \\ \frac{350}{140} &= \frac{140t}{140} \\ 2.5 &= t \end{aligned}$$

It takes the plane 2.5 h to travel 350 mi.

Problem 7

Strategy ▶ Rate of the second plane: r
Rate of the first plane: $r + 30$

	Rate	Time	Distance
First plane	$r + 30$	4	$4(r + 30)$
Second plane	r	4	$4r$

▶ The total distance traveled by the two planes is 1160 mi.

Solution

$$\begin{aligned} 4(r + 30) + 4r &= 1160 \\ 4r + 120 + 4r &= 1160 \\ 8r + 120 &= 1160 \\ 8r &= 1040 \\ r &= 130 \end{aligned}$$

$$r + 30 = 130 + 30 = 160$$

The first plane is traveling 160 mph.
The second plane is traveling 130 mph.

SECTION 2.4

Problem 1

Strategy ▶ Each equal side: s

▶ Use the equation for the perimeter of a square.

Solution

$$\begin{aligned} P &= 4s \\ 52 &= 4s \\ 13 &= s \end{aligned}$$

The length of each side is 13 ft.

Problem 2

Strategy To find the supplement, let x represent the supplement of a 107° angle. Use the fact that supplementary angles are two angles whose sum is 180° to write an equation. Solve for x.

Solution

$$\begin{aligned} 107° + x &= 180° \\ x &= 73° \end{aligned}$$

The supplement of a 107° angle is a 73° angle.

Problem 3

Strategy The angles labeled are adjacent angles of intersecting lines and are therefore supplementary angles. To find x, write an equation and solve for x.

Solution

$$\begin{aligned} x + (3x + 20°) &= 180° \\ 4x + 20° &= 180° \\ 4x &= 160° \\ x &= 40° \end{aligned}$$

Problem 4

Strategy $2x = y$ because alternate exterior angles have the same measure.

$y + (x + 15°) = 180°$ because adjacent angles of intersecting lines are supplementary angles. Substitute $2x$ for y and solve for x.

Solution

$$y + (x + 15°) = 180°$$
$$2x + (x + 15°) = 180°$$
$$3x + 15° = 180°$$
$$3x = 165°$$
$$x = 55°$$

Problem 5

Strategy

- To find the measure of $\angle a$, use the fact that $\angle a$ and $\angle y$ are vertical angles.
- To find the measure of $\angle b$, use the fact that the sum of the interior angles of a triangle is 180°.
- To find the measure of $\angle d$, use the fact that $\angle b$ and $\angle d$ are supplementary angles.

Solution $\angle a = \angle y = 55°$

$$\angle a + \angle b + 90° = 180°$$
$$55° + \angle b + 90° = 180°$$
$$145° + \angle b = 180°$$
$$\angle b = 35°$$

$$\angle b + \angle d = 180°$$
$$35° + \angle d = 180°$$
$$\angle d = 145°$$

Problem 6

Strategy To find the measure of the third angle, use the fact that the sum of the measures of the interior angles of a triangle is 180°. Write an equation using x to represent the measure of the third angle. Solve the equation for x.

Solution

$$x + 90° + 27° = 180°$$
$$x + 117° = 180°$$
$$x = 63°$$

The measure of the third angle is 63°.

SECTION 2.5

Problem 1

Strategy Given: $C = \$120$
$S = \$180$
Unknown markup rate: r
Use the equation $S = C + rC$.

Solution

$$S = C + rC$$
$$180 = 120 + 120r$$
$$60 = 120r$$
$$0.5 = r$$

The markup rate is 50%.

Problem 2

Strategy Given: $r = 40\% = 0.40$
$S = \$266$
Unknown cost: C
Use the equation $S = C + rC$.

Solution

$$S = C + rC$$
$$266 = C + 0.4C$$
$$266 = 1.40C$$
$$190 = C$$

The cost is \$190.

Problem 3

Strategy Given: $R = \$39.80$
$S = \$29.85$
Unknown discount rate: r
Use the equation $S = R - rR$.

Solution

$$S = R - rR$$
$$29.85 = 39.80 - 39.80r$$
$$-9.95 = -39.80r$$
$$0.25 = r$$

The discount rate is 25%.

Problem 4

Strategy Given: $S = \$43.50$
$r = 25\% = 0.25$
Unknown regular price: R
Use the equation $S = R - rR$.

Solution

$$S = R - rR$$
$$43.50 = R - 0.25R$$
$$43.50 = 0.75R$$
$$58 = R$$

The regular price is \$58.

SECTION 2.6

Problem 1

Strategy

- Amount invested at 7.5%: x

	Principal	Rate	Interest
Amount at 5.2%	3500	0.052	0.052(3500)
Amount at 7.5%	x	0.075	$0.075x$

- The sum of the interest earned by the two investments equals the total annual interest earned (\$575).

Solution

$$0.052(3500) + 0.075x = 575$$
$$182 + 0.075x = 575$$
$$0.075x = 393$$
$$x = 5240$$

The amount invested at 7.5% is \$5240.

Problem 2

Strategy

- Pounds of 22% fat hamburger: x
Pounds of 12% fat hamburger: $80 - x$

	Amount	Percent	Quantity
22%	x	0.22	$0.22x$
12%	$80 - x$	0.12	$0.12(80 - x)$
18%	80	0.18	0.18(80)

- The sum of the quantities before mixing is equal to the quantity after mixing.

Solution

$$\begin{aligned} 0.22x + 0.12(80 - x) &= 0.18(80) \\ 0.22x + 9.6 - 0.12x &= 14.4 \\ 0.10x + 9.6 &= 14.4 \\ 0.10x &= 4.8 \\ x &= 48 \end{aligned}$$

$80 - x = 80 - 48 = 32$

The butcher needs 48 lb of the hamburger that is 22% fat and 32 lb of the hamburger that is 12% fat.

SECTION 2.7

Problem 1

$\{x \mid x < 7, x \in \text{real numbers}\}$

Problem 2

$A \cup C = \{-5, -2, -1, 0, 1, 2, 5\}$

Problem 3

$E \cap F = \varnothing$

Problem 4

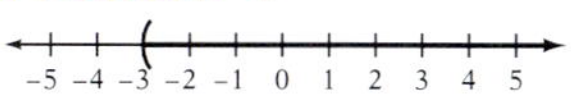

Problem 5

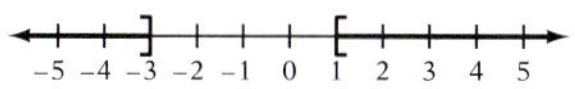

Problem 6

$[-8, -1)$

Problem 7

$\{x \mid x > -12\}$

Problem 8

$$\begin{aligned} 2x - 1 &< 6x + 7 \\ -4x - 1 &< 7 \\ -4x &< 8 \\ \frac{-4x}{-4} &> \frac{8}{-4} \\ x &> -2 \end{aligned}$$

$\{x \mid x > -2\}$

Problem 9

$$\begin{aligned} 5x - 2 &\le 4 - 3(x - 2) \\ 5x - 2 &\le 4 - 3x + 6 \\ 5x - 2 &\le 10 - 3x \\ 8x - 2 &\le 10 \\ 8x &\le 12 \\ \frac{8x}{8} &\le \frac{12}{8} \\ x &\le \frac{3}{2} \end{aligned}$$

$\left(-\infty, \frac{3}{2}\right]$

Problem 10

$$\begin{aligned} -2 &\le 5x + 3 \le 13 \\ -2 - 3 &\le 5x + 3 - 3 \le 13 - 3 \\ -5 &\le 5x \le 10 \\ \frac{-5}{5} &\le \frac{5x}{5} \le \frac{10}{5} \\ -1 &\le x \le 2 \end{aligned}$$

$[-1, 2]$

Problem 11

$5 - 4x > 1$ and $6 - 5x < 11$

$-4x > -4 \qquad -5x < 5$

$x < 1 \qquad x > -1$

$\{x \mid x < 1\} \qquad \{x \mid x > -1\}$

$\{x \mid x < 1\} \cap \{x \mid x > -1\} = (-1, 1)$

Problem 12

$2 - 3x > 11$ or $5 + 2x > 7$

$-3x > 9 \qquad 2x > 2$

$x < -3 \qquad x > 1$

$\{x \mid x < -3\} \qquad \{x \mid x > 1\}$

$\{x \mid x < -3\} \cup \{x \mid x > 1\}$
$= \{x \mid x < -3 \text{ or } x > 1\}$

Problem 13

Strategy To find the maximum height, substitute the given values in the inequality $\frac{1}{2}bh < A$ and solve for x.

Solution

$$\begin{aligned} \frac{1}{2}bh &< A \\ \frac{1}{2}(12)(x + 2) &< 50 \\ 6(x + 2) &< 50 \\ 6x + 12 &< 50 \\ 6x &< 38 \\ x &< \frac{19}{3} \end{aligned}$$

The largest integer less than $\frac{19}{3}$ is 6.

$x + 2 = 6 + 2 = 8$

The maximum height of the triangle is 8 in.

Problem 14

Strategy To find the range of scores, write and solve an inequality using N to represent the score on the fifth exam.

Solution

$$\begin{aligned} 80 &\le \frac{72 + 94 + 83 + 70 + N}{5} \le 89 \\ 80 &\le \frac{319 + N}{5} \le 89 \\ 5(80) &\le 5\left(\frac{319 + N}{5}\right) \le 5(89) \\ 400 &\le 319 + N \le 445 \\ 400 - 319 &\le 319 - 319 + N \le 445 - 319 \\ 81 &\le N \le 126 \end{aligned}$$

Because 100 is a maximum score, the range of scores that will give the student a B for the course is $81 \le N \le 100$.

SECTION 2.8

Problem 1

A. $|x| = 25$

$x = 25 \qquad x = -25$

The solutions are 25 and -25.

B. $|2x - 3| = 5$

$$\begin{aligned} 2x - 3 &= 5 & 2x - 3 &= -5 \\ 2x &= 8 & 2x &= -2 \\ x &= 4 & x &= -1 \end{aligned}$$

The solutions are 4 and -1.

C. $|x - 3| = -2$

The absolute value of a number must be nonnegative.

There is no solution to this equation.

D. $5 - |3x + 5| = 3$

$$\begin{aligned} -|3x + 5| &= -2 \\ |3x + 5| &= 2 \end{aligned}$$

$$\begin{aligned} 3x + 5 &= 2 & 3x + 5 &= -2 \\ 3x &= -3 & 3x &= -7 \\ x &= -1 & x &= -\frac{7}{3} \end{aligned}$$

The solutions are -1 and $-\frac{7}{3}$.

Problem 2

$|3x + 2| < 8$

$$\begin{aligned} -8 < 3x + 2 &< 8 \\ -8 - 2 < 3x + 2 - 2 &< 8 - 2 \\ -10 < 3x &< 6 \\ \frac{-10}{3} < \frac{3x}{3} &< \frac{6}{3} \\ -\frac{10}{3} < x &< 2 \end{aligned}$$

$\left\{x \middle| -\frac{10}{3} < x < 2\right\}$

Problem 3

$|3x - 7| < 0$

The absolute value of a number must be nonnegative. The solution set is the empty set.

$\varnothing$

Problem 4

$|5x + 3| > 8$

$$\begin{aligned} 5x + 3 &< -8 \quad \text{or} & 5x + 3 &> 8 \\ 5x &< -11 & 5x &> 5 \\ x &< -\frac{11}{5} & x &> 1 \end{aligned}$$

$\left\{x \middle| x < -\frac{11}{5}\right\} \qquad \{x \mid x > 1\}$

$$\left\{x \middle| x < -\frac{11}{5}\right\} \cup \{x \mid x > 1\} = \left\{x \middle| x < -\frac{11}{5} \text{ or } x > 1\right\}$$

Problem 5

Strategy Let b represent the desired diameter of the bushing, T the tolerance, and d the actual diameter of the bushing. Solve the absolute value inequality $|d - b| \le T$ for d.

Solution

$$\begin{aligned} |d - b| &\le T \\ |d - 2.55| &\le 0.003 \end{aligned}$$

$$\begin{aligned} -0.003 \le d - 2.55 &\le 0.003 \\ -0.003 + 2.55 \le d - 2.55 + 2.55 &\le 0.003 + 2.55 \\ 2.547 \le d &\le 2.553 \end{aligned}$$

The lower and upper limits of the diameter of the bushing are 2.547 in. and 2.553 in.

Chapter 3

SECTION 3.1

Problem 1

Replace x by -2 and solve for y.

$$y = \frac{3(-2)}{-2 + 1} = \frac{-6}{-1} = 6$$

The ordered-pair solution is $(-2, 6)$.

Problem 2

x	y
-3	2
-2	1
-1	0
0	1
1	2

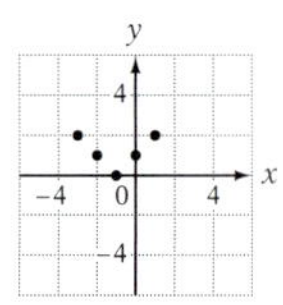

Problem 3

$(x_1, y_1) = (5, -2) \qquad (x_2, y_2) = (-4, 3)$

$$\begin{aligned} d &= \sqrt{(x_2 - x_1)^2 + (y_2 - y_1)^2} \\ &= \sqrt{(-4 - 5)^2 + (3 - (-2))^2} \\ &= \sqrt{(-9)^2 + (5)^2} \\ &= \sqrt{81 + 25} \\ &= \sqrt{106} \end{aligned}$$

Problem 4

$(x_1, y_1) = (-3, -5) \qquad (x_2, y_2) = (-2, 3)$

$$\begin{aligned} x_m &= \frac{x_1 + x_2}{2} & y_m &= \frac{y_1 + y_2}{2} \\ &= \frac{-3 + (-2)}{2} & &= \frac{-5 + 3}{2} \\ &= -\frac{5}{2} & &= -1 \end{aligned}$$

The midpoint is $\left(-\frac{5}{2}, -1\right)$.

Problem 5

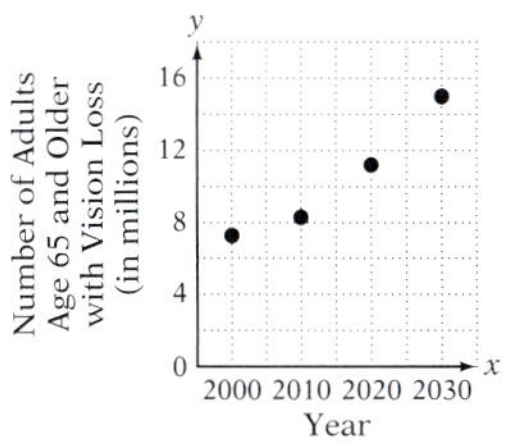

Problem 6

In 1900, the population was 1.5 million: (1900, 1.5)

In 2000, the population was 33.9 million: (2000, 33.9)

Average rate of change

$$= \frac{\text{change in } y}{\text{change in } x}$$

$$= \frac{\text{population in 2000} - \text{population in 1900}}{2000 - 1900}$$

$$= \frac{33.9 - 1.5}{2000 - 1900}$$

$$= \frac{32.4}{100} = 0.324$$

The average rate of change in the population was 324,000 people per year.

Problem 7

A. In 1990, the number of women competing was 26,000: (1900, 26,000)

In 1995, the number of women competing was 39,100: (1995, 39,100)

Average rate of change

$$= \frac{\text{change in } y}{\text{change in } x}$$

$$= \frac{\text{women competing in 1995} - \text{women competing in 1990}}{1995 - 1990}$$

$$= \frac{39{,}100 - 26{,}000}{1995 - 1990}$$

$$= \frac{13{,}100}{5} = 2620$$

The average rate of change was 2620 women per year.

B. In 1995, the number of women competing was 39,100: (1995, 39,100)

In 2000, the number of women competing was 72,000: (2000, 72,000)

Average rate of change

$$= \frac{\text{change in } y}{\text{change in } x}$$

$$= \frac{\text{women competing in 2000} - \text{women competing in 1995}}{2000 - 1995}$$

$$= \frac{72{,}000 - 39{,}100}{2000 - 1995}$$

$$= \frac{32{,}900}{5} = 6580$$

The average rate of change was 6580 women per year.

$2620 < 6580$

The average rate of change in the number of women competing in triathlons in the United States between 1990 and 1995 is less than the average rate of change in the number of women competing in triathlons in the United States between 1995 and 2000.

SECTION 3.2

Problem 1

The domain is $\{-1, 3, 4, 6\}$.

The range is $\{5\}$.

Problem 2

$$G(x) = \frac{2x}{x + 4}$$

$$G(-2) = \frac{2(-2)}{-2 + 4} = \frac{-4}{2} = -2$$

Problem 3

$$\begin{aligned} f(x) &= x^2 - x - 2 \\ f(-3) &= (-3)^2 - (-3) - 2 = 10 \\ f(-2) &= (-2)^2 - (-2) - 2 = 4 \\ f(-1) &= (-1)^2 - (-1) - 2 = 0 \\ f(0) &= 0^2 - 0 - 2 = -2 \\ f(1) &= 1^2 - 1 - 2 = -2 \\ f(2) &= 2^2 - 2 - 2 = 0 \end{aligned}$$

The range is $\{-2, 0, 4, 10\}$.

Problem 4

$$f(x) = \frac{2}{x - 5}$$

For $x = 5$, $5 - 5 = 0$.

$f(5) = \dfrac{2}{5 - 5} = \dfrac{2}{0}$, which is not a real number.

5 is excluded from the domain of $f(x) = \dfrac{2}{x - 5}$.

SECTION 3.3

Problem 1

x	y
−5	−7
0	−4
5	−1

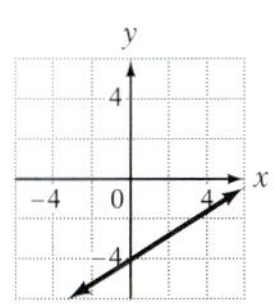

Problem 2

$$\begin{aligned} -3x + 2y &= 4 \\ 2y &= 3x + 4 \\ y &= \frac{3}{2}x + 2 \end{aligned}$$

x	y
-2	-1
0	2
2	5

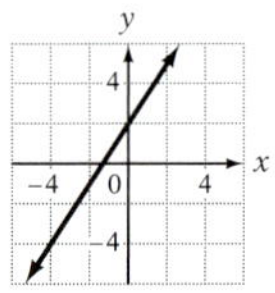

Problem 3

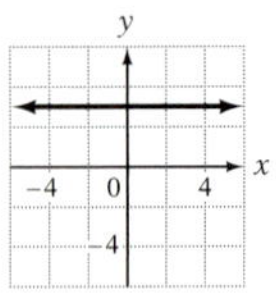

Problem 4

x-intercept:

$$3x - y = 2$$
$$3x - 0 = 2$$
$$3x = 2$$
$$x = \frac{2}{3}$$
$$\left(\frac{2}{3}, 0\right)$$

y-intercept:

$$3x - y = 2$$
$$3(0) - y = 2$$
$$-y = 2$$
$$y = -2$$
$$(0, -2)$$

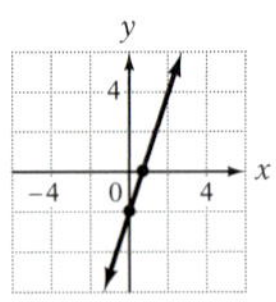

Problem 5

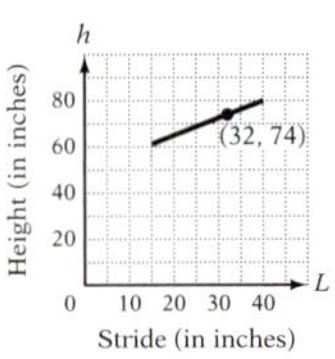

The ordered pair (32, 74) means that a person with a stride of 32 in. is 74 in. tall.

SECTION 3.4

Problem 1

Let $P_1 = (4, -3)$ and $P_2 = (2, 7)$.

$$m = \frac{y_2 - y_1}{x_2 - x_1} = \frac{7 - (-3)}{2 - 4} = \frac{10}{-2} = -5$$

The slope is -5.

Problem 2

$$m = \frac{55{,}000 - 25{,}000}{2 - 5} = \frac{30{,}000}{-3} = -10{,}000$$

A slope of $-10{,}000$ means that the value of the printing press is decreasing by \$10,000 per year.

Problem 3

$$2x + 3y = 6$$
$$3y = -2x + 6$$
$$y = -\frac{2}{3}x + 2$$

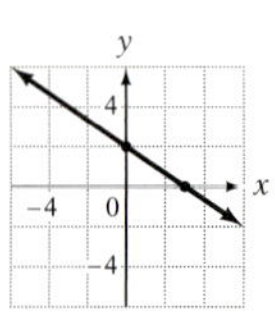

$$m = -\frac{2}{3} = \frac{-2}{3}$$

y-intercept $= (0, 2)$

Problem 4

$(x_1, y_1) = (-3, -2)$

$$m = 3 = \frac{3}{1} = \frac{\text{change in } y}{\text{change in } x}$$

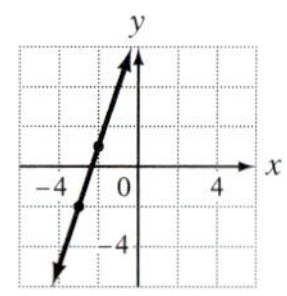

SECTION 3.5

Problem 1

$m = -3 \qquad (x_1, y_1) = (4, -3)$

$$y - y_1 = m(x - x_1)$$
$$y - (-3) = -3(x - 4)$$
$$y + 3 = -3x + 12$$
$$y = -3x + 9$$

The equation of the line is $y = -3x + 9$.

Problem 2

Let $(x_1, y_1) = (2, 0)$ and $(x_2, y_2) = (5, 3)$.

$$m = \frac{y_2 - y_1}{x_2 - x_1} = \frac{3 - 0}{5 - 2} = \frac{3}{3} = 1$$

$$y - y_1 = m(x - x_1)$$
$$y - 0 = 1(x - 2)$$
$$y = x - 2$$

The equation of the line is $y = x - 2$.

Problem 3

Strategy
- Select the independent and dependent variables. Because the function will predict the Celsius temperature, that quantity is the dependent variable, y. The Fahrenheit temperature is the independent variable.
- From the given data, two ordered pairs are (212, 100) and (32, 0). Use these ordered pairs to determine the linear function.

Solution Let $(x_1, y_1) = (32, 0)$ and $(x_2, y_2) = (212, 100)$.

$$m = \frac{y_2 - y_1}{x_2 - x_1} = \frac{100 - 0}{212 - 32} = \frac{100}{180} = \frac{5}{9}$$

$$y - y_1 = m(x - x_1)$$
$$y - 0 = \frac{5}{9}(x - 32)$$
$$y = \frac{5}{9}(x - 32), \text{ or } C = \frac{5}{9}(F - 32).$$

The linear function is $f(F) = \frac{5}{9}(F - 32)$.

SECTION 3.6

Problem 1

$(x_1, y_1) = (-2, -3), (x_2, y_2) = (7, 1)$

$$m_1 = \frac{1 - (-3)}{7 - (-2)} = \frac{4}{9}$$

$(x_1, y_1) = (4, 1), (x_2, y_2) = (6, -5)$

$$m_2 = \frac{-5 - 1}{6 - 4} = \frac{-6}{2} = -3$$

$$m_1 \cdot m_2 = \frac{4}{9} \cdot -3 = -\frac{4}{3}$$

No, the lines are not perpendicular.

Problem 2

$$5x + 2y = 2$$
$$2y = -5x + 2$$
$$y = -\frac{5}{2}x + 1$$

$$m_1 = -\frac{5}{2}$$

$$5x + 2y = -6$$
$$2y = -5x - 6$$
$$y = -\frac{5}{2}x - 3$$

$$m_2 = -\frac{5}{2}$$

$$m_1 = m_2 = -\frac{5}{2}$$

Yes, the lines are parallel.

Problem 3

$$x - 4y = 3$$
$$-4y = -x + 3$$
$$y = \frac{1}{4}x - \frac{3}{4}$$

$$m_1 = \frac{1}{4}$$

$$m_1 \cdot m_2 = -1$$
$$\frac{1}{4} \cdot m_2 = -1$$
$$m_2 = -4$$

$$y - y_1 = m(x - x_1)$$
$$y - 3 = -4[x - (-2)]$$
$$y - 3 = -4(x + 2)$$
$$y - 3 = -4x - 8$$
$$y = -4x - 5$$

The equation of the line is $y = -4x - 5$.

SECTION 3.7

Problem 1

$$x + 3y > 6$$
$$3y > -x + 6$$
$$y > -\frac{1}{3}x + 2$$

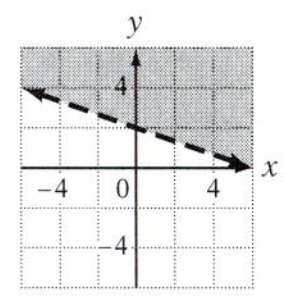

Problem 2

$y < 2$

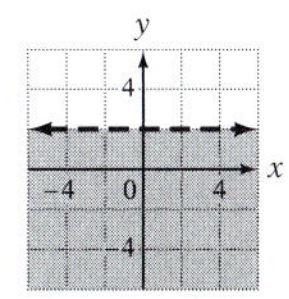

Chapter 4

SECTION 4.1

Problem 1

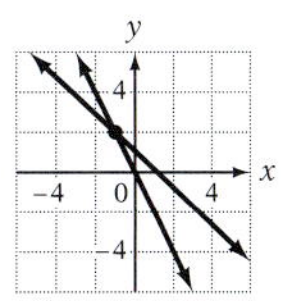

The solution is $(-1, 2)$.

Problem 2

A.

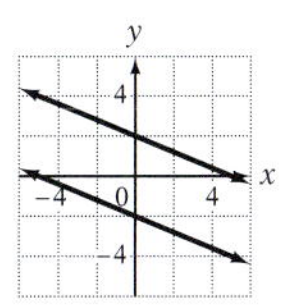

The lines are parallel and therefore do not intersect. The system of equations has no solution. The system of equations is inconsistent.

B.

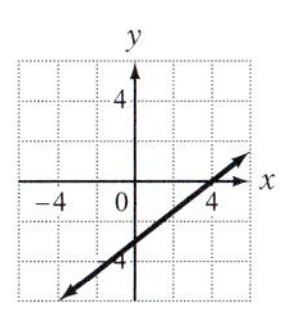

The two equations represent the same line. The system of equations is dependent. The solutions are the ordered pairs $\left(x, \frac{3}{4}x - 3\right)$.

Problem 3

A. (1) $3x - y = 3$
(2) $6x + 3y = -4$

Solve equation (1) for y.

$$3x - y = 3$$
$$-y = -3x + 3$$
$$y = 3x - 3$$

Substitute into equation (2).

$$6x + 3y = -4$$
$$6x + 3(3x - 3) = -4$$
$$6x + 9x - 9 = -4$$
$$15x - 9 = -4$$
$$15x = 5$$
$$x = \frac{5}{15} = \frac{1}{3}$$

Substitute into equation (1).

$$3x - y = 3$$
$$3\left(\frac{1}{3}\right) - y = 3$$
$$1 - y = 3$$
$$-y = 2$$
$$y = -2$$

The solution is $\left(\frac{1}{3}, -2\right)$.

B. (1) $6x - 3y = 6$
(2) $2x - y = 2$

Solve equation (2) for y.

$$2x - y = 2$$
$$-y = -2x + 2$$
$$y = 2x - 2$$

Substitute into equation (1).

$$6x - 3y = 6$$
$$6x - 3(2x - 2) = 6$$
$$6x - 6x + 6 = 6$$
$$6 = 6$$

$6 = 6$ is a true equation. The system of equations is dependent. The solutions are the ordered pairs $(x, 2x - 2)$.

SECTION 4.2

Problem 1

A. (1) $2x + 5y = 6$
(2) $3x - 2y = 6x + 2$

Write equation (2) in the form $Ax + By = C$.

$$3x - 2y = 6x + 2$$
$$-3x - 2y = 2$$

Solve the system: $2x + 5y = 6$
$-3x - 2y = 2$

Eliminate y.

$$2(2x + 5y) = 2(6)$$
$$5(-3x - 2y) = 5(2)$$

$$4x + 10y = 12$$
$$-15x - 10y = 10$$

Add the equations.

$$-11x = 22$$
$$x = -2$$

Replace x in equation (1).

$$2x + 5y = 6$$
$$2(-2) + 5y = 6$$
$$-4 + 5y = 6$$
$$5y = 10$$
$$y = 2$$

The solution is $(-2, 2)$.

B. $2x + y = 5$
$4x + 2y = 6$

Eliminate y.

$$-2(2x + y) = -2(5)$$
$$4x + 2y = 6$$

$$-4x - 2y = -10$$
$$4x + 2y = 6$$

Add the equations.

$$0x + 0y = -4$$
$$0 = -4$$

$0 = -4$ is not a true equation. The system is inconsistent and therefore has no solution.

Problem 2

(1) $x - y + z = 6$
(2) $2x + 3y - z = 1$
(3) $x + 2y + 2z = 5$

Eliminate z. Add equations (1) and (2).

$$x - y + z = 6$$
$$2x + 3y - z = 1$$
(4) $3x + 2y = 7$

Eliminate z. Multiply equation (2) by 2 and add to equation (3).

$$4x + 6y - 2z = 2$$
$$x + 2y + 2z = 5$$
(5) $5x + 8y = 7$

Solve the system of two equations.

(4) $3x + 2y = 7$
(5) $5x + 8y = 7$

Multiply equation (4) by -4 and add to equation (5).

$$-12x - 8y = -28$$
$$5x + 8y = 7$$
$$-7x = -21$$
$$x = 3$$

Replace x by 3 in equation (4).

$$3x + 2y = 7$$
$$3(3) + 2y = 7$$
$$9 + 2y = 7$$
$$2y = -2$$
$$y = -1$$

Replace x by 3 and y by -1 in equation (1).

$$x - y + z = 6$$
$$3 - (-1) + z = 6$$
$$4 + z = 6$$
$$z = 2$$

The solution is $(3, -1, 2)$.

SECTION 4.3

Problem 1

$$\left[\begin{array}{rrr|r} 2 & -3 & 1 & 4 \\ 1 & 0 & -2 & 3 \\ 0 & 1 & 2 & -3 \end{array}\right]$$

$$2x - 3y + z = 4$$
$$x - 2z = 3$$
$$y + 2z = -3$$

Problem 2

A. $\left[\begin{array}{rrr|r} 1 & 8 & -2 & 3 \\ 2 & -3 & 4 & 1 \\ 3 & 5 & -7 & 3 \end{array}\right] \quad R_2 \leftrightarrow R_3 \quad \left[\begin{array}{rrr|r} 1 & 8 & -2 & 3 \\ 3 & 5 & -7 & 3 \\ 2 & -3 & 4 & 1 \end{array}\right]$

B. $\left[\begin{array}{rrr|r} 1 & 8 & -2 & 3 \\ 2 & -3 & 4 & 1 \\ 3 & 5 & -7 & 3 \end{array}\right] \quad 3R_2 \rightarrow \quad \left[\begin{array}{rrr|r} 1 & 8 & -2 & 3 \\ 6 & -9 & 12 & 3 \\ 3 & 5 & -7 & 3 \end{array}\right]$

C. $\left[\begin{array}{rrr|r} 1 & 8 & -2 & 3 \\ 2 & -3 & 4 & 1 \\ 3 & 5 & -7 & 3 \end{array}\right] \quad -3R_1 + R_3 \rightarrow \quad \left[\begin{array}{rrr|r} 1 & 8 & -2 & 3 \\ 2 & -3 & 4 & 1 \\ 0 & -19 & -1 & -6 \end{array}\right]$

Problem 3

$$\left[\begin{array}{rrr|r} 1 & -3 & 2 & 1 \\ -4 & 14 & 0 & -2 \\ 2 & -5 & -3 & 16 \end{array}\right] \quad 4R_1 + R_2 \rightarrow \quad \left[\begin{array}{rrr|r} 1 & -3 & 2 & 1 \\ 0 & 2 & 8 & 2 \\ 2 & -5 & -3 & 16 \end{array}\right]$$

$$\left[\begin{array}{rrr|r} 1 & -3 & 2 & 1 \\ 0 & 2 & 8 & 2 \\ 2 & -5 & -3 & 16 \end{array}\right] \quad -2R_1 + R_3 \rightarrow \quad \left[\begin{array}{rrr|r} 1 & -3 & 2 & 1 \\ 0 & 2 & 8 & 2 \\ 0 & 1 & -7 & 14 \end{array}\right]$$

$$\left[\begin{array}{rrr|r} 1 & -3 & 2 & 1 \\ 0 & 2 & 8 & 2 \\ 0 & 1 & -7 & 14 \end{array}\right] \quad \frac{1}{2}R_2 \rightarrow \quad \left[\begin{array}{rrr|r} 1 & -3 & 2 & 1 \\ 0 & 1 & 4 & 1 \\ 0 & 1 & -7 & 14 \end{array}\right]$$

$$\left[\begin{array}{rrr|r} 1 & -3 & 2 & 1 \\ 0 & 1 & 4 & 1 \\ 0 & 1 & -7 & 14 \end{array}\right] \quad -1R_2 + R_3 \rightarrow \quad \left[\begin{array}{rrr|r} 1 & -3 & 2 & 1 \\ 0 & 1 & 4 & 1 \\ 0 & 0 & -11 & 13 \end{array}\right]$$

$$\left[\begin{array}{rrr|r} 1 & -3 & 2 & 1 \\ 0 & 1 & 4 & 1 \\ 0 & 0 & -11 & 13 \end{array}\right] \quad -\frac{1}{11}R_3 \rightarrow \quad \left[\begin{array}{rrr|r} 1 & -3 & 2 & 1 \\ 0 & 1 & 4 & 1 \\ 0 & 0 & 1 & -\frac{13}{11} \end{array}\right]$$

A row echelon form of the matrix is $\left[\begin{array}{rrr|r} 1 & -3 & 2 & 1 \\ 0 & 1 & 4 & 1 \\ 0 & 0 & 1 & -\frac{13}{11} \end{array}\right]$.

Problem 4

$4x - 5y = 17$
$3x + 2y = 7$

$$\left[\begin{array}{rr|r} 4 & -5 & 17 \\ 3 & 2 & 7 \end{array}\right] \xrightarrow[\frac{1}{4}R_1]{\text{Change } a_{11} \text{ to } 1.} \left[\begin{array}{rr|r} 1 & -\frac{5}{4} & \frac{17}{4} \\ 3 & 2 & 7 \end{array}\right]$$

• **We multiplied R_1 by the reciprocal of a_{11}.**

$$\left[\begin{array}{rr|r} 1 & -\frac{5}{4} & \frac{17}{4} \\ 3 & 2 & 7 \end{array}\right] \xrightarrow[-3R_1 + R_2]{\text{Change } a_{21} \text{ to } 0.} \left[\begin{array}{rr|r} 1 & -\frac{5}{4} & \frac{17}{4} \\ 0 & \frac{23}{4} & -\frac{23}{4} \end{array}\right]$$

• **We multiplied R_1 by the opposite of a_{21}.**

$$\left[\begin{array}{rr|r} 1 & -\frac{5}{4} & \frac{17}{4} \\ 0 & \frac{23}{4} & -\frac{23}{4} \end{array}\right] \xrightarrow[\frac{4}{23}R_2]{\text{Change } a_{22} \text{ to } 1.} \left[\begin{array}{rr|r} 1 & -\frac{5}{4} & \frac{17}{4} \\ 0 & 1 & -1 \end{array}\right]$$

• **This is row echelon form.**

$$x - \frac{5}{4}y = \frac{17}{4}$$
$$y = -1$$

$$x - \frac{5}{4}(-1) = \frac{17}{4}$$
$$x + \frac{5}{4} = \frac{17}{4}$$
$$x = 3$$

The solution is $(3, -1)$.

Problem 5

$2x + 3y + 3z = -2$
$x + 2y - 3z = 9$
$3x - 2y - 4z = 1$

$$\left[\begin{array}{rrr|r} 2 & 3 & 3 & -2 \\ 1 & 2 & -3 & 9 \\ 3 & -2 & -4 & 1 \end{array}\right] \quad R_1 \leftrightarrow R_2 \quad \left[\begin{array}{rrr|r} 1 & 2 & -3 & 9 \\ 2 & 3 & 3 & -2 \\ 3 & -2 & -4 & 1 \end{array}\right]$$

$$\left[\begin{array}{rrr|r} 1 & 2 & -3 & 9 \\ 2 & 3 & 3 & -2 \\ 3 & -2 & -4 & 1 \end{array}\right] \quad -2R_1 + R_2 \rightarrow \quad \left[\begin{array}{rrr|r} 1 & 2 & -3 & 9 \\ 0 & -1 & 9 & -20 \\ 3 & -2 & -4 & 1 \end{array}\right]$$

$$\left[\begin{array}{rrr|r} 1 & 2 & -3 & 9 \\ 0 & -1 & 9 & -20 \\ 3 & -2 & -4 & 1 \end{array}\right] \quad -3R_1 + R_3 \rightarrow \quad \left[\begin{array}{rrr|r} 1 & 2 & -3 & 9 \\ 0 & -1 & 9 & -20 \\ 0 & -8 & 5 & -26 \end{array}\right]$$

$$\left[\begin{array}{rrr|r} 1 & 2 & -3 & 9 \\ 0 & -1 & 9 & -20 \\ 0 & -8 & 5 & -26 \end{array}\right] \quad -1R_2 \rightarrow \quad \left[\begin{array}{rrr|r} 1 & 2 & -3 & 9 \\ 0 & 1 & -9 & 20 \\ 0 & -8 & 5 & -26 \end{array}\right]$$

$$\left[\begin{array}{rrr|r} 1 & 2 & -3 & 9 \\ 0 & 1 & -9 & 20 \\ 0 & -8 & 5 & -26 \end{array}\right] \quad 8R_2 + R_3 \rightarrow \quad \left[\begin{array}{rrr|r} 1 & 2 & -3 & 9 \\ 0 & 1 & -9 & 20 \\ 0 & 0 & -67 & 134 \end{array}\right]$$

$$\left[\begin{array}{rrr|r} 1 & 2 & -3 & 9 \\ 0 & 1 & -9 & 20 \\ 0 & 0 & -67 & 134 \end{array}\right] \quad -\frac{1}{67}R_2 \rightarrow \quad \left[\begin{array}{rrr|r} 1 & 2 & -3 & 9 \\ 0 & 1 & -9 & 20 \\ 0 & 0 & 1 & -2 \end{array}\right]$$

$$x + 2y - 3z = 9$$
$$y - 9z = 20$$
$$z = -2$$

$$y - 9(-2) = 20$$
$$y + 18 = 20$$
$$y = 2$$

$$x + 2(2) - 3(-2) = 9$$
$$x + 10 = 9$$
$$x = -1$$

The solution is $(-1, 2, -2)$.

Problem 6

A. $\begin{vmatrix} -1 & -4 \\ 3 & -5 \end{vmatrix} = -1(-5) - 3(-4)$
$= 5 + 12 = 17$

The value of the determinant is 17.

B. Expand by cofactors of the first column.

$$\begin{vmatrix} 1 & 4 & -2 \\ 3 & 1 & 1 \\ 0 & -2 & 2 \end{vmatrix} = 1\begin{vmatrix} 1 & 1 \\ -2 & 2 \end{vmatrix} - 3\begin{vmatrix} 4 & -2 \\ -2 & 2 \end{vmatrix} + 0$$
$$= 1(2 + 2) - 3(8 - 4)$$
$$= 4 - 12$$
$$= -8$$

The value of the determinant is −8.

Problem 7

$$D = \begin{vmatrix} 6 & -6 \\ 2 & -10 \end{vmatrix} = -48$$
$$D_x = \begin{vmatrix} 5 & -6 \\ -1 & -10 \end{vmatrix} = -56$$
$$D_y = \begin{vmatrix} 6 & 5 \\ 2 & -1 \end{vmatrix} = -16$$
$$x = \frac{D_x}{D} = \frac{-56}{-48} = \frac{7}{6} \qquad y = \frac{D_y}{D} = \frac{-16}{-48} = \frac{1}{3}$$

The solution is $\left(\frac{7}{6}, \frac{1}{3}\right)$.

Problem 8

$$D = \begin{vmatrix} 2 & -1 & 1 \\ 3 & 2 & -1 \\ 1 & 3 & 1 \end{vmatrix} = 21$$
$$D_x = \begin{vmatrix} -1 & -1 & 1 \\ 3 & 2 & -1 \\ -2 & 3 & 1 \end{vmatrix} = 9$$
$$D_y = \begin{vmatrix} 2 & -1 & 1 \\ 3 & 3 & -1 \\ 1 & -2 & 1 \end{vmatrix} = -3$$
$$D_z = \begin{vmatrix} 2 & -1 & -1 \\ 3 & 2 & 3 \\ 1 & 3 & -2 \end{vmatrix} = -42$$
$$x = \frac{D_x}{D} = \frac{9}{21} = \frac{3}{7} \qquad y = \frac{D_y}{D} = \frac{-3}{21} = -\frac{1}{7}$$
$$z = \frac{D_z}{D} = \frac{-42}{21} = -2$$

The solution is $\left(\frac{3}{7}, -\frac{1}{7}, -2\right)$.

SECTION 4.4

Problem 1

Strategy ► Rate of the rowing team in calm water: t
Rate of the current: c

	Rate	Time	Distance
With current	$t + c$	2	$2(t + c)$
Against current	$t - c$	2	$2(t - c)$

► The distance traveled with the current is 18 mi.
The distance traveled against the current is 10 mi.

Solution
$$2(t + c) = 18$$
$$\frac{1}{2} \cdot 2(t + c) = \frac{1}{2} \cdot 18 \rightarrow t + c = 9$$
$$2(t - c) = 10$$
$$\frac{1}{2} \cdot 2(t - c) = \frac{1}{2} \cdot 10 \rightarrow t - c = 5$$
$$2t = 14$$
$$t = 7$$

$$t + c = 9$$
$$7 + c = 9$$
$$c = 2$$

The rate of the rowing team in calm water is 7 mph.
The rate of the current is 2 mph.

Problem 2

Strategy ► Cost of an orange tree: x
Cost of a grapefruit tree: y

First purchase:

	Amount	Unit cost	Value
Orange trees	25	x	$25x$
Grapefruit trees	20	y	$20y$

Second purchase:

	Amount	Unit cost	Value
Orange trees	20	x	$20x$
Grapefruit trees	30	y	$30y$

► The total of the first purchase was \$290.
The total of the second purchase was \$330.

Solution

$$25x + 20y = 290$$
$$20x + 30y = 330$$

$$4(25x + 20y) = 4 \cdot 290$$
$$-5(20x + 30y) = -5 \cdot 330$$

$$100x + 80y = 1160$$
$$-100x - 150y = -1650$$
$$-70y = -490$$
$$y = 7$$

$$25x + 20y = 290$$
$$25x + 20(7) = 290$$
$$25x + 140 = 290$$
$$25x = 150$$
$$x = 6$$

The cost of an orange tree is \$6.
The cost of a grapefruit tree is \$7.

Problem 3

Strategy Number of full-price tickets sold: x
Number of member admission tickets sold: y
Number of student tickets sold: z
There were 750 tickets sold: $x + y + z = 750$
The income was \$5400: $10x + 7y + 5z = 5400$
20 more student tickets were sold than full-price tickets: $z = x + 20$

Solution

(1) $x + y + z = 750$
(2) $10x + 7y + 5z = 5400$
(3) $z = x + 20$

Solve the system of equations by substitution. Replace z in equation (1) and equation (2) by $x + 20$.

$$x + y + (x + 20) = 750$$
$$10x + 7y + 5(x + 20) = 5400$$

(4) $2x + y = 730$
(5) $15x + 7y = 5300$

► Multiply equation (4) by −7 and add to equation (5).

$$-14x - 7y = -5110$$
$$15x + 7y = 5300$$
$$x = 190$$

Substitute the value of x into equation (4) and solve for y.

$$2x + y = 730$$
$$2(190) + y = 730$$
$$380 + y = 730$$
$$y = 350$$

Substitute the values of x and y into equation (1) and solve for z.

$$x + y + z = 750$$
$$190 + 350 + z = 750$$
$$540 + z = 750$$
$$z = 210$$

The museum sold 190 full-priced tickets, 350 member tickets, and 210 student tickets.

SECTION 4.5

Problem 1

A. Shade above the solid line $y = 2x - 3$.
Shade above the dashed line $y = -3x$.

The solution set of the system is the intersection of the solution sets of the individual inequalities.

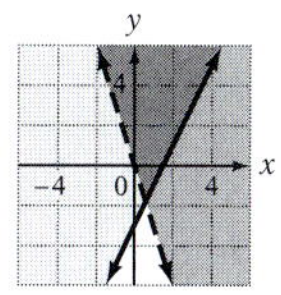

B. $3x + 4y > 12$

$$4y > -3x + 12$$
$$y > -\frac{3}{4}x + 3$$

Shade above the dashed line $y = -\frac{3}{4}x + 3$.

Shade below the dashed line $y = \frac{3}{4}x - 1$.

The solution set of the system is the intersection of the solution sets of the individual inequalities.

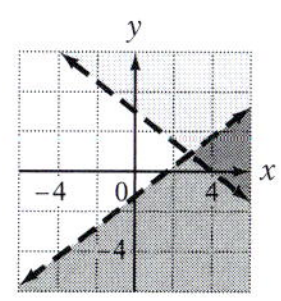

Chapter 5

SECTION 5.1

Problem 1

Strategy Evaluate the function when $r = 0.4$.

Solution

$$v(r) = 600r^2 - 1000r^3$$
$$= 600(0.4)^2 - 1000(0.4)^3$$
$$= 32$$

The velocity will be 32 cm/s.

Problem 2

The leading coefficient is −3, the constant term is −12, and the degree is 4.

Problem 3

A. This is a polynomial function.

B. This is not a polynomial function. A polynomial function does not have a variable expression raised to a negative power.

C. This is not a polynomial function. A polynomial function does not have a variable expression within a radical.

Problem 4

x	y
−1	3
0	0
1	−1
2	0
3	3

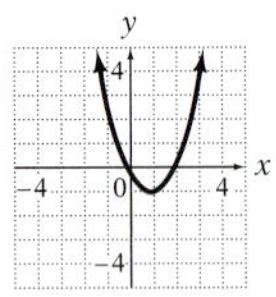

Problem 5

x	y
−3	28
−2	9
−1	2
0	1
1	0
2	−7
3	−26

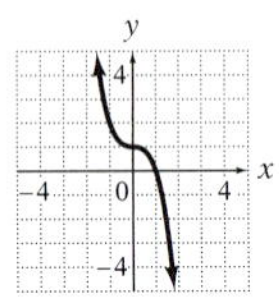

Problem 6

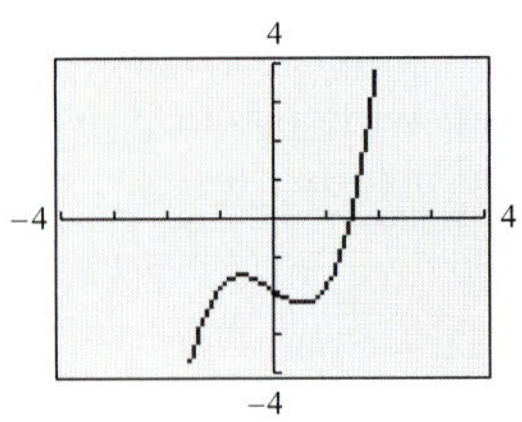

The value of x for which $S(x) = 2$ is approximately 1.8.

Problem 7

$$\begin{array}{r} 5x^2 + 3x - 1 \\ 2x^2 + 4x - 6 \\ x^2 - 7x + 8 \\ \hline 8x^2 \qquad + 1 \end{array}$$

Problem 8

Rewrite the subtraction as the addition of the opposite.

$(-5x^2 + 2x - 3) + (-6x^2 - 3x + 7)$

$$\begin{array}{r} -5x^2 + 2x - 3 \\ -6x^2 - 3x + 7 \\ \hline -11x^2 - x + 4 \end{array}$$

Problem 9

$$\begin{aligned} D(x) &= (4x^3 - 3x^2 + 2) - (-2x^2 + 2x - 3) \\ &= (4x^3 - 3x^2 + 2) + (2x^2 - 2x + 3) \\ &= 4x^3 - x^2 - 2x + 5 \end{aligned}$$

SECTION 5.2

Problem 1

$(3x^2)(6x^3) = (3 \cdot 6)(x^2 \cdot x^3) = 18x^5$

Problem 2

$$\begin{aligned} &(-3xy^2)(-4x^2y^3) \\ &= [(-3)(-4)](x \cdot x^2)(y^2 \cdot y^3) \\ &= 12x^3y^5 \end{aligned}$$

Problem 3

$$\begin{aligned} (3x)(2x^2y)^3 &= (3x)(2^3x^6y^3) = (3x)(8x^6y^3) \\ &= (3 \cdot 8)(x \cdot x^6)y^3 = 24x^7y^3 \end{aligned}$$

SECTION 5.3

Problem 1

A. $(-2y + 3)(-4y) = 8y^2 - 12y$

B. $-a^2(3a^2 + 2a - 7) = -3a^4 - 2a^3 + 7a^2$

Problem 2

$$\begin{array}{r} 2y^3 + 2y^2 \qquad - 3 \\ 3y - 1 \\ \hline -2y^3 - 2y^2 \qquad + 3 \\ 6y^4 + 6y^3 \qquad - 9y \qquad \\ \hline 6y^4 + 4y^3 - 2y^2 - 9y + 3 \end{array}$$

Problem 3

$$\begin{array}{r} 3x^3 - 2x^2 + x - 3 \\ 2x + 5 \\ \hline 15x^3 - 10x^2 + 5x - 15 \\ 6x^4 - 4x^3 + 2x^2 - 6x \qquad \\ \hline 6x^4 + 11x^3 - 8x^2 - x - 15 \end{array}$$

Problem 4

$$\begin{aligned} &(4y - 5)(3y - 3) \\ &= 12y^2 - 12y - 15y + 15 \\ &= 12y^2 - 27y + 15 \end{aligned}$$

Problem 5

$$\begin{aligned} &(3a + 2b)(3a - 5b) \\ &= 9a^2 - 15ab + 6ab - 10b^2 \\ &= 9a^2 - 9ab - 10b^2 \end{aligned}$$

Problem 6

$$\begin{aligned} (2a + 5c)(2a - 5c) &= (2a)^2 - (5c)^2 \\ &= 4a^2 - 25c^2 \end{aligned}$$

Problem 7

$$\begin{aligned} (3x + 2y)^2 &= (3x)^2 + 2(3x)(2y) + (2y)^2 \\ &= 9x^2 + 12xy + 4y^2 \end{aligned}$$

Problem 8

$$\begin{aligned} (6x - y)^2 &= (6x)^2 + 2(6x)(-y) + (-y)^2 \\ &= 36x^2 - 12xy + y^2 \end{aligned}$$

Problem 9

Strategy To find the area in terms of x, replace the variables L and W in the equation $A = LW$ with the given values, and solve for A.

Solution $A = LW$
$A = (x + 7)(x - 4)$
$A = x^2 - 4x + 7x - 28$
$A = x^2 + 3x - 28$

The area is $(x^2 + 3x - 28)$ m^2.

Problem 10

Strategy To find the area of the triangle in terms of x, replace the variables b and h in the equation $A = \frac{1}{2}bh$ with the given values, and solve for A.

Solution
$$A = \frac{1}{2}bh$$
$$A = \frac{1}{2}(x + 3)(4x - 6)$$
$$A = \frac{1}{2}(4x^2 + 6x - 18)$$
$$A = 2x^2 + 3x - 9$$

The area is $(2x^2 + 3x - 9)$ cm^2.

SECTION 5.4

Problem 1

$$\frac{2^{-2}}{2^3} = 2^{-2-3} = 2^{-5} = \frac{1}{2^5} = \frac{1}{32}$$

Problem 2

$$\frac{b^8}{a^{-5}b^6} = a^5b^{8-6} = a^5b^2$$

Problem 3

A. $\dfrac{12x^{-8}y^4}{-16xy^{-3}} = -\dfrac{3x^{-9}y^7}{4} = -\dfrac{3y^7}{4x^9}$

B. $(-3ab)(2a^3b^{-2})^{-3} = (-3ab)(2^{-3}a^{-9}b^6)$
$= -\dfrac{3a^{-8}b^7}{2^3} = -\dfrac{3b^7}{8a^8}$

Problem 4

A. $57{,}000{,}000{,}000 = 5.7 \times 10^{10}$

B. $0.000000017 = 1.7 \times 10^{-8}$

Problem 5

A. $5 \times 10^{12} = 5{,}000{,}000{,}000{,}000$

B. $4.0162 \times 10^{-9} = 0.0000000040162$

Problem 6

$$\frac{5{,}600{,}000 \times 0.000000081}{900 \times 0.000000028}$$
$$= \frac{5.6 \times 10^6 \times 8.1 \times 10^{-8}}{9 \times 10^2 \times 2.8 \times 10^{-8}}$$
$$= \frac{(5.6)(8.1) \times 10^{6+(-8)-2-(-8)}}{(9)(2.8)} = 1.8 \times 10^4$$

Problem 7

Strategy To find the number of arithmetic operations:

- Find the reciprocal of 1×10^{-7}, which is the number of operations performed in 1 s.
- Write the number of seconds in 1 min (60) in scientific notation.
- Multiply the number of arithmetic operations per second by the number of seconds in 1 min.

Solution $\dfrac{1}{1 \times 10^{-7}} = 10^7$
$60 = 6 \times 10$
$6 \times 10 \times 10^7 = 6 \times 10^8$

The computer can perform 6×10^8 operations in 1 min.

SECTION 5.5

Problem 1

$$\frac{4x^3y + 8x^2y^2 - 4xy^3}{2xy} = \frac{4x^3y}{2xy} + \frac{8x^2y^2}{2xy} - \frac{4xy^3}{2xy}$$
$$= 2x^2 + 4xy - 2y^2$$

Problem 2

$$\frac{24x^2y^2 - 18xy + 6y}{6xy} = \frac{24x^2y^2}{6xy} - \frac{18xy}{6xy} + \frac{6y}{6xy}$$
$$= 4xy - 3 + \frac{1}{x}$$

Problem 3

A.
$$\begin{array}{r} 5x - 1 \\ 3x + 4\overline{)15x^2 + 17x - 20} \\ \underline{15x^2 + 20x} \\ -3x - 20 \\ \underline{-3x - 4} \\ -16 \end{array}$$

$$\frac{15x^2 + 17x - 20}{3x + 4} = 5x - 1 - \frac{16}{3x + 4}$$

B.
$$\begin{array}{r} x^2 + 3x - 1 \\ 3x - 1\overline{)3x^3 + 8x^2 - 6x + 2} \\ \underline{3x^3 - x^2} \\ 9x^2 - 6x \\ \underline{9x^2 - 3x} \\ -3x + 2 \\ \underline{-3x + 1} \\ 1 \end{array}$$

$$\frac{3x^3 + 8x^2 - 6x + 2}{3x - 1} = x^2 + 3x - 1 + \frac{1}{3x - 1}$$

Problem 4

A.

$$\begin{array}{r|rrr} -2 & 6 & 8 & -5 \\ & & -12 & 8 \\ \hline & 6 & -4 & 3 \end{array}$$

$$(6x^2 + 8x - 5) \div (x + 2) = 6x - 4 + \frac{3}{x + 2}$$

B.

$$\begin{array}{r|rrrrr} 3 & 2 & -3 & -8 & 0 & -2 \\ & & 6 & 9 & 3 & 9 \\ \hline & 2 & 3 & 1 & 3 & 7 \end{array}$$

$$(2x^4 - 3x^3 - 8x^2 - 2) \div (x - 3)$$
$$= 2x^3 + 3x^2 + x + 3 + \frac{7}{x - 3}$$

Problem 5

$$\begin{array}{r|rrrr} 3 & 2 & 0 & -4 & -5 \\ & & 6 & 18 & 42 \\ \hline & 2 & 6 & 14 & 37 \end{array}$$

By the Remainder Theorem, $P(3) = 37$.

Solutions to Chapter 6 Problems

SECTION 6.1

Problem 1

$4x^6y = 2 \cdot 2 \cdot x^6 \cdot y$
$18x^2y^6 = 2 \cdot 3 \cdot 3 \cdot x^2 \cdot y^6$
$\text{GCF} = 2 \cdot x^2 \cdot y = 2x^2y$

Problem 2

A. $14a^2 = 2 \cdot 7 \cdot a^2$
$21a^4b = 3 \cdot 7 \cdot a^4 \cdot b$
The GCF is $7a^2$.

$$\begin{aligned} 14a^2 - 21a^4b &= 7a^2(2) + 7a^2(-3a^2b) \\ &= 7a^2(2 - 3a^2b) \end{aligned}$$

B. $6x^4y^2 = 2 \cdot 3 \cdot x^4 \cdot y^2$
$9x^3y^2 = 3 \cdot 3 \cdot x^3 \cdot y^2$
$12x^2y^4 = 2 \cdot 2 \cdot 3 \cdot x^2 \cdot y^4$
The GCF is $3x^2y^2$.

$$\begin{aligned} &6x^4y^2 - 9x^3y^2 + 12x^2y^4 \\ &\quad = 3x^2y^2(2x^2) + 3x^2y^2(-3x) + 3x^2y^2(4y^2) \\ &\quad = 3x^2y^2(2x^2 - 3x + 4y^2) \end{aligned}$$

Problem 3

$a(b - 7) + b(b - 7) = (b - 7)(a + b)$

Problem 4

$$\begin{aligned} &3y(5x - 2) - 4(2 - 5x) \\ &\quad = 3y(5x - 2) + 4(5x - 2) \\ &\quad = (5x - 2)(3y + 4) \end{aligned}$$

Problem 5

$$\begin{aligned} &y^5 - 5y^3 + 4y^2 - 20 \\ &\quad = (y^5 - 5y^3) + (4y^2 - 20) \\ &\quad = y^3(y^2 - 5) + 4(y^2 - 5) \\ &\quad = (y^2 - 5)(y^3 + 4) \end{aligned}$$

SECTION 6.2

Problem 1

Factors of 15	Sum
−1, −15	−16
−3, −5	−8

$x^2 - 8x + 15 = (x - 3)(x - 5)$

Problem 2

Factors of −18	Sum
1, −18	−17
−1, 18	17
2, −9	−7
−2, 9	7
3, −6	−3
−3, 6	3

$x^2 + 3x - 18 = (x + 6)(x - 3)$

Problem 3

The GCF is $3b$.

$3a^2b - 18ab - 81b = 3b(a^2 - 6a - 27)$

Factors of −27	Sum
1, −27	−26
−1, 27	26
3, −9	−6
−3, 9	6

$3a^2b - 18ab - 81b = 3b(a + 3)(a - 9)$

Problem 4

The GCF is 4.

$$\begin{aligned} &4x^2 - 40xy + 84y^2 \\ &\quad = 4(x^2 - 10xy + 21y^2) \end{aligned}$$

Factors of 21	Sum
−1, −21	−22
−3, −7	−10

$4x^2 - 40xy + 84y^2 = 4(x - 3y)(x - 7y)$

SECTION 6.3

Problem 1

Factors of 6	Factors of 5
1, 6 2, 3	−1, −5

Trial Factors	Middle Term
$(x-1)(6x-5)$	$-5x-6x=-11x$

$6x^2-11x+5=(x-1)(6x-5)$

Problem 2

Factors of 8	Factors of −15
1, 8 2, 4	1, −15 −1, 15 3, −5 −3, 5

Trial Factors	Middle Term
$(x+1)(8x-15)$	$-15x+8x=-7x$
$(x-1)(8x+15)$	$15x-8x=7x$
$(x+3)(8x-5)$	$-5x+24x=19x$
$(x-3)(8x+5)$	$5x-24x=-19x$
$(2x+1)(4x-15)$	$-30x+4x=-26x$
$(2x-1)(4x+15)$	$30x-4x=26x$
$(2x+3)(4x-5)$	$-10x+12x=2x$
$(2x-3)(4x+5)$	$10x-12x=-2x$
$(8x+1)(x-15)$	$-120x+x=-119x$
$(8x-1)(x+15)$	$120x-x=119x$
$(8x+3)(x-5)$	$-40x+3x=-37x$
$(8x-3)(x+5)$	$40x-3x=37x$
$(4x+1)(2x-15)$	$-60x+2x=-58x$
$(4x-1)(2x+15)$	$60x-2x=58x$
$(4x+3)(2x-5)$	$-20x+6x=-14x$
$(4x-3)(2x+5)$	$20x-6x=14x$

$8x^2+14x-15=(4x-3)(2x+5)$

Problem 3

Factors of 24	Factors of −1
1, 24 2, 12 3, 8 4, 6	1, −1

Trial Factors	Middle Term
$(1-y)(24+y)$	$y-24y=-23y$
$(2-y)(12+y)$	$2y-12y=-10y$
$(3-y)(8+y)$	$3y-8y=-5y$
$(4-y)(6+y)$	$4y-6y=-2y$

$24-2y-y^2=(4-y)(6+y)$

Problem 4

The GCF is $2a^2$.

$4a^2b^2-30a^2b+14a^2$
$=2a^2(2b^2-15b+7)$

Factors of 2	Factors of 7
1, 2	−1, −7

Trial Factors	Middle Term
$(b-1)(2b-7)$	$-7b-2b=-9b$
$(b-7)(2b-1)$	$-b-14b=-15b$

$4a^2b^2-30a^2b+14a^2$
$=2a^2(b-7)(2b-1)$

Problem 5

$a\cdot c=-14$
$-1(14)=-14,\ -1+14=13$

$2a^2+13a-7$
$=2a^2-a+14a-7$
$=(2a^2-a)+(14a-7)$
$=a(2a-1)+7(2a-1)$
$=(2a-1)(a+7)$

Problem 6

$a\cdot c=-12$
$1(-12)=-12,\ 1-12=-11$

$4a^2-11a-3$
$=4a^2+a-12a-3$
$=(4a^2+a)-(12a+3)$
$=a(4a+1)-3(4a+1)$
$=(4a+1)(a-3)$

Problem 7

The GCF is $5x$.

$15x^3+40x^2-80x=5x(3x^2+8x-16)$
$3(-16)=-48$
$-4(12)=-48,\ -4+12=8$

$3x^2+8x-16$
$=3x^2-4x+12x-16$
$=(3x^2-4x)+(12x-16)$
$=x(3x-4)+4(3x-4)$
$=(3x-4)(x+4)$

$15x^3+40x^2-80x=5x(3x^2+8x-16)$
$=5x(3x-4)(x+4)$

SECTION 6.4

Problem 1

$x^2-36y^4=x^2-(6y^2)^2$
$=(x+6y^2)(x-6y^2)$

Problem 2

$9x^2 + 12x + 4 = (3x + 2)^2$

Problem 3

A. $8x^3 + y^3z^3 = (2x)^3 + (yz)^3$
$= (2x + yz)(4x^2 - 2xyz + y^2z^2)$

B. $(x - y)^3 + (x + y)^3$
$= [(x - y) + (x + y)]$
$\times [(x - y)^2 - (x - y)(x + y) + (x + y)^2]$
$= 2x[x^2 - 2xy + y^2 - (x^2 - y^2) + x^2 + 2xy + y^2]$
$= 2x[x^2 - 2xy + y^2 - x^2 + y^2 + x^2 + 2xy + y^2]$
$= 2x(x^2 + 3y^2)$

Problem 4

A. $6x^2y^2 - 19xy + 10$
$= (3xy - 2)(2xy - 5)$

B. $3x^4 + 4x^2 - 4 = (x^2 + 2)(3x^2 - 2)$

Problem 5

A. $4x - 4y - x^3 + x^2y$
$= (4x - 4y) - (x^3 - x^2y)$
$= 4(x - y) - x^2(x - y)$
$= (x - y)(4 - x^2)$
$= (x - y)(2 + x)(2 - x)$

B. $x^{4n} - x^{2n}y^{2n} = x^{2n+2n} - x^{2n}y^{2n}$
$= x^{2n}(x^{2n} - y^{2n})$
$= x^{2n}[(x^n)^2 - (y^n)^2]$
$= x^{2n}(x^n + y^n)(x^n - y^n)$

SECTION 6.5

Problem 1

A. $4x^2 + 11x = 3$
$4x^2 + 11x - 3 = 0$
$(4x - 1)(x + 3) = 0$

$4x - 1 = 0 \quad x + 3 = 0$
$4x = 1 \quad x = -3$
$x = \frac{1}{4}$

The solutions are $\frac{1}{4}$ and -3.

B. $(x - 2)(x + 5) = 8$
$x^2 + 3x - 10 = 8$
$x^2 + 3x - 18 = 0$
$(x + 6)(x - 3) = 0$

$x + 6 = 0 \quad x - 3 = 0$
$x = -6 \quad x = 3$

The solutions are -6 and 3.

C. $x^3 + 4x^2 - 9x - 36 = 0$
$(x^3 + 4x^2) - (9x + 36) = 0$
$x^2(x + 4) - 9(x + 4) = 0$
$(x + 4)(x^2 - 9) = 0$
$(x + 4)(x + 3)(x - 3) = 0$

$x + 4 = 0 \quad x + 3 = 0 \quad x - 3 = 0$
$x = -4 \quad x = -3 \quad x = 3$

The solutions are -4, -3, and 3.

Problem 2

$s(c) = 4$
$c^2 - c - 2 = 4$
$c^2 - c - 6 = 0$
$(c + 2)(c - 3) = 0$

$c + 2 = 0 \quad c - 3 = 0$
$c = -2 \quad c = 3$

The values of c are -2 and 3.

Problem 3

Strategy Replace D by 20 in the equation $D = \frac{n(n - 3)}{2}$ and then solve for n.

Solution $D = \frac{n(n - 3)}{2}$

$20 = \frac{n(n - 3)}{2}$
$40 = n^2 - 3n$
$0 = n^2 - 3n - 40$
$0 = (n - 8)(n + 5)$

$n - 8 = 0 \quad n + 5 = 0$
$n = 8 \quad n = -5$

The answer -5 does not make sense in the context of this problem.

The polygon has 8 sides.

Chapter 7

SECTION 7.1

Problem 1

$f(x) = \frac{3 - 5x}{x^2 + 5x + 6}$

$f(2) = \frac{3 - 5(2)}{2^2 + 5(2) + 6}$

$f(2) = \frac{3 - 10}{4 + 10 + 6}$

$f(2) = \frac{-7}{20}$

$f(2) = -\frac{7}{20}$

Problem 2

$x^2 - 4 = 0$
$(x - 2)(x + 2) = 0$

$x - 2 = 0 \quad x + 2 = 0$
$x = 2 \quad x = -2$

The domain is $\{x \mid x \neq -2, 2\}$.

Problem 3

$x^2 + 4$ is a positive number.

The domain is $\{x \mid x \in \text{real numbers}\}$.

Problem 4

A.
$$\frac{6x^4 - 24x^3}{12x^3 - 48x^2} = \frac{6x^3(x-4)}{12x^2(x-4)}$$
$$= \frac{6x^3\overset{1}{\cancel{(x-4)}}}{12x^2\underset{1}{\cancel{(x-4)}}} = \frac{x}{2}$$

B.
$$\frac{20x - 15x^2}{15x^3 - 5x^2 - 20x} = \frac{5x(4-3x)}{5x(3x^2 - x - 4)}$$
$$= \frac{5x(4-3x)}{5x(3x-4)(x+1)}$$
$$= \frac{5x\overset{-1}{\cancel{(4-3x)}}}{5x\underset{1}{\cancel{(3x-4)}}(x+1)}$$
$$= -\frac{1}{x+1}$$

C.
$$\frac{x^{2n} + x^n - 12}{x^{2n} - 3x^n} = \frac{(x^n+4)(x^n-3)}{x^n(x^n-3)}$$
$$= \frac{(x^n+4)\overset{1}{\cancel{(x^n-3)}}}{x^n\underset{1}{\cancel{(x^n-3)}}} = \frac{x^n+4}{x^n}$$

SECTION 7.2

Problem 1

A.
$$\frac{12 + 5x - 3x^2}{x^2 + 2x - 15} \cdot \frac{2x^2 + x - 45}{3x^2 + 4x}$$
$$= \frac{(4+3x)(3-x)}{(x+5)(x-3)} \cdot \frac{(2x-9)(x+5)}{x(3x+4)}$$
$$= \frac{(4+3x)(3-x)(2x-9)(x+5)}{(x+5)(x-3) \cdot x(3x+4)}$$
$$= \frac{\overset{1}{\cancel{(4+3x)}}\overset{-1}{\cancel{(3-x)}}(2x-9)\overset{1}{\cancel{(x+5)}}}{\underset{1}{\cancel{(x+5)}}\underset{1}{\cancel{(x-3)}} \cdot x\underset{1}{\cancel{(3x+4)}}}$$
$$= -\frac{2x-9}{x}$$

B.
$$\frac{2x^2 - 13x + 20}{x^2 - 16} \cdot \frac{2x^2 + 9x + 4}{6x^2 - 7x - 5}$$
$$= \frac{(2x-5)(x-4)}{(x-4)(x+4)} \cdot \frac{(2x+1)(x+4)}{(3x-5)(2x+1)}$$
$$= \frac{(2x-5)(x-4)(2x+1)(x+4)}{(x-4)(x+4)(3x-5)(2x+1)}$$
$$= \frac{(2x-5)\overset{1}{\cancel{(x-4)}}\overset{1}{\cancel{(2x+1)}}\overset{1}{\cancel{(x+4)}}}{\underset{1}{\cancel{(x-4)}}\underset{1}{\cancel{(x+4)}}(3x-5)\underset{1}{\cancel{(2x+1)}}}$$
$$= \frac{2x-5}{3x-5}$$

Problem 2

A.
$$\frac{6x^2 - 3xy}{10ab^4} \div \frac{16x^2y^2 - 8xy^3}{15a^2b^2}$$
$$= \frac{6x^2 - 3xy}{10ab^4} \cdot \frac{15a^2b^2}{16x^2y^2 - 8xy^3}$$
$$= \frac{3x(2x-y)}{10ab^4} \cdot \frac{15a^2b^2}{8xy^2(2x-y)}$$
$$= \frac{45a^2b^2x(2x-y)}{80ab^4xy^2(2x-y)} = \frac{45a^2b^2x\overset{1}{\cancel{(2x-y)}}}{80ab^4xy^2\underset{1}{\cancel{(2x-y)}}}$$
$$= \frac{9a}{16b^2y^2}$$

B.
$$\frac{6x^2 - 7x + 2}{3x^2 + x - 2} \div \frac{4x^2 - 8x + 3}{5x^2 + x - 4}$$
$$= \frac{6x^2 - 7x + 2}{3x^2 + x - 2} \cdot \frac{5x^2 + x - 4}{4x^2 - 8x + 3}$$
$$= \frac{(2x-1)(3x-2)}{(x+1)(3x-2)} \cdot \frac{(x+1)(5x-4)}{(2x-1)(2x-3)}$$
$$= \frac{(2x-1)(3x-2)(x+1)(5x-4)}{(x+1)(3x-2)(2x-1)(2x-3)}$$
$$= \frac{\overset{1}{\cancel{(2x-1)}}\overset{1}{\cancel{(3x-2)}}\overset{1}{\cancel{(x+1)}}(5x-4)}{\underset{1}{\cancel{(x+1)}}\underset{1}{\cancel{(3x-2)}}\underset{1}{\cancel{(2x-1)}}(2x-3)}$$
$$= \frac{5x-4}{2x-3}$$

Problem 3

The LCM is $a(a-5)(a+5)$.

$$\frac{a-3}{a^2-5a} + \frac{a-9}{a^2-25}$$
$$= \frac{a-3}{a(a-5)} \cdot \frac{a+5}{a+5} + \frac{a-9}{(a-5)(a+5)} \cdot \frac{a}{a}$$
$$= \frac{a^2 + 2a - 15}{a(a-5)(a+5)} + \frac{a^2 - 9a}{a(a-5)(a+5)}$$
$$= \frac{(a^2 + 2a - 15) + (a^2 - 9a)}{a(a-5)(a+5)}$$
$$= \frac{a^2 + 2a - 15 + a^2 - 9a}{a(a-5)(a+5)}$$
$$= \frac{2a^2 - 7a - 15}{a(a-5)(a+5)} = \frac{(2a+3)(a-5)}{a(a-5)(a+5)}$$
$$= \frac{(2a+3)\overset{1}{\cancel{(a-5)}}}{a\underset{1}{\cancel{(a-5)}}(a+5)} = \frac{2a+3}{a(a+5)}$$

Problem 4

The LCM is $(x-2)(2x-3)$.

$$\frac{x-1}{x-2} - \frac{7-6x}{2x^2 - 7x + 6} + \frac{4}{2x-3}$$
$$= \frac{x-1}{x-2} \cdot \frac{2x-3}{2x-3} - \frac{7-6x}{(x-2)(2x-3)}$$
$$+ \frac{4}{2x-3} \cdot \frac{x-2}{x-2}$$

$$= \frac{2x^2 - 5x + 3}{(x-2)(2x-3)} - \frac{7 - 6x}{(x-2)(2x-3)} + \frac{4x - 8}{(x-2)(2x-3)}$$

$$= \frac{(2x^2 - 5x + 3) - (7 - 6x) + (4x - 8)}{(x-2)(2x-3)}$$

$$= \frac{2x^2 - 5x + 3 - 7 + 6x + 4x - 8}{(x-2)(2x-3)}$$

$$= \frac{2x^2 + 5x - 12}{(x-2)(2x-3)} = \frac{(x+4)(2x-3)}{(x-2)(2x-3)}$$

$$= \frac{(x+4)\overset{1}{\cancel{(2x-3)}}}{(x-2)\underset{1}{\cancel{(2x-3)}}} = \frac{x+4}{x-2}$$

SECTION 7.3

Problem 1

A. The LCM of x and x^2 is x^2.

$$\frac{3 + \frac{16}{x} + \frac{16}{x^2}}{6 + \frac{5}{x} - \frac{4}{x^2}} = \frac{3 + \frac{16}{x} + \frac{16}{x^2}}{6 + \frac{5}{x} - \frac{4}{x^2}} \cdot \frac{x^2}{x^2}$$

$$= \frac{3 \cdot x^2 + \frac{16}{x} \cdot x^2 + \frac{16}{x^2} \cdot x^2}{6 \cdot x^2 + \frac{5}{x} \cdot x^2 - \frac{4}{x^2} \cdot x^2}$$

$$= \frac{3x^2 + 16x + 16}{6x^2 + 5x - 4}$$

$$= \frac{(3x+4)(x+4)}{(2x-1)(3x+4)}$$

$$= \frac{\overset{1}{\cancel{(3x+4)}}(x+4)}{(2x-1)\underset{1}{\cancel{(3x+4)}}} = \frac{x+4}{2x-1}$$

B. The LCM is $x - 3$.

$$\frac{2x + 5 + \frac{14}{x-3}}{4x + 16 + \frac{49}{x-3}}$$

$$= \frac{2x + 5 + \frac{14}{x-3}}{4x + 16 + \frac{49}{x-3}} \cdot \frac{x-3}{x-3}$$

$$= \frac{(2x+5)(x-3) + \frac{14}{x-3}(x-3)}{(4x+16)(x-3) + \frac{49}{x-3}(x-3)}$$

$$= \frac{2x^2 - x - 15 + 14}{4x^2 + 4x - 48 + 49} = \frac{2x^2 - x - 1}{4x^2 + 4x + 1}$$

$$= \frac{(2x+1)(x-1)}{(2x+1)(2x+1)} = \frac{\overset{1}{\cancel{(2x+1)}}(x-1)}{\underset{1}{\cancel{(2x+1)}}(2x+1)}$$

$$= \frac{x-1}{2x+1}$$

SECTION 7.4

Problem 1

A.

$$\frac{5}{2x-3} = \frac{-2}{x+1}$$

$$(x+1)(2x-3)\frac{5}{2x-3} = (x+1)(2x-3)\frac{-2}{x+1}$$

$$5(x+1) = -2(2x-3)$$

$$5x + 5 = -4x + 6$$

$$9x + 5 = 6$$

$$9x = 1$$

$$x = \frac{1}{9}$$

The solution is $\frac{1}{9}$.

B.

$$\frac{4x+1}{2x-1} = 2 + \frac{3}{x-3}$$

$$(2x-1)(x-3)\frac{4x+1}{2x-1} = (2x-1)(x-3)\left(2 + \frac{3}{x-3}\right)$$

$$(x-3)(4x+1) = (2x-1)(x-3)2 + (2x-1)3$$

$$4x^2 - 11x - 3 = 4x^2 - 14x + 6 + 6x - 3$$

$$-11x - 3 = -8x + 3$$

$$-3x = 6$$

$$x = -2$$

The solution is -2.

Problem 2

Strategy ▶ Time required for the small pipe to fill the tank: x

	Rate	Time	Part
Large pipe	$\frac{1}{9}$	6	$\frac{6}{9}$
Small pipe	$\frac{1}{x}$	6	$\frac{6}{x}$

▶ The sum of the part of the task completed by the large pipe and the part of the task completed by the small pipe is 1.

Solution

$$\frac{6}{9} + \frac{6}{x} = 1$$

$$\frac{2}{3} + \frac{6}{x} = 1$$

$$3x\left(\frac{2}{3} + \frac{6}{x}\right) = 3x \cdot 1$$

$$2x + 18 = 3x$$
$$18 = x$$

The small pipe working alone will fill the tank in 18 h.

Problem 3

Strategy ▶ Rate of the wind: r

	Distance	Rate	Time
With wind	700	$150 + r$	$\frac{700}{150 + r}$
Against wind	500	$150 - r$	$\frac{500}{150 - r}$

▶ The time flying with the wind equals the time flying against the wind.

Solution

$$\frac{700}{150 + r} = \frac{500}{150 - r}$$
$$(150 + r)(150 - r)\left(\frac{700}{150 + r}\right) = (150 + r)(150 - r)\left(\frac{500}{150 - r}\right)$$
$$(150 - r)700 = (150 + r)500$$
$$105{,}000 - 700r = 75{,}000 + 500r$$
$$30{,}000 = 1200r$$
$$25 = r$$

The rate of the wind is 25 mph.

SECTION 7.5

Problem 1

Strategy To find the cost, write and solve a proportion using x to represent the cost.

Solution

$$\frac{2}{5.80} = \frac{15}{x}$$
$$x(5.80)\frac{2}{5.80} = x(5.80)\frac{15}{x}$$
$$2x = 15(5.80)$$
$$2x = 87$$
$$x = 43.5$$

The cost of 15 lb of cashews is \$43.50.

Problem 2

Strategy To find the distance:

▶ Write the basic direct variation equation, replace the variables by the given values, and solve for k.

▶ Write the direct variation equation, replacing k by its value. Substitute 5 for t, and solve for s.

Solution

$$s = kt^2$$
$$64 = k(2)^2$$
$$64 = k \cdot 4$$
$$16 = k$$

$$s = 16t^2$$
$$s = 16(5)^2 = 400$$

The object will fall 400 ft in 5 s.

Problem 3

Strategy To find the resistance:

▶ Write the basic inverse variation equation, replace the variables by the given values, and solve for k.

▶ Write the inverse variation equation, replacing k by its value. Substitute 0.02 for d, and solve for R.

Solution

$$R = \frac{k}{d^2}$$
$$0.5 = \frac{k}{(0.01)^2}$$
$$0.5 = \frac{k}{0.0001}$$
$$0.00005 = k$$

$$R = \frac{0.00005}{d^2}$$
$$R = \frac{0.00005}{(0.02)^2} = 0.125$$

The resistance is 0.125 ohm.

Problem 4

Strategy To find the strength of the beam:

▶ Write the basic combined variation equation, replace the variables by the given values, and solve for k.

▶ Write the basic combined variation equation, replacing k by its value and substituting 4 for W; 8 for d, and 16 for L. Solve for s.

Solution

$$s = \frac{kWd^2}{L}$$
$$1200 = \frac{k(2)(12)^2}{12}$$
$$1200 = \frac{k(288)}{12}$$
$$14{,}400 = 288k$$
$$50 = k$$

$$s = \frac{50Wd^2}{L}$$
$$s = \frac{50(4)8^2}{16}$$
$$s = \frac{12{,}800}{16}$$
$$s = 800$$

The strength of the beam is 800 lb.

SECTION 7.6

Problem 1

A.
$$\frac{1}{R_1}+\frac{1}{R_2}=\frac{1}{R}$$
$$RR_1R_2\left(\frac{1}{R_1}+\frac{1}{R_2}\right)=RR_1R_2\left(\frac{1}{R}\right)$$
$$RR_1R_2\left(\frac{1}{R_1}\right)+RR_1R_2\left(\frac{1}{R_2}\right)=R_1R_2$$
$$RR_2+RR_1=R_1R_2$$
$$R(R_2+R_1)=R_1R_2$$
$$R=\frac{R_1R_2}{R_2+R_1}$$

B.
$$t=\frac{r}{r+1}$$
$$(r+1)t=(r+1)\left(\frac{r}{r+1}\right)$$
$$tr+t=r$$
$$t=r-tr$$
$$t=r(1-t)$$
$$\frac{t}{1-t}=r$$

Chapter 8

SECTION 8.1

Problem 1

A. $64^{\frac{2}{3}}=(2^6)^{\frac{2}{3}}=2^4=16$

B. $16^{-\frac{3}{4}}=(2^4)^{-\frac{3}{4}}=2^{-3}=\frac{1}{2^3}=\frac{1}{8}$

C. $(-81)^{\frac{3}{4}}$

The base of the exponential expression is negative, and the denominator of the exponent is a positive even number.

Therefore, $(-81)^{\frac{3}{4}}$ is not a real number.

Problem 2

A. $\dfrac{x^{\frac{1}{2}}y^{-\frac{5}{4}}}{x^{-\frac{4}{3}}y^{\frac{1}{3}}}=\dfrac{x^{\frac{1}{2}+\frac{4}{3}}}{y^{\frac{1}{3}+\frac{5}{4}}}=\dfrac{x^{\frac{11}{6}}}{y^{\frac{19}{12}}}$

B. $\left(x^{\frac{3}{4}}y^{\frac{1}{2}}z^{-\frac{2}{3}}\right)^{-\frac{4}{3}}=x^{-1}y^{-\frac{2}{3}}z^{\frac{8}{9}}=\dfrac{z^{\frac{8}{9}}}{xy^{\frac{2}{3}}}$

C. $\left(\dfrac{16a^{-2}b^{\frac{4}{3}}}{9a^4b^{-\frac{2}{3}}}\right)^{-\frac{1}{2}}=\left(\dfrac{2^4a^{-6}b^2}{3^2}\right)^{-\frac{1}{2}}=\dfrac{2^{-2}a^3b^{-1}}{3^{-1}}$

$=\dfrac{3a^3}{2^2b}=\dfrac{3a^3}{4b}$

Problem 3

A. $(2x^3)^{\frac{3}{4}}=\sqrt[4]{(2x^3)^3}=\sqrt[4]{8x^9}$

B. $-5a^{\frac{5}{6}}=-5(a^5)^{\frac{1}{6}}=-5\sqrt[6]{a^5}$

Problem 4

A. $\sqrt[3]{3ab}=(3ab)^{\frac{1}{3}}$

B. $\sqrt[4]{x^4+y^4}=(x^4+y^4)^{\frac{1}{4}}$

Problem 5

A. $\sqrt{121x^{10}y^4}=11x^5y^2$

B. $\sqrt[3]{-8x^{12}y^3}=-2x^4y$

C. $-\sqrt[4]{81x^{12}y^8}=-3x^3y^2$

SECTION 8.2

Problem 1

$$\sqrt[5]{128x^7}=\sqrt[5]{32x^5(4x^2)}$$
$$=\sqrt[5]{32x^5}\sqrt[5]{4x^2}$$
$$=2x\sqrt[5]{4x^2}$$

Problem 2

A. $3xy\sqrt[3]{81x^5y}-\sqrt[3]{192x^8y^4}$

$=3xy\sqrt[3]{27x^3}\sqrt[3]{3x^2y}-\sqrt[3]{64x^6y^3}\sqrt[3]{3x^2y}$

$=3xy\cdot 3x\sqrt[3]{3x^2y}-4x^2y\sqrt[3]{3x^2y}$

$=9x^2y\sqrt[3]{3x^2y}-4x^2y\sqrt[3]{3x^2y}$

$=5x^2y\sqrt[3]{3x^2y}$

B. $4a\sqrt[3]{54a^7b^9}+a^2b\sqrt[3]{128a^4b^6}$

$=4a\sqrt[3]{27a^6b^9}\sqrt[3]{2a}+a^2b\sqrt[3]{64a^3b^6}\sqrt[3]{2a}$

$=4a\cdot 3a^2b^3\sqrt[3]{2a}+a^2b\cdot 4ab^2\sqrt[3]{2a}$

$=12a^3b^3\sqrt[3]{2a}+4a^3b^3\sqrt[3]{2a}$

$=16a^3b^3\sqrt[3]{2a}$

Problem 3

$\sqrt{5b}\left(\sqrt{3b}-\sqrt{10}\right)$

$=\sqrt{15b^2}-\sqrt{50b}$

$=\sqrt{b^2}\sqrt{15}-\sqrt{25}\sqrt{2b}$

$=b\sqrt{15}-5\sqrt{2b}$

Problem 4

A. $\left(2\sqrt[3]{2x}-3\right)\left(\sqrt[3]{2x}-5\right)$

$=2\sqrt[3]{4x^2}-10\sqrt[3]{2x}-3\sqrt[3]{2x}+15$

$=2\sqrt[3]{4x^2}-13\sqrt[3]{2x}+15$

B. $\left(2\sqrt{x}-3\right)\left(2\sqrt{x}+3\right)$

$=\left(2\sqrt{x}\right)^2-9=4x-9$

Problem 5

A. $\dfrac{y}{\sqrt{3y}}=\dfrac{y}{\sqrt{3y}}\cdot\dfrac{\sqrt{3y}}{\sqrt{3y}}=\dfrac{y\sqrt{3y}}{\sqrt{9y^2}}=\dfrac{y\sqrt{3y}}{3y}$

$=\dfrac{\sqrt{3y}}{3}$

B. $\dfrac{3}{\sqrt[3]{3x^2}}=\dfrac{3}{\sqrt[3]{3x^2}}\cdot\dfrac{\sqrt[3]{9x}}{\sqrt[3]{9x}}=\dfrac{3\sqrt[3]{9x}}{\sqrt[3]{27x^3}}=\dfrac{3\sqrt[3]{9x}}{3x}$

$=\dfrac{\sqrt[3]{9x}}{x}$

Problem 6

A. $\dfrac{4+\sqrt{2}}{3-\sqrt{3}} = \dfrac{4+\sqrt{2}}{3-\sqrt{3}} \cdot \dfrac{3+\sqrt{3}}{3+\sqrt{3}}$

$= \dfrac{12+4\sqrt{3}+3\sqrt{2}+\sqrt{6}}{9-(\sqrt{3})^2}$

$= \dfrac{12+4\sqrt{3}+3\sqrt{2}+\sqrt{6}}{6}$

B. $\dfrac{\sqrt{2}+\sqrt{x}}{\sqrt{2}-\sqrt{x}} = \dfrac{\sqrt{2}+\sqrt{x}}{\sqrt{2}-\sqrt{x}} \cdot \dfrac{\sqrt{2}+\sqrt{x}}{\sqrt{2}+\sqrt{x}}$

$= \dfrac{\sqrt{4}+\sqrt{2x}+\sqrt{2x}+\sqrt{x^2}}{(\sqrt{2})^2-(\sqrt{x})^2}$

$= \dfrac{2+2\sqrt{2x}+x}{2-x}$

SECTION 8.3

Problem 1

A. Because Q contains an odd root, there are no restrictions on the radicand.

The domain is $(-\infty, \infty)$.

B. $3x + 9 \geq 0$

$3x \geq -9$

$x \geq -3$

The domain is $[-3, \infty)$.

Problem 2

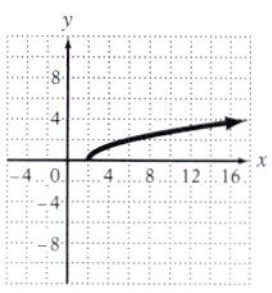

Problem 3

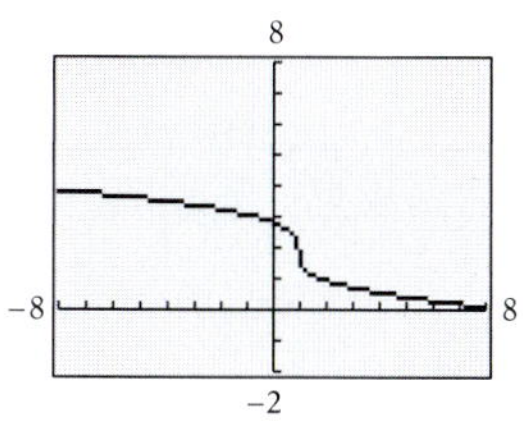

SECTION 8.4

Problem 1

A. $\sqrt{4x+5} - 12 = -5$

$\sqrt{4x+5} = 7$

$(\sqrt{4x+5})^2 = 7^2$

$4x + 5 = 49$

$4x = 44$

$x = 11$

Check: $\sqrt{4x+5} - 12 = -5$

$\sqrt{4 \cdot 11 + 5} - 12 \quad -5$

$\sqrt{44+5} - 12$

$\sqrt{49} - 12$

$7 - 12$

$-5 = -5$

The solution is 11.

B. $\sqrt[4]{x-8} = 3$

$(\sqrt[4]{x-8})^4 = 3^4$

$x - 8 = 81$

$x = 89$

Check: $\sqrt[4]{x-8} = 3$

$\sqrt[4]{89-8} \quad 3$

$\sqrt[4]{81}$

$3 = 3$

The solution is 89.

Problem 2

A. $x + 3\sqrt{x+2} = 8$

$3\sqrt{x+2} = 8 - x$

$(3\sqrt{x+2})^2 = (8-x)^2$

$9(x+2) = 64 - 16x + x^2$

$9x + 18 = 64 - 16x + x^2$

$0 = x^2 - 25x + 46$

$0 = (x-2)(x-23)$

$x - 2 = 0 \qquad x - 23 = 0$

$x = 2 \qquad x = 23$

Check: $x + 3\sqrt{x+2} = 8$

$2 + 3\sqrt{2+2} \quad 8$

$2 + 3\sqrt{4}$

$2 + 3 \cdot 2$

$2 + 6$

$8 = 8$

$x + 3\sqrt{x+2} = 8$

$23 + 3\sqrt{23+2} \quad 8$

$23 + 3\sqrt{25}$

$23 + 3 \cdot 5$

$23 + 15$

$38 \neq 8$

23 does not check as a solution. The solution is 2.

B. $\sqrt{x+5} = 5 - \sqrt{x}$

$(\sqrt{x+5})^2 = (5-\sqrt{x})^2$

$x + 5 = 25 - 10\sqrt{x} + x$

$-20 = -10\sqrt{x}$

$2 = \sqrt{x}$

$2^2 = (\sqrt{x})^2$

$4 = x$

Check: $\sqrt{x+5} = 5 - \sqrt{x}$

$\sqrt{4+5} \quad 5 - \sqrt{4}$

$\sqrt{9} \quad 5 - 2$

$3 = 3$

The solution is 4.

Problem 3

Strategy To find the diagonal, use the Pythagorean Theorem. One leg is the length of the rectangle. The second leg is the width of the rectangle. The hypotenuse is the diagonal of the rectangle.

Solution

$$\begin{aligned} c^2 &= a^2 + b^2 \\ c^2 &= (6)^2 + (3)^2 \\ c^2 &= 36 + 9 \\ c^2 &= 45 \\ (c^2)^{\frac{1}{2}} &= (45)^{\frac{1}{2}} \\ c &= \sqrt{45} \\ c &\approx 6.7 \end{aligned}$$

The diagonal is 6.7 cm.

SECTION 8.5

Problem 1

$\sqrt{-45} = i\sqrt{45} = i\sqrt{9 \cdot 5} = 3i\sqrt{5}$

Problem 2

$$\begin{aligned} &\sqrt{98} - \sqrt{-60} = \sqrt{98} - i\sqrt{60} \\ &= \sqrt{49 \cdot 2} - i\sqrt{4 \cdot 15} \\ &= 7\sqrt{2} - 2i\sqrt{15} \end{aligned}$$

Problem 3

$(-4 + 2i) - (6 - 8i) = -10 + 10i$

Problem 4

$$\begin{aligned} &\left(16 - \sqrt{-45}\right) - \left(3 + \sqrt{-20}\right) \\ &= \left(16 - i\sqrt{45}\right) - \left(3 + i\sqrt{20}\right) \\ &= \left(16 - i\sqrt{9 \cdot 5}\right) - \left(3 + i\sqrt{4 \cdot 5}\right) \\ &= \left(16 - 3i\sqrt{5}\right) - \left(3 + 2i\sqrt{5}\right) \\ &= 13 - 5i\sqrt{5} \end{aligned}$$

Problem 5

$$\begin{aligned} &\sqrt{-3}\left(\sqrt{27} - \sqrt{-6}\right) \\ &= i\sqrt{3}\left(\sqrt{27} - i\sqrt{6}\right) \\ &= i\sqrt{81} - i^2\sqrt{18} \\ &= i\sqrt{81} - (-1)\sqrt{2 \cdot 9} \\ &= 9i + 3\sqrt{2} \\ &= 3\sqrt{2} + 9i \end{aligned}$$

Problem 6

A. $$\begin{aligned} (4 - 3i)(2 - i) &= 8 - 4i - 6i + 3i^2 \\ &= 8 - 10i + 3i^2 = 8 - 10i + 3(-1) \\ &= 5 - 10i \end{aligned}$$

B. $$\begin{aligned} &(3 - i)\left(\frac{3}{10} + \frac{1}{10}i\right) \\ &= \frac{9}{10} + \frac{3}{10}i - \frac{3}{10}i - \frac{1}{10}i^2 = \frac{9}{10} - \frac{1}{10}i^2 \\ &= \frac{9}{10} - \frac{1}{10}(-1) = \frac{9}{10} + \frac{1}{10} = 1 \end{aligned}$$

C. $$\begin{aligned} (3 + 6i)(3 - 6i) &= 3^2 + 6^2 = 9 + 36 \\ &= 45 \end{aligned}$$

D. $$\begin{aligned} (1 + 5i)^2 &= 1 + 10i + 25i^2 \\ &= 1 + 10i + 25(-1) \\ &= 1 + 10i - 25 \\ &= -24 + 10i \end{aligned}$$

Problem 7

$$\begin{aligned} &\frac{2 - 3i}{4i} = \frac{2 - 3i}{4i} \cdot \frac{i}{i} = \frac{2i - 3i^2}{4i^2} = \frac{2i - 3(-1)}{4(-1)} \\ &= \frac{3 + 2i}{-4} = -\frac{3}{4} - \frac{1}{2}i \end{aligned}$$

Problem 8

$$\begin{aligned} &\frac{2 + 5i}{3 - 2i} = \frac{(2 + 5i)}{(3 - 2i)} \cdot \frac{(3 + 2i)}{(3 + 2i)} \\ &= \frac{6 + 4i + 15i + 10i^2}{3^2 + 2^2} = \frac{6 + 19i + 10(-1)}{13} \\ &= \frac{-4 + 19i}{13} = -\frac{4}{13} + \frac{19}{13}i \end{aligned}$$

Chapter 9

SECTION 9.1

Problem 1

$$\begin{aligned} x^2 - 3ax - 4a^2 &= 0 \\ (x + a)(x - 4a) &= 0 \end{aligned}$$

$$\begin{aligned} x + a &= 0 \qquad & x - 4a &= 0 \\ x &= -a \qquad & x &= 4a \end{aligned}$$

The solutions are $-a$ and $4a$.

Problem 2

$$\begin{aligned} (x - r_1)(x - r_2) &= 0 \\ \left[x - \left(-\frac{2}{3}\right)\right]\left(x - \frac{1}{6}\right) &= 0 \\ \left(x + \frac{2}{3}\right)\left(x - \frac{1}{6}\right) &= 0 \\ x^2 + \frac{3}{6}x - \frac{2}{18} &= 0 \\ 18\left(x^2 + \frac{3}{6}x - \frac{2}{18}\right) &= 0 \\ 18x^2 + 9x - 2 &= 0 \end{aligned}$$

Problem 3

$$\begin{aligned} 2(x + 1)^2 + 24 &= 0 \\ 2(x + 1)^2 &= -24 \\ (x + 1)^2 &= -12 \\ \sqrt{(x + 1)^2} &= \sqrt{-12} \\ x + 1 &= \pm\sqrt{-12} \\ x + 1 &= \pm 2i\sqrt{3} \end{aligned}$$

$$\begin{aligned} x + 1 &= 2i\sqrt{3} \qquad & x + 1 &= -2i\sqrt{3} \\ x &= -1 + 2i\sqrt{3} \qquad & x &= -1 - 2i\sqrt{3} \end{aligned}$$

The solutions are $-1 + 2i\sqrt{3}$ and $-1 - 2i\sqrt{3}$.

SECTION 9.2

Problem 1

A. $4x^2 - 4x - 1 = 0$

$$4x^2 - 4x = 1$$
$$\frac{4x^2 - 4x}{4} = \frac{1}{4}$$
$$x^2 - x = \frac{1}{4}$$

Complete the square.

$$x^2 - x + \frac{1}{4} = \frac{1}{4} + \frac{1}{4}$$
$$\left(x - \frac{1}{2}\right)^2 = \frac{2}{4}$$
$$\sqrt{\left(x - \frac{1}{2}\right)^2} = \sqrt{\frac{2}{4}}$$
$$x - \frac{1}{2} = \pm\sqrt{\frac{2}{4}}$$
$$x - \frac{1}{2} = \pm\frac{\sqrt{2}}{2}$$

$$x - \frac{1}{2} = \frac{\sqrt{2}}{2} \qquad x - \frac{1}{2} = -\frac{\sqrt{2}}{2}$$
$$x = \frac{1}{2} + \frac{\sqrt{2}}{2} \qquad x = \frac{1}{2} - \frac{\sqrt{2}}{2}$$

The solutions are $\frac{1 + \sqrt{2}}{2}$ and $\frac{1 - \sqrt{2}}{2}$.

B. $2x^2 + x - 5 = 0$

$$2x^2 + x = 5$$
$$\frac{2x^2 + x}{2} = \frac{5}{2}$$
$$x^2 + \frac{1}{2}x = \frac{5}{2}$$

Complete the square.

$$x^2 + \frac{1}{2}x + \frac{1}{16} = \frac{5}{2} + \frac{1}{16}$$
$$\left(x + \frac{1}{4}\right)^2 = \frac{41}{16}$$
$$\sqrt{\left(x + \frac{1}{4}\right)^2} = \sqrt{\frac{41}{16}}$$
$$x + \frac{1}{4} = \pm\sqrt{\frac{41}{16}}$$
$$x + \frac{1}{4} = \pm\frac{\sqrt{41}}{4}$$

$$x + \frac{1}{4} = \frac{\sqrt{41}}{4} \qquad x + \frac{1}{4} = -\frac{\sqrt{41}}{4}$$
$$x = -\frac{1}{4} + \frac{\sqrt{41}}{4} \qquad x = -\frac{1}{4} - \frac{\sqrt{41}}{4}$$

The solutions are $\frac{-1 + \sqrt{41}}{4}$ and $\frac{-1 - \sqrt{41}}{4}$.

Problem 2

A. $x^2 + 6x - 9 = 0$

$a = 1, b = 6, c = -9$

$$x = \frac{-b \pm \sqrt{b^2 - 4ac}}{2a}$$
$$= \frac{-6 \pm \sqrt{6^2 - 4(1)(-9)}}{2 \cdot 1}$$
$$= \frac{-6 \pm \sqrt{36 + 36}}{2}$$
$$= \frac{-6 \pm \sqrt{72}}{2}$$
$$= \frac{-6 \pm 6\sqrt{2}}{2} = -3 \pm 3\sqrt{2}$$

The solutions are $-3 + 3\sqrt{2}$ and $-3 - 3\sqrt{2}$.

B. $4x^2 = 4x - 1$

$$4x^2 - 4x + 1 = 0$$

$a = 4, b = -4, c = 1$

$$x = \frac{-b \pm \sqrt{b^2 - 4ac}}{2a}$$
$$= \frac{-(-4) \pm \sqrt{(-4)^2 - 4(4)(1)}}{2 \cdot 4}$$
$$= \frac{4 \pm \sqrt{16 - 16}}{8}$$
$$= \frac{4 \pm \sqrt{0}}{8} = \frac{4}{8} = \frac{1}{2}$$

The solution is $\frac{1}{2}$.

Problem 3

$3x^2 - x - 1 = 0$

$a = 3, b = -1, c = -1$

$$b^2 - 4ac = (-1)^2 - 4(3)(-1)$$
$$= 1 + 12 = 13$$

$13 > 0$

Because the discriminant is greater than zero, the equation has two real number solutions.

SECTION 9.3

Problem 1

A. $x - 5x^{\frac{1}{2}} + 6 = 0$

$$\left(x^{\frac{1}{2}}\right)^2 - 5\left(x^{\frac{1}{2}}\right) + 6 = 0$$
$$u^2 - 5u + 6 = 0$$
$$(u - 2)(u - 3) = 0$$

$$u - 2 = 0 \qquad u - 3 = 0$$
$$u = 2 \qquad u = 3$$

Replace u with $x^{\frac{1}{2}}$.

$$x^{\frac{1}{2}} = 2 \qquad x^{\frac{1}{2}} = 3$$
$$\left(x^{\frac{1}{2}}\right)^2 = 2^2 \qquad \left(x^{\frac{1}{2}}\right)^2 = 3^2$$
$$x = 4 \qquad x = 9$$

4 and 9 check as solutions.

The solutions are 4 and 9.

B.
$$4x^4 + 35x^2 - 9 = 0$$
$$4(x^2)^2 + 35(x^2) - 9 = 0$$
$$4u^2 + 35u - 9 = 0$$
$$(4u - 1)(u + 9) = 0$$

$$4u - 1 = 0 \qquad u + 9 = 0$$
$$4u = 1 \qquad u = -9$$
$$u = \frac{1}{4}$$

Replace u with x^2.

$$x^2 = \frac{1}{4} \qquad x^2 = -9$$
$$\sqrt{x^2} = \sqrt{\frac{1}{4}} \qquad \sqrt{x^2} = \sqrt{-9}$$
$$x = \pm\sqrt{\frac{1}{4}} \qquad x = \pm\sqrt{-9}$$
$$x = \pm\frac{1}{2} \qquad x = \pm 3i$$

The solutions are $\frac{1}{2}$, $-\frac{1}{2}$, $3i$, and $-3i$.

Problem 2

$$\sqrt{2x + 1} + x = 7$$
$$\sqrt{2x + 1} = 7 - x$$
$$\left(\sqrt{2x + 1}\right)^2 = (7 - x)^2$$
$$2x + 1 = 49 - 14x + x^2$$
$$0 = x^2 - 16x + 48$$
$$0 = (x - 4)(x - 12)$$

$$x - 4 = 0 \qquad x - 12 = 0$$
$$x = 4 \qquad x = 12$$

4 checks as a solution, but 12 does not check as a solution.

The solution is 4.

Problem 3

$$\sqrt{2x - 1} + \sqrt{x} = 2$$

Solve for one of the radical expressions.

$$\sqrt{2x - 1} = 2 - \sqrt{x}$$
$$\left(\sqrt{2x - 1}\right)^2 = \left(2 - \sqrt{x}\right)^2$$
$$2x - 1 = 4 - 4\sqrt{x} + x$$
$$x - 5 = -4\sqrt{x}$$

Square each side of the equation.

$$(x - 5)^2 = \left(-4\sqrt{x}\right)^2$$
$$x^2 - 10x + 25 = 16x$$
$$x^2 - 26x + 25 = 0$$
$$(x - 1)(x - 25) = 0$$

$$x - 1 = 0 \qquad x - 25 = 0$$
$$x = 1 \qquad x = 25$$

1 checks as a solution, but 25 does not check as a solution.

The solution is 1.

Problem 4

$$3y + \frac{25}{3y - 2} = -8$$
$$(3y - 2)\left(3y + \frac{25}{3y - 2}\right) = (3y - 2)(-8)$$
$$(3y - 2)(3y) + (3y - 2)\left(\frac{25}{3y - 2}\right) = (3y - 2)(-8)$$
$$9y^2 - 6y + 25 = -24y + 16$$
$$9y^2 + 18y + 9 = 0$$
$$9(y^2 + 2y + 1) = 0$$
$$9(y + 1)(y + 1) = 0$$

$$y + 1 = 0 \qquad y + 1 = 0$$
$$y = -1 \qquad y = -1$$

The solution is -1.

SECTION 9.4

Problem 1

Strategy
- ▶ This is a geometry problem.
- ▶ Width of the rectangle: W
  Length of the rectangle: $W + 3$
- ▶ Use the equation $A = L \cdot W$

Solution
$$A = L \cdot W$$
$$54 = (W + 3)(W)$$
$$54 = W^2 + 3W$$
$$0 = W^2 + 3W - 54$$
$$0 = (W + 9)(W - 6)$$

$$W + 9 = 0 \qquad W - 6 = 0$$
$$W = -9 \qquad W = 6$$

The solution -9 is not possible.

Length $= W + 3 = 6 + 3 = 9$

The length is 9 m.

SECTION 9.5

Problem 1

$$2x^2 - x - 10 \le 0$$
$$(2x - 5)(x + 2) \le 0$$

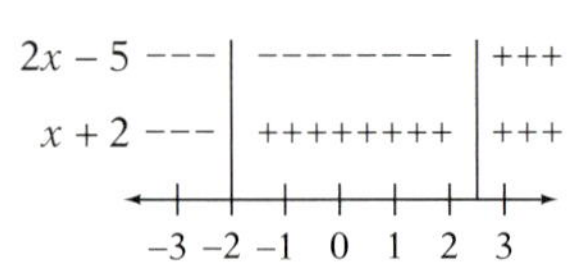

$$\left\{x \middle| -2 \le x \le \frac{5}{2}\right\}$$

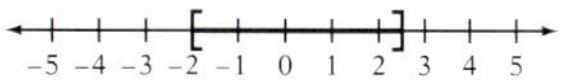

Problem 2

$$\frac{x}{x-2} \le 0$$

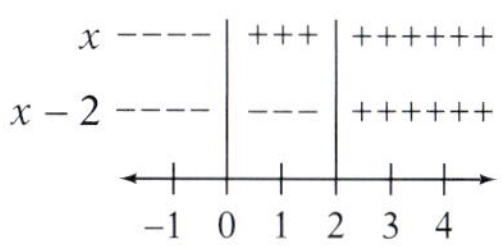

$\{x \mid 0 \le x < 2\}$

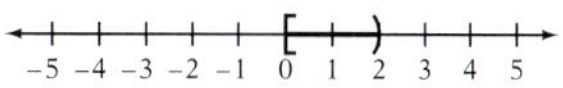

SECTION 9.6

Problem 1

x-coordinate: $-\dfrac{b}{2a} = -\dfrac{0}{2(1)} = 0$

$$\begin{aligned} y &= x^2 - 2 \\ &= 0^2 - 2 \\ &= -2 \end{aligned}$$

The vertex is $(0, -2)$.

The axis of symmetry is the line $x = 0$.

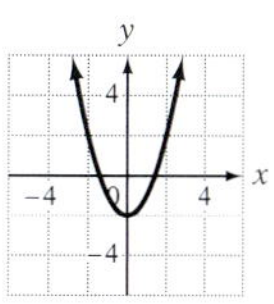

Problem 2

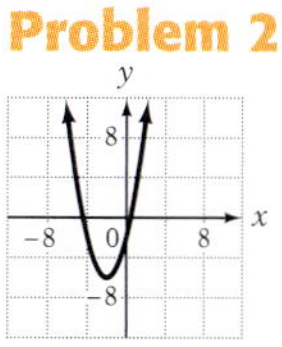

The domain is $\{x \mid x \in \text{real numbers}\}$. The range is $\{y \mid y \ge -6\}$.

Problem 3

A. $y = 2x^2 - 5x + 2$
$0 = 2x^2 - 5x + 2$
$0 = (2x - 1)(x - 2)$

$$\begin{aligned} 2x - 1 &= 0 & \quad x - 2 &= 0 \\ 2x &= 1 & x &= 2 \\ x &= \frac{1}{2} \end{aligned}$$

The x-intercepts are $\left(\dfrac{1}{2}, 0\right)$ and $(2, 0)$.

B. $y = x^2 + 4x + 4$
$0 = x^2 + 4x + 4$
$0 = (x + 2)(x + 2)$

$$\begin{aligned} x + 2 &= 0 & \quad x + 2 &= 0 \\ x &= -2 & x &= -2 \end{aligned}$$

The x-intercept is $(-2, 0)$.

Problem 4

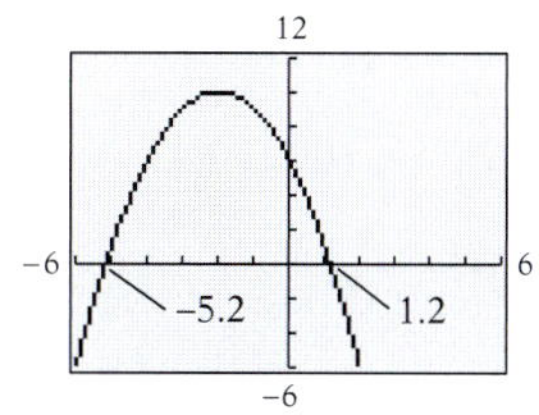

The approximate values of the zeros are -5.2 and 1.2.

Problem 5

$y = x^2 - x - 6$

$a = 1, b = -1, c = -6$

$$\begin{aligned} b^2 - 4ac &= (-1)^2 - 4(1)(-6) \\ &= 1 + 24 = 25 \end{aligned}$$

Because the discriminant is greater than zero, the parabola has two x-intercepts.

SECTION 9.7

Problem 1

$$x = -\frac{b}{2a} = -\frac{-3}{2(2)} = \frac{3}{4}$$

$$f(x) = 2x^2 - 3x + 1$$

$$\begin{aligned} f\left(\frac{3}{4}\right) = 2\left(\frac{3}{4}\right)^2 - 3\left(\frac{3}{4}\right) + 1 &= \frac{9}{8} - \frac{9}{4} + 1 \\ &= -\frac{1}{8} \end{aligned}$$

Because a is positive, the function has a minimum value. The minimum value of the function is $-\frac{1}{8}$.

Problem 2

Strategy
- ▶ To find the time it takes the ball to reach its maximum height, find the t-coordinate of the vertex.
- ▶ To find the maximum height, evaluate the function at the t-coordinate of the vertex.

Solution $t = -\dfrac{b}{2a} = -\dfrac{64}{2(-16)} = 2$

The ball reaches its maximum height in 2 s.

$s(t) = -16t^2 + 64t$
$s(2) = -16(2)^2 + 64(2) = -64 + 128 = 64$

The maximum height is 64 ft.

Problem 3

Strategy

$$P = 2x + y$$
$$100 = 2x + y \qquad \text{• } \mathbf{P = 100}$$
$$100 - 2x = y \qquad \text{• Solve for } \mathbf{y}.$$

Express the area of the rectangle in terms of x.

$$A = xy$$
$$A = x(100 - 2x) \qquad \text{• } \mathbf{y = 100 - 2x}$$
$$A = -2x^2 + 100x$$

► To find the width, find the x-coordinate of the vertex of $f(x) = -2x^2 + 100x$.

► To find the length, replace x in $y = 100 - 2x$ by the x-coordinate of the vertex.

Solution $x = -\dfrac{b}{2a} = -\dfrac{100}{2(-2)} = 25$

The width is 25 ft.

$$100 - 2x = 100 - 2(25)$$
$$= 100 - 50 = 50$$

The length is 50 ft.

Chapter 10

SECTION 10.1

Problem 1

Because any vertical line would intersect the graph at no more than one point, the graph is the graph of a function.

Problem 2

A.

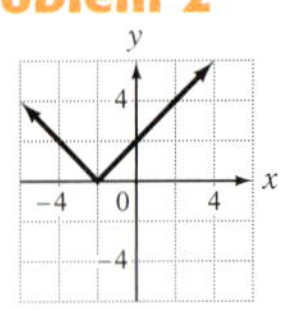

The domain is $\{x \mid x \in \text{real numbers}\}$.
The range is $\{y \mid y \geq 0\}$.

B.

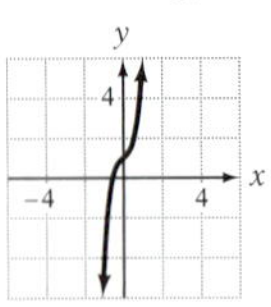

The domain is $\{x \mid x \in \text{real numbers}\}$.
The range is $\{y \mid y \in \text{real numbers}\}$.

C.

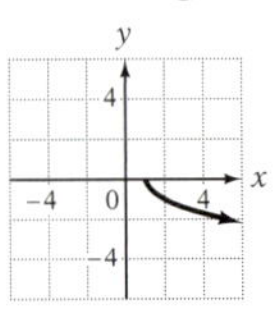

► $x - 1 \geq 0$
$\quad x \geq 1$

The domain is $\{x \mid x \geq 1\}$.
The range is $\{y \mid y \leq 0\}$.

SECTION 10.2

Problem 1

The graph of $g(x) = x^2 - 2x - 3$ is the graph of $f(x) = x^2 - 2x$ shifted 3 units down.

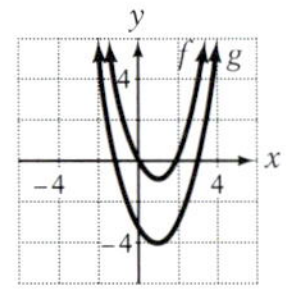

Problem 2

The graph of $g(x) = \sqrt{x - 3}$ is the graph of $f(x) = \sqrt{x}$ shifted 3 units to the right.

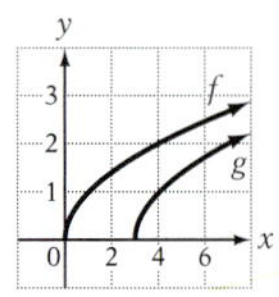

Problem 3

The graph of $V(x) = (x - 2)^3 + 1$ is the graph of $g(x) = x^3$ shifted up 1 unit and shifted 2 units to the right.

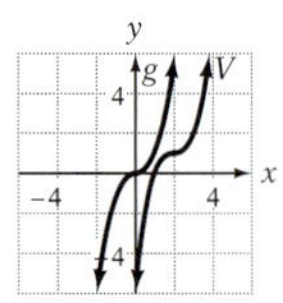

SECTION 10.3

Problem 1

$$\begin{aligned}(f + g)(-2) &= f(-2) + g(-2)\\ &= [(-2)^2 + 2(-2)] + [5(-2) - 2]\\ &= (4 - 4) + (-10 - 2)\\ &= -12\end{aligned}$$

$$(f + g)(-2) = -12$$

Problem 2

$$\begin{aligned}(f \cdot g)(3) &= f(3) \cdot g(3)\\ &= (4 - 3^2) \cdot [3(3) - 4]\\ &= (4 - 9) \cdot (9 - 4)\\ &= (-5)(5)\\ &= -25\end{aligned}$$

$$(f \cdot g)(3) = -25$$

Problem 3

$$\begin{aligned}\left(\frac{f}{g}\right)(4) &= \frac{f(4)}{g(4)}\\ &= \frac{4^2 - 4}{4^2 + 2 \cdot 4 + 1}\\ &= \frac{16 - 4}{16 + 8 + 1}\\ &= \frac{12}{25}\end{aligned}$$

$$\left(\frac{f}{g}\right)(4) = \frac{12}{25}$$

Problem 4

A. $h(x) = x^2 + 1$
$h(0) = 0 + 1 = 1$

$g(x) = 3x - 2$
$g(1) = 3(1) - 2 = 1$

$g[h(0)] = 1$

B. $h[g(x)] = h(3x - 2)$
$= (3x - 2)^2 + 1$
$= 9x^2 - 12x + 4 + 1$
$= 9x^2 - 12x + 5$

SECTION 10.4

Problem 1

A. Because any vertical line will intersect the graph at no more than one point, and any horizontal line will intersect the graph at no more than one point, the graph is the graph of a 1–1 function.

B. Because any vertical line will intersect the graph at no more than one point, and any horizontal line will intersect the graph at no more than one point, the graph is the graph of a 1–1 function.

Problem 2

$$f(x) = 4x + 2$$
$$y = 4x + 2$$
$$x = 4y + 2$$
$$4y = x - 2$$
$$y = \frac{1}{4}x - \frac{1}{2}$$
$$f^{-1}(x) = \frac{1}{4}x - \frac{1}{2}$$

Problem 3

$$h[g(x)] = 4\left(\frac{1}{4}x - \frac{1}{2}\right) + 2$$
$$= x - 2 + 2 = x$$
$$g[h(x)] = \frac{1}{4}(4x + 2) - \frac{1}{2}$$
$$= x + \frac{1}{2} - \frac{1}{2} = x$$

The functions are inverses of each other.

SECTION 10.5

Problem 1

$$-\frac{b}{2a} = -\frac{1}{2(-1)} = \frac{1}{2}$$
$$y = -x^2 + x + 3$$
$$y = -\left(\frac{1}{2}\right)^2 + \frac{1}{2} + 3 = \frac{13}{4}$$

The vertex is $\left(\frac{1}{2}, \frac{13}{4}\right)$.
The axis of symmetry is the line $x = \frac{1}{2}$.

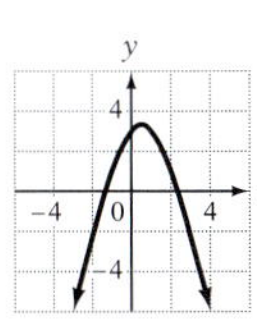

Problem 2

$$y\text{-coordinate: } -\frac{b}{2a} = -\frac{-4}{2(-2)} = -1$$
$$x = -2y^2 - 4y - 3$$
$$x = -2(-1)^2 - 4(-1) - 3$$
$$x = -1$$

The vertex is $(-1, -1)$.
The axis of symmetry is the line $y = -1$.

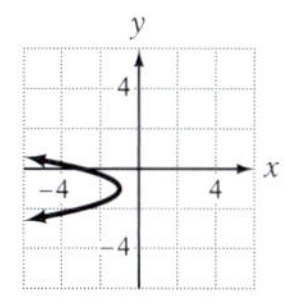

Problem 3

$$(x - h)^2 + (y - k)^2 = r^2$$
$$(x - 2)^2 + [y - (-3)]^2 = 4^2$$
$$(x - 2)^2 + (y + 3)^2 = 16$$

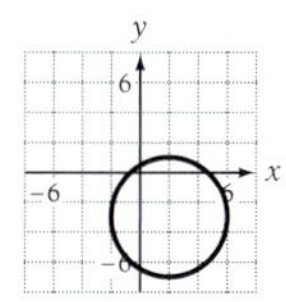

Problem 4

$$x_m = \frac{x_1 + x_2}{2} \qquad y_m = \frac{y_1 + y_2}{2}$$
$$x_m = \frac{-2 + 4}{2} = 1 \qquad y_m = \frac{1 + (-1)}{2} = 0$$

Center: $(1, 0)$

$$r = \sqrt{(x_1 - x_m)^2 + (y_1 - y_m)^2}$$
$$r = \sqrt{(-2 - 1)^2 + (1 - 0)^2} = \sqrt{9 + 1}$$
$$= \sqrt{10}$$

Radius: $\sqrt{10}$

$$(x - h)^2 + (y - k)^2 = r^2$$
$$(x - 1)^2 + (y - 0)^2 = 10$$
$$(x - 1)^2 + y^2 = 10$$

Problem 5

A. x-intercepts:
$(2, 0)$ and $(-2, 0)$

y-intercepts:
$(0, 5)$ and $(0, -5)$

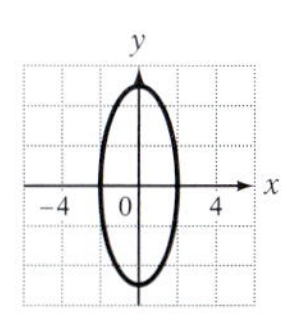

B. x-intercepts:
$(3\sqrt{2}, 0)$ and $(-3\sqrt{2}, 0)$

y-intercepts:
$(0, 3)$ and $(0, -3)$

$\left(3\sqrt{2} \approx 4\frac{1}{4}\right)$

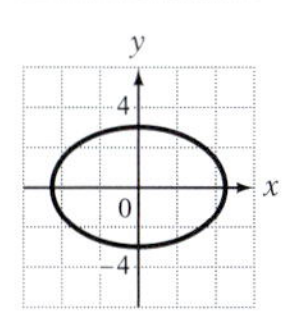

Problem 6

A. Axis of symmetry:
x-axis

Vertices:
$(3, 0)$ and $(-3, 0)$

Asymptotes:
$y = \frac{5}{3}x$ and $y = -\frac{5}{3}x$

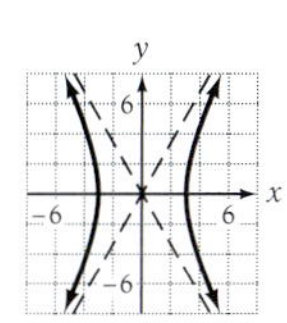

B. Axis of symmetry:
y-axis

Vertices:
(0, 3) and (0, −3)

Asymptotes:
$y = x$ and $y = -x$

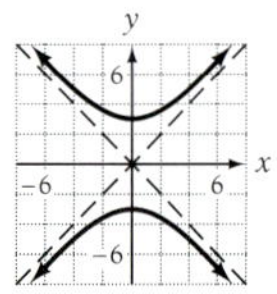

Chapter 11

SECTION 11.1

Problem 1

$$f(x) = \left(\frac{2}{3}\right)^x$$
$$f(3) = \left(\frac{2}{3}\right)^3 = \frac{8}{27}$$
$$f(-2) = \left(\frac{2}{3}\right)^{-2} = \left(\frac{3}{2}\right)^2 = \frac{9}{4}$$

Problem 2

$$f(x) = 2^{2x+1}$$
$$f(0) = 2^{2(0)+1} = 2^1 = 2$$
$$f(-2) = 2^{2(-2)+1} = 2^{-3} = \frac{1}{2^3} = \frac{1}{8}$$

Problem 3

$$f(x) = \pi^x$$
$$f(3) = \pi^3 \approx 31.0063$$
$$f(-2) = \pi^{-2} \approx 0.1013$$
$$f(\pi) = \pi^\pi \approx 36.4622$$

Problem 4

$$f(x) = e^x$$
$$f(1.2) = e^{1.2} \approx 3.3201$$
$$f(-2.5) = e^{-2.5} \approx 0.0821$$
$$f(e) = e^e \approx 15.1543$$

Problem 5

A.

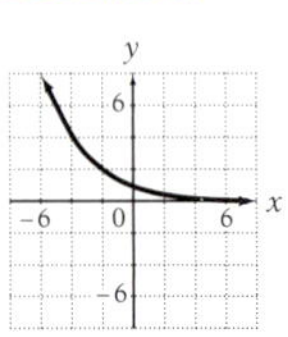

B.

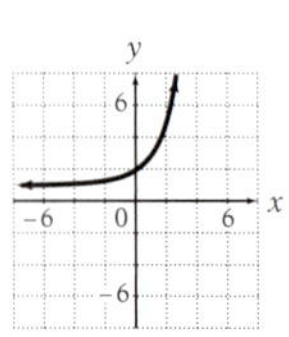

Problem 6

A.

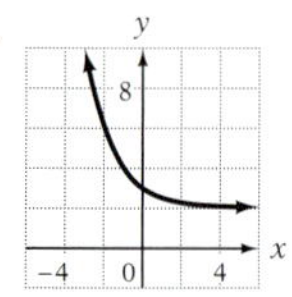

B.

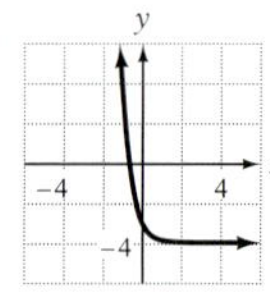

Problem 7

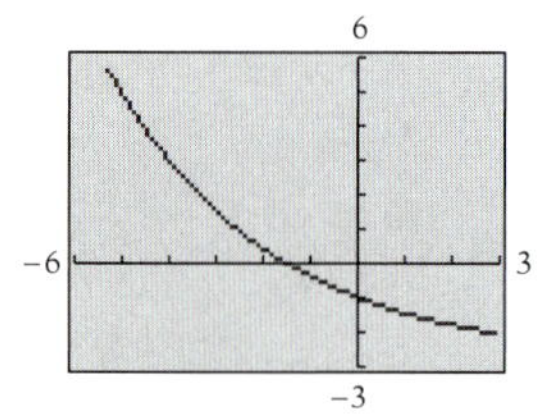

−2.4

SECTION 11.2

Problem 1

$3^{-4} = \frac{1}{81}$ is equivalent to $\log_3 \frac{1}{81} = -4$.

Problem 2

$\log_{10} 0.0001 = -4$ is equivalent to $10^{-4} = 0.0001$.

Problem 3

$$\log_4 64 = x$$
$$64 = 4^x$$
$$4^3 = 4^x$$
$$3 = x$$
$$\log_4 64 = 3$$

Problem 4

$$\log_2 x = -4$$
$$2^{-4} = x$$
$$\frac{1}{2^4} = x$$
$$\frac{1}{16} = x$$

The solution is $\frac{1}{16}$.

Problem 5

$$\log x = 1.5$$
$$10^{1.5} = x$$
$$31.6228 \approx x$$

The solution is 31.6228.

Problem 6

A. $\log_b \frac{x^2}{y} = \log_b x^2 - \log_b y$
$= 2 \log_b x - \log_b y$

B. $\ln y^{\frac{1}{3}}z^3 = \ln y^{\frac{1}{3}} + \ln z^3$

$= \frac{1}{3}\ln y + 3\ln z$

C. $\log_8 \sqrt[3]{xy^2} = \log_8(xy^2)^{\frac{1}{3}}$

$= \frac{1}{3}\log_8 xy^2$

$= \frac{1}{3}(\log_8 x + \log_8 y^2)$

$= \frac{1}{3}(\log_8 x + 2\log_8 y)$

$= \frac{1}{3}\log_8 x + \frac{2}{3}\log_8 y$

Problem 7

A. $2\log_b x - 3\log_b y - \log_b z = \log_b x^2 - \log_b y^3 - \log_b z$

$= \log_b \frac{x^2}{y^3} - \log_b z$

$= \log_b \frac{x^2}{y^3 z}$

B. $\frac{1}{3}(\log_4 x - 2\log_4 y + \log_4 z)$

$= \frac{1}{3}(\log_4 x - \log_4 y^2 + \log_4 z)$

$= \frac{1}{3}\left(\log_4 \frac{x}{y^2} + \log_4 z\right)$

$= \frac{1}{3}\left(\log_4 \frac{xz}{y^2}\right)$

$= \log_4\left(\frac{xz}{y^2}\right)^{\frac{1}{3}} = \log_4 \sqrt[3]{\frac{xz}{y^2}}$

C. $\frac{1}{2}(2\ln x - 5\ln y) = \frac{1}{2}(\ln x^2 - \ln y^5)$

$= \frac{1}{2}\left(\ln \frac{x^2}{y^5}\right)$

$= \ln\left(\frac{x^2}{y^5}\right)^{\frac{1}{2}}$

$= \ln\sqrt{\frac{x^2}{y^5}}$

Problem 8

A. $\log_{16} 1 = 0$

B. $12\log_3 3 = \log_3 3^{12} = 12$

Problem 9

$$\log_4 2.4 = \frac{\ln 2.4}{\ln 4} \approx 0.6315$$

Problem 10

$$f(x) = 4\log_8(3x+4) = 4\frac{\log(3x+4)}{\log 8}$$

$$= \frac{4}{\log 8}\log(3x+4)$$

SECTION 11.3

Problem 1

A. $f(x) = \log_2(x-1)$

$y = \log_2(x-1)$

$y = \log_2(x-1)$ is equivalent to $2^y = x - 1$.

$2^y + 1 = x$

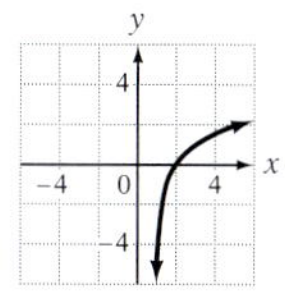

B. $f(x) = \log_3 2x$

$y = \log_3 2x$

$y = \log_3 2x$ is equivalent to $3^y = 2x$.

$\frac{3^y}{2} = x$

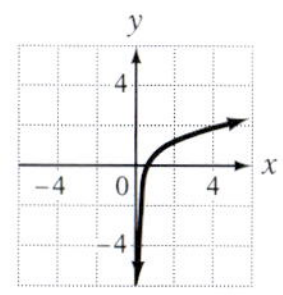

Problem 2

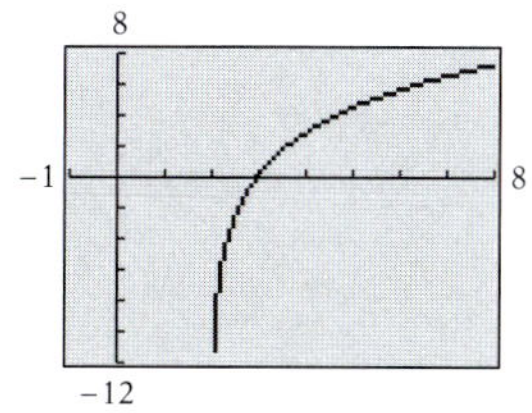

Problem 3

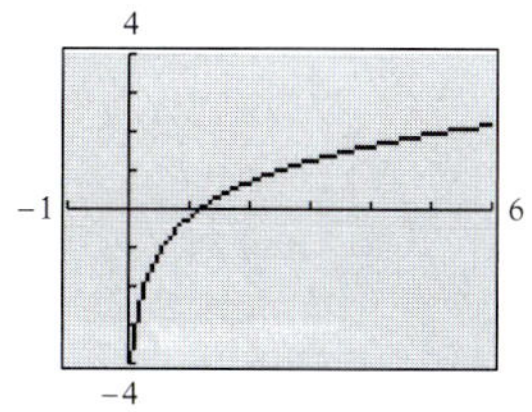

SECTION 11.4

Problem 1

$10^{3x+5} = 10^{x-3}$

$3x + 5 = x - 3$

$2x + 5 = -3$

$2x = -8$

$x = -4$

Check: $10^{3x+5} = 10^{x-3}$

$10^{3(-4)+5} \quad 10^{-4-3}$

$10^{-12+5} \quad 10^{-7}$

$10^{-7} = 10^{-7}$

The solution is −4.

Problem 2

A.
$$4^{3x} = 25$$
$$\log 4^{3x} = \log 25$$
$$3x \log 4 = \log 25$$
$$3x = \frac{\log 25}{\log 4}$$
$$x = \frac{\log 25}{3 \log 4}$$
$$x \approx 0.7740$$

The solution is 0.7740.

B.
$$(1.06)^x = 1.5$$
$$\log(1.06)^x = \log 1.5$$
$$x \log 1.06 = \log 1.5$$
$$x = \frac{\log 1.5}{\log 1.06}$$
$$x \approx 6.9585$$

The solution is 6.9585.

Problem 3

$$e^x = x$$
$$e^x - x = 0$$

Graph $f(x) = e^x - x$.

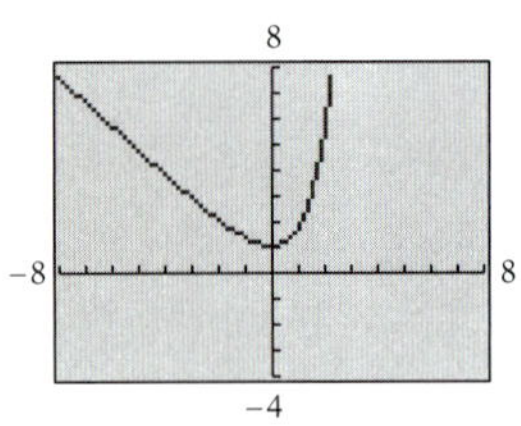

The equation has no real number solutions.

Problem 4

A. $\log_4(x^2 - 3x) = 1$

Rewrite in exponential form.

$$4^1 = x^2 - 3x$$
$$4 = x^2 - 3x$$
$$0 = x^2 - 3x - 4$$
$$0 = (x + 1)(x - 4)$$

$x + 1 = 0 \qquad x - 4 = 0$

$x = -1 \qquad x = 4$

The solutions are −1 and 4.

B. $\log_3 x + \log_3(x + 3) = \log_3 4$

$\log_3[x(x + 3)] = \log_3 4$

Use the One-to-One Property of Logarithms.

$$x(x + 3) = 4$$
$$x^2 + 3x = 4$$
$$x^2 + 3x - 4 = 0$$
$$(x + 4)(x - 1) = 0$$

$x + 4 = 0 \qquad x - 1 = 0$

$x = -4 \qquad x = 1$

−4 does not check as a solution. The solution is 1.

Problem 5

$$\log(3x - 2) = -2x$$
$$\log(3x - 2) + 2x = 0$$

Graph $f(x) = \log(3x - 2) + 2x$.

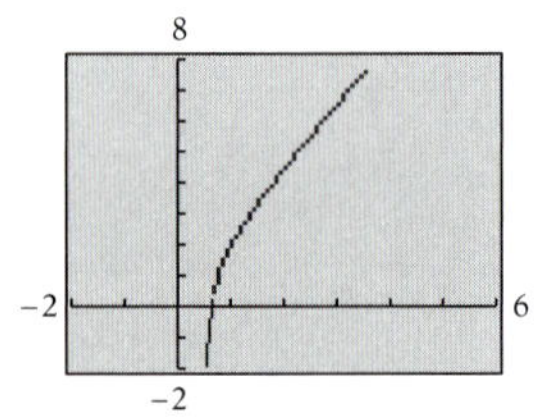

The solution is 0.68.

SECTION 11.5

Problem 1

Strategy ► To find the time, solve the compound interest formula for n. Use $P = 6000$, $A = 3000$, and $i = \frac{12\%}{12} = \frac{0.12}{12} = 0.01$.

Solution

$$P = A(1 + i)^n$$
$$6000 = 3000(1 + 0.01)^n$$
$$6000 = 3000(1.01)^n$$
$$2 = (1.01)^n$$
$$\log 2 = \log(1.01)^n$$
$$\log 2 = n \log 1.01$$
$$\frac{\log 2}{\log 1.01} = n$$
$$70 \approx n$$

70 months ÷ 12 ≈ 6 years

In approximately 6 years, the investment will be worth $6000.

Problem 2

Strategy ► To find the intensity, use the equation for the Richter scale magnitude of an earthquake, $M = \log \frac{I}{I_0}$. Replace M by 5.2 and solve for I.

Solution

$$M = \log \frac{I}{I_0}$$
$$5.2 = \log \frac{I}{I_0}$$ • Replace M by 5.2.
$$10^{5.2} = \frac{I}{I_0}$$ • Write in exponential form.
$$10^{5.2} I_0 = I$$
$$158{,}489 I_0 \approx I$$

The earthquake had an intensity that was approximately 158,489 times the intensity of a zero-level earthquake.

Chapter R

SECTION R.1

Problem 1

$$\begin{aligned}(-4)(6-8)^2-(-12\div 4) &= -4(-2)^2-(-3)\\ &= -4(4)-(-3)\\ &= -16-(-3)\\ &= -16+3\\ &= -13\end{aligned}$$

Problem 2

$3xy^2 - 3x^2y$

$$\begin{aligned}3(-2)(5)^2-3(-2)^2(5) &= 3(-2)(25)-3(4)(5)\\ &= -6(25)-3(4)(5)\\ &= -150-3(4)(5)\\ &= -150-12(5)\\ &= -150-60\\ &= -150+(-60)\\ &= -210\end{aligned}$$

Problem 3

$$\begin{aligned}-5(-3a) &= [-5(-3)]a\\ &= 15a\end{aligned}$$

Problem 4

$$\begin{aligned}2z^2-5z-3z^2+6z &= 2z^2-3z^2-5z+6z\\ &= (2z^2-3z^2)+(-5z+6z)\\ &= -1z^2+z\\ &= -z^2+z\end{aligned}$$

Problem 5

a. $$\begin{aligned}-3(5y-2) &= -3(5y)-(-3)(2)\\ &= -15y+6\end{aligned}$$

b. $$\begin{aligned}-2(4x+2y-6z) &= -2(4x)+(-2)(2y)-(-2)(6z)\\ &= -8x-4y+12z\end{aligned}$$

Problem 6

a. $$\begin{aligned}7(-3x-4y)-3(3x+y) &= -21x-28y-9x-3y\\ &= -30x-31y\end{aligned}$$

b. $$\begin{aligned}2y-3[5-3(3+2y)] &= 2y-3[5-9-6y]\\ &= 2y-3[-4-6y]\\ &= 2y+12+18y\\ &= 20y+12\end{aligned}$$

SECTION R.2

Problem 1

a. $$\begin{aligned}4x+3 &= 7x+9\\ 4x-7x+3 &= 7x-7x+9\\ -3x+3 &= 9\\ -3x+3-3 &= 9-3\\ -3x &= 6\\ \frac{-3x}{-3} &= \frac{6}{-3}\\ x &= -2\end{aligned}$$

The solution is -2.

b. $$\begin{aligned}4-(5x-8) &= 4x+3\\ 4-5x+8 &= 4x+3\\ -5x+12 &= 4x+3\\ -5x-4x+12 &= 4x-4x+3\\ -9x+12 &= 3\\ -9x+12-12 &= 3-12\\ -9x &= -9\\ \frac{-9x}{-9} &= \frac{-9}{-9}\\ x &= 1\end{aligned}$$

The solution is 1.

Problem 2

$$\begin{aligned}3x-1 &\le 5x-7\\ 3x-5x-1 &\le 5x-5x-7\\ -2x-1 &\le -7\\ -2x-1+1 &\le -7+1\\ -2x &\le -6\\ \frac{-2x}{-2} &\ge \frac{-6}{-2}\\ x &\ge 3\end{aligned}$$

$\{x|x \ge 3\}$

Problem 3

$$\begin{aligned}3-2(3x+1) &\le 7-2x\\ 3-6x-2 &\le 7-2x\\ 1-6x &\le 7-2x\\ 1-6x+2x &\le 7-2x+2x\\ 1-4x &\le 7\\ 1-1-4x &\le 7-1\\ -4x &\le 6\\ \frac{-4x}{-4} &\ge \frac{6}{-4}\\ x &\ge -\frac{3}{2}\end{aligned}$$

$\left[-\frac{3}{2}, \infty\right)$

SECTION R.3

Problem 1

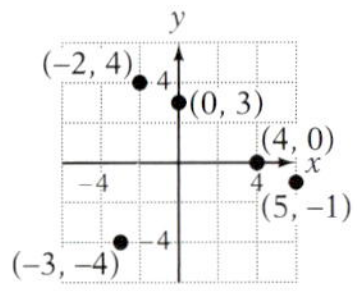

Problem 2

$$y = \frac{3}{5}x - 4$$

x	y
0	−4
−5	−7
5	−1

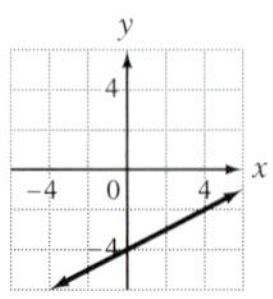

Problem 3

$$-3x + 2y = 4$$
$$2y = 3x + 4$$
$$y = \frac{3}{2}x + 2$$

x	y
0	2
−2	−1
2	5

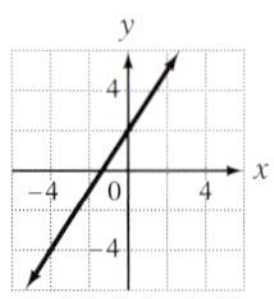

Problem 4

Let $(x_1, y_1) = (-2, 3)$ and $(x_2, y_2) = (1, -3)$.

$$m = \frac{y_2 - y_1}{x_2 - x_1} = \frac{-3 - 3}{1 - (-2)} = \frac{-6}{3} = -2$$

The slope is -2.

Problem 5

$$y = -\frac{2}{3}x + 2$$

$$m = -\frac{2}{3} = \frac{-2}{3}$$

y-intercept = (0, 2)

Place a dot at the y-intercept.

Starting at the y-intercept, move to the right 3 units (the change in x) and down 2 units (the change in y).

Place a dot at that location.

Draw a line through the two points.

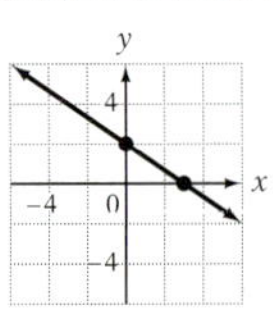

Problem 6

$$f(x) = 4 - 2x$$
$$f(-3) = 4 - 2(-3)$$
$$f(-3) = 4 + 6$$
$$f(-3) = 10$$

An ordered pair of the function is $(-3, 10)$.

Problem 7

$$(x_1, y_1) = (-2, 2),\ m = -\frac{1}{2}$$

$$y - y_1 = m(x - x_1)$$
$$y - 2 = -\frac{1}{2}[x - (-2)]$$
$$y - 2 = -\frac{1}{2}(x + 2)$$
$$y - 2 = -\frac{1}{2}x - 1$$
$$y = -\frac{1}{2}x + 1$$

SECTION R.4

Problem 1

a. $$(-2ab)(2a^3b^{-2})^{-3} = (-2ab)(2^{-3}a^{-9}b^6)$$
$$= -(2^{-2})a^{-8}b^7$$
$$= -\frac{b^7}{2^2a^8}$$
$$= -\frac{b^7}{4a^8}$$

b. $$\left(\frac{2x^2y^{-4}}{4x^{-2}y^{-5}}\right)^{-3} = \left(\frac{x^2y^{-4}}{2x^{-2}y^{-5}}\right)^{-3}$$
$$= \frac{x^{2(-3)}y^{(-4)(-3)}}{2^{1(-3)}x^{(-2)(-3)}y^{(-5)(-3)}}$$
$$= \frac{x^{-6}y^{12}}{2^{-3}x^6y^{15}}$$
$$= 2^3x^{-6-6}y^{12-15}$$
$$= 2^3x^{-12}y^{-3}$$
$$= \frac{8}{x^{12}y^3}$$

Problem 2

$$(-4x^3 + 2x^2 - 8) + (4x^3 + 6x^2 - 7x + 5)$$
$$= (-4x^3 + 4x^3) + (2x^2 + 6x^2) + (-7x) + (-8 + 5)$$
$$= 8x^2 - 7x - 3$$

Problem 3

$$(-4w^3 + 8w - 8) - (3w^3 - 4w^2 + 2w - 1)$$
$$= (-4w^3 + 8w - 8) + (-3w^3 + 4w^2 - 2w + 1)$$
$$= (-4w^3 - 3w^3) + 4w^2 + (8w - 2w) + (-8 + 1)$$
$$= -7w^3 + 4w^2 + 6w - 7$$

Problem 4

$3mn^2(2m^2 - 3mn - 1)$
$= 3mn^2(2m^2) - 3mn^2(3mn) - 3mn^2(1)$
$= 6m^3n^2 - 9m^2n^3 - 3mn^2$

Problem 5

$$\begin{array}{r} 3c^2 - 4c + 5 \\ 2c - 3 \\ \hline -9c^2 + 12c - 15 \\ 6c^3 - 8c^2 + 10c \\ \hline 6c^3 - 17c^2 + 22c - 15 \end{array}$$

Problem 6

$(4y - 7)(3y - 5)$
$= (4y)(3y) + (4y)(-5) + (-7)(3y) + (-7)(-5)$
$= 12y^2 - 20y - 21y + 35$
$= 12y^2 - 41y + 35$

Problem 7

$(x^3 - 7 - 2x) \div (x - 2)$

$$\begin{array}{r} x^2 + 2x + 2 \\ x - 2 \overline{)x^3 + 0x^2 - 2x - 7} \\ \underline{x^3 - 2x^2} \\ 2x^2 - 2x \\ \underline{2x^2 - 4x} \\ 2x - 7 \\ \underline{2x - 4} \\ -3 \end{array}$$

Check: $(x - 2)(x^2 + 2x + 2) - 3$
$= x^3 - 2x - 4 - 3$
$= x^3 - 2x - 7$

$(x^3 - 7 - 2x) \div (x - 2)$
$= x^2 + 2x + 2 - \dfrac{3}{x - 2}$

Problem 8

The GCF is $3x^2y^2$.

$6x^4y^2 - 9x^3y^2 + 12x^2y^4$
$= 3x^2y^2(2x^2 - 3x + 4y^2)$

Problem 9

Two factors of 15 whose sum is -8 are -3 and -5.

$x^2 - 8x + 15 = (x - 3)(x - 5)$

Problem 10

$3x^2 - x - 2$

Positive Factors of 3	Factors of −2
1, 3	1, −2
	−1, 2

Trial Factors	Middle Term
$(x + 1)(3x - 2)$	$-2x + 3x = x$
$(x - 2)(3x + 1)$	$x - 6x = -5x$
$(x - 1)(3x + 2)$	$2x - 3x = -x$
$(x + 2)(3x - 1)$	$-x + 6x = 5x$

$3x^2 - x - 2 = (x - 1)(3x + 2)$

Check: $(x - 1)(3x + 2) = 3x^2 + 2x - 3x - 2$
$= 3x^2 - x - 2$

Problem 11

$4a^3 - 4a^2 - 24a$

There is a common factor, $4a$.
Factor out the GCF.

$$4a^3 - 4a^2 - 24a = 4a(a^2 - a - 6)$$

Factor $a^2 - a - 6$. The two factors of -6 whose sum is -1 are 2 and -3.

$$4a(a^2 - a - 6) = 4a(a + 2)(a - 3)$$

$4a^3 - 4a^2 - 24a = 4a(a + 2)(a - 3)$

Check: $4a(a + 2)(a - 3) = (4a^2 + 8a)(a - 3)$
$= 4a^3 - 12a^2 + 8a^2 - 24a$
$= 4a^3 - 4a^2 - 24a$

Answers to Selected Exercises

Chapter 1

PREP TEST

1. 127.16 **2.** 49,743 **3.** 4517 **4.** 11,396 **5.** 24 **6.** 24 **7.** 4 **8.** $3 \cdot 7$ **9.** $\frac{2}{5}$

CONCEPT REVIEW 1.1

1. Sometimes true **3.** Never true **5.** Sometimes true **7.** Always true **9.** Always true

SECTION 1.1

3. $>$ **5.** $<$ **7.** $>$ **9.** $>$ **11.** $<$ **13.** $>$ **15.** $>$ **17.** $<$ **19.** $\{1, 2, 3, 4, 5, 6, 7, 8\}$ **21.** $\{1, 2, 3, 4, 5, 6, 7, 8\}$ **23.** $\{-6, -5, -4, -3, -2, -1\}$ **25.** 5 **27.** $-23, -18$ **29.** 21, 37 **31.** $-52, -46, 0$ **33.** $-17, 0, 4, 29$ **35.** 5, 6, 7, 8, 9 **37.** $-10, -9, -8, -7, -6, -5$ **41.** -22 **43.** 31 **45.** 168 **47.** -630 **49.** 18 **51.** -49 **53.** 16 **55.** 12 **57.** -29 **59.** -14 **61.** 0 **63.** -34 **65.** **a.** $8, 5, 2, -1, -3$ **b.** 8, 5, 2, 1, 3 **67.** $>$ **69.** $<$ **71.** $>$ **73.** $<$ **75.** $-19, -|-8|, |-5|, 6$ **77.** $-22, -(-3), |-14|, |-25|$ **79.** **a.** 5°F with a 20 mph wind feels colder. **b.** −15°F with a 20 mph wind feels colder. **81.** $-4, 4$ **83.** $-3, 11$ **85.** Negative **87.** 0 **89.** True

CONCEPT REVIEW 1.2

1. Sometimes true **3.** Always true **5.** Always true **7.** Never true **9.** Never true

SECTION 1.2

3. -11 **5.** -9 **7.** -3 **9.** 1 **11.** -5 **13.** -30 **15.** 9 **17.** 1 **19.** -10 **21.** -28 **23.** -392 **27.** 8 **29.** -7 **31.** 7 **33.** -2 **35.** -28 **37.** -13 **39.** 6 **41.** -9 **43.** 2 **45.** -138 **47.** -8 **53.** 42 **55.** -20 **57.** -16 **59.** 25 **61.** 0 **63.** -72 **65.** -102 **67.** 140 **69.** -70 **71.** 162 **73.** 120 **75.** 36 **77.** 192 **79.** -108 **81.** 0 **83.** $3(-12) = -36$ **85.** $-5(11) = -55$ **87.** -2 **89.** 8 **91.** 0 **93.** -9 **95.** -9 **97.** 9 **99.** -24 **101.** -12 **103.** -13 **105.** -18 **107.** 19 **109.** 26 **111.** 11 **113.** -13 **115.** 13 **117.** The temperature is 3°C. **119.** The difference is 14°C. **121.** The difference is 399°C. **123.** The difference in elevation is 5662 m. **125.** The difference in elevation is 6028 m. **127.** The difference in elevation is 9248 m. **129.** The average daily high temperature is −3°C. **131.** The temperature dropped 100°F. **133.** The average temperature at 30,000 ft is 36°F colder than at 20,000 ft. **135.** The student's score is 93. **137.** The 5-day moving average for the Disney stock is $-24, -14, -12, 4, 8, 20$. **139.** 17 **141.** 3 **143.** $-4, -9, -14$ **145.** $-16, 4, -1$ **147.** 5436 **149.** b **151.** a **153.** No. $10 - (-8) = 18$

CONCEPT REVIEW 1.3

1. Never true **3.** Sometimes true **5.** Always true **7.** Never true **9.** Always true

SECTION 1.3

1. $0.\overline{3}$ **3.** 0.25 **5.** 0.4 **7.** $0.1\overline{6}$ **9.** 0.125 **11.** $0.\overline{2}$ **13.** $0.\overline{45}$ **15.** $0.58\overline{3}$ **17.** $0.2\overline{6}$ **19.** 0.4375 **21.** 0.24 **23.** 0.225 **25.** $0.6\overline{81}$ **27.** $0.458\overline{3}$ **29.** $0.1\overline{5}$ **31.** $0.\overline{081}$ **33.** $\frac{13}{12}$ **35.** $-\frac{5}{24}$ **37.** $-\frac{19}{24}$ **39.** $\frac{5}{26}$ **41.** $\frac{7}{24}$ **43.** 0 **45.** $-\frac{7}{16}$ **47.** $\frac{11}{24}$ **49.** 1 **51.** $\frac{11}{8}$ **53.** $\frac{169}{315}$ **55.** −38.8 **57.** −6.192 **59.** 13.355 **61.** 4.676 **63.** −10.03 **65.** −37.19 **67.** −17.5 **69.** 853.2594 **71.** $-\frac{3}{8}$ **73.** $\frac{1}{10}$ **75.** $\frac{15}{64}$ **77.** $\frac{3}{2}$ **79.** $-\frac{8}{9}$ **81.** $\frac{2}{3}$ **83.** 4.164 **85.** 4.347 **87.** −4.028 **89.** −2.22 **91.** 12.26448 **93.** 0.75 **95.** −2060.55 **97.** 0.09 **101.** $\frac{3}{4}$, 0.75 **103.** $\frac{1}{2}$, 0.5 **105.** $\frac{16}{25}$, 0.64 **107.** $1\frac{3}{4}$, 1.75 **109.** $\frac{19}{100}$, 0.19 **111.** $\frac{1}{20}$, 0.05 **113.** $4\frac{1}{2}$, 4.5 **115.** $\frac{2}{25}$, 0.08 **117.** $\frac{1}{9}$ **119.** $\frac{5}{16}$ **121.** $\frac{1}{200}$ **123.** $\frac{1}{16}$ **125.** 0.073 **127.** 0.158 **129.** 0.0915 **131.** 0.1823 **133.** 15% **135.** 5% **137.** 17.5% **139.** 115% **141.** 0.8% **143.** 6.5% **145.** 54% **147.** 33.3% **149.** 44.4% **151.** 250% **153.** $37\frac{1}{2}\%$ **155.** $35\frac{5}{7}\%$ **157.** 125% **159.** $155\frac{5}{9}\%$ **161.** $\frac{1}{25}$ of the respondents found their most recent jobs on the Internet. **163.** Fewer than one-quarter of the respondents found their most recent jobs through a newspaper ad. **165.** Natural number, integer, positive integer, rational number, real number **167.** Rational number, real number **169.** Irrational number, real number **171.** The cost is \$1.29. **173.** The difference is \$138.499 billion. **175.** The deficit in 1985 was 4 times greater than in 1975. **177.** The average deficit for the years 1995 through 2000 was −\$4,579 million. **179.** The new fraction is greater than $\frac{2}{5}$. **181.** $1.06x$ **183.** 53¢ **185.** Yes. For example, $c = \frac{a + b}{2}$.

CONCEPT REVIEW 1.4

1. Never true **3.** Never true **5.** Always true **7.** Never true **9.** Never true

SECTION 1.4

1. 36 **3.** −49 **5.** 9 **7.** 81 **9.** $\frac{1}{4}$ **11.** 0.09 **13.** 12 **15.** 0.216 **17.** −12 **19.** 16 **21.** −864 **23.** −1008 **25.** 3 **27.** −77,760 **31.** 9 **33.** 12 **35.** 1 **37.** 7 **39.** −36 **41.** 13 **43.** 4 **45.** 15 **47.** −1 **49.** 4 **51.** 0.51 **53.** 1.7 **55.** $\frac{17}{48}$ **57.** $<$ **59.** $>$ **61.** 100 **63.** 120 **65.** It will take the computer 17 s. **67.** 6 **69.** 9

CONCEPT REVIEW 1.5

1. Never true **3.** Always true **5.** Always true **7.** Never true

SECTION 1.5

3. -7 **5.** 13 **7.** -15 **9.** 41 **11.** 57 **13.** 5 **15.** 12 **17.** 6 **19.** 10
21. 24 **23.** 4.96 **25.** -5.68 **27.** The volume is 25.8 in^3. **29.** The area is 93.7 cm^2.
31. The volume is 484.9 m^3. **33.** 41 **35.** 2 **37.** 5 **39.** 6 **41.** -8
43. The Inverse Property of Addition **45.** The Commutative Property of Addition
47. The Commutative Property of Multiplication **49.** The Associative Property of Multiplication
53. $3x, 5x$ **55.** $5a$ **57.** $-2ab$ **59.** $5xy$ **61.** 0 **63.** $-\frac{5}{6}x$ **65.** $-\frac{1}{24}x^2$
67. $11x$ **69.** $7a$ **71.** $-14x^2$ **73.** $-x + 3y$ **75.** $17x - 3y$ **77.** $-2a - 6b$
79. $-3x - 8y$ **81.** $-4x^2 - 2x$ **83.** $12x$ **85.** $-21a$ **87.** $6y$ **89.** $8x$ **91.** $-6a$
93. $12b$ **95.** $-15x^2$ **97.** x^2 **99.** x **101.** n **103.** x **105.** n **107.** $2x$
109. $-2x$ **111.** $-15a^2$ **113.** $6y$ **115.** $-x - 2$ **117.** $8x - 6$
119. $-2a - 14$ **121.** $-6y + 24$ **123.** $35 - 21b$ **125.** $-9 + 15x$ **127.** $15x^2 + 6x$
129. $2y - 18$ **131.** $-15x - 30$ **133.** $-6x^2 - 28$ **135.** $-6y^2 + 21$ **137.** $-6a^2 + 7b^2$
139. $4x^2 - 12x + 20$ **141.** $-3y^2 + 9y + 21$ **143.** $-12a^2 - 20a + 28$ **145.** $a - 7$
147. $-11x + 13$ **149.** $-4y - 4$ **151.** $-2x - 16$ **153.** $18y - 51$ **155.** $a + 7b$
157. $6x + 28$ **159.** $5x - 75$ **161.** $4x - 4$ **163.** $38x - 63$ **165.** b **167.** $20x + y$
169. $-3x + y$ **171 a.** 0 **b.** 1 or -1

CONCEPT REVIEW 1.6

1. Never true **3.** Never true **5.** Always true **7.** Never true

SECTION 1.6

1. $19 - d$ **3.** $r - 12$ **5.** $28a$ **7.** $5(n - 7)$ **9.** $y - 3y$ **11.** $-6b$ **13.** $\frac{4}{p - 6}$
15. $\frac{x - 9}{2x}$ **17.** $-4s - 21$ **19.** $\frac{d + 8}{d}$ **21.** $\frac{3}{8}(t + 15)$ **23.** $w + \frac{7}{w}$ **25.** $d + (16d - 3)$
27. $\frac{n}{19}$ **29.** $n + 40$ **31.** $(n - 90)^2$ **33.** $\frac{4}{9}n + 20$ **35.** $n(n + 10)$ **37.** $7n + 14$
39. $\frac{12}{n + 2}$ **41.** $\frac{2}{n + 1}$ **43.** $60 - \frac{n}{50}$ **45.** $n^2 + 3n$ **47.** $(n + 3) + n^3$ **49.** $n^2 - \frac{1}{4}n$
51. $2n - \frac{7}{n}$ **53.** $n^3 - 12n$ **55.** $n + (n + 10); 2n + 10$ **57.** $n - (9 - n); 2n - 9$
59. $\frac{1}{5}n - \frac{3}{8}n; -\frac{7}{40}n$ **61.** $(n + 9) + 4; n + 13$ **63.** $2(3n + 40); 6n + 80$ **65.** $7(5n); 35n$
67. $17n + 2n; 19n$ **69.** $n + 12n;13n$ **71.** $3(n^2 + 4); 3n^2 + 12$ **73.** $\frac{3}{4}(16n + 4); 12n + 3$
75. $16 - (n + 9); -n + 7$ **77.** $\frac{4n}{2} + 5; 2n + 5$ **79.** $6(n + 8); 6n + 48$ **81.** $7 - (n + 2); -n + 5$
83. $\frac{1}{3}(n + 6n); \frac{7n}{3}$ **85.** $(8 + n^3) + 2n^3; 3n^3 + 8$ **87.** $(n - 6) + (n + 12); 2n + 6$
89. $(n - 20) + (n + 9); 2n - 11$ **91.** $14 + (n - 3)10; 10n - 16$ **93.** Let x be one number; x and $18 - x$
95. Let G be the number of genes in the roundworm genome; $G + 11{,}000$ **97.** Let N be the total number of Americans; $\frac{3}{4}N$ **99.** Let s be the number of points awarded for a safety; $3s$ **101.** Let m be the number of moons of Jupiter; $m + 9$ **103.** Let p be the pounds of pecans produced in Texas; $\frac{1}{2}p$ **105.** Let p be the

world population in 1980; $2p$ **107.** Let L be the measure of the largest angle; $\frac{1}{2}L - 10$ **109.** Let n be either the number of nickels or the number of dimes; n and $35 - n$ **111.** Let h be the number of hours of overtime worked; $720 + 27h$ **113.** Let m be the number of minutes of phone calls; $19.95 + 0.08m$ **115.** $\frac{1}{4}x$

CHAPTER REVIEW EXERCISES*

1. $\{1, 2, 3, 4, 5, 6\}$ [1.1.1] **2.** 62.5% [1.3.4] **3.** -4 [1.1.2] **4.** 4 [1.2.2] **5.** 17 [1.2.4] **6.** $0.\overline{7}$ [1.3.1] **7.** -5.3578 [1.3.3] **8.** 8 [1.4.2] **9.** 67.2% [1.3.4] **10.** $\frac{159}{200}$ [1.3.4] **11.** 90 [1.2.3] **12.** -6.881 [1.3.2] **13.** $-5, -3$ [1.1.1] **14.** 0.07 [1.3.4] **15.** 12 [1.4.1] **16.** $>$ [1.1.1] **17.** $\frac{1}{15}$ [1.3.2] **18.** -108 [1.4.1] **19.** $-\frac{7}{6}$ [1.3.3] **20.** 12 [1.4.2] **21.** $288\frac{8}{9}\%$ [1.3.4] **22. a.** $12, 8, 1, -7$ **b.** 12, 8, 1, 7 [1.1.2] **23.** -11.5 [1.3.3] **24.** 8 [1.4.2] **25.** -8 [1.2.1] **26.** $-\frac{11}{24}$ [1.3.2] **27.** 3.561 [1.3.2] **28.** -27 [1.2.2] **29.** $-\frac{1}{10}$ [1.3.3] **30.** $<$ [1.1.1] **31.** y^2 [1.5.3] **32.** $3x$ [1.5.4] **33.** $7x + 38$ [1.5.6] **34.** 16 [1.5.1] **35.** 9 [1.5.2] **36.** $-3x + 21y$ [1.5.6] **37.** $15 - 35b$ [1.5.5] **38.** $-7x + 33$ [1.5.6] **39.** The Commutative Property of Multiplication [1.5.2] **40.** $24 - 6x$ [1.5.5] **41.** 3 [1.5.1] **42.** $36y$ [1.5.4] **43.** $5x - 43$ [1.5.6] **44.** $2x$ [1.5.4] **45.** $-6x^2 + 21y^2$ [1.5.5] **46.** 6 [1.5.1] **47.** $-x + 6$ [1.5.6] **48.** $-5a - 2b$ [1.5.3] **49.** $-10x^2 + 15x - 30$ [1.5.5] **50.** $-9x - 7y$ [1.5.3] **51.** $-2x - 5y$ [1.5.3] **52.** $30b$ [1.5.4] **53.** 21 [1.5.2] **54.** $-2x^2 + 4x$ [1.5.3] **55.** $b - 7b$ [1.6.1] **56.** $\frac{10}{y-2}$ [1.6.1] **57.** The temperature is 8°C. [1.2.5] **58.** The average low temperature was −2°C. [1.2.5] **59.** The difference is 106°F. [1.2.5] **60.** The temperature is −6°C. [1.2.5] **61.** The difference is 714°C. [1.2.5] **62.** $8\left(\frac{2n}{16}\right)$; n [1.6.2] **63.** $4(2 + 5n)$; $8 + 20n$ [1.6.2] **64.** Let h be the height of the triangle; $h + 15$ [1.6.3] **65.** Let b be the amount of either espresso beans or mocha java beans; b and $20 - b$ [1.6.3]

CHAPTER TEST

1. $\frac{11}{20}$ [1.3.4] **2.** $-6, -8$ [1.1.1] **3.** -3 [1.2.2] **4.** 0.15 [1.3.1] **5.** $-\frac{1}{14}$ [1.3.3] **6.** -15 [1.2.4] **7.** $-\frac{8}{3}$ [1.4.1] **8.** 2 [1.2.1] **9.** $\{1, 2, 3, 4, 5, 6\}$ [1.1.1] **10.** 159% [1.3.4] **11.** 29 [1.1.2] **12.** $>$ [1.1.1] **13.** $-\frac{23}{18}$ [1.3.2] **14.** $\frac{5}{6}$ [1.3.3] **15.** 23.1% [1.3.4] **16.** 0.062 [1.3.4] **17.** 14 [1.2.2] **18.** $0.4\overline{3}$ [1.3.1] **19.** -2.43 [1.3.3] **20.** -12 [1.2.1] **21. a.** $17, 6, -5, -9$ **b.** 17, 6, 5, 9 [1.1.2] **22.** -11.384 [1.3.2] **23.** 160 [1.2.3] **24.** $36y$ [1.5.4] **25.** $4x - 3y$ [1.5.3] **26.** $10n - 6$ [1.5.6] **27.** The Multiplication Property of

*The numbers in brackets following the answers in the Chapter Review Exercises refer to the objective that corresponds to that problem. For example, the reference [1.2.1] stands for Section 1.2, Objective 1. This notation will be used for all Pretests, Chapter Review Exercises, Chapter Tests, and Cumulative Review Exercises throughout the text.

One [1.5.2] **28.** $4x - 40$ [1.5.5] **29.** $\frac{1}{12}x^2$ [1.5.3] **30.** $4x$ [1.5.4] **31.** $-24y^2 + 48$ [1.5.5] **32.** 19 [1.5.2] **33.** 6 [1.5.1] **34.** $-3x + 13y$ [1.5.6] **35.** $3x^2 - 15x + 12$ [1.5.5] **36.** -32 [1.5.1] **37.** $37x - 5y$ [1.5.6] **38.** -2 [1.4.2] **39.** The temperature is 4°C. [1.2.5] **40.** The average is -48°F. [1.2.5] **41.** $(a + b) - b^2$ [1.6.1] **42.** $20(n + 9)$; $20n + 180$ [1.6.2] **43.** $(n - 3) + (n + 2)$; $2n - 1$ [1.6.2] **44.** Let d be the distance from Earth to the sun; $30d$ [1.6.3] **45.** Let L be the length of one piece; L and $9 - L$ [1.6.3]

Chapter 2

PREP TEST

1. -4 [1.2.2] **2.** 1 [1.3.3] **3.** -10 [1.3.3] **4.** 0.9 [1.3.4] **5.** 75% [1.3.4] **6.** 63 [1.5.1] **7.** $10x - 5$ [1.5.3] **8.** -9 [1.5.3] **9.** $9x - 18$ [1.5.5]

CONCEPT REVIEW 2.1

1. Sometimes true **3.** Always true **5.** Always true **7.** Always true

SECTION 2.1

3. Yes **5.** No **7.** No **9.** Yes **11.** Yes **13.** Yes **15.** No **17.** No **19.** No **21.** Yes **23.** No **25.** No **31.** 2 **33.** 15 **35.** 6 **37.** -6 **39.** 3 **41.** 0 **43.** -2 **45.** -7 **47.** -7 **49.** -12 **51.** 2 **53.** -5 **55.** 15 **57.** 9 **59.** 14 **61.** -1 **63.** 1 **65.** $-\frac{1}{2}$ **67.** $-\frac{3}{4}$ **69.** $\frac{1}{12}$ **71.** $-\frac{7}{12}$ **73.** $-\frac{5}{7}$ **75.** $-\frac{1}{3}$ **77.** 1.869 **79.** 0.884 **85.** 7 **87.** -7 **89.** -4 **91.** 5 **93.** 9 **95.** -8 **97.** 0 **99.** -7 **101.** 8 **103.** 12 **105.** 12 **107.** -18 **109.** -18 **111.** 6 **113.** -12 **115.** 3 **117.** -24 **119.** 9 **121.** $\frac{1}{3}$ **123.** $\frac{15}{7}$ **125.** 4 **127.** 3 **129.** 4.48 **131.** 2.06 **133.** -2.1 **137.** 24% **139.** 7.2 **141.** 400 **143.** 9 **145.** 25% **147.** 5 **149.** 200% **151.** 400 **153.** 7.7 **155.** 40% of 80 is equal to 80% of 40. **157.** 250 seats are reserved for wheelchair accessibility. **159.** In the average single-family home, 74 gal of water are used per person per day. **161.** 51% of a typical traveler's vacation bill is paid in cash. **163.** There is insufficient information. **165.** 12% of the deaths were not attributed to motor vehicle accidents. **167.** In this country 96.1 billion kilowatt-hours of electricity are used for home lighting per year. **169.** The annual simple interest rate is 9%. **171.** Sal earned $240 in interest from the two accounts. **173.** The interest rate on the combined investment is between 6% and 9%. **175.** The percent concentration of the hydrogen perioxide is 2%. **177.** Apple Dan's has the greater concentration of apple juice. **179.** 12.5 g of the cream are not glycerine. **181.** The percent concentration of salt in the remaining solution is 12.5%. **183.** -21 **185.** -27 **187.** 21 **189.** The cost of the dinner was $80. **191.** The new value is two times the original value. **193.** One possible answer is $5x = -10$.

CONCEPT REVIEW 2.2

1. Always true **3.** Never true **5.** Never true **7.** Sometimes true

SECTION 2.2

1. 3 **3.** 6 **5.** −1 **7.** 1 **9.** −9 **11.** 3 **13.** 3 **15.** 3 **17.** 2 **19.** 2 **21.** 4 **23.** $\frac{6}{7}$ **25.** $\frac{2}{3}$ **27.** −1 **29.** $\frac{3}{4}$ **31.** $\frac{1}{3}$ **33.** $-\frac{3}{4}$ **35.** $\frac{1}{3}$ **37.** $-\frac{1}{6}$ **39.** 1 **41.** 0 **43.** $\frac{2}{5}$ **45.** $-\frac{3}{2}$ **47.** 18 **49.** 8 **51.** −16 **53.** 25 **55.** 21 **57.** 15 **59.** −16 **61.** $\frac{7}{6}$ **63.** $\frac{1}{8}$ **65.** $\frac{15}{2}$ **67.** $-\frac{18}{5}$ **69.** 2 **71.** 3 **73.** 1 **75.** 2 **77.** 3.95 **79.** −0.8 **81.** −11 **83.** 0 **85.** 3 **87.** 8 **89.** 2 **91.** −2 **93.** −3 **95.** 2 **97.** −2 **99.** −2 **101.** −7 **103.** 0 **105.** −2 **107.** −2 **109.** 4 **111.** 10 **113.** 3 **115.** $\frac{3}{4}$ **117.** $\frac{2}{7}$ **119.** $-\frac{3}{4}$ **121.** 3 **123.** 10 **125.** $\frac{4}{3}$ **127.** 3 **129.** −14 **131.** 7 **133.** 3 **135.** 4 **137.** 2 **139.** 2 **141.** 2 **143.** −7 **145.** $\frac{4}{7}$ **147.** $\frac{1}{2}$ **149.** $-\frac{1}{3}$ **151.** $\frac{10}{3}$ **153.** $-\frac{1}{4}$ **155.** 0.5 **157.** 0 **159.** −1 **161.** The car will slide 168 ft. **163.** The depth of the diver is 40 ft. **165.** The initial velocity is 8 ft/s. **167.** The length of the humerus is approximately 31.8 in. **169.** The passenger was driven 6 mi. **171.** He made 8 errors. **173.** The population is approximately 51,000 people. **175.** The break-even point is 350 television sets. **179.** No, the seesaw is not balanced. **181.** A force of 25 lb must be applied to the other end of the lever. **183.** The fulcrum must be placed 10 ft from the 128-pound acrobat. **185.** No solution **187.** $-\frac{11}{4}$ **189.** 62 **191.** One possible answer is $3x - 6 = 2x - 2$.

CONCEPT REVIEW 2.3

1. Always true **3.** Always true **5.** Always true **7.** Never true **9.** Always true

SECTION 2.3

1. $n - 15 = 7; n = 22$ **3.** $7n = -21; n = -3$ **5.** $3n - 4 = 5; n = 3$ **7.** $4(2n + 3) = 12; n = 0$ **9.** $12 = 6(n - 3); n = 5$ **11.** $2(n - 25) = 3n; n = -50$ **13.** $3n = 2(20 - n)$; 8 and 12 **15.** $3n + 2(18 - n) = 44$; 8 and 10 **17.** The original value was \$16,000. **19.** There are 58 calories in a medium-size orange. **21.** The amount of mulch is 15 lb. **23.** The intensity of the sound is 140 decibels. **25.** The length is 80 ft.; the width is 50 ft. **27.** To replace the water pump, 5 h of labor were required. **29.** The customer used the service for 11 min. **31.** The integers are 17, 18, and 19. **33.** The integers are 26, 28, and 30. **35.** The integers are 17, 19, and 21. **37.** The integers are 8 and 10. **39.** The integers are 7 and 9. **41.** The integers are −9, −8, and −7. **43.** The integers are 10, 12, and 14. **45.** No solution **49.** The vegetable medley costs \$1.66 per pound. **51.** There were 320 adult tickets sold. **53.** The mixture must contain 225 L of imitation maple syrup. **55.** The instructor used 8 lb of peanuts. **57.** The cost of the mixture is \$4.04 per pound. **59.** The mixture must contain 37.5 gal of cranberry juice. **61.** The student drives at an average rate of 40 mph. **63.** It would take a Boeing 737-800 jet 3.2 h to make the same trip. **65.** The two cyclists will meet at 2:30 P.M. **67.** The bicyclist overtakes the in-line skater 11.25 mi from the starting point. **69.** The speed of the first plane is 420 mph. The speed of the second plane is 500 mph. **71.** The distance to the island is 43.2 mi. **73.** The rate of the first plane is 255 mph. The rate of the second plane is 305 mph. **75.** The distance between the student's home and the bicycle shop is 2.8 mi.

77. The two trains will pass each other in 2 h. **79.** At 3:39 P.M., $\frac{1}{3}$ of the container is filled. **81.** The integers are −15, −13, −11, and −9. **83.** For any four consecutive integers, the sum of the first and fourth integers equals the sum of the second and third integers. **85.** It will take the runner $\frac{2}{3}$ h. **87.** The grass seed is 18 ft from the 40-foot tower. **89. a.** Rutan and Yeager landed at Edwards Air Force Base on December 23. **b.** The average speed was 116 mph. **c.** The approximate distance of the flight above Earth was 14 mi.

CONCEPT REVIEW 2.4

1. Always true **3.** Always true **5.** Never true

SECTION 2.4

3. The sides measure 50 ft, 50 ft, and 25 ft. **5.** The length is 13 m. The width is 8 m. **7.** The length is 40 ft. The width is 20 ft. **9.** The sides measure 40 cm, 20 cm, and 50 cm. **11.** The length is 130 ft. The width is 39 ft. **13.** The width is 12 ft. **15.** Each side measures 12 in. **17. a.** 90° **b.** 180° **c.** 360° **d.** between 0° and 90° **e.** between 90° and 180° **f.** 90° **g.** 180° **19.** The complement of a 28° angle is 62°. **21.** The supplement of a 73° angle is 107°. **23.** 35° **25.** 20° **27.** 53° **29.** 121° **31.** 15° **33.** 18° **35.** 45° **37.** 49° **39.** 12° **41.** $\angle a = 122°$, $\angle b = 58°$ **43.** $\angle a = 44°$; $\angle b = 136°$ **45.** 20° **47.** 40° **49.** 128° **51.** $\angle x = 160°$, $\angle y = 145°$ **53.** $\angle a = 40°$, $\angle b = 140°$ **55.** $75° - x$ **57.** The measure of the third angle is 45°. **59.** The measure of the third angle is 73°. **61.** The measure of the third angle is 43°. **63.** The angles measure 38°, 38°, and 104°. **65.** The angles measure 60°, 30°, and 90°. **67.** The length is 9 cm. The width is 4 cm.

CONCEPT REVIEW 2.5

1. Always true **3.** Never true **5.** Always true

SECTION 2.5

3. The selling price is \$56. **5.** The selling price is \$565.64. **7.** The markup rate is 75%. **9.** The markup rate is 44.4%. **11.** The markup rate is 60%. **13.** The cost of the compact disc player is \$120. **15.** The cost of the basketball is \$59. **17.** The cost of the computer is \$1750. **21.** The sale price is \$101.25. **23.** The discount price is \$218.50. **25.** The discount rate is 25%. **27.** The discount rate is 23.2%. **29.** The markdown rate is 38%. **31.** The regular price is \$510. **33.** The regular price is \$300. **35.** The regular price is \$275. **37.** The markup is \$18. **39.** The markup rate is 40%. **41.** The cost of the camera is \$230. **43.** The regular price is \$80. **45.** No. The single discount that would give the same sale price is 28%.

CONCEPT REVIEW 2.6

1. Never true **3.** Never true

SECTION 2.6

3. Joseph will earn \$327.50 on the two accounts. **5.** Deon must invest \$1600 in the account that earns 8% interest. **7.** The amount invested in the CD is \$15,000. **9.** An additional \$2000 must be invested at an annual simple interest rate of 10.5%. **11.** The amount invested at 8.5% is \$3000. The amount invested at 6.4% is \$5000. **13.** She must invest an additional \$3000 at an annual interest rate of 10%. **15.** The amount that should be invested at 4.2% is \$8000. The amount that should be invested at 6% is \$5600. **19.** A 40-ounce bottle of Orange Ade contains 10 oz of orange juice. **21.** The 850-milliliter solution contains 12.5 more milliliters of hydrogen peroxide. **23.** The resulting alloy is 30% silver. **25.** The resulting alloy is $33\frac{1}{3}\%$ silver.

27. 500 lb of the 12% aluminum alloy is needed. **29.** 25 L of the 65% solution and 25 L of the 15% solution were used. **31.** 3 qt of water must be added. **33.** 30 oz of pure water must be added. **35.** The result is 6% real fruit juice. **37.** The resulting alloy is 31.1% copper. **39.** The amount invested at 9% was \$5000. The amount invested at 8% was \$6000. The amount invested at 9.5% was \$9000. **41.** The cost per pound of the tea mixture is \$4.84. **43.** 60 g of pure water were in the beaker before the acid was added. **45. a.** 1990 **b.** 2000

CONCEPT REVIEW 2.7

1. Always true **3.** Sometimes true **5.** Sometimes true

SECTION 2.7

3. $\{x \mid x > 4, x \in \text{integers}\}$ **5.** $\{x \mid x \geq -2\}$ **7.** $\{x \mid 0 < x < 1\}$ **9.** $\{x \mid 1 \leq x \leq 4\}$ **11.** $\{1, 2, 4, 6, 9\}$ **13.** $\{2, 3, 5, 8, 9, 10\}$ **15.** $\{-4, -2, 0, 2, 4, 8\}$ **17.** $\{1, 2, 3, 4, 5\}$ **19.** $\{6\}$ **21.** $\{5, 10, 20\}$ **23.** $\varnothing$ **25.** $\{4, 6\}$ **27.** 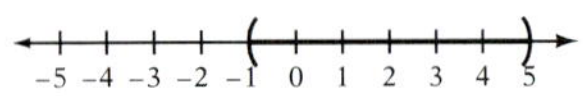

29. 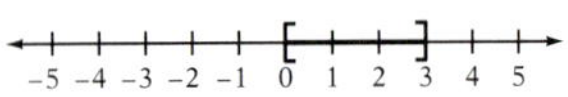 **31.** 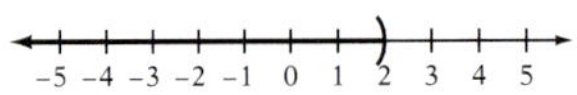 **33.** 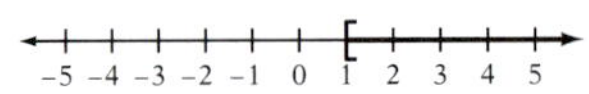

35. 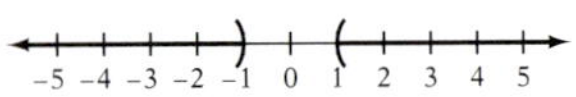 **37.** 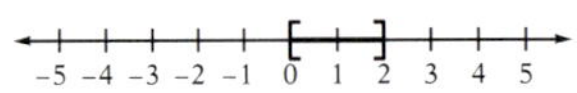 **39.** 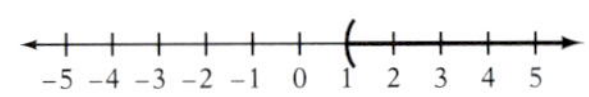

41. 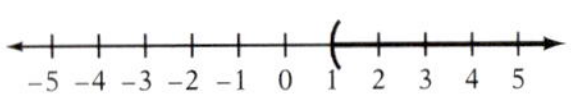 **43.** $\{x \mid 0 < x < 8\}$ **45.** $\{x \mid -5 \leq x \leq 7\}$ **47.** $\{x \mid -3 \leq x < 6\}$ **49.** $\{x \mid x \leq 4\}$ **51.** $\{x \mid x > 5\}$ **53.** $(-2, 4)$ **55.** $[-1, 5]$ **57.** $(-\infty, 1)$ **59.** $[-2, 6)$ **61.** $(-\infty, \infty)$ **65.** a, c **67.** $\{x \mid x < 5\}$ **69.** $\{x \mid x \leq 2\}$ **71.** $\{x \mid x < -4\}$ **73.** $\{x \mid x > 3\}$ **75.** $\{x \mid x > 4\}$ **77.** $\{x \mid x > -2\}$ **79.** $\{x \mid x \geq 2\}$ **81.** $\{x \mid x \leq 3\}$ **83.** $\{x \mid x < -3\}$ **85.** $(-\infty, 5]$ **87.** $[1, \infty)$ **89.** $(-\infty, -5)$ **91.** $\left(-\infty, \frac{23}{16}\right)$ **93.** $\left[\frac{8}{3}, \infty\right)$ **95.** $(-\infty, 1)$ **97.** $(-\infty, 3)$ **101.** $(-1, 2)$ **103.** $(-\infty, 1] \cup [3, \infty)$ **105.** $(-2, 4)$ **107.** $(-\infty, -3) \cup (0, \infty)$ **109.** $[3, \infty)$ **111.** $\varnothing$ **113.** $\varnothing$ **115.** $(-\infty, 1) \cup (3, \infty)$ **117.** $\{x \mid -3 < x < 4\}$ **119.** $\{x \mid 3 < x < 5\}$ **121.** $\left\{x \mid x > 3 \text{ or } x \leq -\frac{5}{2}\right\}$ **123.** $\{x \mid x > 4\}$ **125.** $\varnothing$ **127.** $\{x \mid x \text{ is a real number}\}$ **129.** $\{x \mid -4 < x < -2\}$ **131.** $\{x \mid x < -4 \text{ or } x > 3\}$ **133.** $\{x \mid -4 \leq x < 1\}$ **135.** $\left\{x \mid x > \frac{27}{2} \text{ or } x < \frac{5}{2}\right\}$ **137.** $\left\{x \mid -10 \leq x \leq \frac{11}{2}\right\}$ **139.** The smallest integer is 3. **141.** The maximum width of the rectangle is 2 ft. **143.** The four integers are 15, 16, 17, 18; or 16, 17, 18, 19; or 17, 18, 19, 20. **145.** The length of the second side could be 5 in., 6 in., or 7 in. **147.** A customer can use a cellular phone 160 min before the charges exceed the first option. **149.** The AirTouch Plan is less expensive for more than 460 pages. **151.** A call must be 7 min or less if it is to be cheaper to pay with coins. **153.** If a business chooses the Glendale Federal Bank, then the business writes more than 200 checks per month. **155.** The car can travel between 429 mi and 536.25 mi on a full tank of gasoline. **157.** $\{1, 2\}$ **159.** $\{1, 2, 3, 4, 5, 6\}$ **161.** $\{1, 2, 3, 4\}$ **163.** $\{1, 2\}$ **165.** The temperature is between 25°C and 30°C. **167.** The largest whole number of minutes the call would last is 10 min.

CONCEPT REVIEW 2.8

1. Always true **3.** Never true **5.** Never true

SECTION 2.8

1. Yes **3.** Yes **7.** −7 and 7 **9.** −3 and 3 **11.** No solution **13.** −5 and 1 **15.** 8 and 2 **17.** 2 **19.** No solution **21.** $\frac{1}{2}$ and $\frac{9}{2}$ **23.** 0 and $\frac{4}{5}$ **25.** −1

27. No solution **29.** 4 and 14 **31.** 4 and 12 **33.** $\frac{7}{4}$ **35.** No solution **37.** -3 and $\frac{3}{2}$
39. $\frac{4}{5}$ and 0 **41.** No solution **43.** No solution **45.** -2 and 1 **47.** $-\frac{3}{5}$
51. $\{x|x > 3 \text{ or } x < -3\}$ **53.** $\{x|x > 1 \text{ or } x < -3\}$ **55.** $\{x|4 \le x \le 6\}$ **57.** $\{x|x \le -1 \text{ or } x \ge 5\}$
59. $\{x|-3 < x < 2\}$ **61.** $\left\{x\middle|x > 2 \text{ or } x < -\frac{14}{5}\right\}$ **63.** $\varnothing$ **65.** $\{x|x \in \text{real numbers}\}$
67. $\left\{x\middle|x \le -\frac{1}{3} \text{ or } x \ge 3\right\}$ **69.** $\left\{x\middle|-2 \le x \le \frac{9}{2}\right\}$ **71.** $\{x|x = 2\}$ **73.** $\left\{x\middle|x < -2 \text{ or } x > \frac{22}{9}\right\}$
75. The lower and upper limits of the diameter of the bushing are 1.742 in. and 1.758 in. **77.** The lower and upper limits of the amount of medicine to be given to the patient are 2.3 cc and 2.7 cc. **79.** The lower and upper limits of voltage to the computer are 93.5 volts and 126.5 volts. **81.** The lower and upper limits of the length of the piston rod are $10\frac{11}{32}$ in. and $10\frac{13}{32}$ in. **83.** The lower and upper limits of the resistor are 13,500 ohms and 16,500 ohms. **85.** The lower and upper limits of the resistor are 53.2 ohms and 58.8 ohms. **87.** $2, -\frac{2}{3}$
89. $\{x|-7 \le x \le 8\}$ **91.** $\{y|y \ge -6\}$ **93.** $\{b|b \le 7\}$ **95.** $|x - 2| < 5$ **97.** **a.** $\le$ **b.** $\ge$ **c.** $\le$ **d.** $=$ **e.** $=$

CHAPTER REVIEW EXERCISES

1. No [2.1.1] **2.** 20 [2.1.2] **3.** -7 [2.1.3] **4.** 7 [2.2.1] **5.** 4 [2.2.2]
6. $-\frac{1}{5}$ [2.2.3] **7.** 405 [2.1.4] **8.** 25 [2.1.4] **9.** 67.5% [2.1.4] **10.** $-\frac{9}{19}$ [2.2.3]
11. $\{2, 3\}$ [2.7.1] **12.** 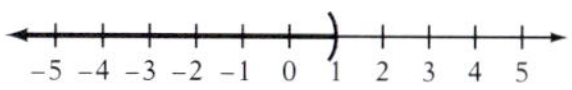

[2.7.1] **13.** $\left(\frac{5}{3}, \infty\right)$ [2.7.2] **14.** $(-4, \infty)$ [2.7.2] **15.** $(-1, 2)$ [2.7.3] **16.** $(-\infty, -2) \cup (2, \infty)$ [2.7.3] **17.** $\left\{x\middle|-3 < x < \frac{4}{3}\right\}$ [2.7.3]
18. $\{x|x \in \text{real numbers}\}$ [2.7.3] **19.** $-\frac{5}{2}, \frac{11}{2}$ [2.8.1] **20.** $-\frac{8}{5}$ [2.8.1] **21.** No solution [2.8.1]
22. $\{x|1 \le x \le 4\}$ [2.8.2] **23.** $\left\{x\middle|x \le \frac{1}{2} \text{ or } x \ge 2\right\}$ [2.8.2] **24.** $\varnothing$ [2.8.2] **25.** A force of 24 lb must be applied to the other end of the lever. [2.2.4] **26.** Approximately 11,065 plants and animals are at risk of extinction on Earth. [2.3.1] **27.** The depth is 80 ft. [2.2.4] **28.** $2x - 7 = x$; 7 [2.3.1]
29. The length of the shorter piece is 14 in. [2.3.1] **30.** The two numbers are 8 and 13. [2.3.1]
31. The discount rate is $33\frac{1}{3}\%$ [2.5.2] **32.** The angles measure 65°, 65°, and 50°. [2.4.3] **33.** The markup rate is 80%. [2.5.1] **34.** The concentration of butterfat is 14%. [2.6.2] **35.** The length is 80 ft. The width is 20 ft. [2.4.1] **36.** The integers are -17, -15, and -13. [2.3.2] **37.** $\angle a = 138°$, $\angle b = 42°$ [2.4.2] **38.** The lower and upper limits of the amount of medicine to be given are 1.75 cc and 2.25 cc. [2.8.3]
39. The integers are -1, 0, and 1. [2.3.2] **40.** The mixture costs $4.25 per ounce. [2.3.3] **41.** The mixture must contain 52 gal of apple juice. [2.3.3] **42.** The speed of the first plane is 440 mph. The speed of the second plane is 520 mph. [2.3.4] **43.** The amount invested at 10.5% was $3000. The amount invested at 6.4% was $5000. [2.6.1] **44.** 375 lb of the 30% tin alloy and 125 lb of the 70% tin alloy were used. [2.6.2]
45. The range of scores to receive a B grade is $82 \le N \le 100$. [2.7.4]

CHAPTER TEST

1. -12 [2.1.3] **2.** $-\frac{1}{2}$ [2.2.2] **3.** -3 [2.2.1] **4.** No [2.1.1] **5.** $-\frac{1}{3}$ [2.2.3] **6.** -5

[2.1.2] **7.** 40 [2.1.4] **8.** $(-1, \infty)$ [2.7.2] **9.** $\{x \mid x > -2\}$ [2.7.3] **10.** 3 and $-\frac{9}{5}$ [2.8.1] **11.** $\{x \mid x > 2 \text{ or } x < -1\}$ [2.8.2] **12.** $3n - 15 = 27; n = 14$ [2.3.1] **13.** The cost is \$200. [2.5.1] **14.** The number of calculators produced was 200. [2.2.4] **15.** The integers are 5, 7, and 9. [2.3.2] **16.** The measures of the three angles are 48°, 33°, and 99°. [2.4.3] **17.** The minimum length is 24 ft. [2.7.4] **18.** The sale price is \$79.20. [2.5.2] **19.** The measures of the three other angles are 45°, 135°, and 45°. [2.4.2] **20.** It costs less to rent from agency A if the car is driven less than 120 mi. [2.7.4] **21.** The lower and upper limits of the amount of medication to be given are 2.9 cc and 3.1 cc. [2.8.3] **22.** The price of the hamburger mixture is \$3.20/lb. [2.3.3] **23.** The jogger ran a total distance of 12 mi. [2.3.4] **24.** The amount invested at 7.8% was \$5000. The amount invested at 9% was \$7000. [2.6.1] **25.** 100 oz of pure water must be added. [2.6.2]

CUMULATIVE REVIEW EXERCISES

1. 6 [1.2.2] **2.** −48 [1.2.3] **3.** $-\frac{19}{48}$ [1.3.2] **4.** −2 [1.3.3] **5.** 54 [1.4.1] **6.** 24 [1.4.2] **7.** 6 [1.5.1] **8.** $-17x$ [1.5.3] **9.** $-5a - 2b$ [1.5.3] **10.** $2x$ [1.5.4] **11.** $36y$ [1.5.4] **12.** $2x^2 + 6x - 4$ [1.5.5] **13.** $-4x + 14$ [1.5.5] **14.** $6x - 34$ [1.5.5] **15.** $\{-7, -6, -5, -4, -3, -2, -1\}$ [1.1.1] **16.** $87\frac{1}{2}\%$ [1.3.4] **17.** 3.42 [1.3.4] **18.** $\frac{5}{8}$ [1.3.4] **19.** Yes [2.1.1] **20.** −5 [2.1.2] **21.** −25 [2.1.3] **22.** 3 [2.2.1] **23.** −3 [2.2.2] **24.** 13 [2.2.3] **25.** $x < -\frac{8}{9}$ [2.7.2] **26.** $x \geq 12$ [2.7.2] **27.** $x > 9$ [2.7.2] **28.** $8 - \frac{n}{12}$ [2.3.1] **29.** $n + (n + 2); 2n + 2$ [2.3.1] **30.** A force of 24 lb must be applied to the other end of the lever. [2.2.4] **31.** The first integer is 1. [2.3.2] **32.** 48 adult tickets were sold. [2.3.3] **33.** The speed of the faster plane is 340 mph. [2.3.4] **34.** 3 L of 12% acid solution must be in the mixture. [2.6.2] **35.** \$6500 was invested at 9.8%. [2.6.1]

Chapter 3

PREP TEST

1. $-4x + 12$ [1.5.5] **2.** 10 [1.4.2] **3.** −2 [1.4.2] **4.** 11 [1.5.1] **5.** 2.5 [1.5.1] **6.** 5 [1.5.1] **7.** 1 [1.5.1] **8.** 4 [2.2.1]

CONCEPT REVIEW 3.1

1. Always true **3.** Never true **5.** Never true

SECTION 3.1

3.

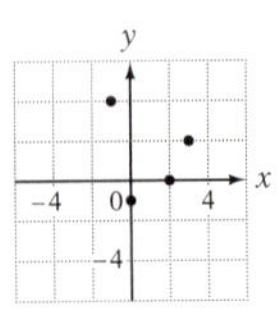

5. A is (0, 3); B is (1, 1); C is (3, −4); D is (−4, 4).

7.

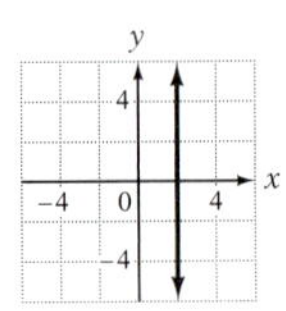

9.

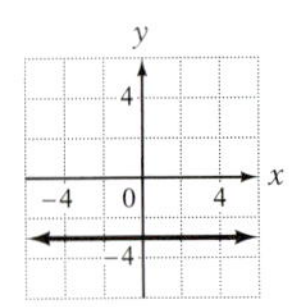

11.

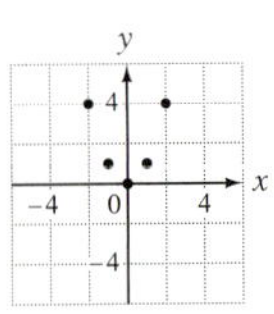

13.

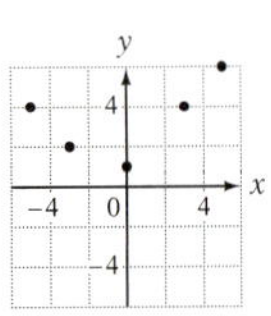

15.

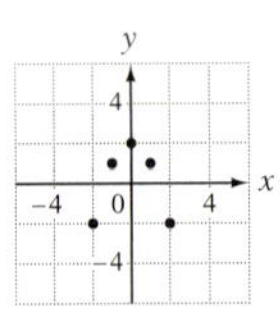

17.

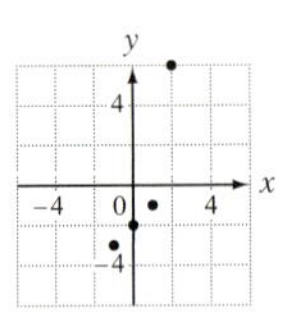

21. Length: $\sqrt{17}$; midpoint: $\left(\frac{7}{2}, 3\right)$ **23.** Length: $\sqrt{5}$; midpoint: $\left(-1, \frac{7}{2}\right)$ **25.** Length: $\sqrt{26}$; midpoint: $\left(-\frac{1}{2}, -\frac{9}{2}\right)$ **27.** Length: $\sqrt{85}$; midpoint: $\left(2, \frac{3}{2}\right)$ **29.** Length: 3; midpoint: $\left(\frac{7}{2}, -5\right)$
31. **a.** There are 275 calories in a hamburger. **b.** There are 1100 mg of sodium in a Big Mac.
33. **a.** There were 8 million bushels of apples in storage in 1977. **b.** The number of bushels was first greater than 12 million in 1999. **c.** One would expect that the number of bushels of apples in storage in 2001 would be greater than the number in storage in 2000. **35.**

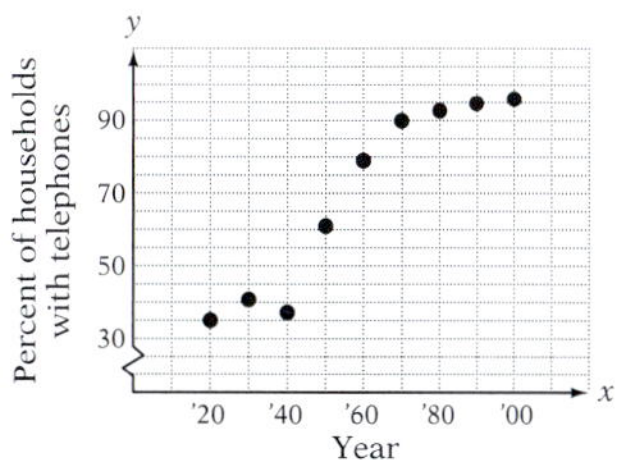

37. **a.** The average rate of change in the population was 179,000 people per year. **b.** The average annual rate of change in the population from 1900 to 1950 was less than the average annual rate of change from 1950 to 2000. **c.** The average annual rate of change in the population of Texas from 1980 to 1990 was less than the average annual rate of change in population of California from 1980 to 1990. **d.** The average annual rate of change was the least from 1860 to 1870. The average annual rate of change during the decade was 20,000 people per year.
39. **a.** The average annual rate of change in the number of pedestrian fatalities from 1990 to 2000 was −175.5 pedestrians per year. **b.** The answer to part **a** can be considered encouraging because it reflects the fact that the number of fatalities is decreasing. **41.** **a.** The average rate of change in the number of applications from 1990 to 2000 was 25,500 applications per year. **b.** The average annual rate of change in the number of applications from 1995 to 2000 was 29,000 applications per year greater than that from 1990 to 1995.

43.

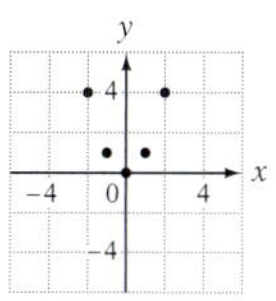

45.

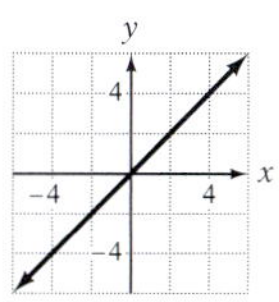

CONCEPT REVIEW 3.2

1. Sometimes true **3.** Sometimes true **5.** Always true

SECTION 3.2

3. Yes. No element of the domain is paired with more than one element of the range. **5.** Yes. No element of the domain is paired with more than one element of the range. **7.** No. The number 6 in the domain is paired with 2 and 3. Therefore, an element of the domain is paired with more than one element of the range.
9. Function **11.** Function **13.** Function **15.** Not a function **17.** **a.** Yes **b.** $29.62
21. 11 **23.** −4 **25.** 4 **27.** 10 **29.** 5 **31.** 0 **33.** 24 **35.** −4 **37.** 1
39. $\frac{3t}{t+2}$ **41.** 6 **43.** $a^3 - 3a + 4$ **45.** **a.** $4.75 per game **b.** $4.00 per game
47. **a.** $3000 **b.** $950 **49.** Domain: {1, 2, 3, 4, 5}; range: {1, 4, 7, 10, 13} **51.** Domain: {0, 2, 4, 6}; range: {1, 2, 3, 4} **53.** Domain: {1, 3, 5, 7, 9}; range: {0} **55.** Domain: {−2, −1, 0, 1, 2}; range: {0, 1, 2}
57. 1 **59.** −8 **61.** None **63.** None **65.** 0 **67.** None **69.** None **71.** −2
73. None **75.** Range: {−3, 1, 5, 9} **77.** Range: {−23, −13, −8, −3, 7} **79.** Range: {0, 1, 4}

81. Range: {2, 14, 26, 42} **83.** Range: $\left\{-5, \frac{5}{3}, 5\right\}$ **85.** Range: $\left\{-1, -\frac{1}{2}, -\frac{1}{3}, 1\right\}$

87. Range: {−38, −8, 2} **89.** $4h$ **91.** $-3h$ **93.** The car will skid for 61.2 ft.

97. a. The parachutist will be falling at a rate of 20 ft/s after 5 s. **b.** The parachutist will be falling at a rate of 28 ft/s after 15 s. **99. a.** The temperature of the cola will be 60°F after 10 h. **b.** The temperature of the cola will be 52°F after 15 h. **101. a.**

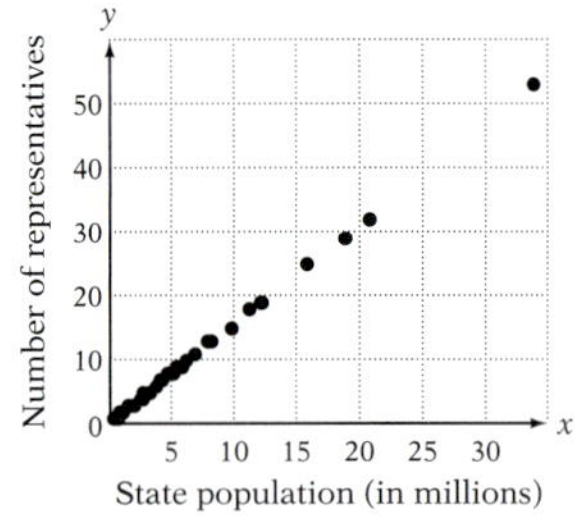

b. The 435 representatives in the House of Representatives are apportioned among the states according to their populations.

CONCEPT REVIEW 3.3

1. Never true **3.** Always true **5.** Sometimes true

SECTION 3.3

3.

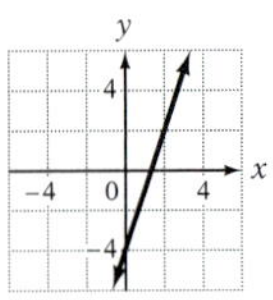

5.

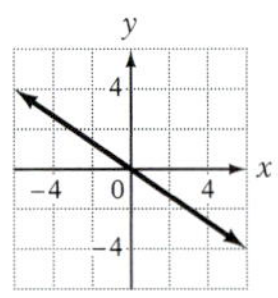

7.

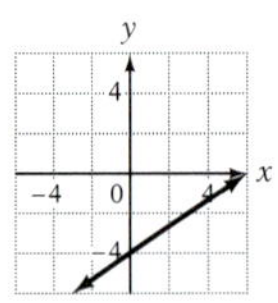

9.

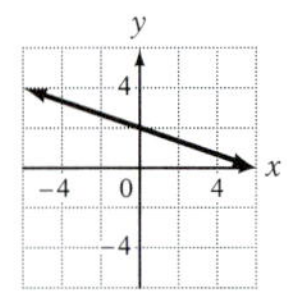

11.

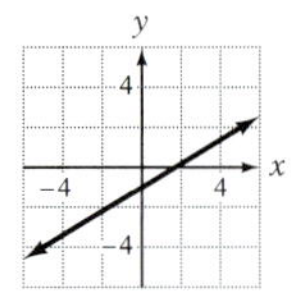

15.

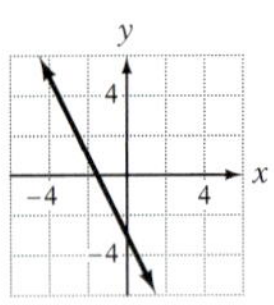

17.

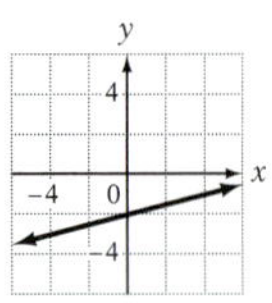

19.

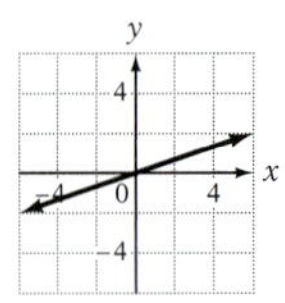

21.

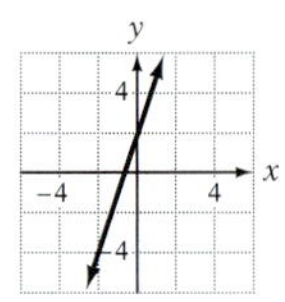

23. x-intercept: (−4, 0)
y-intercept: (0, 2)

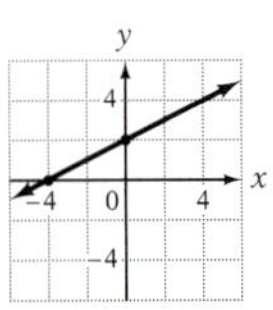

25. x-intercept: $\left(\frac{9}{2}, 0\right)$
y-intercept: (0, −3)

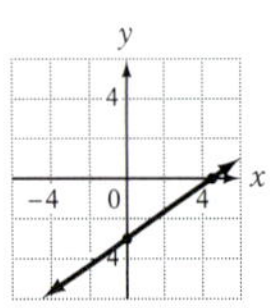

27. x-intercept: (2, 0)
y-intercept: (0, −4)

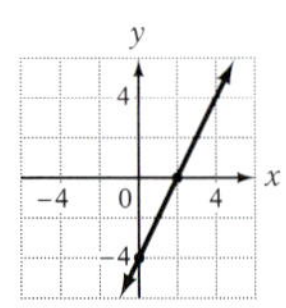

29. x-intercept: $\left(\frac{5}{3}, 0\right)$
y-intercept: $\left(0, \frac{5}{2}\right)$

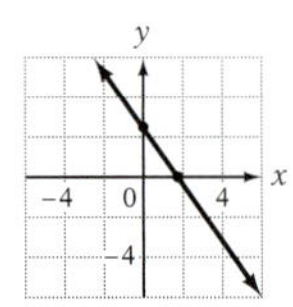

31. x-intercept: $\left(\frac{4}{3}, 0\right)$
y-intercept: (0, 2)

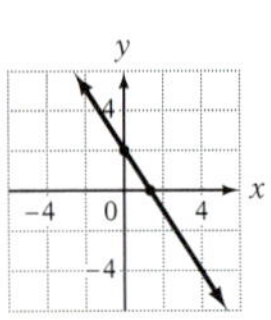

33. x-intercept: (3, 0)
y-intercept: $\left(0, -\frac{9}{5}\right)$

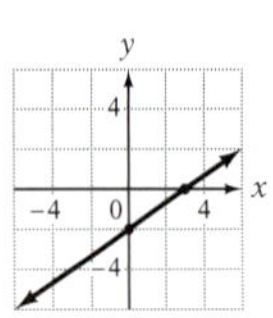

35.

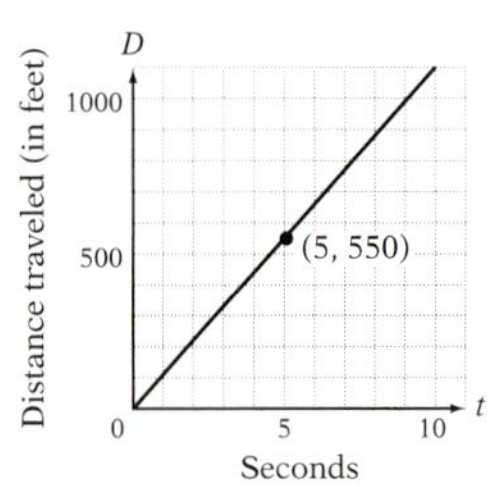

The roller coaster travels 550 ft in 5 s.

37.

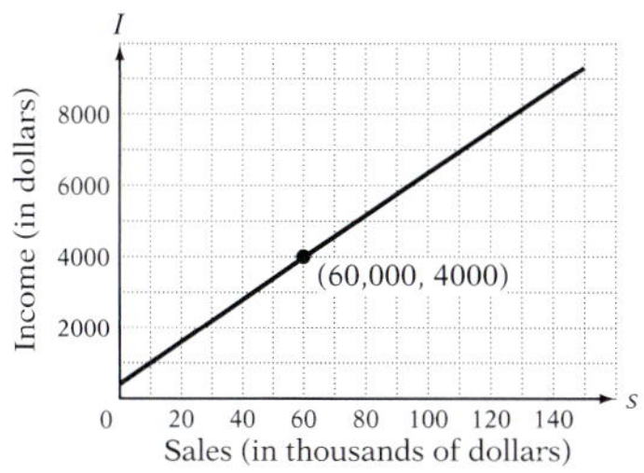

The realtor will earn a monthly income of \$4000 for \$60,000 worth of property.

39.

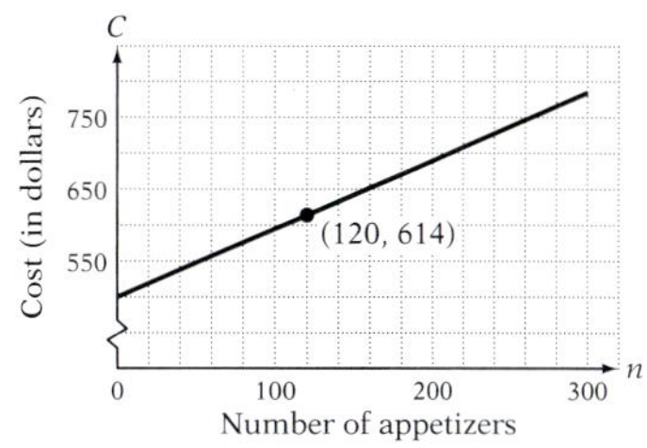

The caterer will charge \$614 for 120 hot appetizers.

41. The sale price is \$160.

43.

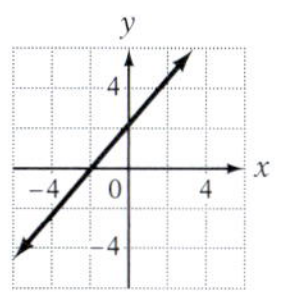

45.

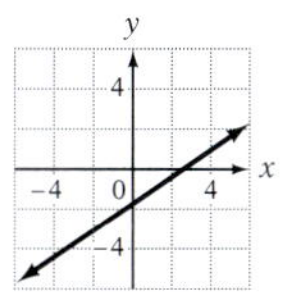

47.

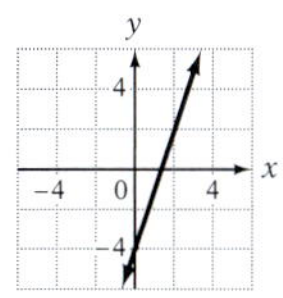

CONCEPT REVIEW 3.4

1. Always true **3.** Never true

SECTION 3.4

3. -1 **5.** $\frac{1}{3}$ **7.** $-\frac{2}{3}$ **9.** $-\frac{3}{4}$ **11.** Undefined **13.** $\frac{7}{5}$ **15.** 0 **17.** $-\frac{1}{2}$

19. Undefined **21.** The slope is 40. The slope is the average speed of the motorist in miles per hour.

23. The slope is -0.05. For each mile the car is driven, approximately 0.05 gal of fuel is used.

25. Line A represents Lois's distance; line B, Tanya's distance; and line C, the distance between.

27. **a.** The ramp does not meet the ANSI requirements. **b.** The ramp does meet the ANSI requirements.

29.

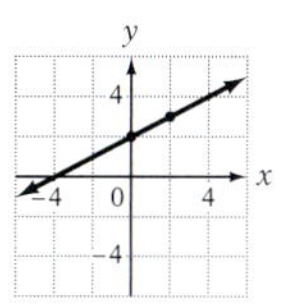

31.

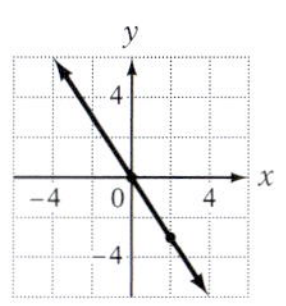

33.

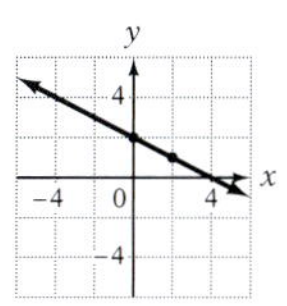

35.

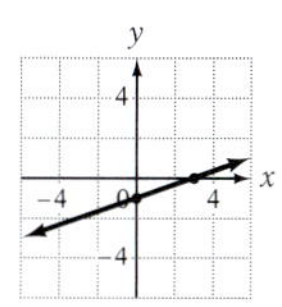

37.

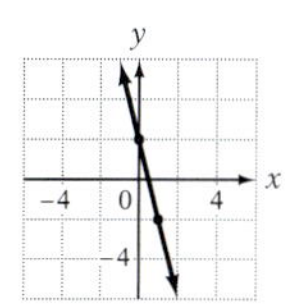

39.

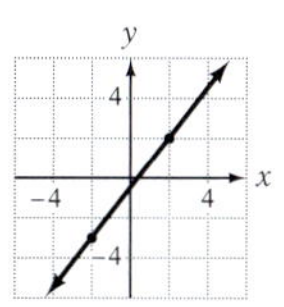

41.

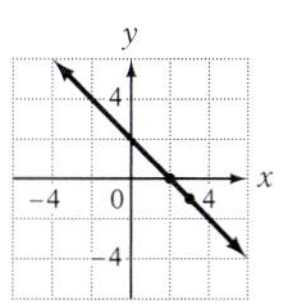

43. increases by 2 **45.** increases by $\frac{1}{2}$ **47.** 10

49. i and D; ii and C; iii and B; iv and F; v and E; vi and A

CONCEPT REVIEW 3.5

1. Always true **3.** Never true **5.** Never true **7.** Never true

SECTION 3.5

3. $y = 2x + 5$ **5.** $y = \frac{1}{2}x + 2$ **7.** $y = -\frac{5}{3}x + 5$ **9.** $y = -3x + 4$ **11.** $y = \frac{1}{2}x$

13. $y = 3x - 9$ **15.** $y = -\frac{2}{3}x + 7$ **17.** $y = -x - 3$ **19.** $x = 3$ **21.** $y = -3$

23. $y = -2x + 3$ **25.** $x = -5$ **27.** $y = x + 2$ **29.** $y = -2x - 3$ **31.** $y = \frac{1}{3}x + \frac{10}{3}$

33. $y = -\frac{3}{2}x + 3$ **35.** $y = x - 1$ **37.** $y = \frac{2}{3}x + \frac{5}{3}$ **39.** $y = \frac{1}{2}x - 1$ **41.** $y = -4$

43. $y = \frac{3}{4}x$ **45.** $x = -2$ **47.** $y = x - 1$ **49.** $y = -x + 3$ **51.** $y = 1200x$; the model predicts the plane will be 13,200 ft above sea level in 11 min. **53.** $y = 0.59x + 4.95$; the model predicts that using a cell phone for 13 min in one month will cost $12.62. **55.** $y = 63x$; the model predicts that there are 315 calories in a 5-ounce serving of lean hamburger. **57.** $y = -0.431x + 864$; the model predicts that in 1967 a plane could travel from New York to Paris in 16.2 h. **59.** $y = -0.032x + 16$; the model predicts that this car would contain 11.2 gal of gas after driving 150 mi. **61.** $\left(-\frac{b}{m}, 0\right)$ **63.** 0 **65.** Changing b moves the graph of the line up or down.

CONCEPT REVIEW 3.6

1. Sometimes true **3.** Never true **5.** Always true **7.** Always true

SECTION 3.6

3. -5 **5.** $-\frac{1}{4}$ **7.** Yes **9.** No **11.** No **13.** Yes **15.** Yes **17.** Yes

19. No **21.** Yes **23.** Yes **25.** $y = \frac{2}{3}x - \frac{8}{3}$ **27.** $y = \frac{1}{3}x - \frac{1}{3}$ **29.** $y = -\frac{5}{3}x - \frac{14}{3}$

31. $\frac{A_1}{B_1} = \frac{A_2}{B_2}$ **33.** $y = -2x + 15$ **35.** Any equation of the form $y = 2x + b$, where $b \neq -13$, or of the form $y = -\frac{3}{2}x + c$, where $c \neq 8$.

CONCEPT REVIEW 3.7

1. Never true **3.** Always true **5.** Always true

SECTION 3.7

3. Yes **5.** No

7.
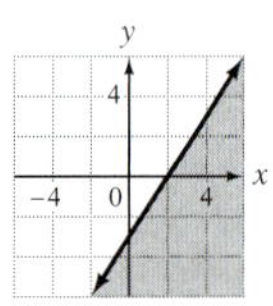

9.
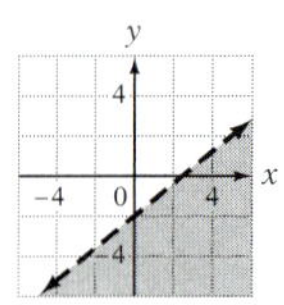

11.
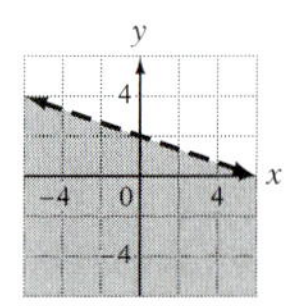

13.
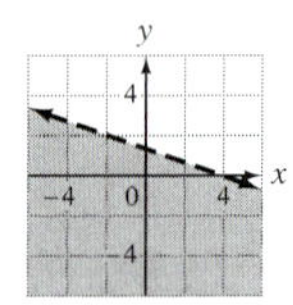

15.
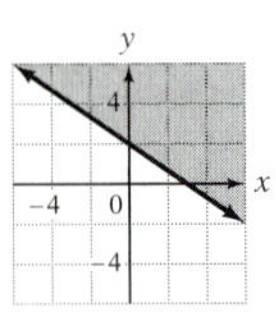

17.
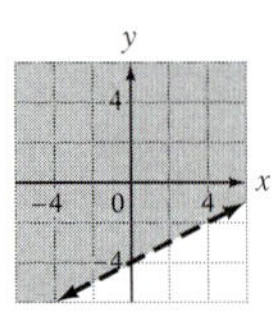

19.
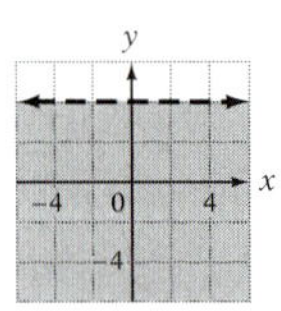

21.
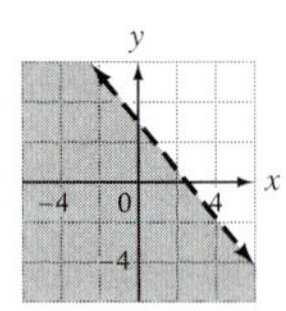

23.
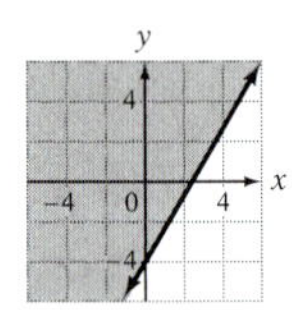

CHAPTER REVIEW EXERCISES

1. (4, 2) [3.2.1] **2.** Midpoint: $\left(\frac{1}{2}, \frac{9}{2}\right)$; length: $\sqrt{26}$ [3.1.2] **3.**
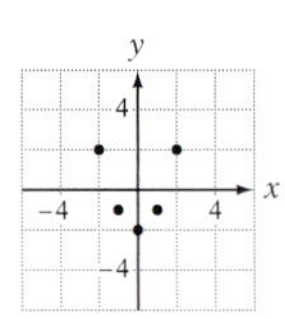
[3.1.1]

4. 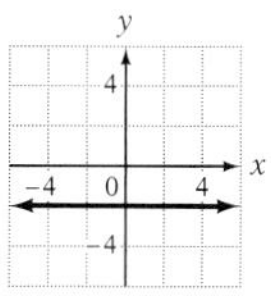 [3.1.1] **5.** $P(-2) = -2$; $P(a) = 3a + 4$ [3.2.1] **6.** Domain: $\{-1, 0, 1, 2, 5\}$; range: $\{0, 2, 3\}$ [3.2.1]

7. Range: $\{-2, -1, 2\}$ [3.2.1] **8.** -4 [3.2.1] **9.** x-intercept: $(3, 0)$ y-intercept: $(0, -2)$ 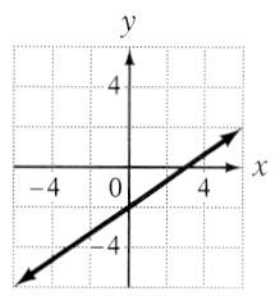 [3.3.2]

10. 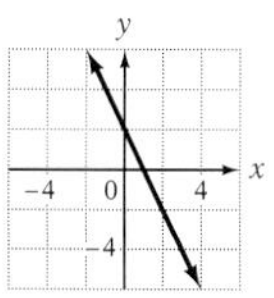 [3.3.1] **11.** 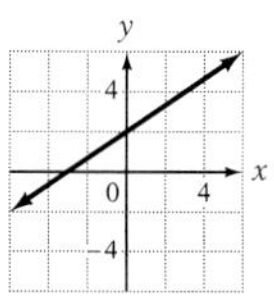 [3.3.2] **12.** -1 [3.4.1]

13. x-intercept: $\left(-\frac{4}{3}, 0\right)$ y-intercept: $(0, -2)$ 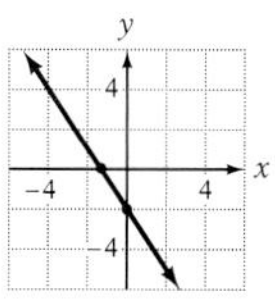 [3.3.2] **14.** 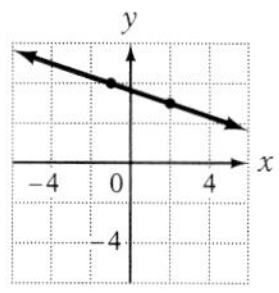 [3.4.2] **15.** $y = \frac{5}{2}x + \frac{23}{2}$ [3.5.1]

16. $y = -\frac{7}{6}x + \frac{5}{3}$ [3.5.2] **17.** $y = -3x + 7$ [3.6.1] **18.** $y = \frac{2}{3}x - \frac{8}{3}$ [3.6.1]

19. $y = \frac{3}{2}x + 2$ [3.6.1] **20.** $y = -\frac{1}{2}x - \frac{5}{2}$ [3.6.1] **21.** 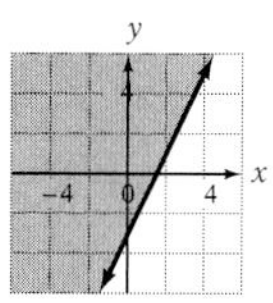 [3.7.1]

22. 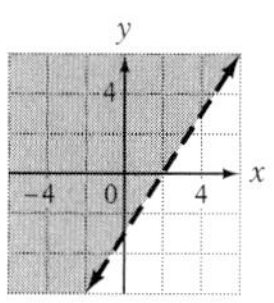 [3.7.1] **23.** 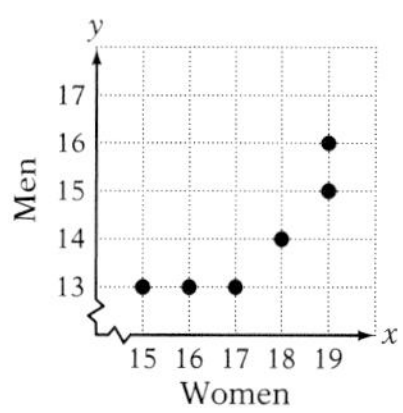 [3.1.3]

24. **a.** The average rate of change in the number of foreign-born people living in the United States from 1980 to 2000 was 715,000 people per year. **b.** The average rate of change in the number of foreign-born people living in the United States from 1970 to 1980 was 450,000 people per year. This was less than the average annual rate of change from 1980 to 1990. [3.1.4] **25.** 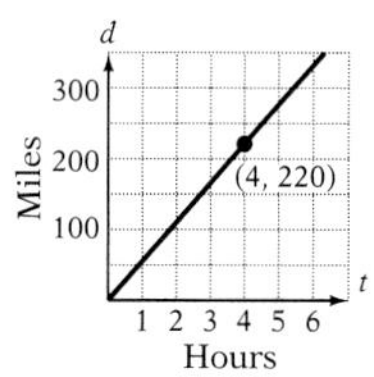 After 4 h, the car has traveled 220 mi. [3.3.3]

26. The slope is 20. The cost of manufacturing one calculator is \$20. [3.4.1] **27.** $y = 80x + 25{,}000$; this model predicts that the cost to build a house that contains 2000 ft^2 will be \$185,000. [3.5.3]

CHAPTER TEST

1. 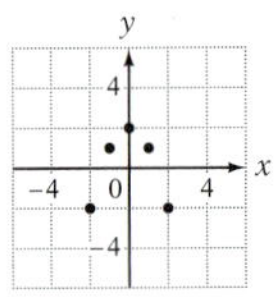

[3.1.1] **2.** (−3, 0) [3.1.1] **3.** 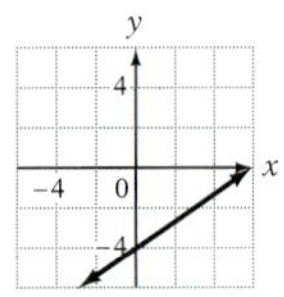

[3.3.1] **4.** 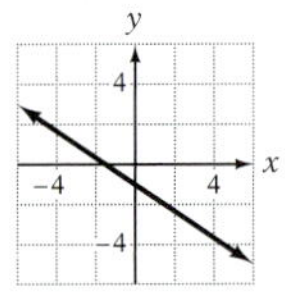

[3.3.2]

5. $x = -2$ [3.5.1] **6.** Midpoint: $\left(-\frac{1}{2}, 5\right)$; length: $\sqrt{117}$ [3.1.2] **7.** $-\frac{1}{6}$ [3.4.1] **8.** 9 [3.2.1]

9. 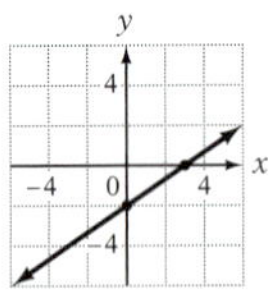

[3.3.2] **10.** 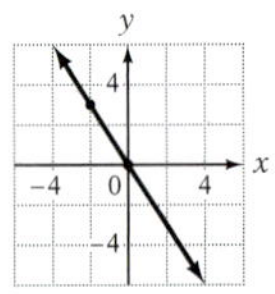

[3.4.2] **11.** $y = \frac{2}{5}x + 4$ [3.5.1] **12.** 0 [3.2.1]

13. $y = -\frac{7}{5}x + \frac{1}{5}$ [3.5.2] **14.** $y = -3$ [3.5.1] **15.** Domain: {−4, −2, 0, 3}; range: {0, 2, 5} [3.2.1]

16. $y = -\frac{3}{2}x + \frac{7}{2}$ [3.6.1] **17.** $y = 2x + 1$ [3.6.1] **18.** 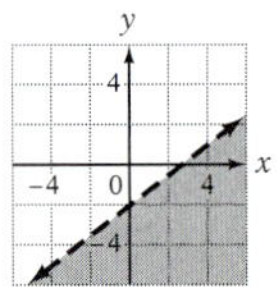

[3.7.1]

19. $y = -\frac{3}{10}x + 175$; this model predicts that 85 students will enroll when the tuition is $300. [3.5.3]

20. The slope is $-\frac{10{,}000}{3}$. The value of the house decreases by $3333.33 each year. [3.4.1]

21. a. The average rate of change in the Consumer Confidence Index from August 2000 to January 2001 was −5.28 points per month. **b.** The average rate of change in the Consumer Confidence Index from January 2001 to October 2001 was $-3.2\overline{1}$ points per month. [3.1.4]

CUMULATIVE REVIEW EXERCISES

1. Commutative Property of Multiplication [1.5.2] **2.** $\frac{9}{2}$ [2.2.3] **3.** $\frac{8}{9}$ [2.2.3] **4.** $-\frac{1}{14}$ [2.2.3]

5. $\left\{x \mid x < -1 \text{ or } x > \frac{1}{2}\right\}$ [2.7.3] **6.** $\frac{5}{2}$ and $-\frac{3}{2}$ [2.8.1] **7.** $\left\{x \mid 0 < x < \frac{10}{3}\right\}$ [2.8.2] **8.** 8 [1.4.2]

9. −18 [1.5.1] **10.** 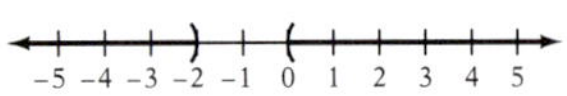

[2.7.1] **11.** ∅ [2.7.3] **12.** 14 [3.2.1]

13. (−8, 13) [3.1.1] **14.** $-\frac{7}{4}$ [3.4.1] **15.** $y = \frac{3}{2}x + \frac{13}{2}$ [3.5.1] **16.** $y = -\frac{5}{4}x + 3$ [3.5.2]

17. $y = -\frac{3}{2}x + 7$ [3.6.1] **18.** $y = -\frac{2}{3}x + \frac{8}{3}$ [3.6.1] **19.** 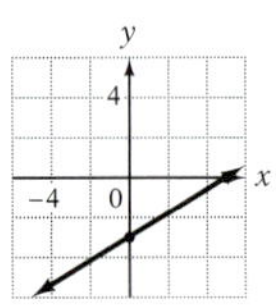

[3.3.2]

20. 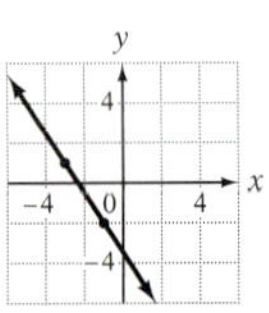

[3.4.2] **21.** 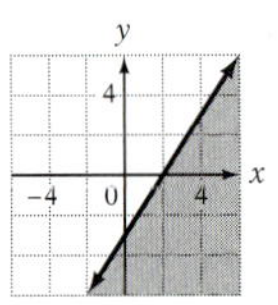

[3.7.1]

22. The cost is \$671.25. [2.5.1] **23.** The rate of the first plane is 200 mph, and the rate of the second plane is 400 mph. [2.3.4] **24.** The mixture consists of 32 lb of \$8 coffee and 48 lb of \$3 coffee. [2.3.3] **25.** $y = -2500x + 15{,}000$; the value of the truck decreases by \$2500 each year. [3.4.1, 3.5.2]

Chapter 4

PREP TEST

1. $6x + 5y$ [1.5.5] **2.** 7 [1.5.1] **3.** 0 [2.1.1] **4.** −3 [2.1.2] **5.** 1000 [2.2.3]

6. 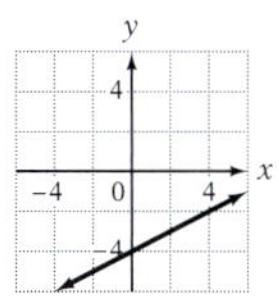 [3.3.1] **7.** 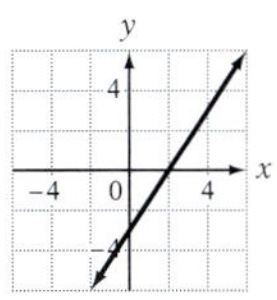 [3.3.2] **8.** 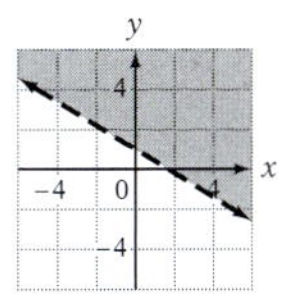 [3.7.1]

CONCEPT REVIEW 4.1

1. Always true **3.** Sometimes true **5.** Always true

SECTION 4.1

1. No **3.** Yes **5.** Independent **7.** Inconsistent

11. 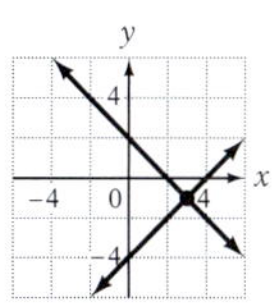 (3, −1) **13.** 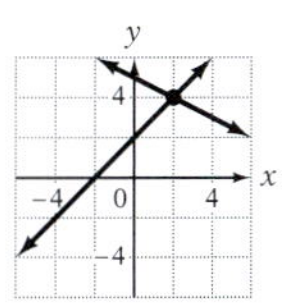 (2, 4) **15.** 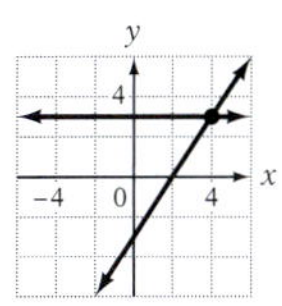 (4, 3) **17.** 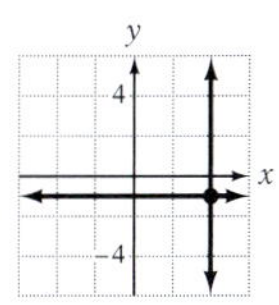 (4, −1) **19.** 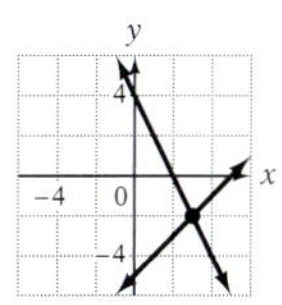 (3, −2)

21. 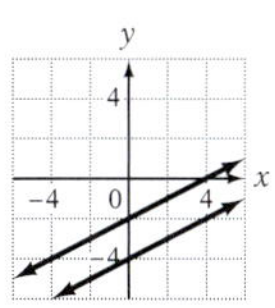 Inconsistent **23.** 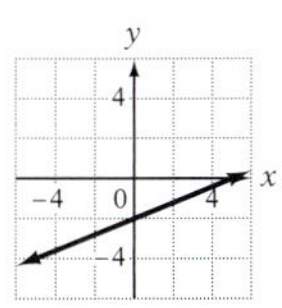 $\left(x, \frac{2}{5}x - 2\right)$ **25.** 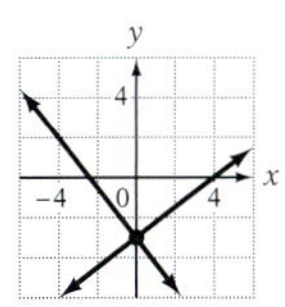 (0, −3) **29.** (2, 1) **31.** (1, 1) **33.** (2, 1)

35. (−2, −3) **37.** (3, −4) **39.** (0, −1) **41.** $\left(\frac{1}{2}, 3\right)$ **43.** (1, 2) **45.** (−1, 2)

47. (−2, 5) **49.** (0, 0) **51.** (−4, −6) **53.** Inconsistent **55.** $\left(x, \frac{1}{2}x - 4\right)$

57. Inconsistent **59.** (1, 5) **61.** (5, 2) **63.** (1, 4) **65.** 2 **67.** $\frac{1}{2}$

69. The two numbers are 26 and 18. **71.** The two numbers are 8 and 11. **73.** (2, 1) **75.** $\left(\frac{13}{11}, \frac{13}{5}\right)$

77. (1.20, 1.40) **79.** (0.64, −0.10) **81. a.** No **b.** 2016 **c.** The population is decreasing at a rate of 3600 people per year. **d.** The population is increasing at a rate of 41,900 people per year.

CONCEPT REVIEW 4.2

1. Always true **3.** Sometimes true **5.** Never true

SECTION 4.2

3. (6, 1) **5.** (1, 1) **7.** (2, 1) **9.** $(x, 3x - 4)$ **11.** $\left(-\frac{1}{2}, 2\right)$ **13.** Inconsistent **15.** (−1, −2) **17.** (2, 5) **19.** $\left(\frac{1}{2}, \frac{3}{4}\right)$ **21.** (0, 0) **23.** $\left(\frac{2}{3}, -\frac{2}{3}\right)$ **25.** (1, −1) **27.** $\left(x, \frac{4}{3}x - 6\right)$ **29.** (5, 3) **31.** $\left(\frac{1}{3}, -1\right)$ **33.** $\left(\frac{5}{3}, \frac{1}{3}\right)$ **35.** Inconsistent **41.** (−1, 2, 1) **43.** (6, 2, 4) **45.** (4, 1, 5) **47.** (3, 1, 0) **49.** (−1, −2, 2) **51.** Inconsistent **53.** (2, 1, −3) **55.** (2, −1, 3) **57.** (6, −2, 2) **59.** (0, −2, 0) **61.** (2, 3, 1) **63.** (1, 1, 3) **65.** (2, 1) **67.** (2, −1) **69.** (2, 0, −1) **71.** $A = 2, B = 3, C = -3$ **73.** The number of nickels is $3z - 5$, the number of dimes is $-4z + 35$, and the number of quarters is z, where $z = 2, 3, 4, 5, 6, 7,$ or 8. **75.** (1, 1) **77.** (3, 2)

CONCEPT REVIEW 4.3

1. Always true **3.** Sometimes true **5.** Always true

SECTION 4.3

3. $\left[\begin{array}{cc|c} 1 & -2 & \frac{1}{2} \\ 0 & 1 & \frac{5}{2} \end{array}\right]$ **5.** $\left[\begin{array}{cc|c} 1 & -\frac{2}{5} & \frac{3}{5} \\ 0 & 1 & 26 \end{array}\right]$ **7.** $\left[\begin{array}{ccc|c} 1 & 4 & 1 & -2 \\ 0 & 1 & 4 & -8 \\ 0 & 0 & 1 & -\frac{32}{19} \end{array}\right]$ **9.** $\left[\begin{array}{ccc|c} 1 & -3 & \frac{1}{2} & -\frac{3}{2} \\ 0 & 1 & \frac{3}{2} & \frac{5}{2} \\ 0 & 0 & 1 & 3 \end{array}\right]$

11. (−7, 4, 3) **13.** (1, 3) **15.** (−1, −3) **17.** (3, 2) **19.** Inconsistent **21.** (−2, 2) **23.** (0, 0, −3) **25.** (1, −1, −1) **27.** Inconsistent **29.** $\left(\frac{1}{3}, \frac{1}{2}, 0\right)$ **31.** $\left(\frac{1}{5}, \frac{2}{5}, -\frac{3}{5}\right)$ **33.** $\left(\frac{1}{4}, 0, -\frac{2}{3}\right)$ **35.** 11 **37.** 18 **39.** 0 **41.** 15 **43.** −30 **45.** 0 **49.** (3, −4) **51.** (4, −1) **53.** $\left(\frac{11}{14}, \frac{17}{21}\right)$ **55.** $\left(\frac{1}{2}, 1\right)$ **57.** Not possible by Cramer's Rule **59.** (−1, 0) **61.** (1, −1, 2) **63.** (2, −2, 3) **65.** Not possible by Cramer's Rule **67.** $\left(\frac{68}{25}, \frac{56}{25}, -\frac{8}{25}\right)$ **69.** $\left(\frac{1}{3}, \frac{2}{5}, \frac{1}{2}\right)$ **71.** −14 **73.** 0 **75.** **a.** 0 **b.** 0 **77.** The area of the polygon is 239 ft^2.

CONCEPT REVIEW 4.4

1. Never true **3.** Always true

SECTION 4.4

1. The rate of the plane in calm air is 150 mph. The rate of the wind is 10 mph. **3.** The rate of the cabin cruiser in calm water is 14 mph. The rate of the current is 2 mph. **5.** The rate of the plane in calm air is 165 mph. The rate of the wind is 15 mph. **7.** The rate of the boat in calm water is 19 km/h. The rate of the current is 3 km/h. **9.** The rate of the plane in calm air is 105 mph. The rate of the wind is 15 mph. **11.** The rate of the boat in calm water is 16.5 mph. The rate of the current is 1.5 mph. **13.** The cost of the redwood is \$.30 per foot. The cost of the pine is \$.18 per foot. **15.** The cost per unit of gas is \$.16. **17.** There are 18 quarters in the bank. **19.** The company plans to manufacture 25 mountain bikes during the week. **21.** The pharmacist should use 200 mg of the first powder and 450 mg of the second powder.

23. The model VI costs \$4000. **25.** \$8000 was invested at 9%, \$6000 at 7%, and \$4000 at 5%. **27.** The distances are $d_1 = 6$ in., $d_2 = 3$ in., and $d_3 = 9$ in. **29.** There were 10 nickels, 5 dimes, and 4 quarters. **31.** There was \$14,000 invested at 12%, \$10,000 invested at 8%, and \$9000 invested at 9%. **33.** Las Vegas has 265 hotels, Orlando has 350 hotels, and Los Angeles–Long Beach has 653 hotels. **35.** The measures of the two angles are 9° and 81°. **37.** There are 25 nickels, 10 dimes, and 5 quarters in the coin bank.

CONCEPT REVIEW 4.5

1. Sometimes true **3.** Always true

SECTION 4.5

1. b **5.** 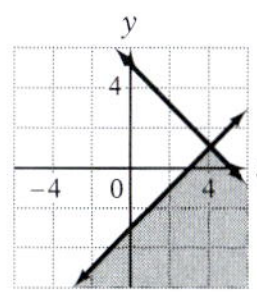 **7.** 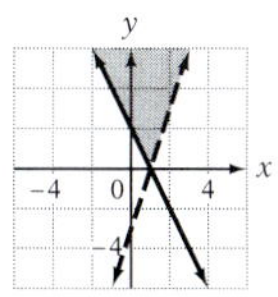 **9.** 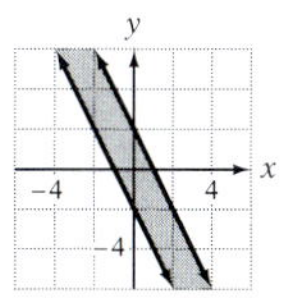 **11.** 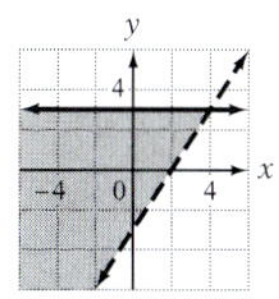

13. 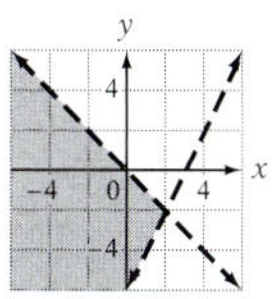 **15.** 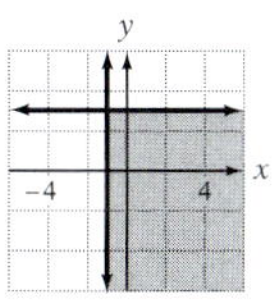 **17.** 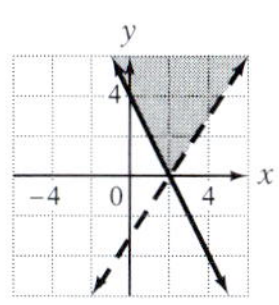 **19.** 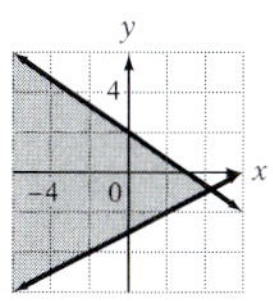 **21.** 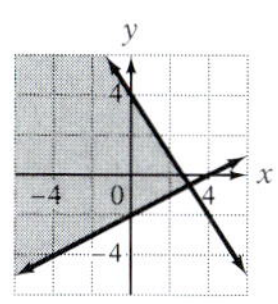

23. 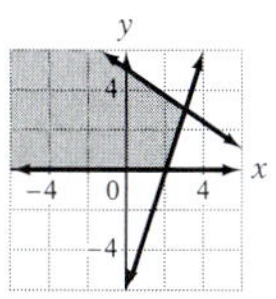 **25.** 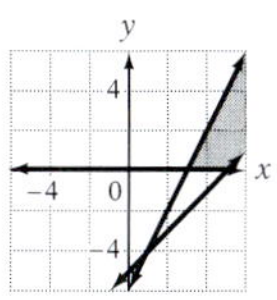 **27.** 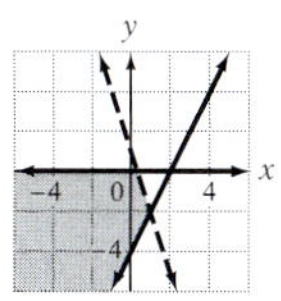

CHAPTER REVIEW EXERCISES

1. Inconsistent [4.1.2] **2.** $\left(x, -\frac{1}{4}x + \frac{3}{2}\right)$ [4.1.2] **3.** $(-4, 7)$ [4.2.1] **4.** $\left(x, \frac{1}{3}x - 2\right)$ [4.2.1]

5. $(5, -2, 3)$ [4.2.2] **6.** $(3, -1, -2)$ [4.2.2] **7.** 28 [4.3.2] **8.** 0 [4.3.2] **9.** $(3, -1)$ [4.3.3]

10. $\left(\frac{110}{23}, \frac{25}{23}\right)$ [4.3.3] **11.** $(-1, -3, 4)$ [4.3.3] **12.** $(2, 3, -5)$ [4.3.3] **13.** $(1, -1, 4)$ [4.2.2]

14. $\left(\frac{8}{5}, \frac{7}{5}\right)$ [4.3.3] **15.** $\left(\frac{1}{2}, -1, \frac{1}{3}\right)$ [4.3.1] **16.** 12 [4.3.2] **17.** $(2, -3)$ [4.3.3]

18. $(2, -3, 1)$ [4.3.1] **19.** 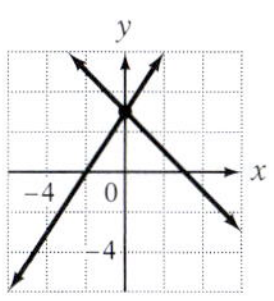 [4.1.1] (0, 3) **20.** 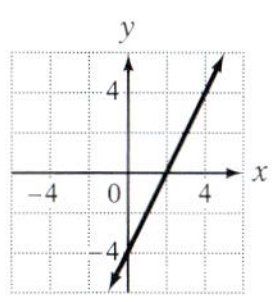 [4.1.1] $(x, 2x - 4)$

21. 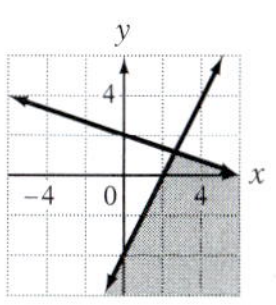 [4.5.1] **22.** 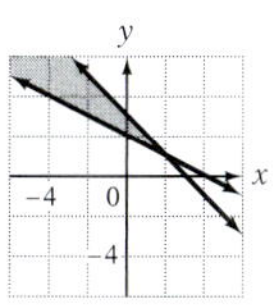 [4.5.1] **23.** The rate of the cabin cruiser in calm water is 16 mph. The rate of the current is 4 mph. [4.4.1]

24. The rate of the plane in calm air is 175 mph. The rate of the wind is 25 mph. [4.4.1] **25.** The number of children attending on Friday was 100. [4.4.2] **26.** The chef must use 3 oz of meat, 5 oz of potatoes, and 4 oz of green beans. [4.2.2]

CHAPTER TEST

1. $\left(\frac{3}{4}, \frac{7}{8}\right)$ [4.1.2] **2.** $(-3, -4)$ [4.1.2] **3.** $(2, -1)$ [4.1.2] **4.** $(-2, 1)$ [4.3.1] **5.** Inconsistent [4.2.1] **6.** $(1, 1)$ [4.2.1] **7.** Inconsistent [4.2.2] **8.** $(2, -1, -2)$ [4.3.1] **9.** 10 [4.3.2] **10.** -32 [4.3.2] **11.** $\left(-\frac{1}{3}, -\frac{10}{3}\right)$ [4.3.3] **12.** $\left(\frac{1}{5}, -\frac{6}{5}, \frac{3}{5}\right)$ [4.3.3] **13.** $(0, -2, 3)$ [4.3.3] **14.**

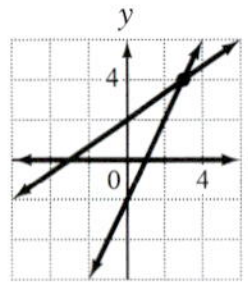

(3, 4) [4.1.1] **15.**

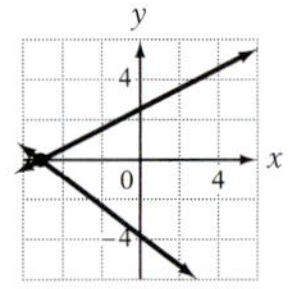

(−5, 0) [4.1.1] **16.**

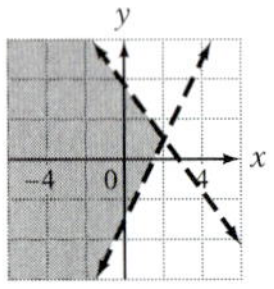

[4.5.1] **17.**

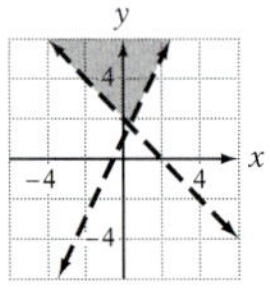

[4.5.1] **18.** $(-0.14, 2.43)$ [4.1.1] **19.** The rate of the plane in calm air is 150 mph. The rate of the wind is 25 mph. [4.4.1] **20.** The cost per yard of cotton is \$9. The cost per yard of wool is \$14. [4.4.2]

CUMULATIVE REVIEW EXERCISES

1. $-\frac{11}{28}$ [2.2.3] **2.** $y = 5x - 11$ [3.5.2] **3.** $3x - 24$ [1.5.6] **4.** -4 [1.5.1] **5.** $\{x \mid x < 6\}$ [2.7.3] **6.** $\{x \mid -4 < x < 8\}$ [2.8.2] **7.** $\{x \mid x > 4 \text{ or } x < -1\}$ [2.8.2] **8.** -98 [3.2.1] **9.** $\{0, 1, 5, 8, 16\}$ [3.2.1] **10.** 1 [3.2.1] **11.** $3h$ [3.2.1] **12.**

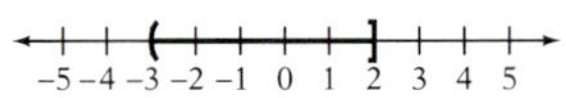

[2.7.1] **13.** $y = -\frac{2}{3}x + \frac{5}{3}$ [3.5.1] **14.** $y = -\frac{3}{2}x + \frac{1}{2}$ [3.6.1] **15.** $\sqrt{37}$ [3.1.2] **16.** $\left(-\frac{1}{2}, 4\right)$ [3.1.2] **17.**

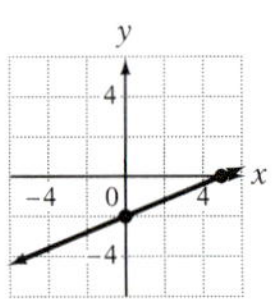

[3.4.2] **18.**

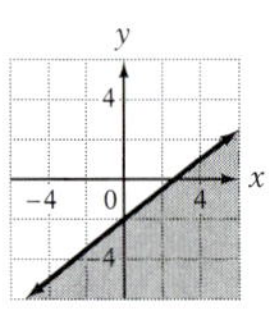

[3.7.1] **19.** $(-5, -11)$ [4.1.2] **20.** $(1, 0, -1)$ [4.2.2] **21.** 3 [4.3.2] **22.** $(2, 0)$

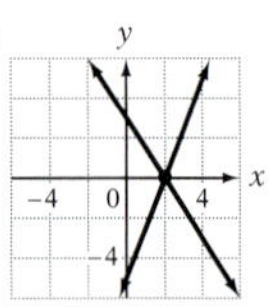

[4.1.1] **23.** $(2, -3)$ [4.3.3] **24.**

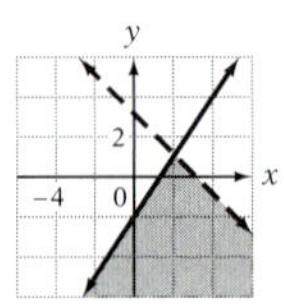

[4.5.1] **25.** The lowest score the student can receive is 72. [2.7.4] **26.** The amount of water that should be added is 60 ml. [2.6.2] **27.** The rate of the wind is 12.5 mph. [4.4.1] **28.** The cost per pound of steak is \$5. [4.4.2] **29.** The lower and upper limits of the resistor are 10,200 ohms and 13,800 ohms. [2.8.3] **30.** The slope is 40. The account executive earns \$40 for each \$1000 of sales. [3.4.1]

Answers to Chapter 5 Selected Exercises

PREP TEST

1. 1 [1.2.2] **2.** -18 [1.2.3] **3.** $\frac{2}{3}$ [1.2.4] **4.** 48 [1.5.1] **5.** 0 [1.2.4]
6. No [1.5.3] **7.** $5x^2 - 9x - 6$ [1.5.3] **8.** 0 [1.5.3] **9.** $-6x + 24$ [1.5.5]
10. $-7xy + 10y$ [1.5.6]

CONCEPT REVIEW 5.1

1. Always true **3.** Always true **5.** Sometimes true

SECTION 5.1

3. 13 **5.** 10 **7.** -11 **9.** The length of the wave is 23.1 m. **11.** There must be 56 games scheduled. **13.** The baked Alaska needs 260 in^3 of meringue. **15. a.** The leading coefficient is -1. **b.** The constant term is 8. **c.** The degree is 2. **17.** The expression is not a polynomial. **19.** The expression is not a polynomial. **21. a.** The leading coefficient is 3. **b.** The constant term is π. **c.** The degree is 5. **23. a.** The leading coefficient is -5. **b.** The constant term is 2. **c.** The degree is 3. **25. a.** The leading coefficient is 14. **b.** The constant term is 14. **c.** The degree is 0.

27.
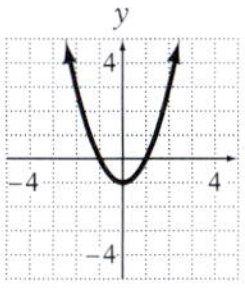

29.
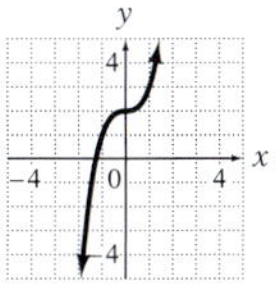

31.
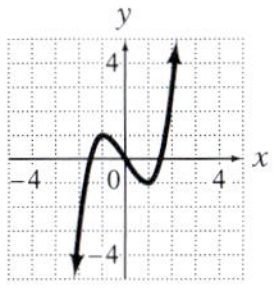

35. $6x^2 - 6x + 5$ **37.** $-x^2 + 1$

39. $5y^2 - 15y + 2$ **41.** $7a^2 - a + 2$ **43.** $5x^3 - 4x^2 + 4x - 7$ **45.** $9x^n + 5$ **47.** $3x^4 - 8x^2 + 2x$
49. $-2x^3 + 4x^2 + 8$ **51.** 1.5 **53.** 3.4, 0.6 **55.** $-1.9, 0.35, 1.5$ **57.** $1.6, -0.6$ **59.** -2
61. $P(x) + Q(x)$ is a fourth-degree polynomial. **63.** The maximum deflection of the beam is 3.125 in. **65.** 2

67.
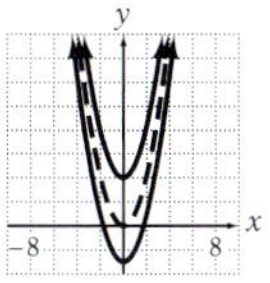

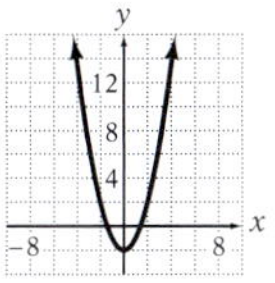

The graph of k is the graph of f moved 2 units down.

CONCEPT REVIEW 5.2

1. Never true **3.** Always true **5.** Never true

SECTION 5.2

3. a. Product of two exponential expressions **b.** Power of an exponential expression
5. a. Power of an exponential expression **b.** Product of two exponential expressions
7. $2x^2$ **9.** $12x^2$ **11.** $6a^7$ **13.** x^3y^5 **15.** $-10x^9y$ **17.** x^7y^8 **19.** $-6x^3y^5$
21. x^4y^5z **23.** $a^3b^5c^4$ **25.** $-a^5b^8$ **27.** $-6a^5b$ **29.** $40y^{10}z^6$ **31.** $-20a^2b^3$
33. $x^3y^5z^3$ **35.** $-12a^{10}b^7$ **37.** 64 **39.** 4 **41.** -64 **43.** x^9 **45.** x^{14} **47.** x^4
49. $4x^2$ **51.** $-8x^6$ **53.** x^4y^6 **55.** $9x^4y^2$ **57.** $27a^8$ **59.** $-8x^7$ **61.** x^8y^4
63. a^4b^6 **65.** $16x^{10}y^3$ **67.** $-18x^3y^4$ **69.** $-8a^7b^5$ **71.** $-54a^9b^3$ **73.** $-72a^5b^5$
75. $32x^3$ **77.** $-3a^3b^3$ **79.** $13x^5y^5$ **81.** $27a^5b^3$ **83.** $-5x^7y^4$ **85.** a^{2n} **87.** a^{2n}
89. True **91.** False; $x^{2 \cdot 5} = x^{10}$ **93.** No; $2^{(3^2)}$ is larger. $(2^3)^2 = 8^2 = 64$, whereas $2^{(3^2)} = 2^9 = 512$.
95. $x = y$ when n is odd.

CONCEPT REVIEW 5.3

1. Always true **3.** Sometimes true **5.** Always true **7.** Sometimes true

SECTION 5.3

1. $x^2 - 2x$ **3.** $-x^2 - 7x$ **5.** $3a^3 - 6a^2$ **7.** $-5x^4 + 5x^3$ **9.** $-3x^5 + 7x^3$ **11.** $12x^3 - 6x^2$ **13.** $6x^2 - 12x$ **15.** $-x^3y + xy^3$ **17.** $2x^4 - 3x^2 + 2x$ **19.** $2a^3 + 3a^2 + 2a$ **21.** $3x^6 - 3x^4 - 2x^2$ **23.** $-6y^4 - 12y^3 + 14y^2$ **25.** $-2a^3 - 6a^2 + 8a$ **27.** $6y^4 - 3y^3 + 6y^2$ **29.** $x^3y - 3x^2y^2 + xy^3$ **31.** $x^3 + 4x^2 + 5x + 2$ **33.** $a^3 - 6a^2 + 13a - 12$ **35.** $-2b^3 + 7b^2 + 19b - 20$ **37.** $-6x^3 + 31x^2 - 41x + 10$ **39.** $x^4 - 4x^3 - 3x^2 + 14x - 8$ **41.** $15y^3 - 16y^2 - 70y + 16$ **43.** $5a^4 - 20a^3 - 15a^2 + 62a - 8$ **45.** $y^4 + 4y^3 + y^2 - 5y + 2$ **47.** **a.** $y, 2y$ **b.** $y, 3$ **c.** $-8, 2y$ **d.** $-8, 3$ **49.** $x^2 + 4x + 3$ **51.** $a^2 + a - 12$ **53.** $y^2 - 5y - 24$ **55.** $y^2 - 10y + 21$ **57.** $2x^2 + 15x + 7$ **59.** $3x^2 + 11x - 4$ **61.** $4x^2 - 31x + 21$ **63.** $3y^2 - 2y - 16$ **65.** $9x^2 + 54x + 77$ **67.** $21a^2 - 83a + 80$ **69.** $15b^2 + 47b - 78$ **71.** $2a^2 + 7ab + 3b^2$ **73.** $6a^2 + ab - 2b^2$ **75.** $2x^2 - 3xy - 2y^2$ **77.** $10x^2 + 29xy + 21y^2$ **79.** $6a^2 - 25ab + 14b^2$ **81.** $2a^2 - 11ab - 63b^2$ **83.** $100a^2 - 100ab + 21b^2$ **85.** $15x^2 + 56xy + 48y^2$ **87.** $14x^2 - 97xy - 60y^2$ **89.** $56x^2 - 61xy + 15y^2$ **91.** $y^2 - 25$ **93.** $4x^2 - 9$ **95.** $x^2 + 2x + 1$ **97.** $9a^2 - 30a + 25$ **99.** $9x^2 - 49$ **101.** $4a^2 + 4ab + b^2$ **103.** $x^2 - 4xy + 4y^2$ **105.** $16 - 9y^2$ **107.** $25x^2 + 20xy + 4y^2$ **109.** The area is $(10x^2 - 35x)$ ft^2. **111.** The area is $(18x^2 + 12x + 2)$ in^2. **113.** The area $(4x^2 + 4x + 1)$ km^2. **115.** The area is $(4x^2 + 10x)$ m^2. **117.** The area is $(\pi x^2 + 8\pi x + 16\pi)$ cm^2. **119.** The area is $(90x + 2025)$ ft^2. **121.** $4ab$ **123.** $9a^4 - 24a^3 + 28a^2 - 16a + 4$ **125.** $24x^3 - 3x^2$ **127.** $x^{2n} + x^n$ **129.** $x^{2n} + 2x^n + 1$ **131.** $x^2 - 1$ **133.** $x^4 - 1$ **135.** $x^6 - 1$ **137.** 1024 **139.** $12x^2 - x - 20$ **141.** $7x^2 - 11x - 8$

CONCEPT REVIEW 5.4

1. Never true **3.** Always true **5.** Never true **7.** Never true

SECTION 5.4

1. $\frac{1}{25}$ **3.** 64 **5.** $\frac{1}{27}$ **7.** 1 **11.** y^4 **13.** a^3 **15.** p^4 **17.** $2x^3$ **19.** $2k$ **21.** m^5n^2 **23.** $\frac{3r^2}{2}$ **25.** $-\frac{2a}{3}$ **27.** $\frac{1}{x^2}$ **29.** a^6 **31.** $\frac{4}{x^7}$ **33.** $5b^8$ **35.** $\frac{x^2}{3}$ **37.** 1 **39.** $\frac{1}{y^5}$ **41.** $\frac{1}{a^6}$ **43.** $\frac{1}{3x^3}$ **45.** $\frac{2}{3x^5}$ **47.** $\frac{y^4}{x^2}$ **49.** $\frac{2}{5m^3n^8}$ **51.** $\frac{1}{p^3q}$ **53.** $\frac{1}{2y^3}$ **55.** $\frac{7xz}{8y^3}$ **57.** $\frac{p^2}{2m^3}$ **59.** $-\frac{8x^3}{y^6}$ **61.** $\frac{9}{x^2y^4}$ **63.** $\frac{2}{x^4}$ **65.** $-\frac{5}{a^8}$ **67.** $-\frac{a^5}{8b^4}$ **69.** $\frac{10y^3}{x^4}$ **71.** $\frac{1}{2x^3}$ **73.** $\frac{3}{x^3}$ **75.** $\frac{1}{2x^2y^6}$ **77.** $-\frac{1}{6x^3}$ **81.** No. 39.4 is not between 1 and 10. **83.** No. 2.4 is not an integer. **85.** 2.37×10^6 **87.** 4.5×10^{-4} **89.** 3.09×10^5 **91.** 6.01×10^{-7} **93.** 710,000 **95.** 0.000043 **97.** 671,000,000 **99.** 0.00000713 **101.** 1.6×10^{10} **103.** 5.98×10^{24} **105.** 1.6×10^{-19} **107.** 1×10^{-12} **109.** 6.65×10^{19} **111.** 3.22×10^{-14} **113.** 3.6×10^5 **115.** 1.8×10^{-18} **117.** The computer can perform 7.2×10^9 operations in 1 h. **119.** Light travels 8.64×10^{12} m in 8 h. **121.** The time for the centrifuge to make one revolution is 1.5×10^{-7} s. **123.** The sun is 3.3898305×10^5 times heavier than Earth. **125.** It takes light 5×10^2 s for light to travel to Earth from the sun. **127.** Each American would have to pay 2.2×10^4 dollars. **129.** The surface area of Neptune is 2.82743339×10^9 mi^2. **131.** The weight of one atom of hydrogen is

$1.66030217 \times 10^{-24}$ g. **133.** $\frac{3}{64}$ **135.** For 2^x: $\frac{1}{4}, \frac{1}{2}, 1, 2, 4$. For 2^{-x}: $4, 2, 1, \frac{1}{2}, \frac{1}{4}$. **137.** 0.0625 **139.** xy^2 **141.** $\frac{x^6z^6}{4y^5}$ **143.** 1 **145.** 0 **147.** False; $(2a)^{-3} = \frac{1}{8a^3}$ **149.** False; $(2 + 3)^{-1} = 5^{-1} = \frac{1}{5}$

CONCEPT REVIEW 5.5

1. Always true **3.** Never true

SECTION 5.5

1. $x + 1$ **3.** $2a - 5$ **5.** $3a + 2$ **7.** $4b^2 - 3$ **9.** $x - 2$ **11.** $-x + 2$ **13.** $x^2 + 3x - 5$ **15.** $x^4 - 3x^2 - 1$ **17.** $xy + 2$ **19.** $-3y^3 + 5$ **21.** $3x - 2 + \frac{1}{x}$ **23.** $-3x + 7 - \frac{6}{x}$ **25.** $4a - 5 + 6b$ **27.** $9x + 6 - 3y$ **29.** $x + 8$ **31.** $x^2 + \frac{2}{x - 3}$ **33.** $3x + 5 + \frac{3}{2x + 1}$ **35.** $5x + 7 + \frac{2}{2x - 1}$ **37.** $4x^2 + 6x + 9 + \frac{18}{2x - 3}$ **39.** $3x^2 + 1 + \frac{1}{2x^2 - 5}$ **41.** $x^2 - 3x - 10$ **43.** $-x^2 + 2x - 1 + \frac{1}{x - 3}$ **45.** $2x^3 - 3x^2 + x - 4$ **47.** $x - 4 + \frac{x + 3}{x^2 + 1}$ **49.** $2x - 3 + \frac{1}{x - 1}$ **51.** $3x + 1 + \frac{10x + 7}{2x^2 - 3}$ **53.** $2x - 8$ **55.** $3x + 3 - \frac{1}{x - 1}$ **57.** $x - 4 + \frac{7}{x + 4}$ **59.** $x - 2 + \frac{16}{x + 2}$ **61.** $2x^2 - 3x + 9$ **63.** $x^2 - 3x + 2$ **65.** $x^2 - 5x + 16 - \frac{41}{x + 2}$ **67.** $x^2 - x + 2 - \frac{4}{x + 1}$ **69.** $4x^2 + 8x + 15 + \frac{12}{x - 2}$ **71.** $2x^3 - 3x^2 + x - 4$ **73.** $3x^3 + 2x^2 + 12x + 19 + \frac{33}{x - 2}$ **75.** $3x^3 - x + 4 - \frac{2}{x + 1}$ **77.** $2x^3 + 6x^2 + 17x + 51 + \frac{155}{x - 3}$ **79.** $x^2 - 5x + 25$ **81.** $P(3) = 8$ **83.** $R(4) = 43$ **85.** $P(-2) = -39$ **87.** $Q(2) = 31$ **89.** $F(-3) = 190$ **91.** $P(5) = 122$ **93.** $R(-3) = 302$ **95.** $Q(2) = 0$ **97.** 3 **99.** -1 **101.** $(x + 2), (x - 3)$ **103.** $3a^2$ **105.** $4x^2 - 6x + 3$

CHAPTER REVIEW EXERCISES

1. $2x^2 - 5x - 2$ [5.1.2] **2.** $4x^2 - 8xy + 5y^2$ [5.1.2] **3.** $70xy^2z^6$ [5.4.1] **4.** $-\frac{48y^9z^{17}}{x}$ [5.4.1] **5.** $-\frac{x^3}{4y^2z^3}$ [5.4.1] **6.** $\frac{c^{10}}{2b^{17}}$ [5.4.1] **7.** 9.3×10^7 [5.4.2] **8.** 0.00254 [5.4.2] **9.** 2×10^{-6} [5.4.2] **10.** -7 [5.1.1] **11.**

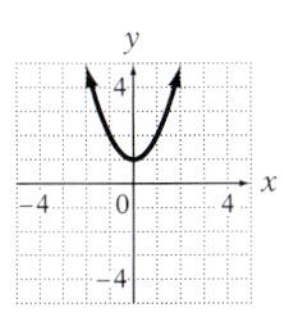

[5.1.1] **12. a.** 3 **b.** 8 **c.** 5 [5.1.1] **13.** 68 [5.5.4] **14.** $5x + 4 + \frac{6}{3x - 2}$ [5.5.2] **15.** $2x - 3 - \frac{4}{6x + 1}$ [5.5.2] **16.** $4x^2 + 3x - 8 + \frac{50}{x + 6}$ [5.5.2/5.5.3] **17.** $x^3 + 4x^2 + 16x + 64 + \frac{252}{x - 4}$ [5.5.2/5.5.3] **18.** $12x^5y^3 + 8x^3y^2 - 28x^2y^4$ [5.3.1] **19.** $a^{3n+3} - 5a^{2n+4} + 2a^{2n+3}$ [5.3.1] **20.** $9x^2 + 8x$ [5.3.1]

21. $x^{3n+1} - 3x^{2n} - x^{n+2} + 3x$ [5.3.3] **22.** $x^4 + 3x^3 - 23x^2 - 29x + 6$ [5.3.2] **23.** $6x^3 - 29x^2 + 14x + 24$ [5.3.2] **24.** $25a^2 - 4b^2$ [5.3.3] **25.** $16x^2 - 24xy + 9y^2$ [5.3.4] **26.** The distance from Earth to the Great Galaxy of Andromeda is 1.137048×10^{23} mi. [5.4.3] **27.** The sun generates 1.09×10^{21} horsepower. [5.4.3] **28.** The area is $(10x^2 - 29x - 21)$ cm². [5.3.5] **29.** The volume is $(27x^3 - 27x^2 + 9x - 1)$ ft³. [5.3.5] **30.** The area is $(5x^2 + 8x - 8)$ in². [5.3.5]

CHAPTER TEST

1. $2x^3 - 4x^2 + 6x - 14$ [5.1.2] **2.** $64a^7b^7$ [5.2.1] **3.** $\frac{2b^7}{a^{10}}$ [5.4.1] **4.** 5.01×10^{-7} [5.4.2] **5.** 6.048×10^5 [5.4.2] **6.** $\frac{x^{12}}{16y^4}$ [5.4.1] **7.** $35x^2 - 55x$ [5.3.1] **8.** $6a^2 - 13ab - 28b^2$ [5.3.3] **9.** $6t^5 - 8t^4 - 15t^3 + 22t^2 - 5$ [5.3.2] **10.** $9z^2 - 30z + 25$ [5.3.4] **11.** $2x^2 + 3x + 5$ [5.5.2] **12.** $x^2 - 2x - 1 + \frac{2}{x - 3}$ [5.5.2/5.5.3] **13.** -3 [5.1.1] **14.** -8 [5.5.4] **15.** $3x^3 + 6x^2 - 8x + 3$ [5.1.2] **16.** $4x^3 - 6x^2$ [5.3.1] **17.** $-4x^6$ [5.4.1] **18.** $-5a^3 + 3a^2 - 4a + 3$ [5.1.2] **19.** $4x^4 - 2x^2 + 5$ [5.5.1] **20.** $-6x^3y^6$ [5.2.1] **21.** $\frac{9}{x^3}$ [5.4.1] **22.** $10x^2 - 43xy + 28y^2$ [5.3.3] **23.** $\frac{a^4}{b^6}$ [5.4.1] **24.** The area of the rectangle is $(10x^2 - 3x - 1)$ ft². [5.3.5] **25.** It takes 12 h for the space vehicle to reach the moon. [5.4.3]

CUMULATIVE REVIEW EXERCISES

1. 4 [1.4.2] **2.** $-\frac{5}{4}$ [1.5.1] **3.** Inverse Property of Addition [1.5.2] **4.** $-18x + 8$ [1.5.6] **5.** $-\frac{1}{6}$ [2.1.2] **6.** $-\frac{11}{4}$ [2.2.2] **7.** $x^2 + 3x + 9 + \frac{24}{x - 3}$ [5.5.2/5.5.3] **8.** -1 and $\frac{7}{3}$ [2.8.1] **9.** 18 [3.2.1] **10.** -2 [3.2.1] **11.** $\{-4, -1, 8\}$ [3.2.1] **12.** $-\frac{1}{6}$ [3.5.2] **13.** $y = -\frac{3}{2}x + \frac{1}{2}$ [3.5.1] **14.** $y = \frac{2}{3}x + \frac{16}{3}$ [3.6.1] **15.** $\left(-\frac{7}{5}, -\frac{8}{5}\right)$ [4.3.3] **16.** $\left(-\frac{9}{7}, \frac{2}{7}, \frac{11}{7}\right)$ [4.2.2] **17.**

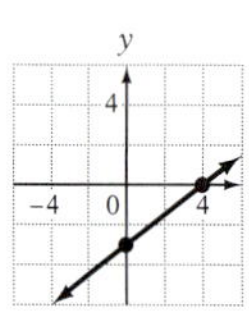

[3.3.2] **18.**

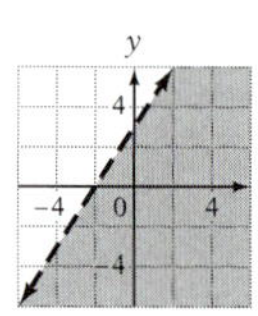

[3.7.1] **19.**

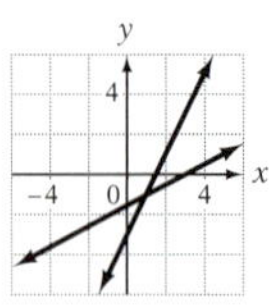

(1, −1) [4.1.1] **20.**

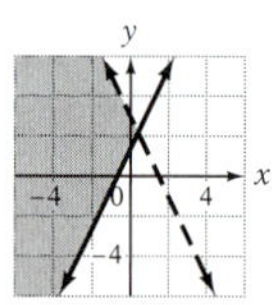

[4.5.1] **21.** $\frac{b^5}{a^8}$ [5.4.1] **22.** $\frac{y^2}{25x^6}$ [5.4.1] **23.** $\frac{21}{8}$ [5.4.1] **24.** $4x^3 - 7x + 3$ [5.3.2] **25.** $4x + 8 + \frac{21}{2x - 3}$ [5.5.2] **26.** a^3b^7 [5.2.1] **27.** $a^2 + ab - 12b^2$ [5.3.3] **28.** 0.000000035 [5.4.2] **29.** The integers are 9 and 15. [2.3.2] **30.** 40 oz of pure gold must be mixed with the alloy. [2.3.3] **31.** The slower cyclist travels at 5 mph, the faster cyclist at 7.5 mph. [2.3.4] **32.** The additional investment is \$4500. [2.6.1] **33.** $m = 50$; a slope of 50 means the average speed was 50 mph. [3.4.1]

Answers to Chapter 6 Selected Exercises

PREP TEST

1. $2 \cdot 3 \cdot 5$ [1.1.4] **2.** $-12y + 15$ [1.5.5] **3.** $-a + b$ [1.5.5] **4.** $-3a + 3b$ [1.5.6] **5.** 0 [2.1.3] **6.** $-\frac{1}{2}$ [2.2.1] **7.** $x^2 - 2x - 24$ [5.3.3] **8.** $6x^2 - 11x - 10$ [5.3.3] **9.** x^3 [5.4.1] **10.** $3x^3y$ [5.4.1]

CONCEPT REVIEW 6.1

1. Always true **3.** Always true **5.** Never true **7.** Never true

SECTION 6.1

3. x^3 **5.** xy^4 **7.** xy^4z^2 **9.** $7a^3$ **11.** 1 **13.** $3a^2b^2$ **15.** ab **17.** $2x$ **19.** $3x$ **21.** $5(a + 1)$ **23.** $8(2 - a^2)$ **25.** $4(2x + 3)$ **27.** $6(5a - 1)$ **29.** $x(7x - 3)$ **31.** $a^2(3 + 5a^3)$ **33.** $y(14y + 11)$ **35.** $2x(x^3 - 2)$ **37.** $2x^2(5x^2 - 6)$ **39.** $xy(x - y^2)$ **41.** $2a^5b + 3xy^3$ **43.** $6b^2(a^2b - 2)$ **45.** $2abc(3a + 2b)$ **47.** $9(2x^2y^2 - a^2b^2)$ **49.** $6x^3y^3(1 - 2x^3y^3)$ **51.** $x(x^2 - 3x - 1)$ **53.** $2(x^2 + 4x - 6)$ **55.** $b(b^2 - 5b - 7)$ **57.** $4(2y^2 - 3y + 8)$ **59.** $5y(y^2 - 4y + 2)$ **61.** $3y^2(y^2 - 3y - 2)$ **63.** $3y(y^2 - 3y + 8)$ **65.** $a^2(6a^3 - 3a - 2)$ **67.** $x^2(8y^2 - 4y + 1)$ **69.** $x^3y^3(4x^2y^2 - 8xy + 1)$ **71.** $(a + b)(x + 2)$ **73.** $(b + 2)(x - y)$ **75.** $(y - 4)(a - b)$ **77.** $a(x - 2) + b(x - 2)$ **79.** $b(a - 7) - 3(a - 7)$ **81.** $x(a - 2b) + y(a - 2b)$ **83.** $(x - 2)(a - 5)$ **85.** $(y - 3)(b - 3)$ **87.** $(x - y)(a + 2)$ **89.** $(c + 5)(z - 8)$ **91.** $(3x - 4)(w + 1)$ **93.** $(x + 4)(x^2 + 3)$ **95.** $(y + 2)(2y^2 + 3)$ **97.** $(a + 3)(b - 2)$ **99.** $(a - 2)(x^2 - 3)$ **101.** $(a - b)(3x - 2y)$ **103.** $(x - 3)(x + 4a)$ **105.** $(x - 5)(y - 2)$ **107.** $(7x + 2y)(3x - 7)$ **109.** $(2r + a)(a - 1)$ **111.** $(4x + 3y)(x - 4)$ **113.** $(2y - 3)(5xy + 3)$ **115. a.** $(x + 3)(2x + 5)$ **b.** $(2x + 5)(x + 3)$ **117. a.** $(a - b)(2a - 3b)$ **b.** $(2a - 3b)(a - b)$ **119.** 28 **121.** P doubles

CONCEPT REVIEW 6.2

1. Never true **3.** Always true **5.** Never true

SECTION 6.2

1.

Factors of 18	Sum
1, 18	19
2, 9	11
3, 6	9

3.

Factors of −21	Sum
1, −21	−20
−1, 21	20
3, −7	−4
−3, 7	4

5.

Factors of −28	Sum
1, −28	−27
−1, 28	27
2, −14	−12
−2, 14	12
4, −7	−3
−4, 7	3

7. $(x + 1)(x + 2)$ **9.** $(x + 1)(x - 2)$ **11.** $(a + 4)(a - 3)$ **13.** $(a - 1)(a - 2)$ **15.** $(a + 2)(a - 1)$ **17.** $(b - 3)(b - 3)$ **19.** $(b + 8)(b - 1)$ **21.** $(y + 11)(y - 5)$ **23.** $(y - 2)(y - 3)$ **25.** $(z - 5)(z - 9)$ **27.** $(z + 8)(z - 20)$ **29.** $(p + 3)(p + 9)$ **31.** $(x + 10)(x + 10)$ **33.** $(b + 4)(b + 5)$ **35.** $(x + 3)(x - 14)$ **37.** $(b + 4)(b - 5)$ **39.** $(y + 3)(y - 17)$ **41.** $(p + 3)(p - 7)$ **43.** Nonfactorable over the integers **45.** $(x - 5)(x - 15)$ **47.** $(x - 7)(x - 8)$ **49.** $(x + 8)(x - 7)$ **51.** $(a + 3)(a - 24)$ **53.** $(a - 3)(a - 12)$ **55.** $(z + 8)(z - 17)$

57. $(c + 9)(c - 10)$ **59.** $2(x + 1)(x + 2)$ **61.** $3(a + 3)(a - 2)$ **63.** $a(b + 5)(b - 3)$
65. $x(y - 2)(y - 3)$ **67.** $z(z - 3)(z - 4)$ **69.** $3y(y - 2)(y - 3)$ **71.** $3(x + 4)(x - 3)$
73. $5(z + 4)(z - 7)$ **75.** $2a(a + 8)(a - 4)$ **77.** $(x - 2y)(x - 3y)$ **79.** $(a - 4b)(a - 5b)$
81. $(x + 4y)(x - 7y)$ **83.** Nonfactorable over the integers **85.** $z^2(z - 5)(z - 7)$
87. $b^2(b - 10)(b - 12)$ **89.** $2y^2(y + 3)(y - 16)$ **91.** $x^2(x + 8)(x - 1)$ **93.** $4y(x + 7)(x - 2)$
95. $8(y - 1)(y - 3)$ **97.** $c(c + 3)(c + 10)$ **99.** $3x(x - 3)(x - 9)$ **101.** $(x - 3y)(x - 5y)$
103. $(a - 6b)(a - 7b)$ **105.** $(y + z)(y + 7z)$ **107.** $3y(x + 21)(x - 1)$ **109.** $3x(x + 4)(x - 3)$
111. $4z(z + 11)(z - 3)$ **113.** $(c + 4)(c + 5)$ **115.** $a^2(b - 5)(b - 9)$ **117.** 36, 12, −12, −36
119. 22, 10, −10, −22 **121.** 3, 4 **123.** 5, 8, 9 **125.** 2

CONCEPT REVIEW 6.3

1. Never true **3.** Sometimes true **5.** Always true

SECTION 6.3

1. $(x + 1)(2x + 1)$ **3.** $(y + 3)(2y + 1)$ **5.** $(a - 1)(2a - 1)$ **7.** $(b - 5)(2b - 1)$
9. $(x + 1)(2x - 1)$ **11.** $(x - 3)(2x + 1)$ **13.** Nonfactorable over the integers **15.** $(2t - 1)(3t - 4)$
17. $(x + 4)(8x + 1)$ **19.** $(b - 4)(3b - 4)$ **21.** $(z - 14)(2z + 1)$ **23.** $(p + 8)(3p - 2)$
25. $(2x - 3)(3x - 4)$ **27.** $(b + 7)(5b - 2)$ **29.** $(2a - 3)(3a + 8)$ **31.** $(3t + 1)(6t - 5)$
33. $(3a + 7)(5a - 3)$ **35.** $(2y - 5)(4y - 3)$ **37.** $(2z + 3)(4z - 5)$ **39.** $(x + 5)(3x - 1)$
41. $(3x + 4)(4x + 3)$ **43.** Nonfactorable over the integers **45.** $(x + y)(3x - 2y)$ **47.** $(4 + z)(7 - z)$
49. $(1 - x)(8 + x)$ **51.** $3(x + 5)(3x - 4)$ **53.** $4(2x - 3)(3x - 2)$ **55.** $a^2(5a + 2)(7a - 1)$
57. $5(b - 7)(3b - 2)$ **59.** $2x(x + 1)(5x + 1)$ **61.** $2y(y - 4)(5y - 2)$ **63.** $yz(z + 2)(4z - 3)$
65. $b^2(4b + 5)(5b + 4)$ **67.** $xy(3x + 2)(3x + 2)$ **69.** $(t + 2)(2t - 5)$ **71.** $(p - 5)(3p - 1)$
73. $(3y - 1)(4y - 1)$ **75.** Nonfactorable over the integers **77.** $(3y + 1)(4y + 5)$ **79.** $(a + 7)(7a - 2)$
81. $(z + 2)(4z + 3)$ **83.** $(2p + 5)(11p - 2)$ **85.** $(y + 1)(8y + 9)$ **87.** $(2b - 3)(3b - 2)$
89. $(3b + 5)(11b - 7)$ **91.** $(3y - 4)(6y - 5)$ **93.** Nonfactorable over the integers
95. $(2z - 5)(5z - 2)$ **97.** $(6z + 5)(6z + 7)$ **99.** $(2y - 3)(7y - 4)$ **101.** $(x + 6)(6x - 1)$
103. $3(x + 3)(4x - 1)$ **105.** $10(y + 1)(3y - 2)$ **107.** $x(x + 1)(2x - 5)$ **109.** $(a - 3b)(2a - 3b)$
111. $(y + z)(2y + 5z)$ **113.** $(1 + x)(18 - x)$ **115.** $4(9y + 1)(10y - 1)$ **117.** $8(t + 4)(2t - 3)$
119. $p(2p + 1)(3p + 1)$ **121.** $3(2z - 5)(5z - 2)$ **123.** $3a(2a + 3)(7a - 3)$ **125.** $y(3x - 5y)(3x - 5y)$
127. $xy(3x - 4y)(3x - 4y)$ **129.** $2y(y - 3)(4y - 1)$ **131.** $ab(a + 4)(a - 6)$ **133.** $5t(3 + 2t)(4 - t)$
135. $(y + 3)(2y + 1)$ **137.** $(2x - 1)(5x + 7)$ **139.** 7, −7, 5, −5 **141.** 7, −7, 5, −5
143. 11, −11, 7, −7 **145.** $(x + 2)(x - 3)(x - 1)$

CONCEPT REVIEW 6.4

1. Never true **3.** Never true **5.** Never true

SECTION 6.4

1. 4; $25x^6$; $100x^4y^4$ **3.** $4z^4$ **5.** $9a^2b^3$ **9.** $(x + 4)(x - 4)$ **11.** $(2x + 1)(2x - 1)$
13. $(b - 1)^2$ **15.** $(4x - 5)^2$ **17.** $(xy + 10)(xy - 10)$ **19.** Nonfactorable **21.** $(x + 3y)^2$
23. $(2x + y)(2x - y)$ **25.** $(a^n + 1)(a^n - 1)$ **27.** $(a + 2)^2$ **29.** $(x - 6)^2$ **31.** $(4x + 11)(4x - 11)$
33. $(1 + 3a)(1 - 3a)$ **35.** Nonfactorable **37.** Nonfactorable **39.** $(5 + ab)(5 - ab)$
41. $(5a - 4b)^2$ **43.** $(x^n + 3)^2$ **45.** 8; x^9; $27c^{15}d^{18}$ **47.** $2x^3$ **49.** $4a^2b^6$
51. $(x - 3)(x^2 + 3x + 9)$ **53.** $(2x - 1)(4x^2 + 2x + 1)$ **55.** $(x - y)(x^2 + xy + y^2)$
57. $(m + n)(m^2 - mn + n^2)$ **59.** $(4x + 1)(16x^2 - 4x + 1)$ **61.** $(3x - 2y)(9x^2 + 6xy + 4y^2)$
63. $(xy + 4)(x^2y^2 - 4xy + 16)$ **65.** Nonfactorable **67.** Nonfactorable **69.** $(a - 2b)(a^2 - ab + b^2)$

71. $(x^{2n} + y^n)(x^{4n} - x^{2n}y^n + y^{2n})$ **75.** $(xy - 3)(xy - 5)$ **77.** $(xy - 5)(xy - 12)$ **79.** $(x^2 - 6)(x^2 - 3)$ **81.** $(b^2 - 18)(b^2 + 5)$ **83.** $(x^2y^2 - 6)(x^2y^2 - 2)$ **85.** $(x^n + 2)(x^n + 1)$ **87.** $(3xy - 5)(xy - 3)$ **89.** $(2ab - 3)(3ab - 7)$ **91.** $(2x^2 - 15)(x^2 + 1)$ **93.** $(2x^n - 1)(x^n - 3)$ **95.** $(2a^n + 5)(3a^n + 2)$ **97.** $3(2x - 3)^2$ **99.** $a(3a - 1)(9a^2 + 3a + 1)$ **101.** $5(2x + 1)(2x - 1)$ **103.** $y^3(y + 11)(y - 5)$ **105.** $(4x^2 + 9)(2x + 3)(2x - 3)$ **107.** $2a(2 - a)(4 + 2a + a^2)$ **109.** $(x + 2)(x + 1)(x - 1)$ **111.** $(x + 2)(2x^2 - 3)$ **113.** $(x + 1)(x + 4)(x - 4)$ **115.** $b^3(ab - 1)(a^2b^2 + ab + 1)$ **117.** $x^2(x - 7)(x + 5)$ **119.** Nonfactorable **121.** $2x^3(3x + 1)(x + 12)$ **123.** $(4a^2 + b^2)(2a + b)(2a - b)$ **125.** Nonfactorable **127.** $3b^2(b - 2)(b^2 + 2b + 4)$ **129.** $x^2y^2(x - 3y)(x - 2y)$ **131.** $2xy(4x + 7y)(2x - 3y)$ **133.** $(x - 2)(x + 1)(x - 1)$ **135.** $4(b - 1)(2x - 1)$ **137.** $(y + 1)(y - 1)(2x + 3)(2x - 3)$ **139.** $(x + 2)(x - 2)^2(x^2 + 2x + 4)$ **141.** $a^2(a^n - 3)^2$ **143.** $x^n(2x - 1)(x - 3)$ **145.** 20, −20 **147.** 8, −8 **149.** $(x - 1)(x^2 + x + 1)(a - b)$ **151.** $(y^{2n} + 1)^2(y^n - 1)^2(y^n + 1)^2$ **153.** The sum of the numbers is 16. **155.** The palindromic perfect squares less than 500 are 1, 4, 9 ,121, and 484. **157.** $(x^2 - 4x + 8)(x^2 + 4x + 8)$

CONCEPT CHECK 6.5

1. Always true **3.** Sometimes true **5.** Sometimes true

SECTION 6.5

3. −4 and −6 **5.** 0 and 7 **7.** 0 and $-\frac{5}{2}$ **9.** $-\frac{3}{2}$ and 7 **11.** 7 and −7 **13.** $\frac{4}{3}$ and $-\frac{4}{3}$ **15.** −5 and 1 **17.** $-\frac{3}{2}$ and 4 **19.** 0 and 9 **21.** 0 and 4 **23.** 7 and −4 **25.** $\frac{2}{3}$ and −5 **27.** $\frac{2}{5}$ and 3 **29.** $\frac{3}{2}$ and $-\frac{1}{4}$ **31.** 7 and −5 **33.** 9 and 3 **35.** $-\frac{4}{3}$ and 2 **37.** $\frac{4}{3}$ and −5 **39.** 5 and −3 **41.** −11 and −4 **43.** 4 and 2 **45.** 2 and 1 **47.** 2 and 3 **49.** −4, −1, and 1 **51.** 1 and 2 **53.** $-\frac{1}{2}$ and 1 **55.** $\frac{1}{2}$ **57.** −9, −1, and 1 **59.** The number is −10 or 9. **61.** The integers are 8 and 9. **63.** The number is 0, 3, or 4. **65.** The width is 8 in. The length is 21 in. **67.** The value of x is 2 m or 3 m. **69.** The velocity of the rocket is 100 m/s. **71.** The height is 4 cm. The base is 12 cm. **73.** The object will hit the ground in 5 s. **75.** The length of the larger rectangle is 11 cm. The width is 8 cm. **77.** 175 or 15 **79.** −16 or 4 **81.** The length is 18 in. The width is 8 in. **83.** For example, $x^3 - 4x^2 + x + 6 = 0$

CHAPTER REVIEW EXERCISES

1. $7y^3(2y^6 - 7y^3 + 1)$ [6.1.1] **2.** $(a - 4)(3a + b)$ [6.1.2] **3.** $(c + 2)(c + 6)$ [6.2.1] **4.** $a(a - 2)(a - 3)$ [6.2.2] **5.** $(2x - 7)(3x - 4)$ [6.3.1/6.3.2] **6.** $(y + 6)(3y - 2)$ [6.3.1/6.3.2] **7.** $(3a + 2)(6a - 5)$ [6.3.1/6.3.2] **8.** $(ab + 1)(ab - 1)$ [6.4.1] **9.** $4(y - 2)^2$ [6.4.2] **10.** $0, -\frac{1}{5}$ [6.5.1] **11.** $3ab(4a + b)$ [6.1.1] **12.** $(b - 3)(b - 10)$ [6.2.1] **13.** $(2x + 5)(5x + 2y)$ [6.1.2] **14.** $3(a + 2)(a - 7)$ [6.2.2] **15.** $n^2(n + 1)(n - 3)$ [6.2.2] **16.** Nonfactorable over the integers [6.3.1/6.3.2] **17.** $(2x - 1)(3x - 2)$ [6.3.1/6.3.2] **18.** Nonfactorable over the integers [6.4.1] **19.** $2, \frac{3}{2}$ [6.5.1] **20.** $7(x + 1)(x - 1)$ [6.4.2] **21.** $x^3(3x^2 - 9x - 4)$ [6.1.1] **22.** $(x - 3)(4x + 5)$ [6.1.2] **23.** $(a + 7)(a - 2)$ [6.2.1] **24.** $(y + 9)(y - 4)$ [6.2.1] **25.** $5(x - 12)(x + 2)$ [6.2.2] **26.** −3, 7 [6.5.1] **27.** $x^2(5x^{n+3} + x^{n+1} + 4)$ [6.1.1] **28.** $(y - 3)(x - 4)$ [6.1.2] **29.** $(a + 2b)(2x - 3y)$ [6.1.2]

30. $(x + 5)(x + 7)$ [6.2.1] **31.** $(3 + x)(4 - x)$ [6.3.1] **32.** $(x - 7)(x - 9)$ [6.2.1] **33.** $(3x - 2)(2x - 9)$ [6.3.1/6.3.2] **34.** $(8x - 1)(3x + 8)$ [6.3.1/6.3.2] **35.** $(xy + 3)(xy - 3)$ [6.4.1] **36.** $(2x + 3y)^2$ [6.4.1] **37.** $(x^n - 6)^2$ [6.4.1] **38.** $(6 + a^n)(6 - a^n)$ [6.4.1] **39.** $(4a - 3b)(16a^2 + 12ab + 9b^2)$ [6.4.2] **40.** $(2 - y^n)(4 + 2y^n + y^{2n})$ [6.4.2] **41.** $(3x^2 + 2)(5x^2 - 3)$ [6.4.3] **42.** $(6x^4 - 5)(6x^4 - 1)$ [6.4.3] **43.** $(3x^2y^2 + 2)(7x^2y^2 + 3)$ [6.4.3] **44.** $3a^2(a^2 - 6)(a^2 + 1)$ [6.4.4] **45.** $(x^n + 2)^2(x^n - 2)^2$ [6.4.4] **46.** $3ab(a - b)(a^2 + ab + b^2)$ [6.4.4] **47.** -2, 0, and 3 [6.5.1] **48.** $\frac{5}{2}$ and 4 [6.5.1] **49.** -4, 0, and 4 [6.5.1] **50.** -1, -6, and 6 [6.5.1] **51.** $(2x - 11)(8x - 3)$ [6.3.1/6.3.2] **52.** $(3x - 5)^2$ [6.4.1] **53.** $(2y + 3)(6y - 1)$ [6.3.1/6.3.2] **54.** $3(x + 6)^2$ [6.4.4] **55.** The length is 100 yd. The width is 50 yd. [6.5.2] **56.** The length is 100 yd. The width is 60 yd. [6.5.2] **57.** The two integers are 4 and 5. [6.5.2] **58.** The distance is 20 ft. [6.5.2] **59.** The width of the larger rectangle is 15 ft. [6.5.2] **60.** The length of a side of the original square was 20 ft. [6.5.2]

CHAPTER TEST

1. $3y^2(2x^2 + 3x + 4)$ [6.1.1] **2.** $2x(3x^2 - 4x + 5)$ [6.1.1] **3.** $(p + 2)(p + 3)$ [6.2.1] **4.** $(x - 2)(a - b)$ [6.1.2] **5.** $\frac{3}{2}, -7$ [6.5.1] **6.** $(a - 3)(a - 16)$ [6.2.1] **7.** $x(x + 5)(x - 3)$ [6.2.2] **8.** $4(x + 4)(2x - 3)$ [6.3.1/6.3.2] **9.** $(b + 6)(a - 3)$ [6.1.2] **10.** $\frac{1}{2}, -\frac{1}{2}$ [6.5.1] **11.** $(2x + 1)(3x + 8)$ [6.3.1/6.3.2] **12.** $(2a^2 - 5)(3a^2 + 1)$ [6.4.3] **13.** $2(b + 4)(b - 4)$ [6.4.4] **14.** $(2a - 3b)^2$ [6.4.1] **15.** $(p + 1)(x - 1)$ [6.1.2] **16.** $(3x - 2)(9x^2 + 6x + 4)$ [6.4.2] **17.** Nonfactorable over the integers [6.3.1/6.3.2] **18.** $(2x + 7y)(2x - 7y)$ [6.4.1] **19.** 3, 5 [6.5.1] **20.** $(p + 6)^2$ [6.4.1] **21.** $2(3x - 4y)^2$ [6.4.4] **22.** $2y^2(y + 1)(y - 8)$ [6.2.2] **23.** The length is 15 cm. The width is 6 cm. [6.5.2] **24.** The length of the base is 12 in. [6.5.2] **25.** The two integers are -13 and -12. [6.5.2]

CUMULATIVE REVIEW EXERCISES

1. -8 [1.2.2] **2.** -8.1 [1.3.3] **3.** 4 [1.4.2] **4.** -7 [1.5.1] **5.** The Associative Property of Addition [1.5.2] **6.** $18x^2$ [1.5.4] **7.** $-6x + 24$ [1.5.6] **8.** $\frac{2}{3}$ [2.1.3] **9.** 5 [2.2.3] **10.** $\frac{7}{4}$ [2.2.2] **11.** 3 [2.2.3] **12.** 35 [2.1.4] **13.** $x \le -3$ [2.7.2] **14.** $x < \frac{1}{5}$ [2.7.2]

15. 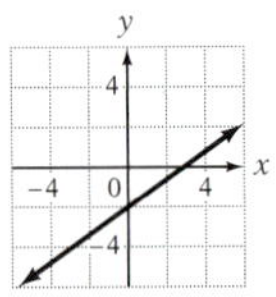

[3.3.1] **16.** 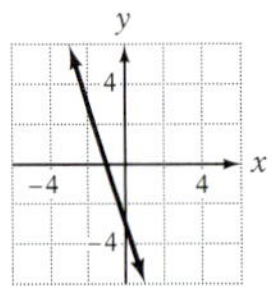

[3.3.1]

17. Domain: $\{-5, -3, -1, 1, 3\}$; range: $\{-4, -2, 0, 2, 4\}$; yes [3.2.1] **18.** 61 [3.2.1]

19. 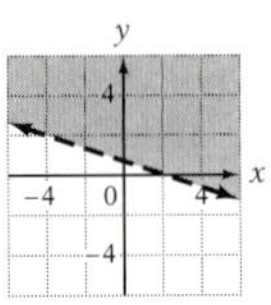

[3.7.1] **20.** $(2, -5)$ [4.1.2] **21.** $(1, 2)$ [4.2.1] **22.** $3y^3 - 3y^2 - 8y - 5$ [5.1.2]

23. $-27a^{12}b^6$ [5.2.1] **24.** $x^3 - 3x^2 - 6x + 8$ [5.3.2] **25.** $4x + 8 + \frac{21}{2x - 3}$ [5.5.2]

26. $\frac{y^6}{x^{12}}$ [5.4.1] **27.** $(a - b)(3 - x)$ [6.1.2] **28.** $(x + 5y)(x - 2y)$ [6.2.2]
29. $2a^2(3a + 2)(a + 3)$ [6.3.1/6.3.2] **30.** $(5a + 6b)(5a - 6b)$ [6.4.1] **31.** $3(2x - 3y)^2$ [6.4.4]
32. $\frac{4}{3}, -5$ [6.5.1] **33.** $\{-11, -7, -3, 1, 5\}$ [3.2.1] **34.** The average daily high temperature was −3°C. [1.2.5] **35.** The length is 15 cm. The width is 6 cm. [2.4.1] **36.** The pieces measure 4 ft and 6 ft. [2.3.1] **37.** The maximum distance you can drive a Company A car is 359 mi. [2.7.4]
38. $6500 must be invested at an annual simple interest rate of 11%. [2.6.1]
39. The discount rate is 40%. [2.5.2] **40.** The three integers are 10, 12, and 14. [2.3.2]

Chapter 7

PREP TEST

1. 50 [1.3.2] **2.** $-\frac{1}{6}$ [1.3.3] **3.** $-\frac{3}{2}$ [1.3.3] **4.** $\frac{1}{24}$ [1.3.2] **5.** $\frac{5}{24}$ [1.3.2]
6. $-\frac{2}{9}$ [1.4.2] **7.** $\frac{1}{3}$ [1.5.1] **8.** −2 [2.2.3] **9.** $\frac{10}{7}$ [2.2.3] **10.** 110 mph, 130 mph [2.3.4]

CONCEPT REVIEW 7.1

1. Never true **3.** Always true **5.** Never true

SECTION 7.1

3. $f(4) = 2$ **5.** $f(-2) = -2$ **7.** $f(-2) = \frac{1}{9}$ **9.** $f(3) = \frac{1}{35}$ **11.** $f(-1) = \frac{3}{4}$ **13.** $\{x \mid x \neq 3\}$
15. $\{x \mid x \neq -4\}$ **17.** $\{x \mid x \neq -3\}$ **19.** $\{x \mid x \neq -2, 4\}$ **21.** $\left\{x \mid x \neq -\frac{5}{2}, 2\right\}$ **23.** $\{x \mid x \neq 0\}$
25. $\{x \mid x \in \text{real numbers}\}$ **27.** $\{x \mid x \neq -3, 2\}$ **29.** $\{x \mid x \neq -6, 4\}$ **31.** $\{x \mid x \in \text{real numbers}\}$
33. $\left\{x \mid x \neq \frac{2}{3}, \frac{3}{2}\right\}$ **35.** $\left\{x \mid x \neq 0, \frac{1}{2}, -5\right\}$ **39.** $1 - 2x$ **41.** $3x - 1$ **43.** $2x$ **45.** $-\frac{2}{x}$
47. $\frac{x}{2}$ **49.** The expression is in simplest form. **51.** $4x^2 - 2x + 3$ **53.** $a^2 + 2a - 3$
55. $\frac{x^n - 3}{4}$ **57.** $\frac{x - 3}{x - 5}$ **59.** $\frac{x + y}{x - y}$ **61.** $-\frac{x + 3}{3x - 4}$ **63.** $-\frac{x + 7}{x - 7}$ **65.** The expression is in simplest form. **67.** $\frac{a - b}{a^2 - ab + b^2}$ **69.** $\frac{x + y}{3x}$ **71.** $\frac{x + 2y}{3x + 4y}$ **73.** $-\frac{2x - 3}{2(2x + 3)}$ **75.** $\frac{x - 2}{a - b}$
77. $\frac{x^2 + 2}{(x + 1)(x - 1)}$ **79.** $\frac{xy + 7}{xy - 7}$ **81.** $\frac{a^n - 2}{a^n + 2}$ **83.** decrease **85. a.**

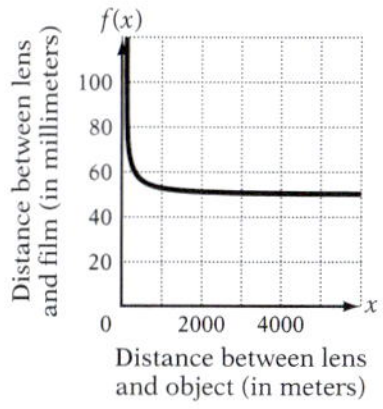

b. The ordered pair (2000, 51) means that when the distance between the object and the lens is 2000 m, the distance between the lens and the film is 51 mm. **c.** For $x = 50$, the expression $\frac{50x}{x - 50}$ is undefined. For $0 < x < 50$, $f(x)$ is negative, and distance cannot be negative. Therefore, the domain is $x > 50$. **d.** For $x > 1000$, $f(x)$ changes very little for large changes in x.

CONCEPT REVIEW 7.2

1. Always true 3. Never true

SECTION 7.2

3. $\frac{15b^4xy}{4}$ 5. 1 7. $\frac{2x-3}{x^2(x+1)}$ 9. $-\frac{x-1}{x-5}$ 11. $\frac{2(x+2)}{x-1}$ 13. $-\frac{x-3}{2x+3}$ 15. $\frac{x^n-6}{x^n-1}$ 17. $x+y$ 19. $\frac{a^2y}{10x}$ 21. $\frac{3}{2}$ 23. $\frac{5(x-y)}{xy(x+y)}$ 25. $\frac{(x-3)^2}{(x+3)(x-4)}$ 27. $-\frac{x+2}{x+5}$ 29. $\frac{x^{2n}}{8}$ 31. $\frac{(x-3)(3x-5)}{(2x+5)(x+4)}$ 33. -1 35. 1 37. $-\frac{2(x^2+x+1)}{(x+1)^2}$ 39. $\frac{9y^3}{12x^2y^4}, \frac{17x}{12x^2y^4}$ 41. $\frac{6x^2+9x}{(2x-3)(2x+3)}, \frac{10x^2-15x}{(2x-3)(2x+3)}$ 43. $\frac{2x}{(x+3)(x-3)}, \frac{x^2+4x+3}{(x+3)(x-3)}$ 45. $\frac{3x^2-3x}{(x+1)(x-1)^2}, \frac{5x^2+5x}{(x+1)(x-1)^2}$ 47. $-\frac{13}{2xy}$ 49. $\frac{1}{x-1}$ 51. $\frac{15-16xy-9x}{10x^2y}$ 53. $-\frac{5y+21}{30xy}$ 55. $-\frac{6x+19}{36x}$ 57. $\frac{10x+6y+7xy}{12x^2y^2}$ 59. $-\frac{x^2+x}{(x-3)(x-5)}$ 61. -1 63. $\frac{5x-8}{(x+5)(x-5)}$ 65. $-\frac{3x^2-4x+8}{x(x-4)}$ 67. $\frac{4x^2-14x+15}{2x(2x-3)}$ 69. $\frac{2x^2+x+3}{(x+1)^2(x-1)}$ 71. $\frac{x+1}{x+3}$ 73. $\frac{x+2}{x+3}$ 75. $-\frac{x^n-2}{(x^n+1)(x^n-1)}$ 77. $\frac{12x+13}{(2x+3)(2x-3)}$ 79. $\frac{2x+3}{2x-3}$ 81. $\frac{2x+5}{(x+3)(x-2)(x+1)}$ 83. $\frac{3x^2+4x+3}{(2x+3)(x+4)(x-3)}$ 85. $\frac{x-2}{x-3}$ 87. $-\frac{2x^2-9x-9}{(x+6)(x-3)}$ 89. $\frac{3x^2-14x+24}{(3x-2)(2x+5)}$ 91. $\frac{2}{x^2+1}$ 93. $-x-1$ 95. $\frac{y-2}{x^4}$ 97. $\frac{4y}{(y-1)^2}$ 99. $\frac{5(x+4)(x-4)}{2(x+2)(x-2)(x-3)}$ 101. $-\frac{(a+3)(a-2)(5a-2)}{(3a+1)(a+2)(2a-3)}$ 103. $\frac{6(x-1)}{x(2x+1)}$ 105. $\frac{8}{15} \neq \frac{1}{8}$ 107. $\frac{3}{y}+\frac{6}{x}$ 109. $\frac{4}{b^2}+\frac{3}{ab}$

111 a. $f(x) = \frac{2x^3+528}{x}$ b.

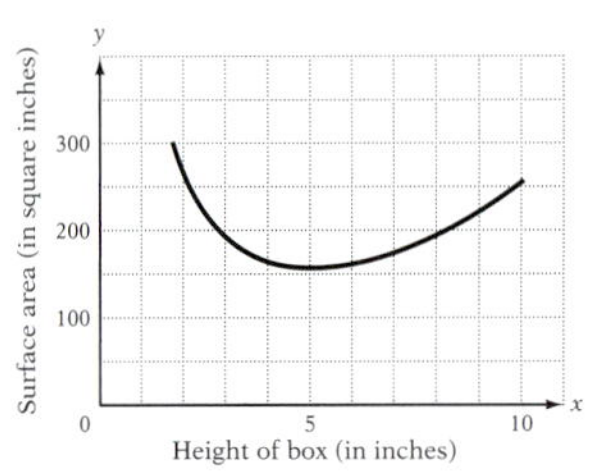

c. When the height of the box is 4 in., 164 in^2 of cardboard is needed.
d. The height of the box is 5.1 in.
e. The minimum amount of cardboard is 155.5 in^2.

CONCEPT REVIEW 7.3

1. Sometimes true 3. Always true

SECTION 7.3

3. $\frac{5}{23}$ 5. $\frac{2}{5}$ 7. $\frac{1-y}{y}$ 9. $\frac{5-a}{a}$ 11. $\frac{b}{2(2-b)}$ 13. $\frac{4}{5}$ 15. $-\frac{1}{2}$ 17. $\frac{2a^2-3a+3}{3a-2}$ 19. $3x+2$ 21. $\frac{x+2}{x-1}$ 23. $-\frac{(x-3)(x+5)}{4x^2-5x+4}$ 25. $\frac{1}{x+4}$ 27. $\frac{x-7}{x+5}$ 29. $-\frac{2(a+1)}{7a-4}$ 31. $\frac{x+y}{x-y}$ 33. $-\frac{2x}{x^2+1}$ 35. $\frac{6(x-2)}{2x-3}$ 37. $\frac{2y}{x}$ 39. $x-1$ 41. $\frac{c-3}{c-2}$ 43. $\frac{2a(2a-3)}{3a-4}$ 45. $-\frac{1}{x(x+h)}$ 47. $\frac{3n^2+12n+8}{n(n+2)(n+4)}$ 49. a. $P(x) = \frac{Cx(x+1)^{60}}{(x+1)^{60}-1}$

b.

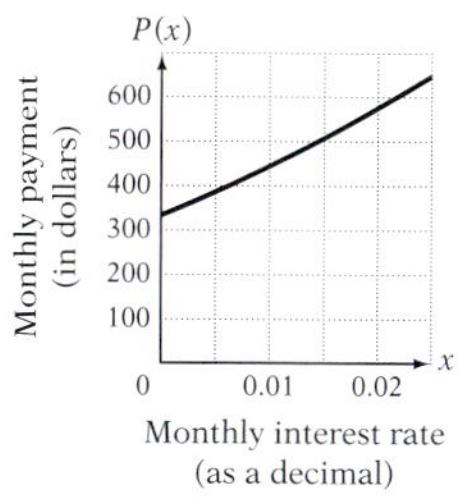

c. The interval of annual interest rates is 0% to 30%. **d.** The ordered pair (0.005, 386.66) means that when the monthly interest rate on a car loan is 0.5%, the monthly payment on the loan is \$386.66. **e.** The monthly payment with a loan amount of \$20,000 and an annual interest rate of 8% is \$406.

CONCEPT REVIEW 7.4

1. Never true **3.** Always true **5.** Always true

SECTION 7.4

3. -5 **5.** -1 **7.** $\frac{5}{2}$ **9.** 0 **11.** -3 and 3 **13.** 8 **15.** 5 **17.** -3 **19.** No solution **21.** 4 and 10 **23.** -1 **25.** -3 and 4 **27.** 2 **29.** $\frac{7}{2}$ **31.** No solution **35.** It would take 1.2 h to process the data with both computers working. **37.** With both panels working, it would take 18 min to raise the temperature 1 degree. **39.** With both members working together, it would take 3 h to wire the telephone lines. **41.** Working alone, the new machine would take 10 h to package the transistors. **43.** Working alone, the smaller printer would take 80 min to print the payroll. **45.** It would take the apprentice 20 h to shingle the roof working alone. **47.** It will take them 40 min to address 140 envelopes. **49.** In 30 min, the bathtub will start to overflow. **51.** With all three pipes open, it would take 30 h to fill the tank. **53.** The rate of the first skater is 15 mph; the rate of the second skater is 12 mph. **55.** The rate of the freight train is 45 mph. The rate of the passenger train is 59 mph. **57.** The cyclist was riding at the rate of 12 mph. **59.** The motorist walks at the rate of 4 mph. **61.** The business executive can walk at the rate of 6 ft/s. **63.** The rate of the single-engine plane is 180 mph. The rate of the jet is 720 mph. **65.** The rate of the Gulf Stream is 6 mph. **67.** The rate of the current is 3 mph. **69.** The original fraction is $\frac{17}{21}$. **71.** With all three printers operating, it would take $5\frac{1}{3}$ min to print the checks. **73.** The bus usually travels at 60 mph. **75.** The total time is $\frac{5}{8}$ h.

CONCEPT REVIEW 7.5

1. Never true **3.** Never true **5.** Never true

SECTION 7.5

3. 3 **5.** -2 **7.** -4 **9.** $-\frac{6}{5}$ **11.** There are 4000 ducks in the preserve. **13.** The caterer will need an additional 3 gal of punch. **15.** The dimensions of the room are 17 ft by 22 ft. **17.** To lose 1 pound, a person would need to walk 21.54 mi. **19.** An additional 0.5 oz of medication is required. **21.** An additional 34 oz of insecticide is required. **23.** The Postal Service delivers 4 billion pieces of mail per week. **25.** The length of the whale is 96 ft. **29.** The pressure is 13.5 lb/in^2. **31.** When the company sells 300 products, the profit is \$37,500. **33.** The length of the person's face is 10.2 in. **35.** The horizon is 24.03 mi from a point that is 800 ft high. **37.** The ball will roll 80 ft in 4 s. **39.** When the volume is 150 ft^3, the pressure is $66\frac{2}{3}$ lb/in^2. **41.** A gear that has 30 teeth will make 32 revolutions per minute. **43.** The repulsive force is 112.5 lb when the distance is 1.2 in. **45.** The resistance is 56.25 ohms.

47. The wind force is 180 lb. **49. a.** 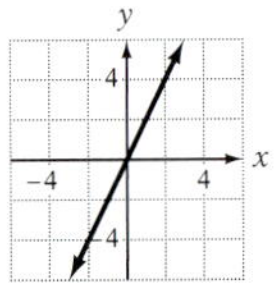 **b.** A linear function

51. a. 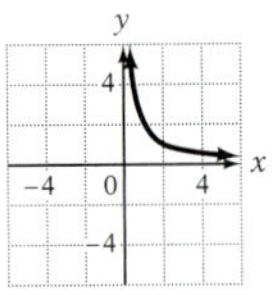 **b.** Yes **53.** x is halved. **55.** directly, inversely **57.** directly

59. a. $\frac{2ab}{a+b}$ **b.** 12

CONCEPT REVIEW 7.6

1. Always true **3.** Never true

SECTION 7.6

1. $W = \frac{P - 2L}{2}$ **3.** $C = \frac{S}{1 - r}$ **5.** $R = \frac{PV}{nT}$ **7.** $m_2 = \frac{Fr^2}{Gm_1}$ **9.** $R = \frac{E - Ir}{I}$ **11.** $b_2 = \frac{2A - hb_1}{h}$ **13.** $R_2 = \frac{RR_1}{R_1 - R}$ **15.** $d = \frac{a_n - a_1}{n - 1}$ **17.** $H = \frac{S - 2WL}{2W + 2L}$ **19.** $x = \frac{-by - c}{a}$ **21.** $x = \frac{d - b}{a - c}$ **23.** $x = \frac{ac}{b}$ **25.** $x = \frac{ab}{a + b}$ **27.** $5 + 2y$ **29.** $0, y$ **31.** $\frac{y}{2}, 5y$ **33.** $f = \frac{w_2 f_2 + w_1 f_1}{w_1 + w_2}$

CHAPTER REVIEW EXERCISES

1. 4 [7.1.1] **2.** $\frac{2}{21}$ [7.1.1] **3.** $\{x \mid x \neq 3\}$ [7.1.1] **4.** $\{x \mid x \neq -3, 2\}$ [7.1.1] **5.** $\{x \mid x \in \text{real numbers}\}$ [7.1.1] **6.** $3a^{2n} + 2a^n - 1$ [7.1.2] **7.** $-\frac{x + 4}{x(x + 2)}$ [7.1.2] **8.** $\frac{x^2 + 3x + 9}{x + 3}$ [7.1.2] **9.** $\frac{1}{a - 1}$ [7.2.1] **10.** x [7.2.1] **11.** $-\frac{(x + 4)(x + 3)(x + 2)}{6(x - 1)(x - 4)}$ [7.2.1] **12.** $\frac{x^n + 6}{x^n - 3}$ [7.2.1] **13.** $\frac{3x + 2}{x}$ [7.2.1] **14.** $-\frac{x - 3}{x + 3}$ [7.2.1] **15.** $\frac{21a + 40b}{24a^2b^4}$ [7.2.2] **16.** $\frac{4x - 1}{x + 2}$ [7.2.2] **17.** $\frac{3x + 26}{(3x - 2)(3x + 2)}$ [7.2.2] **18.** $\frac{9x^2 + x + 4}{(3x - 1)(x - 2)}$ [7.2.2] **19.** $-\frac{5x^2 - 17x - 9}{(x - 3)(x + 2)}$ [7.2.2] **20.** $\frac{x - 4}{x + 5}$ [7.3.1] **21.** $\frac{x - 1}{4}$ [7.3.1] **22.** $\frac{15}{13}$ [7.4.1] **23.** −3 and 5 [7.4.1] **24.** 6 [7.4.1] **25.** No solution [7.4.1] **26.** $R = \frac{V}{I}$ [7.6.1] **27.** $N = \frac{S}{1 - Q}$ [7.6.1] **28.** $r = \frac{S - a}{S}$ [7.6.1] **29.** The number of tanks of fuel is $6\frac{2}{3}$. [7.5.1] **30.** The number of miles is 48. [7.5.1] **31.** Working alone, the apprentice would take 104 min to install the ceiling fan. [7.4.2] **32.** With both pipes open, it would take 40 min to empty the tub. [7.4.2] **33.** It would take 4 h for the three painters to paint the dormitory together. [7.4.2] **34.** The rate of the current is 2 mph. [7.4.3] **35.** The rate of the bus is 45 mph. [7.4.3] **36.** The rate of the tractor is 30 mph. [7.4.3] **37.** The pressure is 40 lb. [7.5.2] **38.** The illumination is 300 lumens 2 ft from the light. [7.5.2] **39.** The resistance of the cable is 0.4 ohm. [7.5.2]

CHAPTER TEST

1. $\frac{v(v+2)}{2v-1}$ [7.1.2] **2.** $-\frac{2(a-2)}{3a+2}$ [7.1.2] **3.** $\frac{6(x-2)}{5}$ [7.2.1] **4.** 2 [7.1.1] **5.** $\frac{x+1}{3x-4}$ [7.2.1] **6.** $\frac{x^n-1}{x^n+2}$ [7.2.1] **7.** $\frac{4y^2+6x^2-5xy}{2x^2y^2}$ [7.2.2] **8.** $\frac{2(5x+2)}{(x-2)(x+2)}$ [7.2.2] **9.** $\{x \mid x \neq -3, 3\}$ [7.1.1] **10.** $-\frac{x^2+5x-2}{(x+1)(x+4)(x-1)}$ [7.2.2] **11.** $\frac{x-4}{x+3}$ [7.3.1] **12.** $\frac{x+4}{x+2}$ [7.3.1] **13.** 2 [7.4.1] **14.** No solution [7.4.1] **15.** $x = \frac{c}{a-b}$ [7.6.1] **16.** It would take 80 min to empty the full tank with both pipes open. [7.4.2] **17.** The office requires 14 rolls of wallpaper. [7.5.1] **18.** Working together, the landscapers would complete the task in 10 min. [7.4.2] **19.** The rate of the cyclist is 10 mph. [7.4.3] **20.** The stopping distance of the car is 61.2 ft. [7.5.2]

CUMULATIVE REVIEW EXERCISES

1. $\frac{36}{5}$ [1.4.2] **2.** $\frac{15}{2}$ [2.2.3] **3.** 7 and 1 [2.8.1] **4.** $\{x \mid x \neq 3\}$ [7.1.1] **5.** $\frac{3}{7}$ [7.1.1] **6.** 3.5×10^{-8} [5.4.2] **7.** No solution [7.4.1] **8.** $\frac{1}{9}$ and 4 [6.5.1] **9.** $\frac{8b^3}{a}$ [5.4.1] **10.** $\{x \mid x \leq 4\}$ [2.7.2] **11.** $-4a^4+6a^3-2a^2$ [5.3.1] **12.** $(2x^n-1)(x^n+2)$ [6.4.3] **13.** $(xy-3)(x^2y^2+3xy+9)$ [6.4.2] **14.** $x+3y$ [7.1.2] **15.** $y = \frac{3}{2}x+2$ [3.6.1] **16.** $\frac{x}{x^2-1}$ [5.4.1] **17.**

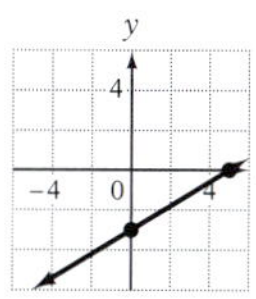

[3.3.2] **18.**

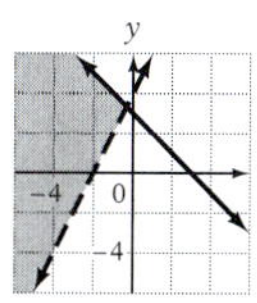

[4.5.1]

19. $4x^2+10x+10+\frac{21}{x-2}$ [5.5.2] **20.** $\frac{3xy}{x-y}$ [7.2.1] **21.** $\{x \mid x \in \text{real numbers}\}$ [7.1.1] **22.** $-\frac{x}{(3x+2)(x+1)}$ [7.2.2] **23.** −28 [4.3.2] **24.** $\frac{x-3}{x+3}$ [7.3.1] **25.** (1, 1, 1) [4.2.2 or 4.3.3] **26.** 1, 2 [6.5.1] **27.** 9 [7.5.1] **28.** 13 [7.4.1] **29.** $a^4+5a^3-3a^2-11a+20$ [5.3.2] **30.** $r = \frac{E-IR}{I}$ [7.6.1] **31.** No [3.6.1] **32.**

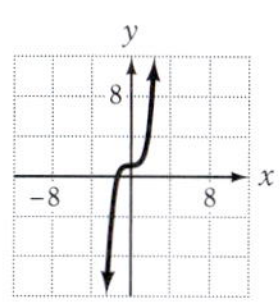

[5.1.1]

33. The smaller integer is 5 and the larger integer is 10. [2.3.1] **34.** The number of pounds of almonds is 50. [2.3.3] **35.** The number of people expected to vote is 75,000. [7.5.1] **36.** It would take the new computer 14 min to do the job working alone. [7.4.2] **37.** The rate of the wind is 60 mph. [7.4.3] **38.** The frequency of vibration of the open pipe organ is 80 vibrations/min. [7.5.2]

Chapter 8

PREP TEST

1. 16 [1.2.4] **2.** 32 [1.4.1] **3.** 9 [1.3.3] **4.** $\frac{1}{12}$ [1.3.2] **5.** $-5x - 1$ [1.5.6]
6. $\frac{xy^5}{4}$ [5.4.1] **7.** $9x^2 - 12x + 4$ [5.3.4] **8.** $-12x^2 + 14x + 10$ [5.3.3] **9.** $36x^2 - 1$ [5.3.4]
10. $-1, 15$ [6.5.1]

CONCEPT REVIEW 8.1

1. Sometimes true **3.** Always true **5.** Always true

SECTION 8.1

3. 2 **5.** 27 **7.** $\frac{1}{9}$ **9.** 4 **11.** Not a real number **13.** $\frac{343}{125}$ **15.** x **17.** $y^{\frac{1}{2}}$
19. $x^{\frac{1}{12}}$ **21.** $a^{\frac{7}{12}}$ **23.** $\frac{1}{a}$ **25.** $\frac{1}{y}$ **27.** $y^{\frac{3}{2}}$ **29.** $\frac{1}{x}$ **31.** $\frac{1}{x^4}$ **33.** a
35. $x^{\frac{3}{10}}$ **37.** a^3 **39.** $\frac{1}{x^{\frac{1}{2}}}$ **41.** $y^{\frac{2}{3}}$ **43.** x^4y **45.** $x^6y^3z^9$ **47.** $\frac{x}{y^2}$ **49.** $\frac{x^{\frac{3}{2}}}{y^{\frac{1}{4}}}$
51. x^2y^8 **53.** $\frac{1}{x^{\frac{11}{12}}}$ **55.** $\frac{1}{y^{\frac{5}{2}}}$ **57.** $\frac{1}{b^{\frac{7}{8}}}$ **59.** a^5b^{13} **61.** $\frac{m^2}{4n^{\frac{3}{2}}}$ **63.** $\frac{y^{\frac{17}{2}}}{x^3}$
65. $\frac{16b^2}{a^{\frac{1}{3}}}$ **67.** $\frac{y^2}{x^3}$ **69.** $a - a^2$ **71.** $y + 1$ **73.** $a - \frac{1}{a}$ **75.** $\frac{1}{a^{10}}$ **77.** $a^{\frac{n}{6}}$
79. $\frac{1}{b^{\frac{2m}{3}}}$ **81.** x^{10n^2} **83.** $x^{3n}y^{2n}$ **85.** $\sqrt{5}$ **87.** $\sqrt[3]{b^4}$ **89.** $\sqrt[3]{9x^2}$ **91.** $-3\sqrt[5]{a^2}$
93. $\sqrt[4]{x^6y^9}$ **95.** $\sqrt{a^9b^{21}}$ **97.** $\sqrt[3]{3x - 2}$ **99.** $14^{\frac{1}{2}}$ **101.** $x^{\frac{1}{3}}$ **103.** $x^{\frac{4}{3}}$ **105.** $b^{\frac{3}{5}}$
107. $(2x^2)^{\frac{1}{3}}$ **109.** $-(3x^5)^{\frac{1}{2}}$ **111.** $3xy^{\frac{2}{3}}$ **113.** $(a^2 + 2)^{\frac{1}{2}}$ **115.** y^7 **117.** $-a^3$
119. a^7b^3 **121.** $11y^6$ **123.** a^2b^4 **125.** $-a^3b^3$ **127.** $5b^5$ **129.** $-a^2b^3$
131. $5x^4y$ **133.** Not a real number **135.** $2a^7b^2$ **137.** $-3ab^5$ **139.** y^3 **141.** $3a^5$
143. $-a^4b$ **145.** ab^5 **147.** $2a^2b^5$ **149.** $-2x^3y^4$ **151.** $\frac{4x}{y^7}$ **153.** $\frac{3b}{a^3}$ **155.** $2x + 3$
157. $x + 1$ **159.** x **161.** b **163.** ab^2 **165.** $\frac{3}{5}$ **167.** $\frac{3}{4}$

CONCEPT REVIEW 8.2

1. Sometimes true **3.** Never true **5.** Always true

SECTION 8.2

3. $3\sqrt{2}$ **5.** $7\sqrt{2}$ **7.** $2\sqrt[3]{9}$ **9.** $2\sqrt[3]{2}$ **11.** $x^2yz^2\sqrt{yz}$ **13.** $2ab^4\sqrt{2a}$ **15.** $3xyz^2\sqrt{5yz}$
17. $-5y\sqrt[3]{x^2y}$ **19.** $-6xy^3\sqrt[3]{x^2}$ **21.** $ab^2\sqrt[3]{a^2b^2}$ **23.** $2\sqrt{2}$ **25.** $3\sqrt[3]{7}$ **27.** $-6\sqrt{x}$
29. $-2\sqrt{2}$ **31.** $\sqrt{2x}$ **33.** $3\sqrt{3a} - 2\sqrt{2a}$ **35.** $10x\sqrt{2x}$ **37.** $2x\sqrt{3xy}$ **39.** $xy\sqrt{2y}$
41. $5a^2b^2\sqrt{ab}$ **43.** $9\sqrt[3]{2}$ **45.** $-7a\sqrt[3]{3a}$ **47.** $-5xy^2\sqrt[3]{x^2y}$ **49.** $16ab\sqrt[4]{ab}$ **51.** $6\sqrt{3} - 6\sqrt{2}$
53. $6x^3y^2\sqrt{xy}$ **55.** $-9a^2\sqrt{3ab}$ **57.** 16 **59.** $2\sqrt[3]{4}$ **61.** $xy^3\sqrt{x}$ **63.** $8xy\sqrt{x}$
65. $2x^2y\sqrt[3]{2}$ **67.** $2ab\sqrt[4]{3a^2b}$ **69.** 6 **71.** $x - \sqrt{2x}$ **73.** $4x - 8\sqrt{x}$ **75.** $x - 6\sqrt{x} + 9$
77. $84 + 16\sqrt{5}$ **79.** $672x^2y^2$ **81.** $4a^3b^3\sqrt[3]{a}$ **83.** $-8\sqrt{5}$ **85.** $x - y^2$ **87.** $12x - y$

89. $x - 3\sqrt{x} - 28$ **91.** $\sqrt[3]{x^2} + \sqrt[3]{x} - 20$ **95.** $4\sqrt{x}$ **97.** $b^2\sqrt{3a}$ **99.** $\frac{\sqrt{5}}{5}$ **101.** $\frac{\sqrt{2x}}{2x}$

103. $\frac{\sqrt{5x}}{x}$ **105.** $\frac{\sqrt{5x}}{5}$ **107.** $\frac{3\sqrt[3]{4}}{2}$ **109.** $\frac{3\sqrt[3]{2x}}{2x}$ **111.** $\frac{\sqrt{2xy}}{2y}$ **113.** $\frac{2\sqrt{3ab}}{3b^2}$

115. $2\sqrt{5} - 4$ **117.** $\frac{3\sqrt{y} + 6}{y - 4}$ **119.** $-5 + 2\sqrt{6}$ **121.** $8 + 4\sqrt{3} - 2\sqrt{2} - \sqrt{6}$

123. $\frac{\sqrt{15} + \sqrt{10} - \sqrt{6} - 5}{3}$ **125.** $\frac{3\sqrt[4]{2x}}{2x}$ **127.** $\frac{2\sqrt[5]{2a^3}}{a}$ **129.** $\frac{\sqrt[5]{16x^2}}{2}$ **131.** $\frac{1 + 2\sqrt{ab} + ab}{1 - ab}$

133. $\frac{5x\sqrt{y} + 5y\sqrt{x}}{x - y}$ **135.** $2\sqrt{2}$ **137.** $29\sqrt{2} - 45$ **139.** $\frac{3\sqrt{y+1} - 3}{y}$ **141.** $\sqrt[6]{x + y}$

143. $\sqrt[4]{2y(x+3)^2}$ **145.** $\sqrt[6]{a^3(a+3)^2}$ **147.** $16^{\frac{1}{4}}, 2$ **149.** $32^{-\frac{1}{5}}, \frac{1}{2}$ **151. a.** $D(x) = 3.50\sqrt[10]{x^2}$ **b.** The model predicts a national debt of approximately \$4.83 trillion in 1995. **c.** The national debt will be approximately \$6.37 trillion in 2010.

CONCEPT REVIEW 8.3

1. Always true **3.** Never true

SECTION 8.3

3. $\{x \mid x \in \text{real numbers}\}$ **5.** $\{x \mid x \geq -1\}$ **7.** $\{x \mid x \geq 0\}$ **9.** $\{x \mid x \geq 0\}$ **11.** $\{x \mid x \geq 2\}$

13. $(-\infty, \infty)$ **15.** $(-\infty, 3]$ **17.** $[2, \infty)$ **19.** $\left(-\infty, \frac{2}{3}\right]$ **21.**

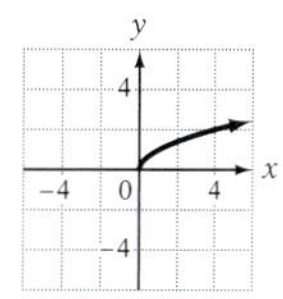

23.

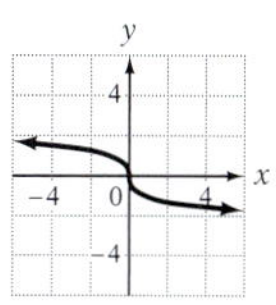

25.

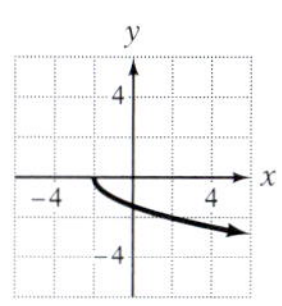

27.

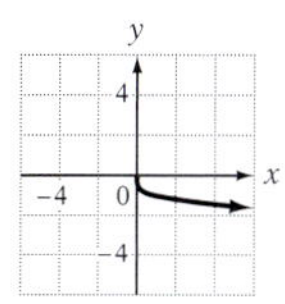

29.

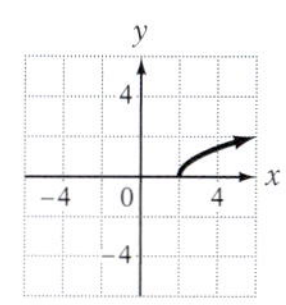

31.

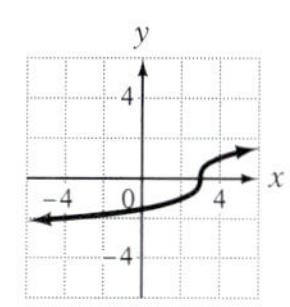

33.

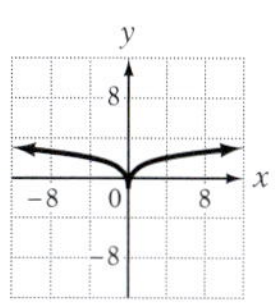

35.

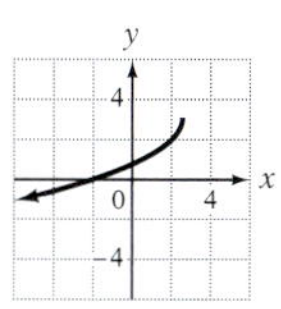

37.

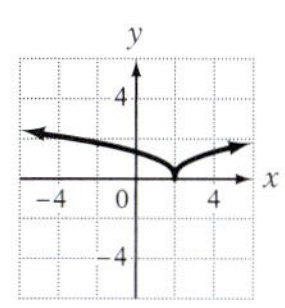

39.

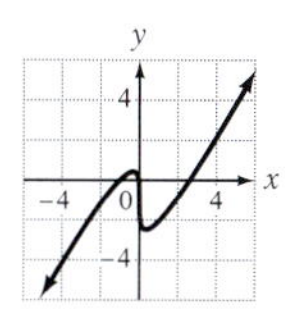

41.

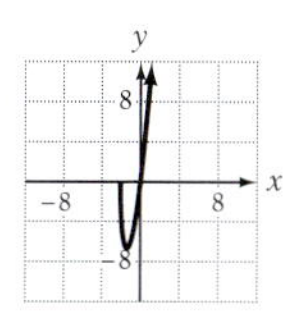

CONCEPT REVIEW 8.4

1. Sometimes true **3.** Sometimes true

SECTION 8.4

3. 25 **5.** 27 **7.** 48 **9.** -2 **11.** No solution **13.** 9 **15.** -23 **17.** -16

19. $\frac{9}{4}$ **21.** 14 **23.** 7 **25.** 7 **27.** -122 **29.** 35 **31.** 45 **33.** 23

35. -4 **37.** 10 **39.** 1 **41.** 3 **43.** 21 **45.** $\frac{25}{72}$ **47.** 5 **49.** $-2, 4$

51. 6 **53.** 0, 1 **55.** $-2, 1$ **57.** 2, 11 **59.** 3 **61.** The air pressure is 909.8 mb. The wind speed increases. **63.** The mean distance is 295,000 km. **65.** The elephant weighs 3700 lb. **67.** The distance is 10 ft. **69.** The distance between the joggers after 3 min is approximately 829.76 m. **71.** The length of the diagonal is 8.9 cm. **73.** The periscope must be 8.17 ft above the water.

75. The distance is 225 ft. **77.** The distance is 96 ft. **79.** The length of the pendulum is approximately 4.67 ft. **81.** 27 **83.** $d = \frac{v^2}{64}$ **85.** $r = \frac{\sqrt{V\pi h}}{\pi h}$ **87.** 2 **89.** The distance between the corners is 7.81 in. **91.** $\sqrt{6}$

CONCEPT REVIEW 8.5

1. Always true **3.** Always true

SECTION 8.5

3. $2i$ **5.** $7i\sqrt{2}$ **7.** $3i\sqrt{3}$ **9.** $4 + 2i$ **11.** $2\sqrt{3} - 3i\sqrt{2}$ **13.** $4\sqrt{10} - 7i\sqrt{3}$ **15.** $2ai$ **17.** $7x^6i$ **19.** $12ab^2i\sqrt{ab}$ **21.** $2\sqrt{a} + 2ai\sqrt{3}$ **23.** $3b^2\sqrt{2b} - 3bi\sqrt{3b}$ **25.** $5x^3y\sqrt{y} + 5xyi\sqrt{2xy}$ **27.** $2a^2bi\sqrt{a}$ **29.** $2a\sqrt{3a} + 3bi\sqrt{3b}$ **31.** $10 - 7i$ **33.** $11 - 7i$ **35.** $-6 + i$ **37.** $-4 - 8i\sqrt{3}$ **39.** $-\sqrt{10} - 11i\sqrt{2}$ **41.** $6 - 4i$ **43.** 0 **45.** $11 + 6i$ **47.** -24 **49.** -15 **51.** $-5\sqrt{2}$ **53.** $-15 - 12i$ **55.** $3\sqrt{2} + 6i$ **57.** $-10i$ **59.** $29 - 2i$ **61.** 1 **63.** 1 **65.** 89 **67.** 50 **69.** $7 - 24i$ **71.** $-2i$ **73.** $-\frac{4}{5}i$ **75.** $-\frac{5}{3} + \frac{16}{3}i$ **77.** $\frac{30}{29} - \frac{12}{29}i$ **79.** $\frac{20}{17} + \frac{5}{17}i$ **81.** $\frac{11}{13} + \frac{29}{13}i$ **83.** $-\frac{1}{5} + \frac{\sqrt{6}}{10}i$ **85.** $-1 + 4i$ **87.** -1 **89.** i **91.** -1 **93.** -1 **95. a.** Yes **b.** Yes **97. a.** 2, −2; integer, rational number, real number **b.** $i, -i$; imaginary **c.** $\frac{16}{15}$; rational number, real number **d.** $2\sqrt{2}, -2\sqrt{2}$; irrational number, real number **e.** 16; integer, rational number, real number **f.** −19,683; integer, rational number, real number

CHAPTER REVIEW EXERCISES

1. $\frac{1}{3}$ [8.1.1] **2.** $\frac{1}{x^5}$ [8.1.1] **3.** $\frac{1}{a^{10}}$ [8.1.1] **4.** $\frac{160y^{11}}{x}$ [8.1.1] **5.** $3\sqrt[4]{x^3}$ [8.1.2] **6.** $7x^{\frac{2}{3}}y$ [8.1.2] **7.** $3a^2b^3$ [8.1.3] **8.** $-7x^3y^8$ [8.1.3] **9.** $-2a^2b^4$ [8.1.3] **10.** $3ab^3\sqrt{2a}$ [8.2.1] **11.** $-2ab^2\sqrt[5]{2a^3b^2}$ [8.2.1] **12.** $xy^2z^2\sqrt[4]{x^2z^2}$ [8.2.1] **13.** $5\sqrt{6}$ [8.2.2] **14.** $4x^2\sqrt{3xy} - 4x^2\sqrt{5xy}$ [8.2.2] **15.** $2a^2b\sqrt{2b}$ [8.2.2] **16.** $6x^2\sqrt{3y}$ [8.2.2] **17.** 40 [8.2.3] **18.** $4xy^2\sqrt[3]{x^2}$ [8.2.3] **19.** $3x + 3\sqrt{3x}$ [8.2.3] **20.** $31 - 10\sqrt{6}$ [8.2.3] **21.** $-13 + 6\sqrt{3}$ [8.2.3] **22.** $5x\sqrt{x}$ [8.2.4] **23.** $\frac{8\sqrt{3y}}{3y}$ [8.2.4] **24.** $\frac{x\sqrt{x} - x\sqrt{2} + 2\sqrt{x} - 2\sqrt{2}}{x - 2}$ [8.2.4] **25.** $\frac{x + 2\sqrt{xy} + y}{x - y}$ [8.2.4] **26.** $6i$ [8.5.1] **27.** $5i\sqrt{2}$ [8.5.1] **28.** $7 - 4i$ [8.5.1] **29.** $10\sqrt{2} + 2i\sqrt{3}$ [8.5.1] **30.** $9 - i$ [8.5.2] **31.** $-12 + 10i$ [8.5.2] **32.** $14 + 2i$ [8.5.2] **33.** $-4\sqrt{2} + 8i\sqrt{2}$ [8.5.2] **34.** $10 - 9i$ [8.5.2] **35.** -16 [8.5.3] **36.** $7 + 3i$ [8.5.3] **37.** $-6\sqrt{2}$ [8.5.3] **38.** $39 - 2i$ [8.5.3] **39.** $6i$ [8.5.4] **40.** $\frac{2}{3} - \frac{5}{3}i$ [8.5.4] **41.** $\frac{14}{5} + \frac{7}{5}i$ [8.5.4] **42.** $1 + i$ [8.5.4] **43.** $-2 + 7i$ [8.5.4] **44.** -24 [8.4.1] **45.** $\left\{x \mid x \geq \frac{2}{3}\right\}$ [8.3.1] **46.** $\{x \mid x \in \text{real numbers}\}$ [8.3.1] **47.**

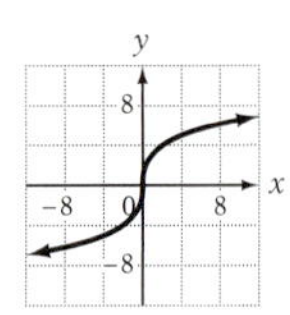

[8.3.2]

48. 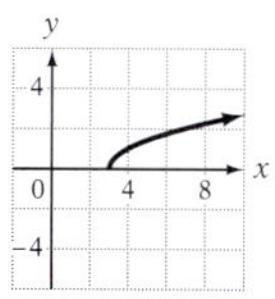 [8.3.2] **49.** $\frac{13}{3}$ [8.4.1] **50.** -2 [8.4.1] **51.** The width is 5 in. [8.4.2]

52. The amount of power is approximately 120 watts. [8.4.2] **53.** The distance required is 242 ft. [8.4.2]
54. The distance is approximately 6.63 ft. [8.4.2]

CHAPTER TEST

1. $r^{\frac{1}{6}}$ [8.1.1] **2.** $\frac{64x^3}{y^6}$ [8.1.1] **3.** $\frac{b^3}{8a^6}$ [8.1.1] **4.** $3\sqrt[5]{y^2}$ [8.1.2] **5.** $\frac{1}{2}x^{\frac{3}{4}}$ [8.1.2]
6. $\{x|x \le 4\}$ [8.3.1] **7.** $(-\infty, \infty)$ [8.3.1] **8.** $3abc^2\sqrt[3]{ac}$ [8.2.1] **9.** $8a\sqrt{2a}$ [8.2.2]
10. $-2x^2y\sqrt[3]{2x}$ [8.2.2] **11.** $-4x\sqrt{3}$ [8.2.3] **12.** $14 + 10\sqrt{3}$ [8.2.3]
13. $2a - \sqrt{ab} - 15b$ [8.2.3] **14.** $4x + 4\sqrt{xy} + y$ [8.2.3] **15.** $\frac{4x^2}{y}$ [8.2.4] **16.** 2 [8.2.4]
17. $\frac{x + \sqrt{xy}}{x - y}$ [8.2.4] **18.** -4 [8.5.3] **19.** $-3 + 2i$ [8.5.2] **20.** $18 + 16i$ [8.5.3]
21. $-\frac{4}{5} + \frac{7}{5}i$ [8.5.4] **22.** 4 [8.5.2] **23.** 4 [8.4.1] **24.** -3 [8.4.1] **25.** The length of the guy wire is approximately 30.6 ft. [8.4.2]

CUMULATIVE REVIEW EXERCISES

1. The Distributive Property [1.5.2] **2.** $-x - 24$ [1.5.6] **3.** $\varnothing$ [2.7.1] **4.** 16 [8.4.1]
5. $\frac{3}{2}$ [2.2.1] **6.** $\frac{2}{3}$ [2.2.3] **7.** $\{x|x \ge -1\}$ [2.7.2] **8.** $\{x|4 < x < 5\}$ [2.7.3]
9. $\left\{x|x < 2 \text{ or } x > \frac{8}{3}\right\}$ [2.8.2] **10.** $(8a + b)(8a - b)$ [6.4.1] **11.** $x(x^2 + 3)(x + 1)(x - 1)$ [6.4.4]
12. $\frac{2}{3}$ and -5 [6.5.1] **13.** 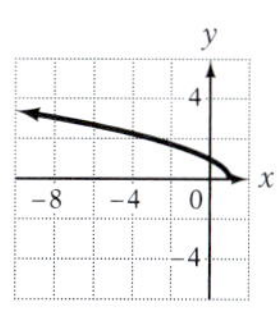 [8.3.2] **14.** Yes [3.6.1] **15.** $\frac{3x^3}{y}$ [5.4.1]
16. $\frac{y^8}{x^2}$ [8.1.1] **17.** $-x\sqrt{5x}$ [8.2.2] **18.** $11 - 5\sqrt{5}$ [8.2.3] **19.** $\frac{x\sqrt[3]{4y^2}}{2y}$ [8.2.4]
20. $-\frac{3}{5} + \frac{6}{5}i$ [8.5.4] **21.** 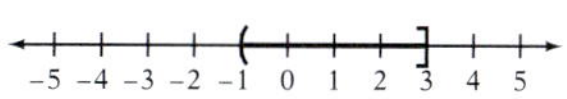 [2.7.1] **22.** $\{x|x \in \text{real numbers}\}$ [8.3.1]
23. 34 [5.1.1] **24.** $y = \frac{1}{3}x + \frac{7}{3}$ [3.5.2] **25.** 3 [4.3.2] **26.** $\left(\frac{17}{4}, \frac{9}{2}\right)$ [4.3.3]
27. 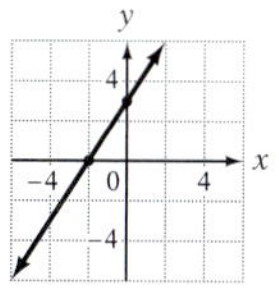 $m = \frac{3}{2}, b = 3$ [3.3.2] **28.** 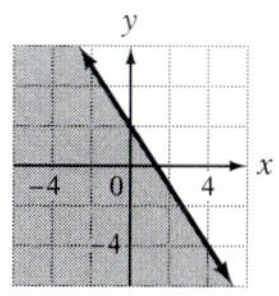 [3.7.1]
29. The additional investment must be $4000. [2.6.1] **30.** The length is 12 ft. The width is 6 ft. [6.5.2]
31. The rate of the plane is 250 mph. [7.4.3] **32.** The time is 1.25 s. [5.4.2] **33.** The slope is 0.08. A slope of 0.08 means that the annual interest income is 8% of the investment. [3.4.1]

Chapter 9

PREP TEST

1. $3\sqrt{2}$ [8.2.1] **2.** $3i$ [8.5.1] **3.** $\frac{2x-1}{x-1}$ [7.2.2] **4.** 8 [1.5.1] **5.** Yes [6.4.1]
6. $(2x-1)^2$ [6.4.1] **7.** $(3x+2)(3x-2)$ [6.4.1] **8.** 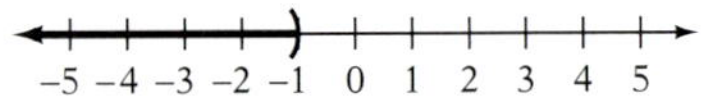

[2.7.1]
9. $-3, 5$ [6.5.1] **10.** 4 [7.4.1]

CONCEPT REVIEW 9.1

1. Sometimes true **3.** Never true

SECTION 9.1

3. $2x^2-4x-5=0$; $a=2, b=-4, c=-5$ **5.** $4x^2-5x-6=0$; $a=4, b=-5, c=-6$ **7.** 0 and 4
9. -5 and 5 **11.** -2 and 3 **13.** 3 **15.** 0 and 2 **17.** -2 and 5 **19.** 2 and 5
21. 6 and $-\frac{3}{2}$ **23.** $\frac{1}{4}$ and 2 **25.** -4 and $\frac{1}{3}$ **27.** $-\frac{2}{3}$ and $\frac{9}{2}$ **29.** -4 and $\frac{1}{4}$ **31.** -2 and 9
33. -2 and $-\frac{3}{4}$ **35.** 2 and -5 **37.** $-\frac{3}{2}$ and -4 **39.** $2b$ and $7b$ **41.** $-c$ and $7c$
43. $-\frac{b}{2}$ and $-b$ **45.** $\frac{2a}{3}$ and $4a$ **47.** $x^2-7x+10=0$ **49.** $x^2+6x+8=0$
51. $x^2-5x-6=0$ **53.** $x^2-9=0$ **55.** $x^2-8x+16=0$ **57.** $x^2-5x=0$ **59.** $x^2-3x=0$
61. $2x^2-7x+3=0$ **63.** $4x^2-5x-6=0$ **65.** $3x^2+11x+10=0$ **67.** $9x^2-4=0$
69. $6x^2-5x+1=0$ **71.** $10x^2-7x-6=0$ **73.** $8x^2+6x+1=0$ **75.** $50x^2-25x-3=0$
77. 7 and -7 **79.** $2i$ and $-2i$ **81.** 2 and -2 **83.** $\frac{9}{2}$ and $-\frac{9}{2}$ **85.** $7i$ and $-7i$
87. $4\sqrt{3}$ and $-4\sqrt{3}$ **89.** $5\sqrt{3}$ and $-5\sqrt{3}$ **91.** $3i\sqrt{2}$ and $-3i\sqrt{2}$ **93.** 7 and -5 **95.** 0 and -6
97. -7 and 3 **99.** $-5+5i$ and $-5-5i$ **101.** $4+2i$ and $4-2i$ **103.** $9+3i$ and $9-3i$
105. 0 and 1 **107.** $-\frac{1}{5}$ and 1 **109.** $-\frac{3}{2}$ and 0 **111.** $\frac{7}{3}$ and 1 **113.** $-5+\sqrt{6}$ and $-5-\sqrt{6}$
115. $2+2\sqrt{6}$ and $2-2\sqrt{6}$ **117.** $-1+2i\sqrt{3}$ and $-1-2i\sqrt{3}$ **119.** $3+3i\sqrt{5}$ and $3-3i\sqrt{5}$
121. $\frac{-2+9\sqrt{2}}{3}$ and $\frac{-2-9\sqrt{2}}{3}$ **123.** $-\frac{1}{2}+2i\sqrt{10}$ and $-\frac{1}{2}-2i\sqrt{10}$ **125.** $x^2-2=0$
127. $x^2+1=0$ **129.** $x^2-8=0$ **131.** $x^2-12=0$ **133.** $x^2+12=0$ **135.** $-\frac{5z}{y}$ and $\frac{5z}{y}$
137. $y+2$ and $y-2$ **139.** 1

CONCEPT REVIEW 9.2

1. Always true **3.** Never true **5.** Never true

SECTION 9.2

3. 5 and -1 **5.** -9 and 1 **7.** 3 **9.** $-2+\sqrt{11}$ and $-2-\sqrt{11}$ **11.** $3+\sqrt{2}$ and $3-\sqrt{2}$
13. $1+i$ and $1-i$ **15.** $\frac{1+\sqrt{5}}{2}$ and $\frac{1-\sqrt{5}}{2}$ **17.** $3+\sqrt{13}$ and $3-\sqrt{13}$ **19.** 3 and 5
21. $2+3i$ and $2-3i$ **23.** $-3+2i$ and $-3-2i$ **25.** $1+3\sqrt{2}$ and $1-3\sqrt{2}$
27. $\frac{1+\sqrt{17}}{2}$ and $\frac{1-\sqrt{17}}{2}$ **29.** $1+2i\sqrt{3}$ and $1-2i\sqrt{3}$ **31.** $-\frac{1}{2}$ and -1 **33.** $\frac{1}{2}$ and $\frac{3}{2}$

35. $-\frac{1}{2}$ and $\frac{4}{3}$ **37.** $\frac{1}{2} + i$ and $\frac{1}{2} - i$ **39.** $\frac{1}{3} + \frac{1}{3}i$ and $\frac{1}{3} - \frac{1}{3}i$ **41.** $\frac{2 + \sqrt{14}}{2}$ and $\frac{2 - \sqrt{14}}{2}$

43. $-\frac{3}{2}$ and 1 **45.** $1 + \sqrt{5}$ and $1 - \sqrt{5}$ **47.** $\frac{1}{2}$ and 5 **49.** $2 + \sqrt{5}$ and $2 - \sqrt{5}$

51. -3.236 and 1.236 **53.** 1.707 and 0.293 **55.** 0.309 and -0.809 **61.** 5 and -2

63. 4 and -9 **65.** $4 + 2\sqrt{22}$ and $4 - 2\sqrt{22}$ **67.** 3 and -8 **69.** $\frac{1}{2}$ and -3 **71.** $-\frac{1}{4}$ and $\frac{3}{2}$

73. $1 + 2\sqrt{2}$ and $1 - 2\sqrt{2}$ **75.** 10 and -2 **77.** $6 + 2\sqrt{3}$ and $6 - 2\sqrt{3}$ **79.** $\frac{1 + 2\sqrt{2}}{2}$ and $\frac{1 - 2\sqrt{2}}{2}$

81. $\frac{1}{2}$ and 1 **83.** $\frac{-5 + \sqrt{7}}{3}$ and $\frac{-5 - \sqrt{7}}{3}$ **85.** $\frac{2}{3}$ and $\frac{5}{2}$ **87.** $2 + i$ and $2 - i$

89. $-3 + 2i$ and $-3 - 2i$ **91.** $3 + i$ and $3 - i$ **93.** $-\frac{3}{2}$ and $-\frac{1}{2}$ **95.** $-\frac{1}{2} + \frac{5}{2}i$ and $-\frac{1}{2} - \frac{5}{2}i$

97. $\frac{-3 + \sqrt{6}}{3}$ and $\frac{-3 - \sqrt{6}}{3}$ **99.** $1 + \frac{\sqrt{6}}{3}i$ and $1 - \frac{\sqrt{6}}{3}i$ **101.** $-\frac{3}{2}$ and -1 **103.** $-\frac{3}{2}$ and 4

105. 7.606 and 0.394 **107.** 0.236 and -4.236 **109.** 1.851 and -1.351 **111.** Two complex

113. One real **115.** Two real **117.** Two real **119.** Quadratic **121.** Linear

123. Quadratic **125.** $\frac{\sqrt{2}}{2}$ and $-2\sqrt{2}$ **127.** $-3\sqrt{2}$ and $\frac{\sqrt{2}}{2}$ **129.** $\frac{\sqrt{3}}{2} + \frac{1}{2}i$ and $\frac{\sqrt{3}}{2} - \frac{1}{2}i$

131. $2a$ and $-a$ **133.** $\frac{a}{2}$ and $-2a$ **135.** $1 + \sqrt{y + 1}$ and $1 - \sqrt{y + 1}$

137. $\{p|p < 9, p \in \text{real numbers}\}$ **139.** $\{p|p > 1, p \in \text{real numbers}\}$ **141.** $i, -2i$ **143.** The ball takes 4.43 s to hit the ground. **145. a.** 1.05: The model predicts that the population will increase. **b.** 0.96: The model predicts that the population will decrease.

CONCEPT REVIEW 9.3

1. Always true **3.** Always true

SECTION 9.3

3. 3, -3, 2, -2 **5.** 2, -2, $\sqrt{2}$, $-\sqrt{2}$ **7.** 1 and 4 **9.** 16 **11.** $2i, -2i, 1, -1$

13. $4i, -4i, 2, -2$ **15.** 16 **17.** 512 and 1 **19.** $2, 1, -1 + i\sqrt{3}, -1 - i\sqrt{3}, -\frac{1}{2} + \frac{\sqrt{3}}{2}i, -\frac{1}{2} - \frac{\sqrt{3}}{2}i$

21. $1, -1, 2, -2, i, -i, 2i, -2i$ **23.** -64 and 8 **25.** $\frac{1}{4}$ and 1 **27.** 3 **29.** 9 **31.** 2 and -1

33. 0 and 2 **35.** $-\frac{1}{2}$ and 2 **37.** -2 **39.** 1 **41.** 1 **43.** -3 **45.** 10 and -1

47. $-\frac{1}{2} + \frac{\sqrt{7}}{2}i$ and $-\frac{1}{2} - \frac{\sqrt{7}}{2}i$ **49.** 1 and -3 **51.** 0 and -1 **53.** $\frac{1}{2}$ and $-\frac{1}{3}$

55. 6 and $-\frac{2}{3}$ **57.** $\frac{4}{3}$ and 3 **59.** $-\frac{1}{4}$ and 3 **61.** $\frac{-5 + \sqrt{33}}{2}$ and $\frac{-5 - \sqrt{33}}{2}$ **63.** 4 and -6

65. 5 and 4 **67.** $3, -3, i, -i$ **69.** $i\sqrt{6}, -i\sqrt{6}, 2, -2$ **71.** $\sqrt{2}$ **73.** 4

CONCEPT REVIEW 9.4

1. Sometimes true **3.** Always true

SECTION 9.4

1. The length is 40 ft. The width is 16 ft. **3.** The length is 13 ft. The width is 5 ft. **5.** The cardboard needs to be 90 cm by 90 cm. **7.** The rate returning was 55 mph. **9.** The projectile takes 12.5 s to return to

Earth. **11.** The maximum speed is 72.5 km/h. **13.** The rocket will be 300 ft above the ground after 1.74 s and after 10.76 s. **15.** Working alone, it would take the smaller pipe 12 min to fill the tank. It would take the larger pipe 6 min. **17.** The rate of the cruise ship for the first 40 mi was 10 mph. **19.** The rate of the wind is 20 mph. **21.** The rate of the crew in calm water is 6 mph. **23.** The fraction is $\frac{3}{6}$. **25.** The width is 8 cm. The length is 16 cm. The height is 2 cm. **27.** The depth will be 10 cm after 22.2 s. **29.** The bottom of the scoop of ice cream is approximately 2.3 in. from the bottom of the cone.

CONCEPT REVIEW 9.5

1. Sometimes true **3.** Sometimes true

SECTION 9.5

3. $\{x|x < -2 \text{ or } x > 4\}$ 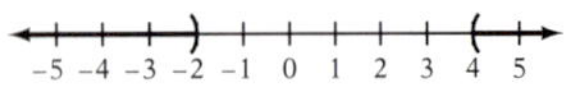
5. $\{x|x \le 1 \text{ or } x \ge 2\}$ 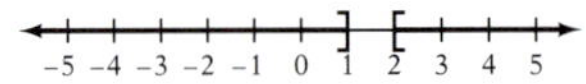
7. $\{x|-3 < x < 4\}$ 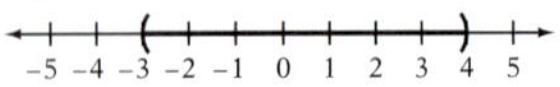
9. $\{x|x < -2 \text{ or } 1 < x < 3\}$ 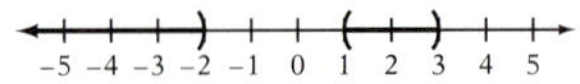
11. $\{x|x < -2 \text{ or } x > 4\}$ 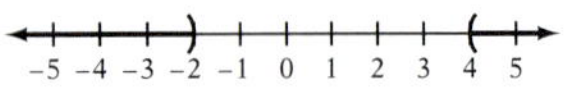
13. $\{x|-1 < x \le 3\}$ 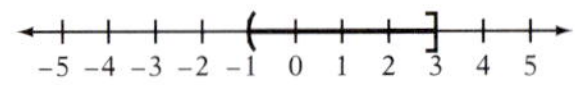
15. $\{x|x \le -2 \text{ or } 1 \le x < 3\}$ 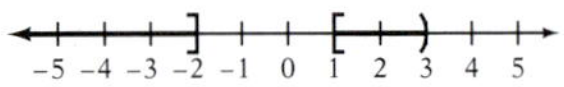
17. $\{x|x < -4 \text{ or } x > 4\}$
19. $\{x|x < 2 \text{ or } x > 2\}$ **21.** $\{x|-3 \le x \le 12\}$ **23.** $\left\{x\middle|x \le \frac{1}{2} \text{ or } x \ge 2\right\}$ **25.** $\left\{x\middle|\frac{1}{2} < x < \frac{3}{2}\right\}$
27. $\{x|x \le -3 \text{ or } 2 \le x \le 6\}$ **29.** $\left\{x\middle|-\frac{3}{2} < x < \frac{1}{2} \text{ or } x > 4\right\}$ **31.** $\{x|x \le -3 \text{ or } -1 \le x \le 1\}$
33. $\{x|-2 \le x \le 1 \text{ or } x \ge 2\}$ **35.** $\{x|x < -1 \text{ or } x > 2\}$ **37.** $\{x|-1 < x \le 0\}$
39. $\{x|-2 < x \le 0 \text{ or } x > 1\}$ **41.** $\left\{x\middle|x < 0 \text{ or } x > \frac{1}{2}\right\}$ **43.** 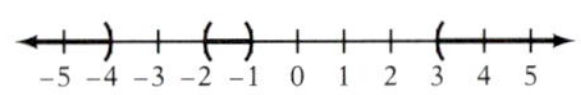
45. 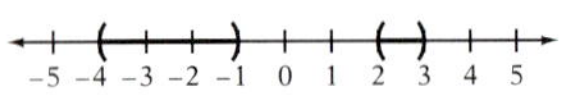 **47.** 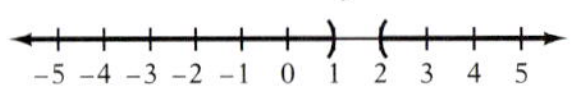 **49.** 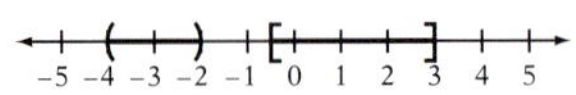
51. 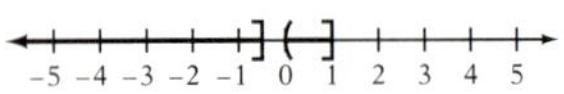

CONCEPT REVIEW 9.6

1. Sometimes true **3.** Sometimes true **5.** Always true

SECTION 9.6

3. -5 **5.** $x = 7$ **7.** Vertex: (0, 0)
Axis of symmetry: $x = 0$

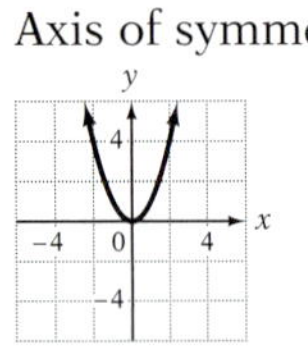

9. Vertex: (0, −2)
Axis of symmetry: $x = 0$

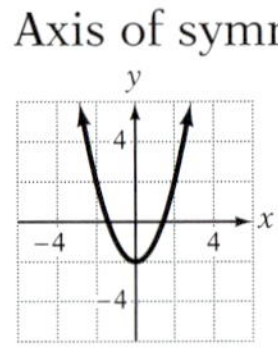

11. Vertex: (0, 3)
Axis of symmetry: $x = 0$

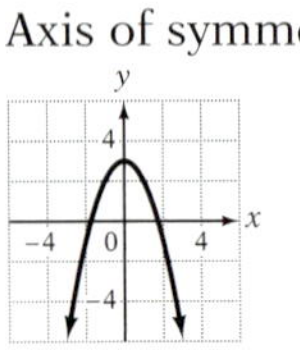

13. Vertex: (0, 0)
Axis of symmetry: $x = 0$

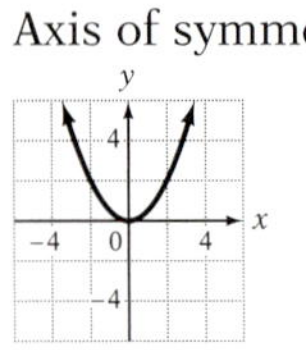

15. Vertex: (0, −1)
Axis of symmetry: $x = 0$

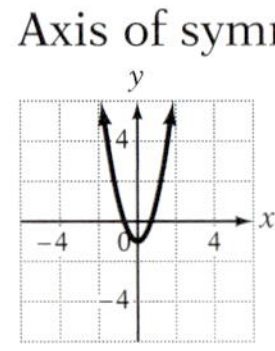

17. Vertex: $(1, -1)$

Axis of symmetry: $x = 1$

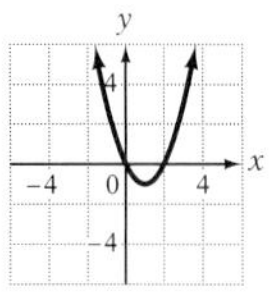

19. Vertex: $(1, 2)$

Axis of symmetry: $x = 1$

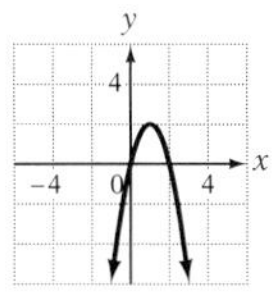

21. Vertex: $\left(\frac{1}{2}, -\frac{9}{4}\right)$

Axis of symmetry: $x = \frac{1}{2}$

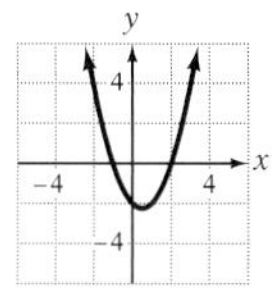

23. Vertex: $\left(\frac{1}{4}, -\frac{41}{8}\right)$

Axis of symmetry: $x = \frac{1}{4}$

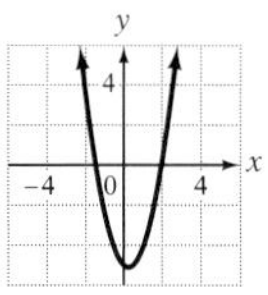

25.

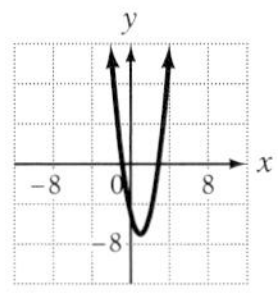

Domain: $\{x | x \in \text{real numbers}\}$
Range: $\{y | y \geq -7\}$

27.

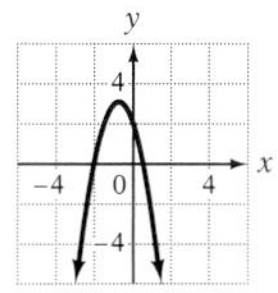

Domain: $\{x | x \in \text{real numbers}\}$
Range: $\left\{y | y \leq \frac{25}{8}\right\}$

29.

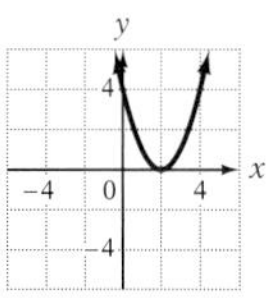

Domain: $\{x | x \in \text{real numbers}\}$
Range: $\{y | y \geq 0\}$

31.

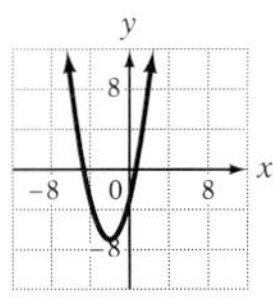

Domain: $\{x | x \in \text{real numbers}\}$
Range: $\{y | y \geq -7\}$

33.

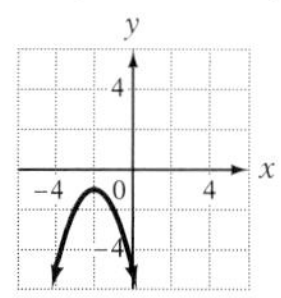

Domain: $\{x | x \in \text{real numbers}\}$
Range: $\{y | y \leq -1\}$

37. $(2, 0)$ and $(-2, 0)$ **39.** $(0, 0)$ and $(2, 0)$ **41.** $(2, 0)$ and $(-1, 0)$ **43.** $(3, 0)$ and $\left(-\frac{1}{2}, 0\right)$

45. $\left(-\frac{2}{3}, 0\right)$ and $(7, 0)$ **47.** $(5, 0)$ and $\left(\frac{4}{3}, 0\right)$ **49.** $\left(\frac{2}{3}, 0\right)$ **51.** $\left(\frac{\sqrt{2}}{3}, 0\right)$ and $\left(-\frac{\sqrt{2}}{3}, 0\right)$

53. $\left(-\frac{3}{2}, 0\right), \left(\frac{5}{2}, 0\right)$ **55.** $(-2 + \sqrt{7}, 0)$ and $(-2 - \sqrt{7}, 0)$ **57.** No x-intercepts

59. $(-1 + \sqrt{2}, 0)$ and $(-1 - \sqrt{2}, 0)$ **61.** -2 and -1 **63.** $2 + i$ and $2 - i$ **65.** 0 and $\frac{3}{2}$

67. $\sqrt{2}$ and $-\sqrt{2}$ **69.** $-\frac{3}{4} + \frac{\sqrt{7}}{4}i$ and $-\frac{3}{4} - \frac{\sqrt{7}}{4}i$ **71.** $\frac{1}{3}$ and 1 **73.**

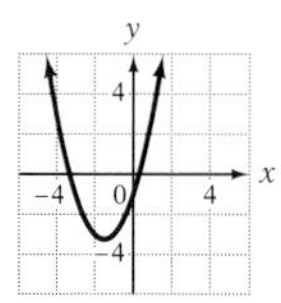

-3.3 and 0.3

75.

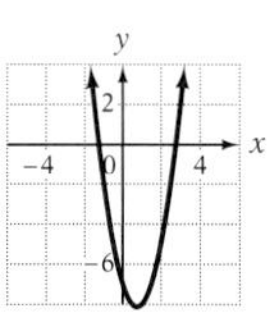

-1.3 and 2.8

77.

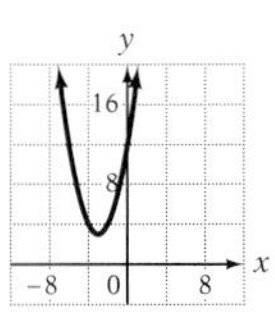

No real zeros

79. No x-intercepts **81.** Two **83.** One

85. No x-intercepts **87.** Two **89.** No x-intercepts **91.** One **93.** Two

95. $(-4, 0)$ and $(5, 0)$ **97.** -2 **99.** 1 **101.** 16 **103.** $\frac{1 + \sqrt{19}}{3}$

105. i and E; ii and F, iii and C; iv and A; v and B; vi and D **107.** When the bicyclist is traveling at 15 mph, the drag force on the bicyclist is 33.975 lb. **109.** $y = (x - 2)^2 + 3$; vertex: (2, 3)

111. $y = \left(x + \frac{1}{2}\right)^2 + \frac{7}{4}$; vertex: $\left(-\frac{1}{2}, \frac{7}{4}\right)$ **113.** $y = 3x^2 - 6x + 5$

CONCEPT REVIEW 9.7

1. Sometimes true **3.** Always true **5.** Sometimes true

SECTION 9.7

3. a. Maximum value **b.** Minimum value **c.** Maximum value

5. Minimum value at 2 **7.** Maximum value at −1 **9.** Maximum value at $\frac{41}{8}$ **11.** Minimum value at $-\frac{73}{8}$ **13.** Maximum value at $-\frac{71}{12}$ **15.** Minimum value at $-\frac{13}{4}$ **17.** a **19.** The function has a minimum value at 7. **21.** The rock will attain a maximum height above the beach of 114 ft. **23.** The maximum profit the tour operator can expect is $1600. **25.** The pool will have the least amount of algae 5 days after treatment. **27.** The minimum thickness of the mirror is 20.21 in. **29.** The water will land $31\frac{2}{3}$ ft up on the building. **31.** The two numbers are 10 and 10. **33.** The dimensions that will maximize the enclosed area are 100 ft by 50 ft. **35.** 3.0 **37.** 0.5

CHAPTER REVIEW EXERCISES

1. 0 and $\frac{3}{2}$ [9.1.1] **2.** $-2c$ and $\frac{c}{2}$ [9.1.1] **3.** $-4\sqrt{3}$ and $4\sqrt{3}$ [9.1.3] **4.** $-\frac{1}{2} - 2i$ and $-\frac{1}{2} + 2i$ [9.1.3] **5.** $-\frac{17}{4}$ [9.7.1] **6.** 3 [9.7.1] **7.** $3x^2 + 8x - 3 = 0$ [9.1.2] **8.** −5 and $\frac{1}{2}$ [9.1.1] **9.** $-1 - 3\sqrt{2}$ and $-1 + 3\sqrt{2}$ [9.1.3] **10.** $-3 - i$ and $-3 + i$ [9.2.1]

11. 2 and 3 [9.3.3] **12.** $-\sqrt{2}, \sqrt{2}, -2, 2$ [9.3.1] **13.** $\frac{25}{18}$ [9.3.2] **14.** −8 and $\frac{1}{8}$ [9.3.1]

15. 2 [9.3.2] **16.** $3 - \sqrt{11}$ and $3 + \sqrt{11}$ [9.2.2] **17.** 5 and $-\frac{4}{9}$ [9.3.3] **18.** $\frac{1 + \sqrt{3}}{2}$ and $\frac{1 - \sqrt{3}}{2}$ [9.2.2] **19.** $\frac{5}{4}$ [9.3.2] **20.** 3 and −1 [9.3.3] **21.** −1 [9.3.3] **22.** $\frac{-3 + \sqrt{249}}{10}$ and $\frac{-3 - \sqrt{249}}{10}$ [9.3.3] **23.** $\frac{-11 - \sqrt{129}}{2}$ and $\frac{-11 + \sqrt{129}}{2}$ [9.3.3] **24.** $x = 3$ [9.6.1]

25. $\left(\frac{3}{2}, \frac{1}{4}\right)$ [9.6.1] **26.** No x-intercepts [9.6.2] **27.** Two [9.6.2] **28.** $\left(\frac{-3 - \sqrt{5}}{2}, 0\right)$ and $\left(\frac{-3 + \sqrt{5}}{2}, 0\right)$ [9.6.2] **29.** (−2, 0) and $\left(\frac{1}{2}, 0\right)$ [9.6.2] **30.** $-\frac{1}{3} + \frac{\sqrt{5}}{3}i$ and $-\frac{1}{3} - \frac{\sqrt{5}}{3}i$ [9.6.2]

31. $\left\{x \mid -3 < x < \frac{5}{2}\right\}$ [9.5.1] **32.** $\left\{x \mid x \le -4 \text{ or } -\frac{3}{2} \le x \le 2\right\}$ [9.5.1] **33.** $\left\{x \mid x < \frac{3}{2} \text{ or } x \ge 2\right\}$

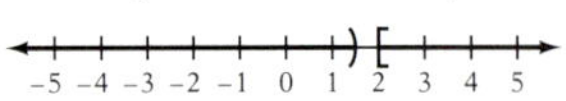

[9.5.1] **34.** $\left\{x \mid x \le -3 \text{ or } \frac{1}{2} \le x < 4\right\}$

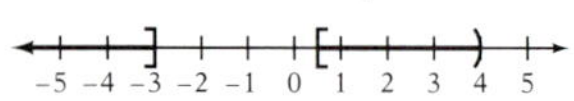

[9.5.1]

35. 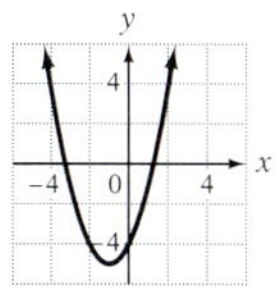 [9.6.1]

Domain: $\{x|x \in \text{real numbers}\}$
Range: $\{y|y \geq -5\}$

36. 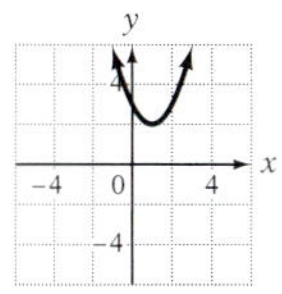 [9.6.1]

Vertex: (1, 2)
Axis of symmetry: $x = 1$

37. The integers are 2, 4, and 6 or −6, −4, −2. [9.4.1] **38.** The width of the rectangle is 5 cm. The length of the rectangle is 12 cm. [9.4.1] **39.** Working alone, the new computer can print the payroll in 12 min. [9.4.1] **40.** The rate of the first car is 40 mph. The rate of the second car is 50 mph. [9.4.1]

CHAPTER TEST

1. $\frac{3}{2}$ and -2 [9.1.1] **2.** $\frac{3}{4}$ and $-\frac{4}{3}$ [9.1.1] **3.** 9 [9.7.1] **4.** $3x^2 - 8x - 3 = 0$ [9.1.2]
5. $-3 + 3\sqrt{2}$ and $-3 - 3\sqrt{2}$ [9.1.3] **6.** $-2 + \sqrt{5}$ and $-2 - \sqrt{5}$ [9.2.2] **7.** $\frac{-3 + \sqrt{41}}{2}$ and $\frac{-3 - \sqrt{41}}{2}$ [9.6.2] **8.** $\frac{1}{6} - \frac{\sqrt{95}}{6}i$ and $\frac{1}{6} + \frac{\sqrt{95}}{6}i$ [9.2.2] **9.** $-3 - \sqrt{10}$ and $-3 + \sqrt{10}$ [9.3.3]
10. $\frac{1}{4}$ [9.3.1] **11.** $-3, 3, -\sqrt{2}, \sqrt{2}$ [9.3.1] **12.** 4 [9.3.2] **13.** No solution [9.3.2]
14. Two [9.6.2] **15.** $(-4, 0)$ and $\left(\frac{3}{2}, 0\right)$ [9.6.2] **16.** $x = -\frac{3}{2}$ [9.6.1]

17. 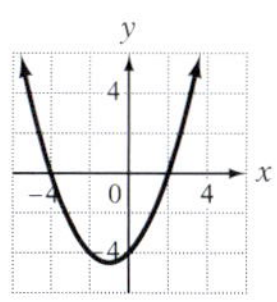 [9.6.1]

Domain: $\{x|x \in \text{real numbers}\}$
Range: $\{y|y \geq -4.5\}$

18. $\left\{x\middle|-4 < x \leq \frac{3}{2}\right\}$ 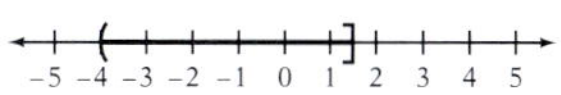 [9.5.1]

19. The height of the triangle is 4 ft. The base of the triangle is 15 ft. [9.4.1] **20.** The rate of the canoe in calm water is 4 mph. [9.4.1]

CUMULATIVE REVIEW EXERCISES

1. 14 [1.5.1] **2.** −28 [2.2.3] **3.** $-\frac{3}{2}$ [3.4.1] **4.** $y = x + 1$ [3.6.1]
5. $-3xy(x^2 - 2xy + 3y^2)$ [6.1.1] **6.** $(2x - 5)(3x + 4)$ [6.3.1/6.3.2] **7.** $(x + y)(a^n - 2)$ [6.1.2]
8. $x^2 - 3x - 4 - \frac{6}{3x - 4}$ [5.5.2] **9.** $\frac{x}{2}$ [7.2.1] **10.** $2\sqrt{5}$ [3.1.2] **11.** $b = \frac{2S - an}{n}$ [7.6.1]
12. $-8 - 14i$ [8.5.3] **13.** $1 - a$ [8.1.1] **14.** $\frac{x\sqrt[3]{4y^2}}{2y}$ [8.2.4] **15.** $-\frac{3}{2}$ and -1 [9.3.3]
16. $-\frac{1}{2}$ [9.3.3] **17.** $2, -2, \sqrt{2}$, and $-\sqrt{2}$ [9.3.1] **18.** 0 and 1 [9.3.2]
19. $\left\{x\middle|-2 < x < \frac{10}{3}\right\}$ [2.8.2] **20.** $\left(\frac{5}{2}, 0\right)$ and $(0, -3)$ [3.3.2] **21.** 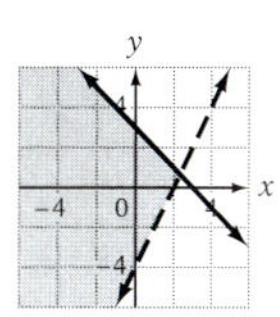 [4.5.1]

22. $(1, -1, 2)$ [4.2.2] **23.** $f(-2) = -\frac{7}{3}$ [7.1.1] **24.** $\{x \mid x \neq 5, -3\}$ [7.1.1]

25. $\{x \mid x < -3 \text{ or } 0 < x < 2\}$ 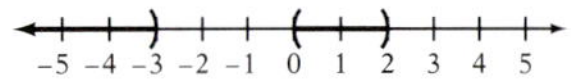

[9.5.1]

26. $\{x \mid -3 < x \leq 1 \text{ or } x \geq 5\}$ 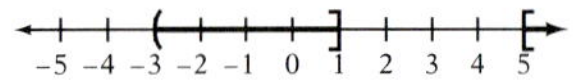

[9.5.1] **27.** The concentration of orange juice is 28%. [2.6.2] **28.** The lower and upper limits of the length of the piston rod are $9\frac{23}{64}$ in. and $9\frac{25}{64}$ in. [2.8.3]

29. The area of the triangle is $(x^2 + 6x - 16)$ ft^2. [5.3.5] **30.** $m = -\frac{25{,}000}{3}$; a slope of $-\frac{25{,}000}{3}$ means that the value of the building decreases \$8333.33 each year. [3.4.1]

Chapter 10

PREP TEST

1. 1 [1.5.1] **2.** −7 [1.5.1] **3.** 30 [3.2.1] **4.** $h^2 + 4h - 1$ [3.2.1] **5.** $-\frac{2}{3}, 3$ [9.1.1]

6. $2 - \sqrt{3}, 2 + \sqrt{3}$ [9.2.1, 9.2.2] **7.** $y = \frac{1}{2}x - 2$ [7.6.1] **8.** D: $\{-2, 3, 4, 6\}$; R: $\{4, 5, 6\}$; Yes [3.2.1]

9. 8 [3.2.1] **10.** 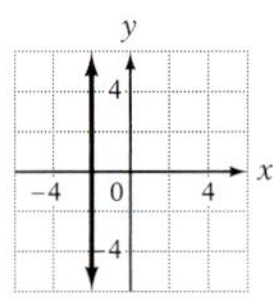

[3.3.2] **11.** 7.21 [3.1.2]

CONCEPT REVIEW 10.1

1. Sometimes true **3.** Sometimes true **5.** Always true

SECTION 10.1

3. Yes **5.** No **7.** Yes

9. 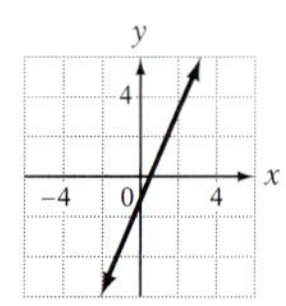

Domain: $\{x \mid x \in \text{real numbers}\}$
Range: $\{y \mid y \in \text{real numbers}\}$

11. 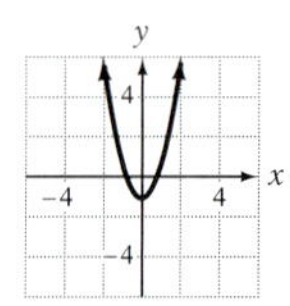

Domain: $\{x \mid x \in \text{real numbers}\}$
Range: $\{y \mid y \geq -1\}$

13. 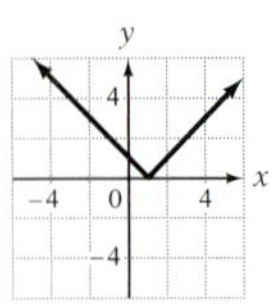

Domain: $\{x \mid x \in \text{real numbers}\}$
Range: $\{y \mid y \geq 0\}$

15. 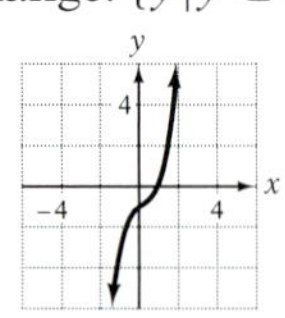

Domain: $\{x \mid x \in \text{real numbers}\}$
Range: $\{y \mid y \in \text{real numbers}\}$

17. 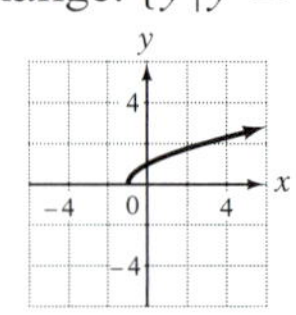

Domain: $\{x \mid x \geq -1\}$
Range: $\{y \mid y \geq 0\}$

19. 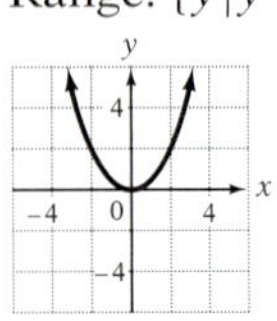

Domain: $\{x \mid x \in \text{real numbers}\}$
Range: $\{y \mid y \geq 0\}$

21. 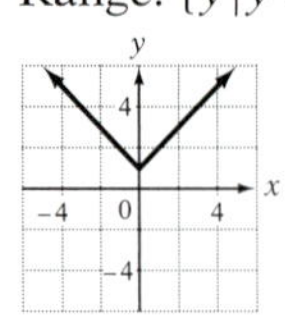

Domain: $\{x \mid x \in \text{real numbers}\}$
Range: $\{y \mid y \geq 1\}$

23. 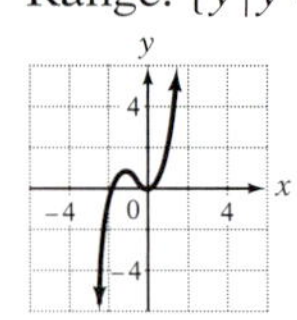

Domain: $\{x \mid x \in \text{real numbers}\}$
Range: $\{y \mid y \in \text{real numbers}\}$

25. 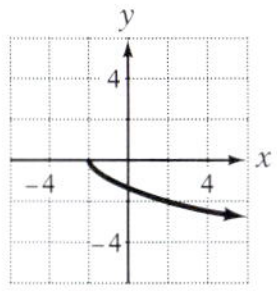

Domain: $\{x \mid x \geq -2\}$
Range: $\{y \mid y \leq 0\}$

27. 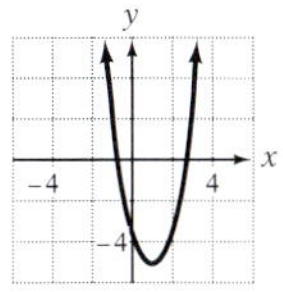

Domain: $\{x \mid x \in \text{real numbers}\}$
Range: $\{y \mid y \geq -5\}$

29. 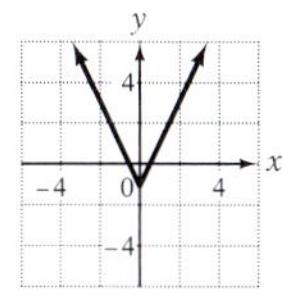

Domain: $\{x \mid x \in \text{real numbers}\}$
Range: $\{y \mid y \geq -1\}$

31. 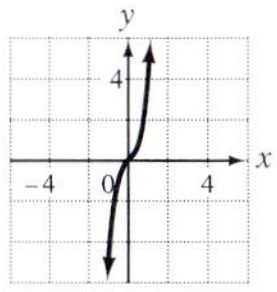

Domain: $\{x \mid x \in \text{real numbers}\}$
Range: $\{y \mid y \in \text{real numbers}\}$

33. 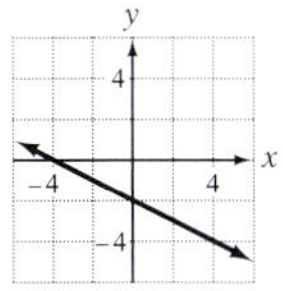

Domain: $\{x \mid x \in \text{real numbers}\}$
Range: $\{y \mid y \in \text{real numbers}\}$

35. 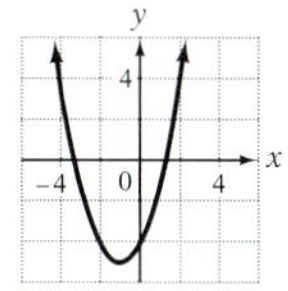

Domain: $\{x \mid x \in \text{real numbers}\}$
Range: $\{y \mid y \geq -5\}$

37. 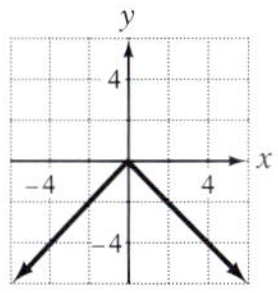

Domain: $\{x \mid x \in \text{real numbers}\}$
Range: $\{y \mid y \leq 0\}$

39. 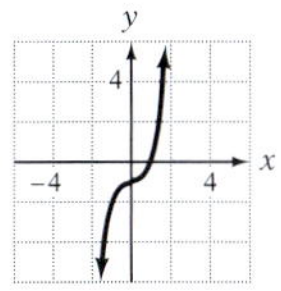

Domain: $\{x \mid x \in \text{real numbers}\}$
Range: $\{y \mid y \in \text{real numbers}\}$

41. c **43.** $a = 18$

45. a and c **47.** 77 **49.** $\{x \mid -1 < x < 1\}$ **51.** -3 **53. a.** 0 **b.** (0, 0) **55.** 1

CONCEPT REVIEW 10.2

1. Always true **3.** Always true **5.** Never true

SECTION 10.2

3. 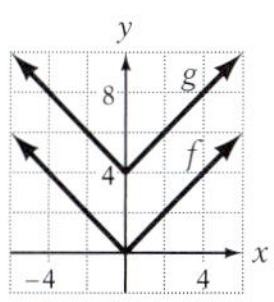

5. 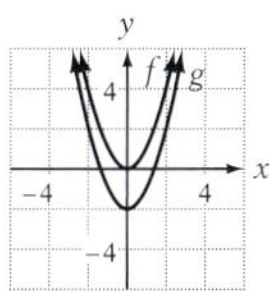

7. 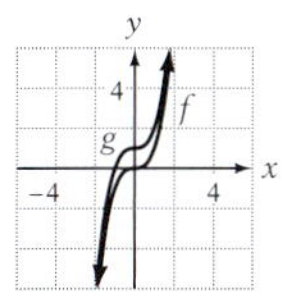

9. 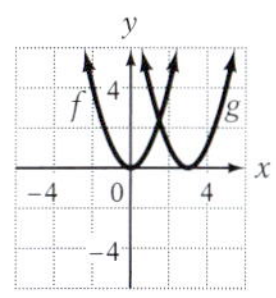

11. 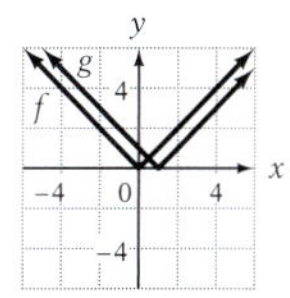

13. 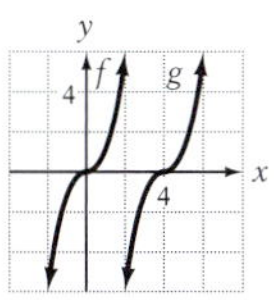

15. 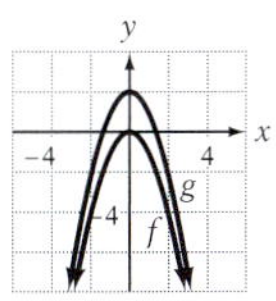

17. 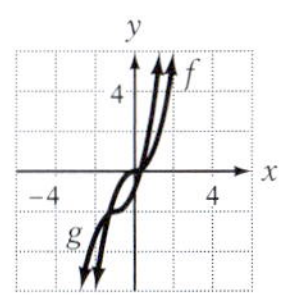

19. 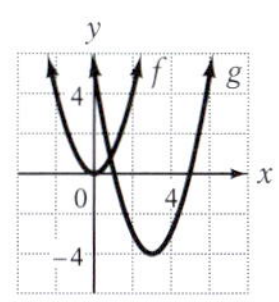

21. 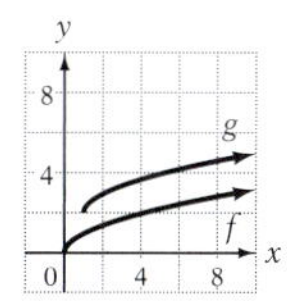

23. 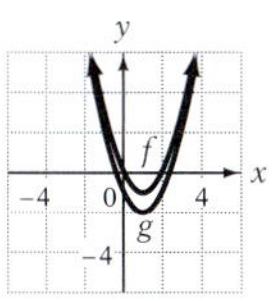

25. 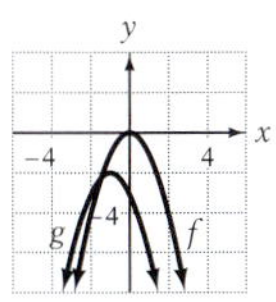

27. 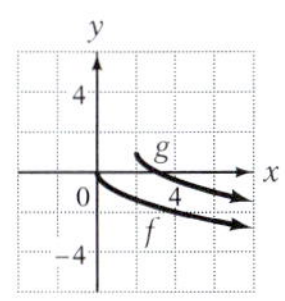

29. 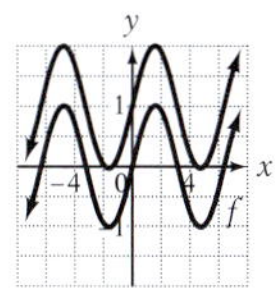

31. 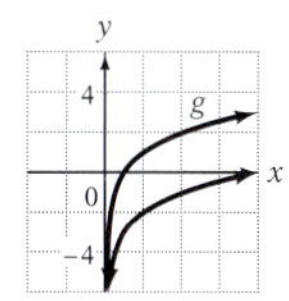

33. 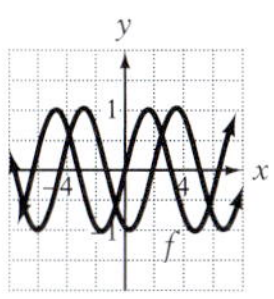

35. 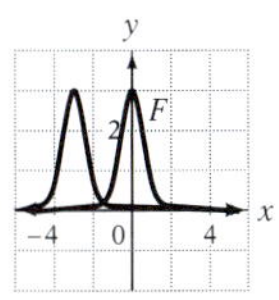

37. 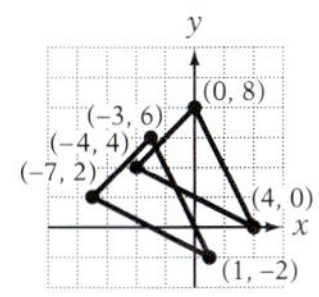

CONCEPT REVIEW 10.3

1. Always true **3.** Never true

SECTION 10.3

1. 5 **3.** 1 **5.** 0 **7.** $-\frac{29}{4}$ **9.** $\frac{2}{3}$ **11.** 2 **13.** 39 **15.** −8 **17.** $-\frac{4}{5}$
19. −2 **21.** 7 **25.** −13 **27.** −29 **29.** $8x - 13$ **31.** 1 **33.** 5 **35.** $x^2 + 1$
37. 5 **39.** 11 **41.** $3x^2 + 3x + 5$ **43.** −3 **45.** −27 **47.** $x^3 - 6x^2 + 12x - 8$
49. $6h + h^2$ **51.** $2 + h$ **53.** $h + 2a$ **55.** −1 **57.** −6 **59.** $6x - 13$ **61.** −2
63. −2 **65.** 0 **67.** 6

CONCEPT REVIEW 10.4

1. Always true **3.** Never true **5.** Sometimes true

SECTION 10.4

3. Yes **5.** No **7.** Yes **9.** No **11.** No **13.** No **17.** {(0, 1), (3, 2), (8, 3), (15, 4)}
19. No inverse **21.** $f^{-1}(x) = \frac{1}{4}x + 2$ **23.** No inverse **25.** $f^{-1}(x) = x + 5$ **27.** $f^{-1}(x) = 3x - 6$
29. $f^{-1}(x) = -\frac{1}{3}x - 3$ **31.** $f^{-1}(x) = \frac{3}{2}x - 6$ **33.** $f^{-1}(x) = -3x + 3$ **35.** $f^{-1}(x) = \frac{1}{2}x + \frac{5}{2}$
37. No inverse **39.** $\frac{5}{3}$ **41.** 3 **43.** Yes; yes **45.** Yes **47.** No **49.** Yes
51. Yes **53.** Range **55.** 3 **57.** 0 **59.** 2 **61.**

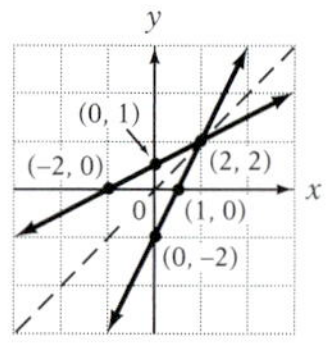

63.

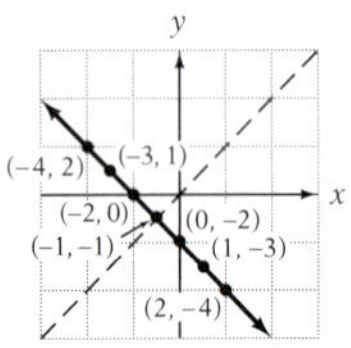

The function and its inverse have the same graph.

65.

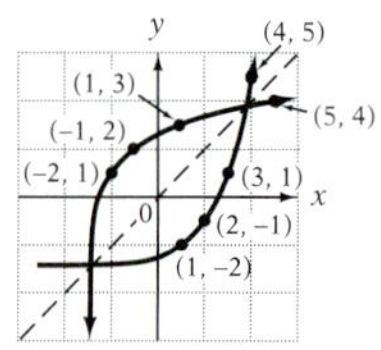

CONCEPT REVIEW 10.5

1. Sometimes true **3.** Sometimes true **5.** Never true **7.** Never true **9.** Always true

SECTION 10.5

1. a. The axis of symmetry is a vertical line. **b.** The parabola opens up. **3. a.** The axis of symmetry is a horizontal line. **b.** The parabola opens right. **5. a.** The axis of symmetry is a horizontal line. **b.** The parabola opens left. **7.** Vertex: (1, −5)
Axis of symmetry: $x = 1$

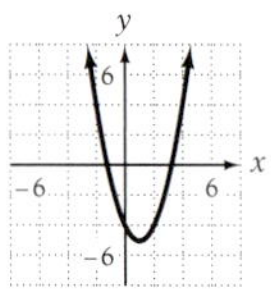

9. Vertex: (1, −2)
Axis of symmetry: $x = 1$

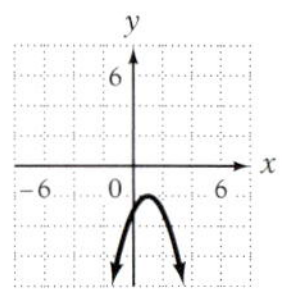

11. Vertex: $(-4, -3)$
Axis of symmetry: $y = -3$

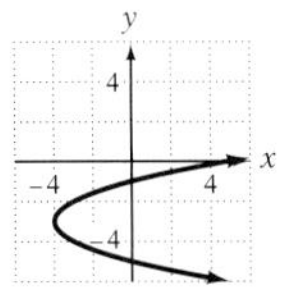

13. Vertex: $(1, -1)$
Axis of symmetry: $x = 1$

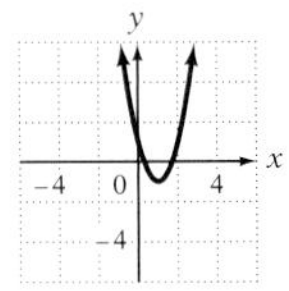

15. Vertex: $\left(\frac{5}{2}, -\frac{9}{4}\right)$
Axis of symmetry: $x = \frac{5}{2}$

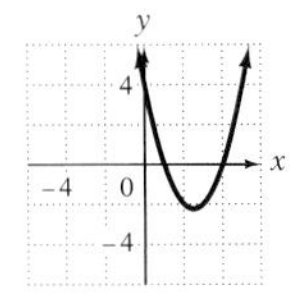

17. Vertex: $(-6, 1)$
Axis of symmetry: $y = 1$

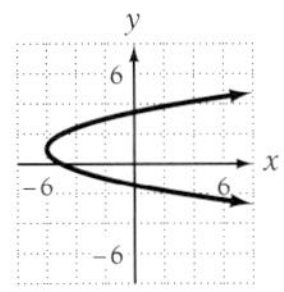

19. Vertex: $\left(-\frac{3}{2}, \frac{27}{4}\right)$
Axis of symmetry: $x = -\frac{3}{2}$

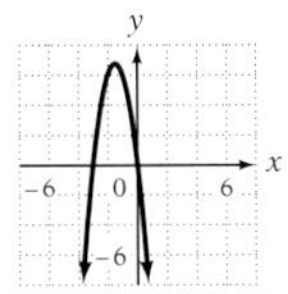

21. Vertex: $(4, 0)$
Axis of symmetry: $y = 0$

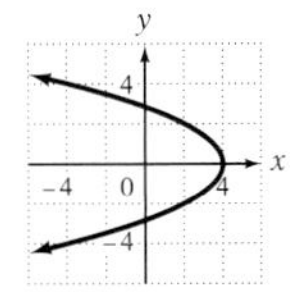

23. Vertex: $\left(\frac{1}{2}, 1\right)$
Axis of symmetry: $y = 1$

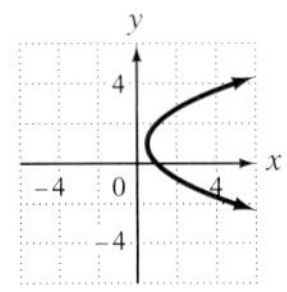

25. Vertex: $(-2, -8)$
Axis of symmetry: $x = -2$

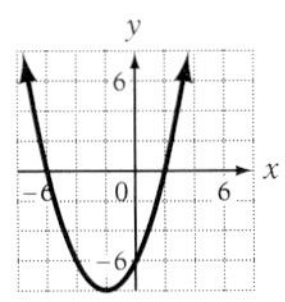

29.

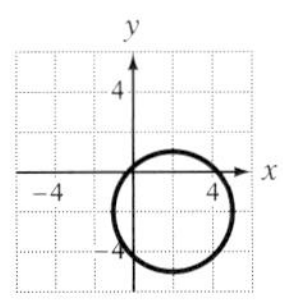

31.

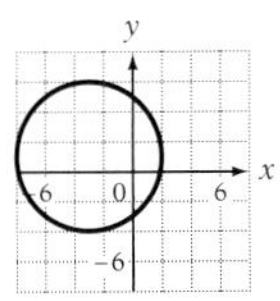

33.

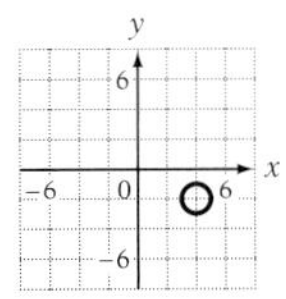

35.

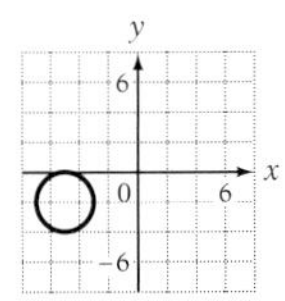

37. $(x - 2)^2 + (y + 1)^2 = 4$

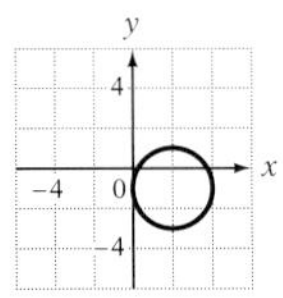

39. $(x + 1)^2 + (y - 1)^2 = 5$

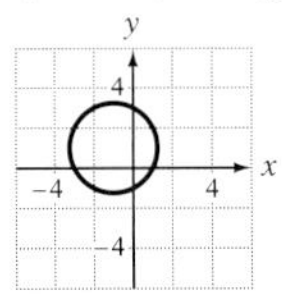

41. $(x + 3)^2 + (y - 6)^2 = 8$

43. $(x + 2)^2 + (y - 1)^2 = 5$

45.

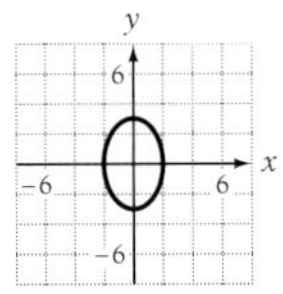

47.

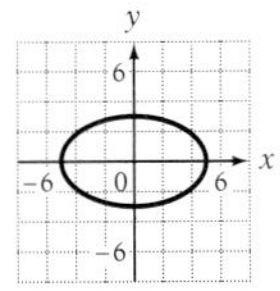

49.

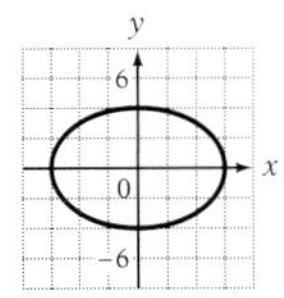

51.

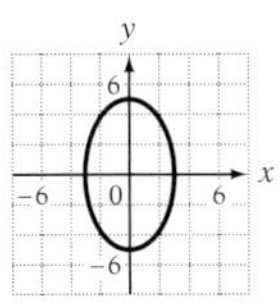

53.

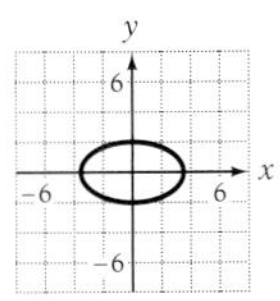

55.

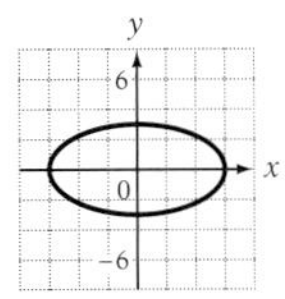

59.

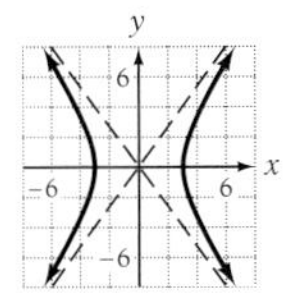

61.

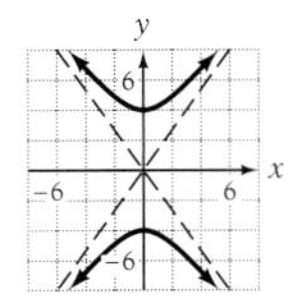

63.

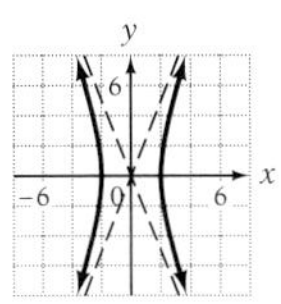

65.

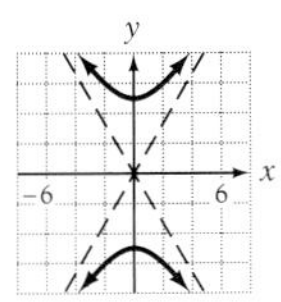

67.

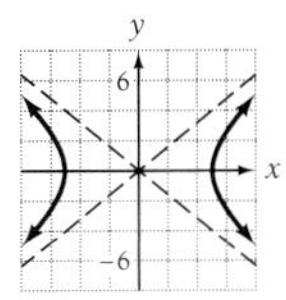

69.

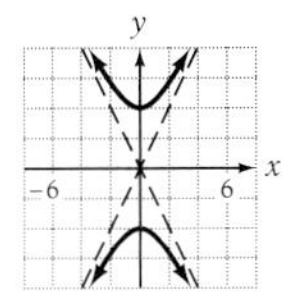

71.

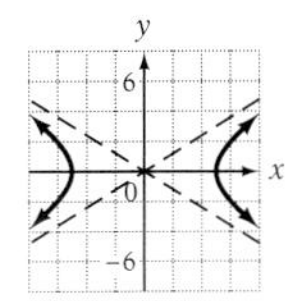

73. Domain: $\{x \mid x \in \text{real numbers}\}$
Range: $\{y \mid y \geq -6\}$

75. Domain: $\{x \mid x \geq -14\}$
Range: $\{y \mid y \in \text{real numbers}\}$

77. $\left(0, \frac{1}{8}\right)$

79. $(x - 3)^2 + y^2 = 9$

81. $6\sqrt{3}$ in.

83.
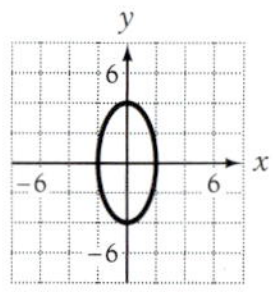

85.
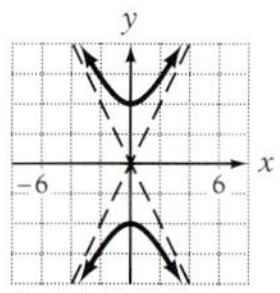

87.
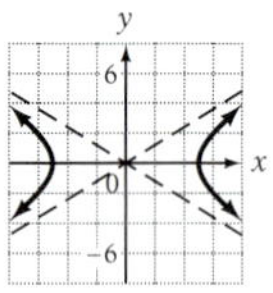

89. **a.** $\frac{x^2}{324} + \frac{y^2}{20.25} = 1$ **b.** The aphelion is 3,293,400,000 mi.
c. The perihelion is approximately 53,100,000 mi.

91. **a.** $\frac{x^2}{2.310} + \frac{x^2}{2.235} = 1$ **b.** The aphelion is approximately 166,800,000 mi.
c. The perihelion is approximately 115,800,000 mi.

CHAPTER REVIEW EXERCISES

1. Yes [10.1.1]

2.
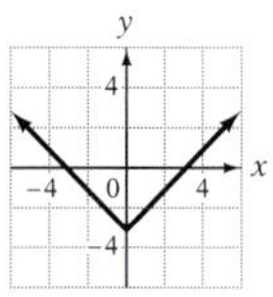

[10.1.1]
Domain: $\{x \mid x \in \text{real numbers}\}$
Range: $\{y \mid y \geq -3\}$

3.
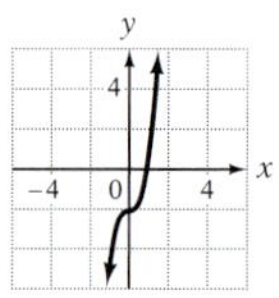

[10.1.1]
Domain: $\{x \mid x \in \text{real numbers}\}$
Range: $\{y \mid y \in \text{real numbers}\}$

4.
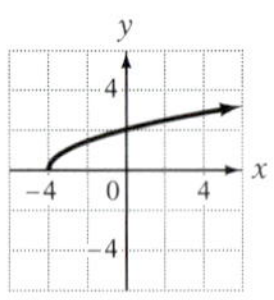

[10.1.1]
Domain: $[-4, \infty)$
Range: $[0, \infty)$

5.
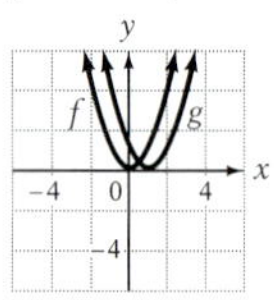

[10.2.1]

6.
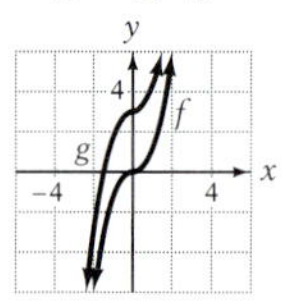

[10.2.1]

7.
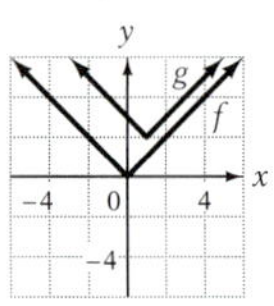

[10.2.1]

8.
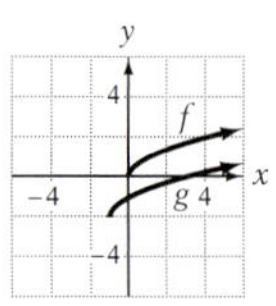

[10.2.1]

9. 7 [10.3.1] **10.** −9 [10.3.1]

11. 70 [10.3.1] **12.** $\frac{12}{7}$ [10.3.1] **13.** $12x^2 + 12x - 1$ [10.3.2] **14.** 5 [10.3.2]

15. 10 [10.3.2] **16.** $6x^2 + 3x - 16$ [10.3.2] **17.** No [10.4.1]

18. $\{(1, -2), (3, 2), (-4, 5), (9, 7)\}$ [10.4.2] **19.** No [10.4.1] **20.** Yes [10.4.1]

21. $f^{-1}(x) = 2x - 16$ [10.4.2] **22.** $f^{-1}(x) = -\frac{1}{6}x + \frac{2}{3}$ [10.4.2] **23.** $f^{-1}(x) = \frac{3}{2}x + 18$ [10.4.2]

24. Yes [10.4.2] **25.** Vertex: (2, 4); axis of symmetry: $x = 2$ [10.5.1]

26. Vertex: $\left(\frac{7}{2}, \frac{17}{4}\right)$; axis of symmetry: $x = \frac{7}{2}$ [10.5.1]

27. 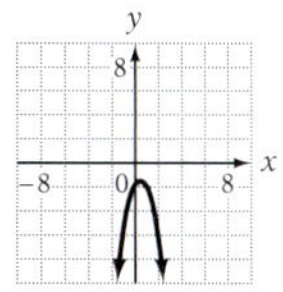 [10.5.1]

28. 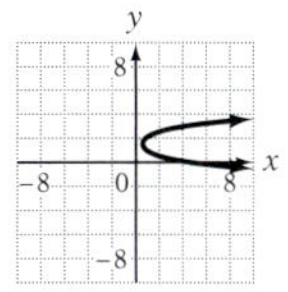 [10.5.1]

29. $(x + 1)^2 + (y - 2)^2 = 18$ [10.5.2]

30. $(x + 1)^2 + (y - 5)^2 = 36$ [10.5.2]

31. 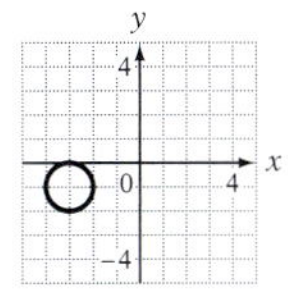 [10.5.2]

32. 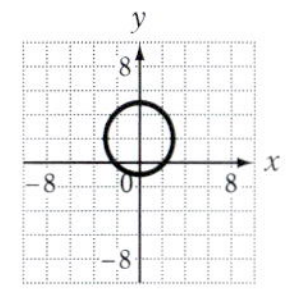 [10.5.2]

33. $x^2 + (y + 3)^2 = 97$ [10.5.2]

34. 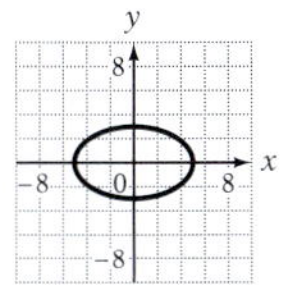 [10.5.3]

35. 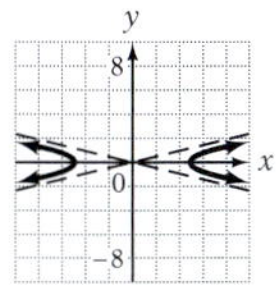 [10.5.4]

36. 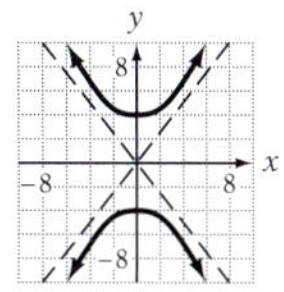 [10.5.4]

CHAPTER TEST

1. No [10.1.1]

2. -2 [10.3.1]

3. 234 [10.3.1]

4. $-\frac{13}{2}$ [10.3.1]

5. -5 [10.3.1]

6. 5 [10.3.2]

7. 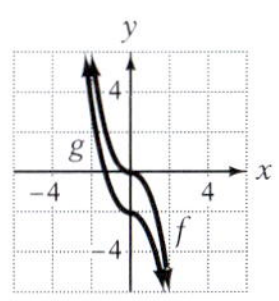 [10.2.1]

8. 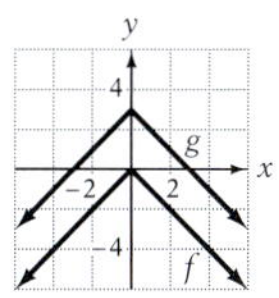 [10.2.1]

9. 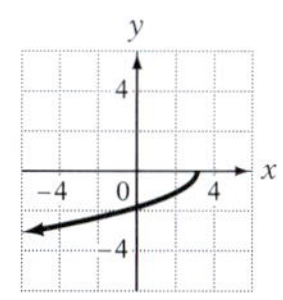 [10.1.1]

Domain: $(-\infty, 3]$
Range: $(-\infty, 0]$

10. 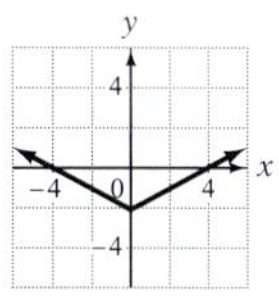 [10.1.1]

Domain: $\{x \mid x \in \text{real numbers}\}$
Range: $\{y \mid y \geq -2\}$

11. 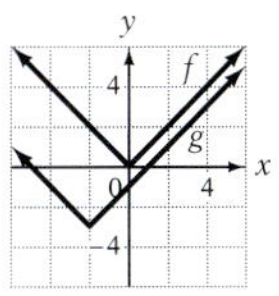 [10.2.1]

12. 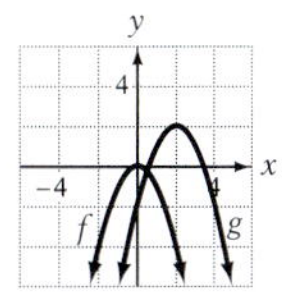 [10.2.1]

13. $f^{-1}(x) = 4x + 16$ [10.4.2]

14. $\{(6, 2), (5, 3), (4, 4), (3, 5)\}$ [10.4.2]

15. Yes [10.4.2]

16. $2x^2 - 4x - 5$ [10.3.2]

17. $f^{-1}(x) = 2x + 6$ [10.4.2]

18. No [10.4.2]

19. 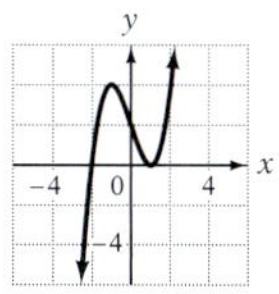 [10.1.1]

Domain: $\{x \mid x \in \text{real numbers}\}$
Range: $\{y \mid y \in \text{real numbers}\}$

20. No [10.4.1]

21. $x = 3$ [10.5.1] **22.** $\left(\frac{3}{2}, \frac{1}{4}\right)$ [10.5.1] **23.** 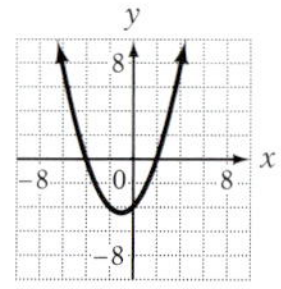 [10.5.1] **24.** 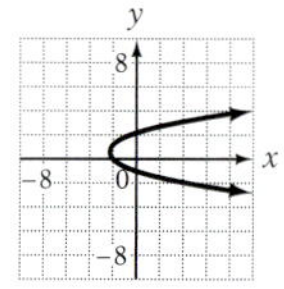 [10.5.1]

25. $(x + 3)^2 + (y + 3)^2 = 16$ [10.5.2] **26.** $(x + 1)^2 + (y + 3)^2 = 58$ [10.5.2]

27. 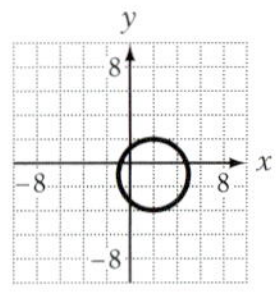 [10.5.2] **28.** 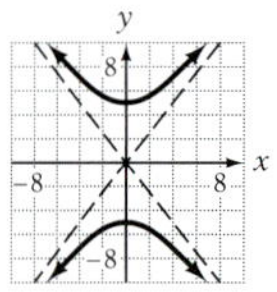 [10.5.4] **29.** 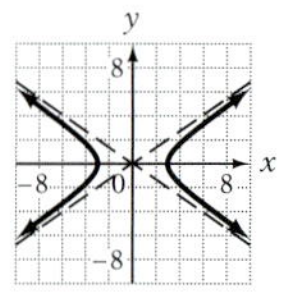 [10.5.4]

30. 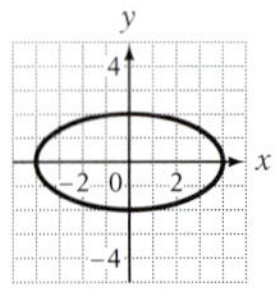 [10.5.3]

CUMULATIVE REVIEW EXERCISES

1. $-\frac{23}{4}$ [1.5.1] **2.** 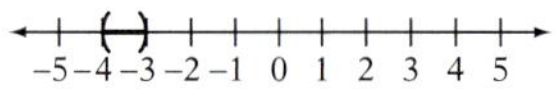 [2.7.1] **3.** 3 [2.2.3]

4. $\{x \mid x < -2 \text{ or } x > 3\}$ [2.7.3] **5.** 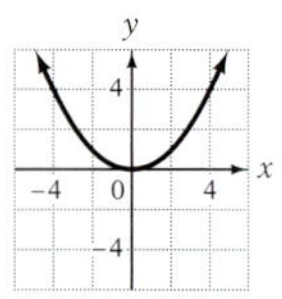 [9.6.1] **6.** 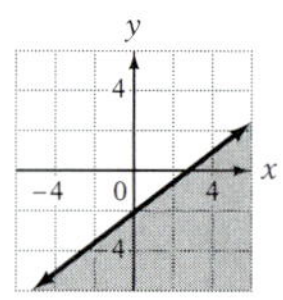 [3.7.1]

Vertex: (0, 0)
Axis of symmetry: $x = 0$

7. $\{x \mid x \in \text{real numbers}\}$ [2.8.2] **8.** $-\frac{a^{10}}{12b^4}$ [5.4.1] **9.** $2x^3 - 4x^2 - 17x + 4$ [5.3.2]

10. $(a + 2)(a - 2)(a^2 + 2)$ [6.4.4] **11.** $xy(x - 2y)(x + 3y)$ [6.4.4] **12.** -3 and 8 [9.1.1]

13. $\{x \mid x < -3 \text{ or } x > 5\}$ [9.5.1] **14.** $\frac{5}{2x - 1}$ [7.2.2] **15.** -2 [7.4.1] **16.** $-3 - 2i$ [8.5.4]

17. $y = -2x - 2$ [3.5.2] **18.** $y = -\frac{3}{2}x - \frac{7}{2}$ [3.6.1] **19.** $\frac{1}{2} + \frac{\sqrt{3}}{6}i$ and $\frac{1}{2} - \frac{\sqrt{3}}{6}i$ [9.2.2]

20. 3 [8.4.1] **21.** -3 [9.7.1] **22.** {1, 2, 4, 5} [3.2.1] **23.** Yes [3.2.1] **24.** 2 [8.4.1]

25. $g[h(2)] = 10$ [10.3.2] **26.** $f^{-1}(x) = -\frac{1}{3}x + 3$ [10.4.2] **27.** The cost per pound of the mixture is $3.96. [2.3.3] **28.** 25 lb of the 80% copper alloy must be used. [2.6.2] **29.** An additional 4.5 oz of insecticide is required. [7.5.1] **30.** It would take the larger pipe 4 min to fill the tank. [7.4.2] **31.** A force of 40 lb will stretch the spring 24 in. [7.5.2] **32.** The car overtakes the cyclist 25 mi from the starting point. [7.4.3] **33.** The frequency is 80 vibrations per minute. [7.5.2]

Chapter 11

PREP TEST

1. $\frac{1}{9}$ [5.4.1] **2.** 16 [5.4.1] **3.** -3 [5.4.1] **4.** 0; 108 [3.2.1] **5.** -6 [2.2.2]

6. $-2, 8$ [9.1.1] **7.** 6326.60 [1.5.1] **8.** 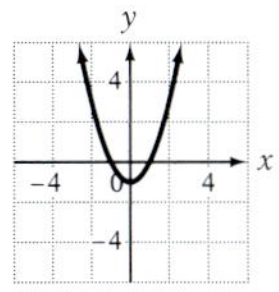 [9.6.1]

CONCEPT REVIEW 11.1

1. Never true **3.** Never true **5.** Never true

SECTION 11.1

3. c **5. a.** 9 **b.** 1 **c.** $\frac{1}{9}$ **7. a.** 16 **b.** 4 **c.** $\frac{1}{4}$ **9. a.** 1 **b.** $\frac{1}{8}$ **c.** 16 **11. a.** 7.3891 **b.** 0.3679 **c.** 1.2840 **13. a.** 54.5982 **b.** 1 **c.** 0.1353 **15. a.** 16 **b.** 16 **c.** 1.4768 **17. a.** 0.1353 **b.** 0.1353 **c.** 0.0111

19. 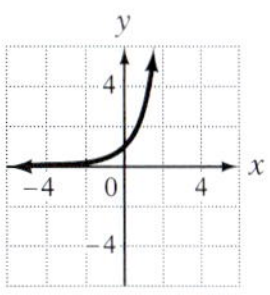 **21.** 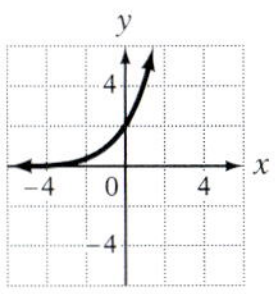 **23.** 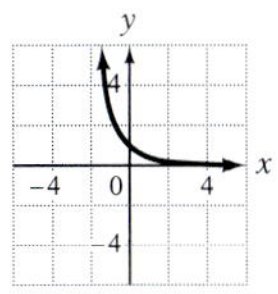 **25.** 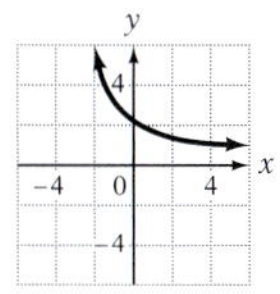 **27.** 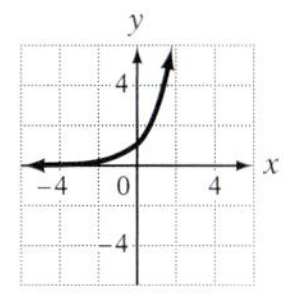

29. 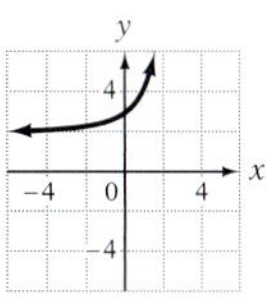 **31.** b and d **33.** (0, 1) **35.** There is no x-intercept. The y-intercept is (0, 1).

37. 1.6 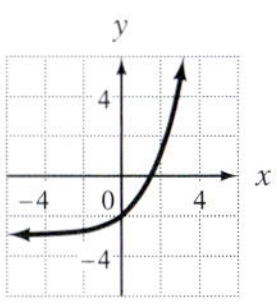 **39.** 1.1 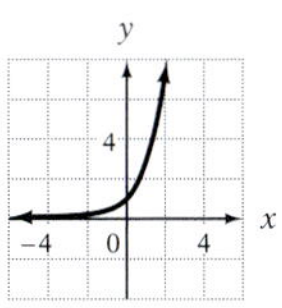

41. 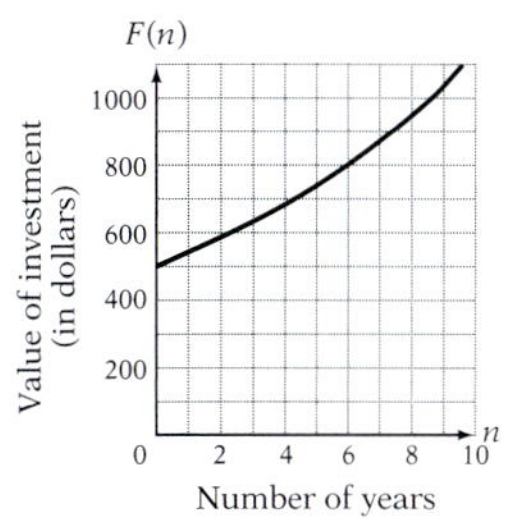

The investment will be worth \$1000 in approximately 9 years.

43. 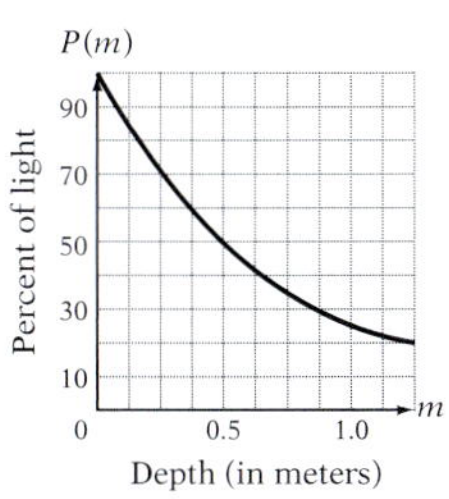

50% of the light will reach to a depth of approximately 0.5 m.

45. a. 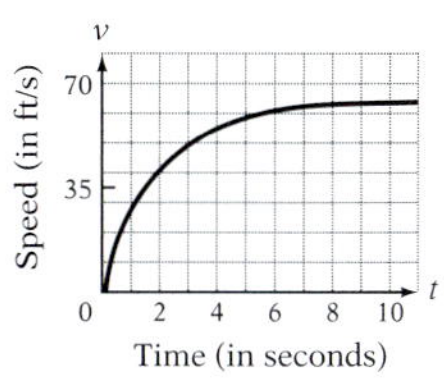

b. The point (4, 55.3) means that after 4 s, the object is moving at a speed of 55.3 ft/s.

CONCEPT REVIEW 11.2

1. Always true **3.** Never true **5.** Always true **7.** Always true

SECTION 11.2

3. $\log_5 25 = 2$ **5.** $\log_4 \frac{1}{16} = -2$ **7.** $\log_{10} x = y$ **9.** $\log_a w = x$ **11.** $3^2 = 9$ **13.** $10^{-2} = 0.01$ **15.** $e^y = x$ **17.** $b^v = u$ **19.** 4 **21.** 7 **23.** 2 **25.** 3 **27.** 0 **29.** 4 **31.** 9 **33.** 64 **35.** $\frac{1}{7}$ **37.** 1 **39.** 316.23 **41.** 0.02 **43.** 7.39

45. 0.61 **49.** $\log_8 x + \log_8 z$ **51.** $5 \log_3 x$ **53.** $\ln r - \ln s$ **55.** $2 \log_3 x + 6 \log_3 y$
57. $3 \log_7 u - 4 \log_7 v$ **59.** $2 \log_2 r + 2 \log_2 s$ **61.** $2 \log_9 x + \log_9 y + \log_9 z$
63. $\ln x + 2 \ln y - 4 \ln z$ **65.** $2 \log_8 x - \log_8 y - 2 \log_8 z$ **67.** $\frac{1}{2} \log_7 x + \frac{1}{2} \log_7 y$
69. $\frac{1}{2} \log_2 x - \frac{1}{2} \log_2 y$ **71.** $\frac{3}{2} \ln x + \frac{1}{2} \ln y$ **73.** $\frac{3}{2} \log_7 x - \frac{1}{2} \log_7 y$ **75.** $\log_3 \frac{x^3}{y}$ **77.** $\log_8 x^4y^2$
79. $\ln x^3$ **81.** $\log_5 x^3y^4$ **83.** $\log_4 \frac{1}{x^2}$ **85.** $\log_3 \frac{x^2z^2}{y}$ **87.** $\log_b \frac{x}{y^2z}$ **89.** $\ln x^2y^2$
91. $\log_6 \sqrt{\frac{x}{y}}$ **93.** $\log_4 \frac{s^2r^2}{t^4}$ **95.** $\log_5 \frac{x}{y^2z^2}$ **97.** $\ln \frac{t^3v^2}{r^2}$ **99.** $\log_4 \sqrt{\frac{x^3z}{y^2}}$ **101.** 1.3863
103. 1.0415 **105.** 0.8617 **107.** 2.1133 **109.** −0.6309 **111.** 0.2727 **113.** 1.6826
115. 1.9266 **117.** $\frac{\log(3x - 2)}{\log 3}$ **119.** $\frac{\log(4 - 9x)}{\log 8}$ **121.** $\frac{5}{\log 9}\log(6x + 7)$ **123.** $\frac{\ln(x + 5)}{\ln 2}$
125. $\frac{\ln(x^2 + 9)}{\ln 3}$ **127.** $\frac{7}{\ln 8}\ln(10x - 7)$ **129.** 8 **131.** 6 **133.** 2 **135.** 3
137. **a.** b **b.** b **139.** 13.12 **141.** −6.59 **143.** 15.1543

CONCEPT REVIEW 11.3

1. Never true **3.** Always true

SECTION 11.3

5. 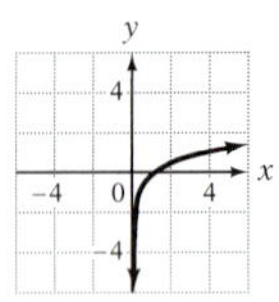
7. 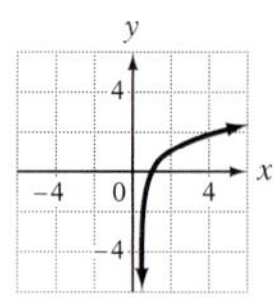
9. 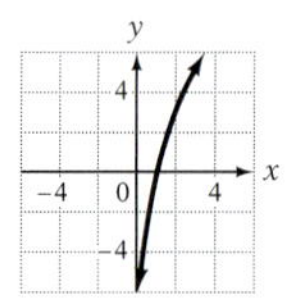
11. 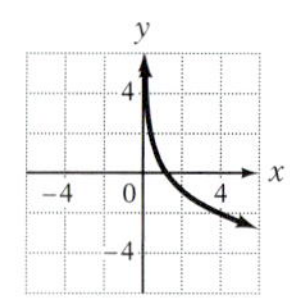
13. 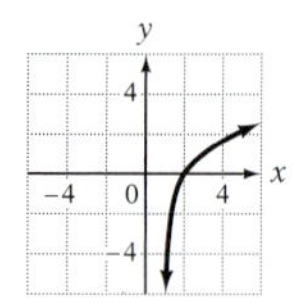

15. 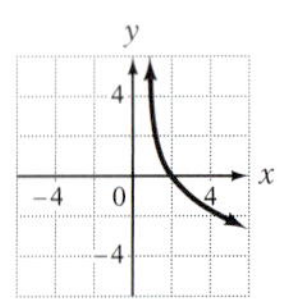
17. 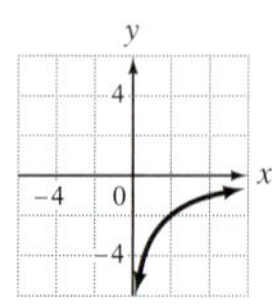
19. 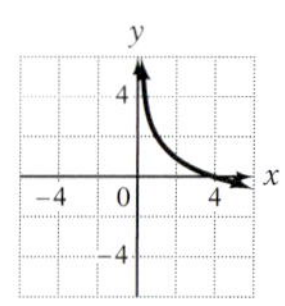
21. 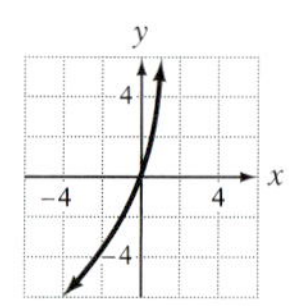
23. 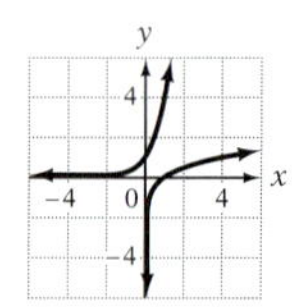

25. 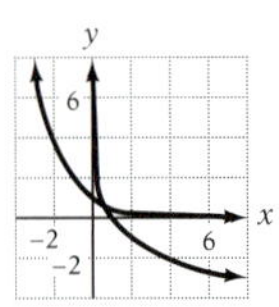

27. **a.** 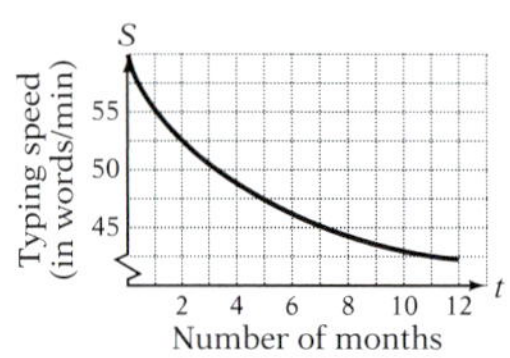

b. The point (4, 49) means that after 4 months, the typist's proficiency has dropped to 49 words per minute.

29. **a.** 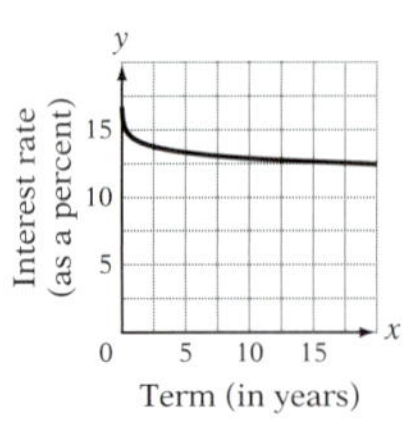

b. A security that has a yield of 13% has a term of 8.5 years.
c. The model predicts an interest rate of 12.2%.

31. A quadratic function **33.** A linear function **35.** An exponential function **37.** A logarithmic function

CONCEPT REVIEW 11.4

1. Always true **3.** Never true **5.** Always true

SECTION 11.4

3. 1 **5.** −3 **7.** 1.1133 **9.** 0.7211 **11.** −1.5850 **13.** 1.7095 **15.** 1.3222 **17.** −2.8074 **19.** 3.5850 **21.** 1.1309 **23.** 1 **25.** 6 **27.** 1.3754 and −1.3754 **29.** 1.5805 **31.** −0.6309 **33.** 1.6610 **35.** −1.5129 **37.** 1.5 **39.** 0.63 **41.** −1.86 and 3.44 **43.** −1.16 **47.** 8 **49.** $\frac{11}{2}$ **51.** 2 and −4 **53.** $\frac{5}{3}$ **55.** $\frac{1}{2}$ **57.** 1,000,000,000 **59.** 4 **61.** 3 **63.** 3 **65.** 2 **67.** No solution **69.** 4 **71.** $\frac{5 \pm \sqrt{33}}{2}$ **73.** 1.76 **75.** 2.42 **77.** −1.51 and 2.10 **79.** 1.7233 **81.** 0.8060 **83.** 5.6910 **85.** 2744 **87. a.**

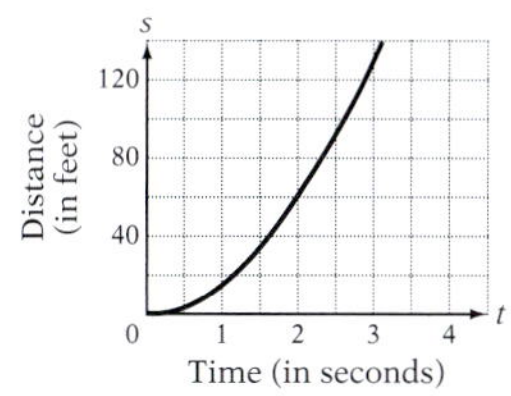

b. It will take the object approximately 2.64 s to fall 100 ft.

CONCEPT REVIEW 11.5

1. Always true **3.** Always true

SECTION 11.5

3. The value of the investment after 2 years is approximately \$1172. **5.** The investment will be worth \$15,000 in approximately 18 years. **7. a.** The amount of technetium will be approximately 21.2 mg after 3 h. **b.** It takes approximately 3.5 h for the amount of technetium to reach 20 mg. **9.** The half-life of promethium-147 is approximately 2.5 years. **11. a.** The approximate pressure at 40 km above Earth is 0.098 newtons per square centimeter. **b.** The approximate pressure on Earth's surface is 10.13 newtons per square centimeter. **c.** The atmospheric pressure decreases as you rise about Earth's surface. **13.** The pH of milk is approximately 6.4. **15.** At a depth of approximately 2.5 m, the percent of light will be 75% of the light at the surface of the pool. **17.** Normal conversation is approximately 65 decibels. **19.** The average time of a major-league baseball game increased approximately 22.5 min during the years from 1981 to 1999. **21.** A thickness of approximately 0.4 cm is needed. **23.** The Richter scale magnitude of the earthquake is approximately 6.8. **25.** The intensity of the earthquake is approximately $794{,}328{,}234I_0$. **27.** The magnitude of the earthquake is approximately 5.3. **29.** The magnitude of the earthquake is approximately 5.6. **31.** Prices will double in approximately 14 years. **33.** The value of the investment after 5 years is \$3210.06. **35. a.** 104 months are required to reduce the loan amount to \$90,000. **b.** 269 months are required to reduce the loan amount to one-half the original amount. **c.** The total interest paid exceeds \$100,000 in month 163.

CHAPTER REVIEW EXERCISES

1. 2 [11.2.1] **2.** $\log_3 \sqrt{\frac{x}{y}}$ [11.2.2] **3.** 1 [11.1.1] **4.** −3 [11.4.1] **5.** 1 [11.1.1] **6.** $\frac{1}{9}$ [11.2.1] **7.** $\log_2 32 = 5$ [11.2.1] **8.** 6 [11.4.2] **9.** $\frac{1}{2}\log_6 x + \frac{3}{2}\log_6 y$ [11.2.2] **10.** 2 [11.4.1] **11.** −2 [11.4.1] **12.** $\frac{1}{3}$ [11.1.1] **13.** 4 [11.2.1] **14.** 2 [11.4.2]

15. 2.3219 [11.2.2] **16.** 1.7251 [11.2.2] **17.** 3 [11.4.1] **18.** 4 [11.1.1]

19. $\frac{1}{2}\log_5 x - \frac{1}{2}\log_5 y$ [11.2.2] **20.** $\frac{1}{4}$ [11.4.2] **21.** 125 [11.2.1] **22.** $\frac{1}{2}$ [11.4.2]

23. $\log_b \frac{x^3}{y^5}$ [11.2.2] **24.** 4 [11.1.1] **25.** 2.6801 [11.2.2] **26.** −0.5350 [11.4.1]

27. 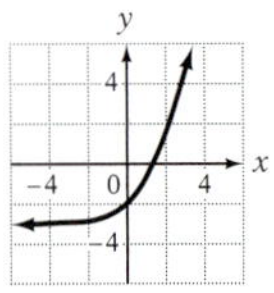 [11.1.2] **28.** 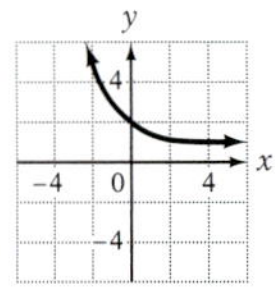 [11.1.2] **29.** 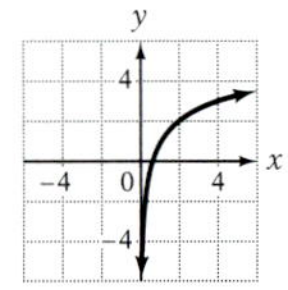 [11.3.1]

30. 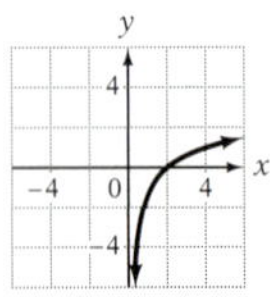 [11.3.1] **31.** 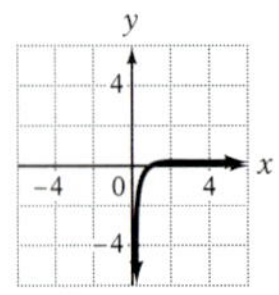 1.0 [11.3.1] **32.** The half-life is approximately 33 h. [11.5.1]

33. The material must be approximately 0.602 cm thick. [11.5.1]

CHAPTER TEST

1. 1 [11.1.1] **2.** $\frac{1}{64}$ [11.1.1] **3.** 3 [11.2.1] **4.** $\frac{1}{16}$ [11.2.1]

5. $\frac{2}{3}\log_6 x + \frac{5}{3}\log_6 y$ [11.2.2] **6.** $\log_5 \sqrt{\frac{x}{y}}$ [11.2.2] **7.** 3 [11.4.2] **8.** $\frac{1}{3}$ [11.1.1]

9. 2.5789 [11.4.1] **10.** 10 [11.4.2] **11.** 3 [11.4.1] **12.** −4 [11.4.1] **13.** $\frac{3}{2}$ [11.4.2]

14. 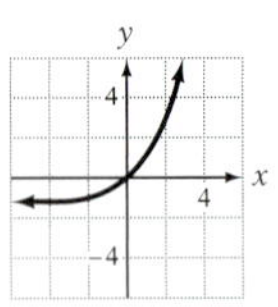 [11.1.2] **15.** 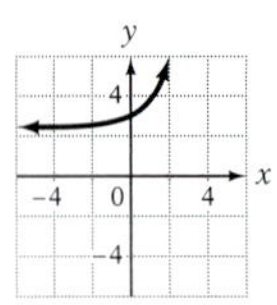 [11.1.2] **16.** 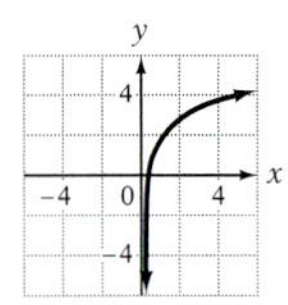 [11.3.1]

17. 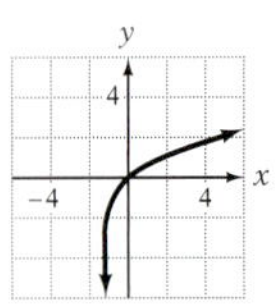 [11.3.1] **18.** 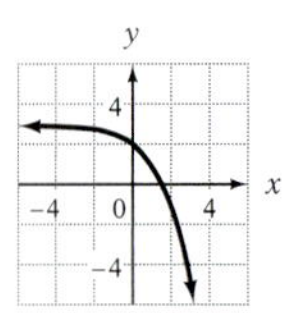 1.6 [11.1.2] **19.** The value of the investment after 6 years is \$15,661. [11.5.1]

20. The half-life is approximately 24 h. [11.5.1]

CUMULATIVE REVIEW EXERCISES

1. $\frac{8}{7}$ [2.2.3] **2.** $L = \frac{S - 2WH}{2W + 2H}$ [7.6.1] **3.** $\{x \mid 1 \le x \le 4\}$ [2.8.2] **4.** $(4x^n + 3)(x^n + 1)$ [6.4.3]

5. $\{x \mid -5 \le x \le 1\}$ [9.5.1] **6.** $\frac{x-3}{x+3}$ [7.3.1] **7.** $\frac{x\sqrt{y} + y\sqrt{x}}{x - y}$ [8.2.4] **8.** $-4x^2y^3\sqrt{2x}$ [8.2.2]

9. $-\frac{1}{5} + \frac{2}{5}i$ [8.5.4] **10.** $y = 2x - 6$ [3.6.1] **11.** $3x^2 + 8x - 3 = 0$ [9.1.2] **12.** $2 + \sqrt{10}$ and $2 - \sqrt{10}$ [9.2.1, 9.2.2] **13.** $\{-6, -4, 0\}$ [3.2.1] **14.** 4 [10.3.2] **15.** $(0, -1, 2)$ [4.2.2]

16. $\left(-\frac{1}{2}, -2\right)$ [4.1.2] **17.** 243 [11.1.1] **18.** 64 [11.2.1] **19.** 8 [11.4.1] **20.** 1 [11.4.2]

21. 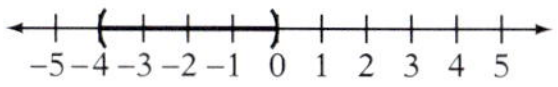

[2.7.1] **22.** 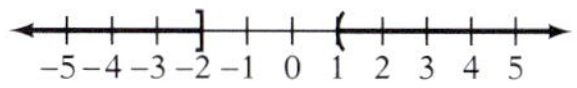

[9.5.1] **23.** 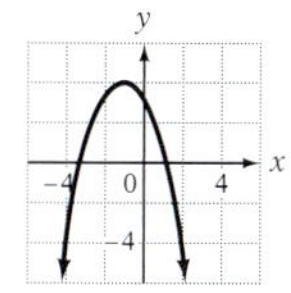

[9.6.1]

24. 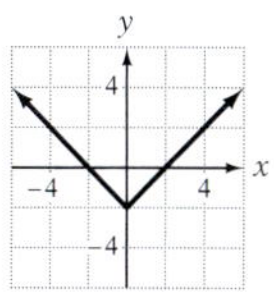

[10.1.1] **25.** 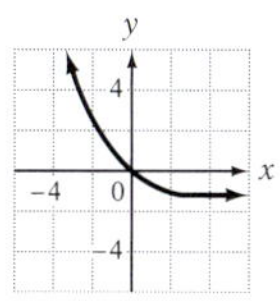

[11.1.2] **26.** 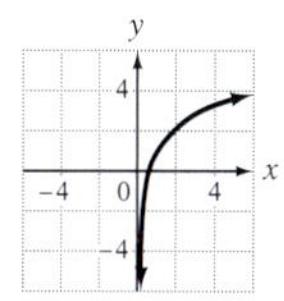

[11.3.1]

27. 800 lb of the alloy containing 25% tin and 1200 lb of the alloy containing 50% tin were used. [2.6.2]

28. To earn \$3000 or more a month, the sales executive must make sales amounting to \$31,250 or more. [2.7.4]

29. 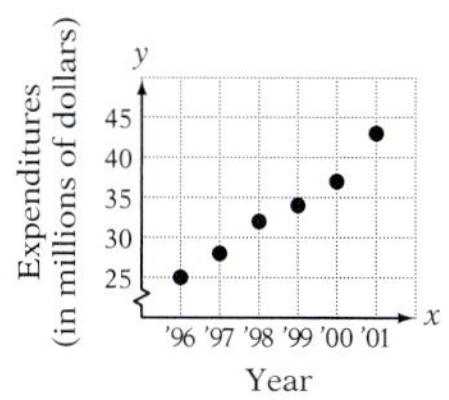

[3.1.3] **30.** With both printers operating, it would take 7.5 min to print the checks. [7.4.2]

31. When the volume is 25 ft^3, the pressure is 500 lb/in^2. [7.5.2] **32.** The cost per yard of the wool carpet is \$24. [4.4.2] **33.** The value of the investment after 5 years is \$15,657. [11.5.1]

FINAL EXAM

1. -31 [1.4.2] **2.** -1 [1.5.1] **3.** $33 - 10x$ [1.5.6] **4.** 8 [2.2.1] **5.** $\frac{2}{3}$ [2.2.3]

6. 4 and $-\frac{2}{3}$ [2.8.1] **7.** $\{x \mid -4 < x < -1\}$ [2.8.2] **8.** $\left\{x \mid x > \frac{3}{2}\right\}$ [2.7.3]

9. $y = -\frac{2}{3}x - \frac{1}{3}$ [3.6.1] **10.** $6a^3 - 5a^2 + 10a$ [5.3.1] **11.** $\frac{6}{5} - \frac{3}{5}i$ [8.5.4]

12. $2x^2 - 3x - 2 = 0$ [9.1.2] **13.** $(2 - xy)(4 + 2xy + x^2y^2)$ [6.4.2] **14.** $(x - y)(1 + x)(1 - x)$ [6.4.4]

15. $x^2 - 2x - 3 - \frac{5}{2x - 3}$ [5.5.2] **16.** $\frac{x(x - 1)}{2x - 5}$ [7.2.1] **17.** $-\frac{10x}{(x + 2)(x - 3)}$ [7.2.2]

18. $\frac{x + 3}{x + 1}$ [7.3.1] **19.** $-\frac{7}{4}$ [7.4.1] **20.** $d = \frac{a_n - a_1}{n - 1}$ [7.6.1] **21.** $\frac{y^4}{162x^3}$ [5.4.1]

22. $\frac{1}{64x^8y^5}$ [8.1.1] **23.** $-2x^2y\sqrt{2y}$ [8.2.2] **24.** $\frac{x^2\sqrt{2y}}{2y^2}$ [8.2.4] **25.** $\frac{3 + \sqrt{17}}{4}$ and $\frac{3 - \sqrt{17}}{4}$ [9.2.2] **26.** 27 and -8 [9.3.1] **27.** $y = -3x + 7$ [3.5.2] **28.** $\frac{3}{2}$ and -2 [9.3.3]

29. $(3, 4)$ [4.2.1] **30.** 10 [4.3.2] **31.** 6 [11.4.2] **32.** $f^{-1}(x) = \frac{3}{2}x + 6$ [10.4.2]

33. $\log_2 \frac{a^2}{b^2}$ [11.2.2] **34.** x-intercept: $\left(\frac{9}{2}, 0\right)$
y-intercept: $(0, -3)$ 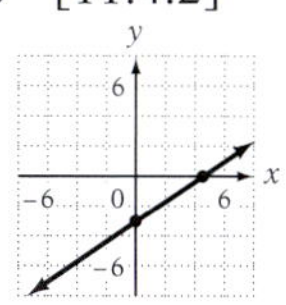

[3.3.2]

35. 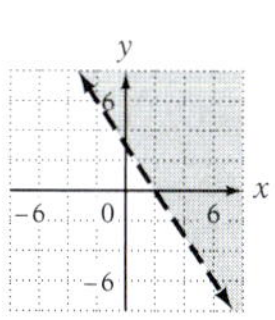

[3.7.1] **36.** 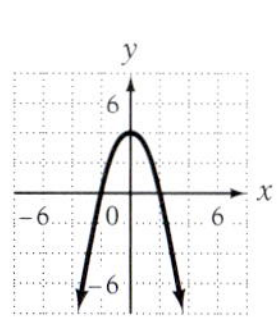

[9.6.1/10.5.1] **37.** 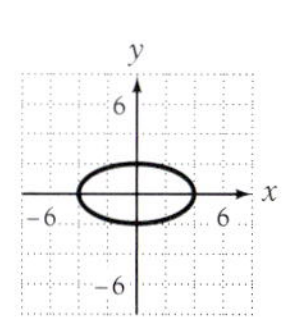

[10.5.3]

38. 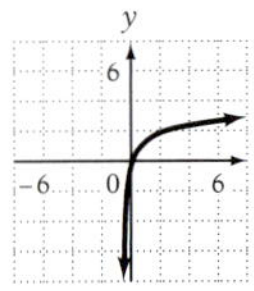 [11.3.1] **39.** 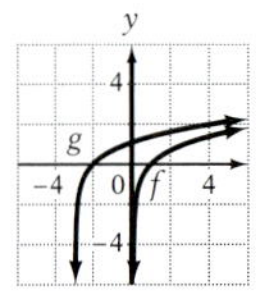 [11.3.1] **40.** The range of scores is 69 or better. [2.7.4]

41. The cyclist traveled 40 mi. [2.3.4] **42.** The amount invested at 8.5% is \$8000. The amount invested at 6.4% is \$4000. [2.6.1] **43.** The width is 7 ft and the length is 20 ft. [9.4.1] **44.** The number of additional shares to be purchased is 200. [7.5.1] **45.** The rate of the plane is 420 mph. [7.4.3] **46.** The distance the object has fallen is 88 ft. [8.4.2] **47.** The rate of the plane for the first 360 mi is 120 mph. [9.4.1] **48.** The intensity is 200 lumens. [7.5.2] **49.** The rate of the boat in calm water is 12.5 mph. The rate of the current is 2.5 mph. [4.4.1] **50.** The value of the investment is \$4785.65. [11.5.1]

Chapter R

SECTION R.1

1. 10 **3.** 21 **5.** -2 **7.** 5 **9.** -3 **11.** -1 **13.** 1 **15.** 18 **17.** -6 **19.** 2 **21.** -40 **23.** 2.01 **25.** $8x$ **27.** $3b$ **29.** $5a$ **31.** $11x$ **33.** $15x - 2y$ **35.** $9a$ **37.** $22y^2$ **39.** $\frac{1}{2}x$ **41.** $-20x$ **43.** $-24a$ **45.** x **47.** $7x$ **49.** $3y$ **51.** $-3a - 15$ **53.** $-16x - 48$ **55.** $-10y^2 + 5$ **57.** $18x^2 - 12xy - 6y^2$ **59.** $-8y - 7$ **61.** $5x - 75$ **63.** $-a + 56$

SECTION R.2

1. -12 **3.** -1 **5.** -8 **7.** -20 **9.** 8 **11.** -1 **13.** 1 **15.** $\frac{3}{4}$ **17.** 3 **19.** 3 **21.** 2 **23.** -1 **25.** $\frac{1}{2}$ **27.** 5 **29.** $-\frac{1}{9}$ **31.** $\{x \mid x > 3\}$ **33.** $\{n \mid n \geq 2\}$ **35.** $\{x \mid x \leq -3\}$ **37.** $\{n \mid n > 0\}$ **39.** $\{x \mid x > 4\}$ **41.** $\{x \mid x > -2\}$ **43.** $\{x \mid x > 3\}$ **45.** $\{x \mid x \geq 2\}$ **47.** $\{x \mid x < 2\}$ **49.** $(3, \infty)$ **51.** $[2, \infty)$ **53.** $(-\infty, 4)$ **55.** $\left(\frac{14}{11}, \infty\right)$ **57.** $\left(-\infty, \frac{3}{8}\right]$ **59.** $\left[-\frac{5}{4}, \infty\right)$

SECTION R.3

1. 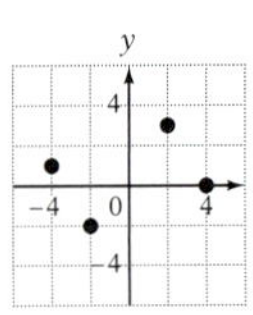 **3.** 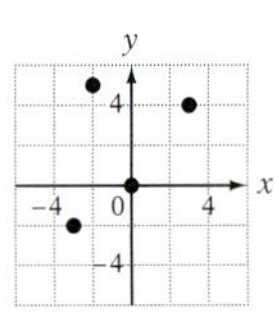 **5.** 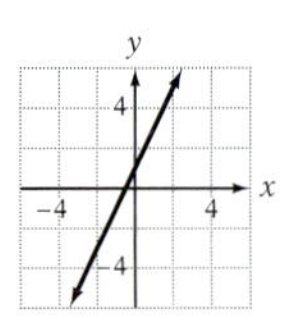 **7.** 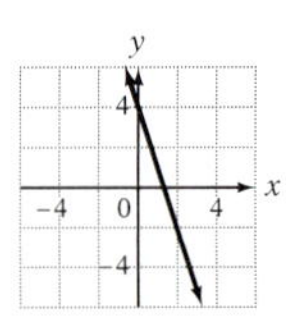 **9.** 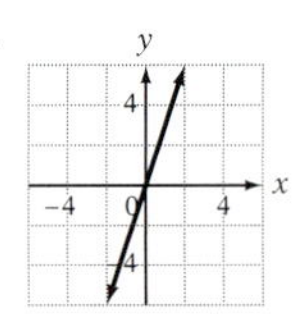

11. 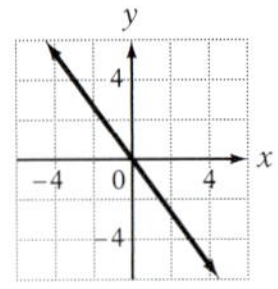 **13.** 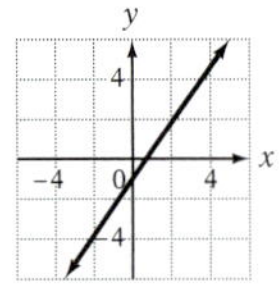 **15.** 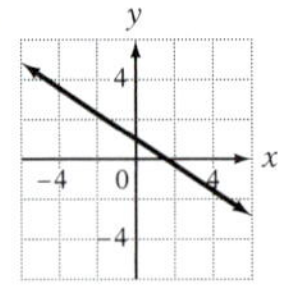 **17.** 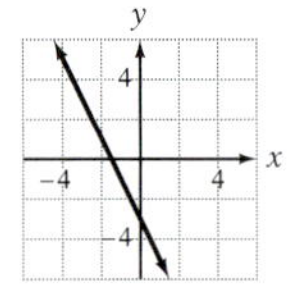 **19.** 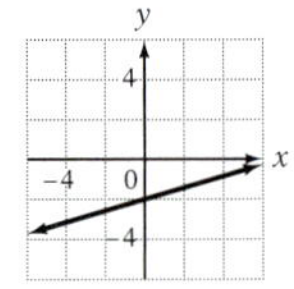

21. 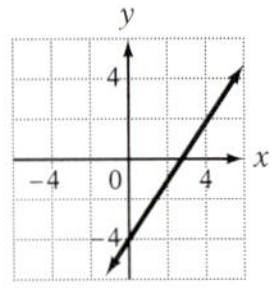 **23.** 3 **25.** $\frac{1}{4}$ **27.** $-\frac{3}{2}$ **29.** undefined **31.** 0 **33.** 1

35. $m = \frac{5}{2}, b = (0, -4)$ **37.** 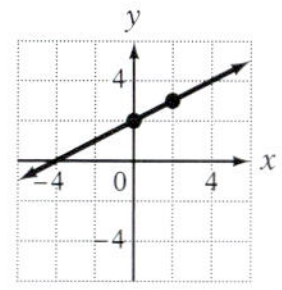

39. 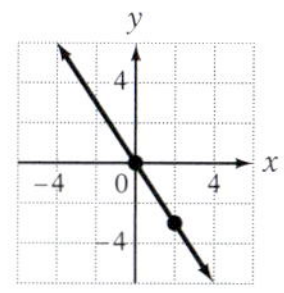

41. 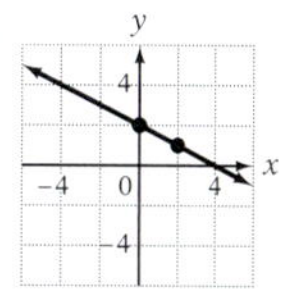

43. 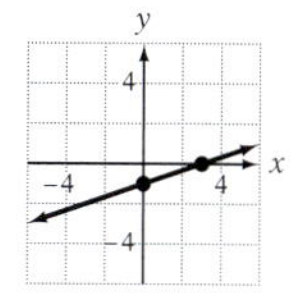

45. 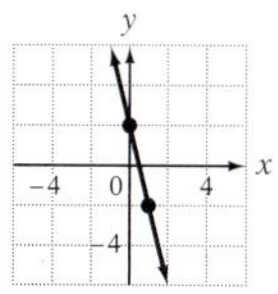

47. 13; (−4, 13) **49.** 14; (−1, 14) **51.** −1; (2, −1) **53.** −15; (−2, −15)

55. $y = -3x - 1$ **57.** $y = -2x + 3$ **59.** $y = -\frac{3}{5}x$ **61.** $y = -\frac{2}{3}x - 4$ **63.** $y = -\frac{1}{2}x + 4$

SECTION R.4

1. z^8 **3.** x^{15} **5.** $x^{12}y^{18}$ **7.** a^6 **9.** $-m^9n^3$ **11.** $-8a^9b^3c^6$ **13.** mn^2 **15.** $-\frac{2a}{3}$
17. $-27m^2n^4p^3$ **19.** $54n^{14}$ **21.** $\frac{7xz}{8y^3}$ **23.** $-8x^{13}y^{14}$ **25.** $\frac{4}{x^7}$ **27.** $\frac{1}{d^{10}}$ **29.** $\frac{1}{x^5}$ **31.** $\frac{x^2}{3}$
33. $\frac{x^6}{y^{12}}$ **35.** $\frac{9}{x^2y^4}$ **37.** $\frac{2}{x^4}$ **39.** $\frac{1}{2x^3}$ **41.** $\frac{1}{2x^2y^6}$ **43.** $\frac{1}{x^6}$ **45.** $\frac{y^8}{x^4}$ **47.** $\frac{b^{10}}{4a^{10}}$
49. $7b^2 + b - 4$ **51.** $3a^2 - 3a + 17$ **53.** $-7x - 7$ **55.** $-2x^3 + x^2 + 2$ **57.** $x^3 + 2x^2 - 6x - 6$
59. $5x^3 + 10x^2 - x - 4$ **61.** $y^3 - y^2 + 6y - 6$ **63.** $4y^3 - 2y^2 + 2y - 4$ **65.** $12b^4 - 48b^3 - 24b$
67. $9b^5 - 9b^3 + 24b$ **69.** $-2x^4y + 6x^3y^2 - 4x^2y^3$ **71.** $x^3 + 4x^2 + 5x + 2$ **73.** $a^3 - 6a^2 + 13a - 12$
75. $-2b^3 + 7b^2 + 19b - 20$ **77.** $x^4 - 4x^3 - 3x^2 + 14x - 8$ **79.** $y^4 + 4y^3 + y^2 - 5y + 2$ **81.** $a^2 + a - 12$
83. $y^2 - 10y + 21$ **85.** $2x^2 + 15x + 7$ **87.** $3x^2 + 11x - 4$ **89.** $4x^2 - 31x + 21$ **91.** $3y^2 - 2y - 16$
93. $21a^2 - 83a + 80$ **95.** $2x^2 + 3xy + y^2$ **97.** $3x^2 - 10xy + 8y^2$ **99.** $10a^2 + 14ab - 12b^2$
101. $16x^2 - 49$ **103.** $x + 2$ **105.** $2y - 7$ **107.** $x - 2 + \frac{8}{x + 2}$ **109.** $3y - 5 + \frac{10}{y + 2}$
111. $b - 5 - \frac{24}{b - 3}$ **113.** $3x + 17 + \frac{64}{x - 4}$ **115.** $5y + 3 + \frac{1}{2y + 3}$ **117.** $x^2 - 5x + 2$
119. $x^2 + 5$ **121.** $y(12y - 5)$ **123.** $5xyz(2xz + 3y^2)$ **125.** $5(x^2 - 3x + 7)$ **127.** $3y^2(y^2 - 3y - 2)$
129. $x^2y^2(x^2y^2 - 3xy + 6)$ **131.** $8x^2y(2 - xy^3 - 6y)$ **133.** $(x + 2)(x - 1)$ **135.** $(a + 4)(a - 3)$
137. $(a - 1)(a - 2)$ **139.** $(b + 8)(b - 1)$ **141.** $(z + 5)(z - 9)$ **143.** $(z - 5)(z - 9)$
145. $(b + 4)(b + 5)$ **147.** Nonfactorable **149.** $(x - 7)(x - 8)$ **151.** $(y + 3)(2y + 1)$
153. $(a - 1)(3a - 1)$ **155.** $(x - 3)(2x + 1)$ **157.** $(2t + 1)(5t + 3)$ **159.** $(2z - 1)(5z + 4)$
161. Nonfactorable **163.** $(t + 2)(2t - 5)$ **165.** $(3y + 1)(4y + 5)$ **167.** $(a - 5)(11a + 1)$
169. $(2b - 3)(3b - 2)$ **171.** $3(x + 2)(x + 3)$ **173.** $a(b + 8)(b - 1)$ **175.** $2y^2(y + 3)(y - 16)$
177. $x(x - 5)(2x - 1)$ **179.** $5(t + 2)(2t - 5)$ **181.** $p(2p + 1)(3p + 1)$

Glossary

abscissa The first number in an ordered pair; it measures a horizontal distance and is also called the first coordinate of an ordered pair. (Section 3.1)

absolute value of a number The distance of the number from zero on the number line. (Sections 1.1; 2.8)

absolute value equation An equation that contains the absolute value symbol. (Section 2.8)

addition method Method of finding an exact solution of a system of linear equations. (Section 4.2)

additive inverses Numbers that are the same distance from zero on the number line but lie on different sides of zero. (Section 1.1)

analytic geometry Geometry in which a coordinate system is used to study relationships between variables. (Section 3.1)

antilogarithm If $\log_b M = N$, then the antilogarithm, base b, of N is M. (Section 11.2)

asymptotes The two straight lines that a hyperbola "approaches." (Section 10.5)

axes The two number lines that form a rectangular coordinate system; also called coordinate axes. (Section 3.1)

axis of symmetry of a parabola A line of symmetry that passes through the vertex of the parabola. (Sections 9.6; 10.5)

base In an exponential expression, the number that is taken as a factor as many times as indicated by the exponent. (Section 1.4)

binomial A polynomial of two terms. (Section 5.1)

center of a circle The central point that is equidistant from all the points that make up a circle. (Section 10.5)

center of a hyperbola The point halfway between the two vertices. (Section 10.5)

center of an ellipse The intersection of the two axes of symmetry of the ellipse. (Section 10.5)

characteristic The integer part of a common logarithm. (Section 11.2)

circle The set of all points (x, y) in the plane that are a fixed distance from a given point (h, k) called the center. (Section 10.5)

clearing denominators Multiplying each side of an equation by the least common multiple of the denominators. (Sections 2.2; 7.4)

closed interval An interval that contains its endpoints. (Section 2.7)

cofactor of an element of a matrix $(-1)^{i+j}$ times the minor of that element, where i is the row number of the element and j is its column number. (Section 4.3)

combined variation A variation in which two or more types of variation occur at the same time. (Section 7.5)

common logarithms Logarithms to the base 10. (Section 11.2)

completing the square Adding to a binomial the constant term that makes it a perfect-square trinomial. (Section 9.2)

complex fraction A fraction whose numerator or denominator contains one or more fractions (Section 7.3)

complex number A number of the form $a + bi$, where a and b are real numbers and $i = \sqrt{-1}$. (Section 8.5)

composition of functions The operation on two functions f and g denoted by $f \circ g$. The value of the composition of f and g is given by $(f \circ g)(x) = f[g(x)]$. (Section 10.3)

compound inequality Two inequalities joined with a connective word such as "and" or "or." (Section 2.7)

compound interest Interest that is computed not only on the original principal but also on the interest already earned. (Section 11.5)

conditional equation An equation that is true if the variable it contains is replaced by the proper value. $x + 2 = 5$ is a conditional equation. (Sections 2.1; 2.3)

conic section A curve that can be constructed from the intersection of a plane and a right circular cone. The four conic sections are the parabola, hyperbola, ellipse, and circle. (Section 10.5)

conjugates Binomial expressions that differ only in the sign of a term. The expressions $a + b$ and $a - b$ are conjugates. (Section 8.2)

consecutive even integers Even integers that follow one another in order. (Section 2.3)

consecutive integers Integers that follow one another in order. (Section 2.3)

consecutive odd integers Odd integers that follow one another in order. (Section 2.3)

constant function A function given by $f(x) = b$, where b is a constant. (Section 3.3)

constant of proportionality k in the direct variation equation $y = kx$; also called the constant of variation. (Section 7.5)

constant of variation k in the direct variation equation $y = kx$; also called the constant of proportionality. (Section 7.5)

constant term A term that contains no variable part. (Section 1.5)

coordinate axes The two number lines that form a rectangular coordinate system; also called simply axes. (Section 3.1)

coordinates of a point The numbers in the ordered pair that is associated with the point. (Section 3.1)

Cramer's Rule A method of solving a system of equations. (Section 4.3)

cube root of a perfect cube One of the three equal factors of the perfect cube. (Section 6.4)

cubic function A third-degree polynomial function. (Section 5.1)

degree of a monomial The sum of the exponents of the variables. (Section 5.1)

degree of a polynomial The greatest of the degrees of any of its terms. (Section 5.1)

dependent system of equations A system of equations whose graphs coincide. (Section 4.1)

dependent variable A variable whose value depends on that of another variable, which is known as the independent variable. (Section 3.2)

descending order The terms of a polynomial in one variable are arranged in descending order when the exponents of the variable decrease from left to right. (Section 5.1)

determinant The number associated with a square matrix. (Section 4.3)

direct variation A special function that can be expressed as the equation $y = kx$, where k is a constant called the constant of variation or the constant of proportionality. (Section 7.5)

discriminant For an equation of the form $ax^2 + bx + c = 0$, the quantity $b^2 - 4ac$ is called the discriminant. (Section 9.2)

domain The set of the first coordinates of all the ordered pairs of a function. (Section 3.2)

double root When a quadratic equation has two solutions that are the same number, the solution is called a double root of the equation. (Section 9.1)

element of a matrix A number in a matrix. (Section 4.3)

elements of a set The objects in the set. (Sections 1.1; 2.7)

ellipse An oval shape that is one of the conic sections. (Section 10.5)

empty set The set that contains no elements. (Section 2.7)

equation A statement of the equality of two mathematical expressions. (Section 2.1)

equivalent equations Equations that have the same solution. (Section 2.1)

evaluating the function Determining $f(x)$ for a given value of x. (Section 3.2)

evaluating a variable expression Replacing the variable in a variable expression by a numerical value and then simplifying. (Section 1.5)

even integer An integer that is divisible by 2. (Section 2.3)

expanding by cofactors A technique for finding the value of a 3×3 or larger determinant. (Section 4.3)

exponent In an exponential expression, the raised number that indicates how many times the factor, or base, occurs in the multiplication. (Section 1.4)

exponential equation An equation in which the variable occurs in the exponent. (Section 11.4)

exponential form The expression 2^6 is in exponential form. Compare factored form. (Section 1.4)

exponential function The exponential function with base b is defined by $f(x) = b^x$, where b is a positive real number not equal to 1. (Section 11.1)

factored form The multiplication $2 \cdot 2 \cdot 2 \cdot 2 \cdot 2$ is in factored form. Compare exponential form. (Section 1.4)

factoring a polynomial Writing the polynomial as a product of other polynomials. (Section 6.1)

factoring a quadratic trinomial Expressing the trinomial as the product of two binomials. (Section 6.2)

finite set A set in which all the elements can be listed. (Section 2.7)

first coordinate of an ordered pair The first number of the ordered pair; it measures a horizontal distance and is also called the abscissa. (Section 3.1)

first-degree equation An equation of the form $ax + b = 0$, where a is not equal to zero. (Section 2.1)

FOIL A method of finding the product of two binomials. The letters stand for First, Outer, Inner, and Last. (Section 5.3)

function A relation in which no two ordered pairs that have the same first coordinate have different second coordinates. (Section 3.2)

functional notation Notation used for those equations that define functions. The letter f is commonly used to name a function. (Section 3.2)

graph of a function A graph of the ordered pairs that belong to the function. (Section 3.3)

graph of an integer A heavy dot placed directly above the number on the number line. (Section 1.1)

graph of an ordered pair The dot drawn at the coordinates of the point in the plane. (Section 3.1)

graphing a point in the plane Placing a dot at the location given by the ordered pair; also called plotting a point in the plane. (Section 3.1)

half-plane The solution set of a linear inequality in two variables. (Section 3.7)

horizontal shift The displacement of a graph to the right or left on the coordinate axes; also known as a horizontal translation. (Section 10.2)

horizontal translation The displacement of a graph to the right or left on the coordinate axes; also known as a horizontal shift. (Section 10.2)

hyperbola A conic section formed by the intersection of a right circular cone and a plane perpendicular to the base of the cone. (Section 10.5)

hypotenuse In a right triangle, the side opposite the 90° angle. (Sections 3.1; 8.4)

imaginary number A number of the form ai, where a is a real number and $i = \sqrt{-1}$. (Section 8.5)

imaginary part of a complex number For the complex number $a + bi$, b is the imaginary part. (Section 8.5)

inconsistent system of equations A system of equations that has no solution. (Section 4.1)

independent system of equations A system of equations that has exactly one solution. (Section 4.1)

independent variable A variable whose value determines that of another variable, which is known as the dependent variable. (Section 3.2)

index In the expression $\sqrt[n]{a}$, n is the index of the radical. (Section 8.1)

infinite set A set in which all the elements cannot be listed. (Section 2.7)

integers The numbers $\ldots, -3, -2, -1, 0, 1, 2, 3, \ldots$. (Section 1.1)

intersection of two sets The set that contains all elements that are common to both of the sets. (Section 2.7)

interval notation A type of set-builder notation in which the property that distinguishes the elements of the set is their location within a specified interval. (Section 2.7)

inverse of a function The set of ordered pairs formed by reversing the coordinates of each ordered pair of the function. (Section 10.4)

inverse variation A function that can be expressed as the equation $y = k/x$, where k is a constant. (Section 7.5)

irrational number The decimal representation of an irrational number never terminates or repeats. (Sections 1.3; 8.2)

joint variation A variation in which a variable varies directly as the product of two or more variables. A joint variation can be expressed as the equation $z = kxy$, where k is a constant. (Section 7.5)

least common multiple (LCM) The LCM of two or more numbers is the smallest number that is a multiple of each of those numbers. (Section 1.3)

least common multiple of two or more polynomials The simplest polynomial of least degree that contains the factors of each polynomial. (Section 7.2)

leg In a right triangle, one of the two sides that are not opposite the 90° angle. (Section 8.4)

like terms Terms of a variable expression that have the same variable part. Having no variable part, constant terms are like terms. (Section 1.5)

linear equation in three variables An equation of the form $Ax + By + Cz = D$, where A, B, and C are coefficients of the variables and D is a constant. (Section 4.2)

linear equation in two variables An equation of the form $y = mx + b$, where m is the coefficient of x and b is a constant, or of the form $Ax + By = C$. Also called a linear function. (Section 3.3)

linear function An equation of the form $y = mx + b$, where m is the coefficient of x and b is a constant; also called a linear equation in two variables. Its graph is a straight line. (Section 3.3)

linear inequality in two variables An inequality of the form $y > mx + b$ or $Ax + By > C$. The symbol $>$ could be replaced by $\geq$, $<$, or $\leq$. (Section 3.7)

literal equation An equation that contains more than one variable. (Sections 3.3; 7.6)

logarithm For b greater than zero and not equal to 1, the statement $y =$ the logarithm of x, base b, is equivalent to $x = b^y$. (Section 11.2)

main diagonal The elements $a_{11}, a_{22}, a_{33}, \ldots, a_{nn}$ of a matrix. (Section 4.3)

mantissa The decimal part of a common logarithm. (Section 11.2)

matrix A rectangular array of numbers. (Section 4.3)

maximum The value of a function at its highest point. (Section 9.7)

minimum The value of a function at its lowest point. (Section 9.7)

minor of an element The minor of an element in a 3×3 determinant is the 2×2 determinant obtained by eliminating the row and column that contain the element. (Section 4.3)

monomial A number, a variable, or a product of a number and variables. (Section 5.1)

monomial factor The GCF of all the terms in a polynomial. (Section 6.1)

multiplicative inverse The multiplicative inverse of a nonzero real number a is $1/a$; also called the reciprocal of a. (Section 1.5)

natural exponential function The function defined by $f(x) = e^x$, where $e \approx 2.71828$. (Section 11.1)

natural logarithm When e (the base of the natural exponential function) is used as the base of a logarithm, the logarithm is referred to as the natural logarithm and is abbreviated $\ln x$. (Section 11.2)

natural numbers The numbers 1, 2, 3,...; also called the positive integers. (Section 1.1)

negative integers The numbers ..., −3, −2, −1. (Section 1.1)

negative slope The slope of a line that slants downward to the right. (Section 3.4)

nonfactorable over the integers A polynomial is nonfactorable over the integers if it does not factor using only integers. (Section 6.2)

n*th root of *a A number b such that $b^n = a$. The nth root of a can be written $a^{\frac{1}{n}}$ or $\sqrt[n]{a}$. (Section 8.1)

null set A set that contains no elements; also called an empty set. (Section 2.7)

numerical coefficient The number part of a variable term. (Section 1.5)

odd integer An integer that is not divisible by 2. (Section 2.3)

one-to-one function In a one-to-one function, given any y, there is only one x that can be paired with the given y. (Section 10.4)

open interval An interval that does not contain its endpoints. (Section 2.7)

opposites Numbers that are the same distance from zero on the number line but lie on different sides of zero; also called additive inverses. (Section 1.1)

order $m \times n$ A matrix of m rows and n columns is of order $m \times n$. (Section 4.3)

Order of Operations Agreement Rules that specify the order in which operations are performed in simplifying numerical expressions. (Section 1.4)

ordered pair A pair of numbers expressed in the form (a, b) and used to locate a point in a rectangular coordinate system. (Section 3.1)

ordinate The second number of an ordered pair; it measures a vertical distance and is also called the second coordinate of an ordered pair. (Section 3.1)

origin The point of intersection of the two number lines that form a rectangular coordinate system. (Section 3.1)

parabola The graph of a quadratic function is called a parabola. (Sections 9.6; 10.5)

parallel lines Lines that have the same slope and thus do not intersect. (Sections 2.4; 3.6)

percent mixture problems A problem that involves combining two ingredients with different percent concentrations into a single mixture. (Section 2.6)

perfect cube The product of the same three factors. (Section 6.4)

perfect square The product of a term and itself. (Section 6.4)

perpendicular lines Lines that intersect at right angles. (Sections 2.4; 3.6)

plotting a point in the plane Placing a dot at the location given by the ordered pair; also called graphing a point in the plane. (Section 3.1)

point-slope formula The equation $y - y_1 = m(x - x_1)$, where m is the slope of a line and (x_1, y_1) is a point on the line. (Section 3.5)

polynomial A variable expression in which the terms are monomials. (Section 5.1)

positive integers The numbers 1, 2, 3,...; also called the natural numbers. (Section 1.1)

positive slope The slope of a line that slants upward to the right. (Section 3.4)

principal square root The positive square root of a number. (Section 8.1)

proportion An equation that states the equality of two ratios or rates. (Section 7.5)

Pythagorean Theorem If a and b are the lengths of the legs of a right triangle and c is the length of the hypotenuse, then $c^2 = a^2 + b^2$. (Section 8.4)

quadrant One of the four regions into which a rectangular coordinate system divides the plane. (Section 3.1)

quadratic equation An equation that can be written in the form $ax^2 + bx + c = 0$, where a and b are coefficients, c is a constant and $a \neq 0$; also called a second-degree equation. (Sections 6.5; 9.1)

quadratic formula A general formula derived by applying the method of completing the square to the standard form of a quadratic equation. (Section 9.2)

quadratic function A function that can be expressed by the equation $f(x) = ax^2 + bx + c$, where a is not equal to zero. (Sections 5.1; 9.6)

quadratic in form An equation that can be written in the form $au^2 + bu + c = 0$. (Section 9.3)

quadratic inequality in one variable An inequality that can be written in the form $ax^2 + bx + c < 0$ or $ax^2 + bx + c > 0$, where a is not equal to zero. The symbols $\leq$ and $\geq$ can also be used. (Section 9.5)

quadratic trinomial A trinomial of the form $ax^2 + bx + c$, where a and b are nonzero coefficients and c is a nonzero constant. (Sections 6.2; 6.3)

radical The symbol $\sqrt{\ }$. (Section 8.1)

radical equation An equation that contains a variable expression in a radical. (Section 8.4)

radical function A function that contains a variable expression in a radical. (Section 8.3)

radicand In a radical expression, the expression within the radical. (Section 8.1)

radius The fixed distance, from the center of a circle, of all points that make up the circle. (Section 10.5)

range The set of the second coordinates of all the ordered pairs of a function. (Section 3.2)

rate The quotient of two quantities that have different units. (Section 7.5)

rate of work The part of a task that is completed in one unit of time. (Section 7.4)

ratio The quotient of two quantities that have the same unit. (Section 7.5)

rational expression A fraction in which the numerator and denominator are polynomials. (Section 7.1)

rational function A function that is written in terms of a rational expression. (Section 7.1)

rational number A number of the form a/b, where a and b are integers and b is not equal to zero. (Section 1.3)

rationalizing the denominator The procedure used to remove a radical from the denominator. (Section 8.2)

real numbers The rational numbers and the irrational numbers taken together. (Sections 1.3; 8.2)

real part of a complex number For the complex number $a + bi$, a is the real part. (Section 8.5)

reciprocal The reciprocal of a nonzero number a is $1/a$; also called the multiplicative inverse. (Section 1.3)

reciprocal of a rational expression The rational expression with the numerator and denominator interchanged. (Section 7.2)

rectangular coordinate system A coordinate system formed by two number lines, one horizontal and one vertical, that intersect at the zero point of each line. (Section 3.1)

relation A set of ordered pairs. (Section 3.2)

repeating decimal A decimal formed when dividing the numerator of a fraction by its denominator, in which a digit or a sequence of digits in the decimal repeat infinitely. (Section 1.3)

roster method A method of designating a set by enclosing a list of its elements in braces. (Sections 1.1; 2.7)

scatter diagram A graph of ordered-pair data. (Section 3.1)

scientific notation Notation in which a number is expressed as the product of a number between 1 and 10 and a power of 10. (Section 5.4)

second coordinate of an ordered pair The second number of the ordered pair; it measures a vertical distance and is also called the ordinate. (Section 3.1)

second-degree equation An equation of the form $ax^2 + bx + c = 0$, where a, b, and c are constants and a is not equal to zero; also called a quadratic equation. (Section 9.1)

set A collection of objects. (Sections 1.1; 2.7)

set-builder notation A method of designating a set that makes use of a variable and a certain property that only elements of that set possess. (Section 2.7)

slope A measure of the slant, or tilt, of a line. (Section 3.4)

slope-intercept form of a straight line The equation $y = mx + b$, where m is the slope of the line and $(0, b)$ is the y-intercept. (Section 3.4)

solution(s) of an equation The replacement value(s) of the variable that will make the equation true; also called the root(s) of the equation. (Section 2.1)

solution of an equation in two variables An ordered pair whose coordinates make the equation a true statement. (Section 3.1)

solution of a system of equations in three variables An ordered triple that is a solution of each equation of the system. (Section 4.2)

solution of a system of equations in two variables An ordered pair that is a solution of each equation of the system. (Section 4.1)

solution set of an inequality A set of numbers, each element of which, when substituted for the variable, results in a true inequality. (Section 2.7)

solution set of a system of inequalities The intersection of the solution sets of the individual inequalities. (Section 4.5)

solving an equation Finding a root, or solution, of the equation. (Section 2.1)

square matrix A matrix that has the same number of rows as columns. (Section 4.3)

square root of a perfect square One of the two equal factors of the perfect square. (Section 6.4)

standard form of a quadratic equation A quadratic equation is in standard form when the polynomial is in descending order and is equal to zero. (Section 9.1)

substitution method Method of finding a solution of a system of linear equations. (Section 4.1)

synthetic division A shorter method of dividing a polynomial by a binomial of the form $x - a$. This method uses only the coefficients of the variable terms. (Section 5.5)

system of equations Two or more equations considered together. (Section 4.1)

system of inequalities Two or more inequalities considered together. (Section 4.5)

terminating decimal A decimal formed when dividing the numerator of a fraction by its denominator and the remainder is zero. (Section 1.3)

terms of a variable expression The addends of the expression. (Section 1.5)

tolerance of a component The acceptable amount by which the component may vary from a given measurement. (Section 2.8)

trinomial A polynomial of three terms. (Section 5.1)

undefined slope The slope of a vertical line is undefined. (Section 3.4)

uniform motion The motion of an object whose speed and direction do not change. (Section 2.3)

union of two sets The set that contains all elements that belong to either of the sets. (Section 2.7)

value mixture problem A problem that involves combining two ingredients that have different prices into a single blend. (Section 2.3)

value of a function The value of the dependent variable for a given value of the independent variable. (Section 3.2)

variable A letter of the alphabet used to stand for some number. (Section 1.1)

variable expression An expression that contains one or more variables. (Section 1.5)

variable term A term composed of a numerical coefficient and a variable part. When the numerical coefficient is 1 or -1, the 1 is not usually written. (Section 1.5)

vertex of a parabola The point on the parabola with the smallest y-coordinate or the largest y-coordinate. (Sections 9.6; 10.5)

vertical shift The displacement of a graph up or down on the coordinate axes; also known as a vertical translation. (Section 10.2)

vertical translation The displacement of a graph up or down on the coordinate axes; also known as a vertical shift. (Section 10.2)

whole numbers The numbers 0, 1, 2, 3,.... The whole numbers include the natural numbers and zero. (Section 1.1)

***x*-coordinate** The abscissa in an ordered pair. (Section 3.1)

***x*-intercept** The point at which a graph crosses the x-axis. (Section 3.3)

***xy*-coordinate system** A rectangular coordinate system in which the horizontal axis is labeled x and the vertical axis is labeled y. (Section 3.1)

***y*-coordinate** The ordinate in an ordered pair. (Section 3.1)

***y*-intercept** The point at which a graph crosses the y-axis. (Section 3.3)

zero slope The slope of a horizontal line. (Section 3.4)

Index

JUST LOOK FOR PATTERNS!

By Joanne Peeples

What makes a mathematics course different from most of the other courses you are taking? Why do you, the student, often seem to have so much trouble with the course – and of course the instructor thinks it's easy? Just what does the instructor know that you don't know? If these questions could just be answered, math courses would be so much easier.

In the front of your text is "AIM for Success", page xxv, which gives you hints on what to do to be successful in this course. The suggestions are good, and it will help make the course easier if you follow the suggestions, but perhaps something is missing! I'd like to call it a "point of view". The way you look at a problem, and the way a mathematician looks at a problem is often quite different. Take a word problem for example, the student often just sees the words (and maybe even has trouble reading and understanding what the problem is saying), while the instructor is probably thinking: "Is this like any other problem I've done before? How is it similar? How is it different?" The instructor will possibly skip most of the words, until the general idea of the problem "sits in her head", only then will she go back and read the problem closely, and start putting the pieces together – like the pieces of a puzzle.

What does a mathematician look for in a problem? Patterns! She looks for similarities and differences to problems she has already worked. Actually a mathematician is selectively lazy, and if possible will try to reduce the problem to one she has already worked. Then the problem is solved, except for the arithmetic.

Look for patterns in real life, you can see them all around. There are patterns in shape (circular, square, triangular); there are patterns in numbers (2, 4, 6, 8, ...); there are patterns in motion (two steps forward, one to the side and then one back – could that be a dance step?); there are patterns in behavior. Sometimes you can actually see the patterns, sometimes the patterns are in your imagination (have you ever seen a cloud that looks like a "dog"?). We depend on patterns in our lives (cars must stop at stop lights – a behavior pattern), and to be successful in mathematics you must learn to look for patterns.

At the beginning of this course, you should know the patterns for combining real numbers – do you remember what a real number is? Could you explain to your best friend just what a real number is? (If not *go back and review* or go to Section 1.3, page 21, of this text and remind yourself of the different names for numbers and what those names mean). If you were told that $m \times n = n \times m$, and that m and n are real numbers, what have you really been told? First, you have been shown some symbols – they make a pretty pattern – and knowing how these symbols are used and what they mean, should be woven into the fabric of your brain, so you can "see" a pattern. You might want to think about specific numbers when looking at the symbols. For example, what does $3 \times 4 = 4 \times 3$ mean? It's an example that shows a pattern for real numbers, and it "says" that order of multiplication does not matter. Actually you use this in real life, suppose you need to multiply $2 \times 4 \times 5$. Don't you look at the numbers and first multiply 2 times 5, which is 10? And then multiply 4 times 10. Doing it this way, because you know that multiplication by 10 just means adding a zero – another pattern.

What about exponents? There's a pattern to be found! Look at 2^x : $2^1 = 2$, $2^2 = 4$, $2^3 = 8$, … As x increases by 1, 2^x doubles. What happens if x decreases by 1? $2^3 = 8$ and $2^2 = 4$. Look for the pattern! This time we get half as much. If $2^1 = 2$, doesn't it then make sense that $2^0 = (1/2) \times 2 = 1$? Patterns!!! And, what about $2^{-1} = (1/2) \times 1 = 1/2$; and $2^{-2} = (1/2) \times (1/2) = 1/4$. Do you see the pattern? $2^{-x} = 1/2^x$.

So – whenever you're stuck working a problem. Take a piece of paper, and start writing all possible patterns you can find in the problem (maybe even drawing some pictures). Pretend the problem is a puzzle, the patterns have to be there, and so is the solution. And, if you can learn to see the patterns, you'll have fewer rules to memorize, and a better understanding of the concepts to take with you to the next class.

Table of Equations and Formulas

Slope of a Straight Line

Slope $= m = \frac{y_2 - y_1}{x_2 - x_1}$, $x_1 \neq x_2$

Slope-intercept Form of a Straight Line

$y = mx + b$

Point-slope Formula for a Line

$y - y_1 = m(x - x_1)$

Quadratic Formula

$$x = \frac{-b \pm \sqrt{b^2 - 4ac}}{2a}$$

Perimeter and Area of a Triangle, and Sum of the Measures of the Angles

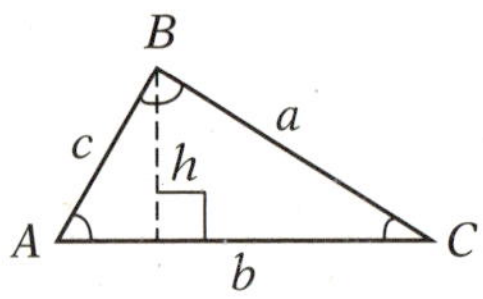

$P = a + b + c$

$A = \frac{1}{2}bh$

$\angle A + \angle B + \angle C = 180°$

Pythagorean Theorem

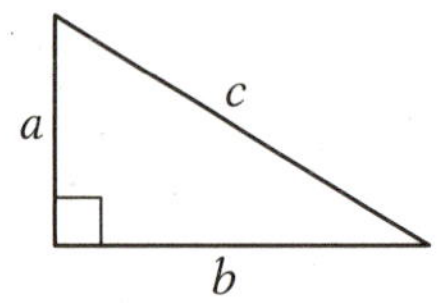

$a^2 + b^2 = c^2$

Perimeter and Area of a Rectangle

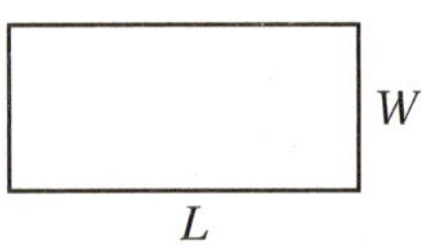

$P = 2L + 2W$

$A = LW$

Perimeter and Area of a Square

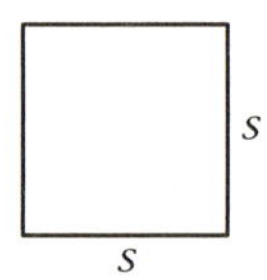

$P = 4s$

$A = s^2$

Circumference and Area of a Circle

$C = 2\pi r$ or $C = \pi d$

$A = \pi r^2$

Area of a Trapezoid

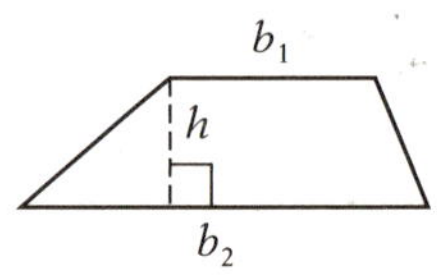

$A = \frac{1}{2}h(b_1 + b_2)$

Volume and Surface Area of a Rectangular Solid

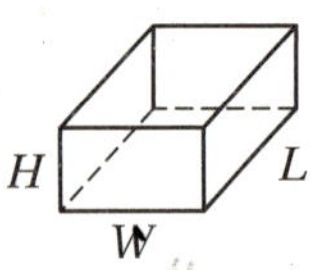

$V = LWH$

$SA = 2LW + 2LH + 2WH$

Volume and Surface Area of a Cube

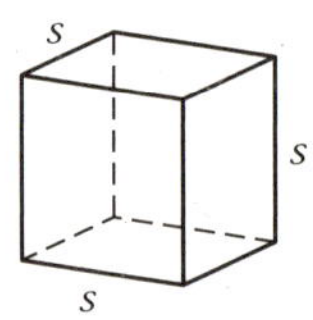

$V = s^3$

$SA = 6s^2$

Volume and Surface Area of a Right Circular Cylinder

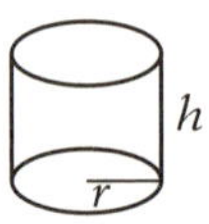

$V = \pi r^2 h$

$SA = 2\pi r^2 + 2\pi rh$

Volume and Surface Area of a Right Circular Cone

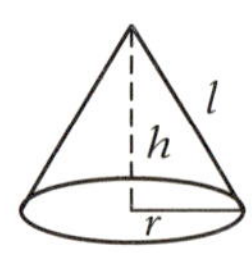

$V = \frac{1}{3}\pi r^2 h$

$SA = \pi r^2 + \pi rl$

Volume and Surface Area of a Sphere

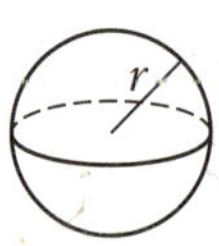

$V = \frac{4}{3}\pi r^3$

$SA = 4\pi r^2$